U.S. Customary Units and Their SI Equivalents

Quantity	U.S. Customary Unit	SI Equivalent
Acceleration	ft/s^2	0.3048 m/s^2
	in./s^2	0.0254 m/s^2
Area	ft^2	0.0929 m^2
	in^2	645.2 mm^2
Energy	ft · lb	1.356 J
Force	kip	4.448 kN
	lb	4.448 N
	oz	0.2780 N
Impulse	lb · s	4.448 N · s
Length	ft	0.3048 m
	in.	25.40 mm
	mi	1.609 km
Mass	oz mass	28.35 g
	lb mass	0.4536 kg
	slug	14.59 kg
	ton	907.2 kg
Moment of a force	lb · ft	1.356 N · m
	lb · in.	0.1130 N · m
Moment of inertia		
Of an area	in^4	0.4162×10^6 mm^4
Of a mass	lb · ft · s^2	1.356 kg · m^2
Momentum	lb · s	4.448 kg · m/s
Power	ft · lb/s	1.356 W
	hp	745.7 W
Pressure or stress	lb/ft^2	47.88 Pa
	lb/in^2 (psi)	6.895 kPa
Velocity	ft/s	0.3048 m/s
	in./s	0.0254 m/s
	mi/h (mph)	0.4470 m/s
	mi/h (mph)	1.609 km/h
Volume, solids	ft^3	0.02832 m^3
	in^3	16.39 cm^3
Liquids	gal	3.785 l
	qt	0.9464 l
Work	ft · lb	1.356 J

E. HABERSTROH

Mechanics for Engineers
STATICS and DYNAMICS

McGRAW-HILL BOOK COMPANY

New York
St. Louis
San Francisco
Auckland
Düsseldorf
Johannesburg
Kuala Lumpur
London
Mexico
Montreal
New Delhi
Panama
Paris
São Paulo
Singapore
Sydney
Tokyo
Toronto

Third Edition

Mechanics for Engineers

STATICS and DYNAMICS

Ferdinand P. Beer

Professor and Chairman
Department of Mechanical Engineering and Mechanics
Lehigh University

E. Russell Johnston, Jr.

Professor and Head
Department of Civil Engineering
University of Connecticut

Mechanics for Engineers
STATICS and DYNAMICS

9 0 DODO 8 9 8 7 6 5

This book was set in Laurel by York Graphic Services, Inc. The editors were B. J. Clark and J. W. Maisel; the designer was Merrill Haber; the production supervisor was Thomas J. LoPinto. The drawings were done by Felix Cooper. Cover photographs were taken by Milton J. Heiberg.
R. R. Donnelley & Sons Company was printer and binder.

Library of Congress Cataloging in Publication Data

Beer, Ferdinand Pierre, date
Mechanics for engineers.

1. Mechanics. I. Johnston, Elwood Russell, date joint author. II. Title.
QA807.B392 1976 531 75-34004
ISBN 0-07-004270-5

CONTENTS

PREFACE

The main objective of a first course in mechanics should be to develop in the engineering student the ability to analyze any problem in a simple and logical manner and to apply to its solution a few, well-understood, basic principles. This text is designed for the first courses in statics and dynamics offered in the sophomore or junior year, and it is hoped that it will help the instructor achieve this goal.†

In this third edition, the vectorial character of mechanics has again been emphasized. The concept of vectors and the laws governing the addition and the resolution of vectors into components are discussed with care at the beginning of statics. Throughout the text, forces, velocities, accelerations, and other vector quantities have been clearly distinguished from scalar quantities through the use of boldface type. Products and derivatives of vectors, however, are not used in this text.‡

One of the characteristics of the approach used in this book is that the mechanics of *particles* has been clearly separated from the mechanics of *rigid bodies*. This approach makes it possible to consider simple practical applications at an early stage and to postpone the introduction of the more difficult concepts. In statics, for example, the statics of particles is treated first (Chap. 2), and the principle of equilibrium is immediately applied to practical situations involving only concurrent forces. The statics of rigid bodies is considered in Chaps. 3 and 4, where the principle of transmissibility and the associated concept of moment of a force are introduced. In dynamics, the same division is observed. The basic concepts of force, mass, and acceleration, of work and energy, and of impulse and momentum are introduced and first applied to problems involving only particles. Thus, the student may familiarize himself with the three basic methods used in

† This text is also available in two separate volumes, *Mechanics for Engineers: Statics*, and *Mechanics for Engineers: Dynamics*.

‡ In a parallel text, *Vector Mechanics for Engineers*, vector and scalar products are introduced at an early stage and used to solve three-dimensional problems in statics. The use of vector analysis in the presentation of dynamics makes it possible also to analyze more advanced problems in three-dimensional kinematics and kinetics.

dynamics and learn their respective advantages before facing the difficulties associated with the motion of rigid bodies.

Although the authors strongly believe in the advantage of the approach they use, they do not wish to impose their choice on the instructor. Consequently, the various chapters in dynamics have been written so that they may be taught in alternate sequences. For example, Chap. 15 may be taken immediately after Chap. 11 if the instructor prefers to keep kinematics apart from kinetics, and Chaps. 13 and 14 may be postponed if he wishes to teach the method of work and energy (Chaps. 13 and 17) and the method of impulse and momentum (Chaps. 14 and 18) as single units. Whatever the sequence used, the very fact that distinct chapters are devoted to the dynamics of particles and to the dynamics of rigid bodies should help the student organize his own thoughts.

Since this text is designed for first courses in statics and dynamics, new concepts have been presented in simple terms and every step explained in detail. On the other hand, by discussing the broader aspects of the problems considered, and by stressing methods of general applicability, a definite maturity of approach has been achieved. For example, the concepts of partial constraints and of statical indeterminacy are introduced early in the text and used throughout statics. In dynamics, the concepts of mass and weight are carefully distinguished, and the limitations of the methods used to analyze plane motion are indicated so that the student will not be tempted to apply them to cases in which they are not valid.

The fact that mechanics is essentially a *deductive* science based on a few fundamental principles has been stressed. Derivations have been presented in their logical sequence and with all the rigor warranted at this level. However, the learning process being largely *inductive,* simple applications have been considered first. Thus the statics of particles precedes the statics of rigid bodies; coplanar forces are introduced before forces in space; and problems involving internal forces are postponed until Chap. 6. Again, the dynamics of particles precedes the dynamics of rigid bodies, and, in the latter, the emphasis has been placed on the study of plane motion.

Free-body diagrams are introduced early, and their importance is emphasized throughout the text. Color has been used in this edition to distinguish forces from other elements of the free-body diagrams. This makes it easier for the students to identify the forces acting on a given particle or rigid body and to follow the discussion of sample problems and other examples given in the text.

Free-body diagrams are used not only to solve equilibrium problems but also to express the equivalence of two systems of forces or, more generally, of two systems of vectors. The advantage of this approach becomes apparent in the study of the dynamics of rigid bodies. By placing the emphasis on "free-body-diagram equations" rather than on the standard algebraic equations of motion, a more intuitive and more complete understanding of the fundamental principles of dynamics may be achieved. This approach, which was first introduced in 1962 in the first edition of *Vector Mechanics for Engineers*, has now gained wide acceptance among mechanics teachers in this country. It is, therefore, being used in preference to the method of dynamic equilibrium in the solution of sample problems in this new edition of *Mechanics for Engineers*.

Because of the current trend among American engineers to adopt the international system of units (SI metric units), the SI units most frequently used in mechanics have been introduced in Chap. 1 of this edition and are used throughout the text. Half the sample problems and problems to be assigned have been stated in these units, while the other half retain U.S. customary units. The authors believe that this approach will best serve the needs of the students, who will be entering the engineering profession during the period of transition from one system of units to the other. It also should be recognized that the passage from one system to the other entails more than the use of conversion factors. Since the SI system of units is an absolute system based on the units of time, length, and mass, whereas the U.S. customary system is a gravitational system based on the units of time, length, and force, different approaches are required for the solution of many problems. For example, when SI units are used, a body is generally specified by its mass expressed in kilograms; in most problems of statics it will be necessary to determine the weight of the body in newtons, and an additional calculation will be required for this purpose. On the other hand, when U.S. customary units are used, a body is specified by its weight in pounds and, in dynamics problems, an additional calculation will be required to determine its mass in slugs (or $\text{lb}\cdot\text{sec}^2/\text{ft}$). The authors, therefore, believe that problems assignments should include both types of units. A sufficient number of problems, however, have been provided so that, if so desired, two complete sets of assignments may be selected from problems stated in SI units only and two others from problems stated in U.S. customary units. Since the answers to all even-numbered problems stated in U.S. customary units have been given in both systems of units, teachers who wish to give special instruction to their students in the conversion of units may assign these problems and ask their students to use SI units in their solutions. This has been illustrated in two sample problems involving, respectively, the computation of mass moments of inertia

(Sample Prob. 9.13) and the kinetics of particles (Sample Prob. 12.2).

A large number of optional sections have been included. These sections are indicated by asterisks and may thus easily be distinguished from those which form the core of the basic mechanics course. They may be omitted without prejudice to the understanding of the rest of the text. Among the topics covered in these additional sections are applications to hydrostatics, shear and bending-moment diagrams for beams, equilibrium of cables, products of inertia and Mohr's circle, the method of virtual work, graphical methods for the solution of rectilinear-motion problems, the trajectory of a particle under a central force, the deflection of fluid streams, problems involving jet and rocket propulsion, Coriolis acceleration, damped mechanical vibrations, and electrical analogues. The sections on beams are especially useful when the course in statics is immediately followed by a course in mechanics of materials, while the additional topics included in dynamics will be found of interest when dynamics is taught in the junior year.

The material presented in the text and most of the problems require no previous mathematical knowledge beyond algebra, trigonometry, and elementary calculus. However, special problems have been included, which make use of a more advanced knowledge of calculus, and certain sections, such as Secs. 19.8 and 19.9 on damped vibrations, should be assigned only if the students possess the proper mathematical background. In the portions of the text using elementary calculus, a greater emphasis has been placed on the correct understanding and application of the concepts of differentiation and integration than on the nimble manipulation of mathematical formulas. In this connection, it should be mentioned that the determination of the centroids of composite areas precedes the calculation of centroids by integration, making it thus possible to establish firmly the concept of moment of area before introducing the use of integration.

Instructions on the use of pocket calculators for the solution of a number of typical statics problems have been added to Chap. 2. These instructions are complemented by the notes on the proper use of the slide rule which were included in the previous editions of *Mechanics for Engineers.*

The text has been divided into units, each consisting of one or several theory sections, one or several sample problems, and a large number of problems to be assigned. Each unit corresponds to a well-defined topic and generally may be covered in one lesson. In a number of cases, however, the instructor will find it desirable to devote more than one lesson to a given topic. The sample problems have been set up in much the same form that a student will use in solving the assigned problems. They thus serve the double purpose of amplifying the text and demonstrating the type of neat and orderly work that the student should cultivate in his own solutions. Most of the problems to be assigned are of a practical nature and should appeal to the engineering student. They are primarily designed, however, to illustrate the material presented in the text and to help the student understand the basic principles of mechanics. The problems have been grouped according to the portions of material they illustrate and have been arranged in order of increasing difficulty. Problems requiring special attention have been indicated by asterisks. Answers to all even-numbered problems are given at the end of the book.

The authors wish to acknowledge gratefully the many helpful comments and suggestions offered by the users of the two previous editions of *Mechanics for Engineers* and of the first and second editions of *Vector Mechanics for Engineers.*

FERDINAND P. BEER
E. RUSSELL JOHNSTON, JR.

LIST OF SYMBOLS

a, a Acceleration
a Constant; radius; distance; semimajor axis of ellipse
$\bar{\mathbf{a}}$, $\bar{a}$ Acceleration of mass center
$\mathbf{a}_{B/A}$ Relative acceleration of B with respect to A
$\mathbf{a}_c$ Coriolis acceleration
A, B, C, . . . Reactions at supports and connections
$A, B, C, \ldots$ Points
A Area
b Width; distance; semiminor axis of ellipse
c Constant; coefficient of viscous damping
C Centroid; instantaneous center of rotation; capacitance
d Distance
e Coefficient of restitution; base of natural logarithms
E Total mechanical energy; voltage
f Frequency; function
F Force; friction force
g Acceleration of gravity
G Center of gravity; mass center; constant of gravitation
h Height; sag of cable; angular momentum per unit mass
$\mathbf{H}_O$ Angular momentum about point O
i Current
$I, I_x, \ldots$ Moment of inertia
$\bar{I}$ Centroidal moment of inertia
J Polar moment of inertia
k Spring constant
k_x, k_y, k_O Radius of gyration
$\bar{k}$ Centroidal radius of gyration
l Length
L Length; span; inductance
m Mass
M Couple
$\mathbf{M}_O$ Moment about point O
M Moment; mass of earth
n Normal direction
N Normal component of reaction
O Origin of coordinates
p Pressure; circular frequency
P Force; vector
$P_{xy}, \ldots$ Product of inertia
q Electric charge
Q Force; vector
r Position vector
r Radius; distance; polar coordinate
R Resultant force; resultant vector; reaction
R Radius of earth; resistance
s Position vector
s Distance; length of arc; length of cable
S Force; vector
t Time; thickness; tangential direction
T Force
T Tension; kinetic energy
u Velocity
u Rectangular coordinate
U Work
v, v Velocity
v Speed; rectangular coordinate
$\bar{\mathbf{v}}$, $\bar{v}$ Velocity of mass center
$\mathbf{v}_{B/A}$ Relative velocity of B with respect to A
V Shearing force
V Volume; potential energy; shear
w Load per unit length
W, W Weight; load
x, y, z Rectangular coordinates; distances
$\dot{x}, \dot{y}, \dot{z}$ Time derivatives of coordinates x, y, z

$\bar{x}, \bar{y}, \bar{z}$ Rectangular coordinates of centroid, center of gravity, or mass center
$\boldsymbol{\alpha}, \alpha$ Angular acceleration
α, β, γ Angles
γ Specific weight
δ Elongation
$\delta \mathbf{s}$ Virtual displacement
δU Virtual work
ε Eccentricity of conic section or of orbit
η Efficiency
θ Angular coordinate; angle; polar coordinate
μ Coefficient of friction
ρ Density; radius of curvature
τ Period; periodic time
ϕ Angle of friction; phase angle; angle
φ Phase difference
$\boldsymbol{\omega}, \omega$ Angular velocity
ω Circular frequency of forced vibration
$\boldsymbol{\Omega}$ Rate of precession

Chapter
1

Introduction

1.1. What Is Mechanics? Mechanics may be defined as that science which describes and predicts the conditions of rest or motion of bodies under the action of forces. It is divided into three parts: mechanics of *rigid bodies,* mechanics of *deformable bodies,* and mechanics of *fluids.*

The mechanics of rigid bodies is subdivided into *statics* and *dynamics,* the former dealing with bodies at rest, the latter with bodies in motion. In this part of the study of mechanics, bodies are assumed to be perfectly rigid. Actual structures and machines, however, are never absolutely rigid and deform under the loads to which they are subjected. But these deformations are usually small and do not appreciably affect the conditions of equilibrium or motion of the structure under consideration. They are important, though, as far as the resistance of the structure to failure is concerned and are studied in mechanics of materials, which is a part of the mechanics of deformable bodies. The third division of mechanics, the mechanics of fluids, is subdivided into the study of *incompressible fluids* and of *compressible fluids.* An important subdivision of the study of incompressible fluids is *hydraulics,* which deals with problems involving liquids.

Mechanics is a physical science, since it deals with the study of physical phenomena. However, some associate mechanics with mathematics, while many consider it as an engineering subject. Both these views are justified in part. Mechanics is the foundation of most engineering sciences and is an indispensable prerequisite to their study. However, it does not have the *empiricism* found in many engineering sciences; by its rigor and the emphasis it places on deductive reasoning it resembles mathematics. But, again, it is not an *abstract* or even a *pure* science; mechanics is an *applied* science. The purpose of mechanics is to explain and predict physical phenomena and thus to lay the foundations for engineering applications.

1.2. Fundamental Concepts and Principles. Although the study of mechanics goes back to the time of Aristotle (384–322 B.C.) and Archimedes (287–212 B.C.), one has to wait until Newton (1642–1727) to find a satisfactory formulation of its fundamental principles. These principles were later expressed in a modified form by D'Alembert, Lagrange, and Hamilton. Their validity remained unchallenged, however, until Einstein formulated his *theory of relativity* (1905). While its limitations have now been recognized, *newtonian mechanics* still remains the basis of today's engineering sciences.

The basic concepts used in mechanics are *space, time, mass,* and *force.* These concepts cannot be truly defined; they should be accepted on the basis of our intuition and experience and used as a mental frame of reference for our study of mechanics.

The concept of *space* is associated with the notion of the position of a point P. The position of P may be defined by three lengths measured from a certain reference point, or *origin,* in three given directions. These lengths are known as the *coordinates* of P.

To define an event, it is not sufficient to indicate its position in space. The *time* of the event should also be given.

The concept of *mass* is used to characterize and compare bodies on the basis of certain fundamental mechanical experiments. Two bodies of the same mass, for example, will be attracted by the earth in the same manner; they will also offer the same resistance to a change in translational motion.

A *force* represents the action of one body on another. It may be exerted by actual contact or at a distance, as in the case of gravitational forces and magnetic forces. A force is characterized by its *point of application,* its *magnitude,* and its *direction;* a force is represented by a *vector* (Sec. 2.2).

In newtonian mechanics, space, time, and mass are absolute concepts, independent of each other. (This is not true in *relativistic mechanics,* where the time of an event depends upon its position, and where the mass of a body varies with its velocity.) On the other hand, the concept of force is not independent of the other three. Indeed, one of the fundamental principles of newtonian mechanics listed below indicates that the resultant force acting on a body is related to the mass of the body and to the manner in which its velocity varies with time.

We shall study the conditions of rest or motion of particles and rigid bodies in terms of the four basic concepts we have introduced. By *particle* we mean a very small amount of matter which may be assumed to occupy a single point in space. A *rigid body* is a combination of a large number of particles occupying fixed positions with respect to each other. The study of the mechanics of particles is obviously a prerequisite to that of rigid bodies. Besides, the results obtained for a particle may be used directly in a large number of problems dealing with the conditions of rest or motion of actual bodies.

The study of elementary mechanics rests on six fundamental principles based on experimental evidence:

The Parallelogram Law for the Addition of Forces. This states that two forces acting on a particle may be replaced by a single force, called their *resultant*, obtained by drawing the diagonal of the parallelogram which has sides equal to the given forces (Sec. 2.1).

The Principle of Transmissibility. This states that the conditions of equilibrium or of motion of a rigid body will remain unchanged if a force acting at a given point of the rigid body is replaced by a force of the same magnitude and same direction, but acting at a different point, provided that the two forces have the same line of action (Sec. 3.2).

Newton's Three Fundamental Laws. Formulated by Sir Isaac Newton in the latter part of the seventeenth century, these laws may be stated as follows:

FIRST LAW. If the resultant force acting on a particle is zero, the particle will remain at rest (if originally at rest) or will move with constant speed in a straight line (if originally in motion) (Sec. 2.9).

SECOND LAW. If the resultant force acting on a particle is not zero, the particle will have an acceleration proportional to the magnitude of the resultant and in the direction of this resultant force.

As we shall see in Sec. 12.1, this law may be stated as

$$\mathbf{F} = m\mathbf{a} \tag{1.1}$$

where $\mathbf{F}$, m, and $\mathbf{a}$ represent, respectively, the resultant force acting on the particle, the mass of the particle, and the acceleration of the particle, expressed in a consistent system of units.

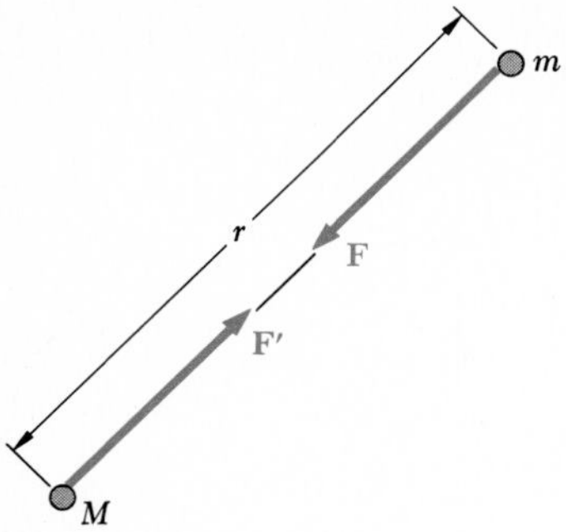

Fig. 1.1

THIRD LAW. The forces of action and reaction between bodies in contact have the same magnitude, same line of action, and opposite sense (Sec. 6.1).

Newton's Law of Gravitation. This states that two particles of mass M and m are mutually attracted with equal and opposite forces $\mathbf{F}$ and $\mathbf{F}'$ (Fig. 1.1) of magnitude F given by the formula

$$F = G\frac{Mm}{r^2} \tag{1.2}$$

where r = distance between the two particles
G = universal constant called the *constant of gravitation*

Newton's law of gravitation introduces the idea of an action exerted at a distance and extends the range of application of Newton's third law: the action $\mathbf{F}$ and the reaction $\mathbf{F}'$ in Fig. 1.1 are equal and opposite, and they have the same line of action.

A particular case of great importance is that of the attraction of the earth on a particle located on its surface. The force $\mathbf{F}$ exerted by the earth on the particle is then defined as the *weight* $\mathbf{W}$ of the particle. Taking M equal to the mass of the earth, m equal to the mass of the particle, and r equal to the radius R of the earth, and introducing the constant

$$g = \frac{GM}{R^2} \tag{1.3}$$

the magnitude W of the weight of a particle of mass m may be expressed as†

$$W = mg \tag{1.4}$$

The value of R in formula (1.3) depends upon the elevation of the point considered; it also depends upon its latitude, since the earth is not truly spherical. The value of g therefore varies with the position of the point considered. As long as the point actually remains on the surface of the earth, it is sufficiently accurate in most engineering computations to assume that g equals 9.81 m/s^2 or 32.2 ft/s^2.

The principles we have just listed will be introduced in the course of our study of mechanics as they are needed. The study of the statics of particles, carried out in Chap. 2, will be based on the parallelogram law of addition and on Newton's first law alone. The principle of transmissibility will be introduced in Chap. 3 as we begin the study of the statics of rigid bodies, and

† A more accurate definition of the weight $\mathbf{W}$ should take into account the rotation of the earth.

Newton's third law in Chap. 6 as we analyze the forces exerted on each other by the various members forming a structure. It should be noted that the four aforementioned principles underlie the entire study of the statics of particles, rigid bodies, and systems of rigid bodies. In the study of dynamics, Newton's second law and Newton's law of gravitation will be introduced. It will then be shown that Newton's first law is a particular case of Newton's second law (Sec. 12.1). It can also be shown that the principle of transmissibility could be derived from the other principles and thus eliminated.† However, Newton's first and third laws, the parallelogram law of addition, and the principle of transmissibility will provide us with the necessary and sufficient foundation for the study of the entire field of statics.

As noted earlier, the six fundamental principles listed above are based on experimental evidence. Except for Newton's first law and the principle of transmissibility, they are independent principles which cannot be derived mathematically from each other or from any other elementary physical principle. On these principles rests most of the intricate structure of newtonian mechanics. For more than two centuries a tremendous number of problems dealing with the conditions of rest and motion of rigid bodies, deformable bodies, and fluids have been solved by applying these fundamental principles. Many of the solutions obtained could be checked experimentally, thus providing a further verification of the principles from which they were derived. It is only recently that Newton's mechanics was found at fault, in the study of the motion of atoms and in the study of the motion of certain planets, where it must be supplemented by the theory of relativity. But on the human or engineering scale, where velocities are small compared with the velocity of light, Newton's mechanics has yet to be disproved.

1.3. Systems of Units. With the four fundamental concepts introduced in the preceding section are associated the so-called *kinetic units*, i.e., the units of *length, time, mass,* and *force*. These units cannot be chosen independently if Eq. (1.1) is to be satisfied. Three of the units may be defined arbitrarily; they are then referred to as *base units*. The fourth unit, however, must be chosen in accordance with Eq. (1.1) and is referred to as a *derived unit*. Kinetic units selected in that way are said to form a *consistent system of units*.

International System of Units (SI Units‡). In this system, which will be in universal use after the United States has com-

†Cf. F. P. Beer and E. R. Johnston, "Vector Mechanics for Engineers," sec. 16.5, McGraw-Hill Book Company, 1972.

‡SI stands for *Système International d'Unités* (French).

pleted its current conversion, the base units are the units of length, mass, and time, and are called, respectively, the *meter* (m), the *kilogram* (kg), and the *second* (s). All three are arbitrarily defined. The second, which is supposed to represent the 1/86,400 part of the mean solar day, is actually defined as the duration of 9,192,631,770 cycles of the radiation associated with a specified transition of the cesium atom. The meter, originally intended to represent one-ten-millionth of the distance from the equator to the pole, is now defined as 1,650,763.73 wavelengths of the orange-red line of krypton 86. The kilogram, which is approximately equal to the mass of 0.001 m^3 of water, is actually defined as the mass of a platinum standard kept at the International Bureau of Weights and Measures at Sèvres, near Paris, France. The unit of force is a derived unit. It is called the *newton* (N) and is defined as the force which gives an acceleration of 1 m/s^2 to a mass of 1 kg (Fig. 1.2). From Eq. (1.1) we write

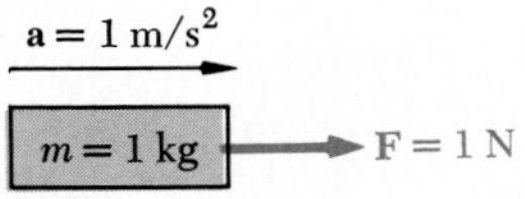

Fig. 1.2

$$1\text{ N} = (1\text{ kg})(1\text{ m/s}^2) = 1\text{ kg}\cdot\text{m/s}^2 \tag{1.5}$$

The SI units are said to form an *absolute* system of units. This means that the three base units chosen are independent of the location where measurements are made. The meter, the kilogram, and the second may be used anywhere on the earth; they may even be used on another planet. They will always have the same significance.

Like any other force, the *weight* of a body should be expressed in newtons. From Eq. (1.4) it follows that the weight of a body of mass 1 kg (Fig. 1.3) is

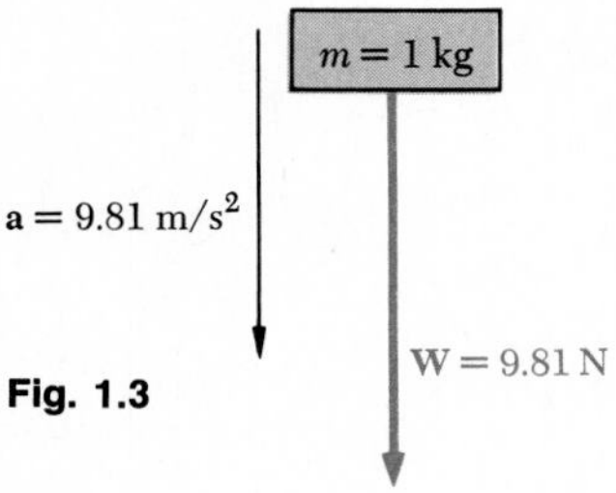

Fig. 1.3

$$\begin{aligned} W &= mg \\ &= (1\text{ kg})(9.81\text{ m/s}^2) \\ &= 9.81\text{ N} \end{aligned}$$

Multiples and submultiples of the fundamental SI units may be obtained through the use of the prefixes defined in Table 1.1. The multiples and submultiples of the units of length, mass, and force most frequently used in engineering are, respectively, the *kilometer* (km) and the *millimeter* (mm); the *megagram*† (Mg) and the *gram* (g); and the *kilonewton* (kN). According to Table 1.1, we have

$$\begin{aligned} 1\text{ km} &= 1000\text{ m} \qquad & 1\text{ mm} &= 0.001\text{ m} \\ 1\text{ Mg} &= 1000\text{ kg} \qquad & 1\text{ g} &= 0.001\text{ kg} \end{aligned}$$
$$1\text{ kN} = 1000\text{ N}$$

The conversion of these units into meters, kilograms, and newtons, respectively, can be effected by simply moving the decimal point three places to the right or to the left. For example, to

† Also known as a *metric ton*.

Table 1.1 SI Prefixes

Multiplication Factor	Prefix†	Symbol
$1\,000\,000\,000\,000 = 10^{12}$	tera	T
$1\,000\,000\,000 = 10^{9}$	giga	G
$1\,000\,000 = 10^{6}$	mega	M
$1\,000 = 10^{3}$	kilo	k
$100 = 10^{2}$	hecto‡	h
$10 = 10^{1}$	deka‡	da
$0.1 = 10^{-1}$	deci‡	d
$0.01 = 10^{-2}$	centi‡	c
$0.001 = 10^{-3}$	milli	m
$0.000\,001 = 10^{-6}$	micro	μ
$0.000\,000\,001 = 10^{-9}$	nano	n
$0.000\,000\,000\,001 = 10^{-12}$	pico	p
$0.000\,000\,000\,000\,001 = 10^{-15}$	femto	f
$0.000\,000\,000\,000\,000\,001 = 10^{-18}$	atto	a

† The first syllable of every prefix is accented so that the prefix will retain its identity. Thus, the preferred pronunciation of kilometer places the accent on the first syllable, not the second.

‡ The use of these prefixes should be avoided, except for the measurement of areas and volumes and for the nontechnical use of centimeter, as for body and clothing measurements.

convert 3.82 km into meters, one moves the decimal point three places to the right:

$$3.82 \text{ km} = 3820 \text{ m}$$

Similarly, 47.2 mm is converted into meters by moving the decimal point three places to the left:

$$47.2 \text{ mm} = 0.0472 \text{ m}$$

Using the scientific notation, one may also write

$$3.82 \text{ km} = 3.82 \times 10^{3} \text{ m}$$
$$47.2 \text{ mm} = 47.2 \times 10^{-3} \text{ m}$$

The multiples of the unit of time are the *minute* (min) and the *hour* (h). Since

$$1 \text{ min} = 60 \text{ s} \qquad \text{and} \qquad 1 \text{ h} = 60 \text{ min} = 3600 \text{ s}$$

these multiples cannot be converted as readily as the others.

Units of Area and Volume. The unit of area is the *square meter* (m^2), which represents the area of a square of side 1 m; the unit of volume is the *cubic meter* (m^3), equal to the volume of a cube of side 1 m. In order to avoid exceedingly small or large numerical values in the computation of areas and volumes, one uses systems of subunits obtained by respectively squaring and cubing, not only the millimeter, but also two intermediate submultiples of the meter, namely, the *decimeter* (dm) and the *centimeter* (cm). Since, by definition,

$$1 \text{ dm} = 0.1 \text{ m} = 10^{-1} \text{ m}$$
$$1 \text{ cm} = 0.01 \text{ m} = 10^{-2} \text{ m}$$
$$1 \text{ mm} = 0.001 \text{ m} = 10^{-3} \text{ m}$$

the submultiples of the unit of area are

$$1 \text{ dm}^2 = (1 \text{ dm})^2 = (10^{-1} \text{ m})^2 = 10^{-2} \text{ m}^2$$
$$1 \text{ cm}^2 = (1 \text{ cm})^2 = (10^{-2} \text{ m})^2 = 10^{-4} \text{ m}^2$$
$$1 \text{ mm}^2 = (1 \text{ mm})^2 = (10^{-3} \text{ m})^2 = 10^{-6} \text{ m}^2$$

and the submultiples of the unit of volume are

$$1 \text{ dm}^3 = (1 \text{ dm})^3 = (10^{-1} \text{ m})^3 = 10^{-3} \text{ m}^3$$
$$1 \text{ cm}^3 = (1 \text{ cm})^3 = (10^{-2} \text{ m})^3 = 10^{-6} \text{ m}^3$$
$$1 \text{ mm}^3 = (1 \text{ mm})^3 = (10^{-3} \text{ m})^3 = 10^{-9} \text{ m}^3$$

It should be noted that when the volume of a liquid is being measured, the cubic decimeter (dm^3) is usually referred to as a *liter* (l).

Other derived SI units used to measure the moment of a force, the work of a force, etc., will be introduced in later chapters as they are needed. These units are shown in Table 1.2.

U.S. Customary Units. Most practicing American engineers still commonly use a system in which the base units are the units of length, force, and time. These units are, respectively, the *foot* (ft), the *pound* (lb), and the *second* (s). The second is the same as the corresponding SI unit. The foot is defined as 0.3048 m. The pound is defined as the *weight* of a platinum standard, called the *standard pound* and kept at the National Bureau of Standards in Washington, the mass of which is 0.453 592 43 kg. Since the weight of a body depends upon the gravitational attraction of the earth, which varies with location, it is specified that the standard pound should be placed at sea level and at the latitude of 45° to properly define a force of 1 lb. Clearly the U.S. customary units do not form an absolute system of units. Because of their dependence upon the gravitational attraction of the earth, they form a *gravitational* system of units.

While the standard pound also serves as the unit of mass in commercial transactions in the United States, it cannot be so used in engineering computations since such a unit would not be consistent with the base units defined in the preceding paragraph. Indeed, when acted upon by a force of 1 lb, that is, when subjected to its own weight, the standard pound receives the acceleration of gravity, $g = 32.2 \text{ ft/s}^2$ (Fig. 1.4), not the unit

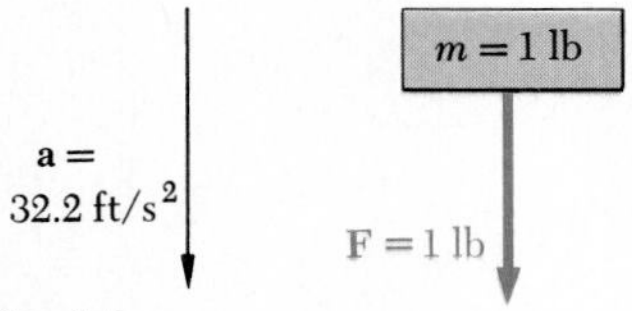

Fig. 1.4

Table 1.2 Principal SI Units Used in Mechanics

Quantity	Unit	Symbol	Formula
Acceleration	Meter per second squared	. . .	m/s^2
Angle	Radian	rad	†
Angular acceleration	Radian per second squared	. . .	rad/s^2
Angular velocity	Radian per second	. . .	rad/s
Area	Square meter	. . .	m^2
Density	Kilogram per cubic meter	. . .	kg/m^3
Energy	Joule	J	$N \cdot m$
Force	Newton	N	$kg \cdot m/s^2$
Frequency	Hertz	Hz	s^{-1}
Length	Meter	m	‡
Mass	Kilogram	kg	‡
Moment of a force	Newton-meter	. . .	$N \cdot m$
Power	Watt	W	J/s
Pressure	Pascal	Pa	N/m^2
Stress	Pascal	Pa	N/m^2
Time	Second	s	‡
Velocity	Meter per second	. . .	m/s
Volume, solids	Cubic meter	. . .	m^3
Liquids	Liter	l	10^{-3} m^3
Work	Joule	J	$N \cdot m$

†Supplementary unit (1 revolution = 2π rad = 360°).
‡Base unit.

acceleration required by Eq. (1.1). The unit of mass consistent with the foot, the pound, and the second is the mass which receives an acceleration of 1 ft/s² when a force of 1 lb is applied to it (Fig. 1.5). This unit, sometimes called a *slug*, can be derived from the equation $F = ma$ after substituting 1 lb and 1 ft/s² for F and a, respectively. We write

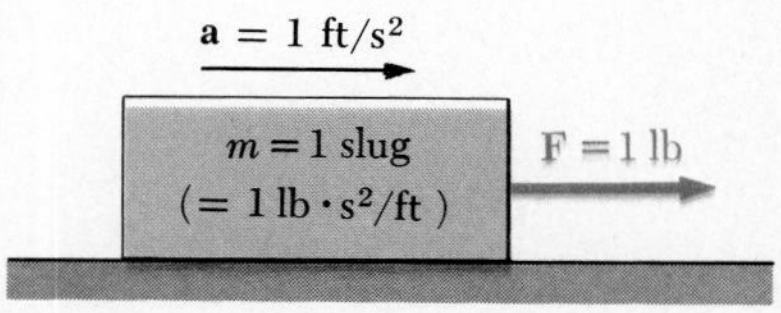

Fig. 1.5

$$F = ma \qquad 1 \text{ lb} = (1 \text{ slug})(1 \text{ ft/s}^2)$$

and obtain

$$1 \text{ slug} = \frac{1 \text{ lb}}{1 \text{ ft/s}^2} = 1 \text{ lb} \cdot \text{s}^2/\text{ft} \tag{1.6}$$

Comparing Figs. 1.4 and 1.5, we conclude that the slug is a mass 32.2 times larger than the mass of the standard pound.

The fact that in the U.S. customary system of units bodies are characterized by their weight in pounds, rather than by their mass in slugs, will be a convenience in the study of statics, where we shall constantly deal with weights and other forces and only seldom with masses. However, in the study of dynamics, where forces, masses, and accelerations are involved, we shall have to express the mass m in slugs of a body of which the weight W has been given in pounds. Recalling Eq. (1.4), we shall write

$$m = \frac{W}{g} \tag{1.7}$$

where g is the acceleration of gravity ($g = 32.2 \text{ ft/s}^2$).

Other U.S. customary units frequently encountered in engineering problems are the *mile* (mi), equal to 5280 ft; the *inch* (in.), equal to $\frac{1}{12}$ ft; and the *kilopound* (kip), equal to a force of 1000 lb. The *ton* is often used to represent a mass of 2000 lb but, like the pound, must be converted into slugs in engineering computations.

The conversion into feet, pounds, and seconds of quantities expressed in other U.S. customary units is generally more involved and requires greater attention than the corresponding operation in SI units. If, for example, the magnitude of a velocity is given as $v = 30$ mi/h, we shall proceed as follows to convert it to ft/s. First we write

$$v = 30\frac{\text{mi}}{\text{h}}$$

Since we want to get rid of the unit miles and introduce instead the unit feet, we should multiply the right-hand member of the equation by an expression containing miles in the denominator and feet in the numerator. But, since we do not want to change the value of the right-hand member, the expression used should have a value equal to unity. The quotient (5280 ft)/(1 mi) is such an expression. Operating in a similar way to transform the unit hour into seconds, we write

$$v = \left(30\frac{\text{mi}}{\text{h}}\right)\left(\frac{5280\text{ ft}}{1\text{ mi}}\right)\left(\frac{1\text{ h}}{3600\text{ s}}\right)$$

Carrying out the numerical computations and canceling out units which appear both in the numerator and the denominator, we obtain

$$v = 44\frac{\text{ft}}{\text{s}} = 44 \text{ ft/s}$$

1.4. Conversion from One System of Units to Another. There are many instances when an engineer wishes to convert into SI units a numerical result obtained in U.S. customary units or vice versa. Because the unit of time is the same in both systems, only two kinetic base units need be converted.

Thus, since all other kinetic units can be derived from these base units, only two conversion factors need be memorized.†

Units of Length. By definition the U.S. customary unit of length is

$$1 \text{ ft} = 0.3048 \text{ m} \tag{1.8}$$

It follows that

$$1 \text{ mi} = 5280 \text{ ft} = 5280(0.3048 \text{ m}) = 1609 \text{ m}$$

or

$$1 \text{ mi} = 1.609 \text{ km} \tag{1.9}$$

Also

$$1 \text{ in.} = \tfrac{1}{12} \text{ ft} = \tfrac{1}{12}(0.3048 \text{ m}) = 0.0254 \text{ m}$$

or

$$1 \text{ in.} = 25.4 \text{ mm} \tag{1.10}$$

Units of Force. Recalling that the U.S. customary unit of force (pound) is defined as the weight of the standard pound (of mass 0.4536 kg) at sea level and at the latitude of 45° (where $g = 9.807 \text{ m/s}^2$), and using Eq. (1.4), we write

$$W = mg$$
$$1 \text{ lb} = (0.4536 \text{ kg})(9.807 \text{ m/s}^2) = 4.448 \text{ kg} \cdot \text{m/s}^2$$

or, recalling (1.5),

$$1 \text{ lb} = 4.448 \text{ N} \tag{1.11}$$

Units of Mass. The U.S. customary unit of mass (slug) is a derived unit. Thus, using (1.6), (1.8), and (1.11), we write

$$1 \text{ slug} = 1 \text{ lb} \cdot \text{s}^2/\text{ft} = \frac{1 \text{ lb}}{1 \text{ ft/s}^2} = \frac{4.448 \text{ N}}{0.3048 \text{ m/s}^2} = 14.59 \text{ N} \cdot \text{s}^2/\text{m}$$

and, recalling (1.5),

$$1 \text{ slug} = 1 \text{ lb} \cdot \text{s}^2/\text{ft} = 14.59 \text{ kg} \tag{1.12}$$

† While all conversion factors are given in this section with four significant figures, it should be noted that such an accuracy is seldom needed in engineering computations (Sec. 1.6). It is suggested that students using a slide rule round off the given values as follows:

$1 \text{ ft} = 0.305 \text{ m}$ $\quad 1 \text{ lb} = 4.45 \text{ N}$ $\quad 1 \text{ slug} = 14.6 \text{ kg}$

Although it cannot be used as a consistent unit of mass, we recall that the mass of the standard pound is, by definition,

$$1 \text{ pound mass} = 0.4536 \text{ kg} \tag{1.13}$$

This constant may be used to determine the *mass* in SI units (kilograms) of a body which has been characterized by its *weight* in U.S. customary units (pounds).

To convert into SI units a derived U.S. customary unit, one simply multiplies or divides the appropriate conversion factors. For example, to convert into SI units the moment of a force which was found to be $M = 46 \text{ lb} \cdot \text{in.}$, we use formulas (1.10) and (1.11) and write

$$\begin{aligned} M = 46 \text{ lb} \cdot \text{in.} &= 46(4.45 \text{ N})(25.4 \text{ mm}) \\ &= 5200 \text{ N} \cdot \text{mm} = 5.20 \text{ N} \cdot \text{m} \end{aligned}$$

The conversion factors given in this section may also be used to convert into U.S. customary units a numerical result obtained in SI units. For example, if the moment of a force was found to be $M = 40 \text{ N} \cdot \text{m}$, we write, following the procedure used in the last paragraph of Sec. 1.3,

$$M = 40 \text{ N} \cdot \text{m} = (40 \text{ N} \cdot \text{m})\left(\frac{1 \text{ lb}}{4.45 \text{ N}}\right)\left(\frac{1 \text{ ft}}{0.305 \text{ m}}\right)$$

Carrying out the numerical computations and canceling out units which appear both in the numerator and the denominator, we obtain

$$M = 29.5 \text{ lb} \cdot \text{ft}$$

The U.S. customary units most frequently used in mechanics are listed in Table 1.3 with their SI equivalents.

1.5. Method of Problem Solution. The student should approach a problem in mechanics as he would approach an actual engineering situation. By drawing on his own experience and on his intuition, he will find it easier to understand and formulate the problem. Once the problem has been clearly stated, however, there is no place in its solution for the student's particular fancy. *The solution must be based on the six fundamental principles stated in Sec. 1.2 or on theorems derived from them.* Every step taken must be justified on that basis. Strict rules must be followed, which lead to the solution in an almost automatic fashion, leaving no room for the student's intuition or "feeling." After an answer has been obtained, it should be checked. Here again, the student may call upon his common sense and personal experience. If not completely satisfied with the result obtained, he should carefully check his formulation of the problem, the validity of the methods used for its solution, and the accuracy of his computations.

Table 1.3 U.S. Customary Units and Their SI Equivalents

Quantity	U.S. Customary Unit	SI Equivalent
Acceleration	ft/s^2	$0.3048 m/s^2$
	$in./s^2$	$0.0254 m/s^2$
Area	ft^2	$0.0929 m^2$
	in^2	$645.2 mm^2$
Energy	ft · lb	1.356 J
Force	kip	4.448 kN
	lb	4.448 N
	oz	0.2780 N
Length	ft	0.3048 m
	in.	25.40 mm
	mi	1.609 km
Mass	oz mass	28.35 g
	lb mass	0.4536 kg
	slug	14.59 kg
	ton	907.2 kg
Moment of a force	lb · ft	1.356 N · m
	lb · in.	0.1130 N · m
Moment of inertia		
Of an area	in^4	$0.4162 \times 10^6 mm^4$
Of a mass	$lb \cdot ft \cdot s^2$	$1.356 kg \cdot m^2$
Power	ft · lb/s	1.356 W
	hp	745.7 W
Pressure or stress	lb/ft^2	47.88 Pa
	lb/in^2 (psi)	6.895 kPa
Velocity	ft/s	0.3048 m/s
	in./s	0.0254 m/s
	mi/h (mph)	0.4470 m/s
	mi/h (mph)	1.609 km/h
Volume, solids	ft^3	$0.02832 m^3$
	in^3	$16.39 cm^3$
Liquids	gal	3.785 l
	qt	0.9464 l
Work	ft · lb	1.356 J

The *statement* of a problem should be clear and precise. It should contain the given data and indicate what information is required. A neat drawing showing all quantities involved should be included. Separate diagrams should be drawn for all bodies involved, indicating clearly the forces acting on each body. These diagrams are known as *free-body diagrams* and are described in detail in Secs. 2.10 and 3.13.

The *fundamental principles* of mechanics listed in Sec. 1.2 *will be used to write equations* expressing the conditions of rest or motion of the bodies considered. Each equation should be clearly related to one of the free-body diagrams. The student will then proceed to solve the problem, observing strictly the usual rules of algebra and recording neatly the various steps taken.

After the answer has been obtained, it should be *carefully checked.* Mistakes in *reasoning* may often be detected by checking the units. For example, to determine the moment of a force of 50 N about a point 0.60 m from its line of action, we would have written (Sec. 3.4)

$$M = Fd = (50\text{ N})(0.60\text{ m}) = 30\text{ N}\cdot\text{m}$$

The unit N · m obtained by multiplying newtons by meters is the correct unit for the moment of a force; if another unit had been obtained, we would have known that some mistake had been made.

Errors in *computation* will usually be found by substituting the numerical values obtained into an equation which has not yet been used and verifying that the equation is satisfied. The importance of correct computations in engineering cannot be overemphasized.

1.6. Numerical Accuracy. The accuracy of the solution of a problem depends upon two items: (1) the accuracy of the given data; (2) the accuracy of the computations performed.

The solution cannot be more accurate than the less accurate of these two items. For example, if the loading of a bridge is known to be 75,000 lb with a possible error of 100 lb either way, the relative error which measures the degree of accuracy of the data is

$$\frac{100\text{ lb}}{75{,}000\text{ lb}} = 0.0013 = 0.13\text{ percent}$$

It would then be meaningless, in computing the reaction at one of the bridge supports, to record it as 14,322 lb. The accuracy of the solution cannot be greater than 0.13 percent, no matter how accurate the computations are, and the possible error in the answer may be as large as $(0.13/100)(14{,}322\text{ lb}) \approx 20$ lb. The solution should be properly recorded as $14{,}320 \pm 20$ lb.

In engineering problems, the data are seldom known with an accuracy greater than 0.2 percent. It is therefore unnecessary to carry out computations with a greater accuracy. In almost all cases, the slide rule will provide the desired accuracy. On looking at the main scales of a 10-in. slide rule, it will be observed that the number 502 may be easily interpolated between the graduations corresponding, respectively, to 500 and 505. It might possibly be mistaken for 501 or 503, but no larger error in reading or in adjusting the rule will be made if a minimum of care is observed. The error involved is thus at most 1 unit out of 500; the relative error is 0.2 percent. For other positions on the rule, the absolute error will be different (for example, 2 units out of 1000); owing to the logarithmic nature of the scales, however, the relative error will remain 0.2 percent.

It is sometimes said that the accuracy obtained on the slide rule is *three significant figures*. This is correct as far as the central portion of the scale is concerned, where numbers such as 298, 299, 300, 301 may be easily read. The third significant figure, however, becomes doubtful toward the right end of the scale, where it is difficult to tell 997 from 998. On the other hand, one should be able, toward the left end of the scale, to estimate the fourth significant figure within a couple of units. The relative error between 996 and 998 is the same as between 1002 and 1004 (0.2 percent in both cases). We should therefore define the accuracy of a computation by the possible relative error involved, rather than by the number of significant figures obtained, and we shall attempt to obtain all answers with an accuracy of about 0.2 percent. A practical rule consists in using four figures to record readings taken on a 10-in. slide rule between 1 and 2 and three figures to record readings between 2 and 10. Unless otherwise indicated, the data given in a problem should be assumed known with a comparable degree of accuracy. A force of 40 lb, for example, should actually be read 40.0 lb, and a force of 15 lb should be read 15.00 lb.

Pocket electronic calculators are widely used by practicing engineers and engineering students. The speed and accuracy of these calculators facilitates the numerical computations in the solution of many problems. However, the student should not record more significant figures than can be justified, merely because they are easily obtained. As noted above, an accuracy greater than 0.2 percent is seldom necessary or meaningful in the solution of practical engineering problems.

Chapter

2

Statics of Particles

FORCES IN A PLANE

2.1. Force on a Particle. Resultant of Two Forces. A force represents the action of one body on another. It is characterized by its *point of application*, its *magnitude*, and its *direction*. In this chapter we shall study the effect of forces on particles. The use of the word "particles" does not imply that we shall restrict our study to that of small corpuscles. It means that the size and shape of the bodies under consideration will not affect the solution of the problems treated in this chapter and that all the forces acting on a given body will be assumed to have the same point of application. Each force will thus be completely defined by its magnitude and direction.

The magnitude of a force is characterized by a certain number of units. As indicated in Chap. 1, the SI units used by engineers to measure the magnitude of a force are the newton (N) and its multiple the kilonewton (kN), equal to 1000 N, while the U.S. customary units used for the same purpose are the pound (lb) and its multiple the kilopound (kip, or k), equal to 1000 lb. The direction of a force is defined by the *line of action* and the *sense* of the force. The line of action is the infinite straight line along which the force acts; it is characterized by the angle it forms with some fixed axis (Fig. 2.1). The force itself is represented by a segment of that line; through the use of an appropriate scale, the length of this segment may be chosen to represent

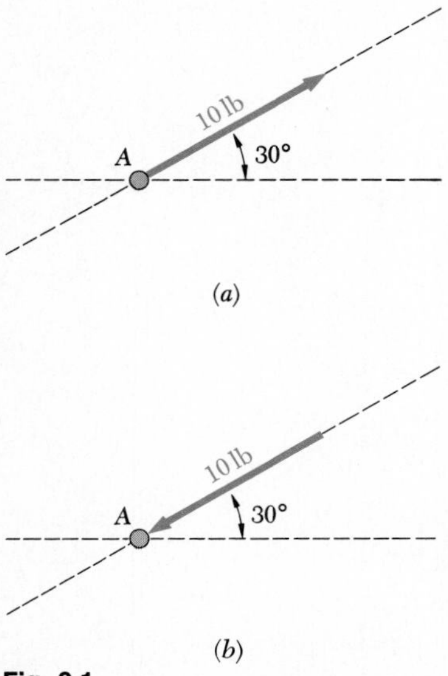

Fig. 2.1

the magnitude of the force. Finally, the sense of the force should be indicated by an arrowhead. It is important in defining a force to indicate its sense. Two forces, such as those shown in Fig. 2.1*a* and *b*, having the same magnitude and the same line of action but different sense, will have directly opposite effects on a particle.

Experimental evidence shows that two forces **P** and **Q** acting on a particle *A* (Fig. 2.2*a*) may be replaced by a single force **R** which has the same effect on the particle (Fig. 2.2*c*). This force is called the *resultant* of the forces **P** and **Q** and may be obtained, as shown in Fig. 2.2*b*, by constructing a parallelogram, using **P** and **Q** as two sides of the parallelogram. *The diagonal that passes through A represents the resultant.* This is known as the *parallelogram law* for the addition of two forces. This law is based on experimental evidence; it cannot be proved or derived mathematically.

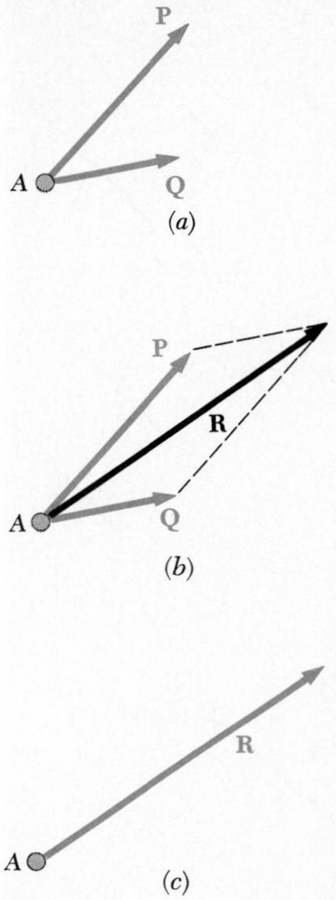

Fig. 2.2

2.2. Vectors. It appears from the above that forces do not obey the rules of addition defined in ordinary arithmetic or algebra. For example, two forces acting at a right angle to each other, one of 4 lb and the other of 3 lb, add up to a force of 5 lb, *not* to a force of 7 lb. Forces are not the only expressions which follow the parallelogram law of addition. As we shall see later, *displacements, velocities, accelerations, momenta* are other examples of physical quantities possessing magnitude and direction and which are added according to the parallelogram law. All these quantities may be represented mathematically by *vectors,* while those physical quantities which do not have direction, such as *volume, mass,* or *energy,* are represented by ordinary numbers or *scalars.*

Vectors are defined as *mathematical expressions possessing magnitude and direction, which add according to the parallelogram law.* Vectors are represented by arrows in the illustrations and will be distinguished from scalar quantities in this text through the use of boldface type (**P**). In longhand writing, a vector may be characterized by drawing a short arrow above the letter used to represent it ($\vec{P}$), or by underlining the letter ($\underline{P}$). The last method is gaining wider acceptance since it can also be used on a typewriter. The magnitude of a vector defines the length of the arrow used to represent the vector. In this text, italic type will be used to denote the magnitude of a vector. Thus, the magnitude of the vector **P** will be referred to as *P.*

A vector used to represent a force acting on a given particle has a well-defined point of application, namely, the particle itself. Such a vector is said to be a *fixed,* or *bound,* vector and cannot be moved without modifying the conditions of the problem. Other physical quantities, however, such as couples (see Chap. 4), are represented by vectors which may be freely moved in space; these vectors are called *free* vectors. Still other physical quantities, such as forces acting on a rigid body (see Chap. 3), are represented by vectors which may be moved, or slid, along their line of action; they are known as *sliding* vectors.†

Two vectors which have the same magnitude and the same direction are said to be *equal*, whether or not they also have the same point of application (Fig. 2.4); equal vectors may be denoted by the same letter. Two vectors which have the same magnitude, parallel lines of action, and opposite sense are said to be *equal and opposite* (Fig. 2.5).

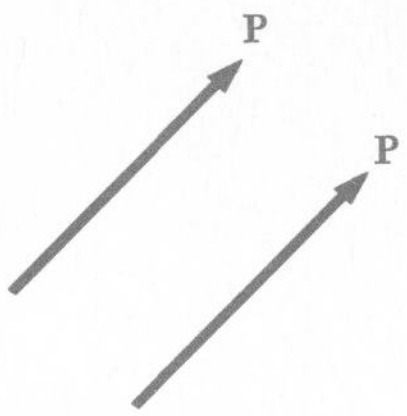

Fig. 2.4

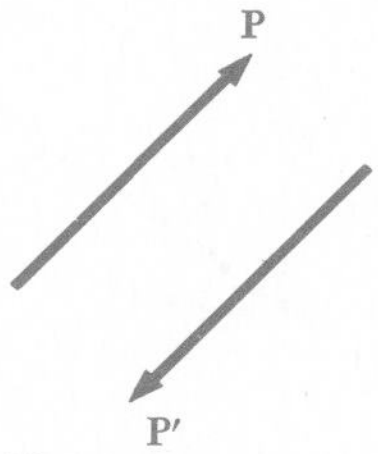

Fig. 2.5

2.3. Addition of Vectors. We saw in the preceding section that, by definition, vectors add according to the parallelogram law. Thus the sum of two vectors **P** and **Q** is obtained by attaching the two vectors to the same point *A* and constructing a parallelogram, using **P** and **Q** as two sides of the parallelogram (Fig. 2.6). The diagonal that passes through *A* represents the sum of the vectors **P** and **Q**, and this sum is denoted by **P** + **Q**. The fact that the sign + is used to denote both vector and scalar addition should not cause any confusion if vector and scalar quantities are always carefully distinguished. Thus, we should note that the magnitude of the vector **P** + **Q** is *not*, in general, equal to the sum $P + Q$ of the magnitudes of the vectors **P** and **Q**.

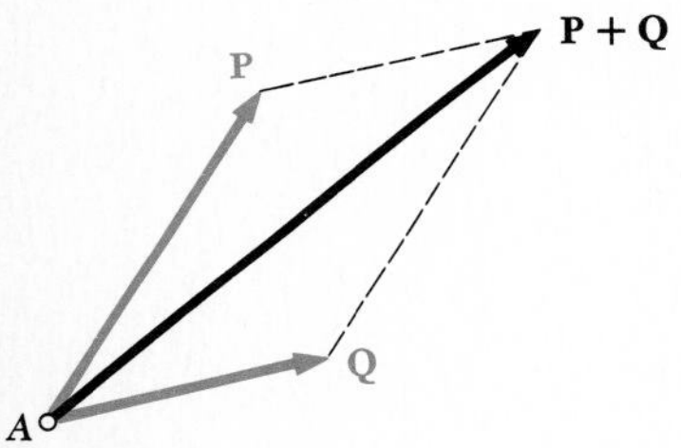

Fig. 2.6

Since the parallelogram constructed on the vectors **P** and **Q** does not depend upon the order in which **P** and **Q** are selected, we conclude that the addition of two vectors is *commutative*,

† Some expressions have magnitude and direction, but do not add according to the parallelogram law. While these expressions may be represented by arrows, they *cannot* be considered as vectors.

A group of such expressions are the finite rotations of a rigid body. Place a closed book on a table in front of you, so that it lies in the usual fashion, with its front cover up and its binding to the left. Now rotate it through 180° about an axis parallel to the binding (Fig. 2.3*a*); this rotation may be represented by an arrow of length equal to 180 units and oriented as shown. Picking up the book as it lies in its new position, rotate it now through 180° about a

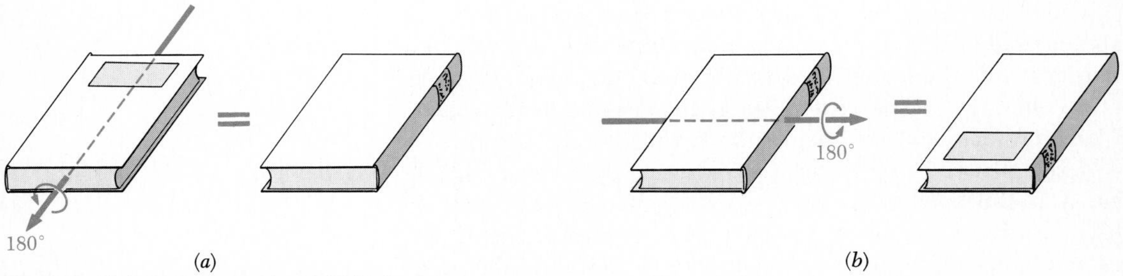

Fig. 2.3 Finite rotations of a rigid body.

and we write

$$\mathbf{P} + \mathbf{Q} = \mathbf{Q} + \mathbf{P} \tag{2.1}$$

From the parallelogram law, we can derive an alternate method for determining the sum of two vectors. This method, known as the *triangle rule,* is derived as follows: Consider Fig. 2.6, where the sum of the vectors **P** and **Q** has been determined by the parallelogram law. Since the side of the parallelogram opposite **Q** is equal to **Q** in magnitude and direction, we could draw only half of the parallelogram (Fig. 2.7*a*). The sum of the two vectors may thus be found by *arranging* **P** *and* **Q** *in tip-to-tail fashion and then connecting the tail of* **P** *with the tip of* **Q**. In Fig. 2.7*b*, the other half of the parallelogram is considered, and the same result is obtained. This confirms the fact that vector addition is commutative.

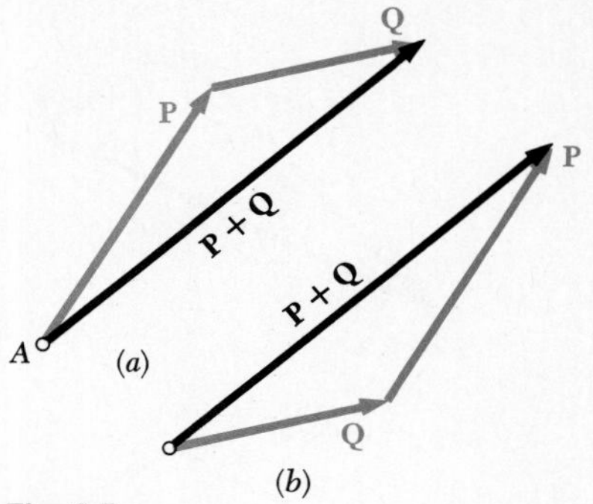

Fig. 2.7

The *subtraction* of a vector is defined as the addition of an equal and opposite vector. Thus, the vector **P** − **Q** representing the difference between the vectors **P** and **Q** is obtained by adding to **P** the vector **Q**′ equal and opposite to **Q** (Fig. 2.8). Here again we should observe that, while the same sign is used to denote both vector and scalar subtraction, confusion will be avoided if care is taken to distinguish between vector and scalar quantities.

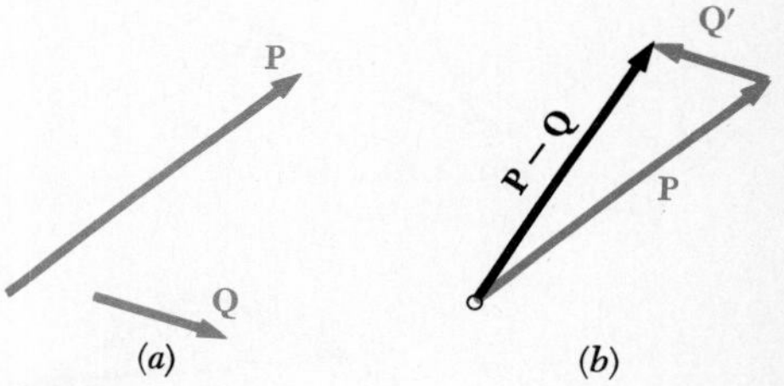

Fig. 2.8

We shall now consider the *sum of three or more vectors*. The sum of three vectors **P**, **Q**, and **S** will, *by definition,* be obtained

horizontal axis perpendicular to the binding (Fig. 2.3*b*); this second rotation may be represented by an arrow 180 units long and oriented as shown. But the book could have been placed in this final position through a single 180° rotation about a vertical axis (Fig. 2.3*c*). We conclude that the sum of the two 180° rotations represented by arrows directed respectively along the z and x axes is a 180° rotation represented by an arrow directed along the y axis (Fig. 2.3*d*). Clearly, the finite rotations of a rigid body *do not* obey the parallelogram law of addition; therefore they *cannot* be represented by vectors.

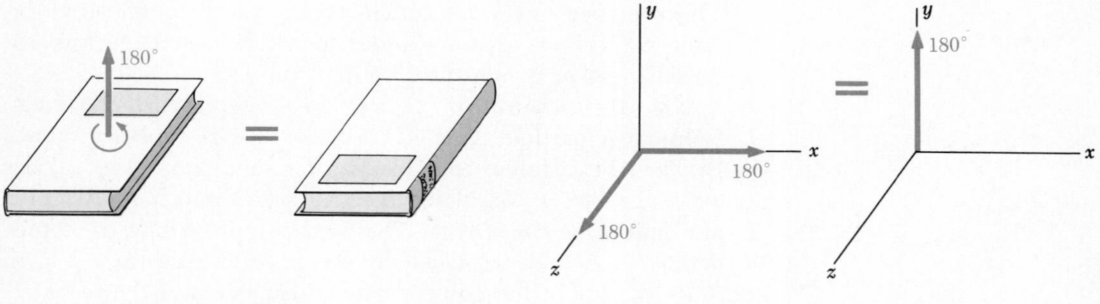

(*c*) (*d*)

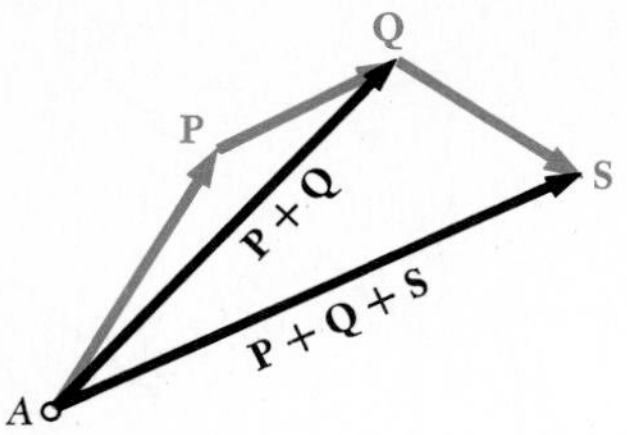

Fig. 2.9

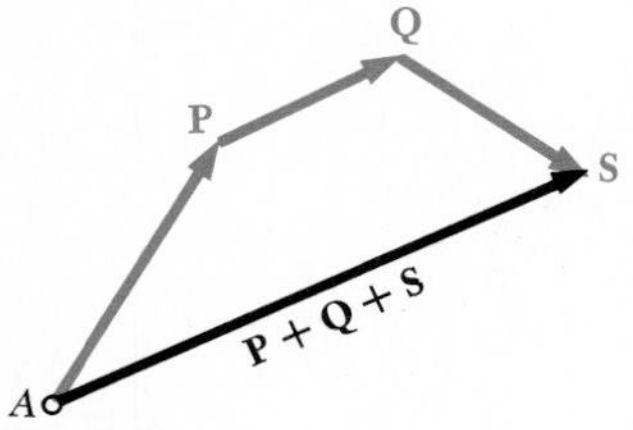

Fig. 2.10

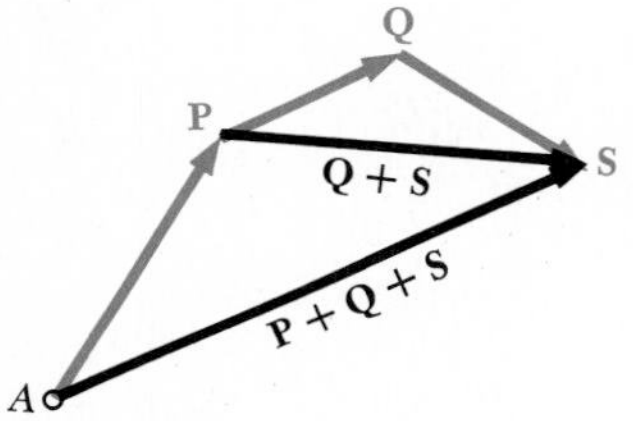

Fig. 2.11

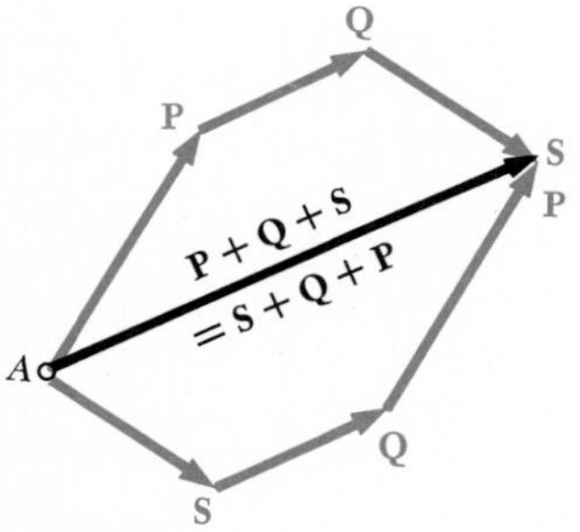

Fig. 2.12

by first adding the vectors **P** and **Q**, and then adding the vector **S** to the vector **P** + **Q**. We thus write

$$\mathbf{P} + \mathbf{Q} + \mathbf{S} = (\mathbf{P} + \mathbf{Q}) + \mathbf{S} \tag{2.2}$$

Similarly, the sum of four vectors will be obtained by adding the fourth vector to the sum of the first three. It follows that the sum of any number of vectors may be obtained by applying repeatedly the parallelogram law to successive pairs of vectors until all the given vectors are replaced by a single vector.

If the given vectors are *coplanar*, i.e., if they are contained in the same plane, their sum may be easily obtained graphically. In that case, the repeated application of the triangle rule will be preferred to the application of the parallelogram law. In Fig. 2.9 the sum of three vectors **P**, **Q**, and **S** was obtained in that manner. The triangle rule was first applied to obtain the sum **P** + **Q** of the vectors **P** and **Q**; it was applied again to obtain the sum of the vectors **P** + **Q** and **S**. The determination of the vector **P** + **Q**, however, could have been omitted and the sum of the three vectors could have been obtained directly, as shown in Fig. 2.10, by *arranging the given vectors in tip-to-tail fashion and connecting the tail of the first vector with the tip of the last one.* This is known as the *polygon rule* for the addition of vectors.

We observe that the result obtained would have been unchanged if, as shown in Fig. 2.11, the vectors **Q** and **S** had been replaced by their sum **Q** + **S**. We may thus write

$$\mathbf{P} + \mathbf{Q} + \mathbf{S} = (\mathbf{P} + \mathbf{Q}) + \mathbf{S} = \mathbf{P} + (\mathbf{Q} + \mathbf{S}) \tag{2.3}$$

which expresses the fact that vector addition is *associative*. Recalling that vector addition has also been shown, in the case of two vectors, to be commutative, we write

$$\begin{aligned} \mathbf{P} + \mathbf{Q} + \mathbf{S} &= (\mathbf{P} + \mathbf{Q}) + \mathbf{S} = \mathbf{S} + (\mathbf{P} + \mathbf{Q}) \\ &= \mathbf{S} + (\mathbf{Q} + \mathbf{P}) = \mathbf{S} + \mathbf{Q} + \mathbf{P} \end{aligned} \tag{2.4}$$

This expression, as well as others which may be obtained in the same way, shows that the order in which several vectors are added together is immaterial (Fig. 2.12).

2.4. Resultant of Several Concurrent Forces. Consider a particle *A* acted upon by several coplanar forces, i.e., by several forces contained in the same plane (Fig. 2.13*a*). Since the forces considered here all pass through *A*, they are also said to be *concurrent.* The vectors representing the forces acting on *A* may be added by the polygon rule (Fig. 2.13*b*). Since the use of the polygon rule is equivalent to the repeated application of the parallelogram law, the vector **R** thus obtained represents the resultant of the given concurrent forces, i.e., the

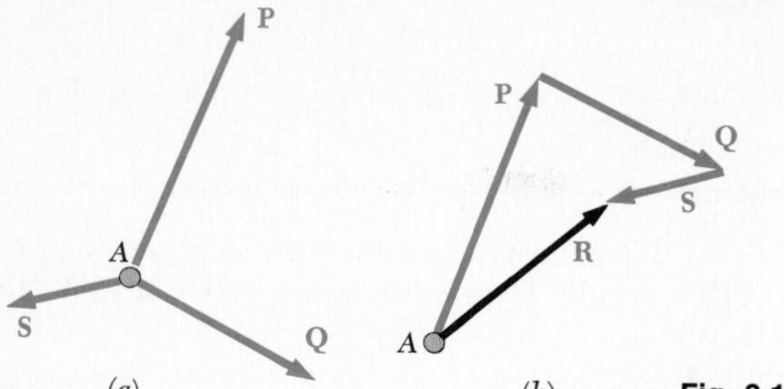

Fig. 2.13

single force which has the same effect on the particle A as the given forces. As indicated above, the order in which the vectors **P**, **Q**, and **S** representing the given forces are added together is immaterial.

2.5. Resolution of a Force into Components. We have seen that two or more forces acting on a particle may be replaced by a single force which has the same effect on the particle. Conversely, a single force **F** acting on a particle may be replaced by two or more forces which, together, have the same effect on the particle. These forces are called the *components* of the original force **F**, and the process of substituting them for **F** is called *resolving the force* **F** *into components*.

Clearly, for each force **F** there exist an infinite number of possible sets of components. Sets of *two components* **P** *and* **Q** are the most important as far as practical applications are concerned. But, even then, the number of ways in which a given force **F** may be resolved into two components is unlimited (Fig. 2.14). Two cases are of particular interest:

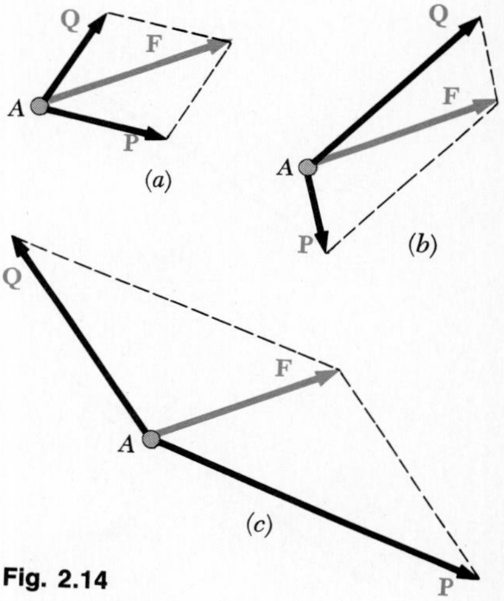

Fig. 2.14

1. *One of the Two Components,* **P**, *Is Known.* The second component, **Q**, is obtained by applying the triangle rule and joining the tip of **P** to the tip of **F** (Fig. 2.15); the magnitude and direction of **Q** are determined graphically or by trigonometry. Once **Q** has been determined, both components **P** and **Q** should be applied at A.
2. *The Line of Action of Each Component Is Known.* The magnitude and sense of the components are obtained by applying the parallelogram law and drawing lines, through the tip of **F**, parallel to the given lines of action (Fig. 2.16). This process leads to two well-defined components, **P** and **Q**, which may be determined graphically or by applying the law of sines.

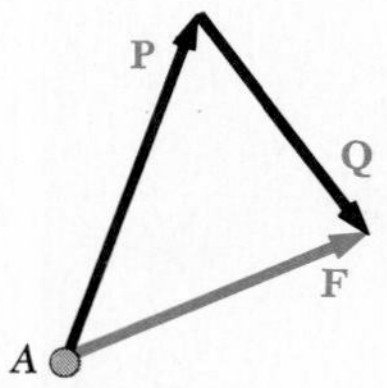

Fig. 2.15

Many other cases may be encountered; for example, the direction of one component may be known while the magnitude of the other component is to be as small as possible (see Sample Prob. 2.2). In all cases the appropriate triangle or parallelogram is drawn, which satisfies the given conditions.

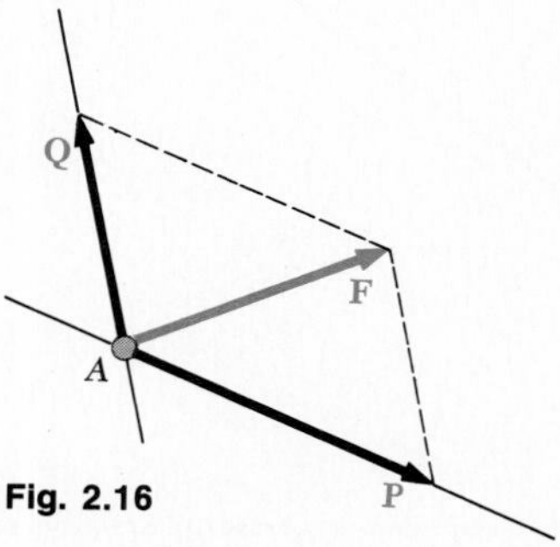

Fig. 2.16

SAMPLE PROBLEM 2.1

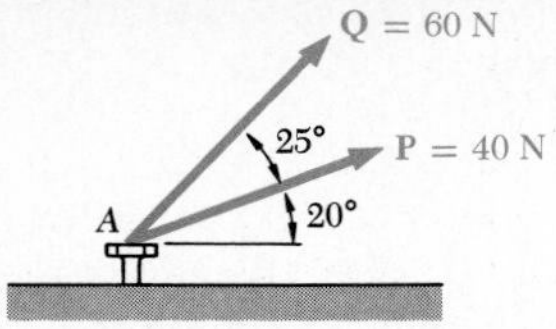

The two forces **P** and **Q** act on a bolt *A*. Determine their resultant.

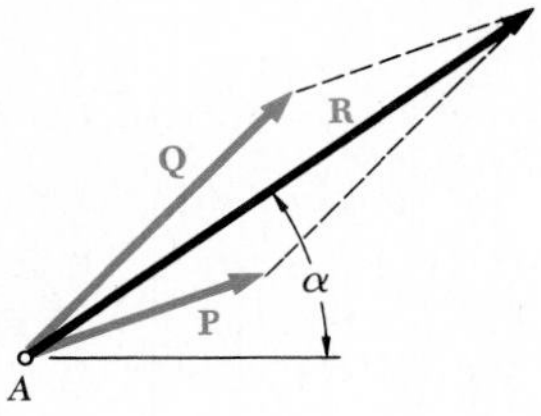

Graphical Solution. A parallelogram with sides equal to **P** and **Q** is drawn to scale. The magnitude and direction of the resultant are measured and found to be

$$R = 98 \text{ N} \qquad \alpha = 35^\circ \qquad \mathbf{R} = 98 \text{ N} \measuredangle 35^\circ \blacktriangleleft$$

The triangle rule may also be used. Forces **P** and **Q** are drawn in tip-to-tail fashion. Again the magnitude and direction of the resultant are measured.

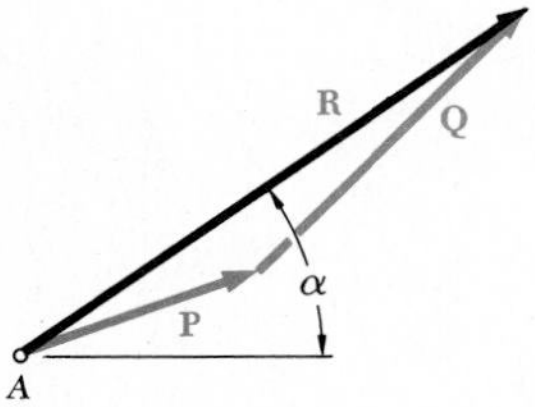

$$R = 98 \text{ N} \qquad \alpha = 35^\circ \qquad \mathbf{R} = 98 \text{ N} \measuredangle 35^\circ \blacktriangleleft$$

Trigonometric Solution. The triangle rule is again used; two sides and the included angle are known. We apply the law of cosines.

$$R^2 = P^2 + Q^2 - 2PQ \cos B$$
$$R^2 = (40 \text{ N})^2 + (60 \text{ N})^2 - 2(40 \text{ N})(60 \text{ N}) \cos 155^\circ$$
$$R = 97.7 \text{ N}$$

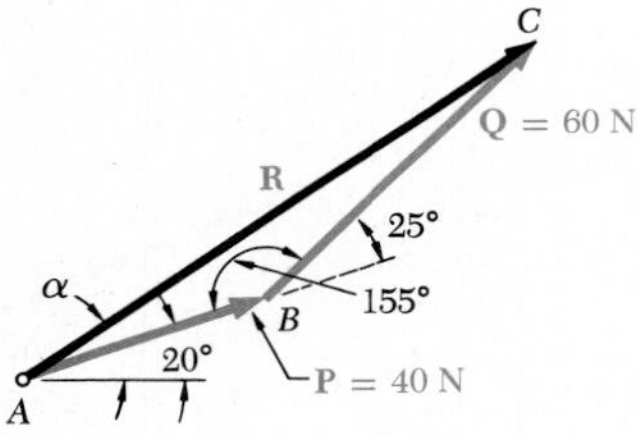

Now, applying the law of sines, we write

$$\frac{\sin A}{Q} = \frac{\sin B}{R} \qquad \frac{\sin A}{60 \text{ N}} = \frac{\sin 155^\circ}{97.7 \text{ N}} \tag{1}$$

With a calculator. Solving Eq. (1) for sin *A*, we have

$$\sin A = \frac{(60 \text{ N}) \sin 155^\circ}{97.7 \text{ N}}$$

Computing first the quotient, and then its arc sine, we obtain

$$A = 15.0^\circ \qquad \alpha = 20^\circ + A = 35.0^\circ$$
$$\mathbf{R} = 97.7 \text{ N} \measuredangle 35.0^\circ \blacktriangleleft$$

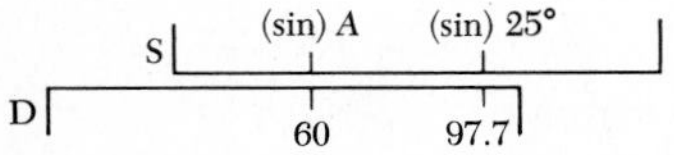

With a slide rule. Noting that sin 155° = sin 25°, and setting the slide rule as shown to reflect Eq. (1), we read $A = 15.0^\circ$ and obtain the same answer as above.

Alternate Trigonometric Solution. We construct the right triangle *BCD* and compute

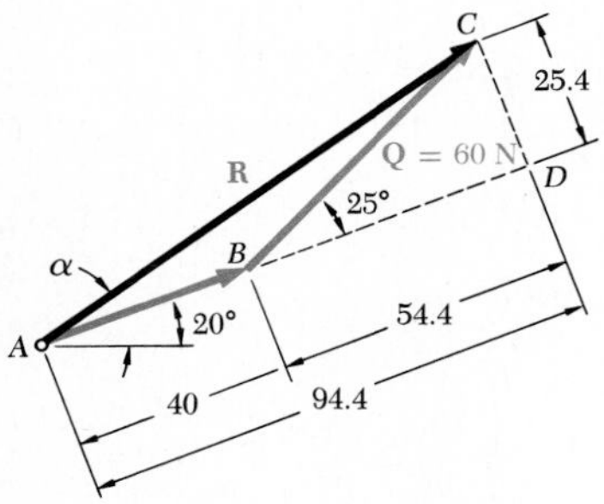

$$CD = (60 \text{ N}) \sin 25^\circ = 25.4 \text{ N}$$
$$BD = (60 \text{ N}) \cos 25^\circ = 54.4 \text{ N}$$

Then, using triangle *ACD*, we obtain

$$\tan A = \frac{25.4 \text{ N}}{94.4 \text{ N}} \qquad A = 15.0^\circ$$
$$R = 25.4/\sin A \qquad R = 97.7 \text{ N}$$

Again,

$$\alpha = 20^\circ + A = 35.0^\circ \qquad \mathbf{R} = 97.7 \text{ N} \measuredangle 35.0^\circ \blacktriangleleft$$

SAMPLE PROBLEM 2.2

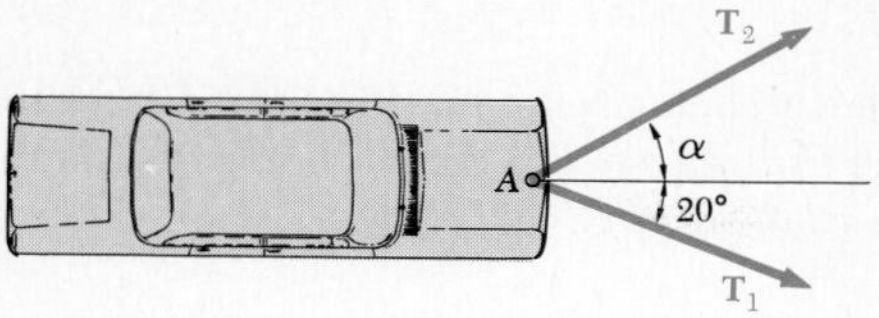

A disabled automobile is pulled by means of two ropes as shown. If the resultant of the two forces exerted by the ropes is a 300-lb force parallel to the axis of the automobile, find (*a*) the tension in each of the ropes, knowing that $\alpha = 30°$, (*b*) the value of α such that the tension in rope *2* is minimum.

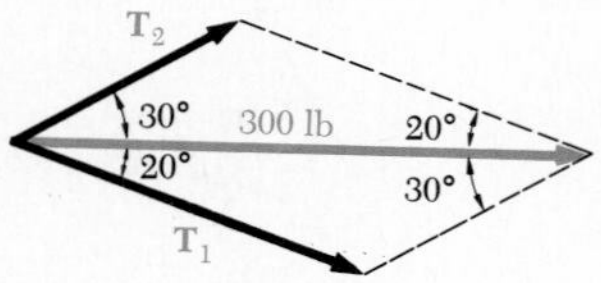

a. **Tension for $\alpha = 30°$.** ***Graphical Solution.*** The parallelogram law is used; the diagonal (resultant) is known to be equal to 300 lb and to be directed to the right. The sides are drawn parallel to the ropes. If the drawing is done to scale, we measure

$$T_1 = 196 \text{ lb} \qquad T_2 = 134 \text{ lb} \blacktriangleleft$$

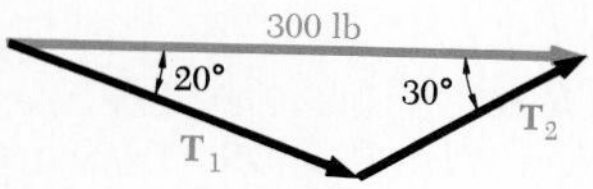

Trigonometric Solution. The triangle rule may be used. We note that the triangle shown represents half of the parallelogram shown above. Using the law of sines, we write

$$\frac{T_1}{\sin 30°} = \frac{T_2}{\sin 20°} = \frac{300 \text{ lb}}{\sin 130°}$$

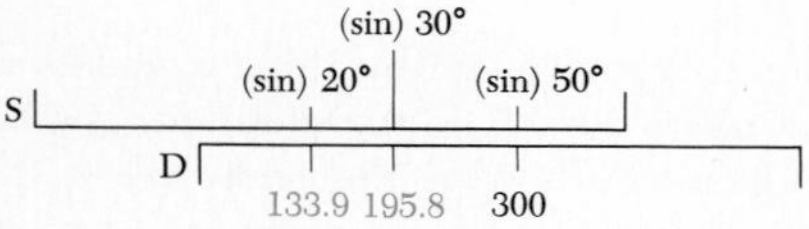

With a calculator. We first compute and store the value of the last quotient. Multiplying this value successively by sin 30° and sin 20°, we obtain

$$T_1 = 195.8 \text{ lb} \qquad T_2 = 133.9 \text{ lb} \blacktriangleleft$$

With a slide rule. Noting that sin 130° = sin 50°, we set the slide rule as shown and read the values of T_1 and T_2.

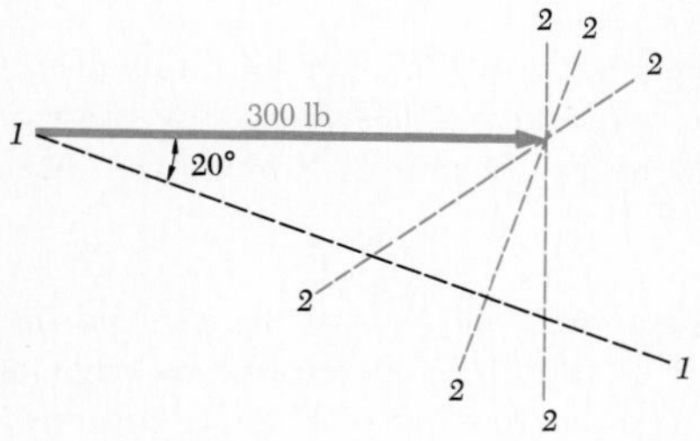

b. **Value of α for Minimum T_2.** To determine the value of α such that the tension in rope *2* is minimum, the triangle rule is again used. In the sketch shown, line *1-1* is the known direction of $\mathbf{T}_1$. Several possible directions of $\mathbf{T}_2$ are shown by the lines *2-2*. We note that the minimum value of T_2 occurs when $\mathbf{T}_1$ and $\mathbf{T}_2$ are perpendicular. The minimum value of T_2 is

$$T_2 = (300 \text{ lb}) \sin 20° = 102.6 \text{ lb}$$

Corresponding values of T_1 and α are

$$T_1 = (300 \text{ lb}) \cos 20° = 282 \text{ lb}$$
$$\alpha = 90° - 20° \qquad \alpha = 70° \blacktriangleleft$$

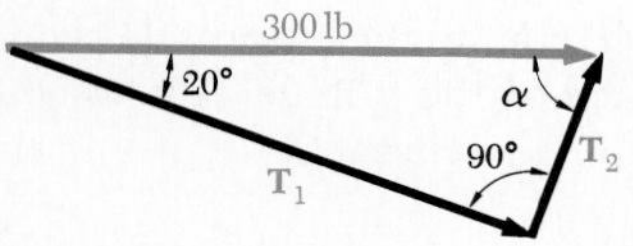

PROBLEMS†

2.1 and 2.2 Determine graphically the magnitude and direction of the resultant of the two forces shown, using in each problem (*a*) the parallelogram law, (*b*) the triangle rule.

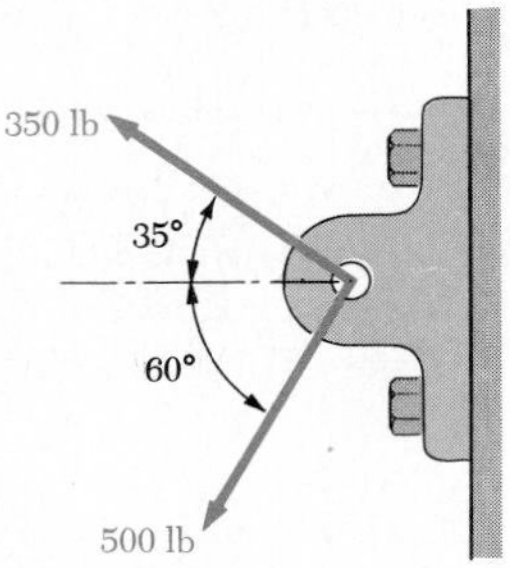

Fig. P2.1

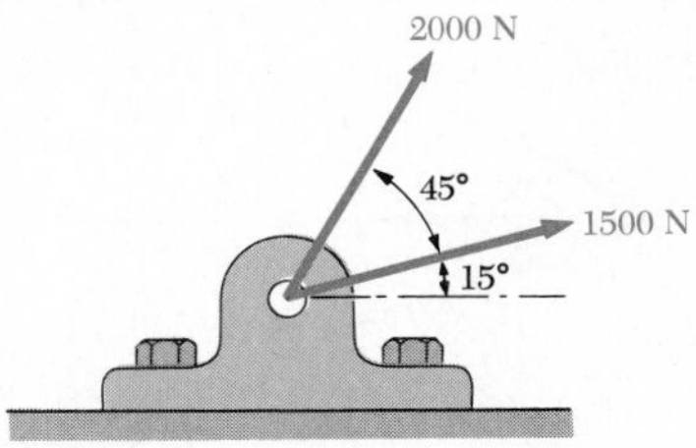

Fig. P2.2

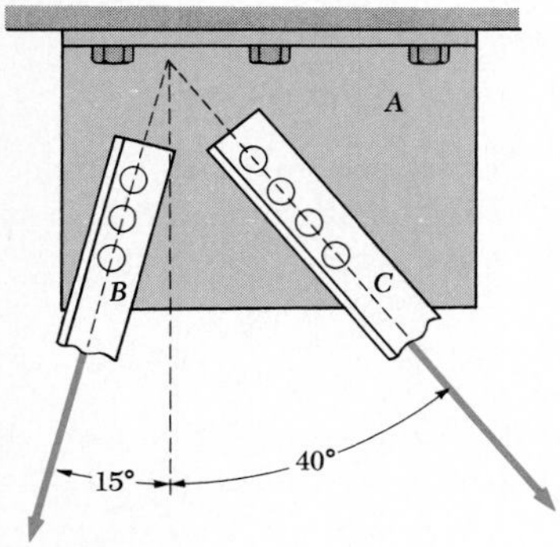

Fig. P2.3

2.3 Two structural members *B* and *C* are riveted to the bracket *A*. Knowing that the tension in member *B* is 2500 lb and that the tension in *C* is 2000 lb, determine graphically the magnitude and direction of the resultant force acting on the bracket.

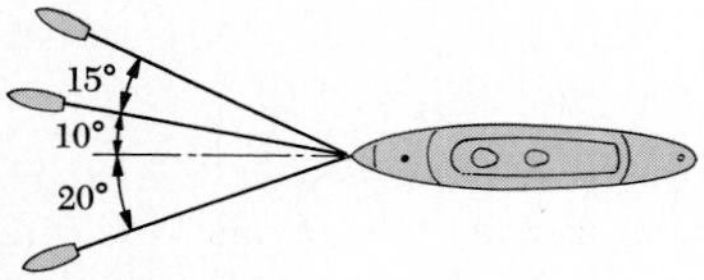

Fig. P2.4

2.4 A disabled ocean liner is being towed by three tugboats as shown. The tension in each cable is 5000 lb. (*a*) Determine graphically the resultant force acting on the bow of the liner. (*b*) If the tugboats cannot operate safely when the angle between any two of the cables is less than 10°, where should the tugboats be located in order to produce the largest resultant force parallel to the axis of the liner? What is the magnitude of this resultant?

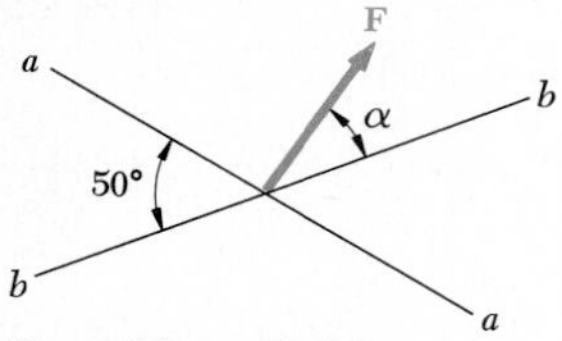

Fig. P2.5 and P2.6

2.5 The force **F** of magnitude 100 lb is to be resolved into two components along the lines *a-a* and *b-b*. Determine by trigonometry the angle α, knowing that the component of **F** along the line *a-a* is to be 70 lb.

2.6 The force **F** of magnitude 800 N is to be resolved into two components along the lines *a-a* and *b-b*. Determine by trigonometry the angle α, knowing that the component of **F** along the line *b-b* is to be 120 N.

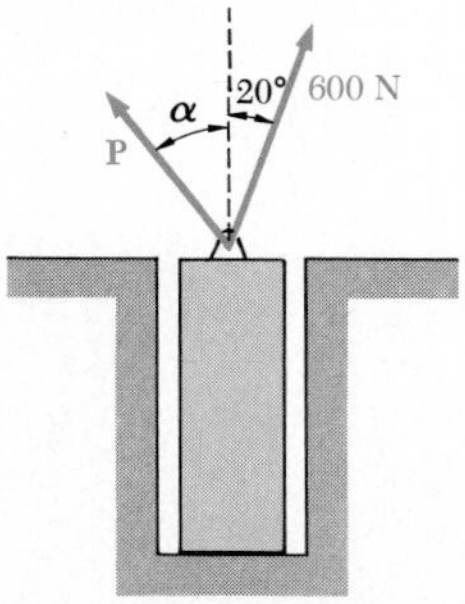

Fig. P2.7 and P2.9

2.7 Knowing that $\alpha = 30°$, determine the magnitude of the force **P** so that the resultant force exerted on the cylinder is vertical. What is the corresponding magnitude of the resultant?

2.8 Determine by trigonometry the magnitude of the force **P** so that the resultant of the two forces applied at *A* is vertical. What is the corresponding magnitude of the resultant?

† Answers to all even-numbered problems are given at the end of the book.

2.9 A cylinder is to be lifted by two cables. Knowing that the tension in one cable is 600 N, determine the magnitude and direction of the force **P** so that the resultant is a vertical force of 900 N.

2.10 Knowing that $P = 30$ lb, determine by trigonometry the resultant of the two forces applied at point A.

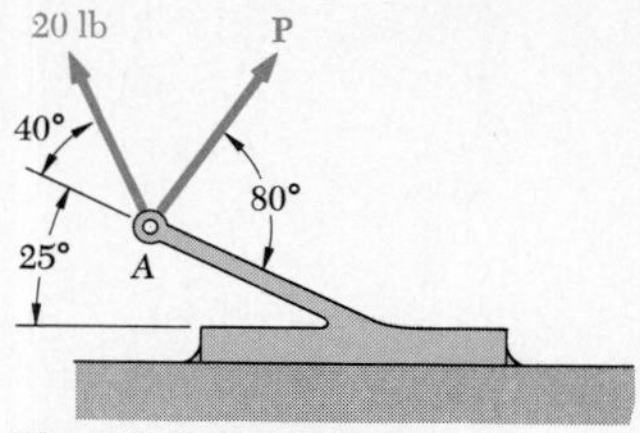

Fig. P2.8 and P2.10

2.11 Solve Prob. 2.3 by trigonometry.

2.12 Determine by trigonometry the magnitude and direction of the force **P** so that the resultant of **P** and the 900-N force is a vertical force of 2700 N directed downward.

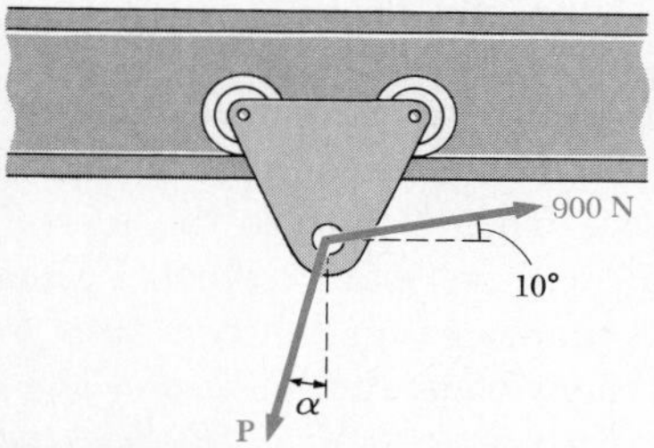

Fig. P2.12

2.13 If the resultant of the two forces acting on the cylinder of Prob. 2.7 is to be vertical, find (*a*) the value of α for which the magnitude of **P** is minimum, (*b*) the corresponding magnitude of **P**.

2.6. Rectangular Components of a Force.†

In many problems it will be found desirable to resolve a force into two components which are perpendicular to each other. In Fig. 2.17, the force **F** has been resolved into a component $\mathbf{F}_x$ along the x axis and a component $\mathbf{F}_y$ along the y axis. The parallelogram drawn to obtain the two components is a *rectangle,* and $\mathbf{F}_x$ and $\mathbf{F}_y$ are called *rectangular components.*

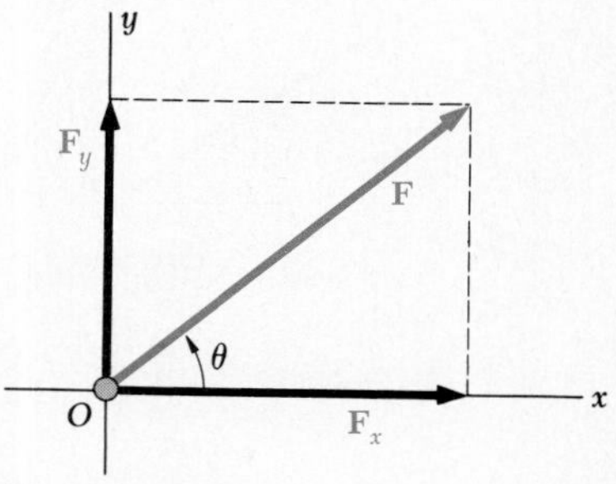

Fig. 2.17

The x and y axes are usually chosen horizontal and vertical, respectively, as in Fig. 2.17; they may, however, be chosen in any two perpendicular directions, as shown in Fig. 2.18. In determining the rectangular components of a force, the student should think of the construction lines shown in Figs. 2.17 and 2.18 as being *parallel* to the x and y axes, rather than *perpendicular* to these axes. This practice will help avoid mistakes in determining *oblique* components as in Sec. 2.5.

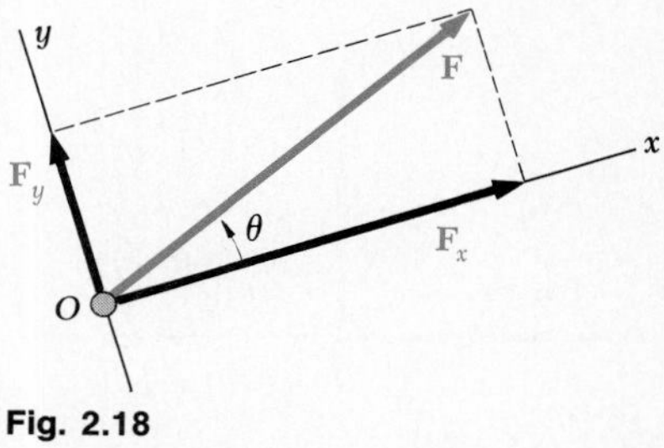

Fig. 2.18

Denoting by F the magnitude of the given force **F**, by θ the

† The properties established in Secs. 2.6 and 2.7 may be readily extended to the rectangular components of any vector quantity.

angle between **F** and the x axis, and by F_x and F_y, respectively, the magnitude of the components $\mathbf{F}_x$ and $\mathbf{F}_y$, we write

$$F_x = F\cos\theta \qquad F_y = F\sin\theta \tag{2.5}$$

The magnitudes F_x and F_y of the components of the force **F** are called the *scalar components* of **F**, while the actual component forces $\mathbf{F}_x$ and $\mathbf{F}_y$ should be referred to as the *vector components* of **F**. However, when there exists no possibility of confusion, the vector as well as the scalar components of **F** may be referred to simply as the *components* of **F**.

We note that, if the angle θ is measured counterclockwise from the positive x axis, it will take values ranging from 0° to 360°. It follows from the relations (2.5) that the scalar components F_x and F_y may have negative as well as positive values. More precisely, F_x will be positive when **F** falls in the first or fourth quadrants, and it will be negative when **F** falls in the second or third quadrants. Similarly, F_y will be positive when **F** falls in the first or second quadrants and negative when **F** falls in the third or fourth quadrants. We conclude that *the scalar component F_x is positive when the vector component $\mathbf{F}_x$ has the same sense as the x axis and negative when $\mathbf{F}_x$ has the opposite sense.* A similar conclusion may be drawn regarding the sign of the scalar component F_y.

It appears from the above that the rectangular vector components of a force acting on a particle are completely defined by the corresponding scalar components. The vector component $\mathbf{F}_x$, for example, has by definition the x axis for its line of action, and both its sense and its magnitude are defined by the scalar component F_x.

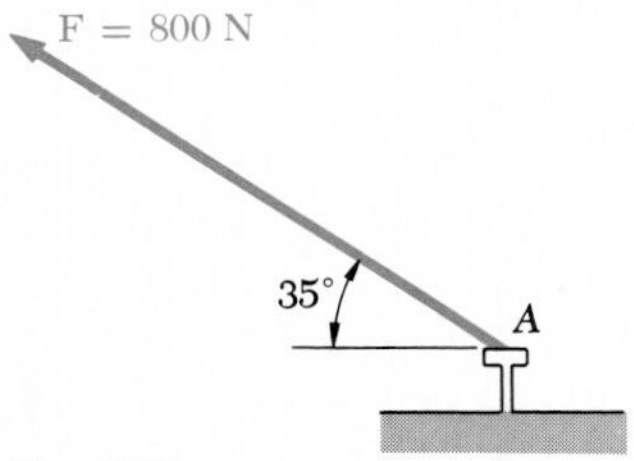

Fig. 2.19

Example 1. A force of 800 N is exerted on a bolt A as shown in Fig. 2.19. Determine the horizontal and vertical components of the force.

In order to obtain the correct sign for the scalar components F_x and F_y, the value $\theta = 180° - 35° = 145°$ should be substituted for θ in the relations (2.5). However, it will be found more practical to determine by inspection the signs of F_x and of F_y (Fig. 2.20) and to use the trigonometric functions of the angle $\alpha = 35°$. We write therefore

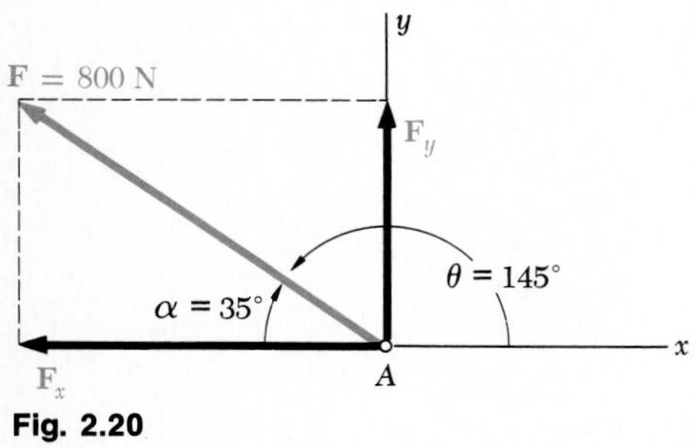

Fig. 2.20

$$F_x = -F\cos\alpha = -(800\text{ N})\cos 35° = -655\text{ N}$$
$$F_y = +F\sin\alpha = +(800\text{ N})\sin 35° = +459\text{ N}$$

The vector components of **F** are thus

$$\mathbf{F}_x = 655\text{ N} \leftarrow \qquad \mathbf{F}_y = 459\text{ N} \uparrow$$

Example 2. A man pulls with a force of 300 N on a rope attached to a building, as shown in Fig. 2.21. What are the horizontal and vertical components of the force exerted by the rope at point A?

It is seen from Fig. 2.22 that

$$F_x = +(300\text{ N})\cos\alpha \qquad F_y = -(300\text{ N})\sin\alpha$$

Observing that $AB = 10$ m, we find from Fig. 2.21

$$\cos\alpha = \frac{8\text{ m}}{AB} = \frac{8\text{ m}}{10\text{ m}} = \frac{4}{5} \qquad \sin\alpha = \frac{6\text{ m}}{AB} = \frac{6\text{ m}}{10\text{ m}} = \frac{3}{5}$$

We thus obtain

$$F_x = +(300\text{ N})\tfrac{4}{5} = +240\text{ N} \qquad F_y = -(300\text{ N})\tfrac{3}{5} = -180\text{ N}$$

and the vector components of **F** are

$$\mathbf{F}_x = 240\text{ N}\rightarrow \qquad \mathbf{F}_y = 180\text{ N}\downarrow$$

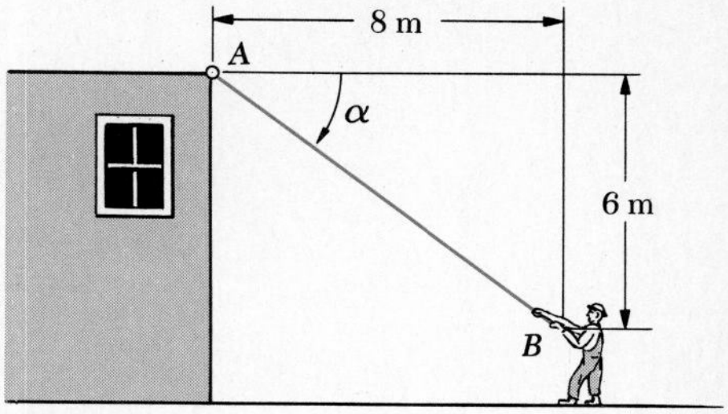

Fig. 2.21

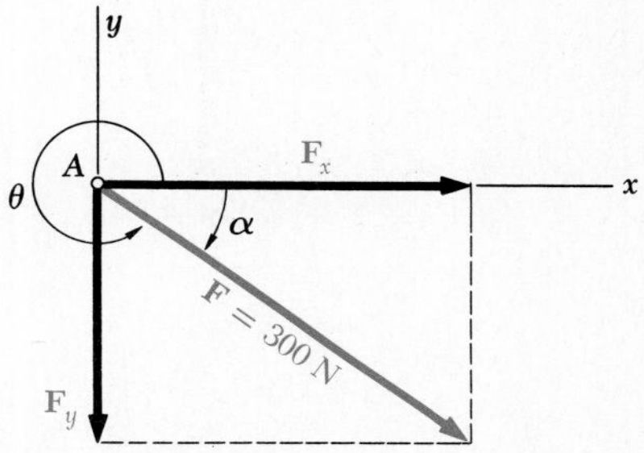

Fig. 2.22

When a force is defined by its rectangular components $\mathbf{F}_x$ and $\mathbf{F}_y$ (see Fig. 2.17 or 2.18), the angle θ defining its direction can be obtained by writing

$$\tan\theta = \frac{F_y}{F_x} \qquad (2.6)$$

The magnitude F of the force may be obtained by applying the Pythagorean theorem and writing

$$F = \sqrt{F_x^2 + F_y^2} \qquad (2.7)$$

However, once θ has been found, it is usually easier to determine the magnitude of the force by the process of solving one of the formulas (2.5) for F.

Example 3. The horizontal and vertical components of a force **F** acting on a bolt A are, respectively, $F_x = +700$ lb and $F_y = +1500$ lb (Fig. 2.23). Determine the magnitude of the force and the angle θ it forms with the horizontal.

With a calculator.† Using Eq. (2.6), we write

$$\tan\theta = \frac{F_y}{F_x} = \frac{1500\text{ lb}}{700\text{ lb}}$$

Entering 1500 lb, dividing by 700 lb, and computing the arc tangent

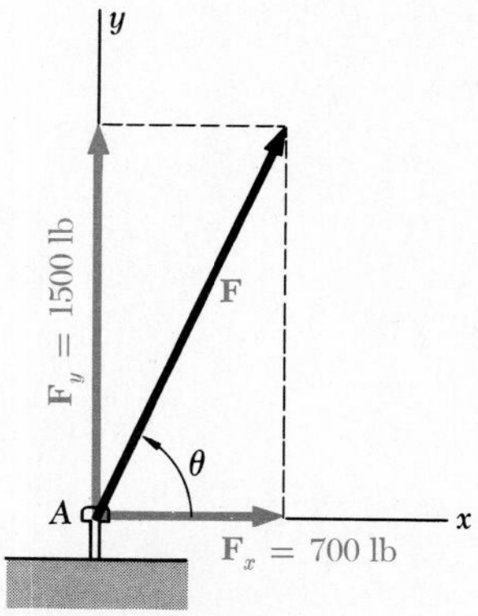

Fig. 2.23

† It is assumed that the calculator used has keys for the computation of trigonometric and inverse trigonometric functions. Some calculators also have keys for the direct conversion of rectangular coordinates into polar coordinates, and vice versa. Such calculators eliminate the need for the computation of trigonometric functions in Examples 1, 2, and 3 and in problems of the same type.

of the quotient, we obtain $\theta = 65.0°$. Solving the second of Eqs. (2.5) for F, we have

$$F = \frac{F_y}{\sin\theta} = \frac{1500 \text{ lb}}{\sin 65.0°} = 1655 \text{ lb}$$

The last calculation is facilitated if the value of F_y is stored when originally entered; it may then be recalled to be divided by $\sin\theta$.

With a slide rule. Rather than θ, we shall determine its complement β (Fig. 2.24) because β being smaller than 45°, its tangent is more easily found on the slide rule. We write

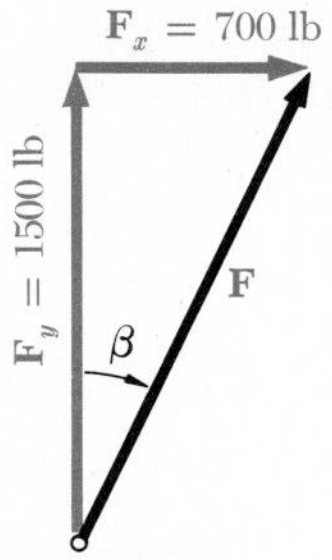

Fig. 2.24

$$\begin{aligned} F_y \tan\beta &= F_x \\ (1500 \text{ lb}) \tan\beta &= 700 \text{ lb} \\ \beta &= 25.0° \end{aligned} \tag{2.8}$$

We then write

$$\begin{aligned} F \sin\beta &= F_x \\ F \sin 25.0° &= 700 \text{ lb} \\ F &= 1656 \text{ lb} \end{aligned} \tag{2.9}$$

The magnitude of the force is thus 1656 lb, and the angle θ it forms with the x axis is $\theta = 90.0° - 25.0° = 65.0°$.

On most slide rules, β and F may be conveniently determined in the following way (Fig. 2.25):

FIRST STEP. Move the slide until one of the extremities of the trigonometric scale (in the present case the left one) coincides with the *larger* of the two components (1500 lb) read on the D scale.

SECOND STEP. Move the indicator until the hairline coincides with the *smaller* of the two components (700 lb). Angle β may then be read *on the* T *scale* under the hairline since the rule is set to perform the product defined by Eq. (2.8).

THIRD STEP. Without touching the indicator, move the slide until angle β (25.0°) is read *on the* S *scale* under the hairline. The value of F (1656 lb) may then be read on the D scale opposite one of the extremities of the trigonometric scale since the rule is set to perform the product defined by Eq. (2.9).

Remark. If the smaller of the two components is less than one-tenth of the larger one, the ST scale should be used in the second step, instead of the T scale, and the angle found will be less than 5.7°. The third step may then be omitted since, within the accuracy of the slide rule, the magnitude F of the force may be assumed equal to the larger of the two components (see Sample Prob. 2.3).

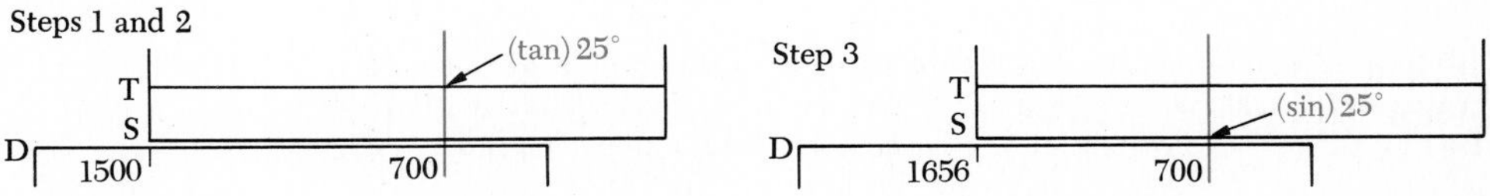

Fig. 2.25

2.7. Addition of Forces by Summing *x* and *y* Components. It was seen in Sec. 2.1 that forces should be added according to the parallelogram law. From this law, two other methods, more readily applicable to the *graphical* solution of problems, were derived in Secs. 2.3 and 2.4: the triangle rule for the addition of two forces and the polygon rule for the addition of three or more forces. It was also seen that the force triangle used to define the resultant of two forces could be used to obtain a *trigonometric* solution.

When three or more forces are to be added, no practical trigonometric solution may be obtained from the force polygon which defines the resultant of the forces. In this case, an *analytic* solution of the problem may be obtained by first resolving each force into two rectangular components. In Fig. 2.26*a*, for example, three forces **P**, **Q**, and **S** are shown acting on a particle A. In Fig. 2.26*b*, each of these forces has been replaced by its horizontal and vertical components; thus the forces of Fig. 2.26*b* produce the same effect on A as the forces of Fig. 2.26*a*. Now the horizontal components may be added into a single force $\mathbf{R}_x$ and the vertical components into a single force $\mathbf{R}_y$. This should be done by applying the parallelogram law; but since, in each case, the forces considered lie in the same line, the actual computation reduces to the algebraic addition of the scalar components. We write

$$R_x = P_x + Q_x + S_x \qquad R_y = P_y + Q_y + S_y$$

or, for short,

$$R_x = \Sigma F_x \qquad R_y = \Sigma F_y \tag{2.10}$$

Thus the three given forces of Fig. 2.26*a* have been replaced by the horizontal force $\mathbf{R}_x$ and the vertical force $\mathbf{R}_y$ shown in Fig. 2.26*c*. These two forces are added vectorially into the resultant **R** of the given system (Fig. 2.26*d*) by the method of Sec. 2.6.

The procedure just described can be carried out most efficiently if the computations are arranged in a table. While it is the only practical analytic method for adding three or more forces, it is also often preferred to the trigonometric solution in the case of the addition of two forces.

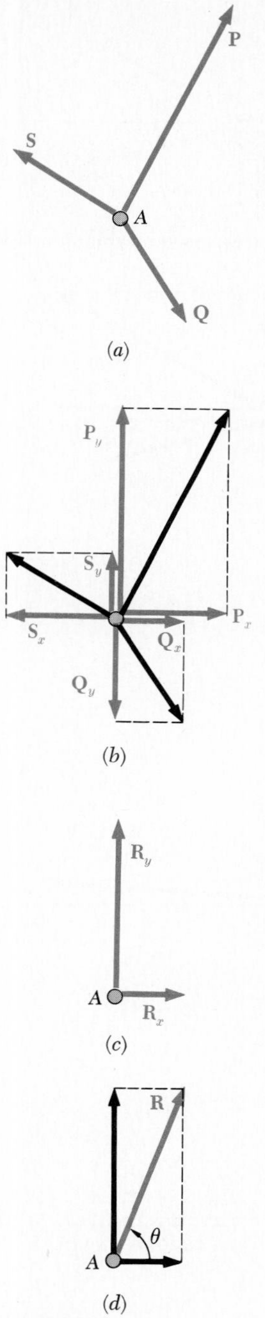

Fig. 2.26

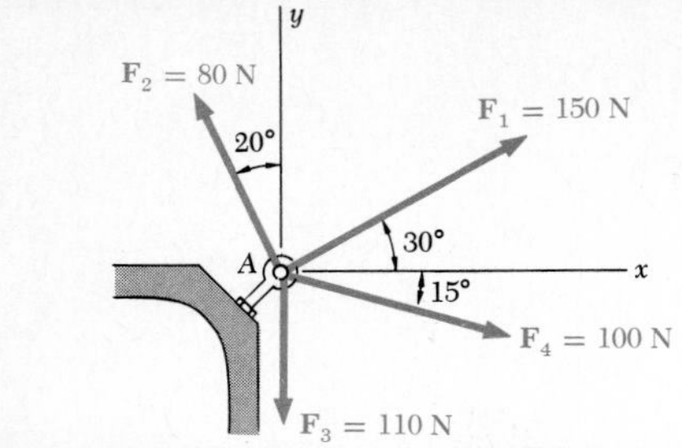

SAMPLE PROBLEM 2.3

Four forces act on bolt A as shown. Determine the resultant of the forces on the bolt.

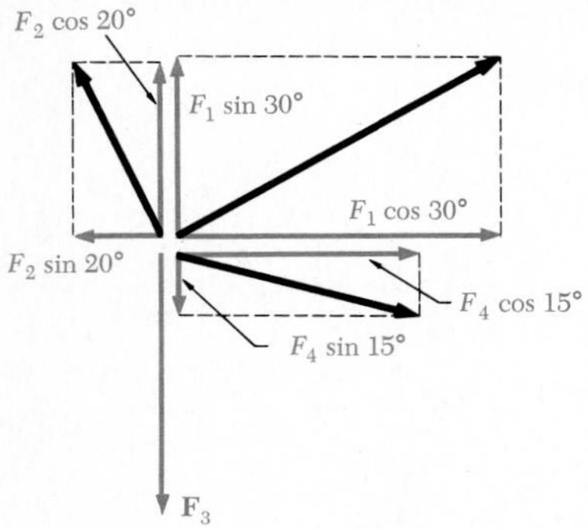

Solution. The x and y components of each force are determined by trigonometry as shown and are entered in the table below. According to the convention adopted in Sec. 2.6, the scalar number representing a force component is positive if the force component has the same sense as the corresponding coordinate axis. Thus, x components acting to the right and y components acting upward are represented by positive numbers.

Force	Magnitude, N	x Component, N	y Component, N
$\mathbf{F}_1$	150	+129.9	+75.0
$\mathbf{F}_2$	80	−27.4	+75.2
$\mathbf{F}_3$	110	0	−110.0
$\mathbf{F}_4$	100	+96.6	−25.9
		$R_x = +199.1$ $\mathbf{R}_x = 199.1 \rightarrow$	$R_y = +14.3$ $\mathbf{R}_y = 14.3 \uparrow$

The magnitude and direction of the resultant may now be determined.

With a calculator. From the triangle shown, we have

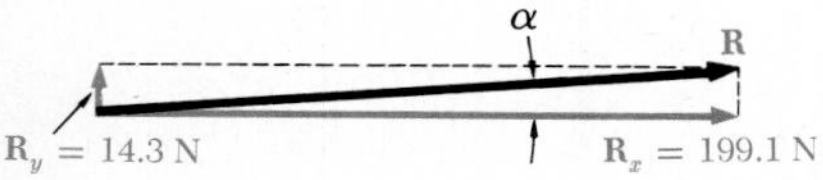

$$\tan \alpha = \frac{R_y}{R_x} = \frac{14.3 \text{ N}}{199.1 \text{ N}} \qquad \alpha = 4.1°$$

$$R = \frac{14.3 \text{ N}}{\sin \alpha} = 199.6 \text{ N} \qquad \mathbf{R} = 199.6 \text{ N} \measuredangle 4.1° \blacktriangleleft$$

The last calculation is facilitated if the value of R_y is stored when originally entered; it may then be recalled to be divided by $\sin \alpha$. (Also see the footnote on p. 27.)

With a slide rule. From the triangle shown, we have

$$R_x \tan \alpha = R_y \qquad (199.1 \text{ N}) \tan \alpha = 14.3 \text{ N}$$

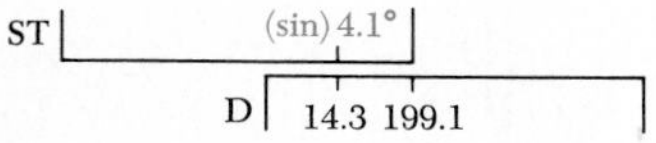

The smaller component (14.3 N) being less than one-tenth of the larger component (199.1 N), we use the ST scale and read $\alpha = 4.1°$. Since within the accuracy of the slide rule the sine and tangent of this angle are equal, the magnitude of **R** may be assumed equal to that of the larger component,

$$\mathbf{R} = 199.1 \text{ N} \measuredangle 4.1° \blacktriangleleft$$

The slide-rule result is accurate to within 0.25 percent, which is quite adequate for this type of engineering problem.

PROBLEMS

2.14 A force of 2.5 kN is applied by a cable attached to the bracket. What are the horizontal and vertical components of this force?

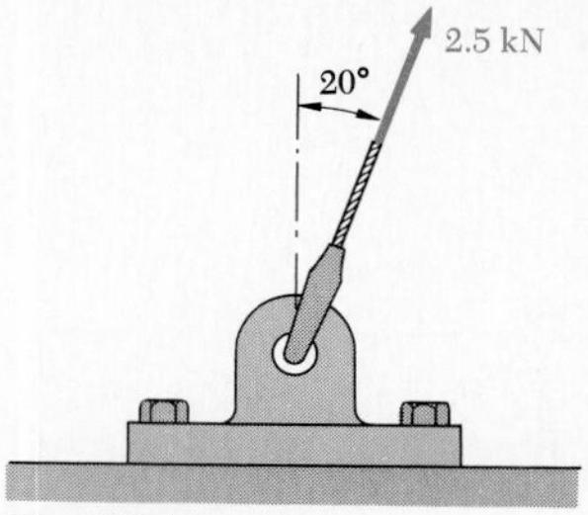

Fig. P2.14

2.15 The force **P** must have a 300-N component acting up the incline. Determine the magnitude of **P** and of its component perpendicular to the incline.

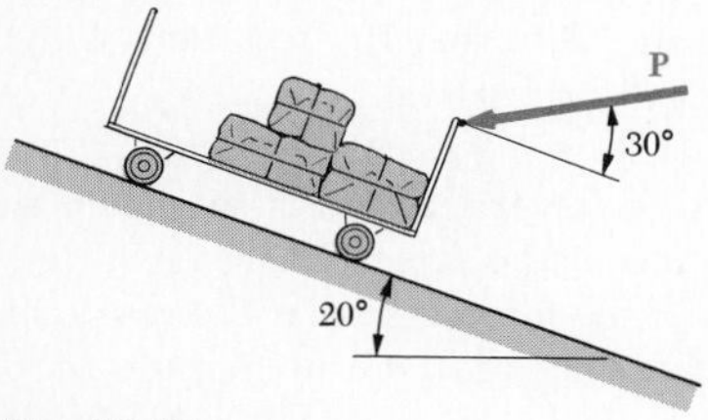

Fig. P2.15

2.16 Determine the x and y components of each of the forces shown.

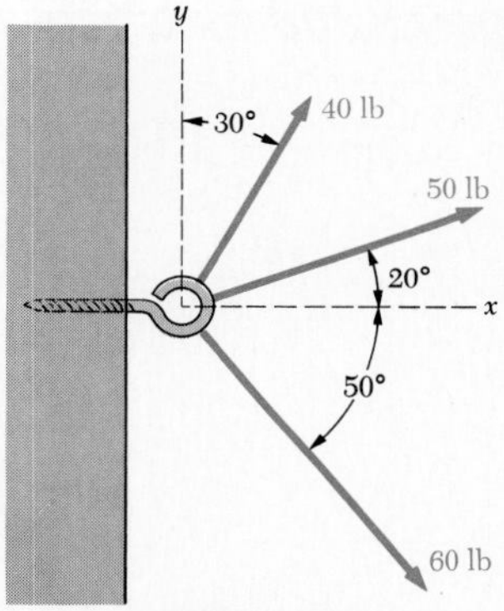

Fig. P2.16 and P2.17

2.17 Determine separately the components of the 40-lb force and of the 60-lb force in directions parallel and perpendicular to the line of action of the 50-lb force.

2.18 The tension in the support wire AB is 650 N. Determine the horizontal and vertical components of the force acting on the pin at A.

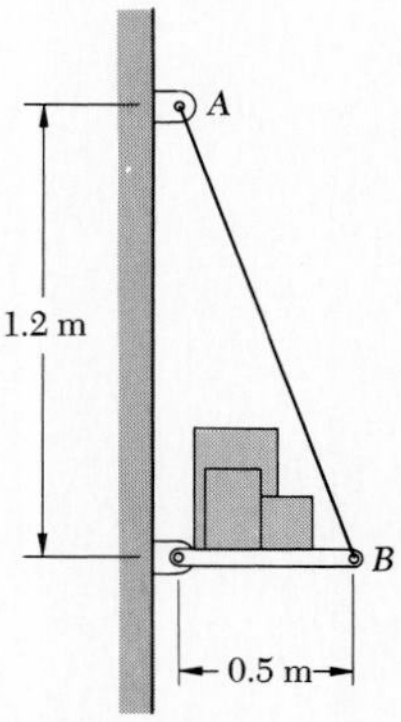

Fig. P2.18

2.19 Determine the x and y components of each of the forces shown.

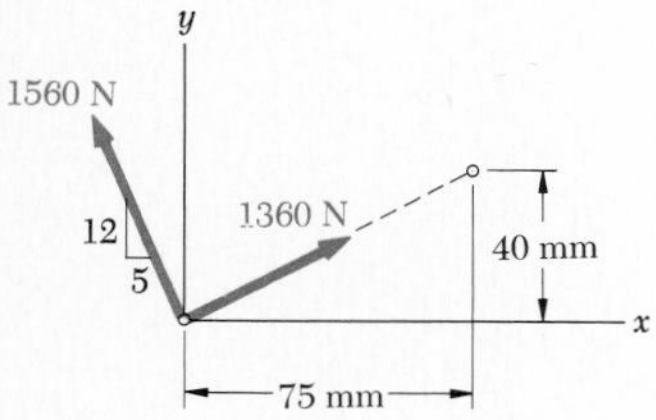

Fig. P2.19

2.20 and 2.21 The x and y components of a force **F** are as shown. Determine the magnitude and direction of the force **F**.

Fig. P2.20

Fig. P2.21

2.22 Determine the resultant of the three forces of Prob. 2.16.

2.23 Determine the resultant of the two forces of Prob. 2.19.

2.24 Using x and y components, solve Prob. 2.2.

2.25 Using x and y components, solve Prob. 2.3.

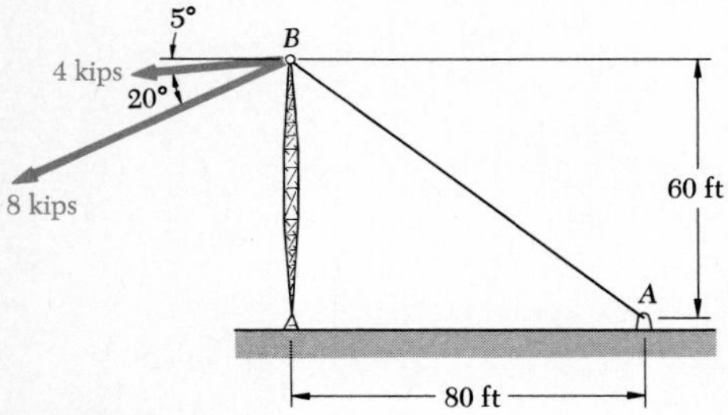

Fig. P2.26

2.26 Two cables which have known tensions are attached at point B. A third cable AB is used as a guy wire and is also attached at B. Determine the required tension in AB so that the resultant of the forces exerted by the three cables will be vertical.

2.27 A collar which may slide on a vertical rod is subjected to the three forces shown. The direction of the force **F** may be varied. If possible, determine the direction of the force **F** so that the resultant of the three forces is horizontal, knowing that the magnitude of **F** is (*a*) 480 lb, (*b*) 280 lb.

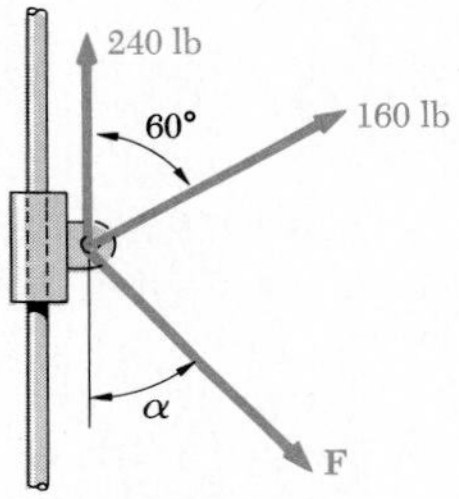

Fig. P2.27

2.28 The directions of the 300-N forces may vary, but the angle between the forces is always 40°. Determine the value of α for which the resultant of the forces acting at A is directed parallel to the plane *b-b*.

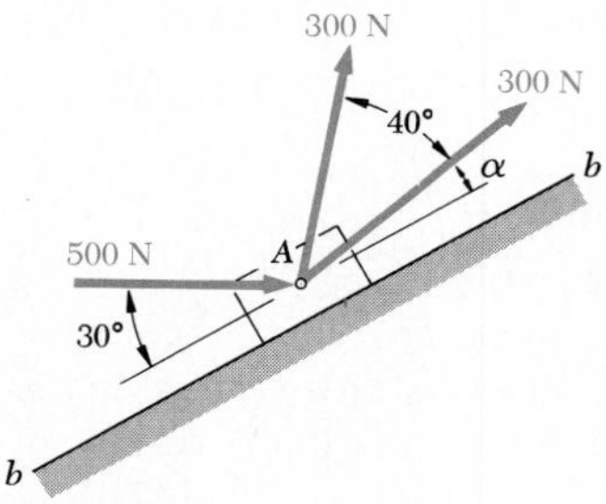

Fig. P2.28

2.29 An additional force **F** is to be attached to the eyebolt of Prob. 2.17. Determine the required magnitude and direction of **F** so that the resultant of all the forces is a 120-lb force directed horizontally to the right.

2.8. Equilibrium of a Particle. In the preceding sections, we discussed the methods for determining the resultant of several forces acting on a particle. Although this has not occurred in any of the problems considered so far, it is quite possible for the resultant to be zero. In such a case, the net effect of the given forces is zero, and the particle is said to be in equilibrium. We thus have the following definition: *When the resultant of all the forces acting on a particle is zero, the particle is in equilibrium.*

A particle which is acted upon by two forces will be in equilibrium if the two forces have the same magnitude, same line of action, and opposite sense. The resultant of the two forces is then zero. Such a case is shown in Fig. 2.27.

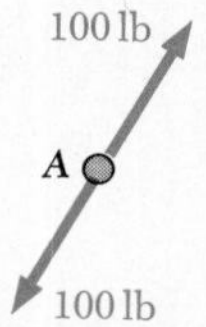

Fig. 2.27

Another case of equilibrium of a particle is represented in Fig. 2.28*a*, where four forces are shown acting on *A*. In Fig. 2.28*b*, the resultant of the given forces is determined by the polygon rule. Starting from point *O* with $\mathbf{F}_1$ and arranging the forces in tip-to-tail fashion, we find that the tip of $\mathbf{F}_4$ coincides with the starting point *O*. Thus the resultant **R** of the given system of forces is zero, and the particle is in equilibrium.

The closed polygon drawn in Fig. 2.28*b* provides a *graphical* expression of the equilibrium of *A*. To express *algebraically* that a particle is in equilibrium, we shall write that the two rectangular components R_x and R_y of the resultant are zero. Since R_x is equal to the sum ΣF_x of the *x* components of the given forces and R_y is equal to ΣF_y, we have the following necessary and sufficient conditions for the equilibrium of a particle:

$$\Sigma F_x = 0 \qquad \Sigma F_y = 0 \tag{2.11}$$

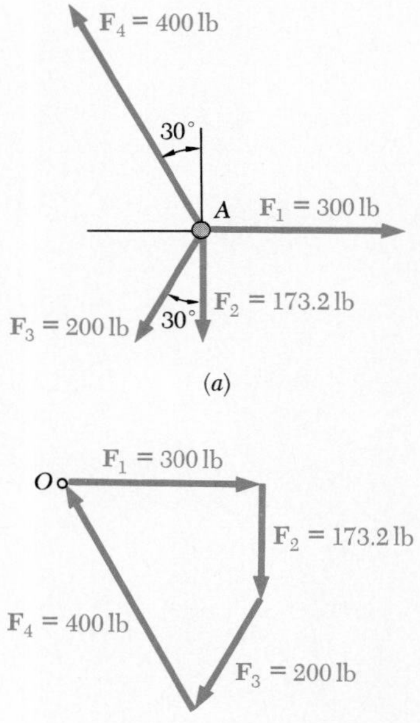

Fig. 2.28

Returning to the particle shown in Fig. 2.28*a* we check that the equilibrium conditions are satisfied. We write

$$\begin{aligned}
\Sigma F_x &= 300 \text{ lb} - (200 \text{ lb}) \sin 30° - (400 \text{ lb}) \sin 30° \\
&= 300 \text{ lb} - 100 \text{ lb} - 200 \text{ lb} = 0 \\
\Sigma F_y &= -173.2 \text{ lb} - (200 \text{ lb}) \cos 30° + (400 \text{ lb}) \cos 30° \\
&= -173.2 \text{ lb} - 173.2 \text{ lb} + 346.4 \text{ lb} = 0
\end{aligned}$$

2.9. Newton's First Law of Motion. In the latter part of the seventeenth century, Sir Isaac Newton formulated three fundamental laws upon which the science of mechanics is based. The first of these laws can be stated as follows:

If the resultant force acting on a particle is zero, the particle will remain at rest (if originally at rest) or will move with constant speed in a straight line (if originally in motion).

From this law and from the definition of equilibrium given in Sec. 2.8, it is seen that a particle in equilibrium either is at rest or is moving in a straight line with constant speed. In the following section, various problems concerning the equilibrium of a particle will be considered.

2.10. Problems Involving the Equilibrium of a Particle. Free-Body Diagram. In practice, a problem in engineering mechanics is derived from an actual physical situation. A sketch showing the physical conditions of the problem is known as a *space diagram.*

The methods of analysis discussed in the preceding sections apply to a system of forces acting on a particle. A large number of problems involving actual structures, however, may be reduced to problems concerning the equilibrium of a particle. This is done by choosing a significant particle and drawing a separate diagram showing this particle and all the forces acting on it. Such a diagram is called a *free-body diagram.*

As an example, consider the 75-kg crate shown in the space diagram of Fig. 2.29*a*. This crate was lying between two buildings, and it is now being lifted onto a truck, which will remove it. The crate is supported by a vertical cable, which is joined at *A* to two ropes which pass over pulleys attached to the buildings at *B* and *C*. It is desired to determine the tension in each of the ropes *AB* and *AC*.

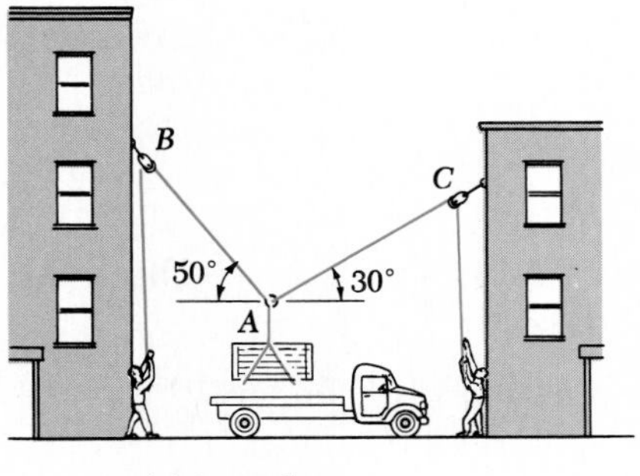

(*a*) Space diagram

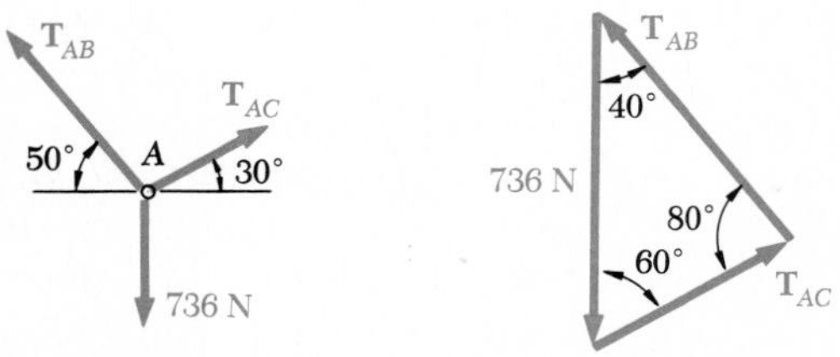

Fig. 2.29 (*b*) Free-body diagram (*c*) Force triangle

In order to solve this problem, a free-body diagram must be drawn, showing a particle in equilibrium. Since we are interested in the rope tensions, the free-body diagram should include at least one of these tensions and, if possible, both tensions. Point A is seen to be a good free body for this problem. The free-body diagram of point A is shown in Fig. 2.29*b*. It shows point A and the forces exerted on A by the vertical cable and the two ropes. The force exerted by the cable is directed downward and is equal to the weight W of the crate. Recalling Eq. (1.4), we write

$$W = mg = (75 \text{ kg})(9.81 \text{ m/s}^2) = 736 \text{ N}$$

and indicate this value in the free-body diagram. The forces exerted by the two ropes are not known. Since they are respectively equal in magnitude to the tension in rope AB and rope AC, we denote them by $\mathbf{T}_{AB}$ and $\mathbf{T}_{AC}$ and draw them away from A in the directions shown in the space diagram. No other detail is included in the free-body diagram.

Since point A is in equilibrium, the three forces acting on it must form a closed triangle when drawn in tip-to-tail fashion. This *force triangle* has been drawn in Fig. 2.29*c*. The values T_{AB} and T_{AC} of the tension in the ropes may be found graphically if the triangle is drawn to scale, or they may be found by trigonometry. If the latter method of solution is chosen, we use the law of sines and write

$$\frac{T_{AB}}{\sin 60°} = \frac{T_{AC}}{\sin 40°} = \frac{736 \text{ N}}{\sin 80°}$$

$$T_{AB} = 647 \text{ N} \qquad T_{AC} = 480 \text{ N}$$

When a particle is in *equilibrium under three forces*, the problem may always be solved by drawing a force triangle. When a particle is in *equilibrium under more than three forces*, the problem may be solved graphically by drawing a force polygon. If an analytic solution is desired, the *equations of equilibrium* given in Sec. 2.8 should be solved:

$$\Sigma F_x = 0 \qquad \Sigma F_y = 0 \tag{2.11}$$

These equations may be solved for no more than *two unknowns*; similarly, the force triangle used in the case of equilibrium under three forces may be solved for two unknowns.

The more common types of problems are those where the two unknowns represent (1) the two components (or the magnitude and direction) of a single force, (2) the magnitude of two forces each of known direction. Problems involving the determination of the maximum or minimum value of the magnitude of a force are also encountered (see Probs. 2.35 and 2.36).

SAMPLE PROBLEM 2.4

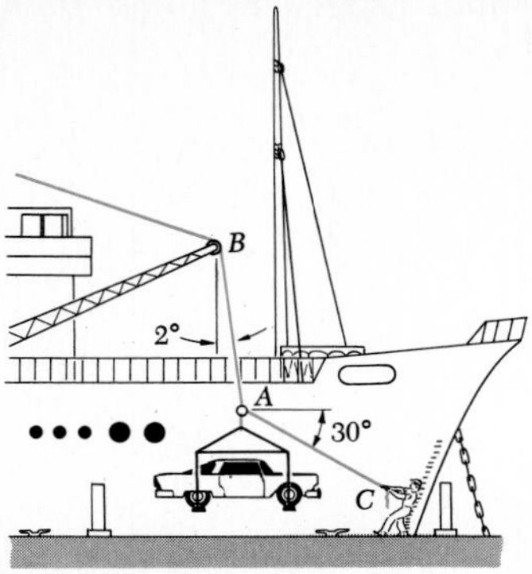

In a ship-unloading operation, a 3500-lb automobile is supported by a cable. A rope is tied to the cable at A and pulled in order to center the automobile over its intended position. The angle between the cable and the vertical is 2°, while the angle between the rope and the horizontal is 30°. What is the tension in the rope?

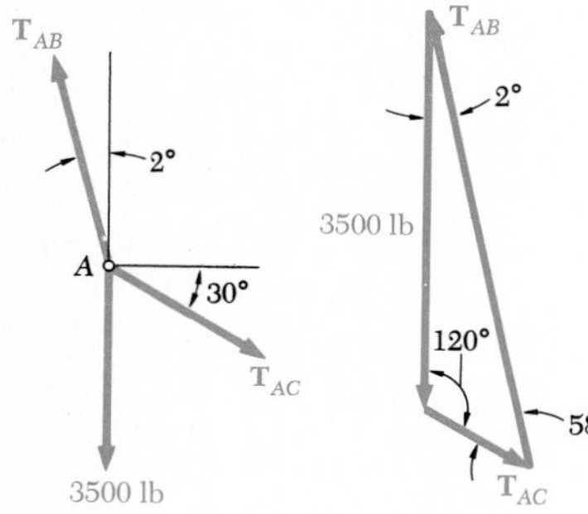

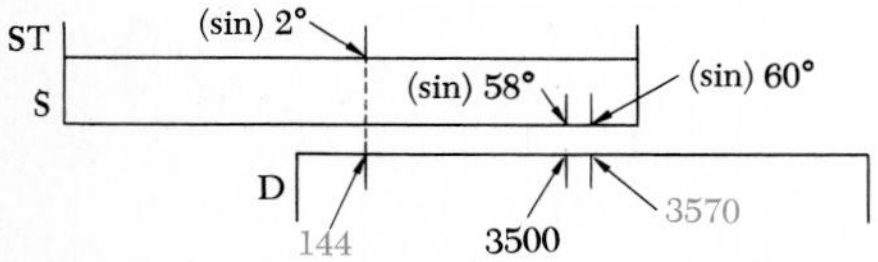

Solution. Point A is chosen as a free body, and the complete free-body diagram is drawn. T_{AB} is the tension in the cable AB, and T_{AC} is the tension in the rope.

Equilibrium Condition. Since only three forces act on the free body, we draw a force triangle to express that it is in equilibrium. Using the law of sines, we write

$$\frac{T_{AB}}{\sin 120°} = \frac{T_{AC}}{\sin 2°} = \frac{3500 \text{ lb}}{\sin 58°}$$

With a calculator. We first compute and store the value of the last quotient. Multiplying this value successively by sin 120° and sin 2°, we obtain

$$T_{AB} = 3570 \text{ lb} \qquad T_{AC} = 144 \text{ lb} \quad ◄$$

With a slide rule. Noting that sin 120° = sin 60° and using both the S and the ST scales, we read the values of T_{AB} and T_{AC}.

SAMPLE PROBLEM 2.5

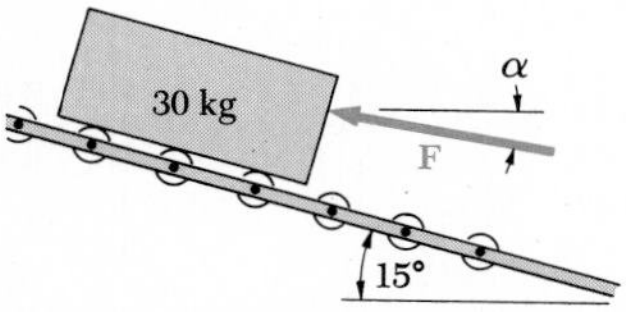

Determine the magnitude and direction of the smallest force **F** which will maintain the package shown in equilibrium. Note that the force exerted by the rollers on the package is perpendicular to the incline.

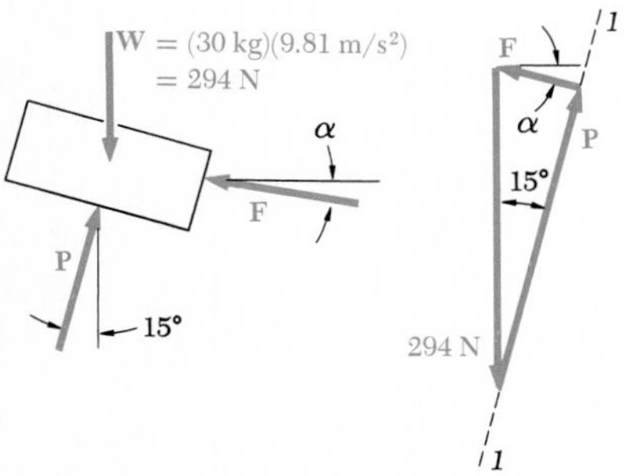

Solution. We choose the package as a free body, assuming that it may be treated as a particle. We draw the corresponding free-body diagram.

Equilibrium Condition. Since only three forces act on the free body, we draw a force triangle to express that it is in equilibrium. Line *1-1* represents the known direction of **P**. In order to obtain the minimum value of the force **F**, we choose the direction of **F** perpendicular to that of **P**. From the geometry of the triangle obtained, we find

$$F = (294 \text{ N}) \sin 15° = 76.1 \text{ N} \qquad \alpha = 15°$$

$$\mathbf{F} = 76.1 \text{ N} ⦪ 15° \quad ◄$$

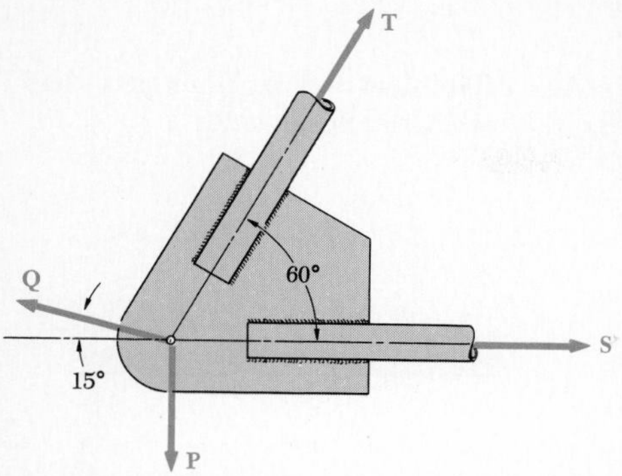

SAMPLE PROBLEM 2.6

Two forces **P** and **Q** of magnitude $P = 1000$ lb and $Q = 1200$ lb are applied to the aircraft connection shown. Knowing that the connection is in equilibrium, determine the tensions S and T.

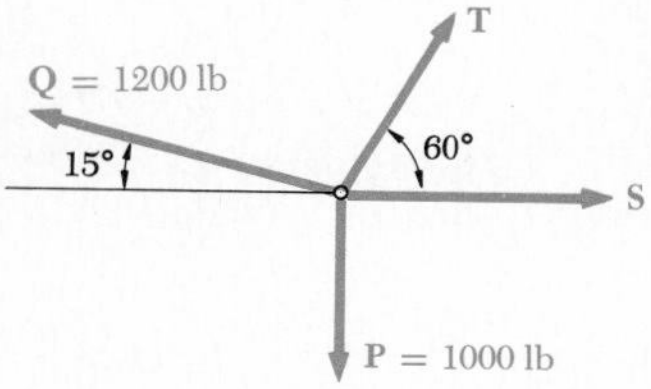

Solution. The connection is considered a particle and taken as a free body. It is acted upon by four forces directed as shown. Each force is resolved into its x and y components. Components directed to the right or upward are assigned a positive sign. We write

$$\begin{aligned} P_x &= 0 & P_y &= -1000 \text{ lb} \\ Q_x &= -(1200 \text{ lb}) \cos 15° & Q_y &= +(1200 \text{ lb}) \sin 15° \\ &= -1159 \text{ lb} & &= +311 \text{ lb} \\ S_x &= +S & S_y &= 0 \\ T_x &= +T \cos 60° & T_y &= +T \sin 60° \\ &= +0.500T & &= +0.866T \end{aligned}$$

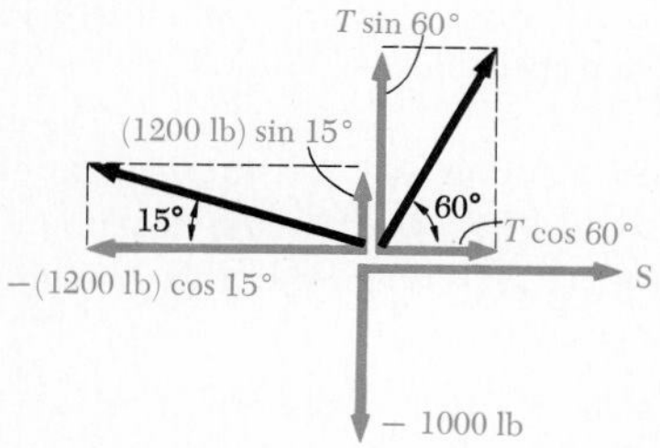

Equilibrium Condition. Since the connection is in equilibrium, the resultant of the forces must be zero. Hence

$$\Sigma F_x = 0 \qquad \Sigma F_y = 0$$

Substituting the values obtained for the components, we write

$$\begin{aligned} \Sigma F_x = 0: &\quad -1159 \text{ lb} + S + 0.500\, T = 0 \\ \Sigma F_y = 0: &\quad -1000 \text{ lb} + 311 \text{ lb} + 0.866\, T = 0 \end{aligned}$$

Solving these equations, we find

$S = 761$ lb $\qquad T = 796$ lb ◄

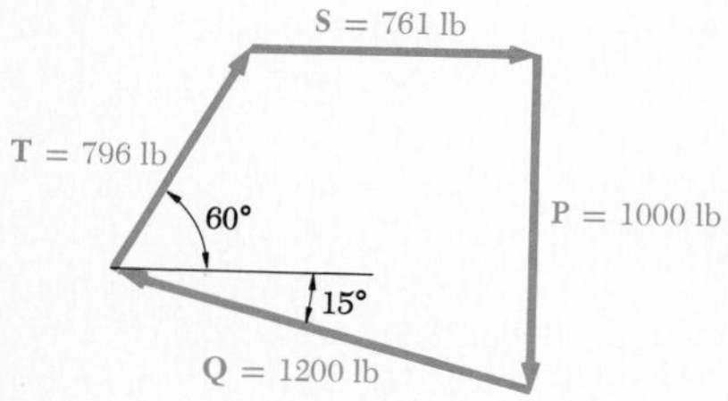

In drawing the free-body diagram, we assumed a sense for each unknown force. A positive sign in the answer indicates that the assumed sense is correct. The complete force polygon may be drawn to check the results.

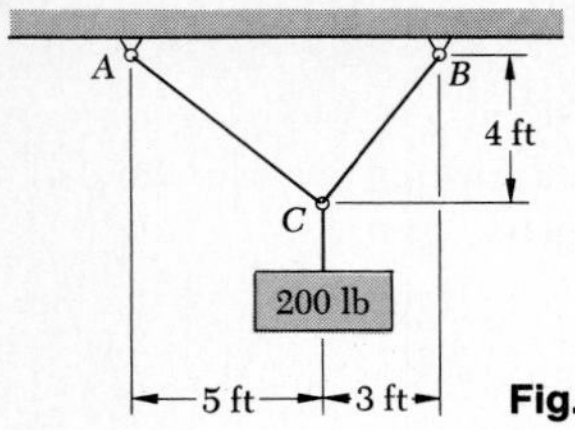

Fig. P2.30

PROBLEMS

2.30 through 2.33 Two cables are tied together at C and loaded as shown. Determine the tension in AC and BC.

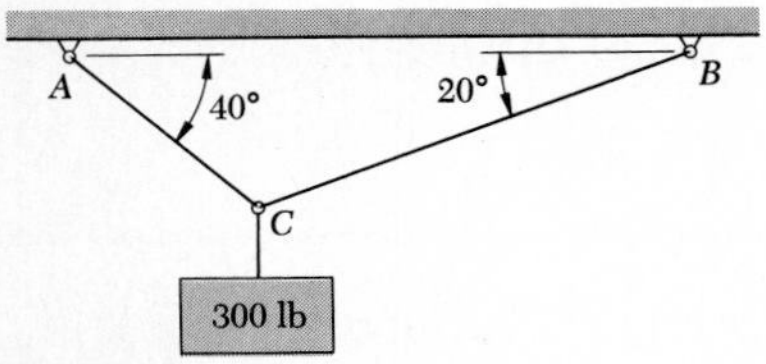

Fig. P2.31

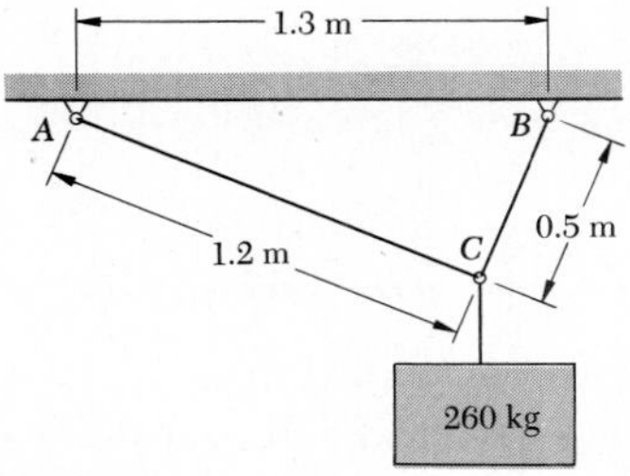

Fig. P2.32

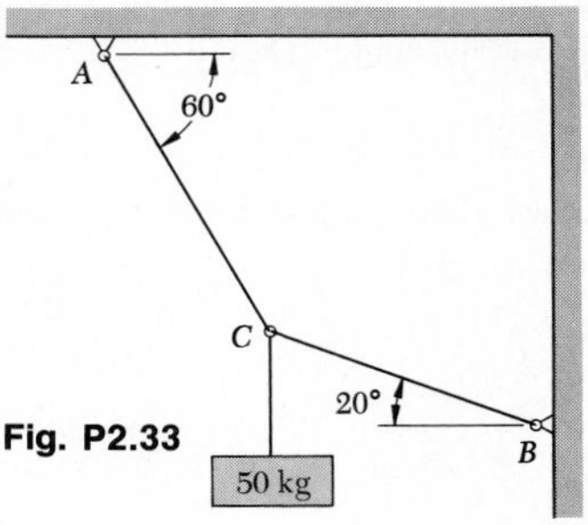

Fig. P2.33

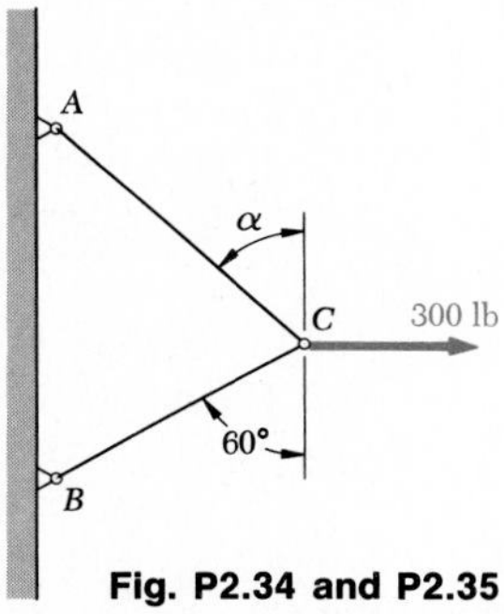

Fig. P2.34 and P2.35

2.34 A 300-lb force is applied at point C. Determine (*a*) the value of α for which the larger of the cable tensions is as small as possible, (*b*) the corresponding values of the tension in cables AC and BC.

2.35 A 300-lb force is applied at point C. (*a*) For what value of α is the tension in cable AC minimum? (*b*) What are the corresponding values of the tension in cables AC and BC?

2.36 Two ropes are tied together at C. If the maximum permissible tension in each rope is 2.5 kN, what is the maximum force **F** that may be applied? In what direction must this maximum force act?

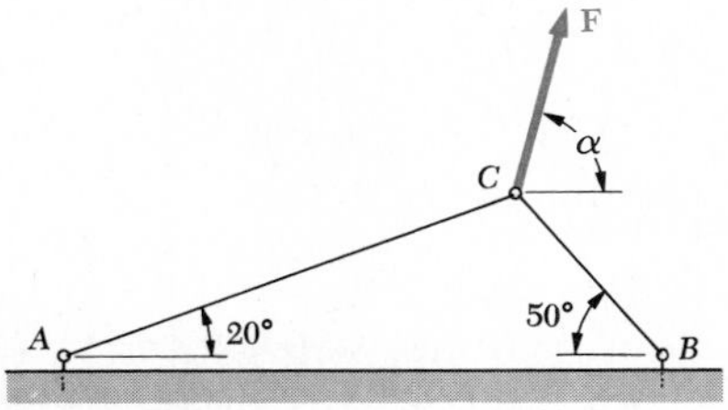

Fig. P2.36

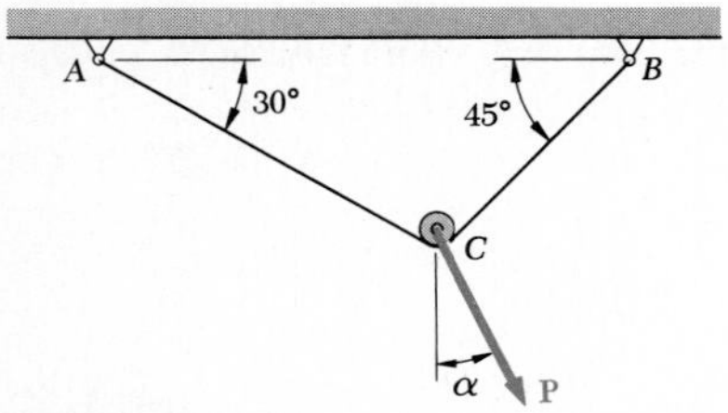

Fig. P2.37

2.37 The force **P** is applied to a small wheel which rolls on the cable ACB. Knowing that the tension in both parts of the cable is 750 N, determine the magnitude and direction of **P**.

2.38 Two forces **A** and **B** of magnitude $A = 5000$ N and $B = 2500$ N are applied to the connection shown. Knowing that the connection is in equilibrium, determine the magnitudes of the forces **C** and **D**.

2.39 Two forces **C** and **D** of magnitude $C = 800$ lb and $D = 1500$ lb are applied to the connection shown. Knowing that the connection is in equilibrium, determine the magnitudes of the forces **A** and **B**.

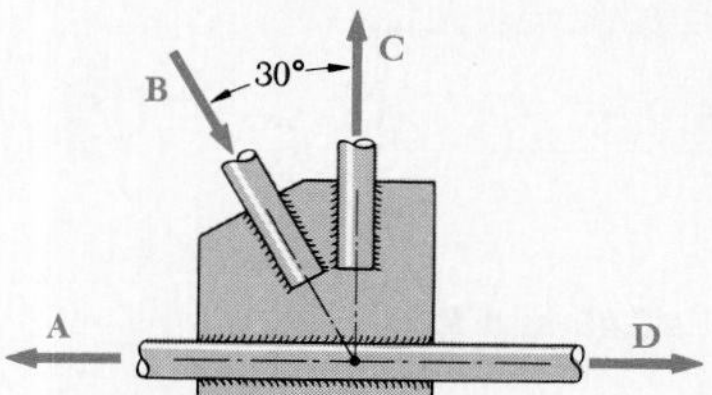

Fig. P2.38 and P2.39

2.40 Knowing that $P = 100$ lb, determine the tension in cables AC and BC.

2.41 Determine the range of values of **P** for which both cables remain taut.

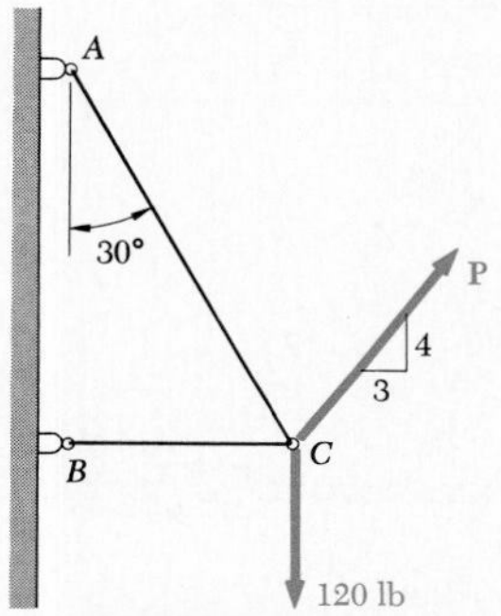

Fig. P2.40 and P2.41

2.42 A portable bin and its contents weigh 1000 lb. Determine the shortest chain sling ACB which may be used to lift the loaded bin if the tension in the chain sling is not to exceed 1300 lb.

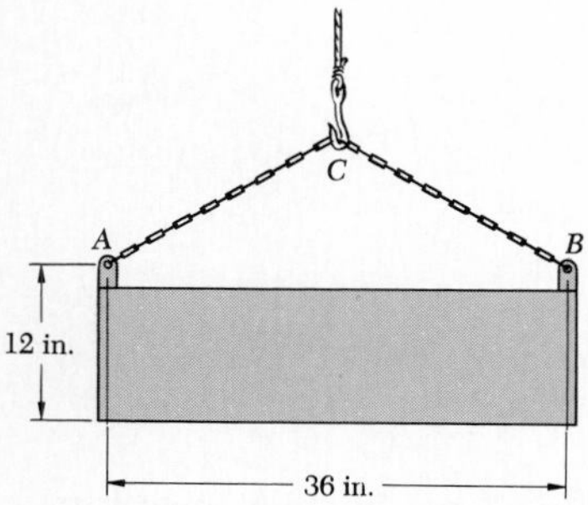

Fig. P2.42

2.43 A 200-kg crate is supported by several rope-and-pulley arrangements as shown. Determine for each arrangement the tension in the rope. (The tension in the rope is the same on each side of a simple pulley. This can be proved by the methods of Chap. 3.)

2.44 Solve parts b and d of Prob. 2.43 assuming that the free end of the rope is attached to the crate.

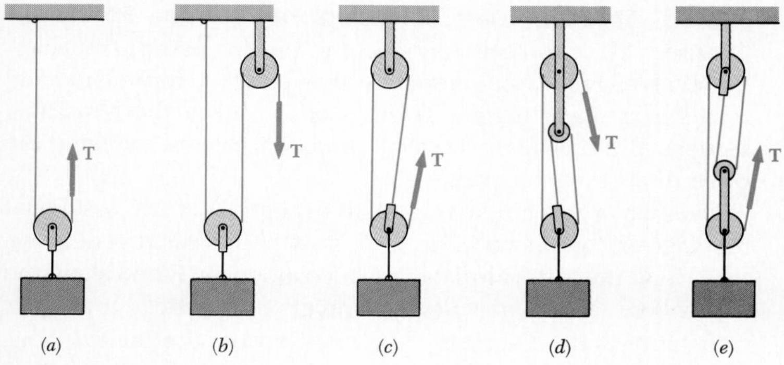

Fig. P2.43

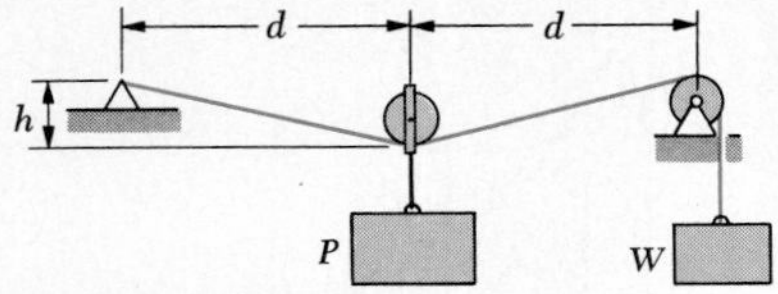

Fig. P2.45 and P2.46

2.45 Express the weight W required to maintain equilibrium in terms of P, d, and h.

2.46 If in the diagram shown $W = 80$ lb, $P = 10$ lb, and $d = 20$ in., determine the value of h consistent with equilibrium.

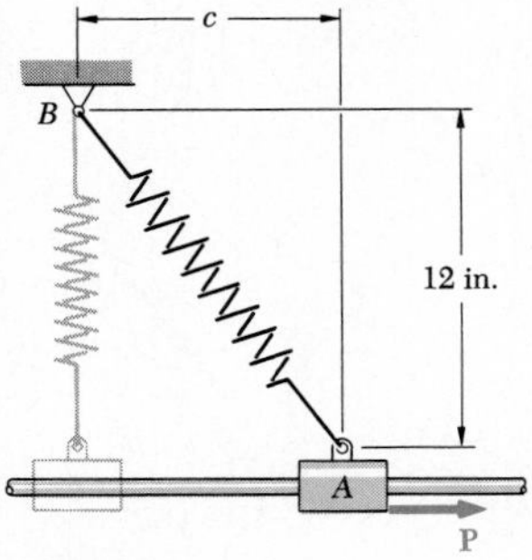

Fig. P2.47

***2.47** The collar A may slide freely on the horizontal smooth rod. The spring attached to the collar has a constant of 10 lb/in. and is undeformed when the collar is directly below support B. Determine the magnitude of the force **P** required to maintain equilibrium when (*a*) $c = 9$ in., (*b*) $c = 16$ in.

2.48 The collar A may slide freely on the horizontal smooth rod. Determine the magnitude of the force **P** required to maintain equilibrium when (*a*) $c = 9$ in., (*b*) $c = 16$ in.

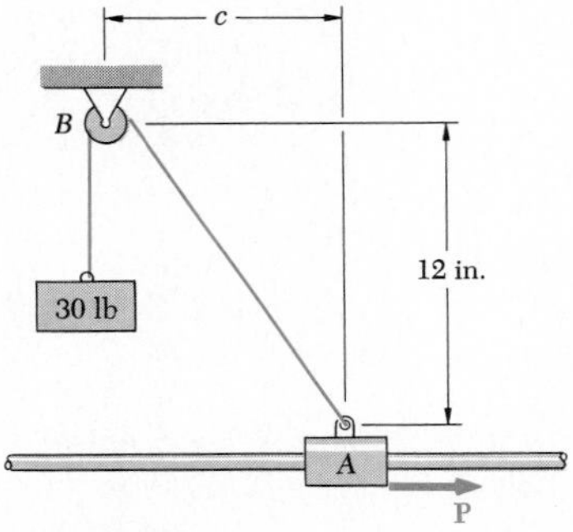

Fig. P2.48

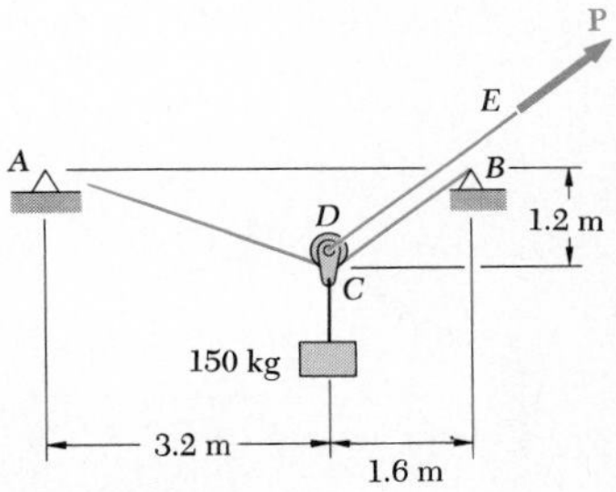

Fig. P2.49

***2.49** A 150-kg block is attached to a small pulley which may roll on the cable ACB. The pulley and load are held in the position shown by a second cable DE which is parallel to the portion CB of the main cable. Determine (*a*) the tension in cable ACB, (*b*) the tension in cable DE. Neglect the radius of the pulleys and the weight of the cables.

FORCES IN SPACE

2.11. Rectangular Components of a Force in Space. The problems considered in the first part of this chapter involved only two dimensions; they could be formulated and solved in a single plane. In this section and in the remaining sections of the chapter, we shall discuss problems involving the three dimensions of space.

Consider a force **F** acting at the origin O of the system of rectangular coordinates x, y, z. To define the direction of **F**, we may draw the vertical plane $OBAC$ containing **F** and shown in Fig. 2.30*a*. This plane passes through the vertical y axis; its orientation is defined by the angle ϕ it forms with the xy plane,

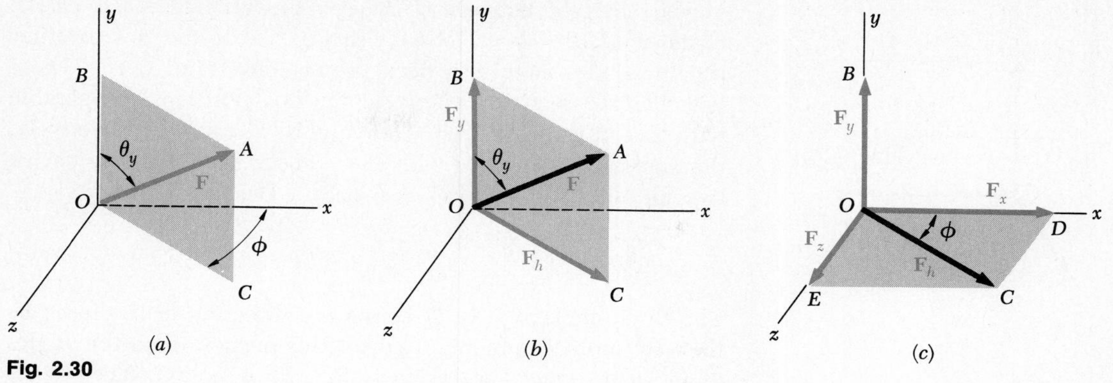

Fig. 2.30

while the direction of **F** within the plane is defined by the angle θ_y that **F** forms with the y axis. The force **F** may be resolved into a vertical component $\mathbf{F}_y$ and a horizontal component $\mathbf{F}_h$; this operation, shown in Fig. 2.30*b*, is carried out inside plane *OBAC* according to the rules developed in the first part of the chapter. The corresponding scalar components are

$$F_y = F\cos\theta_y \qquad F_h = F\sin\theta_y \tag{2.12}$$

But $\mathbf{F}_h$ may be resolved into two rectangular components $\mathbf{F}_x$ and $\mathbf{F}_z$ along the x and z axes, respectively. This operation, shown in Fig. 2.30*c*, is carried out inside the xz plane. We obtain the following expressions for the corresponding scalar components:

$$\begin{aligned} F_x &= F_h\cos\phi = F\sin\theta_y\cos\phi \\ F_z &= F_h\sin\phi = F\sin\theta_y\sin\phi \end{aligned} \tag{2.13}$$

The given force **F** has thus been resolved into three rectangular vector components $\mathbf{F}_x$, $\mathbf{F}_y$, $\mathbf{F}_z$, directed along the three coordinate axes.

Applying the Pythagorean theorem to the triangles *OAB* and *OCD* of Fig. 2.30, we write

$$\begin{aligned} F^2 &= (OA)^2 = (OB)^2 + (BA)^2 = F_y^2 + F_h^2 \\ F_h^2 &= (OC)^2 = (OD)^2 + (DC)^2 = F_x^2 + F_z^2 \end{aligned}$$

Eliminating F_h^2 from these two equations and solving for F, we obtain the following relation between the magnitude of **F** and its rectangular scalar components:

$$F = \sqrt{F_x^2 + F_y^2 + F_z^2} \tag{2.14}$$

The relationship existing between the force **F** and its three components $\mathbf{F}_x$, $\mathbf{F}_y$, $\mathbf{F}_z$ is more easily visualized if a "box" having $\mathbf{F}_x$, $\mathbf{F}_y$, $\mathbf{F}_z$ for edges is drawn as shown in Fig. 2.31. The force **F** is then represented by the diagonal *OA* of this box. Fig. 2.31*b*

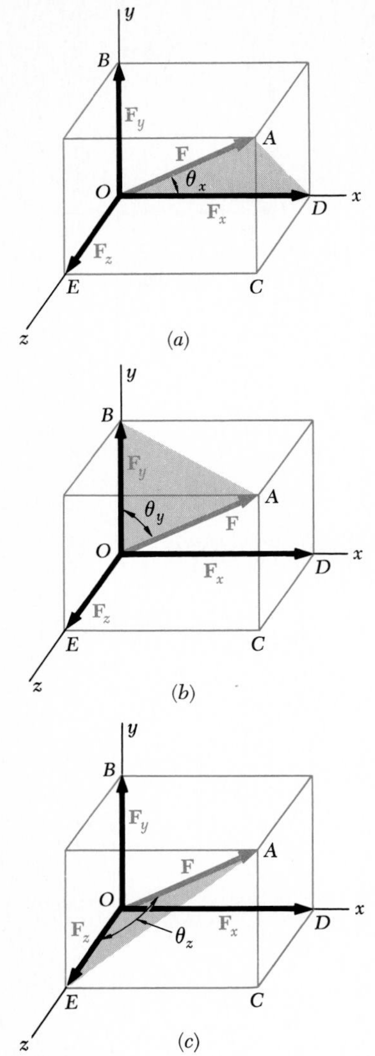

Fig. 2.31

shows the right triangle OAB used to derive the first of the formulas (2.12): $F_y = F\cos\theta_y$. In Fig. 2.31*a* and *c*, two other right triangles have also been drawn: OAD and OAE. These triangles are seen to occupy in the box positions comparable with that of triangle OAB. Denoting by θ_x and θ_z, respectively, the angles that **F** forms with the x and z axes, we may derive two formulas similar to $F_y = F\cos\theta_y$. We thus write

$$F_x = F\cos\theta_x \qquad F_y = F\cos\theta_y \qquad F_z = F\cos\theta_z \tag{2.15}$$

The three angles θ_x, θ_y, θ_z define the direction of the force **F**; they are more commonly used for this purpose than the angles θ_y and ϕ introduced at the beginning of this section. The cosines of θ_x, θ_y, θ_z are known as the direction cosines of the force **F**.

Example 1. A force of 500 N forms angles of 60, 45, and 120°, respectively, with the x, y, and z axes. Find the components F_x, F_y, and F_z of the force.

Substituting $F = 500$ N, $\theta_x = 60°$, $\theta_y = 45°$, $\theta_z = 120°$ into formulas (2.15), we write

$$F_x = (500\text{ N})\cos 60° = +250\text{ N}$$
$$F_y = (500\text{ N})\cos 45° = +354\text{ N}$$
$$F_z = (500\text{ N})\cos 120° = -250\text{ N}$$

As in the case of two-dimensional problems, the plus sign indicates that the component has the same sense as the corresponding axis, and the minus sign that it has the opposite sense.

The angle a force **F** forms with an axis should be measured from the positive side of the axis and will always be comprised between 0 and 180°. An angle θ_x smaller than 90° (acute) indicates that **F** (assumed attached at O) is on the same side of the yz plane as the positive x axis; $\cos\theta_x$ and F_x will then be positive. An angle θ_x larger than 90° (obtuse) would indicate that **F** is on the other side of the yz plane; $\cos\theta_x$ and F_x would then be negative. In Example 1 the angles θ_x and θ_y are acute, while θ_z is obtuse: consequently, F_x and F_y are positive, while F_z is negative.

It should also be observed that the values of the three angles θ_x, θ_y, θ_z are not independent. Substituting for F_x, F_y, F_z from (2.15) into (2.14), we find that the following identity must be satisfied:

$$\cos^2\theta_x + \cos^2\theta_y + \cos^2\theta_z = 1 \tag{2.16}$$

In Example 1, for instance, once the values $\theta_x = 60°$ and $\theta_y = 45°$ have been selected, the value of θ_z *must* be equal to 60° or 120° in order to satisfy identity (2.16).

When the components F_x, F_y, F_z of a force **F** are given, the magnitude F of the force is obtained from (2.14).† The relations (2.15) may then be solved for the direction cosines, and the angles θ_x, θ_y, θ_z characterizing the direction of **F** may be found.‡

Example 2. A force **F** has the components $F_x = 20$ lb, $F_y = -30$ lb, $F_z = 60$ lb. Determine its magnitude F and the angles θ_x, θ_y, θ_z it forms with the axes of coordinates.

From formula (2.14) we obtain†

$$\begin{aligned} F &= \sqrt{F_x^2 + F_y^2 + F_z^2} \\ &= \sqrt{(20 \text{ lb})^2 + (-30 \text{ lb})^2 + (60 \text{ lb})^2} \\ &= \sqrt{4900} \text{ lb} = 70 \text{ lb} \end{aligned}$$

With a calculator, we substitute the values of the components and of the magnitude of **F** into Eqs. (2.15). Solving for the cosines of the angles, we write

$$\cos\theta_x = \frac{F_x}{F} = \frac{20 \text{ lb}}{70 \text{ lb}} \qquad \cos\theta_y = \frac{F_y}{F} = \frac{-30 \text{ lb}}{70 \text{ lb}} \qquad \cos\theta_z = \frac{F_z}{F} = \frac{60 \text{ lb}}{70 \text{ lb}}$$

Calculating successively each quotient and its arc cosine, we obtain

$$\theta_x = 73.4^\circ \qquad \theta_y = 115.4^\circ \qquad \theta_z = 31.0^\circ$$

With a slide rule, we substitute the values of the components and of the magnitude of **F** into (2.17),

$$\frac{\cos\theta_x}{20 \text{ lb}} = \frac{\cos\theta_y}{-30 \text{ lb}} = \frac{\cos\theta_z}{60 \text{ lb}} = \frac{1}{70 \text{ lb}}$$

Bringing one of the extremities of the S scale to coincide with $F = 70$ lb on the D scale, we read the values of θ_x, θ_y, θ_z directly on the S scale (using angles corresponding to cosines) above the values of F_x, F_y, F_z, respectively (Fig. 2.32). Note that, F_y being negative, the value of θ_y must be larger than 90°; thus θ_y is equal not to 64.6° but to its supplement, 115.4°.

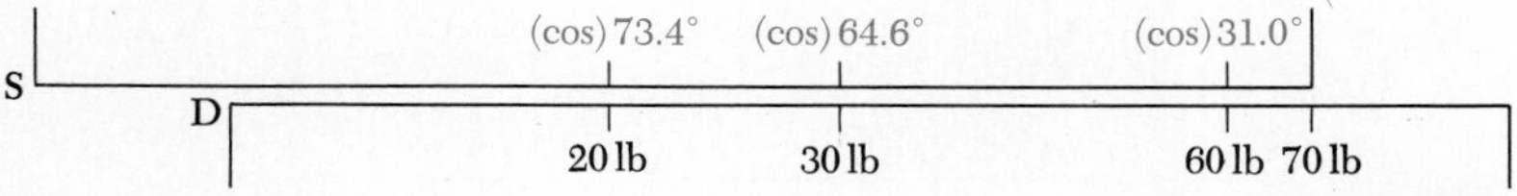

Fig. 2.32

† With a calculator programmed to convert rectangular coordinates into polar coordinates, the following procedure will be found more expeditious for computing F: First determine F_h from its two rectangular components F_x and F_z (Fig. 2.30*c*), then determine F from its two rectangular components F_h and F_y (Fig. 2.30*b*). The actual order in which the three components F_x, F_y, F_z are entered is immaterial.

‡ The relations (2.15) may also be written in the form

$$\frac{\cos\theta_x}{F_x} = \frac{\cos\theta_y}{F_y} = \frac{\cos\theta_z}{F_z} = \frac{1}{F} \tag{2.17}$$

which expresses more clearly the fact that F_x, F_y, F_z, and F are, respectively, proportional to $\cos\theta_x$, $\cos\theta_y$, $\cos\theta_z$, and 1. Formula (2.17) is thus better adapted to the use of the slide rule.

2.12. Force Defined by Its Magnitude and Two Points on Its Line of Action. Another method of defining the direction of a force **F** acting at the origin O of the system of coordinates is to specify a second point N through which the line of action of **F** passes (Fig. 2.33). Through N we may draw lines parallel to the axes of coordinates and thus define a new box. The sides of this box, denoted, respectively, by d_x, d_y, d_z, measure the components of the vector joining O to N in directions parallel to the coordinate axes. This vector is denoted by $\overrightarrow{ON}$. Since, in the case considered, O is the origin of the system of coordinates, the components d_x, d_y, d_z are also the coordinates of point N. Observing that the angles that $\overrightarrow{ON}$ forms with the axes of coordinates are θ_x, θ_y, θ_z, and denoting by d the distance from O to N, we proceed as in Sec. 2.11, and write

$$d_x = d \cos \theta_x \qquad d_y = d \cos \theta_y \qquad d_z = d \cos \theta_z \tag{2.18}$$

$$d = \sqrt{d_x^2 + d_y^2 + d_z^2} \tag{2.19}$$

Dividing member by member the relations (2.15) and (2.18), we obtain

$$\frac{F_x}{d_x} = \frac{F_y}{d_y} = \frac{F_z}{d_z} = \frac{F}{d} \tag{2.20}$$

Formula (2.20) states that the components and magnitude of the force **F** are, respectively, proportional to the components and magnitude of the vector $\overrightarrow{ON}$. This result could have been accepted intuitively by observing that the two boxes shown in Fig. 2.33 have the same shape.

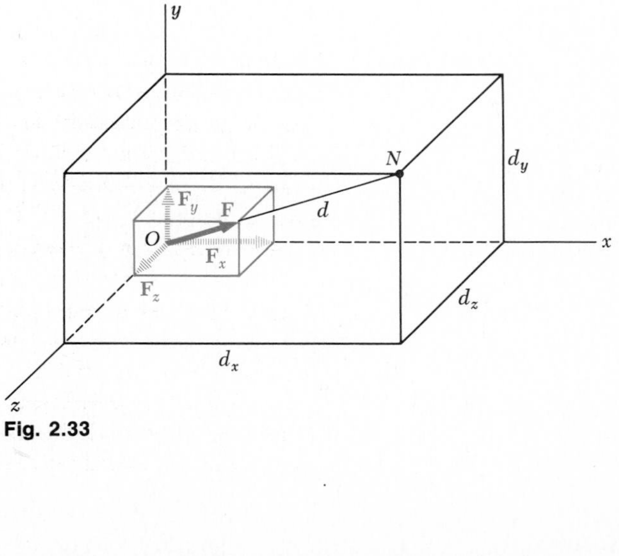

Fig. 2.33

Suppose now that the magnitude F of a force acting at O is known and that the coordinates d_x, d_y, d_z of a point N on its line of action are also given. From (2.19) we may determine the distance d from O to N. Carrying F, d_x, d_y, d_z, and d into (2.20), we obtain the components F_x, F_y, F_z by simple proportions. This computation is easily performed with a calculator or a slide rule, as shown in Sample Prob. 2.7. The angles θ_x, θ_y, θ_z characterizing the direction of **F** may be found by solving Eqs. (2.18) for the direction cosines of the line ON.†

When the line of action of the force **F** is defined by two points M and N, neither of which is at the origin (Fig. 2.34), the components d_x, d_y, d_z are equal to the difference between the coordinates of M and N. These components should have the same sign as the corresponding force components. The sign of d_x, therefore, will be positive if the x coordinate increases when the line MN is described in the sense defined by **F** (from M to N in the case represented in Fig. 2.34); it will be negative if the x coordinate decreases. The signs of d_y and d_z are determined in a similar way.

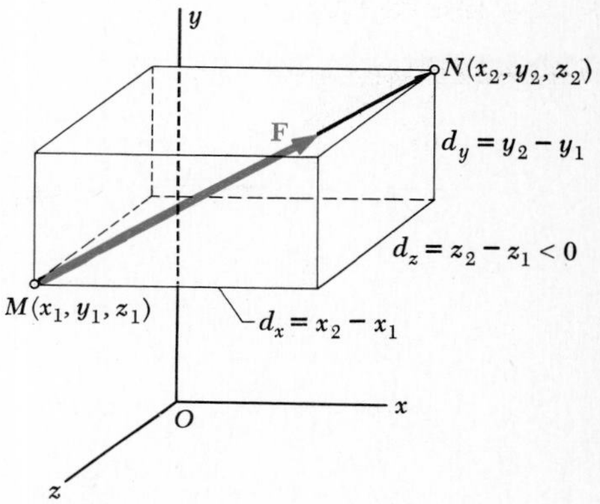

Fig. 2.34

2.13. Addition of Concurrent Forces in Space. We shall determine the resultant **R** of two or more forces in space by summing their rectangular components. Graphical or trigonometric methods are generally not practical in the case of forces in space.

The method followed here is similar to that used in Sec. 2.7 with coplanar forces. First we resolve each of the given forces into rectangular components. Adding algebraically all x components, we obtain the component R_x of the resultant; proceeding in the same manner with the y and z components, we write

$$R_x = \Sigma F_x \qquad R_y = \Sigma F_y \qquad R_z = \Sigma F_z \tag{2.22}$$

The magnitude of the resultant and the angles θ_x, θ_y, θ_z it forms with the axes of coordinates are obtained by the method of Sec. 2.11. We write

$$R = \sqrt{R_x^2 + R_y^2 + R_z^2} \tag{2.23}$$

$$\cos\theta_x = \frac{R_x}{R} \qquad \cos\theta_y = \frac{R_y}{R} \qquad \cos\theta_z = \frac{R_z}{R} \tag{2.24}$$

† The relations (2.18) may also be written in the following form:

$$\frac{\cos\theta_x}{d_x} = \frac{\cos\theta_y}{d_y} = \frac{\cos\theta_z}{d_z} = \frac{1}{d} \tag{2.21}$$

Formula (2.21) may be used to obtain the angles θ_x, θ_y, θ_z on the slide rule, directly from the distance d and the components d_x, d_y, d_z.

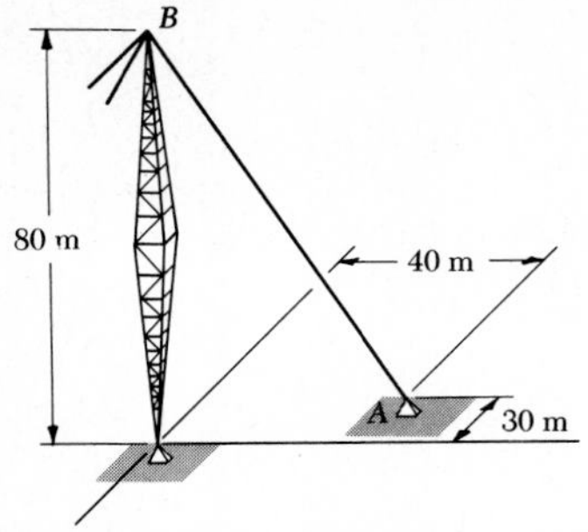

SAMPLE PROBLEM 2.7

A tower guy wire is anchored by means of a bolt at A. The tension in the wire is 2500 N. Determine (*a*) the components F_x, F_y, F_z of the force acting on the bolt, (*b*) the angles θ_x, θ_y, θ_z defining the direction of the force.

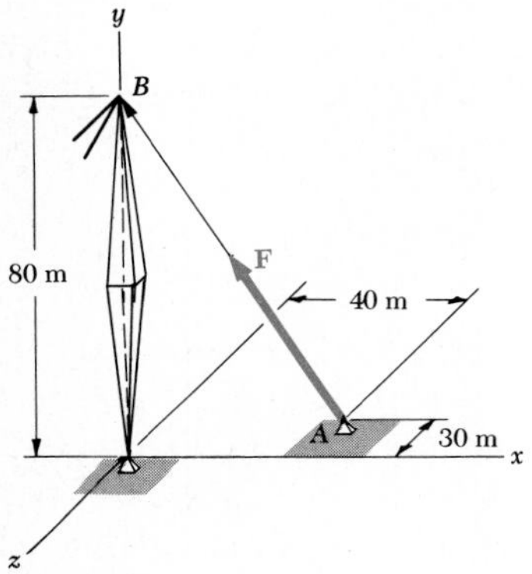

a. **Components of the Force.** The line of action of the force acting on the bolt passes through A and B, and the force is directed from A to B. The components of the vector $\overrightarrow{AB}$, which has the same direction as the force, are

$$d_x = -40 \text{ m} \qquad d_y = +80 \text{ m} \qquad d_z = +30 \text{ m}$$

The total distance from A to B is

$$d = \sqrt{d_x^2 + d_y^2 + d_z^2} = 94.3 \text{ m}$$

Since the components of the force **F** are proportional to the components of the vector $\overrightarrow{AB}$, we write

$$\frac{F_x}{-40 \text{ m}} = \frac{F_y}{+80 \text{ m}} = \frac{F_z}{+30 \text{ m}} = \frac{2500 \text{ N}}{94.3 \text{ m}}$$

With a calculator we first compute and store the value of the last quotient. Multiplying this value successively by -40, $+80$, and $+30$, we obtain

$$F_x = -1060 \text{ N} \qquad F_y = +2120 \text{ N} \qquad F_z = +795 \text{ N} \quad ◄$$

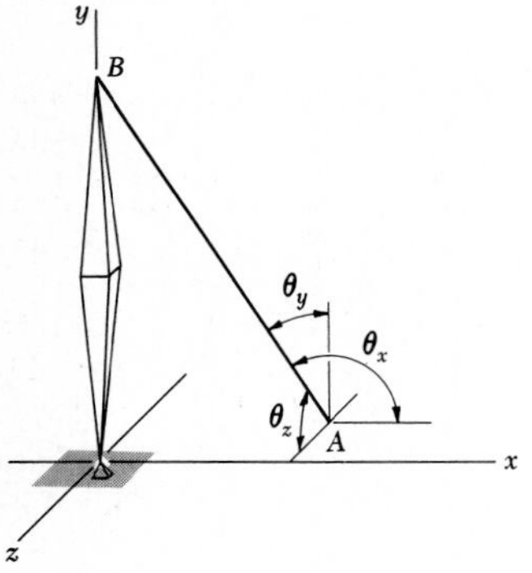

With a slide rule, we set the slide so that 2500 on the C scale is aligned with 94.3 on the D scale. The components of **F** are then read on the C scale, opposite -40, $+80$, and $+30$.

b. **Direction of the Force.** *With a calculator,* we solve Eqs. (2.18) for the direction cosines of the line AB,

$$\cos\theta_x = \frac{d_x}{d} = \frac{-40 \text{ m}}{94.3 \text{ m}} \qquad \cos\theta_y = \frac{d_y}{d} = \frac{+80 \text{ m}}{94.3 \text{ m}}$$

$$\cos\theta_z = \frac{d_z}{d} = \frac{+30 \text{ m}}{94.3 \text{ m}}$$

Calculating successively each quotient and its arc cosine, we obtain

$$\theta_x = 115.1° \qquad \theta_y = 32.0° \qquad \theta_z = 71.5° \quad ◄$$

With a slide rule, we use Eq. (2.21) and write

$$\frac{\cos\theta_x}{-40 \text{ m}} = \frac{\cos\theta_y}{+80 \text{ m}} = \frac{\cos\theta_z}{+30 \text{ m}} = \frac{1}{94.3 \text{ m}}$$

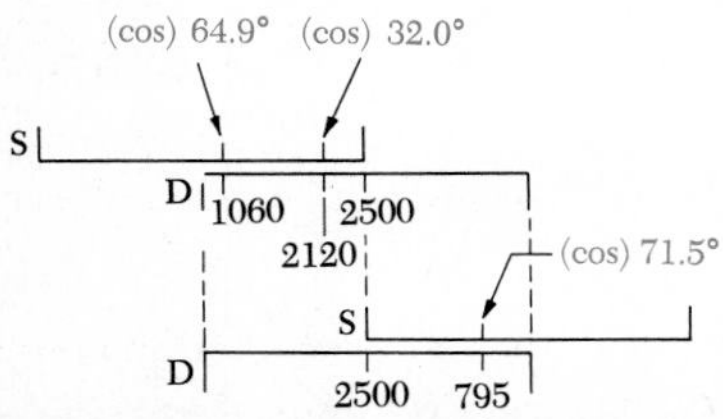

Setting the slide as shown in the adjoining sketch, we obtain

$$\theta_x = 180° - 64.9° = 115.1° \qquad \theta_y = 32.0° \qquad \theta_z = 71.5° \quad ◄$$

Note. The same result would have been obtained by using the components of **F** instead of the components of $\overrightarrow{AB}$.

SAMPLE PROBLEM 2.8

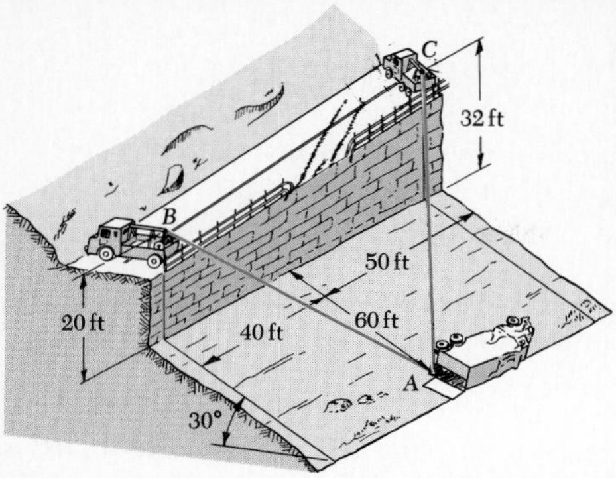

In order to move a wrecked truck, two cables are attached to the truck at A and pulled by winches B and C as shown. Determine the resultant of the forces exerted on the truck by the two cables, knowing that the tension is 2000 lb in cable AB and 1500 lb in cable AC.

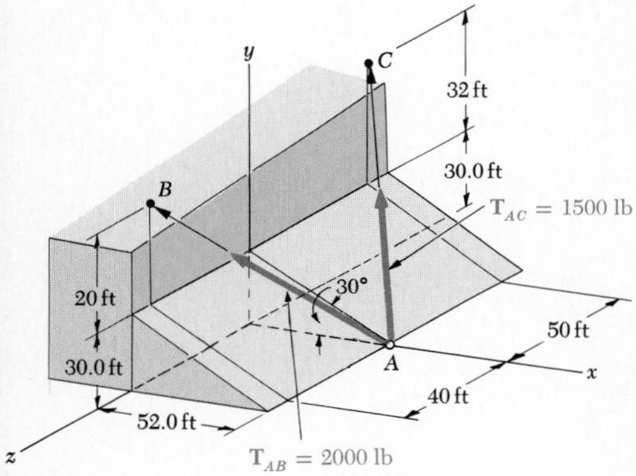

Solution. The force exerted by each cable on the truck will be resolved into x, y, and z components. We first determine the components and magnitudes of the vectors $\overrightarrow{AB}$ and $\overrightarrow{AC}$, measuring them from the truck toward each winch.

Cable AB (A to B): $d_x = -52$ ft $\quad d_y = +50$ ft
$d_z = +40$ ft $\quad d = 82.5$ ft

Cable AC (A to C): $d_x = -52$ ft $\quad d_y = +62$ ft
$d_z = -50$ ft $\quad d = 95.1$ ft

In order to determine the force components, it is necessary to express that force and distance components are proportional. This is most efficiently done in tabular form. The information entered in the table includes the components and the magnitude of the distances as well as the magnitude of the given forces. The force components are then found by proportions and added to obtain the components R_x, R_y, and R_z of the resultant.

Force	Distance Components, ft			d, ft	Force Components, lb			F, lb
	d_x	d_y	d_z		F_x	F_y	F_z	
Cable AB	−52	+50	+40	82.5	−1260	+1212	+970	2000
Cable AC	−52	+62	−50	95.1	−820	+978	−788	1500
					$R_x =$ −2080	$R_y =$ +2190	$R_z =$ +182	

The magnitude and direction of the resultant are now determined.

$$R = \sqrt{R_x^2 + R_y^2 + R_z^2} = \sqrt{(-2080)^2 + (2190)^2 + (182)^2}$$

$$R = 3030 \text{ lb} \quad ◄$$

$$\cos\theta_x = \frac{R_x}{R} = \frac{-2080 \text{ lb}}{3030 \text{ lb}} \qquad \cos\theta_y = \frac{R_y}{R} = \frac{+2190 \text{ lb}}{3030 \text{ lb}}$$

$$\cos\theta_z = \frac{R_z}{R} = \frac{+182 \text{ lb}}{3030 \text{ lb}}$$

$$\theta_x = 133.4° \qquad \theta_y = 43.7° \qquad \theta_z = 86.6° \quad ◄$$

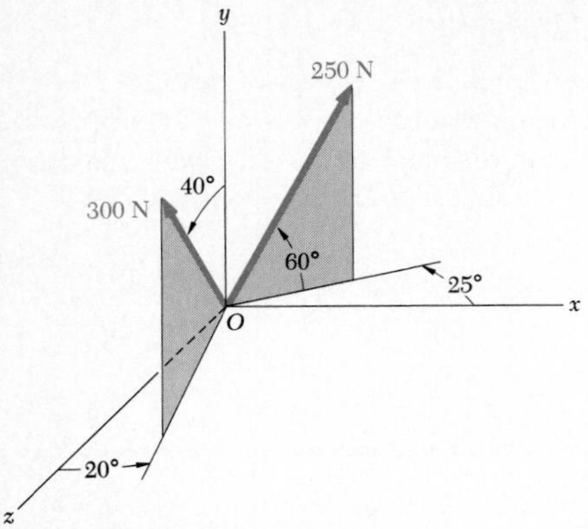

Fig. P2.50 and P2.51

PROBLEMS

2.50 Determine (*a*) the x, y, and z components of the 250-N force, (*b*) the angles θ_x, θ_y, and θ_z that the force forms with the coordinate axes.

2.51 Determine (*a*) the x, y, and z components of the 300-N force, (*b*) the angles θ_x, θ_y, and θ_z that the force forms with the coordinate axes.

2.52 The angle between the guy wire AB and the mast is 20°. Knowing that the tension in AB is 250 lb, determine (*a*) the x, y, and z components of the force exerted on the boat at B, (*b*) the angles θ_x, θ_y, and θ_z defining the direction of the force exerted at B.

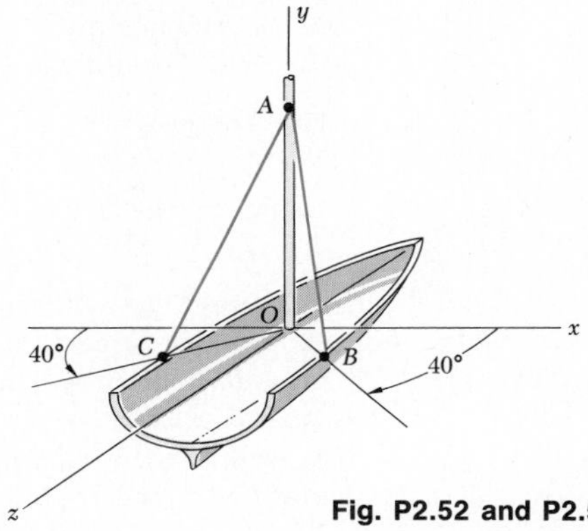

Fig. P2.52 and P2.53

2.53 The angle between the guy wire AC and the mast is 20°. Knowing that the tension in AC is 250 lb, determine (*a*) the x, y, and z components of the force exerted on the boat at C, and (*b*) the angles θ_x, θ_y, and θ_z defining the direction of the force exerted at C.

2.54 A 200-lb force acts at the origin in a direction defined by the angles $\theta_x = 75°$ and $\theta_y = 45°$. It is also known that the z component of the force is positive. Determine the value of θ_z and the components of the force.

2.55 A force acts at the origin in a direction defined by the angles $\theta_y = 110°$ and $\theta_z = 65°$. It is known that the x component of the force is +25 N. Determine the magnitude of the force and the value of θ_x.

2.56 The components of a force are known to be $F_x = +300$ N, $F_y = -360$ N, and $F_z = +520$ N. Determine the magnitude and direction of the force.

2.57 The components of a force are known to be $F_x = -320$ lb, $F_y = -400$ lb, and $F_z = 740$ lb. Determine the magnitude and direction of the force.

2.58 Knowing that the tension in cable AB is 1800 lb, determine the components of the force exerted on the plate at A.

2.59 Knowing that the tension in cable BC is 900 lb, determine the components of the force exerted on the plate at C.

2.60 Determine the angles θ_x, θ_y, and θ_z which define the direction of the force exerted on point A in Prob. 2.58.

2.61 Determine the angles θ_x, θ_y, and θ_z which define the direction of the force exerted on point C in Prob. 2.59.

2.62 Determine the two possible values of θ_y for a force **F**, (*a*) if the force forms equal angles with the positive x, y, and z axes, (*b*) if the force forms equal angles with the positive y and z axes and an angle of 45° with the positive x axis.

2.63 The tension in each of the cables AB and BC is 900 lb. Determine the components of the resultant of the forces exerted on point B.

2.64 Knowing that the tension in AB is 39 kN, determine the required values of the tension in AC and AD so that the resultant of the three forces applied at A is vertical.

2.65 Knowing that the tension in AC is 28 kN, determine the required values of the tension in AB and AD so that the resultant of the three forces applied at A is vertical.

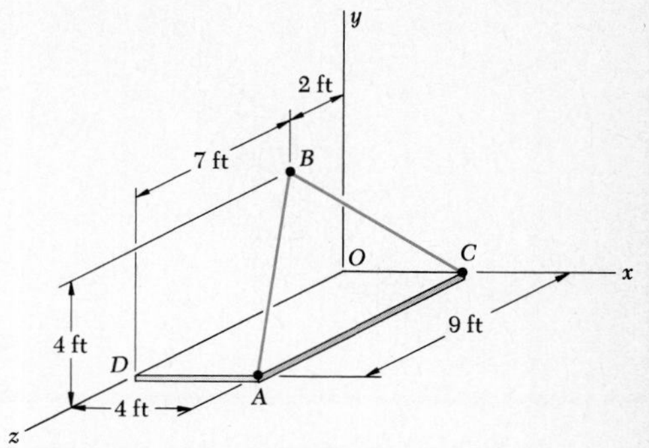

Fig. P2.58, P2.59, and P2.63

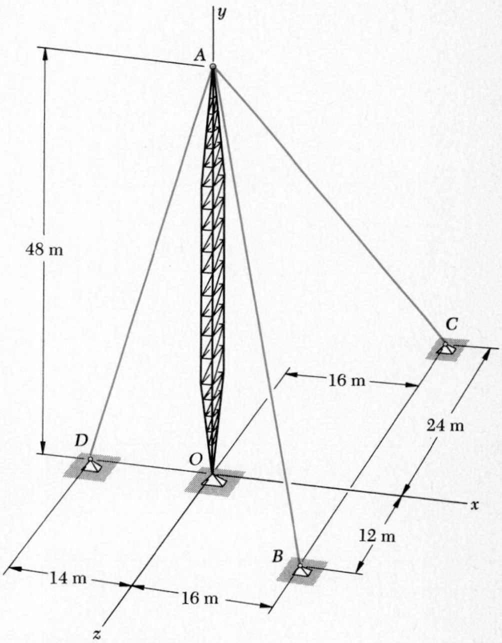

Fig. P2.64 and P2.65

2.14. Equilibrium of a Particle in Space.

According to the definition given in Sec. 2.8, a particle A is in equilibrium if the resultant of all the forces acting on A is zero. The components R_x, R_y, R_z of the resultant are given by the relations (2.22); expressing that the components of the resultant are zero, we write

$$\Sigma F_x = 0 \qquad \Sigma F_y = 0 \qquad \Sigma F_z = 0 \tag{2.25}$$

Equations (2.25) represent the necessary and sufficient conditions for the equilibrium of a particle in space. They may be used to solve problems dealing with the equilibrium of a particle involving no more than three unknowns.

To solve such problems, we first should draw a free-body diagram showing the particle in equilibrium and *all* the forces acting on it. We may then write the equations of equilibrium (2.25) and solve them for three unknowns. In the more common types of problems, these unknowns will represent (1) the three components of a single force or (2) the magnitude of three forces each of known direction.

SAMPLE PROBLEM 2.9

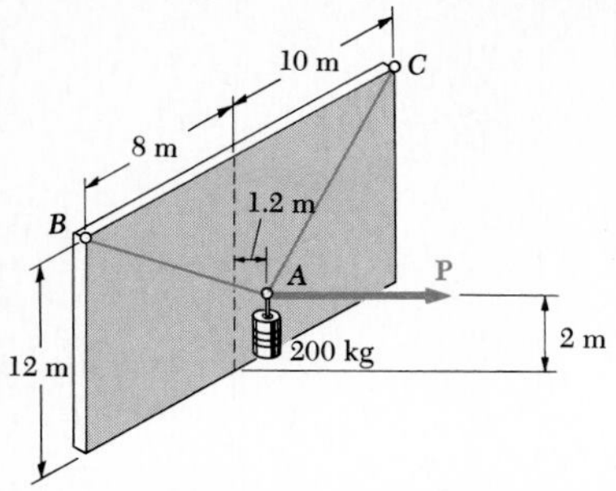

A 200-kg cylinder is hung by means of two cables AB and AC, which are attached to the top of a vertical wall. A horizontal force $\mathbf{P}$ perpendicular to the wall holds the cylinder in the position shown. Determine the magnitude of $\mathbf{P}$ and the tension in each cable.

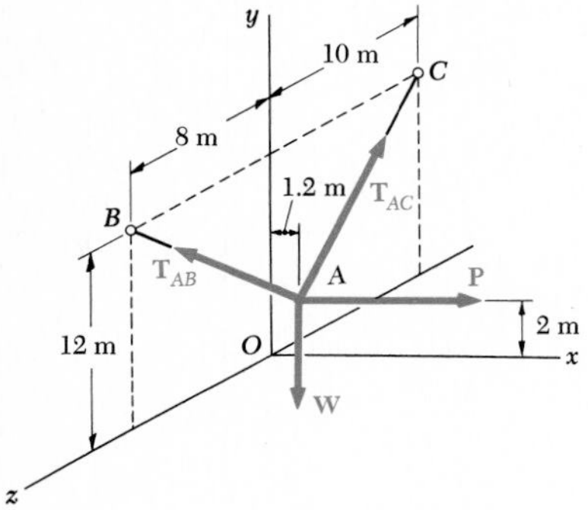

Solution. Point A is chosen as a free body; point A is subjected to four forces as shown. The magnitudes of three of the forces, $\mathbf{P}$, $\mathbf{T}_{AB}$, and $\mathbf{T}_{AC}$, are unknown; the magnitude of the weight $\mathbf{W}$ is found to be

$$W = mg = (200 \text{ kg})(9.81 \text{ m/s}^2) = 1962 \text{ N}$$

The x, y, and z components of each of the unknown forces must be expressed in terms of the unknown magnitudes P, T_{AB}, and T_{AC}. In the case of $\mathbf{T}_{AB}$ and $\mathbf{T}_{AC}$ it is necessary to consider first the corresponding distance components.

Cable AB (A to B): $d_x = -1.2$ m $\quad d_y = 12 - 2 = 10$ m
$\quad d_z = 8$ m $\quad d = 12.86$ m

Cable AC (A to C): $d_x = -1.2$ m $\quad d_y = 12 - 2 = 10$ m
$\quad d_z = -10$ m $\quad d = 14.19$ m

The force components are now obtained by proportions and the results are shown in tabular form.

Force	Distance Components, m			d, m	Force Components, N			Force, N
	d_x	d_y	d_z		F_x	F_y	F_z	
$\mathbf{T}_{AB}$	-1.2	$+10$	$+8$	12.86	$-0.0933T_{AB}$	$+0.778T_{AB}$	$+0.622T_{AB}$	T_{AB}
$\mathbf{T}_{AC}$	-1.2	$+10$	-10	14.19	$-0.0846T_{AC}$	$+0.705T_{AC}$	$-0.705T_{AC}$	T_{AC}
$\mathbf{P}$					$+P$	0	0	P
$\mathbf{W}$					0	-1962	0	-1962

Equilibrium Equations. The equilibrium equations may be written directly by adding successively the force-components columns of the table.

$$\Sigma F_x = 0: \quad -0.0933T_{AB} - 0.0846T_{AC} + P = 0$$
$$\Sigma F_y = 0: \quad +0.778T_{AB} + 0.705T_{AC} - 1962 \text{ N} = 0$$
$$\Sigma F_z = 0: \quad +0.622T_{AB} - 0.705T_{AC} = 0$$

Solving these equations, we obtain

$$P = 235 \text{ N} \qquad T_{AB} = 1401 \text{ N} \qquad T_{AC} = 1236 \text{ N} \blacktriangleleft$$

PROBLEMS

2.66 A load W is supported by three cables as shown. Determine the value of W, knowing that the tension in cable BD is 390 lb.

2.67 A load W is supported by three cables as shown. Determine the value of W, knowing that the tension in cable CD is 750 lb.

2.68 A load W of magnitude 222 lb is supported by three cables as shown. Determine the tension in each cable.

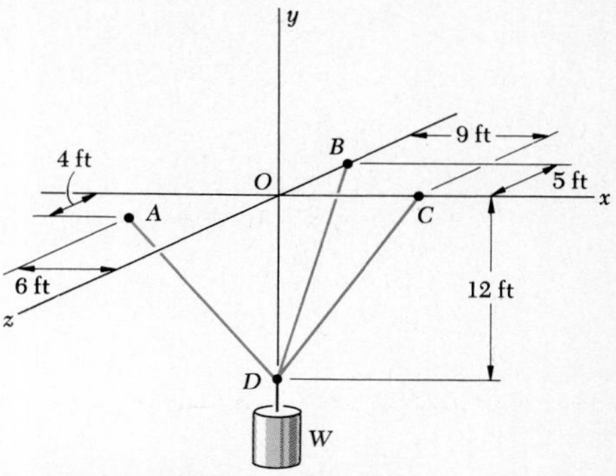

Fig. P2.66, P2.67, and P2.68

2.69 Three cables are joined at D, where two forces **P** and **Q** are applied. If $P = 4$ kN and $Q = 0$, determine the tension in each cable.

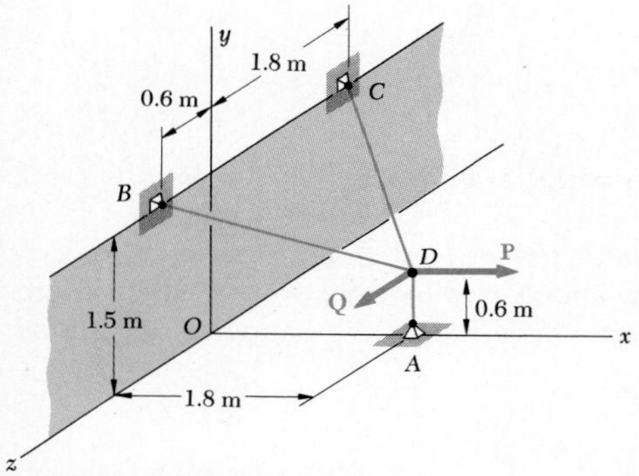

Fig. P2.69 and P2.70

2.70 Three cables are joined at D, where two forces **P** and **Q** are applied. If $P = 3.5$ kN and $Q = 1.5$ kN, determine the tension in each cable.

2.71 A triangular plate of weight 18 lb is supported by three wires as shown. Determine the tension in each wire.

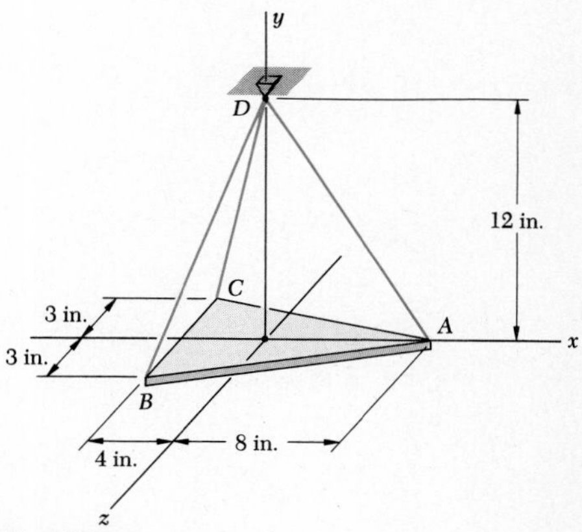

Fig. P2.71

2.72 A 15-kg cylinder is supported by three wires as shown. Determine the tension in each wire when $h = 60$ mm.

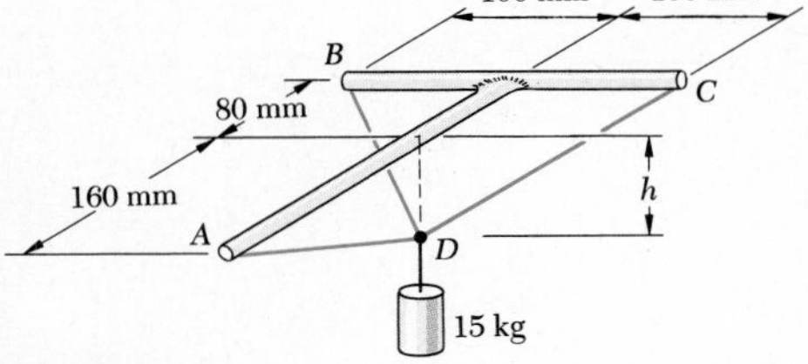

Fig. P2.72

2.73 Solve Prob. 2.72 for the case when $h = 20$ mm. (Explain why this arrangement results in higher tensions.)

2.74 Collar A weighs 5.6 lb and may slide freely on a smooth vertical rod; it is connected to collar B by wire AB. Knowing that the length of wire AB is 18 in., determine the tension in the wire when (a) $c = 2$ in., (b) $c = 8$ in.

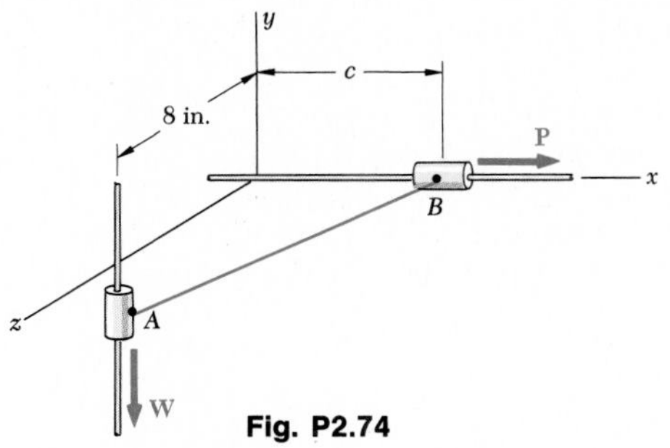

Fig. P2.74

2.75 Solve Prob. 2.74 when (a) $c = 14$ in., (b) $c = 16$ in.

2.76 The uniform circular ring shown has a mass of 20 kg and a diameter of 300 mm. It is supported by three wires each of length 250 mm. If $\alpha = 120°$, $\beta = 150°$, and $\gamma = 90°$, determine the tension in each wire.

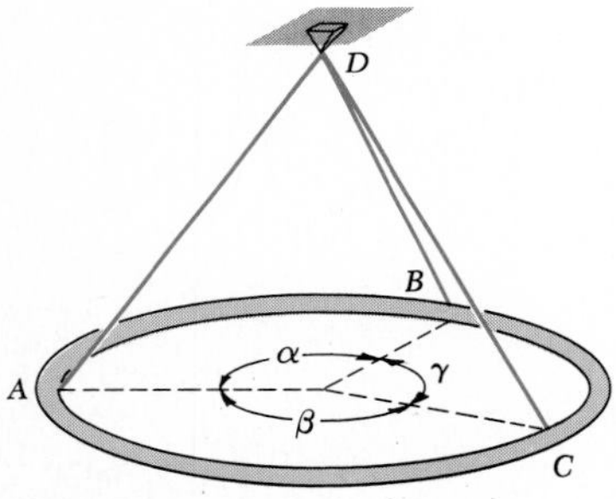

Fig. P2.76

2.77 Solve Prob. 2.76, assuming that $\alpha = \beta = \gamma$.

2.78 Two wires are attached to the top of pole CD. It is known that the force exerted by the pole is vertical and that the 2500-N force applied to point C is horizontal. If the 2500-N force is parallel to the z axis ($\alpha = 90°$), determine the tension in each cable.

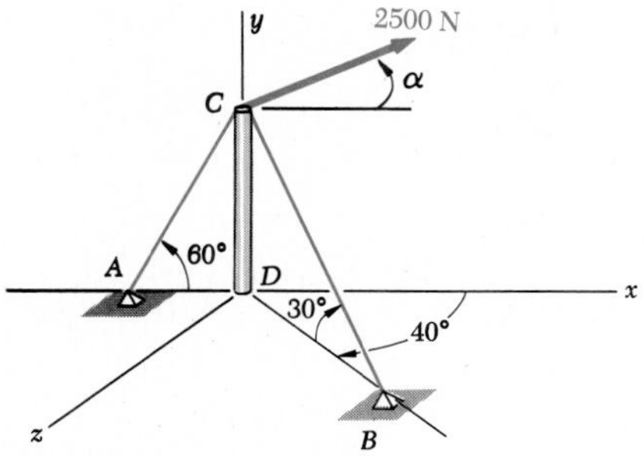

Fig. P2.78

$*$**2.79** In Prob. 2.78, determine (a) the value of the angle α for which the tension in cable AC is maximum, (b) the corresponding tension in each cable.

REVIEW PROBLEMS

2.80 Determine the range of values of P for which the resultant of the three forces applied at A does not exceed 2400 N.

2.81 Knowing that the magnitude of the force $\mathbf{P}$ is 4000 N, determine the resultant of the three forces applied at A.

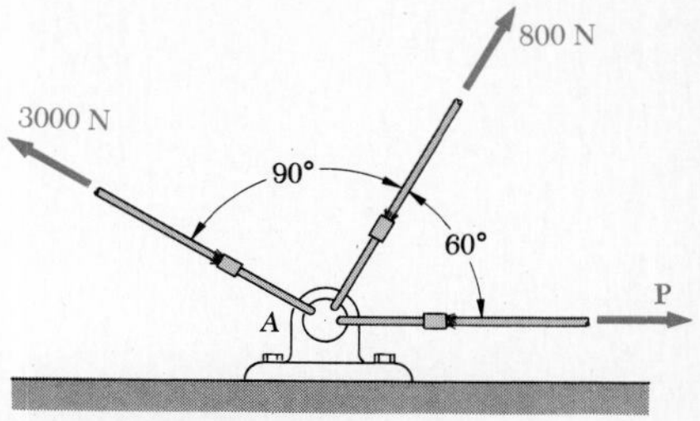

Fig. P2.80 and P2.81

2.82 A 12-ft length of steel pipe weighing 600 lb is lifted by a crane cable CD. Determine the tension in the cable sling ACB, knowing that the length of the sling is (*a*) 15 ft, (*b*) 20 ft.

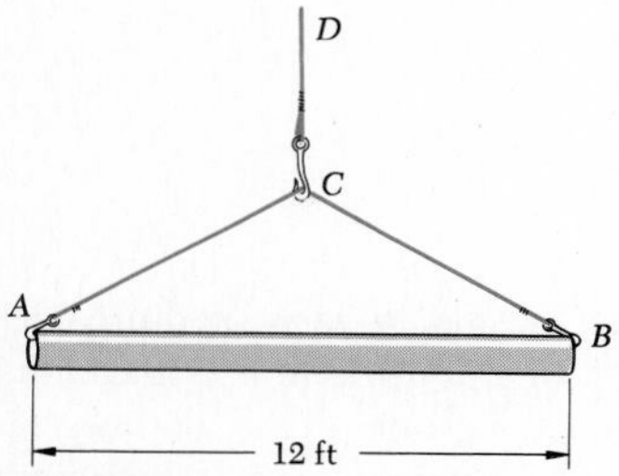

Fig. P2.82

2.83 Three cables are connected at D and support the 3900-lb load shown. Determine the tension in each cable.

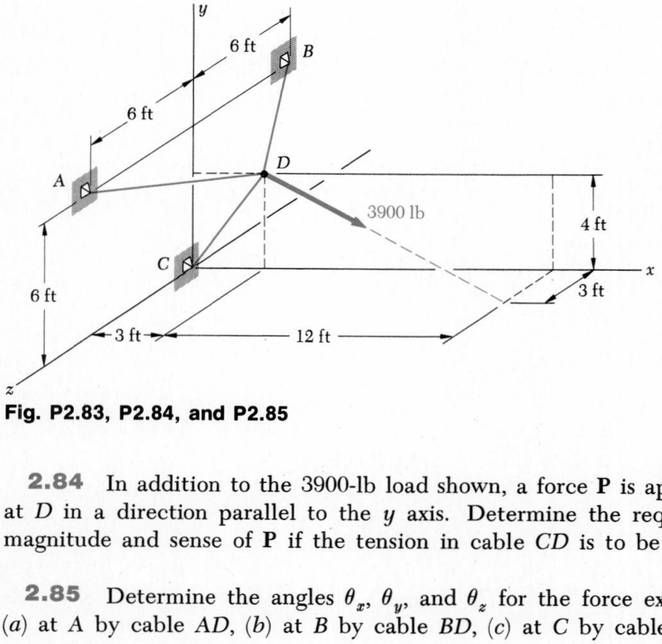

Fig. P2.83, P2.84, and P2.85

2.84 In addition to the 3900-lb load shown, a force $\mathbf{P}$ is applied at D in a direction parallel to the y axis. Determine the required magnitude and sense of $\mathbf{P}$ if the tension in cable CD is to be zero.

2.85 Determine the angles θ_x, θ_y, and θ_z for the force exerted (*a*) at A by cable AD, (*b*) at B by cable BD, (*c*) at C by cable CD.

2.86 Determine the tension in each cable when $F = 800$ N and $\alpha = 90°$.

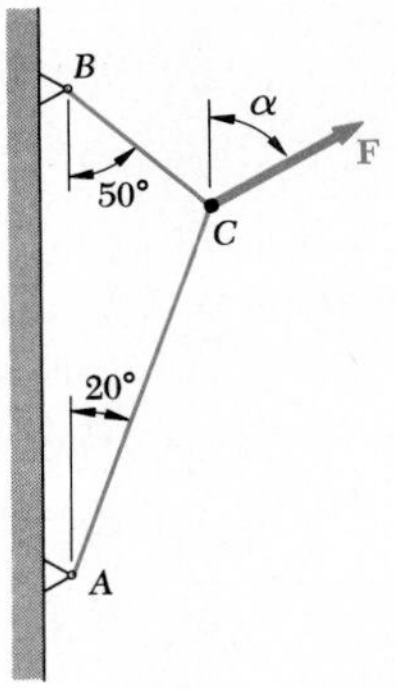

Fig. P2.86 and P2.87

$*$**2.87** The maximum permissible tension in each of the cables AC and CB is 1500 N. For any given value of α, there is a maximum force **F** which may be applied without exceeding the permissible tension in either of the ropes. Considering only cases in which both ropes remain taut, determine the magnitude of the maximum force **F** as a function of the angle α.

2.88 A 100-kg crate is to be supported by the rope-and-pulley arrangement shown. Determine the required magnitude and direction of the force **T**.

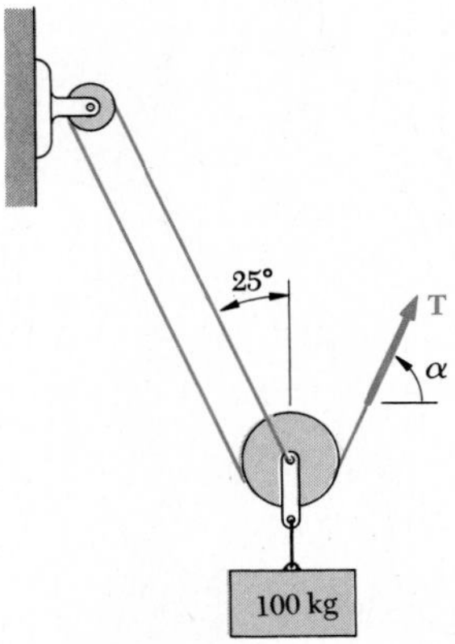

Fig. P2.88

2.89 A 2-kN force acts at the origin in a direction defined by the angles $\theta_y = 46.0°$ and $\theta_z = 80.0°$. It is also known that the x component of the force is positive. Determine the value of θ_x and the components of the force.

2.90 The 68-lb crate is held on the incline by the wire AB and by the horizontal force **P** which is directed parallel to the z axis. Since the crate is mounted on casters, the force exerted by the incline on the crate is perpendicular to the incline. Determine the magnitude of **P** and the tension in wire AB.

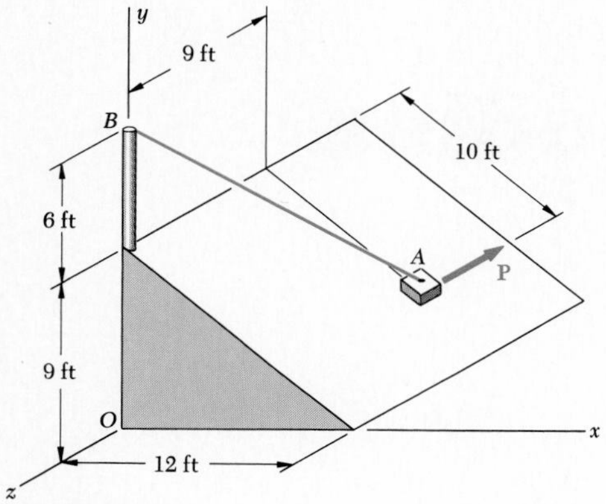

Fig. P2.90

2.91 A painter's scaffold can be supported in the two ways illustrated. If the maximum permissible tension in rope ABC is 100 lb and the maximum permissible tension in rope BD is 150 lb, what is the greatest load that can be carried by each arrangement? (Neglect the effect of the horizontal distance from the building to the scaffold.)

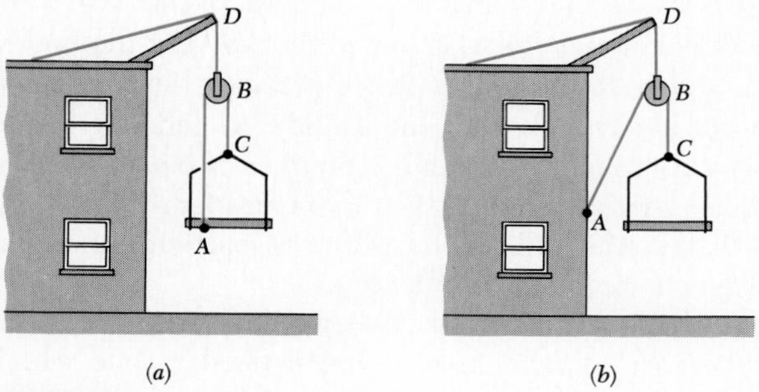

Fig. P2.91

Chapter

3

Statics of Rigid Bodies in Two Dimensions

EQUIVALENT SYSTEMS OF FORCES

3.1. Rigid Bodies. External and Internal Forces. In the preceding chapter it was assumed that each of the bodies considered could be treated as a single particle. Such a view, however, is not always possible, and a body, in general, should be treated as a combination of a large number of particles. The size of the body will have to be taken into consideration, as well as the fact that forces will act on different particles and thus will have different points of application.

Most of the bodies considered in elementary mechanics are assumed to be *rigid*, a *rigid body* being defined as one which does not deform. Actual structures and machines, however, are never absolutely rigid and deform under the loads to which they are subjected. But these deformations are usually small and do not appreciably affect the conditions of equilibrium or motion of the structure under consideration. They are important, though, as far as the resistance of the structure to failure is concerned, and are considered in the study of mechanics of materials.

Forces acting on rigid bodies may be separated into two groups: (1) *external forces;* (2) *internal forces.*

1. The *external forces* represent the action of other bodies on the rigid body under consideration. They are entirely responsible for the external behavior of the rigid body. They will either cause it to move or assure that it remains at rest. We shall be concerned only with external forces in this chapter and in Chaps. 4 and 5.
2. The *internal forces* are the forces which hold together the particles forming the rigid body. If the rigid body is structurally composed of several parts, the forces holding the component parts together are also defined as internal forces. Internal forces will be considered in Chaps. 6 and 7.

As an example of external forces, we shall consider the forces acting on a disabled truck that men are pulling forward by means of a rope attached to the front bumper (Fig. 3.1). The external forces acting on the truck are shown in a *free-body diagram* (Fig. 3.2). Let us first consider the *weight* of the truck. Although it embodies the effect of the earth's pull on each of the particles forming the truck, the weight may be represented by the single force **W**. The *point of application* of this force, i.e., the point at which the force acts, is defined as the *center of gravity* of the truck. It will be seen in Chap. 5 how centers of gravity may be determined. The weight **W** tends to make the truck move vertically downward. In fact, it would actually cause the truck to move downward, i.e., to fall, if it were not for the presence of the ground. The ground opposes the downward motion of the truck by means of the reactions $\mathbf{R}_1$ and $\mathbf{R}_2$. These forces are exerted *by* the ground *on* the truck and must therefore be included among the external forces acting on the truck.

Fig. 3.1

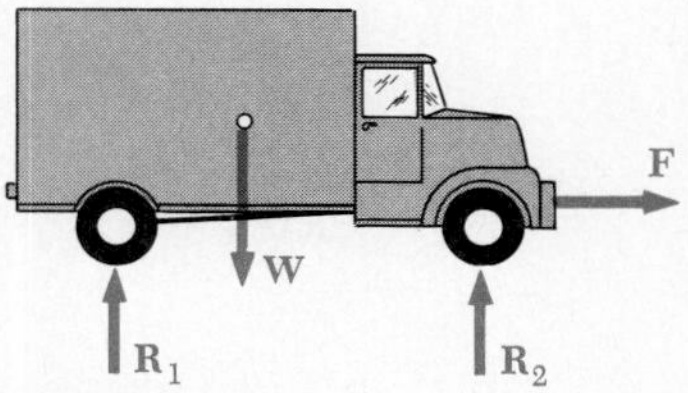

Fig. 3.2

The men pulling on the rope exert the force **F**. The point of application of **F** is on the front bumper. The force **F** tends to make the truck move forward in a straight line and does actually make it move, since no external force opposes this motion. (Rolling resistance has been neglected here for simplicity.) This forward motion of the truck, during which all straight lines remain parallel to themselves (the floor of the truck remains horizontal, and its walls remain vertical), is known as a *translation*. Other forces might cause the truck to move differently. For example, the force exerted by a jack placed under the front axle would cause the truck to pivot about its rear axle. Such a motion is a *rotation*. It may be concluded, therefore, that each of the *external forces* acting on a *rigid body* is capable, if unopposed, of imparting to the rigid body a motion of translation or rotation, or both.

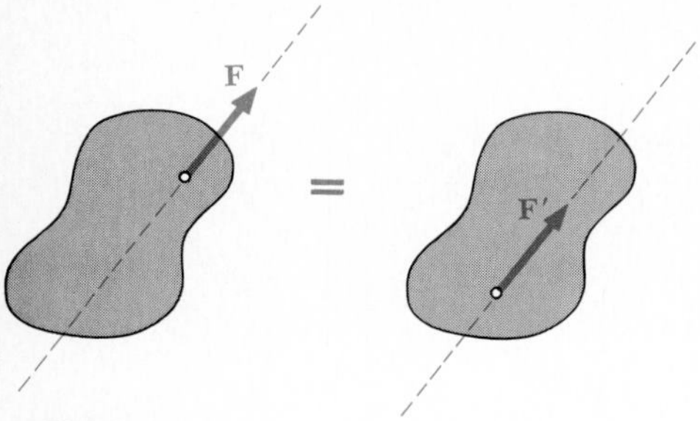

Fig. 3.3

3.2. Principle of Transmissibility. Equivalent Forces. The *principle of transmissibility* states that the conditions of equilibrium or of motion of a rigid body will remain unchanged if a force **F** acting at a given point of the rigid body is replaced by a force **F′** of the same magnitude and same direction, but acting at a different point, *provided that the two forces have the same line of action* (Fig. 3.3). The two forces **F** and **F′** have the same effect on the rigid body and are said to be *equivalent*. This principle, which states in fact that the action of a force may be *transmitted* along its line of action, is based on experimental evidence. It *cannot* be derived from the properties established so far in this text and must therefore be accepted as an experimental law.†

It was indicated in Chap. 2 that the forces acting on a particle could be represented by vectors. These vectors had a well-defined point of application, namely, the particle itself, and were therefore fixed, or bound, vectors. In the case of forces acting on a rigid body, however, the point of application of the force does not matter, as long as the line of action remains unchanged. Thus forces acting on a rigid body must be represented by a different kind of vectors, known as *sliding vectors*, since these vectors may be allowed to slide along their line of action. We should note that all the properties which will be derived in the following sections for the forces acting on a rigid body will be valid more generally for any system of sliding vectors. In order to keep our presentation more intuitive, however, we shall carry it out in terms of physical forces rather than in terms of mathematical sliding vectors.

Returning to the example of the truck, we first observe that the line of action of the force **F** is a horizontal line passing through both the front and the rear bumpers of the truck (Fig. 3.4). Using the principle of transmissibility, we may therefore replace **F** by an *equivalent force* **F′** acting on the rear bumper.

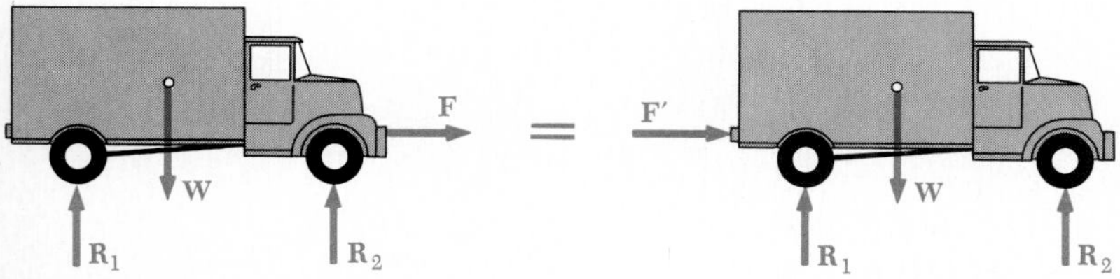

Fig. 3.4

†The principle of transmissibility may be derived from the study of the dynamics of rigid bodies, but this study requires the introduction of all Newton's three laws and of a number of other concepts as well. See F. P. Beer and E. R. Johnston, "Vector Mechanics for Engineers," sec. 16.5, McGraw-Hill Book Company, 1972.

In other words, the conditions of motion are unaffected, and all the other external forces acting on the truck ($\mathbf{W}$, $\mathbf{R}_1$, $\mathbf{R}_2$) remain unchanged if the men push on the rear bumper instead of pulling on the front bumper.

The principle of transmissibility and the concept of equivalent forces have limitations, however. Consider, for example, a short bar AB acted upon by equal and opposite axial forces $\mathbf{P}_1$ and $\mathbf{P}_2$, as shown in Fig. 3.5*a*. According to the principle of transmissibility, the force $\mathbf{P}_2$ may be replaced by a force $\mathbf{P}_2'$ having the same magnitude, same direction, and same line of action, but acting at A instead of B (Fig. 3.5*b*). The forces $\mathbf{P}_1$ and $\mathbf{P}_2'$ acting on the same particle may be added according to the rules of Chap. 2, and, being equal and opposite, their sum is found equal to zero. The original system of forces shown in Fig. 3.5*a* is thus equivalent to no force at all (Fig. 3.5*c*) from the point of view of the external behavior of the bar.

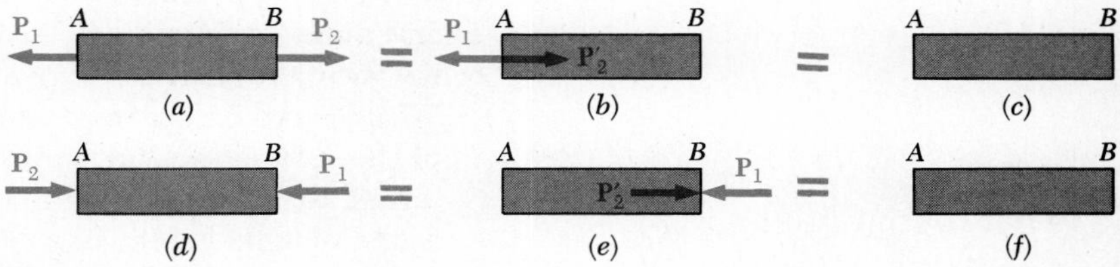

Fig. 3.5

Consider now the two equal and opposite forces $\mathbf{P}_1$ and $\mathbf{P}_2$ acting on the bar AB as shown in Fig. 3.5*d*. The force $\mathbf{P}_2$ may be replaced by a force $\mathbf{P}_2'$ having the same magnitude, same direction, and same line of action, but acting at B instead of A (Fig. 3.5*e*). The forces $\mathbf{P}_1$ and $\mathbf{P}_2'$ may then be added, and their sum is found again to be zero (Fig. 3.5*f*). From the point of view of the mechanics of rigid bodies, the systems shown in Fig. 3.5*a* and *d* are thus equivalent. But the *internal forces* and *deformations* produced by the two systems are clearly different. The bar of Fig. 3.5*a* is in *tension* and, if not absolutely rigid, will increase in length slightly; the bar of Fig. 3.5*d* is in *compression* and, if not absolutely rigid, will decrease in length slightly. Thus, while the principle of transmissibility may be used freely to determine the conditions of motion or equilibrium of rigid bodies and to compute the external forces acting on these bodies, it should be avoided, or at least used with care, in determining internal forces and deformations.

3.3. Two-dimensional Structures. The remarks made in the two preceding sections apply to rigid bodies and to all types of external forces acting on such bodies. For the rest of this chapter, however, we shall be concerned only with two-dimensional structures. By a two-dimensional structure we mean a flat structure which has length and breadth but no depth or, more generally, a structure which contains a plane of symmetry.

All forces acting on these structures will be assumed in the plane of the structure itself, or, more generally, it will be assumed that they may be reduced to forces in the plane of symmetry of the structure. Two-dimensional structures and coplanar forces may be readily represented on a sheet of paper or on a blackboard. Their analysis is therefore considerably simpler than that of three-dimensional structures and forces.

3.4. Moment of a Force about an Axis. It was established in Sec. 3.2 that, from the point of view of the mechanics of rigid bodies, two forces **F** and **F**′ are equivalent if they have the same magnitude, same direction, and same line of action (Fig. 3.6*a*). On the other hand, a force **F**″ of the same magnitude and same direction, but having a different line of action, will not have the same effect as **F** on a given rigid body (Fig. 3.6*b*). While **F** and **F**″ are represented by equal vectors, i.e., by vectors of the same magnitude and same direction (Sec. 2.2), these two forces are *not equivalent*. True, both forces will tend to give to the rigid body the same motion of *translation*, but they would make it *rotate* differently about an axis through A, perpendicular to the plane of the figure.

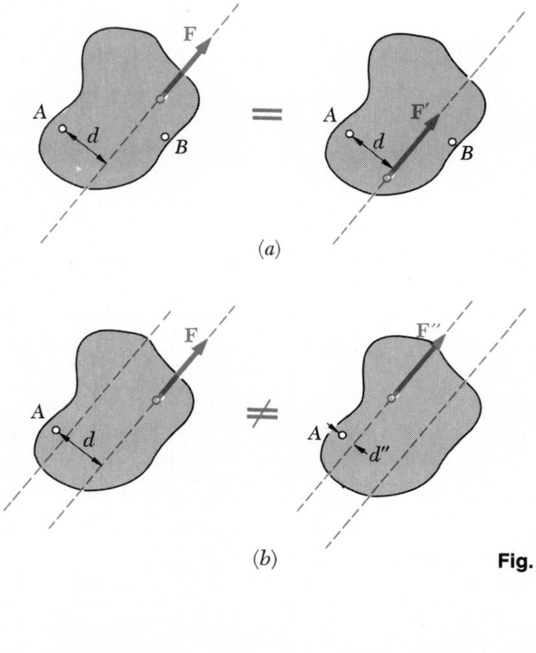

Fig. 3.6

The tendency of a force to make a rigid body rotate about an axis is measured by the moment of the force about that axis. The *moment* M_A *of the force* **F** *about an axis through* A, or, for short, the *moment of* **F** *about* A, is defined as the product of the magnitude F of the force and of the perpendicular distance d from A to the line of action of **F**:

$$M_A = Fd \tag{3.1}$$

In the SI system of units, where a force is expressed in newtons (N) and a distance in meters (m), the moment of a force will be expressed in newton-meters (N · m). In the U.S. customary system of units, where a force is expressed in pounds and a distance in feet or inches, the moment of a force will be expressed in lb · ft or lb · in.

The moment of a force has not only a magnitude but also a sense; depending upon the relative position of the force and of the axis, the moment of the force will be either clockwise ↻ or counterclockwise ↺. It is seen that the moment of **F** about A in Fig. 3.6*a* is counterclockwise, while its moment about B is clockwise. Moments of forces may be added algebraically as scalar quantities if a sign convention is adopted. We shall in this text consider as positive a counterclockwise moment, and as negative a clockwise moment.

Since the forces **F** and **F**′ shown in Fig. 3.6*a* have the same magnitude, same sense, and same line of action, the moment of **F**′ about A will also be Fd and its sense will also be counterclockwise. Both forces have the same moment about A. Clearly they would have *the same moment about any other axis*. It may thus be stated that *two equivalent forces have the same components* F_x *and* F_y *and the same moment about any axis*. On the other hand, the moment of **F**″ about A in Fig. 3.6*b* is different; it is equal to Fd''.

A force acting on a rigid body is completely defined (except for its exact position on its line of action) if its components F_x and F_y and its moment M_A about A are known (in magnitude and sense). The magnitude F of the force and its direction may be determined from the components F_x and F_y by the methods of Chap. 2, while the perpendicular distance d from A to the line of action may be obtained by forming the quotient of M_A over F. The line of action itself will be drawn on one side of A or the other, depending upon the sense of **F** and upon the sense of the moment.

3.5. Varignon's Theorem. An important theorem of statics is due to the French mathematician Varignon (1654–1722). It states that the *moment of a force about any axis is equal to the sum of the moments of its components about that axis.*

To prove this statement, consider a force **F** acting at point A and the components **P** and **Q** of the force **F** in any two directions (Fig. 3.7*a*). The moment of **F** about an axis through an arbitrary point B is Fd, where d is the perpendicular distance from B to the line of action of **F**. Similarly, the moments of **P** and **Q** about the axis through B are Pd_1 and Qd_2, respectively, where d_1 and d_2 denote the perpendicular distances from B to the lines of action of **P** and **Q**. We propose to show that

$$Fd = Pd_1 + Qd_2 \tag{3.2}$$

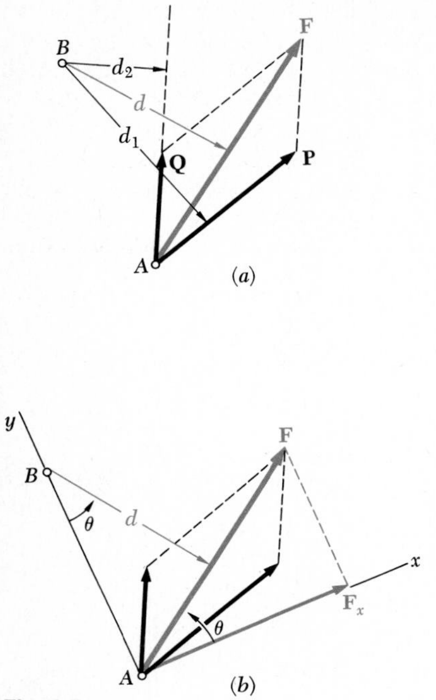

Fig. 3.7

We shall attach at A two rectangular axes as shown in Fig. 3.7*b*; the x axis is perpendicular to AB, while the y axis is directed along AB. Denoting by θ the angle that **F** forms with the x axis, and observing that θ is also the angle between AB and the perpendicular drawn from B onto the line of action of **F**, we write

$$Fd = F(AB)\cos\theta = (AB)F\cos\theta$$
$$Fd = (AB)F_x \tag{3.3}$$

where F_x denotes the x component of the force **F**. Expressing in a similar way the moments Pd_1 and Qd_2 of the forces **P** and **Q**, we write

$$\begin{aligned} Pd_1 + Qd_2 &= (AB)P_x + (AB)Q_x \\ &= (AB)(P_x + Q_x) \end{aligned}$$

But, as we saw in Sec. 2.7, the sum $P_x + Q_x$ of the x components of two forces **P** and **Q** is equal to the x component F_x of their resultant **F**. The expression obtained for the sum $Pd_1 + Qd_2$ of the moments of **P** and **Q** is, therefore, identical with the expression obtained in (3.3) for the moment Fd of **F**. The relation (3.2) is thus verified and Varignon's theorem is proved.

Varignon's theorem is very useful in the computation of moments of forces. Through a judicious choice of components, it is possible to simplify many problems. Two ways in which this may be done are indicated in Sample Probs. 3.2 and 3.3.

It should be noted that, in adding two moments, the sense (counterclockwise or clockwise) of each moment must be taken into account. As indicated in Sec. 3.4, a counterclockwise moment will be considered as positive, and a clockwise moment as negative.

SAMPLE PROBLEM 3.1

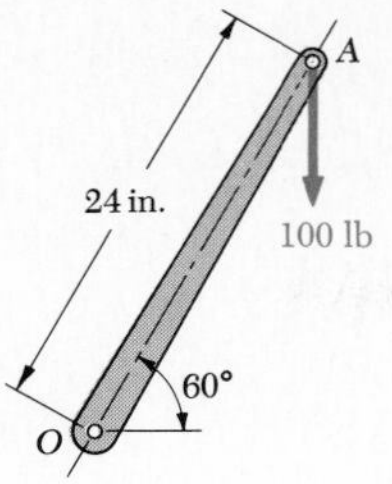

A 100-lb vertical force is applied to the end of a lever which is attached to a shaft at O. Determine (*a*) the moment of the 100-lb force about O; (*b*) the magnitude of the horizontal force applied at A which creates the same moment about O; (*c*) the smallest force applied at A which creates the same moment about O; (*d*) how far from the shaft a 240-lb vertical force must act to create the same moment about O; (*e*) whether any one of the forces obtained in parts *b*, *c*, and *d* is equivalent to the original force.

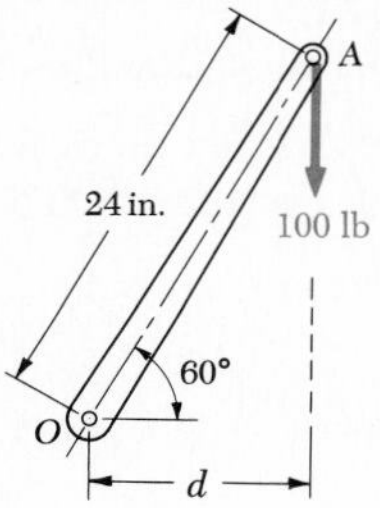

***a.* Moment about *O*.** The perpendicular distance from O to the line of action of the 100-lb force is

$$d = (24 \text{ in.}) \cos 60° = 12 \text{ in.}$$

The moment about O of the 100-lb force is

$$M_O = Fd = (100 \text{ lb})(12 \text{ in.}) = 1200 \text{ lb} \cdot \text{in.}$$

$$M_O = 1200 \text{ lb} \cdot \text{in.} \circlearrowright$$ ◄

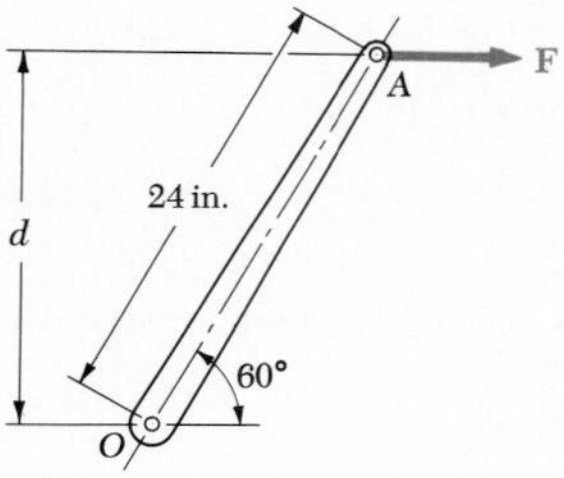

***b.* Horizontal Force.** In this case, we have

$$d = (24 \text{ in.}) \sin 60° = 20.8 \text{ in.}$$

Since the moment about O must be 1200 lb · in., we write

$$M_O = Fd \qquad 1200 \text{ lb} \cdot \text{in.} = F(20.8 \text{ in.})$$
$$F = 57.7 \text{ lb}$$

$$\mathbf{F} = 57.7 \text{ lb} \rightarrow$$ ◄

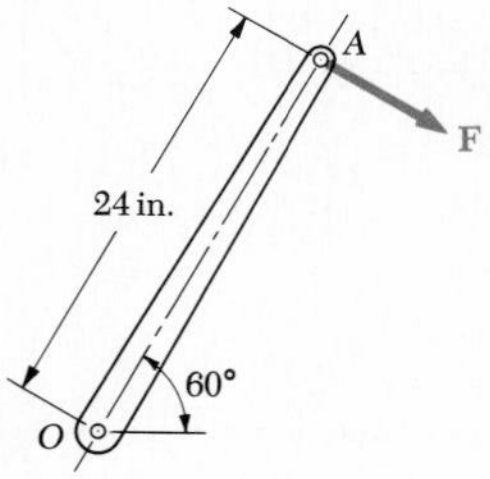

***c.* Smallest Force.** Since $M_O = Fd$, the smallest value of F occurs when d is maximum. We choose the force perpendicular to OA and find $d = 24$ in.; thus

$$M_O = Fd \qquad 1200 \text{ lb} \cdot \text{in.} = F(24 \text{ in.})$$
$$F = 50 \text{ lb}$$

$$\mathbf{F} = 50 \text{ lb} \measuredangle 30°$$ ◄

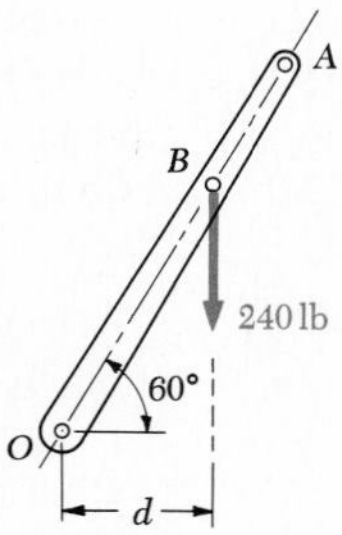

***d.* 240-lb Vertical Force.** In this case $M_O = Fd$ yields

$$1200 \text{ lb} \cdot \text{in.} = (240 \text{ lb})d \qquad d = 5 \text{ in.}$$

but $\quad OB \cos 60° = d \qquad\qquad OB = 10 \text{ in.}$ ◄

e. None of the forces considered in parts *b*, *c*, and *d* is equivalent to the original 100-lb force. Although they have the same moment about O, they have different x and y components. In other words, although each force tends to rotate the shaft in the same manner, each causes the lever to pull on the shaft in a different way.

SAMPLE PROBLEM 3.2

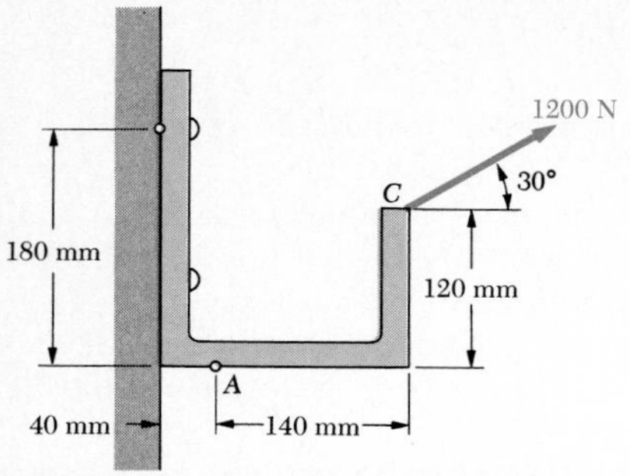

A force of 1200 N acts on a bracket as shown. Determine the moment M_A of the force about A.

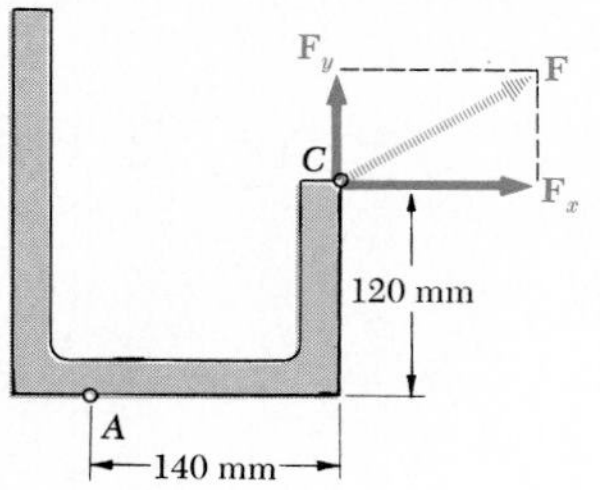

Solution. Resolving the force into its x and y components, we write

$$F_x = (1200\text{ N})\cos 30° = 1039\text{ N}$$
$$F_y = (1200\text{ N})\sin 30° = 600\text{ N}$$

Recalling the sign convention introduced in Sec. 3.4, we find that the moment of $\mathbf{F}_x$ about A is

$$(1039\text{ N})(0.120\text{ m}) = 124.7\text{ N}\cdot\text{m}\ \circlearrowright = -124.7\text{ N}\cdot\text{m}$$

The moment of $\mathbf{F}_y$ about A is

$$(600\text{ N})(0.140\text{ m}) = 84.0\text{ N}\cdot\text{m}\ \circlearrowleft = +84.0\text{ N}\cdot\text{m}$$

Applying Varignon's theorem, we write

$$M_A = +84.0\text{ N}\cdot\text{m} - 124.7\text{ N}\cdot\text{m}$$

$$M_A = -40.7\text{ N}\cdot\text{m} \quad ◄$$

SAMPLE PROBLEM 3.3

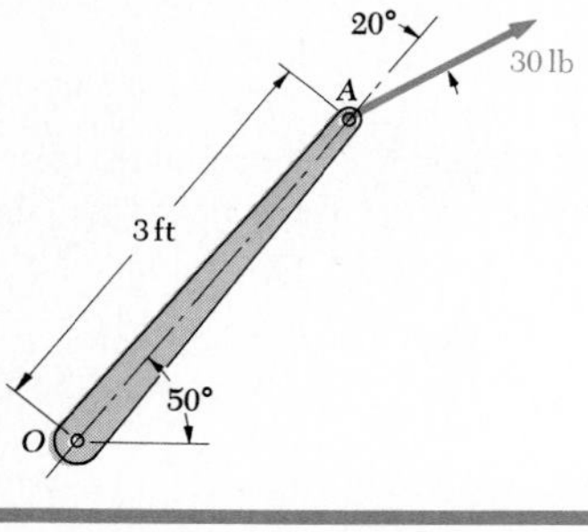

A 30-lb force acts on the end of the 3-ft lever as shown. Determine the moment of the force about O.

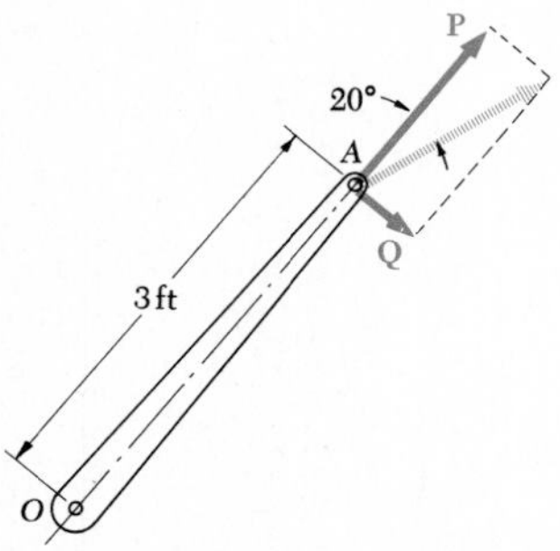

Solution. The force is replaced by two components, one component $\mathbf{P}$ in the direction of OA and one component $\mathbf{Q}$ perpendicular to OA. Since O is on the line of action of $\mathbf{P}$, the moment of $\mathbf{P}$ about O is zero and the moment of the 30-lb force reduces to the moment of $\mathbf{Q}$, which is clockwise and, thus, negative.

$$Q = (30\text{ lb})\sin 20° = 10.26\text{ lb}$$
$$M_O = -Q(3\text{ ft}) = -(10.26\text{ lb})(3\text{ ft})$$

$$M_O = -30.8\text{ lb}\cdot\text{ft} \quad ◄$$

PROBLEMS

3.1 A 150-N force is applied to the control lever at A. Knowing that the distance AB is 250 mm, determine the moment of the force about B when α is 50°.

3.2 Knowing that the distance AB is 250 mm, determine the maximum moment about B which can be caused by the 150-N force. In what direction should the force act?

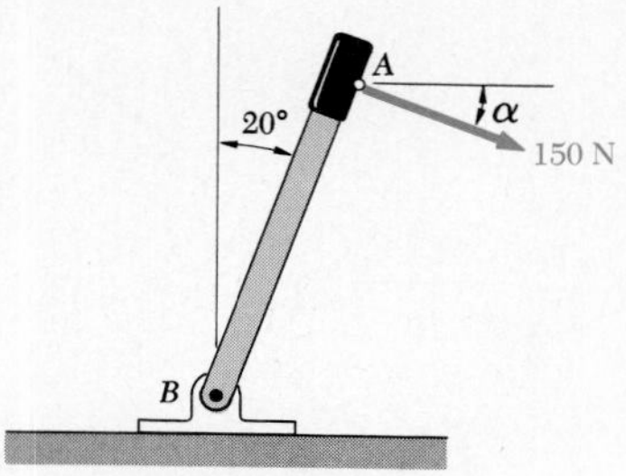

Fig. P3.1 and P3.2

3.3 A 450-N force is applied at A as shown. Determine (*a*) the moment of the 450-N force about D, (*b*) the smallest force applied at B which creates the same moment about D.

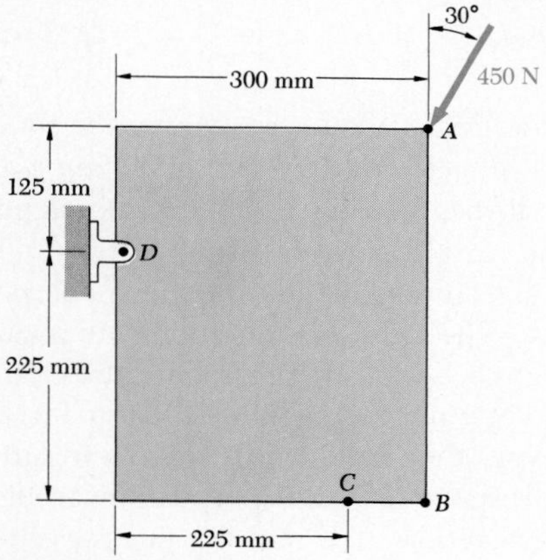

Fig. P3.3 and P3.4

3.4 A 450-N force is applied at A. Determine (*a*) the moment of the 450-N force about D, (*b*) the magnitude and sense of the horizontal force applied at C which creates the same moment about D, (*c*) the smallest force applied at C which creates the same moment about D.

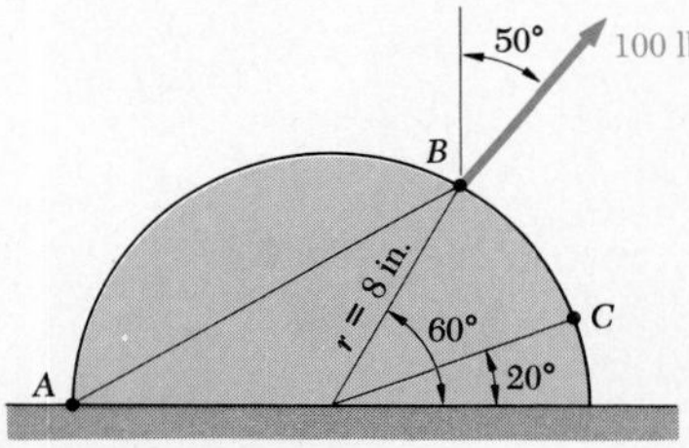

Fig. P3.5 and P3.8

3.5 and 3.6 Compute the moment of the 100-lb force about A, (*a*) by using the definition of the moment of a force, (*b*) by resolving the force into horizontal and vertical components, (*c*) by resolving the force into components along AB and in the direction perpendicular to AB.

3.7 and 3.8 Determine the moment of the 100-lb force about C.

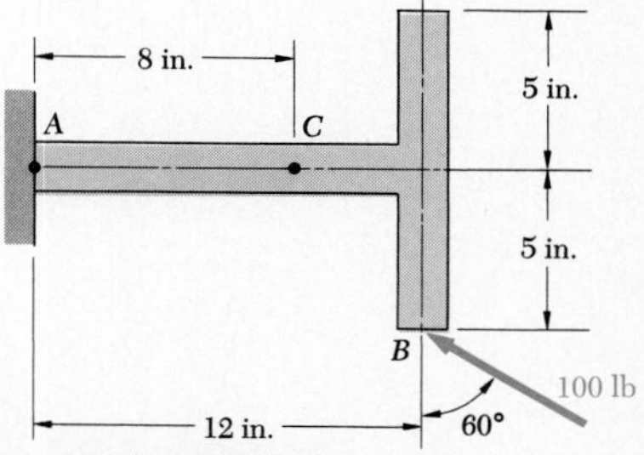

Fig. P3.6 and P3.7

3.9 A force **F** of components F_x and F_y acts at a point of coordinates x and y. Find the expression for the moment of **F** about the origin O of the system of coordinates.

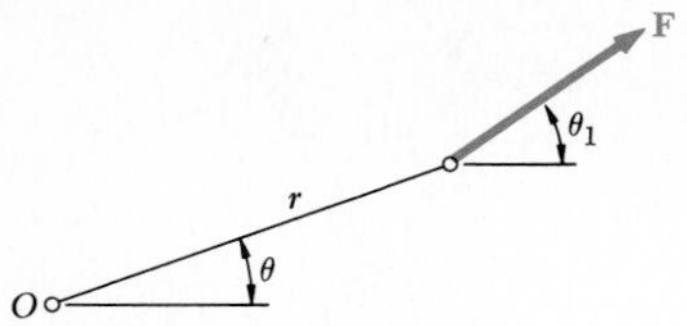

Fig. P3.10

3.10 A force **F** acts at a point of coordinates r and θ as shown. The force forms an angle θ_1 with a line parallel to the horizontal reference axis. Show that the moment of the force about the origin of coordinates O is $Fr \sin(\theta_1 - \theta)$.

3.11 The line of action of a force **P** passes through the two points $A(x_1, y_1)$ and $B(x_2, y_2)$. If the force is directed from A to B, determine the moment of the force about the origin.

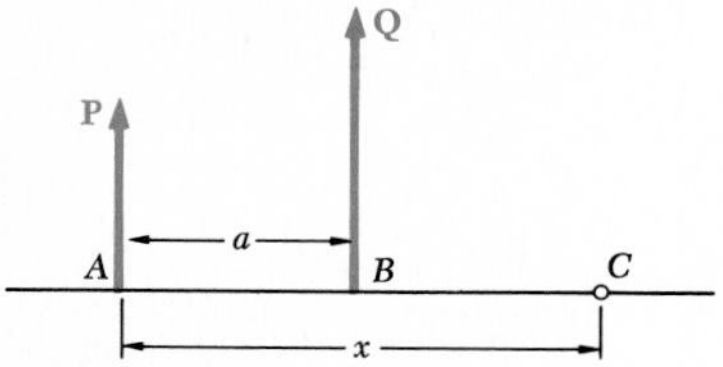

Fig. P3.12

3.12 Two forces **P** and **Q** have parallel lines of action and act at A and B, respectively. The distance between A and B is a. Find the distance x from A to the point C about which both forces have the same moment. Check the formula obtained by assuming $a = 12$ in. and (*a*) $P = 30$ lb up, $Q = 15$ lb up; (*b*) $P = 15$ lb up, $Q = 30$ lb up; (*c*) $P = 30$ lb up, $Q = 15$ lb down; (*d*) $P = 15$ lb up, $Q = 30$ lb down.

3.6. Moment of a Couple. *Two forces having the same magnitude, parallel lines of action, and opposite sense are said to form a couple.* Two such forces, **F** and **F′**, of the same magnitude F, are shown acting on a rigid body in Fig. 3.8.

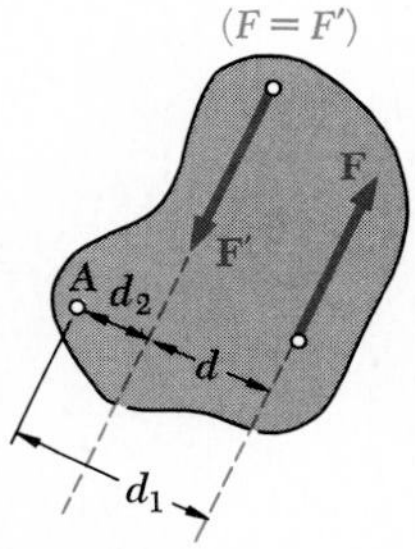

Fig. 3.8

Clearly, the sums of the x components and of the y components of **F** and **F′** are zero. The sum of their moments about an axis through a point A, however, is not zero. Thus, the effect of the couple on the rigid body is not zero. While the two forces will not move the body away, they will tend to make it turn.

Denoting by d_1 and d_2, respectively, the perpendicular distances from A to the lines of action of **F** and **F′**, and recalling that counterclockwise is positive, we write the sum M of the moments of the two forces about A as follows:

$$+\circlearrowleft M = Fd_1 - Fd_2$$
$$= F(d_1 - d_2)$$

or, noting that $d_1 - d_2$ is equal to the distance d between the lines of action of the two forces,

$$M = Fd \circlearrowleft$$

The sum M is called the *moment of the couple.* It may be observed *that M does not depend upon the choice of A; M* will have the same magnitude and same sense regardless of the location of A. This may be checked by repeating the above derivation, choosing A to the left of the couple, and then between the two forces forming the couple. We state therefore:

The moment M of a couple is constant. Its magnitude is equal to the product Fd of the common magnitude F of the two forces and of the distance d between their lines of action. The sense of M (clockwise or counterclockwise) is obtained by direct observation.

3.7. Equivalent Couples. Consider the three couples shown in Fig. 3.9, which are made to act successively on the same corner plate. As seen in the preceding section, the only motion a couple may impart to a rigid body is a rotation. Each of the couples shown will cause the plate to rotate counterclockwise. Since the product Fd is found to be 30 N · m in each case, we may expect the three couples to have the same effect on the corner plate.

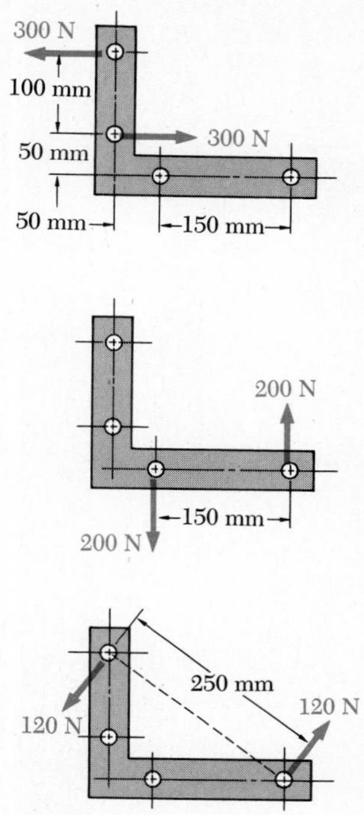

Fig. 3.9

As reasonable as this conclusion may appear, we should not accept it hastily. While intuitive feeling is of great help in the study of mechanics, it should not be accepted as a substitute for logical reasoning. Before stating that two systems (or groups) of forces have the same effect on a rigid body, we should prove that fact on the basis of the experimental evidence introduced so far. This evidence consists of the parallelogram law for the addition of two forces (Sec. 2.1) and of the principle of transmissibility (Sec. 3.2). Therefore, we shall state that *two systems of forces are equivalent* (i.e., they have the same effect on a rigid body) *if we can transform one of them into the other by means of one or several of the following operations:* (1) replacing two forces acting on the same particle by their resultant; (2) resolving a force into two components; (3) canceling two equal and opposite forces acting on the same particle; (4) attaching to the same particle two equal and opposite forces; (5) moving a force along its line of action. Each of these operations is easily justified on the basis of the parallelogram law or the principle of transmissibility.

Let us now prove that *two couples having the same moment (same magnitude and same sense) are equivalent.* Consider the couples shown in Fig. 3.10, consisting, respectively, of the forces $\mathbf{F}_1$ and $\mathbf{F}'_1$, of magnitude F_1 and at a distance d_1 from each other (Fig. 3.10a), and of the forces $\mathbf{F}_2$ and $\mathbf{F}'_2$, of magnitude F_2 and at a distance d_2 from each other (Fig. 3.10d). Both couples are

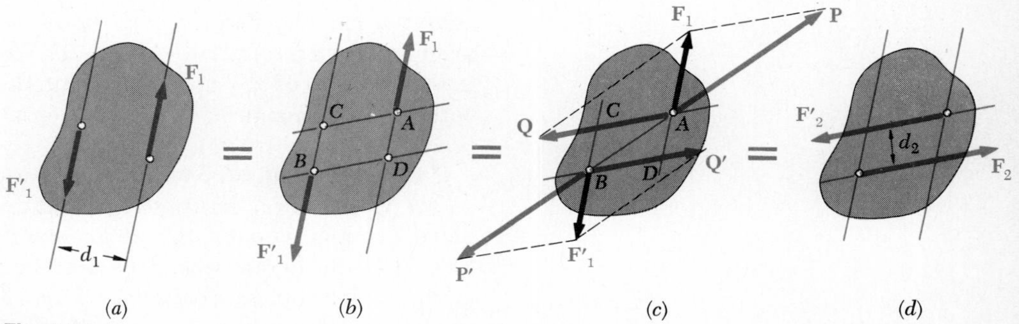

Fig. 3.10

counterclockwise, and we shall assume that they have equal moments:

$$F_1 d_1 = F_2 d_2 \tag{3.4}$$

To prove that they are equivalent, we shall show that the couple $\mathbf{F}_1$, $\mathbf{F}'_1$ may be transformed into the couple $\mathbf{F}_2$, $\mathbf{F}'_2$ by means of the operations listed above.

Denoting by A, B, C, D the points of intersection of the lines of action of the two couples, we first slide the forces $\mathbf{F}_1$ and $\mathbf{F}'_1$ until they are attached, respectively, at A and B, as shown in Fig. 3.10*b*. The force $\mathbf{F}_1$ is then resolved into a component $\mathbf{P}$ along line AB and a component $\mathbf{Q}$ along AC (Fig. 3.10*c*); similarly, the force $\mathbf{F}'_1$ is resolved into $\mathbf{P}'$ along AB and $\mathbf{Q}'$ along BD. The forces $\mathbf{P}$ and $\mathbf{P}'$ have the same magnitude, same line of action, and opposite sense; they may be moved along their common line of action until they are applied at the same point and then canceled. Thus the couple consisting of $\mathbf{F}_1$ and $\mathbf{F}'_1$ reduces to the couple $\mathbf{Q}$, $\mathbf{Q}'$.

We shall now show that the forces $\mathbf{Q}$ and $\mathbf{Q}'$ are respectively equal to the forces $\mathbf{F}'_2$ and $\mathbf{F}_2$. The moment of the couple formed by $\mathbf{Q}$ and $\mathbf{Q}'$ may be obtained by computing the moment of $\mathbf{Q}$ about B; similarly, the moment of the couple formed by $\mathbf{F}_1$ and $\mathbf{F}'_1$ is the moment of $\mathbf{F}_1$ about B. But, by Varignon's theorem, the moment of $\mathbf{F}_1$ is equal to the sum of the moments of its components $\mathbf{P}$ and $\mathbf{Q}$. Since the moment of $\mathbf{P}$ about B is zero, the moment of the couple formed by $\mathbf{Q}$ and $\mathbf{Q}'$ must be equal to the moment of the couple $\mathbf{F}_1$, $\mathbf{F}'_1$. Recalling (3.4), we write

$$Qd_2 = F_1 d_1 = F_2 d_2 \quad \text{and} \quad Q = F_2$$

Thus the forces $\mathbf{Q}$ and $\mathbf{Q}'$ are respectively equal to the forces $\mathbf{F}'_2$ and $\mathbf{F}_2$, and the couple of Fig. 3.10*a* is equivalent to the couple of Fig. 3.10*d*.

The property we have just established is very important for the correct understanding of the mechanics of rigid bodies. It indicates that, when a couple acts on a rigid body, it does not matter where the two forces forming the couple act, or what magnitude and direction they have. The only thing which counts is the *moment* of the couple (magnitude and sense). Couples with the same moment will have the same effect on the rigid body.

3.8. Addition of Couples. Consider two couples $\mathbf{P}$, $\mathbf{P}'$ and $\mathbf{Q}$, $\mathbf{Q}'$ acting on the same rigid body, as shown in Fig. 3.11*a*. Since a couple is completely defined by its moment, the couple formed by the forces $\mathbf{Q}$ and $\mathbf{Q}'$ may be replaced by a couple of the same moment, formed by forces $\mathbf{S}$ and $\mathbf{S}'$ which have, respectively, the same line of action as $\mathbf{P}$ and $\mathbf{P}'$ (Fig. 3.11*b*). The magnitude S of the new forces must be such that $Sp = Qq$. Adding forces along the same line of action, we obtain a single

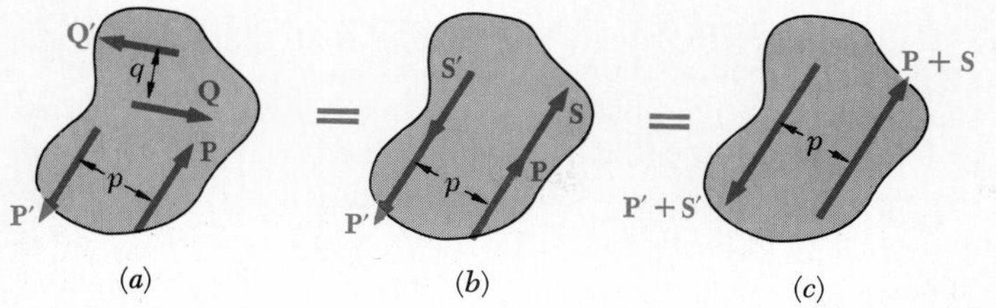

Fig. 3.11

couple consisting of the forces **P** + **S** and **P′** + **S′** (Fig. 3.11*c*). The moment of this couple is

$$M = (P + S)p = Pp + Sp = Pp + Qq$$

The moment M is thus found to be the algebraic sum of the moments Pp and Qq of the two original couples. It should be noted that, if the two couples have opposite senses, their moments will have opposite signs.

We state: *Two couples may be replaced by a single couple of moment equal to the algebraic sum of the moments of the given couples.*

3.9. Resolution of a Given Force into a Force Acting at a Given Point and a Couple. Consider a force **F** acting at point B on a rigid body, as shown in Fig. 3.12*a*. Suppose that for some reason we would rather have it acting at point A. We know that we can move **F** along its line of action (principle of transmissibility); but we cannot move it to a point A away from the original line of action without modifying the action of **F** on the rigid body.

We may attach, however, two forces at point A: a force **F** of the same magnitude and direction as the given force, and a force **F′** equal and opposite (Fig. 3.12*b*). As indicated in Sec. 3.7, this operation does not modify the action of the original force on the rigid body. We observe that the force **F** attached at B and the force **F′** attached at A form a couple of moment $M = Fd$. Since this couple may be replaced by any other couple of the same moment M (Sec. 3.7), we may represent it by the letter **M** and the symbol ↺ placed at any convenient location.†

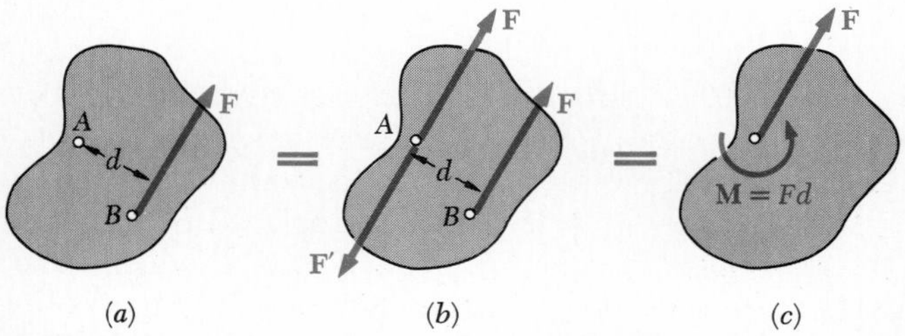

Fig. 3.12

†The vectorial character of couples will be established in Sec. 4.3.

The remaining force **F** attached at A and the couple **M**, however, are usually grouped as shown in Fig. 3.12*c*.

The result we have obtained may be stated as follows: *Any force* **F** *acting on a rigid body may be moved to any given point* A, *provided that a couple* **M** *is added; the moment* M *of the couple must equal the moment of* **F** *(in its original position) about* A. The couple will then impart to the rigid body the same motion of rotation about A that the force **F** produced before it was transferred to A. In the following sections we shall refer to this combination as a *force-couple system*. The location of the couple **M** is immaterial, but the force **F** should not be moved from A. If it were to be moved again, a new couple should be determined.

The transformation we have described may be performed in reverse. That is, *a force* **F** *acting at* A *and a couple* **M** *may be combined into a single resultant force* **F.** This is done by moving the force **F** until its moment about A becomes equal to the moment M of the couple to be eliminated. The magnitude and direction of **F** are unchanged, but its new line of action will be at a distance $d = M/F$ from A. An example of such a transformation is given in Fig. 3.13. Part *a* of the figure shows the head of a bolt which is subjected to the combined action of a 600-lb · in. clockwise couple and of a 50-lb vertical downward force through the center O of the head. The force-couple system may be replaced by a single 50-lb force located 12 in. to the right of O (Fig. 3.13*b*). In this new position, the 50-lb force has a clockwise moment of 600 lb · in. about O. Figure 3.13*a* gives a good representation of the action to which the head of the bolt is subjected, while Fig. 3.13*b* suggests a simple way of exerting this action. The wrench was sketched to help visualize the action exerted on the bolt. This is particularly useful in Fig. 3.13*b*; it was not strictly necessary, however, to suggest an actual contact between the 50-lb force and the bolt.

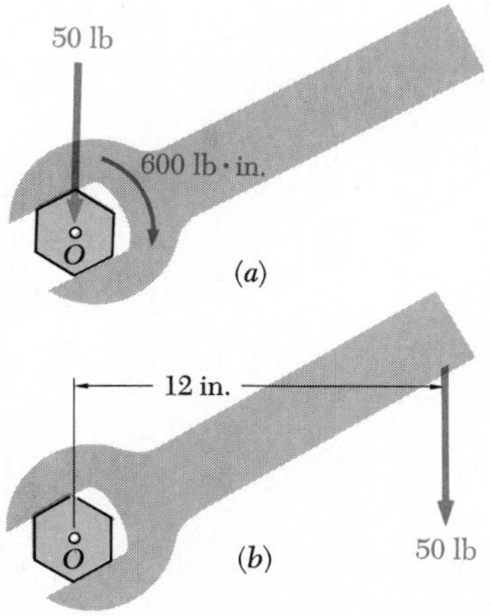

Fig. 3.13

SAMPLE PROBLEM 3.4

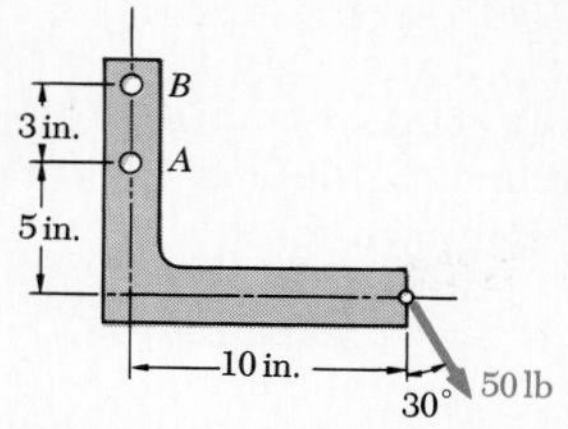

A 50-lb force is applied to a corner plate as shown. Determine (*a*) an equivalent force-couple system at *A*, (*b*) an equivalent system consisting of a 150-lb force at *B* and another force at *A*.

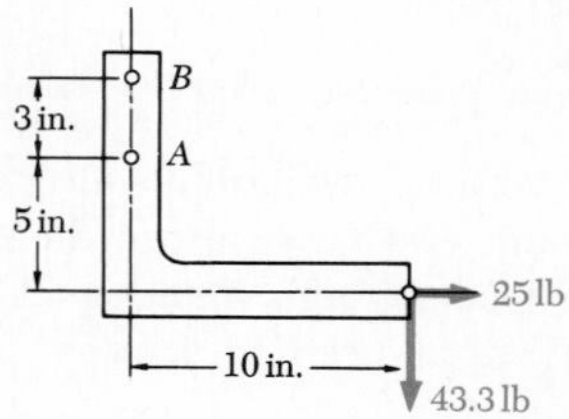

a. Force-Couple System at A. We resolve the 50-lb force into *x* and *y* components

$$F_x = (50 \text{ lb}) \sin 30° \qquad \mathbf{F}_x = 25.0 \text{ lb} \rightarrow$$

$$F_y = -(50 \text{ lb}) \cos 30° \qquad \mathbf{F}_y = 43.3 \text{ lb} \downarrow$$

These components may be moved to *A* if a couple is added, of moment M_A equal to the moment about *A* of the components in their original position. Recalling that counterclockwise is positive, we obtain

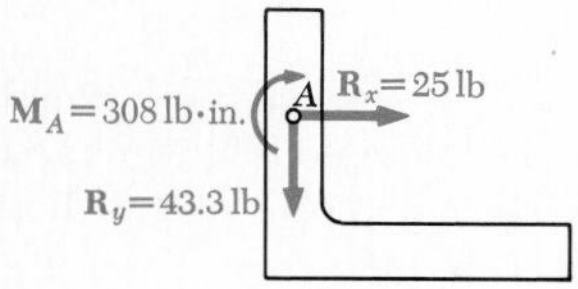

$$M_A = (25.0 \text{ lb})(5 \text{ in.}) - (43.3 \text{ lb})(10 \text{ in.})$$

$$M_A = -308 \text{ lb} \cdot \text{in.} \qquad \mathbf{M}_A = 308 \text{ lb} \cdot \text{in.} \circlearrowright$$

b. Forces at A and B. We shall assume that the couple $\mathbf{M}_A$ found in part *a* consists of two 150-lb forces **P** and **P′** acting respectively at *B* and *A*. The moment of **P** about *A* must be equal to the moment M_A of the couple; denoting by θ the angle that **P** forms with the horizontal, and applying Varignon's theorem, we write

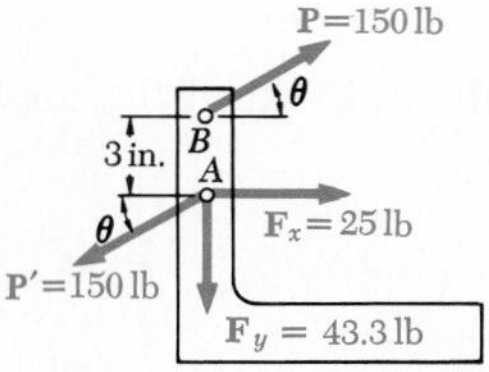

$$M_A = -P \cos \theta \, (3 \text{ in.})$$

$$-308 \text{ lb} \cdot \text{in.} = -(150 \text{ lb}) \cos \theta \, (3 \text{ in.})$$

$$\cos \theta = \frac{308 \text{ lb} \cdot \text{in.}}{450 \text{ lb} \cdot \text{in.}} = 0.684 \qquad \theta = \pm 46.8°$$

Having found the direction of the forces **P** and **P′**, we complete the solution by determining the resultant **Q** of the forces **F** and **P′** acting at *A*.

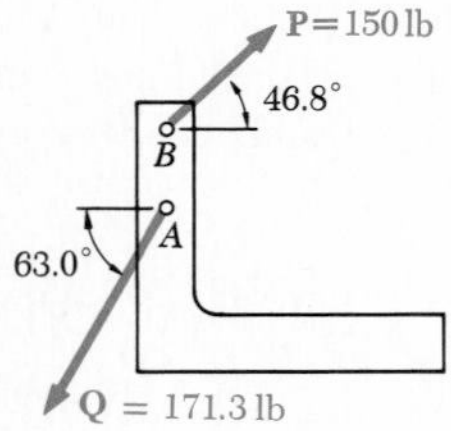

$$Q_x = F_x + P'_x = 25 \text{ lb} - (150 \text{ lb}) \cos \theta$$
$$Q_y = F_y + P'_y = -43.3 \text{ lb} - (150 \text{ lb}) \sin \theta$$

Since we have found two possible values for θ, there will be two pairs of forces **P** and **Q** forming a system equivalent to the original 50-lb force:

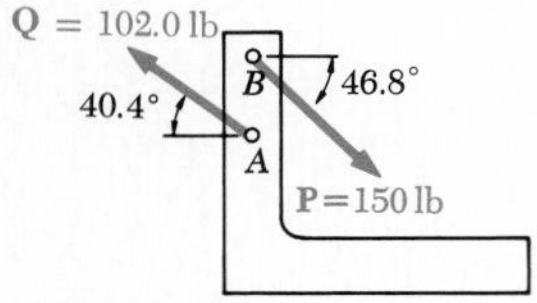

$$\mathbf{P} = 150 \text{ lb} \measuredangle 46.8° \text{ at } B \qquad \mathbf{Q} = 171.3 \text{ lb} \measuredangle 63.0° \text{ at } A$$

or

$$\mathbf{P} = 150 \text{ lb} \measuredangle 46.8° \text{ at } B \qquad \mathbf{Q} = 102.0 \text{ lb} \measuredangle 40.4° \text{ at } A$$

SAMPLE PROBLEM 3.5

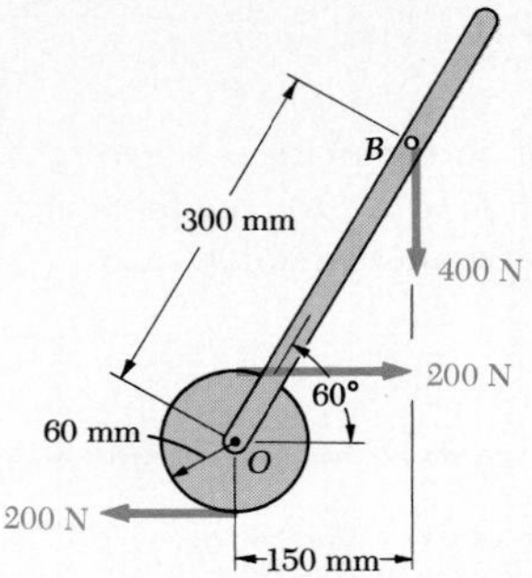

Replace the couple and force shown by an equivalent single force applied to the lever. Determine the distance from the shaft to the point of application of this equivalent force.

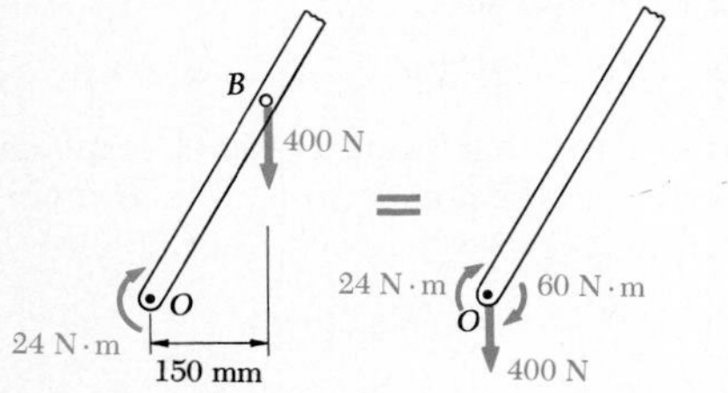

Solution. First the given force and couple are replaced by an equivalent force-couple system at O. We move the 400-N force to O and at the same time add a couple of moment equal to the moment about O of the force in its original position, namely

$$M_O = -(400\text{ N})(0.150\text{ m}) = -60\text{ N}\cdot\text{m} \qquad M_O = 60\text{ N}\cdot\text{m}\,\text{↻}$$

This couple is added to the 24-N · m clockwise couple formed by the two 200-N forces, and an 84-N · m clockwise couple is obtained. We now move the 400-N force to the right a distance d such that the moment of the force about O is 84 N · m clockwise:

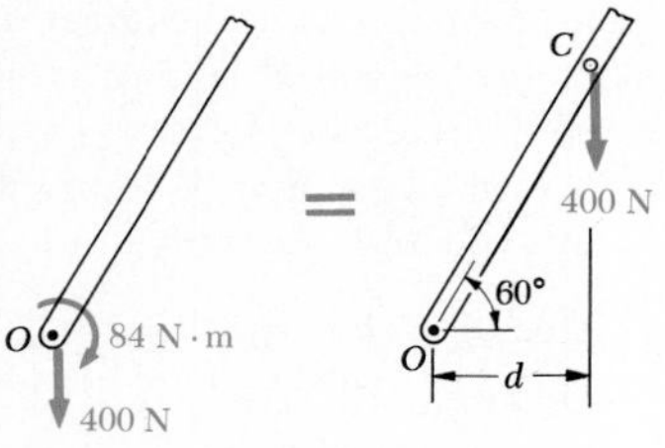

$$84\text{ N}\cdot\text{m} = (400\text{ N})d \qquad d = 0.210\text{ m} = 210\text{ mm}$$

The equivalent single force, or resultant, is attached at point C, where its line of action intersects the lever.

$$(OC)\cos 60° = 210\text{ mm} \qquad OC = 420\text{ mm} \blacktriangleleft$$

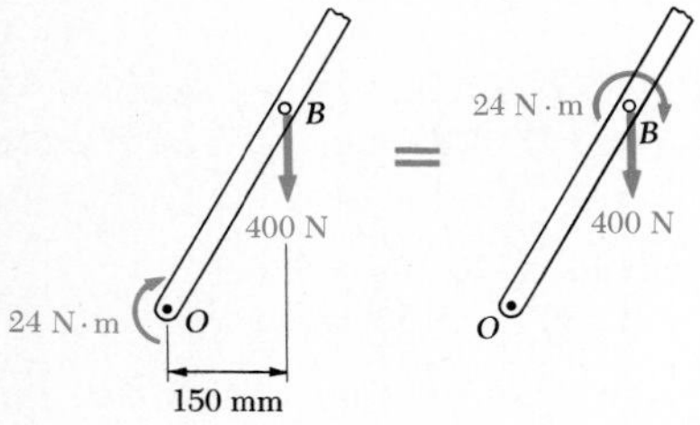

Alternate Solution. Since the effect of a couple does not depend on its location, the 24-N · m clockwise couple may be moved to B; we thus obtain a force-couple system at B. We now move the 400-N force from B to the right a distance d' such that the moment of the force about B is 24 N · m clockwise.

$$24\text{ N}\cdot\text{m} = (400\text{ N})d' \qquad d' = 0.060\text{ m} = 60\text{ mm}$$

Again the point of intersection of the lever and of the line of action is determined.

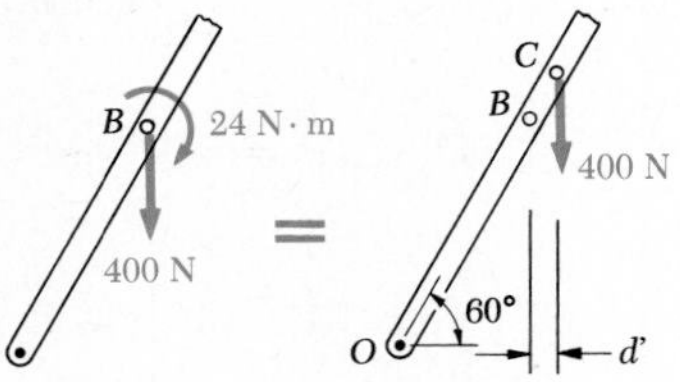

$$(BC)\cos 60° = 60\text{ mm} \qquad BC = 120\text{ mm}$$

$$OC = OB + BC = 300\text{ mm} + 120\text{ mm}$$

$$OC = 420\text{ mm} \blacktriangleleft$$

PROBLEMS

3.13 The two couples shown are applied to a 120- by 160-mm plate. Knowing that $P_1 = P_2 = 150$ N and $Q_1 = Q_2 = 200$ N, prove that their sum is zero (*a*) by adding their moments, (*b*) by combining $\mathbf{P}_1$ and $\mathbf{Q}_1$ into their resultant $\mathbf{R}_1$, combining $\mathbf{P}_2$ and $\mathbf{Q}_2$ into their resultant $\mathbf{R}_2$, and then showing that $\mathbf{R}_1$ and $\mathbf{R}_2$ are equal and opposite and have the same line of action.

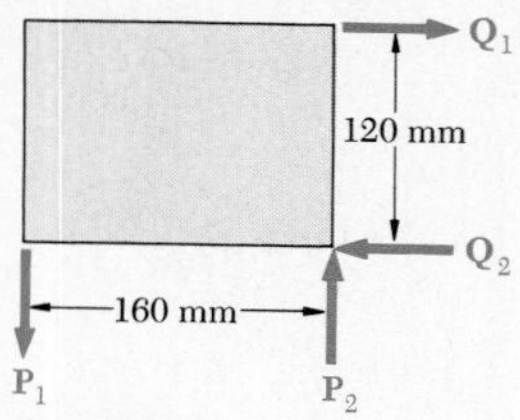

Fig. P3.13

3.14 A couple formed by two 975-N forces is applied to the pulley assembly shown. Determine an equivalent couple which is formed by (*a*) vertical forces acting at *A* and *C*, (*b*) the smallest possible forces acting at *B* and *D*, (*c*) the smallest possible forces which can be attached to the assembly.

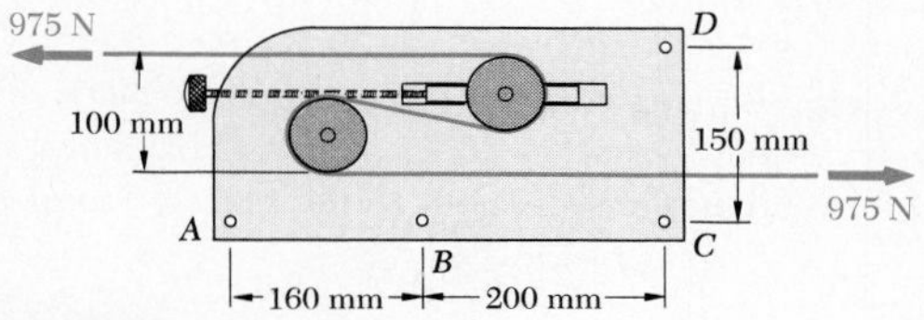

Fig. P3.14

3.15 A multiple-drilling machine is used to drill simultaneously six holes in the steel plate shown. Each drill exerts a clockwise couple of magnitude 40 lb · in. on the plate. Determine an equivalent couple formed by the smallest possible forces acting (*a*) at *A* and *C*, (*b*) at *A* and *D*, (*c*) on the plate.

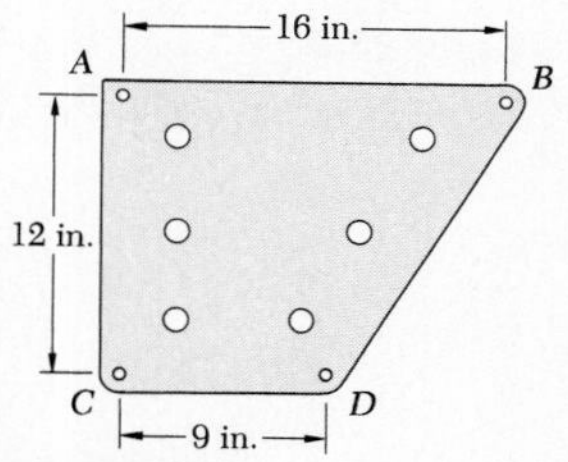

Fig. P3.15

3.16 Four 1-in.-diameter pegs are attached to a board as shown. Two strings are passed around the pegs and pulled with forces of magnitude $P = 20$ lb and $Q = 35$ lb. Determine the resultant couple acting on the board.

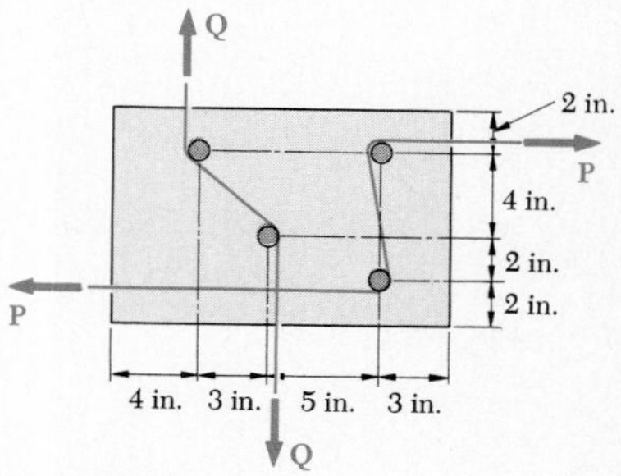

Fig. P3.16

3.17 Replace each of the forces considered in Sample Prob. 3.1 by a force acting on the shaft and a couple.

3.18 A 400-N force is applied to a bent plate as shown. Determine an equivalent force-couple system (*a*) at *A*, (*b*) at *B*.

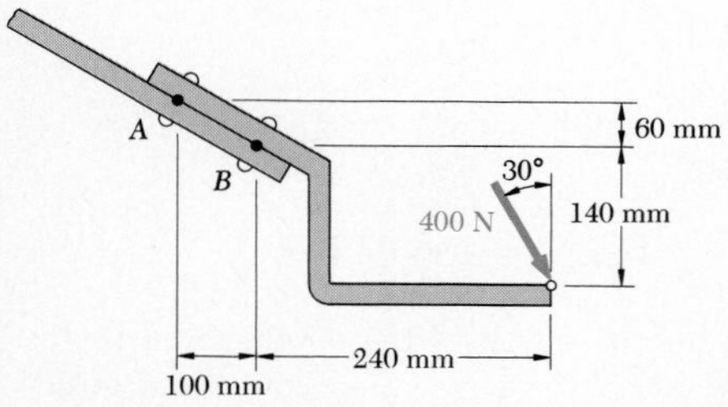

Fig. P3.18

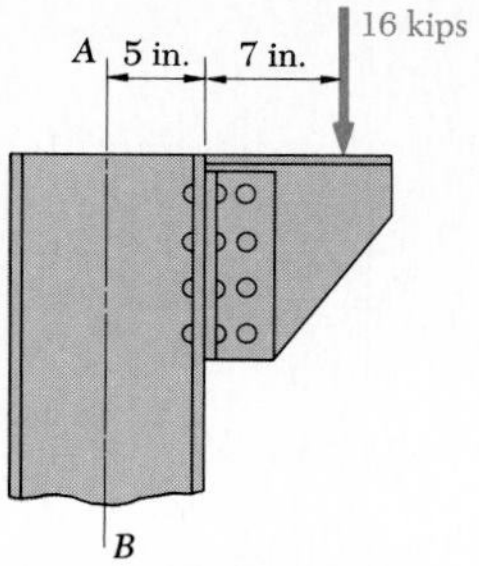

Fig. P3.19 and P3.20

3.19 A crane column supports a 16-kip load as shown. Reduce the load to an axial force along AB and a couple.

3.20 The 16-kip load is to be replaced by an equivalent force-couple system. Knowing that the moment of the couple should be 20 kip · ft clockwise, determine the location of the force-couple system and the magnitude of the force.

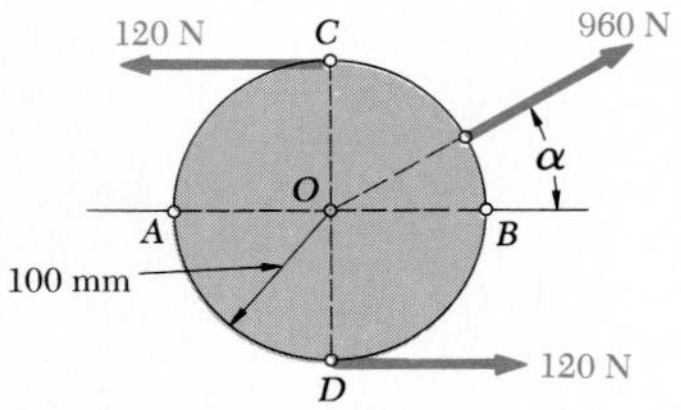

Fig. P3.21 and P3.22

3.21 The force and couple shown are to be replaced by an equivalent single force. Determine the required value of α so that the line of action of the single equivalent force will pass through point B.

3.22 Knowing that $\alpha = 60°$, replace the force and couple shown by a single force applied at a point located (*a*) on line AB, (*b*) on line CD. In each case determine the distance from the center O to the point of application of the force.

3.23 A 50-lb force is applied to a corner plate as shown. Determine (*a*) an equivalent force-couple system at A, (*b*) two horizontal forces at A and B which form a couple equivalent to the couple found in part *a*.

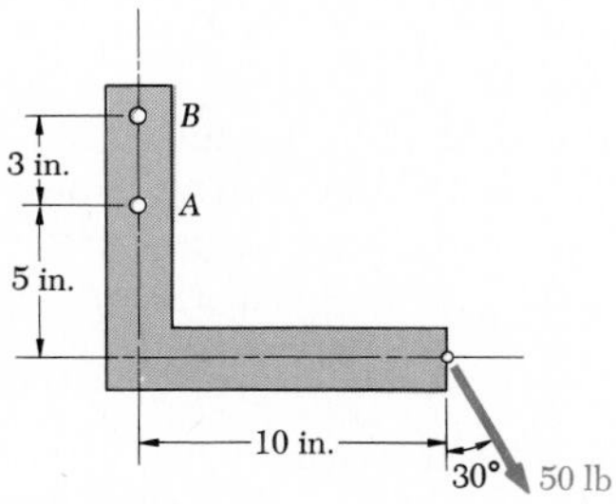

Fig. P3.23

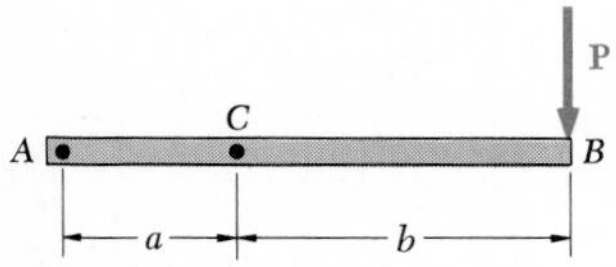

Fig. P3.24

3.24 A force **P** is applied to a beam AB as shown. Find the vertical forces $\mathbf{F}_A$ and $\mathbf{F}_C$ applied, respectively, at A and C, which form a system equivalent to **P**.

3.25 Knowing that $\alpha = 45°$, replace the 600-N force by (*a*) an equivalent force-couple system at C, (*b*) an equivalent system formed by two parallel forces at B and C.

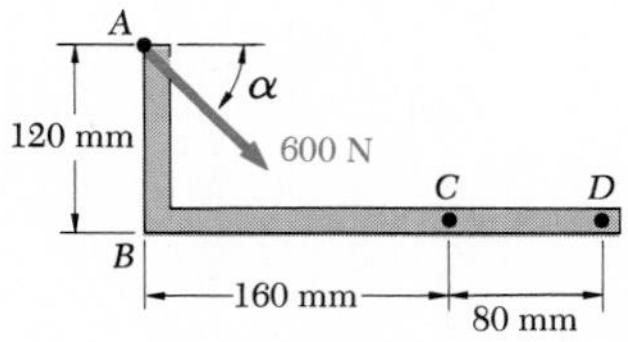

Fig. P3.25 and P3.26

3.26 Knowing that the line of action of the 600-N force passes through point C, replace the 600-N force by (*a*) an equivalent force-couple system at D, (*b*) an equivalent system formed by two parallel forces at B and D.

3.10. Reduction of a System of Coplanar Forces to One Force and One Couple. Resultant of a System of Coplanar Forces. Consider a system of coplanar forces $\mathbf{F}_1$, $\mathbf{F}_2$, $\mathbf{F}_3$, etc., acting on a rigid body as shown in Fig. 3.14*a*. As seen in Sec. 3.9, $\mathbf{F}_1$ may be moved and attached to a given point A if a couple $\mathbf{M}_1$, of moment M_1 equal to the moment of $\mathbf{F}_1$ about A, is added to the original system of forces. Repeating this procedure with $\mathbf{F}_2$, $\mathbf{F}_3$, etc., we obtain the system shown in Fig. 3.14*b*. Since the forces act now on the same particle A, they may be added vectorially and replaced by their resultant $\mathbf{R}$ as shown in Fig. 3.14*c*. On the other hand, the couples may be replaced by a single couple $\mathbf{M}$ of moment M equal to the algebraic sum of the moments M_1, M_2, M_3, etc. Any system of coplanar forces, however complex, may thus be reduced to an *equivalent force-couple system acting at a given point A*.

In the general case, when the force $\mathbf{R}$ is different from zero, the system obtained may be reduced further to a *single force* $\mathbf{R}$, called the *resultant* of the given system (Fig. 3.14*d*). This is done, as indicated in Sec. 3.9, by moving $\mathbf{R}$ until its moment about A becomes equal to M, thus eliminating the need for the couple $\mathbf{M}$. The distance from A to the line of action of $\mathbf{R}$ is $d = M/R$.

When $\mathbf{R}$ is equal to zero, the force-couple system reduces to the couple $\mathbf{M}$. Thus, the given system of forces is reduced to a *single couple* $\mathbf{M}$, called the *resultant couple* of the system.

To summarize, *any given system of coplanar forces may be reduced to a single force or a single couple,* as the case may be. The only exception is when both $\mathbf{R}$ and $\mathbf{M}$ are zero. Then the system exerts no action on the rigid body, and the rigid body is said to be in *equilibrium*. This very particular case is of great practical interest and will be considered under Equilibrium of Rigid Bodies later in this chapter.

In practice, the reduction of a system of coplanar forces to a force $\mathbf{R}$ at A and a couple $\mathbf{M}$ will be considerably simplified if the given forces $\mathbf{F}_1$, $\mathbf{F}_2$, $\mathbf{F}_3$, etc., are resolved into their x and y components as shown in Fig. 3.15*a*. Adding these components, we shall obtain, respectively, the components $\mathbf{R}_x$ and $\mathbf{R}_y$ of the force $\mathbf{R}$. Besides, by using Varignon's theorem, the moments of the given forces about A may be determined more conveniently from the component forces $\mathbf{F}_x$ and $\mathbf{F}_y$ than from the forces $\mathbf{F}$ themselves; the sum ΣM_A of these moments is equal to the moment of the couple $\mathbf{M}$. It is advisable to arrange the computations in tabular form as shown in Sample Prob. 3.7 and to sketch the force $\mathbf{R}_x$, the force $\mathbf{R}_y$, and the couple $\mathbf{M}$ to which the given system of forces is reduced (Fig. 3.15*b*). The component forces $\mathbf{R}_x$ and $\mathbf{R}_y$ may then be added vectorially, and a force-couple system at A is obtained.

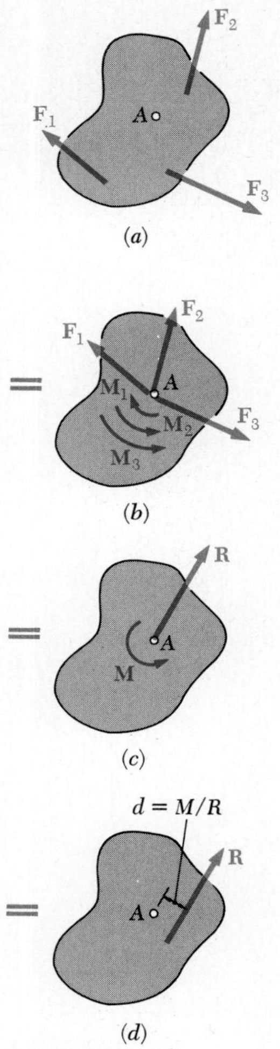

Fig. 3.14

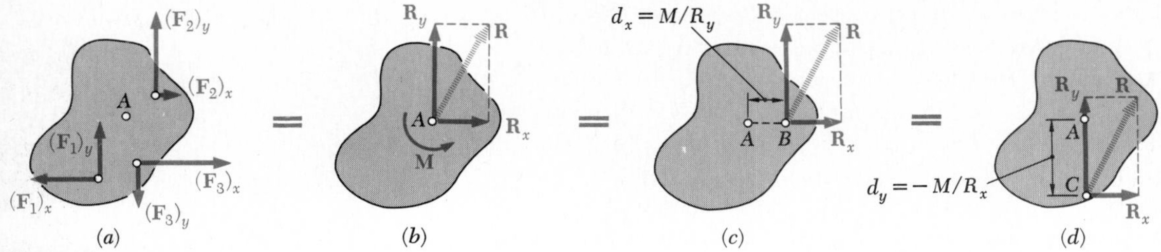

Fig. 3.15

If the given system is to be reduced to a single force (i.e., to its resultant), the following method may be used: The forces $\mathbf{R}_x$ and $\mathbf{R}_y$ are moved to a point B located on the line of action of $\mathbf{R}_x$, as shown in Fig. 3.15*c*; the displacement from A to B is defined by $d_x = M/R_y$. The moment of $\mathbf{R}_x$ about A is still zero, but the moment of $\mathbf{R}_y$ about A is now equal to M. Thus the couple is eliminated, and the given system reduces to the two forces $\mathbf{R}_x$ and $\mathbf{R}_y$ attached at B. These component forces may then be added vectorially, and the resultant $\mathbf{R}$ is obtained.

The forces $\mathbf{R}_x$ and $\mathbf{R}_y$ could also have been moved to point C, located on the line of action of $\mathbf{R}_y$, as shown in Fig. 3.15*d*; the displacement from A to C is defined by $d_y = -M/R_x$. The couple is eliminated, since the moment of $\mathbf{R}_x$ about A is equal to M, and the given system reduces to the forces $\mathbf{R}_x$ and $\mathbf{R}_y$ at C, which may again be added vectorially.

3.11. Equivalent Systems of Coplanar Forces. We have seen in the preceding section that any system of coplanar forces acting on a rigid body may be reduced to a force-couple system at a given point A. This equivalent force-couple system characterizes completely the action exerted by the given system on the rigid body. *Two systems of coplanar forces are equivalent, therefore, if they may be reduced to the same force-couple system at a given point* A. Referring to Fig. 3.15*b* and recalling that $\mathbf{R}_x$, $\mathbf{R}_y$, and $\mathbf{M}$ are obtained, respectively, by summing the x components, the y components, and the moments about A of the forces of the system considered, we state: *Two systems of coplanar forces are equivalent if, and only if, the sums of the x components, of the y components, and of the moments about A of their forces are, respectively, equal.*

The above statement has a simple physical significance. It means that two systems of coplanar forces are equivalent if they tend to impart to a given rigid body (1) the same translation in the x direction, (2) the same translation in the y direction, and (3) the same rotation about a fixed point A. Note that it is sufficient to establish the last property with respect to *one point only*. The property, however, holds with respect to *any point* if the two systems are equivalent.

SAMPLE PROBLEM 3.6

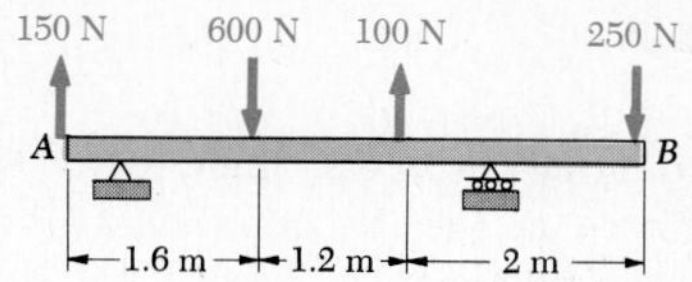

A 4.80-m beam is subjected to the forces shown. Reduce the given system of forces to (*a*) an equivalent force-couple system at *A*, (*b*) an equivalent force-couple system at *B*, (*c*) a single force or resultant.

Note. Since the reactions at the supports are not included in the given system of forces, the given system will not maintain the beam in equilibrium.

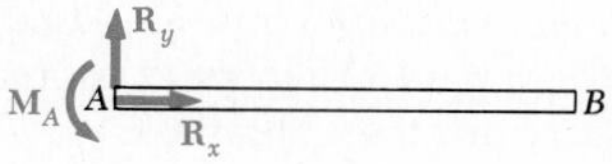

***a.* Force-Couple System at *A*.** Positive senses for the force-couple system are as shown. Adding the components of the given forces and the moments about *A* of the given forces, we obtain

$$\overset{+}{\rightarrow} R_x = 0$$
$$+\uparrow R_y = +150\text{ N} - 600\text{ N} + 100\text{ N} - 250\text{ N} = -600\text{ N}$$
$$+\circlearrowleft M_A = -(600\text{ N})(1.6\text{ m}) + (100\text{ N})(2.8\text{ m}) - (250\text{ N})(4.8\text{ m})$$
$$= -1880\text{ N}\cdot\text{m}$$

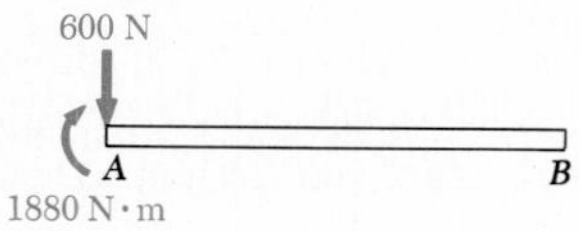

The negative signs obtained indicate that the 600-N force acts downward and that the couple is clockwise. The equivalent force-couple system at *A* is thus

$$\mathbf{R} = 600\text{ N}\downarrow \qquad \mathbf{M}_A = 1880\text{ N}\cdot\text{m}\circlearrowright \quad ◄$$

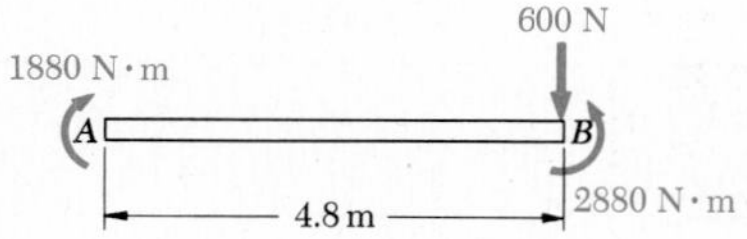

***b.* Force-Couple System at *B*.** We shall find a force-couple system at *B* equivalent to the force-couple system at *A* determined in part *a*. The 600-N force may be moved to *B* provided we add a couple of moment equal to the moment about *B* of the force in its original position, namely,

$$+(600\text{ N})(4.8\text{ m}) = +2880\text{ N}\cdot\text{m}$$

The 1880-N · m clockwise couple may be moved to *B* without changing its effect on the beam. Adding the two couples now at *B*, we obtain a couple of moment

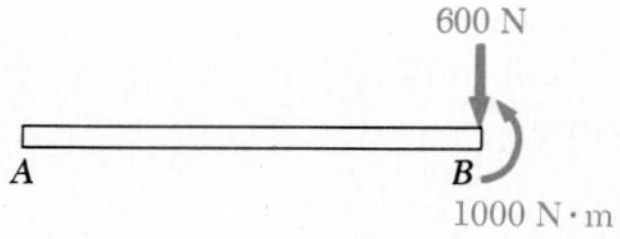

$$+2880\text{ N}\cdot\text{m} - 1880\text{ N}\cdot\text{m} = +1000\text{ N}\cdot\text{m}$$

The equivalent force-couple system at *B* is thus

$$\mathbf{R} = 600\text{ N}\downarrow \qquad \mathbf{M}_B = 1000\text{ N}\cdot\text{m}\circlearrowleft \quad ◄$$

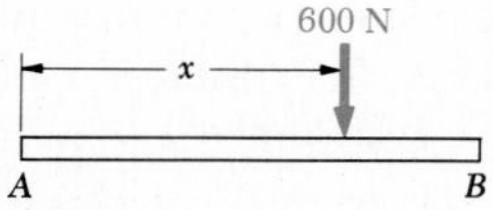

***c.* Single Force or Resultant.** Using the results of part *a*, we move the 600-N force to the right through a distance *x* chosen so that the moment of the force about *A* is -1880 N · m. We write

$$-1880\text{ N}\cdot\text{m} = -(600\text{ N})x \qquad x = 3.13\text{ m}$$
$$\mathbf{R} = 600\text{ N}\downarrow \qquad x = 3.13\text{ m} \quad ◄$$

SAMPLE PROBLEM 3.7

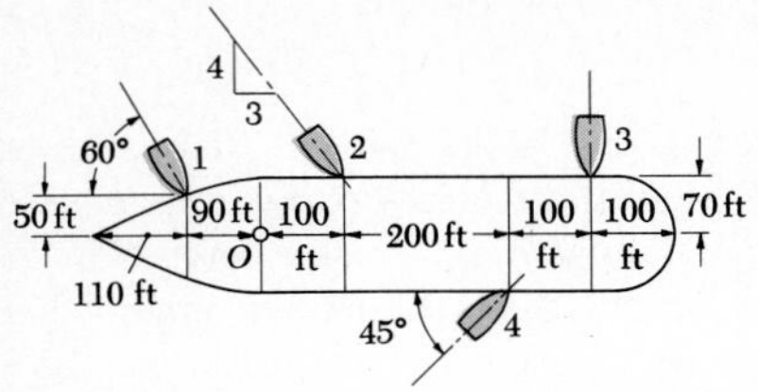

Four tugboats are used to bring an ocean liner to its pier. Each tugboat exerts a 5000-lb force in the direction shown. Determine (*a*) the equivalent force-couple system at the foremast *O*, (*b*) the point on the hull where a single, more powerful tugboat should push to produce the same effect as the original four tugboats.

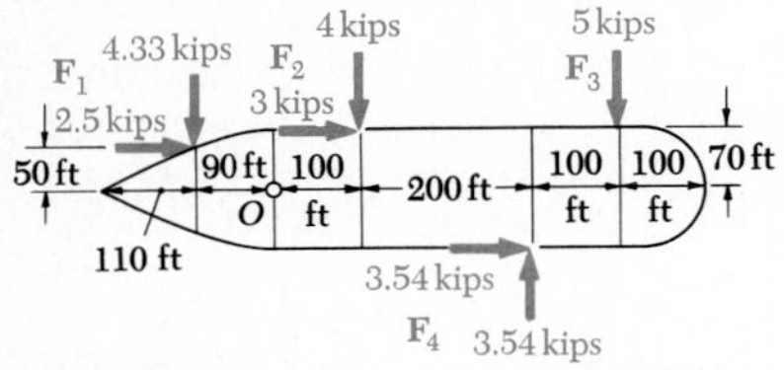

a. **Force-Couple System at *O*.** Each of the given forces is resolved into components in the diagram shown (kip units are used). For convenience the components and their moments about *O* are recorded in a table, from which R_x, R_y, and M are then obtained.

Force	F_x $\overset{+}{\rightarrow}$, kips	$F_y + \uparrow$ kips	M_O + ↺ kip · ft
$\mathbf{F}_1$	+2.50	−4.33	−125 +390
$\mathbf{F}_2$	+3.00	−4.00	−210 −400
$\mathbf{F}_3$	0	−5.00	0 −2000
$\mathbf{F}_4$	+3.54	+3.54	+248 +1062
	$R_x = \Sigma F_x$ $= +9.04$ $= 9.04 \rightarrow$	$R_y = \Sigma F_y$ $= -9.79$ $= 9.79\downarrow$	$M = \Sigma M_O$ $= -1035$ $= 1035$ ↻

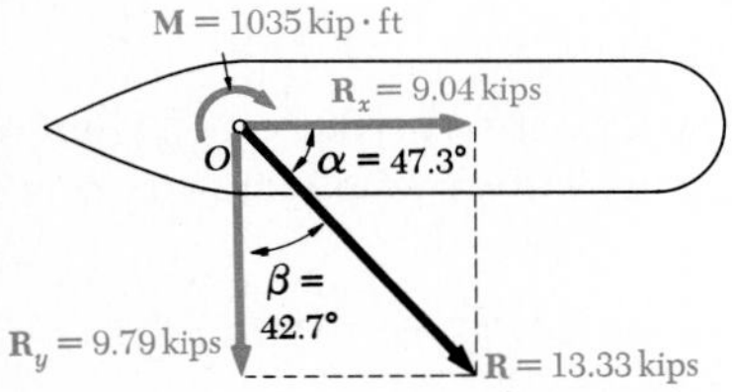

The force components are combined to yield a force-couple system equivalent to the original system.

R = 13.33 kips ⦬47.3° M = 1035 kip · ft ↻ ◄

b. **Single Tugboat.** The force exerted by a single tugboat must be equivalent to the force-couple system found in part *a* and thus have a magnitude of 13.33 kips. Since this force acts on the hull at a point *A*, we resolve it into components at *A*. The distance *x* must be such that the total moment about *O* is −1035 kip · ft.

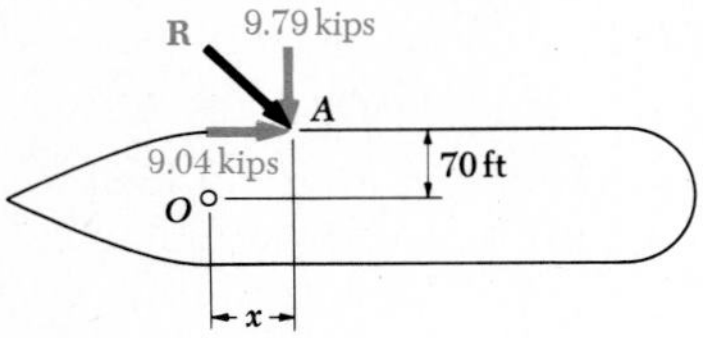

$$-1035 \text{ kip} \cdot \text{ft} = -(9.04 \text{ kips})(70 \text{ ft}) - (9.79 \text{ kips})x$$
$$x = 41.1 \text{ ft}$$

R = 13.33 kips ⦬47.3° $x = 41.1$ ft ◄

PROBLEMS

3.27 A 12-ft beam is loaded in the various ways represented in the figure. Find two loadings which are equivalent.

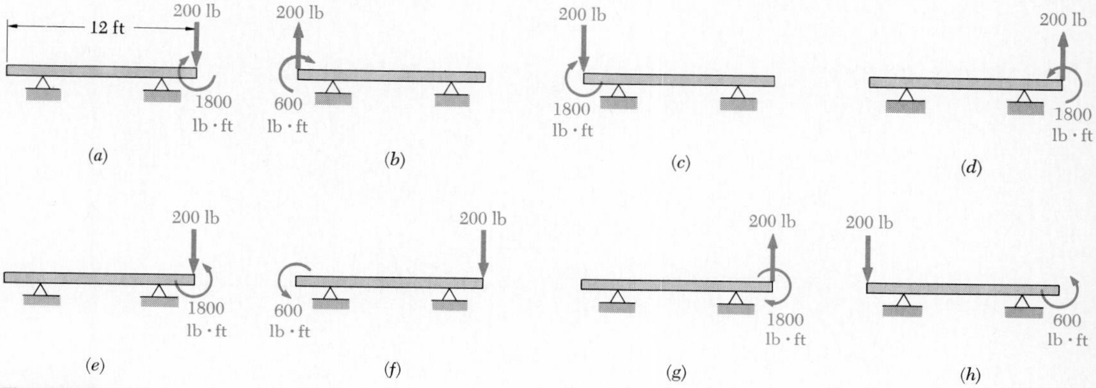

Fig. P3.27

3.28 A 12-ft beam is loaded as shown. Determine the loading of Prob. 3.27 which is equivalent to this loading.

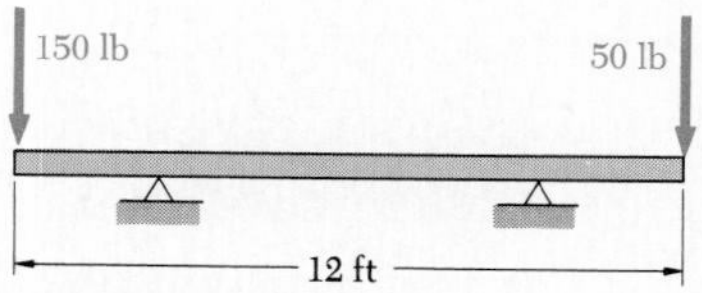

Fig. P3.28

3.29 Two parallel forces **P** and **Q** are applied at the ends of a beam AB of length L. Find the distance x from A to the line of action of their resultant. Check the formula obtained by assuming $L = 8$ in. and (*a*) $P = 10$ lb down, $Q = 30$ lb down; (*b*) $P = 10$ lb down, $Q = 30$ lb up.

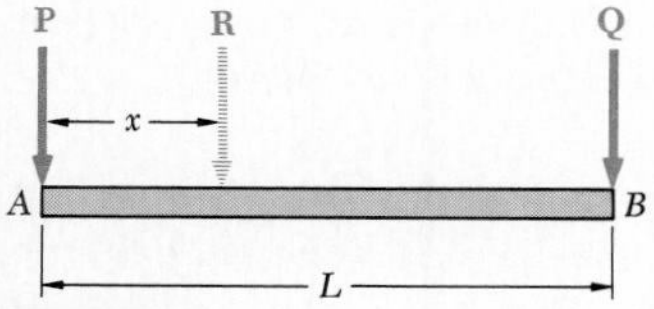

Fig. P3.29

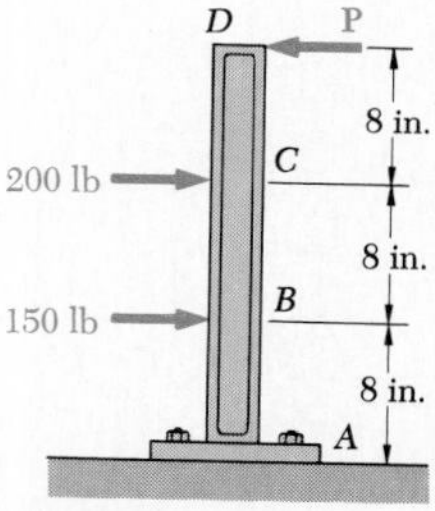

Fig. P3.30

3.30 Three horizontal forces are applied as shown to a machine arm. Determine the resultant of the loads and the point where its line of action intersects AD if the magnitude of **P** is (*a*) $P = 50$ lb, (*b*) $P = 500$ lb, (*c*) $P = 350$ lb.

3.31 Determine the distance from point A to the line of action of the resultant of the three forces shown when (*a*) $a = 1$ m, (*b*) $a = 1.5$ m, (*c*) $a = 2.5$ m.

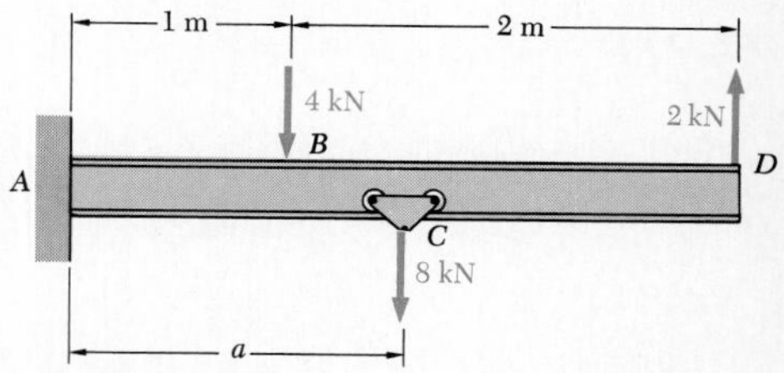

Fig. P3.31

3.32 Solve Prob. 3.31, assuming that the load at C is reduced to 6 kN.

3.33 Four packages are transported at constant speed from A to B by the conveyor. At the instant shown, determine the resultant of the loading and the location of its line of action.

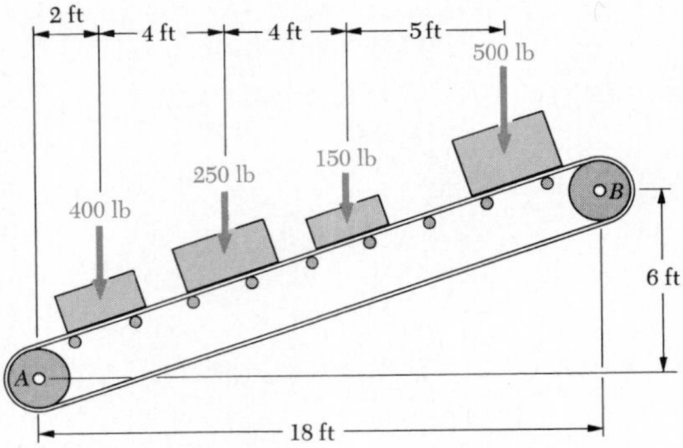

Fig. P3.33

3.34 A 150- by 300-mm plate is subjected to four loads. Find the resultant of the four loads and the two points at which the line of action of the resultant intersects the edge of the plate.

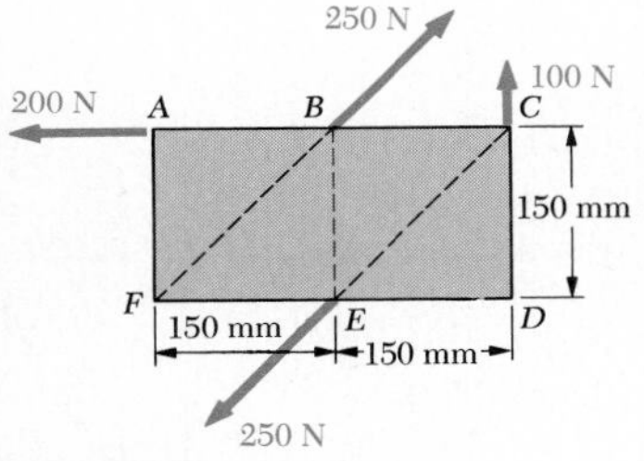

Fig. P3.34

3.35 Solve Prob. 3.34 assuming that the 200-N force applied at A is directed horizontally to the right.

3.36 Solve Prob. 3.33 assuming that the 250-lb package has been removed.

3.37 A bracket is subjected to the system of forces and couples shown. Find the resultant of the system and the point of intersection of its line of action with (*a*) line AB, (*b*) line BC, (*c*) line CD.

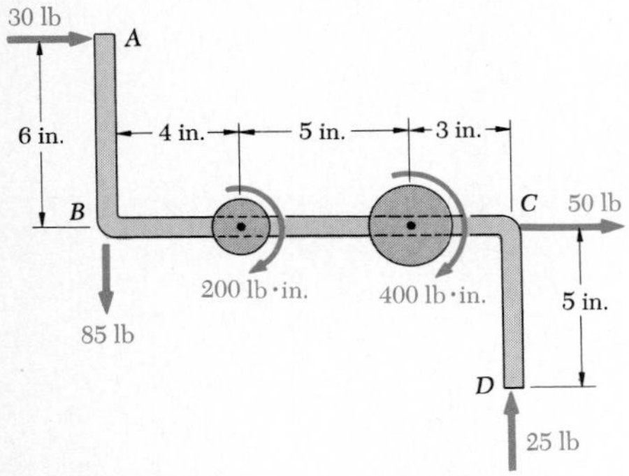

Fig. P3.37

3.38 The roof of a building frame is subjected to the wind loading shown. Determine (*a*) the equivalent force-couple system at D, (*b*) the resultant of the loading and its line of action.

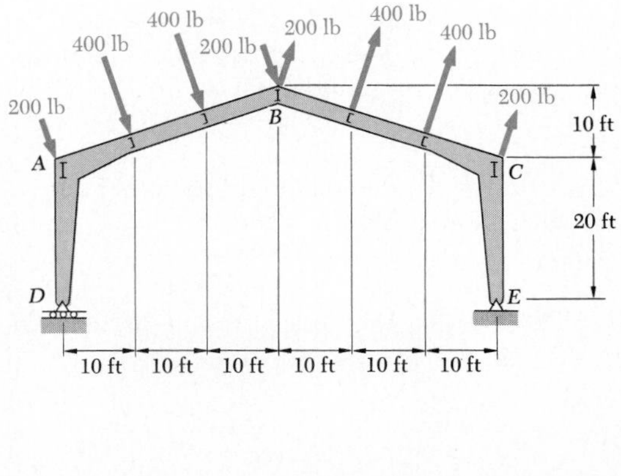

Fig. P3.38

3.39 Three forces act as shown on a traffic-signal pole. Determine (*a*) the equivalent force-couple system at *A*, (*b*) the resultant of the system and the point of intersection of its line of action with the pole.

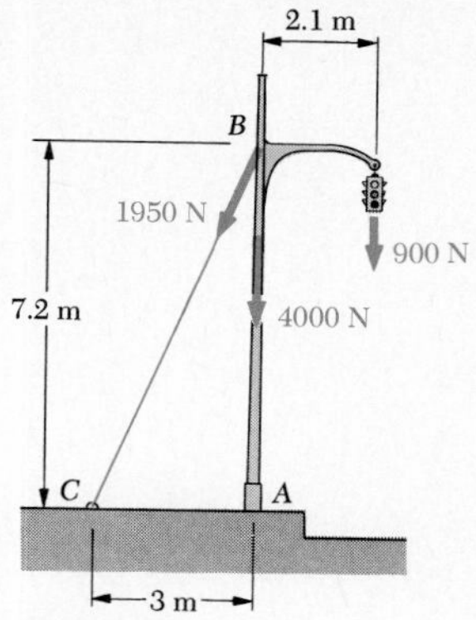

Fig. P3.39

3.40 A force **P** and a couple **M** of moment 20 N · m are applied at *A* to a plate cut in the shape of a regular hexagon of side $a = 125$ mm. Determine the force **P** for which the resultant of the system is directed along (*a*) side *BC*, (*b*) side *CD*, (*c*) diagonal *FC*.

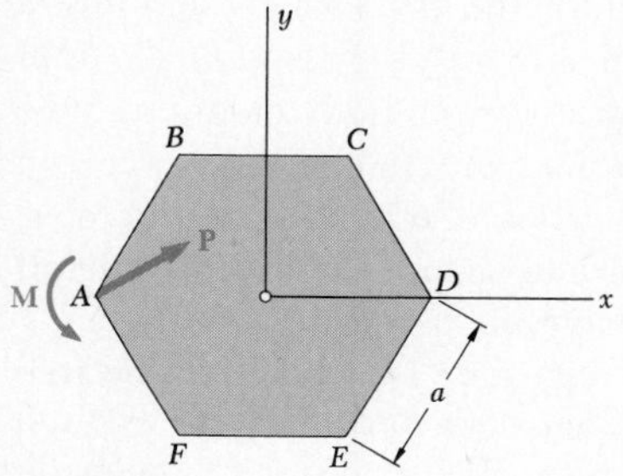

Fig. P3.40

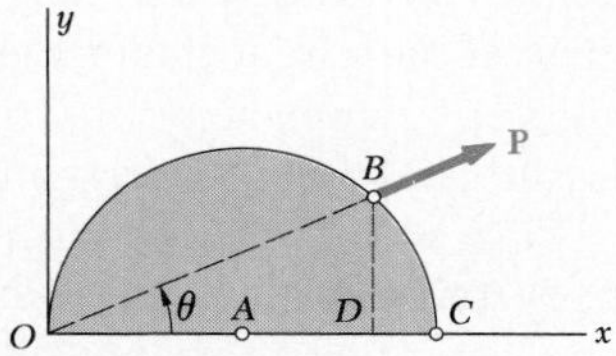

Fig. P3.41

3.41 A force **P** of given magnitude *P* is applied to the edge of a semicircular plate of radius *a* as shown. (*a*) Replace **P** by an equivalent force-couple system at the point *D* obtained by drawing the perpendicular from *B* to the *x* axis. (*b*) Determine the value of θ for which the moment of the equivalent force-couple system at *D* is maximum.

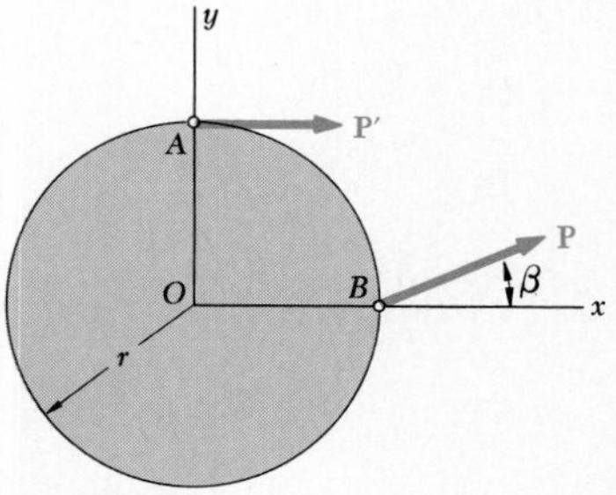

Fig. P3.42

∗**3.42** Two forces, each of magnitude *P*, are applied to the edge of a circular disk as shown. Knowing that the line of action of the resultant of the two forces is tangent to the edge of the disk, determine the required value of β.

EQUILIBRIUM OF RIGID BODIES

3.12. Rigid Body in Equilibrium. *A rigid body is said to be in equilibrium when the external forces acting on it form a system of forces equivalent to zero,* i.e., a system which has no resultant and no resultant couple. The necessary and sufficient conditions for the equilibrium of a rigid body may thus be expressed analytically by writing

$$\Sigma F_x = 0 \qquad \Sigma F_y = 0 \qquad \Sigma M_A = 0 \tag{3.5}$$

Since the direction of the axes of coordinates and the location of point *A* may be chosen arbitrarily, Eqs. (3.5) indicate that the external forces acting on the rigid body do not impart to the body any motion of translation and do not make it rotate about any point. The action of any individual external force is thus canceled by the action of the other forces of the system, and the external forces are said to be *balanced.*

3.13. Free-Body Diagram. In solving a problem concerning the equilibrium of a rigid body, it is essential to consider *all* the forces acting on the body; it is equally important to exclude any force which is not directly applied on the body. Omitting a force or adding an extraneous one would destroy the conditions of equilibrium. Therefore, the first step in the solution of the problem should consist in drawing a *free-body diagram* of the rigid body under consideration. Free-body diagrams have already been used on many occasions in Chap. 2. However, in view of their importance to the solution of equilibrium problems, we shall summarize here the various steps which must be followed in drawing a free-body diagram.

First, a clear decision is made regarding the choice of the free body to be used. This body is then detached from the ground and separated from any other body. The contour of the body thus isolated is sketched.

All external forces are then indicated. These forces represent the action exerted *on* the free body *by* the ground and the bodies which have been detached; they should be applied at the various points where the free body was supported by the ground or connected to the other bodies. The *weight* of the free body should also be included among the external forces, since it represents the attraction exerted by the earth on the various particles forming the free body. As will be seen in Chap. 5, the weight should be applied at the center of gravity of the body. When the free body is made of several parts, the forces the various parts exert on each other should *not* be included among the external forces. These forces are internal forces as far as the free body is concerned.

The magnitude and direction of the *known external forces* should be clearly marked on the free-body diagram. Care should be taken to indicate the sense of the force exerted *on* the free body, not that of the force exerted *by* the free body. Known external forces generally include the *weight* of the free body and *forces applied* for a given purpose.

Unknown external forces usually consist of the *reactions*—also sometimes called *constraining forces*—through which the ground and other bodies oppose a possible motion of the free body and thus constrain it to remain in the same position. Reactions are exerted at the points where the free body is *supported* or *connected* to other bodies. They may be divided into three groups, corresponding to three types of *supports,* or *connections:*

1. *Reactions Equivalent to a Force with Known Line of Action.* Supports and connections causing reactions of this group include *rollers, rockers, frictionless surfaces, short links and cables, collars on frictionless rods,* and *frictionless pins in slots.* Each of these supports and connections can prevent motion in one direction only. They are shown in Fig. 3.16, together with the reaction they produce. Reactions of this group involve *one unknown,* namely, the magnitude of the

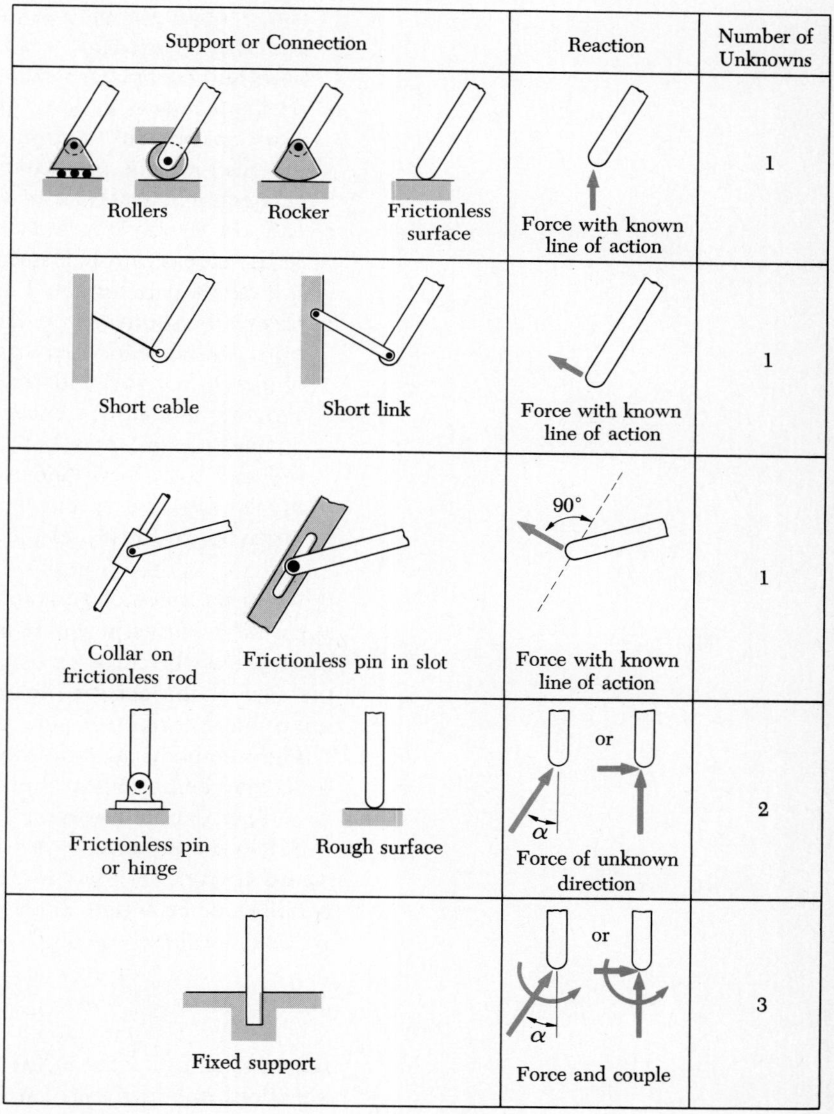

Fig. 3.16 Reactions at supports and connections

reaction; this magnitude should be denoted by an appropriate letter. The line of action of the reaction is known and should be indicated clearly in the free-body diagram. The sense of the reaction must be as shown in Fig. 3.16 in the case of a frictionless surface (away from the surface) or of a cable (tension in the direction of the cable). The reaction may be directed either way in the case of double-track rollers, links, collars on rods, and pins in slots. Single-track rollers and rockers are generally assumed to be reversible, and thus the corresponding reactions may also be directed either way.

2. *Reactions Equivalent to a Force of Unknown Direction.* Supports and connections causing reactions of this group include *frictionless pins in fitted holes, hinges,* and *rough surfaces.* They can prevent translation of the free body in all directions, but they cannot prevent the body from rotating about the connection. Reactions of this group involve *two unknowns* and are usually represented by their x and y components. In the case of a rough surface, the component normal to the surface must be directed away from the surface.
3. *Reactions Equivalent to a Force and a Couple.* These reactions are caused by *fixed supports* which oppose any motion of the free body and thus constrain it completely. Fixed supports actually produce forces over the entire surface of contact; these forces, however, form a system which may be reduced to a force and a couple. Reactions of this group involve *three unknowns,* consisting usually of the two components of the force and the moment of the couple.

When the sense of an unknown force or couple is not clearly apparent, no attempt should be made at determining it. Instead, the sense of the force or couple should be arbitrarily assumed; the sign of the answer obtained will indicate whether the assumption is correct or not.

The free-body diagram should also include dimensions, since these may be needed in the computation of moments of forces. Any other detail, however, should be omitted.

3.14. Problems Involving the Equilibrium of a Rigid Body. We saw in Sec. 3.12 that the conditions for the equilibrium of a rigid body may be expressed analytically by writing the three equations

$$\Sigma F_x = 0 \qquad \Sigma F_y = 0 \qquad \Sigma M_A = 0 \tag{3.5}$$

where A is any point in the plane of the structure. The three equations obtained may be solved for no more than *three unknowns.*

We saw in the preceding section that unknown forces usually consist of reactions, and that the number of unknowns corresponding to a given reaction depends upon the type of support or connection causing that reaction. Referring to Sec. 3.13, we check that the equilibrium equations (3.5) may be used to determine the reactions of two rollers and one cable, or of one fixed support, or of one roller and one pin in a fitted hole, etc.

Consider, for instance, the truss shown in Fig. 3.17*a*, which is subjected to the given forces **P**, **Q**, and **S**. The truss is held in place by a pin at A and a roller at B. The pin prevents point A from moving by exerting on the truss a force which may be resolved into the components $\mathbf{A}_x$ and $\mathbf{A}_y$; the roller keeps the truss from rotating about A by exerting the vertical force **B**. The free-body diagram of the truss is shown in Fig. 3.17*b*; it includes the reactions $\mathbf{A}_x$, $\mathbf{A}_y$, and **B** as well as the applied forces **P**, **Q**, **S**, and the weight **W** of the truss. Expressing that the sum of the moments about A of all the forces shown in Fig. 3.17*b* is zero, we write the equation $\Sigma M_A = 0$, which may be solved for the magnitude B since it does not contain A_x or A_y. Expressing, then, that the sum of the x components and the sum of the y components of the forces are zero, we write the equations $\Sigma F_x = 0$ and $\Sigma F_y = 0$, which may be solved for the components A_x and A_y, respectively.

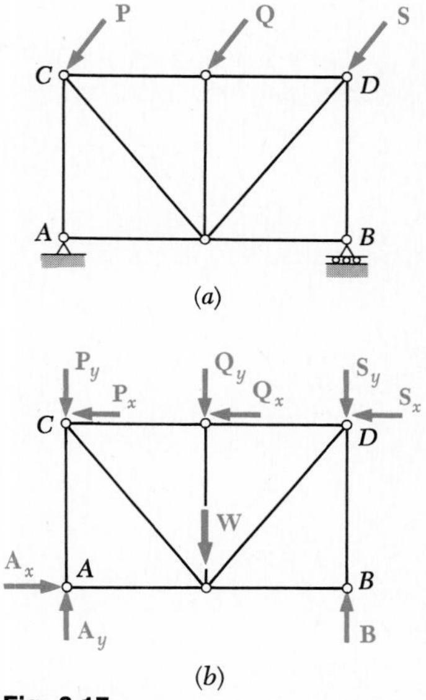

Fig. 3.17

Additional equations could be obtained by expressing that the sum of the moments of the external forces about points other than A is zero. We could write, for instance, $\Sigma M_B = 0$. Such a statement, however, does not contain any new information, since it has already been established that the system of the forces shown in Fig. 3.17*b* is equivalent to zero. The additional equation *is not independent* and cannot be used to determine a fourth unknown. It will be useful, however, for checking the solution obtained from the original three equations of equilibrium.

While the three equations of equilibrium cannot be *augmented* by additional equations, any of them may be *replaced* by another equation. Therefore, an alternate system of equations of equilibrium is

$$\Sigma F_x = 0 \qquad \Sigma M_A = 0 \qquad \Sigma M_B = 0 \tag{3.6}$$

where the line AB is chosen in a direction different from the y direction (Fig. 3.17*b*). These equations are sufficient conditions for the equilibrium of the truss. The first two equations indicate that the external forces must reduce to a single vertical force at A. Since the third equation requires that the moment of this force be zero about a point B which is not on its line of action, the force must be zero and the rigid body is in equilibrium.

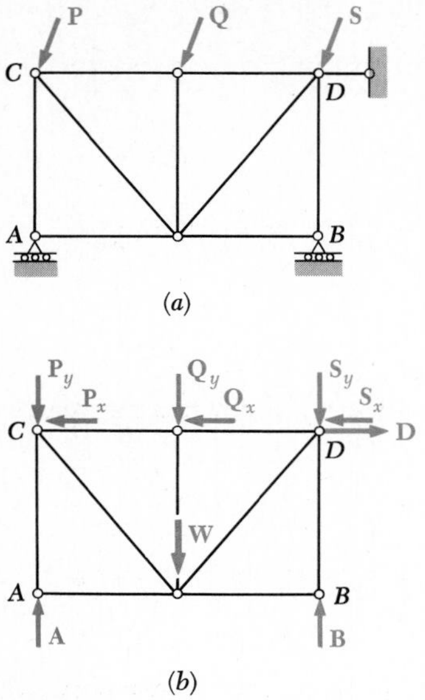

Fig. 3.18

A third possible set of equations of equilibrium is

$$\Sigma M_A = 0 \qquad \Sigma M_B = 0 \qquad \Sigma M_C = 0 \tag{3.7}$$

where the points A, B, and C are not in a straight line (Fig. 3.17*b*). The first equation requires that the external forces reduce to a single force at A; the second equation requires that this force pass through B; the third, that it pass through C. Since the points A, B, C are not in a straight line, the force must be zero, and the rigid body is in equilibrium.

The equation $\Sigma M_A = 0$, which expresses that the sum of the moments of the forces about pin A is zero, possesses a more definite physical meaning than either of the other two equations (3.7). These two equations express a similar idea of balance, but with respect to points about which the rigid body is not actually hinged. They are, however, as useful as the first equation, and our choice of equilibrium equations should not be unduly influenced by the physical meaning of these equations. Indeed, it will be desirable in practice to choose equations of equilibrium containing only one unknown, since this eliminates the necessity of solving simultaneous equations. Equations containing only one unknown may be obtained by summing moments about the point of intersection of the lines of action of two unknown forces or, if these forces are parallel, by summing components in a direction perpendicular to their common direction. In the case of the truss of Fig. 3.18, for example, which is held by rollers at A and B and a short link at D, the reactions at A and B may be eliminated by summing x components. The reactions at A and D will be eliminated by summing moments about C and the reactions at B and D by summing moments about D. The equations obtained are

$$\Sigma F_x = 0 \qquad \Sigma M_C = 0 \qquad \Sigma M_D = 0$$

Each of these equations contains only one unknown.

3.15. Statically Indeterminate Reactions. Partial Constraints. In each of the two examples considered in the preceding section (Figs. 3.17 and 3.18), the reactions to be determined involved *three unknowns;* these unknowns were obtained by solving the three equations of equilibrium. Besides, the types of supports used were such that the rigid body could not possibly move under the given loads or under any other loading conditions. In such cases, the reactions are said to be *statically determinate,* and the rigid body is said to be *completely constrained.*

Consider now the truss shown in Fig. 3.19*a*, which is held by pins at A and B. We note from the free-body diagram of Fig. 3.19*b* that the reactions involve *four unknowns*. Since, as was pointed out in Sec. 3.14, only three independent equilib-

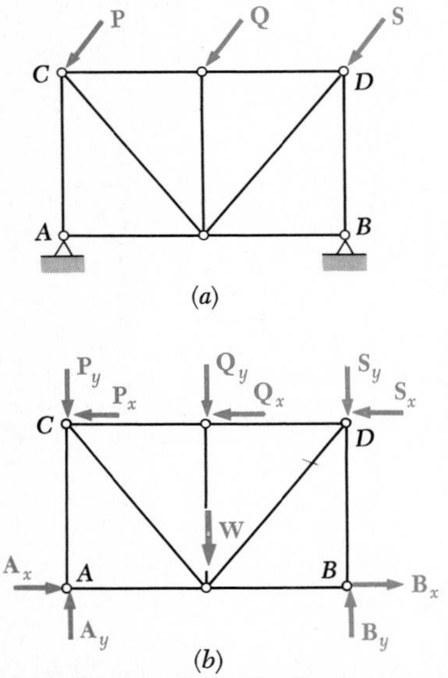

Fig. 3.19 Statically indeterminate reactions

rium equations are available, there are *more unknowns than equations,* and all the unknowns cannot be determined. While the equations $\Sigma M_A = 0$ and $\Sigma M_B = 0$ yield the vertical components B_y and A_y, respectively, the equation $\Sigma F_x = 0$ gives only the sum $A_x + B_x$ of the horizontal components of the reactions at A and B. The components A_x and B_x are said to be *statically indeterminate.* They could be determined by considering the deformations produced in the truss by the given loading, but this method is beyond the scope of statics and belongs to the study of mechanics of materials.

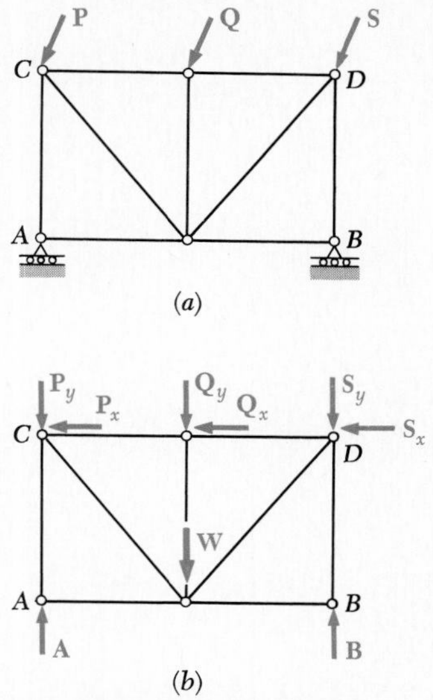

Fig. 3.20 Partial constraints

The supports used to hold the truss shown in Fig. 3.20*a* consist of rollers at A and B. The corresponding reactions, shown in Fig. 3.20*b*, involve *two unknowns.* Since three equations of equilibrium must still be satisfied, there are *fewer unknowns than equations,* and one of the equilibrium equations will not be satisfied. While the equations $\Sigma M_A = 0$ and $\Sigma M_B = 0$ can be satisfied by a proper choice of reactions at A and B, the equation $\Sigma F_x = 0$ will not be satisfied unless the sum of the horizontal components of the applied forces happens to be zero. The physical significance of this result is clear: Equilibrium cannot be maintained under general loading conditions; while any vertical motion is prevented, the truss is free to move horizontally. The truss is said to be only *partially constrained.*†

It appears from the above that, if a rigid body is to be completely constrained and if the reactions at its supports are to be statically determinate, *there must be as many unknowns as there are equations of equilibrium.* We should note, however, that, while *necessary,* this condition is *not sufficient.* Consider, for example, the truss shown in Fig. 3.21*a*, which is held by rollers at A, B, and E. While there are three unknown reactions, **A**, **B**, and **E** (Fig. 3.21*b*), we find that the equation $\Sigma F_x = 0$ will not be satisfied unless the sum of the horizontal components of the applied forces happens to be zero. There is a sufficient number of constraints, but these constraints are not properly arranged; we say that the truss is *improperly constrained.* Since only two equilibrium equations are left for determining the three unknowns, the reactions will be statically indeterminate. Thus, improper constraints also produce static indeterminacy.

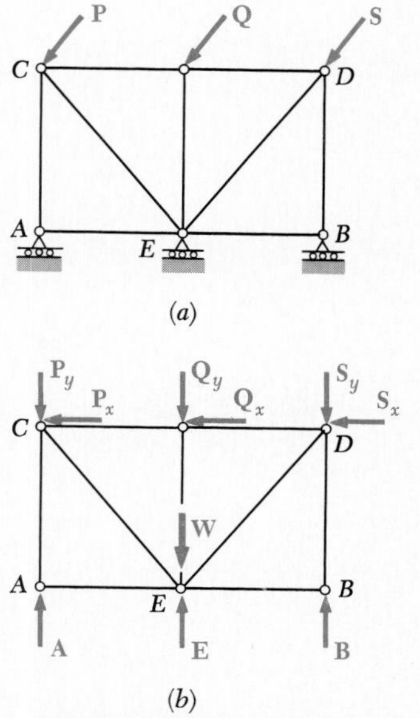

Fig. 3.21 Improper constraints

Another example of improper constraints—and of the accompanying static indeterminacy—is provided by the truss shown in Fig. 3.22. This truss is held by a pin at A and by rollers at B and C, which altogether involve four unknowns. Choosing the

† Partially constrained bodies are often referred to as *unstable.* However, in order to avoid any confusion between this type of instability, due to insufficient constraints, and the type of instability considered in Chap. 10, which relates to the behavior of a rigid body when its equilibrium is disturbed, we shall restrict the use of the words "stable" and "unstable" to the latter case.

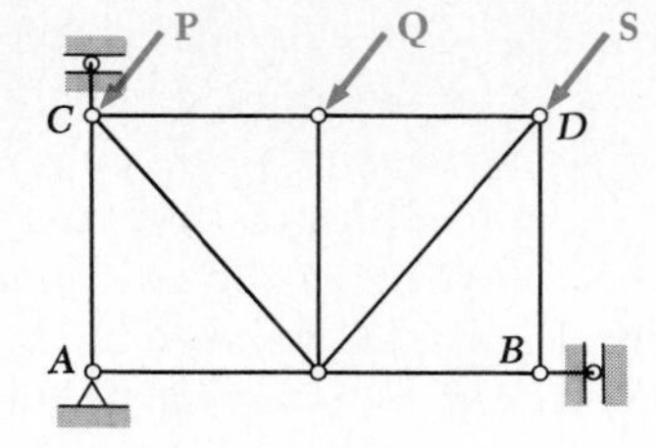

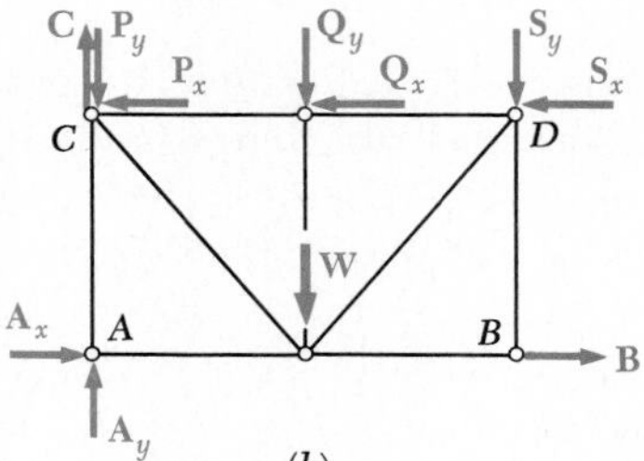

Fig. 3.22 Improper constraints

equilibrium equations $\Sigma M_A = 0$, $\Sigma F_x = 0$, and $\Sigma F_y = 0$, we find that the first equation cannot be satisfied under general loading conditions, while the other two yield only the sums $A_x + B$ and $A_y + C$. The examples of Figs. 3.21 and 3.22 lead us to conclude that *a rigid body is improperly constrained whenever the supports,* even though they may provide a sufficient number of reactions, *are arranged in such a way that the reactions must be either concurrent or parallel.*†

Supports involving statically indeterminate reactions should be used with care in the *design* of structures, and only with a full knowledge of the problems they may cause. On the other hand, the *analysis* of structures possessing statically indeterminate reactions often may be partially carried out by the methods of statics. In the case of the truss of Fig. 3.19, for example, the vertical components of the reactions at A and B were obtained from the equilibrium equations.

For obvious reasons, supports producing partial or improper constraints should be avoided in the design of stationary structures. However, a partially or improperly constrained structure will not necessarily collapse; under particular loading conditions, equilibrium may be maintained. For example, the trusses of Figs. 3.20 and 3.21 will be in equilibrium if the applied forces **P**, **Q**, and **S** are vertical. Besides, structures which are designed to move *should* be only partially constrained. A railroad car, for instance, would be of little use if it were completely constrained by having its brakes applied permanently.

† Because this situation arises from an inadequate arrangement or *geometry* of the supports, it is often referred to as *geometric instability.*

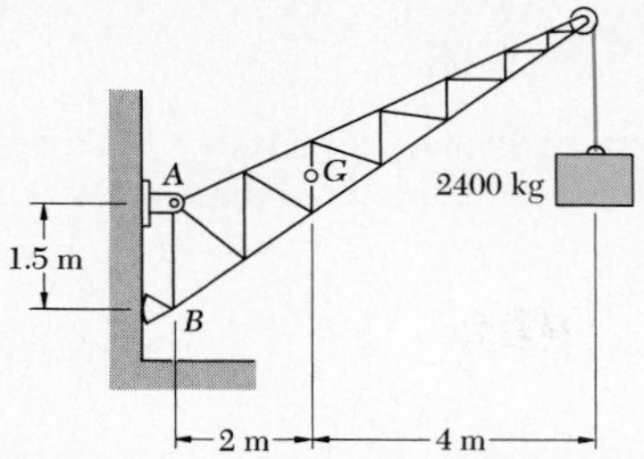

SAMPLE PROBLEM 3.8

A fixed crane has a mass of 1000 kg and is used to lift a 2400-kg crate. It is held in place by a pin at A and a rocker at B. The center of gravity of the crane is located at G. Determine the components of the reactions at A and B.

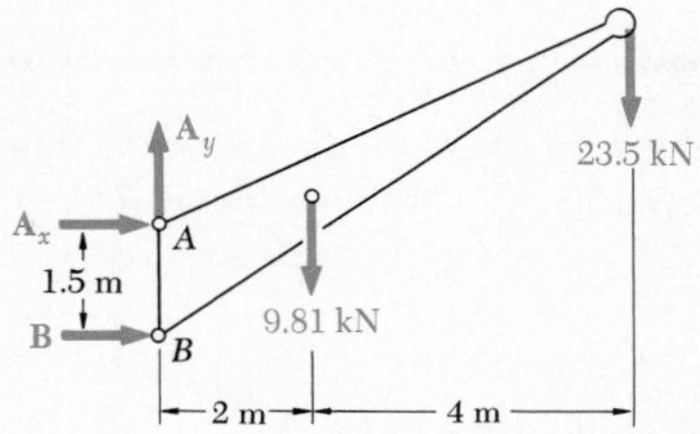

Solution. A free-body diagram of the crane is drawn. Multiplying the masses of the crane and of the crate by $g = 9.81 \text{ m/s}^2$, we obtain the corresponding weights, that is, 9810 N or 9.81 kN, and 23 500 N or 23.5 kN. The reaction at pin A is a force of unknown direction, represented by its components $\mathbf{A}_x$ and $\mathbf{A}_y$. The reaction at the rocker B is perpendicular to the rocker surface; thus it is horizontal. We assume that $\mathbf{A}_x$, $\mathbf{A}_y$, and $\mathbf{B}$ act in the directions shown.

Determination of **B**. We express that the sum of the moments of all external forces about point A is zero. The equation obtained will contain neither A_x nor A_y since the moments of $\mathbf{A}_x$ and $\mathbf{A}_y$ about A are zero. Multiplying the magnitude of each force by its perpendicular distance from B, we write

$$+\circlearrowleft \Sigma M_A = 0: \quad +B(1.5 \text{ m}) - (9.81 \text{ kN})(2 \text{ m}) - (23.5 \text{ kN})(6 \text{ m}) = 0$$
$$B = +107.1 \text{ kN} \qquad \mathbf{B} = 107.1 \text{ kN} \rightarrow$$

Since the result is positive, the reaction is directed as assumed.

Determination of $\mathbf{A}_x$. The magnitude A_x is determined by expressing that the sum of the horizontal components of all external forces is zero.

$$\overset{+}{\rightarrow} \Sigma F_x = 0: \quad A_x + B = 0 \qquad A_x + 107.1 \text{ kN} = 0$$
$$A_x = -107.1 \text{ kN} \qquad \mathbf{A}_x = 107.1 \text{ kN} \leftarrow$$

Since the result is negative, the sense of $\mathbf{A}_x$ is opposite to that assumed originally.

Determination of $\mathbf{A}_y$. The sum of the vertical components must also equal zero.

$$+\uparrow \Sigma F_y = 0: \quad A_y - 9.81 \text{ kN} - 23.5 \text{ kN} = 0$$
$$A_y = +33.3 \text{ kN} \qquad \mathbf{A}_y = 33.3 \text{ kN} \uparrow$$

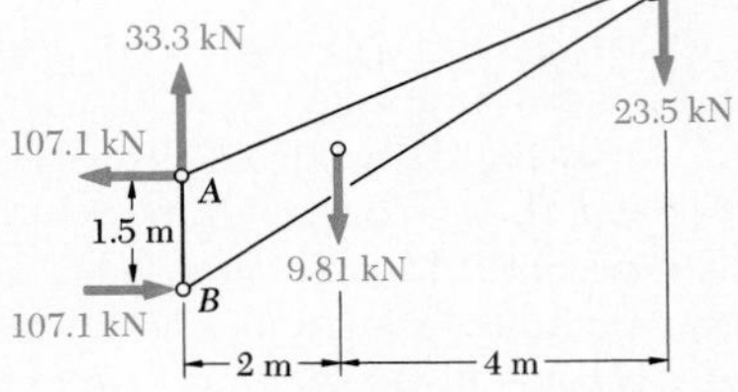

Adding vectorially the components $\mathbf{A}_x$ and $\mathbf{A}_y$, we find that the reaction at A is 112.2 kN $\nwarrow$17.3°.

Check. The values obtained for the reactions may be checked by recalling that the sum of the moments of all external forces about any point must be zero. For example, considering point B, we write

$$+\circlearrowleft \Sigma M_B = -(9.81 \text{ kN})(2 \text{ m}) - (23.5 \text{ kN})(6 \text{ m}) + (107.1 \text{ kN})(1.5 \text{ m}) = 0$$

SAMPLE PROBLEM 3.9

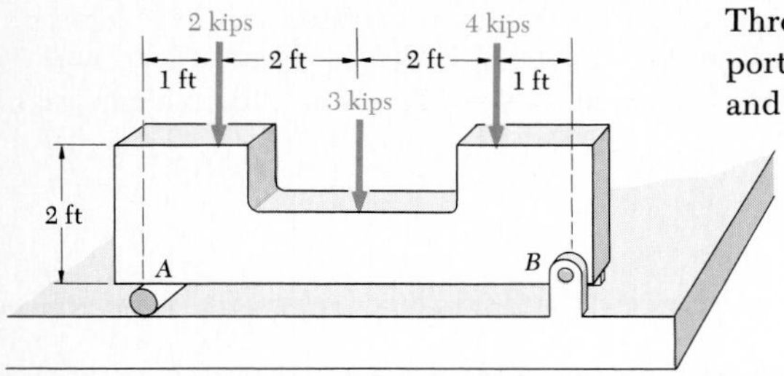

Three loads are applied to a steel plate as shown. The plate is supported by a roller at A and by a pin at B. Determine the reactions at A and B.

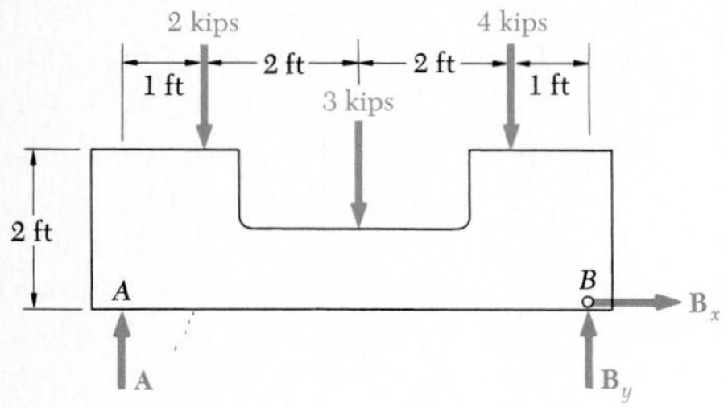

Solution. A free-body diagram of the plate is drawn. The reaction at A is vertical and is denoted by $\mathbf{A}$. The reaction at B is represented by components $\mathbf{B}_x$ and $\mathbf{B}_y$. Each component is assumed to act in the direction shown.

Equilibrium Equations. We write the following three equilibrium equations and solve for the reactions indicated:

$\overset{+}{\rightarrow} \Sigma F_x = 0$: $\quad B_x = 0 \qquad \mathbf{B}_x = 0$ ◄

$+\circlearrowleft \Sigma M_A = 0$:

$$-(2 \text{ kips})(1 \text{ ft}) - (3 \text{ kips})(3 \text{ ft}) - (4 \text{ kips})(5 \text{ ft}) + B_y(6 \text{ ft}) = 0$$

$$B_y = +5.17 \text{ kips} \qquad \mathbf{B}_y = 5.17 \text{ kips} \uparrow$$ ◄

$+\circlearrowleft \Sigma M_B = 0$:

$$-A(6 \text{ ft}) + (2 \text{ kips})(5 \text{ ft}) + (3 \text{ kips})(3 \text{ ft}) + (4 \text{ kips})(1 \text{ ft}) = 0$$

$$A = +3.83 \text{ kips} \qquad \mathbf{A} = 3.83 \text{ kips} \uparrow$$ ◄

Check. The results are checked by adding the vertical components of all the external forces.

$$+\uparrow \Sigma F_y = +5.17 \text{ kips} + 3.83 \text{ kips} - 2 \text{ kips} - 3 \text{ kips} - 4 \text{ kips} = 0$$

Remark. In this problem the reactions at both A and B are vertical; however, these reactions are vertical for different reasons. At A, the plate is supported by a roller; hence the reaction cannot have any horizontal component. At B, the horizontal component of the reaction is zero because it must satisfy the equilibrium equation $\Sigma F_x = 0$ and because none of the other forces acting on the plate has a horizontal component.

We could have noticed at first glance that the reaction at B was vertical and dispensed with the horizontal component $\mathbf{B}_x$. This, however, is a bad practice. In following it, we would run the risk of forgetting the component $\mathbf{B}_x$ when the loading conditions require such a component (i.e., when a horizontal load is included). Also, the component $\mathbf{B}_x$ was found to be zero by using and solving an equilibrium equation, $\Sigma F_x = 0$. By setting $\mathbf{B}_x$ equal to zero immediately, we might not realize that we actually make use of this equation and thus might lose track of the number of equations available for solving the problem.

SAMPLE PROBLEM 3.10

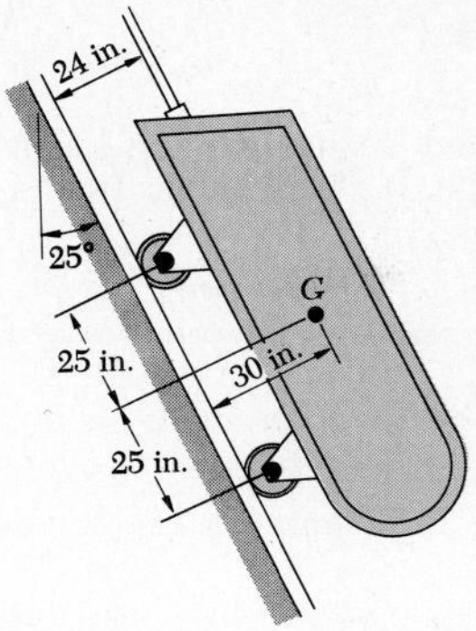

A loading car is at rest on a track forming an angle of 25° with the vertical. The gross weight of the car and its load is 5500 lb, and it is applied at a point 30 in. from the track, halfway between the two axles. The car is held by a cable attached 24 in. from the track. Determine the tension in the cable and the reaction at each pair of wheels.

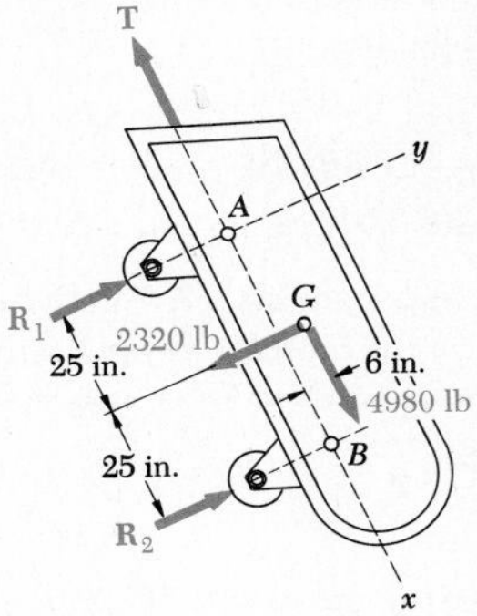

Solution. A free-body diagram of the car is drawn. The reaction at each wheel is perpendicular to the track, and the tension force **T** is parallel to the track. For convenience, we choose the x axis parallel to the track and the y axis perpendicular to the track. The 5500-lb weight is then resolved into x and y components.

$$W_x = +(5500 \text{ lb}) \cos 25° = +4980 \text{ lb}$$
$$W_y = -(5500 \text{ lb}) \sin 25° = -2320 \text{ lb}$$

Equilibrium Equations. We take moments about A to eliminate **T** and $\mathbf{R}_1$ from the computation.

$+\circlearrowleft \Sigma M_A = 0$:

$$-(2320 \text{ lb})(25 \text{ in.}) - (4980 \text{ lb})(6 \text{ in.}) + R_2(50 \text{ in.}) = 0$$
$$R_2 = +1758 \text{ lb} \qquad \mathbf{R}_2 = 1758 \text{ lb} \nearrow \blacktriangleleft$$

Now, taking moments about B to eliminate **T** and $\mathbf{R}_2$ from the computation, we write

$$+\circlearrowleft \Sigma M_B = 0: \quad (2320 \text{ lb})(25 \text{ in.}) - (4980 \text{ lb})(6 \text{ in.}) - R_1(50 \text{ in.}) = 0$$
$$R_1 = +562 \text{ lb} \qquad \mathbf{R}_1 = 562 \text{ lb} \nearrow \blacktriangleleft$$

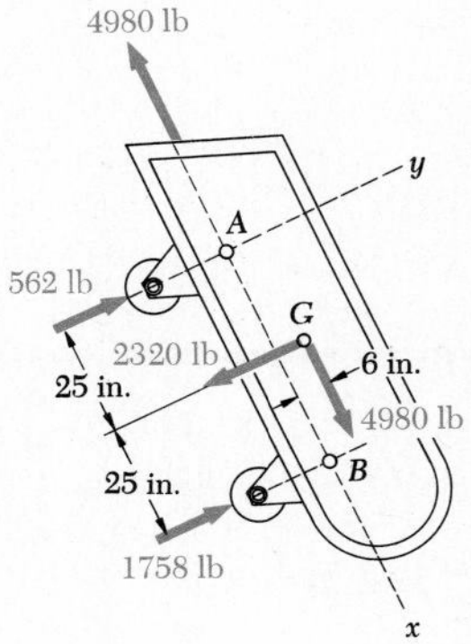

The value of T is found by writing

$$\searrow +\Sigma F_x = 0: \quad +4980 \text{ lb} - T = 0$$
$$T = +4980 \text{ lb} \qquad \mathbf{T} = 4980 \text{ lb} \nwarrow \blacktriangleleft$$

The computed values of the reactions are shown in the adjacent sketch.

Check. The computations are verified by writing

$$\nearrow +\Sigma F_y = +562 \text{ lb} + 1758 \text{ lb} - 2320 \text{ lb} = 0$$

A check could also have been obtained by computing moments about any point except A or B.

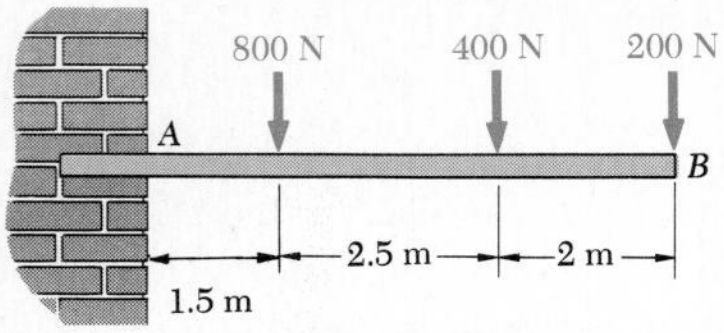

SAMPLE PROBLEM 3.11

A cantilever beam is loaded as shown. The beam is fixed at the left end and free at the right end. Determine the reaction at the fixed end.

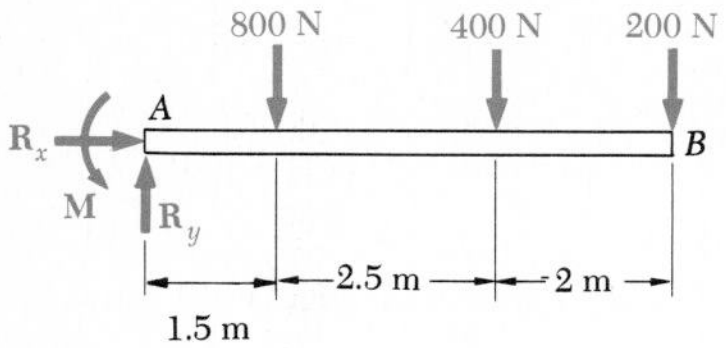

Solution. The portion of the beam which is embedded in the wall is subjected to a large number of forces. These forces, however, are equivalent to a force of components $\mathbf{R}_x$ and $\mathbf{R}_y$ and a couple $\mathbf{M}$.

Equilibrium Equations

$\overset{+}{\rightarrow} \Sigma F_x = 0$: $\quad R_x = 0$ $\qquad \mathbf{R}_x = 0$ ◄

$+\uparrow \Sigma F_y = 0$: $\quad R_y - 800\text{ N} - 400\text{ N} - 200\text{ N} = 0$

$R_y = +1400\text{ N}$ $\qquad \mathbf{R}_y = 1400\text{ N}\uparrow$ ◄

$+\circlearrowleft \Sigma M_A = 0$:

$$-(800\text{ N})(1.5\text{ m}) - (400\text{ N})(4\text{ m}) - (200\text{ N})(6\text{ m}) + M = 0$$

$M = +4000\text{ N}\cdot\text{m}$ $\qquad \mathbf{M} = 4000\text{ N}\cdot\text{m}\circlearrowleft$ ◄

The reaction at the fixed end consists of a vertical upward force of 1400 N and of a 4000-N · m counterclockwise couple.

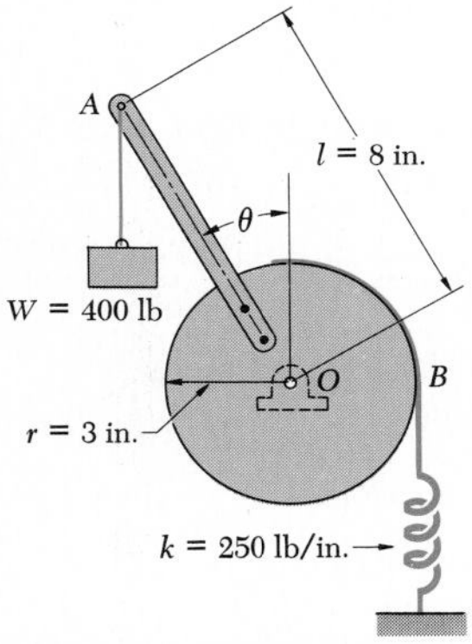

SAMPLE PROBLEM 3.12

A 400-lb weight is attached to the lever AO as shown. The constant of the spring BC is $k = 250$ lb/in., and the spring is unstretched when $\theta = 0$. Determine the position or positions of equilibrium.

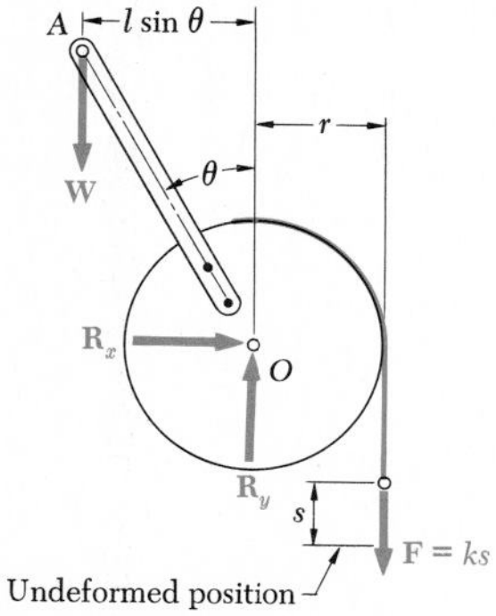

Solution. *Force Exerted by Spring.* Denoting by s the deflection of the spring from its undeformed position, and noting that $s = r\theta$, we write

$$F = ks = kr\theta$$

Equilibrium Equation. Summing the moments of $\mathbf{W}$ and $\mathbf{F}$ about O, we write

$+\circlearrowleft \Sigma M_O = 0$: $\quad Wl \sin\theta - r(kr\theta) = 0 \qquad \sin\theta = \dfrac{kr^2}{Wl}\theta$

Substituting the given data, we obtain

$$\sin\theta = \frac{(250\text{ lb/in.})\,(3\text{ in.})^2}{(400\text{ lb})\,(8\text{ in.})}\theta \qquad \sin\theta = 0.703\theta$$

Solving by trial and error, we find $\qquad \theta = 0 \qquad \theta = 80.3°$ ◄

PROBLEMS

3.43 The 40-ft boom AB weighs 2 kips; the distance from the axle A to the center of gravity G of the boom is 20 ft. For the position shown, determine the tension T in the cable and the reaction at A.

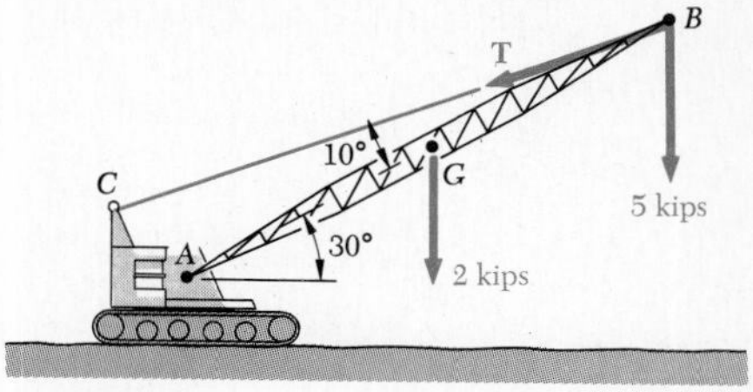

Fig. P3.43

3.44 Knowing that the magnitude of the vertical force **P** is 400 N, determine (*a*) the tension in the cable CD, (*b*) the reaction at B.

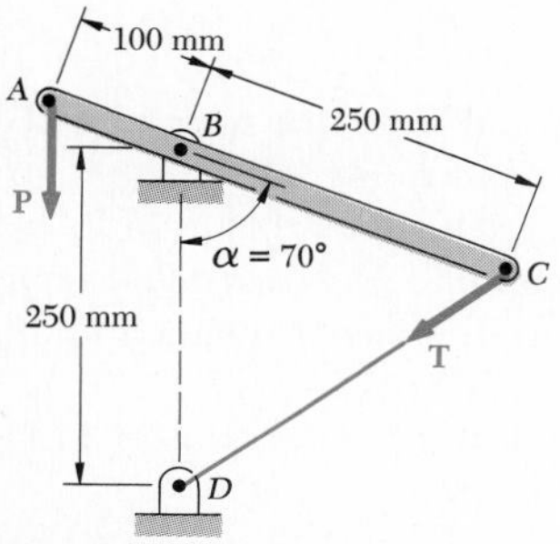

Fig. P3.44

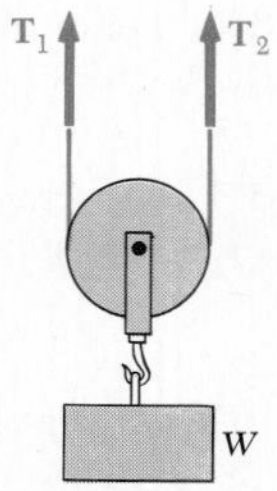

Fig. P3.45

3.45 A load W is supported by the pulley as shown. Prove that, if the pulley is in equilibrium, the tensions T_1 and T_2 are both equal to $W/2$.

3.46 The ladder AB, of length L and weight W, can be raised by the cable BC. Determine the tension T required to raise end B just off the floor (*a*) in terms of W and θ, (*b*) if $h = 8$ ft, $L = 10$ ft, and $W = 35$ lb.

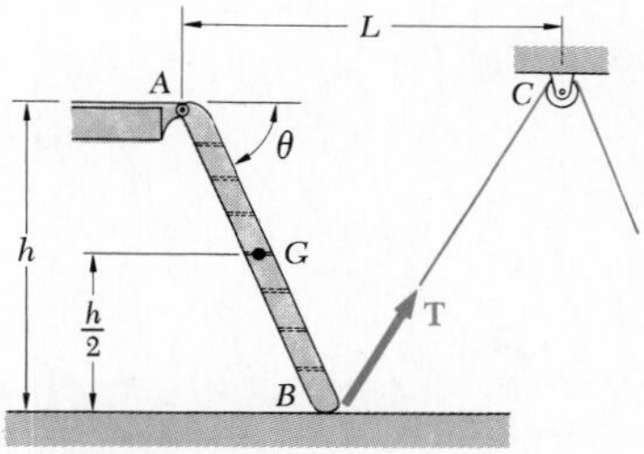

Fig. P3.46

3.47 A block of weight W is to be supported by the winch shown. Determine the required magnitude of the force **P** (*a*) in terms of W, r, l, and θ, (*b*) if $W = 100$ lb, $r = 3$ in., $l = 15$ in., and $\theta = 60°$.

3.48 Determine the values of θ for which the winch is in equilibrium when $W = 125$ lb, $P = 50$ lb, $r = 3$ in., and $l = 15$ in.

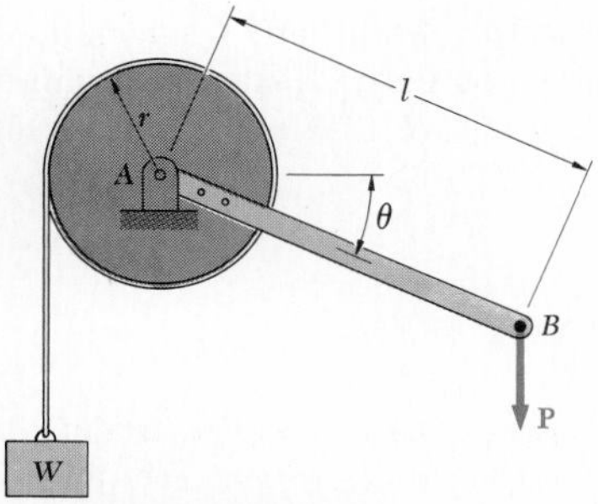

Fig. P3.47 and P3.48

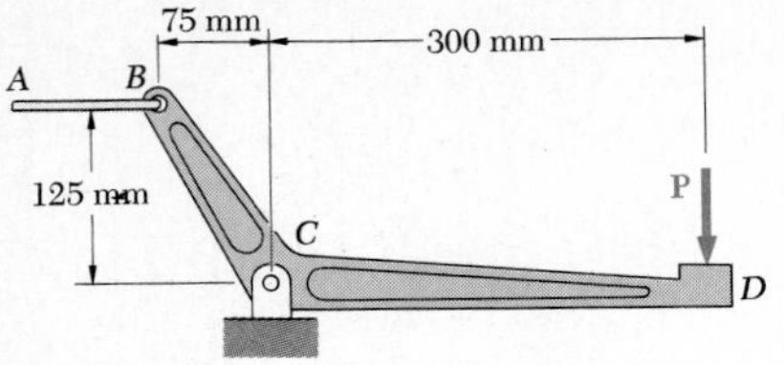

Fig. P3.49 and P3.50

3.49 The required tension in cable AB is 1200 N. Determine (a) the vertical force **P** which must be applied to the pedal, (b) the corresponding reaction at C.

3.50 Determine the maximum tension which may be developed in cable AB if the maximum allowable magnitude of the reaction at C is 2.6 kN.

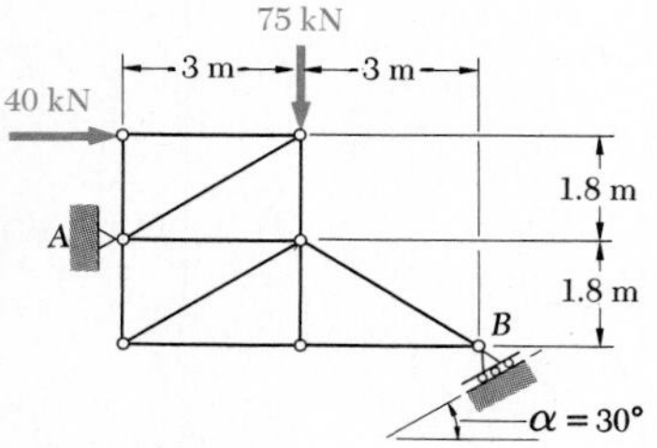

Fig. P3.51

3.51 Determine the reactions at A and B for the truss and loading shown.

3.52 Determine the reactions at A and B for the truss of Prob. 3.51, if the horizontal 40-kN load is directed to the left.

3.53 Determine the reactions at A and B for the truss and loading of Prob. 3.51, if the roller at B rests on a horizontal plane ($\alpha = 0$).

3.54 Determine the reactions at A and B for the truss of Prob. 3.51, if the horizontal 40-kN load is removed.

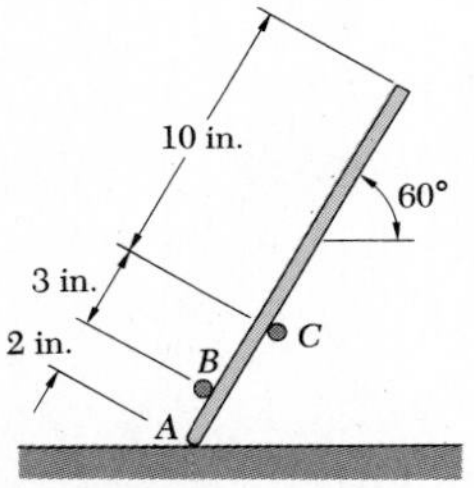

Fig. P3.55

3.55 A slender rod, 15 in. long, weighs 5 lb and is lodged between two frictionless pegs as shown. If the surface at A is also frictionless, determine the reactions at A, B, and C.

3.56 A movable bracket is held at rest by a cable attached at C and by frictionless rollers at A and B. For the loading shown, determine the tension in the cable and the reactions at A and B.

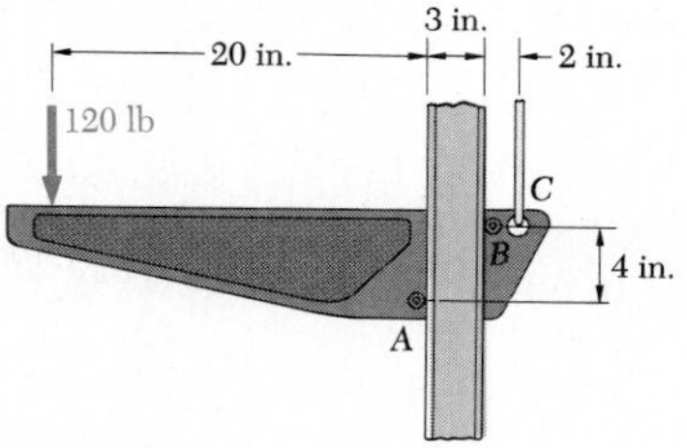

Fig. P3.56

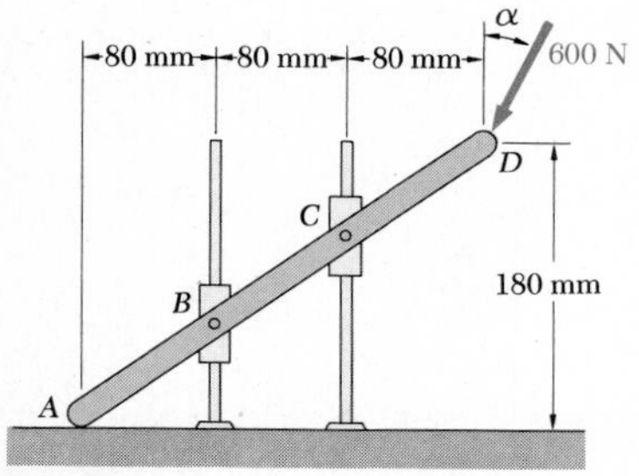

Fig. P3.57

3.57 The light bar AD is attached to collars B and C which may move freely on vertical rods. If the surface at A is frictionless, determine the reactions at A, B, and C (a) if $\alpha = 60°$, (b) if $\alpha = 90°$.

3.58 Solve Prob. 3.57 (a) if $\alpha = 0°$, (b) if $\alpha = 30°$.

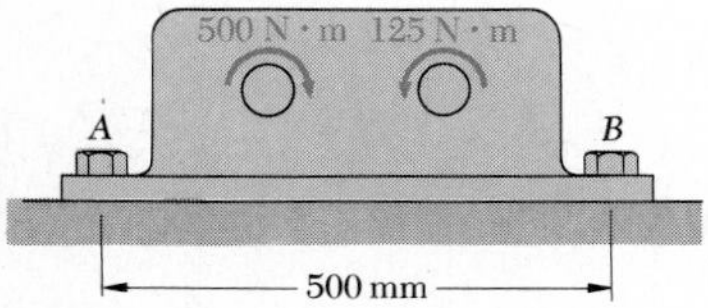

Fig. P3.59

3.59 Two external shafts of a gearbox carry torques as shown. Determine the vertical components of the forces which must be exerted by the bolts at A and B to maintain the gearbox in equilibrium.

3.60 Determine the reactions at A and B for the loading shown.

3.61 If the pin support at A is removed and replaced by a pin support at C, determine the reactions at B and C.

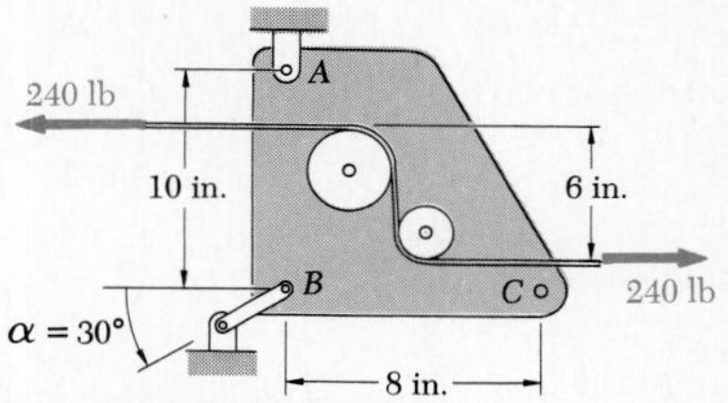

Fig. P3.60 and P3.61

3.62 A couple **M** is applied to a bent rod AB which may be supported in four different ways as shown. In each case determine the reactions at the supports.

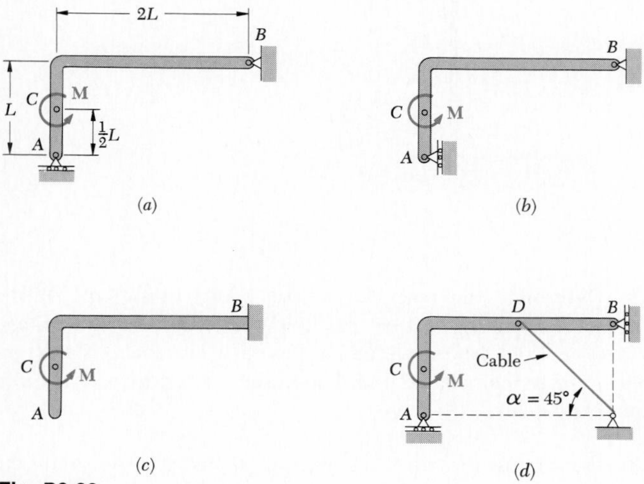

Fig. P3.62

3.63 The maximum allowable value for each of the reactions is 150 kN and the reaction at A must be directed upward. Neglecting the weight of the beam, determine the range of values of P for which the beam is safe.

3.64 Determine the reactions at A and B for the beam and loading shown when (a) $P = 75$ kN, (b) $P = 150$ kN.

3.65 The 10-ft beam AB rests upon, but is not attached to, supports at C and D. Neglecting the weight of the beam, determine the range of values of P for which the beam will remain in equilibrium.

3.66 The 10-ft uniform beam AB weighs 100 lb; it rests upon, but is not attached to, supports at C and D. Determine the range of values of P for which the beam will remain in equilibrium.

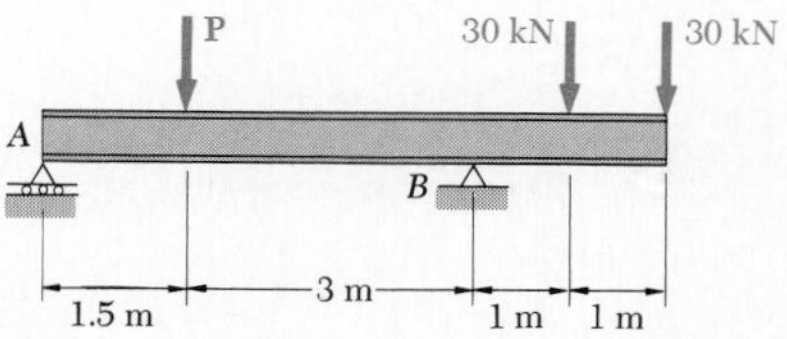

Fig. P3.63 and P3.64

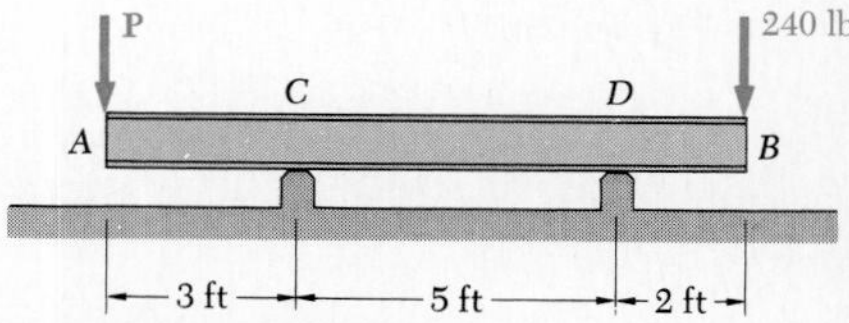

Fig. P3.65 and P3.66

3.67 A traffic-signal pole may be supported in the three ways shown; in part *c* the tension in cable *BC* is known to be 1950 N. Determine the reactions for each type of support shown.

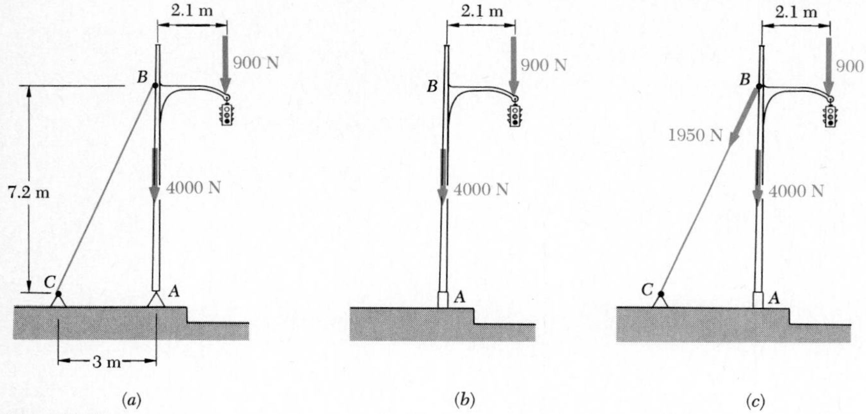

Fig. P3.67

3.68 Determine the reactions at the points of support of the bracket shown, assuming in turn the following types of connections: (*a*) a pin in a fitted hole at *A* and a pin in a horizontal slot at *B*, (*b*) a pin in a vertical slot at *A* and a pin in a fitted hole at *B*, (*c*) a firmly clinched rivet at *A*.

3.69 Determine the reactions at the points of support of the bracket shown, assuming in turn the following types of connections: (*a*) a pin in a fitted hole at *A* and a pin in a vertical slot at *B*, (*b*) a pin in a horizontal slot at *A* and a pin in a fitted hole at *B*, (*c*) a firmly clinched rivet at *B*.

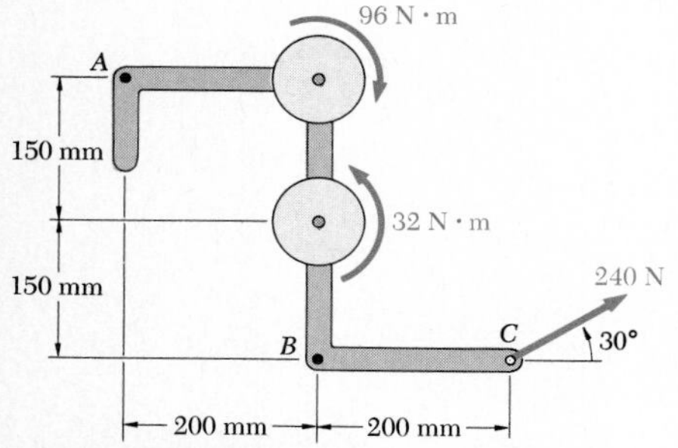

Fig. P3.68 and P3.69

3.70 A 16-ft telephone pole weighing 300 lb is used to support the ends of two wires. The tension in the wire to the left is 80 lb and, at the point of support, the wire forms an angle of 10° with the horizontal. (*a*) If the tension T_2 is zero, determine the reaction at the base *A*. (*b*) Determine the largest and smallest allowable tension T_2, if the magnitude of the couple at *A* may not exceed 600 lb · ft.

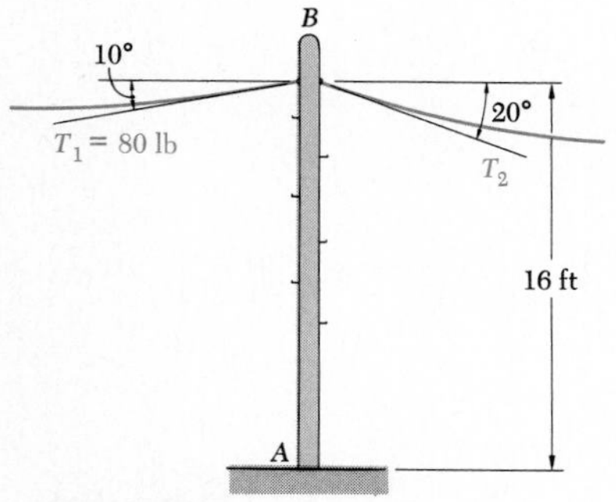

Fig. P3.70

3.71 The 300-lb beam *AB* carries a 500-lb load at *B*. The beam is held by a fixed support at *A* and by the cable *CD* which is attached to the counterweight *W*. (*a*) If *W* = 1300 lb, determine the reaction at *A*. (*b*) Determine the range of values of *W* for which the moment of the couple at *A* does not exceed 1500 lb · ft.

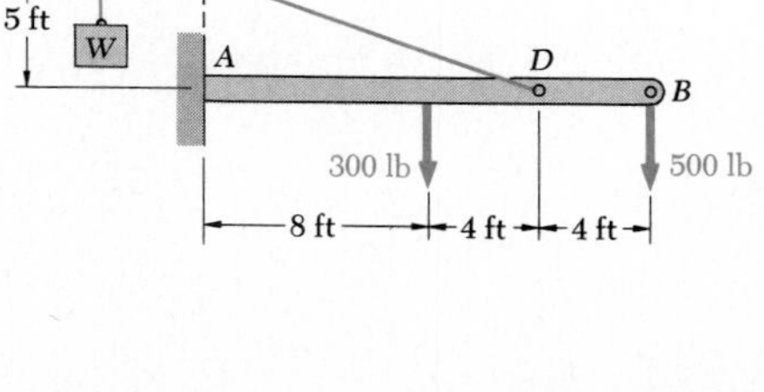

Fig. P3.71

3.72 In the pivoted motor mount, or Rockwood drive, the weight of the motor is used to maintain tension in the drive belt. When the motor is at rest, the tensions T_1 and T_2 may be assumed equal. The mass of the motor is 90 kg, and the diameter of the drive pulley is 150 mm. Assuming that the mass of the platform AB is negligible, determine (*a*) the tension in the belt, (*b*) the reaction at C when the motor is at rest.

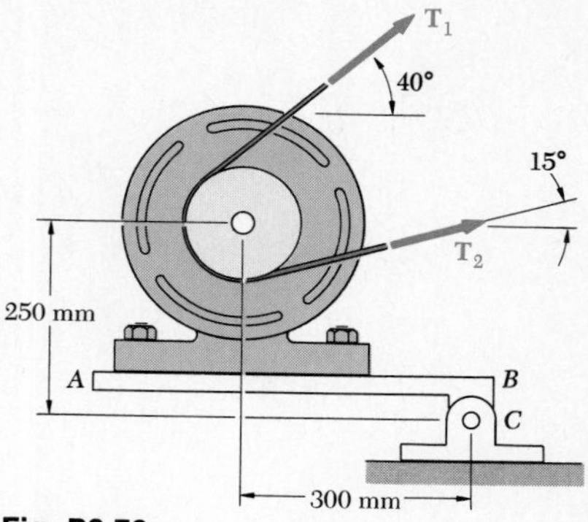

Fig. P3.72

3.73 A force **P** of magnitude 90 lb is applied to the angle AB which is supported by a smooth pin at D and by the cable ACB. Since the cable passes over a pulley at C, the tension may be assumed to be the same in the portions AC and BC of the cable. Determine the tension in the cable and the reaction at D when $a = 3$ in.

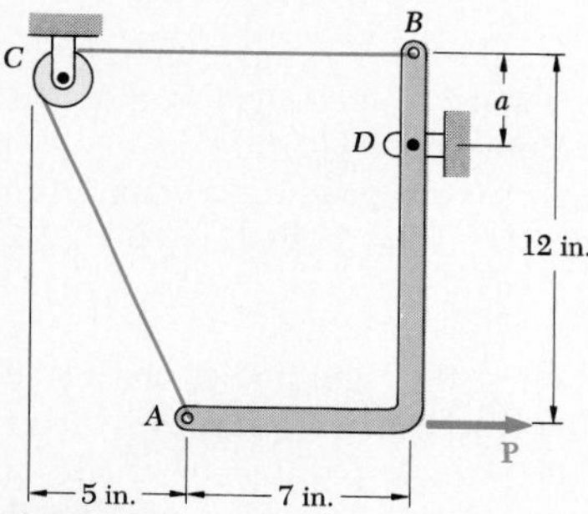

Fig. P3.73

***3.74** Two wheels A and B, of weight $2W$ and W, respectively, are connected by a rod of negligible weight and are free to roll on the surface shown. If $\theta = 45°$, determine the angle that the rod forms with the horizontal when the system is in equilibrium.

***3.75** Two wheels A and B, of weight $2W$ and W, respectively, are connected by a rod of negligible weight and are free to roll on the surface shown. Determine the angle θ for which the rod connecting the wheels is horizontal when the system is in equilibrium.

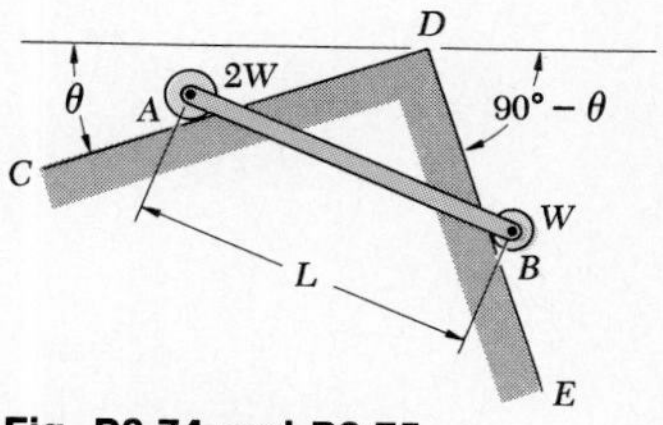

Fig. P3.74 and P3.75

3.76 A slender rod AB, of weight W, is attached to blocks A and B which move freely in the guides shown. The blocks are connected by an elastic cord which passes over a pulley at C. (*a*) Express the tension in the cord in terms of W and θ. (*b*) Determine the value of θ for which the tension in the cord is equal to $2W$.

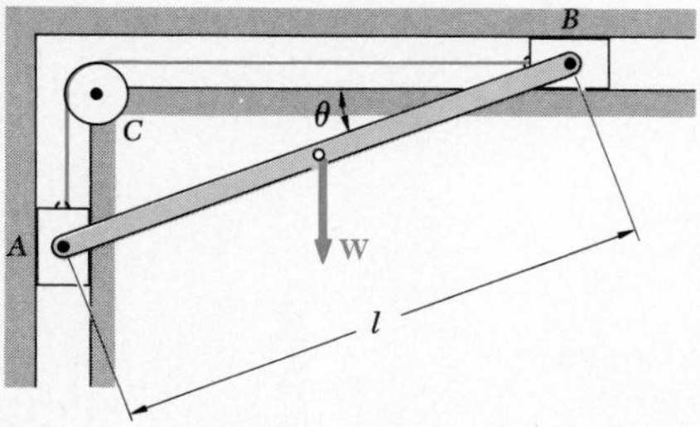

Fig. P3.76

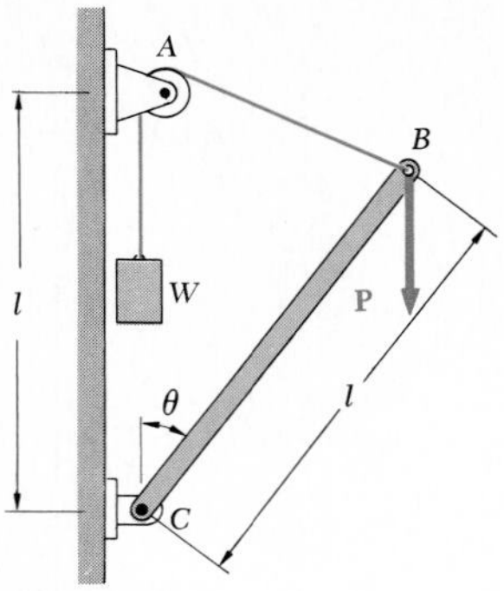

Fig. P3.77

3.77 and 3.78 (*a*) Derive an expression for the angle θ corresponding to equilibrium. (*b*) Determine the value of θ corresponding to the equilibrium position if $W = \frac{1}{2}P$.

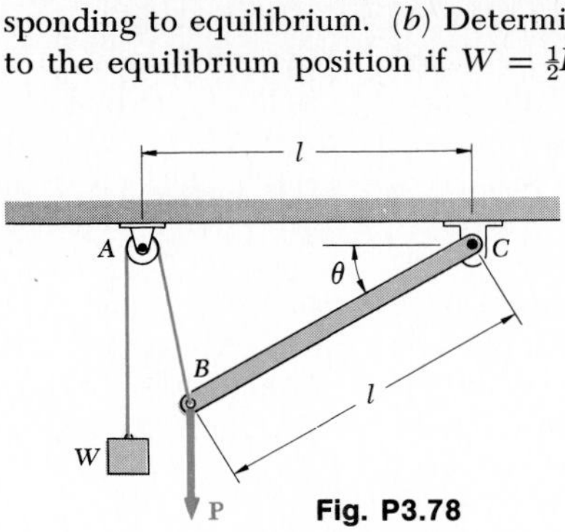

Fig. P3.78

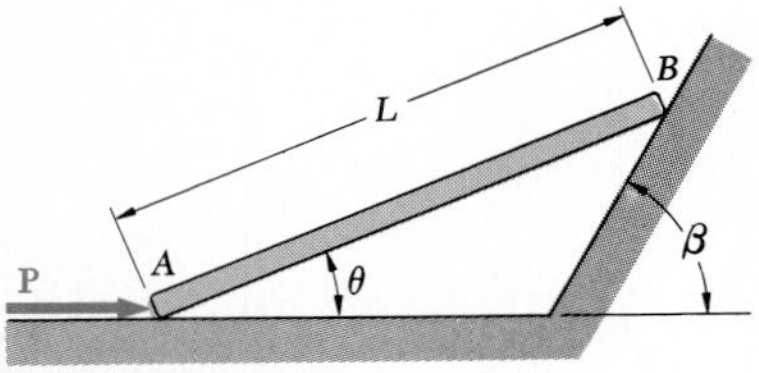

Fig. P3.79

***3.79** A uniform, slender rod of length L and weight W is held in the position shown by the horizontal force **P**. Neglecting the effect of friction at A and B, determine the angle θ corresponding to equilibrium (*a*) in terms of P, W, L, and β, (*b*) if $P = 10$ lb, $W = 20$ lb, $L = 30$ in., and $\beta = 60°$.

***3.80** In the problems listed below, the rigid bodies considered were completely constrained, and the reactions were statically determinate. For each of these rigid bodies it is possible to create an improper set of constraints by changing either a dimension of the body or the direction of a reaction. In each problem determine the value of α or of a which results in improper constraints. (*a*) Prob. 3.44, (*b*) Prob. 3.51, (*c*) Prob. 3.60, (*d*) Prob. 3.62*d*, (*e*) Prob. 3.73.

3.81 The bracket ABC is supported in eight different ways as shown. All connections consist of frictionless pins, rollers, and short links. In each case, determine whether (*a*) the bracket is completely, partially, or improperly constrained, (*b*) the reactions are statically determinate or indeterminate, (*c*) the equilibrium of the bracket is maintained in the position shown. Also, wherever possible, compute the reactions, assuming that the magnitude of the force **P** is 400 N.

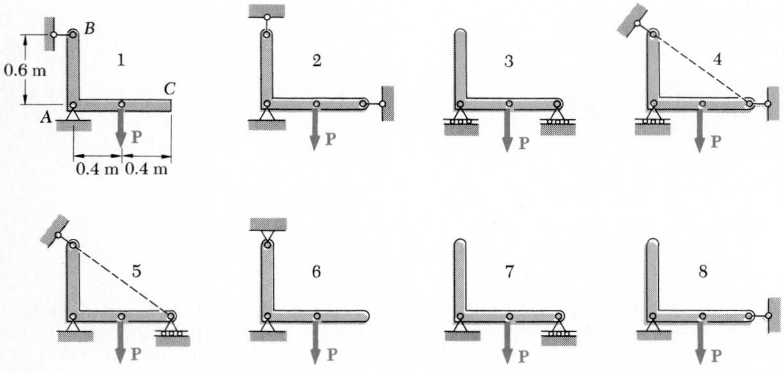

Fig. P3.81

3.82 Nine identical rectangular plates, 2 by 3 ft, weighing 100 lb each are held in a vertical plane as shown. All connections consist of frictionless pins, rollers, or short links. For each case, answer the questions listed in Prob. 3.81, and, wherever possible, compute the reactions.

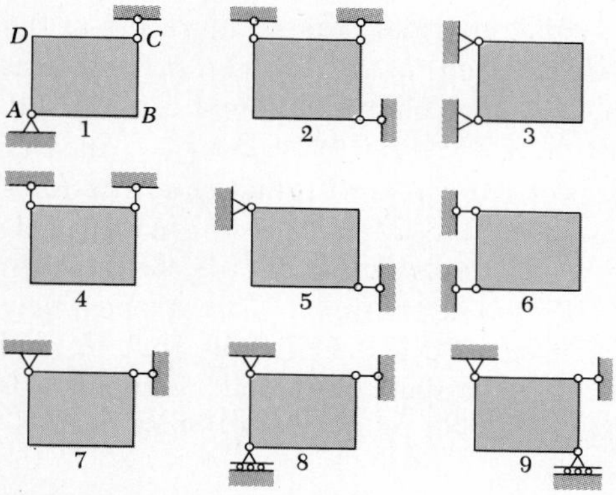

Fig. P3.82

3.16. Equilibrium of a Two-Force Body.

A particular case of equilibrium which is of considerable interest is that of a rigid body subjected to two forces. Such a body is commonly called a *two-force body*. We shall show that, *if a two-force body is in equilibrium, the two forces must have the same magnitude, same line of action, and opposite sense.*

Consider a corner plate subjected to two forces $\mathbf{F}_1$ and $\mathbf{F}_2$ acting at A and B, respectively (Fig. 3.23*a*). If the plate is to be in equilibrium, the sum of the moments of $\mathbf{F}_1$ and $\mathbf{F}_2$ about any axis must be zero. First, we sum moments about A: Since the moment of $\mathbf{F}_1$ is obviously zero, the moment of $\mathbf{F}_2$ must also be zero and the line of action of $\mathbf{F}_2$ must pass through A (Fig. 3.23*b*). Summing moments about B, we prove similarly that the line of action of $\mathbf{F}_1$ must pass through B (Fig. 3.23*c*). Both forces have the same line of action (line AB). From the equation $\Sigma F_x = 0$ or $\Sigma F_y = 0$, it is seen that they must have also the same magnitude but opposite sense.

If several forces act at two points A and B, the forces acting at A may be replaced by their resultant $\mathbf{F}_1$ and those acting at

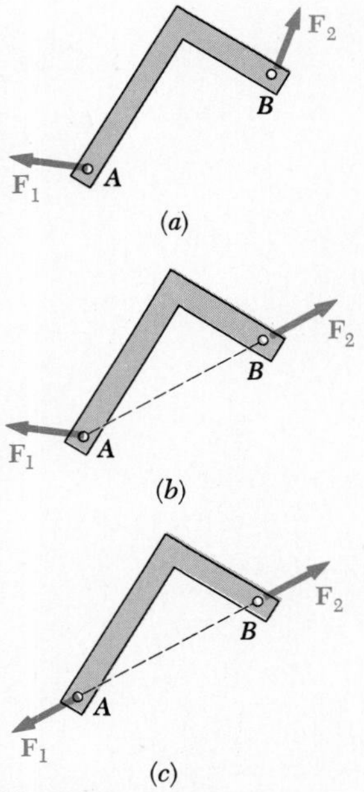

Fig. 3.23

B by their resultant $\mathbf{F}_2$. Thus a two-force body may be more generally defined as *a rigid body subjected to forces acting at only two points*. The resultants $\mathbf{F}_1$ and $\mathbf{F}_2$ then must have the same line of action, same magnitude, and opposite sense.

Although problems dealing with the equilibrium of two-force bodies may be solved by the general methods studied in the preceding sections, it is sometimes desirable to make use of the property we have just established to simplify certain problems so that simple trigonometric or geometric relations can be used.

3.17. Equilibrium of a Three-Force Body. Another case of equilibrium that is of great interest is that of a *three-force body*, i.e., a rigid body subjected to three forces or, more generally, *a rigid body subjected to forces acting at only three points*. Consider a rigid body subjected to a system of forces which may be reduced to three forces $\mathbf{F}_1$, $\mathbf{F}_2$, and $\mathbf{F}_3$ acting at A, B, and C, respectively (Fig. 3.24*a*). We shall show that, if the body is in equilibrium, *the lines of action of the three forces must be either concurrent or parallel.*

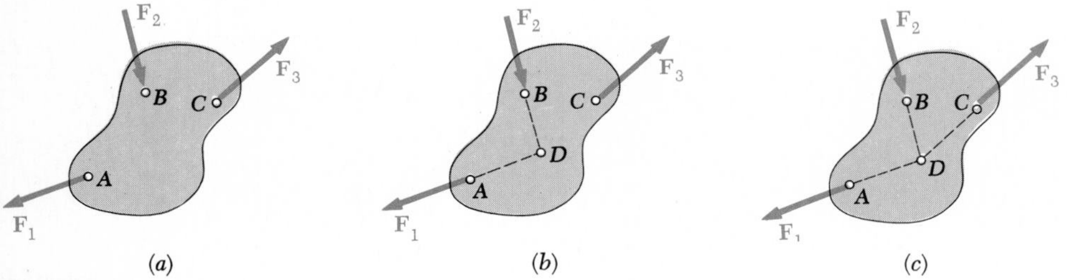

Fig. 3.24

Since the rigid body is in equilibrium, the sum of the moments of $\mathbf{F}_1$, $\mathbf{F}_2$, and $\mathbf{F}_3$ about any axis must be zero. Assuming that the lines of action of $\mathbf{F}_1$ and $\mathbf{F}_2$ intersect, and denoting their point of intersection by D, we sum moments about D (Fig. 3.24*b*); since the moments of $\mathbf{F}_1$ and $\mathbf{F}_2$ about D are zero, the moment of $\mathbf{F}_3$ about D must also be zero and the line of action of $\mathbf{F}_3$ must pass through D (Fig. 3.24*c*). The three lines of action are concurrent. The only exception occurs when none of the lines intersect; the lines of action are then parallel.

Although problems concerning three-force bodies may be solved by the general methods of Secs. 3.12 to 3.15, the property just established may be used to solve them either graphically or from simple trigonometric or geometric relations.

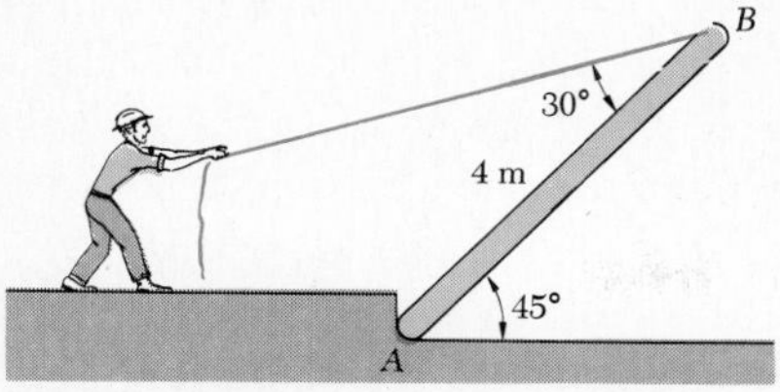

SAMPLE PROBLEM 3.13

A man raises a 10-kg joist, of length 4 m, by pulling on a rope. Find the tension T in the rope and the reaction at A.

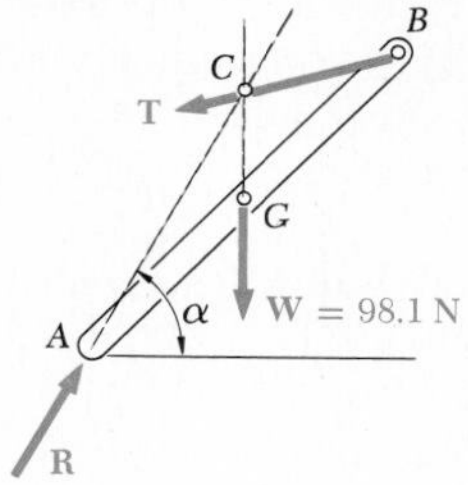

Solution. The joist is a three-force body since it is acted upon by three forces: its weight **W**, the force **T** exerted by the rope, and the reaction **R** of the ground at A. We note that

$$W = mg = (10\text{ kg})(9.81\text{ m/s}^2) = 98.1\text{ N}$$

Three-Force Body. Since the joist is a three-force body, the forces acting on it must be concurrent. The reaction **R**, therefore, will pass through the point of intersection C of the lines of action of the weight **W** and of the tension force **T**. This fact will be used to determine the angle α that **R** forms with the horizontal.

Drawing the vertical BF through B and the horizontal CD through C, we note that

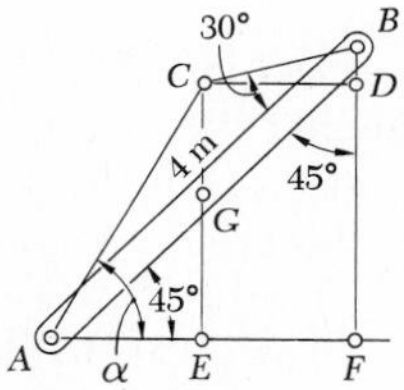

$$AF = BF = (AB)\cos 45^\circ = (4\text{ m})\cos 45^\circ = 2.83\text{ m}$$
$$CD = EF = AE = \tfrac{1}{2}(AF) = 1.415\text{ m}$$
$$BD = (CD)\cot(45^\circ + 30^\circ) = (1.415\text{ m})\tan 15^\circ = 0.379\text{ m}$$
$$CE = DF = BF - BD = 2.83\text{ m} - 0.379\text{ m} = 2.45\text{ m}$$

We write

$$\tan\alpha = \frac{CE}{AE} = \frac{2.45\text{ m}}{1.415\text{ m}} = 1.732$$

$\alpha = 60.0^\circ$ ◄

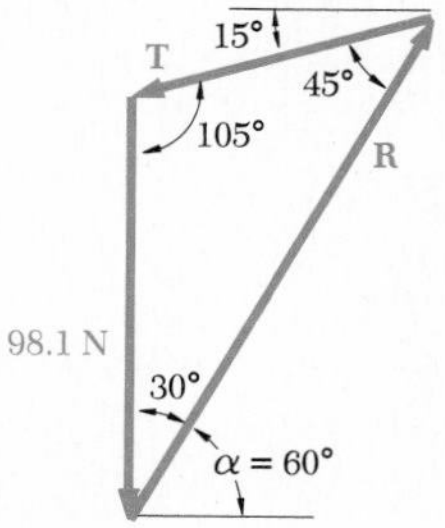

We now know the direction of all the forces acting on the joist.

Force Triangle. A force triangle is drawn as shown, and its interior angles are computed from the known directions of the forces. Using the law of sines, we write

$$\frac{T}{\sin 30^\circ} = \frac{R}{\sin 105^\circ} = \frac{98.1\text{ N}}{\sin 45^\circ}$$

$T = 69.4\text{ N}$ ◄

$\mathbf{R} = 134.0\text{ N} \angle 60.0^\circ$ ◄

PROBLEMS

3.83 Using the method of Sec. 3.17, solve Prob. 3.49.

3.84 Using the method of Sec. 3.17, solve Prob. 3.44.

3.85 Using the method of Sec. 3.17, solve Prob. 3.50.

3.86 In Prob. 3.57, determine (*a*) the value of α for which the reaction at B is zero, (*b*) the corresponding reactions at A and C.

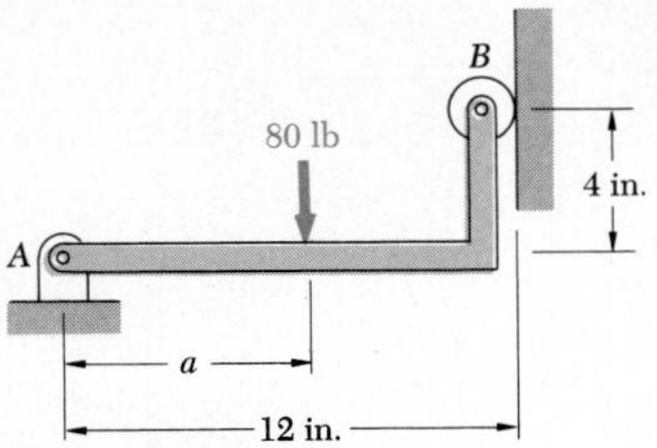

Fig. P3.87 and P3.88

3.87 Determine the reactions at A and B when $a = 7.5$ in.

3.88 Determine the value of a for which the magnitude of the reaction **B** is equal to 200 lb.

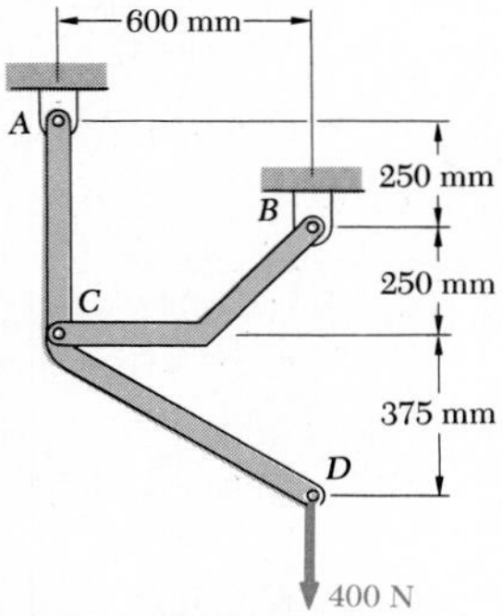

Fig. P3.89 and P3.90

3.89 Determine the reactions at A and B.

3.90 Solve Prob. 3.89, assuming that the 400-N force applied at D is directed horizontally to the right.

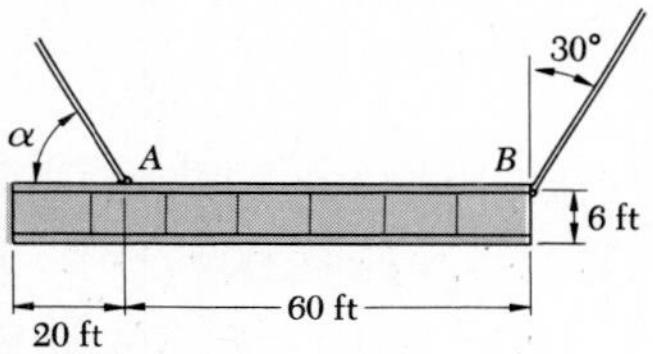

Fig. P3.91

3.91 A uniform plate girder weighs 6000 lb. It is held by two crane cables as shown; the cable attached at B forms an angle of 30° with the vertical. If the girder is to be held in a horizontal position, determine the direction of the cable attached at A and the tension in each cable.

3.92 A slender rod BC of length L and weight W is held by a cable AB and by a pin at C which may slide in a vertical slot. Knowing that $\theta = 30°$, determine (*a*) the value of β for which the rod is in equilibrium, (*b*) the corresponding tension in the cable.

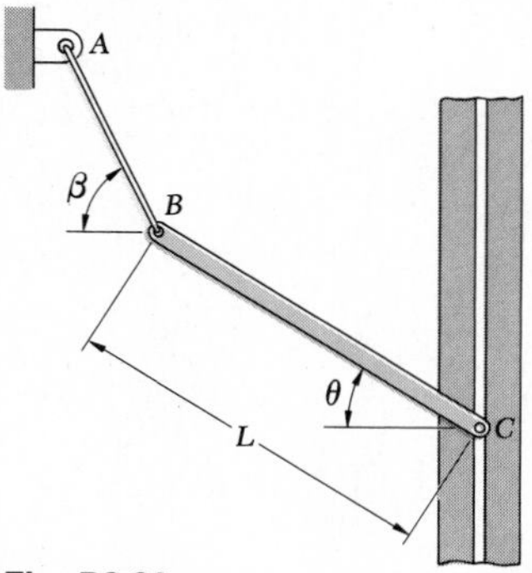

Fig. P3.92

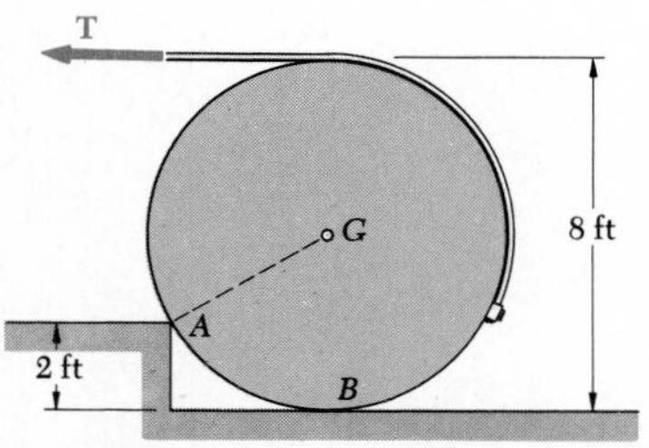

Fig. P3.93

3.93 A 500-lb cylindrical tank, 8 ft in diameter, is to be raised over a 2-ft obstruction. A cable is wrapped around the tank and pulled horizontally as shown. Knowing that the corner of the obstruction at A is rough, find the required tension in the cable and the reaction at A.

3.94 A 2-m rod, of uniform cross section, is held in equilibrium as shown, with one end against a frictionless, vertical wall and the other end attached to a cord. Determine the length of the cord.

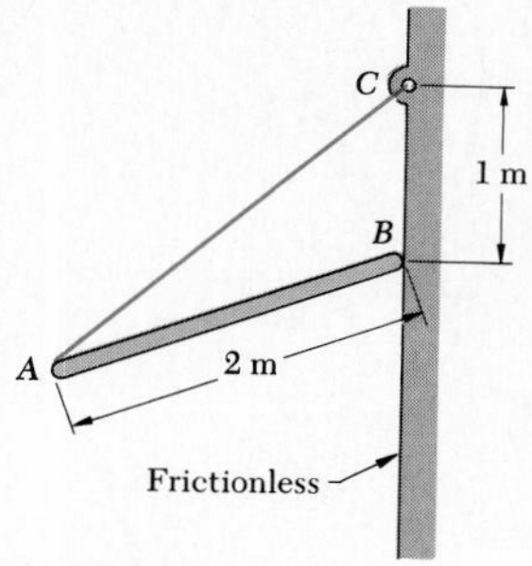

Fig. P3.94

3.95 The slender rod AB of length L and weight W is attached to a collar at A and rests on a small wheel at C. Neglecting the effect of friction and the weight of the collar, determine the angle θ corresponding to equilibrium.

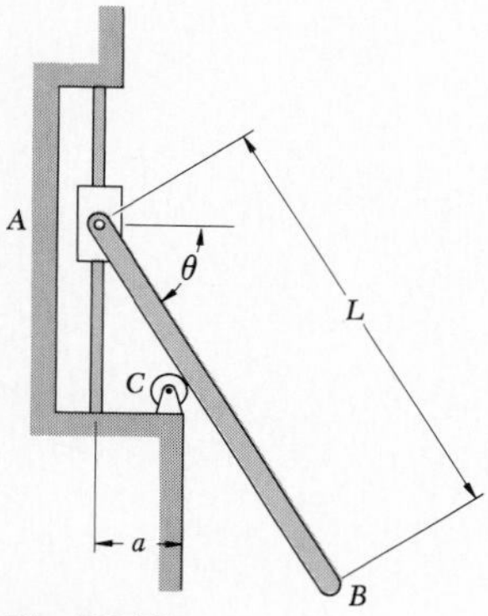

Fig. P3.95

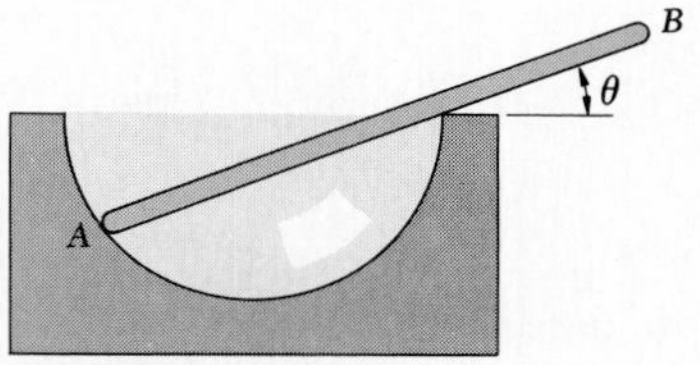

Fig. P3.96

***3.96** A uniform rod AB of length $3R$ rests inside a hemispherical bowl of radius R as shown. Neglecting friction, determine the angle θ corresponding to equilibrium.

***3.97** A slender rod of length $2r$ and weight W is attached to a collar at B and rests on a circular cylinder of radius r. Knowing that the collar may slide freely along a vertical guide and neglecting friction, determine the value of θ corresponding to equilibrium.

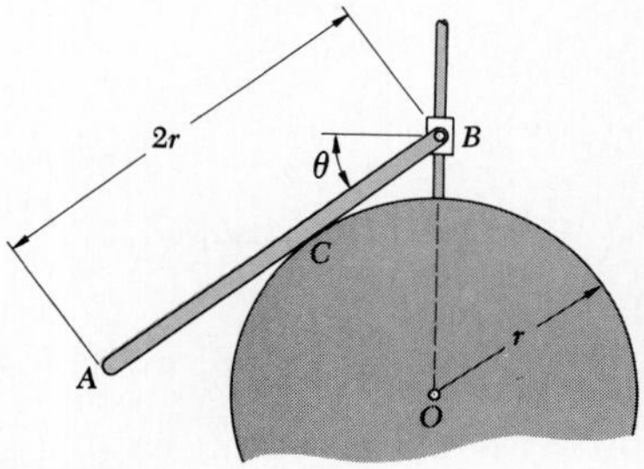

Fig. P3.97

REVIEW PROBLEMS

3.98 In order to move a 173-lb crate, two men push on it while two other men pull on it by means of ropes. The force exerted by man A is 150 lb, and that exerted by man B is 50 lb; both forces are horizontal. Man C pulls with a force equal to 80 lb and man D with a force equal to 120 lb. Both cables form an angle of 30° with the vertical. Determine the resultant of all forces acting on the crate.

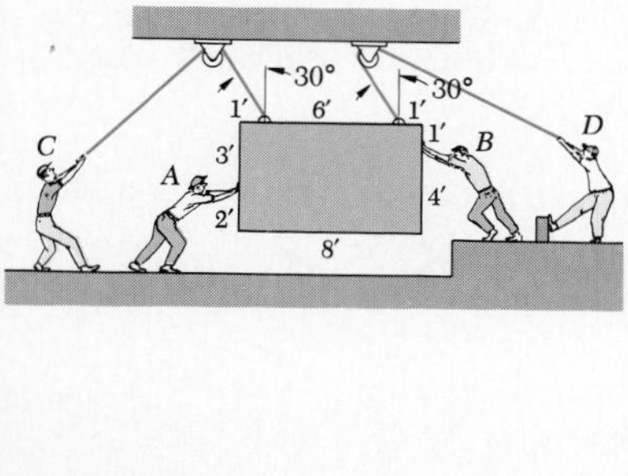

Fig. P3.98

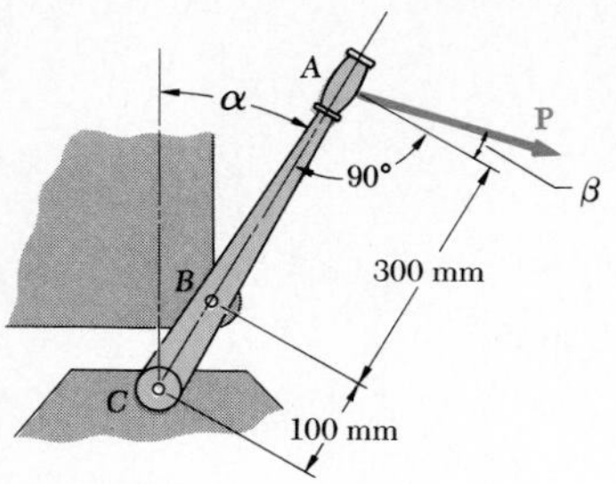

Fig. P3.99

3.99 Replace the force **P** by an equivalent system formed by two parallel forces at *B* and *C*. Show (*a*) that these forces are parallel to force **P**, (*b*) that the magnitude of these forces is independent of both α and β.

3.100 A light rod, supported by rollers at *B*, *C*, and *D*, is subjected to a 200-lb force applied at *A*. If $\beta = 0$, determine (*a*) the reactions at *B*, *C*, and *D*, (*b*) the rollers which may safely be removed for this loading.

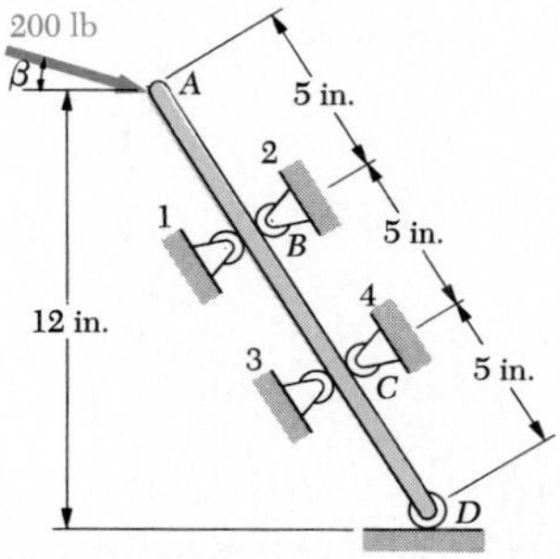

Fig. P3.100

3.101 Solve Prob. 3.100 when the 200-lb force is directed vertically downward, i.e., when $\beta = 90°$.

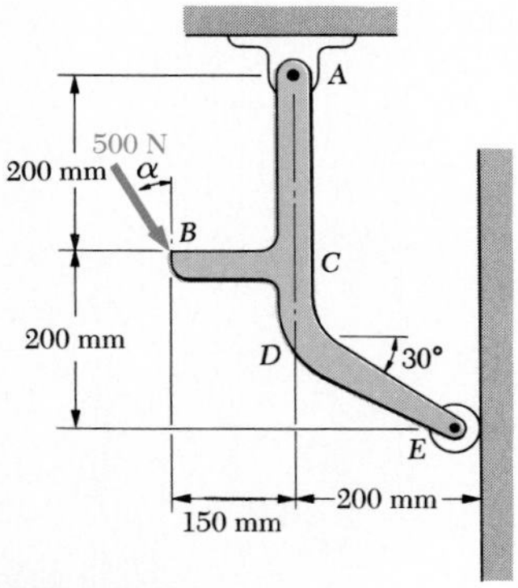

Fig. P3.102

3.102 (*a*) Determine the reactions at *A* and *E*, if $\alpha = 0$. (*b*) Determine the values of α for which the magnitude of the reaction at *E* is 300 N.

3.103 Find the resultant of the system shown and the point of intersection of its line of action with (*a*) line *AC*, (*b*) line *CD*.

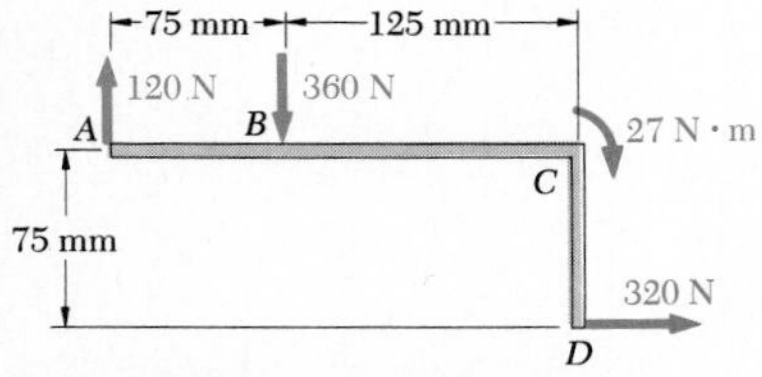

Fig. P3.103, P3.104, and P3.105

3.104 The 27-N · m couple applied at *C* is removed and replaced by a couple of unknown moment *M*. Determine the range of values of *M* for which the line of action of the resultant of the system will intersect the bracket.

3.105 Determine the magnitude and sense of the single vertical force **P** which must be applied at *D* if the line of action of the resultant of the entire system is to pass through point *B*.

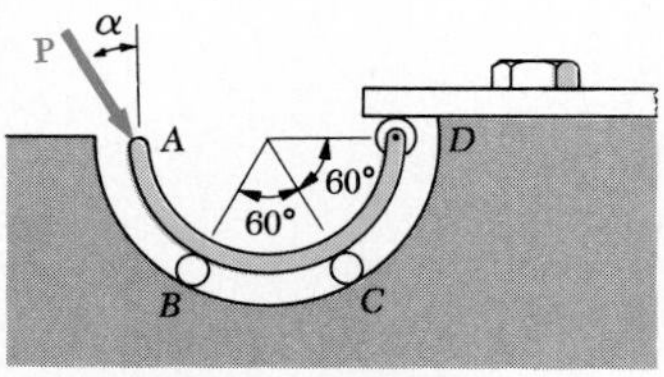

Fig. P3.106 and P3.107

3.106 Determine the range of values of α for which the semicircular rod can be maintained in equilibrium by the small wheel at *D* and the rollers at *B* and *C*.

3.107 Determine the reactions at *B*, *C*, and *D* (*a*) if $\alpha = 0$, (*b*) if $\alpha = 30°$.

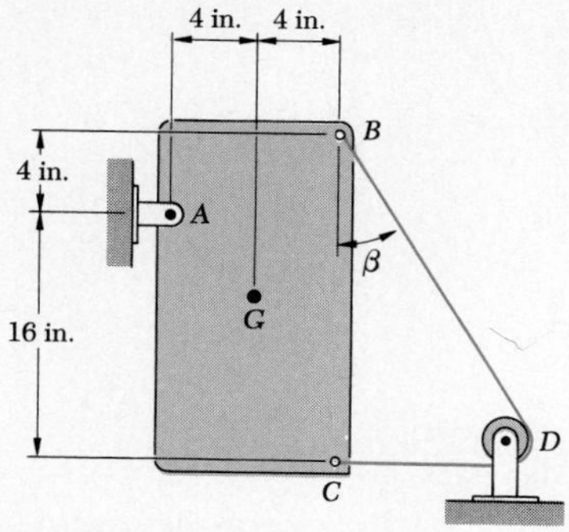

Fig. P3.108 and P3.109

3.108 The plate shown weighs 80 lb and is held by a pin at *A* and by a wire *BDC* which passes over a frictionless pulley at *D*. Determine the components of the reaction at *A* and the tension in the wire when $\beta = 30°$.

3.109 The plate shown weighs 80 lb and is held by a pin at *A* and by a wire *BDC* which passes over a frictionless pulley at *D*. Determine (*a*) the value of β for which the tension in the cable is maximum, (*b*) the corresponding components of the reaction at *A* and the tension in the cable.

Chapter
4

Statics of Rigid Bodies in Three Dimensions

EQUIVALENT SYSTEMS OF FORCES

4.1. Moment of a Force about an Axis. When the moment of a force about an axis was defined in Chap. 3, all the forces considered were contained in the plane of the figure and their moments were computed about axes perpendicular to that plane. In space, however, the forces generally are not contained in a plane perpendicular to the axis about which their moment is to be determined. The definition of the moment of a force given in Sec. 3.4 needs therefore to be extended.

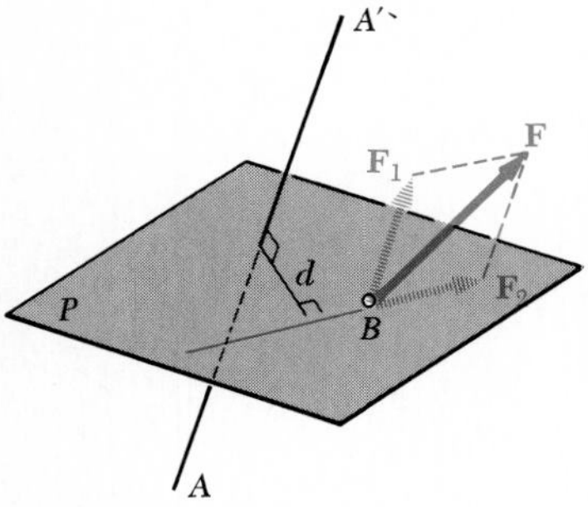

Fig. 4.1

Consider a force **F** applied at B and an axis AA' (Fig. 4.1). We resolve **F** into two perpendicular components $\mathbf{F}_1$ and $\mathbf{F}_2$; the component $\mathbf{F}_1$ is chosen in a direction parallel to AA', and the component $\mathbf{F}_2$ lies in a plane P perpendicular to AA', and passing through B. (We say that $\mathbf{F}_2$ is the *projection* of **F** on P.) If **F** acted on a rigid body, only its component $\mathbf{F}_2$ would tend to make the body rotate about the axis AA'. Therefore, *the moment of* **F** *about the axis* AA' *is defined as the moment of* $\mathbf{F}_2$ *about that axis*. It is equal to the product F_2d, where d is the perpendicular distance from the axis AA' to the line of action of $\mathbf{F}_2$. The sense of the moment (clockwise or counterclockwise) is determined from the point of view of an observer located at A' and looking toward A. The moment of **F** about the axis AA' (directed from A to A') shown in Fig. 4.1 is thus counterclockwise.

It should be noted that the moment of a force **F** about an axis AA' is zero if the line of action of **F** either intersects the axis or is parallel to it. In the first case, the line of action of the projection $\mathbf{F}_2$ of **F** also intersects the axis (Fig. 4.2), and, in the second case, the magnitude of $\mathbf{F}_2$ is zero (Fig. 4.3). In both cases the moment of $\mathbf{F}_2$, and thus the moment of **F**, is zero.

Consider now a force **F** acting on the corner *B* of a rectangular box of sides *a*, *b*, and *c* (Fig. 4.4). The force may impart to the box a motion of translation in the *x*, *y*, or *z* directions, or a motion of rotation about the *x*, *y*, or *z* axes, or a combination of all these motions. The ability of the force to produce a translation is measured by its rectangular components F_x, F_y, F_z; its ability to produce a rotation about any of the coordinate axes is measured by its moment about that axis. The moments of **F** about the *x*, *y*, and *z* axes are denoted, respectively, by M_x, M_y, and M_z.

According to the definition given above for the moment of a force, the moment M_x of **F** about the *x* axis will be obtained by projecting **F** on the *yz* plane and computing the moment of its projection about the *x* axis. This is done in Fig. 4.5*a*, which shows the *yz* plane as an observer located along the positive half of the *x* axis and looking toward *O* would see it. It will

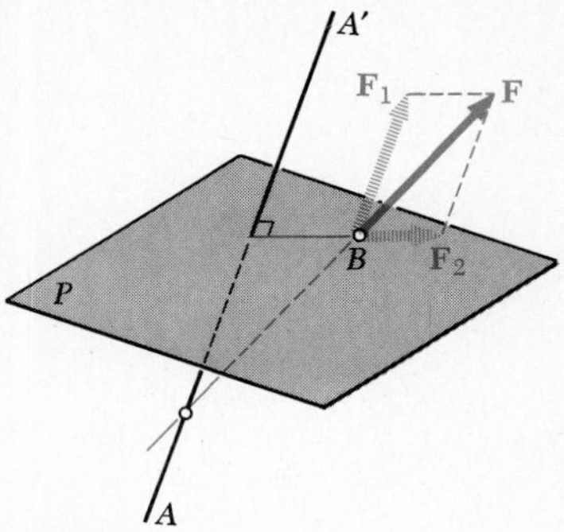

Fig. 4.2

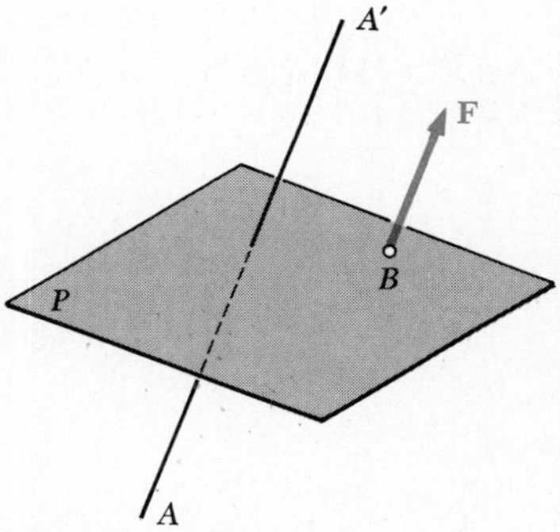

Fig. 4.3

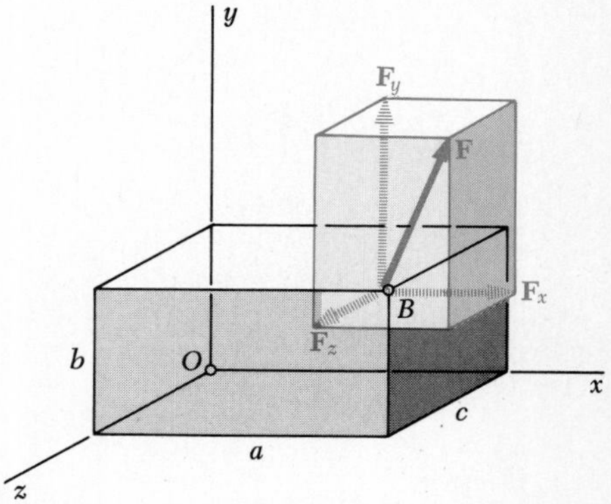

Fig. 4.4

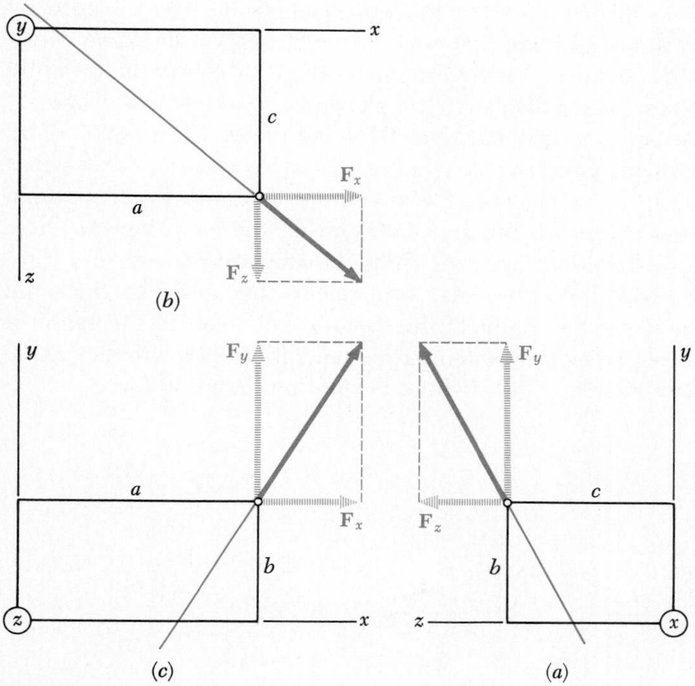

Fig. 4.5

generally be found convenient to carry out the actual computation of M_x by using Varignon's theorem and computing the moments of the component forces $\mathbf{F}_y$ and $\mathbf{F}_z$ about O. Figure 4.5*b* shows the projection of **F** on the *zx* plane, as seen by an observer located along the positive half of the *y* axis and looking toward *O*. The moment of this projection about the *y* axis is the moment M_y of **F** about that axis. Figure 4.5*c* shows the projection of **F** on the *xy* plane, as seen by an observer located along the positive half of the *z* axis and looking toward *O*. The moment of this projection about the *z* axis is the moment M_z of **F** about that axis. In the case considered here, the moments M_x, M_y, and M_z are, respectively, clockwise, clockwise, and counterclockwise. Recalling the sign convention adopted in Sec. 3.4, we may also state that the moments M_x, M_y, and M_z are, respectively, negative, negative, and positive.

4.2 Couples in Space. A *couple* was defined in Sec. 3.6 as the combination of two forces **F** and **F′** having the same magnitude *F*, parallel lines of action, and opposite sense. It was shown in Sec. 3.7 that the action a couple exerts on a rigid body depends only upon the moment *M* of the couple, which is equal to the product *Fd* of the magnitude *F* of the forces and the distance *d* between their lines of action. In the case of a couple in space, however, *the plane in which the couple acts* should be specified (Fig. 4.6*a*). Since the couple may be replaced by any other couple of the same moment acting in the same plane, it will be found convenient to represent a couple in space by a *vector* perpendicular to the plane of the couple and of magnitude equal to the moment *M* of the couple (Fig. 4.6*b*). This vector, denoted by **M**, is called a *couple vector*, and its sense is such that a man located at the tip of **M** will see the couple acting counterclockwise. The sense of the couple vector may also be found by applying the *right-hand rule:* Close your right hand, and hold it so that your fingers are curled in the sense defined by the couple; your thumb will indicate the sense of the couple vector. Note that the symbol ↺ is added to the couple vector to distinguish it from vectors representing forces.

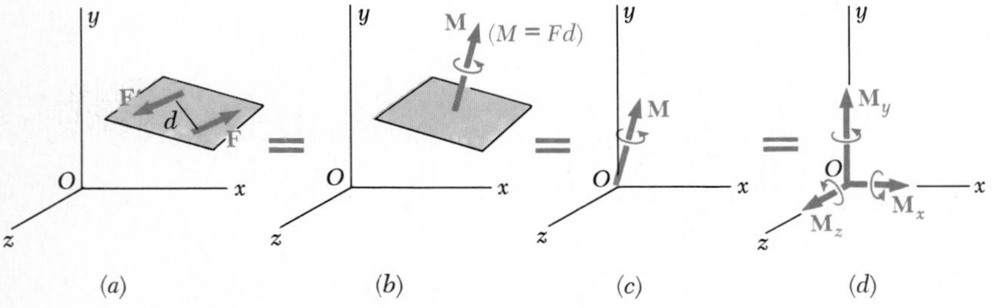

Fig. 4.6

The point of application of the couple vector may be chosen anywhere in the plane of the couple. In fact, it may be chosen anywhere in space since, as will be seen below, only the *orientation* of the plane of the couple matters. (In other words, two couples of the same moment and acting in parallel planes are equivalent.) Because of this fact, couple vectors are *free vectors,* as opposed to bound vectors, which have well-defined points of application, and sliding vectors, which have well-defined lines of action. The point of application of the couple vector, therefore, may be chosen at the origin of the system of coordinates (Fig. 4.6*c*). Furthermore, the couple vector **M** may be resolved into component vectors $\mathbf{M}_x$, $\mathbf{M}_y$, $\mathbf{M}_z$, directed along the coordinate axes (Fig. 4.6*d*). The vector $\mathbf{M}_x$, of magnitude M_x, directed along the x axis, represents a couple of moment M_x acting in the yz plane; similarly, the other two vectors represent couples of moment M_y and M_z acting, respectively, in the zx and xy planes.

4.3. Couples Are Vectors. The representation of a couple by means of a vector is convenient. This representation, however, should be *justified,* i.e., we should show that couples possess all the characteristics of vectors (see Sec. 2.2). We already know that couples have magnitude and direction; we should also show that couples may be added by adding the corresponding couple vectors according to the parallelogram law.

Consider two couples of moment M_1 and M_2 acting, respectively, in planes P_1 and P_2 (Fig. 4.7*a*). Each of these couples may be represented by two forces **F** and **F′** of arbitrary magnitude F. The distance d_1 between the forces forming the first couple and the distance d_2 between the forces forming the second couple should satisfy the equations

$$M_1 = Fd_1 \qquad M_2 = Fd_2 \tag{4.1}$$

One force in each couple may be chosen along the line of intersection AA' of the two planes, one in the sense AA', the other in the opposite sense. The forces along AA' cancel each other,

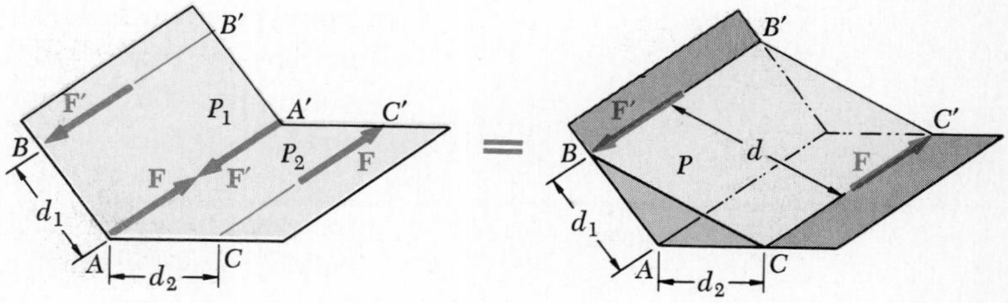

Fig. 4.7 (*a*) (*b*)

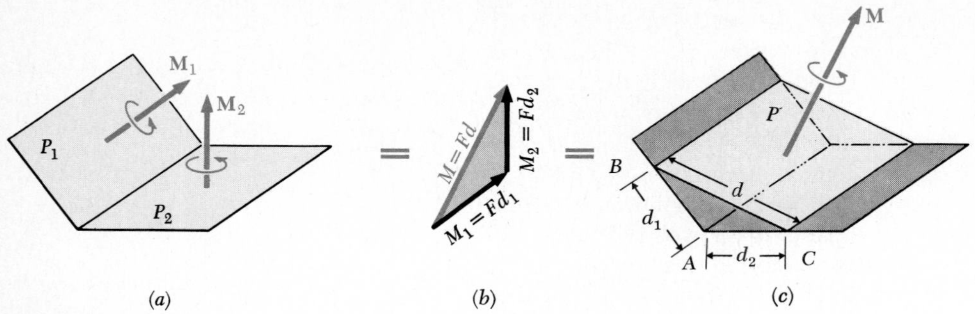

Fig. 4.8

and the given couples thus reduce to the forces along BB' and CC'. These two forces form a couple of moment $M = Fd$ acting in the plane P defined by BB' and CC' (Fig. 4.7*b*).

Let us now represent the two given couples of Fig. 4.7*a* by arrows $\mathbf{M}_1$ and $\mathbf{M}_2$, of magnitude M_1 and M_2, respectively perpendicular to the planes P_1 and P_2 (Fig. 4.8*a*). Drawing $\mathbf{M}_1$ and $\mathbf{M}_2$ in tip-to-tail fashion, we define a plane parallel to plane ABC of Fig. 4.8*c*. Joining the tail of $\mathbf{M}_1$ to the tip of $\mathbf{M}_2$ (Fig. 4.8*b*), we obtain a triangle similar to triangle ABC since the sides $\mathbf{M}_1$ and $\mathbf{M}_2$ are, respectively, perpendicular to the sides AB and AC and since, according to (4.1), they are proportional to d_1 and d_2. The third side of the triangle of Fig. 4.8*b* must therefore be perpendicular to BC and be proportional to d; its magnitude is therefore $Fd = M$. Thus, the third side of the triangle constructed on $\mathbf{M}_1$ and $\mathbf{M}_2$ is equal to the arrow $\mathbf{M}$ which represents the resultant couple (Fig. 4.8*c*). We conclude that $\mathbf{M}$ may be obtained by adding $\mathbf{M}_1$ and $\mathbf{M}_2$ vectorially and, thus, that the arrows $\mathbf{M}_1$, $\mathbf{M}_2$, and $\mathbf{M}$ representing the couples *are truly vectors*.

We note that, if forces $\mathbf{Q}$ and $\mathbf{Q}'$, of magnitude Q different from F, are used to represent the couples in Fig. 4.7*a*, new distances d'_1 and d'_2 are obtained and the resultant couple, which has the same moment M, is found to act in a plane P' parallel to the plane P of Fig. 4.7*b*. Clearly, this new couple is equivalent to the couple of Fig. 4.7*b*. Since, by an appropriate choice of the forces $\mathbf{Q}$ and $\mathbf{Q}'$, the plane P' may be chosen at any distance from P, we conclude that *couples of the same moment and acting in parallel planes are equivalent*. Couples acting in parallel planes may thus be moved freely from one plane to another and added as if they were coplanar.

Summarizing the results obtained, we conclude that *couples in space may be truly represented by vectors*. These vectors, called couple vectors, may be added or resolved according to the parallelogram law; they are *free vectors* and may be applied at any point.

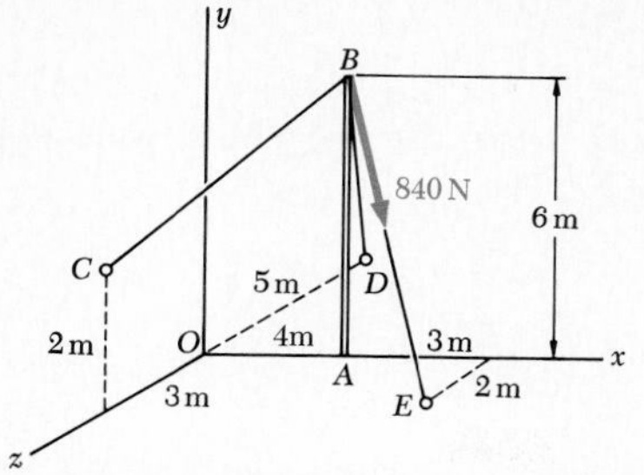

SAMPLE PROBLEM 4.1

A pole AB, 6 m long, is held by three guy wires as shown. Determine the moment about each of the coordinate axes of the force exerted by wire BE on point B. The tension T in wire BE is known to be 840 N.

Solution. The force $\mathbf{T}$ exerted by BE is first resolved into components. The components and magnitude of the vector $\overrightarrow{BE}$, which joins points B and E, are

$$d_x = +3\text{ m} \qquad d_y = -6\text{ m} \qquad d_z = +2\text{ m} \qquad d = 7\text{ m}$$

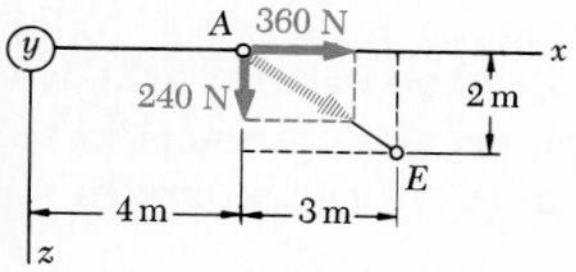

Therefore,

$$\frac{T_x}{+3\text{ m}} = \frac{T_y}{-6\text{ m}} = \frac{T_z}{+2\text{ m}} = \frac{840\text{ N}}{7\text{ m}}$$

$$T_x = +360\text{ N} \qquad T_y = -720\text{ N} \qquad T_z = +240\text{ N}$$

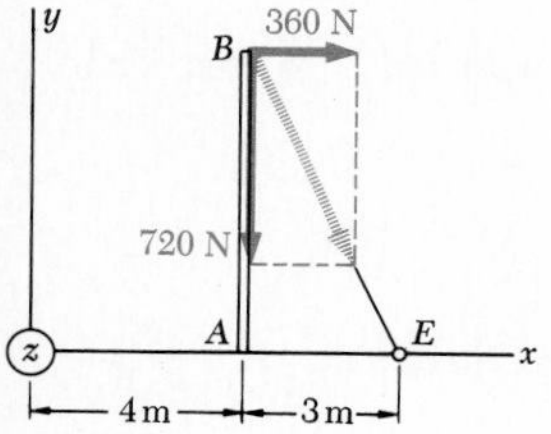

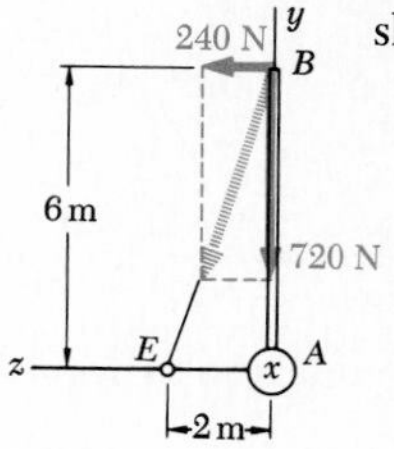

The projections of $\mathbf{T}$ on the three coordinate planes are now sketched. Using Varignon's theorem, we write

$$+\circlearrowleft M_x = +(240\text{ N})(6\text{ m}) = +1440\text{ N}\cdot\text{m}$$
$$+\circlearrowleft M_y = -(240\text{ N})(4\text{ m}) = -960\text{ N}\cdot\text{m}$$
$$+\circlearrowleft M_z = -(720\text{ N})(4\text{ m}) - (360\text{ N})(6\text{ m}) = -5040\text{ N}\cdot\text{m}$$

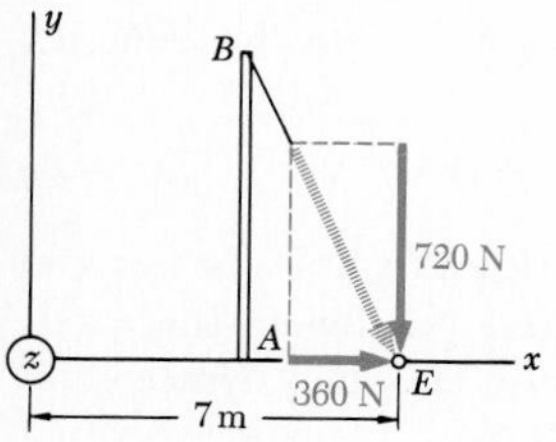

Alternate Solution. The computation of M_z may be simplified by moving the projection of $\mathbf{T}$ along its line of action to the point where it intersects the x axis; the moment of $\mathbf{T}_x$ about the z axis is now zero. We obtain

$$+\circlearrowleft M_z = -(720\text{ N})(7\text{ m}) = -5040\text{ N}\cdot\text{m}$$

SAMPLE PROBLEM 4.2

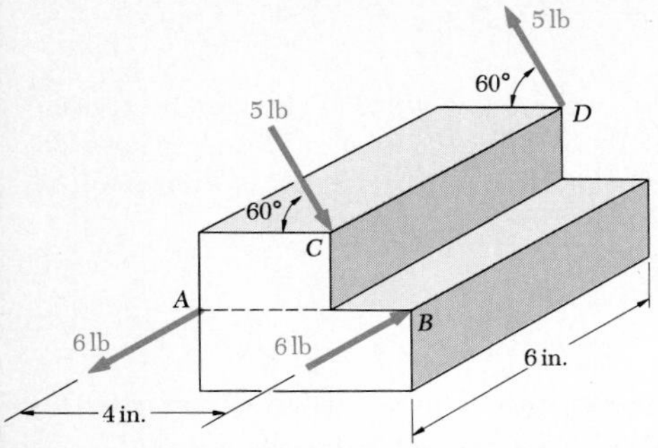

Two couples act on a block as shown. Replace these two couples with a single equivalent couple.

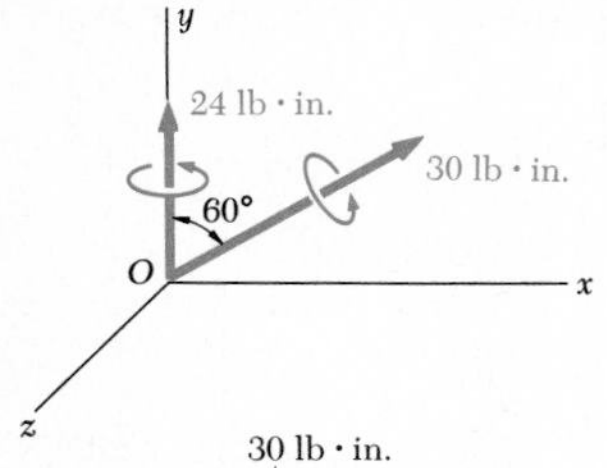

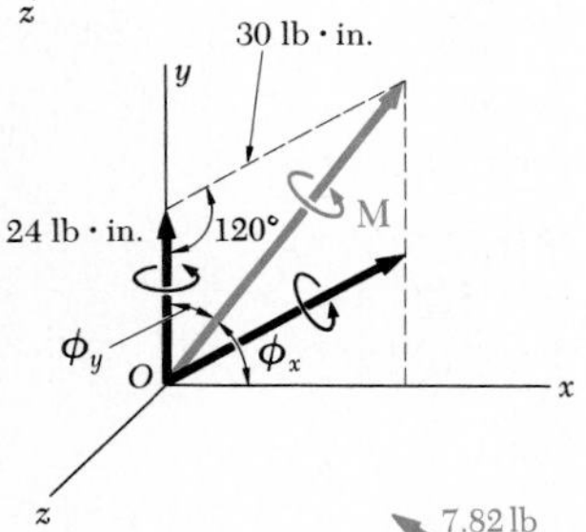

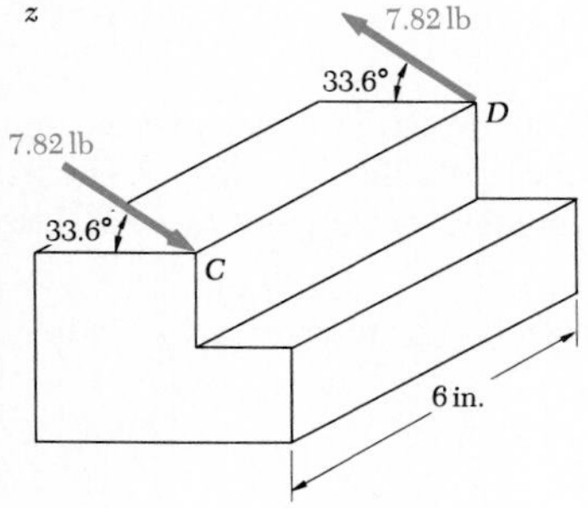

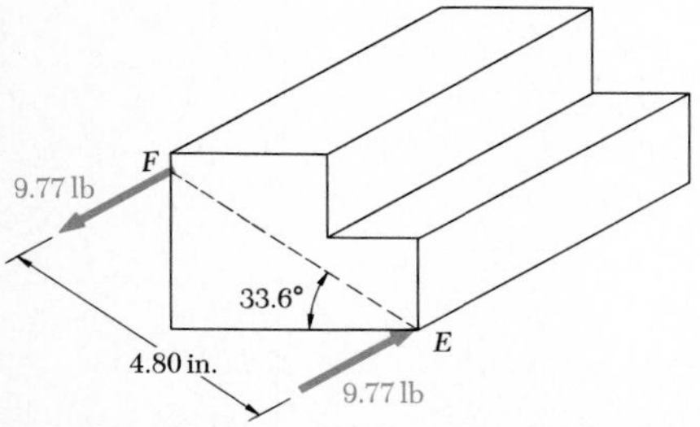

Solution. Each of the given couples is represented by a couple vector which is perpendicular to the plane of the couple and of a magnitude equal to the moment of the couple. The sense of each vector is obtained by applying the right-hand rule, and for convenience both couple vectors are attached at the origin.

The single couple equivalent to the two given couples will be represented by the resultant of the two couple vectors. The moment of the resultant couple vector **M** is obtained from the law of cosines.

$$M^2 = (24)^2 + (30)^2 - (2)(24)(30)\cos 120°$$
$$M = 46.9 \text{ lb} \cdot \text{in.}$$

The angle ϕ_y that the resultant couple vector forms with the vertical is obtained from the law of sines.

$$\frac{\sin \phi_y}{30 \text{ lb} \cdot \text{in.}} = \frac{\sin 120°}{46.9 \text{ lb} \cdot \text{in.}} \qquad \phi_y = 33.6°$$

The angle that the couple vector forms with the x axis is

$$\phi_x = 90° - 33.6° = 56.4°$$

and the angle it forms with the z axis is $\phi_z = 90°$. Thus the resultant couple vector **M** is defined by

$$M = 46.9 \text{ lb} \cdot \text{in.} \qquad \phi_x = 56.4° \qquad \phi_y = 33.6° \qquad \phi_z = 90° \quad ◄$$

The single couple equivalent to the two original couples is a couple of moment $M = 46.9$ lb · in., acting in a plane parallel to the z axis and forming an angle of 33.6° with the horizontal plane. This couple may be formed in many ways, for example, by two 7.82-lb forces acting at corners C and D in the directions shown, or by two 9.77-lb forces parallel to the z axis and applied at E and F as shown.

PROBLEMS

4.1 Knowing that the tension in cable AB is 1800 lb, determine the moment about each of the coordinate axes of the force exerted on the plate at A.

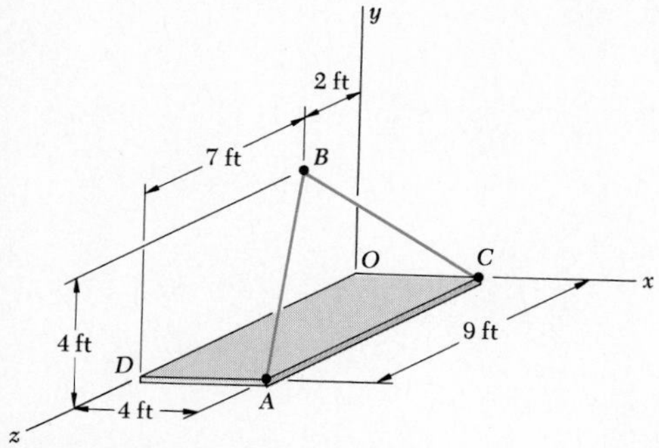

Fig. P4.1 and P4.2

4.2 Knowing that the tension in cable BC is 900 lb, determine the moment about each of the coordinate axes of the force exerted on the plate at C.

4.3 A force **P** of magnitude 360 N is applied at point B as shown. Determine the moment of **P** about each of the coordinate axes.

4.4 A force **Q** of magnitude 450 N is applied at point C as shown. Determine the moment of **Q** about each of the coordinate axes.

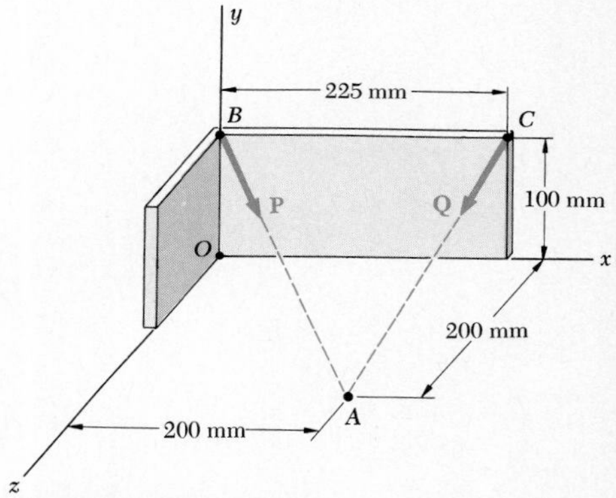

Fig. P4.3 and P4.4

4.5 A vertical force **P** of magnitude 60 lb is applied to the crank at A. Knowing that $\theta = 75°$, determine the moment of **P** about each of the coordinate axes.

4.6 Assuming that the force **P** keeps acting vertically downward as the crank turns, determine the moment of **P** about each of the coordinate axes (*a*) in terms of θ, (*b*) when $\theta = 120°$, (*c*) when $\theta = 240°$.

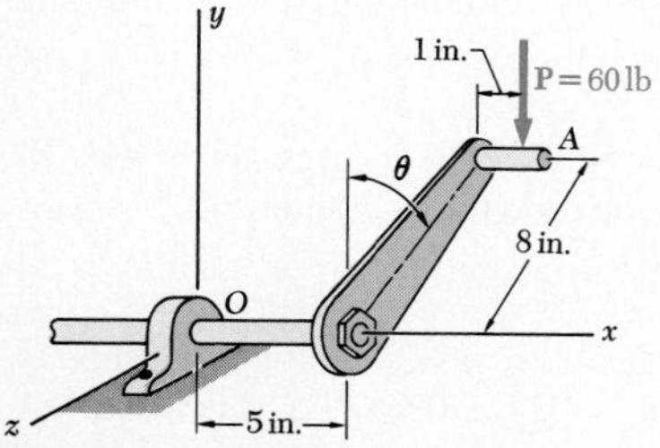

Fig. P4.5 and P4.6

4.7 The jib crane is oriented so that the boom DA is parallel to the x axis. At the instant shown the tension in cable AB is 13 kN. Determine the moment about each of the coordinate axes of the force exerted on A by the cable AB.

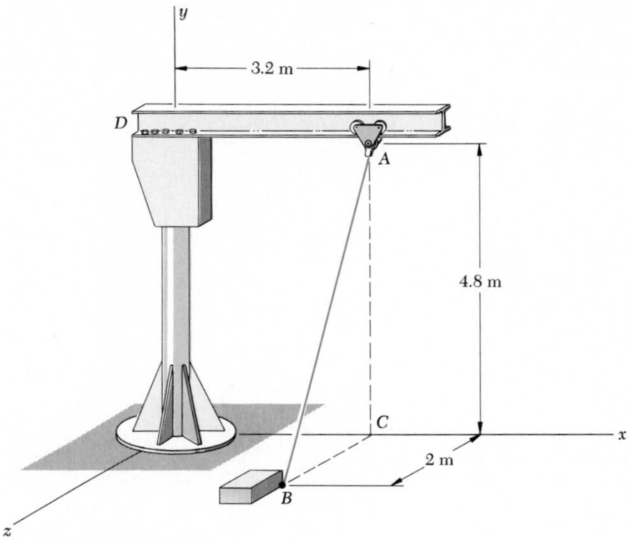

Fig. P4.7 and P4.8

4.8 The jib crane is oriented so that the boom DA is parallel to the x axis. Determine the maximum permissible tension in the cable AB if the absolute value of the moments about the coordinate axes of the force exerted on A must be as follows: $M_x \leq 10\ \text{kN} \cdot \text{m}$, $M_y \leq 6\ \text{kN} \cdot \text{m}$, $M_z \leq 16\ \text{kN} \cdot \text{m}$.

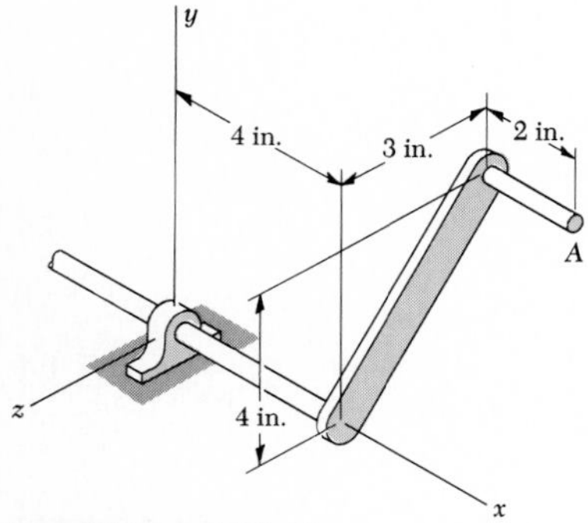

Fig. P4.9 and P4.10

4.9 The primary purpose of the crank shown is, of course, to produce a moment about the x axis. Show that a single force acting at A and having a moment M_x different from zero about the x axis must also have a moment different from zero about at least one of the other coordinate axes.

4.10 A single force $\mathbf{F}$ of unknown magnitude and direction acts at point A of the crank shown. Determine the moment M_x of $\mathbf{F}$ about the x axis, knowing that $M_y = +180\ \text{lb} \cdot \text{in.}$ and $M_z = -320\ \text{lb} \cdot \text{in.}$

4.11 A force $\mathbf{F}$ of components F_x, F_y, and F_z acts at a point of coordinates x, y, and z. Find expressions for the moment of $\mathbf{F}$ about each of the coordinate axes.

4.12 A single force $\mathbf{F}$ acts at a point A of coordinates $x = y = z = a$. Show that $M_x + M_y + M_z = 0$; that is, show that the *algebraic* sum of the moments of $\mathbf{F}$ about the coordinate axes is zero.

4.13 The gearbox shown is acted upon by three couples. Replace the three couples by a single equivalent couple.

4.14 Determine the components of a single couple equivalent to the two couples shown. Check the result obtained by adding the moments of the individual forces about the coordinate axes.

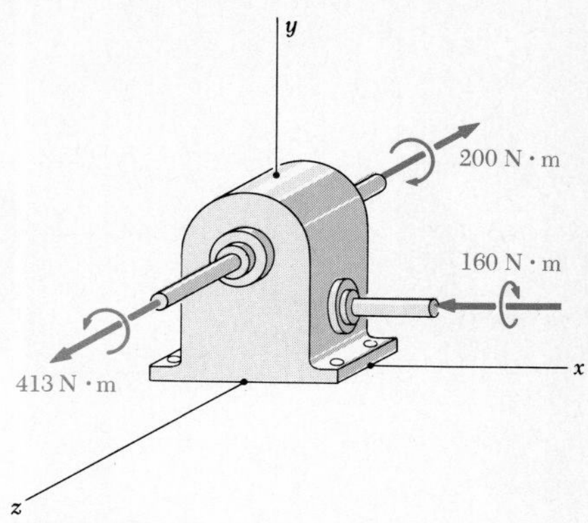

Fig. P4.13

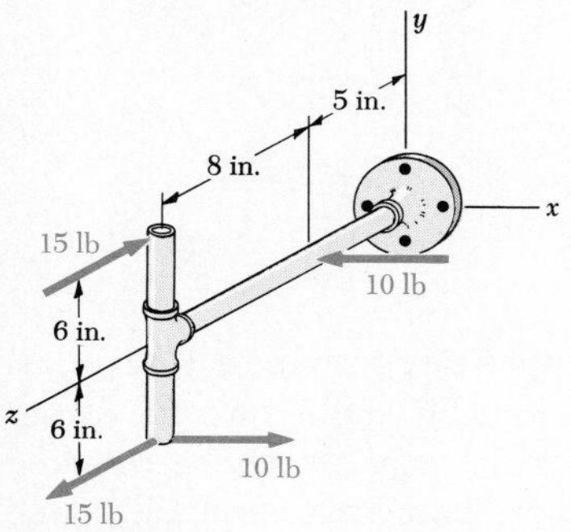

Fig. P4.14

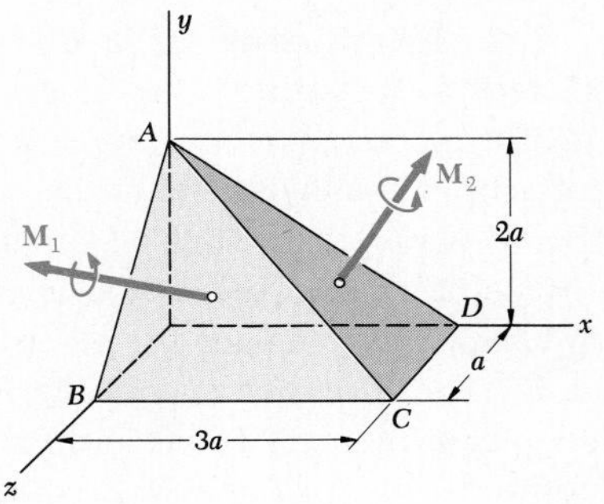

Fig. P4.15

4.15 The couple vectors $\mathbf{M}_1$ and $\mathbf{M}_2$ represent couples which are contained in the planes *ABC* and *ACD*, respectively. Assuming that $M_1 = M_2 = M$, determine a single couple equivalent to the two given couples.

4.16 Three shafts are connected to a gearbox as shown. Shaft *A* is horizontal and shafts *B* and *C* lie in the vertical *yz* plane. Determine the components of the resultant couple exerted on the gearbox.

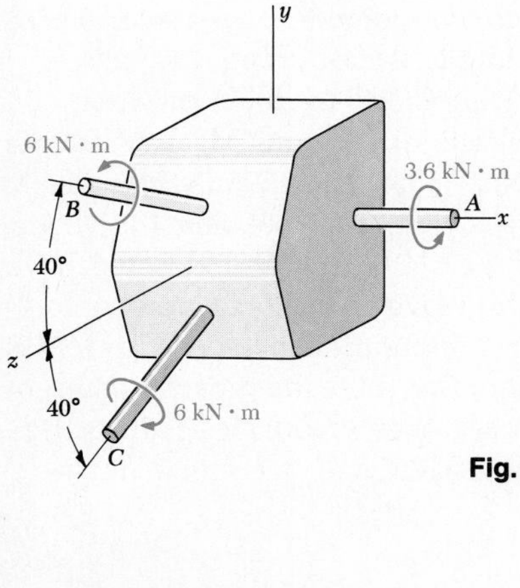

Fig. P4.16

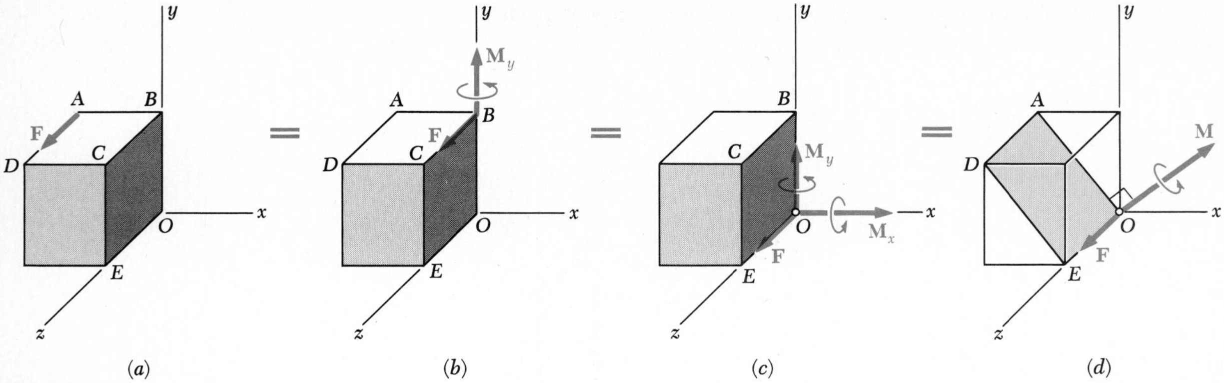

Fig. 4.9

4.4. Resolution of a Given Force into a Force at *O* and a Couple. Consider a force **F** applied at the corner A of a rectangular box. Since a force may always be resolved into components parallel to the coordinate axes, we shall assume that **F** is parallel to one of the axes, say, the z axis (Fig. 4.9*a*).

Operating in plane $ABCD$ according to the rules established in Chap. 3, we replace the given force by a force-couple system at B. The force, of magnitude F, is parallel to the z axis; the couple, of moment equal to the moment M_y of **F** (in its original position) about the y axis, is contained in plane $ABCD$ and may therefore be represented by a vector $\mathbf{M}_y$ directed along the y axis (Fig. 4.9*b*). Operating now in plane $BCEO$, we move the couple vector to O and replace the force **F** acting at B by a force-couple system at O; the new couple, of moment equal to the moment M_x of **F** (acting at B or at A) about the x axis, may be represented by a vector $\mathbf{M}_x$ directed along the x axis (Fig. 4.9*c*). We conclude therefore that *the force* **F** *acting at A may be moved to O, provided that couple vectors are added along the coordinate axes; the magnitude of each couple vector must equal the moment of* **F** *(in its original position) about the corresponding axis.* Since, in the case considered, **F** has zero moment about the z axis, there is no couple vector along that axis.

The couple vectors may be replaced by their resultant. In the case considered, the resultant of $\mathbf{M}_x$ and $\mathbf{M}_y$ is a couple vector **M** lying in the xy plane (Fig. 4.9*d*). Thus, *the couple vector* **M** *is perpendicular to the force* **F**. Note that the vector **M** represents a couple acting in plane $ADEO$ and could have been obtained more directly by moving the force from A to O within that plane. While the representation given in Fig. 4.9*d* is more compact than that given in Fig. 4.9*c*, the representation of Fig. 4.9*c* is usually preferred since it shows more clearly the action of the original force on the rigid body.

4.5. Reduction of a System of Forces to One Force and One Couple. Consider a system of forces $\mathbf{F}_1$, $\mathbf{F}_2$, $\mathbf{F}_3$, etc., acting on a rigid body (Fig. 4.10*a*). We first resolve each force into rectangular components; according to Sec. 4.4, these components may be replaced by equivalent forces and couples at O. Adding all x, y, and z components, we obtain, respectively, the forces $\mathbf{R}_x$, $\mathbf{R}_y$, and $\mathbf{R}_z$, directed along the coordinate axes; similarly, adding all moments about the x, y, and z axes, we obtain, respectively, the couple vectors $\mathbf{M}_x$, $\mathbf{M}_y$, and $\mathbf{M}_z$ (Fig. 4.10*b*). It is advisable to arrange the computations in tabular form as shown in Sample Probs. 4.3 and 4.4. The forces and couple vectors obtained are together equivalent to the given system; they give a clear representation of the action exerted by that system on the rigid body. Note that different couple vectors would be obtained if a different origin O were chosen.

We shall observe that, to be equivalent, two given systems of forces should reduce to the same forces and couple vectors at O. The sums of their x, y, and z components and the sums of their moments about the x, y, and z axes should therefore be, respectively, equal.

If desired, the three forces $\mathbf{R}_x$, $\mathbf{R}_y$, $\mathbf{R}_z$ of Fig. 4.10*b* may be replaced by their resultant $\mathbf{R}$; similarly, the couple vectors $\mathbf{M}_x$, $\mathbf{M}_y$, $\mathbf{M}_z$ may be replaced by a single couple vector $\mathbf{M}$ (Fig. 4.10*c*). The magnitude and direction of the force $\mathbf{R}$ and of the couple vector $\mathbf{M}$ may be obtained by the method of Sec. 2.11. Denoting by θ_x, θ_y, θ_z and ϕ_x, ϕ_y, ϕ_z the angles formed, respectively, by $\mathbf{R}$ and $\mathbf{M}$ with the coordinate axes, we write

$$R = \sqrt{R_x^2 + R_y^2 + R_z^2}$$
$$\cos\theta_x = \frac{R_x}{R} \qquad \cos\theta_y = \frac{R_y}{R} \qquad \cos\theta_z = \frac{R_z}{R} \tag{4.2}$$

$$M = \sqrt{M_x^2 + M_y^2 + M_z^2}$$
$$\cos\phi_x = \frac{M_x}{M} \qquad \cos\phi_y = \frac{M_y}{M} \qquad \cos\phi_z = \frac{M_z}{M} \tag{4.3}$$

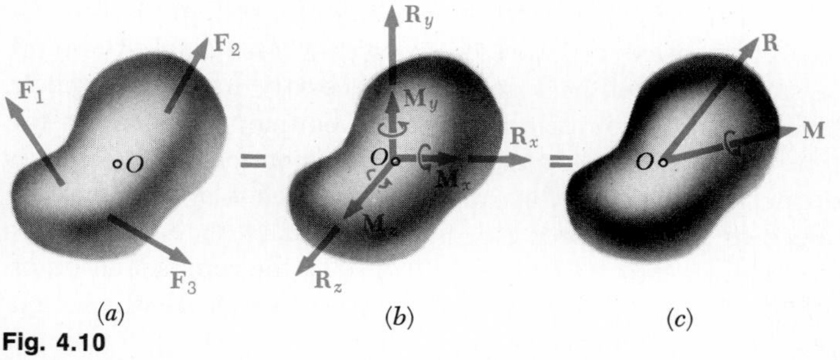

Fig. 4.10

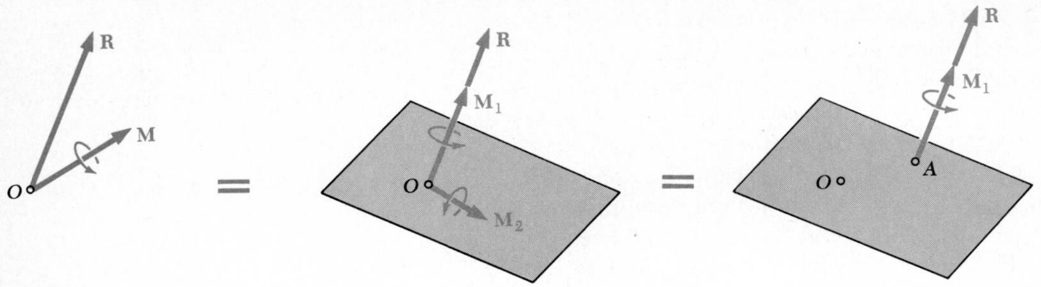

(a) (b) (c)

Fig. 4.11

An important difference between the reduction of a system of coplanar forces and the reduction of a system of space forces should be noted. It was shown in Sec. 3.10 that the force-couple system obtained by reducing a system of coplanar forces may be further reduced to a single force. In the case of a system of space forces, this will generally not be possible. A force-couple system may be reduced to a single force only if the force and the couple act in the same plane, i.e., if the force and the couple vector are mutually perpendicular (see Sec. 4.4).

If, as is generally the case, the force and couple vector are not at a right angle (Fig. 4.11*a*), the couple vector may be replaced by two other couple vectors obtained by resolving $\mathbf{M}$ into a component $\mathbf{M}_1$ along $\mathbf{R}$ and a component $\mathbf{M}_2$ in a direction perpendicular to $\mathbf{R}$ (Fig. 4.11*b*). The couple vector $\mathbf{M}_2$ and the force $\mathbf{R}$ may then be replaced by a single force $\mathbf{R}$ acting along a new line of action; the original system of forces reduces to $\mathbf{R}$ and to a couple of moment M_1 acting in a plane perpendicular to $\mathbf{R}$ (Fig. 4.11*c*). This particular force-couple combination is called a *wrench*. The force $\mathbf{R}$ and the couple of moment M_1 tend to move the rigid body along the line of action of $\mathbf{R}$ and, at the same time, tend to rotate it about that line. The line of action of $\mathbf{R}$ is known as the *axis of the wrench* and the ratio M_1/R is called the *pitch of the wrench.*

Thus, it is seen that, while the action of a system of coplanar forces may always be represented by that of a single force (or of a single couple), it is generally necessary to use a wrench, i.e., the combination of a force and a couple, to represent the action of space forces. An important exception to this rule is the case of a system of parallel forces; such a system may be reduced to a single force (see Sample Prob. 4.4).

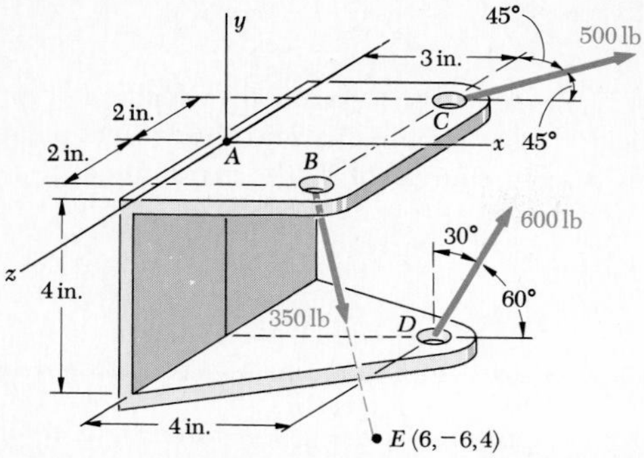

SAMPLE PROBLEM 4.3

Three cables are attached to a bracket as shown. Replace the forces exerted by the cables by three forces at A and three couples directed along the coordinate axes.

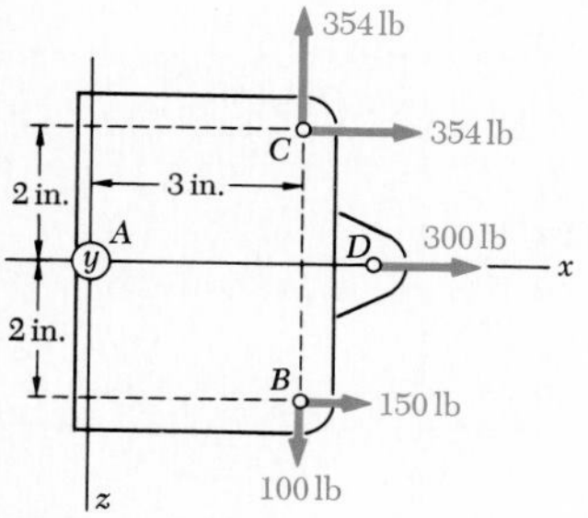

Solution. Because of the many computations involved, this problem is most efficiently solved in tabular fashion. First the components of the forces are determined and their values entered in the table. We note that the components of the force applied at B are computed from the following proportion:

$$\frac{B_x}{+3 \text{ in.}} = \frac{B_y}{-6 \text{ in.}} = \frac{B_z}{+2 \text{ in.}} = \frac{350 \text{ lb}}{7 \text{ in.}}$$

The components are also plotted on the three projection planes, and their moments are computed about each axis and entered in the table. The components and moments are then added.

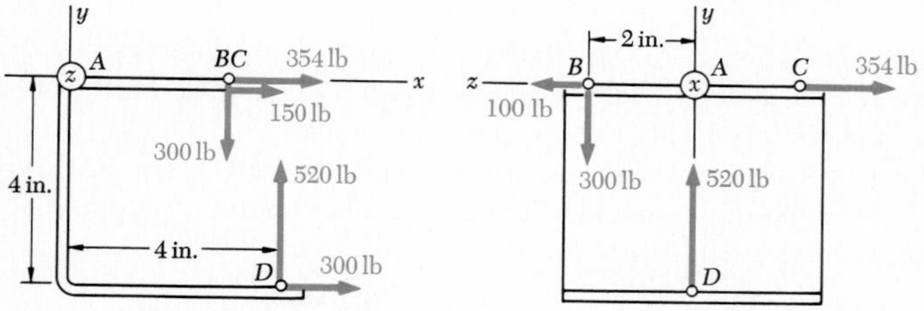

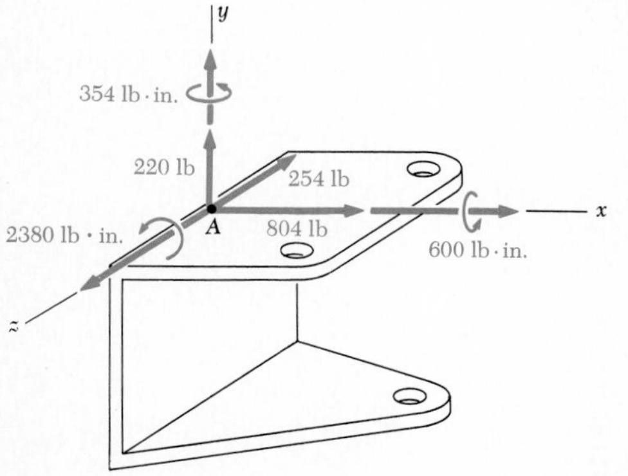

Force	F_x, lb	F_y, lb	F_z, lb	M_x, lb · in.	M_y, lb · in.	M_z, lb · in.
B	+150	−300	+100	+600	+300 −300	−900
C	+354	0	−354	0	+1062 −708	0
D	+300	+520	0	0	0	+2080 +1200
	$R_x =$ +804	$R_y =$ +220	$R_z =$ −254	$M_x =$ +600	$M_y =$ +354	$M_z =$ +2380

The given system may thus be replaced by three forces and three couple vectors, as shown. The three forces could be replaced by a single force **R** of components R_x, R_y, and R_z and the three couple vectors by a single couple vector **M** of components M_x, M_y, and M_z.

SAMPLE PROBLEM 4.4

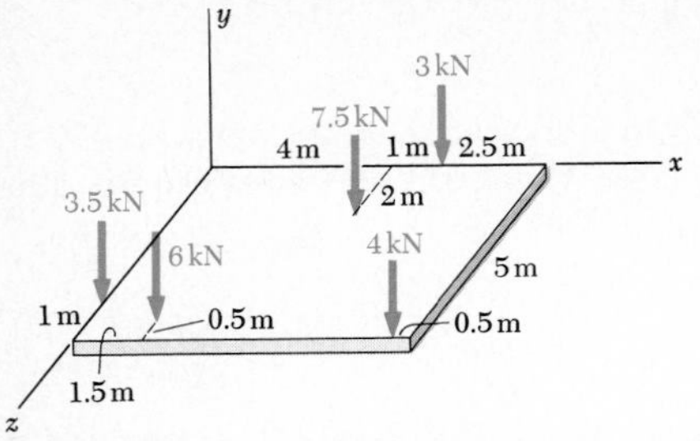

A rectangular slab, 5 by 7.5 m, supports five columns which exert on the slab the forces indicated. Determine the magnitude and point of application of the single force equivalent to the given forces.

Solution. The x and z components of the given forces and their moments about the y axis are obviously zero. Therefore only the y components and the moments M_x and M_z are entered in the table.

F_y, kN	M_x, kN · m	M_z, kN · m
−3	0	−15
−3.5	+14	0
−4	+20	−28
−6	+27	−9
−7.5	+15	−30
$R_y = -24$	$M_x = +76$	$M_z = -82$

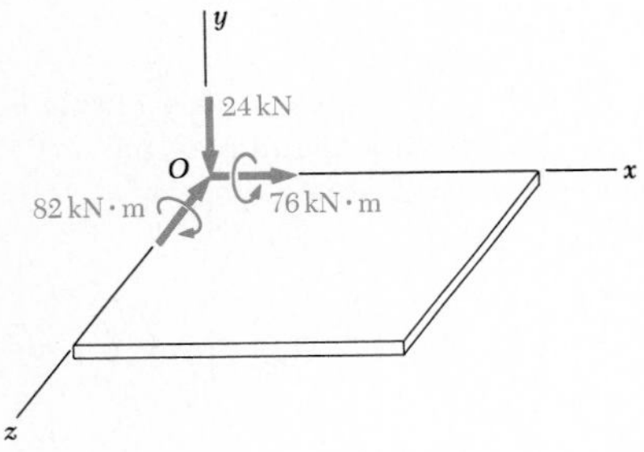

From the table, it is seen that the given system of forces may be replaced by a force of 24 kN acting at O and two couples represented by couple vectors located along the x and z axes.

The two couple vectors are perpendicular to the force and thus may be eliminated by moving the force to a new point of application. The coordinates of this point are

$$x = \frac{82\ \text{kN}\cdot\text{m}}{24\ \text{kN}} = 3.42\ \text{m} \qquad z = \frac{76\ \text{kN}\cdot\text{m}}{24\ \text{kN}} = 3.17\ \text{m}$$

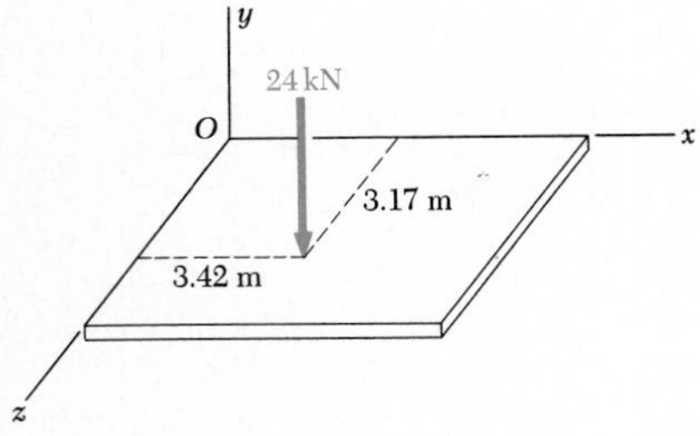

Thus, the single force equivalent to the given system is

$\mathbf{R} = 24\ \text{kN} \downarrow \quad$ at $x = 3.42$ m, $z = 3.17$ m ◄

PROBLEMS

4.17 A 4-kip force is applied on the outside face of the flange of a steel channel. Determine the components of the force and couple at G which are equivalent to the 4-kip load.

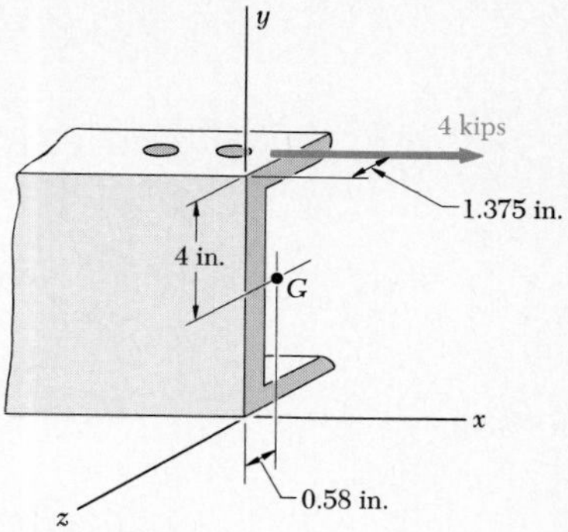

Fig. P4.17

4.18 A 100-kN load is applied eccentrically on a column. Determine the components of the force and couple at G which are equivalent to the 100-kN load.

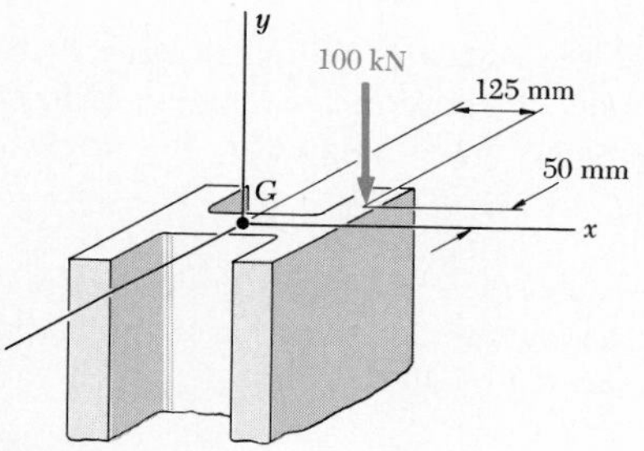

Fig. P4.18

4.19 The tension in cable AB is 1800 lb. Replace the force exerted on the plate at A by a force-couple system located (*a*) at the origin of coordinates O, (*b*) at point D.

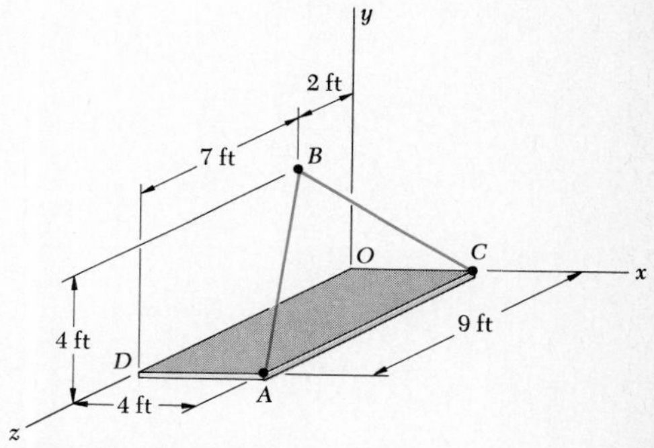

Fig. P4.19 and P4.20

4.20 The tension in cable BC is 900 lb. Replace the force exerted on the plate at C by a force-couple system located (*a*) at the origin of coordinates O, (*b*) at point D.

4.21 Two forces are applied to the vertical pole as shown. Determine the components of the force and couple at O equivalent to the two forces.

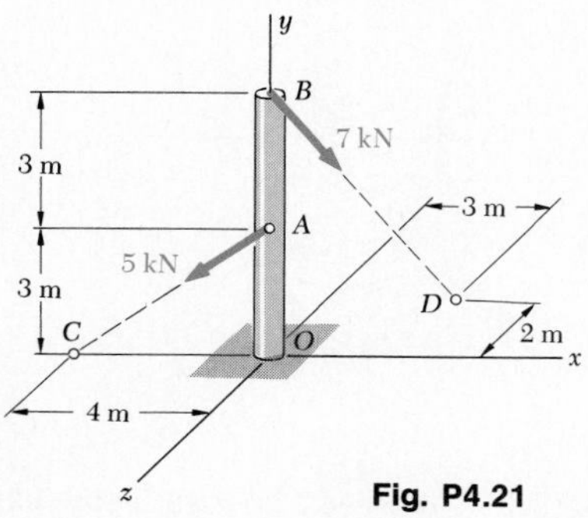

Fig. P4.21

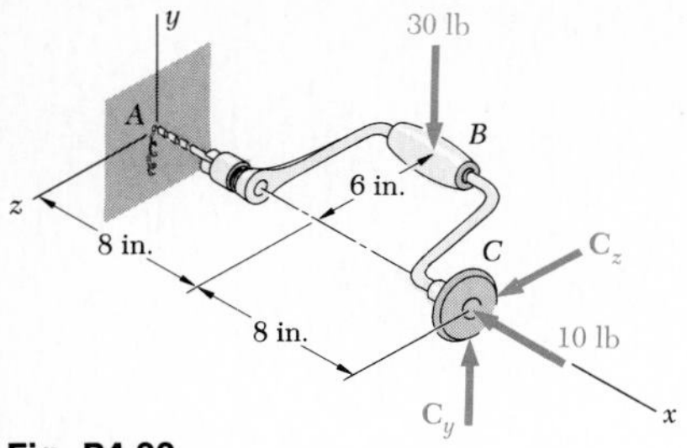

Fig. P4.22

4.22 In drilling a hole in a wall, a man applies a vertical 30-lb force at B on the brace and bit, while pushing at C with a 10-lb force. The brace lies in the horizontal xz plane. (*a*) Determine the other components of the total force which should be exerted at C if the bit is not to be bent about the y and z axes (i.e., if the system of forces applied on the brace is to have zero moment about both the y and z axes). (*b*) Reduce the 30-lb force and the total force at C to an equivalent force and couple at A.

4.23 Solve Prob. 4.21, assuming that the magnitude of the 5-kN force is increased to 10 kN.

4.24 In Prob. 4.21 what is the smallest additional force **P** that can be applied at point B in order to transform the system of forces into a system equivalent to a single force applied at the origin of coordinates?

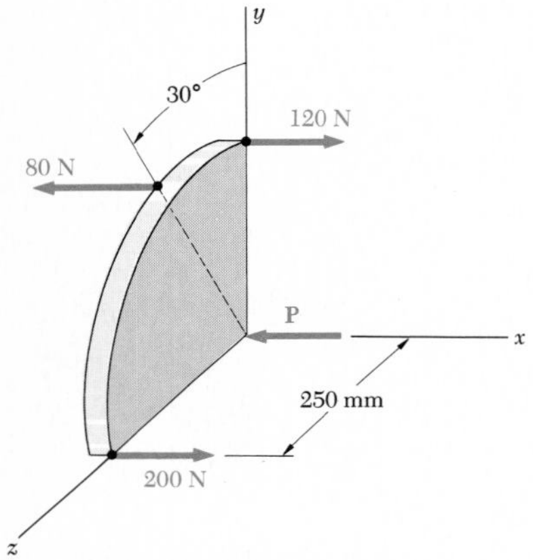

Fig. P4.25 and P4.26

4.25 Four horizontal forces act on a vertical quarter-circular plate of radius 250 mm. Determine the magnitude and point of application of the resultant of the four forces if $P = 40$ N.

4.26 Determine the magnitude of the force **P** for which the resultant of the four forces acts on the rim of the plate.

4.27 Four column loads act on a square foundation mat as shown. Determine the magnitude and point of application of the resultant of the four loads.

4.28 Four column loads act on a square foundation mat as shown. Determine the required additional vertical loads at A and B if the resultant of the six loads is to pass through the center of the mat.

4.29 Determine the magnitude and point of application of the smallest additional load which must be applied to the foundation mat if the resultant of the five loads is to pass through the center of the mat.

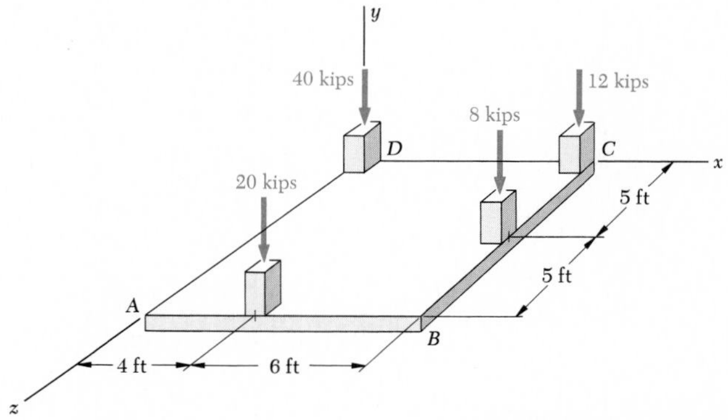

Fig. P4.27, P4.28, and P4.29

4.30 Three vertical loads $\mathbf{P}_1$, $\mathbf{P}_2$, and $\mathbf{P}_3$ are applied at points A, B, and C, respectively, on the perimeter of a horizontal circular plate of radius r. Determine the angles AOB and AOC if the resultant of the three loads is to pass through the center of the plate.

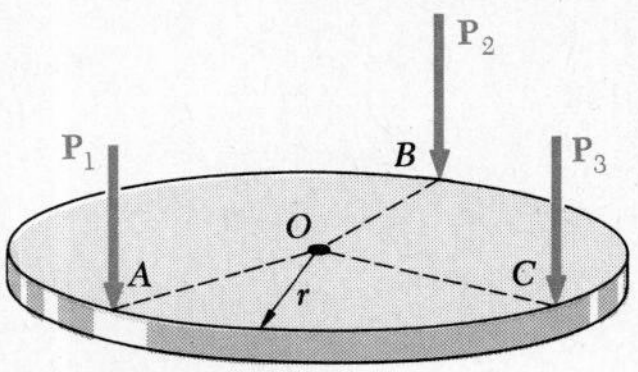

Fig. P4.30

4.31 The worm-gear speed reducer shown weighs 300 N; the center of gravity is located on the x axis at $x = 200$ mm. Replace the weight and couples shown by a wrench. (Specify the axis and pitch of the wrench.)

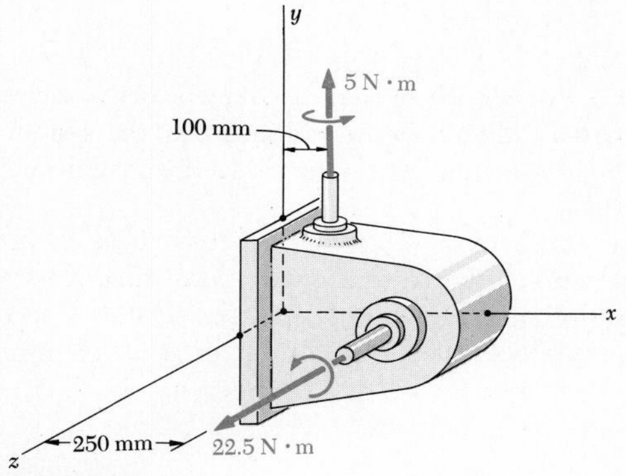

Fig. P4.31

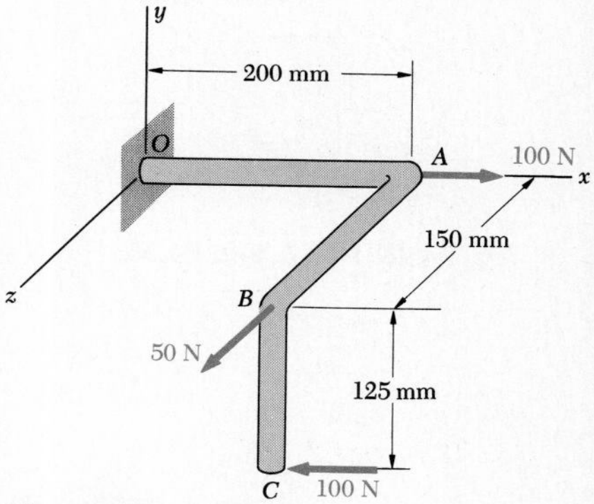

Fig. P4.32

4.32 Replace the three forces shown by (a) a force-couple system at the origin, (b) a wrench. (Specify the axis and pitch of the wrench.)

4.33 Reduce the system of forces shown to a wrench. (Specify the axis and pitch of the wrench.)

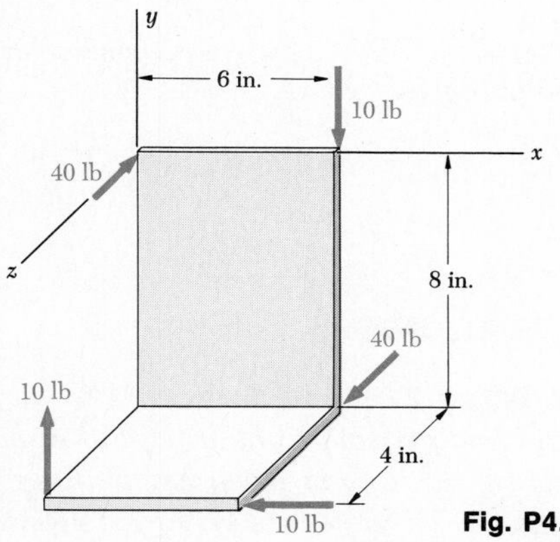

Fig. P4.33

4.34 Solve Prob. 4.33 if the 40-lb forces are replaced by 20-lb forces.

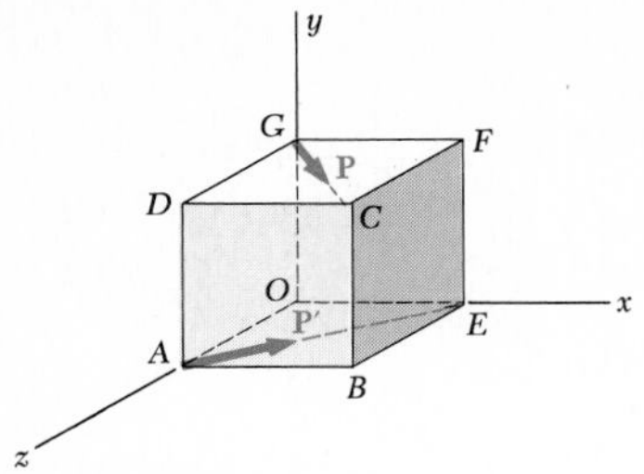

Fig. P4.35

4.35 Two forces of magnitude P act along the diagonals of the faces of a cube of side a as shown. Replace the two forces by a system consisting of (*a*) a single force at O and a couple, (*b*) a wrench. (Specify the axis and pitch of the wrench.)

4.36 Three forces act on a cube of side a as shown. Determine the magnitude and axis of the wrench equivalent to the forces, if $F_1 = F_2 = P$ and $F_3 = 0$.

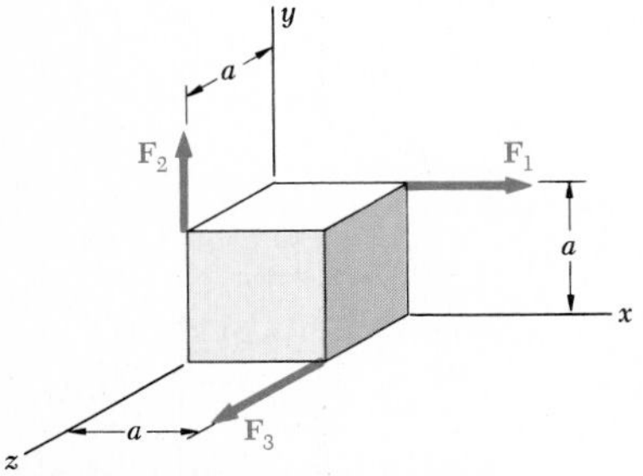

Fig. P4.36, P4.37, and P4.38

4.37 Three forces act on a cube of side a as shown. Determine the magnitude and axis of the wrench equivalent to the three forces, if $F_1 = F_2 = F_3 = P$.

***4.38** Three forces act on a cube of side a as shown. Determine the magnitude of $\mathbf{F}_3$, knowing that $F_1 = F_2 = P$ and that the system may be reduced to a single force **R**. Also determine the magnitude of **R** and its line of action.

4.39 (*a*) Reduce the wrench shown to a system consisting of two forces perpendicular to the y axis and applied respectively at A and B. (*b*) Solve part *a* assuming $R = 240$ N, $M = 40$ N · m, $a = 125$ mm, and $b = 250$ mm.

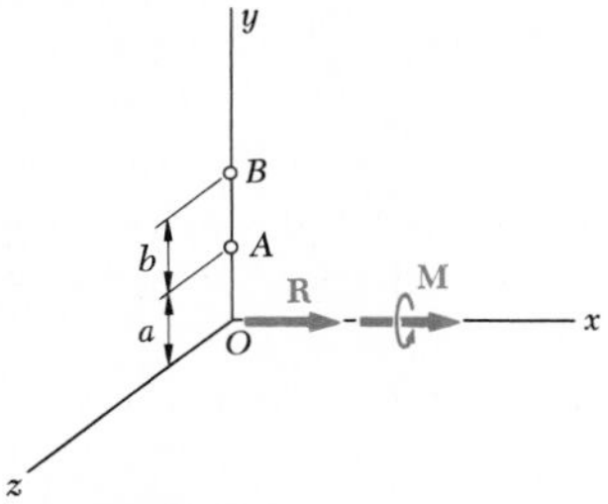

Fig. P4.39

***4.40** Show that a wrench may be replaced by two perpendicular forces, one of which is applied at a given point.

EQUILIBRIUM OF RIGID BODIES

4.6. Equilibrium of a Rigid Body in Space. As was indicated in Sec. 3.12, *a rigid body is said to be in equilibrium when the external forces acting on it form a system of forces equivalent to zero.* In the case of space forces, a system of forces is equivalent to zero when the three forces and three couple vectors shown in Fig. 4.10*b* are all zero. The necessary and sufficient conditions for the equilibrium of a rigid body in space may thus be expressed analytically by writing the six equations

$$\Sigma F_x = 0 \qquad \Sigma F_y = 0 \qquad \Sigma F_z = 0 \tag{4.4}$$
$$\Sigma M_x = 0 \qquad \Sigma M_y = 0 \qquad \Sigma M_z = 0 \tag{4.5}$$

Equations (4.4) express the fact that the components of the external forces in the x, y, and z directions are balanced; Eqs. (4.5) express the fact that the moments of the external forces about the x, y, and z axes are balanced. The system of the external forces, therefore, will impart no motion of translation or rotation to the rigid body considered.

After a free-body diagram has been drawn, showing all the external forces acting on the body, the equations of equilibrium (4.4) and (4.5) may be solved for no more than *six unknowns*. These unknowns generally represent reactions at supports and connections. In practice, it will be found convenient to solve the moment equations (4.5) first, choosing axes which either intersect or are parallel to the lines of action of several unknown forces or force components. In this way, equations containing only two unknowns, possibly one, will be obtained. Once these unknowns have been determined, their values are carried into Eqs. (4.4) in order to reduce the number of unknowns in these equations.

While the six equations of equilibrium cannot be augmented by additional equations, any of them may be replaced by another equation. Thus, moment equations, written with respect to lines other than the axes of coordinates, may sometimes be substituted advantageously for one or several of Eqs. (4.4) and (4.5).

4.7. Reactions at Supports and Connections. The number of unknowns associated with the reaction at a support or connection may vary from one to six in problems involving the equilibrium of a three-dimensional structure. Various types of supports and connections are shown in Fig. 4.12 with the corresponding reactions. A simple way of determining the type of reaction corresponding to a given support or connection and the number of unknowns involved is to find which of the six fundamental motions (translation in x, y, and z directions, rotation about the x, y, and z axes) are allowed and which motions are prevented.

Ball supports, frictionless surfaces, and cables, for example, prevent translation in one direction only and thus exert a single force of known line of action; they each involve one unknown, namely, the magnitude of the reaction. Rollers on rough surfaces and wheels on rails prevent translation in two directions; the corresponding reactions consist of two unknown force components. Rough surfaces in direct contact and ball-and-socket supports prevent translation in three directions; these supports involve three unknown force components.

Some supports and connections may prevent rotation as well as translation; the corresponding reactions include, then, couples as well as forces. The reaction at a fixed support, for example, which prevents any motion (rotation as well as translation), consists of three unknown forces and three unknown couples. A universal joint, which is designed to allow rotation about two axes, will exert a reaction consisting of three unknown force components and one unknown couple.

Other supports and connections are primarily intended to prevent translation; their design, however, is such that they also prevent some rotations. The corresponding reactions consist essentially of force components but *may* also include couples. One group of supports of this type includes hinges and bearings designed to support radial loads only (for example, journal bearings, roller bearings). The corresponding reactions consist of two force components but may also include two couples. Another group includes pin-and-bracket supports, hinges, and bearings designed to support an axial thrust as well as a radial load (for example, ball bearings). The corresponding reactions consist of three force components but may include two couples. However, these supports will not exert any appreciable couples under normal conditions of use. Therefore, *only* force components should be included in their analysis, *unless* it is found that couples are necessary to maintain the equilibrium of the rigid body, or unless the support is known to have been specifically designed to exert a couple.

If the reactions involve more than six unknowns, there are more unknowns than equations, and some of the reactions are *statically indeterminate* (see Probs. 4.58 and 4.59). If the reactions involve less than six unknowns, there are more equations than unknowns, and some of the equations of equilibrium cannot be satisfied under general loading conditions; the rigid body is only *partially constrained*. Under the particular loading conditions corresponding to a given problem, however, the extra equations often reduce to trivial identities such as $0 = 0$ and may be disregarded; although only partially constrained, the rigid body remains in equilibrium (see Sample Probs. 4.5 and 4.6). Even with six or more unknowns, it is possible that some equations of equilibrium will not be satisfied. This may occur when the supports are such that the reactions are forces which must either be parallel or intersect the same line; the rigid body is then *improperly constrained*.

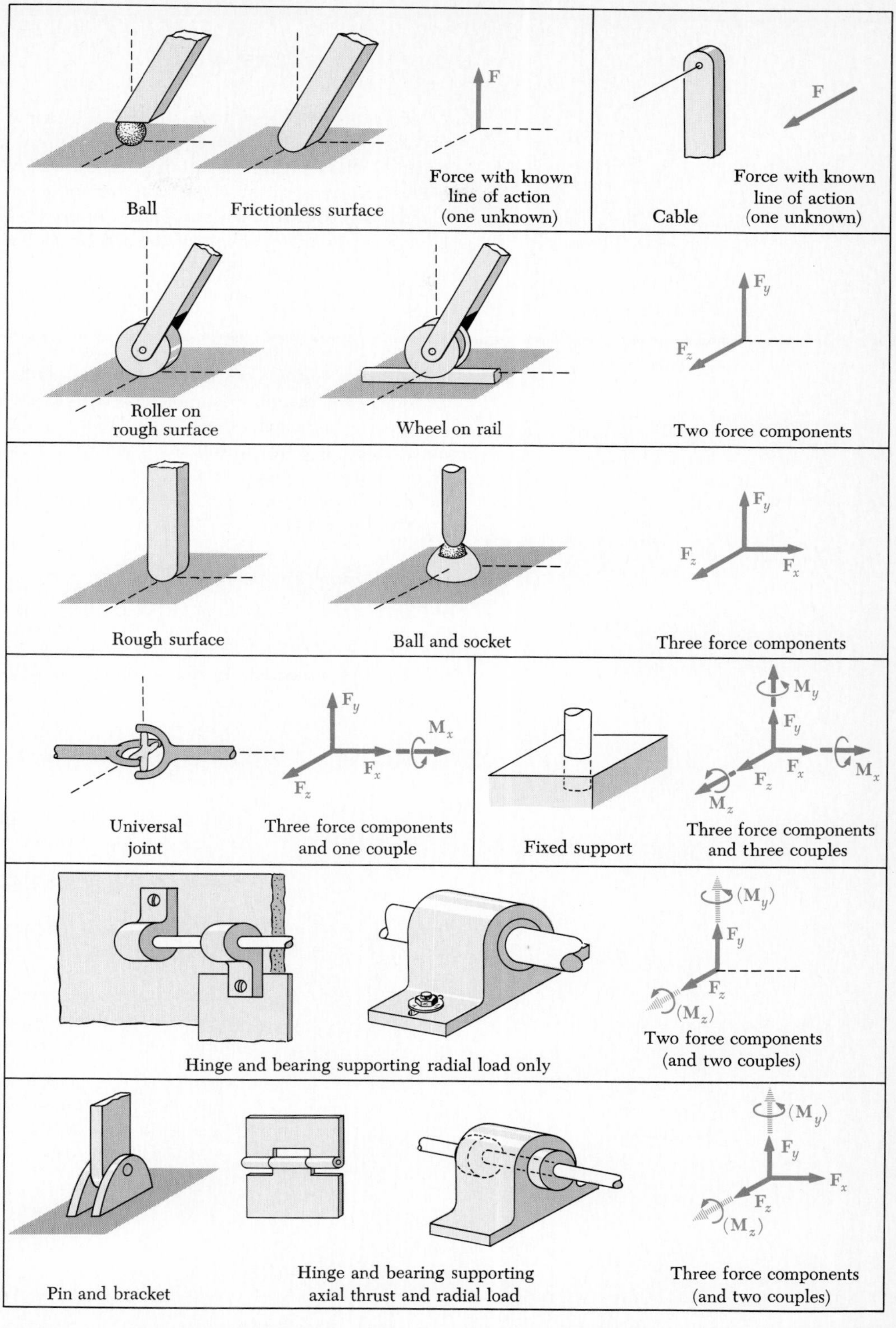

Fig. 4.12 Reactions at supports and connections.

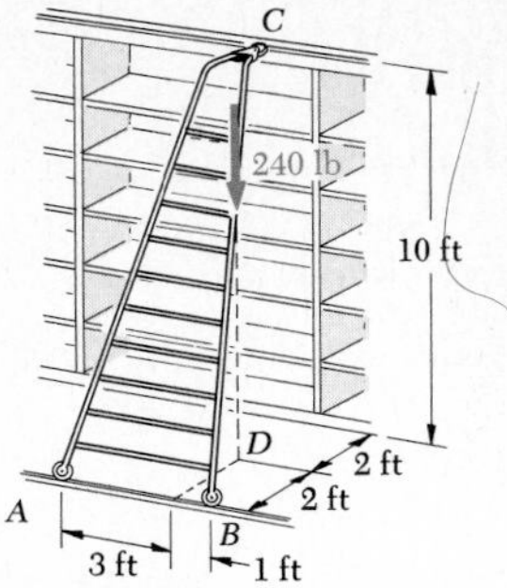

SAMPLE PROBLEM 4.5

A ladder used to reach high shelves in a storeroom is supported by two flanged wheels A and B mounted on a rail and by an unflanged wheel C resting against a rail fixed to the wall. A man stands on the ladder and leans to the right. The line of action of the 240-lb combined weight of the man and ladder intersects the floor at point D. Determine the components of the reactions at A, B, and C.

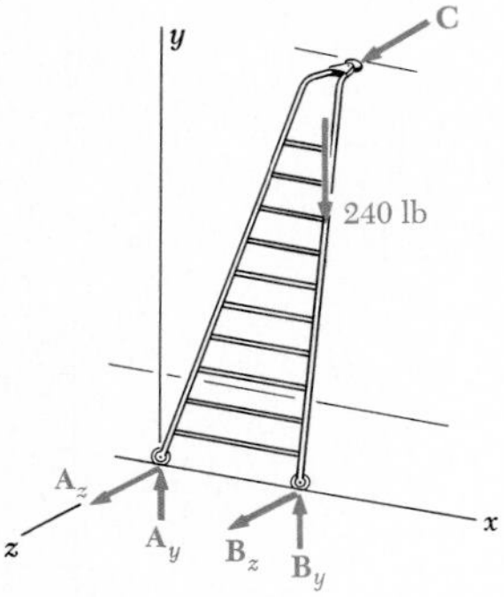

Solution. A free-body diagram of the ladder is drawn; there are five unknown reaction components, two at each flanged wheel and one at the unflanged wheel. The ladder is therefore only partially constrained; it is free to roll along the rails. It is, however, in equilibrium under the given vertical load since the equation $\Sigma F_x = 0$ is satisfied. Three projections of the ladder are drawn, and the following equilibrium equations are written:

Equilibrium Equations

$+\circlearrowleft \Sigma M_x = 0$: $\quad -(240 \text{ lb})(2 \text{ ft}) + C(10 \text{ ft}) = 0$

$C = +48 \text{ lb}$ ◄

Using the projection on the xy plane, we write

$+\circlearrowleft \Sigma M_A = 0$: $\quad -(240 \text{ lb})(3 \text{ ft}) + B_y(4 \text{ ft}) = 0$

$B_y = +180 \text{ lb}$ ◄

$+\circlearrowleft \Sigma M_B = 0$: $\quad +(240 \text{ lb})(1 \text{ ft}) - A_y(4 \text{ ft}) = 0$

$A_y = +60 \text{ lb}$ ◄

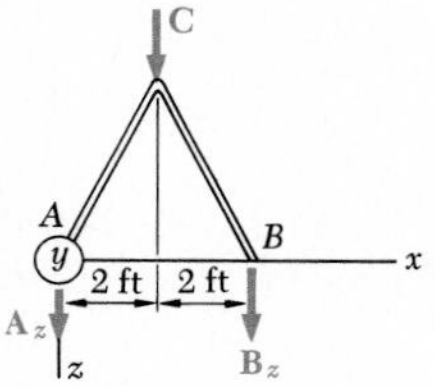

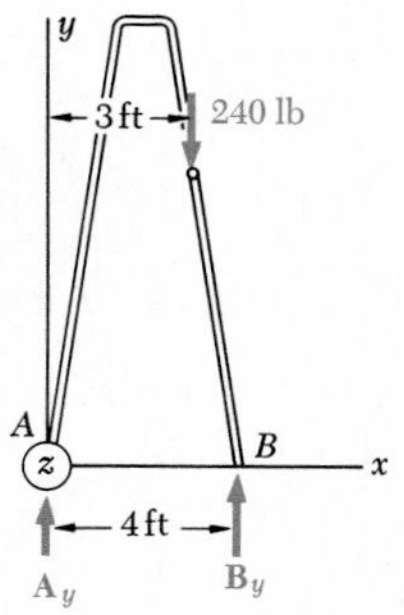

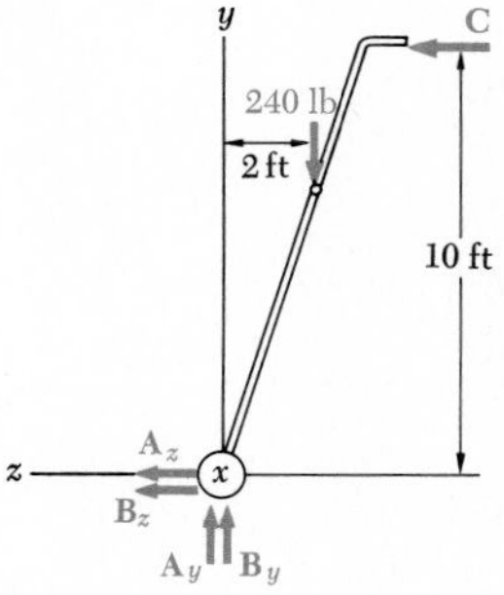

Using the projection on the xz plane and the computed value of C, we write

$+\circlearrowleft \Sigma M_A = 0$: $\quad -(48 \text{ lb})(2 \text{ ft}) - B_z(4 \text{ ft}) = 0$

$B_z = -24 \text{ lb}$ ◄

$+\circlearrowleft \Sigma M_B = 0$: $\quad +(48 \text{ lb})(2 \text{ ft}) + A_z(4 \text{ ft}) = 0$

$A_z = -24 \text{ lb}$ ◄

Check. We may check the results by verifying that equilibrium equations not used in the solution are satisfied; for example,

$$\Sigma F_y = A_y + B_y - 240 \text{ lb}$$
$$= 60 \text{ lb} + 180 \text{ lb} - 240 \text{ lb} = 0$$
$$\Sigma F_z = A_z + B_z + C$$
$$= -24 \text{ lb} - 24 \text{ lb} + 48 \text{ lb} = 0$$

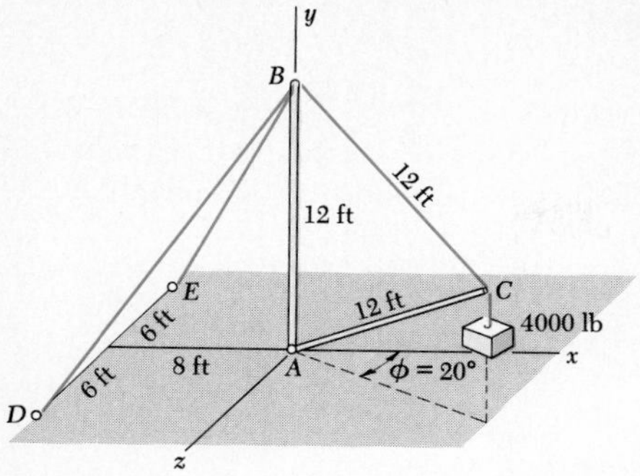

SAMPLE PROBLEM 4.6

The derrick shown supports a 4000-lb load. It is held by a ball and socket at A and by two cables attached at points D and E. In the position shown, the derrick stands in a vertical plane forming an angle $\phi = 20°$ with the xy plane. Determine the tension in each cable and the components of the reaction at A.

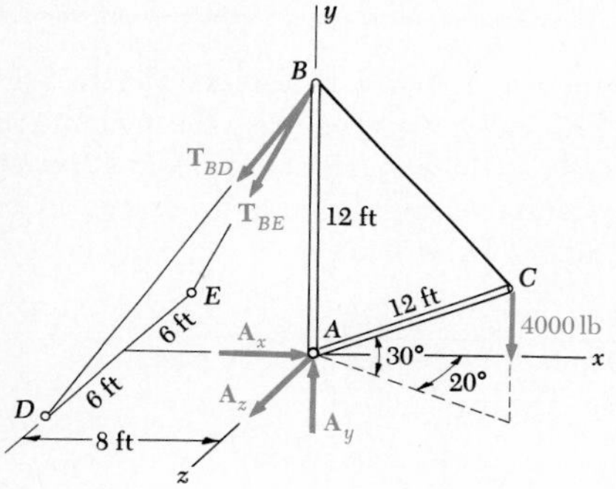

Solution. A free-body diagram of the derrick is drawn. Since the directions of the forces exerted at B by the cables are known, these forces involve one unknown each, namely, the magnitudes T_{BD} and T_{BE}. The reaction at A is a force of unknown direction and is represented by three unknown components $\mathbf{A}_x$, $\mathbf{A}_y$, and $\mathbf{A}_z$. Since there are only five unknowns, the derrick is partially constrained. It may rotate freely about the y axis; it is, however, in equilibrium under the given loading since the equation $\Sigma M_y = 0$ is satisfied.

Recalling that the components and magnitude of the forces $\mathbf{T}_{BD}$ and $\mathbf{T}_{BE}$ are respectively proportional to the components and magnitude of the vectors $\overrightarrow{BD}$ and $\overrightarrow{BE}$, we express the components of these forces in terms of T_{BD} and T_{BE}. On the other hand, the distances a and b are obtained as follows:

$$a = (12 \text{ ft}) \cos 30° \cos 20° = 9.77 \text{ ft}$$
$$b = (12 \text{ ft}) \cos 30° \sin 20° = 3.55 \text{ ft}$$

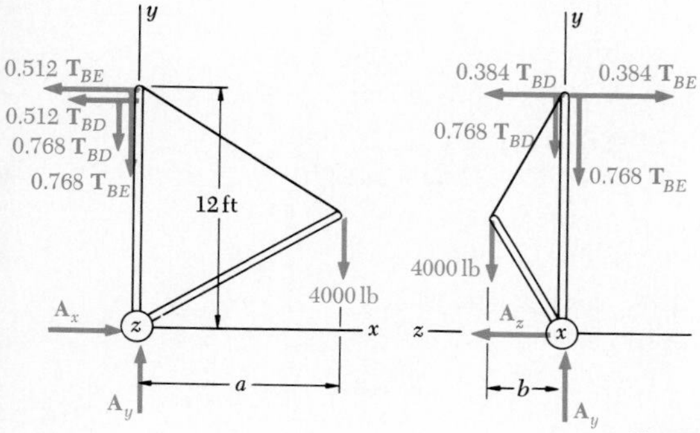

Equilibrium Equations. Using the projections shown, we write

$$+\circlearrowleft \Sigma M_x = 0: \quad 0.384 T_{BD}(12 \text{ ft}) + (4000 \text{ lb})(3.55 \text{ ft}) - 0.384 T_{BE}(12 \text{ ft}) = 0$$

$$+\circlearrowleft \Sigma M_z = 0: \quad 0.512 T_{BD}(12 \text{ ft}) + 0.512 T_{BE}(12 \text{ ft}) - (4000 \text{ lb})(9.77 \text{ ft}) = 0$$

Solving these two equations simultaneously, we obtain

$$T_{BD} = 1640 \text{ lb} \qquad T_{BE} = 4720 \text{ lb} \quad \blacktriangleleft$$

Using the computed magnitudes of T_{BD} and T_{BE}, we write

$\Sigma F_x = 0$: $\quad A_x - 0.512 T_{BD} - 0.512 T_{BE} = 0$

$A_x - 0.512(1640 \text{ lb}) - 0.512(4720 \text{ lb}) = 0$

$$A_x = +3260 \text{ lb} \quad \blacktriangleleft$$

$\Sigma F_y = 0$: $\quad A_y - 0.768 T_{BD} - 0.768 T_{BE} - 4000 \text{ lb} = 0$

$A_y - 0.768(1640 \text{ lb}) - 0.768(4720 \text{ lb}) - 4000 \text{ lb} = 0$

$$A_y = +8880 \text{ lb} \quad \blacktriangleleft$$

$\Sigma F_z = 0$: $\quad A_z + 0.384 T_{BD} - 0.384 T_{BE} = 0$

$A_z + 0.384(1640 \text{ lb}) - 0.384(4720 \text{ lb}) = 0$

$$A_z = +1183 \text{ lb} \quad \blacktriangleleft$$

SAMPLE PROBLEM 4.7

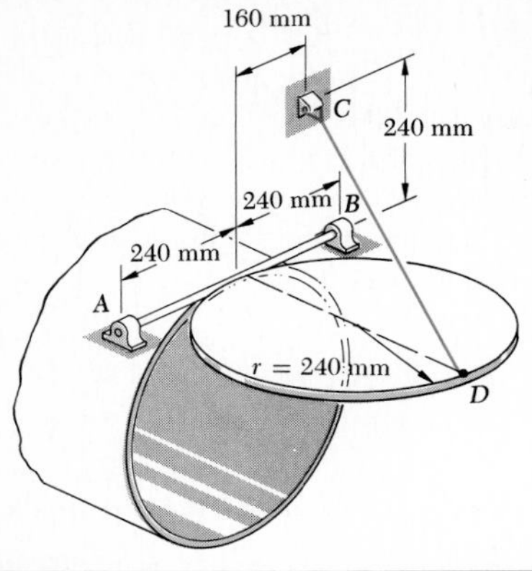

A uniform pipe cover of radius $r = 240$ mm and mass 30 kg is held in a horizontal position by the cable CD. Assuming that the bearing at B does not exert any axial thrust, determine the tension in the cable and the components of the reactions at A and B.

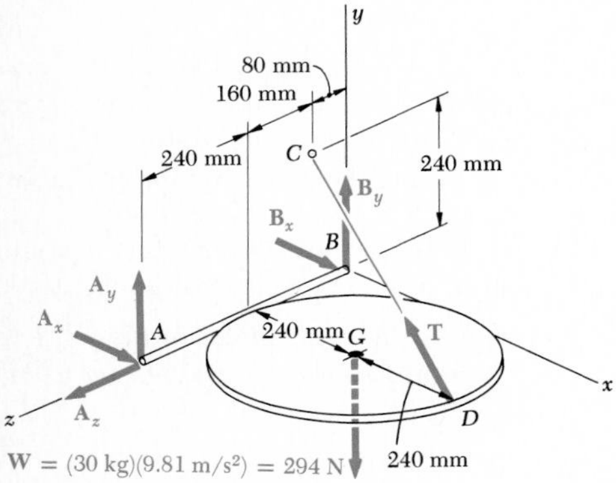

Solution. A free-body diagram is drawn with the coordinate axes shown. The reactions involve six unknowns, namely, the magnitude of the force **T** exerted by the cable, three force components at hinge A, and two at hinge B. The components of **T** may be expressed in terms of the unknown magnitude T by writing

$$\frac{T_x}{-480\text{ mm}} = \frac{T_y}{+240\text{ mm}} = \frac{T_z}{-160\text{ mm}} = \frac{T}{560\text{ mm}}$$

$$T_x = -\tfrac{6}{7}T \qquad T_y = +\tfrac{3}{7}T \qquad T_z = -\tfrac{2}{7}T$$

Equilibrium Equations. Using the projections shown, we write

$+\circlearrowleft\Sigma M_z = 0$: $\quad (\tfrac{3}{7}T)(2r) - (294\text{ N})r = 0 \qquad T = 343\text{ N}$

$+\circlearrowleft\Sigma M_x = 0$: $\quad (294\text{ N})r - (\tfrac{3}{7}T)r - A_y(2r) = 0$

$2A_y = 294\text{ N} - \tfrac{3}{7}(343\text{ N}) \qquad A_y = +73.5\text{ N}$

$+\circlearrowleft\Sigma M_y = 0$: $\quad (\tfrac{2}{7}T)(2r) - (\tfrac{6}{7}T)r + A_x(2r) = 0$

$A_x = -\tfrac{2}{7}T + \tfrac{3}{7}T = \tfrac{1}{7}(343\text{ N}) \qquad A_x = +49.0\text{ N}$

$\Sigma F_x = 0$: $\quad A_x + B_x - \tfrac{6}{7}T = 0$

$B_x = \tfrac{6}{7}(343\text{ N}) - 49\text{ N} \qquad B_x = +245\text{ N}$

$\Sigma F_y = 0$: $\quad A_y + B_y + \tfrac{3}{7}T - 294\text{ N} = 0$

$B_y = 294\text{ N} - 73.5\text{ N} - \tfrac{3}{7}(343\text{ N}) \qquad B_y = +73.5\text{ N}$

$\Sigma F_z = 0$: $\quad A_z - \tfrac{2}{7}T = 0 \qquad A_z = \tfrac{2}{7}(343\text{ N}) \qquad A_z = +98.0\text{ N}$

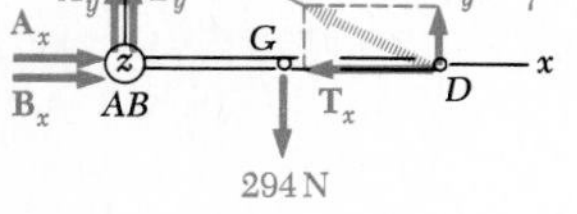

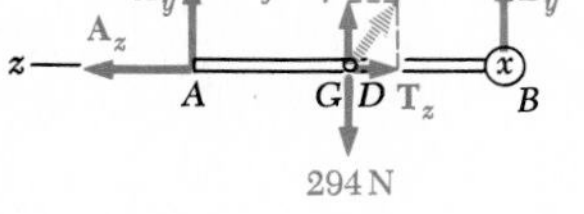

PROBLEMS

4.41 The 10-m pole is acted upon by an 8.4-kN force as shown. It is held by a ball and socket at *A* and by the two cables *BD* and *BE*. Neglecting the weight of the pole, determine the tension in each cable and the reaction at *A*.

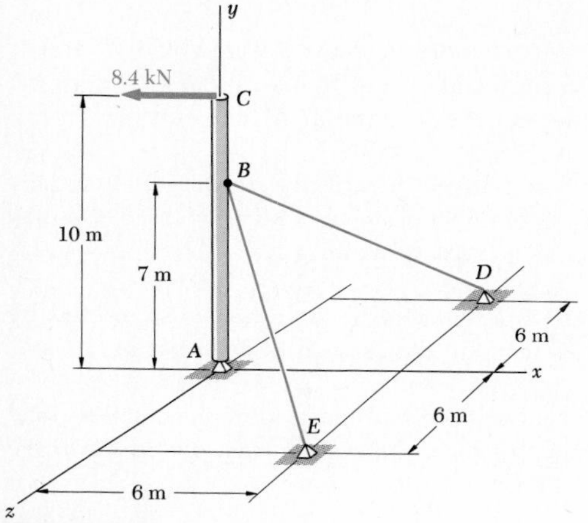

Fig. P4.41

4.42 The 3-m flagpole *AC* forms an angle of 60° with the vertical; it is held by a ball and socket at *A* and by two braces. Knowing that the distance *AB* is 1 m, determine the components of the reaction at *A* and the tension in each brace caused by the 400-N load.

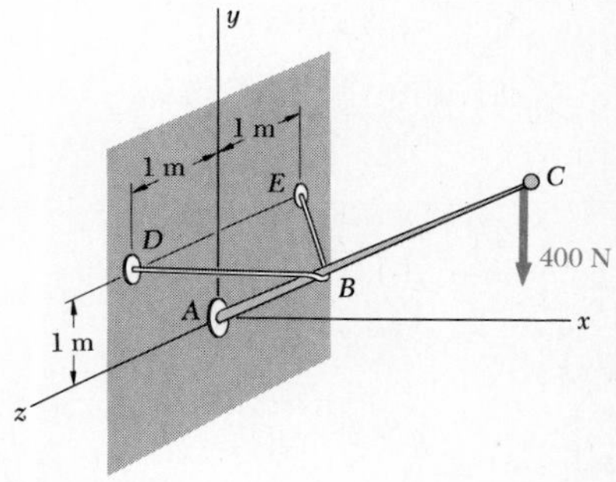

Fig. P4.42

4.43 The 12-ft boom *AB* is acted upon by the 340-lb force shown. Determine the tension in each cable and the reaction at the ball and socket at *A*.

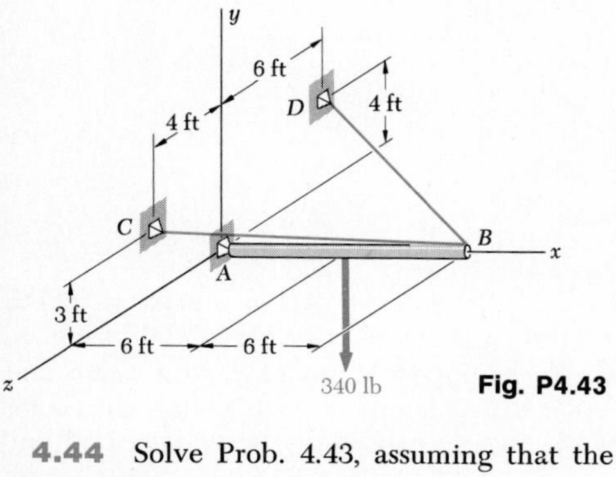

Fig. P4.43

4.44 Solve Prob. 4.43, assuming that the 340-lb load is applied at point *B*.

4.45 A 2- by 3-m plate weighing 600 kg is lifted by three cables which are joined at point *D* directly above the center of the plate. Determine the tension in each cable.

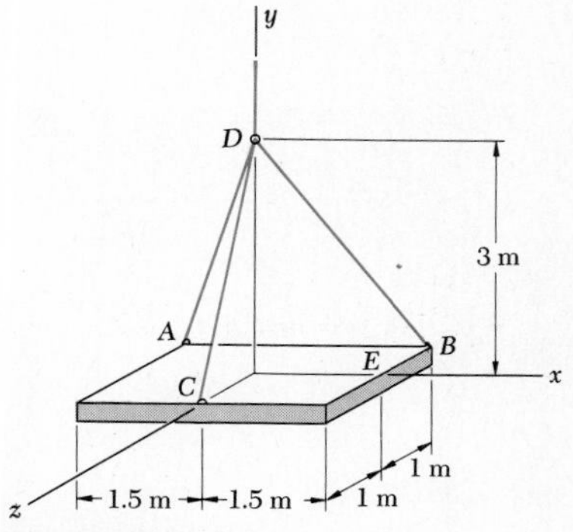

Fig. P4.45

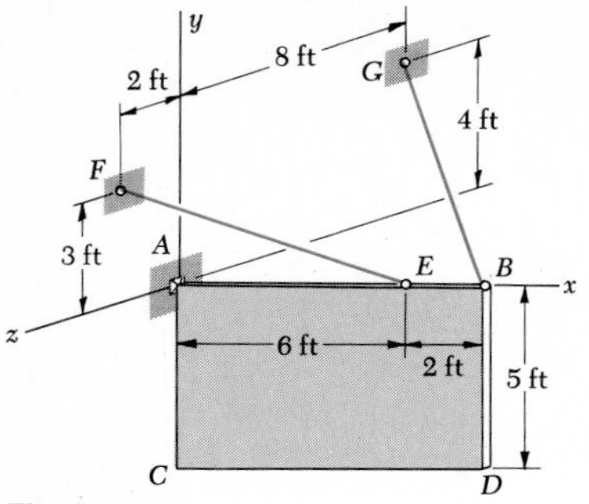

Fig. P4.48

4.46 Solve Prob. 4.45, assuming that cable *BD* is replaced by a cable connecting points *D* and *E*.

4.47 In Sample Prob. 4.6, determine the horizontal angle ϕ for which the tension in *BE* is maximum (*a*) if neither cable may become slack, (*b*) if cables *BE* and *BD* are replaced by rods which can withstand either a tension or a compression force.

4.48 A 5- by 8-ft sign of uniform density weighs 270 lb and is supported by a ball and socket at *A* and by two cables. Determine the tension in each cable and the reaction at *A*.

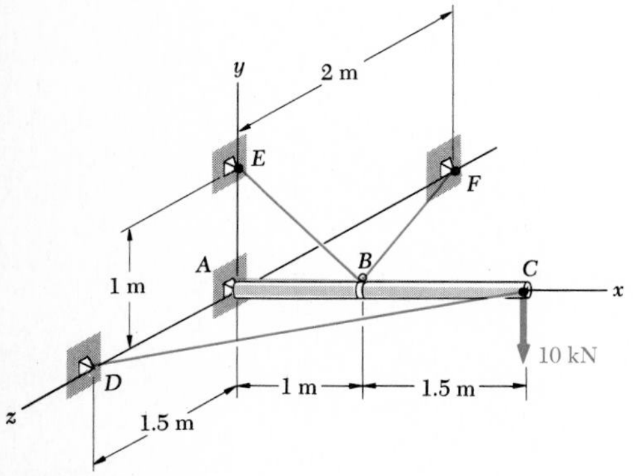

Fig. P4.49

4.49 A 2.5-m boom is held by a ball and socket at *A* and by two cables *EBF* and *DC*; cable *EBF* passes around a frictionless pulley at *B*. Determine the tension in each cable.

4.50 The square plate shown weighs 40 lb and is supported by three wires as shown. Determine the tension in each wire.

4.51 Determine the magnitude and location of the smallest load which must be placed on the 40-lb plate if the tensions in the three wires are to be equal.

4.52 A load *W* is to be placed on the plate shown. Knowing that the weight of the plate is 40 lb, determine the magnitude of the load *W* and the point where the load should be placed if the tension is to be 30 lb in each of the three wires.

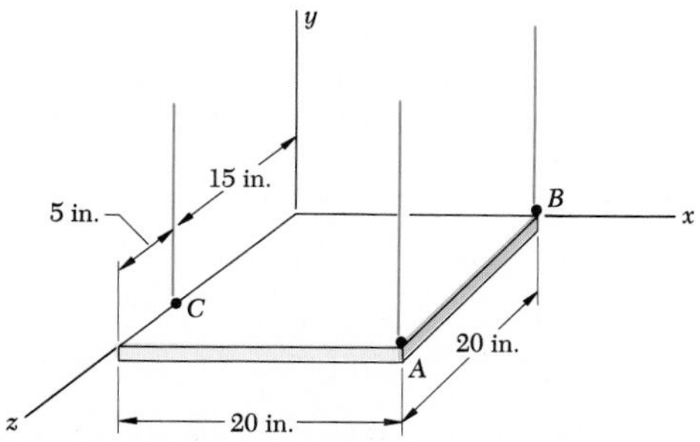

Fig. P4.50, P4.51, and P4.52

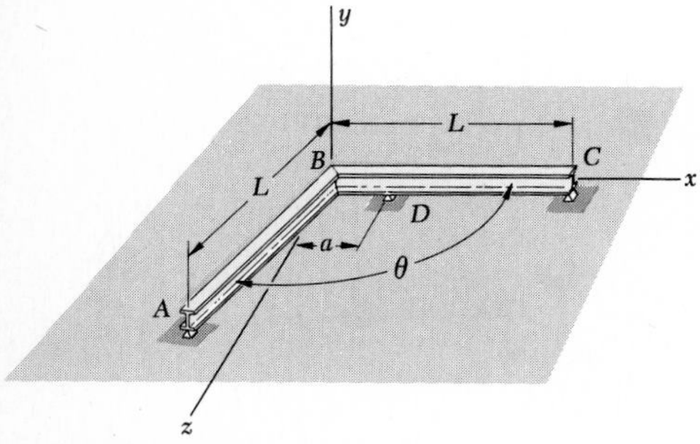

Fig. P4.53

4.53 Two steel wide-flange beams *AB* and *BC*, each of length *L*, are welded together at *B* and rest in the horizontal plane on supports at *A*, *D*, and *C*. Each support is mounted on a caster and can exert only an upward vertical force. Knowing that the beams weigh *w* lb/ft, determine the reactions at the supports if $\theta = 90°$ and $a = \frac{1}{4}L$.

4.54 Knowing that $\theta = 90°$, determine the largest permissible value of *a* if the beams of Prob. 4.53 are not to tip.

4.55 Knowing that $a = \frac{1}{4}L$, determine the reactions in Prob. 4.53 in terms of w, L, and θ. Explain why the results obtained are not valid for $\theta = 0°$ and $\theta = 180°$.

4.56 The winch shown is held in equilibrium by a vertical force **P** applied at point E. Knowing that $\theta = 60°$, determine the magnitude of **P** and the reactions at A and B. It is assumed that the bearing at B does not exert any axial thrust.

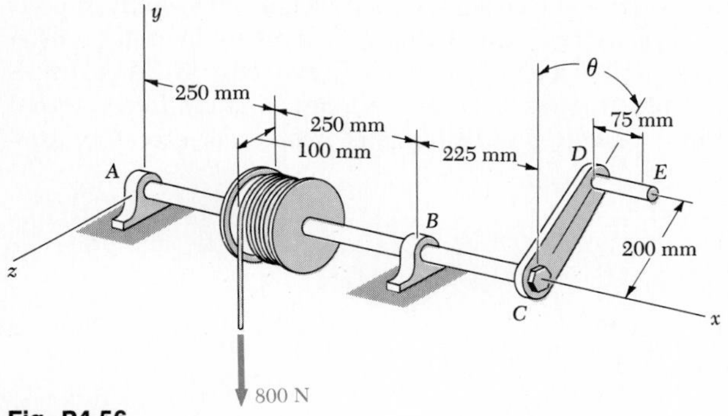

Fig. P4.56

4.57 Solve Prob. 4.56, assuming that the winch is held in equilibrium by a force **P** applied at E in a direction perpendicular to the plane CDE.

4.58 The door of a bank vault weighs 12,000 lb and is supported by two hinges as shown. Determine the components of the reaction at each hinge.

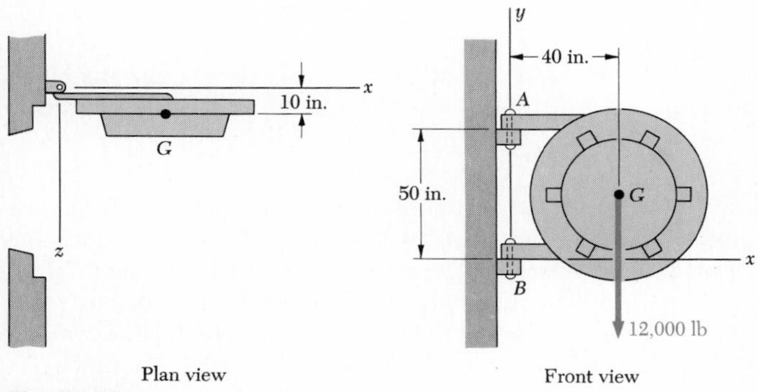

Fig. P4.58

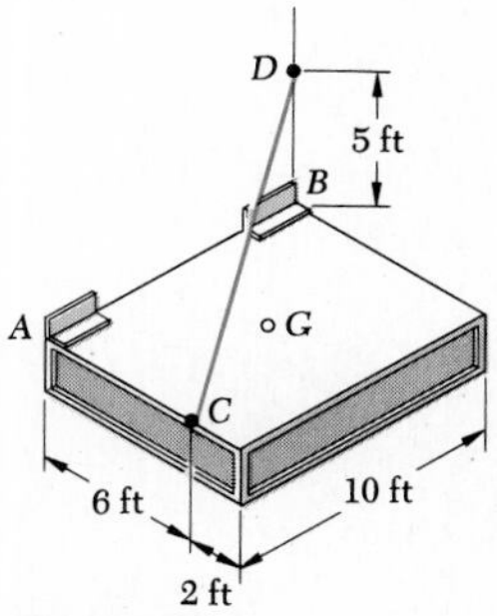

Fig. P4.59

4.59 A 500-lb marquee, 8 by 10 ft, is held in a horizontal position by two horizontal hinges at A and B and by a cable CD attached to a point D located 5 ft directly above B. Determine the tension in the cable and the components of the reactions at the hinges.

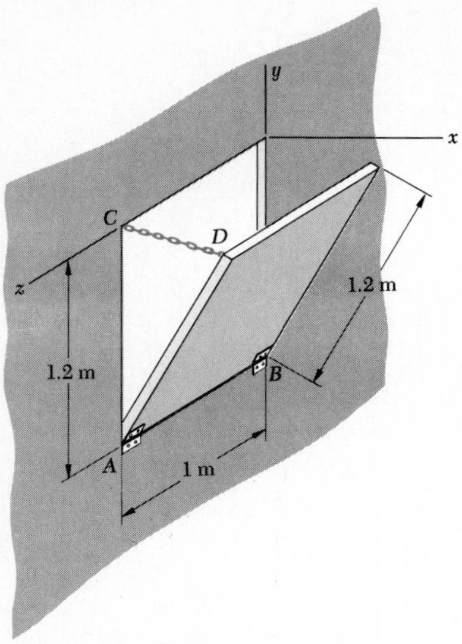

Fig. P4.60

4.60 The 75-kg plate shown is supported by hinges at A and B. It is held by the chain CD of length 0.6 m. Assuming that the hinge at A does not exert any axial thrust, determine the tension in the chain and the reactions at A and B.

4.61 Solve Prob. 4.60, assuming that the hinge at A is removed.

4.62 The horizontal platform $ABCD$ weighs 50 lb and supports a 200-lb load at its center. The platform is normally held in position by hinges at A and B and by braces CE and DE. If brace DE is removed, determine the reactions at the hinges and the force exerted by the remaining brace CE. The hinge at A does not exert any axial thrust.

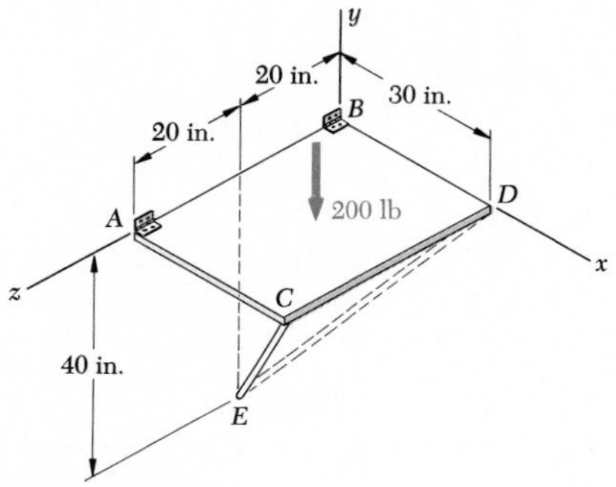

Fig. P4.62

4.63 Solve Prob. 4.62, assuming that the 200-lb load is placed at corner D.

4.64 Solve Prob. 4.62, assuming that the brace DE and the hinge at A are removed.

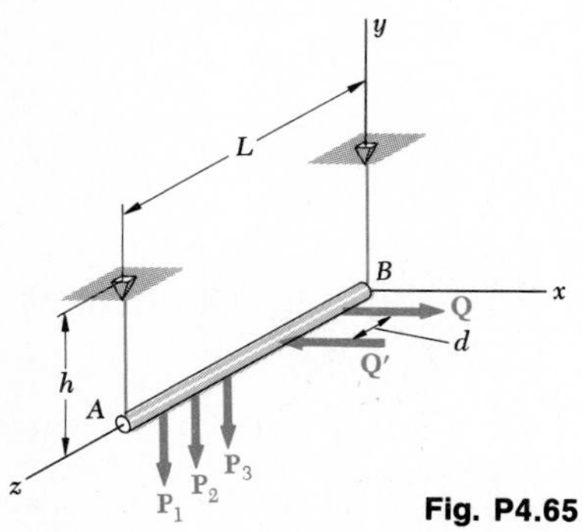

Fig. P4.65

***4.65** A bar AB, of negligible weight, is suspended from two wires and supports several vertical loads $\mathbf{P}_1$, $\mathbf{P}_2$, and $\mathbf{P}_3$. Determine the angle θ through which the bar rotates when two horizontal forces forming a couple of magnitude $M_O = Qd$ are applied to the bar. Also show that the axis of rotation coincides with the axis of the wrench which is equivalent to the total loading. Assume that θ is small.

4.66 The rigid L-shaped member ABC is supported by a ball and socket at A and by three cables. Determine the tension in each cable and the reaction at A caused by the 5-kN load applied at G.

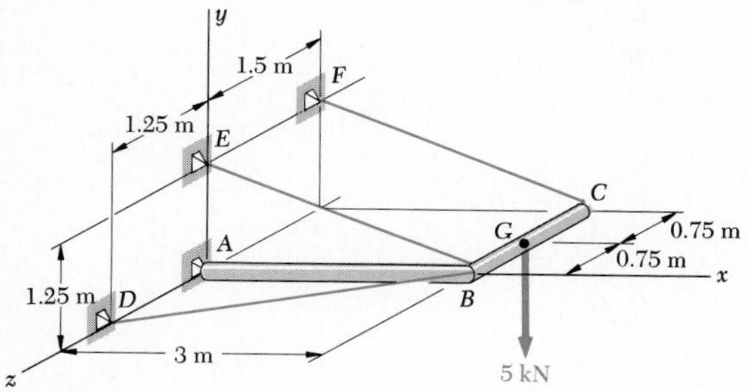

Fig. P4.66

4.67 Solve Prob. 4.66, assuming that cable BD is removed and replaced by a cable joining points E and C.

4.68 Three rods are welded together to form the "corner" shown. The corner is supported by three smooth eyebolts. Determine the reactions at A, B, and C when $P = 800$ N, $a = 150$ mm, $b = 100$ mm, and $c = 125$ mm.

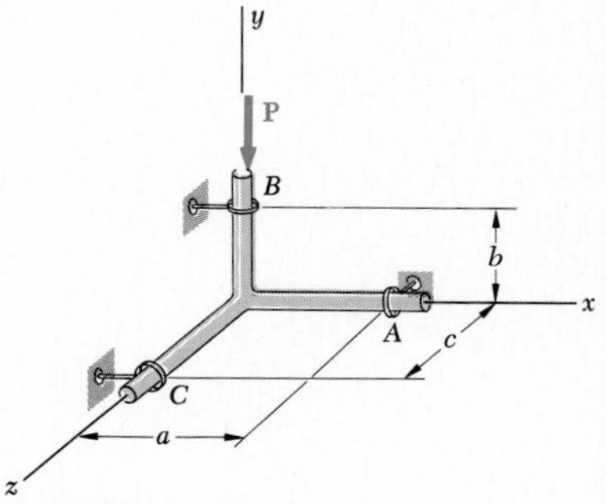

Fig. P4.68

REVIEW PROBLEMS

4.69 The table shown weighs 30 lb and has a diameter of 4 ft. It is supported by three legs equally spaced around the edge. A vertical load **P** of magnitude 75 lb is applied to the top of the table at D. Determine the maximum value of a if the table is not to tip over. Show, on a sketch, the area of the table over which **P** can act without tipping the table.

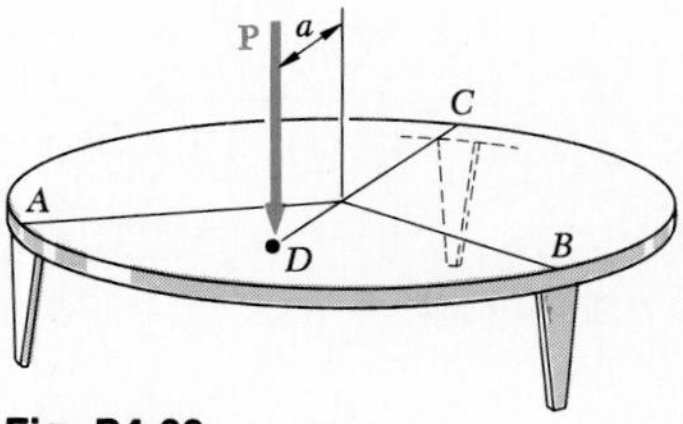

Fig. P4.69

4.70 A force **P** of magnitude 350 N acts along a line joining points A and B as shown. Determine the moment of **P** about each of the coordinate axes.

4.71 Determine (a) the magnitude of the force **P** for which the moment of **P** about the x axis is 50 N · m, (b) the corresponding moment of **P** about the y and z axes.

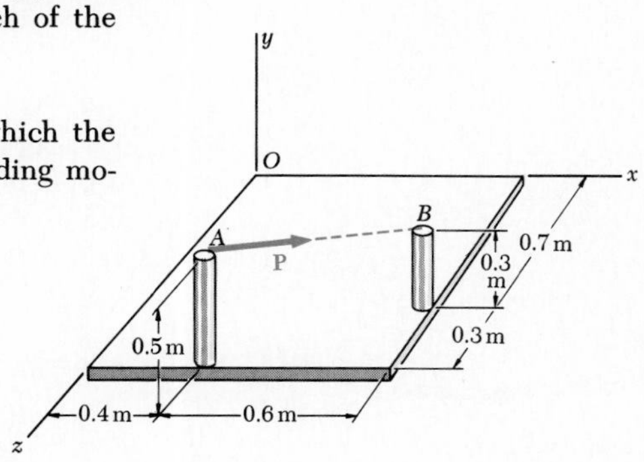

Fig. P4.70 and P4.71

4.72 The uniform rod AB weighs 11.20 lb and is of length 18 in. It is connected by ball-and-socket joints to collars A and B, which slide freely along the two rods shown. Determine the magnitude of the force **P** required to maintain equilibrium when (*a*) $c = 2$ in., (*b*) $c = 8$ in.

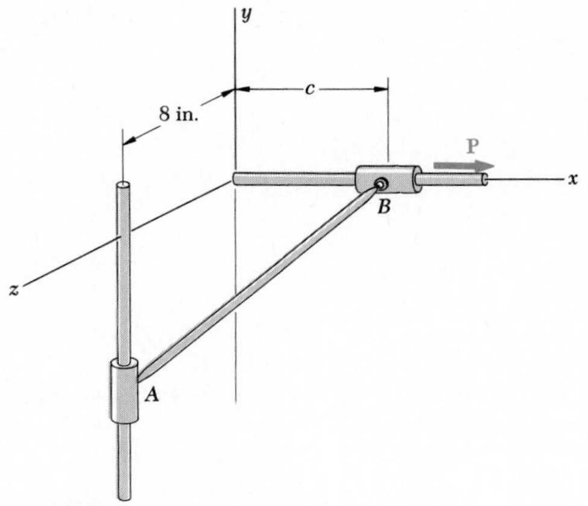

Fig. P4.72

4.73 Solve Prob. 4.72 when (*a*) $c = 14$ in., (*b*) $c = 16$ in.

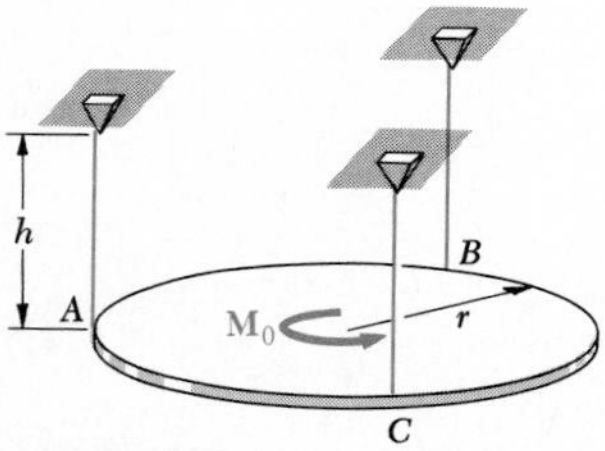

Fig. P4.74

4.74 A uniform circular plate of radius r and weight W is supported by three vertical wires, each of length h, equally spaced around the edge. Determine the horizontal angle θ through which the plate rotates when a couple $\mathbf{M}_0$ is applied in the plane of the plate. Assume that θ is small.

4.75 In Prob. 4.74 show that the angle θ is independent of the location of the points where the vertical wires are attached to the edge of the plate, provided that A, B, and C do not lie on the same side of a diameter. Assume that θ is small.

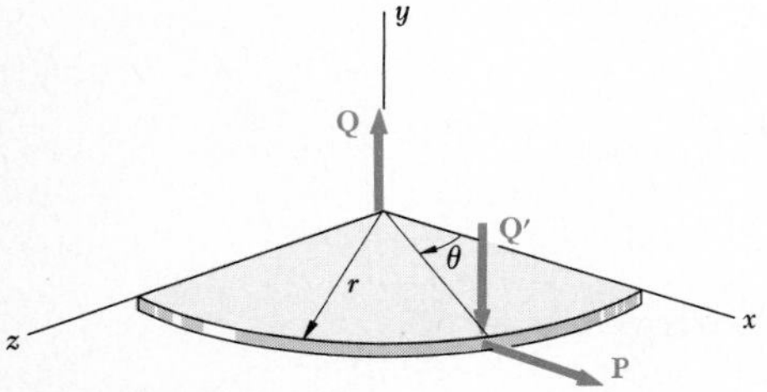

Fig. P4.76 and P4.77

4.76 Three forces act on the plate shown; the horizontal force **P** is parallel to the x axis and the vertical forces **Q** and **Q′** form a couple of moment $M = Qr$. Determine (*a*) the wrench equivalent to the applied system, (*b*) the locus of the points of intersection of the axis of the wrench with the yz plane as θ varies.

4.77 Given $P = 400$ N, $Q = 800$ N, and $r = 200$ mm, determine the wrench equivalent to the given system when (*a*) $\theta = 30°$, (*b*) $\theta = 90°$. (In each case specify the axis and pitch of the wrench.)

4.78 A square plate of side a supports four loads as shown. Determine the magnitude and point of application of the resultant of the four loads.

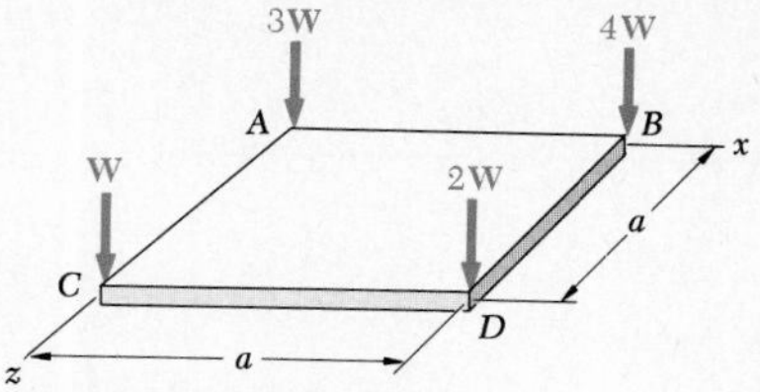

Fig. P4.78 and P4.79

4.79 A square plate of side a supports four loads as shown. Determine the magnitude and point of application of the smallest additional load which must be applied to the plate if the resultant of the five loads is to pass through the center of the plate.

4.80 A gearbox is acted upon by three couples and a horizontal force as shown. Determine (*a*) the equivalent force-couple system at the origin, (*b*) the equivalent wrench. (Specify the axis and pitch of the wrench.)

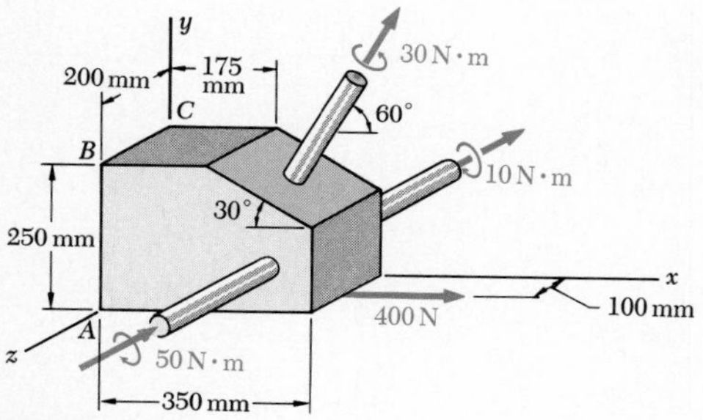

Fig. P4.80

Chapter 5

Distributed Forces: Centroids and Centers of Gravity

AREAS AND LINES

5.1. Center of Gravity of a Two-dimensional Body. We have assumed so far that the attraction exerted by the earth on a rigid body could be represented by a single force **W**. This force, called the weight of the body, was to be applied at the *center of gravity* of the body (Sec. 3.1). Actually, the earth exerts a force on each of the particles forming the body. The action of the earth on a rigid body should thus be represented by a large number of small forces distributed over the entire body. We shall see in this chapter, however, that all these small forces may be replaced by a single equivalent force **W**. We shall also learn to determine the center of gravity, i.e., the point of application of the resultant **W**, for various shapes of bodies.

Let us first consider a flat horizontal plate (Fig. 5.1). We may divide the plate into n small elements. The coordinates of the first element are denoted by x_1 and y_1, those of the second element by x_2 and y_2, etc. The forces exerted by the earth on the elements of plate will be denoted, respectively, by $\Delta\mathbf{W}_1$, $\Delta\mathbf{W}_2, \ldots, \Delta\mathbf{W}_n$. These forces or weights are directed toward the center of the earth; however, for all practical purposes they may be assumed parallel. Their resultant is therefore a single force in the same direction. The magnitude W of this force is

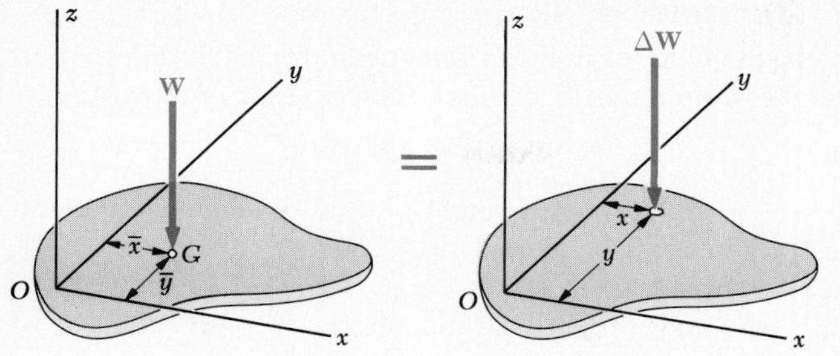

ΣM_y: $\bar{x}W = \Sigma x\,\Delta W$
ΣM_x: $\bar{y}W = \Sigma y\,\Delta W$

Fig. 5.1 Center of gravity of a plate.

obtained by adding the magnitudes of the elementary weights,

$$\Sigma F_z: \qquad W = \Delta W_1 + \Delta W_2 + \cdots + \Delta W_n \tag{5.1}$$

To obtain the coordinates $\bar{x}$ and $\bar{y}$ of the point G where the resultant **W** should be applied, we write that the moments of **W** about the y and x axes are equal to the sum of the corresponding moments of the elementary weights,

$$\begin{aligned} \Sigma M_y: \qquad \bar{x}W &= x_1\,\Delta W_1 + x_2\,\Delta W_2 + \cdots + x_n\,\Delta W_n \\ \Sigma M_x: \qquad \bar{y}W &= y_1\,\Delta W_1 + y_2\,\Delta W_2 + \cdots + y_n\,\Delta W_n \end{aligned} \tag{5.2}$$

If we now increase the number of elements into which the plate is divided and simultaneously decrease the size of each element, we obtain at the limit the following expressions:

$$W = \int dW \qquad \bar{x}W = \int x\,dW \qquad \bar{y}W = \int y\,dW \tag{5.3}$$

These equations define the weight **W** and the coordinates $\bar{x}$ and $\bar{y}$ of the center of gravity G of a flat plate. The same equations may be derived for a wire lying in the xy plane (Fig. 5.2). We shall observe, in the latter case, that the center of gravity G will generally not be located on the wire.

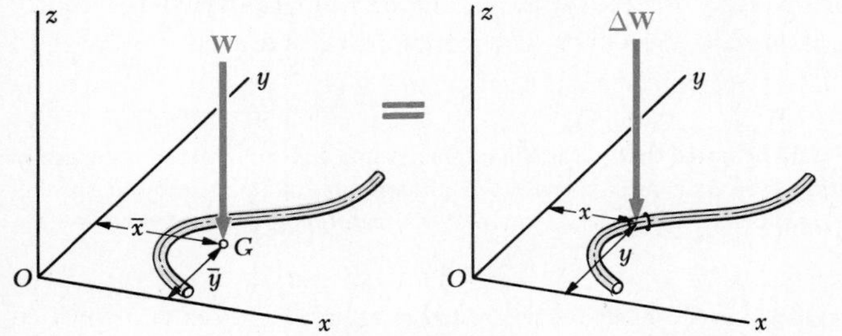

ΣM_y: $\bar{x}W = \Sigma x\,\Delta W$
ΣM_x: $\bar{y}W = \Sigma y\,\Delta W$

Fig. 5.2 Center of gravity of a wire.

5.2. Centroids of Areas and Lines. In the case of a homogeneous plate of uniform thickness, the magnitude ΔW of the weight of an element of plate may be expressed as

$$\Delta W = \gamma t \, \Delta A$$

where γ = specific weight (weight per unit volume) of material
t = thickness of plate
ΔA = area of element

Similarly, we may express the magnitude W of the weight of the entire plate in the form

$$W = \gamma t A$$

where A is the total area of the plate.

If U.S. customary units are used, the specific weight γ should be expressed in lb/ft^3, the thickness t in feet, and the areas ΔA and A in square feet. We check that ΔW and W will then be expressed in pounds. If SI units are used, γ should be expressed in N/m^3, t in meters, and the areas ΔA and A in square meters; the weights ΔW and W will then be expressed in newtons.†

Substituting for ΔW and W in the moment equations (5.2) and dividing throughout by γt, we write

$$\begin{aligned} \Sigma M_y: &\quad \bar{x}A = x_1 \, \Delta A_1 + x_2 \, \Delta A_2 + \cdots + x_n \, \Delta A_n \\ \Sigma M_x: &\quad \bar{y}A = y_1 \, \Delta A_1 + y_2 \, \Delta A_2 + \cdots + y_n \, \Delta A_n \end{aligned} \tag{5.4}$$

If we increase the number of elements into which the area A is divided and simultaneously decrease the size of each element, we obtain at the limit

$$\bar{x}A = \int x \, dA \qquad \bar{y}A = \int y \, dA \tag{5.5}$$

These equations define the coordinates $\bar{x}$ and $\bar{y}$ of the center of gravity of a homogeneous plate. The point of coordinates $\bar{x}$ and $\bar{y}$ is also known as the *centroid C of the area A* of the plate (Fig. 5.3). If the plate is not homogeneous, the equations cannot be used to determine the center of gravity of the plate; they still define, however, the centroid of the area.

† It should be noted that in the SI system of units a given material is generally characterized by its density ρ (mass per unit volume) rather than by its specific weight γ. The specific weight of the material can then be obtained by writing

$$\gamma = \rho g$$

where $g = 9.81 \text{ m/s}^2$. Since ρ is expressed in kg/m^3, we check that γ will be expressed in $(\text{kg/m}^3)(\text{m/s}^2)$, that is, in N/m^3.

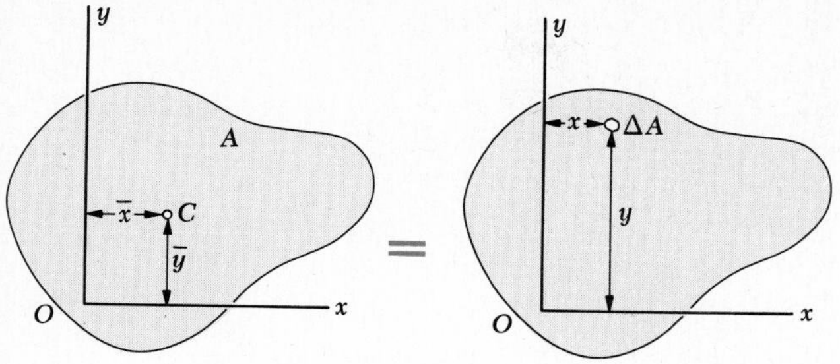

ΣM_y: $\bar{x}A = \Sigma x\, \Delta A$
ΣM_x: $\bar{y}A = \Sigma y\, \Delta A$

Fig. 5.3 Centroid of an area.

The integral $\int x\, dA$ is known as the *first moment of the area A with respect to the y axis.* Similarly, the integral $\int y\, dA$ defines the *first moment of A with respect to the x axis.* It is seen from Eqs. (5.5) that, if the centroid of an area is located on a coordinate axis, the first moment of the area with respect to that axis is zero.

In the case of a homogeneous wire of uniform cross section, the magnitude ΔW of the weight of an element of wire may be expressed as

$$\Delta W = \gamma a\, \Delta L$$

where γ = specific weight of material
a = cross-sectional area of wire
ΔL = length of element

The center of gravity of the wire then coincides with the *centroid C of the line L* defining the shape of the wire (Fig. 5.4). The coordinates $\bar{x}$ and $\bar{y}$ of the centroid of the line L are obtained from the equations

$$\bar{x}L = \int x\, dL \qquad \bar{y}L = \int y\, dL \tag{5.6}$$

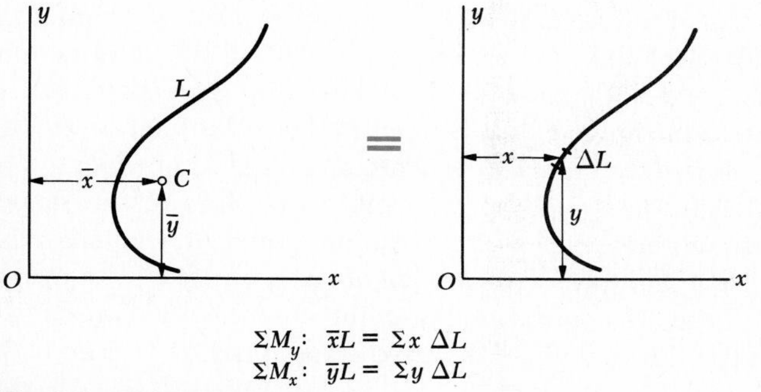

ΣM_y: $\bar{x}L = \Sigma x\, \Delta L$
ΣM_x: $\bar{y}L = \Sigma y\, \Delta L$

Fig. 5.4 Centroid of a line.

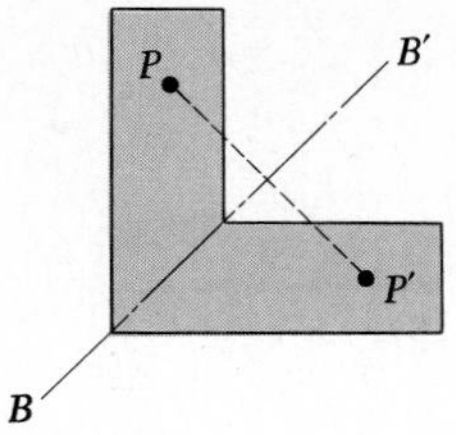

Fig. 5.5

An area A is said to be *symmetrical about an axis BB′* if to every point P of the area corresponds a point P' of the same area such that the line PP' is perpendicular to BB' and is divided into two equal parts by that axis (Fig. 5.5). A line L is said to be symmetrical about BB' if it satisfies similar conditions. When an area A or a line L possesses an axis of symmetry BB', the centroid of the area or line must be located on that axis. Indeed, if the axis of symmetry is chosen as the y axis, the coordinate $\bar{x}$ of the centroid is found to be zero, since to every product $x\,dA$ or $x\,dL$ in the first integral in Eqs. (5.5) or (5.6) will correspond a product of equal magnitude but of opposite sign. It

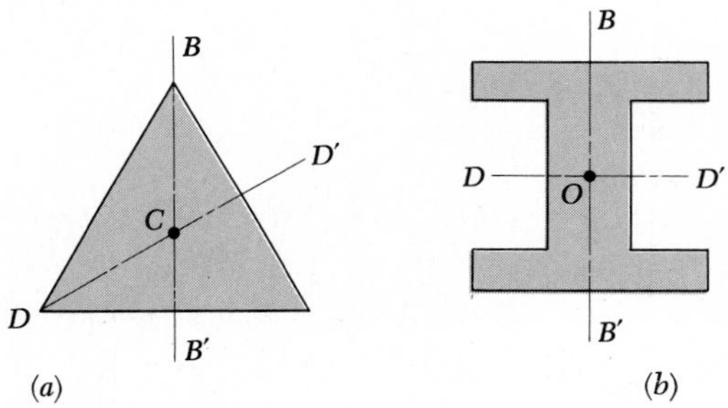

Fig. 5.6

follows that, if an area or line possesses two axes of symmetry, the centroid of the area or line is located at the intersection of the two axes of symmetry (Fig. 5.6). This property enables us to determine immediately the centroid of areas such as circles, ellipses, squares, rectangles, equilateral triangles, or other symmetrical figures, as well as the centroid of lines in the shape of the circumference of a circle, the perimeter of a square, etc.

An area is said to be *symmetrical about a center O* if to every point P of the area corresponds a point P' of the same area such that the line PP' is divided into two equal parts by O (Fig. 5.7).

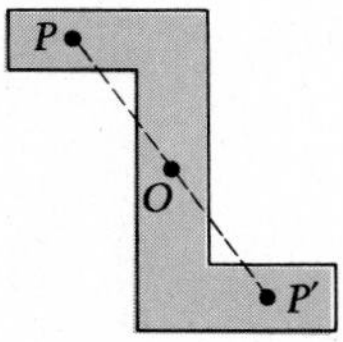

Fig. 5.7

Shape		$\bar{x}$	$\bar{y}$	Area
Triangular area	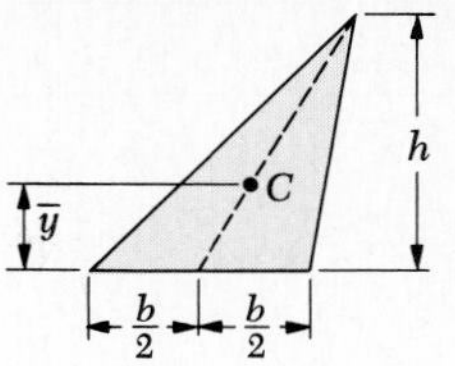		$\dfrac{h}{3}$	$\dfrac{bh}{2}$
Quarter-circular area	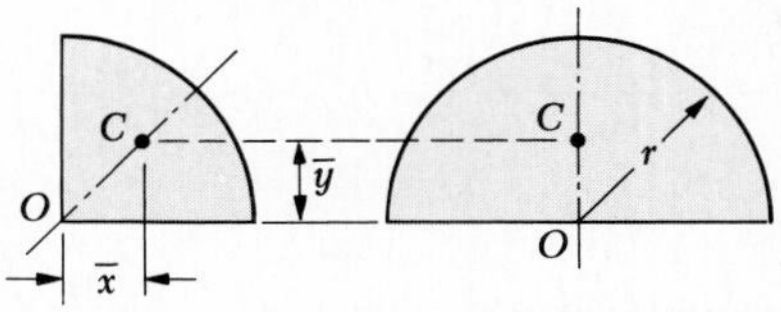	$\dfrac{4r}{3\pi}$	$\dfrac{4r}{3\pi}$	$\dfrac{\pi r^2}{4}$
Semicircular area		0	$\dfrac{4r}{3\pi}$	$\dfrac{\pi r^2}{2}$
Quarter-elliptical area	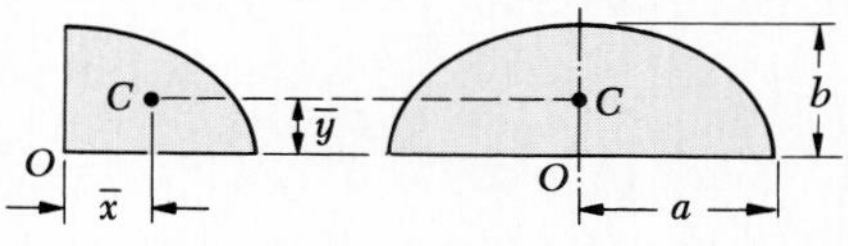	$\dfrac{4a}{3\pi}$	$\dfrac{4b}{3\pi}$	$\dfrac{\pi ab}{4}$
Semielliptical area		0	$\dfrac{4b}{3\pi}$	$\dfrac{\pi ab}{2}$
Semiparabolic area	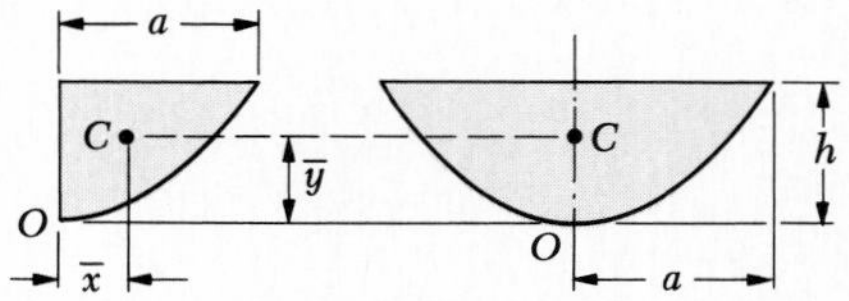	$\dfrac{3a}{8}$	$\dfrac{3h}{5}$	$\dfrac{2ah}{3}$
Parabolic area		0	$\dfrac{3h}{5}$	$\dfrac{4ah}{3}$
Parabolic spandrel	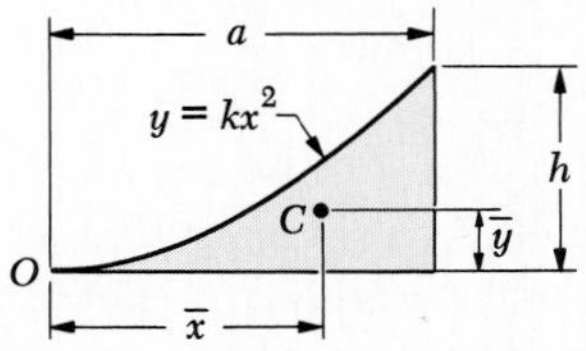	$\dfrac{3a}{4}$	$\dfrac{3h}{10}$	$\dfrac{ah}{3}$
General spandrel	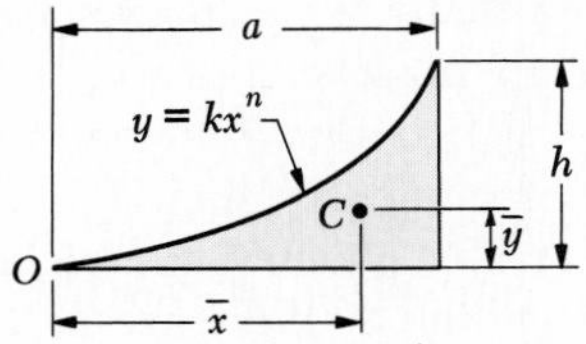	$\dfrac{n+1}{n+2}a$	$\dfrac{n+1}{4n+2}h$	$\dfrac{ah}{n+1}$
Circular sector	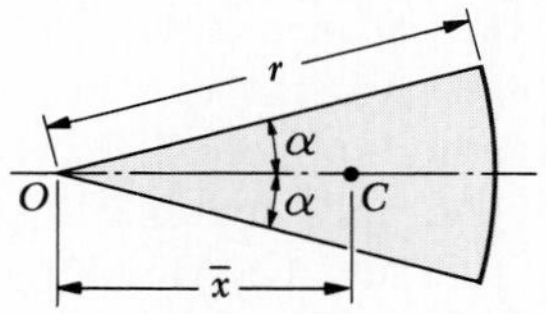	$\dfrac{2r \sin \alpha}{3\alpha}$	0	αr^2

Fig. 5.8A Centroids of common shapes of areas.

Shape		$\bar{x}$	$\bar{y}$	Length
Quarter-circular arc	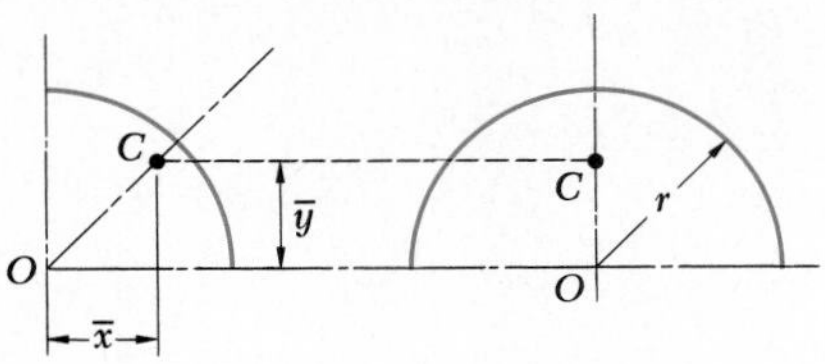	$\frac{2r}{\pi}$	$\frac{2r}{\pi}$	$\frac{\pi r}{2}$
Semicircular arc		0	$\frac{2r}{\pi}$	πr
Arc of circle	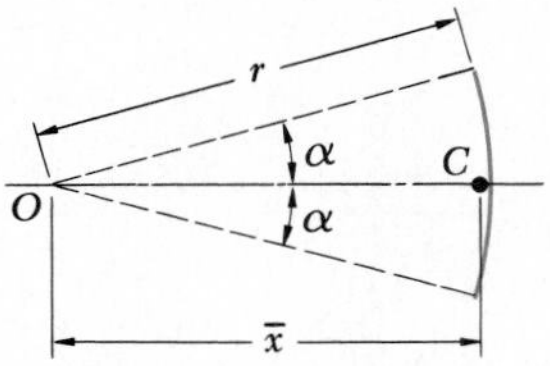	$\frac{r \sin \alpha}{\alpha}$	0	$2\alpha r$

Fig. 5.8B Centroids of common shapes of lines.

A line L is said to be symmetrical about O if it satisfies similar conditions. A reasoning similar to that used above would show that, when an area A or line L possesses a center of symmetry O, the point O must be the centroid of the area or line. It should be noted that a figure possessing a center of symmetry does not necessarily possess an axis of symmetry (Fig. 5.7), while a figure possessing two axes of symmetry does not necessarily possess a center of symmetry (Fig. 5.6*a*). However, if a figure possesses two axes of symmetry at a right angle to each other, the point of intersection of these axes will be a center of symmetry (Fig. 5.6*b*).

Centroids of unsymmetrical areas and lines and of areas and lines possessing only one axis of symmetry will be determined by the methods of Secs. 5.4 and 5.5. Centroids of common shapes of areas and lines are shown in Fig. 5.8*A* and *B*. The formulas defining these centroids will be derived in the Sample Problems and Problems following Secs. 5.4 and 5.5.

5.3. Composite Plates and Wires. In many instances, a flat plate may be divided into rectangles, triangles, or other common shapes shown in Fig. 5.8*A*. The abscissa $\bar{X}$ of its center of gravity G may be determined from the abscissas $\bar{x}_1, \bar{x}_2, \ldots$ of the centers of gravity of the various parts by expressing that the moment of the weight of the whole plate about the y axis is equal to the sum of the moments of the weights of the various parts about the same axis (Fig. 5.9). The ordinate $\bar{Y}$ of the center

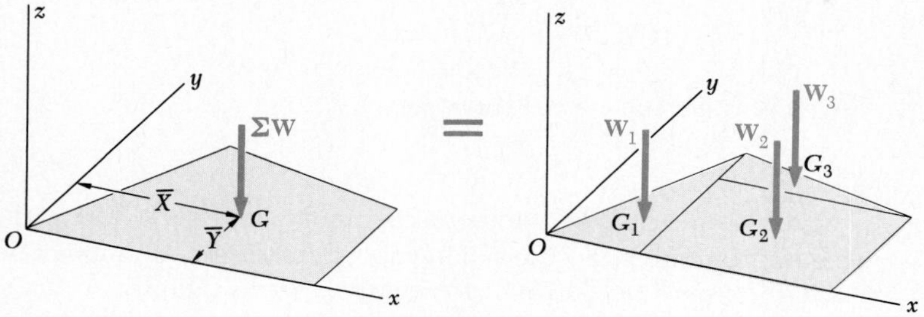

$$\Sigma M_y: \ \bar{X}\,\Sigma W = \Sigma \bar{x} W$$
$$\Sigma M_x: \ \bar{Y}\,\Sigma W = \Sigma \bar{y} W$$

Fig. 5.9 Center of gravity of a composite plate.

of gravity of the plate is found in a similar way by equating moments about the x axis.

$$
\begin{aligned}
\Sigma M_y: \quad & \bar{X}(W_1 + W_2 + \cdots + W_n) \\
& \quad = \bar{x}_1 W_1 + \bar{x}_2 W_2 + \cdots + \bar{x}_n W_n \\
\Sigma M_x: \quad & \bar{Y}(W_1 + W_2 + \cdots + W_n) \\
& \quad = \bar{y}_1 W_1 + \bar{y}_2 W_2 + \cdots + \bar{y}_n W_n
\end{aligned} \tag{5.7}
$$

If the plate is homogeneous and of uniform thickness, the center of gravity coincides with the centroid C of its area. The abscissa $\bar{X}$ of the centroid of the area may then be determined by expressing that the first moment of the composite area with respect to the y axis is equal to the sum of the first moments of the elementary areas with respect to the same axis (Fig. 5.10). The ordinate $\bar{Y}$ of the centroid is found in a similar way by equating first moments of areas with respect to the x axis.

$$
\begin{aligned}
\Sigma M_y: \quad & \bar{X}(A_1 + A_2 + \cdots + A_n) \\
& \quad = \bar{x}_1 A_1 + \bar{x}_2 A_2 + \cdots + \bar{x}_n A_n \\
\Sigma M_x: \quad & \bar{Y}(A_1 + A_2 + \cdots + A_n) \\
& \quad = \bar{y}_1 A_1 + \bar{y}_2 A_2 + \cdots + \bar{y}_n A_n
\end{aligned} \tag{5.8}
$$

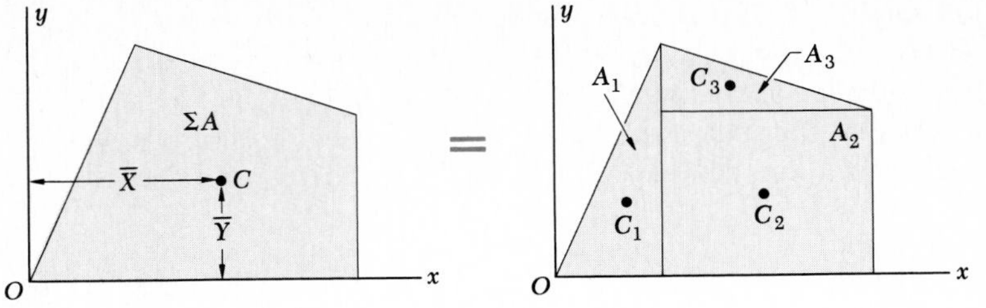

$$\Sigma M_y:\quad \overline{X}\,\Sigma A = \Sigma \bar{x}A$$
$$\Sigma M_x:\quad \overline{Y}\,\Sigma A = \Sigma \bar{y}A$$

Fig. 5.10 Centroid of a composite area.

Care should be taken to record the moment of each area with the appropriate sign. First moments of areas, just like moments of forces, may be positive or negative. For example, an area whose centroid is located to the left of the y axis will have a negative first moment with respect to that axis. Also, the area of a hole should be recorded with a negative sign (Fig. 5.11).

Similarly, it is possible in many cases to determine the center of gravity of a composite wire or the centroid of a composite line by dividing the wire or line into simpler elements (Sample Prob. 5.2).

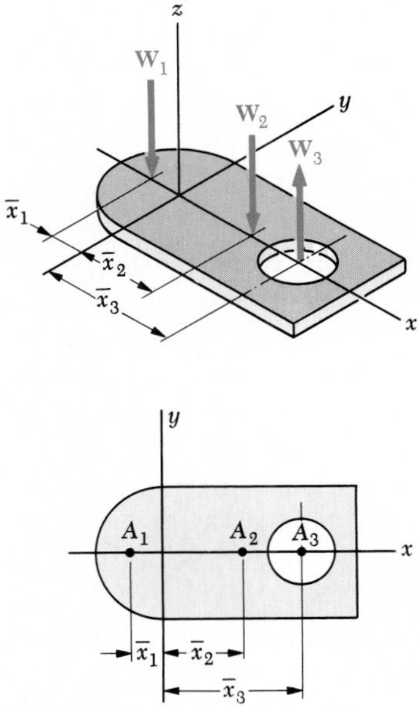

	$\bar{x}$	A	$\bar{x}A$
A_1 Semicircle	−	+	−
A_2 Full rectangle	+	+	+
A_3 Circular hole	+	−	−

Fig. 5.11

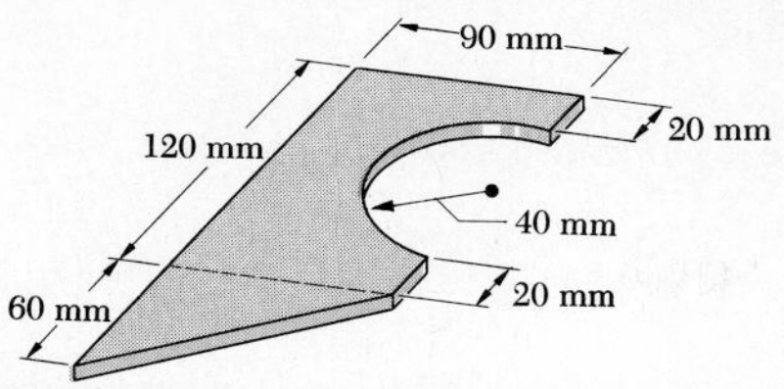

SAMPLE PROBLEM 5.1

Determine the center of gravity of the thin homogeneous plate shown.

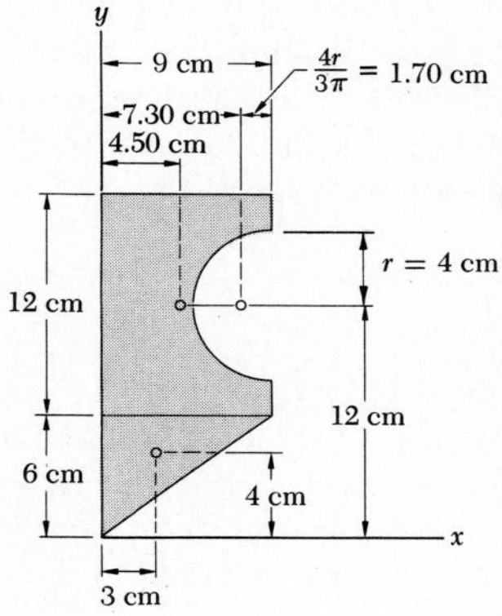

Solution. Since the plate is homogeneous, we may locate its center of gravity by determining the centroid of its area. The area is obtained by adding a rectangle and a triangle and subtracting a semicircle. Coordinate axes are chosen with the origin at the lower left corner of the plate. The area and the coordinates of the centroid of each plate component are determined and entered in the table below. The area of the semicircle is indicated as negative, since it is to be subtracted from the other areas. The moments of the component areas with respect to the coordinate axes are then computed and entered in the table. Note that all dimensions have been converted into *centimeters* (cm) in order to avoid excessively large numerical values in the computation of the areas and moments of areas (1 cm = 10 mm).

Component	A, cm^2	$\bar{x}$, cm	$\bar{y}$, cm	$\bar{x}A$, cm^3	$\bar{y}A$, cm^3
Rectangle	+108	4.50	12	+486	+1296
Triangle	+27	3.00	4	+81	+108
Semicircle	−25.1	7.30	12	−183	−301
	$\Sigma A = 109.9$	. . .	. . .	$\Sigma \bar{x}A = 384$	$\Sigma \bar{y}A = 1103$

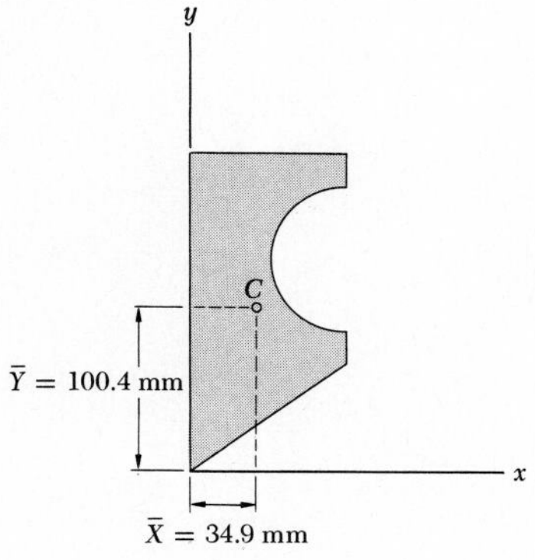

Substituting the values obtained from the table into the equations defining the centroid of a composite area, we obtain

$\bar{X}\Sigma A = \Sigma \bar{x}A$: $\quad \bar{X}(109.9\text{ cm}^2) = 384\text{ cm}^3$

$\bar{X} = 3.49$ cm $\qquad \bar{X} = 34.9$ mm ◄

$\bar{Y}\Sigma A = \Sigma \bar{y}A$: $\quad \bar{Y}(109.9\text{ cm}^2) = 1103\text{ cm}^3$

$\bar{Y} = 10.04$ cm $\qquad \bar{Y} = 100.4$ mm ◄

The above values of $\bar{X}$ and $\bar{Y}$ define the centroid of the area and also the center of gravity of the plate. The center of gravity, of course, is actually located halfway between the upper and lower faces of the plate.

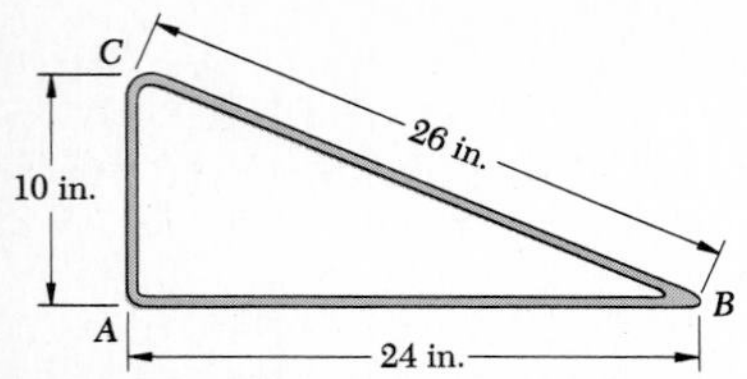

SAMPLE PROBLEM 5.2

The figure shown is made of a thin homogeneous wire. Determine its center of gravity.

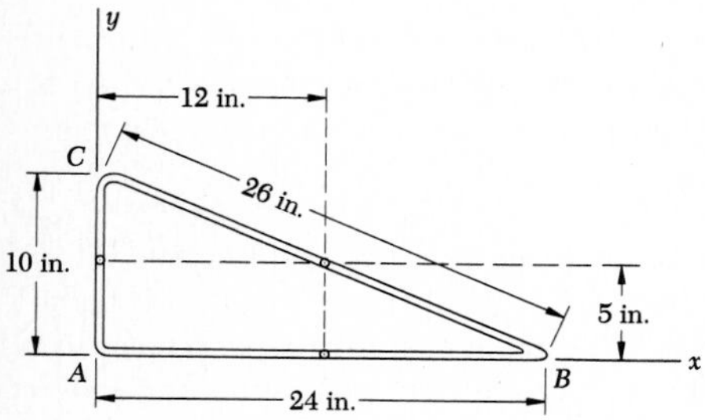

Solution. Since the figure is formed of homogeneous wire, its center of gravity may be located by determining the centroid of the corresponding line. Choosing the coordinate axes shown, with origin at A, we determine the coordinates of the centroid of each segment of line and compute its moments with respect to the coordinate axes.

Segment	L, in.	$\bar{x}$, in.	$\bar{y}$, in.	$\bar{x}L$, in^2	$\bar{y}L$, in^2
AB	24	12	0	288	0
BC	26	12	5	312	130
CA	10	0	5	0	50
	$\Sigma L = 60$	. . .	. . .	$\Sigma\bar{x}L = 600$	$\Sigma\bar{y}L = 180$

Substituting the values obtained from the table into the equations defining the centroid of a composite line, we obtain

$\bar{X}\Sigma L = \Sigma\bar{x}L$: $\quad \bar{X}(60 \text{ in.}) = 600 \text{ in}^2 \qquad \bar{X} = 10 \text{ in.}$ ◄

$\bar{Y}\Sigma L = \Sigma\bar{y}L$: $\quad \bar{Y}(60 \text{ in.}) = 180 \text{ in}^2 \qquad \bar{Y} = 3 \text{ in.}$ ◄

SAMPLE PROBLEM 5.3

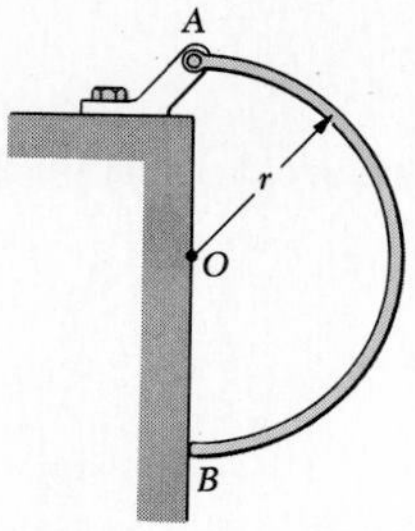

A uniform semicircular rod of weight W and radius r is attached to a pin at A and bears against a frictionless surface at B. Determine the reactions at A and B.

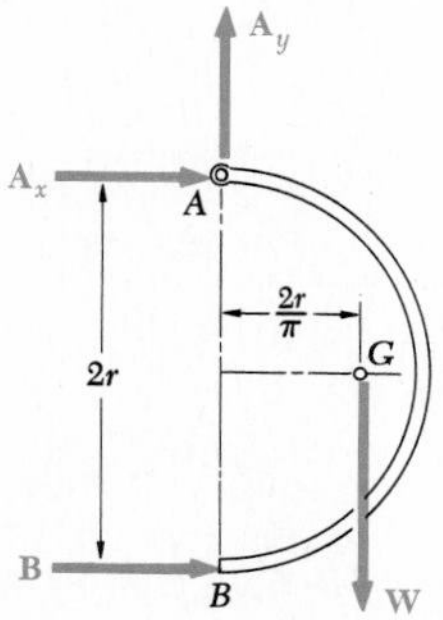

Solution. A free-body diagram of the rod is drawn. The forces acting on the rod consist of its weight $\mathbf{W}$ applied at the center of gravity G, whose position is obtained from Fig. 5.8B, of a reaction at A represented by its components $\mathbf{A}_x$ and $\mathbf{A}_y$, and of a horizontal reaction at B.

$$+\circlearrowleft \Sigma M_A = 0: \qquad B(2r) - W\left(\frac{2r}{\pi}\right) = 0$$

$$B = +\frac{W}{\pi} \qquad\qquad \mathbf{B} = \frac{W}{\pi} \rightarrow \blacktriangleleft$$

$$\xrightarrow{+} \Sigma F_x = 0: \qquad A_x + B = 0$$

$$A_x = -B = -\frac{W}{\pi} \qquad \mathbf{A}_x = \frac{W}{\pi} \leftarrow$$

$$+\uparrow \Sigma F_y = 0: \qquad A_y - W = 0 \qquad \mathbf{A}_y = W\uparrow$$

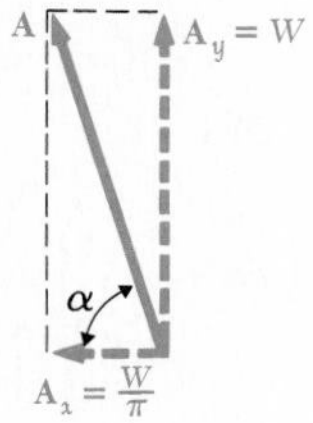

Adding the two components of the reaction at A:

$$A = \left[W^2 + \left(\frac{W}{\pi}\right)^2\right]^{1/2} \qquad\qquad A = W\left(1 + \frac{1}{\pi^2}\right)^{1/2} \blacktriangleleft$$

$$\tan \alpha = \frac{W}{W/\pi} = \pi \qquad\qquad \alpha = \tan^{-1} \pi \blacktriangleleft$$

The answers may also be expressed as follows:

$$\mathbf{A} = 1.049W \measuredangle 72.3° \qquad \mathbf{B} = 0.318W \rightarrow \blacktriangleleft$$

PROBLEMS

5.1 through 5.12 Locate the centroid of the plane area shown.

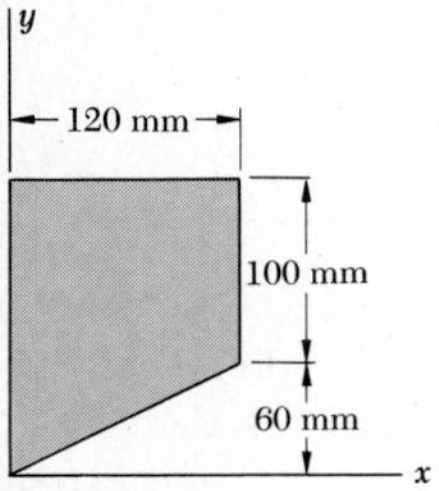

Fig. P5.1

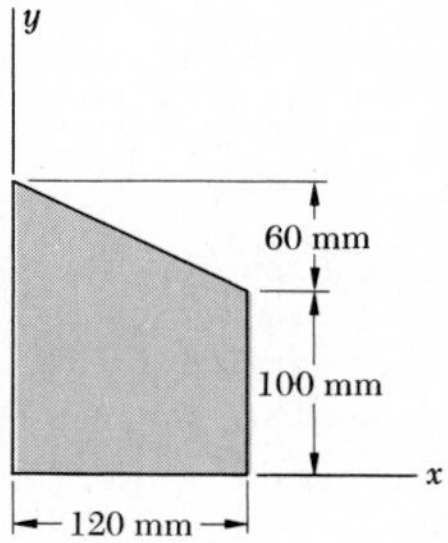

Fig. P5.2

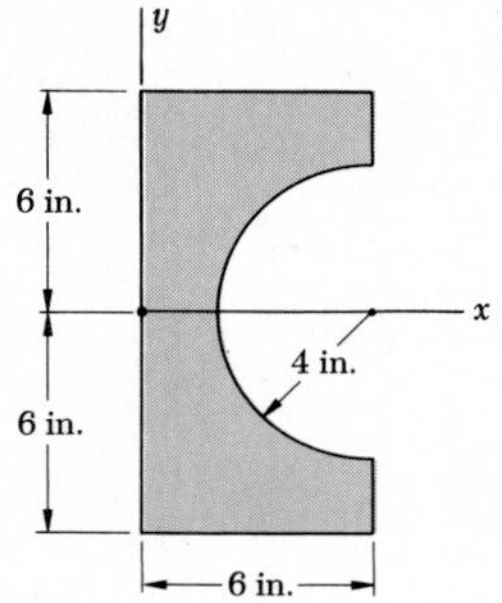

Fig. P5.3

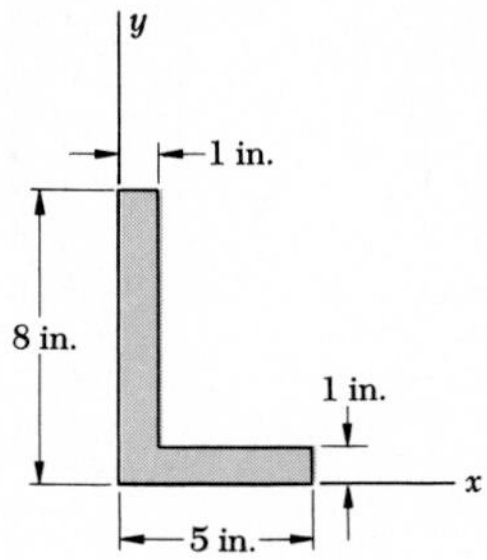

Fig. P5.4

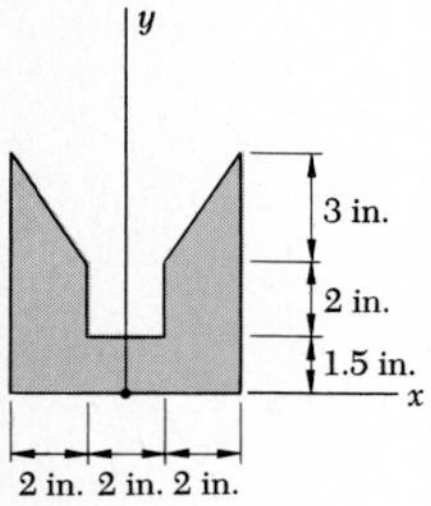

Fig. P5.5

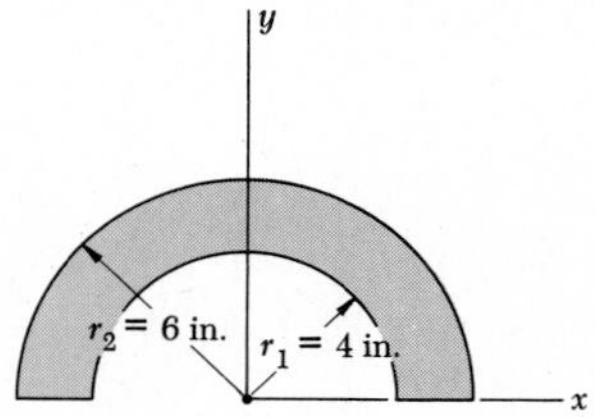

Fig. P5.6

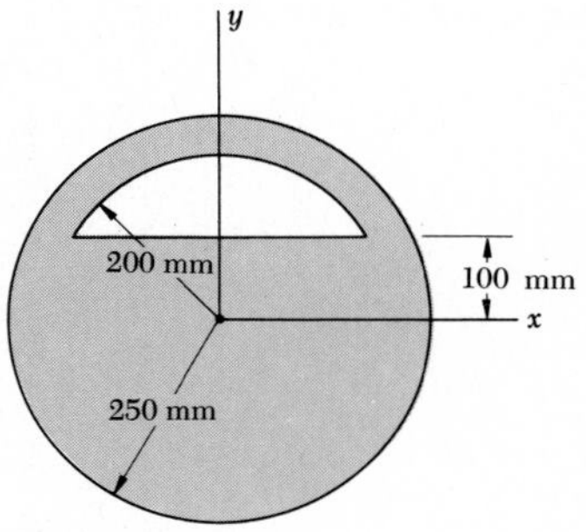

Fig. P5.7

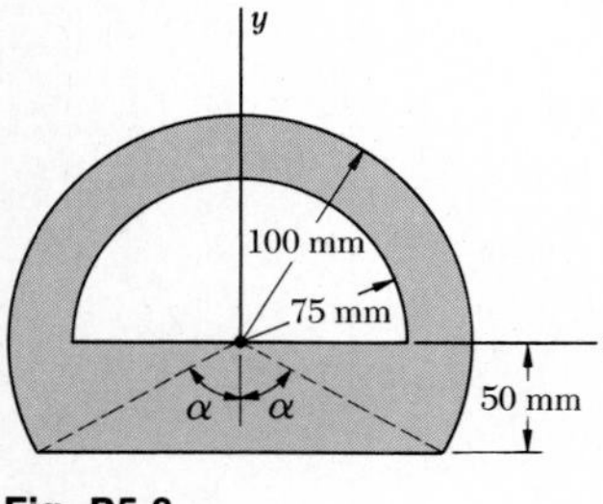

Fig. P5.8

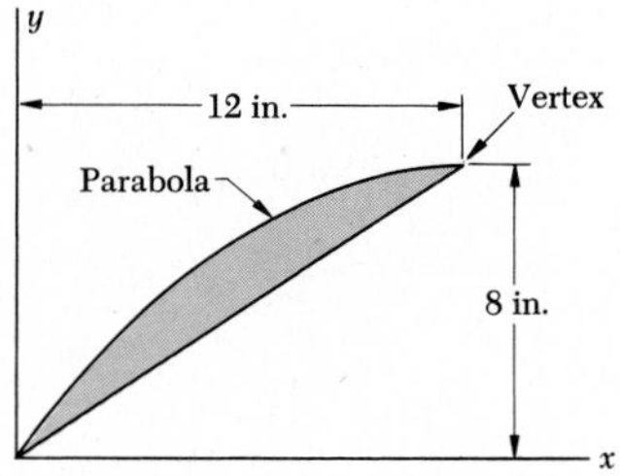

Fig. P5.9

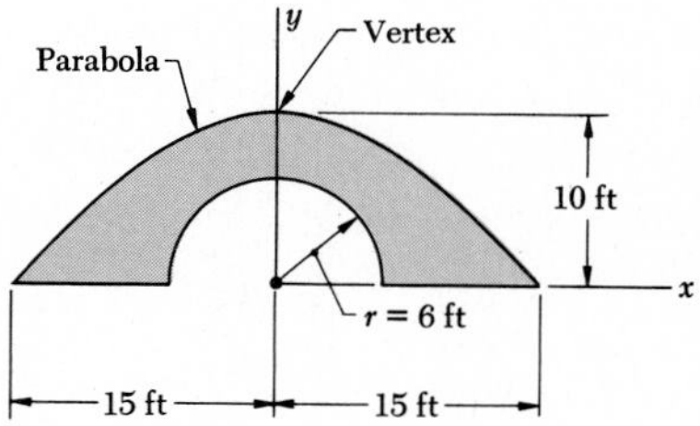

Fig. P5.10

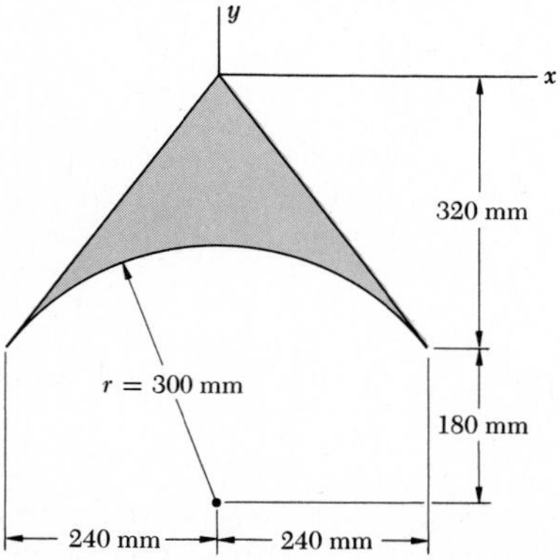

Fig. P5.11

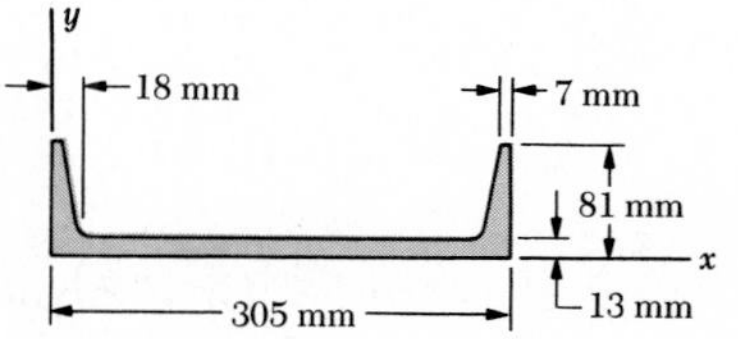

Fig. P5.12

5.13 Determine the abscissa of the centroid of the circular segment in terms of r and α.

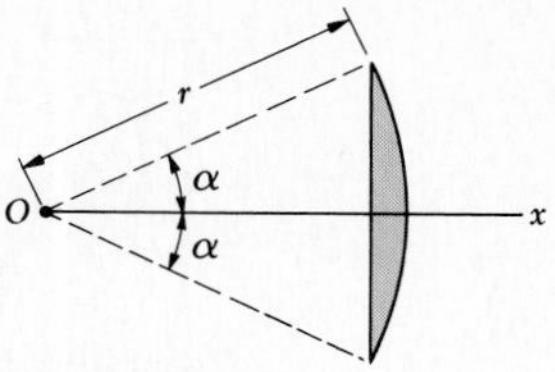

Fig. P5.13

5.14 Determine the abscissa of the centroid of the shaded area in terms of a, b, and α.

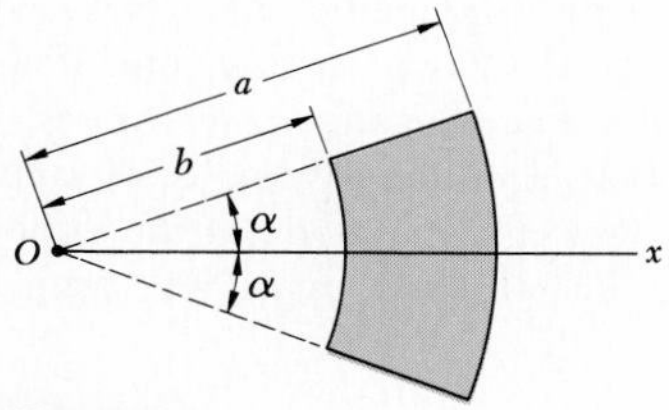

Fig. P5.14

5.15 through 5.18 A thin homogeneous wire is bent to form the *perimeter* of the figure indicated. Locate the center of gravity of the wire figure thus formed.

5.15 Fig. P5.1.
5.16 Fig. P5.2.
5.17 Fig. P5.3.
5.18 Fig. P5.6.

5.19 For the semiannular area of Prob. 5.6, determine the ratio of r_1 to r_2 for which the centroid of the area is located at the point of intersection of the inner circle and the y axis.

5.20 For the semiannular area of Prob. 5.6, determine the ratio of r_1 to r_2 for which the centroid of the area is located at $x = 0$ and $y = \frac{1}{2}r_2$.

$\ast$**5.21** Locate the centroid C in Prob. 5.6 in terms of r_1 and r_2 and show that, as r_1 approaches r_2, the location of C approaches that for a semicircular arc of radius $\frac{1}{2}(r_1 + r_2)$.

5.22 The figure shown is formed of a thin homogeneous wire. Determine the length a of the straight portion and the angle α for which the center of gravity of the entire figure is located at the origin O.

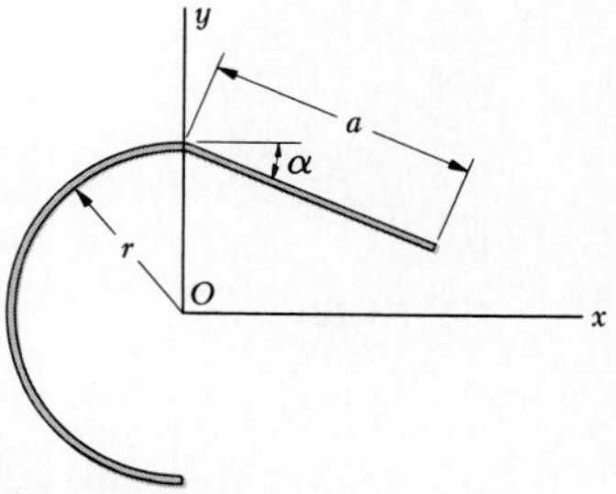

Fig. P5.22

5.23 The homogeneous wire ABC is bent as shown and is attached to a hinge at C. Determine the length L for which portion AB of the wire is horizontal.

5.24 The homogeneous wire ABC is bent as shown and is attached to a hinge at C. Determine the length L for which portion BC of the wire is horizontal.

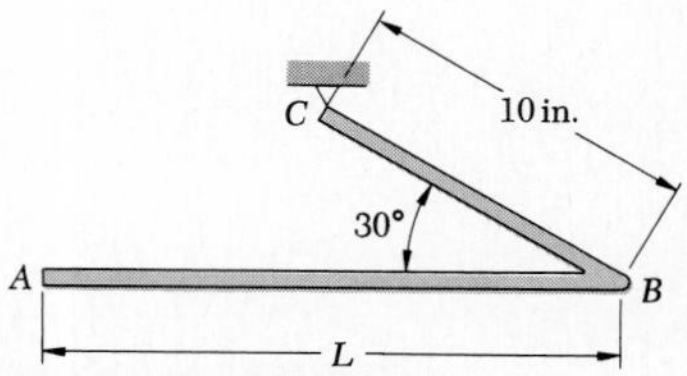

Fig. P5.23 and P5.24

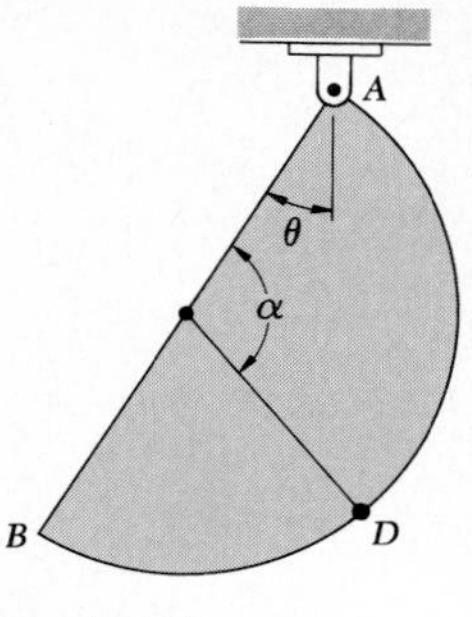

Fig. P5.25

5.25 A uniform semicircular disk of weight W and radius r is suspended from a hinge at A. A weight W of negligible dimensions is attached to the rim of the disk at D. Determine the value of θ when (a) $\alpha = 180°$, (b) $\alpha = 90°$.

5.26 The production-line balancing of an automobile speedometer cup is done as follows. The unbalanced cup is placed on a frictionless shaft at O and is allowed to come to rest. A hole is then punched at A; the cup rotates through an angle θ and again comes to rest. The balancing is completed by punching additional holes at B and C. Knowing that the three holes are equal in size and are at the same distance from the shaft O, determine the required angle α in terms of the angle θ.

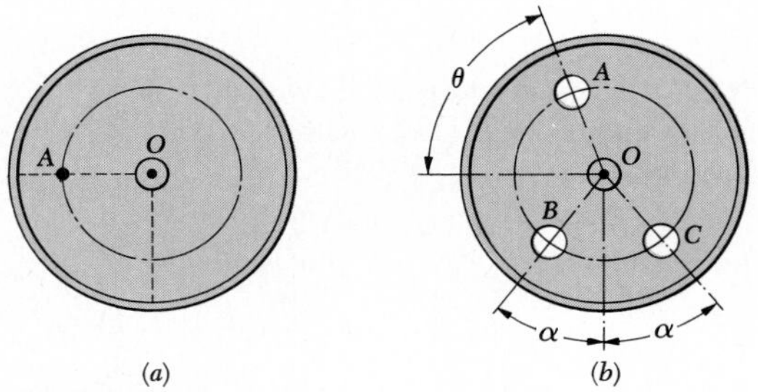

Fig. P5.26

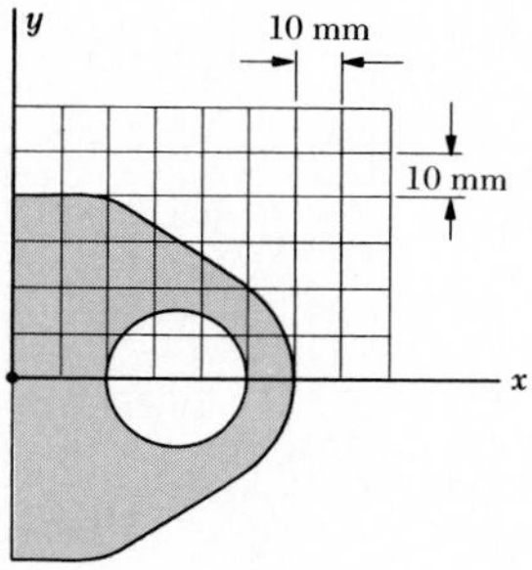

Fig. P5.28

5.27 In Prob. 5.25 determine the angle α for which the angle θ is maximum. Also determine the corresponding maximum value of θ.

5.28 A plate of uniform thickness is cut as shown. Locate by approximate means the center of gravity of the plate.

5.29 Divide the parabolic spandrel shown into five vertical sections and determine by approximate means the x coordinate of its centroid; approximate the spandrel by rectangles of the form $bcc'b'$. What is the percentage error in the answer obtained? (See Fig. 5.8A for exact answer.)

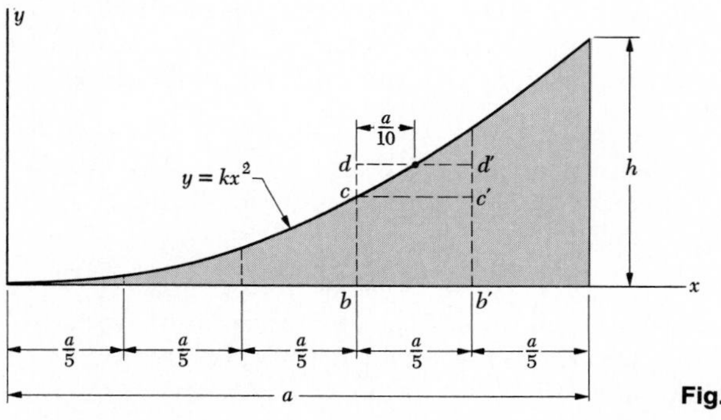

Fig. P5.29

5.30 Solve Prob. 5.29, using rectangles of the form $bdd'b'$.

5.31 Determine by approximate means the x coordinate of the centroid of the area shown.

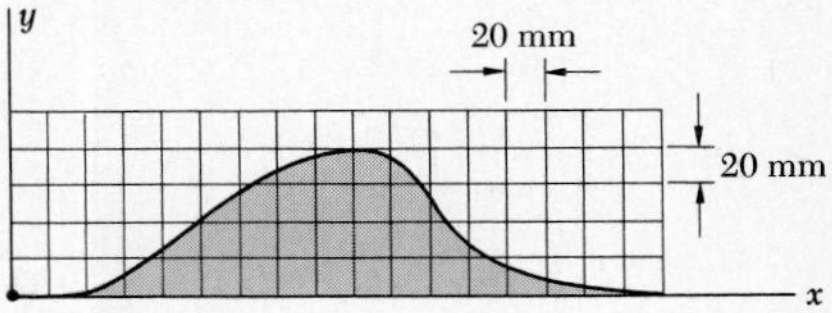

Fig. P5.31

5.4. Determination of Centroids by Integration.

The centroid of an area bounded by analytical curves (i.e., curves defined by algebraic equations) is usually determined by computing the integrals in Eqs. (5.5) of Sec. 5.2.

$$\bar{x}A = \int x\,dA \qquad \bar{y}A = \int y\,dA \tag{5.5}$$

If the element of area dA is chosen equal to a small square of sides dx and dy, the determination of each of these integrals requires a *double integration* in x and y. A double integration is also necessary if polar coordinates are used and if dA is chosen equal to a small square of sides dr and $r\,d\theta$.

In most cases, however, it is possible to determine the coordinates of the centroid of an area by performing a single integration. This is achieved by choosing for dA a thin rectangle or strip, or a thin sector or pie-shaped element (Fig. 5.12). The coordinates of the centroid of the area under consideration are then obtained by expressing that the first moment of the entire area with respect to each of the coordinate axes is equal to the sum (or integral) of the corresponding moments of the elements

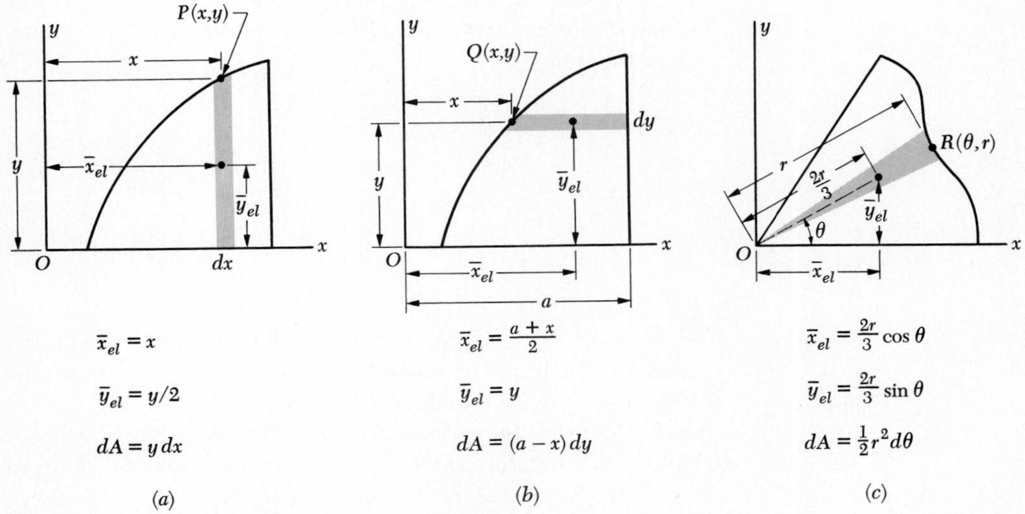

Fig. 5.12 Centroids and areas of differential elements.

of area. Denoting by $\bar{x}_{el}$ and $\bar{y}_{el}$ the coordinates of the centroid of the element dA, we write

$$\begin{aligned} \Sigma M_y: \quad & \bar{x}A = \int \bar{x}_{el}\, dA \\ \Sigma M_x: \quad & \bar{y}A = \int \bar{y}_{el}\, dA \end{aligned} \tag{5.9}$$

If the area itself is not already known, it may also be computed from these elements.

The coordinates $\bar{x}_{el}$ and $\bar{y}_{el}$ of the centroid of the element of area should be expressed in terms of the coordinates of a point located on the curve bounding the area under consideration. Also, the element of area dA should be expressed in terms of the coordinates of the point and their differentials. This has been done in Fig. 5.12 for three common types of elements; the pie-shaped element of part *c* should be used when the equation of the curve bounding the area is given in polar coordinates. The appropriate expressions should be substituted in formulas (5.9), and the equation of the curve should be used to express one of the coordinates in terms of the other. The integration is thus reduced to a single integration which may be performed according to the usual rules of calculus.

The centroid of a line defined by an algebraic equation may be determined by computing the integrals in Eqs. (5.6) of Sec. 5.2.

$$\bar{x}L = \int x\, dL \qquad \bar{y}L = \int y\, dL \tag{5.6}$$

The element dL should be replaced by one of the following expressions, depending upon the type of equation used to define the line (these expressions may be derived by using the Pythagorean theorem).

$$dL = \sqrt{1 + \left(\frac{dy}{dx}\right)^2}\, dx$$

$$dL = \sqrt{1 + \left(\frac{dx}{dy}\right)^2}\, dy$$

$$dL = \sqrt{r^2 + \left(\frac{dr}{d\theta}\right)^2}\, d\theta$$

The equation of the line is then used to express one of the coordinates in terms of the other, and the integration may be performed by the methods of calculus.

5.5. Theorems of Pappus-Guldinus. These theorems, which were first formulated by the Greek geometer Pappus during the third century A.D. and later restated by the Swiss mathematician Guldinus, or Guldin (1577–1643), deal with surfaces and bodies of revolution.

A *surface of revolution* is a surface which may be generated by rotating a plane curve about a fixed axis. For example (Fig. 5.13), the surface of a sphere may be obtained by rotating a

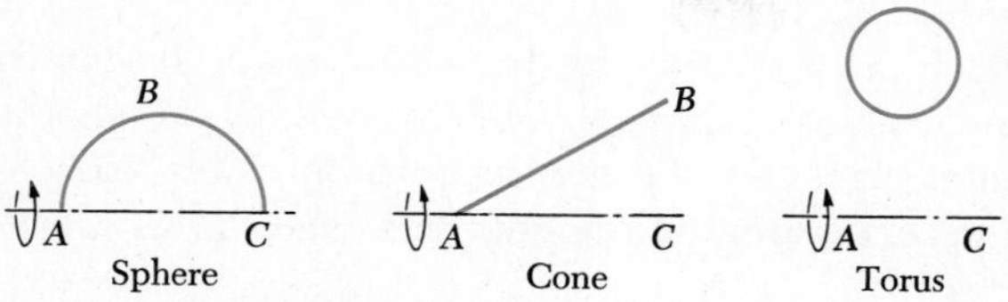

Fig. 5.13 Generating a surface of revolution.

semicircular arc ABC about the diameter AC; the surface of a cone by rotating a straight line AB about an axis AC; the surface of a torus or ring by rotating the circumference of a circle about a nonintersecting axis. A *body of revolution* is a body which may be generated by rotating a plane area about a fixed axis. A solid sphere may be obtained by rotating a semicircular area, a cone by rotating a triangular area, and a solid torus by rotating a full circular area (Fig. 5.14).

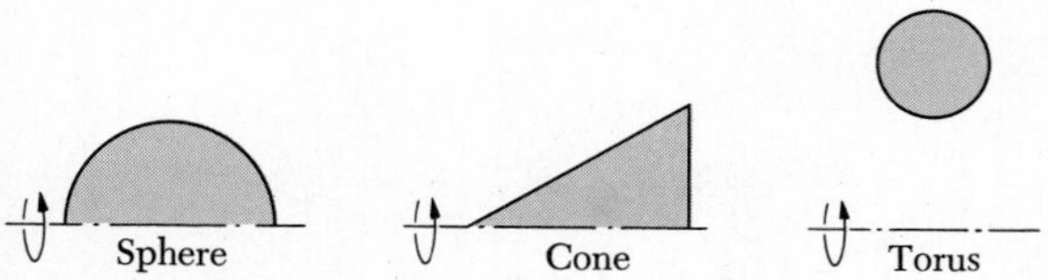

Fig. 5.14 Generating a body of revolution.

THEOREM I. *The area of a surface of revolution is equal to the length of the generating curve times the distance traveled by the centroid of the curve while the surface is being generated.*

Proof. Consider an element dL of the line L (Fig. 5.15) which is revolved about the x axis. The area dA generated by the element dL is equal to $2\pi y\, dL$. Thus, the entire area generated

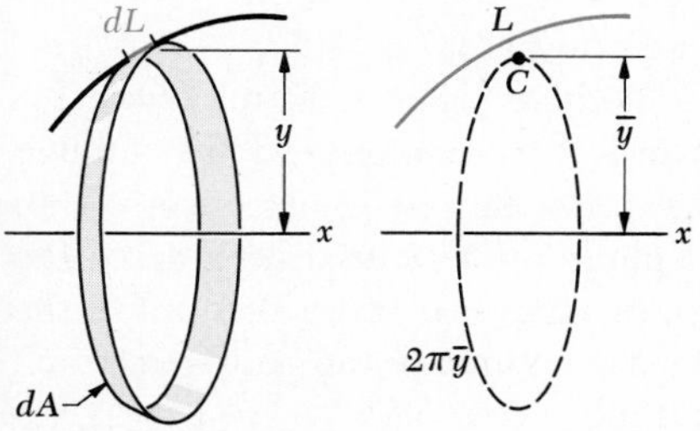

Fig. 5.15

by L is $A = \int 2\pi y\, dL$. But we saw in Sec. 5.2 that the integral $\int y\, dL$ is equal to $\bar{y}L$. We have therefore

$$A = 2\pi\bar{y}L \tag{5.10}$$

where $2\pi\bar{y}$ is the distance traveled by the centroid of L. It should be noted that the generating curve should not cross the axis about which it is rotated; if it did, the two sections on either side of the axis would generate areas of opposite signs and the theorem would not apply.

THEOREM II. *The volume of a body of revolution is equal to the generating area times the distance traveled by the centroid of the area while the body is being generated.*

Proof. Consider an element dA of the area A which is revolved about the x axis (Fig. 5.16). The volume dV generated by the element dA is equal to $2\pi y\, dA$. Thus, the entire volume

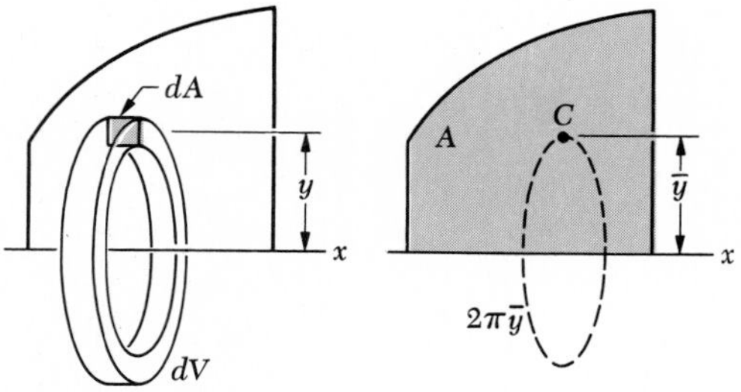

Fig. 5.16

generated by A is $V = \int 2\pi y\, dA$. But since the integral $\int y\, dA$ is equal to $\bar{y}A$ (Sec. 5.2), we have

$$V = 2\pi\bar{y}A \tag{5.11}$$

where $2\pi\bar{y}$ is the distance traveled by the centroid of A. Again, it should be noted that the theorem does not apply if the axis of rotation intersects the generating area.

The theorems of Pappus-Guldinus offer a simple way for computing the area of surfaces of revolution and the volume of bodies of revolution. They may also be used conversely to determine the centroid of a plane curve when the area of the surface generated by the curve is known or to determine the centroid of a plane area when the volume of the body generated by the area is known (see Sample Prob. 5.8).

SAMPLE PROBLEM 5.4

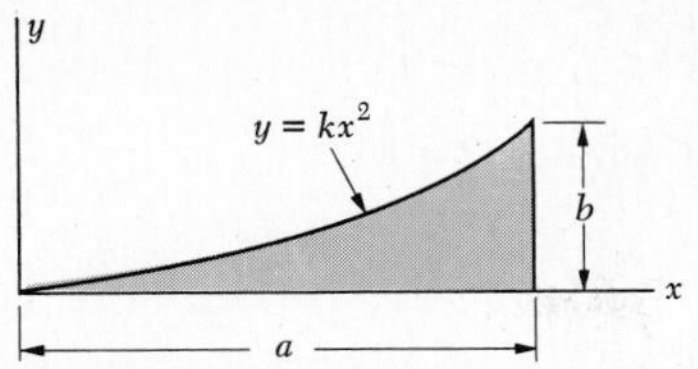

Determine by direct integration the centroid of a parabolic spandrel.

Solution. The value of k is determined by substituting $x = a$ and $y = b$ in the given equation. We have $b = ka^2$ and $k = b/a^2$. The equation of the curve is thus

$$y = \frac{b}{a^2}x^2 \qquad \text{or} \qquad x = \frac{a}{b^{1/2}}y^{1/2}$$

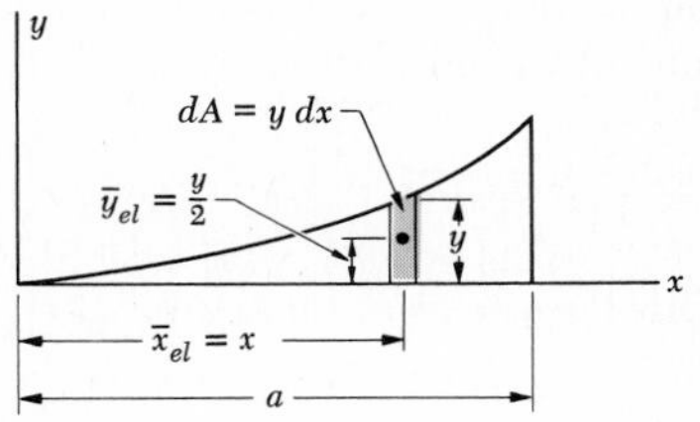

Vertical Differential Element. We choose the differential element shown and find the total area of the figure.

$$A = \int dA = \int y\,dx = \int_0^a \frac{b}{a^2}x^2\,dx = \left[\frac{b}{a^2}\frac{x^3}{3}\right]_0^a = \frac{ab}{3}$$

The moment of the differential element with respect to the y axis is $\bar{x}_{el}\,dA$; hence, the moment of the entire area with respect to this axis is

$$\int \bar{x}_{el}\,dA = \int xy\,dx = \int_0^a x\left(\frac{b}{a^2}x^2\right)dx = \left[\frac{b}{a^2}\frac{x^4}{4}\right]_0^a = \frac{a^2b}{4}$$

Thus,

$$\bar{x}A = \int \bar{x}_{el}\,dA \qquad \bar{x}\frac{ab}{3} = \frac{a^2b}{4} \qquad \bar{x} = \tfrac{3}{4}a \blacktriangleleft$$

Likewise, the moment of the differential element with respect to the x axis is $\bar{y}_{el}\,dA$, and the moment of the entire area is

$$\int \bar{y}_{el}\,dA = \int \frac{y}{2}y\,dx = \int_0^a \frac{1}{2}\left(\frac{b}{a^2}x^2\right)^2 dx = \left[\frac{b^2}{2a^4}\frac{x^5}{5}\right]_0^a = \frac{ab^2}{10}$$

Thus,

$$\bar{y}A = \int \bar{y}_{el}\,dA \qquad \bar{y}\frac{ab}{3} = \frac{ab^2}{10} \qquad \bar{y} = \tfrac{3}{10}b \blacktriangleleft$$

Horizontal Differential Element. The same result may be obtained by considering a horizontal element. The moments of the area are

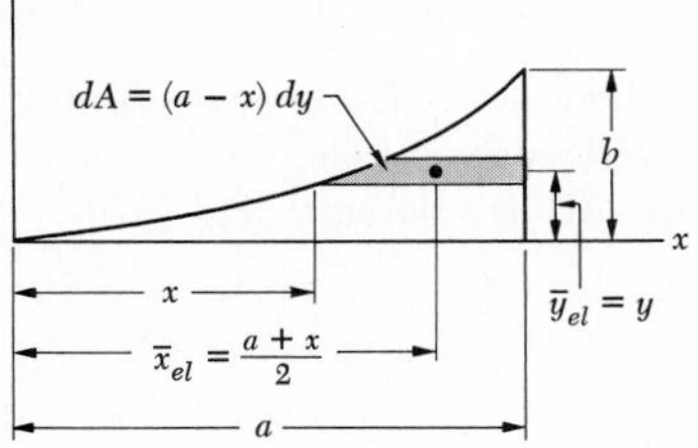

$$\int \bar{x}_{el}\,dA = \int \frac{a+x}{2}(a-x)\,dy = \int_0^b \frac{a^2-x^2}{2}\,dy$$

$$= \frac{1}{2}\int_0^b \left(a^2 - \frac{a^2}{b}y\right)dy = \frac{a^2b}{4}$$

$$\int \bar{y}_{el}\,dA = \int y(a-x)\,dy = \int y\left(a - \frac{a}{b^{1/2}}y^{1/2}\right)dy$$

$$= \int_0^b \left(ay - \frac{a}{b^{1/2}}y^{3/2}\right)dy = \frac{ab^2}{10}$$

These moments are again substituted in the equations defining the centroid of the area to obtain $\bar{x}$ and $\bar{y}$.

SAMPLE PROBLEM 5.5

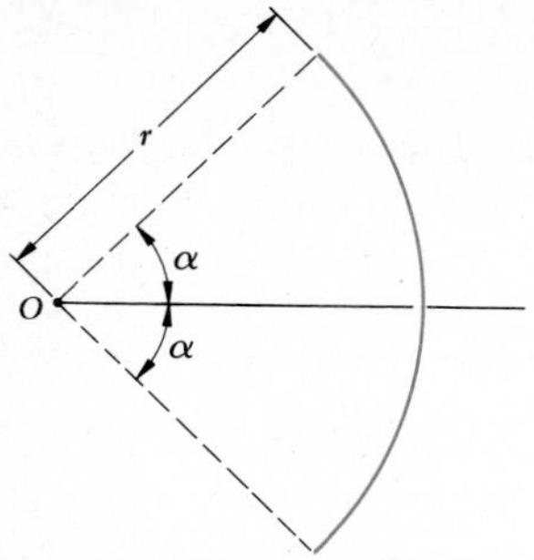

Determine the centroid of the arc of circle shown.

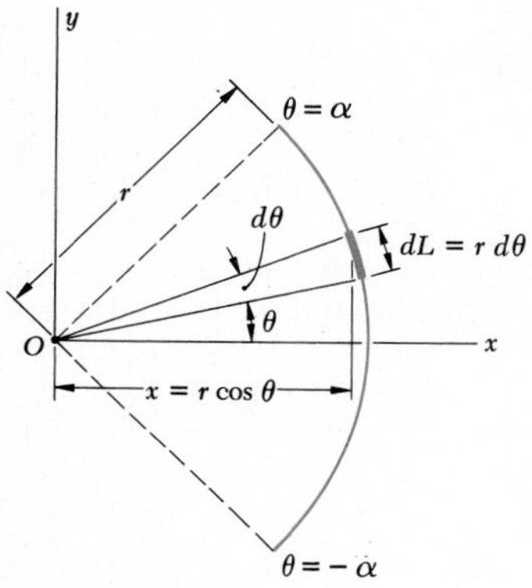

Solution. Since the arc is symmetrical with respect to the x axis, we have $\bar{y} = 0$. A differential element is chosen as shown, and the length of the arc is determined by integration.

$$L = \int dL = \int_{-\alpha}^{\alpha} r\,d\theta = r \int_{-\alpha}^{\alpha} d\theta = 2r\alpha$$

The moment of the arc with respect to the y axis is

$$\int x\,dL = \int_{-\alpha}^{\alpha} (r\cos\theta)(r\,d\theta) = r^2 \int_{-\alpha}^{\alpha} \cos\theta\,d\theta$$

$$= r^2 \Big[\sin\theta\Big]_{-\alpha}^{\alpha} = 2r^2 \sin\alpha$$

Thus, $\quad \bar{x}L = \int x\,dL \qquad \bar{x}(2r\alpha) = 2r^2\sin\alpha \qquad \bar{x} = \dfrac{r\sin\alpha}{\alpha}$ ◄

SAMPLE PROBLEM 5.6

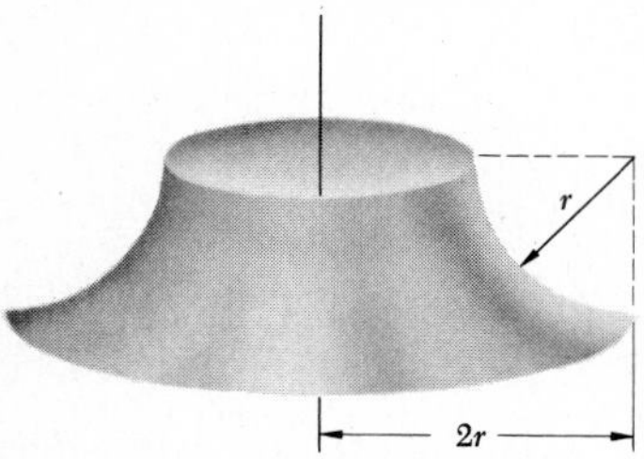

Determine the area of the surface of revolution shown, obtained by rotating a quarter-circular arc about a vertical axis.

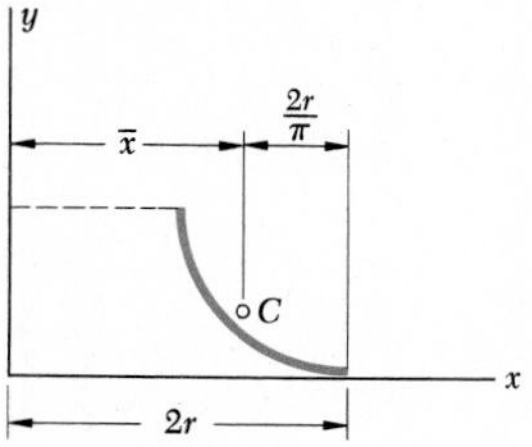

Solution. According to Theorem I of Pappus-Guldinus, the area generated is equal to the product of the length of the arc and the distance traveled by its centroid. Referring to Fig. 5.8B, we have

$$\bar{x} = 2r - \frac{2r}{\pi} = 2r\left(1 - \frac{1}{\pi}\right)$$

$$A = 2\pi\bar{x}L = 2\pi 2r\left(1 - \frac{1}{\pi}\right)\frac{\pi r}{2}$$

$$A = 2\pi r^2(\pi - 1)$$ ◄

SAMPLE PROBLEM 5.7

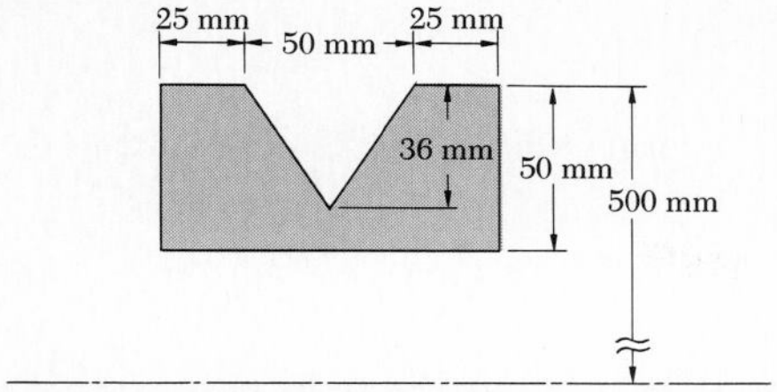

The outside diameter of a V-belt pulley is 1 m, and the cross section of its rim is as shown. Determine the mass and weight of the rim, knowing that the pulley is made of steel and that the density of steel is $\rho = 7.85 \times 10^3 \text{ kg/m}^3$.

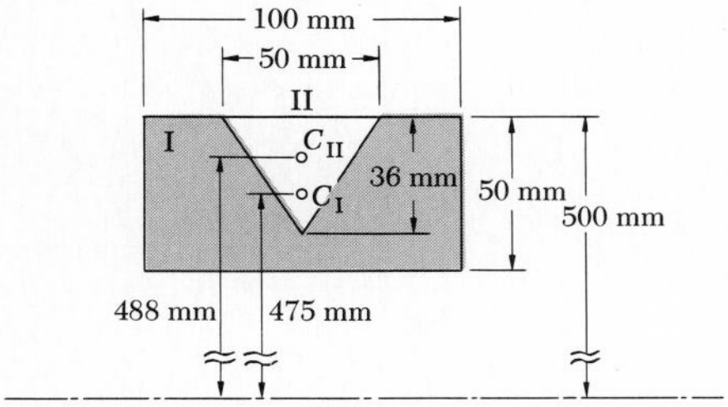

Solution. The volume of the rim may be found by applying Theorem II of Pappus-Guldinus, which states that the volume equals the product of the given cross-sectional area and of the distance traveled by its centroid in one complete revolution. However, the volume will be more easily obtained if we observe that the cross section consists of a rectangle (I) with a positive area and a triangle (II) with a negative area.

Component	Area, mm^2	$\bar{y}$, mm	Distance Traveled by C, mm	Volume, mm^3
I	+5000	475	$2\pi(475) = 2985$	$(5000)(2985) = 14.92 \times 10^6$
II	−900	488	$2\pi(488) = 3066$	$(-900)(3066) = -2.76 \times 10^6$
				Volume of rim $= 12.16 \times 10^6$

Since $1 \text{ mm} = 10^{-3} \text{ m}$, we have $1 \text{ mm}^3 = (10^{-3} \text{ m})^3 = 10^{-9} \text{ m}^3$, and $V = 12.16 \times 10^6 \text{ mm}^3 = (12.16 \times 10^6)(10^{-9} \text{ m}^3) = 12.16 \times 10^{-3} \text{ m}^3$

$$m = \rho V = (7.85 \times 10^3 \text{ kg/m}^3)(12.16 \times 10^{-3} \text{ m}^3) \qquad m = 95.4 \text{ kg} \blacktriangleleft$$
$$W = mg = (95.4 \text{ kg})(9.81 \text{ m/s}^2) = 936 \text{ kg}\cdot\text{m/s}^2 \qquad W = 936 \text{ N} \blacktriangleleft$$

SAMPLE PROBLEM 5.8

Using the theorems of Pappus-Guldinus, determine (*a*) the centroid of a semicircular area, (*b*) the centroid of a semicircular arc. We recall that the volume of a sphere is $\frac{4}{3}\pi r^3$ and that its surface area is $4\pi r^2$.

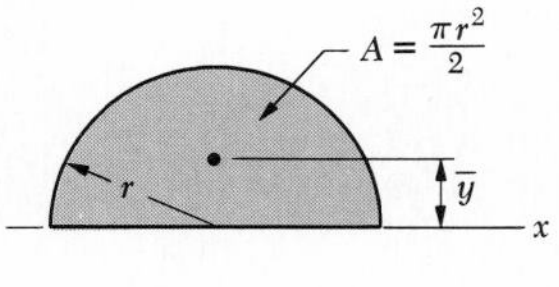

Solution. The volume of a sphere is equal to the product of the area of a semicircle and of the distance traveled by the centroid of the semicircle in one revolution about the x axis.

$$V = 2\pi\bar{y}A \qquad \tfrac{4}{3}\pi r^3 = 2\pi\bar{y}(\tfrac{1}{2}\pi r^2) \qquad \bar{y} = \frac{4r}{3\pi} \blacktriangleleft$$

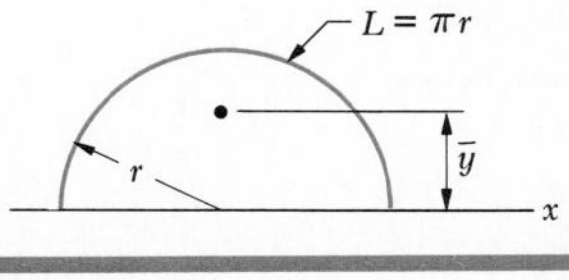

Likewise, the area of a sphere is equal to the product of the length of the generating semicircle and of the distance traveled by its centroid in one revolution.

$$A = 2\pi\bar{y}L \qquad 4\pi r^2 = 2\pi\bar{y}(\pi r) \qquad \bar{y} = \frac{2r}{\pi} \blacktriangleleft$$

PROBLEMS

5.32 through 5.35 Determine by direct integration the centroid of the area shown.

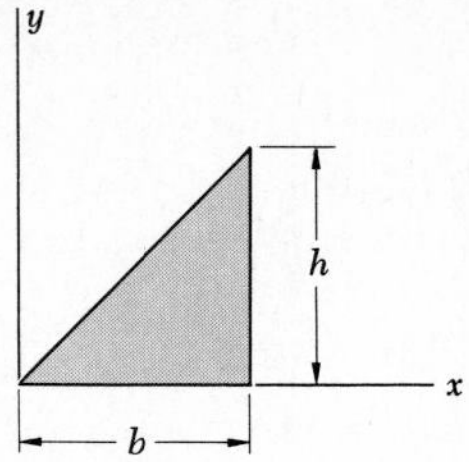

Fig. P5.32

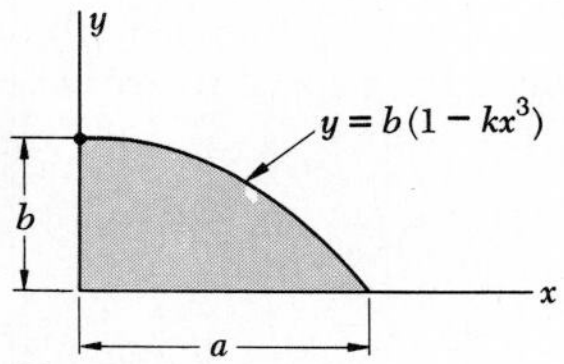

Fig. P5.33

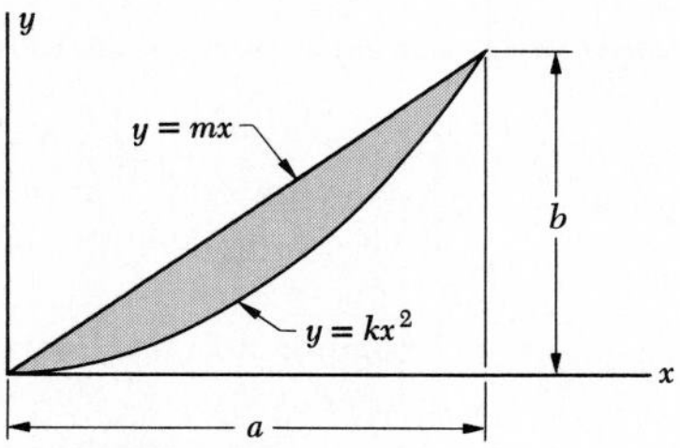

Fig. P5.34

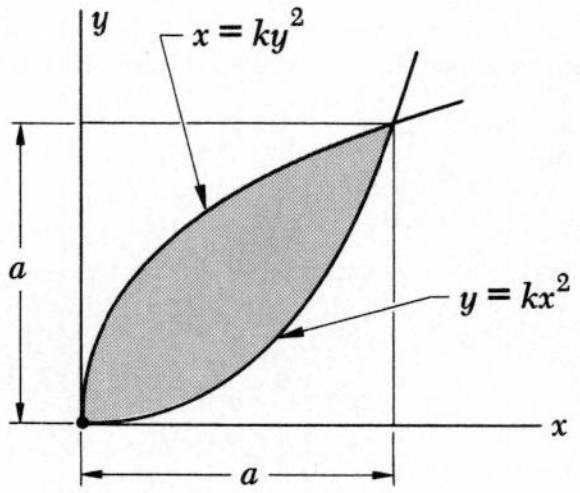

Fig. P5.35

5.36 through 5.41 Derive by direct integration the expressions for $\bar{x}$ and $\bar{y}$ given in Fig. 5.8 for:

5.36 A quarter-elliptical area.
5.37 A semicircular area.
5.38 A semiparabolic area.
5.39 A circular sector.
5.40 A quarter-circular arc.
5.41 A general spandrel ($y = kx^n$).

5.42 Determine by direct integration the x coordinate of the centroid of the area shown.

5.43 Determine by direct integration the y coordinate of the centroid of the area shown.

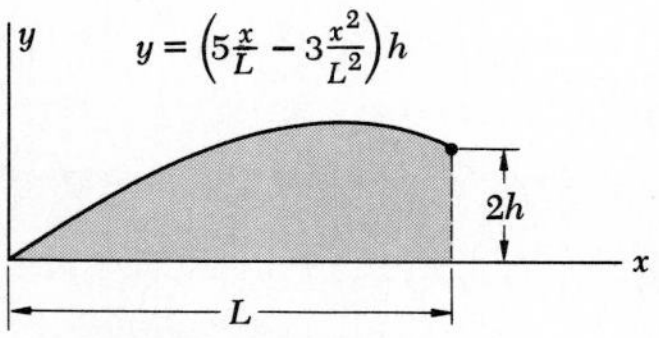

Fig. P5.42 and P5.43

***5.44** Determine by direct integration the centroid of the area shown.

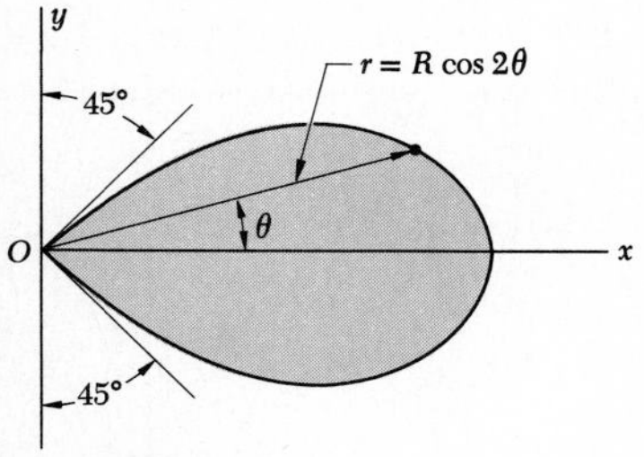

Fig. P5.44

5.45 Determine by direct integration the centroid of the area shown.

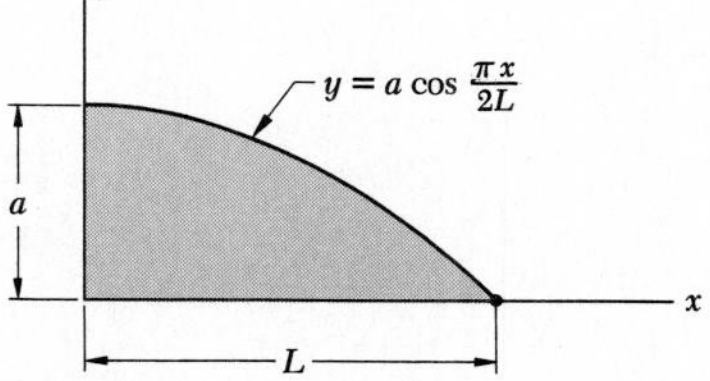

Fig. P5.45

5.46 Determine the centroid of the area shown.

5.47 Determine the centroid of the area of Prob. 5.46 in terms of a.

5.48 Determine by direct integration the centroid of the area located in the first quadrant and bounded by the curves $y = x^n$ and $x = y^n$, for $n > 1$.

5.49 Determine the volume of the solid obtained by rotating the trapezoid of Prob. 5.2 about (*a*) the x axis, (*b*) the y axis.

5.50 Determine the volume of the solid obtained by rotating the semiparabolic area shown about (*a*) the y axis, (*b*) the x axis.

5.51 Determine the surface area and the volume of the half-torus shown.

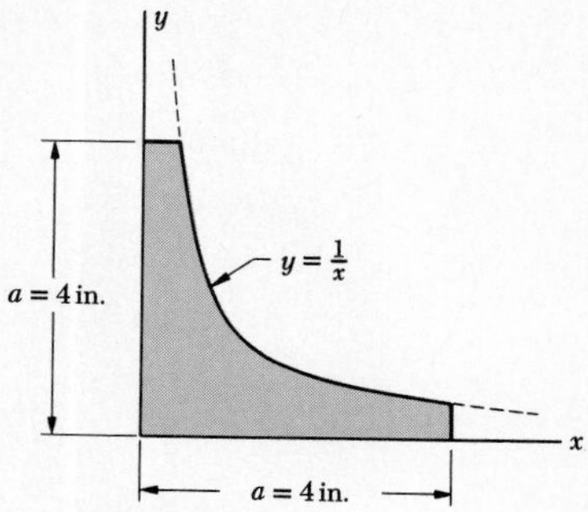

Fig. P5.46

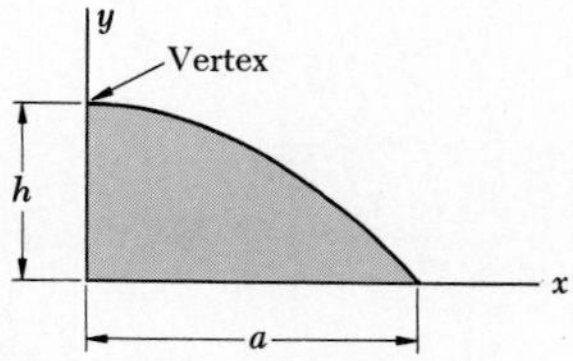

Fig. P5.50

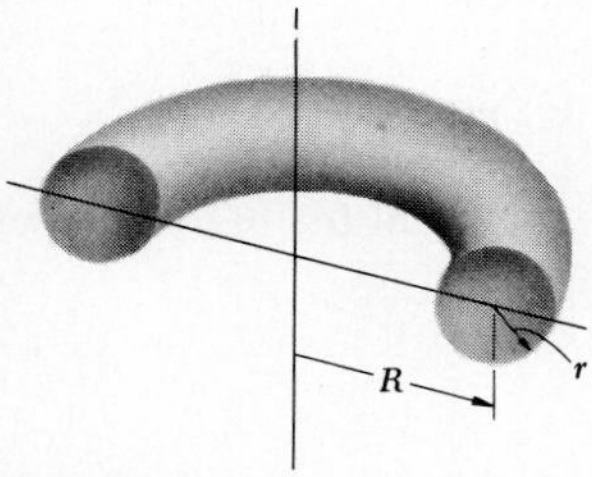

Fig. P5.51

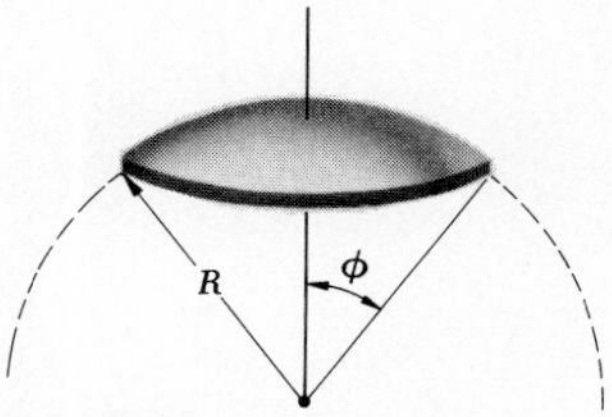

Fig. P5.52

5.52 The spherical cap shown is formed by passing a horizontal plane through a hollow sphere of radius R. Determine the area of the outside surface of the cap in terms of R and ϕ.

5.53 Determine the total surface area of the solid obtained by rotating the area of Prob. 5.3 about the y axis.

5.54 The inside diameter of a spherical tank is 2 m. What volume of liquid is required to fill the tank to a depth of 0.5 m?

5.55 The geometry of the end portion of a proposed embankment is shown. For the portion of the embankment in front of the vertical plane *ABCD*, determine the additional fill material required if the final dimensions are to be $a = 19$ ft, $b = 37$ ft, and $h = 8$ ft.

5.56 Solve Prob. 5.55, assuming that the final dimensions are (*a*) $a = 19$ ft, $b = 34$ ft, $h = 8$ ft, (*b*) $a = 16$ ft, $b = 37$ ft, $h = 8$ ft.

Fig. P5.55

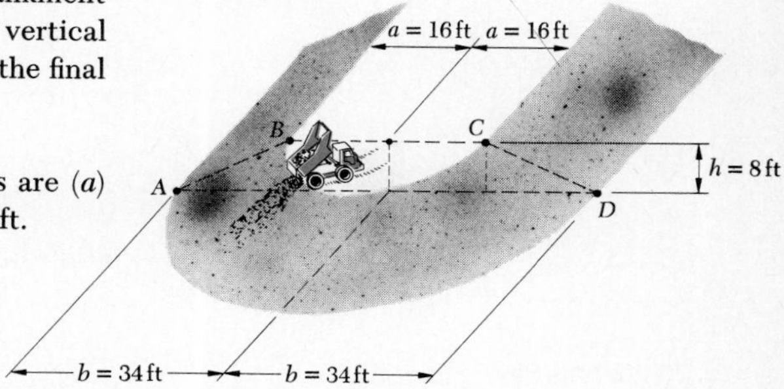

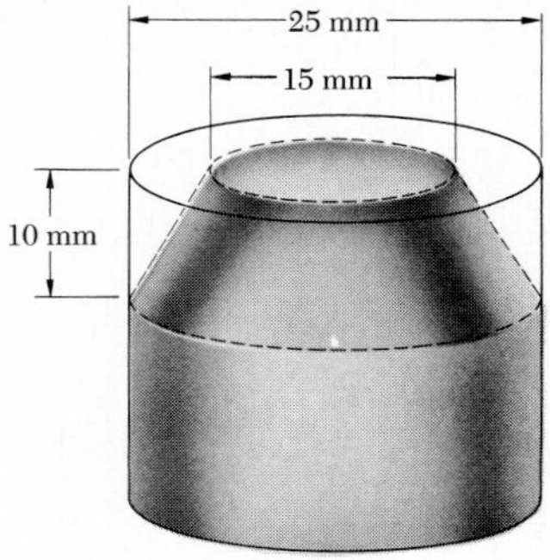

Fig. P5.57

5.57 Determine the volume of material removed as the rod is sharpened to the shape indicated by the dashed lines.

5.58 An automobile tire weighs 22 lb and has a cross-sectional area of 7 in^2. The specific weight of the rubber used is 80 lb/ft^3; determine the location of the centroid of the cross-sectional area.

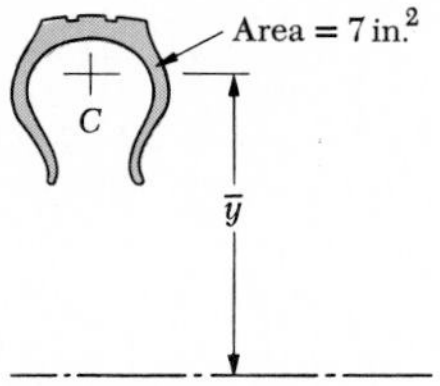

Fig. P5.58

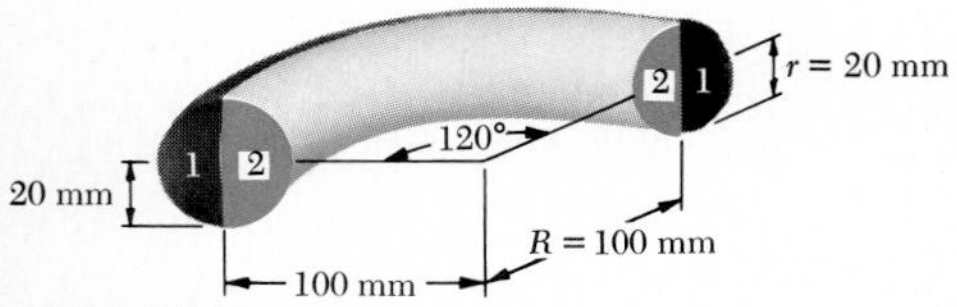

Fig. P5.59

5.59 A section of ring is cut into the two portions shown; the cross section of each portion is a semicircle. Determine the volume and the surface area, including the vertical cut, of portion 1.

5.60 Solve Prob. 5.59 for portion 2 of the ring.

5.61 Determine (*a*) the volume, (*b*) the area of the inside curved surface of the body shown.

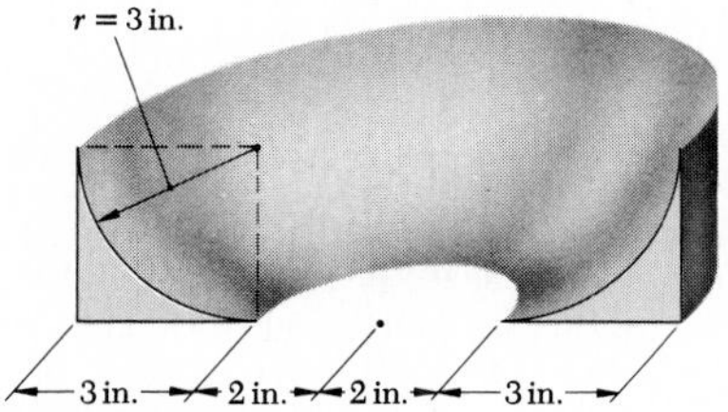

Fig. P5.61

5.62 Determine the volume of the solid of revolution formed by revolving each of the plane areas shown about its vertical edge *AB*. Show that the volumes of the solids formed are in the ratio 6:4:3:2:1.

5.63 Determine the volume of the solid formed by revolving each of the plane areas shown about its base *BC*. Show that the volumes of the solids formed are in the ratio 15:10:8:5:3.

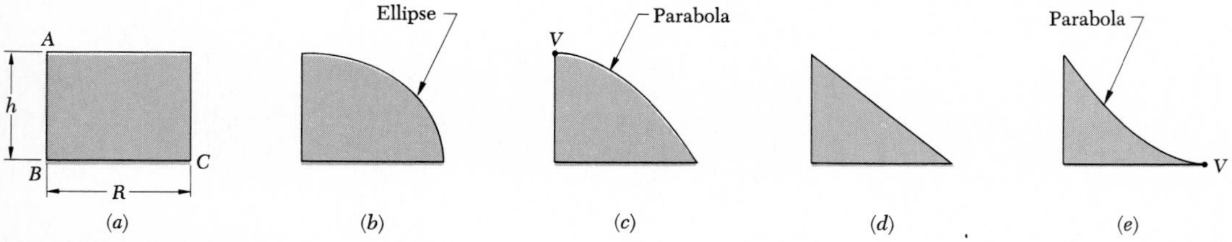

Fig. P5.62 and P5.63

***5.64** A cylindrical hole is drilled through the center of a steel ball bearing shown here in cross section. The length of the *hole* is L; show that the volume of the steel remaining is $\frac{1}{6}\pi L^3$.

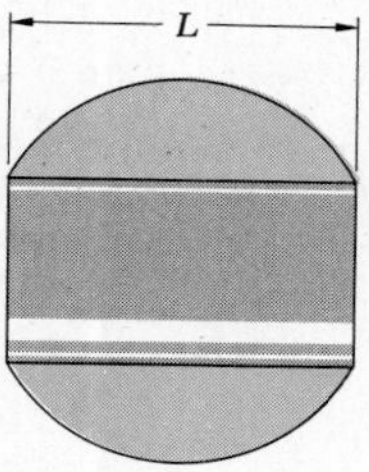

Fig. P5.64

5.65 An oil-storage tank in the form of a shell of constant strength has the cross section shown. (This is also the cross section of a drop of water on an unwetted surface.) Determine by approximate means (*a*) the volume of the tank, (*b*) the surface area of the tank.

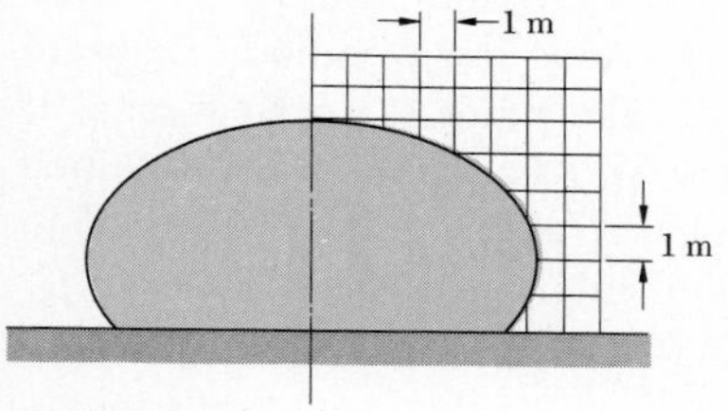

Fig. P5.65

*5.6. Distributed Loads on Beams.

The concept of centroid of an area may be used to solve other problems besides those dealing with the weight of flat plates. Consider, for example, a beam supporting a *distributed load;* this load may consist of the weight of materials supported directly or indirectly by the beam, or it may be caused by wind or hydrostatic pressure. The distributed load may be represented by plotting the load w supported per unit length (Fig. 5.17); this load will be expressed in N/m or in lb/ft. The magnitude of the force exerted on an element of beam of length dx is $dW = w\,dx$, and the total load supported by the beam is

$$W = \int_0^L w\,dx$$

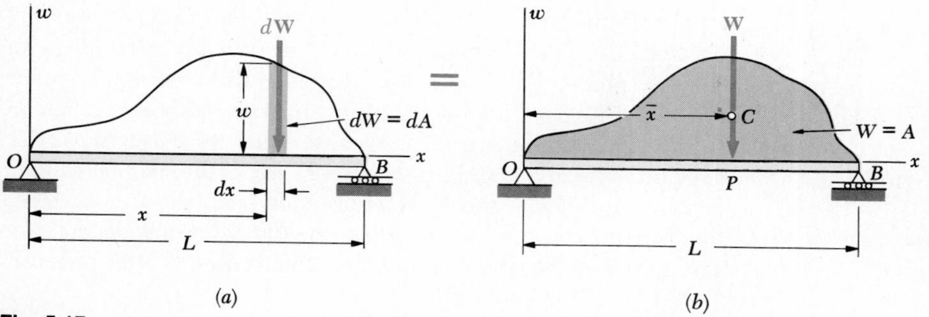

Fig. 5.17

But the product $w\,dx$ is equal in magnitude to the element of area dA shown in Fig. 5.17*a*, and W is thus equal in magnitude to the total area A under the load curve,

$$W = \int dA = A$$

We shall now determine where a *single concentrated load* **W**, of the same magnitude W as the total distributed load, should be applied on the beam if it is to produce the same reactions at the supports (Fig. 5.17*b*). This concentrated load **W**, which represents the resultant of the given distributed loading, should be equivalent to this loading as far as the free-body diagram of the entire beam is concerned. The point of application P of the equivalent concentrated load **W** will therefore be obtained by expressing that the moment of **W** about point O is equal to the sum of the moments of the elementary loads $d\mathbf{W}$ about O:

$$(OP)W = \int x\,dW$$

or, since $dW = w\,dx = dA$ and $W = A$,

$$(OP)A = \int_0^L x\,dA \tag{5.12}$$

Since the integral represents the first moment with respect to the w axis of the area under the load curve, it may be replaced by the product $\bar{x}A$. We have therefore $OP = \bar{x}$, where $\bar{x}$ is the distance from the w axis to the centroid C of the area A (this is *not* the centroid of the beam).

A distributed load on a beam may thus be replaced by a concentrated load; the magnitude of this single load is equal to the area under the load curve, and its line of action passes through the centroid of that area. It should be noted, however, that the concentrated load is equivalent to the given loading only as far as external forces are concerned. It may be used to determine reactions but should not be used to compute internal forces and deflections.

***5.7. Forces on Submerged Surfaces.** Another example of the use of first moments and centroids of areas is obtained by considering the forces exerted on a *rectangular surface* submerged in a liquid. Consider the rectangular plate shown in Fig. 5.18; it has a length L, and its width, perpendicular to the plane of the figure, is assumed equal to unity. Since the gage pressure in a liquid is $p = \gamma h$, where γ is the specific weight of the liquid and h the vertical distance from the free surface, the pressure on the plate varies linearly with the distance x. The width of the plate being taken equal to unity, the pressure

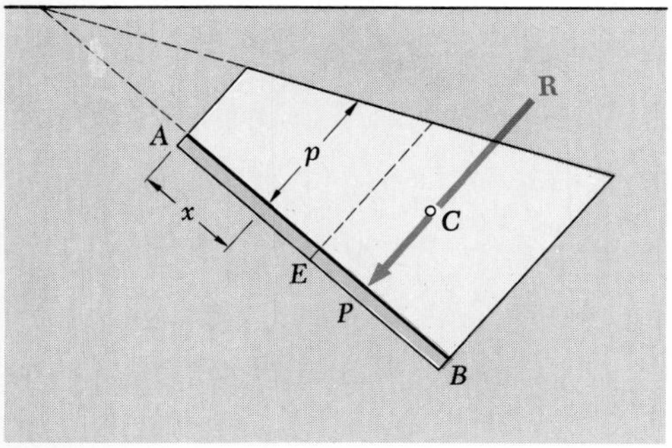

Fig. 5.18

p is equal in magnitude to the load w per unit length used in Sec. 5.6.† The results obtained in that section may thus be used here, and we find that the magnitude of the resultant **R** of the forces exerted on one face of the plate is equal to the area under the pressure curve; we also find that the line of action of **R** passes through the centroid C of that area.

Noting that the area under the pressure curve is equal to $p_E L$, where p_E is the pressure at the center E of the plate and L the length (or area) of the plate, we find that the magnitude R of the resultant may be obtained by multiplying the area of the plate by the pressure at the center E of the plate. The resultant **R**, however, *should not* be applied at E; as indicated above, its line of action passes through the centroid C of the area under the pressure curve. The point of application P of the resultant **R** is known as the *center of pressure*.

We shall consider next the forces exerted by a liquid on a curved surface of constant width (Fig. 5.19*a*). Since the determination of the resultant **R** of these forces by direct integration would not be easy, we shall consider the free body obtained by detaching the volume of liquid ABD bounded by the curved surface AB and by the two plane surfaces AD and DB shown in Fig. 5.19*b*. The forces acting on the free body ABD consist of the weight **W** of the volume of liquid detached, the resultant $\mathbf{R}_1$ of the forces exerted on AD, the resultant $\mathbf{R}_2$ of the forces exerted on BD, and the resultant $\mathbf{R}'$ of the forces exerted *by the curved surface on the liquid*. The resultant $\mathbf{R}'$ is equal and opposite to, and has the same line of action as, the resultant **R** of the forces exerted *by the liquid on the curved surface*. The forces **W**, $\mathbf{R}_1$, and $\mathbf{R}_2$ may be determined by standard methods; after their values have been found, the force $\mathbf{R}'$ will be obtained by solving the equations of equilibrium for the free body of Fig. 5.19*b*. The resultant **R** of the hydrostatic forces exerted on the curved surface will then be obtained by reversing the sense of $\mathbf{R}'$.

The methods outlined in this section may be used to determine the resultant of the hydrostatic forces exerted on the surface of dams or on rectangular gates and vanes. Resultants of forces on submerged surfaces of variable width should be determined by the methods of Chap. 9.

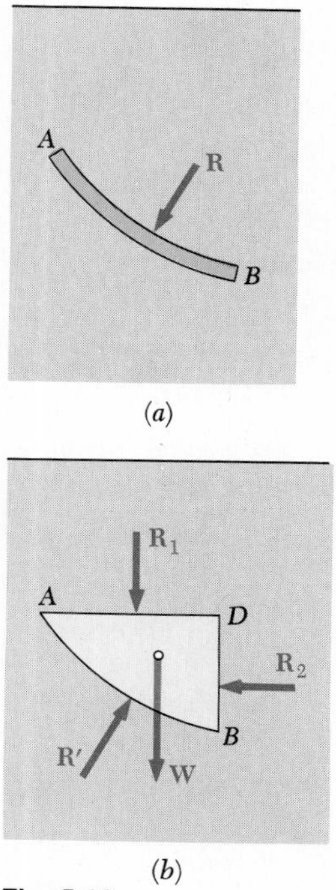

Fig. 5.19

† We should note that, while the load w per unit length is expressed in N/m or in lb/ft, the pressure p, which represents a load per unit area, is expressed in N/m^2 or in lb/ft^2. The derived SI unit N/m^2 is called a *pascal* (Pa).

SAMPLE PROBLEM 5.9

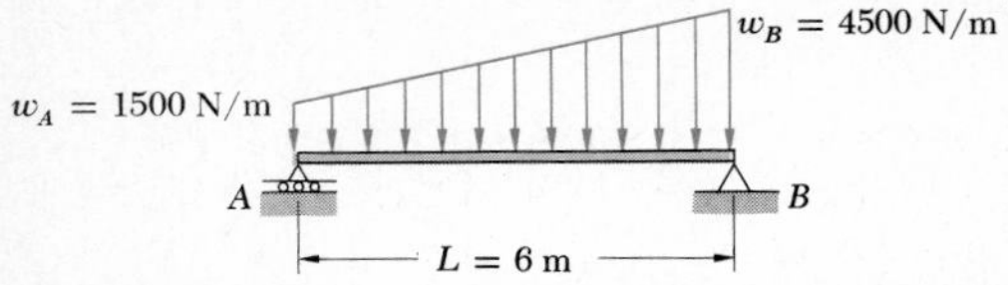

A beam supports a distributed load as shown. (*a*) Determine the equivalent concentrated load. (*b*) Determine the reactions at the supports.

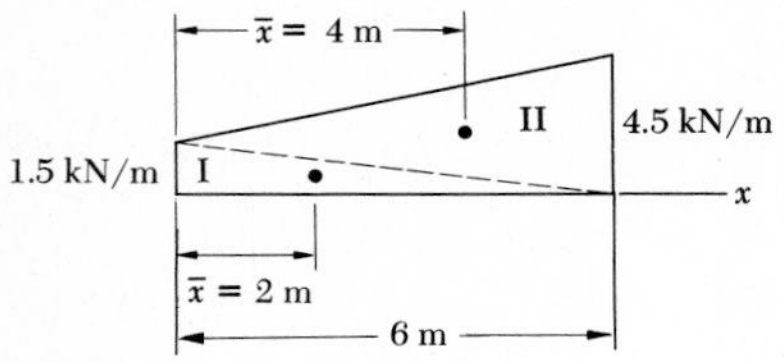

a. Equivalent Concentrated Load. The magnitude of the resultant of the load is equal to the area under the load curve, and the line of action of the resultant passes through the centroid of the same area. We divide the area under the load curve into two triangles and construct the table below. To simplify the computations and tabulation, the given loads per unit length have been converted into kN/m.

Component	A, kN	$\bar{x}$, m	$\bar{x}A$, kN · m
Triangle I	4.5	2	9
Triangle II	13.5	4	54
	$\Sigma A = 18.0$	. . .	$\Sigma \bar{x}A = 63$

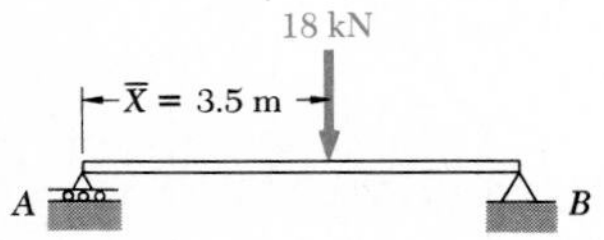

Thus, $\bar{X}\Sigma A = \Sigma \bar{x}A$: $\quad \bar{X}(18 \text{ kN}) = 63 \text{ kN} \cdot \text{m} \quad \bar{X} = 3.5 \text{ m}$

The equivalent concentrated load is

$$\mathbf{W} = 18 \text{ kN} \downarrow \quad ◀$$

and its line of action is located at a distance

$$\bar{X} = 3.5 \text{ m to the right of } A \quad ◀$$

b. Reactions. The reaction at A is vertical and is denoted by **A**; the reaction at B is represented by its components $\mathbf{B}_x$ and $\mathbf{B}_y$. The given load may be considered as the sum of two triangular loads as shown. The resultant of each triangular load is equal to the area of the triangle and acts at its centroid. We write the following equilibrium equations for the free body shown:

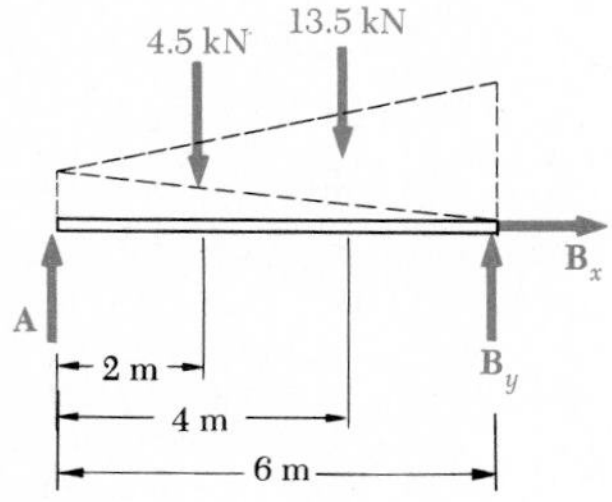

$\xrightarrow{+} \Sigma F_x = 0$: $\qquad \mathbf{B}_x = 0 \quad ◀$

$+\circlearrowleft \Sigma M_A = 0$: $\qquad -(4.5 \text{ kN})(2 \text{ m}) - (13.5 \text{ kN})(4 \text{ m}) + B_y(6 \text{ m}) = 0$

$$\mathbf{B}_y = 10.5 \text{ kN} \uparrow \quad ◀$$

$+\circlearrowleft \Sigma M_B = 0$: $\qquad +(4.5 \text{ kN})(4 \text{ m}) + (13.5 \text{ kN})(2 \text{ m}) - A(6 \text{ m}) = 0$

$$\mathbf{A} = 7.5 \text{ kN} \uparrow \quad ◀$$

Alternate Solution. The given distributed load may be replaced by its resultant, which was found in part *a*. The reactions may be determined by writing the equilibrium equations $\Sigma F_x = 0$, $\Sigma M_A = 0$, and $\Sigma M_B = 0$. We again obtain

$$\mathbf{B}_x = 0 \qquad \mathbf{B}_y = 10.5 \text{ kN} \uparrow \qquad \mathbf{A} = 7.5 \text{ kN} \uparrow \quad ◀$$

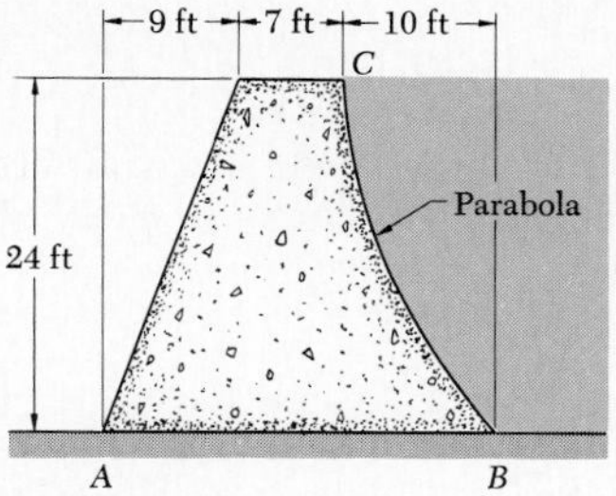

SAMPLE PROBLEM 5.10

The cross section of a concrete dam is as shown. Consider a section of the dam 1 ft thick, and determine (*a*) the resultant of the reaction forces exerted by the ground on the base of the dam *AB*, (*b*) the resultant of the pressure forces exerted by the water on the face *BC* of the dam. Specific weight of concrete = 150 lb/ft³; of water = 62.4 lb/ft³.

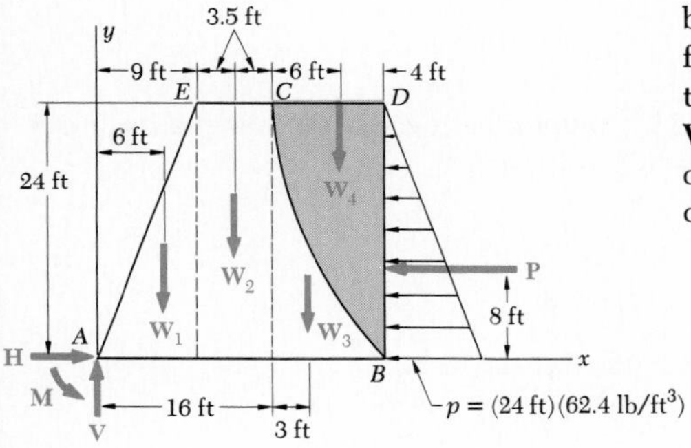

a. **Ground Reaction.** As a free body we choose a section *ABDEA*, 1 ft thick, of the dam and water as shown. The reaction forces exerted by the ground on the base *AB* are represented by an equivalent force-couple system at *A*. Other forces acting on the free body are the weight of the dam, represented by the weights of its components $\mathbf{W}_1$, $\mathbf{W}_2$, and $\mathbf{W}_3$, the weight of the water $\mathbf{W}_4$, and the resultant **P** of the pressure forces exerted on section *BD* by the water to the right of section *BD*. We have

$$
\begin{aligned}
W_1 &= \tfrac{1}{2}(9\text{ ft})(24\text{ ft})(1\text{ ft})(150\text{ lb/ft}^3) = 16{,}200\text{ lb}\\
W_2 &= (7\text{ ft})(24\text{ ft})(1\text{ ft})(150\text{ lb/ft}^3) = 25{,}200\text{ lb}\\
W_3 &= \tfrac{1}{3}(10\text{ ft})(24\text{ ft})(1\text{ ft})(150\text{ lb/ft}^3) = 12{,}000\text{ lb}\\
W_4 &= \tfrac{2}{3}(10\text{ ft})(24\text{ ft})(1\text{ ft})(62.4\text{ lb/ft}^3) = 9{,}980\text{ lb}\\
P &= \tfrac{1}{2}(24\text{ ft})(1\text{ ft})(24\text{ ft})(62.4\text{ lb/ft}^3) = 17{,}970\text{ lb}
\end{aligned}
$$

Equilibrium Equations

$\Sigma F_x = 0$: $\quad H - 17{,}970\text{ lb} = 0 \qquad \mathbf{H} = 17{,}970\text{ lb} \rightarrow$ ◄

$\Sigma F_y = 0$: $\quad V - 16{,}200\text{ lb} - 25{,}200\text{ lb} - 12{,}000\text{ lb} - 9{,}980\text{ lb} = 0$

$\mathbf{V} = 63{,}400\text{ lb} \uparrow$ ◄

$$+\circlearrowleft \Sigma M_A = 0: \quad -(16{,}200\text{ lb})(6\text{ ft}) - (25{,}200\text{ lb})(12.5\text{ ft})$$
$$-(12{,}000\text{ lb})(19\text{ ft}) - (9{,}980\text{ lb})(22\text{ ft}) + (17{,}970\text{ lb})(8\text{ ft}) + M = 0$$

$\mathbf{M} = 716{,}000\text{ lb}\cdot\text{ft} \circlearrowleft$ ◄

We may replace the force-couple system obtained by a single force acting at a distance *d* to the right of *A*, where

$$d = \frac{716{,}000\text{ lb}\cdot\text{ft}}{63{,}400\text{ lb}} \qquad d = 11.29\text{ ft}$$ ◄

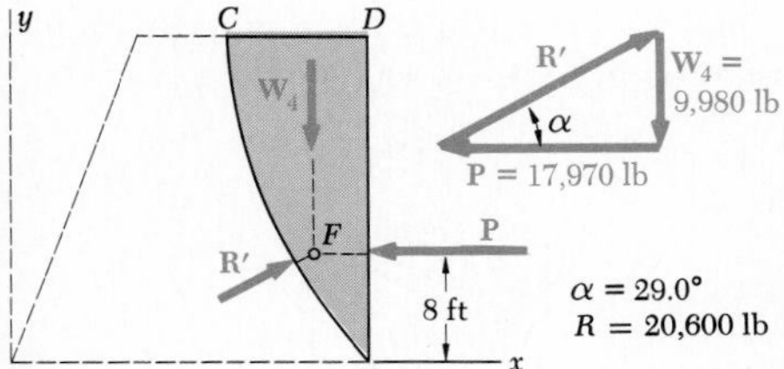

b. **Resultant R of Water Forces.** The parabolic section of water *BCD* is chosen as a free body. The forces involved are the resultant **R′** of the forces exerted by the dam on the water, the weight $\mathbf{W}_4$, and the force **P**. Since these forces must be concurrent, **R′** passes through the point of intersection *F* of $\mathbf{W}_4$ and **P**. A force triangle is drawn from which the magnitude and direction of **R′** are determined. The resultant **R** of the forces exerted by the water on the face *BC* is equal and opposite:

$\mathbf{R} = 20{,}600\text{ lb} \searrow 29.0°$ ◄

PROBLEMS

5.66 and 5.67 Determine the magnitude and location of the resultant of the distributed load shown. Also calculate the reactions at A and B.

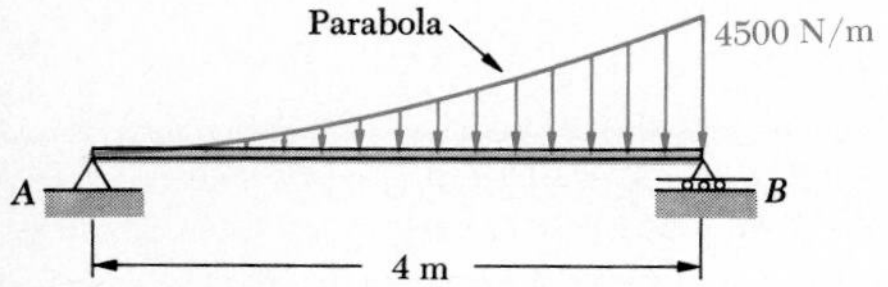

Fig. P5.66

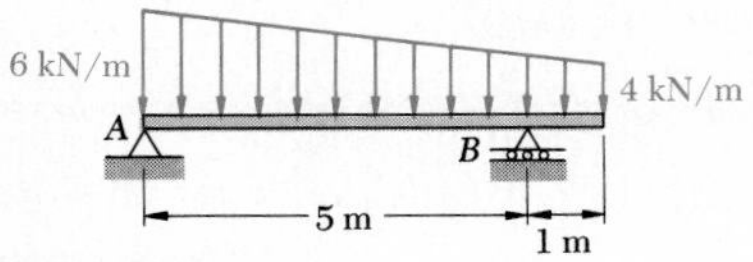

Fig. P5.67

5.68 through 5.71 Determine the reactions at the beam supports for the given loading condition.

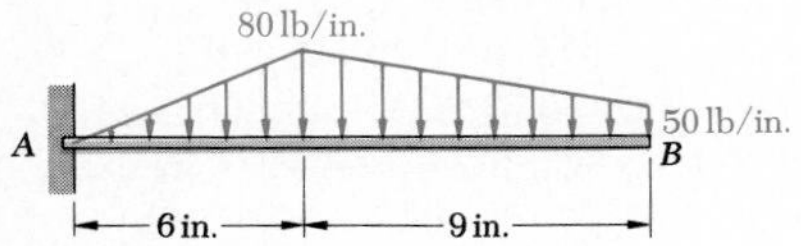

Fig. P5.68

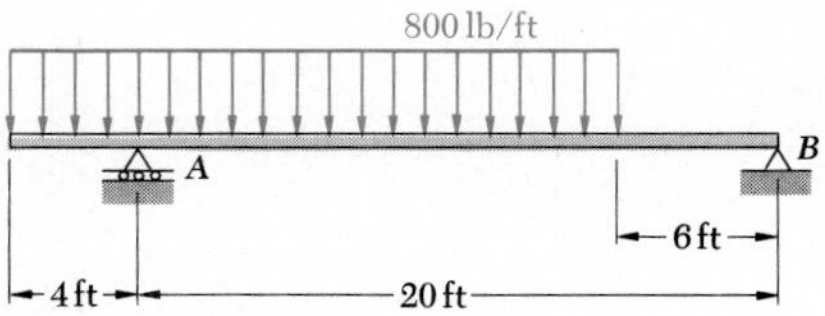

Fig. P5.69

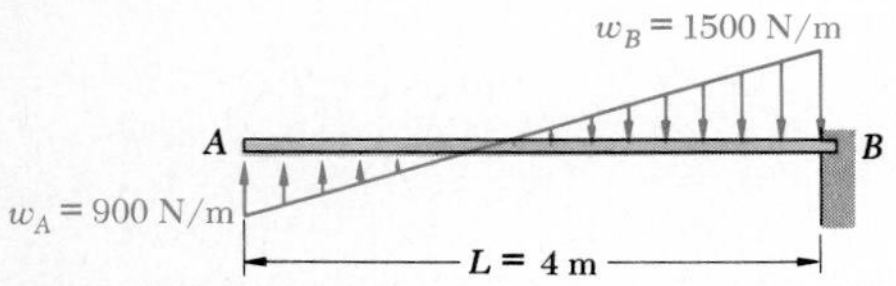

Fig. P5.70

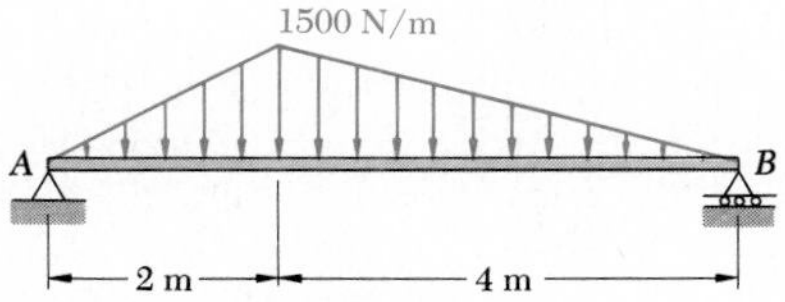

Fig. P5.71

5.72 Determine the ratio of w_A to w_B for which the reaction at A is equal to (*a*) a couple and no force, (*b*) a force and no couple. In each case express the reaction in terms of w_A and L.

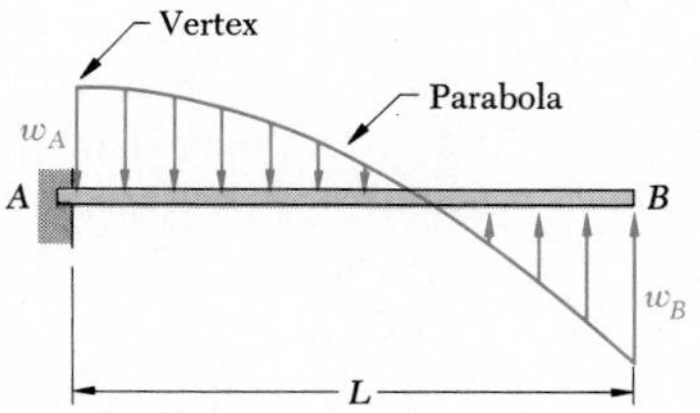

Fig. P5.72

5.73 In Prob. 5.70, determine the ratio of w_A to w_B for which the reaction at B is equal to (*a*) a force and no couple, (*b*) a couple and no force. In each case express the reaction in terms of w_B and L.

5.74 A beam supports a uniformly distributed load w_1 and rests on soil which exerts a uniformly varying upward load as shown. (*a*) Determine w_2 and w_3, corresponding to equilibrium. (*b*) Knowing that at any point the soil can exert only an upward loading on the beam, state for what range of values of a/L the results obtained are valid.

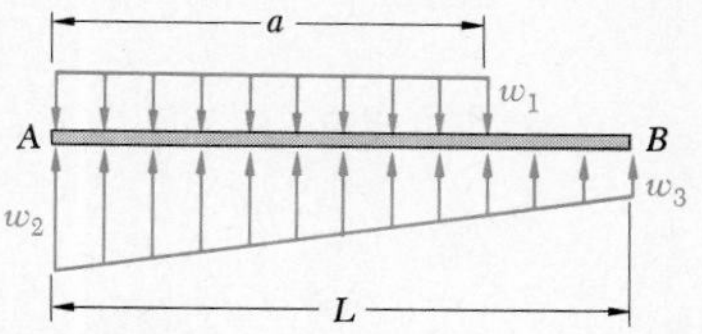

Fig. P5.74

5.75 Solve Sample Prob. 5.9 in terms of the letter quantities w_A, w_B, and L.

In the following problems, use $\gamma = 62.4\ \text{lb/ft}^3$ for the specific weight of fresh water and $\gamma_c = 150\ \text{lb/ft}^3$ for the specific weight of concrete if U.S. customary units are used. With SI units, use $\rho = 10^3\ \text{kg/m}^3$ for the density of fresh water and $\rho_c = 2.40 \times 10^3\ \text{kg/m}^3$ for the density of concrete. (See footnote page 140 for the determination of the specific weight of a material from its density.)

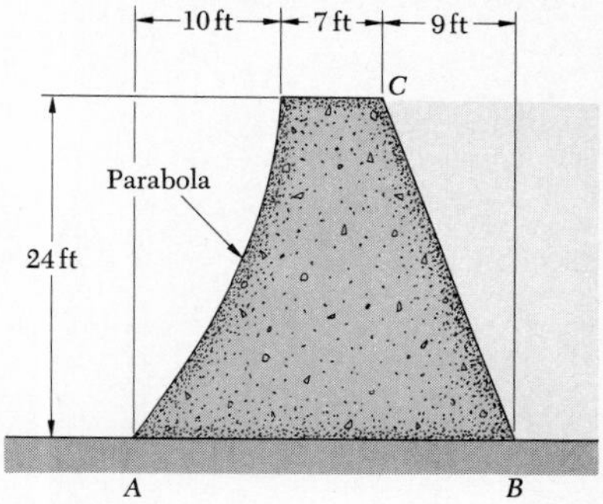

Fig. P5.76

5.76 The cross section of a concrete dam is as shown. Consider a section of the dam 1 ft thick, and determine (*a*) the resultant of the reaction forces exerted by the ground on the base of the dam AB, (*b*) the resultant of the pressure forces exerted by the water on the face BC of the dam.

5.77 A 3- by 3-ft gate is placed in a wall below water level as shown. (*a*) Determine the magnitude and location of the resultant of the forces exerted by the water on the gate. (*b*) If the gate is hinged at A, determine the force exerted by the sill on the gate at B.

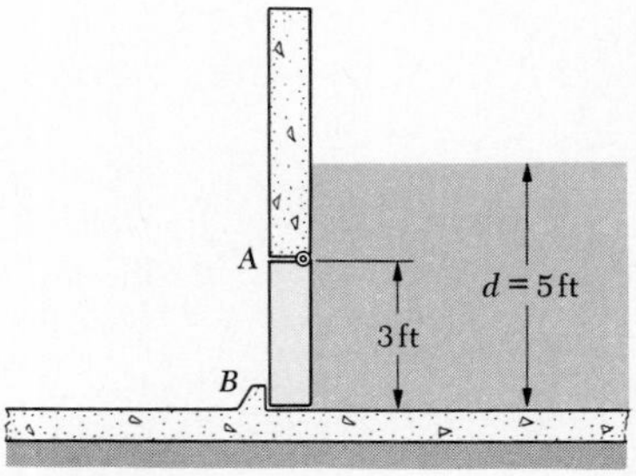

Fig. P5.77

5.78 In Prob. 5.77, determine the depth of water d for which the force exerted by the sill on the gate at B is 2000 lb.

5.79 An automatic valve consists of a square plate, 225 by 225 mm, which is pivoted about a horizontal axis through A located at a distance $h = 100$ mm above the lower edge. Determine the depth of water d for which the valve will open.

5.80 If the valve shown is to open when the depth of water is $d = 300$ mm, determine the distance h from the bottom of the valve to the pivot A.

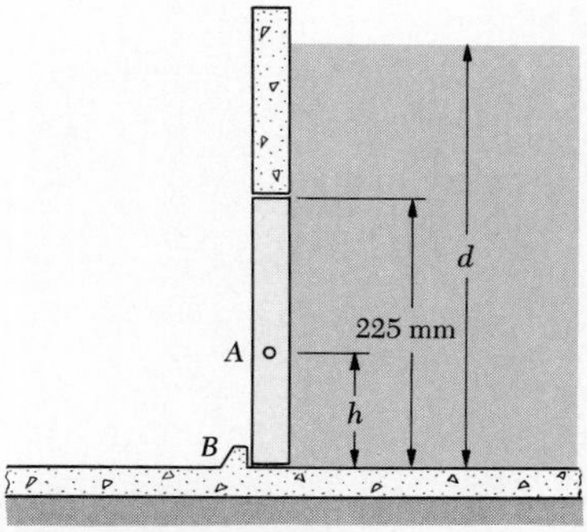

Fig. P5.79 and P5.80

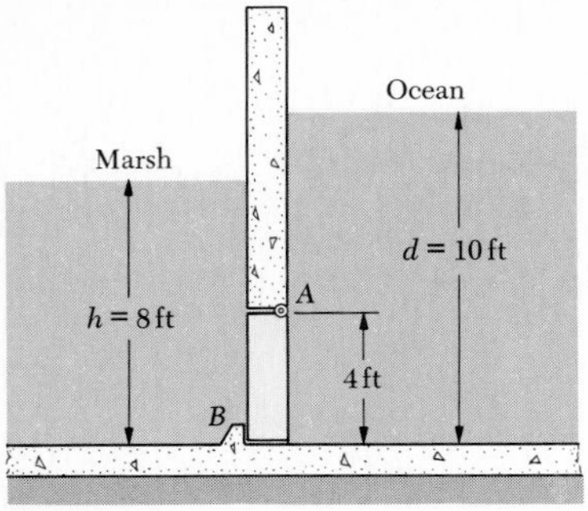

Fig. P5.81

5.81 A freshwater marsh is drained to the ocean through an automatic tide gate which is 5 ft wide and 4 ft high. The gate is held by hinges located along its top edge at A and bears on a sill at B. At a given time, the water level in the marsh is $h = 8$ ft and in the ocean $d = 10$ ft. Determine the force exerted by the sill on the gate at B and the hinge reaction at A. (Specific weight of salt water $= 64\ \text{lb/ft}^3$.)

5.82 The automatic tide gate described in Prob. 5.81 is used to drain a freshwater marsh into the ocean. If the water level in the marsh is $h = 8$ ft, determine the ocean level d for which the gate will open.

5.83 The bent plate $ABCD$ is 2 m wide and is hinged at A. Determine the reactions at A and D for the water level shown.

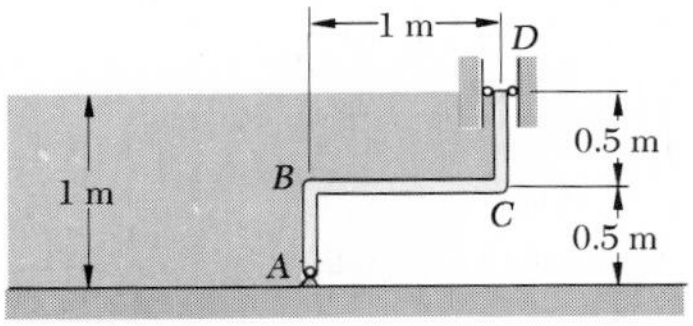

Fig. P5.83

5.84 Solve Prob. 5.83, assuming that the water shown is removed and is replaced by a 1-m depth of water to the right of the plate.

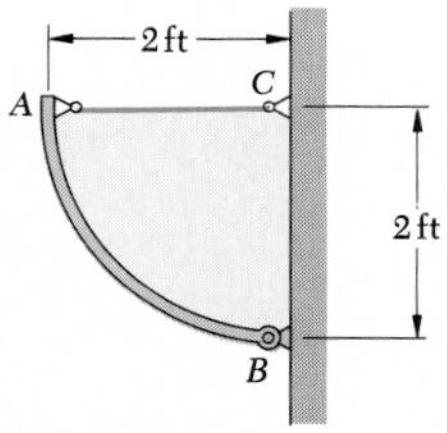

Fig. P5.85

5.85 The metal trough AB is of 2-ft radius and is supported by a hinge at B and cables AC spaced every 3 ft along the trough. Determine the tension in each cable caused by water which fills the trough to the level AC.

5.86 Determine the minimum allowable value of the width a of the rectangular concrete dam if the dam is not to overturn about point A when $d = h = 4$ m.

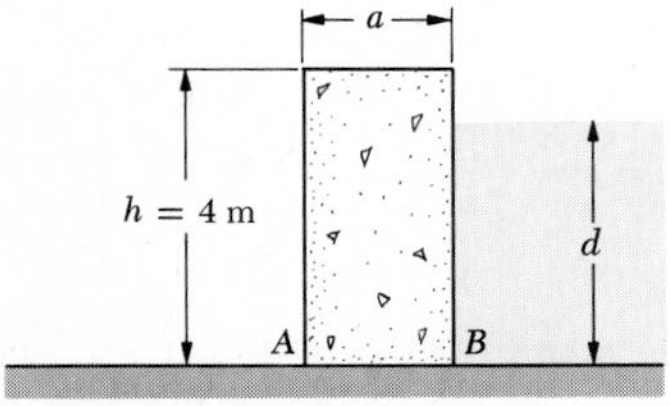

Fig. P5.86, P5.87, and P5.88

5.87 Knowing that the width of the rectangular concrete dam is $a = 1.25$ m and that its height is $h = 4$ m, determine the maximum allowable value of the depth d of water if the dam is not to overturn about A.

5.88 Concrete is a material which is weak in tension. In order to eliminate tension, the line of action of the resultant of the hydrostatic forces and of the weight of the dam must pass through the middle third of the base. Determine the minimum width a for which no tension will occur in the rectangular concrete dam shown when $d = h = 4$ m.

5.89 At the time of construction no silt was present behind the dam shown. At a later time silt has accumulated to a depth $d_s = 9$ ft. Knowing that $d_w = 24$ ft and assuming that the silt is equivalent to a liquid of specific weight $\gamma_s = 110\ \text{lb/ft}^3$, determine the magnitude and location of the resultant force exerted on the face BC. Consider a section of dam 1 ft thick.

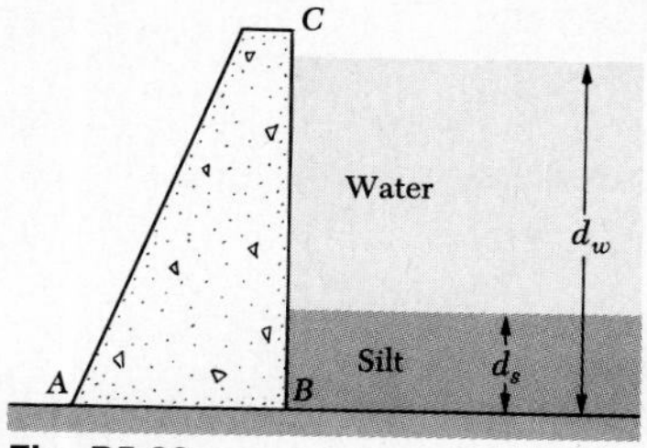

Fig. P5.89

5.90 The gate shown is supported by hinges located along its top edge at B and rests on the floor of a channel at A. Knowing that the gate is 3 ft wide, determine the magnitude and direction of the resultant of the hydrostatic forces exerted on the gate.

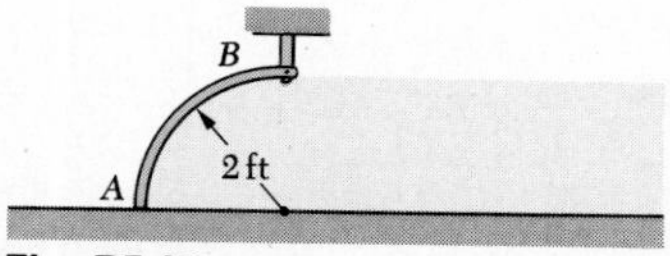

Fig. P5.90

5.91 The end of a freshwater channel consists of a plate ABC which is hinged at B and is 3 ft wide. Knowing that $b = 2$ ft and $h = 1.5$ ft, determine the reactions at A and B.

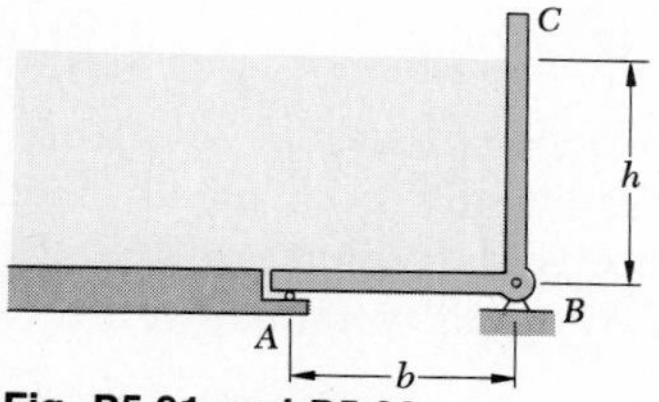

Fig. P5.91 and P5.92

5.92 The end of a freshwater channel consists of a plate ABC which is hinged at B and is 3 ft wide. Determine the ratio h/b for which the reaction at A is zero.

VOLUMES

5.8. Center of Gravity of a Three-dimensional Body. Centroid of a Volume. The *center of gravity* G of a three-dimensional body is obtained by dividing the body into small elements and expressing that the weight $\mathbf{W}$ of the body attached at G is equivalent to the system of distributed forces $\Delta\mathbf{W}$ representing the weights of the small elements. This has been done in Figs. 5.20 and 5.21 for two positions of the body. In Fig. 5.20, the y axis is vertical; in Fig. 5.21, the body and the axes have been rotated so that the z axis is vertical. Three independent equations are obtained, which may be used to determine the coordinates $\bar{x}$, $\bar{y}$, and $\bar{z}$ of the center of gravity G. Written in terms of infinitesimal elements, these equations are

$$\bar{x}W = \int x\,dW \qquad \bar{y}W = \int y\,dW \qquad \bar{z}W = \int z\,dW \qquad (5.13)$$

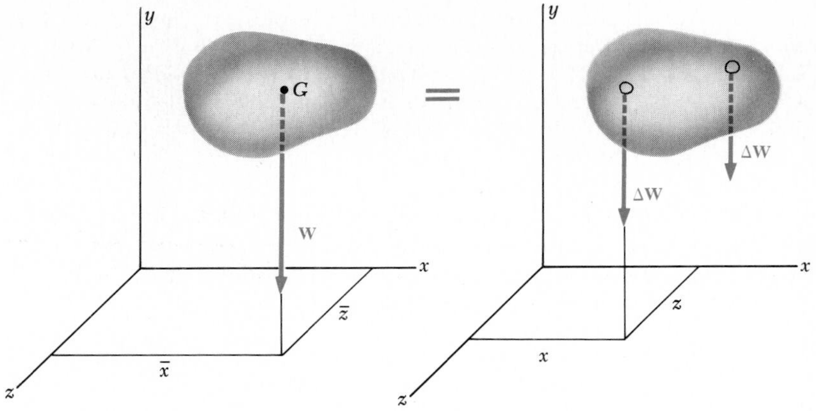

$$\Sigma M_z : \quad \bar{x}W = \Sigma\, x\, \Delta W$$
$$\Sigma M_x : \quad \bar{z}W = \Sigma\, z\, \Delta W$$

Fig. 5.20

If the body is made of a homogeneous material of specific weight γ, the magnitude dW of the weight of an infinitesimal element may be expressed in terms of the volume dV of the element, and the magnitude W of the total weight in terms of the total volume V. We write

$$dW = \gamma\, dV \qquad W = \gamma V$$

Substituting for dW and W in Eqs. (5.13), we obtain

$$\bar{x}V = \int x\, dV \qquad \bar{y}V = \int y\, dV \qquad \bar{z}V = \int z\, dV \quad (5.14)$$

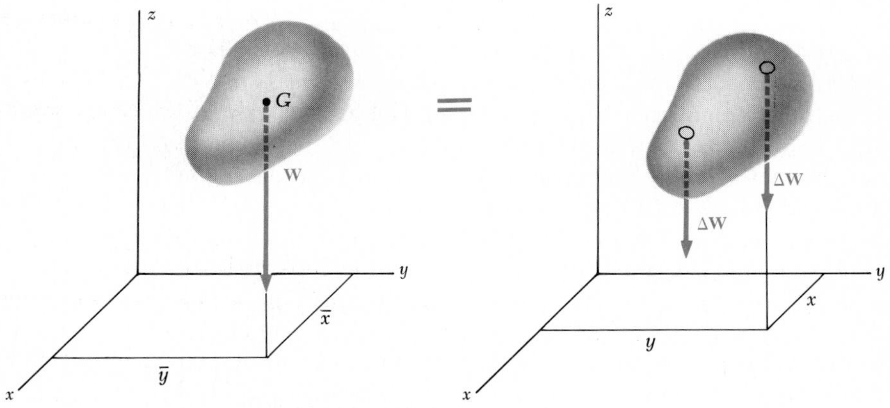

$$\Sigma M_x : \quad \bar{y}W = \Sigma\, y\, \Delta W$$
$$\Sigma M_y : \quad \bar{x}W = \Sigma\, x\, \Delta W$$

Fig. 5.21

The point of coordinates $\bar{x}$, $\bar{y}$, $\bar{z}$ is also known as the *centroid C of the volume V* of the body. If the body is not homogeneous, Eqs. (5.14) cannot be used to determine the center of gravity of the body; they still define, however, the centroid of the volume.

The integral $\int x\,dV$ is known as the *first moment of the volume with respect to the yz plane*. Similarly, the integrals $\int y\,dV$ and $\int z\,dV$ define the first moments of the volume with respect to the zx plane and the xy plane, respectively. It is seen from Eqs. (5.14) that, if the centroid of a volume is located in a coordinate plane, the first moment of the volume with respect to that plane is zero.

A volume is said to be symmetrical with respect to a given plane if to every point P of the volume corresponds a point P' of the same volume, such that the line PP' is perpendicular to the given plane and divided into two equal parts by that plane. The plane is said to be a *plane of symmetry* for the given volume. When a volume V possesses a plane of symmetry, the centroid of the volume must be located in that plane. When a volume possesses two planes of symmetry, the centroid of the volume must be located on the line of intersection of the two planes. Finally, when a volume possesses three planes of symmetry which intersect in a well-defined point (i.e., not along a common line), the point of intersection of the three planes must coincide with the centroid of the volume. This property enables us to determine immediately the centroid of the volume of spheres, ellipsoids, cubes, rectangular parallelepipeds, etc.

Centroids of unsymmetrical volumes or of volumes possessing only one or two planes of symmetry should be determined by integration (Sec. 5.10). Centroids of common shapes of volumes are shown in Fig. 5.22. It should be observed that the centroid of a volume of revolution in general *does not coincide* with the centroid of its cross section. Thus, the centroid of a hemisphere is different from that of a semicircular area, and the centroid of a cone is different from that of a triangle.

5.9. Composite Bodies. If a body can be divided into several of the common shapes shown in Fig. 5.22, the coordinates $\bar{X}$, $\bar{Y}$, $\bar{Z}$ of its center of gravity G may be obtained by expressing that the moment of the total weight is equal to the sum of the moments of the weights of the various component parts. Moments should be taken about three axes for two positions of the body (see Figs. 5.20 and 5.21). The equations obtained are

$$\bar{X}\Sigma W = \Sigma\bar{x}W \qquad \bar{Y}\Sigma W = \Sigma\bar{y}W \qquad \bar{Z}\Sigma W = \Sigma\bar{z}W \tag{5.15}$$

If the body is made of a homogeneous material, its center of

Shape		$\bar{x}$	Volume
Hemisphere	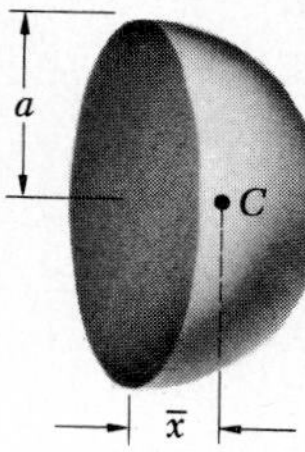	$\frac{3a}{8}$	$\frac{2}{3}\pi a^3$
Semiellipsoid of revolution	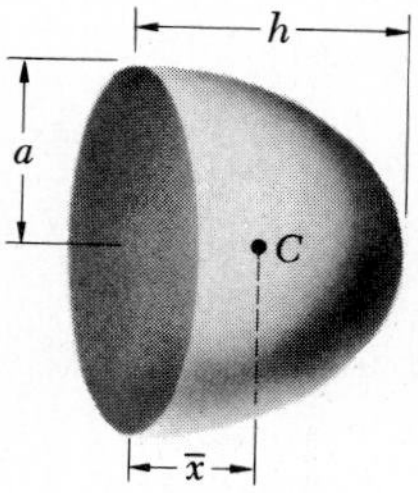	$\frac{3h}{8}$	$\frac{2}{3}\pi a^2 h$
Paraboloid of revolution	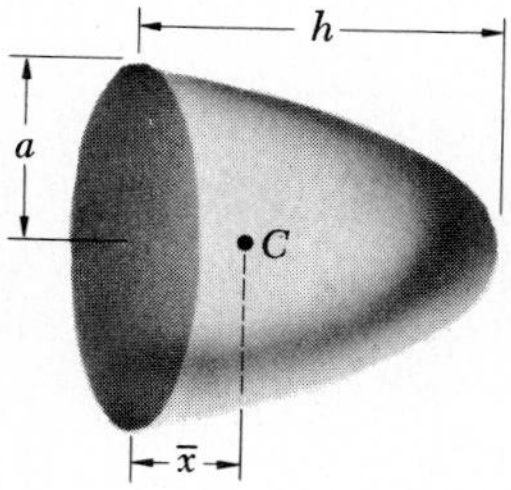	$\frac{h}{3}$	$\frac{1}{2}\pi a^2 h$
Cone	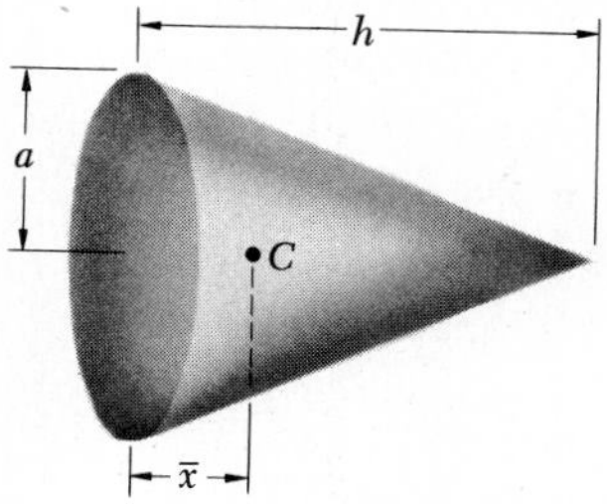	$\frac{h}{4}$	$\frac{1}{3}\pi a^2 h$
Pyramid	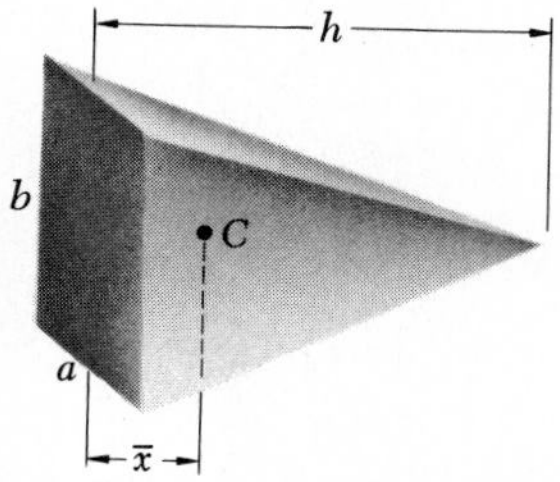	$\frac{h}{4}$	$\frac{1}{3}abh$

Fig. 5.22 Centroids of common shapes of volumes.

gravity coincides with the centroid of its volume and the following equations may be used:

$$\bar{X}\Sigma V = \Sigma \bar{x} V \qquad \bar{Y}\Sigma V = \Sigma \bar{y} V \qquad \bar{Z}\Sigma V = \Sigma \bar{z} V \tag{5.16}$$

5.10. Determination of Centroids of Volumes by Integration. The centroid of a volume bounded by analytical surfaces may be determined by computing the integrals given in Sec. 5.8.

$$\bar{x}V = \int x\,dV \qquad \bar{y}V = \int y\,dV \qquad \bar{z}V = \int z\,dV \tag{5.17}$$

If the element of volume dV is chosen equal to a small cube of sides dx, dy, and dz, the determination of each of these integrals requires a *triple integration* in x, y, and z. However, it is possible to determine the coordinates of the centroid of most volumes by *double integration* if dV is chosen equal to the volume of a thin filament as shown in Fig. 5.23. The coordinates of the centroid of the volume are then obtained by writing

$$\bar{x}V = \int \bar{x}_{el}\,dV \qquad \bar{y}V = \int \bar{y}_{el}\,dV \qquad \bar{z}V = \int \bar{z}_{el}\,dV \tag{5.18}$$

and substituting for the volume dV and the coordinates $\bar{x}_{el}$, $\bar{y}_{el}$, $\bar{z}_{el}$ the expressions given in Fig. 5.23. Using the equation of the surface to express z in terms of x and y, the integration is reduced to a double integration in x and y.

If the volume under consideration possesses *two planes of symmetry*, its centroid must be located on their line of intersection. Choosing the x axis along this line, we have

$$\bar{y} = \bar{z} = 0$$

and the only coordinate to determine is $\bar{x}$. This will be done most conveniently by dividing the given volume into thin slabs parallel to the yz plane. In the particular case of a body of revolution, these slabs are circular; their volume dV is given in Fig. 5.24. Substituting for $\bar{x}_{el}$ and dV into the equation

$$\bar{x}V = \int \bar{x}_{el}\,dV \tag{5.19}$$

and expressing the radius r of the slab in terms of x, we may determine $\bar{x}$ by a single integration.

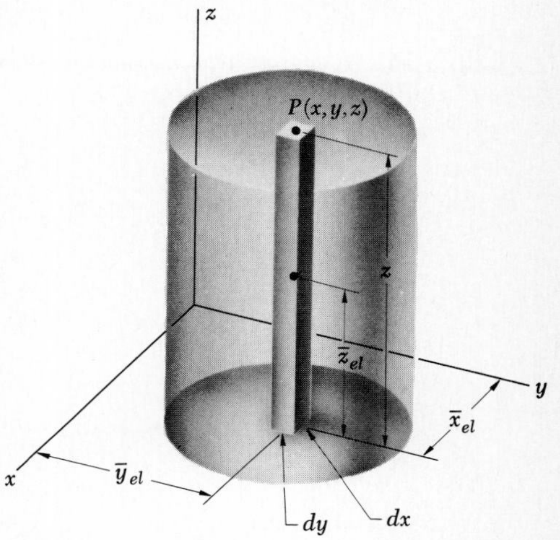

$\bar{x}_{el} = x,\ \bar{y}_{el} = y,\ \bar{z}_{el} = \frac{z}{2}$
$dV = z\,dx\,dy$

Fig. 5.23 Determination of the centroid of a volume by double integration.

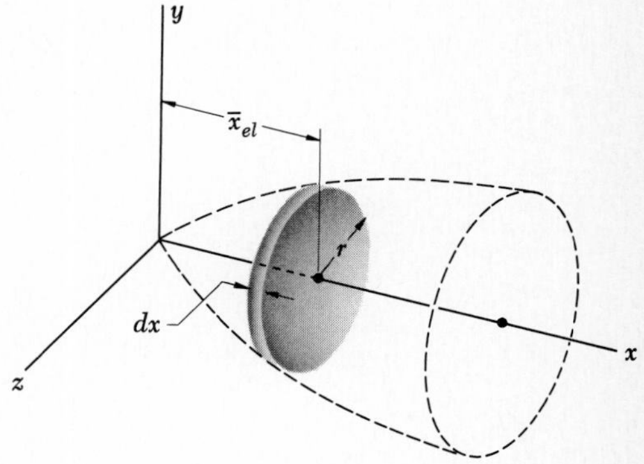

$\bar{x}_{el} = x$
$dV = \pi r^2\,dx$

Fig. 5.24 Determination of the centroid of a body of revolution.

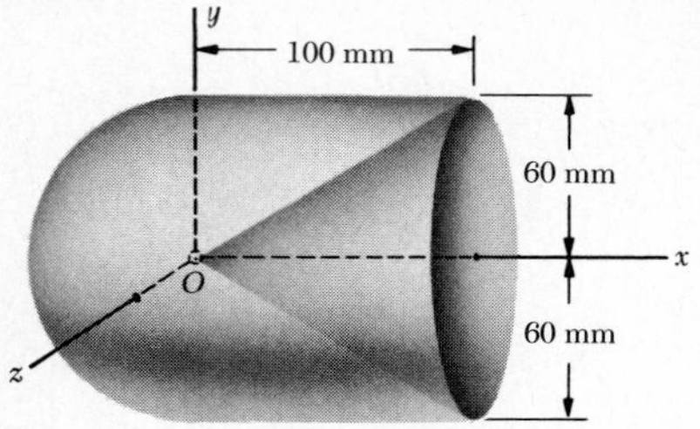

SAMPLE PROBLEM 5.11

Determine the center of gravity of the homogeneous body of revolution shown.

Solution. Because of symmetry, the center of gravity lies on the x axis. The body is seen to consist of a hemisphere, plus a cylinder, minus a cone, as shown. The volume and the abscissa of the centroid of each of these components are obtained from Fig. 5.22 and entered in the table below. All dimensions have been converted into centimeters to avoid excessively large numerical values in the computation of volumes and moments of volumes.

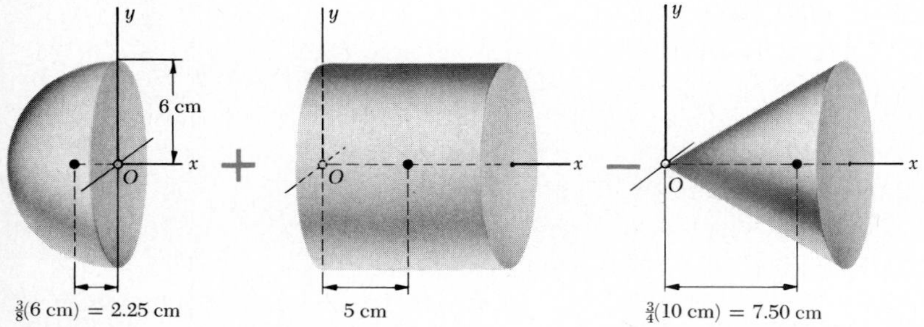

Component	Volume, cm^3	$\bar{x}$, cm	$\bar{x}V$, cm^4
Hemisphere	$\frac{1}{2}\frac{4\pi}{3}(6)^3 = \quad 452$	-2.25	-1017
Cylinder	$\pi(6)^2(10) = \quad 1131$	$+5.00$	$+5655$
Cone	$-\frac{\pi}{3}(6)^2(10) = -377$	$+7.50$	-2828
	$\Sigma V = 1206$	...	$\Sigma \bar{x}V = +1810$

Thus,

$\bar{X}\Sigma V = \Sigma \bar{x}V$: $\quad \bar{X}(1206\ \text{cm}^3) = 1810\ \text{cm}^4$

$$\bar{X} = 1.5\ \text{cm}$$

$\bar{X} = 15\ \text{mm}$ ◄

SAMPLE PROBLEM 5.12

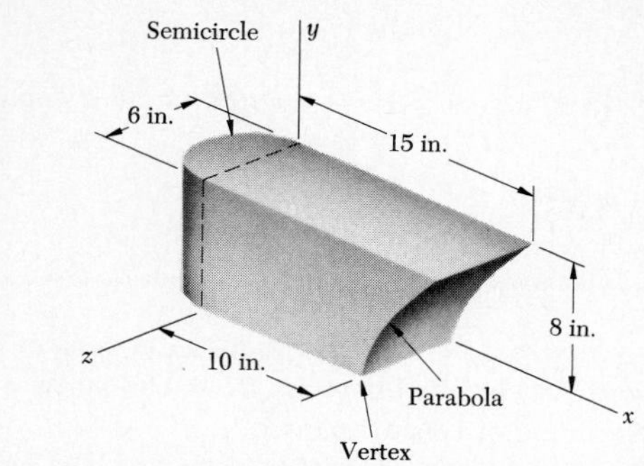

Determine the center of gravity of the homogeneous body shown.

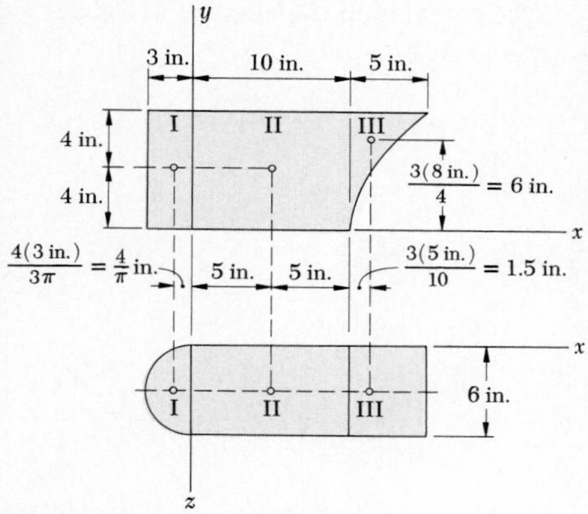

Solution. Because of symmetry, the center of gravity lies in a plane parallel to the xy plane, with $\bar{Z} = 3$ in. The body is seen to consist of a half cylinder (I), a rectangular parallelepiped (II), and a cylindrical shape having for a cross section a parabolic spandrel (III). Using the information given in Fig. 5.8*A* for a semicircle and a parabolic spandrel, we determine the coordinates of the centroids of the various components of the body. These coordinates are shown in the adjoining sketch. The total volume and the moments of the volume with respect to the zx and yz planes are determined in the table below.

Component	V, in^3	$\bar{x}$, in.	$\bar{y}$, in.	$\bar{x}V$, in^4	$\bar{y}V$, in^4
I	$(8)\dfrac{\pi(3^2)}{2} = 113$	$-\dfrac{4}{\pi}$	4	-144	452
II	$(10)(8)(6) = 480$	5	4	2400	1920
III	$\frac{1}{3}(5)(8)(6) = 80$	11.5	6	920	480
	$\Sigma V = 673$	...	...	$\Sigma\bar{x}V = 3176$	$\Sigma\bar{y}V = 2852$

Thus,

$\bar{X}\Sigma V = \Sigma\bar{x}V$: $\quad \bar{X}(673 \text{ in}^3) = 3176 \text{ in}^4 \qquad \bar{X} = 4.72$ in. ◄

$\bar{Y}\Sigma V = \Sigma\bar{y}V$: $\quad \bar{Y}(673 \text{ in}^3) = 2852 \text{ in}^4 \qquad \bar{Y} = 4.24$ in. ◄

$\bar{Z} = 3.00$ in. ◄

SAMPLE PROBLEM 5.13

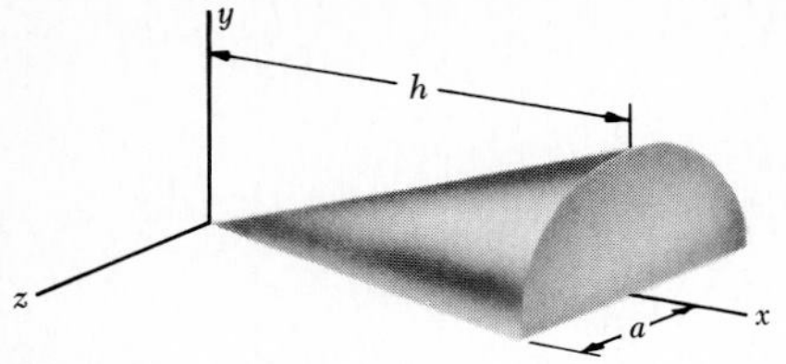

Determine the location of the centroid of the half right circular cone shown.

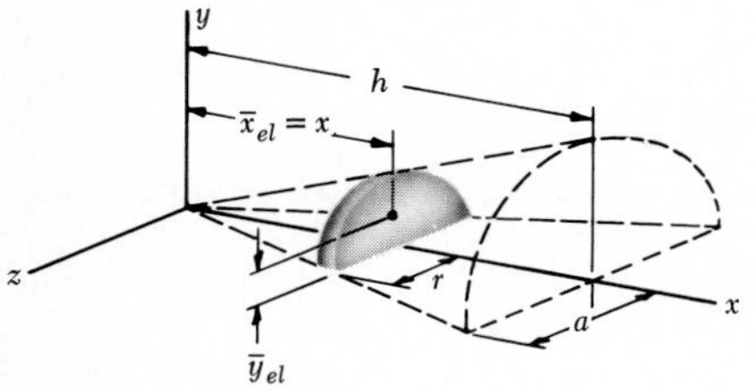

Solution. Since the xy plane is a plane of symmetry, the centroid lies in this plane and $\bar{z} = 0$. A slab of thickness dx is chosen as a differential element. The volume of this element is

$$dV = \tfrac{1}{2}\pi r^2\, dx$$

The coordinates $\bar{x}_{el}$ and $\bar{y}_{el}$ of the centroid of the element are obtained from Fig. 5.8 (semicircular area).

$$\bar{x}_{el} = x \qquad \bar{y}_{el} = \frac{4r}{3\pi}$$

We observe that r is proportional to x and write

$$\frac{r}{x} = \frac{a}{h} \qquad r = \frac{a}{h}x$$

The volume of the body is

$$V = \int dV = \int_0^h \tfrac{1}{2}\pi r^2\, dx = \int_0^h \tfrac{1}{2}\pi\left(\frac{a}{h}x\right)^2 dx = \frac{\pi a^2 h}{6}$$

The moment of the differential element with respect to the yz plane is $\bar{x}_{el}\, dV$; and the total moment of the body with respect to this plane is

$$\int \bar{x}_{el}\, dV = \int_0^h x(\tfrac{1}{2}\pi r^2)\, dx = \int_0^h x(\tfrac{1}{2}\pi)\left(\frac{a}{h}x\right)^2 dx = \frac{\pi a^2 h^2}{8}$$

Thus, $\quad \bar{x}V = \int \bar{x}_{el}\, dV \qquad \bar{x}\frac{\pi a^2 h}{6} = \frac{\pi a^2 h^2}{8} \qquad \bar{x} = \frac{3}{4}h$ ◄

Likewise, the moment of the differential element with respect to the zx plane is $\bar{y}_{el}\, dV$; and the total moment is

$$\int \bar{y}_{el}\, dV = \int_0^h \frac{4r}{3\pi}(\tfrac{1}{2}\pi r^2)\, dx = \frac{2}{3}\int_0^h \left(\frac{a}{h}x\right)^3 dx = \frac{a^3 h}{6}$$

Thus, $\quad \bar{y}V = \int \bar{y}_{el}\, dV \qquad \bar{y}\frac{\pi a^2 h}{6} = \frac{a^3 h}{6} \qquad \bar{y} = \frac{a}{\pi}$ ◄

PROBLEMS

5.93 A cone and a cylinder of the same radius a and height h are attached as shown. Determine the location of the centroid of the composite body.

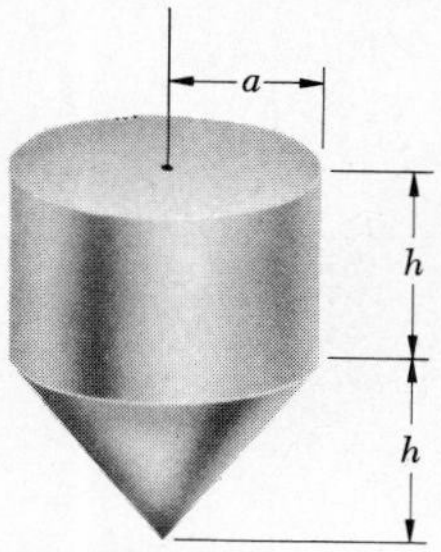

Fig. P5.93

5.94 A hemisphere and a cone are attached as shown. Determine the ratio h/a for which the centroid of the composite body is located in the plane between the hemisphere and the cone.

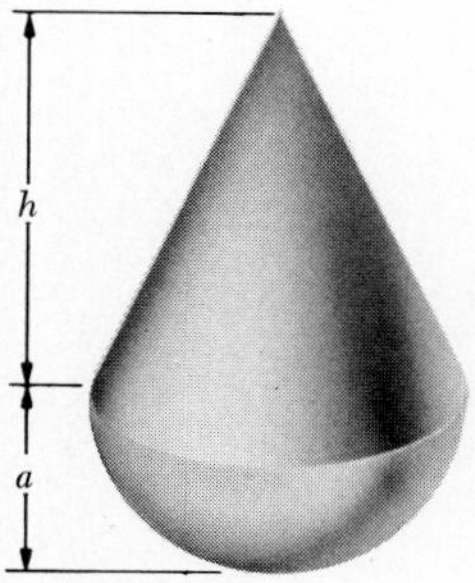

Fig. P5.94

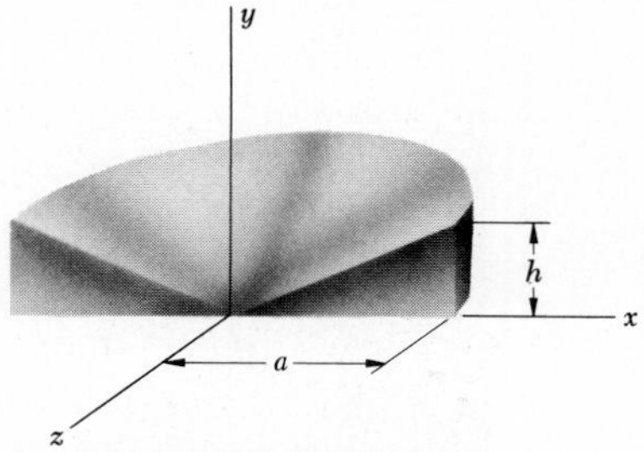

Fig. P5.95 and P5.97

5.95 Determine the y coordinate of the centroid of the solid shown.

5.96 Locate the centroid of the frustum of a right circular cone when $r_1 = 100$ mm, $r_2 = 125$ mm, and $h = 150$ mm.

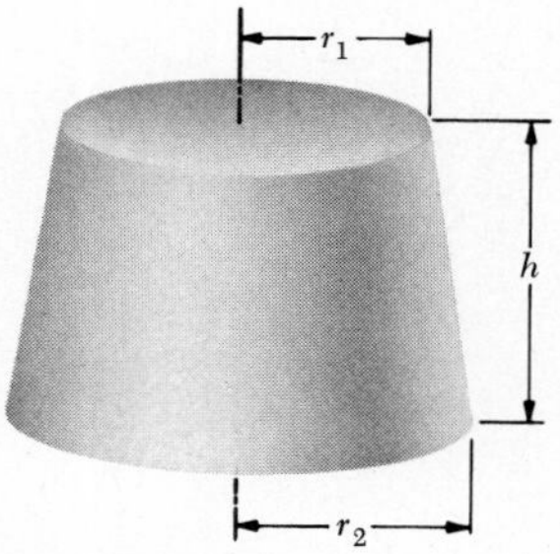

Fig. P5.96

5.97 Determine the z coordinate of the solid shown. (*Hint:* Use the result of Sample Prob. 5.13.)

5.98 and 5.99 Locate the center of gravity of the link shown.

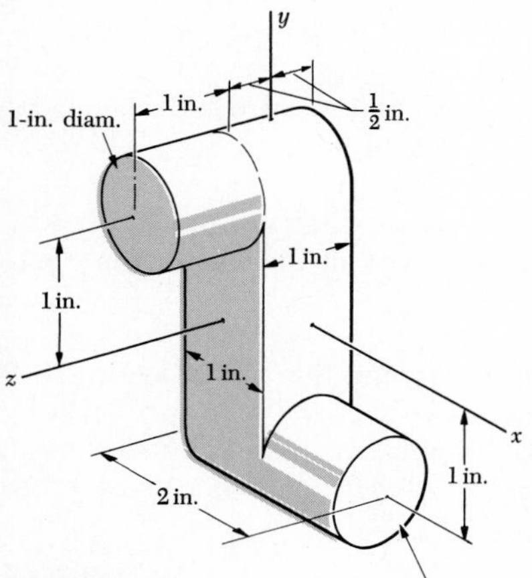

Fig. P5.98

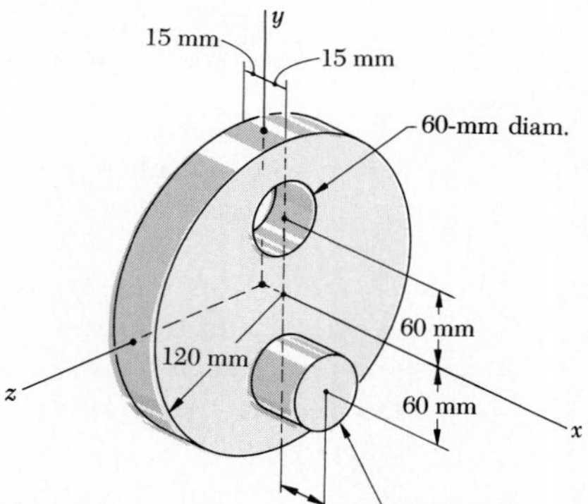

Fig. P5.99

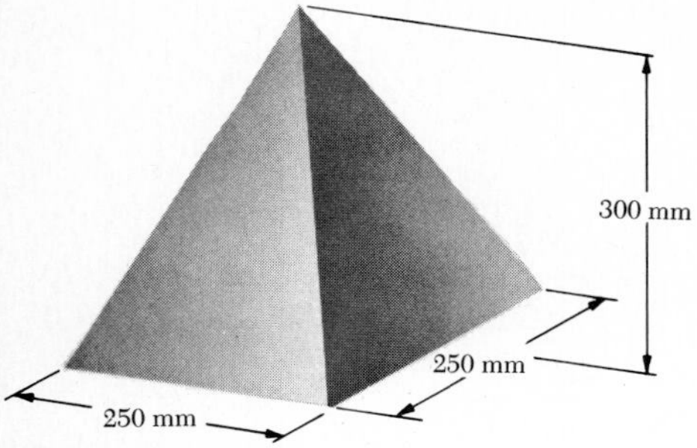

Fig. P5.100

5.100 A regular pyramid 300 mm high, with a square base of side 250 mm, is made of wood. Its four triangular faces are covered with steel sheets 1 mm thick. Locate the center of gravity of the composite body. (Densities: steel = 7850 kg/m^3; wood = 500 kg/m^3.)

5.101 A brass collar, of length 2 in., is mounted on an aluminum bar of length 6 in. Locate the center of gravity of the composite body. (Specific weights: brass = 530 lb/ft^3; aluminum = 170 lb/ft^3.)

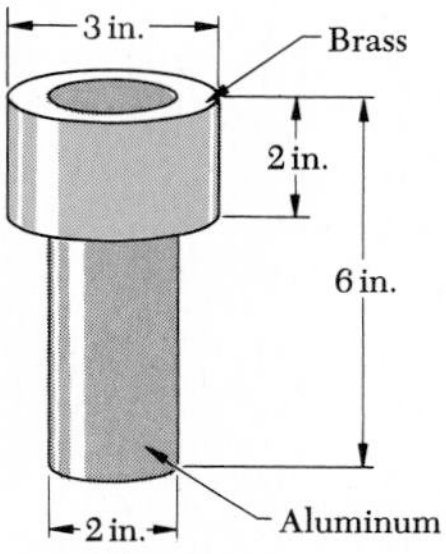

Fig. P5.101

5.102 and 5.103 Locate the center of gravity of the sheet-metal form shown.

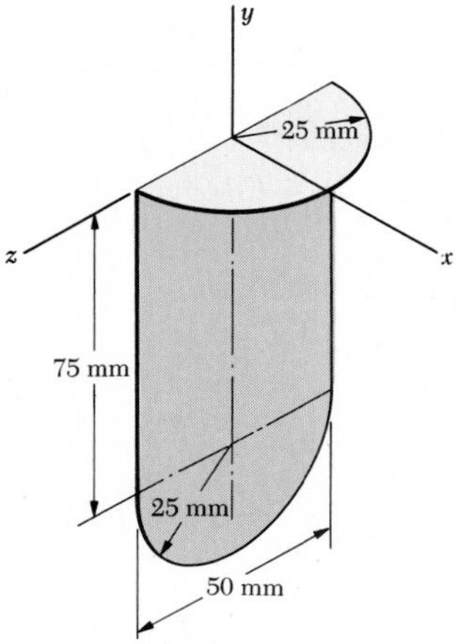

Fig. P5.102

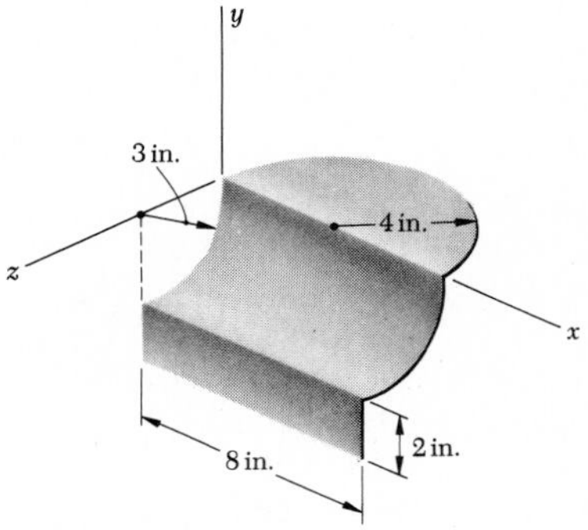

Fig. P5.103

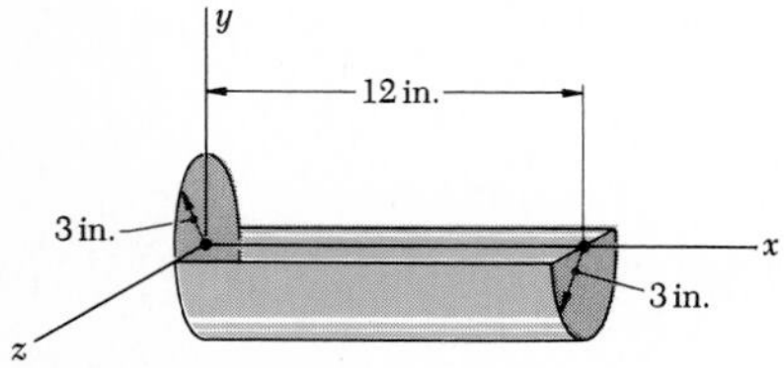

Fig. P5.104

5.104 Locate the center of gravity of the bin shown, knowing that it is made of sheet metal of uniform thickness.

5.105 Locate the centroid of the frustum of the right circular cone of Prob. 5.96, expressing the result in terms of r_1, r_2, and h.

5.106 Sheet metal of thickness t is used to form the half-conical shell shown. Locate the center of gravity of the shell. (*Hint.* Use the differential element shown.)

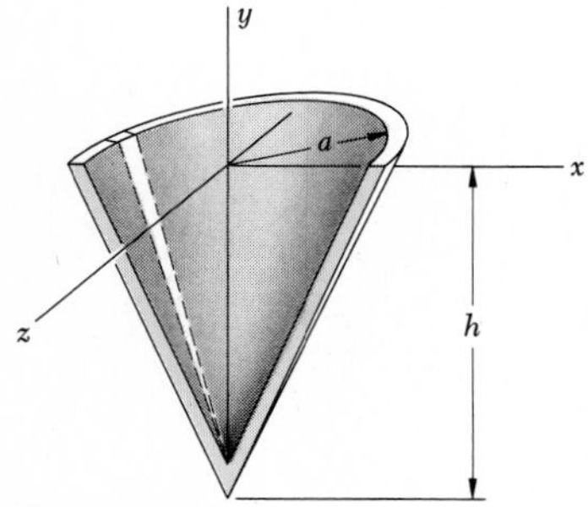

Fig. P5.106

5.107 Locate the center of gravity of a thin hemispherical shell of radius r and thickness t. (*Hint.* Consider the shell as formed by removing a hemisphere of radius r from a hemisphere of radius $r + t$; then neglect the terms containing t^2 and t^3, and keep those terms containing t.)

5.108 Derive by direct integration the expression given for $\bar{x}$ in Fig. 5.22 for a semiellipsoid of revolution.

5.109 Locate the centroid of the volume obtained by rotating the area shown about the x axis. The expression obtained may be used to confirm the values given in Fig. 5.22 for the paraboloid (with $n = \frac{1}{2}$) and the cone (with $n = 1$).

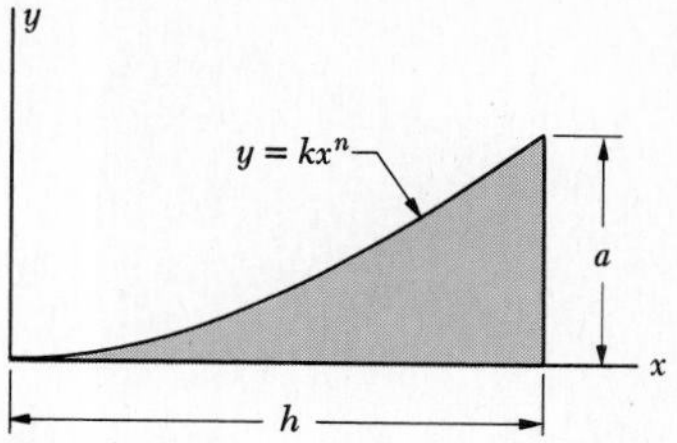

Fig. P5.109

5.110 Locate the centroid of the semiparaboloid of revolution shown.

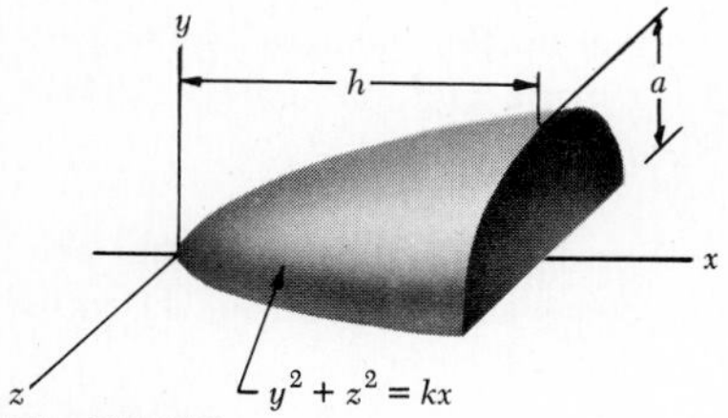

Fig. P5.110

5.111 The volume shown was obtained by rotating the area of Prob. 5.109, with $n = 2$, through 180° about the x axis. Locate its centroid by direct integration.

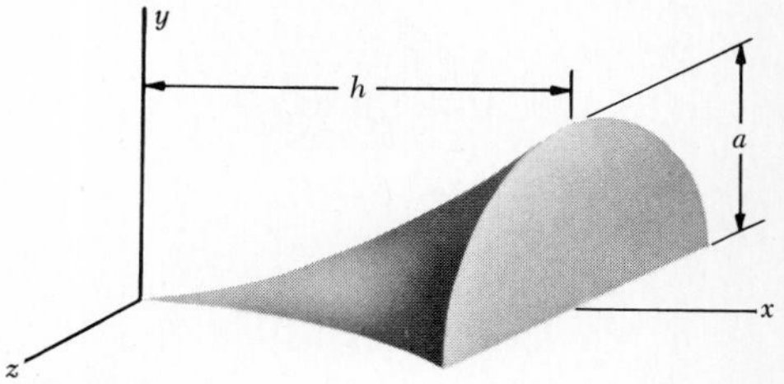

Fig. P5.111

5.112 Locate the centroid of the volume of the irregular pyramid shown.

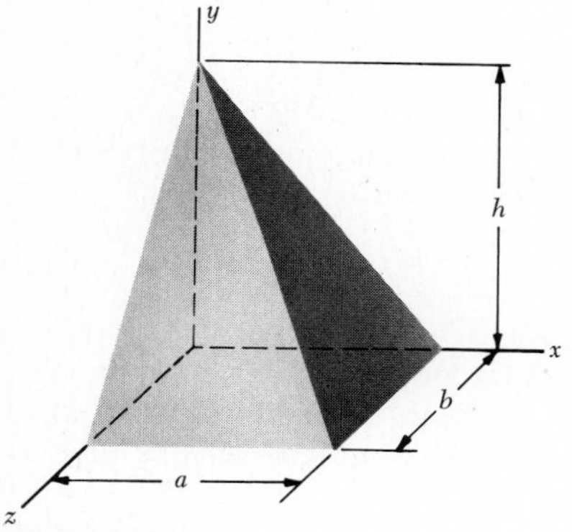

Fig. P5.112

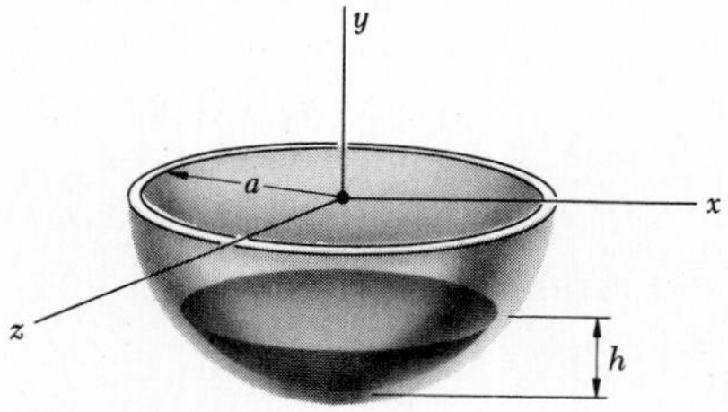

Fig. P5.113

***5.113** A hemispherical tank of radius a is filled with water to a depth h. Determine by direct integration the center of gravity of the water in the tank.

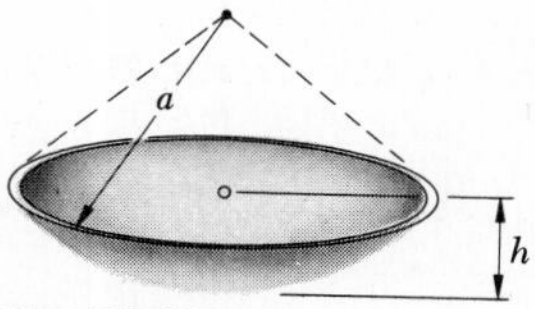

Fig. P5.114

***5.114** A thin spherical cup is of radius a and constant thickness t. Show by direct integration that the center of gravity of the cup is located at a distance $\frac{1}{2}h$ above the base of the cup.

***5.115** A spherical tank is 2 m in diameter and is filled with water to a depth of 1.5 m. Determine by direct integration the center of gravity of the water in the tank.

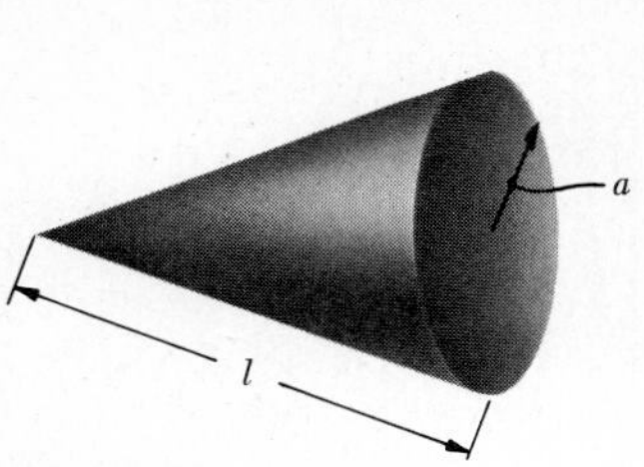

Fig. P5.116

5.116 A right circular cone is made by welding a circular disk of sheet metal of radius a to a conical shell of the same gage and of generatrix l. For what ratio l/a will the center of gravity of this hollow cone coincide with the center of gravity of a solid cone of the same dimensions? (*Hint.* Consider the conical shell as the limiting case of the lateral surface of a regular pyramid when the number of faces of the pyramid approaches infinity.)

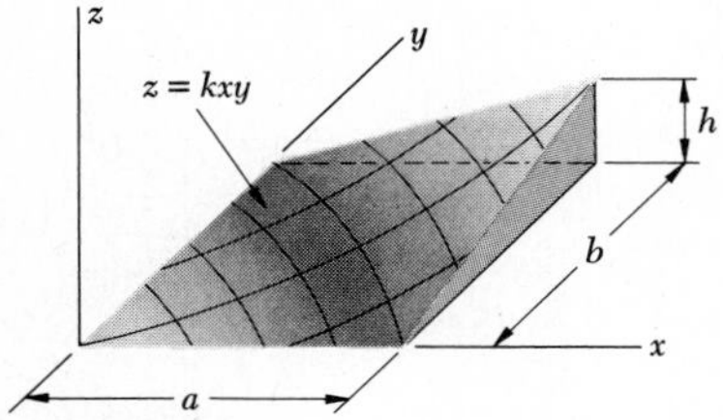

Fig. P5.117

5.117 Determine by direct integration the location of the centroid of the volume between the xy plane and the portion of the hyperbolic paraboloid shown.

5.118 Determine by direct integration the location of the centroid of the volume between the xz plane and the portion shown of the surface $y = h \sin (\pi x/a) \sin (\pi z/b)$.

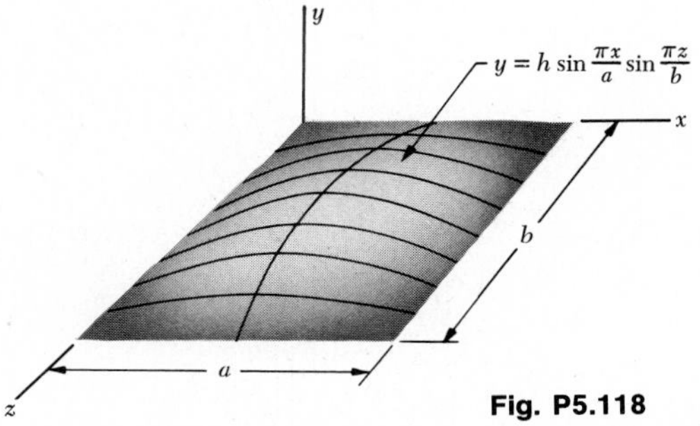

Fig. P5.118

5.119 Locate the centroid of the section shown which was cut from a thin, circular pipe by the oblique plane ABC.

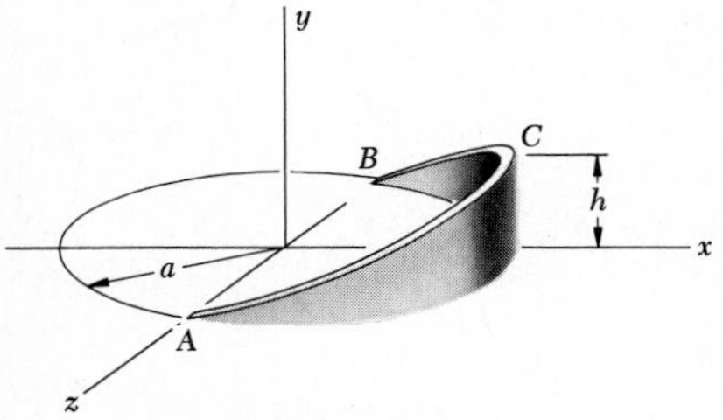

Fig. P5.119

REVIEW PROBLEMS

5.120 Locate the centroid of the plane area shown.

5.121 Determine the volume of the solid generated by rotating the area shown about the y axis.

5.122 Determine the total surface area of the solid generated by rotating the area shown about the y axis.

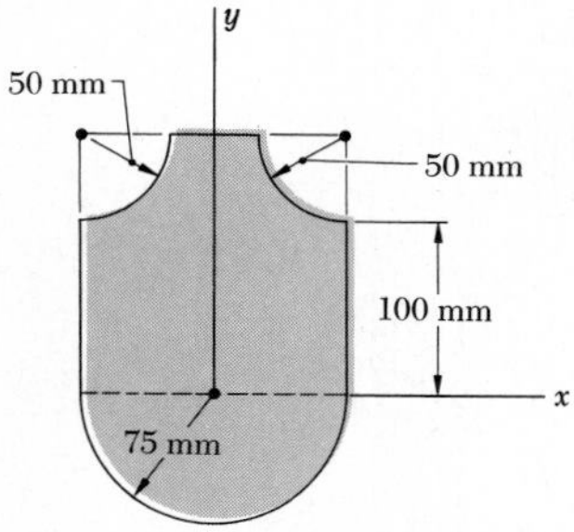

Fig. P5.120, P5.121, and P5.122

5.123 Locate the center of gravity of the sheet-metal form shown.

Fig. P5.123

5.124 Locate the centroid of the irregular pyramid shown. (*Hint.* See Fig. 5.22.)

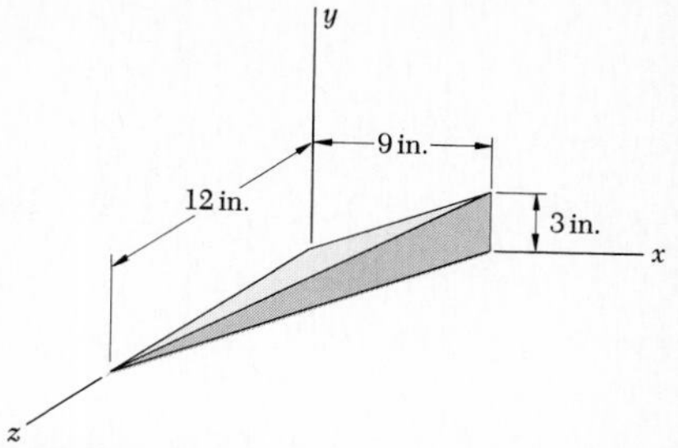

Fig. P5.124

5.125 The gate shown is supported by hinges along its bottom edge at A and rests against a wall at B. Knowing that the gate is 2 m wide, determine the reactions at A and B due to the hydrostatic forces acting on the gate.

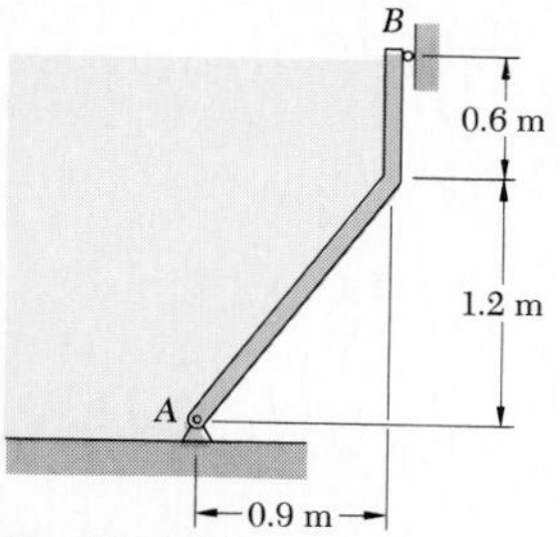

Fig. P5.125

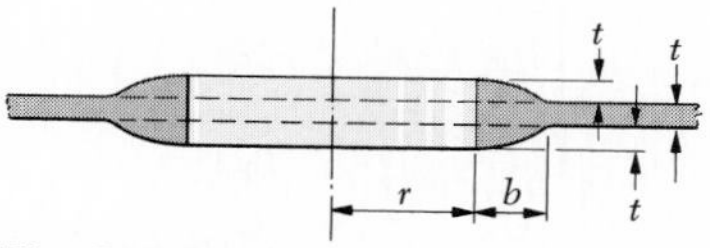

Fig. P5.126

5.126 A hole of radius r is drilled in a flat plate of thickness t. In order to reinforce the plate, a metal "crack stopper" is welded around the hole as shown. Assuming that the cross section of the crack stopper is parabolic, determine the ratio b/r for which the total volume of metal used is the same as that of a plate without a hole.

5.127 Determine the latitude of the parallel which divides the area of the northern hemisphere into two equal parts.

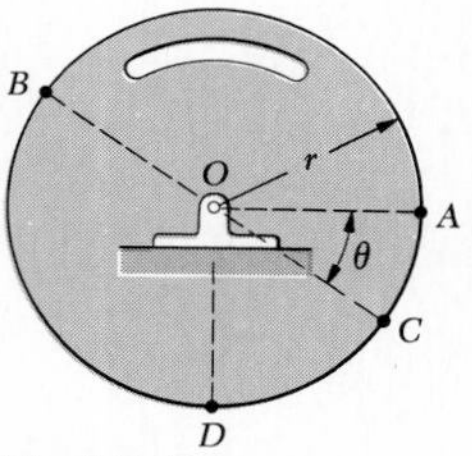

Fig. P5.128

***5.128** A curved slot is cut in a uniform disk of radius r. The disk is mounted on a frictionless shaft at O and is in equilibrium in the position shown. After a weight W_1 has been attached to the rim of the disk at A, the disk takes a new position of equilibrium in which line BOC is vertical. At what point on the rim should a second weight W_2 be attached, and how large should W_2 be, if the center of gravity of the disk and the two weights is to be located at O?

5.129 Concrete is poured into a steel pipe of length 2 ft, inside diameter 1 ft, and weight 100 lb. To what depth should concrete be poured if the combined center of gravity of the pipe and concrete is to be as low as possible? (Specific weight of concrete = 150 lb/ft³.)

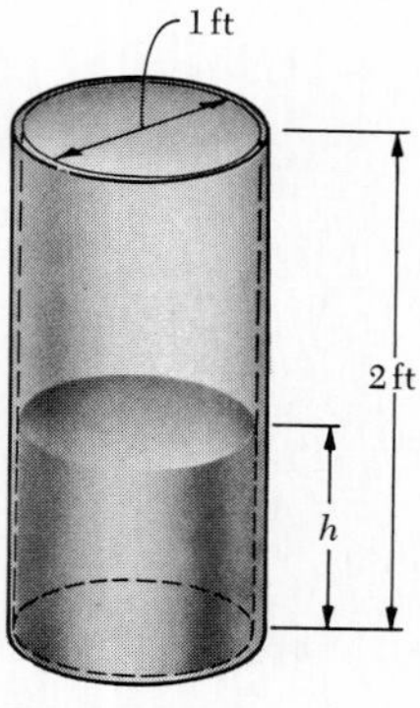

Fig. P5.129

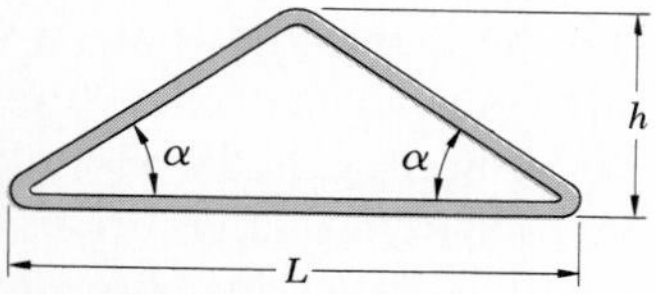

Fig. P5.130

5.130 A homogeneous wire is bent to form an isosceles triangle. Determine the angle α for which the center of gravity of the wire coincides with the centroid of the area enclosed by the wire.

5.131 Determine the reactions at A and B for the loading shown.

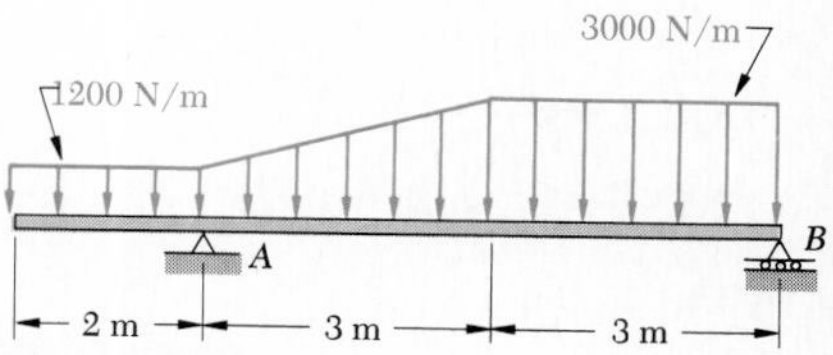

Fig. P5.131

Chapter

6

Analysis of Structures

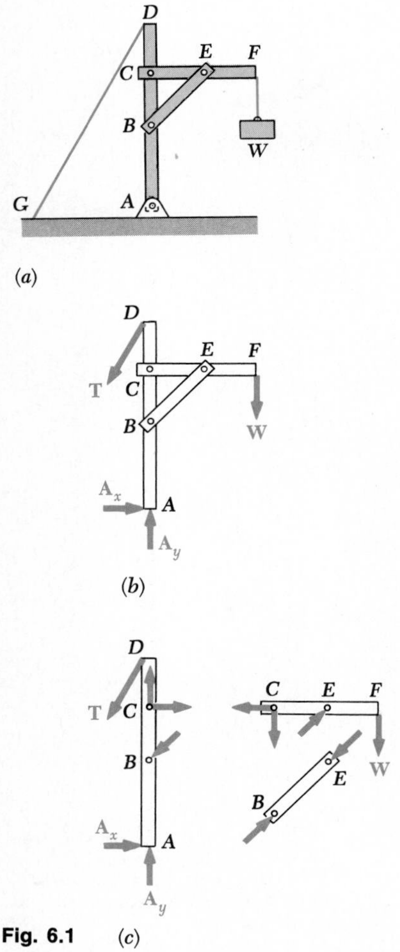

Fig. 6.1 (c)

6.1. Internal Forces. Newton's Third Law. The problems considered in the preceding chapters concerned the equilibrium of a single rigid body, and all forces involved were external to the rigid body. We shall now consider problems dealing with the equilibrium of structures made of several connected parts. These problems call not only for the determination of the external forces acting on the structure but also for the determination of the forces which hold together the various parts of the structure. From the point of view of the structure as a whole, these forces are *internal forces*.

Consider, for example, the crane shown in Fig. 6.1*a*, which carries a load *W*. The crane consists of three beams *AD*, *CF*, and *BE* connected by frictionless pins; it is supported by a pin at *A* and by a cable *DG*. The free-body diagram of the crane has been drawn in Fig. 6.1*b*. The external forces are shown in the diagram and include the weight **W**, the two components $\mathbf{A}_x$ and $\mathbf{A}_y$ of the reaction at *A*, and the force **T** exerted by the cable at *D*. The internal forces holding the various parts of the crane together do not appear in the diagram. If, however, the crane is dismembered and if a free-body diagram is drawn for each of its component parts, the forces holding the three beams together must also be represented, since these forces are external forces from the point of view of each component part (Fig. 6.1*c*).

It will be noted that the force exerted at *B* by member *BE* on member *AD* has been represented as equal and opposite to the force exerted at the same point by member *AD* on member

BE; similarly, the force exerted at *E* by *BE* on *CF* is shown equal and opposite to the force exerted by *CF* on *BE*; and the components of the force exerted at *C* by *CF* on *AD* are shown equal and opposite to the components of the force exerted by *AD* on *CF*. This is in conformity with Newton's third law, which states that *the forces of action and reaction between bodies in contact have the same magnitude, same line of action, and opposite sense*. As pointed out in Chap. 1, this law is one of the six fundamental principles of elementary mechanics and is based on experimental evidence. Its application is essential to the solution of problems involving connected bodies.

TRUSSES

6.2. Definition of a Truss. The truss is one of the major types of engineering structures. It provides both a practical and an economical solution to many engineering situations, especially in the design of bridges and buildings. A truss consists of straight members connected at joints; a typical truss is shown in Fig. 6.2*a*. Truss members are connected at their extremities only; thus no member is continuous through a joint. In Fig. 6.2*a*, for example, there is no member *AB*; there are instead two distinct members *AD* and *DB*. Actual structures are made of several trusses joined together to form a space framework. Each truss is designed to carry those loads which act in its plane and thus may be treated as a two-dimensional structure.

In general, the members of a truss are slender and can support little lateral load; all loads, therefore, must be applied to the various joints, and not to the members themselves. When a concentrated load is to be applied between two joints, or when a distributed load is to be supported by the truss, as in the case of a bridge truss, a floor system must be provided which, through the use of stringers and floor beams, transmits the load to the joints (Fig. 6.3).

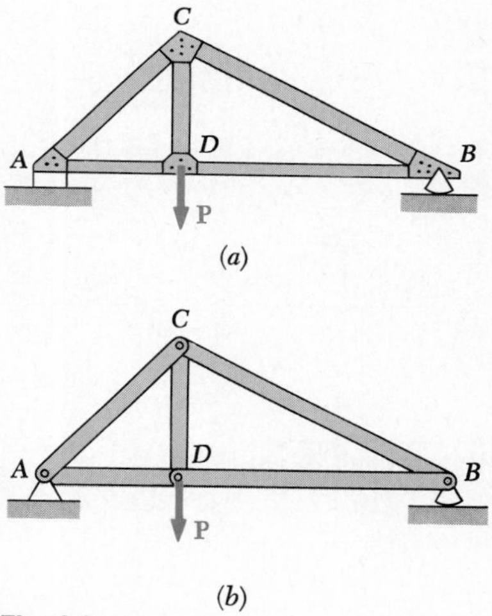

Fig. 6.2

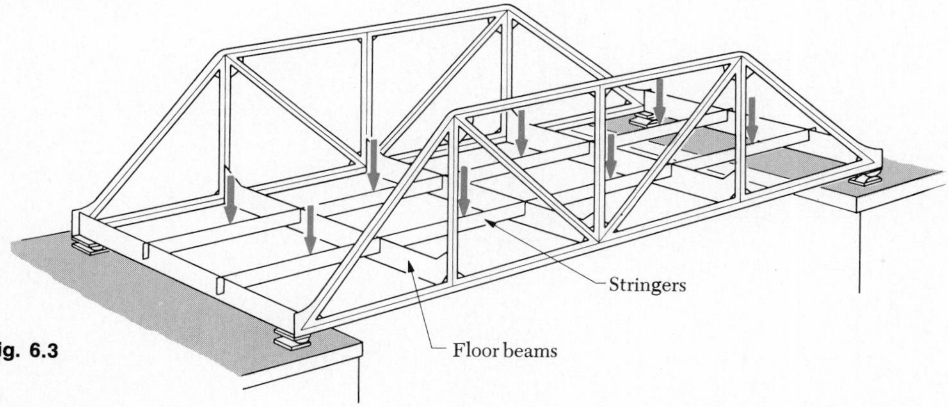

Fig. 6.3

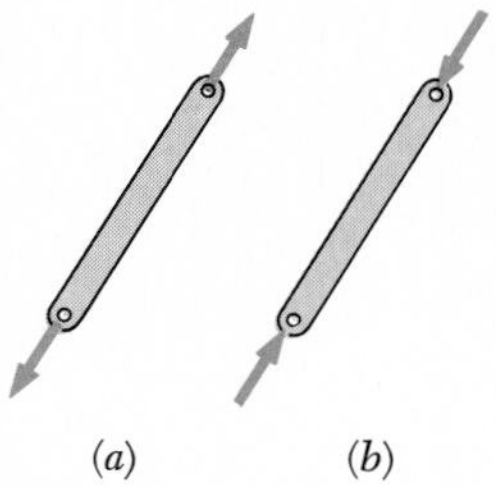

Fig. 6.4

The weights of the members of the truss are also assumed to be applied to the joints, half of the weight of each member being applied to each of the two joints the member connects. Although the members are actually joined together by means of riveted and welded connections, it is customary to assume that the members are pinned together; therefore, the forces acting at each end of a member reduce to a single force and no couple. Thus, the only forces assumed to be applied to a truss member are a single force at each end of the member. Each member may then be treated as a two-force member, and the entire truss may be considered as a group of pins and two-force members (Fig. 6.2*b*). An individual member may be acted upon as shown in either of the two sketches of Fig. 6.4. In the first sketch, the forces tend to pull the member apart, and the member is in tension, while, in the second sketch, the forces tend to compress the member, and the member is in compression. Several typical trusses are shown in Fig. 6.5.

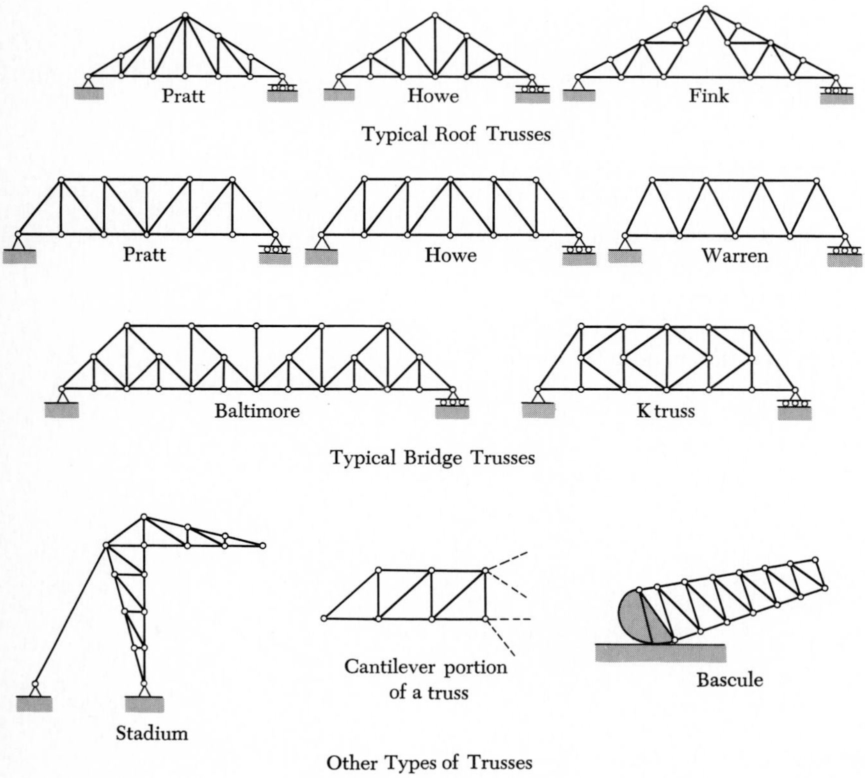

Fig. 6.5

6.3. Simple Trusses. Consider the truss of Fig. 6.6*a*, which is made of four members connected by pins at A, B, C, and D. If a load is applied at B, the truss will greatly deform and lose completely its original shape. On the other hand, the truss of Fig. 6.6*b*, which is made of three members connected by pins at A, B, and C, will deform only slightly under a load applied at B. The only possible deformation for this truss is one involving small changes in the length of its members. The truss of Fig. 6.6*b* is said to be a *rigid truss*, the term rigid being used here to indicate that the truss *will not collapse*.

As shown in Fig. 6.6*c*, a larger rigid truss may be obtained by adding two members BD and CD to the basic triangular truss of Fig. 6.6*b*. This procedure may be repeated as many times as desired, and the resulting truss will be rigid if, each time we add two new members, we attach them to separate existing joints and connect them at a new joint.† A truss which may be constructed in this manner is called a *simple truss*.

It should be noted that a simple truss is not necessarily made only of triangles. The truss of Fig. 6.6*d*, for example, is a simple truss which was constructed from triangle ABC by adding successively the joints D, E, F, and G. On the other hand, rigid trusses are not always simple trusses, even when they appear to be made of triangles. The Fink and Baltimore trusses shown in Fig. 6.5, for instance, are not simple trusses, since they cannot be constructed from a single triangle in the manner described above. All the other trusses shown in Fig. 6.5 are simple trusses, as may be easily checked. (For the K truss, start with one of the central triangles.)

Returning to the basic triangular truss of Fig. 6.6*b*, we note that this truss has three members and three joints. The truss of Fig. 6.6*c* has two more members and one more joint, i.e., altogether five members and four joints. Observing that every time two new members are added, the number of joints is increased by one, we find that in a simple truss the total number of members is $m = 2n - 3$, where n is the total number of joints.

6.4. Analysis of Trusses by the Method of Joints. We saw in Sec. 6.2 that a truss may be considered as a group of pins and two-force members. The truss of Fig. 6.2, whose free-body diagram is shown in Fig. 6.7*a*, may thus be dismembered, and a free-body diagram can be drawn for each pin and each member (Fig. 6.7*b*). Each member is acted upon by two forces, one at each end; these forces have the same magnitude, same line of action, and opposite sense (Sec. 3.16). Besides, Newton's third law indicates that the forces of action and reac-

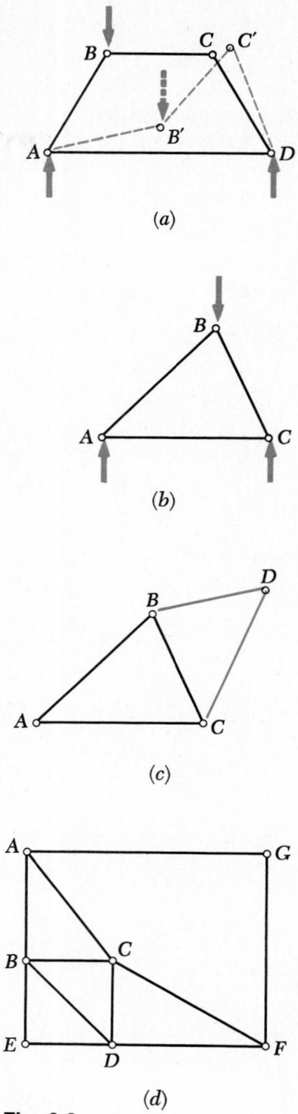

Fig. 6.6

† The three joints must not be in a straight line.

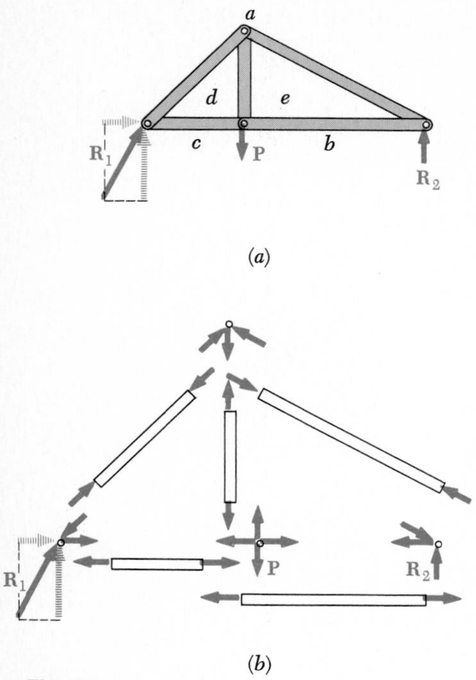

Fig. 6.7

tion between a member and a pin are equal and opposite. Therefore, the forces exerted by a member on the two pins it connects must be directed along that member and be equal and opposite. The common magnitude of the forces exerted by a member on the two pins it connects is commonly referred to as the *force in the member* considered, even though this quantity is actually a scalar. Since the lines of action of all the internal forces in a truss are known, the analysis of a truss reduces to the computation of the forces in its various members and to the determination of whether each of its members is in tension or in compression.

Since the entire truss is in equilibrium, each pin must be in equilibrium. The fact that a pin is in equilibrium may be expressed by drawing its free-body diagram and writing two equilibrium equations (Sec. 2.8). If the truss contains n pins, there will be therefore $2n$ equations available, which may be solved for $2n$ unknowns. In the case of a simple truss, we have $m = 2n - 3$, that is, $2n = m + 3$, and the number of unknowns which may be determined from the free-body diagrams of the pins is thus $m + 3$. This means that the forces in all the members, as well as the two components of the reaction $\mathbf{R}_1$, and the reaction $\mathbf{R}_2$ may be found by considering the free-body diagrams of the pins.

The fact that the entire truss is a rigid body in equilibrium may be used to write three more equations involving the forces shown in the free-body diagram of Fig. 6.7*a*. Since they do not contain any new information, these equations are not independent from the equations associated with the free-body diagrams of the pins. Nevertheless, they may be used to determine immediately the components of the reactions at the supports. The arrangement of pins and members in a simple truss is such that it will then always be possible to find a joint involving only two unknown forces. These forces may be determined by the methods of Sec. 2.10 and their values transferred to the adjacent joints and treated as known quantities at these joints. This procedure may be repeated until all unknown forces have been determined.

In order to expedite the analysis of trusses, it is desirable to establish a uniform method for denoting the various joints, members, loads, and forces. This is done by using the following notation, known as *Bow's notation*. A lowercase letter is assigned to every region between loads and reactions, moving clockwise around the truss, and to every area inside the truss (Fig. 6.7*a*). Each joint, member, load, and force may then be denoted as follows:

1. *Joints.* A joint is specified by naming *in clockwise order* the letters corresponding to all the areas adjacent to the joint. For example, the joint at the left support (Fig. 6.7*a*) is joint *adc*, or joint *dca*, or joint *cad*. For convenience, however, joints will be denoted sometimes by a number.
2. *Members.* A member is specified by naming the letters of the two adjacent areas. For example, the vertical member above the load **P** is member *de* or member *ed*.
3. *Loads.* A load is specified by reading *in clockwise order* the letters of the two areas adjacent to the load. The name of the load is then recorded with capital letters surmounted by an arrow. For example, at joint *debc*, the load **P** is denoted by $\overrightarrow{BC}$.
4. *Forces Exerted by Members on Pins.* As noted above, the forces exerted by a member on the two pins it connects must be directed along the member and be equal and opposite. In considering the action of the member on one of the two pins, we denote the force it exerts by reading the letters of the two areas adjacent to the member *in clockwise order* with respect to the joint. For example, the force exerted by member *ad* on pin *adc* is $\overrightarrow{AD}$, while the force exerted by the same member on pin *aed* is $\overrightarrow{DA}$. The common magnitude of these forces, commonly referred to as the force in member *ad*, may be denoted by either *AD* or *DA*.

We shall now proceed to analyze the truss of Fig. 6.7 by considering successively the equilibrium of each pin, starting with a joint at which only two forces are unknown. In the truss considered, all pins are subjected to at least three unknown forces. Therefore, the reactions at the supports must first be determined by considering the entire truss as a free body and using the equations of equilibrium of a rigid body. We find in this way that $\mathbf{R}_1$ is vertical and determine the magnitudes of $\mathbf{R}_1$ and $\mathbf{R}_2$.

The number of unknown forces at joint *adc* is thus reduced to two; and these forces may be determined by considering the equilibrium of pin *adc*. The magnitude and sense of $\overrightarrow{AD}$ and $\overrightarrow{DC}$ are obtained from the corresponding force triangle (Fig. 6.8).

We may now proceed to joint *debc*, where only two forces, $\overrightarrow{DE}$ and $\overrightarrow{EB}$, are still unknown. $\overrightarrow{BC}$ is the given load **P** and hence is known; $\overrightarrow{CD}$ is the force exerted on the pin by the member *cd* and, as indicated above, is equal and opposite to the force $\overrightarrow{DC}$ exerted by the same member on pin *adc*.

Next, joint *aed* is considered; its free-body diagram is shown in Fig. 6.8. It is noted that both $\overrightarrow{ED}$ and $\overrightarrow{DA}$ are known from the analysis of the preceding joints and that only $\overrightarrow{AE}$ is unknown. Since the equilibrium of each pin provides sufficient information to determine two unknowns, a check of our analysis is obtained at this joint. The force triangle is drawn, and the magnitude and sense of $\overrightarrow{AE}$ are determined. The check is obtained by verifying that the force $\overrightarrow{AE}$ and the member *ae* are parallel.

At joint *eab*, all the forces are known. Since the corresponding

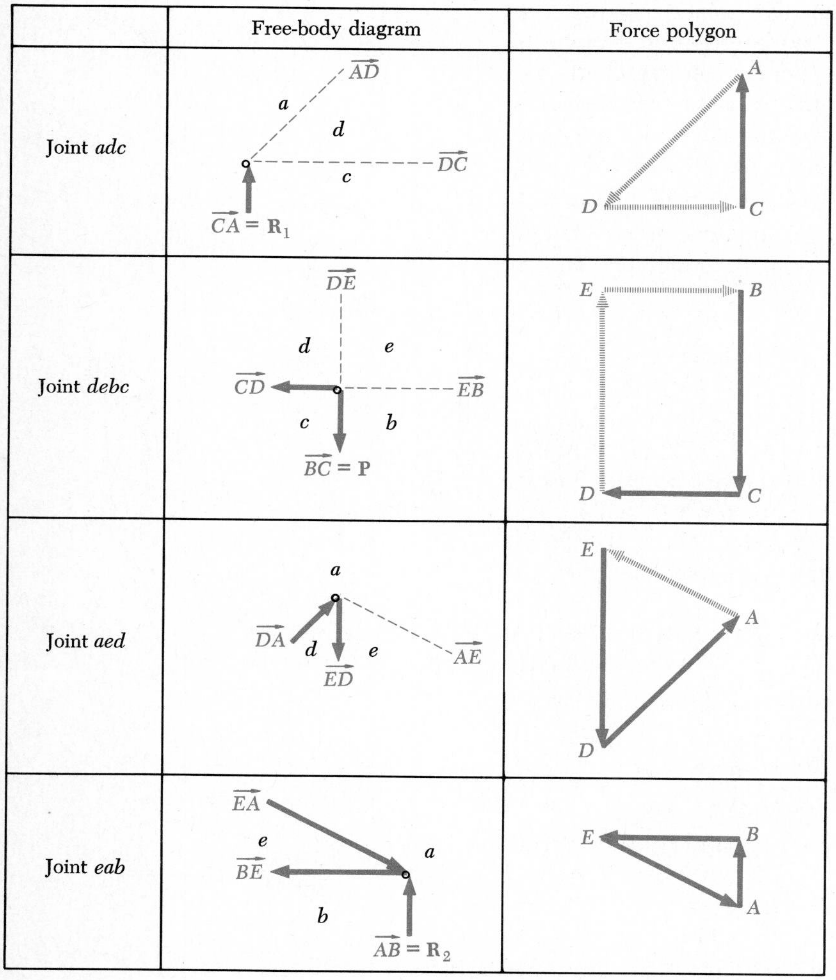

Fig. 6.8

pin is in equilibrium, the force triangle must close and an additional check of the analysis is obtained.

From the free-body diagrams shown in Fig. 6.8, it is seen that some forces act away from a given joint and others toward it. If the force acts away from the joint, the corresponding member pulls on the pin and the member is in tension; if the force acts toward the joint, the corresponding member pushes on the pin and the member is in compression. For example, at joint *adc*, it is seen from the free-body diagram and the force triangle that force $\overrightarrow{AD}$ acts toward the joint; hence, member *ad* pushes on the pin and is in compression. If we consider joint *aed*, it is seen that $\overrightarrow{DA}$ also acts toward the joint; hence, member *ad* is again found to push and thus to be in compression.

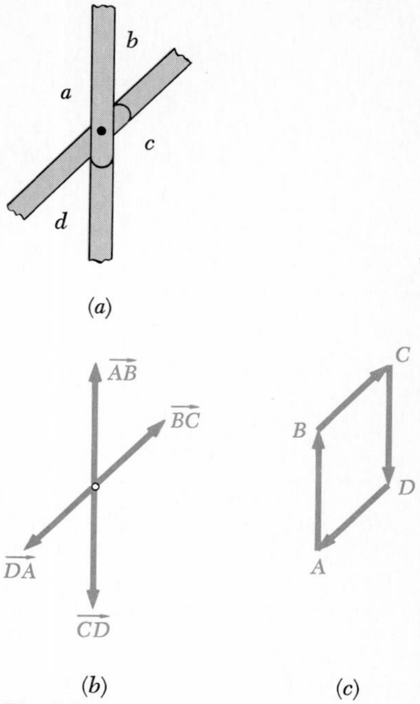

Fig. 6.9

*6.5. Joints under Special Loading Conditions.

Consider the joint shown in Fig. 6.9*a*, which connects four members lying in two intersecting straight lines. The free-body diagram of Fig. 6.9*b* shows that the pin is subjected to two pairs of directly opposite forces. The corresponding force polygon, therefore, must be a parallelogram (Fig. 6.9*c*), and *the forces in opposite members must be equal.*

Consider next the joint shown in Fig. 6.10*a*, which connects three members and supports a load **P**. Two of the members lie in the same line, and the load **P** acts along the third member. The free-body diagram of the pin and the corresponding force polygon again will be as shown in Fig. 6.9*b* and *c*. Thus, *the forces in the two opposite members must be equal, and the force in the other member must equal P.* A particular case of special interest is shown in Fig. 6.10*b*. Since, in this case, no external load is applied to the joint, we have $P = 0$, and the force in member *cd* is zero. Member *cd* is said to be a *zero-force member.*

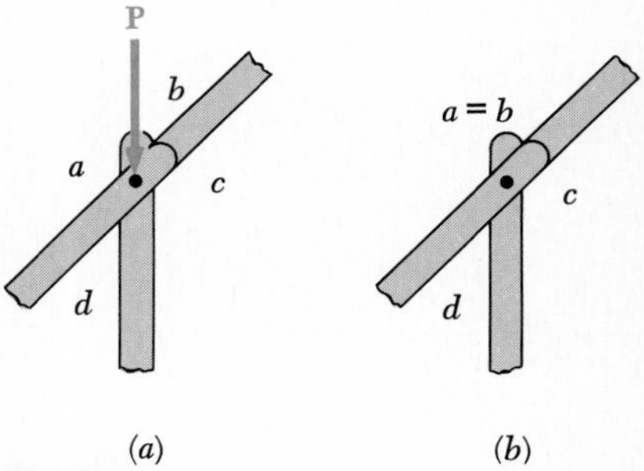

Fig. 6.10

Consider now a joint connecting two members only. From Sec. 2.8, we know that a particle which is acted upon by two forces will be in equilibrium if the two forces have the same magnitude, same line of action, and opposite sense. In the case of the joint of Fig. 6.11*a*, which connects two members lying in the same line, the equilibrium of the pin requires therefore that *the forces in the two members be equal.* In the case of the joint of Fig. 6.11*b*, equilibrium is impossible unless the forces in both members are zero. Members connected as shown in Fig. 6.11*b*, therefore, must be *zero-force members.*

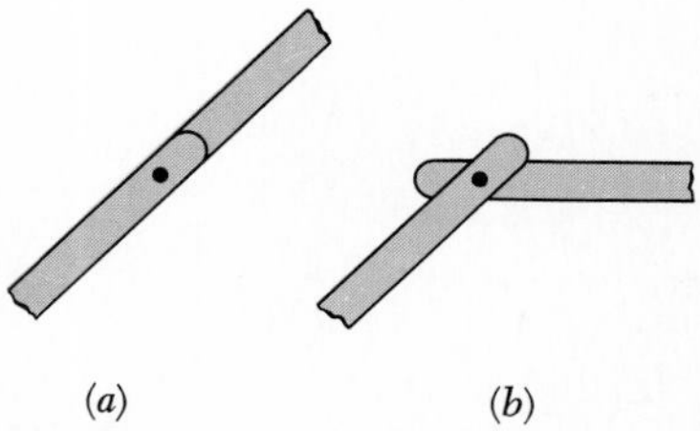

Fig. 6.11

Spotting the joints which are under the special loading conditions listed above will expedite the analysis of a truss. Consider, for example, a Howe truss loaded as shown in Fig. 6.12. All the members represented by colored lines will be recognized as zero-force members. Joint *3* connects three members, two

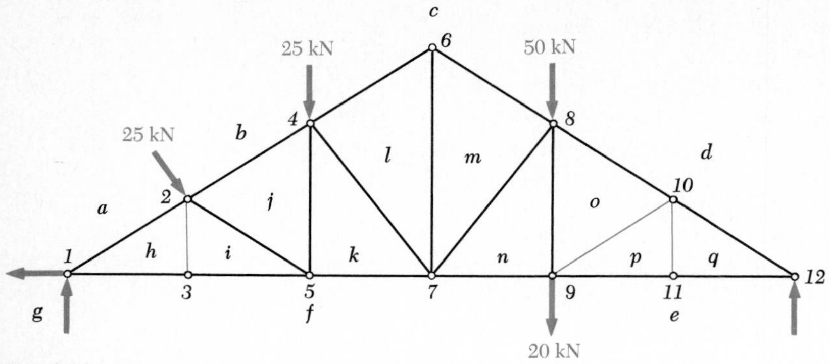

Fig. 6.12

of which lie in the same line, and is not subjected to any external load; member *hi* is thus a zero-force member. Applying the same reasoning to joint *11*, we find that member *pq* is also a zero-force member. But joint *10* is now in the same situation as joints *3* and *11*, and member *po* must be a zero-force member. The examination of joints *3*, *10*, and *11* also shows that the forces in members *fh* and *if* are equal, that the forces in members *od* and *dq* are equal, and that the forces in members *ep* and *qe* are equal. Furthermore, now turning our attention to joint *9*, where the 20-kN load and member *no* are collinear, we note that the force in member *no* is 20 kN (tension) and that the forces in members *fn* and *pe* are equal. Hence, the forces in members *ep*, *eq*, and *fn* are equal.

Students, however, should be warned against misusing the rules established in this section. For example, it would be wrong to assume that the force in member *kj* is 25 kN or that the forces in members *bj* and *ha* are equal. The conditions discussed above do not apply to joints *2* and *4*. The forces in these members and in all remaining members should be found by carrying out the analysis of joints *1*, *2*, *4*, *5*, *6*, *7*, *8*, and *12* in the usual manner. Until they have become thoroughly familiar with the conditions of application of the rules established in this section, students would be well advised to draw the free-body diagrams of all pins and to write the corresponding equilibrium equations (or draw the corresponding force polygons), whether or not the joints considered fall into the categories listed above.

A final remark concerning zero-force members: These members are not useless. While they do not carry any load under the particular loading conditions shown, the zero-force members of Fig. 6.12 will probably carry loads if the loading conditions are changed. Besides, even in the case considered, these members are needed to support the weight of the truss and to maintain the truss in the desired shape.

∗6.6. Space Trusses. When several straight members are joined together at their extremities to form a three-dimensional configuration, the structure obtained is called a *space truss*.

We recall from Sec. 6.3 that the most elementary two-dimensional rigid truss consisted of three members joined at their extremities to form the sides of a triangle; by adding two members at a time to this basic configuration, and connecting them at a new joint, it was possible to obtain a larger rigid structure which was defined as a simple truss. Similarly, the most elementary rigid space truss consists of six members joined at their extremities to form the edges of a tetrahedron $ABCD$ (Fig. 6.13*a*). By adding three members at a time to this basic configuration, such as AE, BE, and CE, attaching them at separate existing joints, and connecting them at a new joint, we can obtain a larger rigid structure which is defined as a *simple space truss* (Fig. 6.13*b*).† Observing that the basic tetrahedron has six members and four joints, and that, every time three members are added, the number of joints is increased by one, we conclude that in a simple space truss the total number of members is $m = 3n - 6$, where n is the total number of joints.

If a space truss is to be completely constrained and if the reactions at its supports are to be statically determinate, the supports should consist of a combination of balls, rollers, and balls and sockets which provides six unknown reactions (see Sec. 4.7). These unknown reactions may be readily determined by solving the six equations expressing that the three-dimensional truss is in equilibrium.

Although the members of a space truss are actually joined together by means of riveted or welded connections, it is assumed that each joint consists of a ball-and-socket connection. Thus, no couple will be applied to the members of the truss, and each member may be treated as a two-force member. The conditions of equilibrium for each joint will be expressed by the three equations $\Sigma F_x = 0$, $\Sigma F_y = 0$, and $\Sigma F_z = 0$. In the case of a simple space truss containing n joints, writing the conditions of equilibrium for each joint will thus yield $3n$ equations. Since $m = 3n - 6$, these equations suffice to determine all unknown forces (forces in m members and six reactions at the supports). However, to avoid solving many simultaneous equations, care should be taken to select joints in such an order that no selected joint will involve more than three unknown forces.

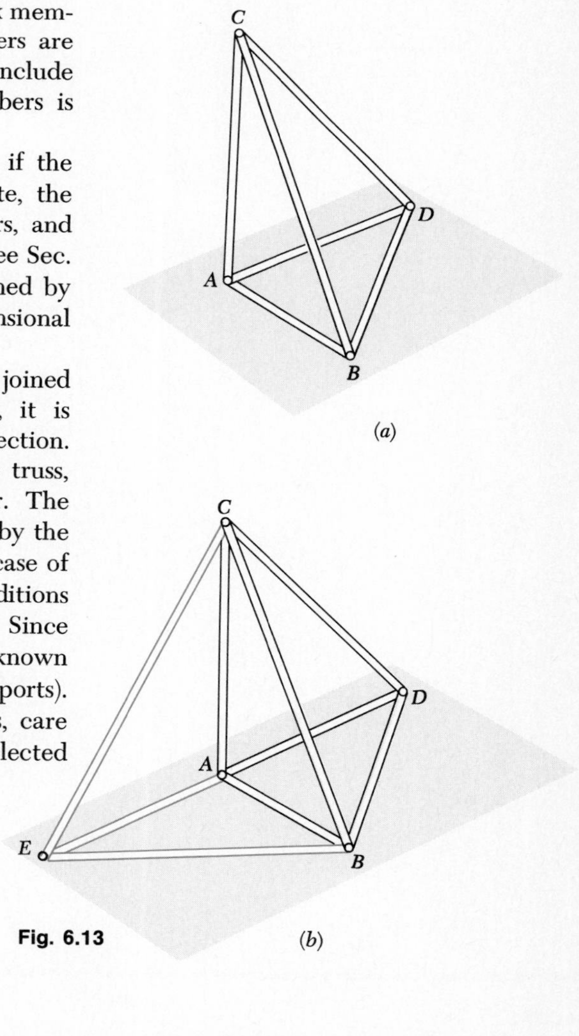

Fig. 6.13

† The four joints must not lie in a plane.

SAMPLE PROBLEM 6.1

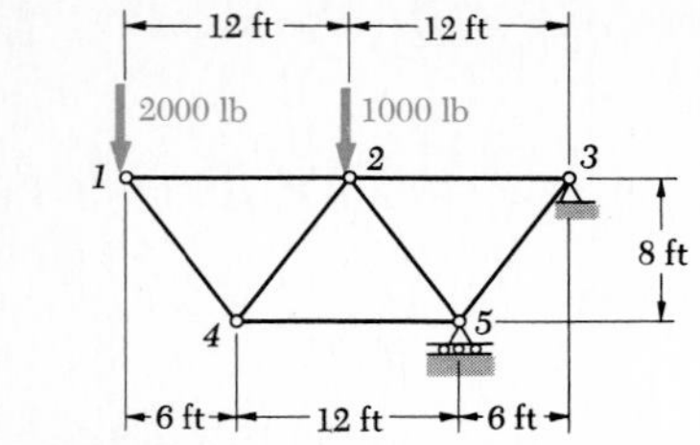

Using the method of joints, determine the force in each member of the truss shown.

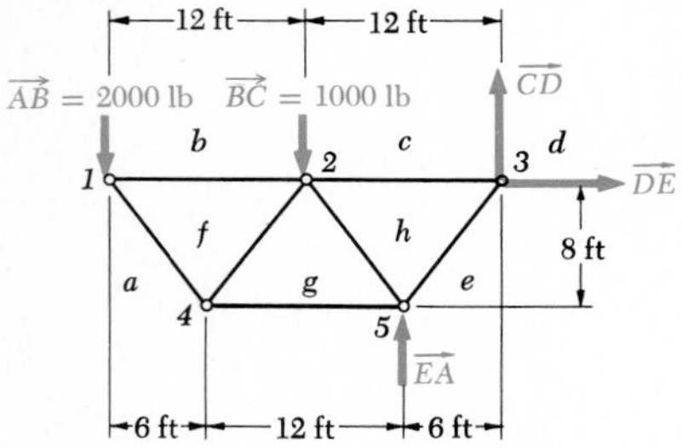

Solution. A free-body diagram of the entire truss is drawn; external forces acting on this free body consist of the applied loads and the reactions. The truss is then lettered, using Bow's notation. The applied loads are thus $\overrightarrow{AB}$ and $\overrightarrow{BC}$; the reactions are $\overrightarrow{CD}$, $\overrightarrow{DE}$, and $\overrightarrow{EA}$.

Equilibrium of Entire Truss.

$+\circlearrowleft \Sigma M_3 = 0$: $\quad (2000 \text{ lb})(24 \text{ ft}) + (1000 \text{ lb})(12 \text{ ft}) - (EA)(6 \text{ ft}) = 0$

$EA = +10{,}000 \text{ lb} \qquad \overrightarrow{EA} = 10{,}000 \text{ lb} \uparrow$

$\xrightarrow{+} \Sigma F_x = 0$: $\qquad \overrightarrow{DE} = 0$

$+\uparrow \Sigma F_y = 0$: $\quad -2000 \text{ lb} - 1000 \text{ lb} + 10{,}000 \text{ lb} + CD = 0$

$CD = -7000 \text{ lb} \qquad \overrightarrow{CD} = 7000 \text{ lb} \downarrow$

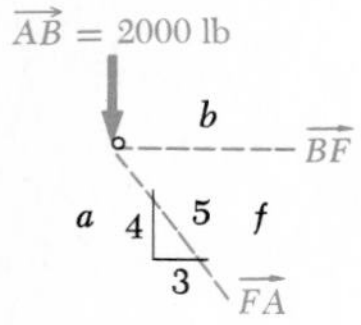

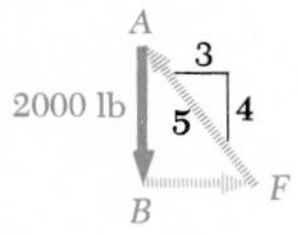

Joint 1. This joint is subjected to only two unknown forces, namely, the forces exerted by members *bf* and *fa*. A force triangle is used to determine $\overrightarrow{BF}$ and $\overrightarrow{FA}$. We note that member *bf* pulls on the joint and thus is in tension and that member *fa* pushes on the joint and thus is in compression. The magnitudes of the two forces are obtained from the proportion

$$\frac{2000 \text{ lb}}{4} = \frac{BF}{3} = \frac{FA}{5}$$

$BF = 1500 \text{ lb } T$ ◄

$FA = 2500 \text{ lb } C$ ◄

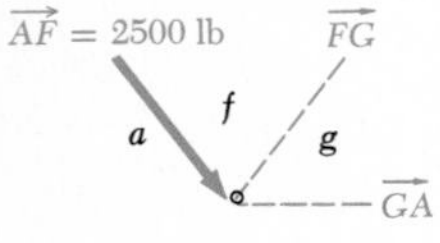

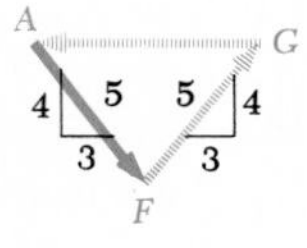

Joint 4. Since the force exerted by member *af* has been determined, only two unknown forces are now involved at this joint. Again, a force triangle is used to determine the unknown forces in members *fg* and *ga*.

$FG = AF \qquad FG = 2500 \text{ lb } T$ ◄

$GA = 2(\tfrac{3}{5})(AF) \qquad GA = 3000 \text{ lb } C$ ◄

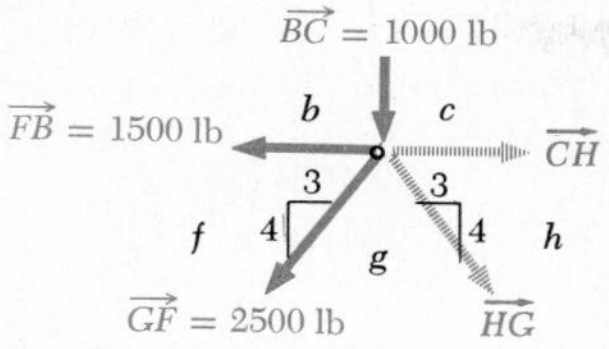

Joint 2. Since more than three forces act at this joint, we determine the two unknown forces $\vec{CH}$ and $\vec{HG}$ by solving the equilibrium equations $\Sigma F_x = 0$ and $\Sigma F_y = 0$. We arbitrarily assume that both unknown forces act away from the joint, i.e., that the members are in tension. The positive value obtained for CH indicates that our assumption was correct; member ch is in tension. The negative value of HG indicates that our assumption was wrong; member hg is in compression.

$$+\uparrow \Sigma F_y = 0: \quad -1000 - \tfrac{4}{5}(2500) - \tfrac{4}{5}(HG) = 0$$
$$HG = -3750 \text{ lb} \qquad HG = 3750 \text{ lb } C$$

$$\overset{+}{\rightarrow} \Sigma F_x = 0: \quad CH - 1500 - \tfrac{3}{5}(2500) - \tfrac{3}{5}(3750) = 0$$
$$CH = +5250 \text{ lb} \qquad CH = 5250 \text{ lb } T$$

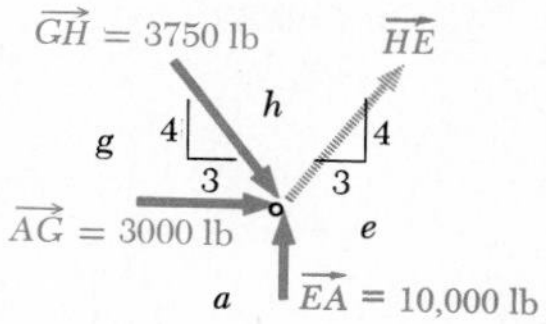

Joint 5. The unknown force $\vec{HE}$ is assumed to act away from the joint. Summing x components, we write

$$\overset{+}{\rightarrow} \Sigma F_x = 0: \quad \tfrac{3}{5}(HE) + 3000 + \tfrac{3}{5}(3750) = 0$$
$$HE = -8750 \text{ lb} \qquad HE = 8750 \text{ lb } C$$

Summing y components, we obtain a check of our computations:

$$+\uparrow \Sigma F_y = 10{,}000 - \tfrac{4}{5}(3750) - \tfrac{4}{5}(8750)$$
$$= 10{,}000 - 3000 - 7000 = 0 \qquad \text{(checks)}$$

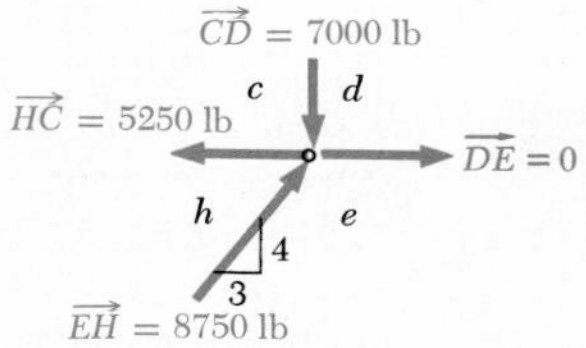

Joint 3. Using the computed values of $\vec{HC}$ and $\vec{EH}$, we may determine the reactions $\vec{CD}$ and $\vec{DE}$ by considering the equilibrium of this joint. Since these reactions have already been determined from the equilibrium of the entire truss, we will obtain two checks of our computations. We may also merely use the computed values of all forces acting on the joint (forces in members and reactions) and check that the joint is in equilibrium

$$\overset{+}{\rightarrow} \Sigma F_x = -5250 + \tfrac{3}{5}(8750) = -5250 + 5250 = 0 \qquad \text{(checks)}$$
$$+\uparrow \Sigma F_y = -7000 + \tfrac{4}{5}(8750) = -7000 + 7000 = 0 \qquad \text{(checks)}$$

PROBLEMS

6.1 through 6.12 Using the method of joints, determine the force in each member of the truss shown. State whether each member is in tension or compression.

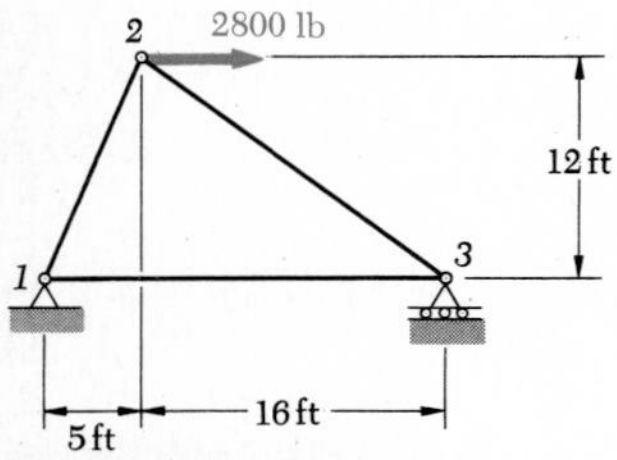

Fig. P6.1 and P6.21

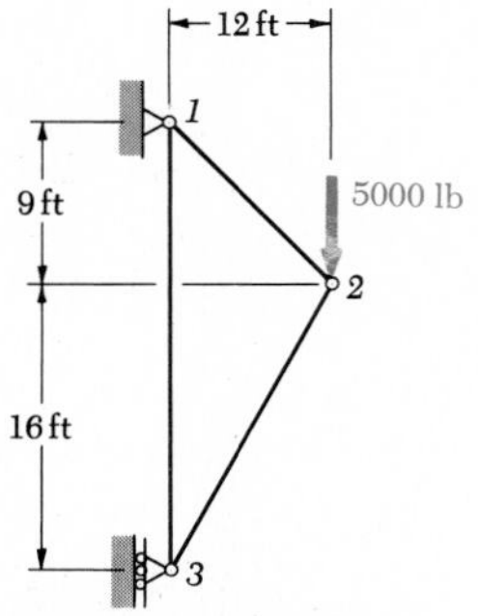

Fig. P6.2 and P6.22

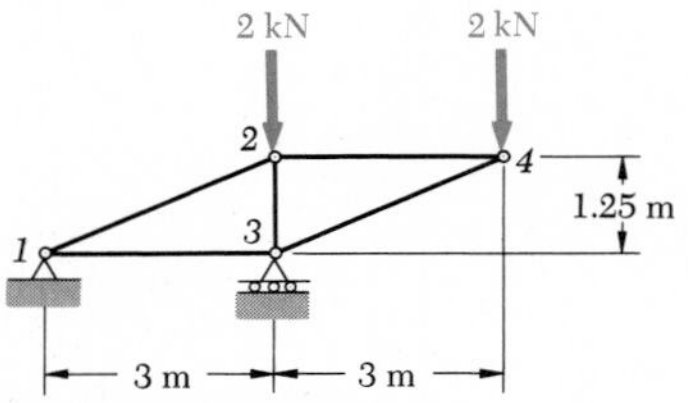

Fig. P6.3 and P6.23

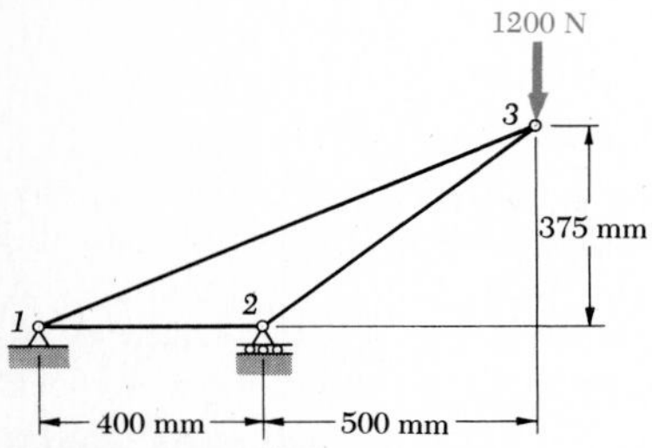

Fig. P6.4 and P6.24

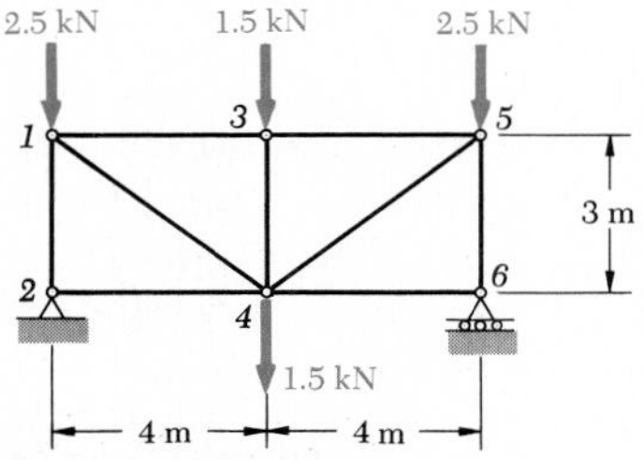

Fig. P6.5 and P6.25

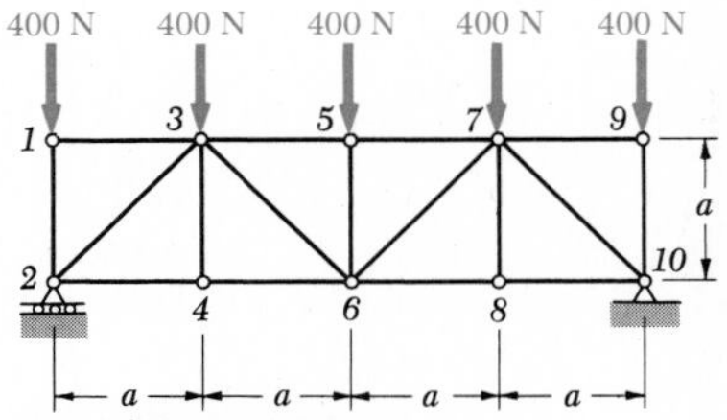

Fig. P6.6 and P6.26

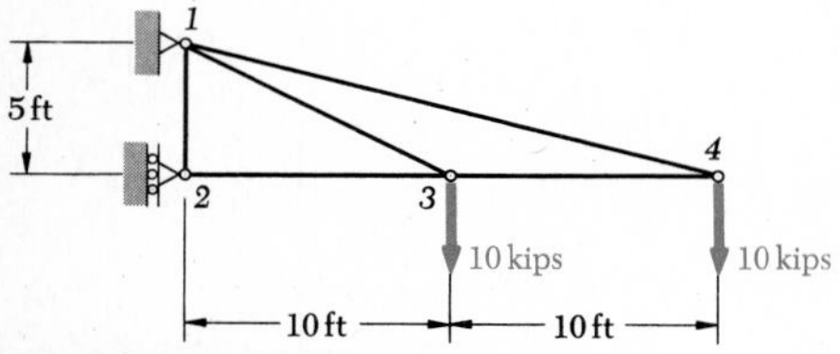

Fig. P6.7 and P6.27

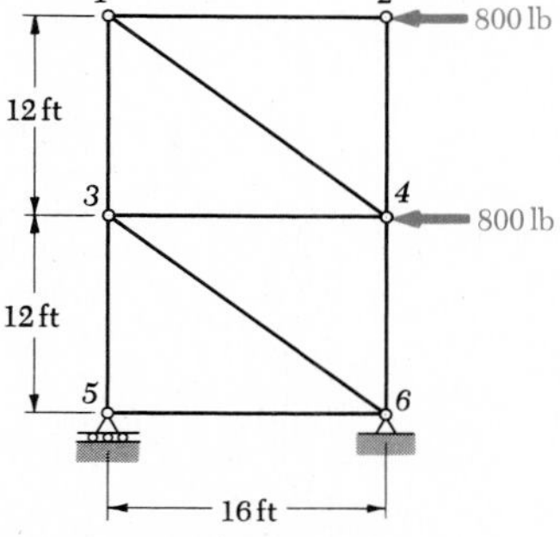

Fig. P6.8 and P6.28

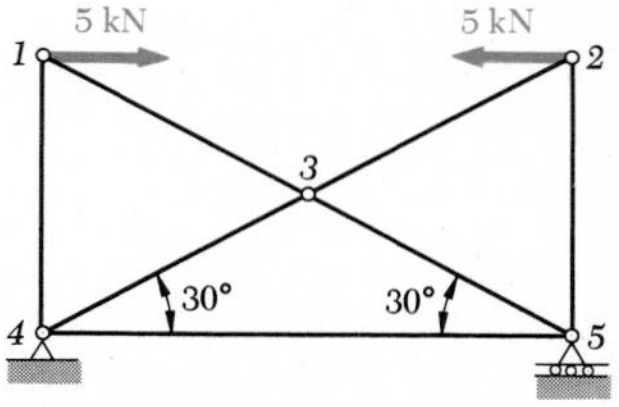

Fig. P6.9 and P6.29

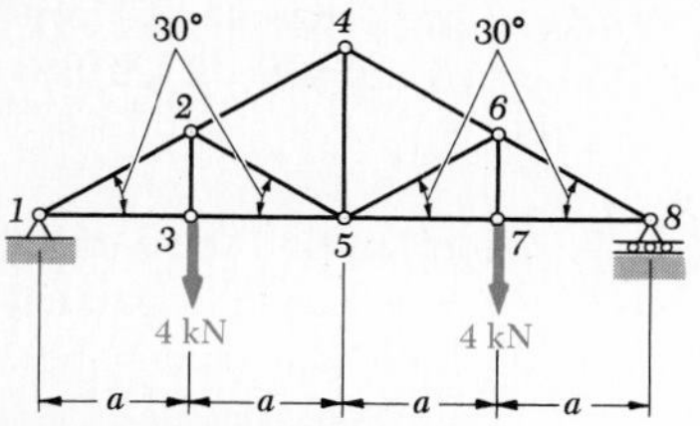

Fig. P6.10 and P6.30

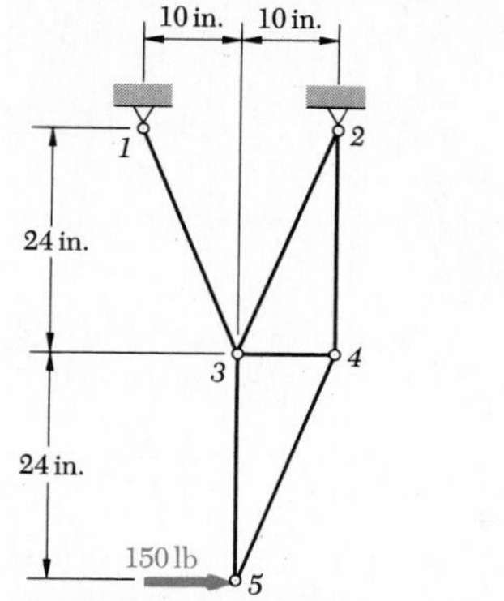

Fig. P6.11 and P6.31

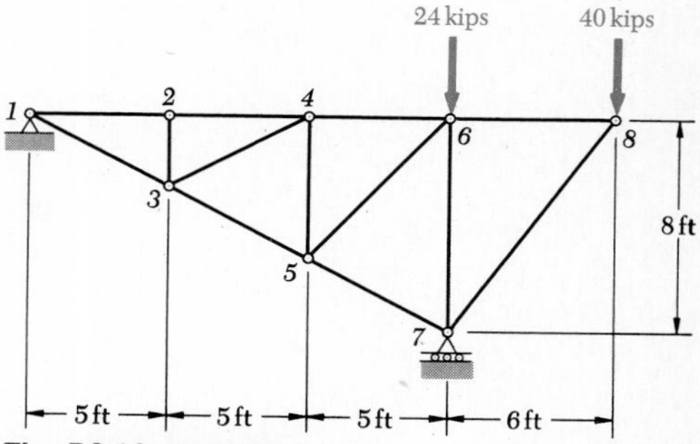

Fig. P6.12 and P6.32

6.13 Determine whether the trusses given in Probs. 6.7, 6.9, 6.14, 6.15, and 6.16 are simple trusses.

6.14 through 6.16 Determine the zero-force members in the truss shown for the given loading.

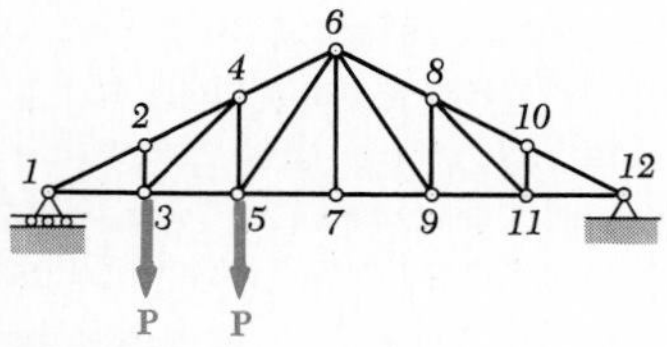

Fig. P6.14

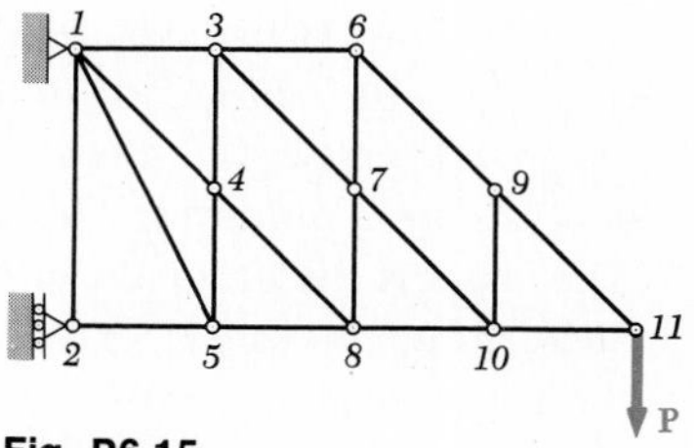

Fig. P6.15

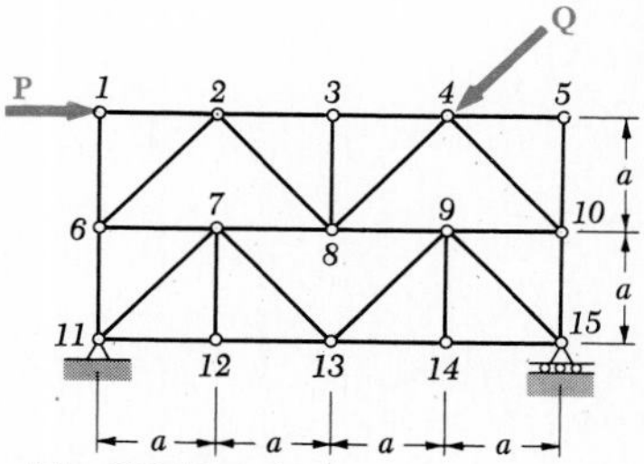

Fig. P6.16

***6.17** Six bars, each of length L, are connected to form a regular tetrahedron which rests on a smooth horizontal surface. Determine the force in each of the six members when a vertical force **P** is applied at A.

Fig. P6.17

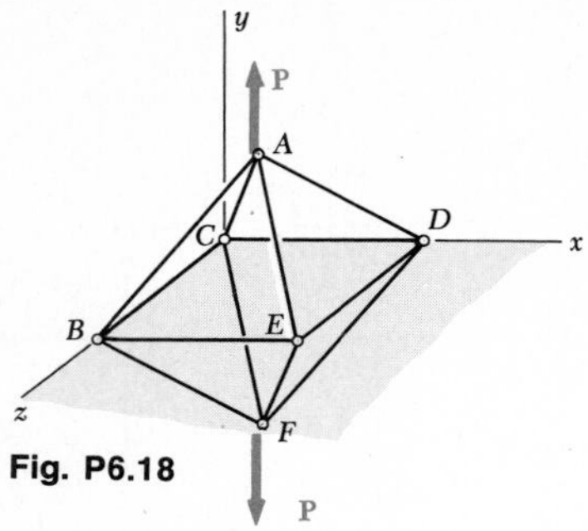

Fig. P6.18

***6.18** Twelve members, each of length L, are connected to form a regular octahedron. Determine the force in each member if two vertical loads are applied as shown.

***6.19** The three-dimensional truss is supported by the six reactions shown. If an 800-lb load is applied at E in a direction parallel to the y axis, determine (*a*) the reactions, (*b*) the force in each member.

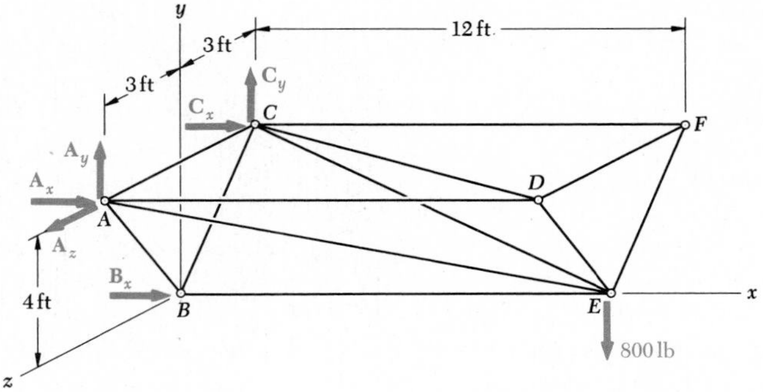

Fig. P6.19

***6.20** Solve Prob. 6.19, assuming that the 800-lb force is applied at E in a direction parallel to the z axis (with positive sense).

6.7. Graphical Analysis of Trusses: Maxwell's Diagram. The method of joints may be used as the basis for a graphical analysis of trusses. We shall develop this graphical analysis by considering the truss already discussed in Sec. 6.4. This truss is shown again in Fig. 6.14*a*, and a force polygon has been drawn to scale for each joint in Fig. 6.14*b*; the force in each member may now be measured from one of these force polygons. The number of lines which have to be drawn can be greatly reduced, however, if the various force polygons are superimposed. The resulting diagram is shown in Fig. 6.14*c* and is known as the *Maxwell diagram* of the truss.

In order to draw the Maxwell diagram directly, we shall proceed as follows:

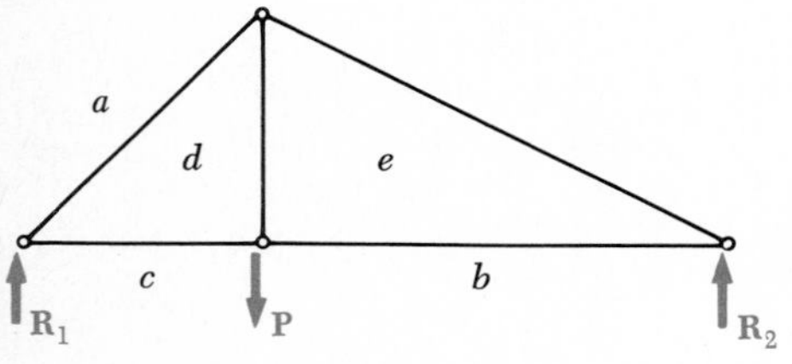

(*a*)

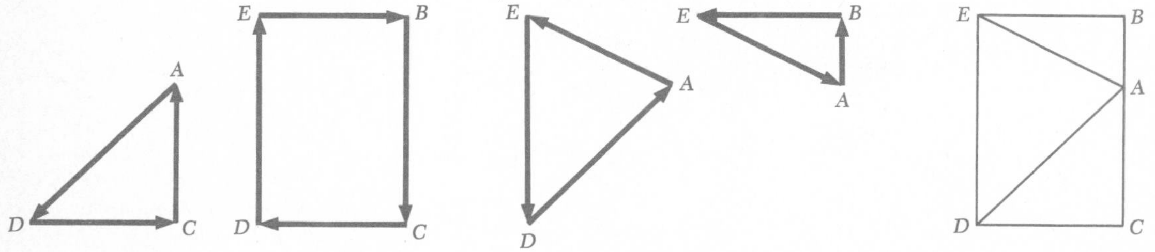

Fig. 6.14 (*b*) (*c*)

1. A lowercase letter is assigned to every region outside the truss, moving clockwise around the truss, and also to every area inside the truss (Fig. 6.14*a*).
2. The reactions $\mathbf{R}_1$ and $\mathbf{R}_2$ are determined by solving the equations of equilibrium for the entire truss.
3. Selecting the external forces in clockwise order around the truss, we draw to scale the force polygon for the entire truss (line *ABCA* in Fig. 6.15*a*). If a joint may be found which is acted upon by only two forces, steps 1 and 2 may be omitted or at least postponed.
4. A force polygon is then drawn for each joint by treating successively joints acted upon by only two unknown forces. All lines previously drawn may be used; thus, it is necessary to draw only two additional lines to complete each new force polygon. For example, starting at joint *adc*, we draw a line parallel to *ad* through point *A* and a line parallel to *dc* through point *C*; this determines point *D* and thus completes the force triangle *ADC*, which corresponds to joint *adc* (Fig. 6.15*a*). Considering next joint *debc*, we draw a line parallel to *de* through *D* and a line parallel to *eb* through *B*; we obtain point *E* and thus complete the force polygon *DEBC*, which corresponds to joint *debc* (Fig. 6.15*b*). Next, we consider joint *aed* and draw a line parallel to *ae* through point *A* (Fig. 6.15*c*); we check that this line passes through the point *E* previously obtained. This completes the force triangle *AED* corresponding to joint *aed* and also completes the Maxwell diagram. At joint *eab*, all the forces are now known; we simply check that they form a closed triangle *EAB* (Fig. 6.15*d*).
5. The magnitude of the force in each member may now be measured on the Maxwell diagram. Thus, the magnitude of the force in member *ad* is *AD*. To determine whether a member is in tension or in compression, we shall determine whether it pulls or pushes on either of the two joints it connects. For example, consider member *ad*, which connects joints *adc* and *aed*. (*a*) We select one of these two joints, say, *adc*. (*b*) We read the names of the areas adjacent to the member in clockwise order around the joint in Fig. 6.14*a*; we read *ad*. (*c*) The direction of the force exerted on the joint is found by reading the corresponding letters of the Maxwell diagram in the same order (Fig. 6.14*c*). Since force $\overrightarrow{AD}$ is directed down and to the left, member *ad* pushes on the joint and must be in compression. The same result may be found by considering joint *aed*. The member is now read *da* and the corresponding force $\overrightarrow{DA}$. Since force $\overrightarrow{DA}$ is directed up and to the right, member *da* pushes on joint *aed* and is again found to be in compression.

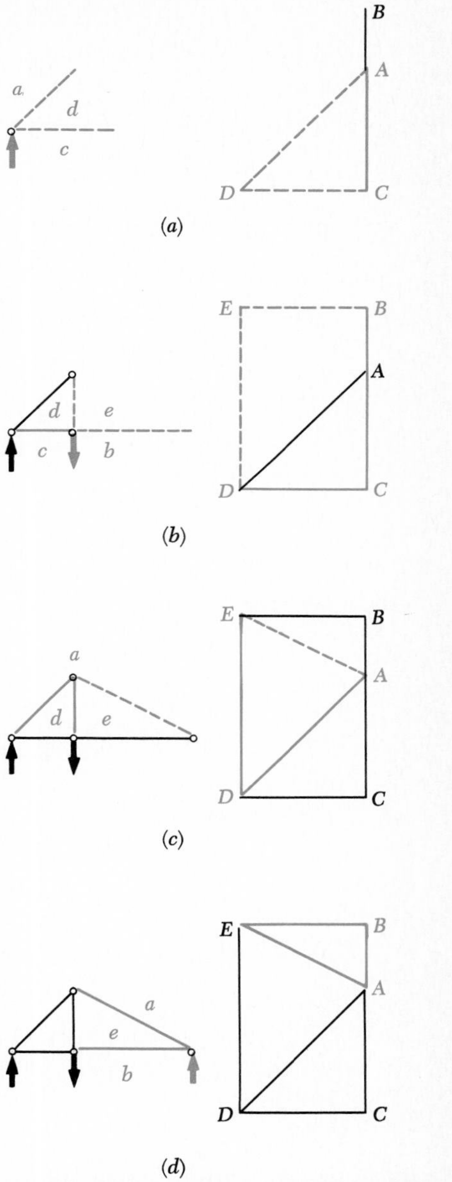

Fig. 6.15 Step-by-step construction of Maxwell's diagram.

SAMPLE PROBLEM 6.2

By drawing Maxwell's diagram, determine the force in each member of the truss considered in Sample Prob. 6.1.

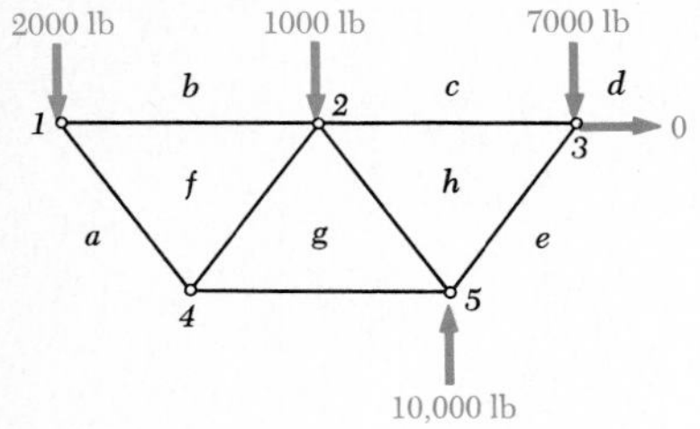

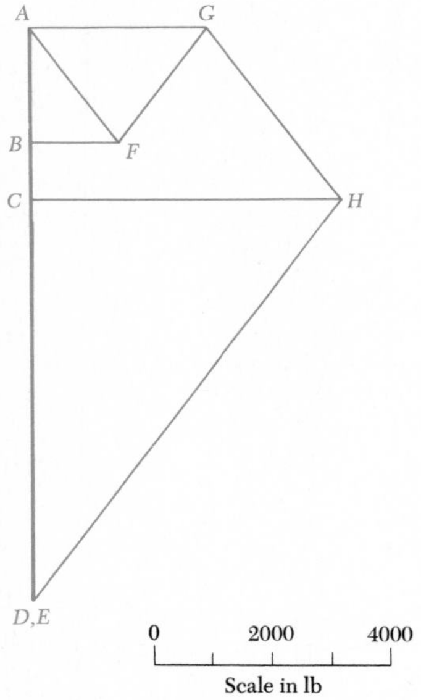

Solution. 1. The truss is drawn to scale; both the applied loads and the reactions are indicated. Using Bow's notation, a lowercase letter is assigned to each region outside the truss and to each individual area inside the truss.

2. By considering the entire truss as a free body, we compute the reactions (see Sample Prob. 6.1).

3. Selecting the external forces in clockwise order around the truss, we draw to scale the force polygon for the entire truss in tip-to-tail fashion; points *A*, *B*, *C*, *D*, and *E* are thus located on the Maxwell diagram.

4. Considering joint *1*, where only two unknown forces exist, we now locate point *F* by drawing lines *BF* and *AF*, which are parallel to members *bf* and *fa*, respectively. We next consider joint *4*, where there are now only two unknown forces and hence only one unknown point in the corresponding triangle of the Maxwell diagram. Lines *FG* and *AG* are drawn parallel to *fg* and *ga*, respectively; point *G* is located at the intersection of these lines. We next consider joint *2* and locate point *H* at the intersection of lines *CH* and *GH* which are drawn parallel to *ch* and *hg*, respectively. All the points of the Maxwell diagram have now been located. Since *HE* represents the force in member *he*, we may draw a line through point *H* parallel to member *he*; this line should pass through point *E*, which is already located. This provides a check on the accuracy of the drawing of the Maxwell diagram.

5. The magnitude and sense of the force in each member are determined from the Maxwell diagram; the magnitude is measured directly, and the sense is found as indicated in Sec. 6.7. The results are given in the table.

Member	Force, lb
bf	1500 *T*
fa	2500 *C*
fg	2500 *T*
ga	3000 *C*
hg	3750 *C*
ch	5250 *T*
he	8750 *C*

PROBLEMS

6.21 through 6.32 By drawing Maxwell's diagram, determine the force in each member of the truss shown. Indicate whether the member is in tension or compression. (The trusses appear on pages 196 and 197.)

6.33 through 6.37 By drawing Maxwell's diagram, determine the force in each member of the truss shown. Indicate whether the member is in tension or compression.

6.38 Solve Prob. 6.35, assuming that the 3000-lb force applied at joint *1* is horizontal and directed to the right.

6.39 Solve Prob. 6.36, assuming that the 2000-N vertical force is applied at joint *5*.

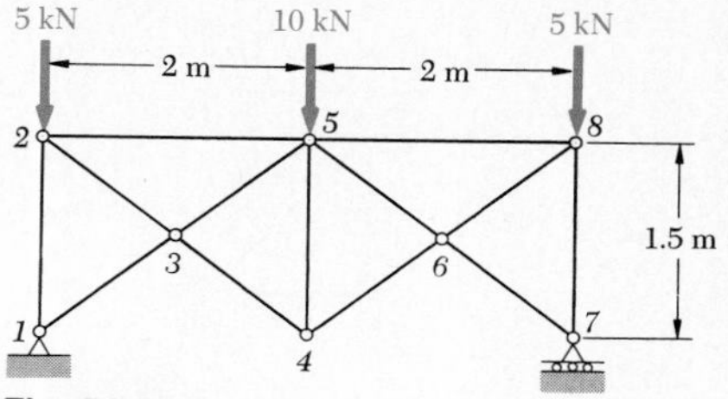

Fig. P6.33

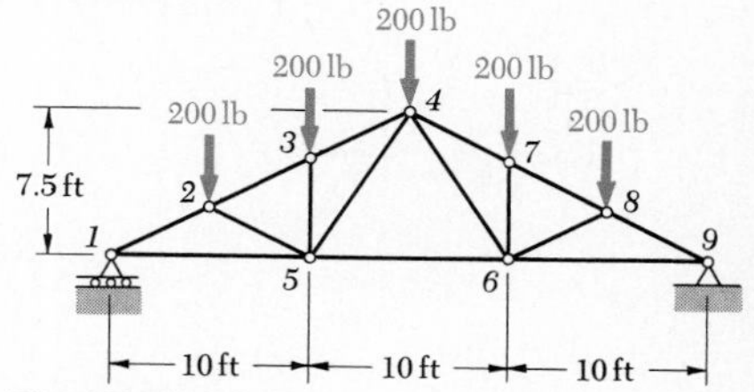

Fig. P6.34

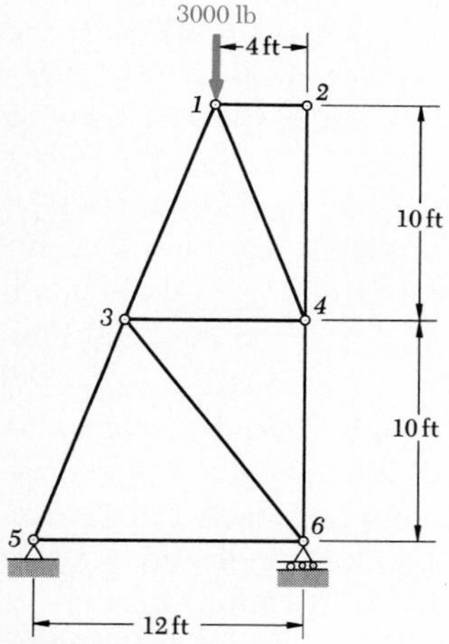

Fig. P6.35

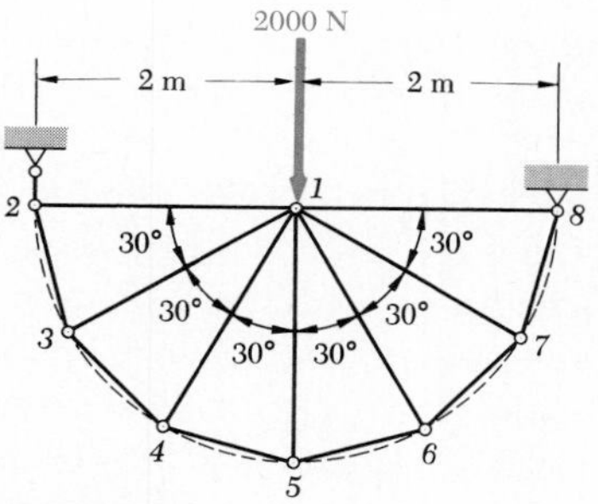

Fig. P6.36

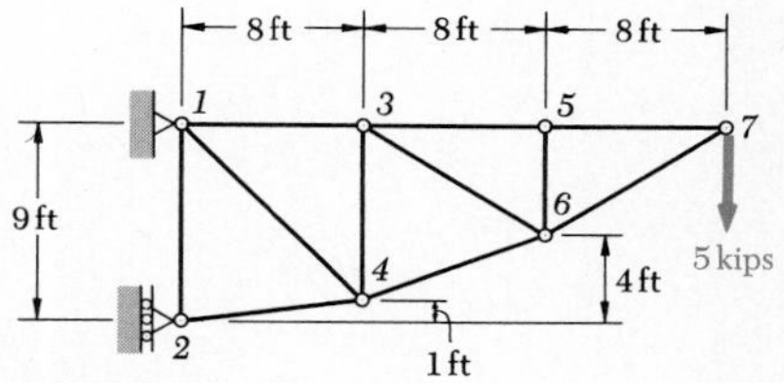

Fig. P6.37

6.8. Analysis of Trusses by the Method of Sections. The method of joints and Maxwell's diagram are most effective when the forces in all the members of a truss are to be determined. If, however, the force in only one member or the forces in a very few members are desired, a third method, the method of sections, will prove more efficient.

Assume, for example, that we want to determine the force in member *BD* of the truss shown in Fig. 6.16*a*. To do this, we must determine the force with which member *BD* acts on either joint *B* or joint *D*. If we were to use the method of joints, we would choose either joint *B* or joint *D* as a free body. However, we may also choose as a free body a larger portion

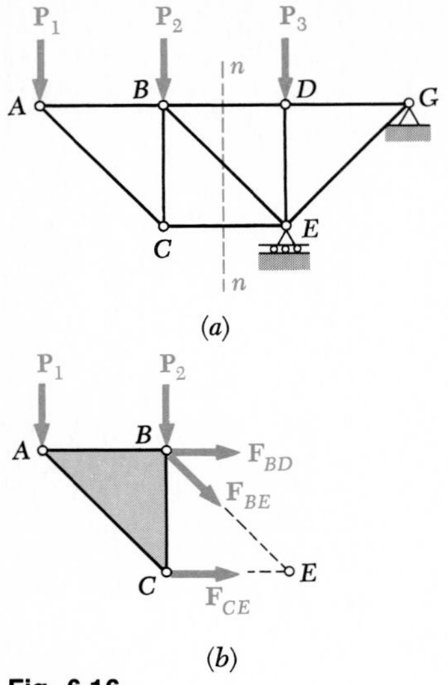

Fig. 6.16

of the truss, composed of several joints and members, provided that the desired force is one of the external forces acting on that portion. If, in addition, the portion of the truss is chosen so that there is a total of only three unknown forces acting upon it, the desired force may be obtained by solving the equations of equilibrium for this portion of the truss. In practice, the portion of the truss to be utilized is obtained by *passing a section* through three members of the truss, one of which is the desired member, i.e., by drawing a line which divides the truss into two completely separate parts but does not intersect more than three members. Either of the two portions of the truss obtained after the intersected members have been removed may then be used as a free body.†

In Fig. 6.16*a*, the section *nn* has been passed through members *BD*, *BE*, and *CE*, and the portion *ABC* of the truss is chosen as the free body (Fig. 6.16*b*). The forces acting on the free body are the loads $\mathbf{P}_1$ and $\mathbf{P}_2$ at points *A* and *B* and the three unknown forces $\mathbf{F}_{BD}$, $\mathbf{F}_{BE}$, and $\mathbf{F}_{CE}$. Since it is not known whether the members removed were in tension or compression, the three forces have been arbitrarily drawn away from the free body as if the members were in tension.

The fact that the rigid body *ABC* is in equilibrium can be expressed by writing three equations which may be solved for the three unknown forces. If only the force $\mathbf{F}_{BD}$ is desired, we need write only one equation, provided that the equation does not contain the other unknowns. Thus the equation $\Sigma M_E = 0$ yields the value of the magnitude F_{BD} of the force $\mathbf{F}_{BD}$. A positive sign in the answer will indicate that our original assumption regarding the sense of $\mathbf{F}_{BD}$ was correct and that member *BD* is in tension; a negative sign will indicate that our assumption was incorrect and that *BD* is in compression.

On the other hand, if only the force $\mathbf{F}_{CE}$ is desired, an equation which does not involve $\mathbf{F}_{BD}$ or $\mathbf{F}_{BE}$ should be written; the appropriate equation is $\Sigma M_B = 0$. Again a positive sign for the magnitude F_{CE} of the desired force indicates a correct assumption, hence tension; and a negative sign indicates an incorrect assumption, hence compression.

If only the force $\mathbf{F}_{BE}$ is desired, the appropriate equation is $\Sigma F_y = 0$. Whether the member is in tension or compression is again determined from the sign of the answer.

When the force in only one member is determined, no independent check of the computation is available. However, when

† In the analysis of certain trusses, sections are passed which intersect more than three members; the forces in one, or possibly two, of the intersected members may be obtained if equilibrium equations can be found, each of which involves only one unknown (see Probs. 6.52 to 6.54).

all the unknown forces acting on the free body are determined, the computations can be checked by writing an additional equation. For instance, if $\mathbf{F}_{BD}$, $\mathbf{F}_{BE}$, and $\mathbf{F}_{CE}$ are determined as indicated above, the computation can be checked by verifying that $\Sigma F_x = 0$.

***6.9. Trusses Made of Several Simple Trusses.** Consider two simple trusses *ABC* and *DEF*. If they are connected by three bars *BD*, *BE*, and *CE* as shown in Fig. 6.17*a*, they will form together a rigid truss *ABDF*. The trusses *ABC* and *DEF* can also be combined into a single rigid truss by joining joints *B* and *D* into a single joint *B* and by connecting joints *C* and *E* by a bar *CE* (Fig. 6.17*b*). The truss thus obtained is known as a Fink truss. It should be noted that the trusses of Fig. 6.17*a* and *b* are *not* simple trusses; they cannot be constructed from a triangular truss by adding successive pairs of members as prescribed in Sec. 6.3. They are rigid trusses, however, as we may check by comparing the systems of connections used to hold the simple trusses *ABC* and *DEF* together (three bars in Fig. 6.17*a*, one pin and one bar in Fig. 6.17*b*) with the systems of supports discussed in Secs. 3.14 and 3.15. Trusses made of several simple trusses rigidly connected are known as *compound trusses*.

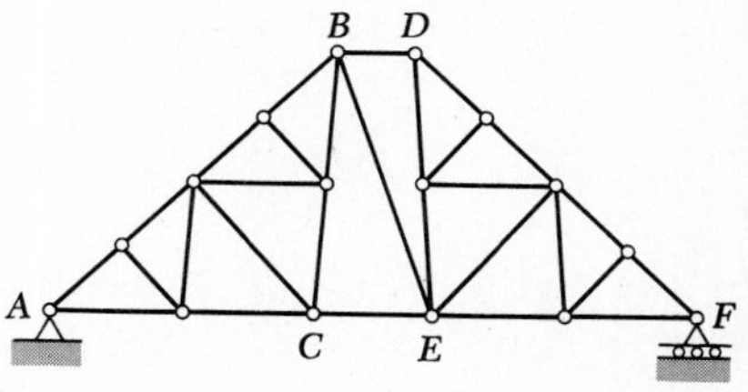

(*a*)

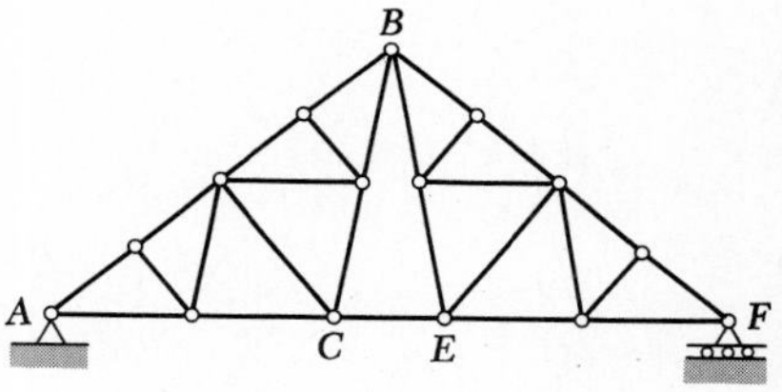

(*b*)

Fig. 6.17

It may be checked that in a compound truss the number of members m and the number of joints n are still related by the formula $m = 2n - 3$. If a compound truss is supported by a frictionless pin and a roller (involving three unknown reactions), the total number of unknowns is $m + 3$ and this number is therefore equal to the number $2n$ of equations obtained by expressing that the n pins are in equilibrium. Compound trusses supported by a pin and a roller, or by an equivalent system of supports, are *statically determinate, rigid,* and *completely constrained.* This means that all unknown reactions and forces in members can be determined by the methods of statics and that, all equilibrium equations being satisfied, the truss will neither collapse nor move. The forces in the members, however, cannot all be determined by the method of joints, except by solving a large number of simultaneous equations. In the case of the compound truss of Fig. 6.17*a*, for example, it will be found more expeditious to pass a section through members *BD*, *BE*, and *CE* to determine their forces.

Suppose, now, that the simple trusses *ABC* and *DEF* are connected by *four* bars *BD*, *BE*, *CD*, and *CE* (Fig. 6.18). The number of members m is now larger than $2n - 3$; the truss obtained is *overrigid,* and one of the four members *BD*, *BE*, *CD*, or *CE* is said to be *redundant.* If the truss is supported by a pin at *A* and a roller at *F*, the total number of unknowns is $m + 3$.

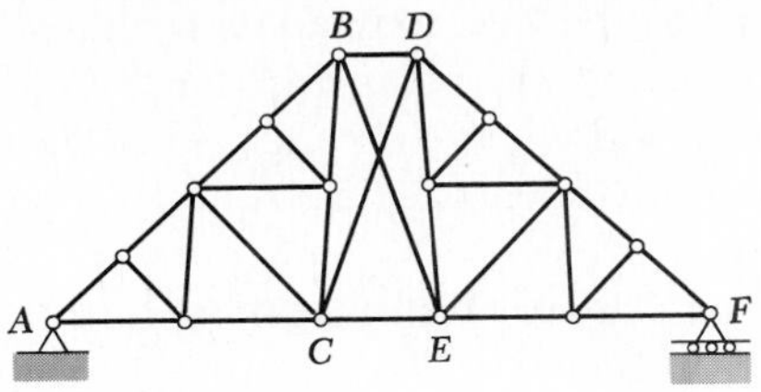

Fig. 6.18

This number is now larger than the number $2n$ of available independent equations; the truss is *statically indeterminate*.

Finally, we shall assume that the two simple trusses ABC and DEF are joined by a pin as shown in Fig. 6.19*a*. The number of members m is smaller than $2n - 3$. If the truss is supported by a pin at A and a roller at F, the total number of unknowns is $m + 3$. This number is now smaller than the number $2n$ of equilibrium equations which should be satisfied; the truss is *nonrigid* and will collapse under its own weight. However, if two pins are used to support it, the truss becomes *rigid* and will not collapse (Fig. 6.19*b*). We note that the total number of unknowns is now $m + 4$ and is thus equal to the number of equations. While necessary, this condition, however, is not sufficient for the equilibrium of a structure which ceases to be rigid when detached from its supports (see Sec. 6.12).

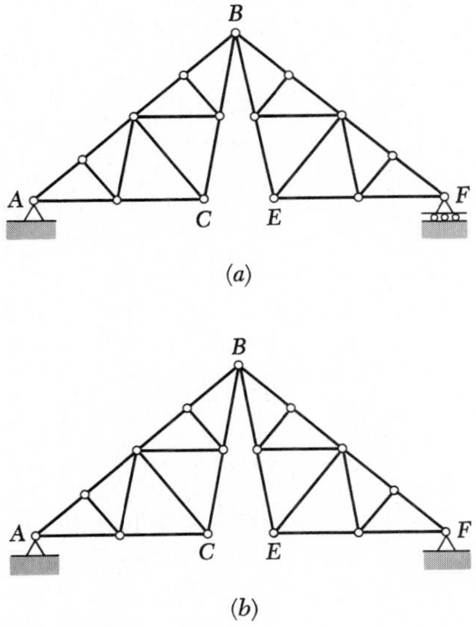

Fig. 6.19

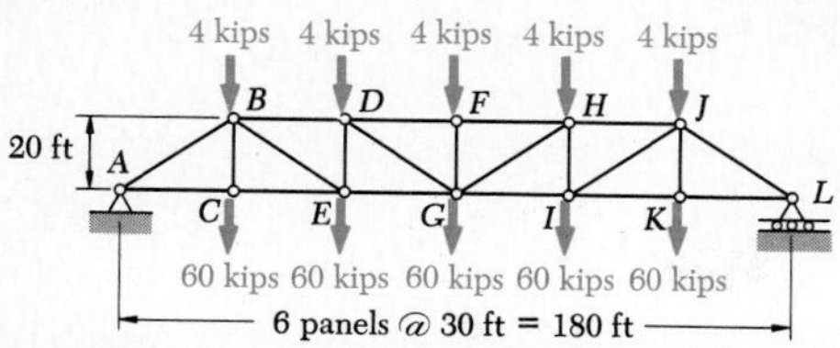

SAMPLE PROBLEM 6.3

Determine the forces in members DE and HJ of the truss shown.

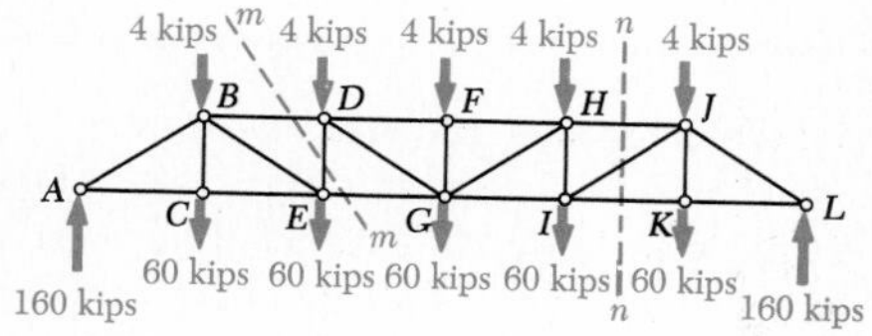

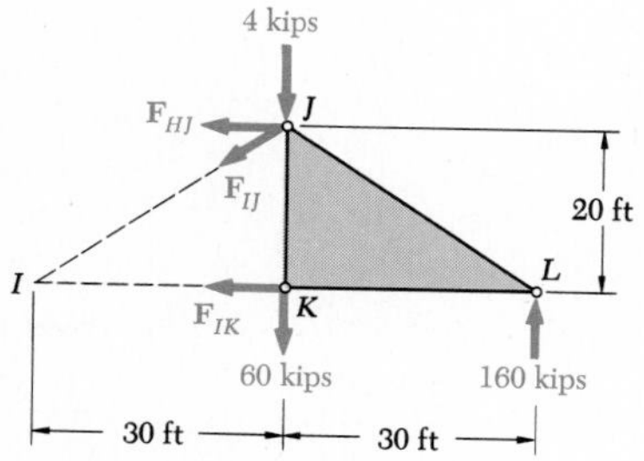

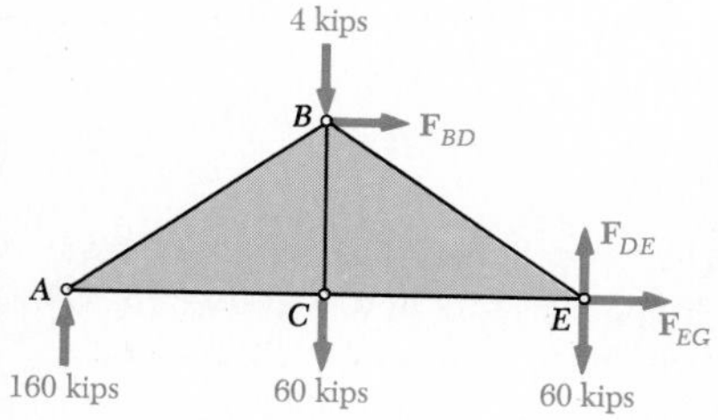

Solution. Considering the entire truss as a free body, we determine the reactions at A and L.

$$\mathbf{A} = 160 \text{ kips} \uparrow$$
$$\mathbf{L} = 160 \text{ kips} \uparrow$$

Force in Member HJ. Section nn is passed through the truss so that it intersects member HJ and only two additional members. After the intersected members have been removed, we choose the right-hand portion of the truss as a free body. Three unknown forces are involved; to eliminate the two forces passing through point I, we write

$+\circlearrowleft \Sigma M_I = 0$:

$$(160 \text{ kips})(60 \text{ ft}) - (60 \text{ kips})(30 \text{ ft}) - (4 \text{ kips})(30 \text{ ft}) + F_{HJ}(20 \text{ ft}) = 0$$
$$F_{HJ} = -384 \text{ kips}$$

The sense of $\mathbf{F}_{HJ}$ was chosen assuming member HJ to be in tension; the negative sign obtained indicates that the member is in compression.

$$F_{HJ} = 384 \text{ kips } C \blacktriangleleft$$

Force in Member DE. Section mm is passed through the truss so that it intersects member DE and only two additional members. After the intersected members have been removed, the left-hand portion of the truss is chosen as a free body. Three unknown forces are again involved; since the equation $\Sigma F_y = 0$ involves only F_{DE} as an unknown, we write

$+\uparrow \Sigma F_y = 0$:

$$+160 \text{ kips} - 60 \text{ kips} - 4 \text{ kips} - 60 \text{ kips} + F_{DE} = 0$$
$$F_{DE} = -36 \text{ kips} \qquad F_{DE} = 36 \text{ kips } C \blacktriangleleft$$

SAMPLE PROBLEM 6.4

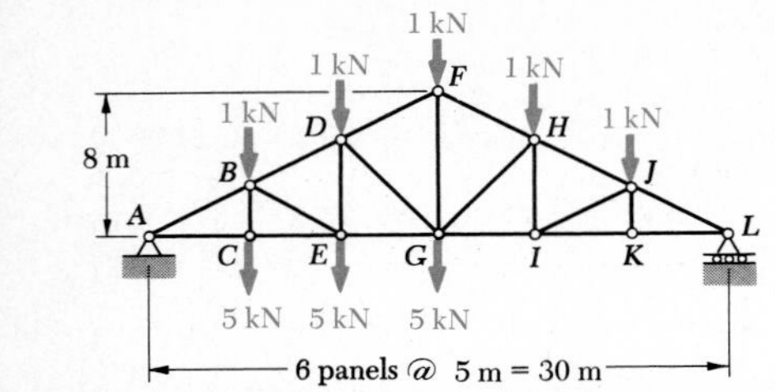

Determine the forces in members *FH*, *GH*, and *GI* of the roof truss shown.

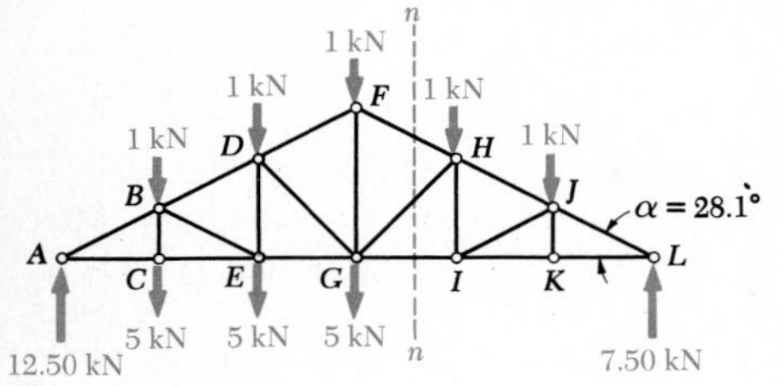

Solution. Section *nn* is passed through the truss as shown. The right-hand portion of the truss will be taken as a free body. Since the reaction at *L* acts on this free body, the value of **L** must be calculated separately, using the entire truss as a free body; the equation $\Sigma M_A = 0$ yields $\mathbf{L} = 7.50 \text{ kN} \uparrow$.

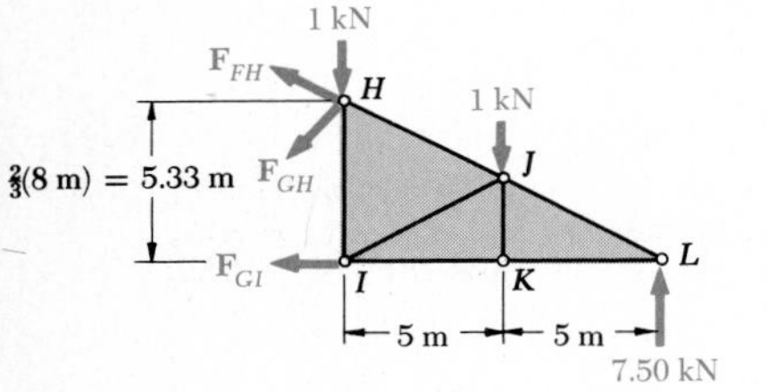

Force in Member GI. Using the portion *HLI* of the truss as a free body, the value of F_{GI} is obtained by writing

$$+\circlearrowleft \Sigma M_H = 0: \quad (7.50 \text{ kN})(10 \text{ m}) - (1 \text{ kN})(5 \text{ m}) - F_{GI}(5.33 \text{ m}) = 0$$

$$F_{GI} = +13.13 \text{ kN} \qquad F_{GI} = 13.13 \text{ kN } T$$

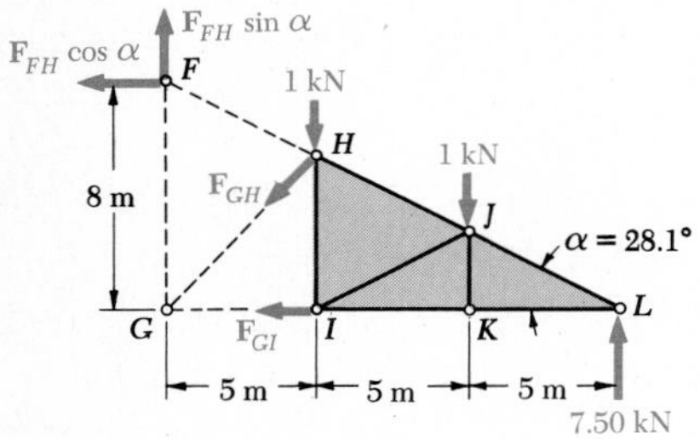

Force in Member FH. The value of F_{FH} is obtained from the equation $\Sigma M_G = 0$. We move $\mathbf{F}_{FH}$ along its line of action until it acts at point *F*, where it is resolved into its *x* and *y* components. The moment of $\mathbf{F}_{FH}$ with respect to point *G* is now equal to $(F_{FH} \cos \alpha)(8 \text{ m})$.

$$+\circlearrowleft \Sigma M_G = 0:$$

$$(7.50 \text{ kN})(15 \text{ m}) - (1 \text{ kN})(10 \text{ m}) - (1 \text{ kN})(5 \text{ m}) + (F_{FH} \cos \alpha)(8 \text{ m}) = 0$$

$$F_{FH} = -13.82 \text{ kN} \qquad F_{FH} = 13.82 \text{ kN } C$$

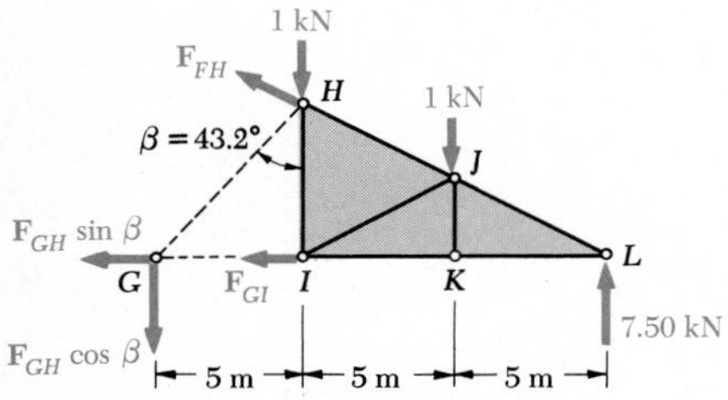

Force in Member GH. The value of F_{GH} is determined by first resolving the force $\mathbf{F}_{GH}$ into *x* and *y* components at point *G* and then solving the equation $\Sigma M_L = 0$.

$$+\circlearrowleft \Sigma M_L = 0: \quad (1 \text{ kN})(10 \text{ m}) + (1 \text{ kN})(5 \text{ m}) + (F_{GH} \cos \beta)(15 \text{ m}) = 0$$

$$F_{GH} = -1.372 \text{ kN} \qquad F_{GH} = 1.372 \text{ kN } C$$

PROBLEMS

6.40 Determine the force in members CE and CF of the truss shown.

6.41 Determine the force in members FG and FH of the truss shown.

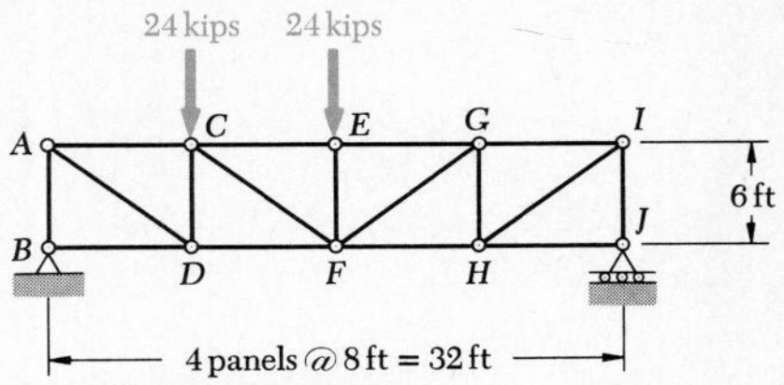

Fig. P6.40 and P6.41

6.42 Determine the force in members DF and DE of the truss shown.

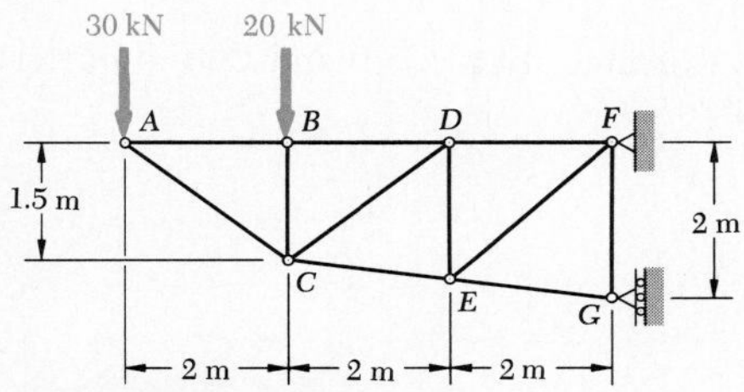

Fig. P6.42 and P6.43

6.43 Determine the force in members CD and CE of the truss shown.

6.44 Determine the force in member CD of the Fink roof truss shown.

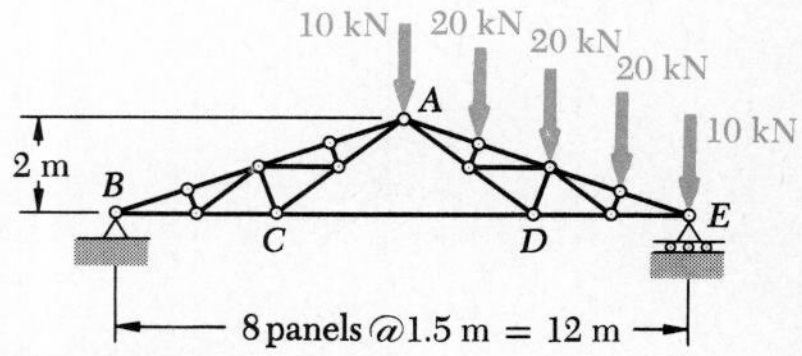

Fig. P6.44

6.45 Determine the force in members CE, CD, and BD of the truss shown.

6.46 Determine the force in members EG, EF, and DF of the truss shown.

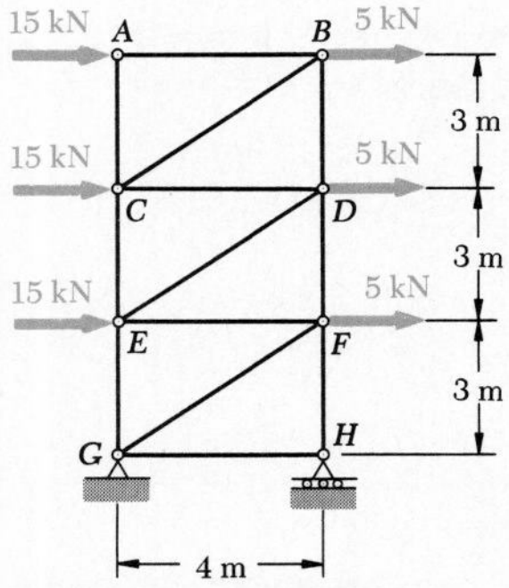

Fig. P6.45 and P6.46

6.47 Determine the force in members EF, FG, and GI of the truss shown.

6.48 Determine the force in members CE, CD, and CB of the truss shown.

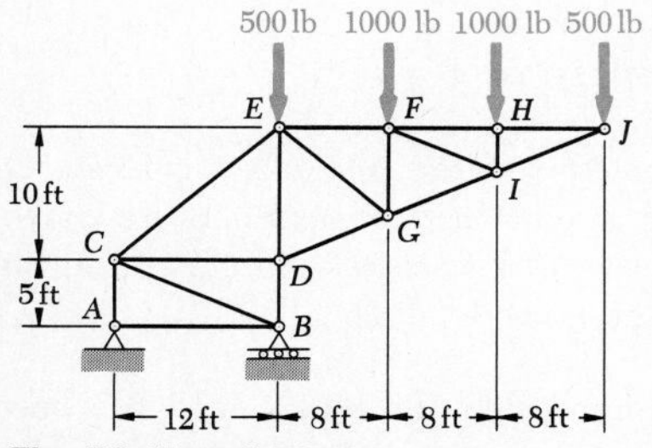

Fig. P6.47 and P6.48

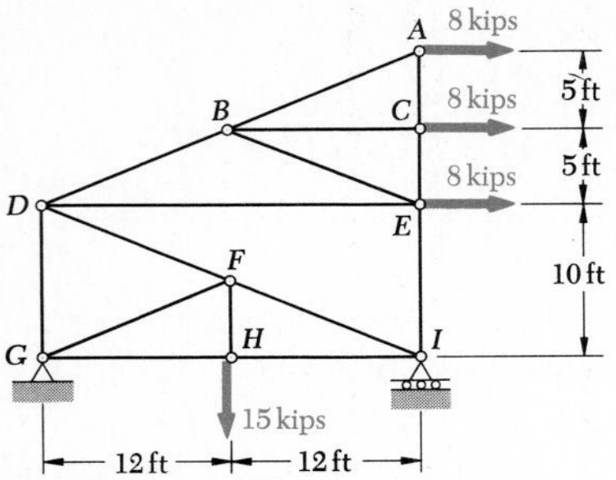

Fig. P6.49, P6.50, and P6.51

6.49 Determine the force in members *BD* and *DE* of the truss shown.

6.50 Determine the force in members *DF* and *DG* of the truss shown.

6.51 Determine the force in members *BE* and *DE* of the truss shown.

6.52 Determine the force in members *AB* and *KL* of the truss shown. (*Hint.* Use section *a-a*.)

6.53 Solve Prob. 6.52, assuming that **P** = 0 and that a vertical load **Q** is applied at joint *B*.

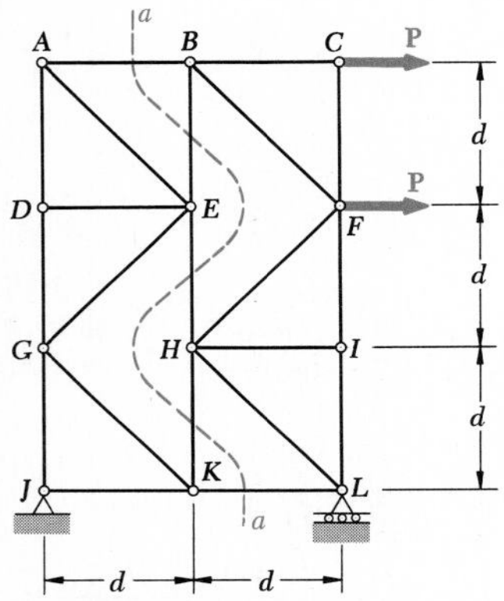

Fig. P6.52

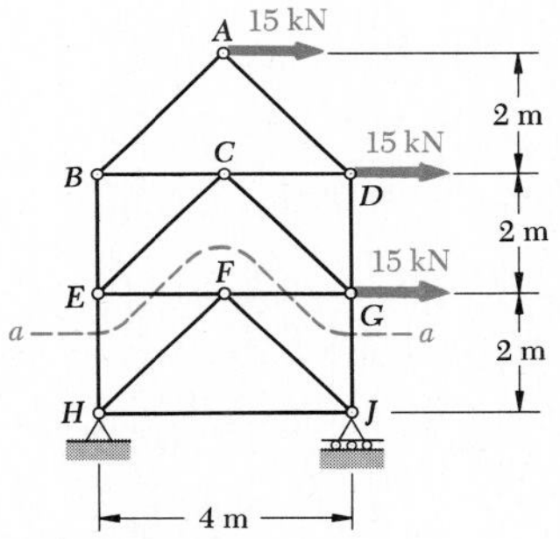

Fig. P6.54

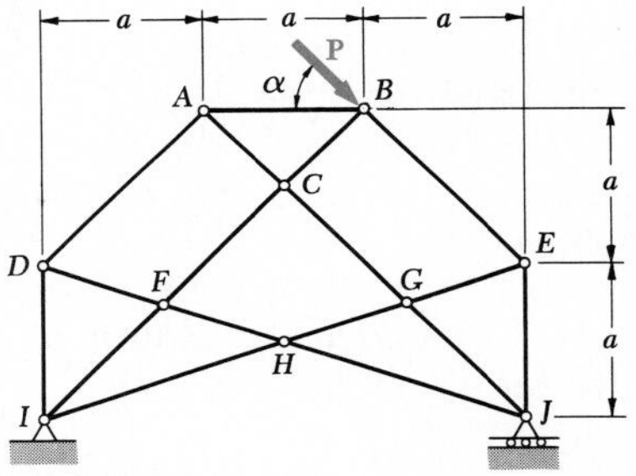

Fig. P6.55

6.54 Determine the force in member *GJ* of the truss shown. (*Hint.* Use section *a-a*.)

6.55 Determine the force in members *AB* and *EJ* of the truss shown, if $\alpha = 0°$. (*Hint.* Use portion *IBE* of the truss as a free body and apply to joints *C*, *F*, *G*, and *H* the results obtained in Sec. 6.5.)

6.56 Solve Prob. 6.55, assuming that $\alpha = 90°$.

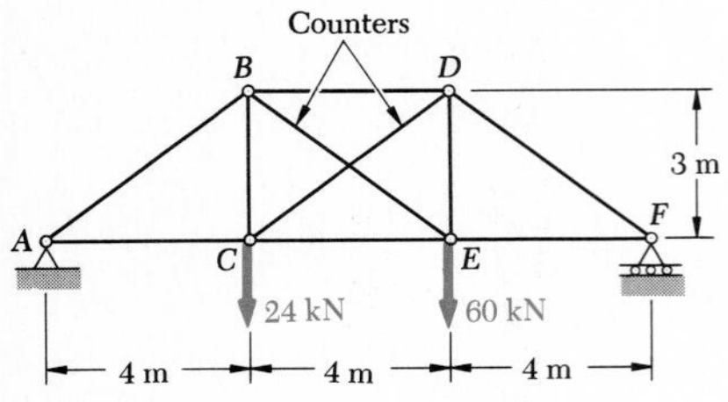

Fig. P6.57

6.57 The diagonal members in the center panel of the truss shown are very slender and can act only in tension; such members are known as *counters*. Determine the force in members *BD* and *CE* and in the counter which is acting under the given loading.

6.58 Solve Prob. 6.57, assuming that the 60-kN load has been removed.

6.59 Determine the force in members BE and CG and in the counter which is acting under the given loading. (See Prob. 6.57 for the definition of a counter.)

6.60 Solve Prob. 6.59, assuming that the 15-kip load has been removed.

***6.61 through 6.64** Classify each of the given structures as completely, partially, or improperly constrained; if completely constrained, further classify as determinate or indeterminate. (All members can act both in tension and in compression.)

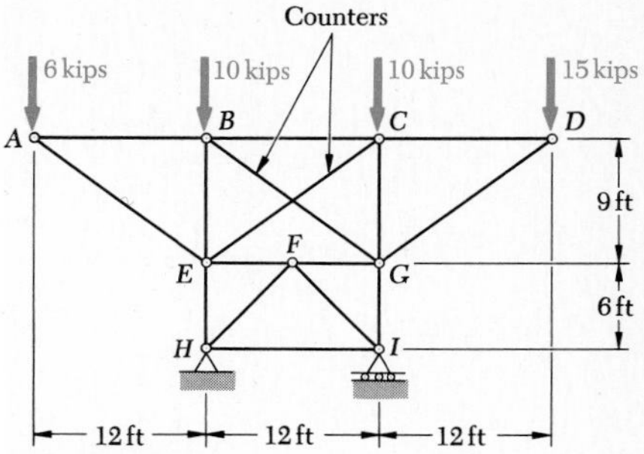

Fig. P6.59

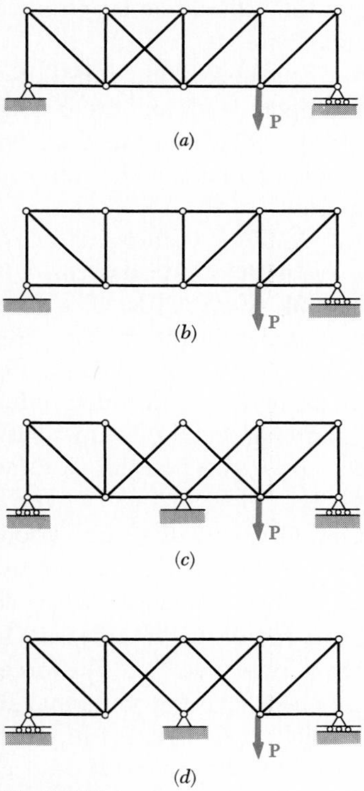

Fig. P6.61

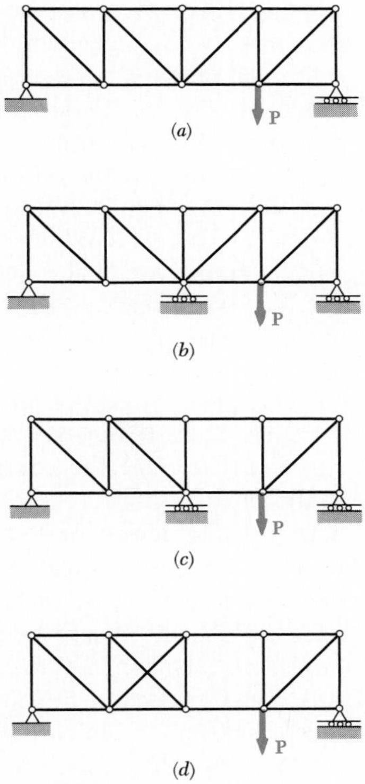

Fig. P6.62

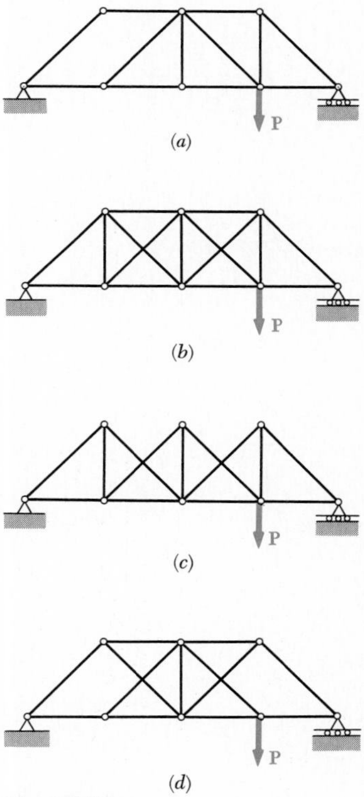

Fig. P6.63

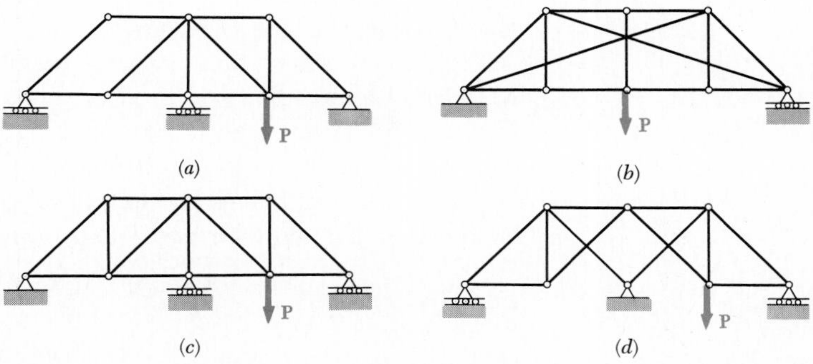

Fig. P6.64

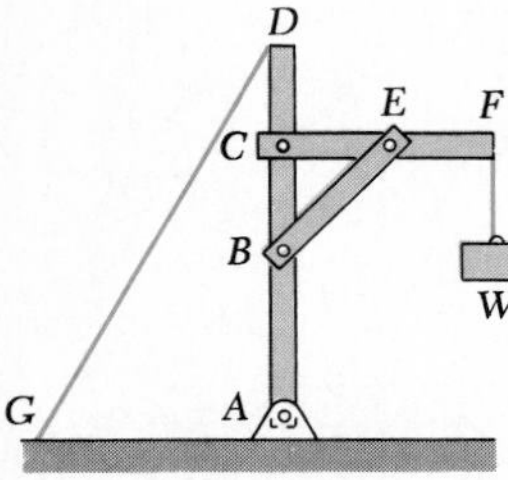

(a)

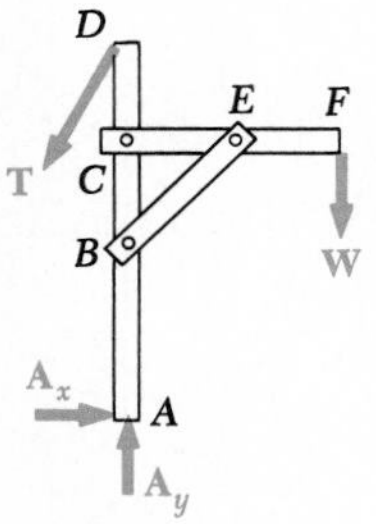

(b)

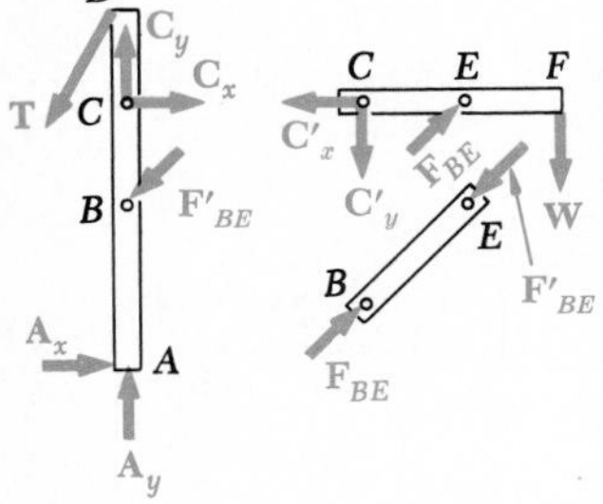

(c)

Fig. 6.20

FRAMES AND MACHINES

6.10. Structures Containing Multiforce Members. Under Trusses, we have considered structures consisting entirely of pins and of straight two-force members. The forces acting on the two-force members were known to be directed along the members themselves. We shall now consider structures in which at least one of the members is a *multiforce* member, i.e., a member acted upon by three or more forces. These forces will generally not be directed along the members on which they act; their direction is unknown, and they should be represented therefore by two unknown components.

Frames and machines are structures containing multiforce members. *Frames* are designed to support loads and are usually stationary, fully constrained structures. *Machines* are designed to transmit and modify forces; they may or may not be stationary and will always contain moving parts.

6.11. Analysis of a Frame. As a first example of analysis of a frame, we shall consider again the crane described in Sec. 6.1, which carries a given load W (Fig. 6.20*a*). The free-body diagram of the entire frame is shown in Fig. 6.20*b*. This diagram may be used to determine the external forces acting on the frame. Summing moments about A, we first determine the force $\mathbf{T}$ exerted by the cable; summing x and y components, we then determine the components $\mathbf{A}_x$ and $\mathbf{A}_y$ of the reaction at the pin A.

In order to determine the internal forces holding the various parts of a frame together, we must dismember the frame and draw a free-body diagram for each of its component parts (Fig. 6.20*c*). First, the two-force members should be considered. In this frame, member BE is the only two-force member. The forces acting at each end of this member must have the same magnitude, same line of action, and opposite sense (Sec. 3.16). They are therefore directed along BE and will be denoted, respectively, by $\mathbf{F}_{BE}$ and $\mathbf{F}'_{BE}$. Their sense will be arbitrarily assumed as shown in Fig. 6.20*c*, and the correctness of this assumption will be checked later by the sign obtained for the common magnitude F_{BE} of the two forces.

Next, we consider the multiforce members, i.e., the members which are acted upon by three or more forces. According to Newton's third law, the force exerted at B by member BE on member AD must be equal and opposite to the force $\mathbf{F}_{BE}$ exerted by AD on BE. Similarly, the force exerted at E by member BE on member CF must be equal and opposite to the force $\mathbf{F}'_{BE}$

exerted by CF on BE. The forces that the two-force member BE exerts on AD and CF are therefore respectively equal to $\mathbf{F}'_{BE}$ and $\mathbf{F}_{BE}$; they have the same magnitude F_{BE} and opposite sense, and should be directed as shown in Fig. 6.20c.

At C two multiforce members are connected. Since neither the direction nor the magnitude of the forces acting at C is known, these forces will be represented by their x and y components. The components $\mathbf{C}_x$ and $\mathbf{C}_y$ of the force acting on member AD will be arbitrarily directed to the right and upward. Since, according to Newton's third law, the forces exerted by member CF on AD and by member AD on CF are equal and opposite, the components $\mathbf{C}'_x$ and $\mathbf{C}'_y$ of the force acting on member CF *must* be directed to the left and downward. Whether the force $\mathbf{C}_x$ is actually directed to the right and the force $\mathbf{C}'_x$ is actually directed to the left will be determined later from the sign of their common magnitude C_x, a plus sign indicating that the assumption made was correct, and a minus sign that it was wrong. The free-body diagrams of the multiforce members are completed by showing the external forces acting at A, D, and F.†

The internal forces may now be determined by considering the free-body diagram of either of the two multiforce members. Choosing the free-body diagram of CF, for example, we write the equations $\Sigma M_C = 0$, $\Sigma M_E = 0$, and $\Sigma F_x = 0$, which yield the values of the magnitudes F_{BE}, C_y, and C_x, respectively. These values may be checked by verifying that member AD is also in equilibrium.

It should be noted that the free-body diagrams of the pins were not shown in Fig. 6.20c. This was because the pins were assumed to form an integral part of one of the two members they connected. This assumption can always be used to simplify the analysis of frames and machines. When a pin connects three or more members, however, or when a pin connects a support and two or more members, or when a load is applied to a pin, a clear decision must be made in choosing the member to which the pin will be assumed to belong. (If multiforce members are involved, the pin should be attached to one of these members.) The various forces exerted on the pin should then be clearly identified. This is illustrated in Sample Prob. 6.7.

†The use of *primes* (′) to distinguish the force exerted by one member on another from the equal and opposite force exerted by the second member on the first is not strictly necessary, since the two forces belong to different free-body diagrams and thus cannot easily be confused. In the Sample Problems, we shall represent by the same symbol equal and opposite forces which are applied to different free bodies.

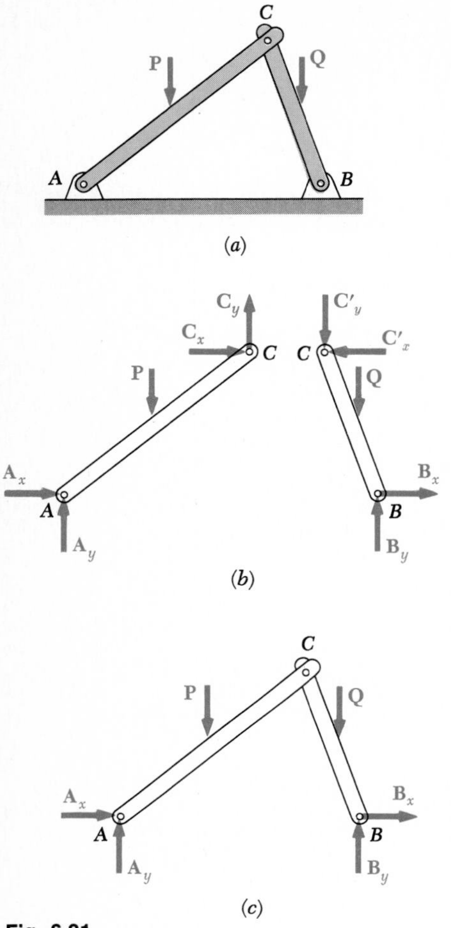

Fig. 6.21

6.12. Frames Which Cease to Be Rigid When Detached from Their Supports. The crane analyzed in Sec. 6.11 was so constructed that it could keep the same shape without the help of its supports; it was therefore considered as a rigid body. Many frames, however, will collapse if detached from their supports; such frames cannot be considered as rigid bodies. Consider, for example, the frame shown in Fig. 6.21*a*, which consists of two members *AC* and *CB* carrying loads **P** and **Q** at their midpoints; the members are supported by pins at *A* and *B* and are connected by a pin at *C*. If detached from its supports, this frame will not maintain its shape; it should therefore be considered as made of *two distinct rigid parts AC* and *CB*.

The equations $\Sigma F_x = 0$, $\Sigma F_y = 0$, $\Sigma M = 0$ (about any given point) express the conditions for the *equilibrium of a rigid body* (Chap. 3); we should use them, therefore, in connection with the free-body diagrams of rigid bodies, namely, the free-body diagrams of members *AC* and *CB* (Fig. 6.21*b*). Since these members are multiforce members, and since pins are used at the supports and at the connection, the reactions at *A* and *B* and the forces at *C* will each be represented by two components. In accordance with Newton's third law, the components $\mathbf{C}_x$ and $\mathbf{C}_y$ of the force exerted by *CB* on *AC* and the components $\mathbf{C}'_x$ and $\mathbf{C}'_y$ of the force exerted by *AC* on *CB* will be represented by vectors of the same magnitude and opposite sense. We note that four unknown force components act on free body *AC*, while only three independent equations may be used to express that the body is in equilibrium; similarly, four unknowns, but only three equations, are associated with *CB*. However, only six different unknowns are involved in the analysis of the two members, and altogether six equations are available to express that the members are in equilibrium. Writing $\Sigma M_A = 0$ for free body *AC* and $\Sigma M_B = 0$ for *CB*, we obtain two simultaneous equations which may be solved for the common magnitude C_x of the components $\mathbf{C}_x$ and $\mathbf{C}'_x$, and for the common magnitude C_y of the components $\mathbf{C}_y$ and $\mathbf{C}'_y$. Writing, then, $\Sigma F_x = 0$ and $\Sigma F_y = 0$ for each of the two free bodies, we obtain successively the magnitudes A_x, A_y, B_x, and B_y.

We shall observe now that, since the equations of equilibrium $\Sigma F_x = 0$, $\Sigma F_y = 0$, $\Sigma M = 0$ (about any given point) are satisfied by the forces acting on free body *AC*, and since they are also satisfied by the forces acting on free body *CB*, they must be satisfied when the forces acting on the two free bodies are considered simultaneously. Since the internal forces at *C* cancel each other, we find that the equations of equilibrium must be satisfied by the external forces shown on the free-body diagram of the frame *ACB* itself (Fig. 6.21*c*), although the frame is not a rigid body. These equations may be used to determine some of the components of the reactions at A and B. We shall note, however, that *the reactions cannot be completely determined from the free-body diagram of the whole frame.* It is thus necessary to dismember the frame and to consider the free-body diagrams of its component parts (Fig. 6.21*b*), even when we are interested only in finding external reactions. This may be explained by the fact that the equilibrium equations obtained for free body ACB are *necessary conditions* for the equilibrium of a nonrigid structure, *but not sufficient conditions.*

The method of solution outlined in the second paragraph of this section involved simultaneous equations. We shall now discuss a more expeditious method, which utilizes the free body ACB as well as the free bodies AC and CB. Writing $\Sigma M_A = 0$ and $\Sigma M_B = 0$ for free body ACB, we obtain B_y and A_y. Writing $\Sigma M_C = 0$, $\Sigma F_x = 0$, and $\Sigma F_y = 0$ for free body AC, we obtain successively A_x, C_x, and C_y. Finally, writing $\Sigma F_x = 0$ for ACB, we obtain B_x.

We noted above that the analysis of the frame of Fig. 6.21 involves six unknown force components and six independent equilibrium equations (the equilibrium equations for the whole frame were obtained from the original six equations and, therefore, are not independent). Moreover, we checked that all unknowns could be actually determined and that all equations could be satisfied. The frame considered is *statically determinate and rigid.*† In general, to determine whether a structure is statically determinate and rigid, we should draw a free-body diagram for each of its component parts and count the reactions and internal forces involved. We should also determine the number of independent equilibrium equations (excluding equations expressing the equilibrium of the whole structure or of groups of component parts already analyzed). If there are more unknowns than equations, the structure is *statically indeterminate.* If there are fewer unknowns than equations, the structure is *nonrigid.* If there are as many unknowns as equations, *and if all unknowns may be determined and all equations satisfied* under general loading conditions, the structure is *statically determinate and rigid;* if, however, due to an *improper arrangement* of members and supports, all unknowns cannot be determined and all equations cannot be satisfied, the structure is *statically indeterminate and nonrigid.*

† The word "rigid" is used here to indicate that the frame will maintain its shape as long as it remains attached to its supports.

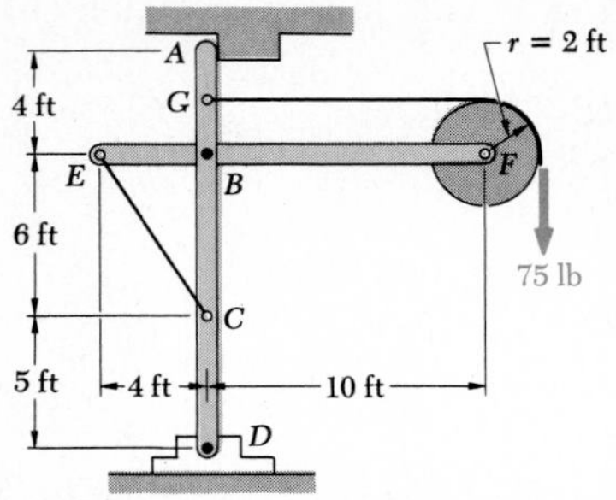

SAMPLE PROBLEM 6.5

In the small frame shown, members *EBF* and *ABCD* are connected by a pin at *B* and by the cable *EC*. A 75-lb load is supported by a second cable which passes over a pulley at *F* and is attached to the vertical member at *G*. Determine the tension in cable *EC* and the components of the pin reaction at *B*.

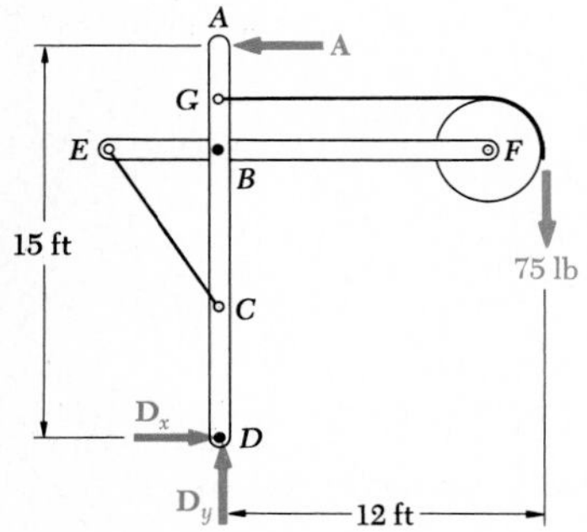

Entire Frame. The external reactions on the frame involve three unknowns; these reactions are determined by taking the entire frame (including the cables) as a free body.

$$+\uparrow\Sigma F_y = 0: \qquad D_y - 75\text{ lb} = 0$$
$$D_y = +75\text{ lb} \qquad \mathbf{D}_y = 75\text{ lb}\uparrow$$

$$+\circlearrowleft\Sigma M_D = 0: \qquad -(75\text{ lb})(12\text{ ft}) + A(15\text{ ft}) = 0$$
$$A = +60\text{ lb} \qquad \mathbf{A} = 60\text{ lb}\leftarrow$$

$$\overset{+}{\rightarrow}\Sigma F_x = 0: \qquad -60\text{ lb} + D_x = 0$$
$$D_x = +60\text{ lb} \qquad \mathbf{D}_x = 60\text{ lb}\rightarrow$$

Since the values obtained are positive, the forces are directed as assumed in the diagram, that is, $\mathbf{D}_x$ to the right, $\mathbf{D}_y$ up, and **A** to the left.

Members. The frame is dismembered; since only two members are connected at *B*, the components of the unknown forces acting on *EBF* and *ABCD* at *B* are, respectively, equal and opposite. The forces exerted at *E* and *C* by the cable *EC* are equal and opposite, and their direction is known. From the free-body diagram of the pulley, it is seen that the force exerted at *F* by the pulley on member *EBF* may be resolved into two 75-lb components as shown. The cable also exerts a 75-lb force on *ABCD* at point *G*.

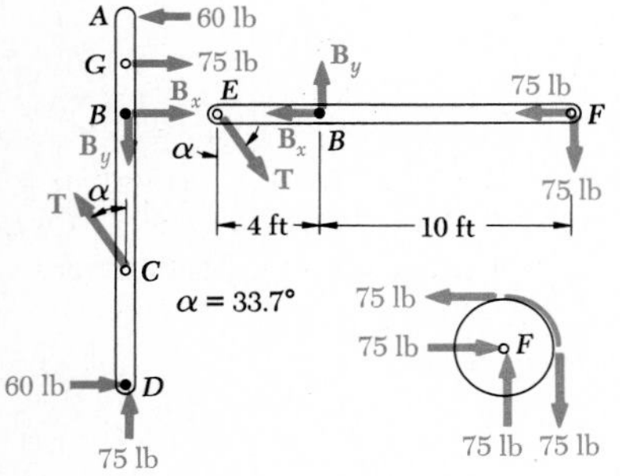

Member *EBF*. Using the free body *EBF*, we write

$$+\circlearrowleft\Sigma M_E = 0: \qquad B_y(4\text{ ft}) - (75\text{ lb})(14\text{ ft}) = 0 \qquad B_y = +263\text{ lb}$$

$$+\circlearrowleft\Sigma M_B = 0: \qquad (T\cos\alpha)(4\text{ ft}) - (75\text{ lb})(10\text{ ft}) = 0$$
$$T = +225\text{ lb}$$

$$\overset{+}{\rightarrow}\Sigma F_x = 0: \qquad +T\sin\alpha - B_x - 75\text{ lb} = 0 \qquad B_x = +50.0\text{ lb}$$

Since the values obtained are positive, the forces are directed as shown on the diagram: the forces $\mathbf{B}_x$ and $\mathbf{B}_y$ acting on member *ABCD* are directed, respectively, to the right and down, while the forces $\mathbf{B}_x$ and $\mathbf{B}_y$ acting on member *EBF* are directed, respectively, to the left and up.

Member *ABCD* (Check). The computations are checked by considering the free body *ABCD*. For example,

$$\overset{+}{\rightarrow}\Sigma F_x = -60\text{ lb} + 75\text{ lb} + B_x - T\sin\alpha + 60\text{ lb}$$
$$= -60\text{ lb} + 75\text{ lb} + 50\text{ lb} - (225\text{ lb})\sin 33.7° + 60\text{ lb} = 0$$

(checks)

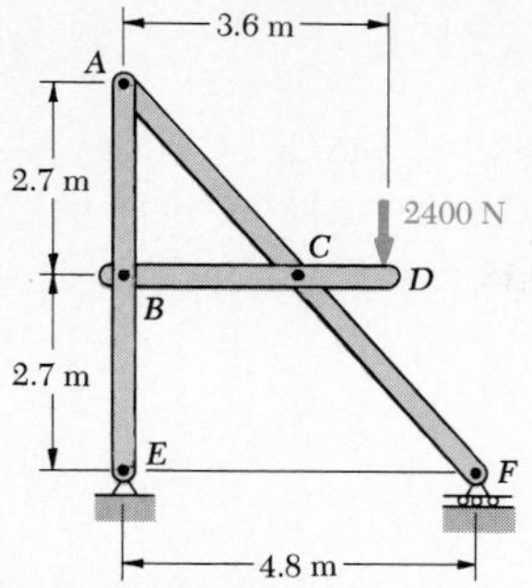

SAMPLE PROBLEM 6.6

Determine the components of the forces acting on each member of the frame shown.

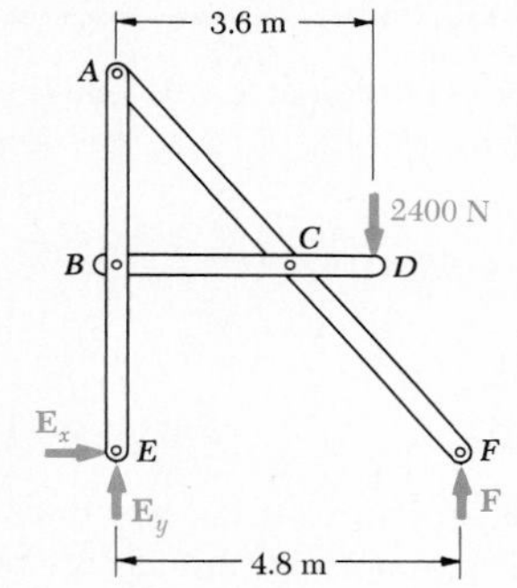

Entire Frame. Since the external reactions involve only three unknowns, we compute the reactions by considering the free-body diagram of the entire frame.

$+\circlearrowleft \Sigma M_E = 0$: $\quad -(2400\text{ N})(3.6\text{ m}) + F(4.8\text{ m}) = 0$

$F = +1800\text{ N}$ $\qquad$ $\mathbf{F} = 1800\text{ N}\uparrow$ ◀

$+\uparrow \Sigma F_y = 0$: $\quad -2400\text{ N} + 1800\text{ N} + E_y = 0$

$E_y = +600\text{ N}$ $\qquad$ $\mathbf{E}_y = 600\text{ N}\uparrow$ ◀

$\xrightarrow{+} \Sigma F_x = 0$: $\qquad$ $E_x = 0$ ◀

The frame is now dismembered; since only two members are connected at each joint, equal and opposite components are shown on each member at each joint.

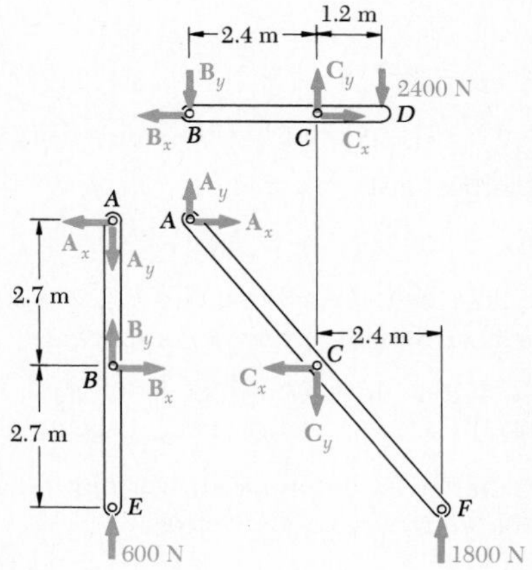

Member *BCD*

$+\circlearrowleft \Sigma M_B = 0$: $-(2400\text{ N})(3.6\text{ m}) + C_y(2.4\text{ m}) = 0$ $\quad C_y = +3600\text{ N}$ ◀

$+\circlearrowleft \Sigma M_C = 0$: $-(2400\text{ N})(1.2\text{ m}) + B_y(2.4\text{ m}) = 0$ $\quad B_y = +1200\text{ N}$ ◀

$\xrightarrow{+} \Sigma F_x = 0$: $\quad -B_x + C_x = 0$

We note that neither B_x nor C_x can be obtained by considering only member *BCD*. The positive values obtained for B_y and C_y indicate that the force components $\mathbf{B}_y$ and $\mathbf{C}_y$ are directed as assumed.

Member *ABE*

$+\circlearrowleft \Sigma M_A = 0$: $\quad B_x(2.7\text{ m}) = 0$ $\qquad B_x = 0$ ◀

$\xrightarrow{+} \Sigma F_x = 0$: $\quad +B_x - A_x = 0$ $\qquad A_x = 0$ ◀

$+\uparrow \Sigma F_y = 0$: $\quad -A_y + B_y + 600\text{ N} = 0$

$-A_y + 1200\text{ N} + 600\text{ N} = 0$ $\quad A_y = +1800\text{ N}$ ◀

Member *BCD*. Returning now to member *BCD*, we write

$\xrightarrow{+} \Sigma F_x = 0$: $\quad -B_x + C_x = 0 \qquad 0 + C_x = 0$ $\qquad C_x = 0$ ◀

Member *ACF* (Check). All unknown components have now been found; to check the results, verify that member *ACF* is in equilibrium.

$$+\circlearrowleft \Sigma M_C = (1800\text{ N})(2.4\text{ m}) - A_y(2.4\text{ m}) - A_x(2.7\text{ m})$$
$$= (1800\text{ N})(2.4\text{ m}) - (1800\text{ N})(2.4\text{ m}) - 0 = 0 \qquad \text{(checks)}$$

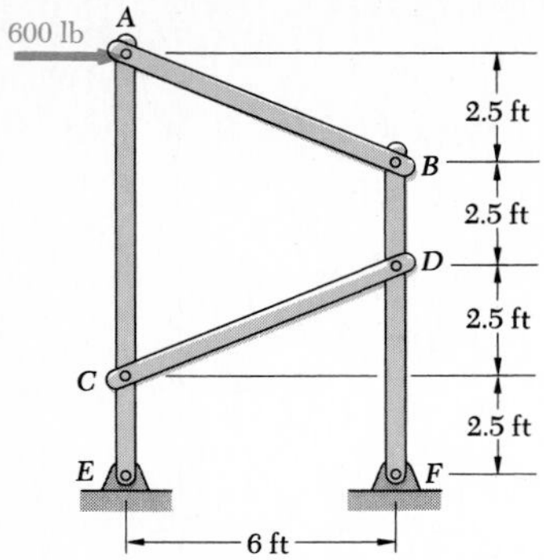

SAMPLE PROBLEM 6.7

A 600-lb horizontal force is applied to pin A of the frame shown. Determine the forces acting on the two vertical members of the frame.

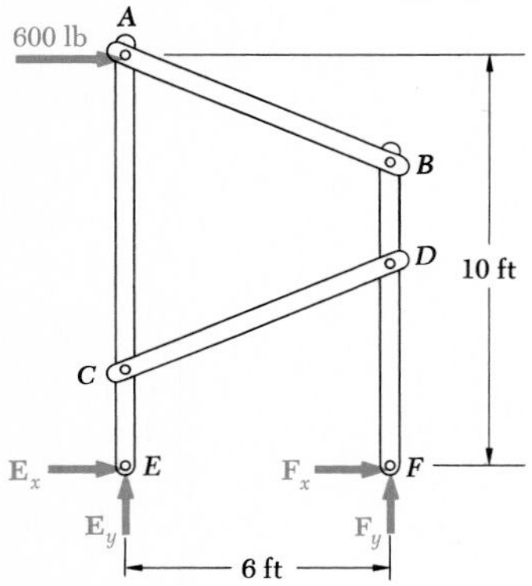

Entire Frame. The entire frame is chosen as a free body; although the reactions involve four unknowns, $\mathbf{E}_y$ and $\mathbf{F}_y$ may be determined by writing

$$+\circlearrowleft \Sigma M_E = 0: \quad -(600 \text{ lb})(10 \text{ ft}) + F_y(6 \text{ ft}) = 0$$
$$F_y = +1000 \text{ lb} \qquad \mathbf{F}_y = 1000 \text{ lb} \uparrow$$

$$+\uparrow \Sigma F_y = 0: \quad E_y + F_y = 0$$
$$E_y = -1000 \text{ lb} \qquad \mathbf{E}_y = 1000 \text{ lb} \downarrow$$

The equations of equilibrium of the entire frame are not sufficient to determine $\mathbf{E}_x$ and $\mathbf{F}_x$. The equilibrium of the various members must now be considered in order to proceed with the solution. In dismembering the frame we shall assume that pin A is attached to the multiforce member ACE and, thus, that the 600-lb force is applied to that member. We also note that AB and CD are two-force members.

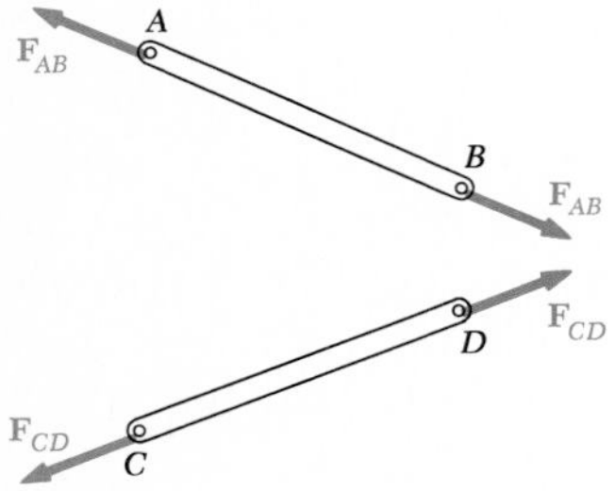

Member *ACE*

$$+\uparrow \Sigma F_y = 0: \quad -\tfrac{5}{13}F_{AB} + \tfrac{5}{13}F_{CD} - 1000 \text{ lb} = 0$$

$$+\circlearrowleft \Sigma M_E = 0: \quad -(600 \text{ lb})(10 \text{ ft}) - (\tfrac{12}{13}F_{AB})(10 \text{ ft}) - (\tfrac{12}{13}F_{CD})(2.5 \text{ ft}) = 0$$

Solving these equations simultaneously, we find

$$F_{AB} = -1040 \text{ lb} \qquad F_{CD} = +1560 \text{ lb}$$

The signs obtained indicate that the sense assumed for F_{CD} was correct and the sense for F_{AB} incorrect. Summing now x components:

$$\xrightarrow{+} \Sigma F_x = 0: \quad 600 \text{ lb} + \tfrac{12}{13}(-1040 \text{ lb}) + \tfrac{12}{13}(+1560 \text{ lb}) + E_x = 0$$
$$E_x = -1080 \text{ lb} \qquad \mathbf{E}_x = 1080 \text{ lb} \leftarrow$$

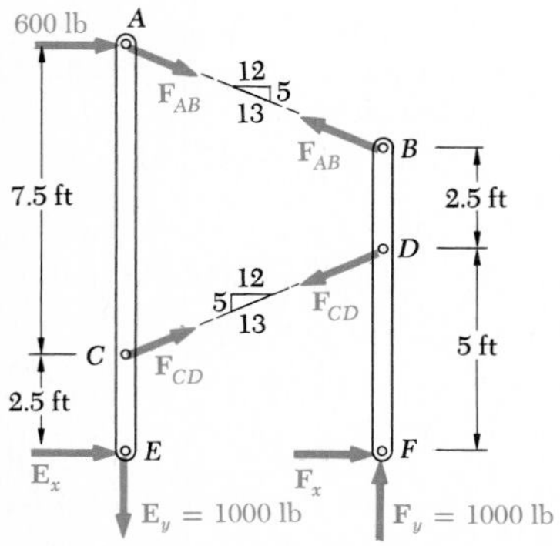

Entire Frame. Since $\mathbf{E}_x$ has been determined, we may return to the free-body diagram of the entire frame and write

$$\xrightarrow{+} \Sigma F_x = 0: \quad 600 \text{ lb} - 1080 \text{ lb} + F_x = 0$$
$$F_x = +480 \text{ lb} \qquad \mathbf{F}_x = 480 \text{ lb} \rightarrow$$

Member *BDF* (Check). We may check our computations by verifying that the equation $\Sigma M_B = 0$ is satisfied by the forces acting on member BDF.

$$+\circlearrowleft \Sigma M_B = -(\tfrac{12}{13}F_{CD})(2.5 \text{ ft}) + (F_x)(7.5 \text{ ft})$$
$$= -\tfrac{12}{13}(1560 \text{ lb})(2.5 \text{ ft}) + (480 \text{ lb})(7.5 \text{ ft})$$
$$= -3600 \text{ lb} \cdot \text{ft} + 3600 \text{ lb} \cdot \text{ft} = 0 \qquad \text{(checks)}$$

PROBLEMS

6.65 and 6.66 Determine the force in member *BD* and the components of the reaction at *C*.

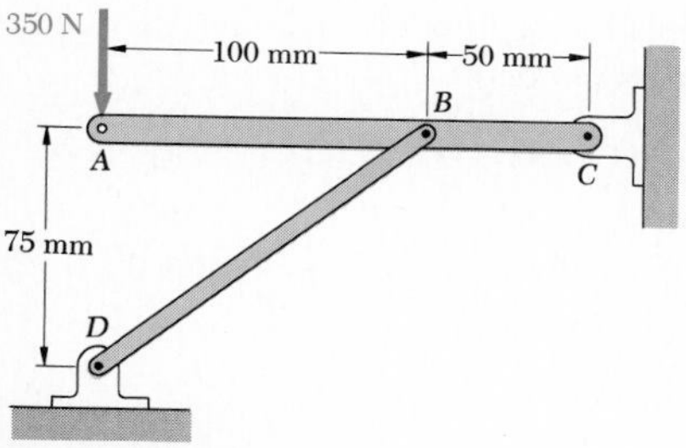

Fig. P6.65

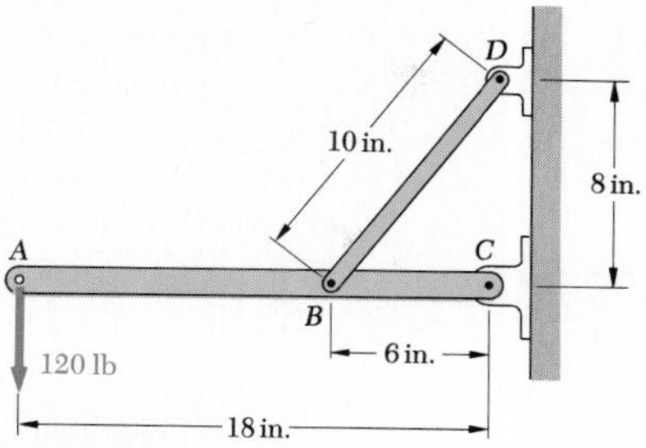

Fig. P6.66

6.67 Determine the components of all forces acting on member *AE* of the frame shown.

6.68 Determine the components of all forces acting on member *ABCD* of the assembly shown.

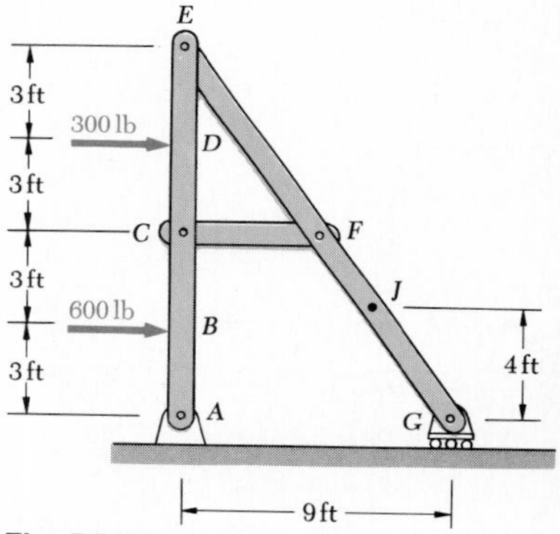

Fig. P6.67

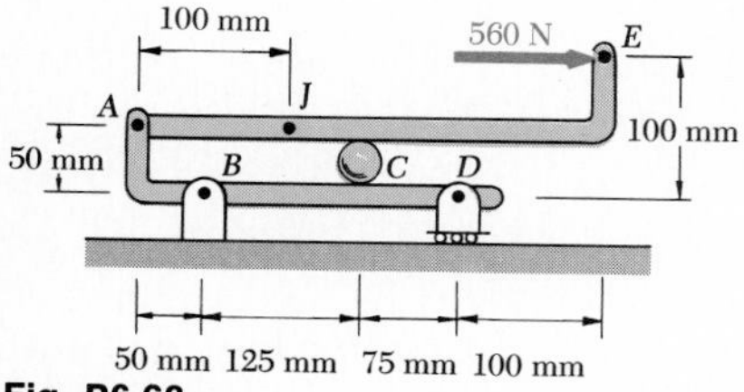

Fig. P6.68

6.69 Solve Prob. 6.68, assuming that the 560-N force applied at *E* is directed vertically downward.

6.70 A 140-lb man stands at *C* on the step stool shown. Half the man's weight is carried by the legs shown. Determine the components of the force exerted at *E* on leg *BE*, assuming that the bottom of the legs is not quite parallel to the floor and that bearing occurs only at points *A* and *B*. Neglect the weight of the stool, and assume the floor to be frictionless.

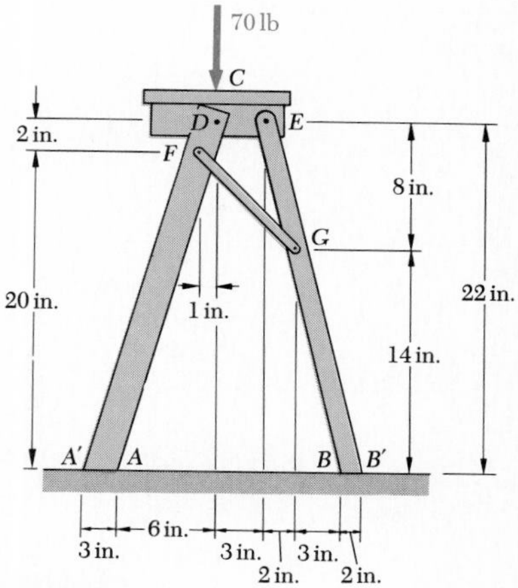

Fig. P6.70

6.71 In Prob. 6.70, bearing between the legs of the step stool and the floor can occur in four ways: at *A* and *B*, *A* and *B'*, *A'* and *B*, or *A'* and *B'*. Determine (*a*) for which combination of bearing points the force in member *FG* is maximum, (*b*) the corresponding value of the force in *FG*.

6.72 Determine the components of the force exerted at B on member BE (a) if the 400-N load is applied as shown, (b) if the 400-N load is moved along its line of action and is applied at point F.

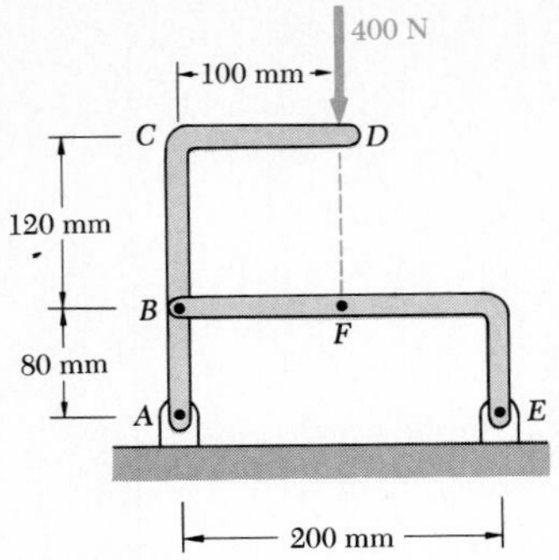

Fig. P6.72

6.73 Determine the forces exerted on member AB if the frame is loaded by a clockwise couple of moment 6 N · m applied (a) at point D, (b) at point E. (c) Determine the forces exerted on member AB if the frame is loaded by vertical forces applied at D and E which are equivalent to a 6-N · m clockwise couple.

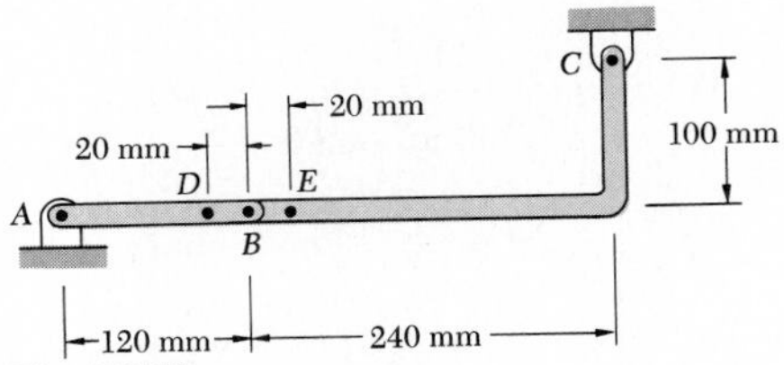

Fig. P6.73

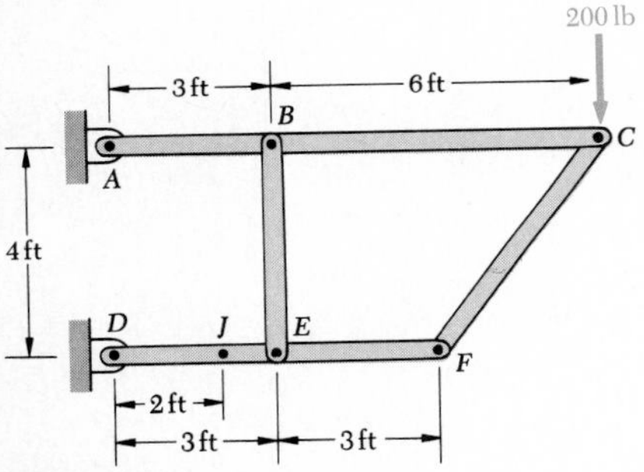

Fig. P6.74

6.74 Determine the components of all forces acting on member DEF of the frame shown.

6.75 Determine the reactions at E and F and the force exerted on pin C for the frame shown.

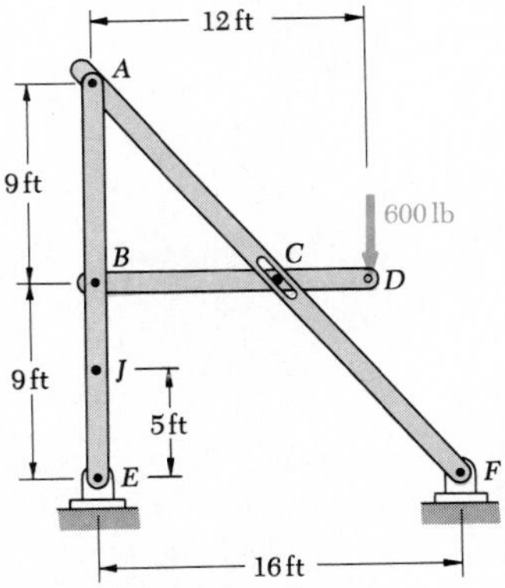

Fig. P6.75

6.76 Determine the reactions at E and F and the force exerted on pin C for the frame of Prob. 6.75, assuming now that pin C is attached to member ACF and may slide in a horizontal slot in member BD.

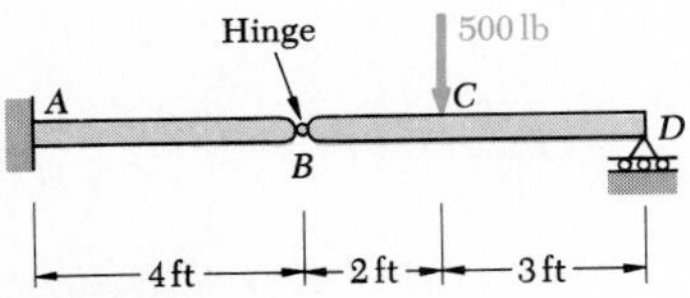

Fig. P6.77

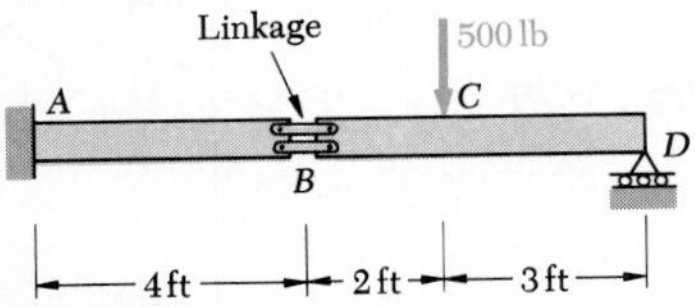

Fig. P6.78

6.77 and 6.78 Determine the reactions at the supports for the beam shown.

6.79 A 3-ft-diameter 500-lb cylinder is supported as shown. Assuming that $\alpha = 30°$, determine all forces acting on member AB. The cylinder is assumed to be frictionless.

6.80 Determine (*a*) the angle α at which the cable in Prob. 6.79 must be attached so that the reaction at A has no vertical component, (*b*) the corresponding reactions at A and B.

6.81 The bent rod DEF fits into the bent pipe ABC as shown. Neglecting the effect of friction, determine the reactions at A and F due to the 360-N force applied at B.

6.82 The tractor and scraper units shown are connected by a vertical pin located 2 ft behind the tractor wheels. The distance from C to D is 30 in. The center of gravity of the 16,000-lb tractor unit is located at G_t. The scraper unit and load together weigh 84,000 lb and have a combined center of gravity located at G_s. Knowing that the machine is at rest, with its brakes released, determine (*a*) the reactions at each of the four wheels, (*b*) the forces exerted on the tractor unit at C and D.

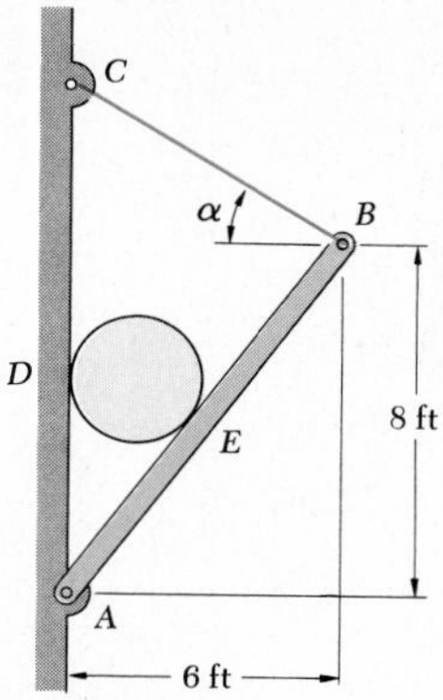

Fig. P6.79

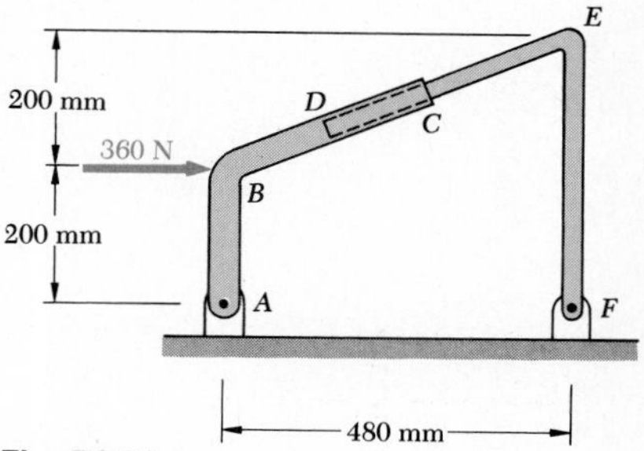

Fig. P6.81

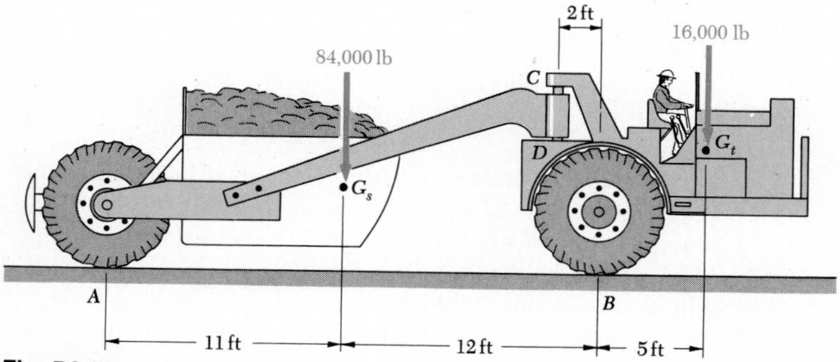

Fig. P6.82

6.83 A trailer weighing 2750 lb is attached to a 3200-lb automobile by a ball-and-socket trailer hitch at D. Determine (*a*) the reactions at each of the six wheels when the automobile and trailer are at rest, (*b*) the additional load on each of the automobile wheels due to the trailer.

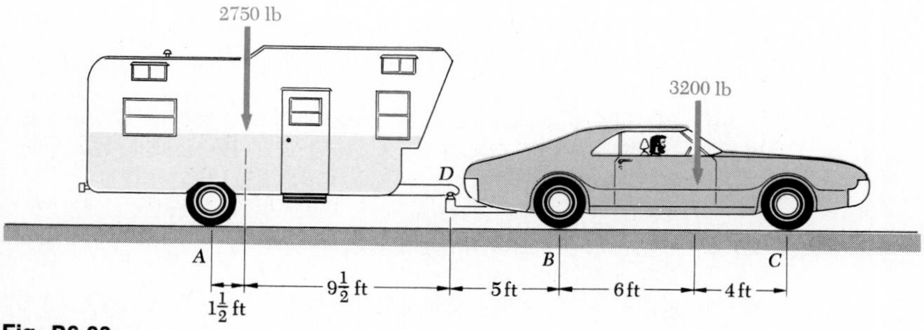

Fig. P6.83

6.84 In order to obtain a better weight distribution over the four wheels of the automobile of Prob. 6.83, a compensating hitch of the type shown is used to attach the trailer to the automobile. This hitch consists of two bar springs (only one is shown in the figure) which fit into bearings inside a support rigidly attached to the automobile. The springs are also connected by chains to the trailer frame, and specially designed hooks make it possible to place both chains under a tension T. Solve Prob. 6.83 assuming that such a compensating hitch is used and that the tension T in each chain is 440 lb.

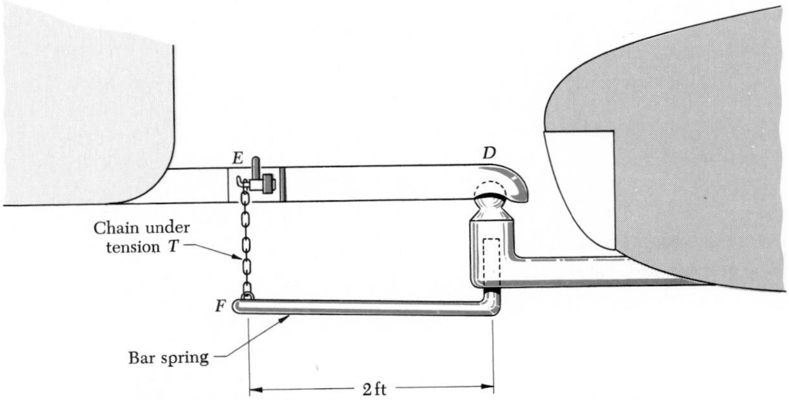

Fig. P6.84 and P6.85

6.85 In order to obtain a better weight distribution over the wheels of the automobile of Prob. 6.83, a compensating hitch of the type described in Prob. 6.84 is used to attach the trailer to the automobile. (*a*) Determine the tension T required in each of the two chains if the additional load due to the trailer is to be evenly distributed over the four wheels of the automobile. (*b*) What are the corresponding reactions at each of the six wheels of the trailer-automobile combination?

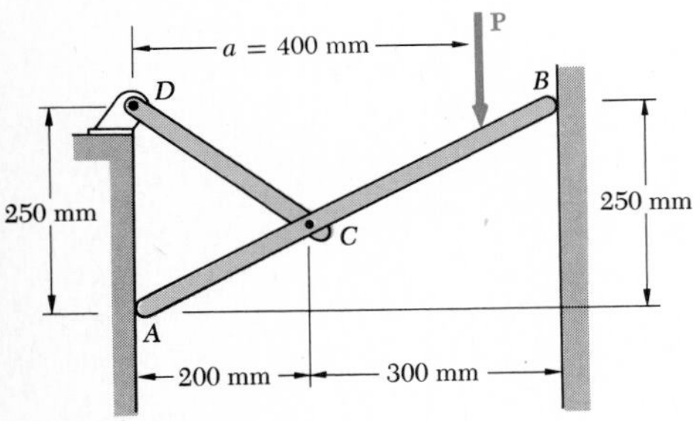

Fig. P6.86

6.86 A vertical load **P** of magnitude 600 N is applied to member AB. Member AB is placed between two frictionless walls and is pin-connected at C to a link CD. Determine all forces exerted on member AB.

6.87 In Prob. 6.86, determine the range of values of the distance a for which the load **P** can be supported.

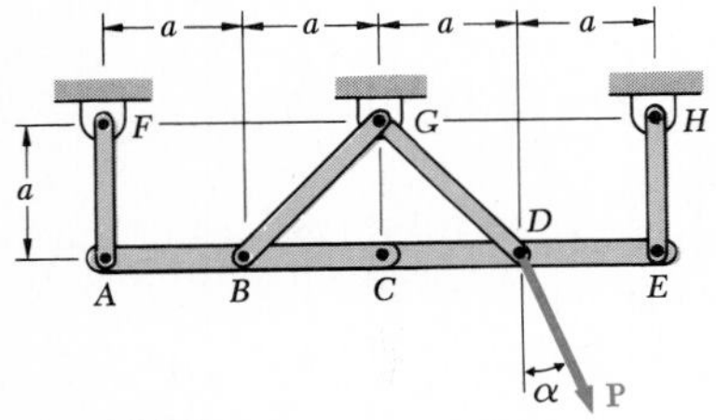

Fig. P6.88

6.88 Members ABC and CDE are pin-connected at C and supported by the four links AF, BG, GD, and EH. Determine the force in each link when $\alpha = 0°$.

6.89 Solve Prob. 6.88 when $\alpha = 90°$.

6.90 Solve Prob. 6.88, assuming that the force **P** is replaced by a clockwise couple of moment M_0 applied to member CDE at D.

6.91 (*a*) Show that, when a frame supports a pulley at *A*, an equivalent loading of the frame and of each of its component parts may be obtained by removing the pulley and applying at *A* two forces equal and parallel to the forces of tension in the cable. (*b*) Further, show that, if one end of the cable is attached to the frame at a point *B*, a force of magnitude equal to the tension should also be applied at *B*.

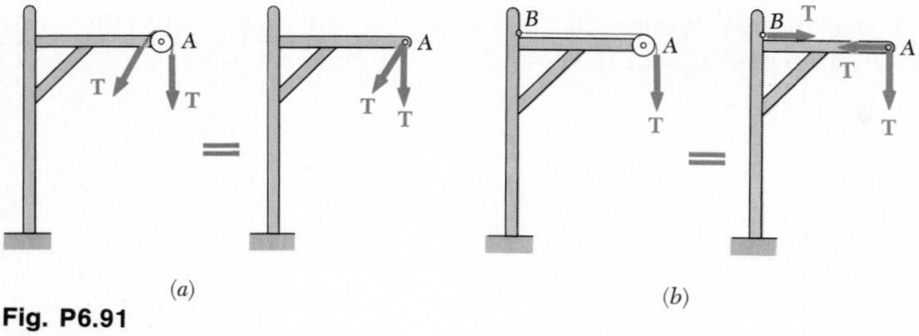

Fig. P6.91

6.92 through 6.94 Determine the reactions at the supports for each of the trusses shown. Indicate whether the truss is rigid. The height of each truss is 4 m; the length of each panel is 4 m; and the magnitude of **P** is 40 kN.

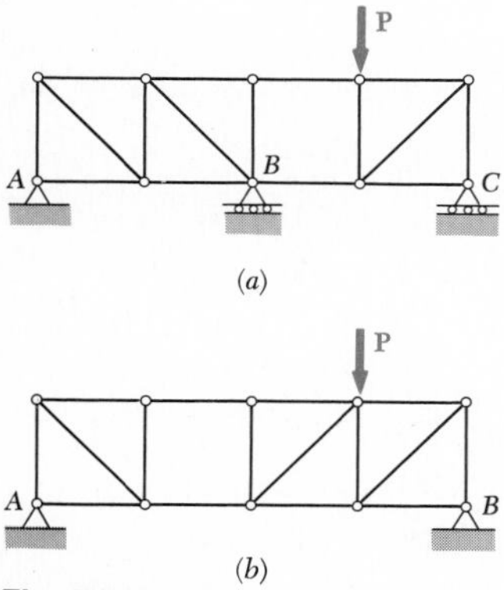

Fig. P6.92

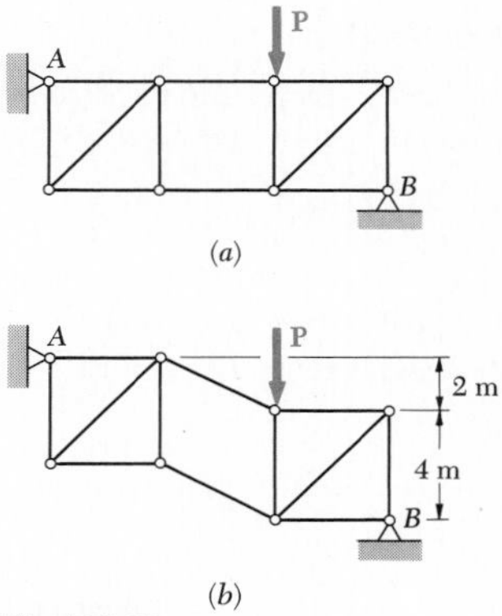

Fig. P6.93

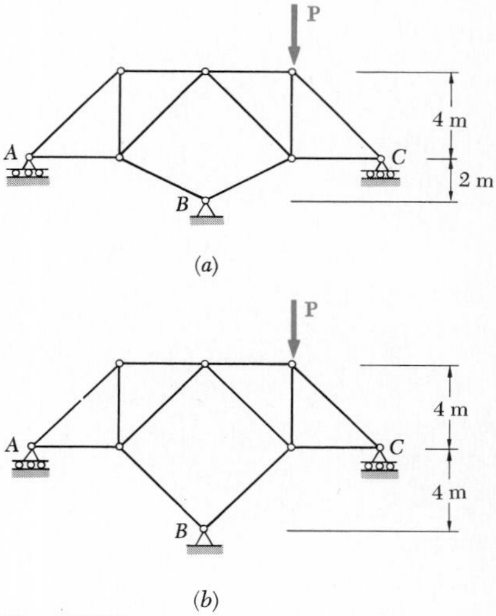

Fig. P6.94

6.95 The frame shown consists of two members *ABC* and *DEF* connected by the links *AE* and *BF*. Determine the reactions at *C* and *D* and the force in links *AE* and *BF*.

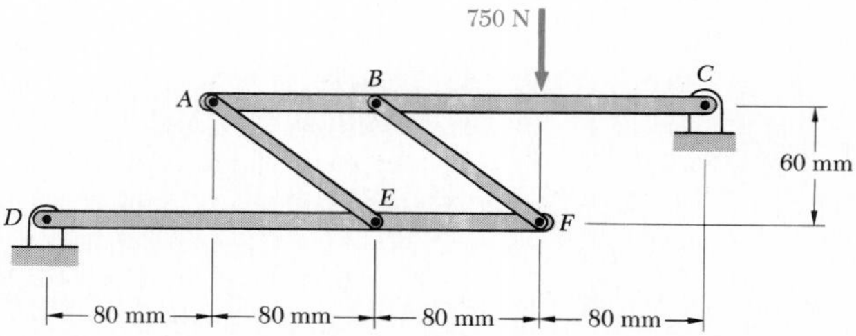

Fig. P6.95

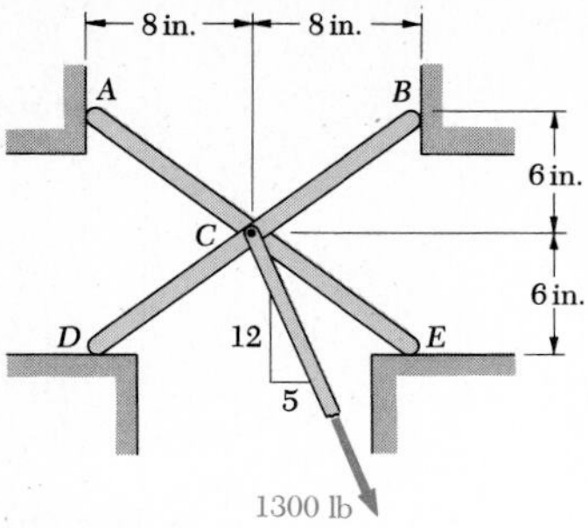

Fig. P6.96

6.96 Knowing that the surfaces at A, B, D, and E are frictionless, determine (a) the reactions, (b) the components of the force exerted on member ACE at point C.

6.97 Solve Prob. 6.95, assuming that the 750-N load is applied at F.

6.98 Determine the vertical reactions at points A, B, C, and D of the runway girders for the overhead traveling crane shown if $P = 12$ kips, $a = 15$ ft, $L = 40$ ft, $b = 10$ ft, and $l = 30$ ft.

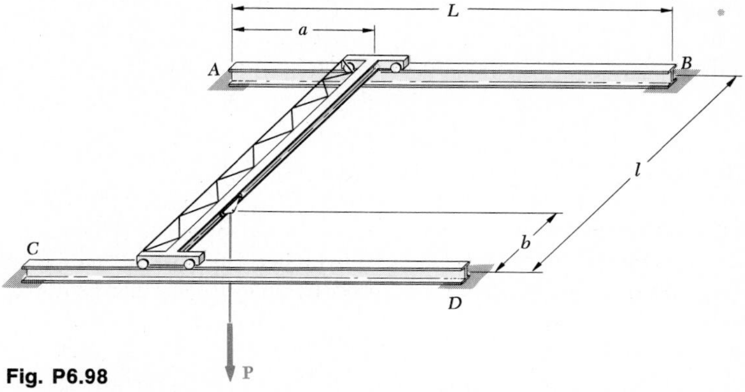

Fig. P6.98

6.99 Determine the vertical reactions at points A and B of the runway girder AB of Prob. 6.98 in terms of P, a, b, l, and L.

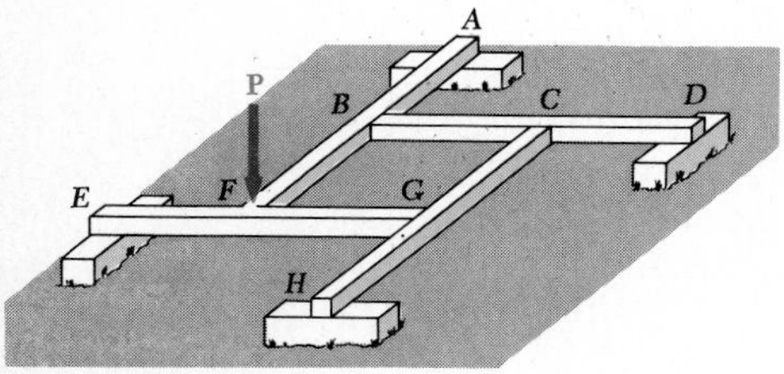

Fig. P6.100

6.100 Four beams, each of length $2a$, are nailed together at their midpoints to form the support system shown. Assuming that only vertical forces are exerted at the connections, determine the vertical reactions at A, D, E, and H.

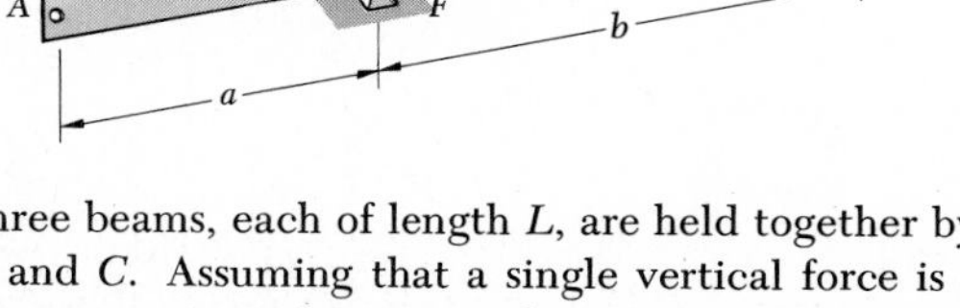

Fig. P6.101

6.101 Three beams, each of length L, are held together by single nails at A, B, and C. Assuming that a single vertical force is exerted at each of the connections, determine the vertical reactions at D, E, and F.

6.102 Solve Prob. 6.101 if $a = 0.6$ m, $b = 1.2$ m, and $P = 600$ N.

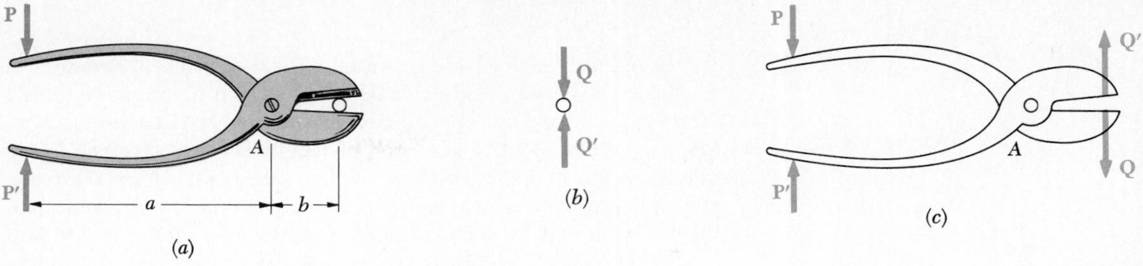

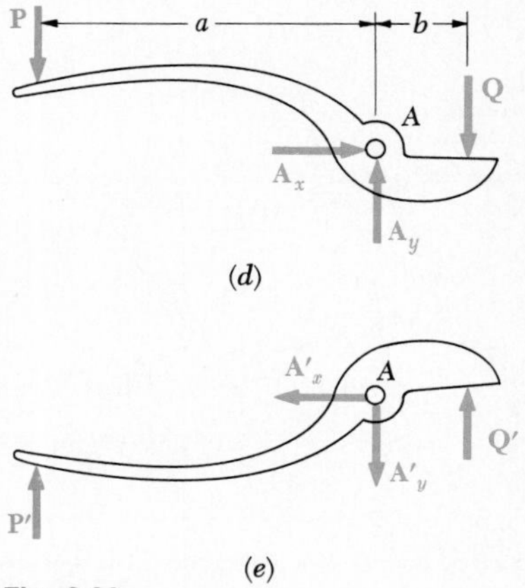

Fig. 6.22

6.13. Machines. Machines are structures designed to transmit and modify forces. Whether they are simple tools or include complicated mechanisms, their main purpose is to transform *input forces* into *output forces*. Consider, for example, a pair of cutting pliers used to cut a wire (Fig. 6.22*a*). If we apply two equal and opposite forces **P** and **P′** on their handles, they will exert two equal and opposite forces **Q** and **Q′** on the wire (Fig. 6.22*b*).

To determine the magnitude Q of the output forces when the magnitude P of the input forces is known (or, conversely, to determine P when Q is known), we draw a free-body diagram of the pliers *alone*, showing the input forces **P** and **P′** and the *reactions* **Q′** and **Q** that the wire exerts on the pliers (Fig. 6.22*c*). However, since a pair of pliers forms a nonrigid structure, we must use one of the component parts as a free body in order to determine the unknown forces. Considering Fig. 6.22*d*, for example, and taking moments about A, we obtain the relation $Pa = Qb$, which defines the magnitude Q in terms of P or P in terms of Q. The same free-body diagram may be used to determine the components of the internal force at A; we find $A_x = 0$ and $A_y = P + Q$.

In the case of more complicated machines, it generally will be necessary to use several free-body diagrams and, possibly, to solve simultaneous equations involving various internal forces. The free bodies should be chosen to include the input forces and the reactions to the output forces, and the total number of unknown force components involved should not exceed the number of available independent equations. While it is advisable to check whether the problem is determinate before attempting to solve it, there is no point in discussing the rigidity of a machine. A machine includes moving parts and thus must be nonrigid.

SAMPLE PROBLEM 6.8

A hydraulic-lift table is used to raise a 1000-kg crate. It consists of a platform and two identical linkages on which hydraulic cylinders exert equal forces. (Only one linkage and one cylinder are shown.) Members *EDB* and *CG* are each of length $2a$, and member *AD* is pinned to the midpoint of *EDB*. If the crate is placed on the table, so that half of its weight is supported by the system shown, determine the force exerted by each cylinder in raising the crate for $\theta = 60°$, $a = 0.70$ m, and $L = 3.20$ m. Show that the result obtained is independent of the distance d.

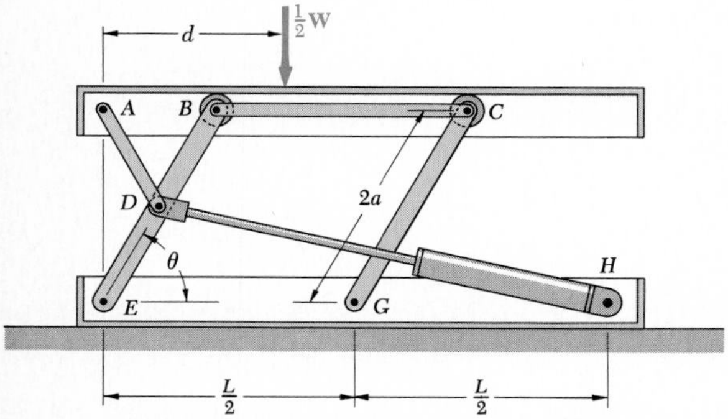

Solution. The machine considered consists of the platform and of the linkage, with an input force $\mathbf{F}_{DH}$ exerted by the cylinder and an output force equal and opposite to $\frac{1}{2}\mathbf{W}$. Since more than three unknowns would be involved, the entire mechanism is not used as a free body. The mechanism is dismembered and free-body diagrams are drawn for platform *ABC*, for roller *C*, and for member *EDB*. Since *AD*, *BC*, and *CG* are two-force members, their free-body diagrams have been omitted, and the forces they exert on the other parts of the mechanism have been drawn parallel to these members.

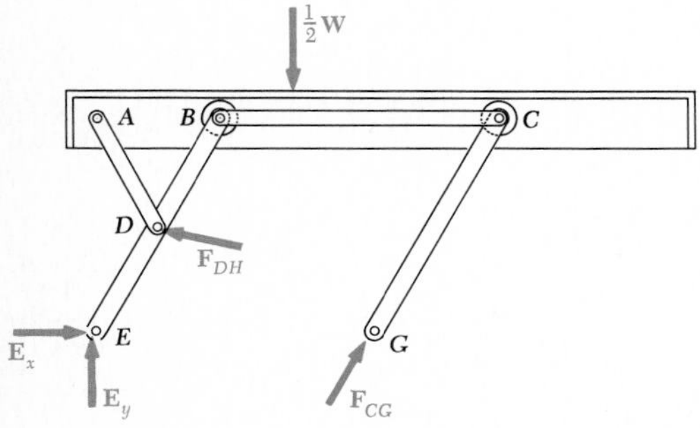

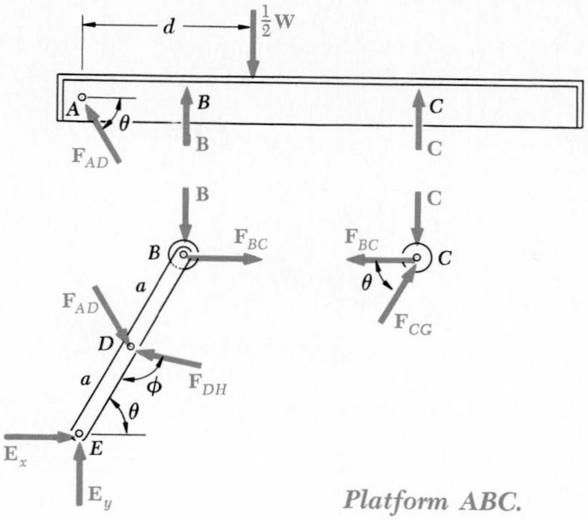

Platform ABC.

$$\overset{+}{\rightarrow}\Sigma F_x = 0: \quad -F_{AD}\cos\theta = 0 \qquad F_{AD} = 0$$
$$+\uparrow\Sigma F_y = 0: \quad B + C - \tfrac{1}{2}W = 0 \qquad B + C = \tfrac{1}{2}W \tag{1}$$

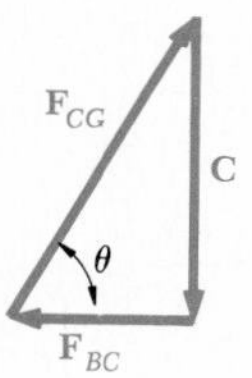

Roller C. We draw a force triangle and obtain $F_{BC} = C\cot\theta$.

Member EDB. Recalling that $F_{AD} = 0$,

$$+\circlearrowleft\Sigma M_E = 0: \quad F_{DH}\cos(\phi - 90°)a - B(2a\cos\theta) - F_{BC}(2a\sin\theta) = 0$$
$$F_{DH}a\sin\phi - B(2a\cos\theta) - (C\cot\theta)(2a\sin\theta) = 0$$
$$F_{DH}\sin\phi - 2(B + C)\cos\theta = 0$$

Recalling Eq. (1), we have

$$F_{DH} = W\frac{\cos\theta}{\sin\phi} \tag{2}$$

and we observe that the result obtained is independent of d. ◀

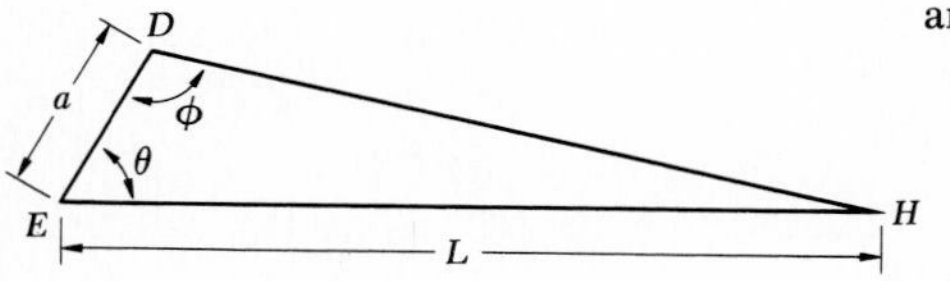

Applying first the law of sines to triangle EDH, we write

$$\frac{\sin\phi}{EH} = \frac{\sin\theta}{DH} \qquad \sin\phi = \frac{EH}{DH}\sin\theta \tag{3}$$

Using now the law of cosines, we have

$$(DH)^2 = a^2 + L^2 - 2aL\cos\theta$$
$$= (0.70)^2 + (3.20)^2 - 2(0.70)(3.20)\cos 60°$$
$$(DH)^2 = 8.49 \qquad DH = 2.91\text{ m}$$

We also note that

$$W = mg = (1000\text{ kg})(9.81\text{ m/s}^2) = 9810\text{ N} = 9.81\text{ kN}$$

Substituting for $\sin\phi$ from (3) into (2) and using the numerical data, we write

$$F_{DH} = W\frac{DH}{EH}\cot\theta = (9.81\text{ kN})\frac{2.91\text{ m}}{3.20\text{ m}}\cot 60°$$

$$F_{DH} = 5.15\text{ kN} \quad ◀$$

PROBLEMS

6.103 A 360-N force is applied to the toggle vise at C. Determine (a) the horizontal force exerted on the block at D, (b) the force exerted on member ABC at B.

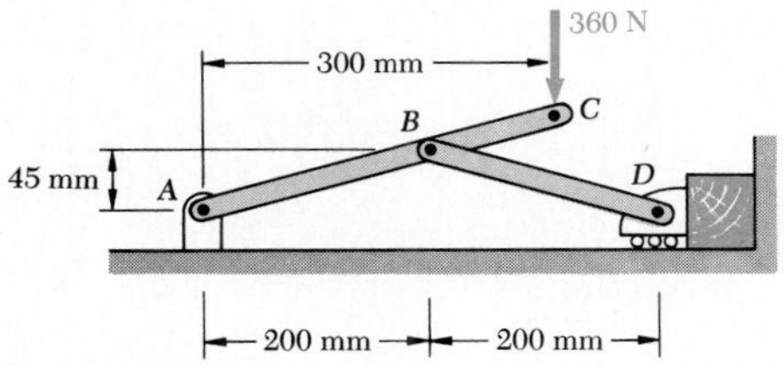

Fig. P6.103

6.104 A cylinder weighs 2000 lb and is lifted by a pair of tongs as shown. Determine the forces exerted at D and C on the tong BCD.

6.105 If the toggle shown is added to the tongs of Prob. 6.104 and the load is lifted by applying a single force at G, determine the forces exerted at D and C on the tong BCD.

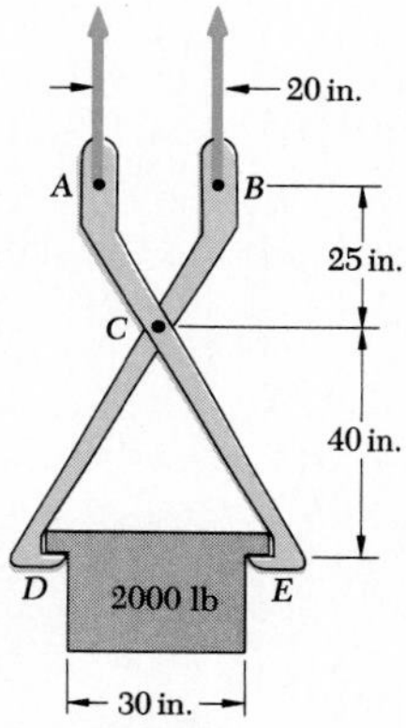

Fig. P6.104

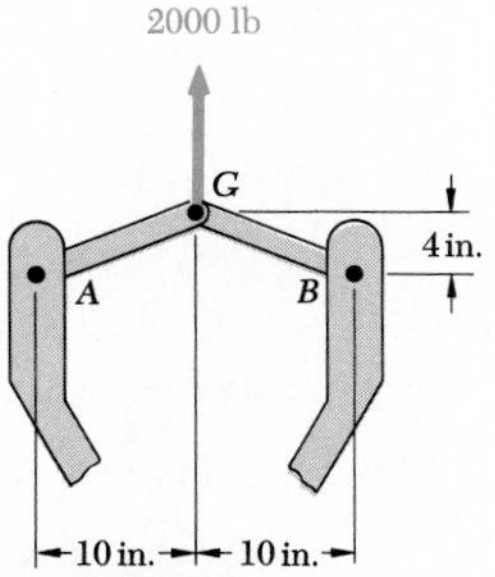

Fig. P6.105

6.106 Two 300-N forces are applied to the handles of the pliers as shown. Determine (a) the magnitude of the forces exerted on the rod, (b) the force exerted by the pin at A on portion AB of the pliers.

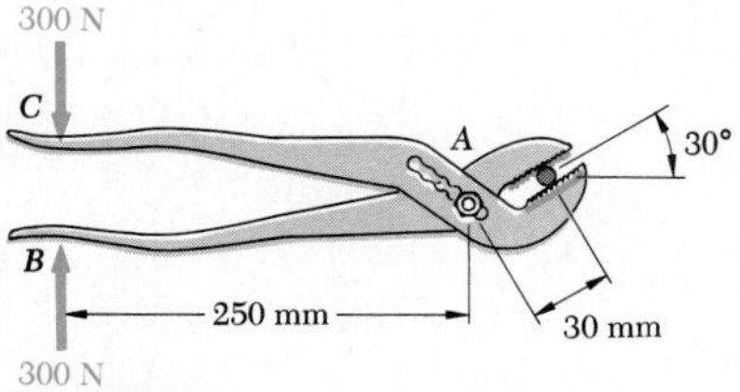

Fig. P6.106

6.107 A force **P** of magnitude 2160 lb is applied to the piston of the engine system shown. For each of the two positions shown, determine the couple **M** required to hold the system in equilibrium.

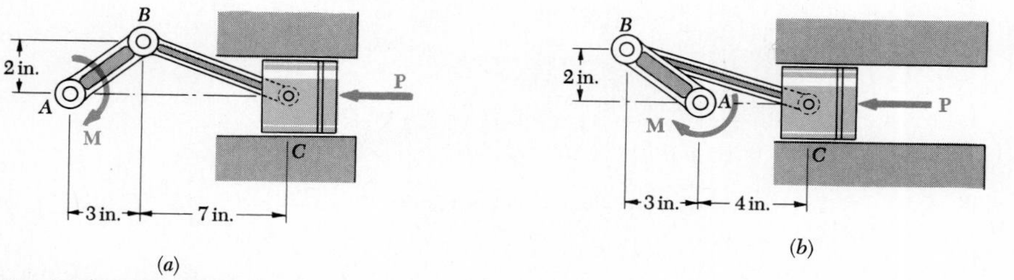

Fig. P6.107 and P6.108

6.108 A couple **M** of moment 400 lb · ft is applied to the crank of the engine system shown. For each of the two positions shown, determine the force **P** required to hold the system in equilibrium.

6.109 and 6.110 Two rods are connected by a frictionless collar B. Knowing that the moment of the couple $\mathbf{M}_A$ is 20 N · m, determine (*a*) the couple $\mathbf{M}_C$ required for equilibrium, (*b*) the corresponding components of the reaction at C.

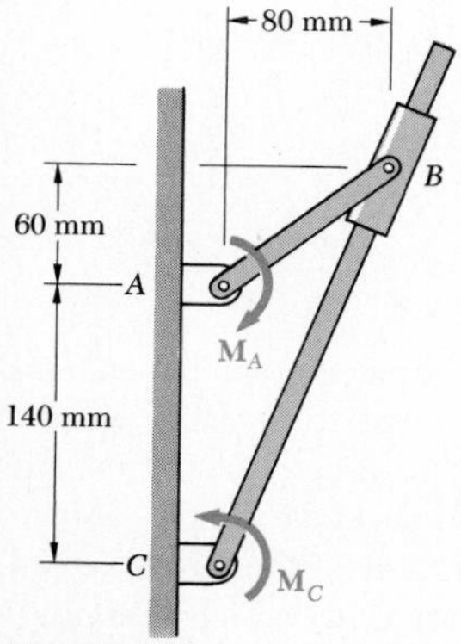

Fig. P6.109

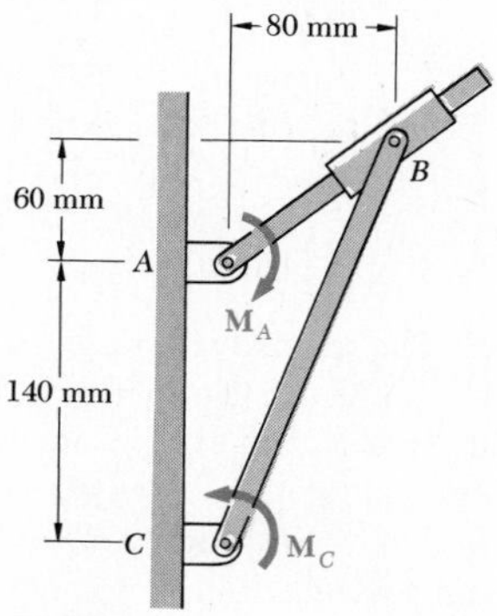

Fig. P6.110

6.111 In using the boltcutter shown, a worker applies two 100-lb forces to the handles. Determine the magnitude of the forces exerted by the cutter on the bolt.

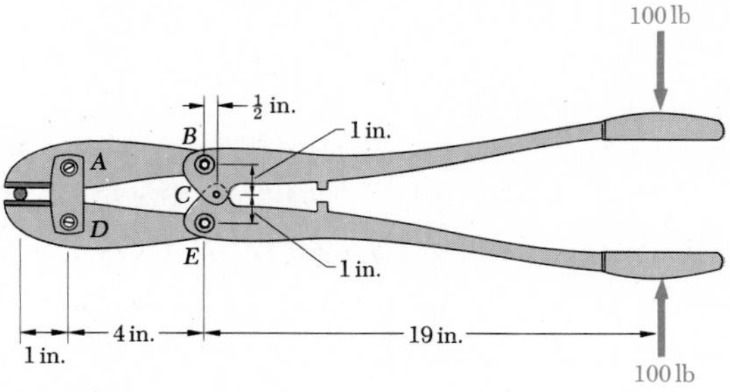

Fig. P6.111

6.112 Determine the magnitude of the gripping forces produced when two 60-lb forces are applied as shown.

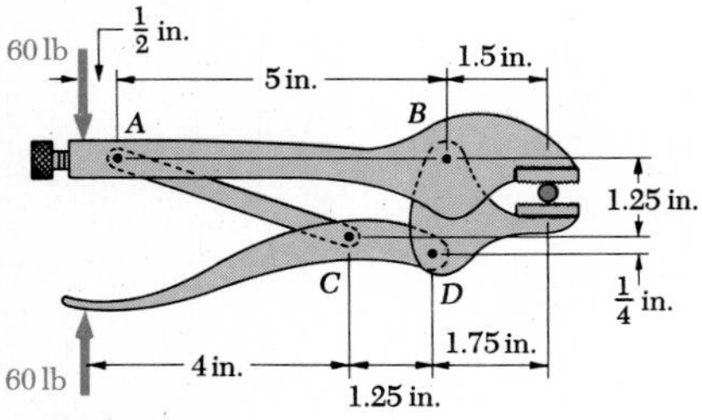

ig. P6.112

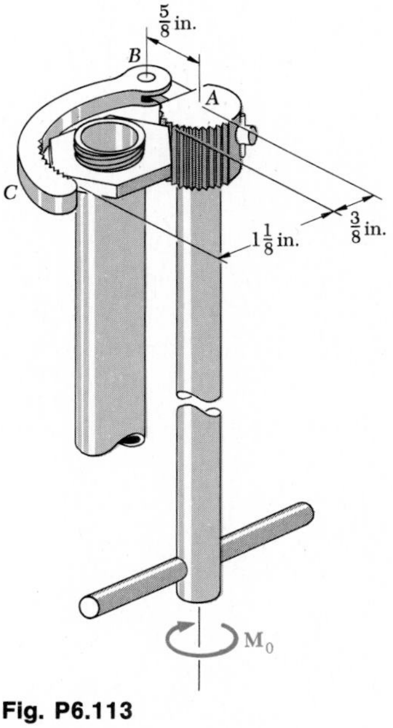

Fig. P6.113

6.113 The specialized plumbing wrench shown is used in confined areas (e.g., under a basin or sink). It consists essentially of a jaw BC pinned at B to a long rod. Knowing that the forces exerted on the nut are equivalent to a clockwise couple (when viewed from above) of moment 135 lb · in., determine (*a*) the magnitude of the force exerted by pin B on jaw BC, (*b*) the moment of the couple $\mathbf{M}_0$ which is applied to the wrench.

6.114 A pipe of diameter $2\frac{1}{2}$ in. is gripped by the Stillson wrench shown. Portions *AB* and *DE* of the wrench are rigidly attached to each other and portion *CF* is connected by a pin at *D*. Assuming that no slipping occurs between the pipe and the wrench, determine the components of the forces exerted on the pipe at *A* and at *C*.

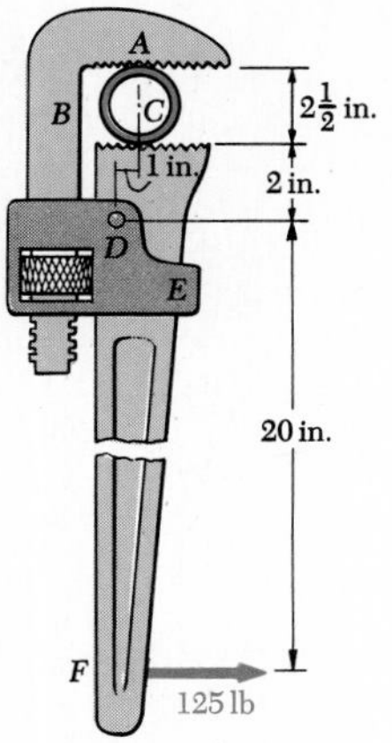

Fig. P6.114

6.115 A 20-kg shelf is held horizontally by a self-locking brace which consists of two parts *EDC* and *CDB* hinged at *C* and bearing against each other at *D*. Determine the force **P** required to release the brace. (*Hint.* To release the brace, the forces of contact at *D* must be zero.)

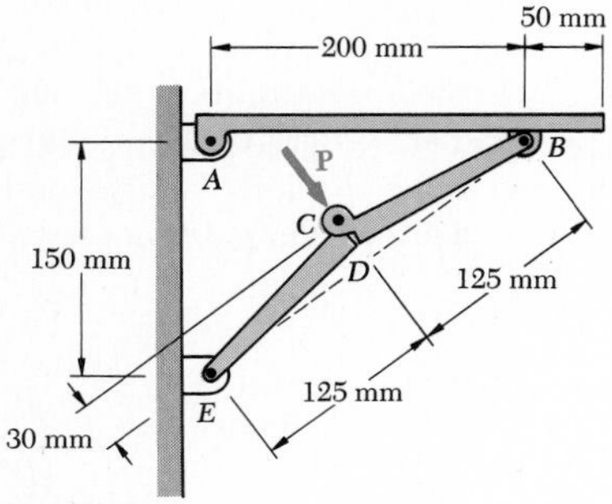

Fig. P6.115

6.116 Since the brace shown must remain in position even when the magnitude of **P** is very small, a single safety spring is attached at *D* and *E*. The spring *DE* has a constant of 10 kN/m and an unstretched length of 140 mm. Knowing that $l = 200$ mm and that the magnitude of **P** is 4000 N, determine the force **Q** required to release the brace.

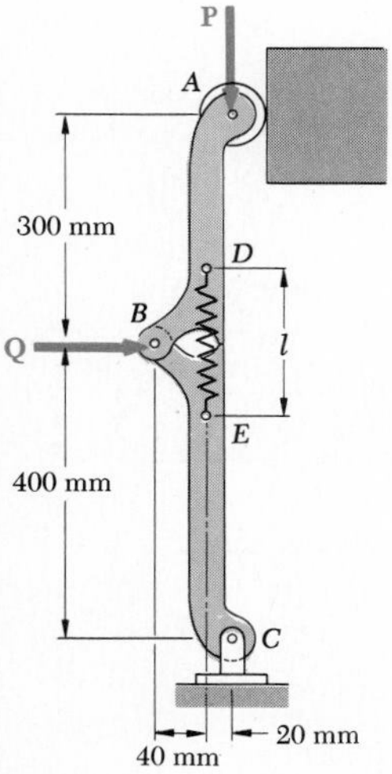

Fig. P6.116

6.117 The action of the backhoe bucket is controlled by the three hydraulic cylinders shown. Determine the force exerted by each cylinder in supporting the 3000-lb load shown.

6.118 The truck shown is used to facilitate work on overhead wires by raising a pair of buckets to the required elevation. Two tubular members *AB* and *BC*, each of length 12 ft, form the main supporting mechanism; the position of *AB* is controlled by means of the hydraulic cylinder *DE*. Two 2-ft-diameter sheaves, one on each side, are rigidly attached to member *BC* at *B*. Cables are fastened to the sheaves, pass over pulleys at *H*, and are fastened to a common movable block at *F*. The position of block *F* is controlled by a second hydraulic cylinder *EF*. Knowing that the workers, buckets, and equipment attached to the buckets together weigh 700 lb and have a combined center of gravity at *G*, determine the force which must be exerted by each hydraulic cylinder to maintain the position shown. Neglect the weight of the mechanism.

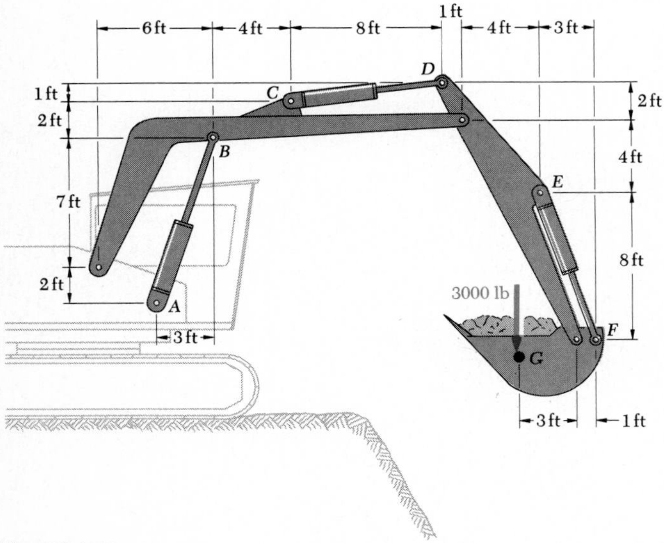

Fig. P6.117

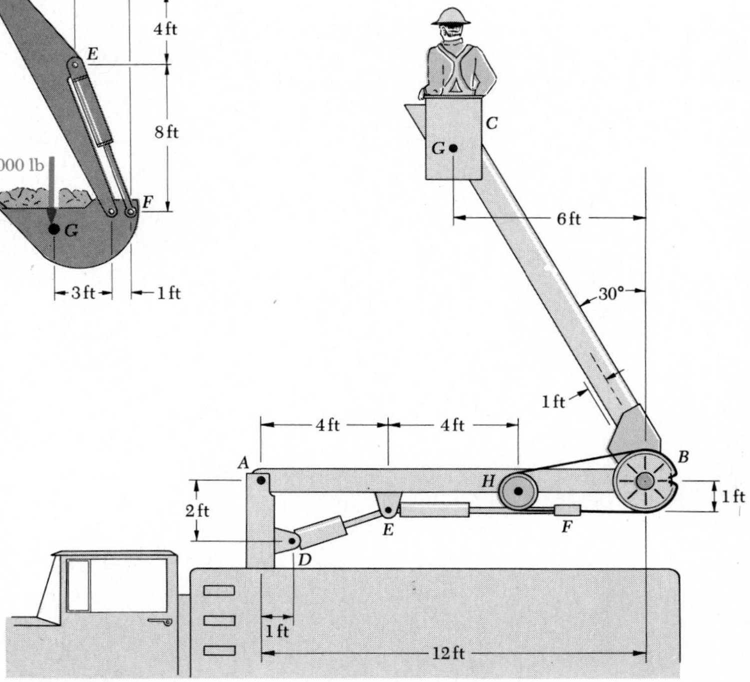

Fig. P6.118

6.119 In order to provide a mobile overhead working platform *JK*, the system shown is mounted on the bed of a truck. The elevation of the platform is controlled by two identical mechanisms, only one of which is shown. A load of 5 kN is applied to the mechanism shown. Knowing that the pin at *C* can transmit only a horizontal force, determine (*a*) the force in link *BE*, (*b*) the components of the force exerted by the hydraulic cylinder on pin *H*.

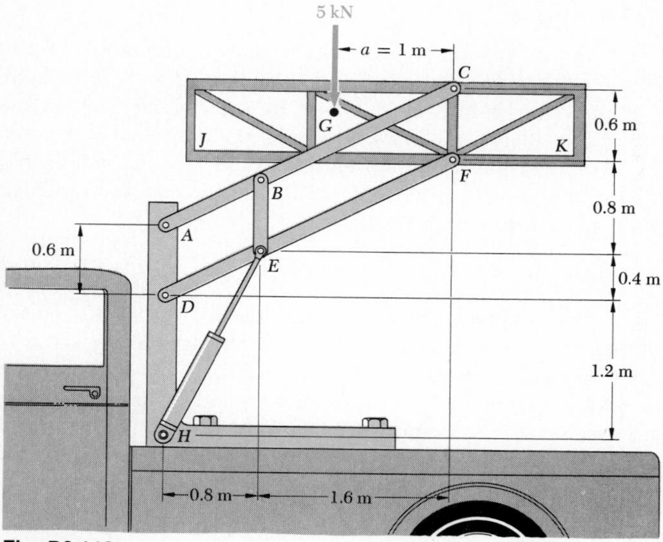

Fig. P6.119

6.120 Solve Prob. 6.119 when the distance from *C* to the point of application of the load is $a = 1.5$ m.

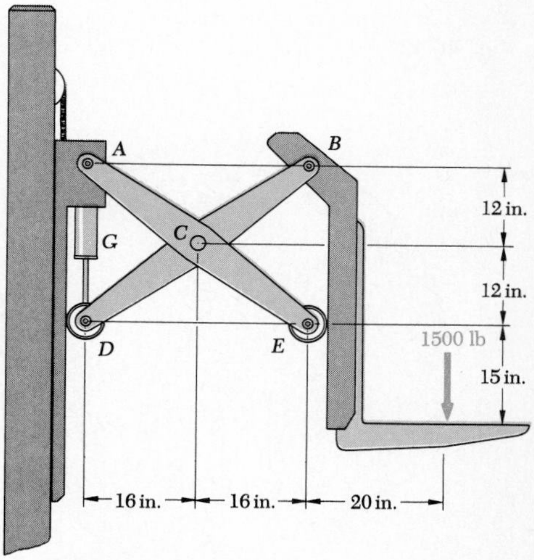

Fig. P6.121

6.121 Two identical linkage-and-hydraulic-cylinder systems control the position of the forks of a fork-lift truck; only one system is shown. Knowing that the load supported by the one system shown is 1500 lb, determine (*a*) the force exerted by the hydraulic cylinder on point *D*, (*b*) the components of the force exerted on member *ACE* at point *C*.

6.122 In the pliers shown, the clamping jaws remain parallel as objects of various sizes are held. If gripping forces of magnitude $Q = 450$ lb are desired, determine the magnitude *P* of the forces which must be applied. Assume that pins *B* and *E* slide freely in the slots cut in the jaws.

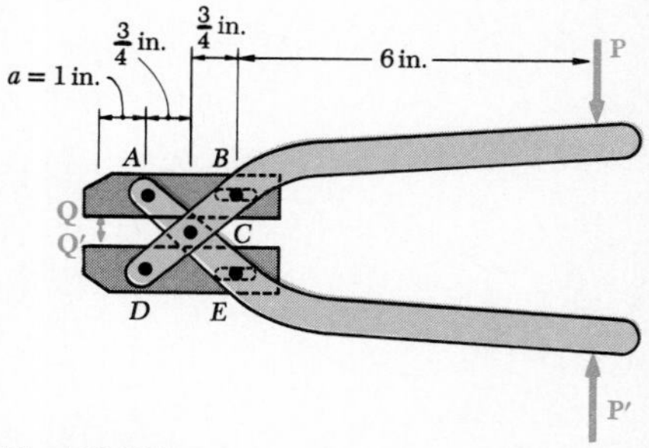

Fig. P6.122

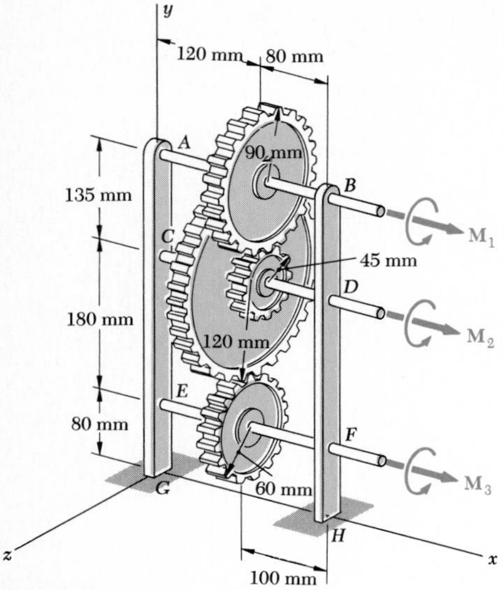

Fig. P6.123

6.123 The four gears are rigidly attached to shafts which are held by frictionless bearings. If $M_1 = 24\ \text{N}\cdot\text{m}$ and $M_2 = 0$, determine (*a*) the couple $\mathbf{M}_3$ which must be applied for equilibrium, (*b*) the reactions at *G* and *H*.

6.124 Solve Prob. 6.123 if $M_1 = 48\ \text{N}\cdot\text{m}$ and $M_2 = -18\ \text{N}\cdot\text{m}$.

6.125 In the planetary-gear system shown, the radius of the central gear *A* is *a*, the radius of each planetary gear is *b*, and the radius of the outer gear *E* is $(a + 2b)$. In a particular gear system where $a = b = 40$ mm, a clockwise couple $\mathbf{M}_A$ is applied to gear *A*. If the system is to be in equilibrium, determine (*a*) the couple $\mathbf{M}_S$ which must be applied to the spider *BCD*, (*b*) the couple $\mathbf{M}_E$ which must be applied to the outer gear *E*.

Fig. P6.125 and P6.126

6.126 In the planetary-gear system shown, the radius of the central gear *A* is *a*, the radius of each of the planetary gears is *b*, and the radius of the outer gear *E* is $(a + 2b)$. A clockwise couple of moment M_A is applied to the central gear *A* and a counterclockwise couple of moment $5M_A$ is applied to the spider *BCD*. If the system is to be in equilibrium, determine (*a*) the required ratio b/a, (*b*) the couple $\mathbf{M}_E$ which must be applied to the outer gear *E*.

***6.127** Two shafts *AC* and *CF*, which lie in the vertical *xy* plane, are connected by a universal joint at *C*. The bearings at *B* and *D* do not exert any axial force. A couple of moment 50 N · m (clockwise when viewed from the positive *x* axis) is applied to shaft *CF* at *F*. At a time when the arm of the crosspiece attached to shaft *CF* is horizontal, determine (*a*) the moment of the couple which must be applied to shaft *AC* at *A* to maintain equilibrium, (*b*) the reactions at *B*, *D*, and *E*. (*Hint.* The sum of the couples exerted on the crosspiece must be zero.)

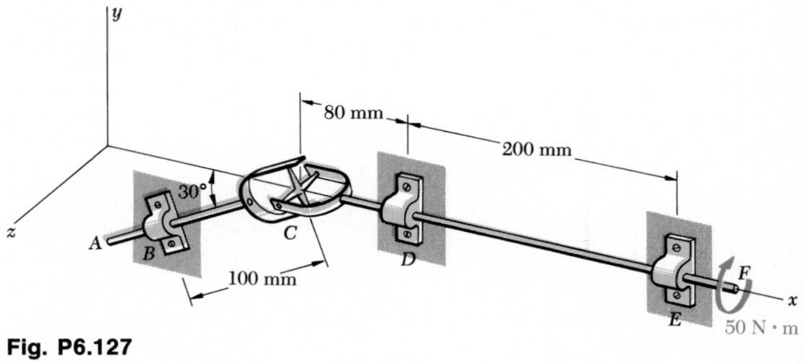

Fig. P6.127

***6.128** Solve Prob. 6.127, assuming that the arm of the crosspiece attached to shaft *CF* is vertical.

6.129 The weight, in ounces, of letters placed on the postal scale shown is indicated on the moving dial by the stationary pointer *P*. The scale is shown in its unloaded position. It is known that the dial, the arm *AB*, and a counterweight together weigh 10 oz and have a combined center of gravity at *G*. The distance *AG* is $1\frac{1}{4}$ in., and the length of the arm *AB* and of the link *CD* is 1.00 in. At what angle θ should the 2-oz number be painted on the dial? (*Hint.* The weights of the tray, of *BC*, and of *CD* are unknown, but their effect must be considered.)

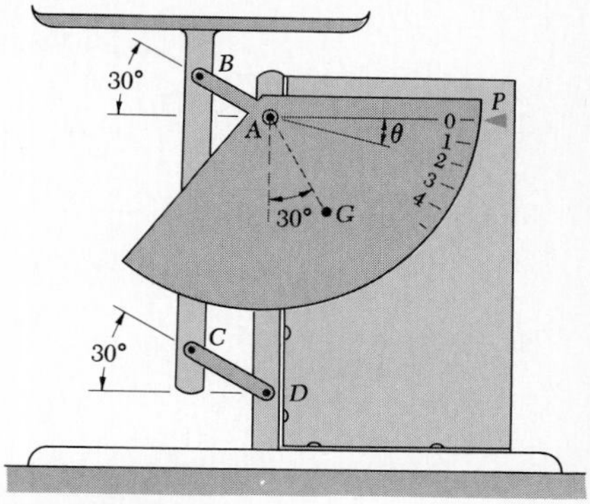

Fig. P6.129

6.130 A letter of unknown weight is placed on the postal scale of Prob. 6.129. Knowing that the dial rotates counterclockwise through 45° before coming to rest, determine the weight of the letter. (See hint of Prob. 6.129.)

REVIEW PROBLEMS

6.131 The retractable shelf is maintained in the position shown by two identical linkage-and-spring systems; only one of the systems is shown. A 20-kg machine is placed on the shelf so that half of its weight is supported by the system shown. Determine (*a*) the force in link *AB*, (*b*) the tension in the spring.

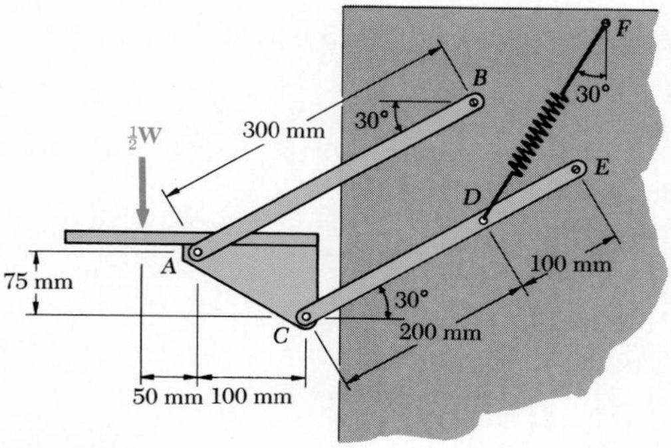

Fig. P6.131

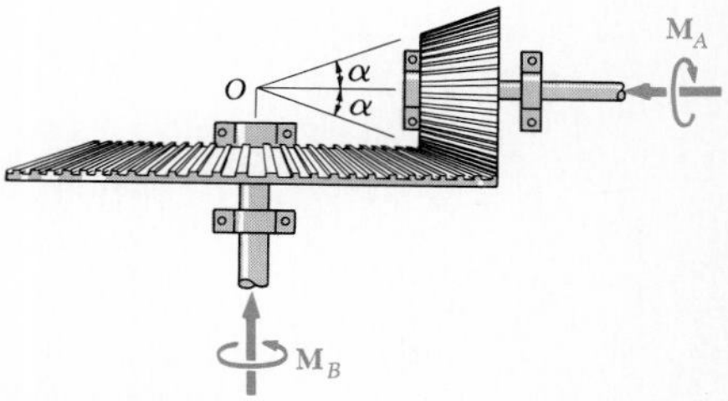

Fig. P6.132

6.132 For the bevel-gear system shown, determine the required value of α if the ratio of M_B to M_A is to be 3.

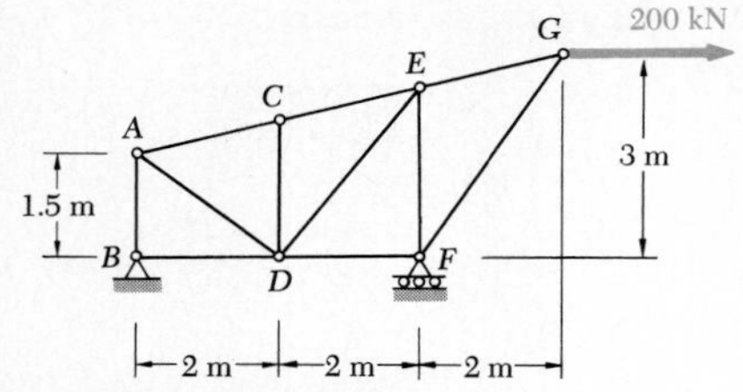

Fig. P6.133 and P6.134

6.133 Determine the force in each of the vertical members of the truss shown.

6.134 Determine the force in members *DF*, *EF*, and *EG*.

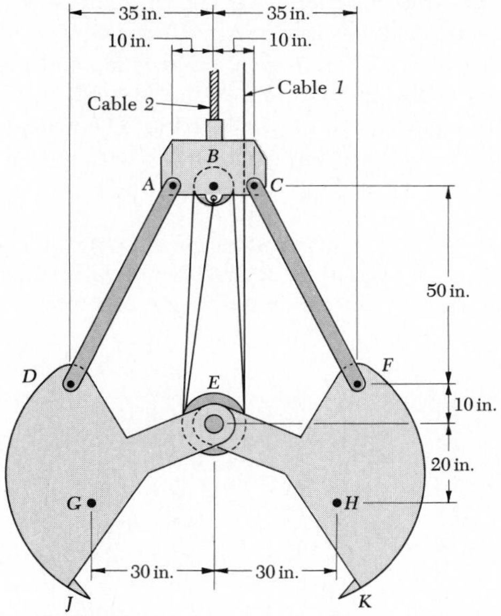

Fig. P6.135

6.135 The total weight of the 1-yd clamshell bucket shown is 4500 lb. The centers of gravity of sections *DEJ* and *EFK*, which weigh 2000 lb each, are located at *G* and *H*, respectively. The double-sheave pulley *E* and a counterweight located at *E* together weigh 400 lb. Determine the tension in cable *1* and cable *2* for the position shown. (Neglect the effect of the horizontal distance between the cables.)

6.136 An automobile front-wheel assembly supports 750 lb. Determine the force exerted by the spring and the components of the forces exerted on the frame at points *A* and *D*.

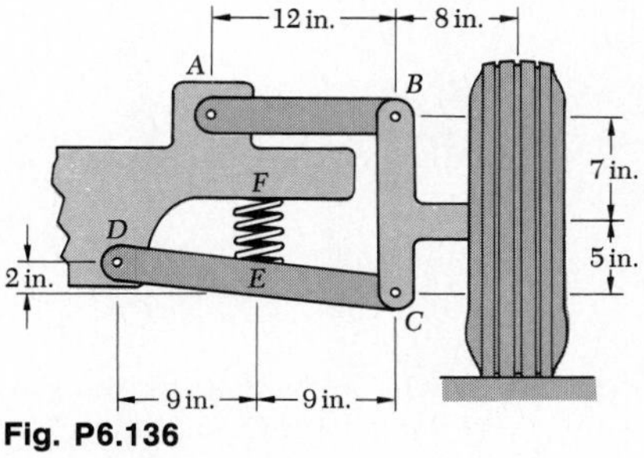

Fig. P6.136

6.137 The rigid bar *EFG* is supported by the truss system shown. Determine the force in each of the two-force members.

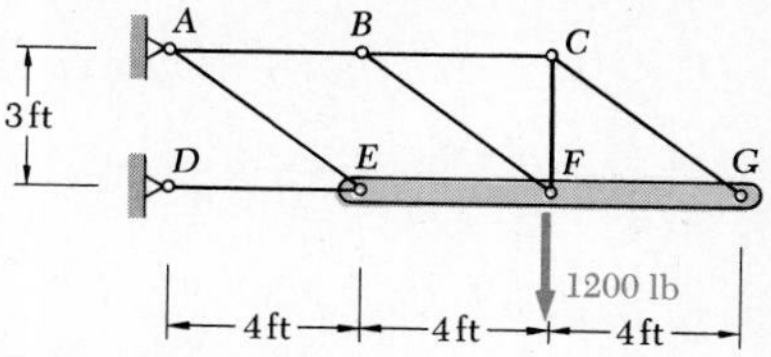

Fig. P6.137

6.138 Solve Prob. 6.137, assuming that the 1200-lb load is replaced by a clockwise couple of moment 4800 lb · ft applied to member *EFG* at *F*.

6.139 A 400-kg block may be supported by a small frame in each of the four ways shown. The diameter of the pulley is 300 mm. For each case, determine the force components and the couple representing the reaction at *A* and also the force exerted at *D* on the vertical member.

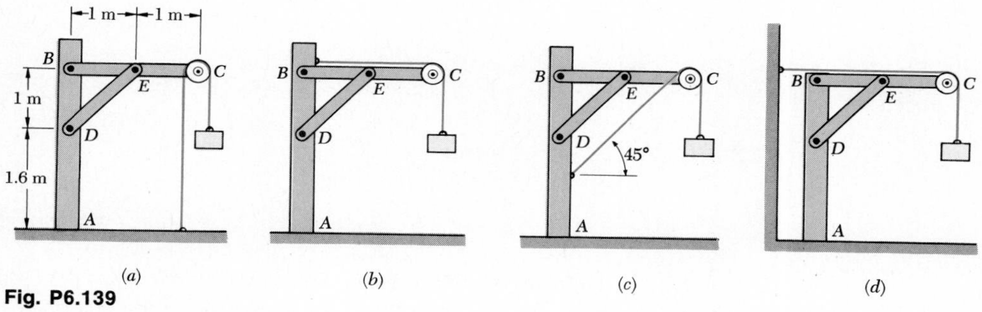

Fig. P6.139

6.140 Determine the couple **M** which must be applied to the crank *CD* to hold the mechanism in equilibrium. The block at *D* is pinned to the crank *CD* and is free to slide in a slot cut in member *AB*.

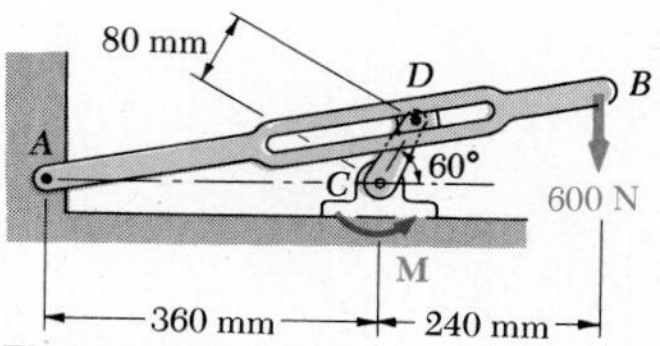

Fig. P6.140

6.141 Determine the reaction at *D* and the force in members *BF* and *EC*.

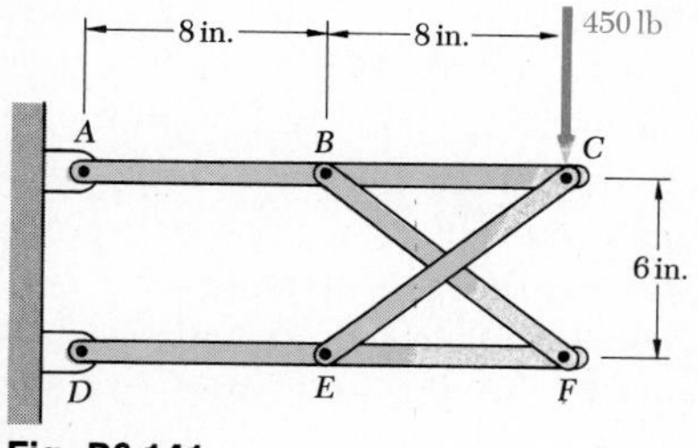

Fig. P6.141

6.142 Solve Prob. 6.141 if the 450-lb load is applied at *F*.

Chapter

7

Forces in Beams and Cables

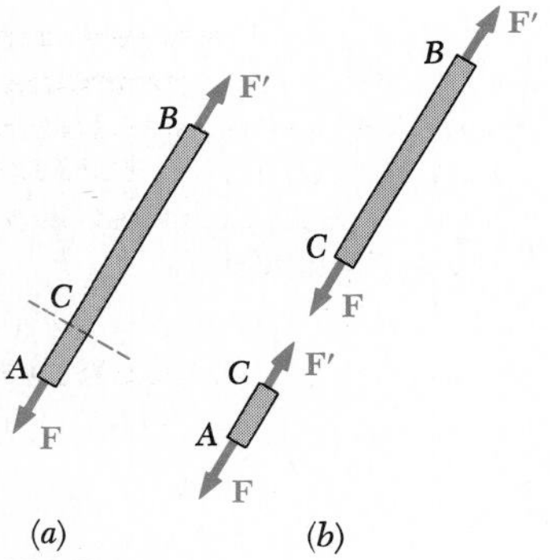

(a) (b)

Fig. 7.1

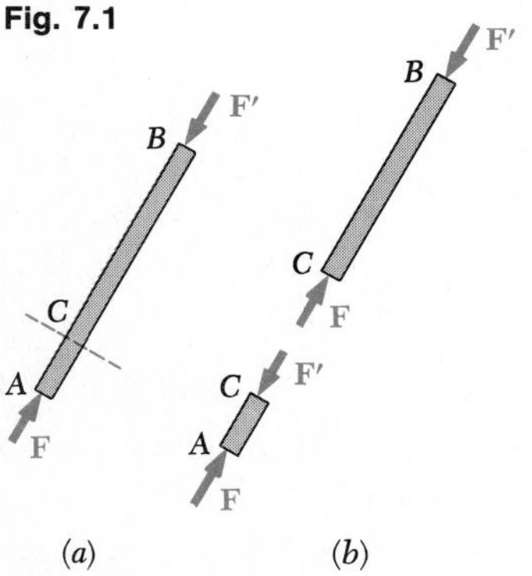

(a) (b)

Fig. 7.2

∗7.1. Introduction. Internal Forces in Members.

In preceding chapters, two basic problems involving structures were considered, (1) the determination of the external forces acting on a structure (Chaps. 3 and 4) and (2) the determination of the forces which hold together the various members forming a structure (Chap. 6). We shall now consider the problem of determining the internal forces which hold together the various parts of a given member.

We shall first consider a *straight two-force member AB* (Fig. 7.1*a*). From Sec. 3.16, we know that the forces **F** and **F**′ acting at *A* and *B*, respectively, must be directed along *AB* in opposite sense and have the same magnitude *F*. Now, let us cut the member at *C*. To maintain the equilibrium of the free bodies *AC* and *CB* thus obtained, we must apply to *AC* a force **F**′ equal and opposite to **F**, and to *CB* a force **F** equal and opposite to **F**′ (Fig. 7.1*b*). These new forces are directed along *AB* in opposite sense and have the same magnitude *F*. Since the two parts *AC* and *CB* were in equilibrium before the member was cut, *internal forces* equivalent to these new forces must have existed in the member itself. We see that, in the case of a straight two-force member, the internal forces acting on each part of the member are equivalent to an axial force. The magnitude *F* of this force does not depend upon the location of the section *C* and is referred to as the *force in member AB*. In the case considered, the member is in tension and will elongate under

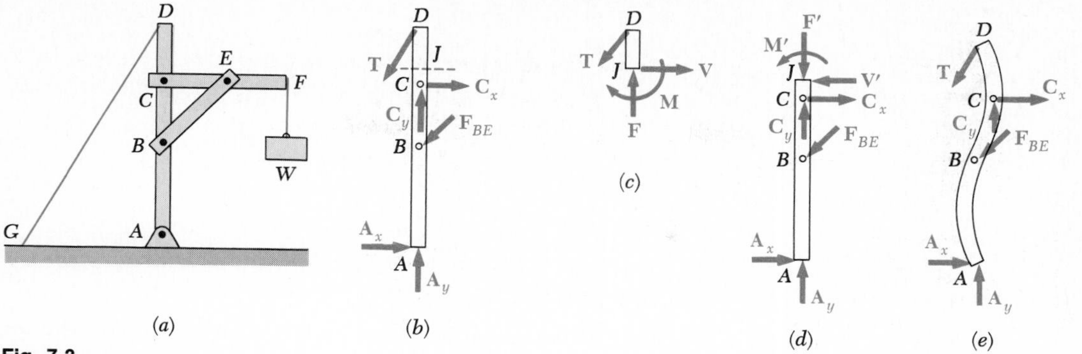

Fig. 7.3

the action of the internal forces. In the case represented in Fig. 7.2, the member is in compression and will decrease in length under the action of the internal forces.

Next we shall consider a *multiforce member*. Take, for instance, member *AD* of the crane analyzed in Sec. 6.11. This crane is shown again in Fig. 7.3*a*, and the free-body diagram of member *AD* is drawn in Fig. 7.3*b*. We now cut member *AD* at *J* and draw a free-body diagram for each of the portions *JD* and *AJ* of the member (Fig. 7.3*c* and *d*). Considering the free body *JD*, we find that its equilibrium will be maintained if we apply at *J* a force **F** to balance the vertical component of **T**, a force **V** to balance the horizontal component of **T**, and a couple **M** to balance the moment of **T** about *J*. Again we conclude that internal forces must have existed at *J* before member *AD* was cut. The internal forces acting on the portion *JD* of member *AD* are equivalent to the force-couple system shown in Fig. 7.3*c*. According to Newton's third law, the internal forces acting on *AJ* must be equivalent to an equal and opposite force-couple system, as shown in Fig. 7.3*d*. It clearly appears that the action of the internal forces in member *AD is not limited to producing tension or compression* as in the case of straight two-force members; the internal forces *also produce shear and bending*. The force **F** is again in this case called an *axial force;* the force **V** is called a *shearing force;* and the moment *M* of the couple is known as the *bending moment* at *J*. We note that, when determining internal forces in a member, we should clearly indicate on which portion of the member the forces are supposed to act. The deformation which will occur in member *AD* is sketched in Fig. 7.3*e*. The actual analysis of such a deformation is part of the study of mechanics of materials.

It should be noted that, in a *two-force member which is not straight*, the internal forces are also equivalent to a force-couple system. This is shown in Fig. 7.4, where the two-force member *ABC* has been cut at *D*.

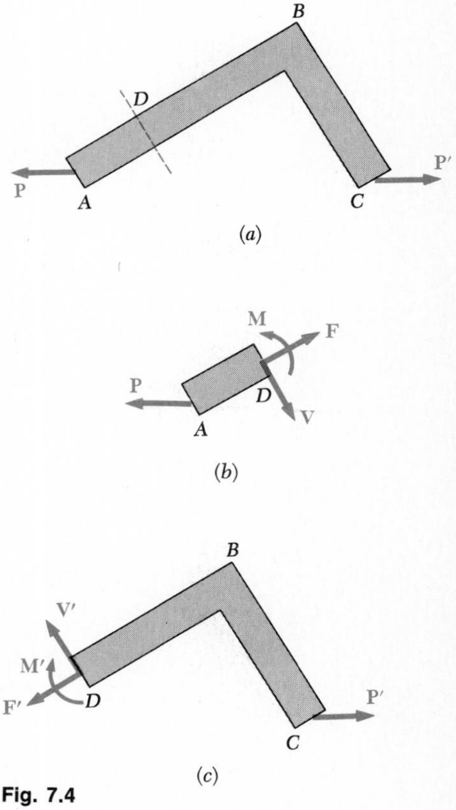

Fig. 7.4

SAMPLE PROBLEM 7.1

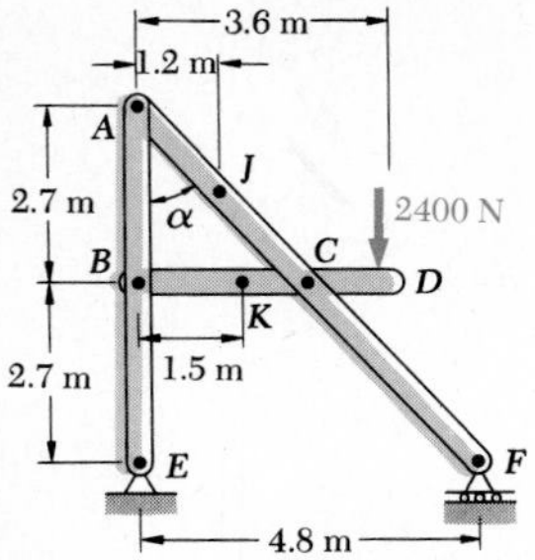

In the frame shown, determine the internal forces (*a*) in member *ACF* at point *J* and (*b*) in member *BCD* at point *K*. This frame has been previously considered in Sample Prob. 6.6.

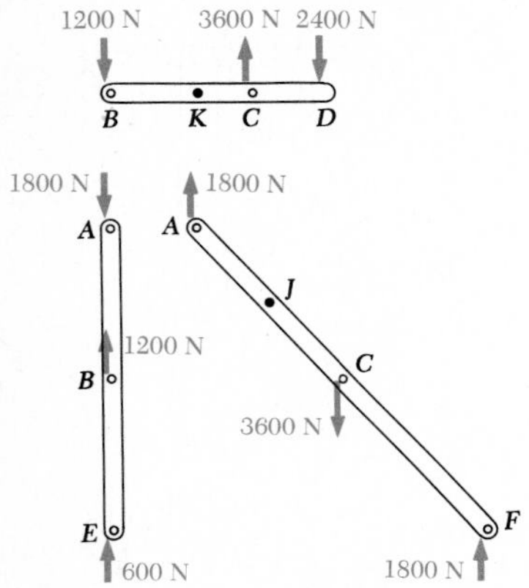

Solution. The reactions and the forces acting on each member of the frame are determined; this has been previously done in Sample Prob. 6.6, and the results are repeated here.

a. Internal Forces at J. Member *ACF* is cut at point *J*, and the two parts shown are obtained. The internal forces at *J* are represented by an equivalent force-couple system and may be determined by considering the equilibrium of either part. Considering the *free body AJ*, we write

$+\circlearrowleft \Sigma M_J = 0$: $\quad -(1800\text{ N})(1.2\text{ m}) + M = 0$

$M = +2160\text{ N}\cdot\text{m} \qquad \mathbf{M} = 2160\text{ N}\cdot\text{m} \circlearrowleft$ ◄

$+\searrow \Sigma F_x = 0$: $\quad F - (1800\text{ N})\cos 41.7° = 0$

$F = +1344\text{ N} \qquad \mathbf{F} = 1344\text{ N} \searrow$ ◄

$+\nearrow \Sigma F_y = 0$: $\quad -V + (1800\text{ N})\sin 41.7° = 0$

$V = +1197\text{ N} \qquad \mathbf{V} = 1197\text{ N} \swarrow$ ◄

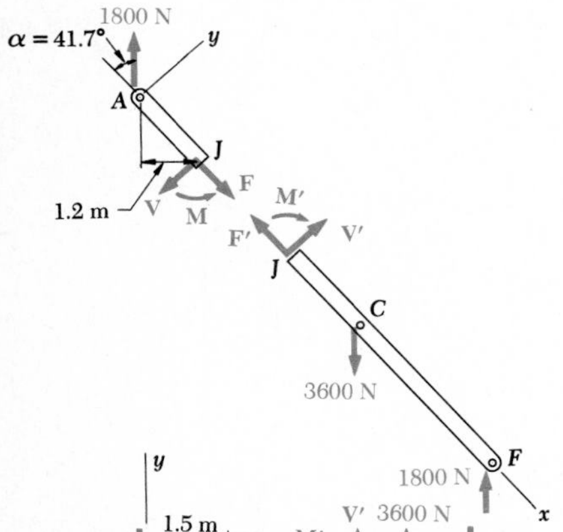

The internal forces at *J* are therefore equivalent to a couple **M**, an axial force **F**, and a shearing force **V**. The internal force-couple system acting on part *JCF* is equal and opposite.

b. Internal Forces at K. We cut member *BCD* at *K* and obtain the two parts shown. Considering the *free body BK*, we write

$+\circlearrowleft \Sigma M_K = 0$: $\quad (1200\text{ N})(1.5\text{ m}) + M = 0$

$M = -1800\text{ N}\cdot\text{m} \qquad \mathbf{M} = 1800\text{ N}\cdot\text{m} \circlearrowright$ ◄

$\xrightarrow{+} \Sigma F_x = 0$: $\quad F = 0 \qquad \mathbf{F} = 0$ ◄

$+\uparrow \Sigma F_y = 0$: $\quad -1200\text{ N} - V = 0$

$V = -1200\text{ N} \qquad \mathbf{V} = 1200\text{ N} \uparrow$ ◄

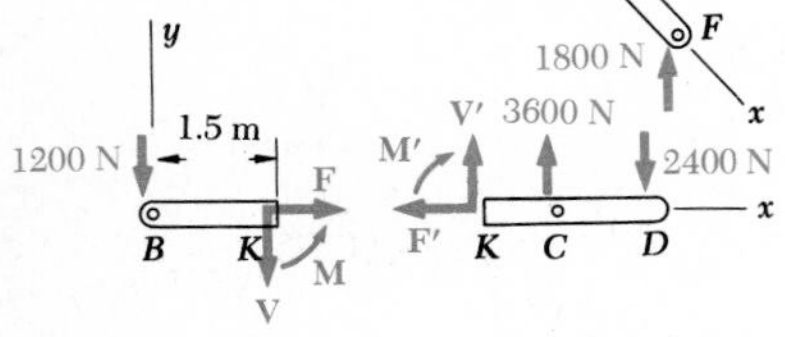

PROBLEMS

7.1 through 7.5 Determine the internal forces (axial force, shearing force, and bending moment) at point J of the structure indicated:

7.1 Frame and loading of Prob. 6.67.
7.2 Frame and loading of Prob. 6.68.
7.3 Frame and loading of Prob. 6.69.
7.4 Frame and loading of Prob. 6.74.
7.5 Frame and loading of Prob. 6.75.

7.6 The bracket AD is supported by a pin at A and by the cable DE. Determine the internal forces just to the left of the load if $a = 200$ mm.

7.7 The bracket AD is supported by a pin at A and by the cable DE. Determine the distance a for which the bending moment at B is equal to the bending moment at C.

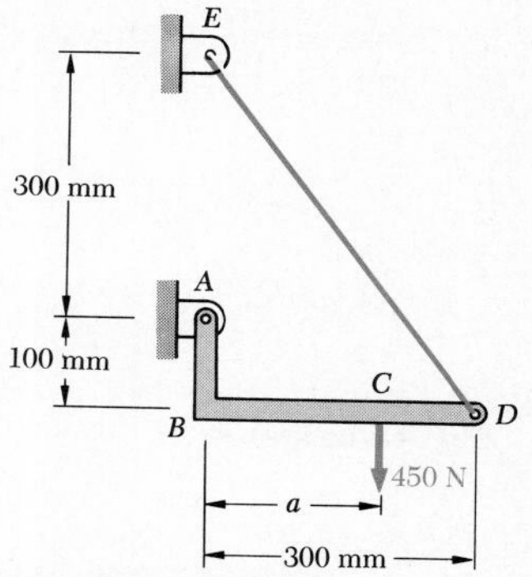

Fig. P7.6 and P7.7

7.8 and 7.9 A steel channel forms one side of a flight of stairs. If one channel weighs w lb/ft, determine the internal forces at the center of one channel due to its own weight, (a) in terms of w, L, and θ, (b) if $w = 20$ lb/ft, $l = 12$ ft, and $h = 9$ ft.

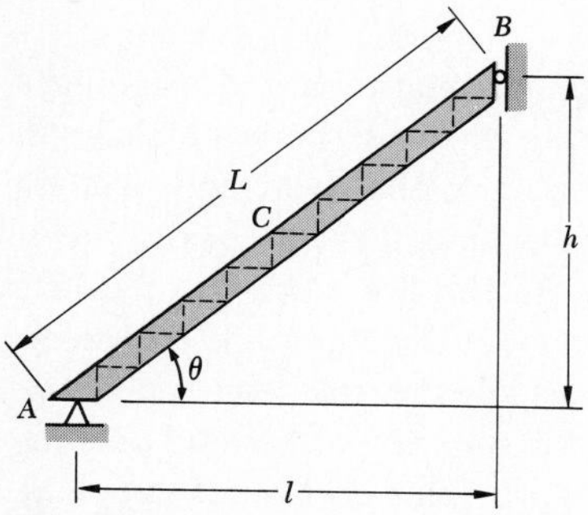

Fig. P7.8

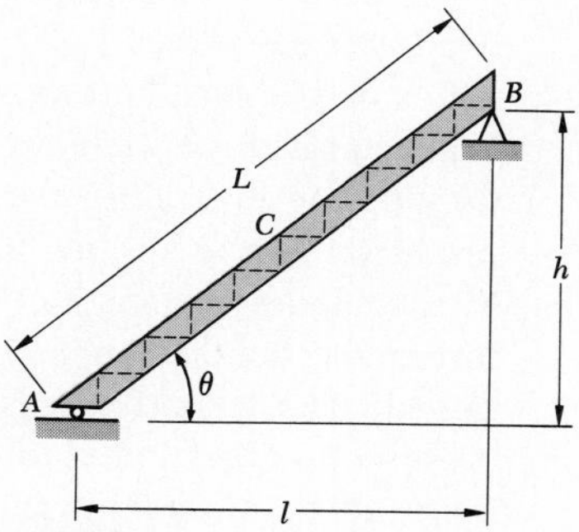

Fig. P7.9

7.10 The axis of the curved member ABC is parabolic. If the tension in the cable AC is 1000 N and the magnitude of the force $\mathbf{P}$ is 500 N, determine the internal forces at a distance $\frac{1}{4}L$ from the left end. Assume $h = 80$ mm and $L = 480$ mm.

7.11 Denoting by T the tension in the cable AC, determine (a) the internal forces just to the left of point B, (b) the value of T for which the bending moment at B is zero.

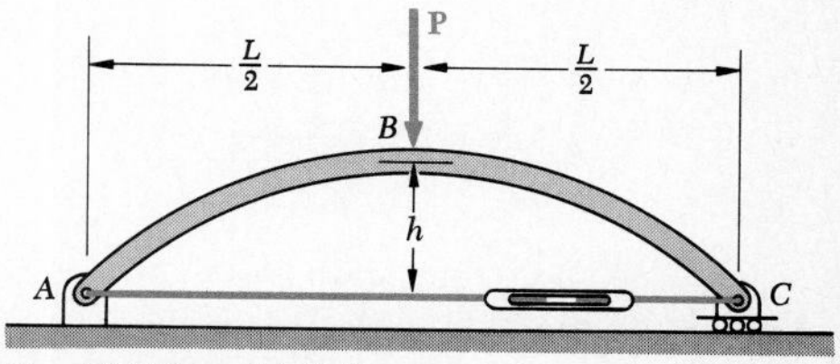

Fig. P7.10 and P7.11

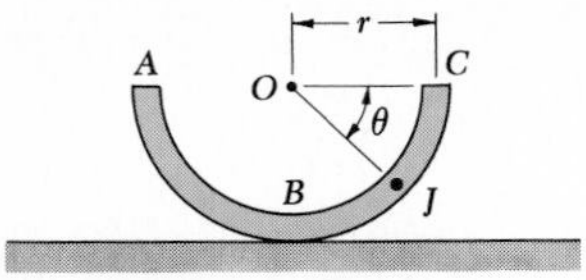

Fig. P7.12 and P7.14

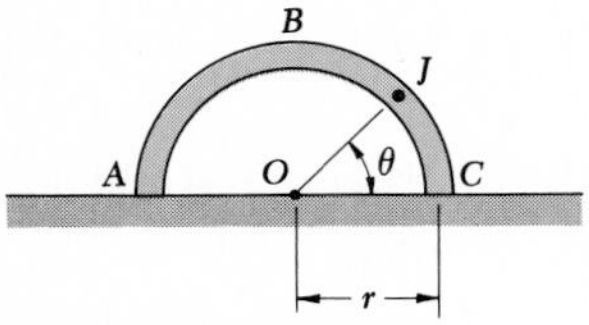

Fig. P7.13 and P7.15

7.12 and 7.13 A half section of pipe, 1 ft long, rests on a frictionless horizontal surface as shown. If the half section of pipe weighs 30 lb and has a diameter of 20 in., determine the bending moment at point J when $\theta = 90°$.

***7.14 and 7.15** A half section of pipe, of weight W and of unit length, rests on a frictionless horizontal surface. Determine the internal forces at a point J in terms of W, r, and θ.

***7.16** In Prob. 7.15, determine the magnitude and location of the maximum internal shearing force.

BEAMS

***7.2. Various Types of Loading and Support.** A structural member designed to support loads applied at various points along the member is known as a *beam*. In most cases, the loads are perpendicular to the axis of the beam and will cause only shear and bending in the beam. When the loads are not at a right angle to the beam, they will also produce axial forces in the beam. Axial forces, however, may usually be neglected in the design of beams, since the ability of a beam to resist shear and especially bending is more critical than its ability to resist axial forces.

Beams are usually long, straight prismatic bars. Designing a beam consists essentially in selecting the cross section which will provide the most effective resistance to the shear and bending produced by the applied loads. The design of the beam, therefore, includes two distinct parts. In the first part, the shearing forces and bending moments produced by the loads are determined. The second part is concerned with the selection of the cross section best suited to resist the shearing forces and bending moments determined in the first part. This portion of the chapter, Beams, deals with the first part of the problem of beam design, namely, the determination of the shearing forces and bending moments in beams subjected to various loading conditions and supported in various ways. The second part of the problem belongs to the study of mechanics of materials.

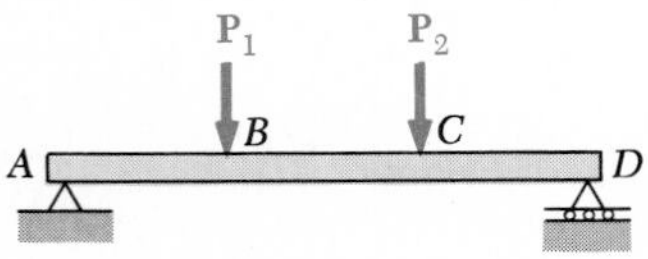

(*a*) Concentrated loads

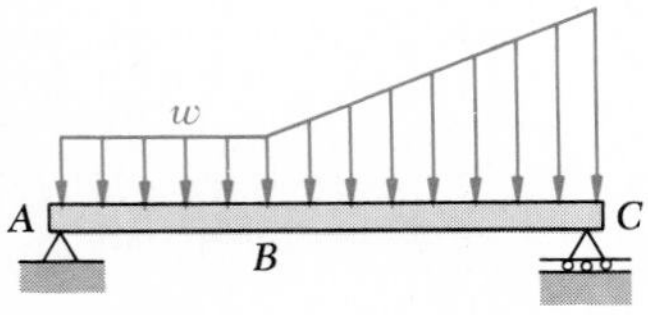

(*b*) Distributed load

Fig. 7.5 Types of loadings

A beam may be subjected to *concentrated loads* $\mathbf{P}_1$, $\mathbf{P}_2$, . . . , expressed in newtons, pounds, or their multiples kilonewtons and kips (Fig. 7.5*a*), to a *distributed load* w, expressed in N/m, kN/m, lb/ft, or kips/ft (Fig. 7.5*b*), or to a combination of both. When the load w per unit length has a constant value over part of the beam (as between A and B in Fig. 7.5*b*), the load is said to be *uniformly distributed* over that part of the beam. The determination of the reactions at the supports may be consid-

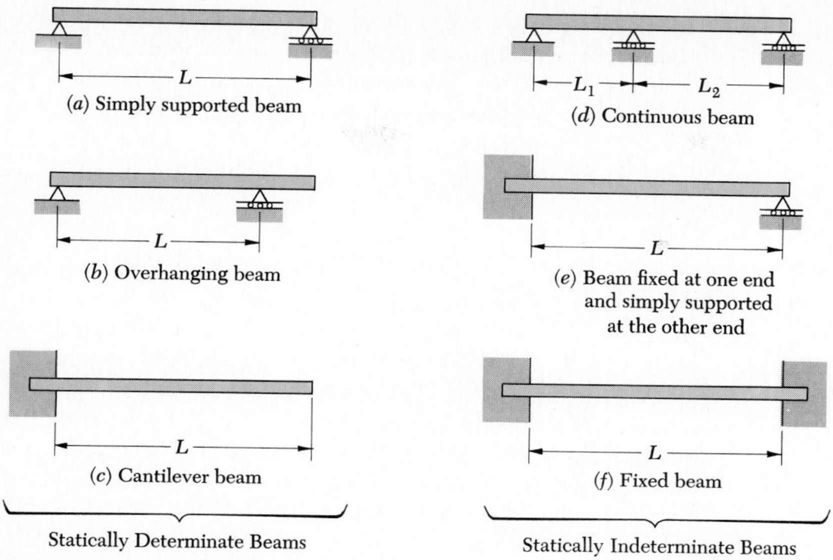

Fig. 7.6 Types of beams

erably simplified if distributed loads are replaced by equivalent concentrated loads, as explained in Sec. 5.6. This substitution, however, should not be performed, or at least should be performed with care, when internal forces are being computed (see Sample Prob. 7.3).

Beams are classified according to the way in which they are supported. Several types of beams frequently used are shown in Fig. 7.6. The distance L between supports is called the *span*. It should be noted that the reactions will be determinate if the supports involve only three unknowns. The reactions will be statically indeterminate if more unknowns are involved; the methods of statics are not sufficient then to determine the reactions, and the properties of the beam with regard to its resistance to bending must be taken into consideration. Beams supported by two rollers are not shown here; such beams are only partially constrained and will move under certain loading conditions.

Sometimes two or more beams are connected by hinges to form a single continuous structure. Two examples of beams hinged at a point H are shown in Fig. 7.7. It will be noted that the reactions at the supports involve four unknowns and cannot be determined from the free-body diagram of the two-beam system. They can be determined, however, by considering the free-body diagram of each beam separately; six unknowns are involved (including two force components at the hinge), and six equations are available.

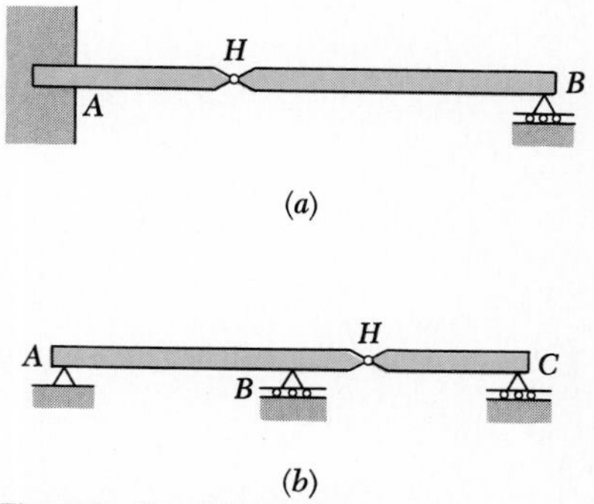

Fig. 7.7 Combined beams

$*$7.3. Shear and Bending Moment in a Beam. Consider a beam AB subjected to various concentrated and distributed loads (Fig. 7.8a). We propose to determine the shearing force and bending moment at any point of the beam. In the example considered here, the beam is simply supported, but the method used could be applied to any type of statically determinate beam.

First we determine the reactions at A and B by choosing the entire beam as a free body (Fig. 7.8b); writing $\Sigma M_A = 0$ and $\Sigma M_B = 0$, we obtain, respectively, $\mathbf{R}_B$ and $\mathbf{R}_A$.

To determine the internal forces at C, we cut the beam at C and draw the free-body diagrams of the portions AC and CB of the beam (Fig. 7.8c). Using the free-body diagram of AC, we may determine the shearing force $\mathbf{V}$ at C by equating to zero the sum of the vertical components of all forces acting on AC. Similarly, the bending moment M at C may be found by equating to zero the sum of the moments about C of all forces and couples acting on AC. We could have used just as well, however, the free-body diagram of CB and determined the shearing force $\mathbf{V}'$ and the bending moment M' by equating to zero the sum of the vertical components and the sum of the moments about C of all forces and couples acting on CB. While this possible choice of alternate free bodies may facilitate the

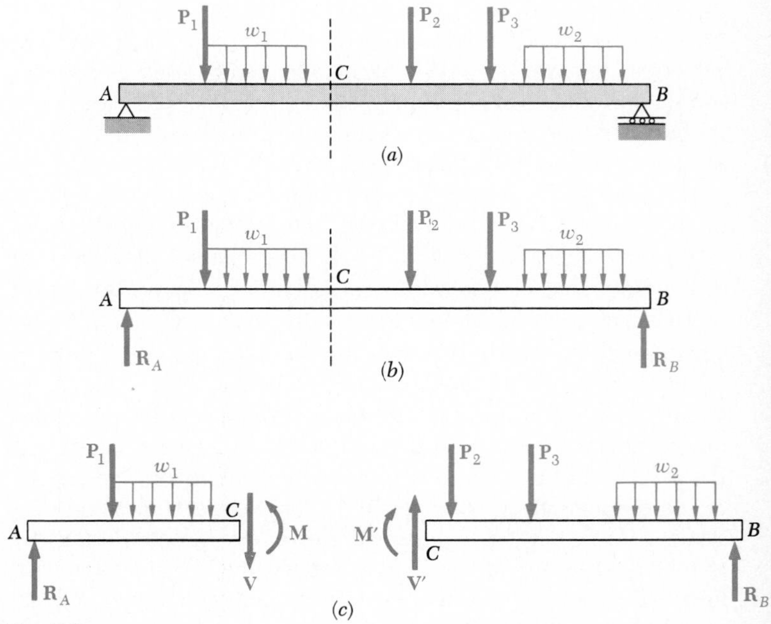

Fig. 7.8

computation of the numerical values of the shearing force and bending moment, it makes it necessary to indicate on which portion of the beam the internal forces considered are acting. If the shearing force and bending moment, however, are to be computed at every point of the beam and efficiently recorded, we should not have to specify every time which portion of the beam is used as a free body. We shall adopt, therefore, the following convention:

To determine the shearing force in a beam, *we shall always assume* that the internal forces **V** and **V′** are directed as shown in Fig. 7.8*c*. A positive value obtained for their common magnitude V will indicate that this assumption was correct and that the shearing forces are actually directed as shown. A negative value obtained for V will indicate that the assumption was wrong and that the shearing forces are directed in the opposite way. Thus, only the magnitude V, together with a plus or minus sign, needs to be recorded to define completely the shearing forces at a given point of the beam. The scalar V is commonly referred to as the *shear* at the given point of the beam.

Similarly, *we shall always assume* that the internal couples **M** and **M′** are directed as shown in Fig. 7.8*c*. A positive value obtained for their magnitude M, commonly referred to as the bending moment, will indicate that this assumption was correct, and a negative value that it was wrong. Summarizing the sign convention we have presented, we state:

The shear V and the bending moment M at a given point of a beam are said to be positive when the internal forces and couples acting on each portion of the beam are directed as shown in Fig. 7.9a.

This convention may be more easily remembered if we note that:

1. *The shear at C is positive when the* **external** *forces (loads and reactions) acting on the beam tend to shear off the beam at C as indicated in Fig. 7.9b.*
2. *The bending moment at C is positive when the* **external** *forces acting on the beam tend to bend the beam at C as indicated in Fig. 7.9c.*

It may also help to note that the situation described in Fig. 7.9, and corresponding to positive values of the shear and of the bending moment, is precisely the situation which occurs in the left half of a simply supported beam carrying a single concentrated load at its midpoint. This particular example is fully discussed in the following section.

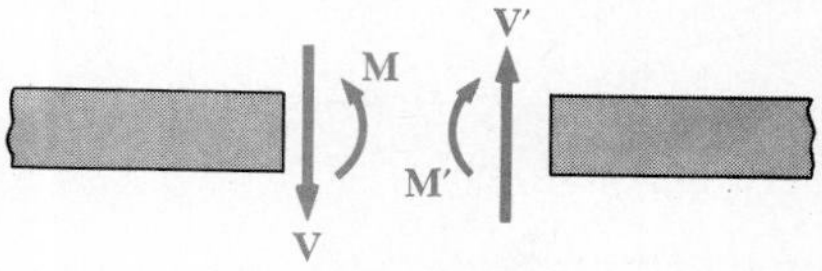

(*a*) Internal forces at section
(positive shear and positive bending moment)

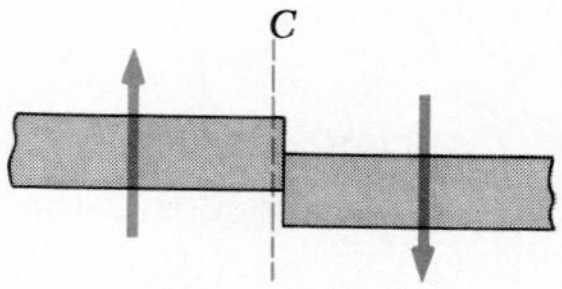

(*b*) Effect of external forces
(positive shear)

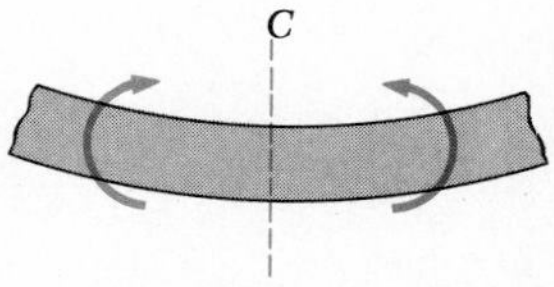

(*c*) Effect of external forces
(positive bending moment)

Fig. 7.9

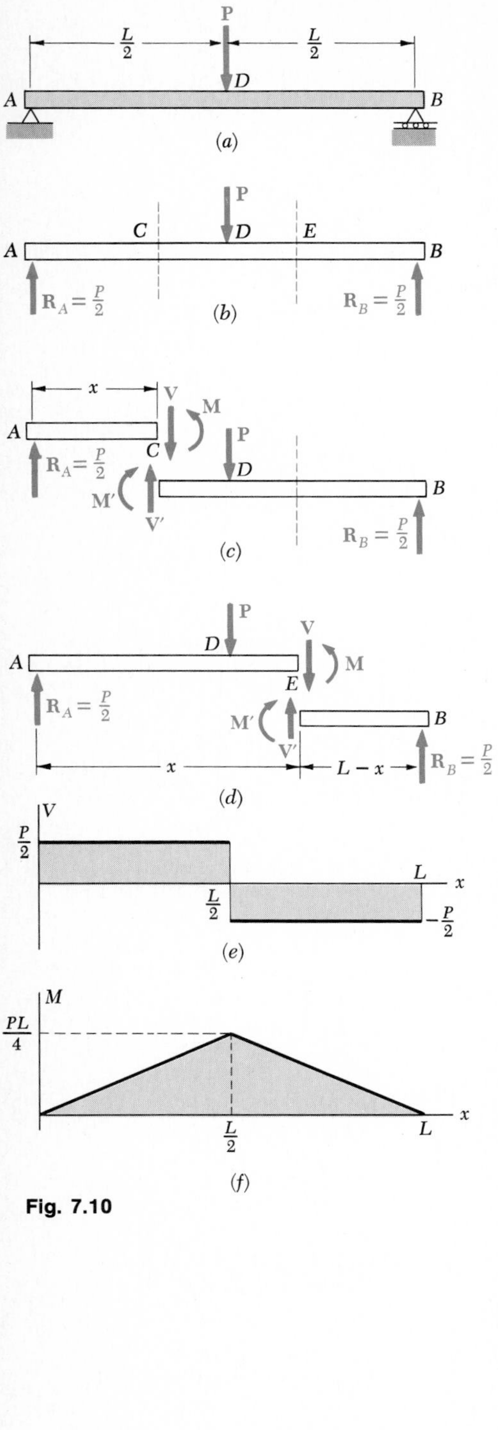

Fig. 7.10

∗7.4. Shear and Bending-Moment Diagrams. Now that shear and bending moment have been clearly defined in sense as well as in magnitude, we may easily record their values at any point of a beam by plotting these values against the distance x measured from one end of the beam. The graphs obtained in this way are called, respectively, the *shear diagram* and the *bending-moment diagram*. As an example, consider a simply supported beam AB of span L subjected to a single concentrated load **P** applied at its midpoint D (Fig. 7.10*a*). We first determine the reactions at the supports from the free-body diagram of the entire beam (Fig. 7.10*b*); we find that the magnitude of each reaction is equal to $P/2$.

Next we cut the beam at a point C between A and D and draw the free-body diagrams of AC and CB (Fig. 7.10*c*). *Assuming that shear and bending moment are positive,* we direct the internal forces **V** and **V′** and the internal couples **M** and **M′** as indicated in Fig. 7.9*a*. Considering the free body AC and writing that the sum of the vertical components and the sum of the moments about C of the forces acting on the free body are zero, we find $V = +P/2$ and $M = +Px/2$. Both the shear and the bending moment are therefore positive; this may be checked by observing that the reaction at A tends to shear off and to bend the beam at C as indicated in Fig. 7.9*b* and *c*. We may plot V and M between A and D (Fig. 7.10*e* and *f*); the shear has a constant value $V = P/2$, while the bending moment increases linearly from $M = 0$ at $x = 0$ to $M = PL/4$ at $x = L/2$.

Cutting, now, the beam at a point E between D and B and considering the free body EB (Fig. 7.10*d*), we write that the sum of the vertical components and the sum of the moments about E of the forces acting on the free body are zero. We obtain $V = -P/2$ and $M = P(L - x)/2$. The shear is therefore negative and the bending moment positive; this may be checked by observing that the reaction at B bends the beam at E as indicated in Fig. 7.9*c* but tends to shear it off in a manner opposite to that shown in Fig. 7.9*b*. We can complete, now, the shear and bending-moment diagrams of Fig. 7.10*e* and *f*; the shear has a constant value $V = -P/2$ between D and B, while the bending moment decreases linearly from $M = PL/4$ at $x = L/2$ to $M = 0$ at $x = L$.

We shall note that, when a beam is subjected only to concentrated loads, the shear is of constant value between loads and the bending moment varies linearly between loads. On the other hand, when a beam is subjected to distributed loads, the shear and bending moment vary quite differently (see Sample Prob. 7.3).

SAMPLE PROBLEM 7.2

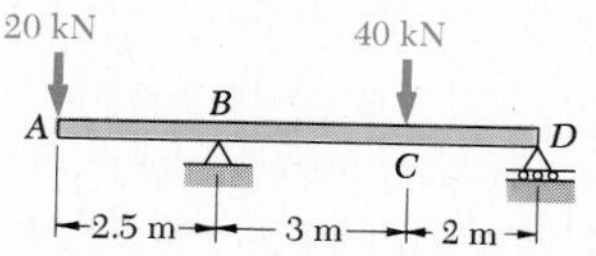

Draw the shear and bending-moment diagram for the beam and loading shown.

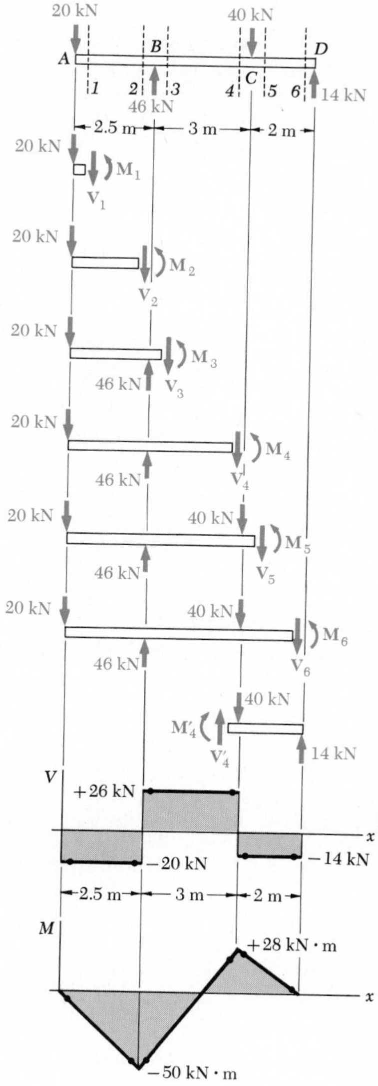

Solution. The reactions are determined by considering the entire beam as a free body; they are

$$\mathbf{R}_B = 46 \text{ kN}\uparrow \qquad \mathbf{R}_D = 14 \text{ kN}\uparrow$$

We first determine the internal forces just to the right of the 20-kN load at A. Considering the stub of beam to the left of section *1* as a free body and assuming V and M to be positive (according to the standard convention), we write

$$+\uparrow\Sigma F_y = 0: \qquad -20 \text{ kN} - V_1 = 0 \qquad V_1 = -20 \text{ kN}$$

$$+\circlearrowleft\Sigma M_1 = 0: \qquad (20 \text{ kN})(0 \text{ m}) + M_1 = 0 \qquad M_1 = 0$$

We next consider as a free body the portion of beam to the left of section *2* and write

$$+\uparrow\Sigma F_y = 0: \qquad -20 \text{ kN} - V_2 = 0 \qquad V_2 = -20 \text{ kN}$$

$$+\circlearrowleft\Sigma M_2 = 0: \qquad (20 \text{ kN})(2.5 \text{ m}) + M_2 = 0 \qquad M_2 = -50 \text{ kN}\cdot\text{m}$$

The shear and bending moment at sections *3, 4, 5,* and *6* are determined in a similar way from the free-body diagrams shown. We obtain

$$V_3 = +26 \text{ kN} \qquad M_3 = -50 \text{ kN}\cdot\text{m}$$

$$V_4 = +26 \text{ kN} \qquad M_4 = +28 \text{ kN}\cdot\text{m}$$

$$V_5 = -14 \text{ kN} \qquad M_5 = +28 \text{ kN}\cdot\text{m}$$

$$V_6 = -14 \text{ kN} \qquad M_6 = 0$$

For several of the latter sections, the results may be more easily obtained by considering as a free body the portion of the beam to the right of the section. For example, considering the portion of the beam to the right of section *4,* we write

$$+\uparrow\Sigma F_y = 0: \qquad V_4 - 40 \text{ kN} + 14 \text{ kN} = 0 \qquad V_4 = +26 \text{ kN}$$

$$+\circlearrowleft\Sigma M_4 = 0: \qquad -M_4 + (14 \text{ kN})(2 \text{ m}) = 0 \qquad M_4 = +28 \text{ kN}\cdot\text{m}$$

We may now plot the six points shown on the shear and bending-moment diagrams. As indicated in Sec. 7.4, the shear is of constant value between concentrated loads, and the bending moment varies linearly; we obtain therefore the shear and bending-moment diagrams shown.

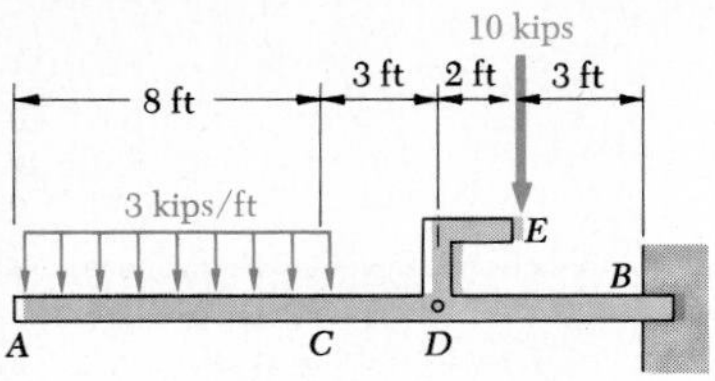

SAMPLE PROBLEM 7.3

Draw the shear and bending-moment diagrams for the cantilever beam AB. The distributed load of 3 kips/ft extends over 8 ft of the beam and the 10-kip load is applied at E.

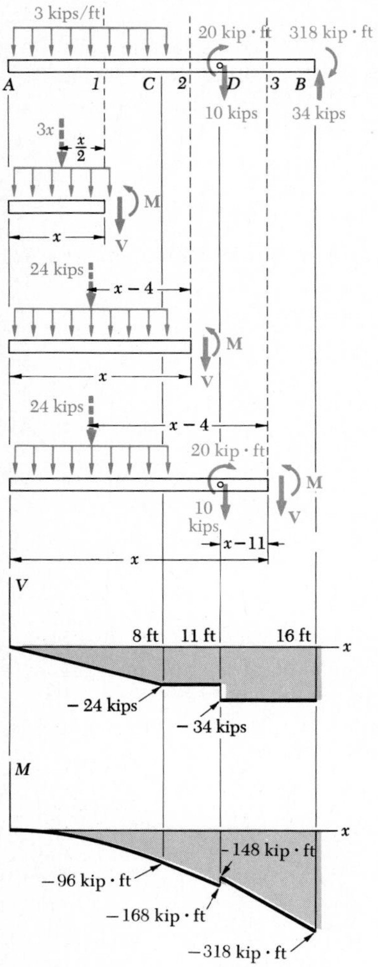

Solution. The 10-kip load is replaced by an equivalent force-couple system acting on the beam at point D. The reaction at B is determined by considering the entire beam as a free body.

From A to C. We determine the internal forces at a distance x from point A by considering the portion of beam to the left of section *1*. That part of the distributed load acting on the free body is replaced by its resultant, and we write

$$+\uparrow\Sigma F_y = 0: \qquad -3x - V = 0 \qquad V = -3x \text{ kips}$$

$$+\circlearrowleft\Sigma M_1 = 0: \qquad 3x(\tfrac{1}{2}x) + M = 0 \qquad M = -1.5x^2 \text{ kip}\cdot\text{ft}$$

Since the free-body diagram shown may be used for all values of x smaller than 8 ft, the expressions obtained for V and M are valid in the region $0 < x < 8$ ft.

From C to D. Considering the portion of beam to the left of section 2 and again replacing the distributed load by its resultant, we obtain

$$+\uparrow\Sigma F_y = 0: \qquad -24 - V = 0 \qquad V = -24 \text{ kips}$$

$$+\circlearrowleft\Sigma M_2 = 0: \qquad 24(x-4) + M = 0 \qquad M = 96 - 24x \quad \text{kip}\cdot\text{ft}$$

These expressions are valid in the region 8 ft $< x <$ 11 ft.

From D to B. Using the portion of beam to the left of section 3, we obtain for the region 11 ft $< x <$ 16 ft

$$V = -34 \text{ kips} \qquad M = 226 - 34x \quad \text{kip}\cdot\text{ft}$$

The shear and bending-moment diagrams for the entire beam may now be plotted. We note that the couple of moment 20 kip · ft applied at point D introduces a discontinuity into the bending-moment diagram.

PROBLEMS

7.17 through 7.22 Draw the shear and bending-moment diagrams for the beam and loading shown.

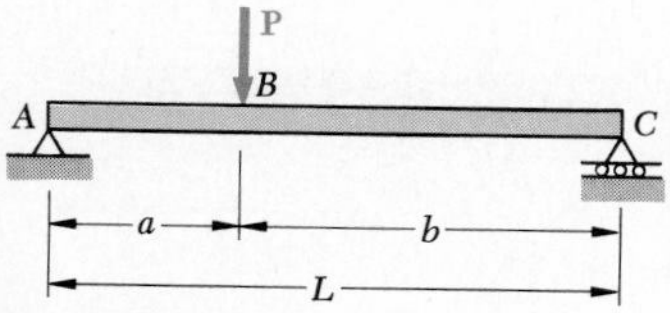

Fig. P7.17

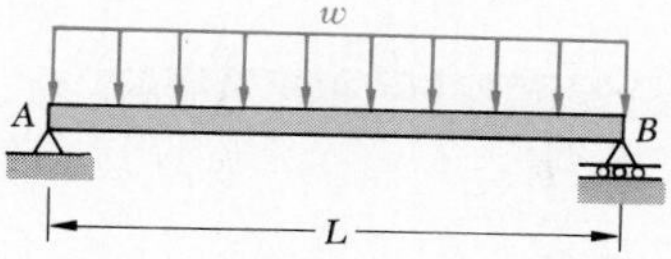

Fig. P7.18

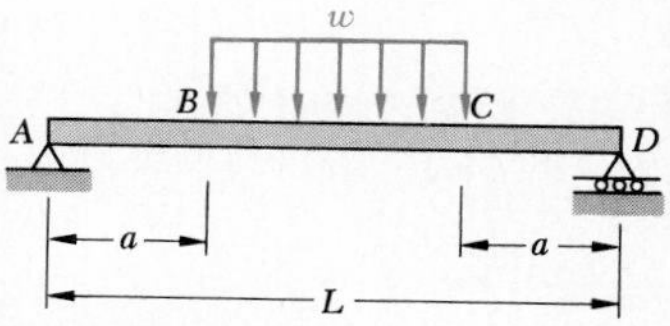

Fig. P7.19

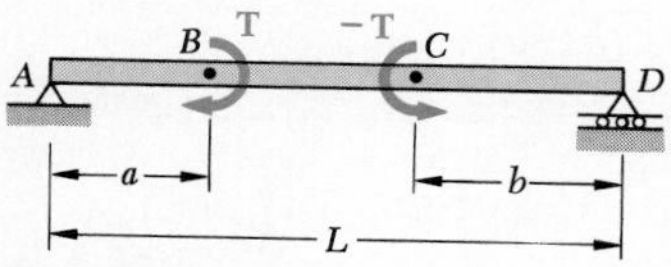

Fig. P7.20

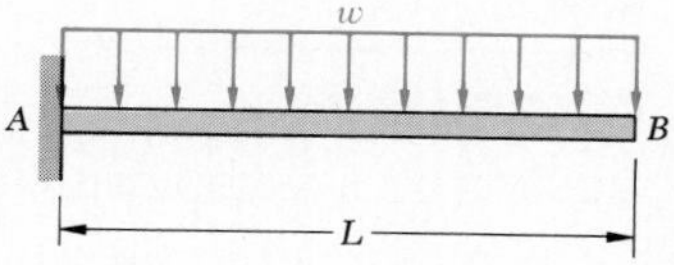

Fig. P7.21

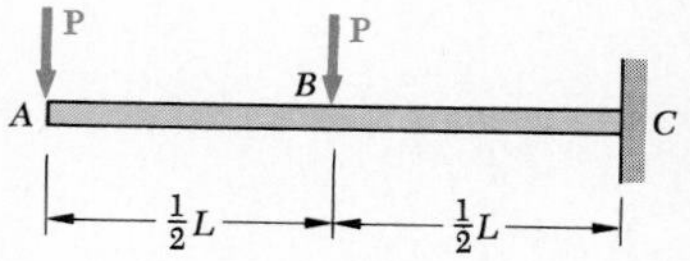

Fig. P7.22

7.23 and 7.24 Draw the shear and bending-moment diagrams for the beam AB.

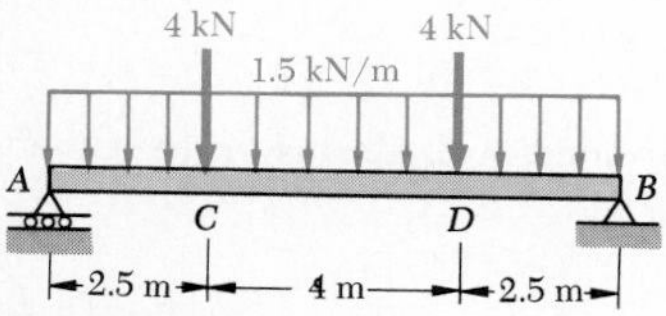

Fig. P7.23

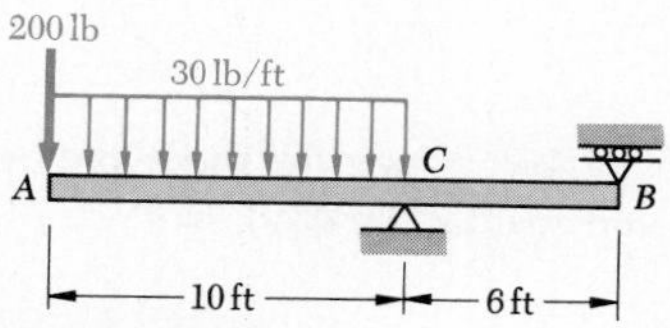

Fig. P7.24

7.25 Draw the shear and bending-moment diagrams for the beam AB if $a = 6$ ft.

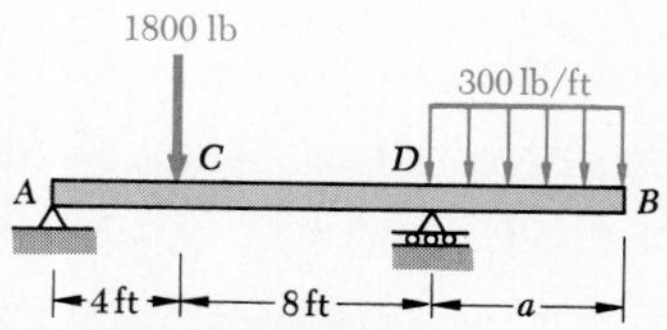

Fig. P7.25 and P7.27

7.26 Draw the shear and bending-moment diagrams for the beam AB if the magnitude of the upward force **P** is 4 kN.

***7.27** Determine the distance a for which the maximum absolute value of the bending moment in the beam is as small as possible.

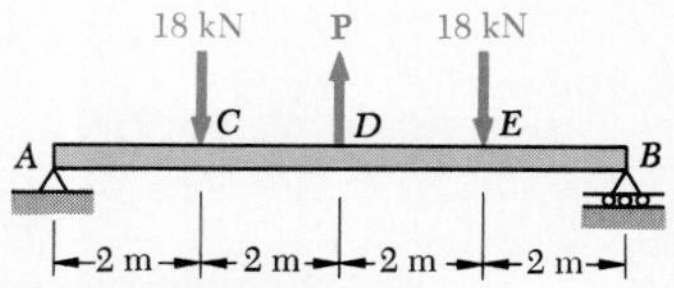

Fig. P7.26 and P7.28

***7.28** Determine the magnitude of the upward force **P** for which the maximum absolute value of the bending moment in the beam is as small as possible.

7.29 and 7.30 Assuming the upward reaction of the ground to be uniformly distributed, draw the shear and bending-moment diagrams for the beam *AB*.

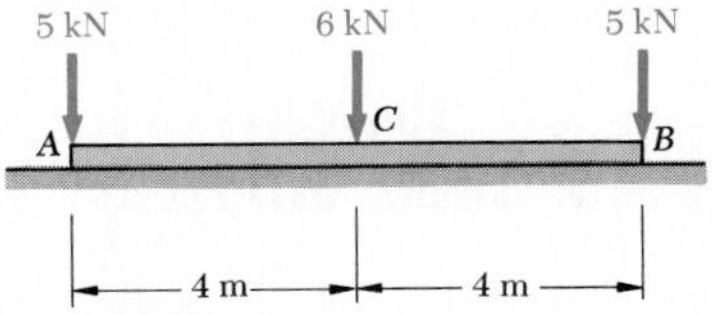

Fig. P7.29

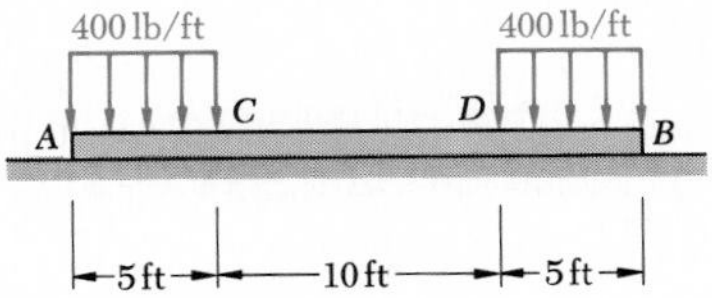

Fig. P7.30

7.31 and 7.32 Draw the shear and bending-moment diagrams for the beam *AB*.

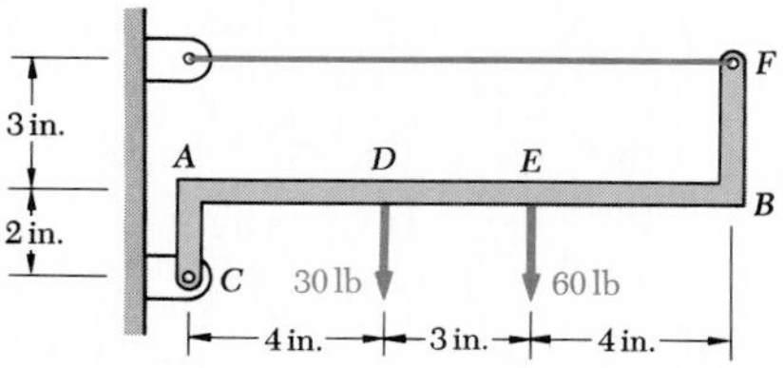

Fig. P7.31

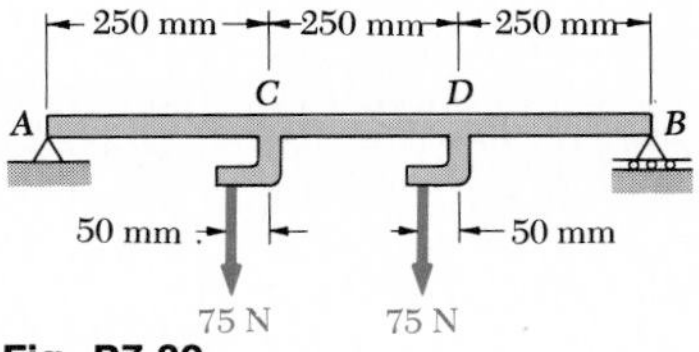

Fig. P7.32

7.33 Draw the shear and bending-moment diagrams for the beam and loading of Prob. 6.77.

7.34 Draw the shear and bending-moment diagrams for the beam and loading of Prob. 6.78.

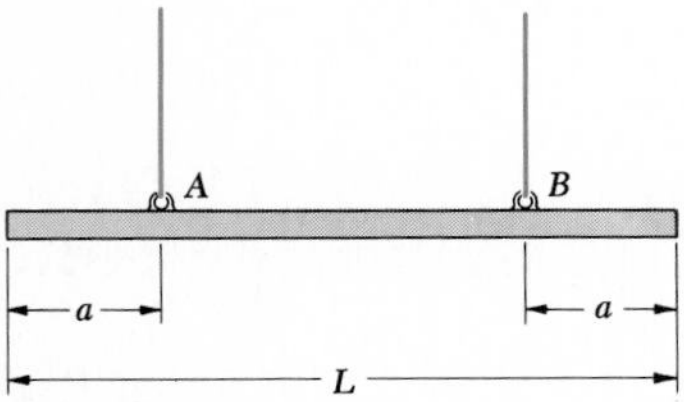

Fig. P7.35

***7.35** A uniform beam is to be picked up by crane cables attached at *A* and *B*. Determine the distance *a* from the ends of the beam to the points where the cables should be attached if the maximum absolute value of the bending moment in the beam is to be as small as possible. (*Hint.* Draw the bending-moment diagram in terms of *a*, *L*, and the weight *w* per unit length, and then equate the maximum positive and negative bending moments obtained.)

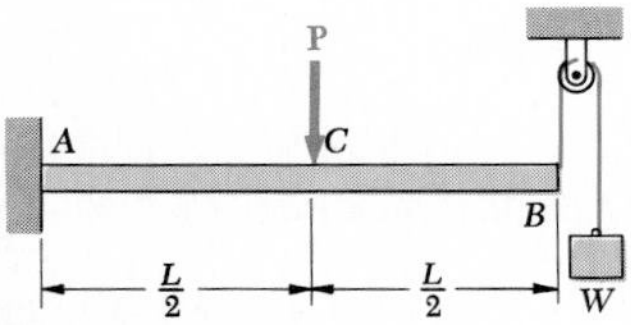

Fig. P7.36

***7.36** In order to reduce the bending moment in the cantilever beam *AB*, a cable and counterweight are permanently attached at end *B*. Determine the magnitude of the counterweight for which the maximum absolute value of the bending moment in the beam is as small as possible. (*a*) Consider only the case when the force **P** is actually applied at *C*. (*b*) Consider the more general case when the force **P** may either be applied at *C* or removed.

*7.5. Relations between Load, Shear, and Bending Moment.

When a beam carries more than two or three concentrated loads, or when it carries distributed loads, the method outlined in Sec. 7.4 for plotting shear and bending moment may prove quite cumbersome. The construction of the shear diagram and, especially, of the bending-moment diagram will be greatly facilitated if certain relations existing between load, shear, and bending moment are taken into consideration.

Let us consider a simply supported beam AB carrying a distributed load w per unit length (Fig. 7.11*a*), and let C and C' be two points of the beam at a distance Δx from each other. The shear and bending moment at C will be denoted by V and M, respectively, and will be assumed positive; the shear and bending moment at C' will be denoted by $V + \Delta V$ and $M + \Delta M$.

We now shall detach the portion of beam CC' and draw its free-body diagram (Fig. 7.11*b*). The forces exerted on the free body include a load of magnitude $w\,\Delta x$ and internal forces and couples at C and C'. Since shear and bending moment have been assumed positive, the forces and couples will be directed as shown in the figure.

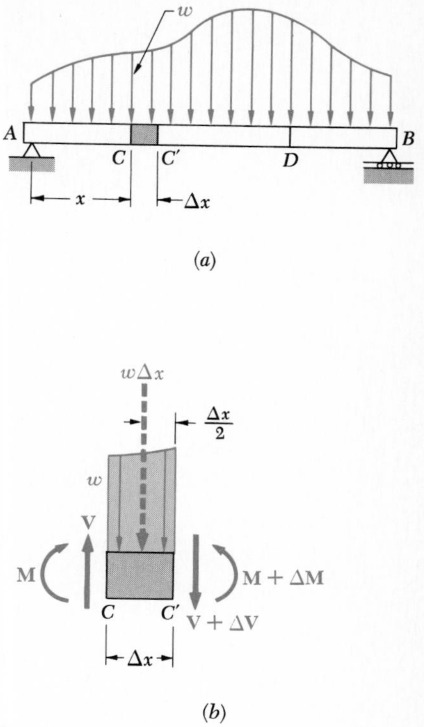

Fig. 7.11

Relations between Load and Shear. Writing that the sum of the vertical components of the forces acting on the free body CC' is zero, we obtain

$$V - (V + \Delta V) - w\,\Delta x = 0$$
$$\Delta V = -w\,\Delta x$$

Dividing both members of the equation by Δx and then letting Δx approach zero, we obtain

$$\frac{dV}{dx} = -w \qquad (7.1)$$

Formula (7.1) indicates that, for a beam loaded as shown in Fig. 7.11*a*, the slope dV/dx of the shear curve is negative; the numerical value of the slope at any point is equal to the load per unit length at that point.

Integrating (7.1) between points C and D, we obtain

$$V_D - V_C = -\int_{x_C}^{x_D} w\,dx \qquad (7.2)$$

$$V_D - V_C = -(\text{area under load curve between } C \text{ and } D) \qquad (7.2')$$

Note that this result could also have been obtained by considering the equilibrium of the portion of beam CD, since the area under the load curve represents the total load applied between C and D.

It should be observed that formula (7.1) *is not valid* at a point where a concentrated load is applied; the shear curve is discontinuous at such a point, as seen in Sec. 7.4. Similarly, formulas (7.2) and (7.2′) cease to be valid when concentrated loads are applied between C and D, since they do not take into account the sudden change in shear caused by a concentrated load. Formulas (7.2) and (7.2′), therefore, should be applied only between successive concentrated loads.

Relations between Shear and Bending Moment. Returning to the free-body diagram of Fig. 7.11*b*, and writing now that the sum of the moments about C' is zero, we obtain

$$(M + \Delta M) - M - V\,\Delta x + w\,\Delta x \frac{\Delta x}{2} = 0$$

$$\Delta M = V\,\Delta x - \tfrac{1}{2}w(\Delta x)^2$$

Dividing both members of the equation by Δx and then letting Δx approach zero, we obtain

$$\frac{dM}{dx} = V \tag{7.3}$$

Formula (7.3) indicates that the slope dM/dx of the bending-moment curve is equal to the value of the shear. This is true at any point where the shear has a well-defined value, i.e., at any point where no concentrated load is applied. Formula (7.3) also shows that the shear is zero at points where the bending moment is maximum. This property facilitates the determination of the points where the beam is likely to fail under bending.

Integrating (7.3) between points C and D, we obtain

$$M_D - M_C = \int_{x_C}^{x_D} V\,dx \tag{7.4}$$

$$M_D - M_C = \text{area under shear curve between } C \text{ and } D \tag{7.4′}$$

Note that the area under the shear curve should be considered positive where the shear is positive and negative where the shear is negative. Formulas (7.4) and (7.4′) are valid even when concentrated loads are applied between C and D, as long as the shear curve has been correctly drawn. The formulas cease to be valid, however, if a *couple* is applied at a point between C and D, since they do not take into account the sudden change in bending moment caused by a couple (see Sample Prob. 7.7).

Example. Let us consider a simply supported beam AB of span L carrying a uniformly distributed load w (Fig. 7.12*a*). From the free-body diagram of the entire beam we determine the magnitude of the reactions at the supports: $R_A = R_B = wL/2$ (Fig. 7.12*b*). Next, we draw the shear diagram. Close to the end A of the beam, the shear is equal to R_A, that is, to $wL/2$, as we may check by considering as a free body a very small portion of the beam. Using formula (7.2), we may then determine the shear V at any distance x from A; we write

$$V - V_A = -\int_0^x w\,dx = -wx$$

$$V = V_A - wx = \frac{wL}{2} - wx = w\left(\frac{L}{2} - x\right)$$

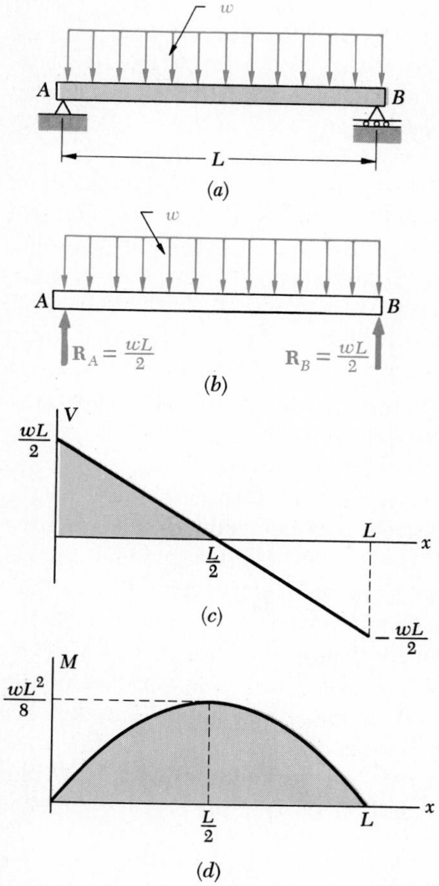

Fig. 7.12

The shear curve is thus an oblique straight line which crosses the x axis at $x = L/2$ (Fig. 7.12c). Considering, now, the bending moment, we first observe that $M_A = 0$. The value M of the bending moment at any distance x from A may then be obtained from formula (7.4); we have

$$M - M_A = \int_0^x V\,dx$$

$$M = \int_0^x w\left(\frac{L}{2} - x\right) dx = \frac{w}{2}(Lx - x^2)$$

The bending-moment curve is a parabola. The maximum value of the bending moment occurs when $x = L/2$, since V (and thus dM/dx) is zero for that value of x. Substituting $x = L/2$ in the last equation, we obtain $M_{\max} = wL^2/8$.

In most engineering applications, the value of the bending moment needs to be known only at a few specific points. Once the shear diagram has been drawn, and after M has been determined at one of the ends of the beam, the value of the bending moment may then be obtained at any given point by computing the area under the shear curve and using formula (7.4′). For instance, since $M_A = 0$ for the beam of Fig. 7.12a, the maximum value of the bending moment for that beam may be obtained simply by measuring the area of the shaded triangle in the shear diagram of Fig. 7.12c. We have

$$M_{\max} = \frac{1}{2}\frac{L}{2}\frac{wL}{2} = \frac{wL^2}{8}$$

We note that, in this example, the load curve is a horizontal straight line, the shear curve an oblique straight line, and the bending-moment curve a parabola. If the load curve had been an oblique straight line (first degree), the shear curve would have been a parabola (second degree) and the bending-moment curve a cubic (third degree). The shear and bending-moment curves will always be, respectively, one and two degrees higher than the load curve. With this in mind, we should be able to sketch the shear and bending-moment diagrams without actually determining the functions $V(x)$ and $M(x)$, once a few values of the shear and bending moment have been computed. The sketches obtained will be more accurate if we make use of the fact that, at any point where the curves are continuous, the slope of the shear curve is equal to $-w$ and the slope of the bending-moment curve is equal to V.

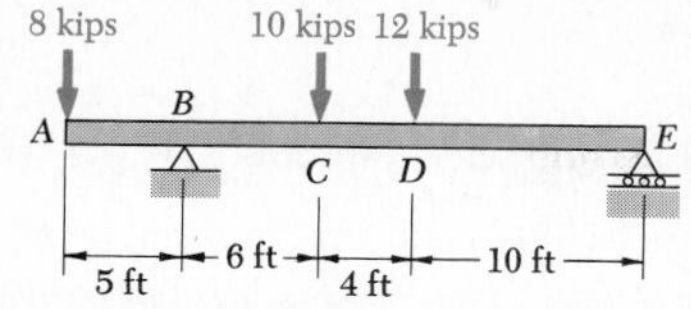

SAMPLE PROBLEM 7.4

Draw the shear and bending-moment diagrams for the beam and loading shown.

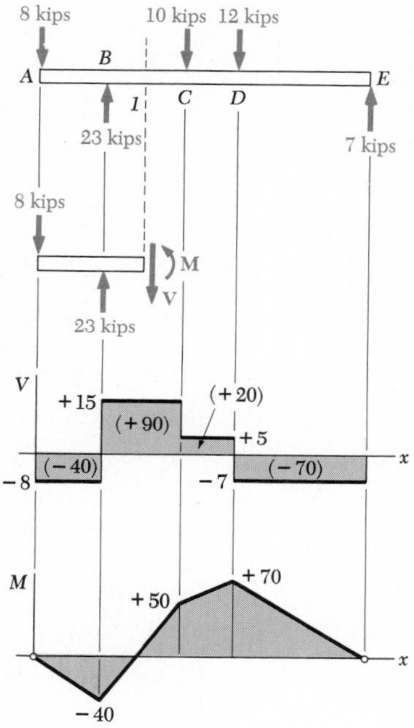

Solution. Considering the entire beam as a free body, we obtain the reactions

$$\mathbf{R}_B = 23 \text{ kips} \uparrow \qquad \mathbf{R}_E = 7 \text{ kips} \uparrow$$

We also note that at both A and E the bending moment is zero; thus two points (indicated by dots) are obtained on the bending-moment diagram.

Shear Diagram. Since $dV/dx = -w$, we find that between loads the slope of the shear diagram is zero (i.e., the shear is constant). The shear at any point is determined by dividing the beam into two parts and considering either part as a free body. For example, using the portion of beam to the left of section *1*, we obtain

$$+\uparrow\Sigma F_y = 0: \qquad -8 + 23 - V = 0 \qquad V = +15 \text{ kips}$$

Bending-Moment Diagram. We recall that the area under the shear curve between two points is equal to the change in bending moment between the same two points. For convenience, the area of each portion of the shear diagram is computed and is indicated on the diagram. Since the bending moment M_A at the free end is known to be zero, we write

$$\begin{aligned} M_B - M_A &= -40 & M_B &= -40 \text{ kip}\cdot\text{ft} \\ M_C - M_B &= +90 & M_C &= +50 \text{ kip}\cdot\text{ft} \\ M_D - M_C &= +20 & M_D &= +70 \text{ kip}\cdot\text{ft} \\ M_E - M_D &= -70 & M_E &= 0 \end{aligned}$$

Since M_E is known to be zero, a check of the computations is obtained.

The shear being constant between successive loads, the slope dM/dx is constant and the bending-moment diagram is obtained by connecting the known points with straight lines. From the V and M diagrams we note that $V_{max} = 15$ kips, and $M_{max} = 70$ kip · ft.

SAMPLE PROBLEM 7.5

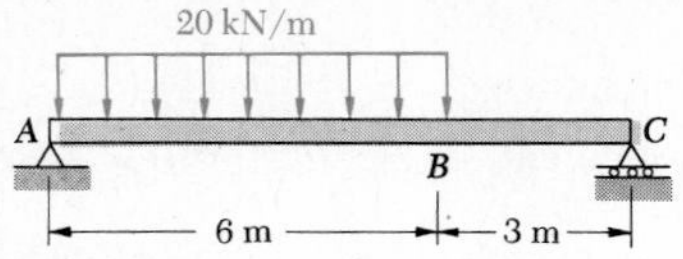

Draw the shear and bending-moment diagrams for the beam and loading shown.

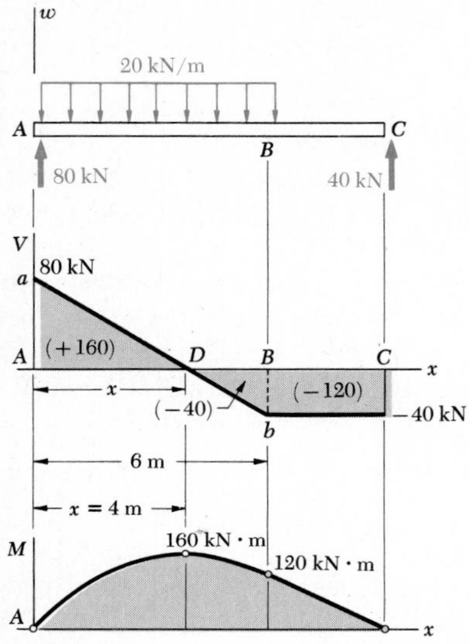

Solution. Considering the entire beam as a free body, we obtain the reactions

$$\mathbf{R}_A = 80 \text{ kN} \uparrow \qquad \mathbf{R}_C = 40 \text{ kN} \uparrow$$

Shear Diagram. The shear just to the right of A is $V_A = +80$ kN. Since the change in shear between two points is equal to *minus* the area under the load curve between the same two points, we obtain V_B by writing

$$V_B - V_A = -(20 \text{ kN/m})(6 \text{ m}) = -120 \text{ kN}$$
$$V_B = -120 + V_A = -120 + 80 = -40 \text{ kN}$$

The slope $dV/dx = -w$ being constant between A and B, the shear diagram between these two points is represented by a straight line. Between B and C, the area under the load curve is zero; therefore,

$$V_C - V_B = 0 \qquad V_C = V_B = -40 \text{ kN}$$

and the shear is constant between B and C.

Bending-Moment Diagram. We note that the bending moment at each end of the beam is zero. In order to determine the maximum bending moment, we locate the section D of the beam where $V = 0$. Considering the portion of the shear diagram between A and B, we note that the triangles DAa and DBb are similar; thus,

$$\frac{x}{80 \text{ kN}} = \frac{6 - x}{40 \text{ kN}} \qquad x = 4 \text{ m}$$

The maximum bending moment occurs at point D, where we have $dM/dx = V = 0$. The areas of the various portions of the shear diagram are computed and are given (in parentheses) on the diagram. Since the area of the shear diagram between two points is equal to the change in bending moment between the same two points, we write

$$M_D - M_A = +160 \text{ kN} \cdot \text{m} \qquad M_D = +160 \text{ kN} \cdot \text{m}$$
$$M_B - M_D = -\ 40 \text{ kN} \cdot \text{m} \qquad M_B = +120 \text{ kN} \cdot \text{m}$$
$$M_C - M_B = -120 \text{ kN} \cdot \text{m} \qquad M_C = 0$$

The bending-moment diagram consists of an arc of parabola followed by a segment of straight line; the slope of the parabola at A is equal to the value of V at that point.

SAMPLE PROBLEM 7.6

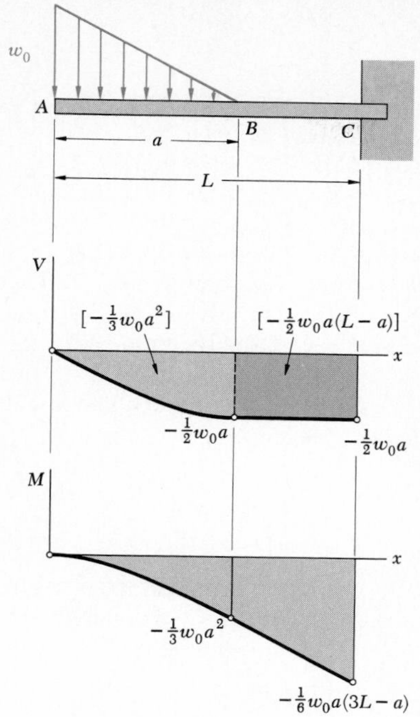

Sketch the shear and bending-moment diagrams for the cantilever beam shown.

Solution. *Shear Diagram.* At the free end of the beam, we find $V_A = 0$. Between A and B, the area under the load curve is $\frac{1}{2}w_0a$; we find V_B by writing

$$V_B - V_A = -\tfrac{1}{2}w_0a \qquad V_B = -\tfrac{1}{2}w_0a$$

Between B and C, the beam is not loaded; thus $V_C = V_B$. At A, we have $w = w_0$ and, according to Eq. (7.1), the slope of the shear curve is $dV/dx = -w_0$, while at B the slope is $dV/dx = 0$. Between A and B, the loading decreases linearly, and the shear diagram is parabolic. Between B and C, $w = 0$, and the shear diagram is a horizontal line.

Bending-Moment Diagram. The bending moment M_A at the free end of the beam is zero. We compute the area under the shear curve and write

$$M_B - M_A = -\tfrac{1}{3}w_0a^2 \qquad M_B = -\tfrac{1}{3}w_0a^2$$
$$M_C - M_B = -\tfrac{1}{2}w_0a(L - a)$$
$$M_C = -\tfrac{1}{6}w_0a(3L - a)$$

The sketch of the bending-moment diagram is completed by recalling that $dM/dx = V$. We find that between A and B the diagram is represented by a cubic curve with zero slope at A , and between B and C by a straight line.

SAMPLE PROBLEM 7.7

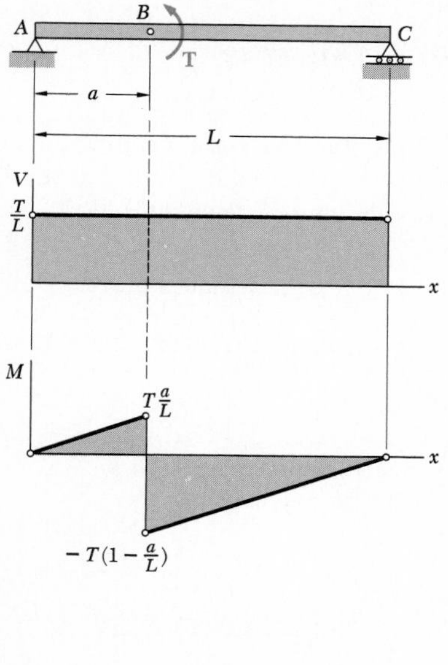

The simple beam AC is loaded by a couple of moment T applied at point B. Draw the shear and bending-moment diagrams of the beam.

Solution. The entire beam is taken as a free body, and we obtain

$$\mathbf{R}_A = \frac{T}{L}\uparrow \qquad \mathbf{R}_C = \frac{T}{L}\downarrow$$

The shear at any section is constant and equal to T/L. Since a couple is applied at B, the bending-moment diagram is discontinuous at B; the bending moment decreases suddenly by an amount equal to T.

PROBLEMS

7.37 Using the methods of Sec. 7.5, solve Prob. 7.17.
7.38 Using the methods of Sec. 7.5, solve Prob. 7.18.
7.39 Using the methods of Sec. 7.5, solve Prob. 7.19.
7.40 Using the methods of Sec. 7.5, solve Prob. 7.22.
7.41 Using the methods of Sec. 7.5, solve Prob. 7.23.
7.42 Using the methods of Sec. 7.5, solve Prob. 7.26.
7.43 Using the methods of Sec. 7.5, solve Prob. 7.25.
7.44 Using the methods of Sec. 7.5, solve Prob. 7.30.
7.45 Using the methods of Sec. 7.5, solve Prob. 7.29.

7.46 through 7.49 Draw the shear and bending-moment diagrams for the beam and loading shown.

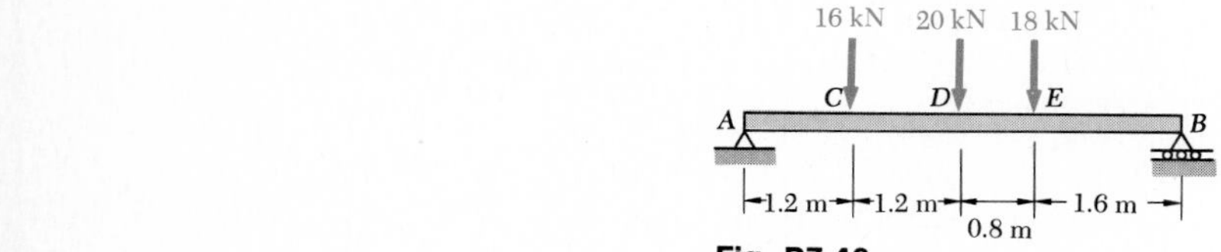

Fig. P7.46

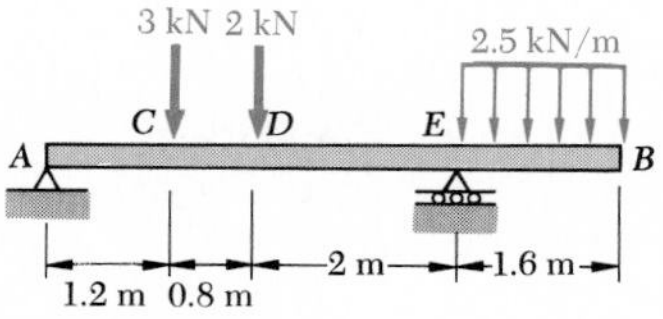

Fig. P7.47

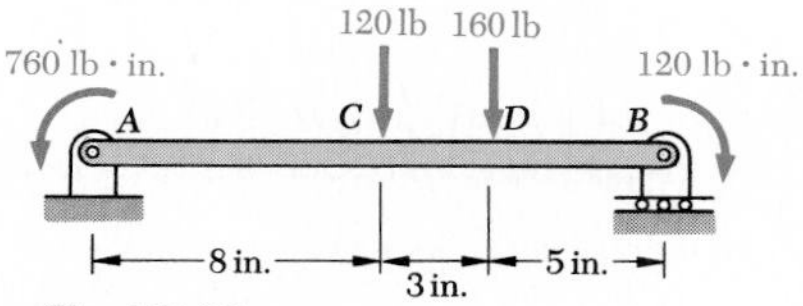

Fig. P7.48

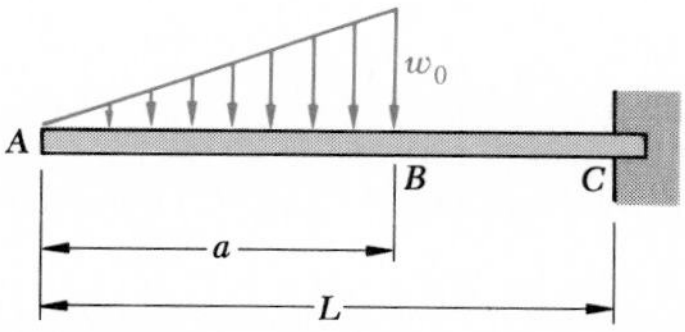

Fig. P7.49

7.50 through 7.53 Draw the shear and bending-moment diagrams for the beam and loading shown, and determine the location and magnitude of the maximum bending moment.

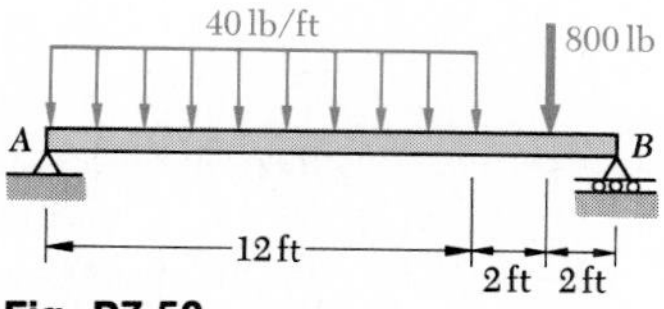

Fig. P7.50

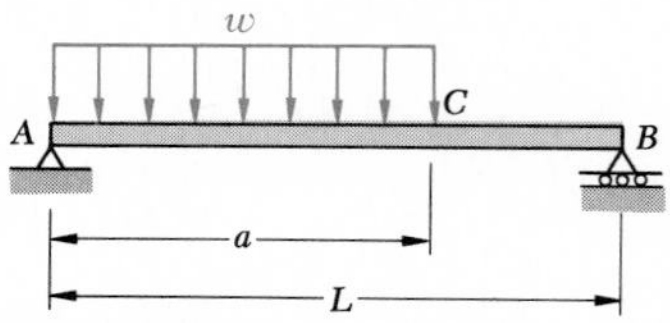

Fig. P7.51

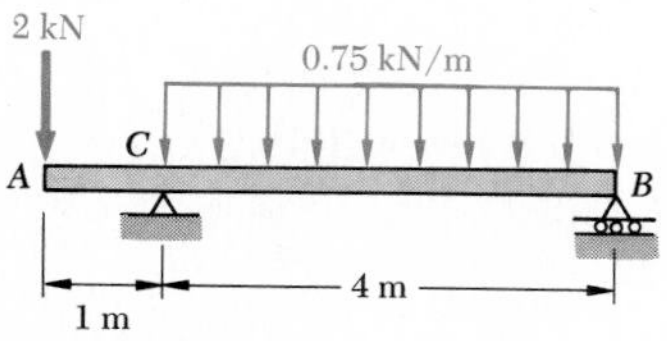

Fig. P7.52

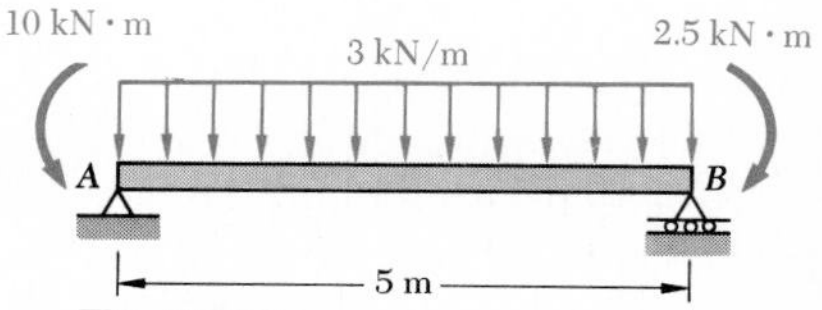

Fig. P7.53

7.54 Determine the equations of the shear and bending-moment curves for the beam and loading of Prob. 7.18. (Place the origin at point *A*.)

7.55 Determine the equations of the shear and bending-moment curves for the beam and loading of Prob. 7.21. (Place the origin at point *A*.)

7.56 and 7.57 Determine the equations of the shear and bending-moment curves for the given beam and loading. Also determine the magnitude and location of the maximum bending moment in the beam.

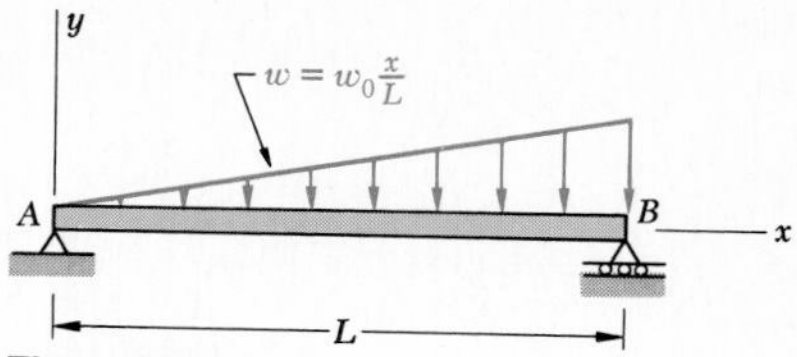

Fig. P7.56

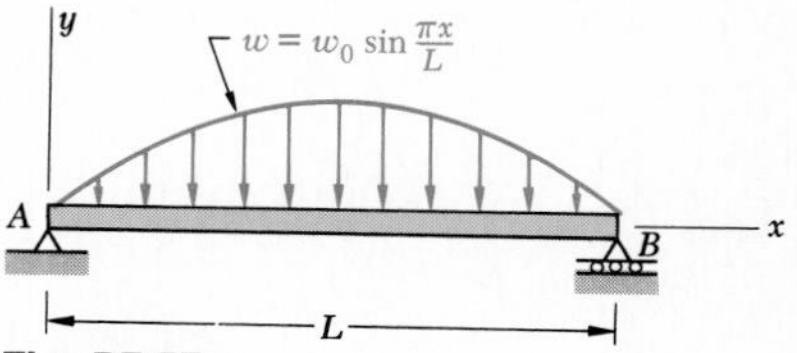

Fig. P7.57

7.58 The rod *AB* is attached to a hinge at *A* and is free at *B*. Determine (*a*) the ratio of w_2 to w_1 for which the rod is in equilibrium, (*b*) the equations of the shear and bending-moment curves for the rod, (*c*) the magnitude and location of the maximum bending moment.

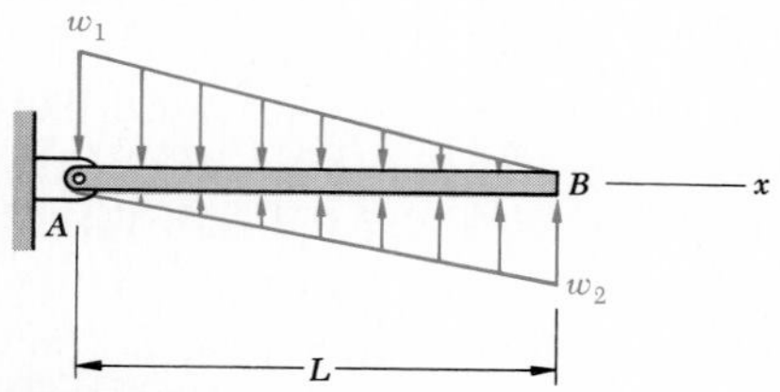

Fig. P7.58

***7.59** The beam *AB* is acted upon by the uniformly distributed load of 100 lb/ft and by two forces **P** and **Q**. It has been experimentally determined that the bending moment is +400 lb · ft at point *D* and +150 lb · ft at point *E*. Draw the shear and bending-moment diagrams for the beam.

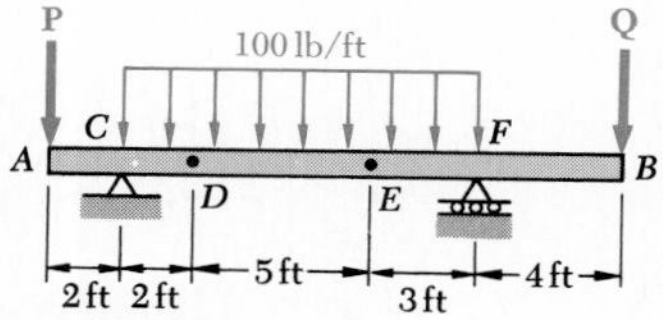

Fig. P7.59

***7.60** The beam *AB* supports the uniformly distributed load of 10 lb/in. and two concentrated loads **P** and **Q**. The bending moment is known to be −1770 lb · in. at point *A* and −930 lb · in. at point *C*. Draw the shear and bending-moment diagrams for the beam.

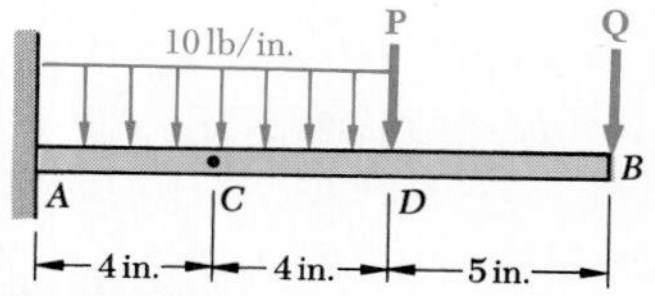

Fig. P7.60

7.61 A beam *AB* is loaded by couples spaced uniformly along its length. Assuming that the couples may be represented by a uniformly distributed couple loading of *m* lb · ft/ft, draw the shear and bending-moment diagrams for the beam when it is supported (*a*) as shown, (*b*) as a cantilever with a fixed support at *A* and no support at *B*.

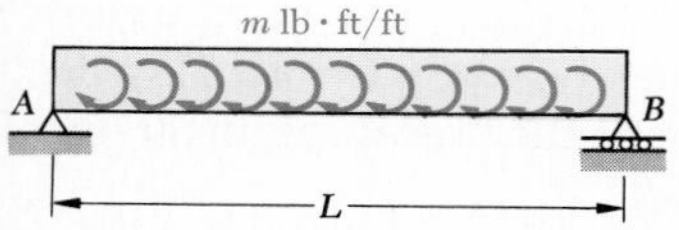

Fig. P7.61

CABLES

***7.6. Cables with Concentrated Loads.** Cables are used in many engineering applications, such as suspension bridges, transmission lines, aerial tramways, guy wires for high towers, etc. Cables may be divided into two categories, according to their loading: (1) cables supporting concentrated loads, (2) cables supporting distributed loads. In this section, we shall examine cables of the first category.

Consider a cable attached to two fixed points A and B and supporting n given vertical concentrated loads $\mathbf{P}_1, \mathbf{P}_2, \ldots, \mathbf{P}_n$ (Fig. 7.13*a*). We assume that the cable is *flexible*, i.e., that its resistance to bending is small and may be neglected. We further assume that the *weight of the cable is negligible* compared with the loads supported by the cable. Any portion of cable between successive loads may therefore be considered as a two-force member, and the internal forces at any point in the cable reduce to a *force of tension directed along the cable.*

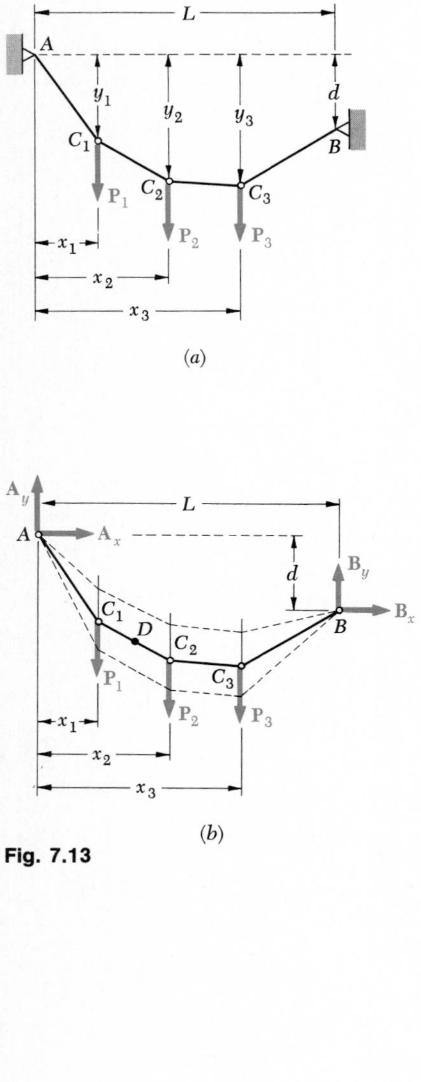

Fig. 7.13

We assume that each of the loads lies in a given vertical line, i.e., that the horizontal distance from support A to each of the loads is known; we also assume that the horizontal and vertical distances between the supports are known. We propose to determine the shape of the cable, i.e., the vertical distance from A to each of the points $C_1, C_2, \ldots, C_n$, and also the tension T in each portion of the cable.

We first draw the free-body diagram of the entire cable (Fig. 7.13*b*). Since the slope of the portions of cable attached at A and B is not known, the reactions at A and B must be represented by two components each. Thus, four unknowns are involved, and the three equations of equilibrium are not sufficient to determine the reactions at A and B.† We must therefore obtain an additional equation by considering the equilibrium of a portion of the cable. This is possible if we know the coordinates x and y of a point D of the cable. Drawing the free-body diagram of the portion of cable AD (Fig. 7.14*a*) and writing $\Sigma M_D = 0$, we obtain an additional relation between the scalar components A_x and A_y and may determine the reactions at A and B. The problem would remain indeterminate, however, if we did not know the coordinates of D, unless some other relation between A_x and A_y (or between B_x and B_y) were given. The cable might hang in any of various possible ways, as indicated by the dashed lines in Fig. 7.13*b*.

Once A_x and A_y have been determined, the vertical distance from A to any point of the cable may be easily found. Considering point C_2, for example, we draw the free-body diagram of

† Clearly, the cable is not a rigid body; the equilibrium equations represent, therefore, *necessary but not sufficient conditions* (see Sec. 6.12).

the portion of cable AC_2 (Fig. 7.14*b*). Writing $\Sigma M_{C_2} = 0$, we obtain an equation which may be solved for y_2. Writing $\Sigma F_x = 0$ and $\Sigma F_y = 0$, we obtain the components of the force **T** representing the tension in the portion of cable to the right of C_2. We observe that $T \cos \theta = -A_x$; *the horizontal component of the tension force is the same at any point of the cable.* It follows that the tension T is maximum when $\cos \theta$ is minimum, i.e., in the portion of cable which has the largest angle of inclination θ. Clearly, this portion of cable must be adjacent to one of the two supports of the cable.

***7.7. Cables with Distributed Loads.** Consider a cable attached to two fixed points A and B and carrying a *distributed load* (Fig. 7.15*a*). We saw in the preceding section that, for a cable supporting concentrated loads, the internal force at any point is a force of tension directed along the cable. In the case of a cable carrying a distributed load, the cable hangs in the shape of a curve, and the internal force at a point D is a force of tension **T** *directed along the tangent to the curve.* Given a certain distributed load, we propose in this section to determine the tension at any point of the cable. We shall also see in the following sections how the shape of the cable may be determined for two particular types of distributed loads.

Considering the most general case of distributed load, we draw the free-body diagram of the portion of cable extending from the lowest point C to a given point D of the cable (Fig. 7.15*b*). The forces acting on the free body are the tension force $\mathbf{T}_0$ at C, which is horizontal, the tension force **T** at D, directed along the tangent to the cable at D, and the resultant **W** of the distributed load supported by the portion of cable CD. Drawing the corresponding force triangle (Fig. 7.15*c*), we obtain the following relations:

$$T \cos \theta = T_0 \qquad T \sin \theta = W \tag{7.5}$$

$$T = \sqrt{T_0^2 + W^2} \qquad \tan \theta = \frac{W}{T_0} \tag{7.6}$$

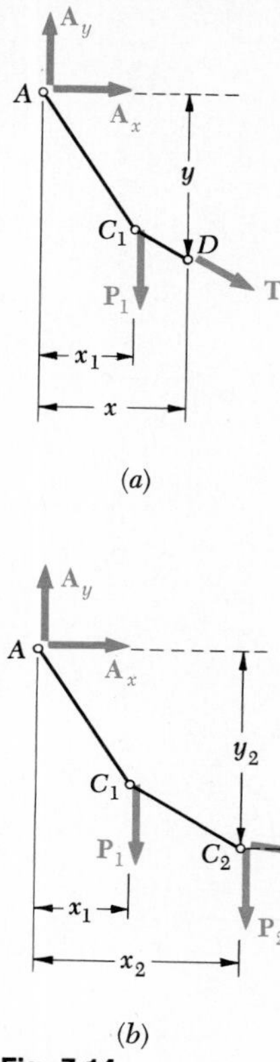

Fig. 7.14

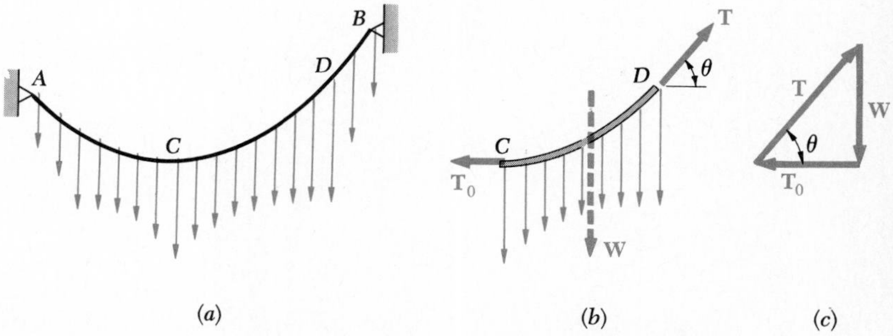

Fig. 7.15

From the relations (7.5), it appears that the horizontal component of the tension force **T** is the same at any point and that the vertical component of **T** is equal to the magnitude W of the load measured from the lowest point. Relations (7.6) show that the tension T is minimum at the lowest point and maximum at one of the two points of support.

***7.8. Parabolic Cable.** Let us assume, now, that the cable AB carries a load *uniformly distributed along the horizontal* (Fig. 7.16*a*). Cables of suspension bridges may be assumed loaded in this way, since the weight of the cables is small compared with the weight of the roadway. We denote by w the load per unit length (*measured horizontally*) and express it in N/m or in lb/ft. Choosing coordinate axes with origin at the lowest point C of the cable, we find that the magnitude W of the total load carried by the portion of cable extending from C to the point D of coordinates x and y is $W = wx$. The relations (7.6) defining the magnitude and direction of the tension force at D become

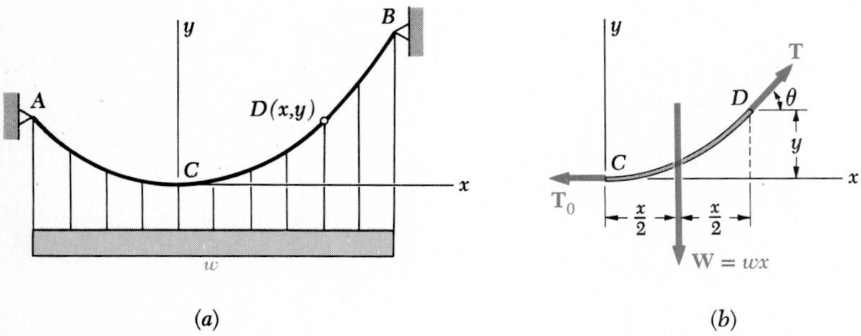

Fig. 7.16

$$T = \sqrt{T_0^2 + w^2x^2} \qquad \tan\theta = \frac{wx}{T_0} \tag{7.7}$$

Moreover, the distance from D to the line of action of the resultant **W** is equal to half the horizontal distance from C to D (Fig. 7.16*b*). Summing moments about D, we write

$$+\circlearrowleft\Sigma M_D = 0: \qquad wx\frac{x}{2} - T_0 y = 0$$

$$y = \frac{wx^2}{2T_0} \tag{7.8}$$

This is the equation of a *parabola* with a vertical axis and its vertex at the origin of coordinates. The curve formed by cables loaded uniformly along the horizontal is thus a parabola.†

When the supports A and B of the cable have the same elevation, the distance L between the supports is called the *span* of the cable and the vertical distance h from the supports to the lowest point is called the *sag* of the cable (Fig. 7.17*a*). If the span and sag of a cable are known, and if the load w per unit horizontal length is given, the minimum tension T_0 may be found by substituting $x = L/2$ and $y = h$ in formula (7.8). Formulas (7.7) and (7.8) will then define the tension at any point and the shape of the cable.

When the supports have different elevations, the position of the lowest point of the cable is not known and the coordinates x_A, y_A and x_B, y_B of the supports must be determined. To this effect, we express that the coordinates of A and B satisfy Eq. (7.8) and that $x_B - x_A = L$, $y_B - y_A = d$, where L and d denote, respectively, the horizontal and vertical distances between the two supports (Fig. 7.17*b* and *c*).

The length of the cable from its lowest point C to its support B may be obtained from the formula

$$s_B = \int_0^{x_B} \sqrt{1 + \left(\frac{dy}{dx}\right)^2}\, dx \tag{7.9}$$

Differentiating (7.8), we obtain the derivative $dy/dx = wx/T_0$; substituting into (7.9) and using the binomial theorem to expand the radical in an infinite series, we have

$$\begin{aligned} s_B &= \int_0^{x_B} \sqrt{1 + \frac{w^2x^2}{T_0^2}}\, dx \\ &= \int_0^{x_B} \left(1 + \frac{w^2x^2}{2T_0^2} - \frac{w^4x^4}{8T_0^4} + \cdots\right) dx \\ &= x_B\left(1 + \frac{w^2x_B^2}{6T_0^2} - \frac{w^4x_B^4}{40T_0^4} + \cdots\right) \end{aligned}$$

and, since $wx_B^2/2T_0 = y_B$,

$$s_B = x_B\left[1 + \frac{2}{3}\left(\frac{y_B}{x_B}\right)^2 - \frac{2}{5}\left(\frac{y_B}{x_B}\right)^4 + \cdots\right] \tag{7.10}$$

The series converges for values of the ratio y_B/x_B less than 0.5; in most cases, this ratio is much smaller, and only the first two terms of the series need be computed.

† Cables hanging under their own weight are not loaded uniformly along the horizontal, and they do not form a parabola. The error introduced by assuming a parabolic shape for cables hanging under their own weight, however, is small when the cable is sufficiently taut. A complete discussion of cables hanging under their own weight is given in the next section.

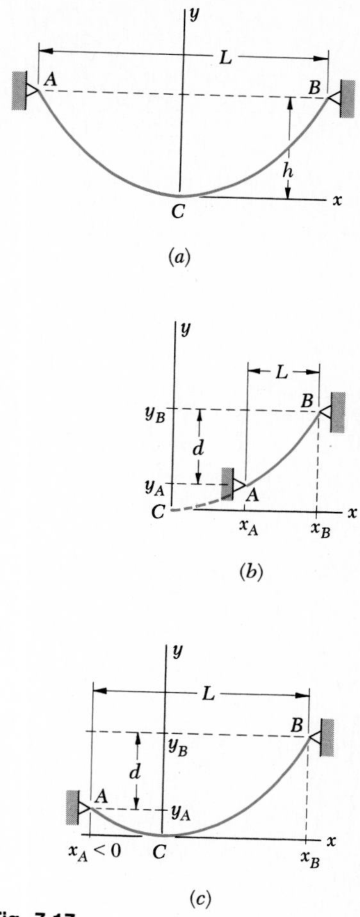

Fig. 7.17

SAMPLE PROBLEM 7.8

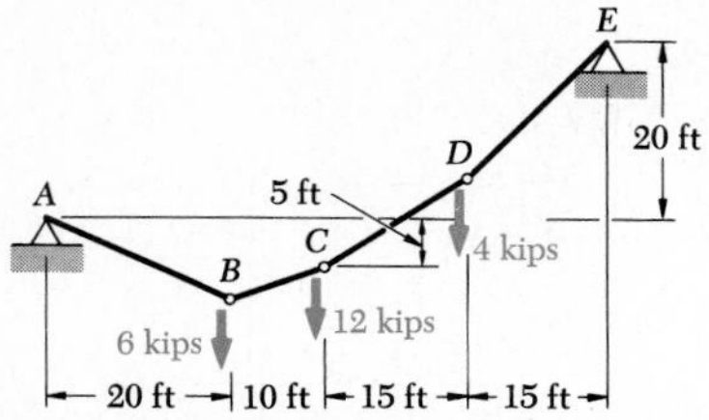

The cable AE supports three vertical loads from the points indicated. If point C is 5 ft below the left support, determine (*a*) the elevations of points B and D, (*b*) the maximum slope and the maximum tension in the cable.

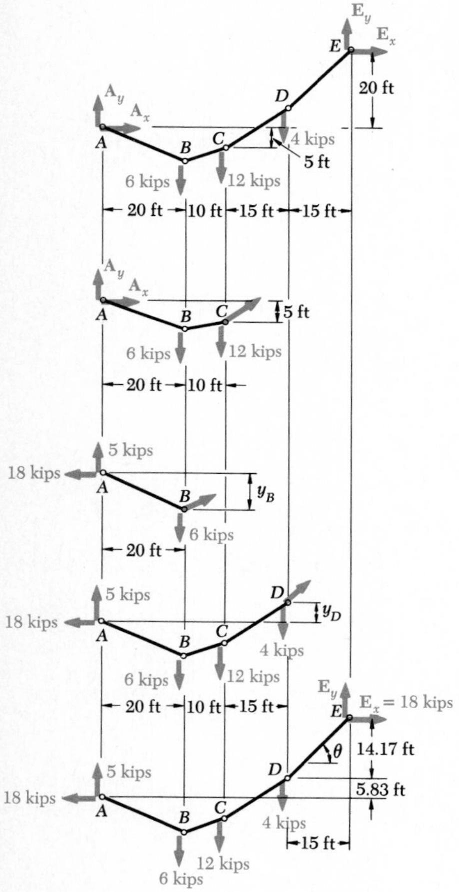

Solution. The reaction components $\mathbf{A}_x$ and $\mathbf{A}_y$ are determined as follows:

Free Body: Entire Cable

$+\circlearrowleft \Sigma M_E = 0$:

$$A_x(20\text{ ft}) - A_y(60\text{ ft}) + (6\text{ kips})(40\text{ ft}) + (12\text{ kips})(30\text{ ft}) + (4\text{ kips})(15\text{ ft}) = 0$$

$$20A_x - 60A_y + 660 = 0$$

Free Body: ABC

$$+\circlearrowleft \Sigma M_C = 0: \qquad -A_x(5\text{ ft}) - A_y(30\text{ ft}) + (6\text{ kips})(10\text{ ft}) = 0$$

$$-5A_x - 30A_y + 60 = 0$$

Solving the two equations simultaneously, we obtain

$$A_x = -18\text{ kips} \qquad \mathbf{A}_x = 18\text{ kips} \leftarrow$$

$$A_y = +5\text{ kips} \qquad \mathbf{A}_y = 5\text{ kips} \uparrow$$

a. Elevation of Point B. Considering the portion of cable AB *as a free body,* we write

$$+\circlearrowleft \Sigma M_B = 0: \qquad (18\text{ kips})y_B - (5\text{ kips})(20\text{ ft}) = 0$$

$$y_B = 5.56\text{ ft below } A \quad ◀$$

Elevation of Point D. Using the portion of cable $ABCD$ *as a free body,* we write

$+\circlearrowleft \Sigma M_D = 0$:

$$-(18\text{ kips})y_D - (5\text{ kips})(45\text{ ft}) + (6\text{ kips})(25\text{ ft}) + (12\text{ kips})(15\text{ ft}) = 0$$

$$y_D = 5.83\text{ ft above } A \quad ◀$$

b. Maximum Slope and Maximum Tension. We observe that the maximum slope occurs in portion DE. Since the horizontal component of the tension is constant and equal to 18 kips, we write

$$\tan\theta = \frac{14.17\text{ ft}}{15\text{ ft}} \qquad \theta = 43.4° \quad ◀$$

$$T_{\max} = \frac{18\text{ kips}}{\cos\theta} \qquad T_{\max} = 24.8\text{ kips} \quad ◀$$

SAMPLE PROBLEM 7.9

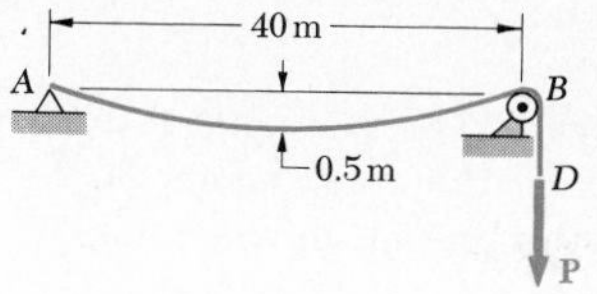

A light cable is attached to a support at A, passes over a small pulley at B, and supports a load **P**. Knowing that the sag of the cable is 0.5 m and that the mass per unit length of the cable is 0.75 kg/m, determine (*a*) the magnitude of the load **P**, (*b*) the slope of the cable at B, and (*c*) the total length of the cable from A to B. Since the ratio of the sag to the span is small, assume the cable to be parabolic. Also, neglect the weight of the portion of cable from B to D.

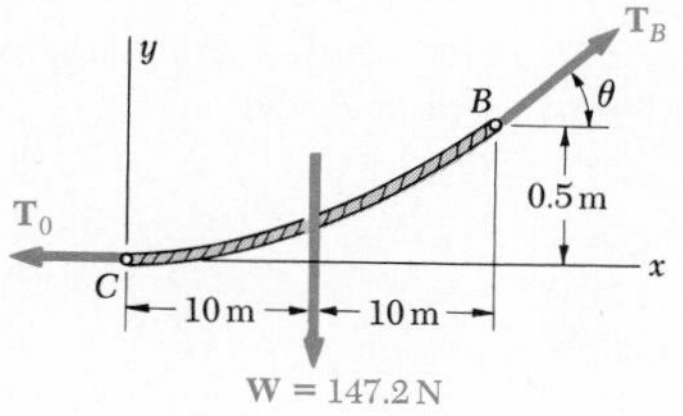

a. **Load P.** We denote by C the lowest point of the cable and draw the free-body diagram of the portion CB of cable. Assuming the load to be uniformly distributed along the horizontal, we write

$$w = (0.75 \text{ kg/m})(9.81 \text{ m/s}^2) = 7.36 \text{ N/m}$$

The total load for the portion CB of the cable is

$$W = wx_B = (7.36 \text{ N/m})(20 \text{ m}) = 147.2 \text{ N}$$

and is applied halfway between C and B. Summing moments about B, we write

$$+\circlearrowleft \Sigma M_B = 0: \qquad (147.2 \text{ N})(10 \text{ m}) - T_0(0.5 \text{ m}) = 0 \qquad T_0 = 2944 \text{ N}$$

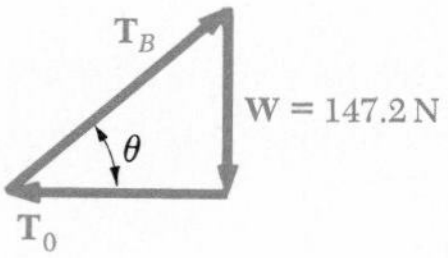

From the force triangle we obtain

$$T_B = \sqrt{T_0^2 + W^2}$$
$$= \sqrt{(2944 \text{ N})^2 + (147.2 \text{ N})^2} = 2948 \text{ N}$$

Since the tension on each side of the pulley is the same, we find

$$P = T_B = 2948 \text{ N} \quad ◄$$

b. **Slope of Cable at B.** We also obtain from the force triangle

$$\tan\theta = \frac{W}{T_0} = \frac{147.2 \text{ N}}{2944 \text{ N}} = 0.05$$

$$\theta = 2.9° \quad ◄$$

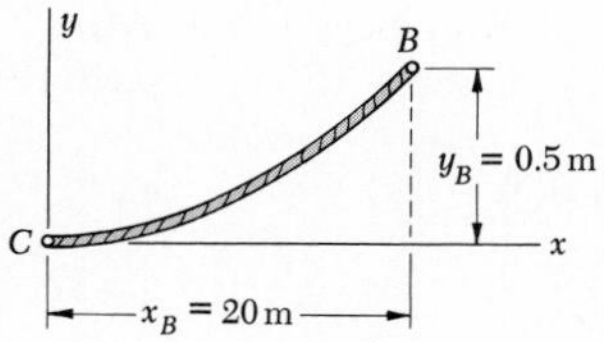

c. **Length of Cable.** Applying Eq. (7.10) between C and B, we write

$$s_B = x_B\left[1 + \frac{2}{3}\left(\frac{y_B}{x_B}\right)^2 + \cdots\right]$$

$$= (20 \text{ m})\left[1 + \frac{2}{3}\left(\frac{0.5 \text{ m}}{20 \text{ m}}\right)^2 + \cdots\right] = 20.00833 \text{ m}$$

The total length of the cable between A and B is twice this value,

$$\text{Length} = 2s_B = 40.0167 \text{ m} \quad ◄$$

PROBLEMS

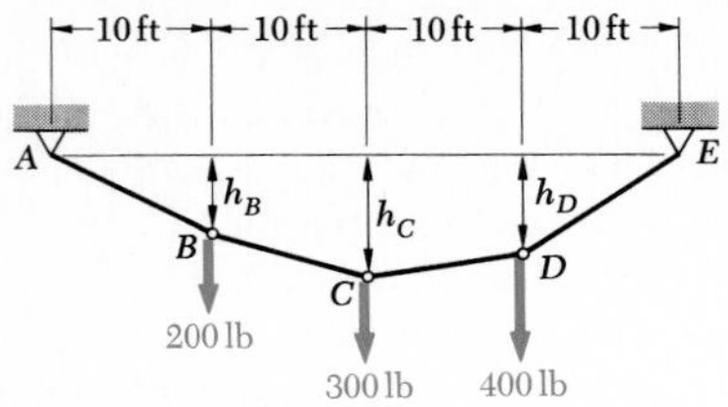

Fig. P7.62 and P7.63

7.62 Three loads are suspended as shown from the cable. Knowing that $h_C = 8$ ft, determine (*a*) the components of the reaction at *E*, (*b*) the maximum value of the tension in the cable.

7.63 Determine the sag at point *C* if the maximum tension in the cable is 1300 lb.

7.64 If $a = 2$ m and $b = 2.25$ m, determine the components of the reaction at *E* for the cable and loading shown.

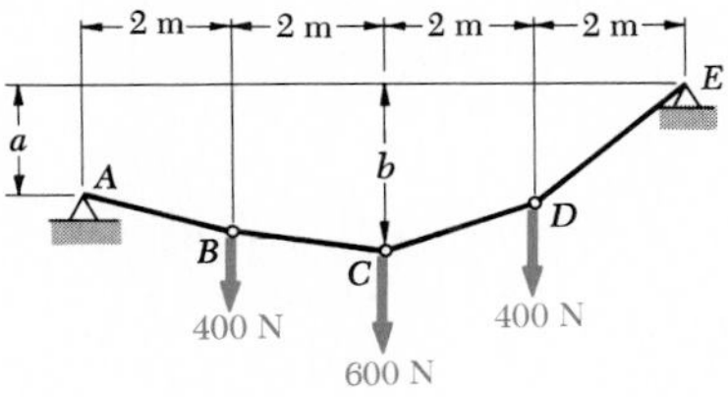

Fig. P7.64, P7.65, and P7.66

7.65 If $a = b = 1.25$ m, determine the components of the reaction at *E* and the maximum tension in the cable.

7.66 Determine the distance *a* if the portion *BC* of the cable is horizontal and if the maximum tension in the cable is 2600 N.

7.67 The center span of the George Washington Bridge, as originally constructed, consisted of a uniform roadway suspended from four cables. The uniform load supported by each cable was $w = 9.75$ kips/ft along the horizontal. Knowing that the span *L* is 3500 ft and that the sag *h* is 316 ft, determine the maximum and minimum tension in each cable.

7.68 Determine the length of each of the cables used in the center span of the George Washington Bridge. (See Prob. 7.67 for data.)

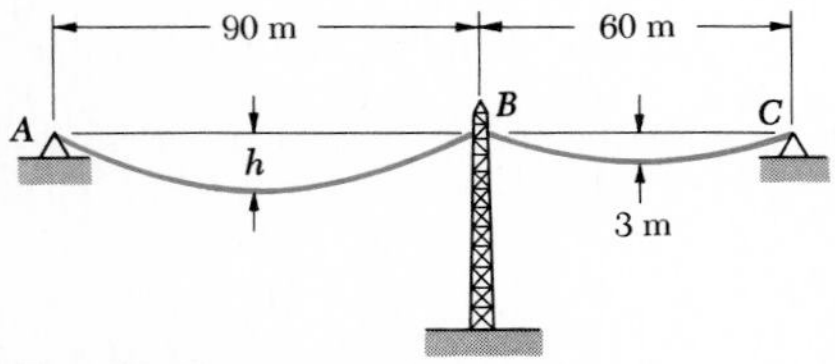

Fig. P7.69

7.69 Two cables of the same gage are attached to a transmission tower at *B*. Since the tower is slender, the horizontal component of the resultant of the forces exerted by the cables at *B* is to be zero. Assuming the cables to be parabolic, determine the required sag *h* of cable *AB*.

7.70 An electric wire having a mass per unit length of 0.3 kg/m is strung between two insulators at the same elevation and 30 m apart. If the maximum tension in the wire is to be 400 N, determine the smallest value of the sag which may be used. (Assume the wire to be parabolic.)

7.71 Knowing that a 51-m length of wire was used in spanning a horizontal distance of 50 m, determine the approximate sag of the wire. Assume the wire to be parabolic. [*Hint.* Use only the first two terms of Eq. (7.10).]

7.72 A cable of length $L + \Delta$ is suspended between two points which are at the same elevation and a distance L apart. (*a*) Assuming that Δ is small compared to L and that the cable is parabolic, determine the approximate sag in terms of L and Δ. (*b*) If $L = 100$ ft and $\Delta = 4$ ft, determine the approximate sag. [*Hint.* Use only the first two terms of Eq. (7.10).]

7.73 Before being fed into a printing press located to the right of D, a continuous sheet of paper weighing 0.20 lb/ft passes over rollers at A and B. Assuming that the curve formed by the sheet is parabolic, determine the location of the lowest point C and the maximum tension in the sheet.

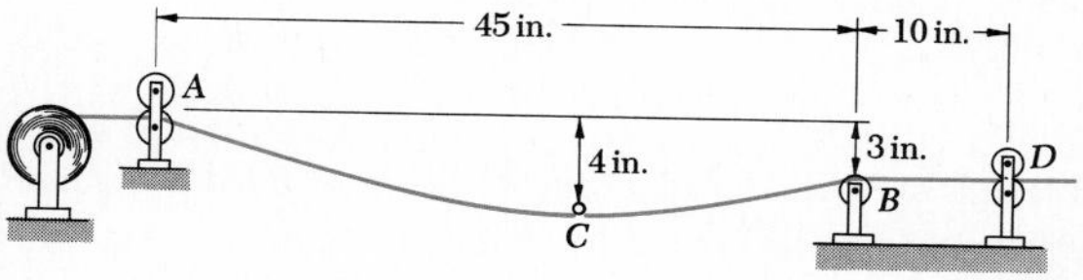

Fig. P7.73

7.74 A steam pipe has a mass per unit length of 70 kg/m. It passes between two buildings 20 m apart and is supported by a system of cables as shown. Assuming that the cable system causes the same loading as a single uniform cable having a mass per unit length of 5 kg/m, determine the location of the lowest point C of the cable and the maximum cable tension.

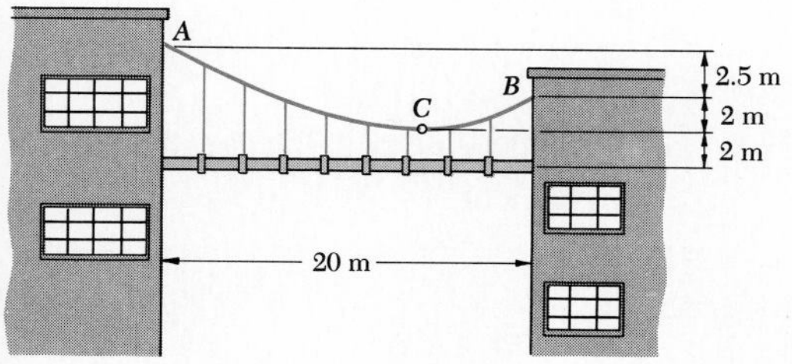

Fig. P7.74

***7.75** The total weight of cable AC is 60 lb. Assuming that the weight of the cable is distributed uniformly along the horizontal, determine the sag h and the slope of the cable at A and C.

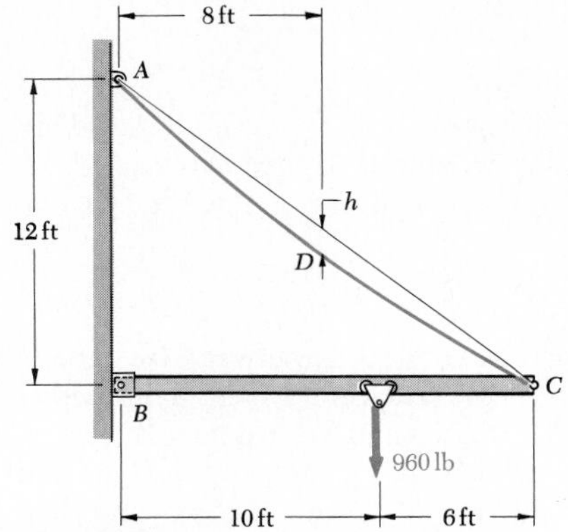

Fig. P7.75

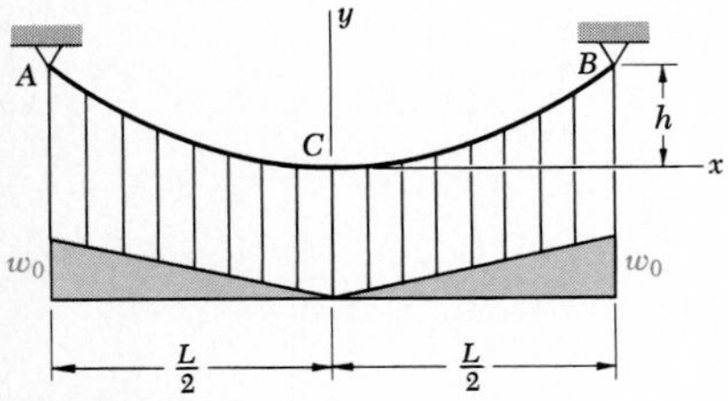

Fig. P7.76

7.76 The loading of a cable varies linearly from zero at the lowest point to w_0 at each support. Determine the equation of the curve assumed by each half of the cable and the tension at the center line.

7.77 Solve Prob. 7.76, assuming that the distributed loading varies uniformly from zero at each end to a maximum value of w_0 at the center line.

***7.78** A cable AB of span L and a simple beam $A'B'$ of the same span are subjected to identical vertical loadings as shown. Show that the magnitude of the bending moment at a point C' in the beam is equal to the product T_0h, where T_0 is the magnitude of the horizontal component of the tension force in the cable and h is the vertical distance between point C of the cable and the chord joining the points of support A and B.

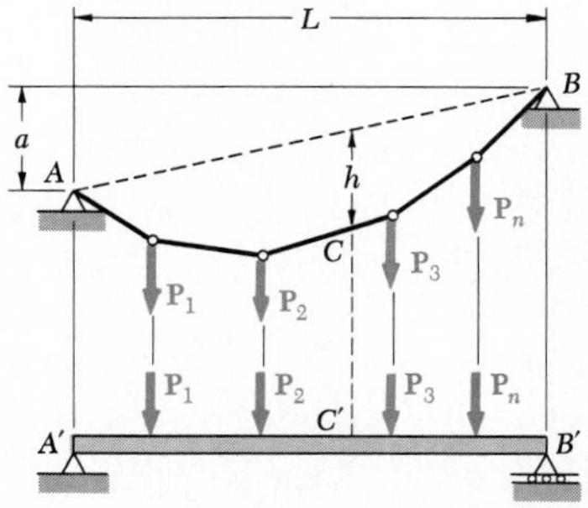

Fig. P7.78

***7.79** Show that the curve assumed by a cable which carries a distributed load $w(x)$ is defined by the differential equation $d^2y/dx^2 = w(x)/T_0$, where T_0 is the tension at the lowest point.

***7.80** Using the property indicated in Prob. 7.79, determine the curve assumed by a cable of span L and sag h carrying a distributed load $w = w_0 \cos(\pi x/L)$, where x is measured from mid-span. Also determine the maximum and minimum values of the tension.

***7.81** A large number of ropes are tied to a light wire which is suspended from two points A and B at the same level. Show that, if the lower ends of the ropes have been cut so that they lie in the same horizontal line and if the ropes are kept uniformly spaced, the curve assumed by the wire ACB is

$$y + d = d \cosh \left(\frac{w_r}{T_0 e}\right)^{1/2} x$$

where w_r is the weight per unit length of the ropes, d is the length of the shortest rope, and e is the horizontal distance between adjacent ropes. (*Hint.* Use the property indicated in Prob. 7.79.)

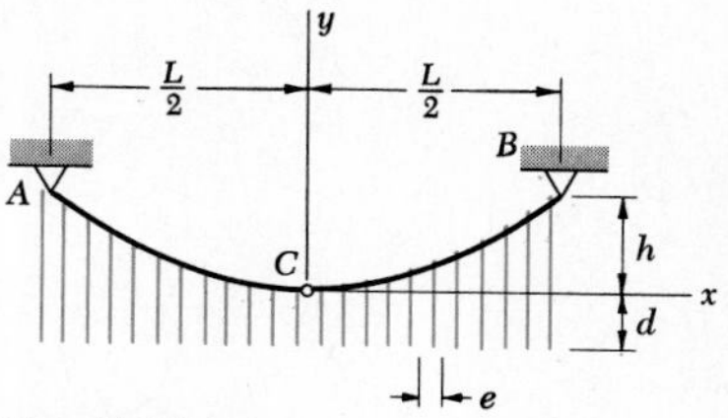

Fig. P7.81

***7.82** If the weight per unit length of the cable AB is $w_0/\cos^2\theta$, prove that the curve formed by the cable is a circular arc. (*Hint.* Use the property indicated in Prob. 7.79.)

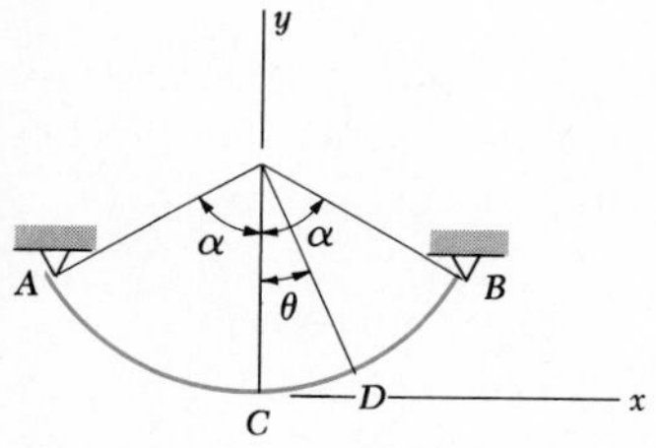

Fig. P7.82

*7.9. Catenary.

We shall consider now a cable AB carrying a load *uniformly distributed along the cable itself* (Fig. 7.18*a*). Cables hanging under their own weight are loaded in this way. We denote by w the load per unit length (*measured along the cable*) and express it in N/m or in lb/ft. The magnitude W of the total load carried by a portion of cable of length s extending from the lowest point C to a point D is $W = ws$. Substituting this value for W in formula (7.6), we obtain the tension at D,

$$T = \sqrt{T_0^2 + w^2 s^2}$$

In order to simplify the subsequent computations, we shall introduce the constant $c = T_0/w$. We thus write

$$T_0 = wc \qquad W = ws \qquad T = w\sqrt{c^2 + s^2} \qquad (7.11)$$

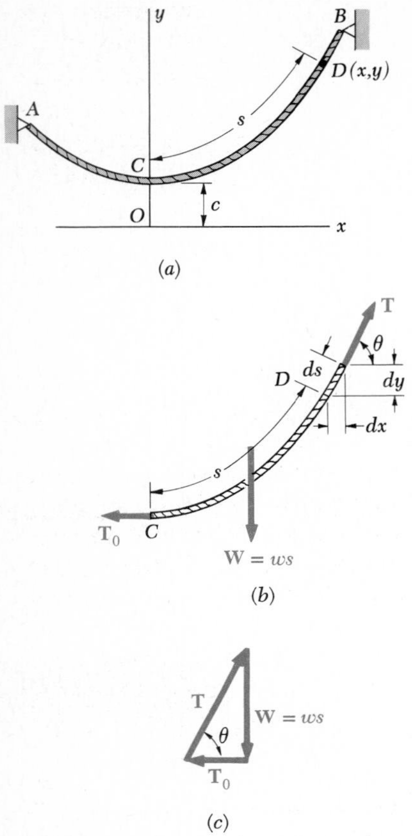

Fig. 7.18

The free-body diagram of the portion of cable CD is shown in Fig. 7.18*b*. This diagram, however, cannot be used to obtain directly the equation of the curve assumed by the cable, since we do not know the horizontal distance from D to the line of action of the resultant $\mathbf{W}$ of the load. To obtain this equation, we shall write first that the horizontal projection of a small element of cable of length ds is $dx = ds \cos\theta$. Observing from Fig. 7.18*c* that $\cos\theta = T_0/T$ and using (7.11), we write

$$dx = ds \cos\theta = \frac{T_0}{T}\, ds = \frac{wc\, ds}{w\sqrt{c^2 + s^2}} = \frac{ds}{\sqrt{1 + s^2/c^2}}$$

Selecting the origin O of the coordinates at a distance c directly below C (Fig. 7.18*a*) and integrating from $C(0,c)$ to $D(x,y)$, we obtain†

$$x = \int_0^s \frac{ds}{\sqrt{1 + s^2/c^2}} = c\left[\sinh^{-1}\frac{s}{c}\right]_0^s = c \sinh^{-1}\frac{s}{c}$$

This equation, which relates the length s of the portion of cable CD and the horizontal distance x, may be written in the form

$$s = c \sinh\frac{x}{c} \tag{7.15}$$

The relation between the coordinates x and y may now be obtained by writing $dy = dx \tan\theta$. Observing from Fig. 7.18*c* that $\tan\theta = W/T_0$ and using (7.11) and (7.15), we write

$$dy = dx \tan\theta = \frac{W}{T_0}\, dx = \frac{s}{c}\, dx = \sinh\frac{x}{c}\, dx$$

Integrating from $C(0,c)$ to $D(x,y)$ and using (7.12) and (7.13), we obtain

† This integral may be found in all standard integral tables. The function

$$z = \sinh^{-1} u$$

(read "arc hyperbolic sine u") is the *inverse* of the function $u = \sinh z$ (read "hyperbolic sine z"). This function and the function $v = \cosh z$ (read "hyperbolic cosine z") are defined as follows:

$$u = \sinh z = \tfrac{1}{2}(e^z - e^{-z}) \qquad v = \cosh z = \tfrac{1}{2}(e^z + e^{-z})$$

Numerical values of the functions $\sinh z$ and $\cosh z$ are found in *tables of hyperbolic functions*. They may also be computed on most calculators either directly or from the above definitions. The student is referred to any calculus text for a complete description of the properties of these functions. In this section, we shall make use only of the following properties, which may be easily derived from the above definitions:

$$\frac{d \sinh z}{dz} = \cosh z \qquad \frac{d \cosh z}{dz} = \sinh z \tag{7.12}$$

$$\sinh 0 = 0 \qquad \cosh 0 = 1 \tag{7.13}$$

$$\cosh^2 z - \sinh^2 z = 1 \tag{7.14}$$

$$y - c = \int_0^x \sinh \frac{x}{c}\, dx = c\left[\cosh \frac{x}{c}\right]_0^x = c\left(\cosh \frac{x}{c} - 1\right)$$

$$y = c \cosh \frac{x}{c} \tag{7.16}$$

This is the equation of a *catenary* with vertical axis. The ordinate c of the lowest point C is called the *parameter* of the catenary. Squaring both sides of Eqs. (7.15) and (7.16), subtracting, and taking (7.14) into account, we obtain the following relation between y and s:

$$y^2 - s^2 = c^2 \tag{7.17}$$

Solving (7.17) for s^2 and carrying into the last of the relations (7.11), we write these relations as follows:

$$T_0 = wc \qquad W = ws \qquad T = wy \tag{7.18}$$

The last relation indicates that the tension at any point D of the cable is proportional to the vertical distance from D to the horizontal line representing the x axis.

When the supports A and B of the cable have the same elevation, the distance L between the supports is called the *span* of the cable and the vertical distance h from the supports to the lowest point C is called the *sag* of the cable. These definitions are the same that were given in the case of parabolic cables, but it should be noted that, because of our choice of coordinate axes, the sag h is now

$$h = y_A - c \tag{7.19}$$

It should also be observed that certain catenary problems involve transcendental equations which must be solved by successive approximations (see Sample Prob. 7.10). When the cable is fairly taut, however, the load may be assumed uniformly distributed *along the horizontal* and the catenary may be replaced by a parabola. The solution of the problem is thus greatly simplified, while the error introduced is small.

When the supports A and B have different elevations, the position of the lowest point of the cable is not known. The problem may be solved then in a manner similar to that indicated for parabolic cables, by expressing that the cable must pass through the supports and that $x_B - x_A = L$, $y_B - y_A = d$, where L and d denote, respectively, the horizontal and vertical distances between the two supports.

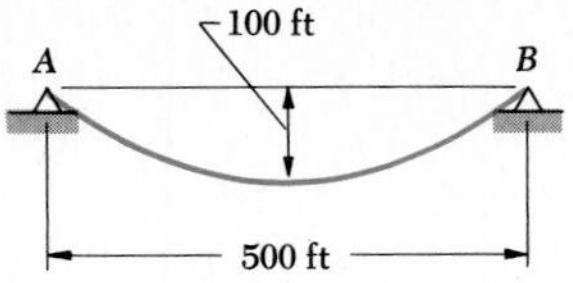

SAMPLE PROBLEM 7.10

A uniform cable weighing 3 lb/ft is suspended between two points A and B as shown. Determine (a) the maximum and minimum values of the tension in the cable, (b) the length of the cable.

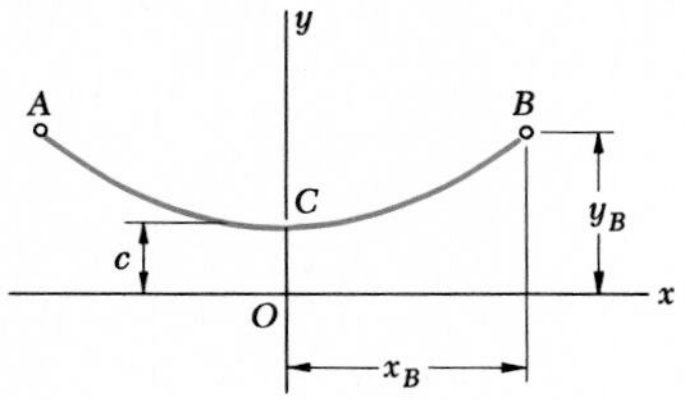

Solution. *Equation of Cable.* The origin of coordinates is placed at a distance c below the lowest point of the cable. The equation of the cable is given by Eq. (7.16),

$$y = c \cosh \frac{x}{c}$$

The coordinates of point B are

$$x_B = 250 \text{ ft} \qquad y_B = 100 + c$$

Substituting these coordinates into the equation of the cable, we obtain

$$100 + c = c \cosh \frac{250}{c}$$

$$\frac{100}{c} + 1 = \cosh \frac{250}{c}$$

The value of c is determined by assuming successive trial values, as shown in the following table:

c	$\frac{250}{c}$	$\frac{100}{c}$	$\frac{100}{c} + 1$	$\cosh \frac{250}{c}$
300	0.833	0.333	1.333	1.367
350	0.714	0.286	1.286	1.266
330	0.758	0.303	1.303	1.301
328	0.762	0.305	1.305	1.305

Taking $c = 328$, we have

$$y_B = 100 + c = 428 \text{ ft}$$

a. Maximum and Minimum Values of the Tension. Using Eqs. (7.18), we obtain

$$T_{min} = T_0 = wc = (3 \text{ lb/ft})(328 \text{ ft}) \qquad T_{min} = 984 \text{ lb} \blacktriangleleft$$
$$T_{max} = T_B = wy_B = (3 \text{ lb/ft})(428 \text{ ft}) \qquad T_{max} = 1284 \text{ lb} \blacktriangleleft$$

b. Length of Cable. One-half the length of the cable is found by solving Eq. (7.17),

$$y_B^2 - s_{CB}^2 = c^2 \qquad s_{CB}^2 = y_B^2 - c^2 = (428)^2 - (328)^2 \qquad s_{CB} = 275 \text{ ft}$$

The total length of the cable is therefore

$$s_{AB} = 2s_{CB} = 2(275 \text{ ft}) \qquad s_{AB} = 550 \text{ ft} \blacktriangleleft$$

PROBLEMS

7.83 An aerial tramway cable of length 200 m and mass 1000 kg is suspended between two points at the same elevation. Knowing that the sag is 50 m, find the horizontal distance between supports and the maximum tension.

7.84 A 100-ft rope is strung between the roofs of two buildings, each of height 30 ft. The maximum tension is found to be 50 lb and the lowest point of the cable is observed to be 10 ft above the ground. Determine the horizontal distance between the buildings and the total weight of the rope.

7.85 A 200-ft steel surveying tape weighs 4 lb. If the tape is stretched between two points at the same elevation and pulled until the tension at each end is 16 lb, determine the horizontal distance between the ends of the tape. Neglect the elongation of the tape due to the tension.

7.86 A copper transmission wire having a mass per unit length of 0.75 kg/m is attached to two insulators at the same elevation and 100 m apart. It has been established that the horizontal component of the tension in the wire must be 900 N if the insulators are not to be bent. Determine (*a*) the length of cable which should be used, (*b*) the resulting sag, (*c*) the resulting maximum tension.

7.87 A tramway wire rope weighing 3.4 lb/ft is suspended across a canyon; it is attached to two supports at the same elevation and 600 ft apart. Knowing that the sag is 80 ft, determine the total length of the cable and the maximum tension.

7.88 A counterweight of mass 200 kg is attached to a cable which passes over a small pulley at *A* and is attached to a support at *B*. If the sag is 3 m, determine (*a*) the length of the cable from *A* to *B*, (*b*) the mass per unit length of the cable. Neglect the mass of the cable from *A* to *D*.

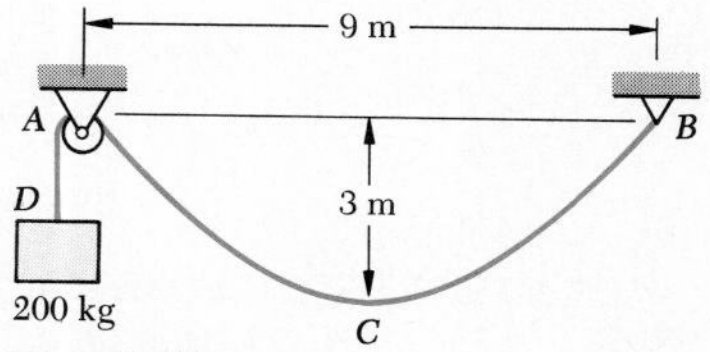

Fig. P7.88

7.89 A chain of length 10 m and total mass 20 kg is suspended between two points at the same elevation and 5 m apart. Determine the sag and the maximum tension.

7.90 A 150-ft wire is suspended between two points at the same elevation and 100 ft apart. Knowing that the maximum tension is 30 lb, determine the sag and the total weight of the wire.

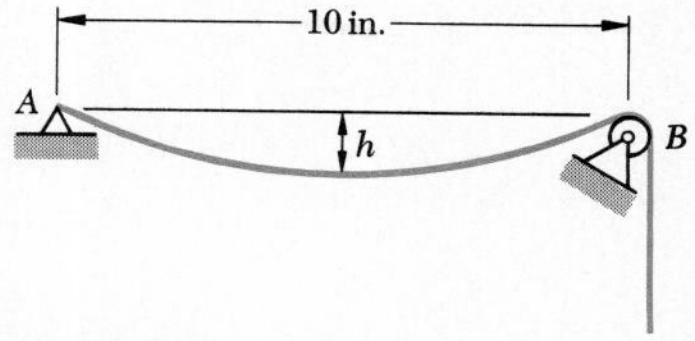

Fig. P7.91

7.91 A uniform cord 30 in. long passes over a frictionless pulley at *B* and is attached to a rigid support at *A*. Determine the smaller of the two values of *h* for which the cord is in equilibrium.

***7.92** The 10-ft cable *AB* weighs 20 lb and is attached to collars at *A* and *B* which may slide freely on the rods shown. Neglecting the weight of the collars, determine (*a*) the magnitude of the horizontal force **F** so that $h = a$, (*b*) the corresponding value of *h* and *a*, (*c*) the maximum tension in the cable.

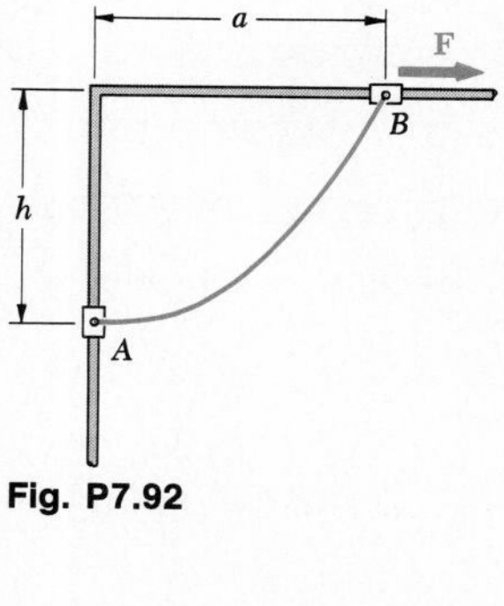

Fig. P7.92

7.93 Denoting by θ the angle formed by a uniform cable and the horizontal, show that at any point $y = c/\cos\theta$.

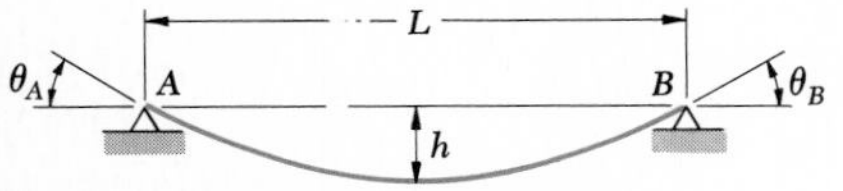

Fig. P7.95, P7.96, and P7.97

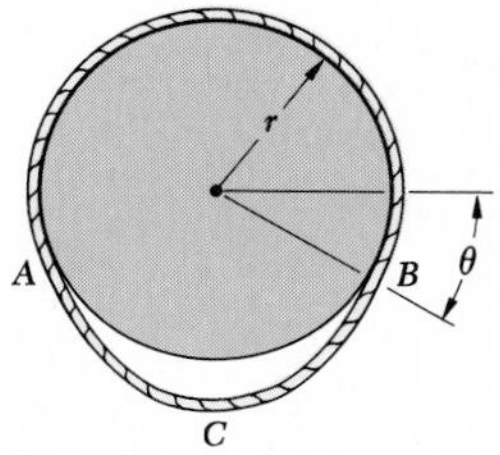

Fig. P7.98

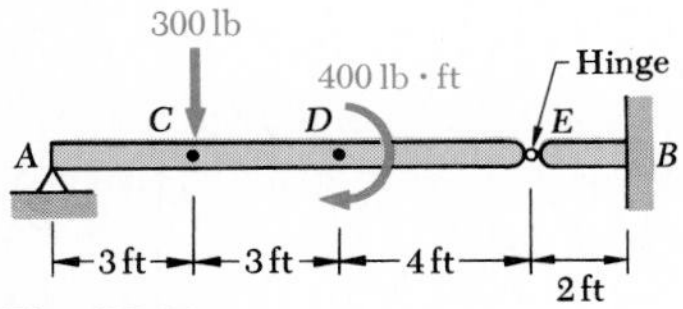

Fig. P7.99

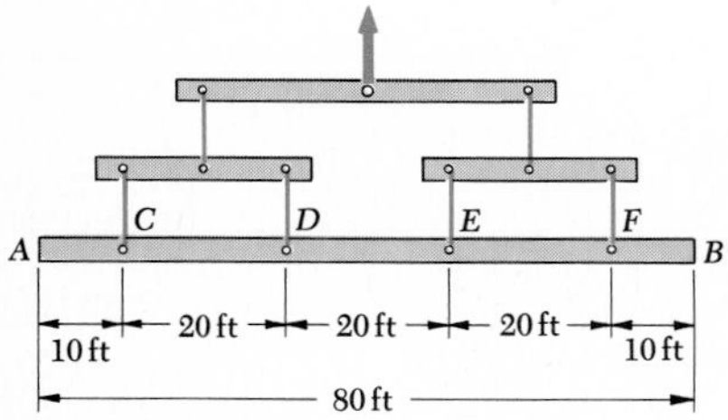

Fig. P7.102

***7.94** (*a*) Determine the maximum allowable horizontal span for a uniform cable of weight w per unit length if the tension in the cable is not to exceed the value T_m. (*b*) Using the result of part *a*, find the maximum span of a drawn-steel wire for which $w = 0.2$ lb/ft and $T_m = 6000$ lb.

***7.95** A chain has a mass per unit length of 6 kg/m and is supported as shown. Knowing that the span L is 8 m, determine the *two* values of the sag h for which the maximum tension is 750 N.

***7.96** Determine the sag-to-span ratio for which the maximum tension in the cable is equal to the total weight of the entire cable AB.

***7.97** A cable, of weight w per unit length, is suspended between two points at the same elevation and a distance L apart. Determine the sag-to-span ratio for which the maximum tension is as small as possible. What are the corresponding values of θ_B and T_m?

***7.98** A cable, of weight w per unit length, is looped over a cylinder and is in contact with the cylinder above points A and B. Knowing that $\theta = 30°$, determine (*a*) the length of the cable, (*b*) the tension in the cable at C. (*Hint.* Use the property indicated in Prob. 7.93.)

REVIEW PROBLEMS

7.99 Draw the shear and bending-moment diagrams for the beam and loading shown.

7.100 Draw the shear and bending-moment diagrams for the beam and loading of Prob. 7.99 if the 400 lb · ft clockwise couple applied at D is replaced by a 400 lb · ft counterclockwise couple applied at the same point.

7.101 A $\frac{3}{4}$-in.-diameter wire rope weighing 0.90 lb/ft is suspended from two supports at the same elevation and 300 ft apart. If the sag is 75 ft, determine (*a*) the total length of the cable, (*b*) the maximum tension.

7.102 Concrete piles are designed primarily to resist axial loads, and their bending resistance is relatively small. To lessen the possibility of breaking them, piles are often lifted at several points along their length. By using the arrangement shown, the concrete pile AB is lifted by four equal forces applied at points C, D, E, and F. If the total weight of the pile is 20,000 lb, draw the shear and bending-moment diagrams of the pile. Assume the weight of the pile to be uniformly distributed.

7.103 The central portion of cable AB is supported by a frictionless horizontal surface. Knowing that the mass per unit length of the cable is 3 kg/m and assuming the cable to be parabolic, determine the magnitude of the load **P** when $a = 8$ m.

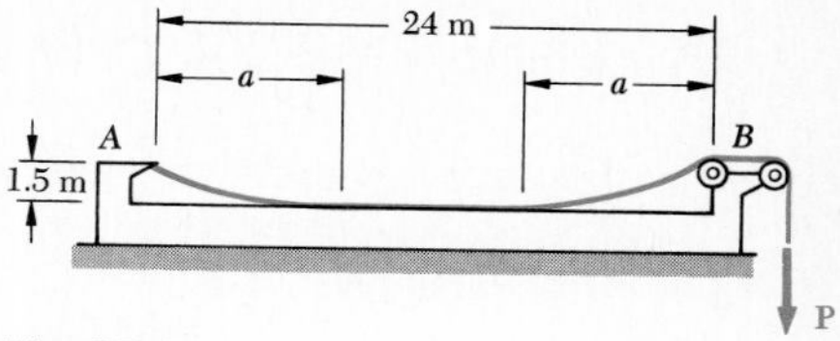

Fig. P7.103

7.104 Solve Prob. 7.103 when (*a*) $a = 6$ m, (*b*) $a = 12$ m.

7.105 Cable ABC supports two loads as shown. Determine (*a*) the required magnitude of the force **P**, (*b*) the corresponding distance a.

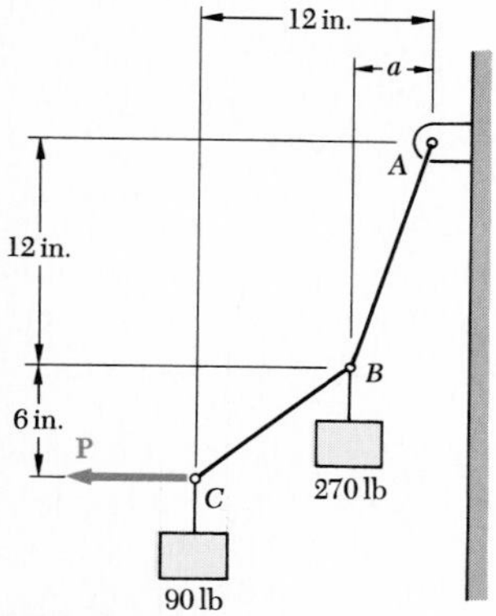

Fig. P7.105

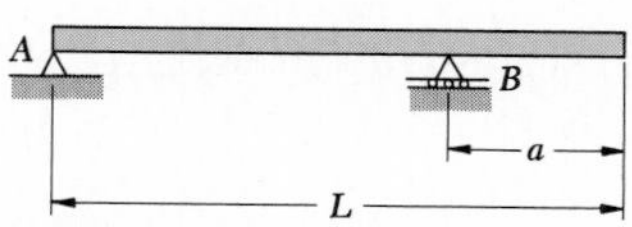

Fig. P7.106

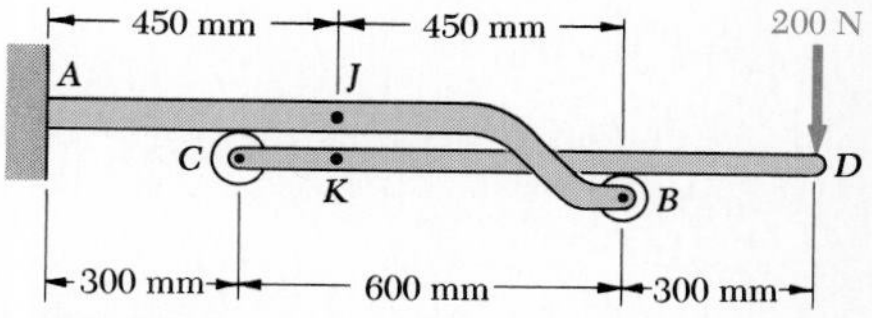

Fig. P7.107

***7.106** A uniform beam of weight w per unit length is supported as shown. Determine the distance a if the maximum absolute value of the bending moment is to be as small as possible. (See hint of Prob. 7.35.)

7.107 Determine the internal forces at points J and K of the adjustable hanger shown.

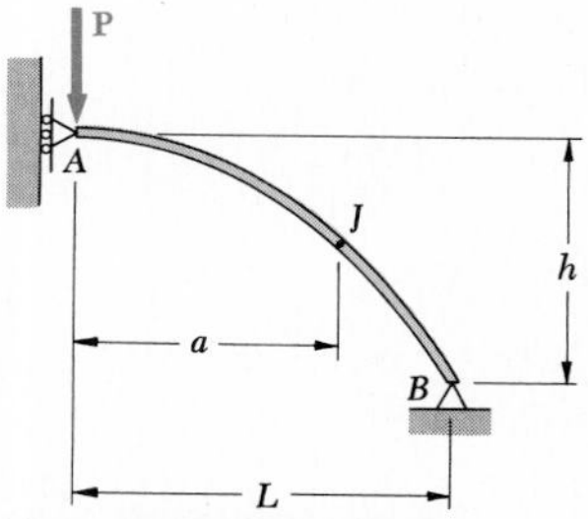

Fig. P7.108 and P7.109

7.108 The axis of the curved member AB is a parabola with vertex at A. If a vertical load **P** of magnitude 3 kN is applied at A, determine the internal forces at J when $h = 225$ mm, $L = 750$ mm, and $a = 500$ mm.

7.109 The axis of the curved member AB is a parabola with vertex at A. (*a*) Determine the magnitude and location of the maximum bending moment. (*b*) Noting that AB is a two-force member, determine the maximum perpendicular distance from the chord AB to the curved member.

***7.110** A uniform semicircular rod of weight W is attached to a pin at A and bears against a frictionless surface at B. Determine (*a*) the bending moment at point J when $\theta = 90°$, (*b*) the magnitude and location of the maximum bending moment.

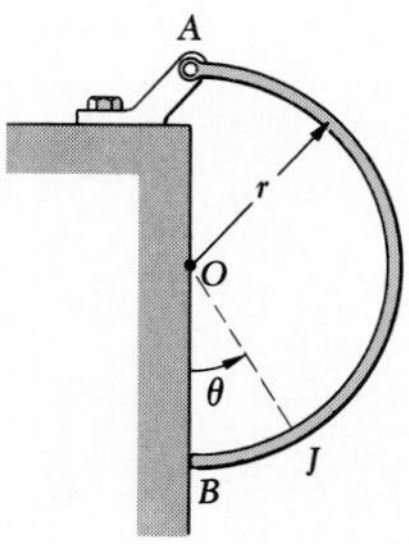

Fig. P7.110

Chapter
8

Friction

8.1. Introduction. In the preceding chapters, it was assumed that surfaces in contact were either frictionless or *rough*. If they were frictionless, the force each surface exerted on the other was normal to the surfaces and the two surfaces could move freely with respect to each other. If they were rough, it was assumed that tangential forces could develop to prevent the motion of one surface with respect to the other.

This view was a simplified one. Actually, no perfectly frictionless surface exists. When two surfaces are in contact, tangential forces, called *friction forces,* will always develop if one attempts to move one surface with respect to the other. On the other hand, these friction forces are limited in magnitude and will not prevent motion if sufficiently large forces are applied. The distinction between frictionless and rough surfaces is thus a matter of degree. This will be seen more clearly in the present chapter, which is devoted to the study of friction and of its applications to common engineering situations.

There are two types of friction: *dry friction,* sometimes called *Coulomb friction,* and *fluid friction.* Fluid friction develops between layers of fluid moving at different velocities. Fluid friction is of great importance in problems involving the flow of fluids through pipes and orifices or dealing with bodies immersed in moving fluids. It is also basic in the analysis of the motion of *lubricated mechanisms*. Such problems are considered in texts on fluid mechanics. We shall limit our present study to dry friction, i.e., to problems involving rigid bodies which are in contact along *nonlubricated* surfaces.

8.2. The Laws of Dry Friction. Coefficients of Friction. The laws of dry friction are best understood by the following experiment. A block of weight **W** is placed on a horizontal plane surface (Fig. 8.1*a*). The forces acting on the block are its weight **W** and the reaction of the surface. Since the weight has no horizontal component, the reaction of the surface also has no horizontal component; the reaction is therefore *normal* to the surface and is represented by **N** in Fig. 8.1*a*. Suppose, now, that a horizontal force **P** is applied to the block (Fig. 8.1*b*). If **P** is small, the block will not move; some other horizontal force must therefore exist, which balances **P**. This other force is the *static-friction force* **F**, which is actually the resultant of a great number of forces acting over the entire surface of contact between the block and the plane. The nature of these forces is not known exactly, but it is generally assumed that these forces are due to the irregularities of the surfaces in contact and also, to a certain extent, to molecular attraction.

If the force **P** is increased, the friction force **F** also increases, continuing to oppose **P**, until its magnitude reaches a certain *maximum value* F_m (Fig. 8.1*c*). If **P** is further increased, the friction force cannot balance it any more and the block starts sliding. As soon as the block has been set in motion, the magnitude of **F** drops from F_m to a lower value F_k. This is because there is less interpenetration between the irregularities of the surfaces in contact when these surfaces move with respect to each other. From then on, the block keeps sliding with increasing velocity while the friction force, denoted by $\mathbf{F}_k$ and called the *kinetic-friction force*, remains approximately constant.

Experimental evidence shows that the maximum value F_m of the static-friction force is proportional to the normal component N of the reaction of the surface. We have

$$F_m = \mu_s N \tag{8.1}$$

where μ_s is a constant called the *coefficient of static friction*. Similarly, the magnitude F_k of the kinetic-friction force may be put in the form

$$F_k = \mu_k N \tag{8.2}$$

where μ_k is a constant called the *coefficient of kinetic friction*.

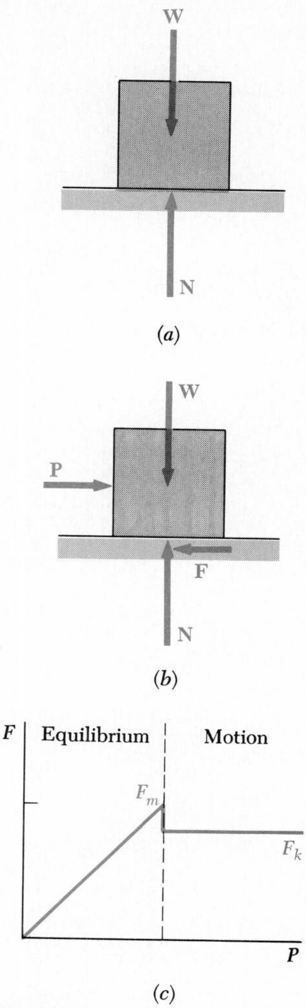

Fig. 8.1

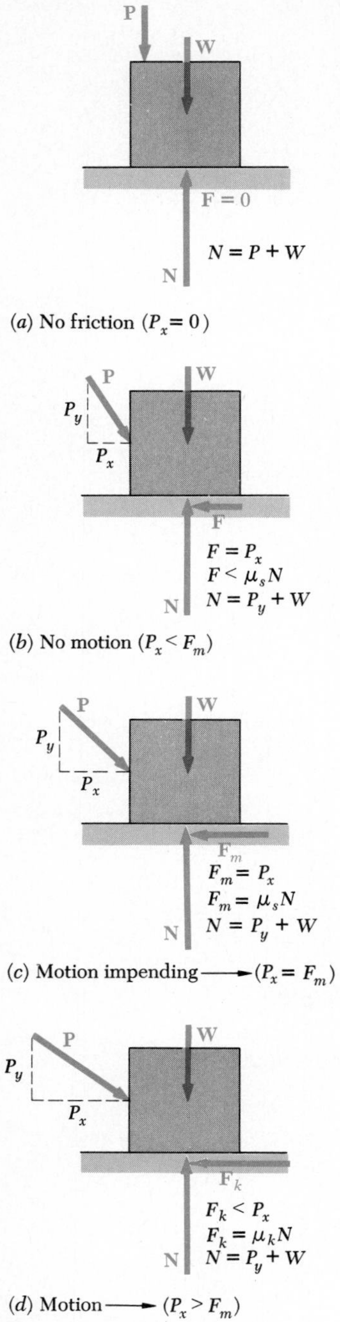

Fig. 8.2

The coefficients of friction μ_s and μ_k do not depend upon the area of the surfaces in contact. Both coefficients, however, depend strongly on the *nature* of the surfaces in contact. Since they also depend upon the exact condition of the surfaces, their value is seldom known with an accuracy greater than 5 percent. Approximate values of coefficients of static friction are given in Table 8.1 for various dry surfaces. The corresponding values

Table 8.1. Approximate Values of Coefficient of Static Friction for Dry Surfaces

Metal on metal	0.15–0.60
Metal on wood	0.20–0.60
Metal on stone	0.30–0.70
Metal on leather	0.30–0.60
Wood on wood	0.25–0.50
Wood on leather	0.25–0.50
Stone on stone	0.40–0.70
Earth on earth	0.20–1.00
Rubber on concrete	0.60–0.90

of the coefficient of kinetic friction would be about 25 percent smaller. Since coefficients of friction are dimensionless quantities, the values given in Table 8.1 may be used with both SI units and U.S. customary units.

From the description given above, it appears that four different situations may occur when a rigid body is in contact with a horizontal surface:

1. The forces applied to the body do not tend to move it along the surface of contact; there is no friction force (Fig. 8.2*a*).
2. The applied forces tend to move the body along the surface of contact but are not large enough to set it in motion. The friction force **F** which has developed may be found by solving the equations of equilibrium for the body. Since there is no evidence that the maximum value of the static-friction force has been reached, the equation $F_m = \mu_s N$ *cannot be used* to determine the friction force (Fig. 8.2*b*).

3. The applied forces are such that the body is just about to slide. We say that *motion is impending*. The friction force **F** has reached its maximum value F_m and, together with the normal force **N**, balances the applied forces. Both the equations of equilibrium and the equation $F_m = \mu_s N$ *may be used*. We also note that the friction force has a sense opposite to the sense of impending motion (Fig. 8.2*c*).
4. The body is sliding under the action of the applied forces, and the equations of equilibrium do not apply any more. However, **F** is now equal to the kinetic-friction force $\mathbf{F}_k$ and the equation $F_k = \mu_k N$ may be used. The sense of $\mathbf{F}_k$ is opposite to the sense of motion (Fig. 8.2*d*).

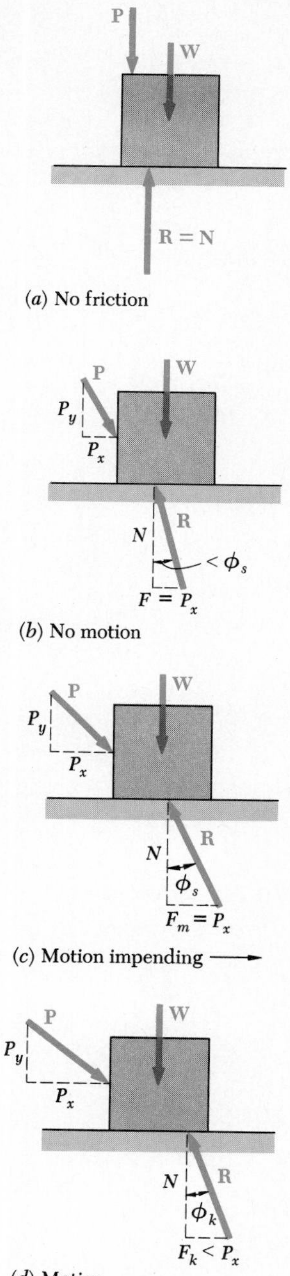

(*a*) No friction

(*b*) No motion

(*c*) Motion impending ⟶

(*d*) Motion ⟶

Fig. 8.3

8.3. Angles of Friction. It is sometimes found convenient to replace the normal force **N** and the friction force **F** by their resultant **R**. Let us consider again a block of weight **W** resting on a horizontal plane surface. If no horizontal force is applied to the block, the resultant **R** reduces to the normal force **N** (Fig. 8.3*a*). However, if the applied force **P** has a horizontal component $\mathbf{P}_x$ which tends to move the block, the force **R** will have a horizontal component **F** and, thus, will form a certain angle with the vertical (Fig. 8.3*b*). If $\mathbf{P}_x$ is increased until motion becomes impending, the angle between **R** and the vertical grows and reaches a maximum value (Fig. 8.3*c*). This value is called the *angle of static friction* and is denoted by ϕ_s. From the geometry of Fig. 8.3*c*, we note that

$$\tan \phi_s = \frac{F_m}{N} = \frac{\mu_s N}{N}$$

$$\tan \phi_s = \mu_s \tag{8.3}$$

If motion actually takes place, the magnitude of the friction force drops to F_k; similarly, the angle between **R** and **N** drops to a lower value ϕ_k, called the *angle of kinetic friction* (Fig. 8.3*d*). From the geometry of Fig. 8.3*d*, we write

$$\tan \phi_k = \frac{F_k}{N} = \frac{\mu_k N}{N}$$

$$\tan \phi_k = \mu_k \tag{8.4}$$

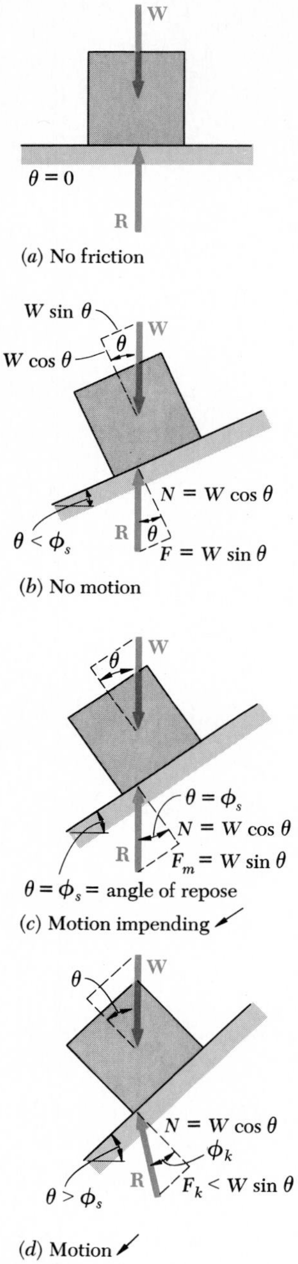

(*a*) No friction

(*b*) No motion

(*c*) Motion impending

(*d*) Motion

Fig. 8.4

Another example will show how the angle of friction may be used to advantage in the analysis of certain types of problems. Consider a block resting on a board which may be given any desired inclination; the block is subjected to no other force than its weight **W** and the reaction **R** of the board. If the board is horizontal, the force **R** exerted by the board on the block is perpendicular to the board and balances the weight **W** (Fig. 8.4*a*). If the board is given a small angle of inclination θ, the force **R** will deviate from the perpendicular to the board by the angle θ and will keep balancing **W** (Fig. 8.4*b*); it will then have a normal component **N** of magnitude $N = W \cos \theta$ and a tangential component **F** of magnitude $F = W \sin \theta$.

If we keep increasing the angle of inclination, motion will soon become impending. At that time, the angle between **R** and the normal will have reached its maximum value ϕ_s (Fig. 8.4*c*). The value of the angle of inclination corresponding to impending motion is called the *angle of repose*. Clearly, the angle of repose is equal to the angle of static friction ϕ_s. If the angle of inclination θ is further increased, motion starts and the angle between **R** and the normal drops to the lower value ϕ_k (Fig. 8.4*d*). The reaction **R** is not vertical any more, and the forces acting on the block are unbalanced.

8.4. Problems Involving Dry Friction. Problems involving dry friction are found in many engineering applications. Some deal with simple situations such as the block sliding on a plane described in the preceding sections. Others involve more complicated situations as in Sample Prob. 8.3; many deal with the stability of rigid bodies in accelerated motion and will be studied in dynamics. Also, a number of common machines and mechanisms may be analyzed by applying the laws of dry friction. These include wedges, screws, journal and thrust bearings, and belt transmissions. They will be studied in the following sections.

The *methods* which should be used to solve problems involving dry friction are the same that were used in the preceding chapters. If a problem involves only a motion of translation, with no possible rotation, the body under consideration may usually be treated as a particle, and the methods of Chap. 2 may be used. If the problem involves a possible rotation, the body must be considered as a rigid body and the methods of Chap. 3 should be used. If the structure considered is made of several parts, the principle of action and reaction must be used as was done in Chap. 6.

If the body considered is acted upon by more than three forces (including the reactions at the surfaces of contact), the reaction

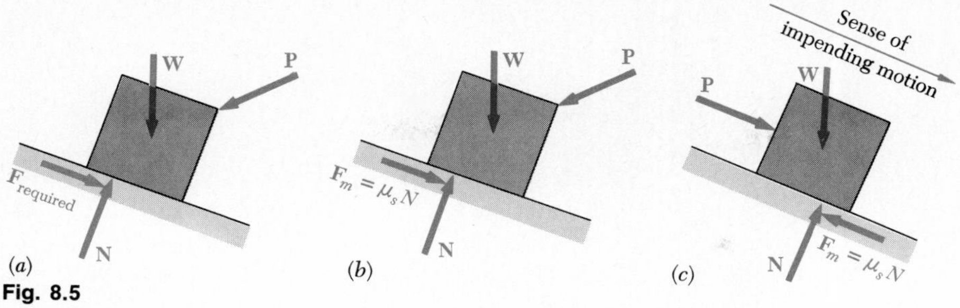

Fig. 8.5

at each surface will be represented by its components **N** and **F** and the problem will be solved from the equations of equilibrium. If only three forces act on the body under consideration, it may be found more convenient to represent each reaction by the single force **R** and to solve the problem by drawing a force triangle.

Most problems involving friction fall into one of the following *three groups:* In the *first group* of problems, all applied forces are given, and the coefficients of friction are known; we are to determine whether the body considered will remain at rest or slide. The friction force **F** *required to maintain equilibrium* is unknown (its magnitude is *not* equal to $\mu_s N$) and should be determined, together with the normal force **N**, by drawing a free-body diagram and *solving the equations of equilibrium* (Fig. 8.5*a*). The value found for the magnitude F of the friction force is then compared with the maximum value $F_m = \mu_s N$. If F is smaller than or equal to F_m, the body remains at rest. If the value found for F is larger than F_m, equilibrium cannot be maintained and motion takes place; the actual magnitude of the friction force is then $F_k = \mu_k N$.

In problems of the *second group,* all applied forces are given, and the motion is known to be impending; we are to determine the value of the coefficient of static friction. Here again, we determine the friction force and the normal force by drawing a free-body diagram and solving the equations of equilibrium (Fig. 8.5*b*). Since we know that the value found for F is the maximum value F_m, the coefficient of friction may be found by writing and solving the equation $F_m = \mu_s N$.

In problems of the *third group,* the coefficient of static friction is given, and it is known that motion is impending in a given direction; we are to determine the magnitude or the direction of one of the applied forces. The friction force should be shown in the free-body diagram with a *sense opposite to that of the impending motion* and with a magnitude $F_m = \mu_s N$ (Fig. 8.5*c*).

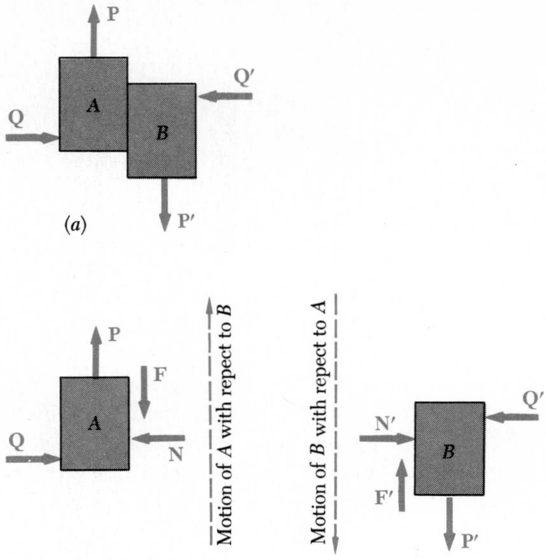

(*a*)

(*b*) (*c*)

Fig. 8.6

The equations of equilibrium may then be written, and the desired force may be determined.

As noted above, it may be more convenient, when only three forces are involved, to represent the reaction of the surface by a single force **R** and to solve the problem by drawing a force triangle. Such a solution is used in Sample Prob. 8.2.

When two bodies A and B are in contact (Fig. 8.6*a*), the forces of friction exerted, respectively, by A on B and by B on A are equal and opposite (Newton's third law). It is important, in drawing the free-body diagram of one of the bodies, to include the appropriate friction force with its correct sense. The following rule should then be observed: *The sense of the friction force acting on A is opposite to that of the motion (or impending motion) of A as observed from B* (Fig. 8.6*b*).† The sense of the friction force acting on B is determined in a similar way (Fig. 8.6*c*). Note that the motion of A as observed from B is a *relative motion*. Body A may be fixed; yet it will have a relative motion with respect to B if B itself moves. Also, A may actually move down yet be observed from B to move up if B moves down faster than A.

† It is therefore *the same as that of the motion of B as observed from A.*

SAMPLE PROBLEM 8.1

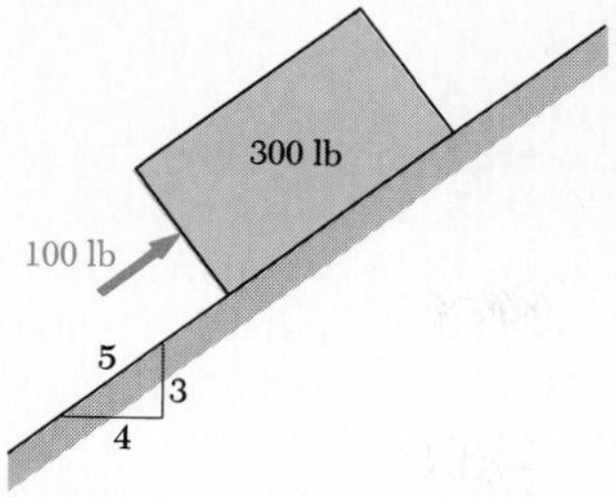

A 100-lb force acts as shown on a 300-lb block placed on an inclined plane. The coefficients of friction between the block and the plane are $\mu_s = 0.25$ and $\mu_k = 0.20$. Determine whether the block is in equilibrium, and find the value of the friction force.

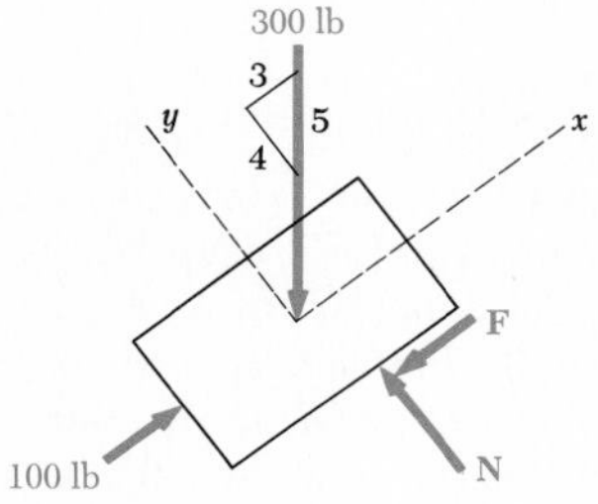

Force Required for Equilibrium. We first determine the value of the friction force *required to maintain equilibrium.* Assuming that **F** is directed down and to the left, we draw the free-body diagram of the block and write

$$+\nearrow \Sigma F_x = 0: \quad 100 \text{ lb} - \tfrac{3}{5}(300 \text{ lb}) - F = 0$$
$$F = -80 \text{ lb} \qquad \mathbf{F} = 80 \text{ lb} \nearrow$$

$$+\nwarrow \Sigma F_y = 0: \quad N - \tfrac{4}{5}(300 \text{ lb}) = 0$$
$$N = +240 \text{ lb} \qquad \mathbf{N} = 240 \text{ lb} \nwarrow$$

The force **F** required to maintain equilibrium is an 80-lb force directed up and to the right; the tendency of the block is thus to move down the plane.

Maximum Friction Force. The magnitude of the maximum friction force which may be developed is

$$F_{\text{max}} = \mu_s N \qquad F_{\text{max}} = 0.25(240 \text{ lb}) = 60 \text{ lb}$$

Since the value of the force required to maintain equilibrium (80 lb) is larger than the maximum value which may be obtained (60 lb), equilibrium will not be maintained and *the block will slide down the plane.*

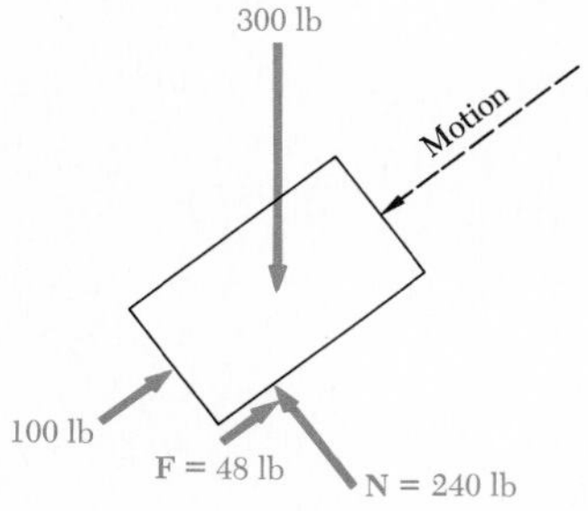

Actual Value of Friction Force. The magnitude of the actual friction force is obtained as follows:

$$F_{\text{actual}} = F_k = \mu_k N$$
$$= 0.20(240 \text{ lb}) = 48 \text{ lb}$$

The sense of this force is opposite to the sense of motion; the force is thus directed up and to the right,

$$\mathbf{F}_{\text{actual}} = 48 \text{ lb} \nearrow$$ ◄

It should be noted that the forces acting on the block are not balanced; the resultant is

$$\tfrac{3}{5}(300 \text{ lb}) - 100 \text{ lb} - 48 \text{ lb} = 32 \text{ lb} \swarrow$$

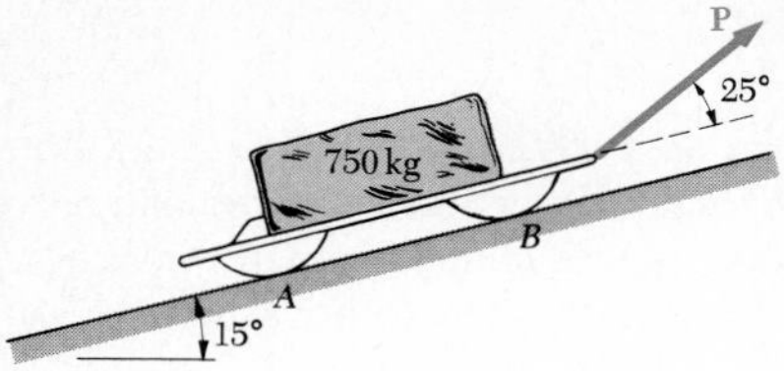

SAMPLE PROBLEM 8.2

A wooden sled supporting a large stone is pulled up a track inclined at 15°. The combined mass of the sled and stone is 750 kg, and the coefficients of friction between the sled runners and the track are $\mu_s = 0.40$ and $\mu_k = 0.30$. Determine the force **P** required (*a*) to start the sled up the track, (*b*) to keep the sled moving up after it has been started, (*c*) to keep the sled from sliding down.

Solution. The magnitude of the combined weight **W** of the sled and stone is

$$W = mg = (750\ \text{kg})(9.81\ \text{m/s}^2) = 7360\ \text{N} = 7.36\ \text{kN}$$

Noting that $\mathbf{R}_A$ and $\mathbf{R}_B$ have the same direction (same angle of friction ϕ), we draw a force triangle including the weight **W**, the force **P**, and the sum $\mathbf{R} = \mathbf{R}_A + \mathbf{R}_B$. The direction of **R** must be redetermined in each part of the problem. The law of sines is used to determine the magnitude of **P** in each part of the problem. If required, we could then determine the individual reactions $\mathbf{R}_A$ and $\mathbf{R}_B$ by summing moments about *B* and *A*, respectively.

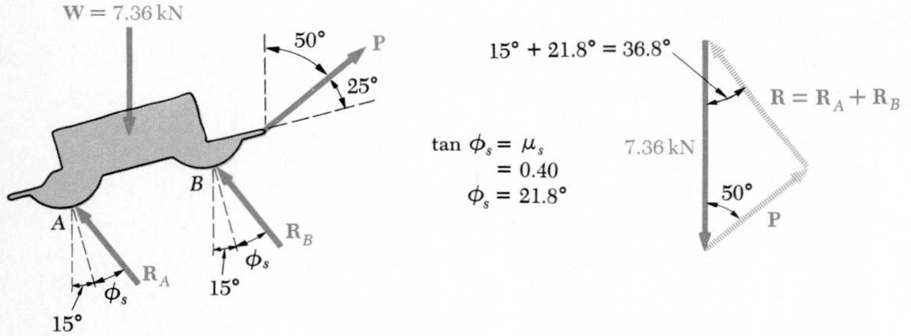

***a.* Force P to Start Sled Moving Up**

$$\frac{P}{\sin 36.8^\circ} = \frac{7.36\ \text{kN}}{\sin\,[180^\circ - (50^\circ + 36.8^\circ)]}$$

$$P = 4.42\ \text{kN} \nearrow$$ ◄

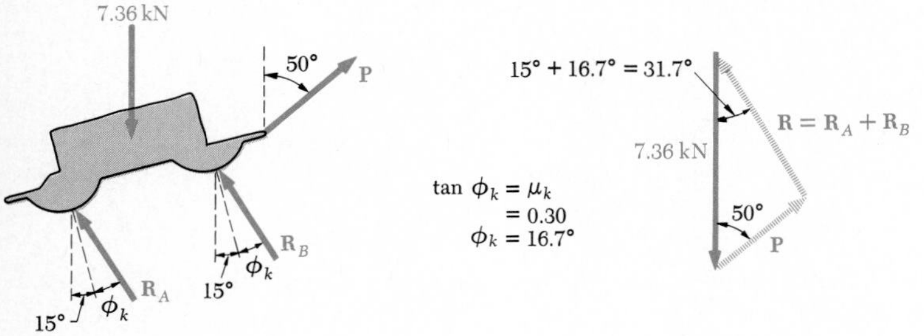

***b.* Force P to Keep Sled Moving**

$$\frac{P}{\sin 31.7^\circ} = \frac{7.36\ \text{kN}}{\sin\,[180^\circ - (50^\circ + 31.7^\circ)]}$$

$$P = 3.91\ \text{kN} \nearrow$$ ◄

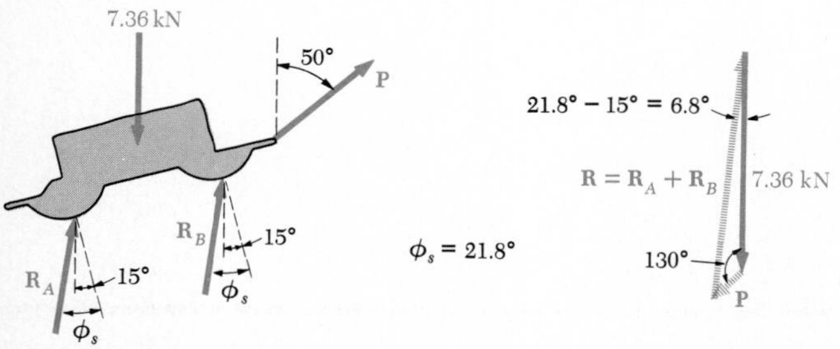

***c.* Force P to Keep Sled from Sliding Down**

$$\frac{P}{\sin 6.8^\circ} = \frac{7.36\ \text{kN}}{\sin\,[180^\circ - (130^\circ + 6.8^\circ)]}$$

$$P = 1.273\ \text{kN} \swarrow$$ ◄

Since the force **P** is directed downward, the sled will not slide down under its own weight.

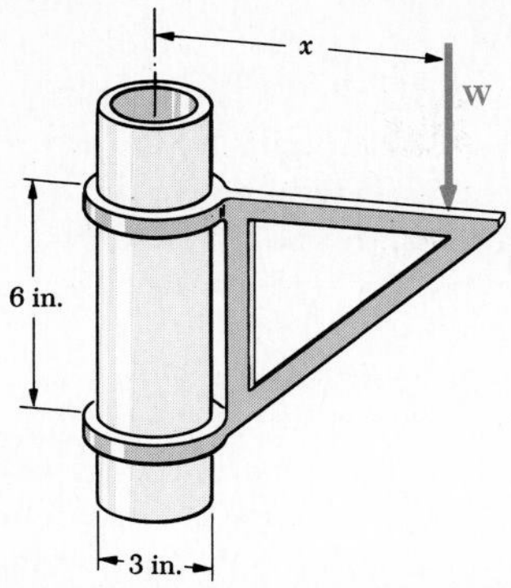

SAMPLE PROBLEM 8.3

The movable bracket shown may be placed at any height on the 3-in.-diameter pipe. If the coefficient of static friction between the pipe and bracket is 0.25, determine the minimum distance x at which the load **W** can be supported. Neglect the weight of the bracket.

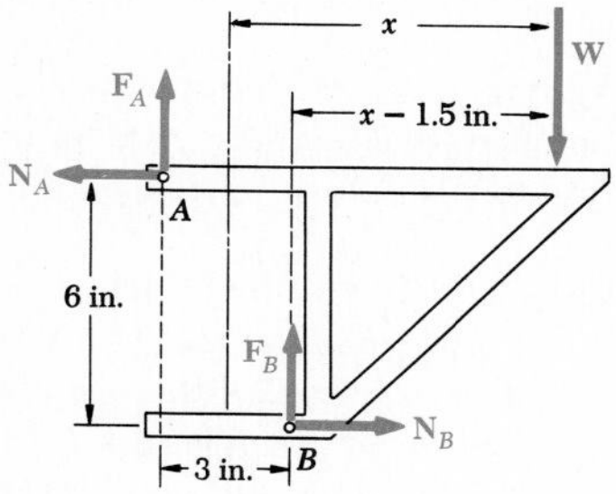

Solution. We draw the free-body diagram of the bracket. When **W** is placed at the minimum distance x from the axis of the pipe, the bracket is just about to slip, and the forces of friction at A and B have reached their maximum values:

$$F_A = \mu N_A = 0.25 N_A$$
$$F_B = \mu N_B = 0.25 N_B$$

Equilibrium Equations

$\xrightarrow{+} \Sigma F_x = 0$: $\quad N_B - N_A = 0$

$$N_B = N_A$$

$+\uparrow \Sigma F_y = 0$: $\quad F_A + F_B - W = 0$

$$0.25 N_A + 0.25 N_B = W$$

And, since N_B has been found equal to N_A,

$$0.50 N_A = W$$
$$N_A = 2W$$

$+\circlearrowleft \Sigma M_B = 0$: $\quad N_A(6 \text{ in.}) - F_A(3 \text{ in.}) - W(x - 1.5 \text{ in.}) = 0$

$$6N_A - 3(0.25 N_A) - Wx + 1.5W = 0$$
$$6(2W) - 0.75(2W) - Wx + 1.5W = 0$$

Dividing through by W and solving for x:

$$x = 12 \text{ in.} \blacktriangleleft$$

PROBLEMS

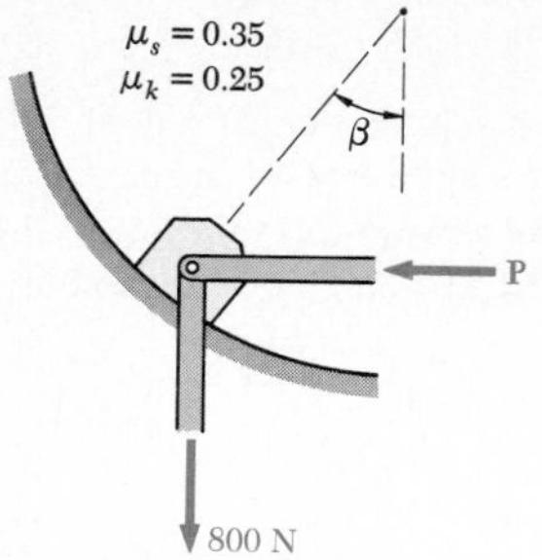

Fig. P8.1 and P8.2

8.1 The support block is acted upon by the two forces shown. For the position when $\beta = 25°$, determine (*a*) the force **P** required to start the block up the surface, (*b*) the smallest force **P** which will prevent the block from moving down the surface.

8.2 Solve Prob. 8.1 for the position when $\beta = 30°$.

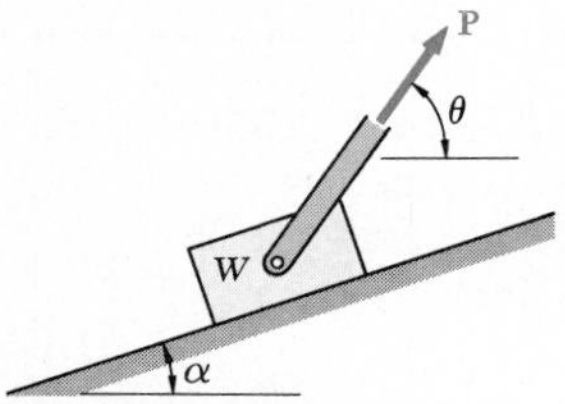

Fig. P8.3 and P8.4

8.3 A block of weight $W = 100$ lb rests on a rough plane as shown. Knowing that $\alpha = 25°$ and $\mu_s = 0.20$, determine the magnitude and direction of the smallest force **P** required (*a*) to start the block up the plane, (*b*) to prevent the block from moving down the plane.

8.4 Denoting by ϕ_s the angle of static friction between the block and the plane, determine the magnitude and direction of the smallest force **P** which will cause the block to move up the plane.

8.5 The two 25-lb boxes *A* and *B* may be attached to the lever by means of either one or two horizontal links. The coefficient of friction between all surfaces is 0.30. Determine the magnitude of the force **P** required to move the lever (*a*) if a single link *EF* is used, (*b*) if a single link *CD* is used, (*c*) if two links *EF* and *CD* are used at the same time.

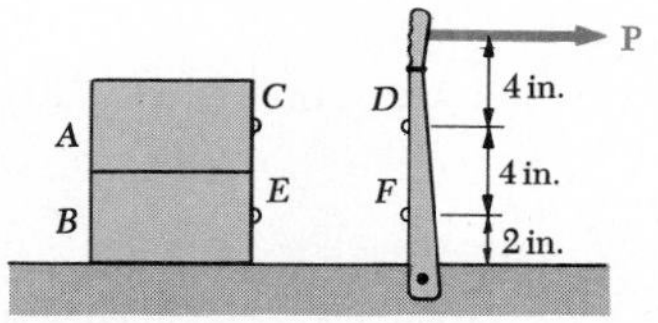

Fig. P8.5

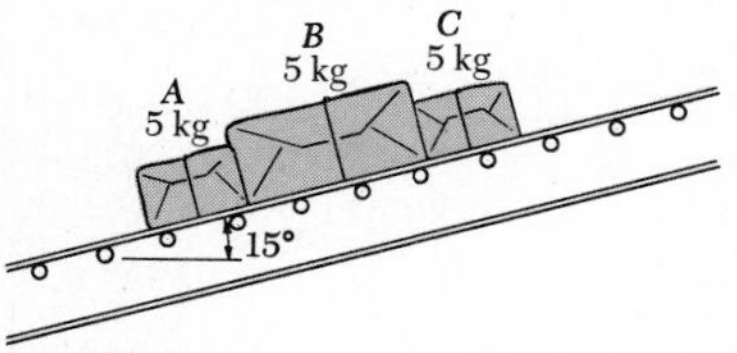

Fig. P8.6

8.6 Three packages *A*, *B*, and *C*, each of mass 5 kg, are placed on a conveyor belt which is at rest. Between the belt and both packages *A* and *C* the coefficients of friction are $\mu_s = 0.30$ and $\mu_k = 0.20$; between package *B* and the belt the coefficients are $\mu_s = 0.10$ and $\mu_k = 0.08$. The packages are placed on the belt so that they are in contact with each other and at rest. Determine which, if any, of the packages will move and the friction force acting on each package.

8.7 Solve Prob. 8.6, assuming that package *B* is placed to the right of both packages *A* and *C*.

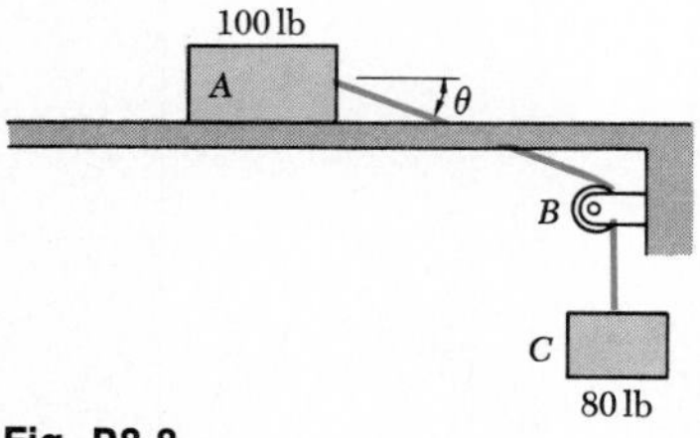

Fig. P8.8

8.8 Two blocks are connected by a cable as shown. Knowing that the coefficient of static friction between block *A* and the horizontal surface is 0.50, determine the range of values of θ for which the blocks will not move.

8.9 A 120-lb cabinet is mounted on casters which can be locked to prevent their rotation. The coefficient of friction is 0.30. If $h = 32$ in., determine the magnitude of the force $\mathbf{P}$ required to move the cabinet to the right (*a*) if all casters are locked, (*b*) if the casters at *B* are locked and the casters at *A* are free to rotate, (*c*) if casters *A* are locked and casters *B* are free to rotate.

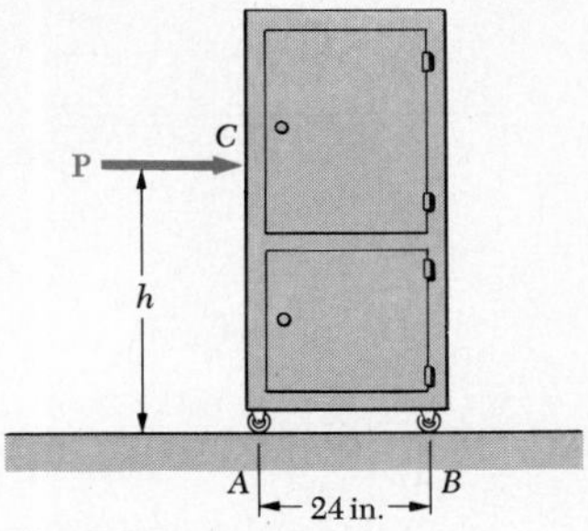

Fig. P8.9 and P8.10

8.10 A 120-lb cabinet is mounted on casters which can be locked to prevent their rotation. The coefficient of friction between the floor and each caster is 0.30. Assuming that the casters at both *A* and *B* are locked, determine (*a*) the force $\mathbf{P}$ required to move the cabinet to the right, (*b*) the largest allowable value of *h* if the cabinet is not to tip over.

8.11 A half section of pipe, of mass 100 kg, is pulled by a cable as shown. The coefficient of friction between the pipe and the floor is 0.40. If $\alpha = 30°$, determine (*a*) the tension *T* required to move the pipe, (*b*) whether the pipe will slide or tip.

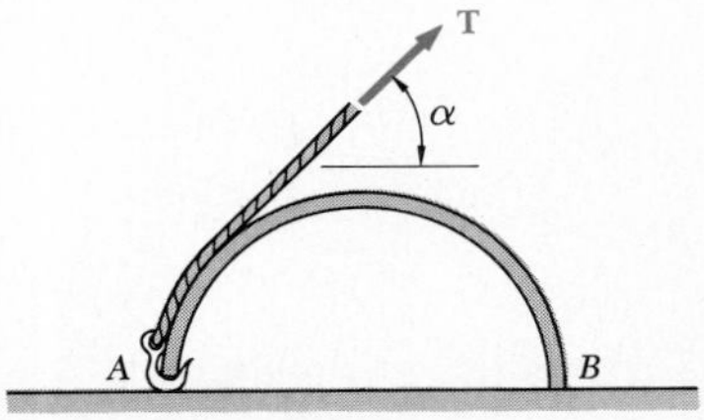

Fig. P8.11 and P8.12

8.12 A half section of pipe, of mass 100 kg, is to be moved to the right along the floor without tipping. Knowing that the coefficient of friction between the pipe and the floor is 0.40, determine (*a*) the largest allowable value of α, (*b*) the corresponding tension *T*.

8.13 A 180-lb sliding door is mounted on a horizontal rail as shown. The coefficients of static friction between the rail and the door at *A* and *B* are 0.20 and 0.30, respectively. Determine the horizontal force which must be applied to the handle *C* in order to move the door to the left.

8.14 Solve Prob. 8.13, assuming that the door is to be moved to the right.

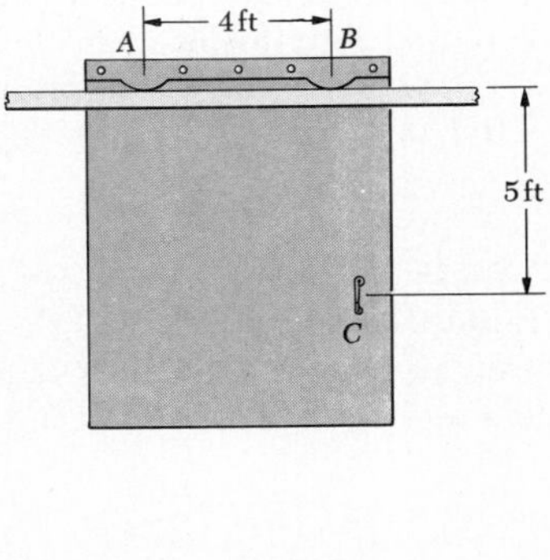

Fig. P8.13

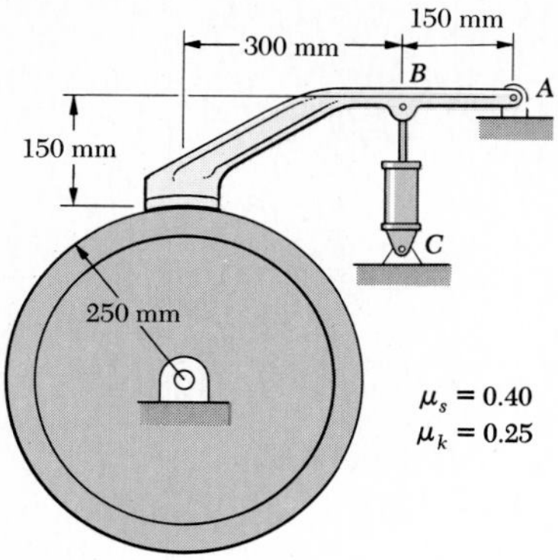

Fig. P8.15 and P8.16

8.15 A couple of moment 90 N · m is applied to the drum. Determine the smallest force which must be exerted by the hydraulic cylinder if the drum is not to rotate, when the applied couple is directed (*a*) clockwise, (*b*) counterclockwise.

8.16 The hydraulic cylinder exerts on point *B* a force of 2400 N directed downward. Determine the moment of the friction force about the axle of the drum when the drum is rotating (*a*) clockwise, (*b*) counterclockwise.

8.17 A uniform beam of weight *W* and length *L* is initially in position *AB*. As the cable is pulled over the pulley *C*, the beam first slides on the floor and is then raised, with its end *A* still sliding. Denoting by μ the coefficient of friction between the beam and the floor, determine the distance *x* that the beam moves before it begins to rise.

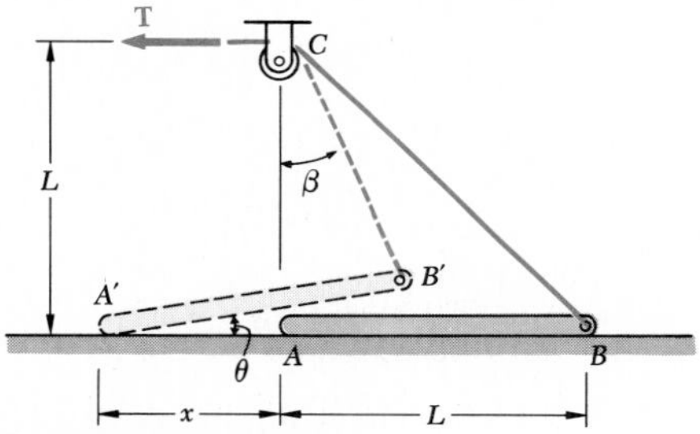

Fig. P8.17 and P8.18

8.18 The coefficient of friction between the beam of Prob. 8.17 and the floor is 0.40. The beam is slowly raised to a position *A′B′* where $\theta = 30°$. Determine the corresponding value of β.

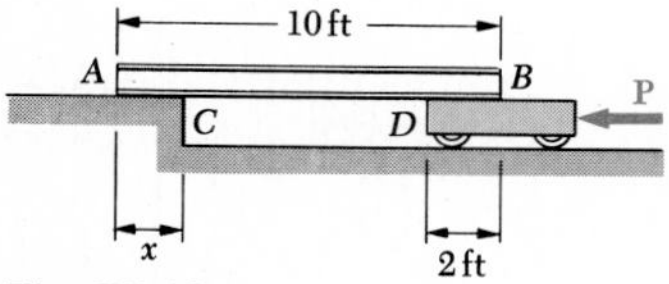

Fig. P8.19

8.19 A 10-ft beam, weighing 1200 lb, is to be moved to the left onto the platform. A horizontal force **P** is applied to the dolly, which is mounted on frictionless wheels. The coefficient of friction between all surfaces is 0.30. Knowing that the horizontal top surface of the dolly is slightly higher than the platform, determine the magnitude of **P** required to move the beam. (*Hint.* The beam is supported at *A* and *D*.)

8.20 (*a*) Show that the beam of Prob. 8.19 cannot be moved if the horizontal top surface of the dolly is slightly *lower* than the platform. (*b*) How far can the beam be moved to the left if two 175-lb men stand on the beam at *B*?

8.21 Two links, of negligible weight, are connected by frictionless pins to each other and to two blocks at A and C. The coefficient of friction is 0.25 at A and C. If neither block is to slip, determine the magnitude of the largest force **P** which can be applied at B.

8.22 In Prob. 8.21, determine the magnitude of the smallest force **P** which must be applied at B if neither block is to slip.

8.23 Two collars, each of weight W, are connected by a cord which passes over a frictionless pulley at C. Determine the smallest value of μ between the collars and the vertical rods for which the system will remain in equilibrium in the position shown.

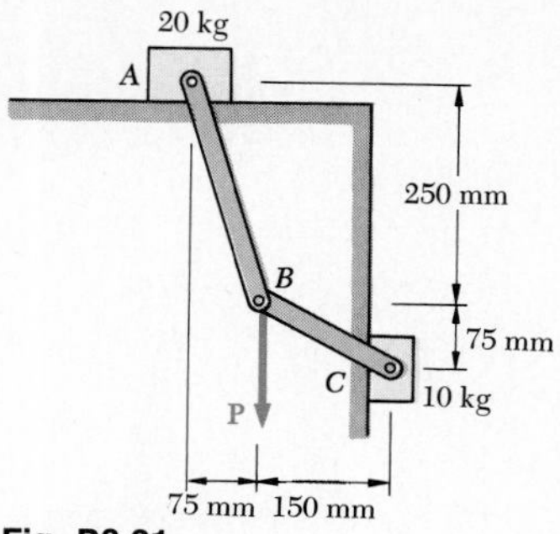

Fig. P8.21

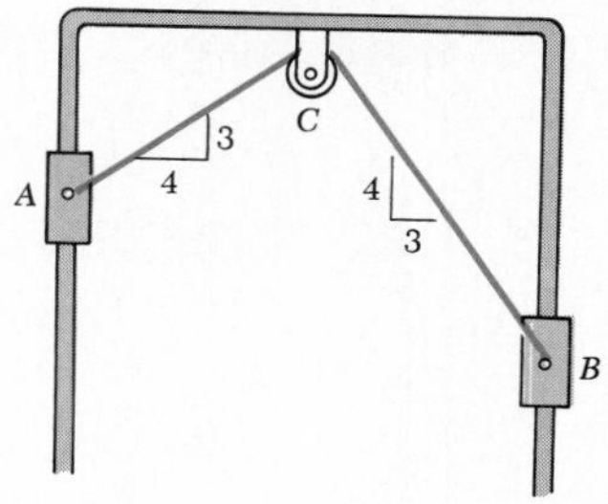

Fig. P8.23

8.24 Two 10-ft beams are pin-connected at D and support two loads **P** and **Q** as shown. Knowing that the coefficient of friction is zero at A, 0.25 at B, and 0.50 at C, determine the largest value of Q for which equilibrium is maintained when $P = 100$ lb. (*Hint.* Note that C moves up when B moves to the right.)

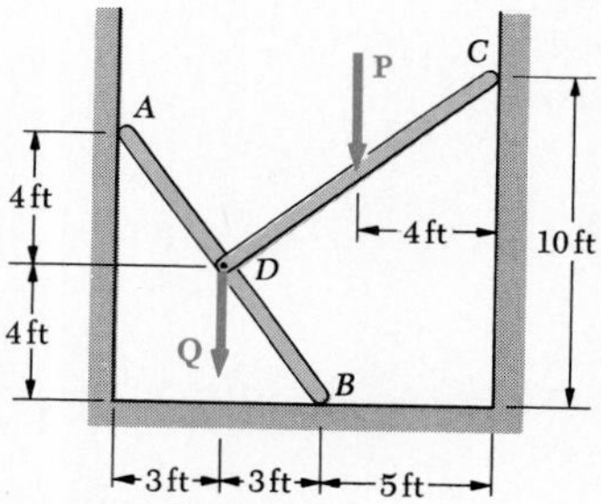

Fig. P8.24

8.25 The rod AB rests on a horizontal surface at A and against a sloping surface at B. Knowing that the coefficient of friction is 0.25 at both A and B, determine the maximum distance a at which the load **W** can be supported. Neglect the weight of the rod.

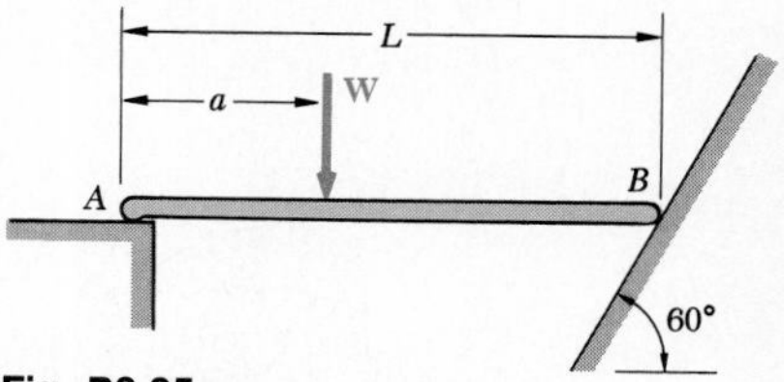

Fig. P8.25

8.26 Two large cylinders, each of radius $r = 600$ mm, rotate in opposite directions and form the main elements of a crusher for stone aggregate. The distance d is set equal to the maximum desired size of the crushed aggregate. If $d = 25$ mm and $\mu = 0.35$, determine the size s of the largest stones which will be pulled through the crusher by friction alone.

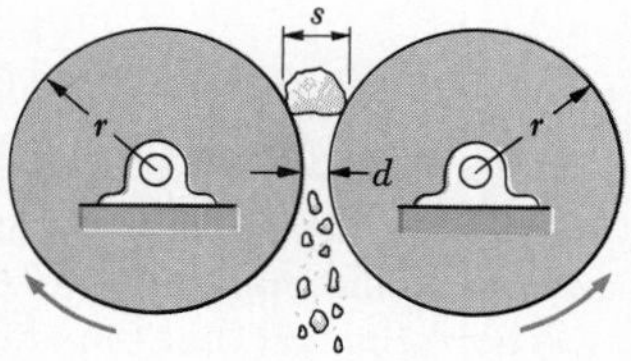

Fig. P8.26

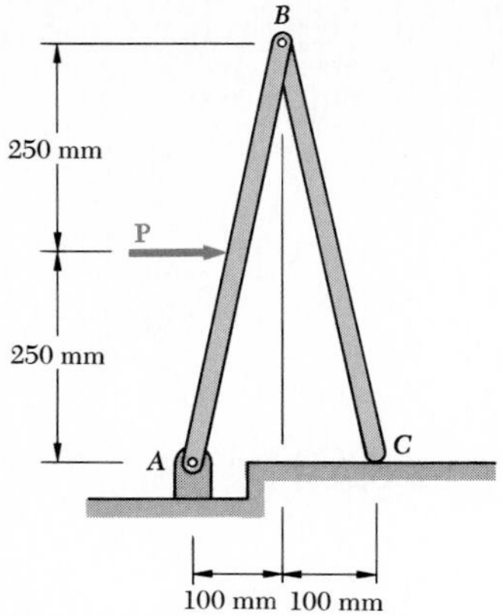

Fig. P8.27

8.27 Two uniform rods, each of mass 5 kg, are held by frictionless pins A and B. It is observed that if the magnitude of $\mathbf{P}$ is larger than 120 N, the rods will collapse. Determine the coefficient of friction at C.

8.28 A window sash of mass 6 kg is normally supported by two 3-kg sash weights. It is observed that the window remains open after one sash cord has broken. What is the smallest possible value of the coefficient of static friction? (Assume that the sash is slightly smaller than the frame and will bind only at points A and D.)

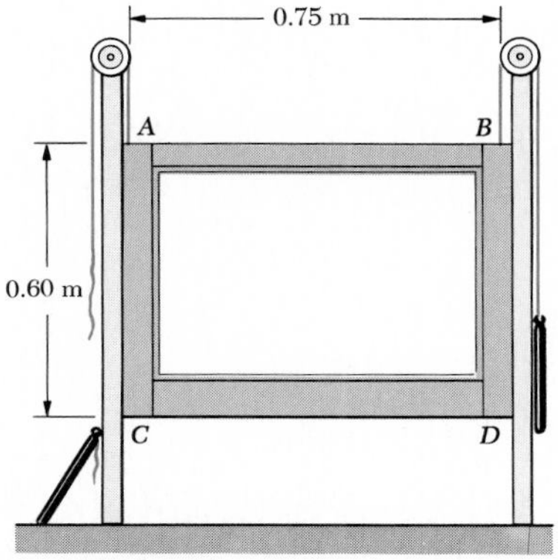

Fig. P8.28

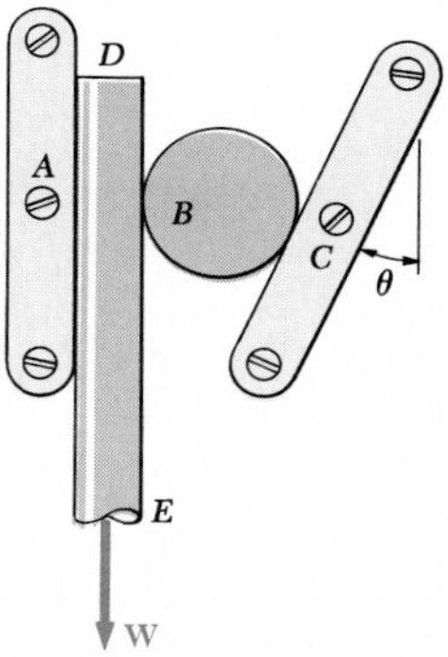

Fig. P8.29

8.29 A rod DE and a small cylinder are placed between two guides as shown. The rod is not to slip downward, however large the force $\mathbf{W}$ may be; i.e., the arrangement is to be *self-locking*. Determine the minimum allowable coefficients of friction at A, B, and C.

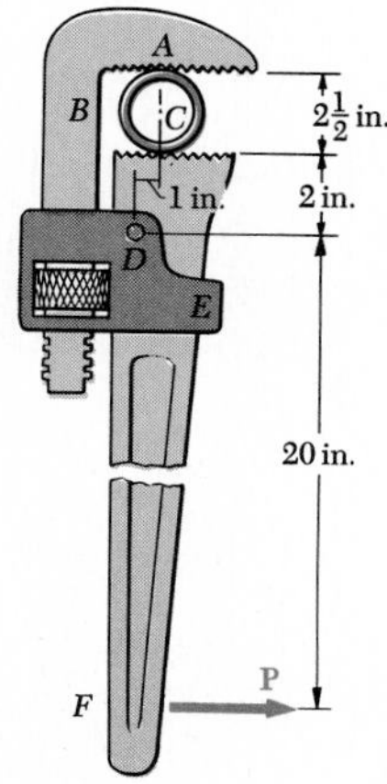

Fig. P8.30

8.30 A pipe of diameter $2\frac{1}{2}$ in. is gripped by the stillson wrench shown. Portions AB and DE of the wrench are rigidly attached to each other and portion CF is connected by a pin at D. If the wrench is to grip the pipe and be self-locking, determine the required minimum coefficients of friction at A and at C.

8.31 If the plumbing wrench of Prob. 6.113 is not to slip, determine the minimum allowable value of the coefficient of friction between (*a*) the nut and the jaw BC, (*b*) the nut and the rod A.

8.32 A uniform plank of weight $W = 90$ lb rests on two joists as shown. The coefficient of friction between the joists and the plank is $\mu = 0.30$. Determine the magnitude of the horizontal force **P** required to move the plank (*a*) when $a = 9$ ft, (*b*) when $a = 10$ ft.

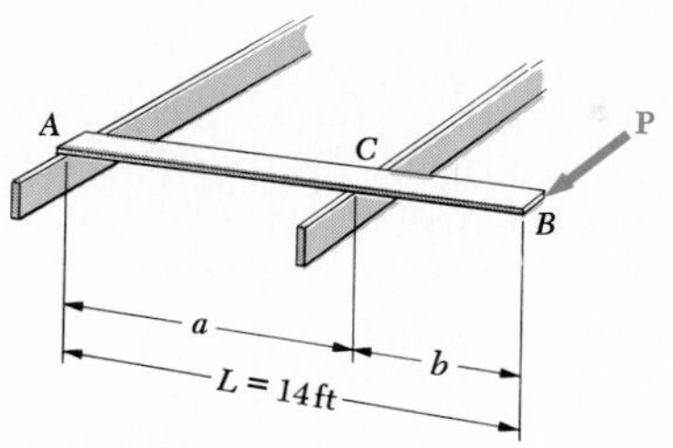

Fig. P8.32

8.33 A uniform plank of weight $W = 90$ lb rests on two joists as shown. The coefficient of friction between the joists and the plank is 0.30. Determine the magnitude of the horizontal force **P** required to move the plank (*a*) when $a = 5$ ft, (*b*) when $a = 3$ ft. (Neglect the angle that the plank forms with the horizontal.)

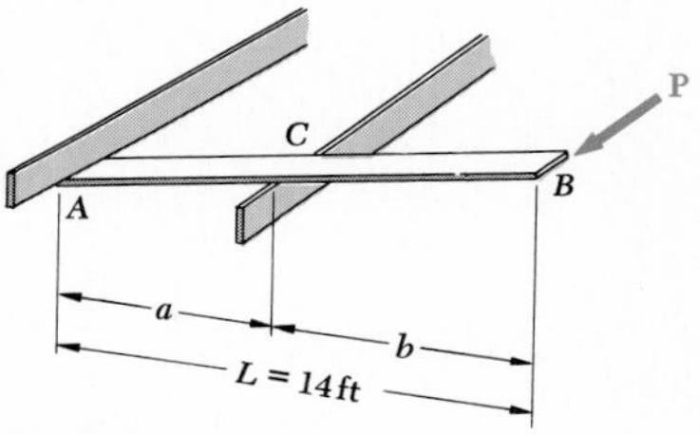

Fig. P8.33

8.34 (*a*) In Prob. 8.32, determine the smallest distance *a* for which the plank will slip at *C*. (*b*) In Prob. 8.33, show that as *a* approaches zero, the magnitude of **P** approaches $\frac{1}{2}\mu W$.

8.35 The mathematical model shown has been developed for the analysis of a certain structure. It consists of 1-lb blocks connected in series by springs, each of constant $k = 0.20$ lb/in. The coefficient of friction between the base and each block is 0.40. Knowing that the initial tension in each spring is zero, construct a graph showing the magnitude of the force **P** versus the position of block *A* as *P* increases from zero to 1.00 lb and then decreases to zero.

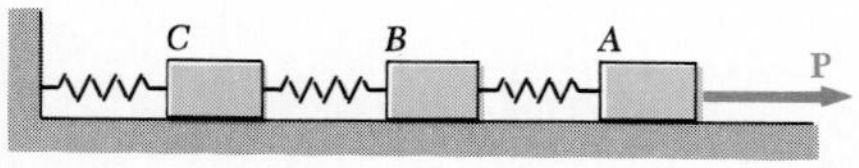

Fig. P8.35

8.36 A cylinder of weight *W* is placed in a V block as shown. Denoting by ϕ the angle of friction between the cylinder and the block and assuming $\phi < \theta$, determine (*a*) the axial force **P** required to move the cylinder, (*b*) the couple **M**, applied in the plane of the cross section of the cylinder, required to rotate the cylinder.

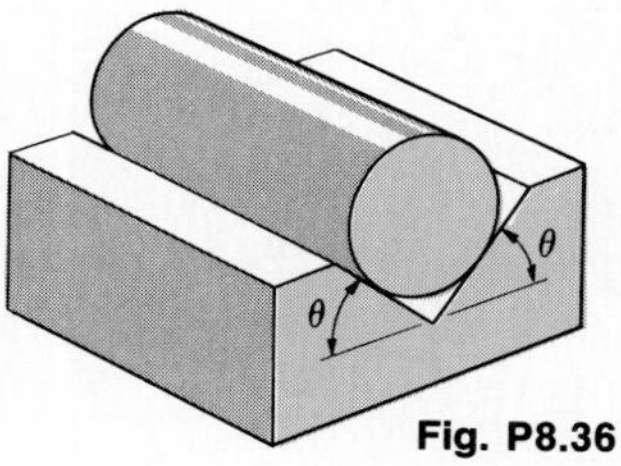

Fig. P8.36

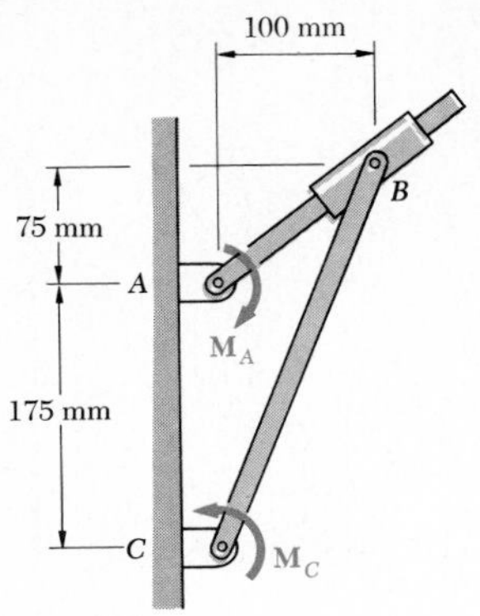

Fig. P8.37

8.37 Two rods are connected by a collar at B; a couple $\mathbf{M}_A$ of moment 40 N · m is applied to rod AB. Knowing that $\mu = 0.30$ between the collar and rod AB, determine the *maximum* couple $\mathbf{M}_C$ for which equilibrium will exist.

8.38 In Prob. 8.37, determine the *minimum* couple $\mathbf{M}_C$ required for equilibrium.

8.39 A 10-kg rod AB is placed between a movable block and a wall as shown. The coefficient of static friction at A and B is 0.25. If a 500-N force acts on the block as shown, determine the angle θ for which slipping impends.

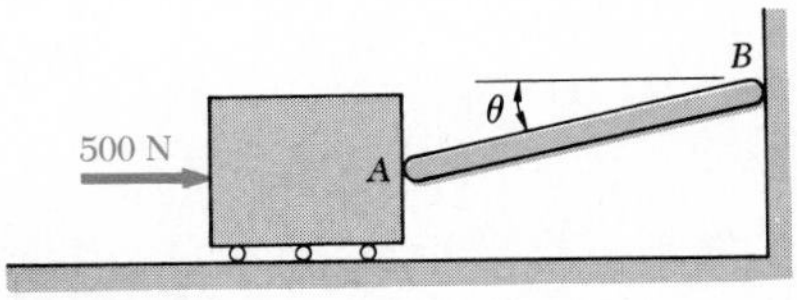

Fig. P8.39

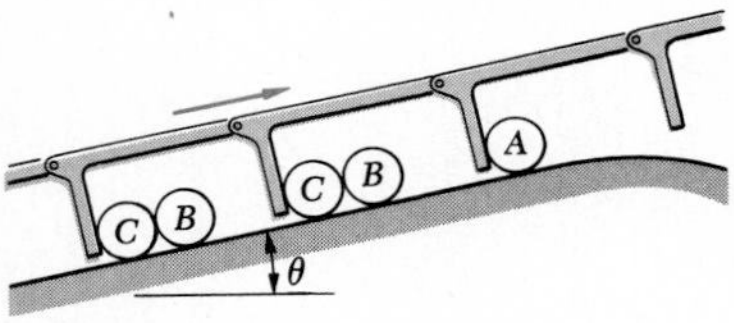

Fig. P8.40

8.40 Identical cylindrical cans, each of weight W, are raised to the top of an incline by a series of moving arms. Either one or two cans are moved by each arm. The coefficient of friction between all surfaces is $\mu = 0.20$. If $W = 2.00$ lb and $\theta = 12°$, determine the force parallel to the incline which the arm must exert on can A to move it. Does can A roll or slide?

***8.41** Solve Prob. 8.40, considering can C instead of can A.

8.42 In Prob. 8.40, determine the range of values of θ for which can A will roll.

***8.43** In Prob. 8.40, determine the range of values of θ for which can C will roll.

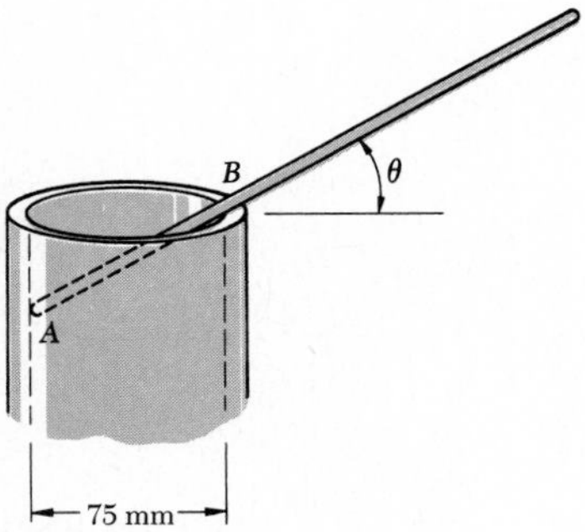

Fig. P8.44

***8.44** A small steel rod, of length 500 mm, is placed inside a pipe as shown. The coefficient of static friction between the rod and the pipe is 0.20. Determine the largest value of θ for which the rod will not fall into the pipe.

***8.45** In Prob. 8.44, determine the smallest value of θ for which the rod will not fall out of the pipe.

8.5. Wedges. Wedges are simple machines used to raise large stone blocks and other heavy loads. These loads may be raised by applying to the wedge a force usually considerably smaller than the weight of the load. Besides, because of the friction existing between the surfaces in contact, a wedge, if properly shaped, will remain in place after being forced under the load. Wedges may thus be used advantageously to make small adjustments in the position of heavy pieces of machinery.

Consider the block *A* shown in Fig. 8.7*a*. This block rests against a vertical wall *B* and is to be raised slightly by forcing a wedge *C* between block *A* and a second wedge *D*. We want to find the minimum value of the force **P** which must be applied to the wedge *C* to move the block. We shall assume that the weight **W** of the block is known. It may have been given in pounds or determined in newtons from the mass of the block expressed in kilograms.

The free-body diagrams of block *A* and of wedge *C* have been drawn in Fig. 8.7*b* and *c*. The forces acting on the block include its weight and the normal and friction forces at the surfaces of contact with wall *B* and wedge *C*. The magnitudes of the friction forces $\mathbf{F}_1$ and $\mathbf{F}_2$ are equal, respectively, to $\mu_s N_1$ and $\mu_s N_2$ since the motion of the block must be started. It is important to show the friction forces with their correct sense. Since the block will move upward, the force $\mathbf{F}_1$ exerted by the wall on the block must be directed downward. On the other hand, since the wedge *C* moves to the right, the relative motion of *A* with respect to *C* is to the left and the force $\mathbf{F}_2$ exerted by *C* on *A* must be directed to the right.

Considering now the free body *C* in Fig. 8.7*c*, we note that the forces acting on *C* include the applied force **P** and the normal and friction forces at the surfaces of contact with *A* and *D*. The weight of the wedge is small compared with the other forces involved and may be neglected. The forces $\mathbf{N}_2'$ and $\mathbf{F}_2'$ acting on *C* are equal and opposite to the forces $\mathbf{N}_2$ and $\mathbf{F}_2$ acting on *A*; the friction force $\mathbf{F}_2'$ must therefore be directed to the left. We check that the force $\mathbf{F}_3$ is also directed to the left.

The total number of unknowns involved in the two free-body diagrams may be reduced to four if the friction forces are expressed in terms of the normal forces. Expressing that block *A* and wedge *C* are in equilibrium will provide four equations which may be solved to obtain the magnitude of **P**. It should be noted that, in the example considered here, it will be more convenient to replace each pair of normal and friction forces by their resultant. Each free body is then subjected to only three forces, and the problem may be solved by drawing the corresponding force triangles (see Sample Prob. 8.4).

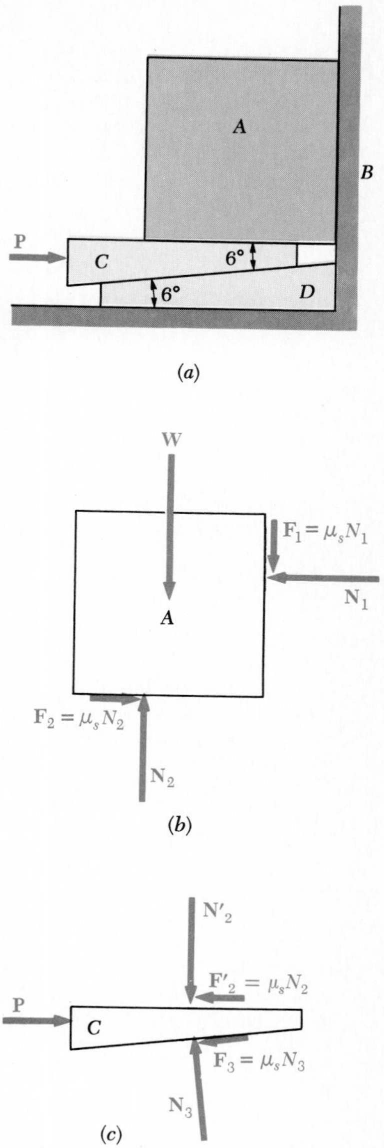

Fig. 8.7

8.6. Square-threaded Screws. Square-threaded screws are frequently used in jacks, presses, and other mechanisms. Their analysis is similar to that of a block sliding along an inclined plane.

Consider the jack shown in Fig. 8.8. The screw carries a load **W** and is supported by the base of the jack. Contact between screw and base takes place along a portion of their threads. By applying a force **P** on the handle, the screw may be made to turn and to raise the load **W**.

The thread of the base has been unwrapped and shown as a straight line in Fig. 8.9*a*. The correct slope was obtained by plotting horizontally the product $2\pi r$, where r is the mean radius of the thread, and vertically the *lead* L of the screw, i.e., the distance through which the screw advances in one turn. The angle θ this line forms with the horizontal is the *lead angle*. Since the force of friction between two surfaces in contact does not depend upon the area of contact, the two threads may be assumed to be in contact over a much smaller area than they actually are and the screw may be represented by the block shown in Fig. 8.9*a*. It should be noted, however, that, in this analysis of the jack, the friction between cap and screw is neglected.

The free-body diagram of the block should include the load **W**, the reaction **R** of the base thread, and a horizontal force **Q** having the same effect as the force **P** exerted on the handle. The force **Q** should have the same moment as **P** about the axis of the screw and its magnitude should thus be $Q = Pa/r$. The force **Q**, and thus the force **P** required to raise the load **W**, may be obtained from the free-body diagram shown in Fig. 8.9*a*. The friction angle is taken equal to ϕ_s since the load will presumably be raised through a succession of short strokes. In mechanisms providing for the continuous rotation of a screw, it may be desirable to distinguish between the force required to start motion (using ϕ_s) and that required to maintain motion (using ϕ_k).

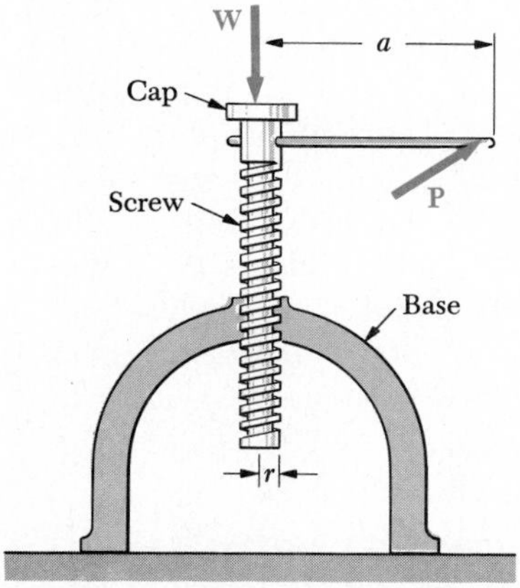

Fig. 8.8

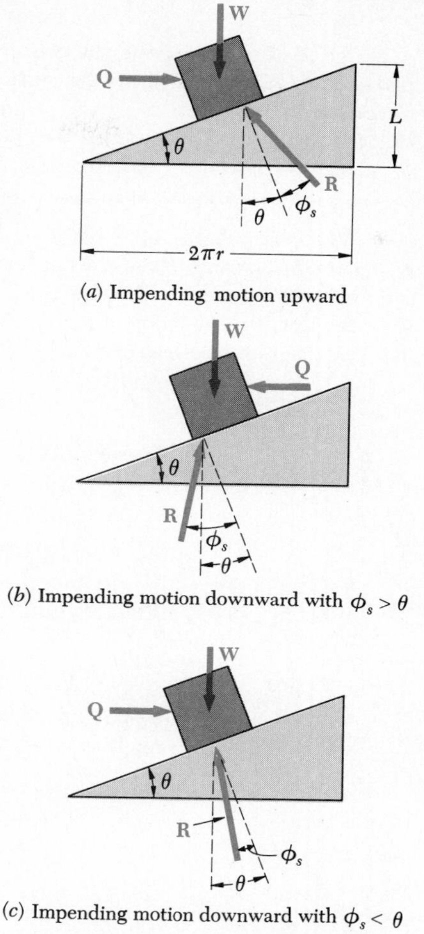

Fig. 8.9 Block-and-incline analysis of a screw.

If the friction angle ϕ_s is larger than the lead angle θ, the screw is said to be *self-locking;* it will remain in place under the load. To lower the load, we must then apply the force shown in Fig. 8.9*b*. If ϕ_s is smaller than θ, the screw will unwind under the load; it is then necessary to apply the force shown in Fig. 8.9*c* to maintain equilibrium.

The lead of a screw should not be confused with its *pitch*. The lead was defined as the distance through which the screw advances in one turn; the pitch is the distance measured between two consecutive threads. While lead and pitch are equal in the case of *single-threaded* screws, they are different in the case of *multiple-threaded* screws, i.e., screws having several independent threads. It is easily verified that, for double-threaded screws, the lead is twice as large as the pitch; for triple-threaded screws, it is three times as large as the pitch; etc.

SAMPLE PROBLEM 8.4

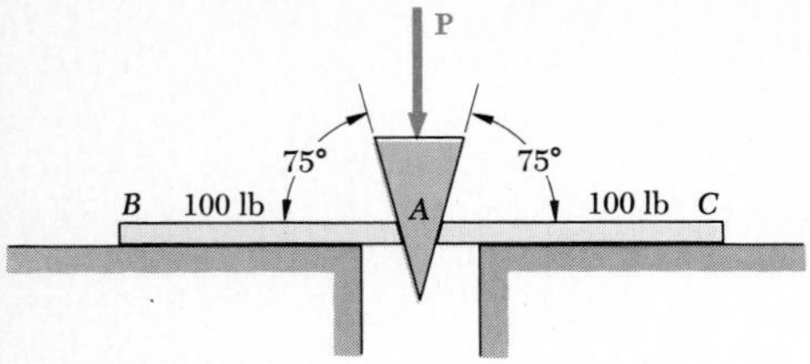

A wedge A of negligible weight is to be driven between two 100-lb plates B and C. The coefficient of static friction between all surfaces of contact is 0.35. Determine the magnitude of the force **P** required to start moving the wedge (a) if the plates are equally free to move, (b) if plate C is securely bolted to the surface.

a. **With Both Plates Free to Move.** The free-body diagrams of wedge A and of plate B are shown, together with the corresponding force triangles. Because of symmetry it will not be necessary to consider the free-body diagram of plate C in the solution of the problem, and this diagram has been omitted.

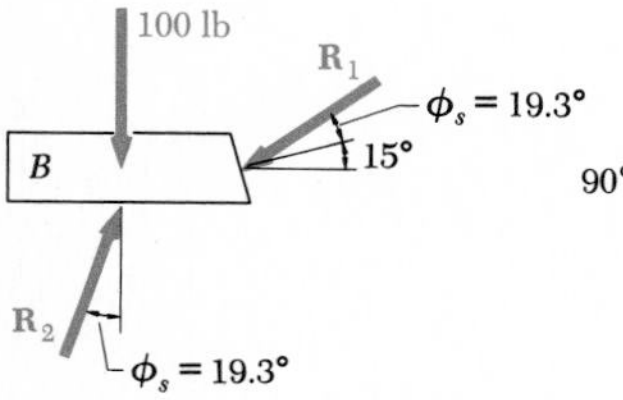

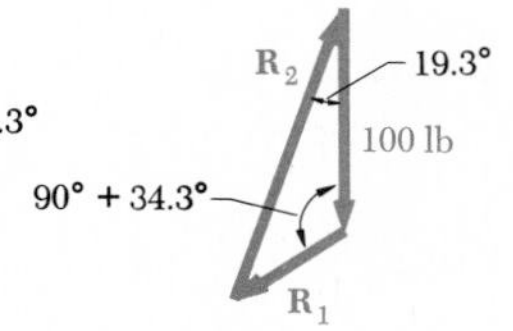

Free Body: Plate B

$$\phi_s = \tan^{-1} 0.35 = 19.3°$$

$$\frac{R_1}{\sin 19.3°} = \frac{100 \text{ lb}}{\sin [180° - (90° + 34.3°) - 19.3°]}$$

$$\frac{R_1}{\sin 19.3°} = \frac{100 \text{ lb}}{\sin 36.4°}$$

$$R_1 = 55.7 \text{ lb}$$

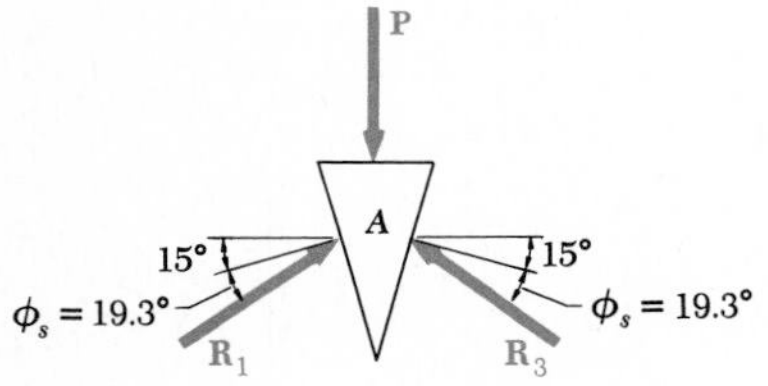

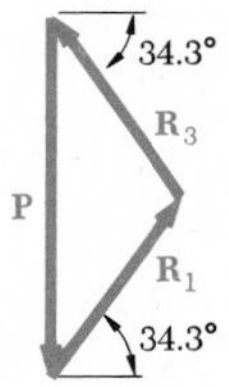

Free Body: Wedge A. We note that the force triangle is an isosceles triangle; thus

$$R_3 = R_1 = 55.7 \text{ lb}$$
$$P = 2R_1 \sin 34.3°$$
$$= 2(55.7 \text{ lb}) \sin 34.3°$$

P $= 62.8$ lb ↓ ◄

b. **With Plate C Bolted.** The only free bodies remaining are those of the wedge A and of the plate B, and the corresponding diagrams will be the same as in part a. Thus, the answer is still

P $= 62.8$ lb ↓ ◄

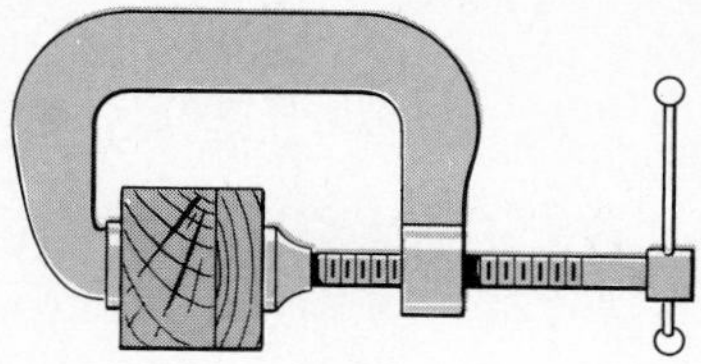

SAMPLE PROBLEM 8.5

A clamp is used to hold two pieces of wood together as shown. The clamp has a double square thread of mean diameter equal to 10 mm and with a pitch of 2 mm. The coefficient of friction between threads is $\mu_s = 0.30$. If a maximum torque of 40 N · m is applied in tightening the clamp, determine (*a*) the force exerted on the pieces of wood, (*b*) the torque required to loosen the clamp.

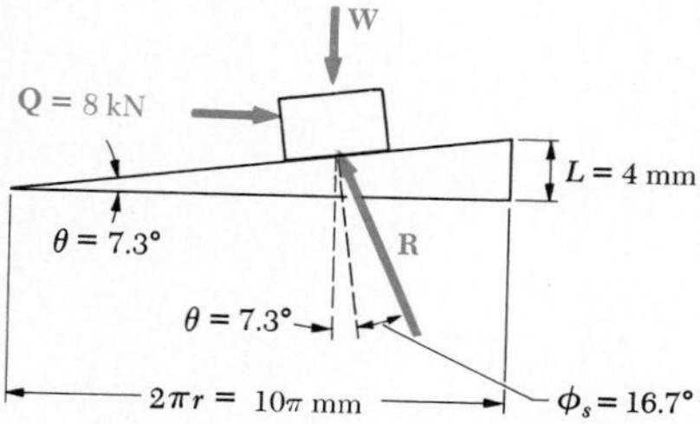

a. Force Exerted by Clamp. The mean radius of the screw is $r = 5$ mm. Since the screw is double-threaded, the lead L is equal to twice the pitch: $L = 2(2 \text{ mm}) = 4$ mm. The lead angle θ and the friction angle ϕ_s are obtained by writing

$$\tan\theta = \frac{L}{2\pi r} = \frac{4 \text{ mm}}{10\pi \text{ mm}} = 0.1273 \qquad \theta = 7.3°$$

$$\tan\phi_s = \mu_s = 0.30 \qquad \phi_s = 16.7°$$

The force **Q** which should be applied to the block representing the screw is obtained by expressing that its moment Qr about the axis of the screw is equal to the applied torque.

$$Q(5 \text{ mm}) = 40 \text{ N}\cdot\text{m}$$

$$Q = \frac{40 \text{ N}\cdot\text{m}}{5 \text{ mm}} = \frac{40 \text{ N}\cdot\text{m}}{5\times 10^{-3}\text{ m}} = 8000 \text{ N} = 8 \text{ kN}$$

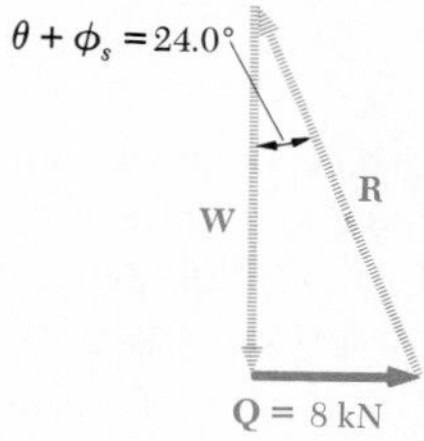

The free-body diagram and the corresponding force triangle may now be drawn for the block; the magnitude of the force **W** exerted on the pieces of wood is obtained by solving the triangle.

$$W = \frac{Q}{\tan(\theta + \phi_s)} = \frac{8 \text{ kN}}{\tan 24.0°}$$

$$W = 17.97 \text{ kN} \quad ◄$$

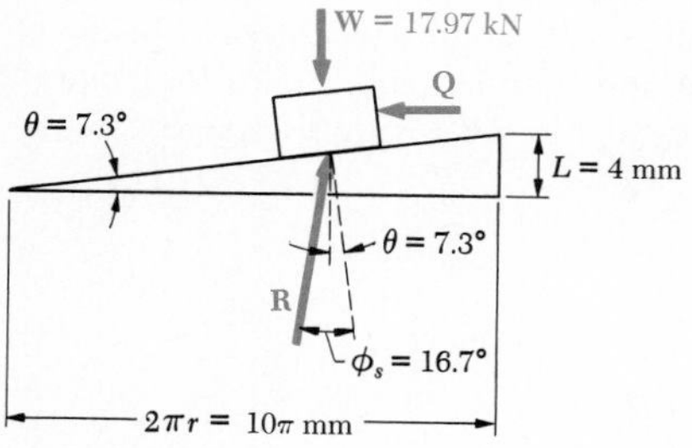

b. Torque Required to Loosen Clamp. The force **Q** required to loosen the clamp and the corresponding torque are obtained from the free-body diagram and force triangle shown.

$$Q = W\tan(\phi_s - \theta) = (17.97 \text{ kN})\tan 9.4°$$
$$= 2.975 \text{ kN}$$

$$\text{Torque} = Qr = (2.975 \text{ kN})(5 \text{ mm})$$
$$= (2.975\times 10^3 \text{ N})(5\times 10^{-3}\text{ m}) = 14.87 \text{ N}\cdot\text{m}$$

$$\text{Torque} = 14.87 \text{ N}\cdot\text{m} \quad ◄$$

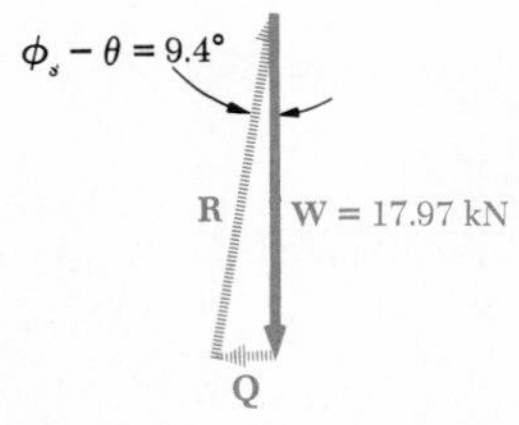

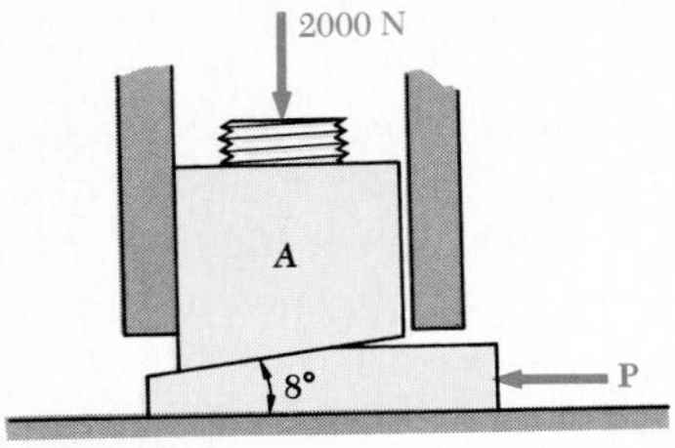

Fig. P8.46

PROBLEMS

8.46 The position of machine block A is adjusted by moving the wedge shown. Knowing that the coefficient of friction is 0.25 at all surfaces of contact, determine the force **P** required to raise the block.

8.47 In Prob. 8.46, determine the force **P** required to lower machine block A.

8.48 and 8.49 Determine the minimum value of **P** which must be applied to the wedge in order to move the 2000-lb block. The coefficient of static friction is 0.30 at all surfaces of contact.

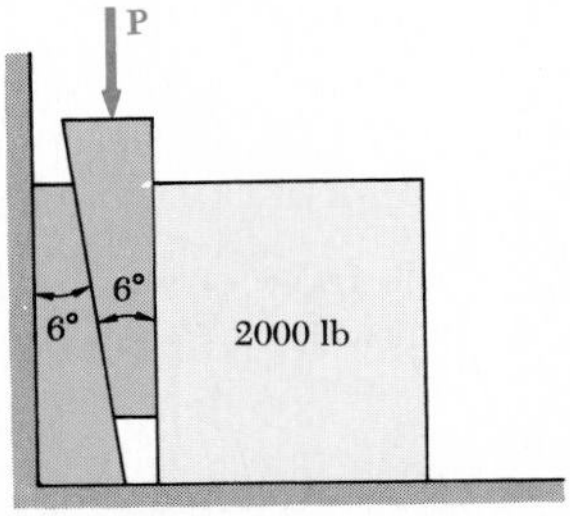

Fig. P8.48

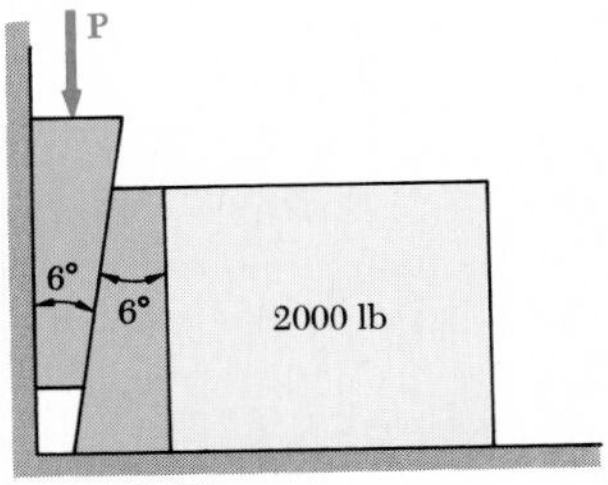

Fig. P8.49

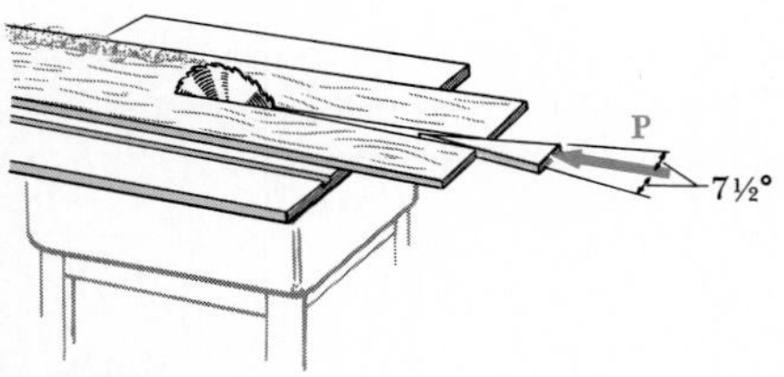

Fig. P8.50

8.50 A 15° wedge is forced into a saw cut to prevent binding of the circular saw. The coefficient of friction between the wedge and the wood is 0.25. Knowing that a horizontal force **P** of magnitude 150 N was required to insert the wedge, determine the magnitude of the forces exerted on the board by the wedge after it has been inserted.

8.51 The spring of the door latch has a constant of 1.5 lb/in. and in the position shown exerts a force of $\frac{1}{2}$ lb on the bolt. The coefficient of friction between the bolt and the guide plate is 0.40; all other surfaces are well lubricated and may be assumed to be frictionless. Determine the magnitude of the force **P** required to start closing the door.

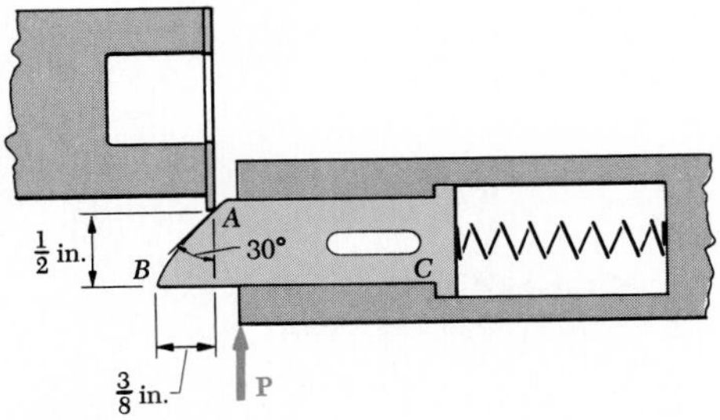

Fig. P8.51

8.52 In Prob. 8.51, determine the angle which the face of the bolt should form with the line *BC* if the force **P** required to close the door is to be the same for both the position shown and the position when *B* is almost at the guide plate.

8.53 The elevation of the end of the steel floor beam is adjusted by the steel wedges *E* and *F*. The steel base plate *CD* has been welded to the lower flange of the beam; the end reaction of the beam is known to be 100 kN. The coefficient of friction between two steel surfaces is 0.30 and between steel and concrete is 0.60. If horizontal motion of the base plate is prevented by the force **Q**, determine (*a*) the force **P** required to raise the beam, (*b*) the corresponding force **Q**.

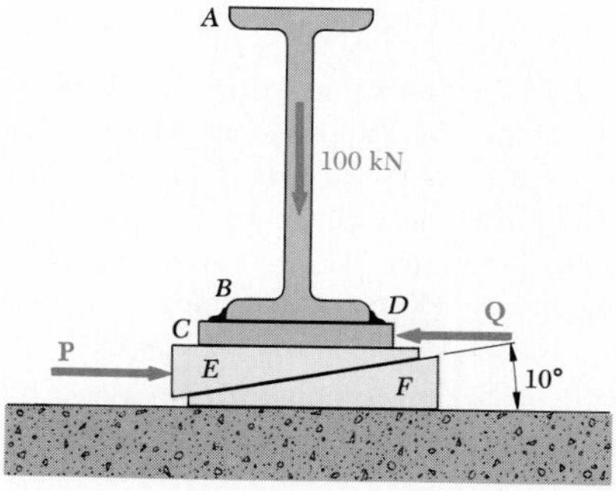

Fig. P8.53

8.54 Solve Prob. 8.53, assuming that the end of the beam is to be lowered.

8.55 A 5° wedge is to be forced under a machine base at *A*. Knowing that $\mu = 0.15$ at all surfaces, (*a*) determine the force **P** required to move the wedge, (*b*) indicate whether the machine will move.

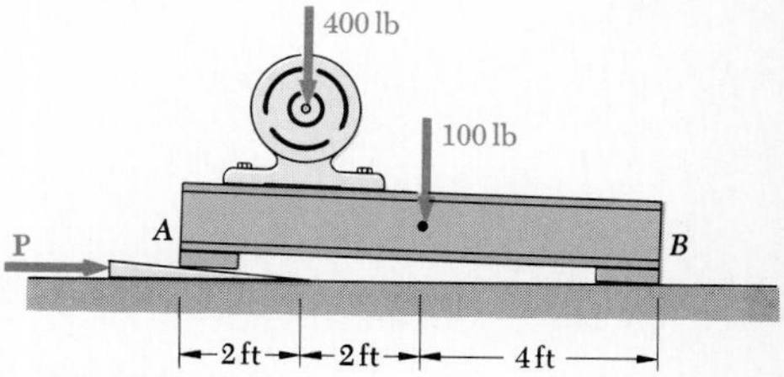

Fig. P8.55

8.56 Solve Prob. 8.55, assuming that the wedge is to be forced under the machine base at *B* instead of *A*.

8.57 A conical wedge is placed between two horizontal plates which are then slowly moved toward each other. Indicate what will happen to the wedge (*a*) if $\mu = 0.15$, (*b*) if $\mu = 0.25$.

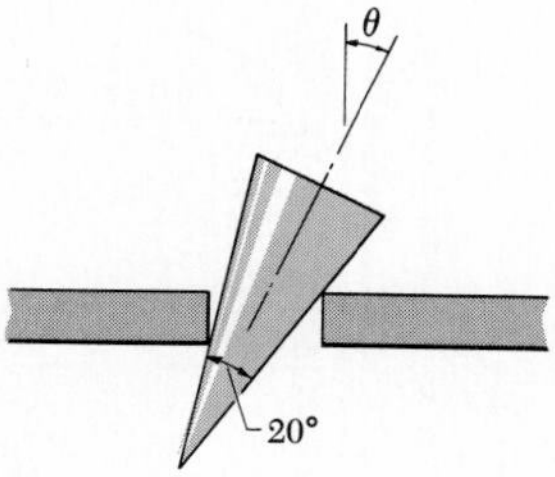

Fig. P8.57

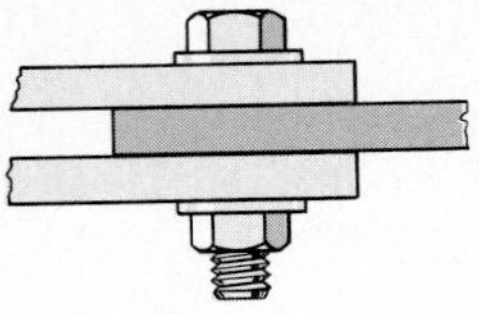

Fig. P8.58

8.58 High-strength bolts are used in the construction of many modern steel structures. For a 1-in. nominal-diameter bolt the required minimum bolt tension is 47,250 lb. Assuming the coefficient of friction to be 0.40, determine the required torque which must be applied to the bolt and nut. The mean diameter of the thread is 0.94 in., and the lead is 0.125 in. Neglect friction between the nut and washer, and assume the bolt to be square-threaded.

8.59 Derive the following formulas relating the load W and the force $\mathbf{P}$ exerted on the handle of the jack discussed in Sec. 8.6: (*a*) $P = (Wr/a)\tan(\theta + \phi_s)$, to raise the load; (*b*) $P = (Wr/a)\tan(\phi_s - \theta)$, to lower the load if the screw is self-locking; (*c*) $P = (Wr/a)\tan(\theta - \phi_s)$, to hold the load if the screw is not self-locking.

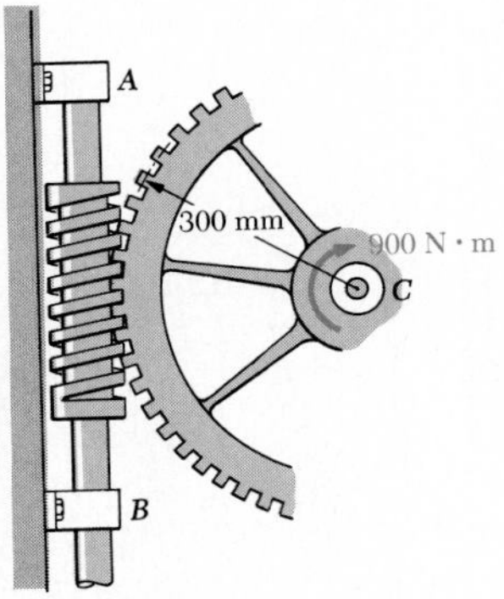

Fig. P8.60

8.60 The square-threaded worm gear shown has a mean radius of 40 mm and a pitch of 10 mm. The large gear is subjected to a constant clockwise torque of 900 N · m. Knowing that the coefficient of friction between gear teeth is 0.10, determine the torque which must be applied to shaft AB in order to rotate the large gear counterclockwise. Neglect friction in the bearings at A, B, and C.

8.61 In Prob. 8.60, determine the torque which must be applied to shaft AB in order to rotate the large gear clockwise.

8.62 The main features of a screw-luffing crane are shown. Distances AD and CD are each 25 ft. The position of the boom CDE is controlled by the screw ABC which is single-threaded at each end (left-handed thread at A, right-handed thread at C). Each thread has a pitch of $1\frac{1}{4}$ in. and a mean diameter of 8 in. If $\mu = 0.08$, determine the moment of the couple which must be applied to the screw (*a*) to raise the boom, (*b*) to lower the boom.

Fig. P8.62

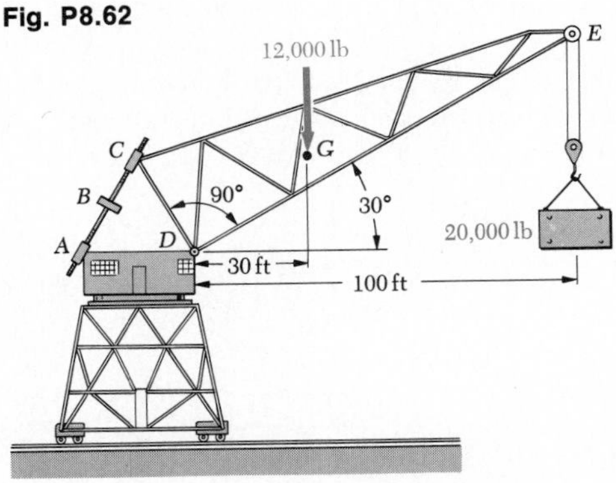

8.63 The ends of two fixed rods A and B are each made in the form of a single-threaded screw of mean radius 6 mm and pitch 2 mm. The coefficient of friction between the rods and the threaded sleeve is 0.15. Determine the moment of the couple which must be applied to the sleeve in order to draw the rods closer together. Rod A has a left-handed thread and rod B a right-handed thread.

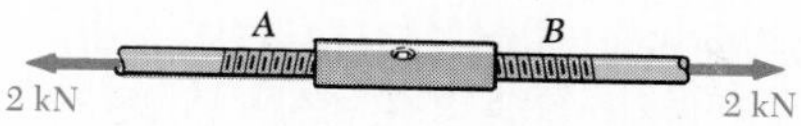

Fig. P8.63

8.64 In Prob. 8.63, a right-handed thread is used on *both* rods A and B. Determine the moment of the couple which must be applied to the sleeve in order to rotate it.

8.65 In the machinist's vise shown, the movable jaw D is rigidly attached to the tongue AB which fits loosely into the fixed body of the vise. The screw is single-threaded into the fixed base and has a mean diameter of 0.50 in. and a pitch of 0.20 in. The coefficient of friction is 0.25 between the threads and also between the tongue and the body. Neglecting bearing friction between the screw and the movable head, determine the torque which must be applied to the handle in order to produce a clamping force of 800 lb.

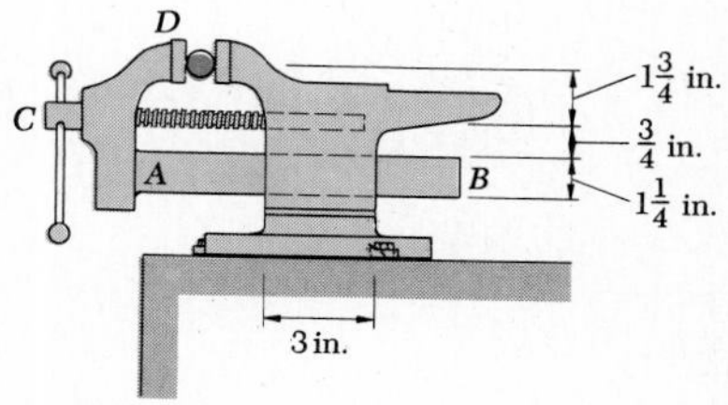

Fig. P8.65

8.66 In Prob. 8.65, a clamping force of 800 lb was obtained by tightening the vise. Determine the torque which must be applied to the screw to loosen the vise.

8.67 The vise shown consists of two members connected by two double-threaded screws of mean radius 5 mm and pitch 1 mm. The lower member is threaded at A and B ($\mu_s = 0.25$), but the upper member is not threaded. It is desired to apply two equal and opposite forces of 500 N on the blocks held between the jaws. (*a*) What screw should be adjusted first? (*b*) What is the maximum torque applied in tightening the second screw?

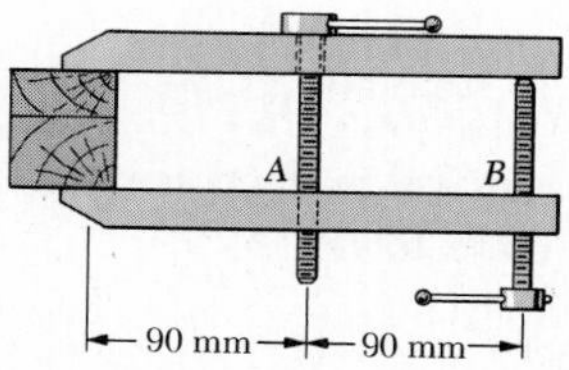

Fig. P8.67

*** 8.7. Journal Bearings. Axle Friction.** Journal bearings are used to provide lateral support to rotating shafts and axles. Thrust bearings, which will be studied in the next section, are used to provide axial support to shafts and axles. If the journal bearing is fully lubricated, the frictional resistance depends upon the speed of rotation, the clearance between axle and bearing, and the viscosity of the lubricant. As indicated in Sec. 8.1, such problems are studied in fluid mechanics. The methods of this chapter, however, may be applied to the study of axle friction when the bearing is not lubricated or only partially lubricated. We may then assume that the axle and the bearing are in direct contact along a single straight line.

Consider two wheels, each of weight **W**, rigidly mounted on an axle supported symmetrically by two journal bearings (Fig. 8.10*a*). If the wheels rotate, we find that, to keep them rotating at constant speed, it is necessary to apply to each of them a couple **M**. A free-body diagram has been drawn in Fig. 8.10*c*, which represents one of the wheels and the corresponding half axle in projection on a plane perpendicular to the axle. The forces acting on the free body include the weight **W** of the wheel, the couple **M** required to maintain its motion, and a force **R** representing the reaction of the bearing. This force is vertical, equal, and opposite to **W** but does not pass through the center O of the axle; **R** is located to the right of O at a distance such that its moment about O balances the moment **M** of the couple. Contact between axle and bearing, therefore, does not take place at the lowest point A when the axle rotates. It takes place at point B (Fig. 8.10*b*) or, rather, along a straight line intersecting the plane of the figure at B. Physically, this is explained by the fact that, when the wheels are set in motion, the axle "climbs" in the bearings until slippage occurs. After sliding back slightly, the axle settles more or less in the position shown. This position is such that the angle between the reaction **R** and the normal to the surface of the bearing is equal to the angle of kinetic friction ϕ_k. The distance from O to the line of action of **R** is thus $r \sin \phi_k$, where r is the radius of the axle. Writing that $\Sigma M_O = 0$ for the forces acting on the free body considered, we obtain the moment M of the couple required to overcome the frictional resistance of one of the bearings:

$$M = Rr \sin \phi_k \tag{8.5}$$

Observing that, for small values of the angle of friction, $\sin \phi_k$ may be replaced by $\tan \phi_k$, that is, by μ_k, we write the approximate formula

$$M \approx Rr\,\mu_k \tag{8.6}$$

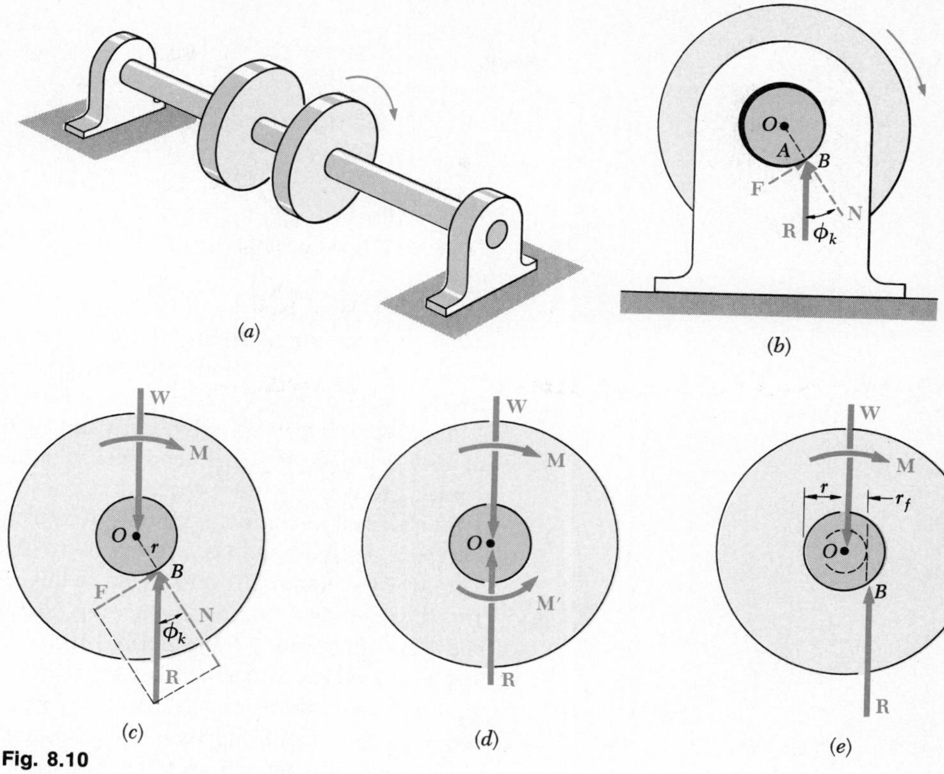

Fig. 8.10

In the solution of certain problems, it may be more convenient to let the line of action of **R** pass through O, as it does when the axle does not rotate. A couple **M′** of the same moment M as the couple **M** but of opposite sense must then be added to the reaction **R** (Fig. 8.10*d*). This couple represents the frictional resistance of the bearing.

In case a graphical solution is preferred, the line of action of **R** may be readily drawn (Fig. 8.10*e*) if we note that it must be tangent to a circle centered at O and of radius

$$r_f = r \sin \phi_k \approx r \mu_k \tag{8.7}$$

This circle is called the *circle of friction* of the axle and bearing and is independent of the loading conditions of the axle.

***8.8. Thrust Bearings. Disk Friction.** Thrust bearings are used to provide axial support to rotating shafts and axles. They are of two types: (1) *end bearings* and (2) *collar bearings* (Fig. 8.11). In the case of collar bearings, friction forces develop between the two ring-shaped areas which are in contact. In the case of end bearings, friction takes place over full

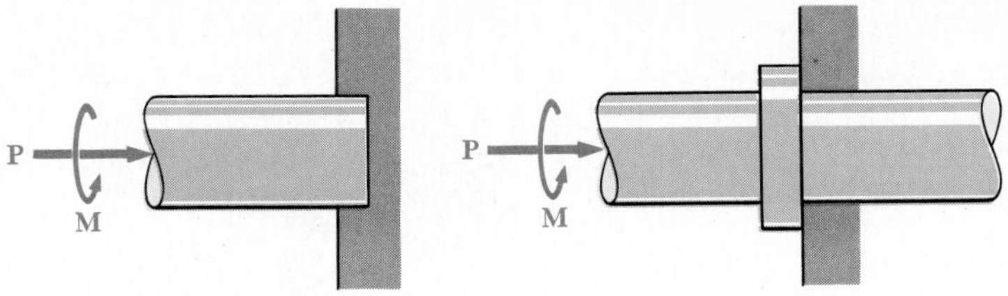

(*a*) End bearing (*b*) Collar bearing

Fig. 8.11 Thrust bearings

circular areas, or over ring-shaped areas when the end of the shaft is hollow. Friction between circular areas, called *disk friction*, also occurs in other mechanisms, such as *disk clutches*.

To obtain a formula which is valid in the most general case of disk friction, we shall consider a rotating hollow shaft. A couple **M** keeps the shaft rotating at constant speed while a force **P** maintains it in contact with a fixed bearing (Fig. 8.12). Contact between the shaft and the bearing takes place over a ring-shaped area of inner radius R_1 and outer radius R_2. Assuming that the pressure between the two surfaces in contact is uniform, we find that the magnitude of the normal force $\Delta\mathbf{N}$ exerted on an element of area ΔA is $\Delta N = P\,\Delta A/A$, where $A = \pi(R_2^2 - R_1^2)$, and that the magnitude of the friction force $\Delta\mathbf{F}$ acting on ΔA is $\Delta F = \mu_k\,\Delta N$. Denoting by r the distance from the axis of the shaft to the element of area ΔA, we express as follows the moment ΔM of $\Delta\mathbf{F}$ about the axis of the shaft:

$$\Delta M = r\,\Delta F = \frac{r\,\mu_k P\,\Delta A}{\pi(R_2^2 - R_1^2)}$$

The equilibrium of the shaft requires that the moment M of the couple applied to the shaft be equal in magnitude to the sum

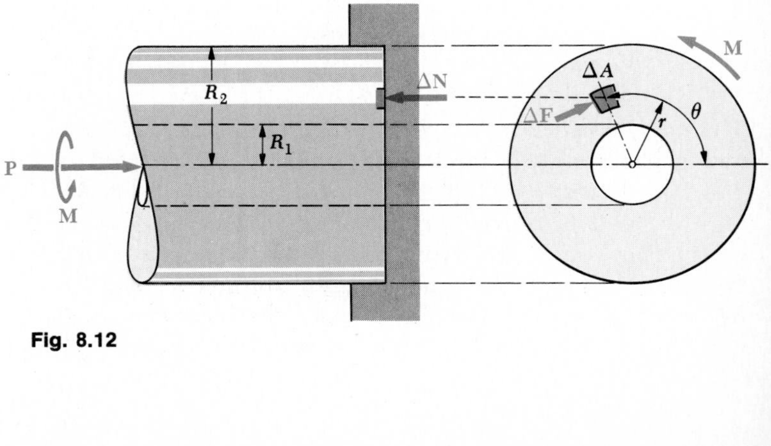

Fig. 8.12

of the moments of the friction forces $\Delta\mathbf{F}$. Replacing ΔA by the infinitesimal element $dA = r\,d\theta\,dr$ used with polar coordinates, and integrating over the area of contact, we thus obtain the following expression for the moment M of the couple required to overcome the frictional resistance of the bearing:

$$M = \frac{\mu_k P}{\pi(R_2^2 - R_1^2)} \int_0^{2\pi} \int_{R_1}^{R_2} r^2\,dr\,d\theta$$

$$= \frac{\mu_k P}{\pi(R_2^2 - R_1^2)} \int_0^{2\pi} \tfrac{1}{3}(R_2^3 - R_1^3)\,d\theta$$

$$M = \tfrac{2}{3}\mu_k P \frac{R_2^3 - R_1^3}{R_2^2 - R_1^2} \tag{8.8}$$

When contact takes place over a full circle of radius R, formula (8.8) reduces to

$$M = \tfrac{2}{3}\mu_k PR \tag{8.9}$$

The value of M is then the same as would be obtained if contact between shaft and bearing took place at a single point located at a distance $2R/3$ from the axis of the shaft.

The largest torque which may be transmitted by a disk clutch without causing slippage is given by a formula similar to (8.9), where μ_k has been replaced by the coefficient of static friction μ_s.

***8.9. Wheel Friction. Rolling Resistance.** The wheel is one of the most important inventions of our civilization. Its use makes it possible to move heavy loads with relatively little effort. Because the point of the wheel in contact with the ground at any given instant has no relative motion with respect to the ground, the wheel eliminates the large friction forces which would arise if the load were in direct contact with the ground. In practice, however, the wheel is not perfect, and some resistance to its motion exists. This resistance has two distinct causes. It is due (1) to a combined effect of axle friction and friction at the rim and (2) to the fact that the wheel and the ground deform, with the result that contact between wheel and ground takes place, not at a single point, but over a certain area.

To understand better the first cause of resistance to the motion of a wheel, we shall consider a railroad car supported by eight wheels mounted on axles and bearings. The car is assumed to be moving to the right at constant speed along a straight horizontal track. The free-body diagram of one of the wheels is shown in Fig. 8.13*a*. The forces acting on the free body include the load $\mathbf{W}$ supported by the wheel and the normal reaction $\mathbf{N}$ of the track. Since $\mathbf{W}$ is drawn through the center O of the

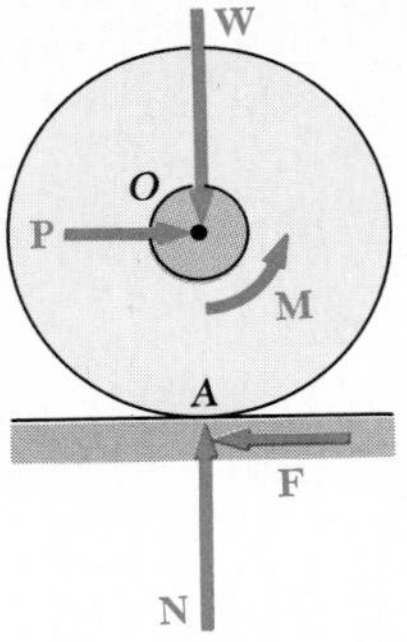

(*a*) Effect of axle friction

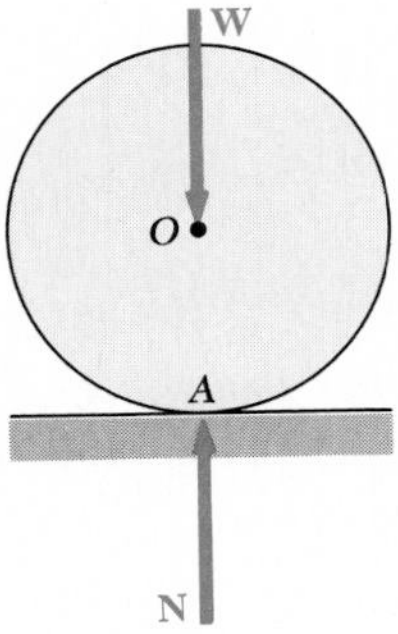

(*b*) Free wheel

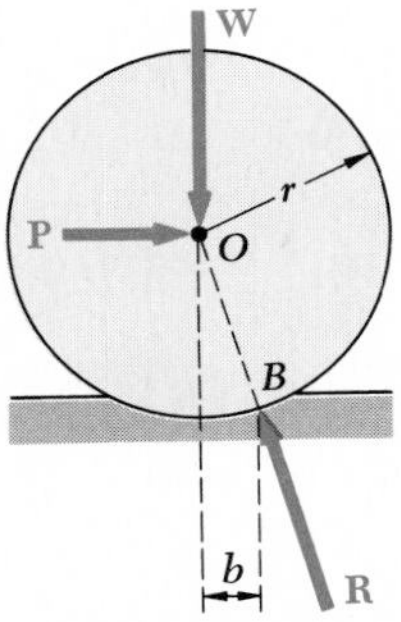

(*c*) Rolling resistance

Fig. 8.13

axle, the frictional resistance of the bearing should be represented by a counterclockwise couple **M** (see Sec. 8.7). To keep the free body in equilibrium, we must add two equal and opposite forces **P** and **F**, forming a clockwise couple of moment M. The force **F** is the friction force exerted by the track on the wheel, and **P** represents the force which should be applied to the wheel to keep it rolling at constant speed. Note that the forces **P** and **F** would not exist if there were no friction between wheel and track. The couple **M** representing the axle friction would then be zero; the wheel would slide on the track without turning in its bearing.

The couple **M** and the forces **P** and **F** also reduce to zero when there is no axle friction. For example, a wheel which is not held in bearings and rolls freely and at constant speed on horizontal ground (Fig. 8.13*b*) will be subjected to only two forces: its own weight **W** and the normal reaction **N** of the ground. No friction force will act on the wheel, regardless of the value of the coefficient of friction between wheel and ground. A wheel rolling freely on horizontal ground should thus keep rolling indefinitely.

Experience, however, indicates that the wheel will slow down and eventually come to rest. This is due to the second type of resistance mentioned at the beginning of this section, known as the *rolling resistance*. Under the load **W**, both the wheel and the ground deform slightly, causing the contact between wheel and ground to take place over a certain area. Experimental evidence shows that the resultant of the forces exerted by the ground on the wheel over this area is a force **R** applied at a point B, which is not located directly under the center O of the wheel, but slightly in front of it (Fig. 8.13*c*). To balance the moment of **W** about B and to keep the wheel rolling at constant speed, it is necessary to apply a horizontal force **P** at the center of the wheel. Writing $\Sigma M_B = 0$, we obtain

$$Pr = Wb \tag{8.10}$$

where r = radius of wheel
b = horizontal distance between O and B

The distance b is commonly called the *coefficient of rolling resistance*. It should be noted that b is not a dimensionless coefficient since it represents a length; b is usually expressed in inches or in millimeters. The value of b depends upon several parameters in a manner which has not yet been clearly established. Values of the coefficient of rolling resistance vary from about 0.01 in. or 0.25 mm for a steel wheel on a steel rail to 5.0 in. or 125 mm for the same wheel on soft ground.

SAMPLE PROBLEM 8.6

A pulley of diameter 4 in. can rotate about a fixed shaft of diameter 2 in. The coefficients of static and kinetic friction between the pulley and shaft are both assumed equal to 0.20. Determine (*a*) the smallest vertical force **P** required to raise a 500-lb load, (*b*) the smallest vertical force **P** required to hold the load, (*c*) the smallest horizontal force **P** required to raise the same load.

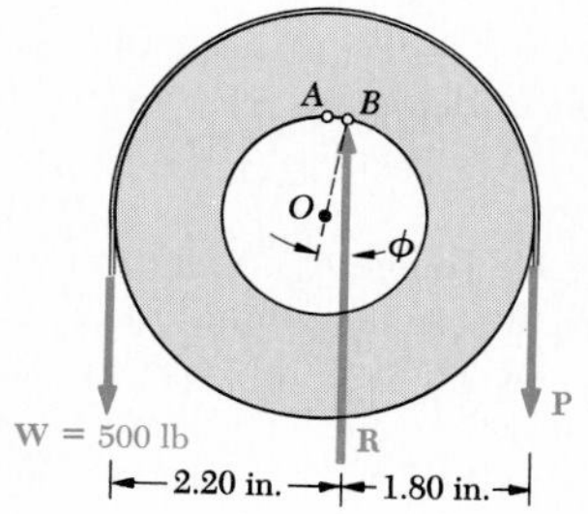

a. **Vertical Force P Required to Raise the Load.** When the forces in both parts of the rope are equal, contact between the pulley and shaft takes place at *A*. When **P** is increased, the pulley rolls around the shaft slightly and contact takes place at *B*. The free-body diagram of the pulley when motion is impending is drawn. The perpendicular distance from the center *O* of the pulley to the line of action of **R** is

$$r_f = r \sin \phi \approx r\mu \qquad r_f \approx (1 \text{ in.})0.20 = 0.20 \text{ in.}$$

Summing moments about *B*, we write

$$+\circlearrowleft \Sigma M_B = 0: \qquad (2.20 \text{ in.})(500 \text{ lb}) - (1.80 \text{ in.})P = 0$$
$$P = 611 \text{ lb}$$

P = 611 lb ↓ ◄

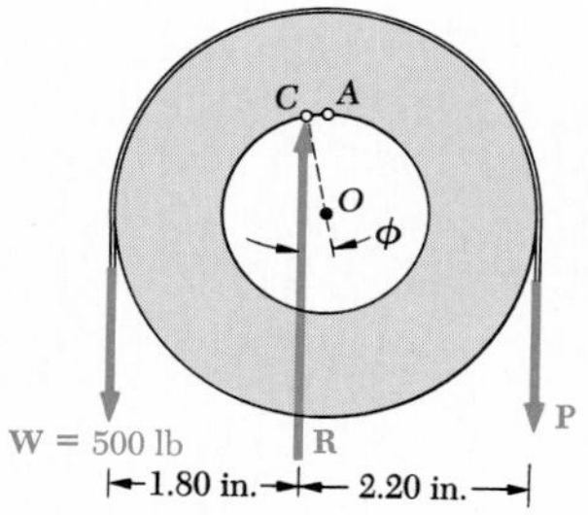

b. **Vertical Force P to Hold the Load.** As the force **P** is decreased, the pulley rolls around the shaft and contact takes place at *C*. Considering the pulley as a free body and summing moments about *C*, we write

$$+\circlearrowleft \Sigma M_C = 0: \qquad (1.80 \text{ in.})(500 \text{ lb}) - (2.20 \text{ in.})P = 0$$
$$P = 409 \text{ lb}$$

P = 409 lb ↓ ◄

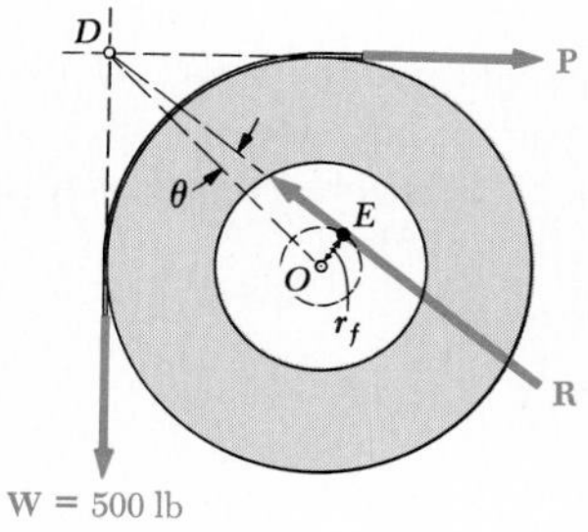

c. **Horizontal Force P to Raise the Load.** Since the three forces **W**, **P**, and **R** are not parallel, they must be concurrent. The direction of **R** is thus determined from the fact that its line of action must pass through the point of intersection *D* of **W** and **P**, and must be tangent to the circle of friction. Recalling that the radius of the circle of friction is $r_f = 0.20$ in., we write

$$\sin \theta = \frac{OE}{OD} = \frac{0.20 \text{ in.}}{(2 \text{ in.})\sqrt{2}} = 0.0707 \qquad \theta = 4.1°$$

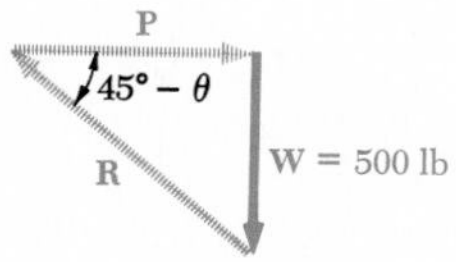

From the force triangle, we obtain

$$P = W \cot (45° - \theta) = (500 \text{ lb}) \cot 40.9°$$
$$= 577 \text{ lb}$$

P = 577 lb → ◄

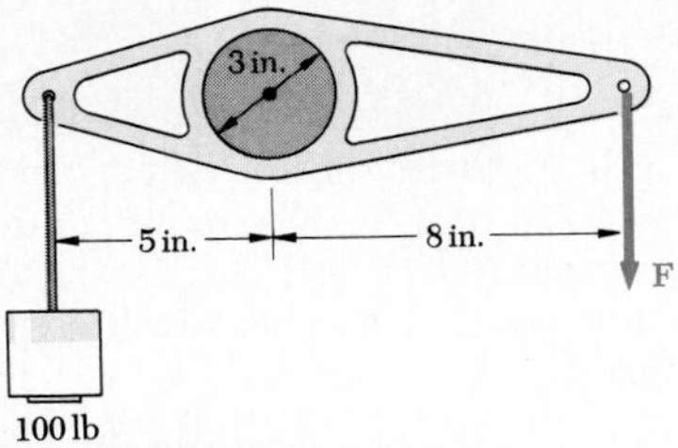

Fig. P8.68

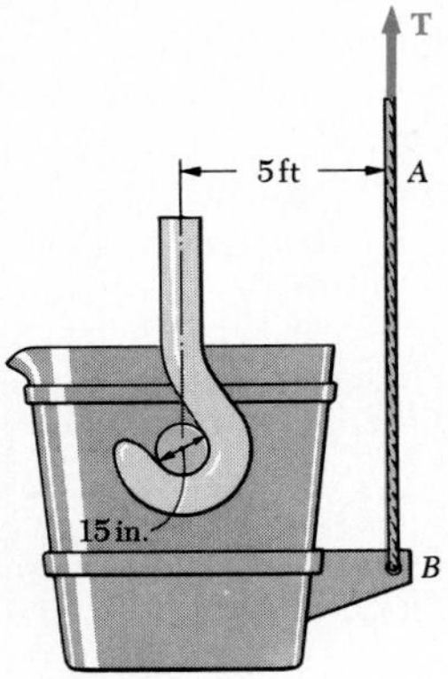

Fig. P8.69

PROBLEMS

8.68 A lever of negligible weight is loosely fitted onto a 3-in.-diameter fixed shaft as shown. It is observed that a force **F** of magnitude 70 lb will just start rotating the lever clockwise. Determine (*a*) the coefficient of friction between the shaft and the lever, (*b*) the smallest force **F** for which the lever does not start rotating counterclockwise.

8.69 A hot-metal ladle and its contents weigh 60 tons. Knowing that the coefficient of friction between the hooks and the pinion is 0.30, determine the tension in the cable *AB* required to start tipping the ladle.

8.70 A bushing of outside diameter 75 mm fits loosely on a horizontal 50-mm-diameter shaft. A horizontal force **P** of magnitude 550 N is required to pull one end of the rope to the right when a 500-N force is applied to its other end. Determine the coefficient of friction between the shaft and the bushing. Assume that the rope does not slip on the bushing.

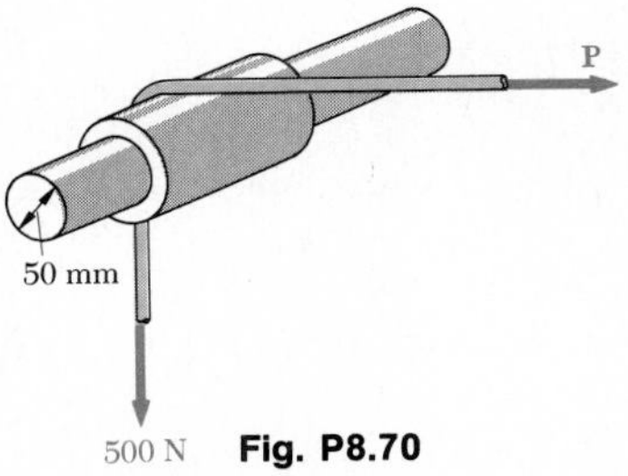

Fig. P8.70

8.71 A 600-N load is to be raised by the block and tackle shown. Each of the 60-mm-diameter pulleys rotates on a 10-mm-diameter axle. Knowing that $\mu = 0.20$, determine the tension in each portion of the rope as the load is slowly raised.

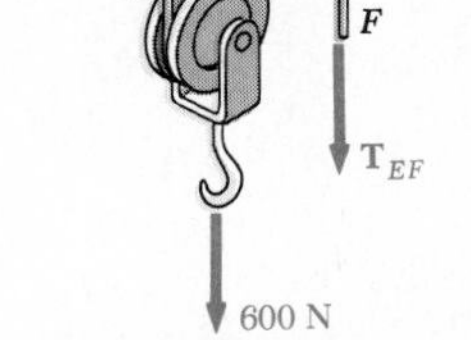

Fig. P8.71

8.72 In Prob. 8.71, determine the tension in each portion of the rope as the 600-N load is slowly lowered.

8.73 A certain railroad freight car has eight steel wheels of 32-in. diameter which are supported on 5-in.-diameter axles. Assuming $\mu = 0.015$, determine the horizontal force per ton of load required to move the car at constant velocity.

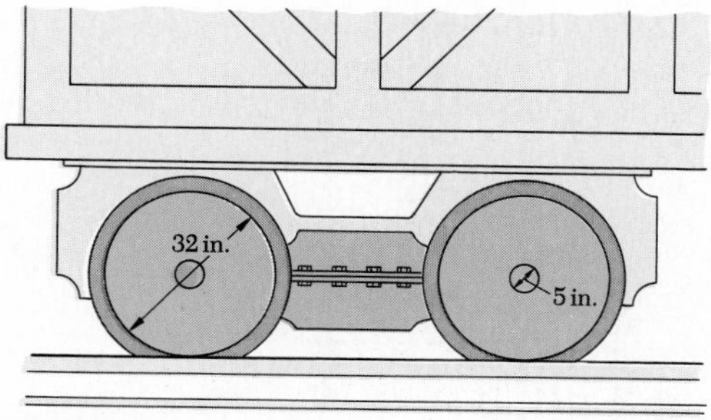

Fig. P8.73

8.74 A scooter is to be designed to roll down a 2 percent slope at constant speed. Assuming that the coefficient of kinetic friction between the 1-in. axles and the bearings is 0.10, determine the required diameter of the wheels. Neglect the rolling resistance between the wheels and the ground.

8.75 A couple of moment 150 lb · ft is required to start the vertical shaft rotating. Determine the coefficient of static friction.

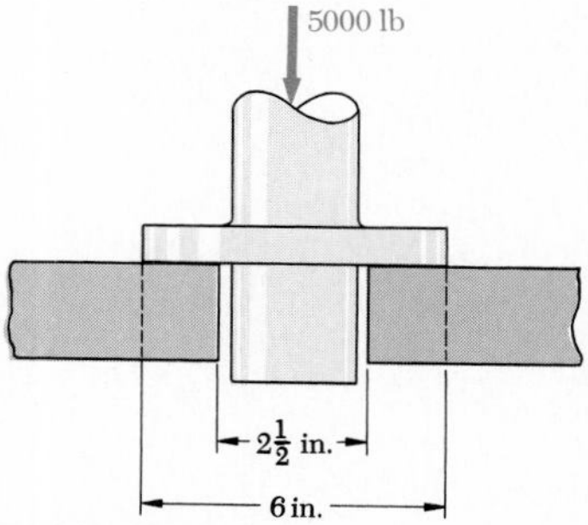

Fig. P8.75

8.76 A 50-lb electric floor polisher is operated on a surface for which the coefficient of friction is 0.25. Assuming the normal force per unit area between the disk and the floor to be uniform, determine the magnitude Q of the horizontal forces required to prevent motion of the machine.

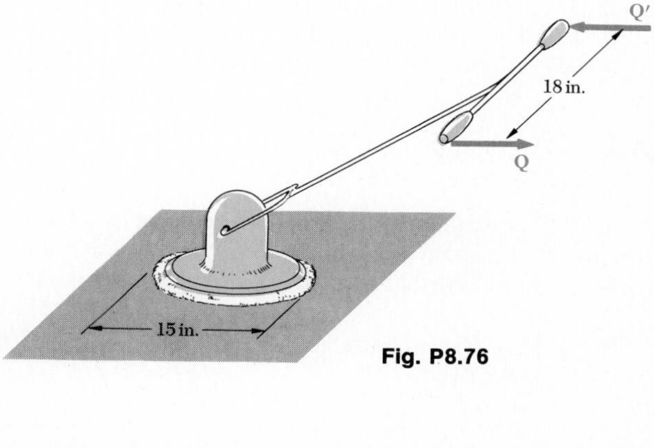

Fig. P8.76

8.77 Four springs, each of constant 6 kN/m, are used to press plate AB against the head on the vertical rod. In the position shown each spring has been compressed 50 mm. The plate AB may move vertically, but it is constrained against rotation by bolts which pass through holes cut in the plate. Knowing that $\mu = 0.30$ and $\mathbf{P} = 0$, determine the moment of the couple $\mathbf{M}$ required to rotate the rod.

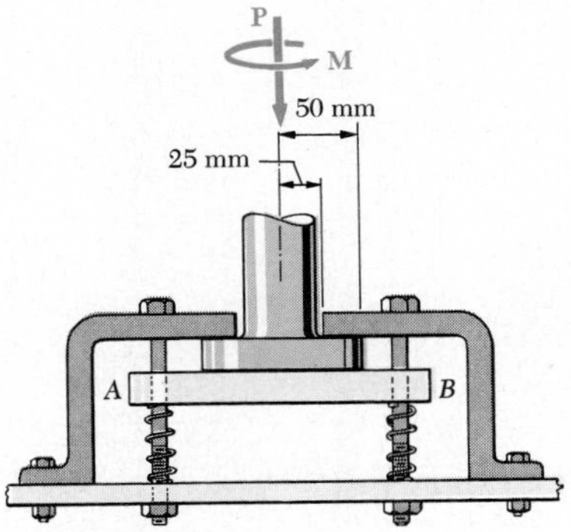

Fig. P8.77

8.78 In Prob. 8.77, determine the required magnitude of the force $\mathbf{P}$ if the rod is to just rotate when the moment of the couple $\mathbf{M}$ is 20 N · m. For what magnitude of the force $\mathbf{P}$ is the rod most easily rotated?

***8.79** As the surfaces of shaft and bearing wear out, the frictional resistance of a thrust bearing decreases. It is generally assumed that the wear is directly proportional to the distance traveled by any given point of the shaft, and thus to the distance r from the point to the axis of the shaft. Assuming, then, that the normal force per unit area is inversely proportional to r, show that the moment M of the couple required to overcome the frictional resistance of a worn-out end bearing (with contact over the full circular area) is equal to 75 percent of the value given by formula (8.9) for a new bearing.

***8.80** Assuming that bearings wear out as indicated in Prob. 8.79, show that the moment M of the couple required to overcome the frictional resistance of a worn-out collar bearing is

$$M = \tfrac{1}{2}\mu_k P(R_1 + R_2)$$

where P = magnitude of the total axial force
R_1, R_2 = inner and outer radii of collar

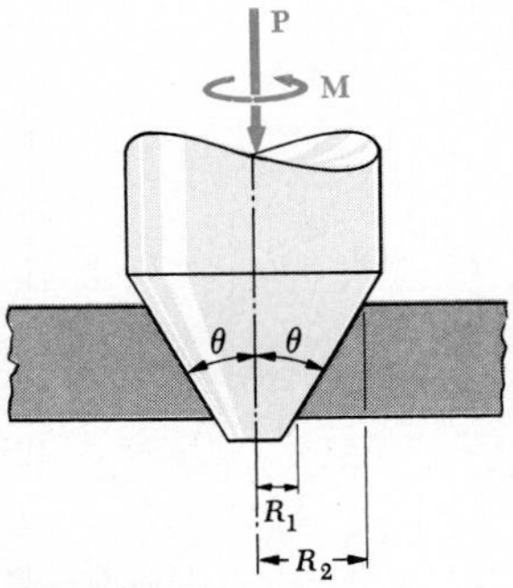

Fig. P8.81

***8.81** Assuming that the pressure between the surfaces of contact is uniform, show that the moment M of the couple required to overcome frictional resistance for the conical pivot shown is

$$M = \frac{2}{3}\frac{\mu_k P}{\sin\theta}\frac{R_2^3 - R_1^3}{R_2^2 - R_1^2}$$

8.82 Solve Prob. 8.76, assuming that between the disk and the surface the normal force per unit area varies uniformly from a maximum at the center to zero at the perimeter of the disk.

8.83 A circular disk of diameter 200 mm rolls at constant velocity down a 2 percent incline. Determine the coefficient of rolling resistance.

8.84 Determine the horizontal force required to move a 1200 kg automobile along a horizontal road at constant velocity. Neglect all forms of friction except rolling resistance, and assume the coefficient of rolling resistance to be 1.5 mm. The diameter of each tire is 600 mm.

8.85 Solve Prob. 8.73, including the effect of a coefficient of rolling resistance of 0.02 in.

8.86 Solve Prob. 8.74, including the effect of a coefficient of rolling resistance of 0.07 in.

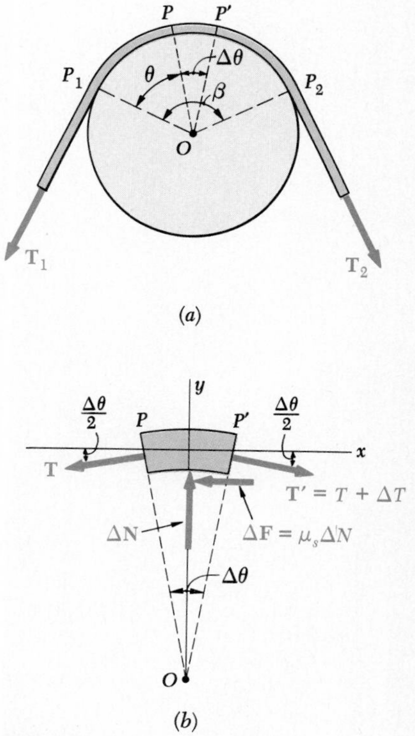

Fig. 8.14

8.10. Belt Friction.

Consider a flat belt passing over a fixed cylindrical drum (Fig. 8.14*a*). We propose to determine the relation existing between the values T_1 and T_2 of the tension in the two parts of the belt when the belt is just about to slide toward the right.

Let us detach from the belt a small element PP' subtending an angle $\Delta\theta$. Denoting by T the tension at P and by $T + \Delta T$ the tension at P', we draw the free-body diagram of the element of the belt (Fig. 8.14*b*). Besides the two forces of tension, the forces acting on the free body are the normal component $\Delta\mathbf{N}$ of the reaction of the drum and the friction force $\Delta\mathbf{F}$. Since motion is assumed to be impending, we have $\Delta F = \mu_s \Delta N$. It should be noted that if $\Delta\theta$ is made to approach zero, the magnitudes ΔN, ΔF, and the *difference* ΔT between the tension at P and the tension at P' will also approach zero; the value T of the tension at P, however, will remain unchanged. This observation helps in understanding our choice of notations.

Choosing the coordinate axes shown in Fig. 8.14*b*, we write the equations of equilibrium for the element PP':

$$\Sigma F_x = 0: \qquad (T + \Delta T)\cos\frac{\Delta\theta}{2} - T\cos\frac{\Delta\theta}{2} - \mu_s\,\Delta N = 0 \quad (8.11)$$

$$\Sigma F_y = 0: \qquad \Delta N - (T + \Delta T)\sin\frac{\Delta\theta}{2} - T\sin\frac{\Delta\theta}{2} = 0 \quad (8.12)$$

Solving Eq. (8.12) for ΔN and substituting into (8.11), we obtain after reductions

$$\Delta T \cos \frac{\Delta\theta}{2} - \mu_s(2T + \Delta T) \sin \frac{\Delta\theta}{2} = 0$$

We shall now divide both terms by $\Delta\theta$; as far as the first term is concerned, this will be simply done by dividing ΔT by $\Delta\theta$. The division of the second term is carried out by dividing the terms in the parentheses by 2 and the sine by $\Delta\theta/2$. We write

$$\frac{\Delta T}{\Delta\theta} \cos \frac{\Delta\theta}{2} - \mu_s \left(T + \frac{\Delta T}{2}\right) \frac{\sin (\Delta\theta/2)}{\Delta\theta/2} = 0$$

If we now let $\Delta\theta$ approach 0, the cosine approaches 1 and $\Delta T/2$ approaches zero as noted above. On the other hand, the quotient of $\sin (\Delta\theta/2)$ over $\Delta\theta/2$ approaches 1, according to a lemma derived in all calculus textbooks. Since the limit of $\Delta T/\Delta\theta$ is by definition equal to the derivative $dT/d\theta$, we write

$$\frac{dT}{d\theta} - \mu_s T = 0 \qquad \frac{dT}{T} = \mu_s \, d\theta$$

We shall now integrate both members of the last equation obtained from P_1 to P_2 (Fig. 8.14*a*). At P_1, we have $\theta = 0$ and $T = T_1$; at P_2, we have $\theta = \beta$ and $T = T_2$. Integrating between these limits, we write

$$\int_{T_1}^{T_2} \frac{dT}{T} = \int_0^{\beta} \mu_s \, d\theta$$

$$\ln T_2 - \ln T_1 = \mu_s \beta$$

$$\ln \frac{T_2}{T_1} = \mu_s \beta \qquad (8.13)$$

This relation may also be written in the form

$$\frac{T_2}{T_1} = e^{\mu_s \beta} \qquad (8.14)$$

The formulas we have derived apply equally well to problems involving flat belts passing over fixed cylindrical drums and to problems involving ropes wrapped around a post or capstan. They may also be used to solve problems involving band brakes. In such problems, it is the drum which is about to rotate, while the band remains fixed. The formulas may also be applied to problems involving belt drives. In these problems, both the pulley and the belt rotate; our concern is then to find whether the belt will slip, i.e., whether it will move *with respect* to the pulley.

Formulas (8.13) and (8.14) should be used only if the belt, rope, or brake is *about to slip*. Formula (8.14) will be used if T_1 or T_2 is desired; formula (8.13) will be preferred if either μ_s or the angle of contact β is desired.† We should note that T_2 is always larger than T_1; T_2 therefore represents the tension in that part of the belt or rope which *pulls*, while T_1 is the tension in the part which *resists*. We should also observe that the angle of contact β must be expressed in *radians*. The angle β may be larger than 2π; for example, if a rope is wrapped n times around a post, β is equal to $2\pi n$.

If the belt, rope, or brake is actually slipping, formulas similar to (8.13) and (8.14), but involving the coefficient of kinetic friction μ_k, should be used. If the belt, rope, or brake does not slip and is not about to slip, none of these formulas may be used.

The belts used in belt drives are often V-shaped. Such a belt, called a *V belt*, is shown in Fig. 8.15*a*. It is seen that contact between belt and pulley takes place along the sides of the groove. The relation existing between the values T_1 and T_2 of the tension in the two parts of the belt when the belt is just about to slip may again be obtained by drawing the free-body diagram of an element of belt (Fig. 8.15*b* and *c*). Equations similar to (8.11) and (8.12) are derived, but the magnitude of the total friction force acting on the element is now $2\,\Delta F$, and the sum of the y components of the normal forces is $2\,\Delta N \sin(\alpha/2)$. Proceeding as above, we obtain

$$\frac{T_2}{T_1} = e^{\mu_s \beta / \sin(\alpha/2)} \tag{8.15}$$

† Since the determination of a power of e and that of a logarithm to the base e (ln) involves the use of the same scales on the slide rule, there is no need for distinguishing between formulas (8.13) and (8.14) in carrying out computations on the slide rule. In every case, the product $\mu_s\beta$ should be read on the D scale and the quotient T_2/T_1 on one of the LL scales.

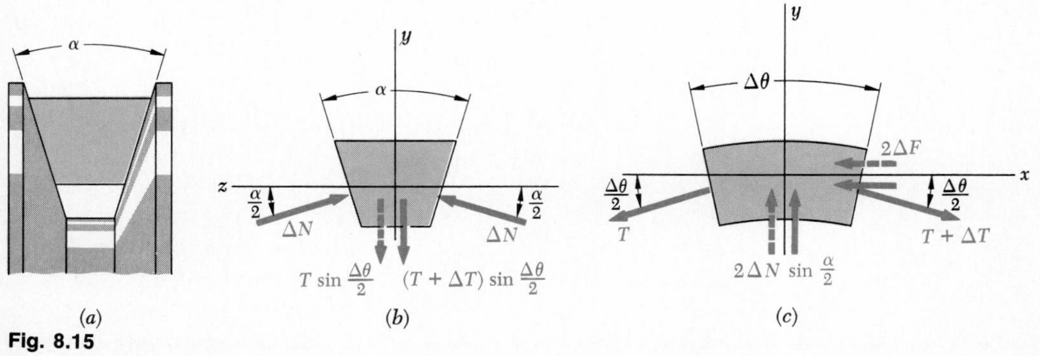

Fig. 8.15

SAMPLE PROBLEM 8.7

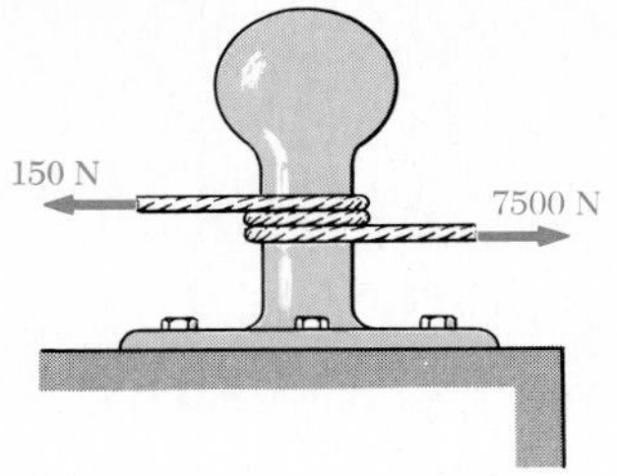

A hawser thrown from a ship to a pier is wrapped two full turns around a capstan. The tension in the hawser is 7500 N; by exerting a force of 150 N on its free end, a longshoreman can just keep the hawser from slipping. (*a*) Determine the coefficient of friction between the hawser and the capstan. (*b*) Determine the tension in the hawser that could be resisted by the 150-N force if the hawser were wrapped three full turns around the capstan.

a. **Coefficient of Friction.** Since slipping of the hawser is impending, we use Eq. (8.13),

$$\ln \frac{T_2}{T_1} = \mu_s \beta$$

Since the hawser is wrapped two full turns around the capstan, we have

$$\beta = 2(2\pi \text{ rad}) = 12.6 \text{ rad}$$
$$T_1 = 150 \text{ N} \qquad T_2 = 7500 \text{ N}$$

Therefore,

$$\mu_s \beta = \ln \frac{T_2}{T_1}$$

$$\mu_s(12.6 \text{ rad}) = \ln \frac{7500 \text{ N}}{150 \text{ N}} = \ln 50 = 3.91$$

$$\mu_s = 0.31 \blacktriangleleft$$

b. **Hawser Wrapped Three Turns around Capstan.** Using the value of μ_s obtained in part *a*, we have now

$$\beta = 3(2\pi \text{ rad}) = 18.9 \text{ rad}$$
$$T_1 = 150 \text{ N} \qquad \mu_s = 0.31$$

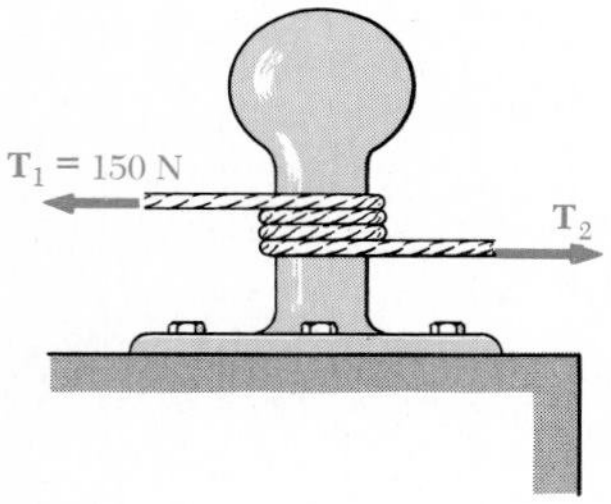

Substituting these values into Eq. (8.14), we obtain

$$\frac{T_2}{T_1} = e^{\mu_s \beta}$$

$$\frac{T_2}{150 \text{ N}} = e^{(0.31)(18.9)} = e^{5.86} = 350$$

$$T_2 = 52\,500 \text{ N} \qquad T_2 = 52.5 \text{ kN} \blacktriangleleft$$

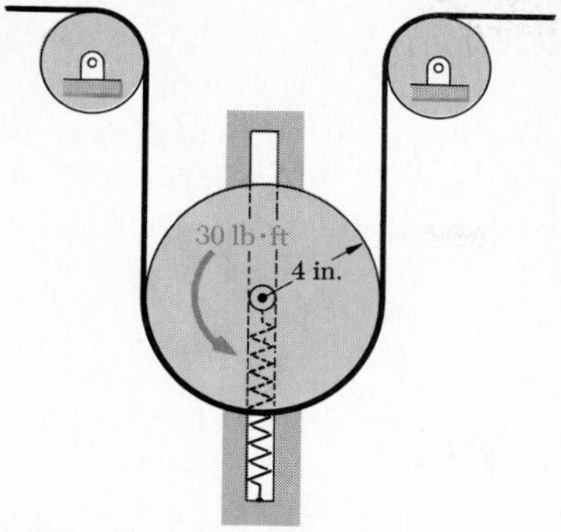

SAMPLE PROBLEM 8.8

A flat belt passes over two idler pulleys and under a rotating drum of diameter 8 in. The axle of the drum is free to move vertically in a slot, and a spring keeps the drum in contact with the belt. What is the minimum force which should be exerted by the spring if slippage is not to occur when a 30-lb · ft torque is applied to the drum? The coefficient of static friction between belt and drum is 0.30.

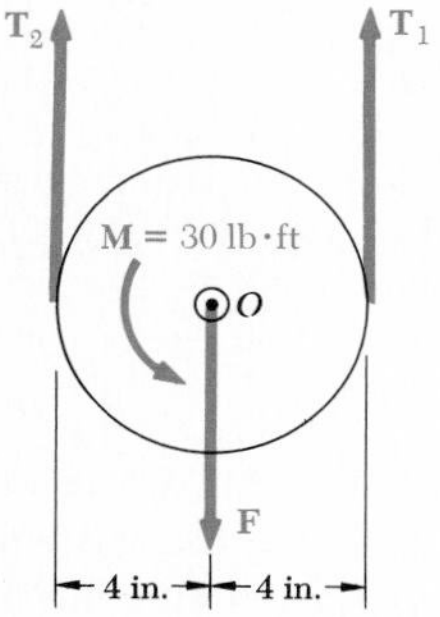

Solution. The free-body diagram of the drum is shown. The forces exerted on the drum by the spring and by each side of the belt are denoted by **F**, $\mathbf{T}_1$, and $\mathbf{T}_2$, respectively. The force $\mathbf{T}_2$ is shown on the left since, if slippage occurs, the belt will be observed as moving clockwise *with respect to the drum.*

$$+\circlearrowleft \Sigma M_O = 0: \qquad T_1(4 \text{ in.}) - T_2(4 \text{ in.}) + 30 \text{ lb} \cdot \text{ft} = 0$$
$$(T_2 - T_1)(4 \text{ in.}) = 360 \text{ lb} \cdot \text{in.}$$
$$T_2 - T_1 = 90 \text{ lb} \qquad (1)$$

The angle of contact between belt and drum is

$$\beta = 180° = 3.14 \text{ rad}$$

Since the values of T_1 and T_2 correspond to impending slippage, we use Eq. (8.14):

$$\frac{T_2}{T_1} = e^{\mu_s \beta}$$
$$= e^{(0.30)(3.14)} = 2.57$$
$$T_2 = 2.57T_1 \qquad (2)$$

Substituting T_2 from Eq. (2) into Eq. (1), we obtain

$$2.57T_1 - T_1 = 90 \text{ lb}$$
$$1.57T_1 = 90 \text{ lb}$$
$$T_1 = 57.3 \text{ lb}$$

$$T_2 = 2.57(57.3 \text{ lb}) = 147.3 \text{ lb}$$

Expressing that the sum of the vertical components of all forces must be zero:

$$+\uparrow \Sigma F_y = 0: \qquad T_1 + T_2 - F = 0$$
$$F = T_1 + T_2 = 57.3 \text{ lb} + 147.3 \text{ lb} = 204.6 \text{ lb}$$

$$\mathbf{F} = 205 \text{ lb} \downarrow \quad ◄$$

PROBLEMS

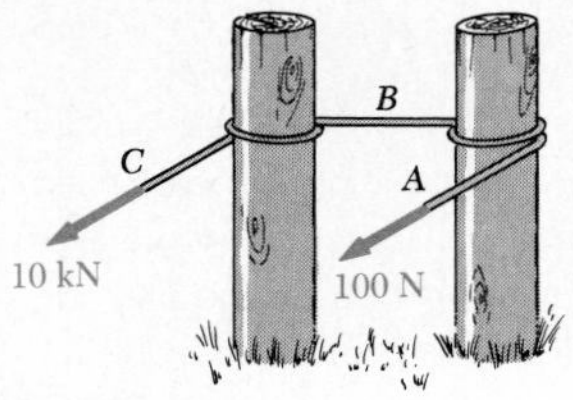

Fig. P8.87

8.87 A rope is wrapped around two posts as shown. If a 100-N force must be exerted at A to resist a 10-kN force at C, determine (*a*) the coefficient of static friction between the rope and the posts, (*b*) the corresponding tension in portion B of the rope.

8.88 A hawser is wrapped two full turns around a capstan head. By exerting a force of 160 lb on the free end of the hawser, a seaman can resist a force of 10,000 lb on the other end of the hawser. Determine (*a*) the coefficient of friction, (*b*) the number of times the hawser should be wrapped around the capstan if a 40,000-lb force is to be resisted by the same 160-lb force.

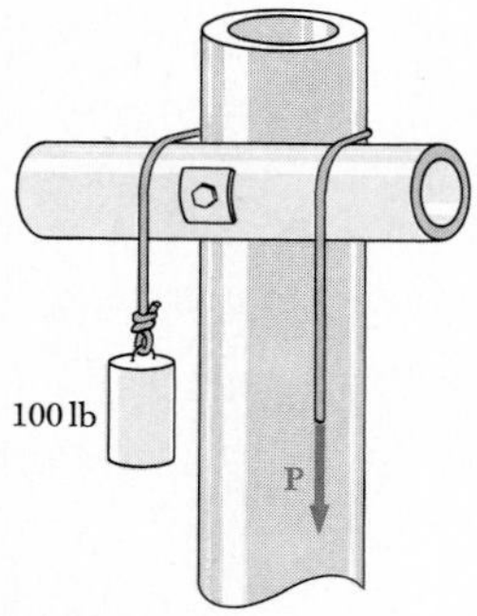

Fig. P8.89

8.89 Knowing that the coefficient of friction between the rope and the rods is 0.30, determine the range of values of P for which the rope will not slip.

8.90 Assume that the bushing of Prob. 8.70 has become frozen to the shaft and cannot rotate. Determine the coefficient of friction between the bushing and the rope if a force **P** of magnitude 600 N is required to pull the rope to the right.

8.91 A flat belt is used to deliver a torque to pulley A. Determine the largest torque which can be delivered to pulley A if the maximum allowable belt tension is 3 kN. Will the belt slip first on pulley A or on pulley B if the torque is increased?

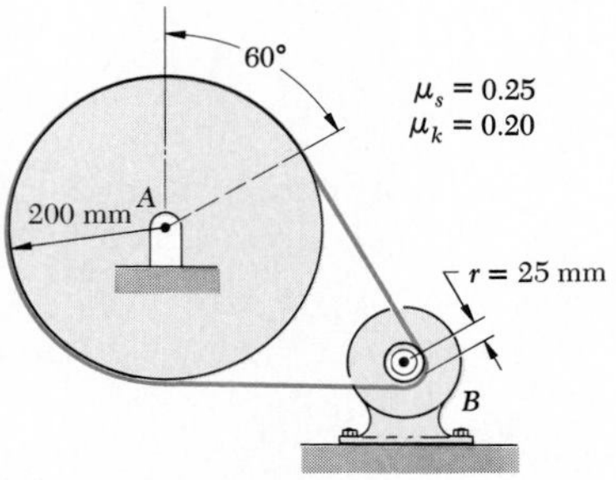

Fig. P8.91

8.92 A band brake is used to control the speed of a flywheel as shown. What torque must be applied to the flywheel in order to keep it rotating at a constant speed, when $P = 8$ lb? Assume that the flywheel rotates clockwise.

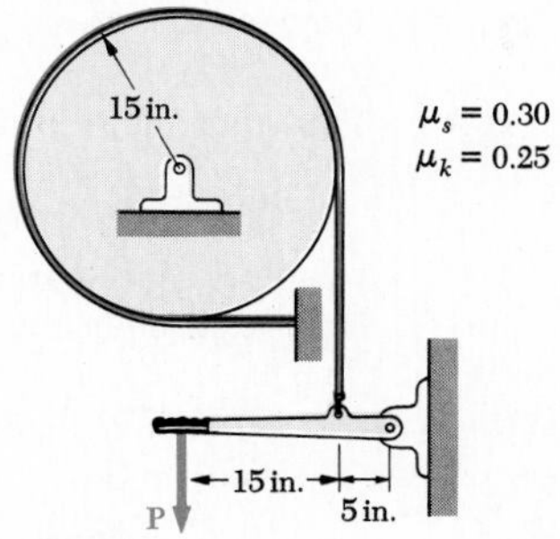

Fig. P8.92

8.93 Solve Prob. 8.92, assuming that the flywheel rotates counterclockwise.

8.94 A brake drum of radius $r = 150$ mm is rotating clockwise when a force **P** of magnitude 75 N is applied at D. Knowing that $\mu = 0.25$, determine the moment about O of the friction forces applied to the drum when $a = 75$ mm and $b = 400$ mm.

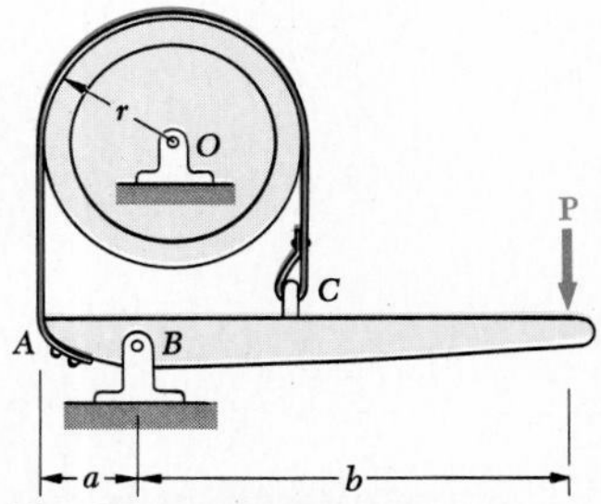

Fig. P8.94 and P8.95

8.95 Knowing that $r = 150$ mm and $a = 75$ mm, determine the maximum value of μ for which the brake is not self-locking. The brake drum rotates clockwise.

8.96 A cord is placed over two cylinders, each of 75-mm diameter. Knowing that $\mu = 0.25$, determine the largest mass m which can be raised if cylinder B is rotated slowly and cylinder A is fixed.

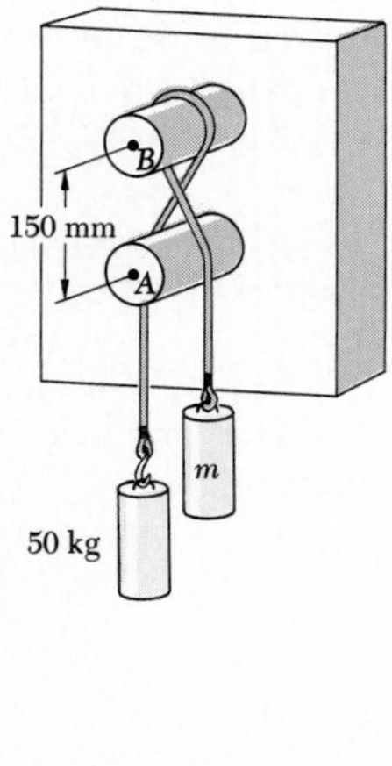

Fig. P8.96

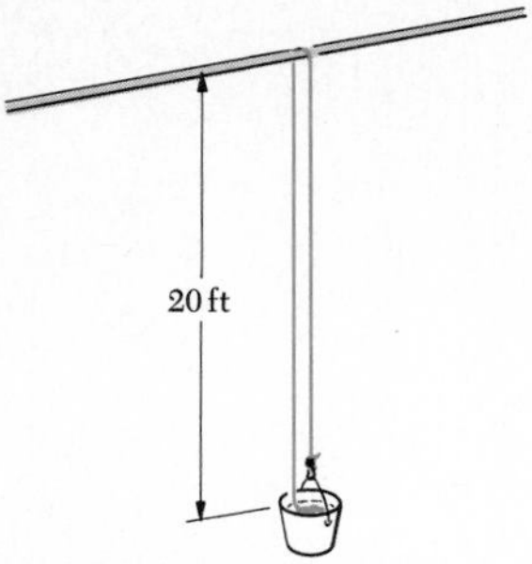

Fig. P8.97

8.97 A 50-ft rope passes over a small horizontal shaft; one end of the rope is attached to a bucket which weighs 5 lb. The excess rope is coiled inside the bucket. The coefficient of static friction between the rope and the shaft is 0.30, and the rope weighs 0.50 lb/ft. (*a*) If the shaft is held fixed, show that the system is in equilibrium. (*b*) If the shaft is slowly rotated, how far will the bucket rise before slipping? Neglect the diameter of the shaft.

8.98 The shaft of Prob. 8.97 is slowly rotated. Determine how far the bucket can be lowered before the rope slips on the shaft.

8.99 A cable is placed around three pipes, each of 100-mm outside diameter, located in the same horizontal plane. Two of the pipes are fixed and do not rotate; the third pipe is rotated slowly. Knowing that $\mu = 0.30$ for each pipe, determine the largest mass m which can be raised (*a*) if only pipe A is rotated, (*b*) if only pipe B is rotated, (*c*) if only pipe C is rotated.

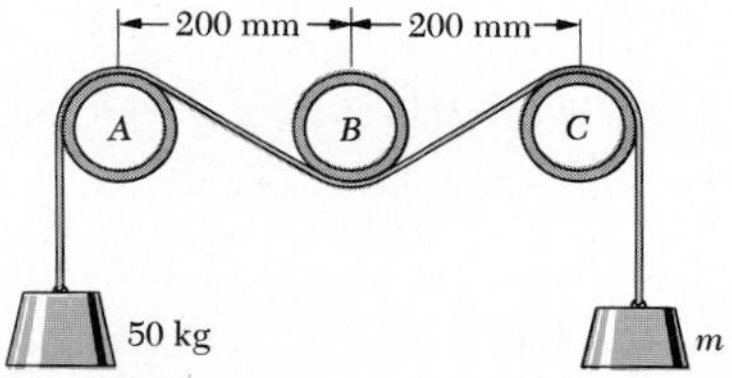

Fig. P8.99

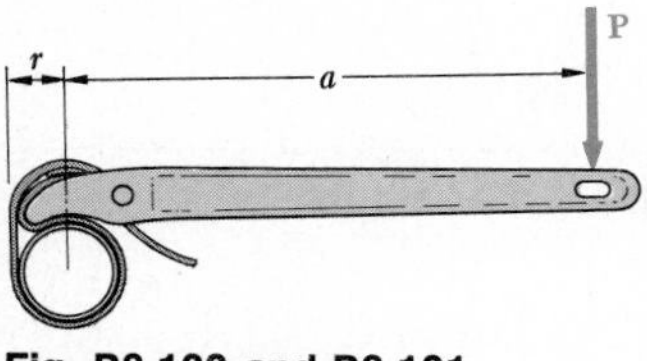

Fig. P8.100 and P8.101

8.100 The strap wrench shown is used to grip the pipe firmly and at the same time not mar the external surface of the pipe. Knowing that $a = 10$ in. and $r = 2$ in., determine the minimum coefficient of friction for which the wrench will be self-locking.

8.101 Denoting by μ the coefficient of friction between the strap and the pipe, determine the minimum value of the ratio r/a for which the wrench will be self-locking.

8.102 Solve Prob. 8.91, assuming that the flat belt and pulley are replaced by a V belt and V pulley with $\alpha = 28°$. (The angle α is as shown in Fig. 8.15*a*.)

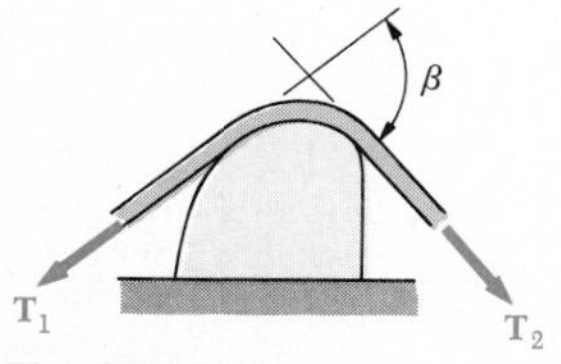

Fig. P8.104

8.103 Complete the derivation of Eq. (8.15), which relates the tension in both parts of a V belt.

8.104 Prove that Eqs. (8.13) and (8.14) are valid for any shape of surface provided that the coefficient of friction is the same at all points of contact.

REVIEW PROBLEMS

8.105 Two uniform rods, each of weight W, are held by frictionless pins A and B. It is observed that if the value of θ is greater than 10°, the rods will not remain in equilibrium. Determine the coefficient of friction at C.

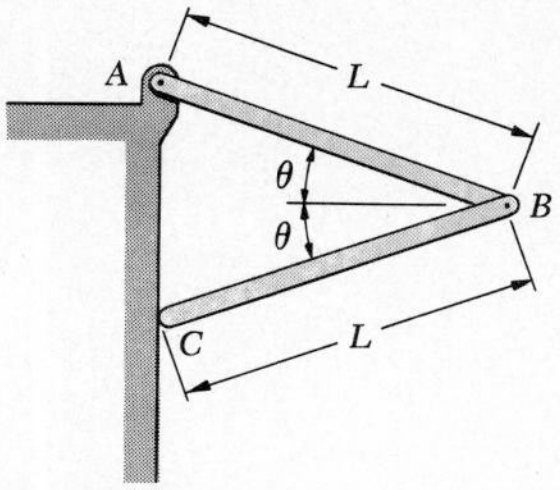

Fig. P8.105

8.106 Determine the smallest value of the coefficient of friction for which three identical cylindrical rods may be placed as shown.

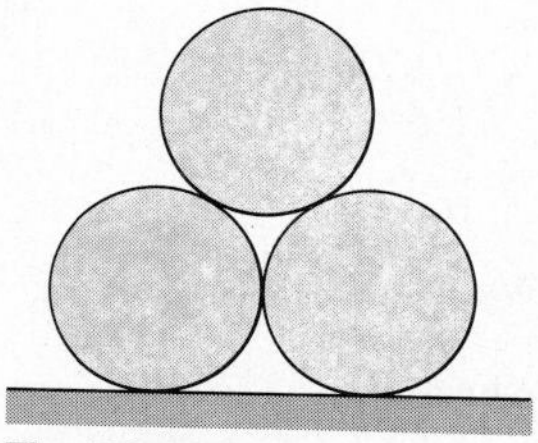
Fig. P8.106

8.107 A rope $ABCD$ is looped over two pipes as shown. Knowing that $\mu = 0.30$, determine (*a*) the smallest weight W for which equilibrium is possible, (*b*) the angle θ for which the corresponding tension in portion BC of the rope is 50 lb.

8.108 A rope $ABCD$ is looped over two pipes as shown. Pipe C is fixed and does not rotate; pipe B is rotated slowly counterclockwise. Knowing that $\mu = 0.30$ and $\theta = 45°$, determine (*a*) the largest weight W which can be raised, (*b*) the corresponding tension in portion BC of the rope.

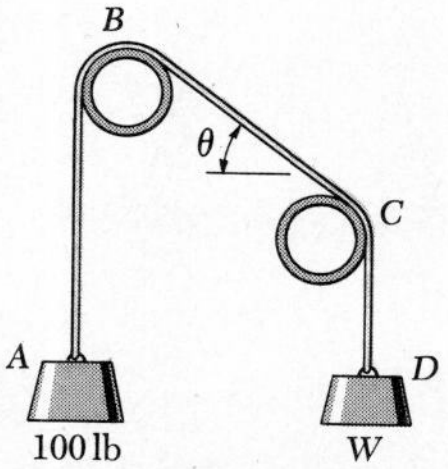

Fig. P8.107 and P8.108

8.109 The drive disk of the Scotch crosshead mechanism shown is slowly rotated counterclockwise by the couple **M**. A constant force **P** is applied to rod BC which slides in the frictionless bearings D and E. Denoting by μ the coefficient of friction between the pin A and the crosshead, determine the moment of the couple **M** in terms of P, r, μ, and θ.

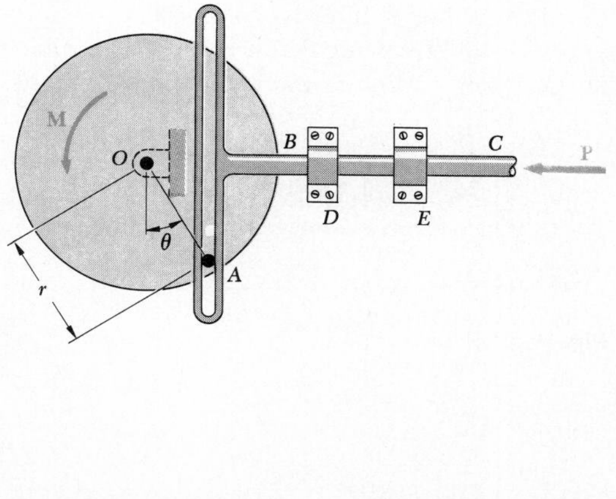

Fig. P8.109

8.110 A windlass, of diameter 150 mm, is used to raise a 40-kg load. It is supported by two axles and bearings of diameter 50 mm, poorly lubricated ($\mu = 0.40$). Determine the magnitude of the force **P** required to raise the load for each of the two positions shown.

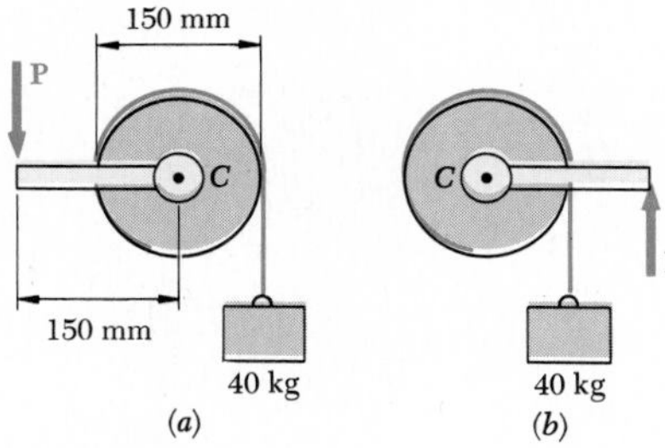

Fig. P8.110

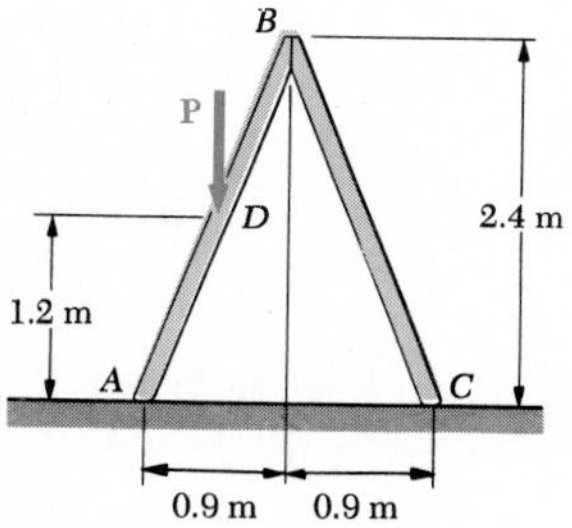

Fig. P8.111

8.111 Two identical uniform boards, each of mass 20 kg, are temporarily leaned against each other as shown. Knowing that the coefficient of friction between all surfaces is 0.40, determine (*a*) the largest magnitude of the force **P** for which equilibrium will be maintained, (*b*) the surface at which motion will impend.

8.112 Solve Prob. 8.111, assuming that the force **P** is applied at *D* and is directed horizontally to the right.

8.113 Knowing that $\mu = 0.20$ at all surfaces of contact, determine the magnitude of the force **P** required to move the 10-kg plate *B* to the left. (Neglect bearing friction in the pulley.)

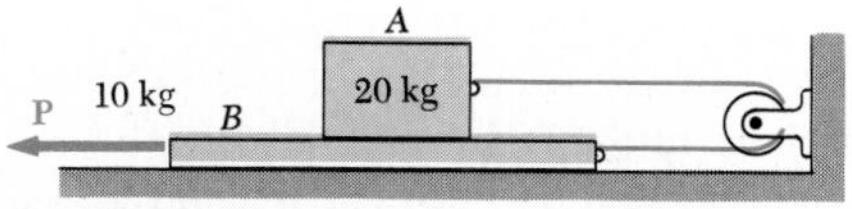

Fig. P8.113

8.114 A couple $\mathbf{M}_0$, of moment 150 lb · ft, must be applied to the drive pulley *B* in order to maintain constant motion in the conveyor shown. Knowing that $\mu = 0.20$, determine the minimun tension in the lower portion of the belt if no slipping is to occur.

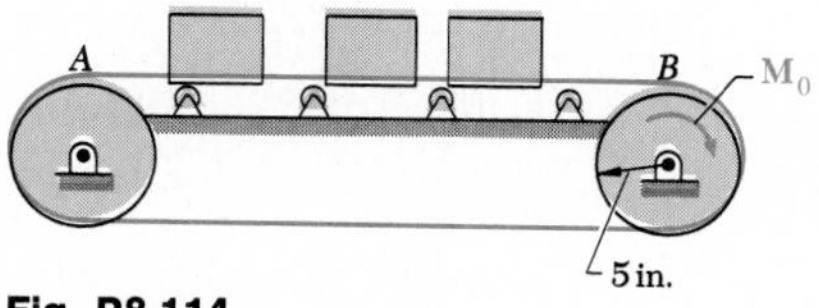

Fig. P8.114

8.115 A block of weight W rests on an incline which forms an angle θ with the horizontal plane. A horizontal force **P**, parallel to the incline, is applied to the block. Denoting by μ the coefficient of friction, determine (*a*) the magnitude of **P** required to move the block, (*b*) the direction in which the block moves. (*c*) Obtain numerical values for parts *a* and *b*, when $\theta = 15°$, $\mu = 0.30$, and $W = 50$ lb.

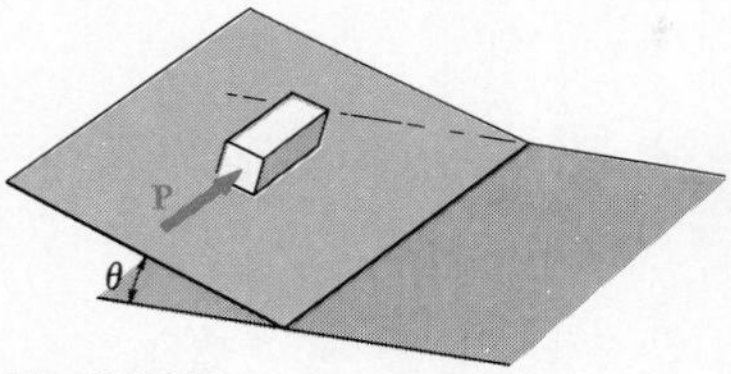

Fig. P8.115

8.116 A uniform slender rod is attached to a collar at B and rests on a circular cylindrical surface of radius r. The collar may slide without friction along a vertical guide. Denoting by ϕ the angle of friction between the rod and the surface, show that the largest and smallest values of θ corresponding to equilibrium must satisfy the equation

$$\tan\theta + \tan^2\theta \tan(\theta \mp \phi) = 1$$

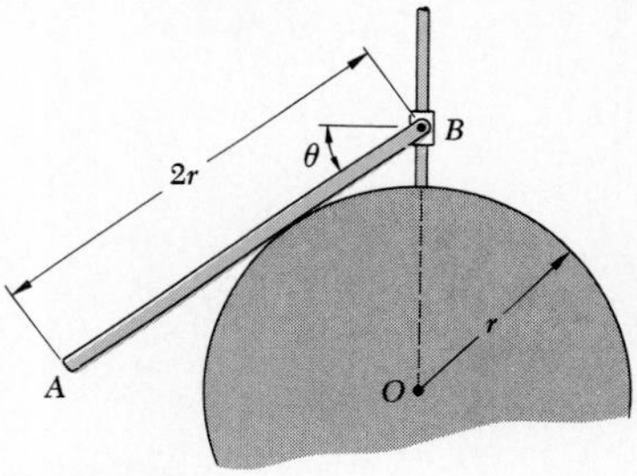

Fig. P8.116

Chapter
9

Distributed Forces: Moments of Inertia

MOMENTS OF INERTIA OF AREAS

9.1. Second Moment, or Moment of Inertia, of an Area. In Chap. 5, we analyzed various systems of forces distributed over an area. The two main types of forces considered were (1) weights of homogeneous plates of uniform thickness (Secs. 5.2 to 5.4) and (2) distributed loads on beams and hydrostatic forces (Secs. 5.6 and 5.7). In the case of homogeneous plates, the magnitude ΔW of the weight of an element of plate was proportional to the area ΔA of the element. In the case of distributed loads on beams, the magnitude ΔW of each elementary weight was represented by an element of area $\Delta A = \Delta W$ under the load curve; in the case of hydrostatic forces on submerged rectangular surfaces, a similar procedure was followed. Thus, in all cases considered in Chap. 5, the distributed forces were proportional to the elementary areas associated with them. The resultant of these forces, therefore, could be obtained by summing the corresponding areas, and the moment of the resultant about any given axis could be determined by computing the first moments of the areas about that axis.

In this chapter, we shall consider distributed forces $\Delta\mathbf{F}$ whose magnitudes depend not only upon the element of area ΔA on which they act but also upon the distance from ΔA to some given axis. More precisely, the magnitude of the force per unit area

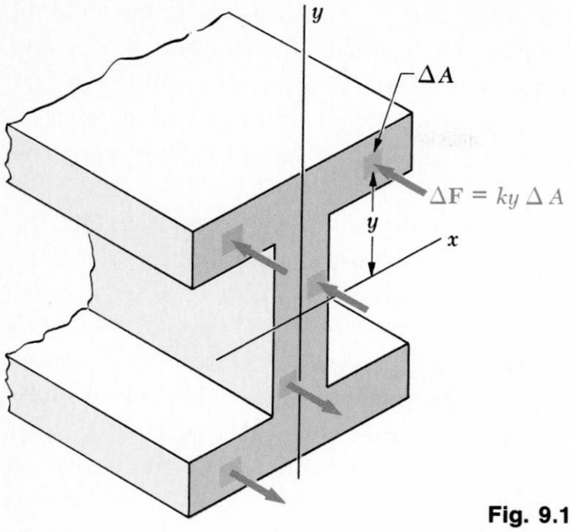

Fig. 9.1

$\Delta F/\Delta A$ will vary linearly with the distance to the axis.

Consider, for example, a beam of uniform cross section, subjected to two equal and opposite couples applied at each end of the beam. Such a beam is said to be in *pure bending*, and it is shown in mechanics of materials that the internal forces in any section of the beam are distributed forces whose magnitudes $\Delta F = ky\,\Delta A$ vary linearly with the distance y from an axis passing through the centroid of the section. This axis, represented by the x axis in Fig. 9.1, is known as the *neutral axis* of the section. The forces on one side of the neutral axis are forces of compression, and on the other side forces of tension, while on the neutral axis itself the forces are zero.

The magnitude of the resultant **R** of the elementary forces $\mathbf{\Delta F}$ over the entire section is

$$R = \int ky\,dA = k\int y\,dA$$

The last integral obtained is recognized as the *first moment* of the section about the x axis; it is equal to $\bar{y}A$ and to zero, since the centroid of the section is located on the x axis. The system of the forces $\mathbf{\Delta F}$ thus reduces to a couple. The moment M of this couple (bending moment) must be equal to the sum of the moments $\Delta M_x = y\,\Delta F = ky^2\,\Delta A$ of the elementary forces. Integrating over the entire section, we obtain

MOMENT OF AN AREA

$$M = \int ky^2\,dA = k\int y^2\,dA$$

The last integral is known as the *second moment*, or *moment of inertia*,† of the beam section with respect to the x axis and

MOMENT OF INERTIA

† The term second moment is more proper than the term moment of inertia, since, logically, the latter should be used only to denote integrals of mass (see Sec. 9.10). In common engineering practice, however, moment of inertia is used in connection with areas as well as masses.

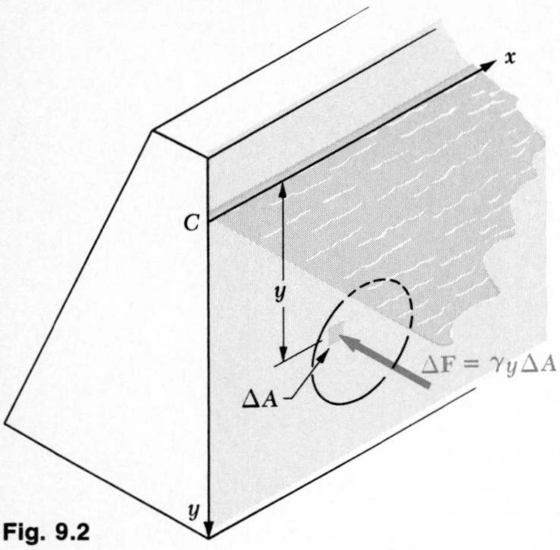

Fig. 9.2

is denoted by I_x. It is obtained by multiplying each element of area dA by the *square of its distance* from the x axis and integrating over the beam section. Since each product $y^2\,dA$ is positive, whether y is itself positive or negative (or zero if y is zero), the integral I_x will always be different from zero and positive.

Another example of second moment, or moment of inertia, of an area is provided by the following problem of hydrostatics: A vertical circular gate used to close the outlet of a large reservoir is submerged under water as shown in Fig. 9.2. What is the resultant of the forces exerted by the water on the gate, and what is the moment of the resultant about the line of intersection of the plane of the gate with the water surface (x axis)?

If the gate were rectangular, the resultant of the forces of pressure could be determined from the pressure curve, as was done in Sec. 5.7. Since the gate is circular, however, a more general method must be used. Denoting by y the depth of an element of area ΔA and by γ the specific weight of water, the pressure at the element is $p = \gamma y$, and the magnitude of the elementary force exerted on ΔA is $\Delta F = p\,\Delta A = \gamma y\,\Delta A$. The magnitude of the resultant of the elementary forces is thus

$$R = \int \gamma y\,dA = \gamma \int y\,dA$$

and may be obtained by computing the first moment of the area of the gate with respect to the x axis. The moment M_x of the resultant must be equal to the sum of the moments $\Delta M_x = y\,\Delta F = \gamma y^2\,\Delta A$ of the elementary forces. Integrating over the area of the gate, we have

$$M_x = \int \gamma y^2\,dA = \gamma \int y^2\,dA$$

Here again, the integral obtained represents the second moment, or moment of inertia, I_x of the area with respect to the x axis.

9.2. Determination of the Moment of Inertia of an Area by Integration. We have defined in the preceding section the second moment, or moment of inertia, of an area A with respect to the x axis. Defining in a similar way the moment of inertia I_y of the area A with respect to the y axis, we write (Fig. 9.3*a*)

$$I_x = \int y^2\,dA \qquad I_y = \int x^2\,dA \tag{9.1}$$

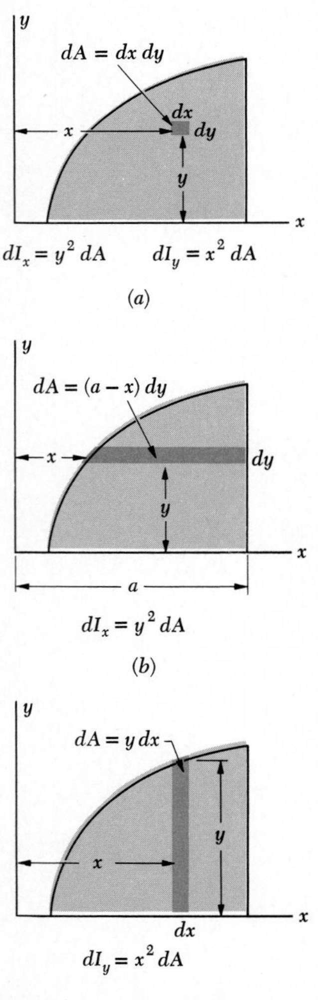

Fig. 9.3 (*c*)

These integrals, known as the *rectangular moments of inertia* of the area A, may be more easily computed if we choose for dA a thin strip parallel to one of the axes of coordinates. To compute I_x, the strip is chosen parallel to the x axis, so that all the points forming the strip are at the same distance y from the x axis (Fig. 9.3*b*); the moment of inertia dI_x of the strip is then obtained by multiplying the area dA of the strip by y^2. To compute I_y, the strip is chosen parallel to the y axis so that all the points forming the strip are at the same distance x from the y axis (Fig. 9.3c); the moment of inertia dI_y of the strip is $x^2\,dA$.

Moment of Inertia of a Rectangular Area. As an example, we shall determine the moment of inertia of a rectangle with respect to its base (Fig. 9.4). Dividing the rectangle into strips parallel to the x axis, we obtain

$$dA = b\,dy \qquad dI_x = y^2 b\,dy \qquad I_x = \int_0^h by^2\,dy = \tfrac{1}{3}bh^3 \qquad (9.2)$$

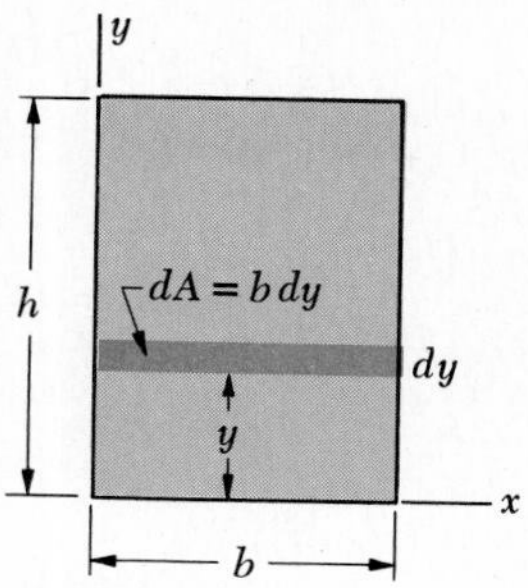

Fig. 9.4

Computing I_x and I_y from the Same Elementary Strips. The formula just derived may be used to determine the moment of inertia dI_x with respect to the x axis of a rectangular strip parallel to the y axis such as the one shown in Fig. 9.3c. Making $b = dx$ and $h = y$ in formula (9.2), we write

$$dI_x = \tfrac{1}{3}y^3\,dx$$

On the other hand, we have

$$dI_y = x^2\,dA = x^2 y\,dx$$

The same element may thus be used to compute the moments of inertia I_x and I_y of a given area (Fig. 9.5).

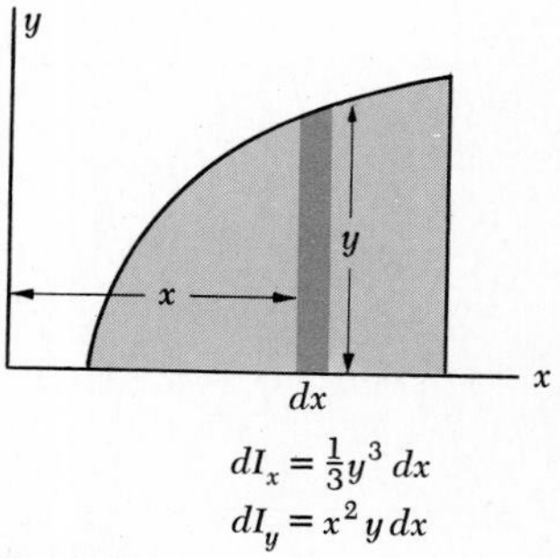

Fig. 9.5

9.3. Polar Moment of Inertia. An integral of great importance in problems concerning the torsion of cylindrical shafts and in problems dealing with the rotation of slabs is

$$J_O = \int r^2\,dA \qquad (9.3)$$

where r is the distance from the element of area dA to the pole O (Fig. 9.6). This integral is the *polar moment of inertia* of the area A with respect to O.

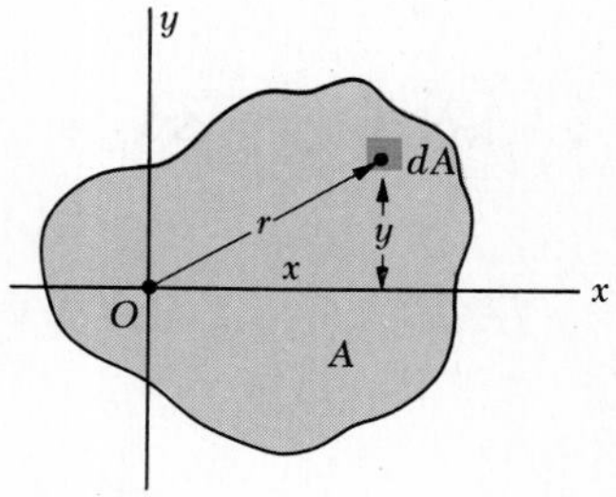

Fig. 9.6

The polar moment of inertia of a given area may be computed from the rectangular moments of inertia I_x and I_y of the area if these integrals are already known. Indeed, noting that $r^2 = x^2 + y^2$, we write

$$J_O = \int r^2\,dA = \int(x^2 + y^2)\,dA = \int y^2\,dA + \int x^2\,dA$$

that is,

$$J_O = I_x + I_y \qquad (9.4)$$

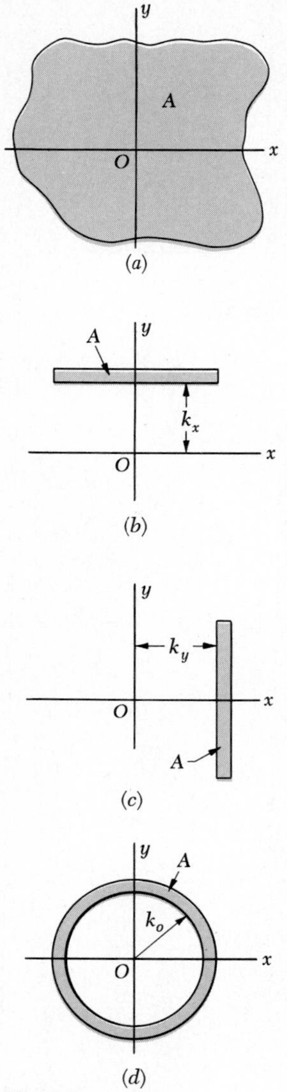

Fig. 9.7

9.4. Radius of Gyration of an Area. Consider an area A which has a moment of inertia I_x with respect to the x axis (Fig. 9.7*a*). Let us imagine that we concentrate this area into a thin strip parallel to the x axis (Fig. 9.7*b*). If the area A, thus concentrated, is to have the same moment of inertia with respect to the x axis, the strip should be placed at a distance k_x from the x axis, defined by the relation

$$I_x = k_x^2 A$$

Solving for k_x, we write

$$k_x = \sqrt{\frac{I_x}{A}} \tag{9.5}$$

The distance k_x is referred to as the *radius of gyration* of the area with respect to the x axis. We may define in a similar way the radii of gyration k_y and k_O; we write

$$I_y = k_y^2 A \qquad k_y = \sqrt{\frac{I_y}{A}} \tag{9.6}$$

$$J_O = k_O^2 A \qquad k_O = \sqrt{\frac{J_O}{A}} \tag{9.7}$$

Substituting for J_O, I_x, and I_y in terms of the radii of gyration in the relation (9.4), we observe that

$$k_O^2 = k_x^2 + k_y^2 \tag{9.8}$$

Example. As an example, let us compute the radius of gyration k_x of the rectangle shown in Fig. 9.4. Using formulas (9.5) and (9.2), we write

$$k_x^2 = \frac{I_x}{A} = \frac{\frac{1}{3}bh^3}{bh} = \frac{h^2}{3} \qquad k_x = \frac{h}{\sqrt{3}}$$

The radius of gyration k_x of the rectangle is shown in Fig. 9.8. It should not be confused with the ordinate $\bar{y} = h/2$ of the centroid of the area. While k_x depends upon the *second moment,* or moment of inertia, of the area, the ordinate $\bar{y}$ is related to the *first moment* of the area.

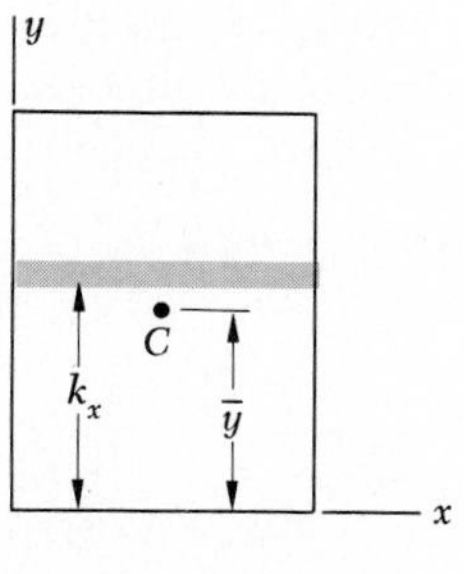

Fig. 9.8

SAMPLE PROBLEM 9.1

Determine the moment of inertia of a triangle with respect to its base.

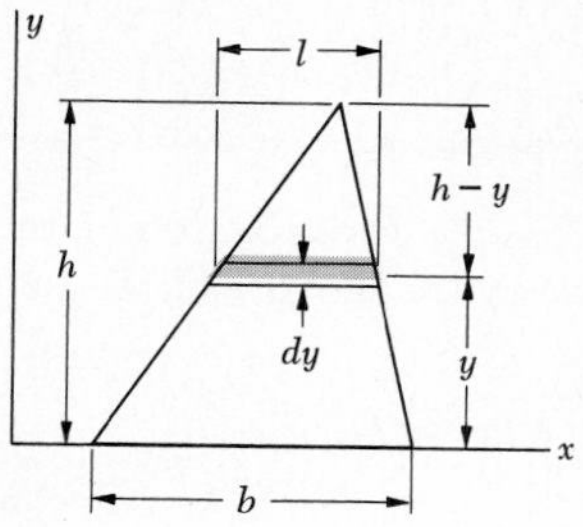

Solution. A triangle of base b and height h is drawn; the x axis is chosen to coincide with the base. A differential strip parallel to the x axis is chosen. Since all portions of the strip are at the same distance from the x axis, we write

$$dI_x = y^2\,dA \qquad dA = l\,dy$$

From the similar triangles, we have

$$\frac{l}{b} = \frac{h-y}{h} \qquad l = b\frac{h-y}{h} \qquad dA = b\frac{h-y}{h}\,dy$$

Integrating dI_x from $y = 0$ to $y = h$, we obtain

$$I_x = \int y^2\,dA = \int_0^h y^2 b\frac{h-y}{h}\,dy = \frac{b}{h}\int_0^h (hy^2 - y^3)\,dy$$

$$= \frac{b}{h}\left[h\frac{y^3}{3} - \frac{y^4}{4}\right]_0^h \qquad I_x = \frac{bh^3}{12} ◄$$

SAMPLE PROBLEM 9.2

(*a*) Determine the centroidal polar moment of inertia of a circular area by direct integration. (*b*) Using the result of part *a*, determine the moment of inertia of a circular area with respect to a diameter.

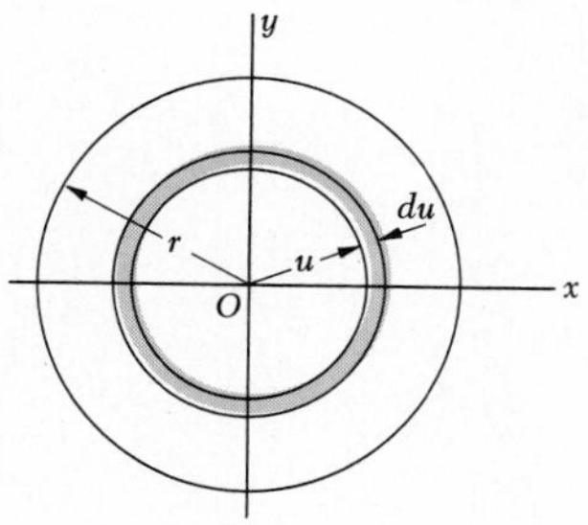

a. Polar Moment of Inertia. An annular differential element of area is chosen. Since all portions of this differential area are at the same distance from the origin, we write

$$dJ_O = u^2\,dA \qquad dA = 2\pi u\,du$$

$$J_O = \int dJ_O = \int_0^r u^2(2\pi u\,du) = 2\pi\int_0^r u^3\,du$$

$$J_O = \frac{\pi}{2}r^4 ◄$$

b. Moment of Inertia. Because of the symmetry of the circular area we have $I_x = I_y$. We then write

$$J_O = I_x + I_y = 2I_x \qquad \frac{\pi}{2}r^4 = 2I_x \qquad I_{\text{diameter}} = I_x = \frac{\pi}{4}r^4 ◄$$

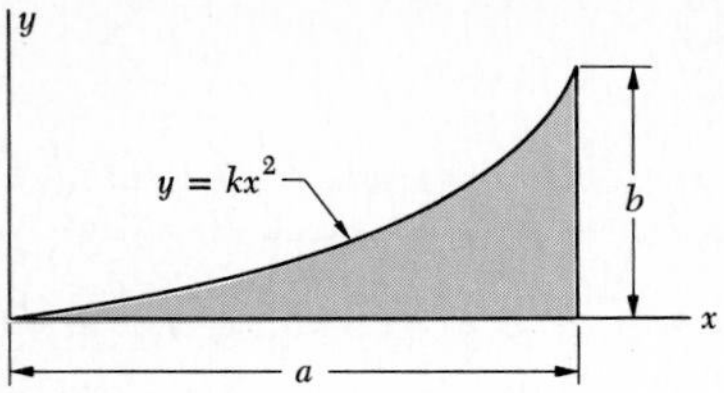

SAMPLE PROBLEM 9.3

(*a*) Determine the moment of inertia of the shaded area shown with respect to each of the coordinate axes. This area has also been considered in Sample Prob. 5.4. (*b*) Using the results of part *a*, determine the radius of gyration of the shaded area with respect to each of the coordinate axes.

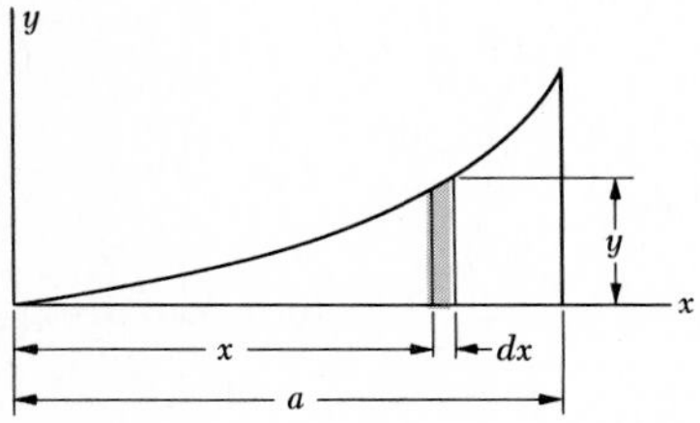

Solution. Referring to Sample Prob. 5.4, we obtain the following expressions for the equation of the curve and the total area:

$$y = \frac{b}{a^2}x^2 \qquad A = \tfrac{1}{3}ab$$

Moment of Inertia I_x. A vertical differential element of area is chosen. Since all portions of this element are *not* at the same distance from the x axis, we must treat the element as a thin rectangle. The moment of inertia of the element with respect to the x axis is

$$dI_x = \tfrac{1}{3}y^3\,dx = \frac{1}{3}\left(\frac{b}{a^2}x^2\right)^3 dx = \frac{1}{3}\frac{b^3}{a^6}x^6\,dx$$

$$I_x = \int dI_x = \int_0^a \frac{1}{3}\frac{b^3}{a^6}x^6\,dx = \left[\frac{1}{3}\frac{b^3}{a^6}\frac{x^7}{7}\right]_0^a$$

$$I_x = \frac{ab^3}{21} \quad ◄$$

Moment of Inertia I_y. The same vertical differential element of area is used. Since all portions of the element are at the same distance from the y axis, we write

$$dI_y = x^2\,dA = x^2(y\,dx) = x^2\left(\frac{b}{a^2}x^2\right)dx = \frac{b}{a^2}x^4\,dx$$

$$I_y = \int dI_y = \int_0^a \frac{b}{a^2}x^4\,dx = \left[\frac{b}{a^2}\frac{x^5}{5}\right]_0^a$$

$$I_y = \frac{a^3b}{5} \quad ◄$$

Radii of Gyration k_x and k_y

$$k_x^2 = \frac{I_x}{A} = \frac{ab^3/21}{ab/3} = \frac{b^2}{7} \qquad k_x = \sqrt{\tfrac{1}{7}}\,b \quad ◄$$

$$k_y^2 = \frac{I_y}{A} = \frac{a^3b/5}{ab/3} = \tfrac{3}{5}a^2 \qquad k_y = \sqrt{\tfrac{3}{5}}\,a \quad ◄$$

PROBLEMS

9.1 through 9.4 Determine by direct integration the moment of inertia of the shaded area with respect to the y axis.

9.5 through 9.8 Determine by direct integration the moment of inertia of the shaded area with respect to the x axis.

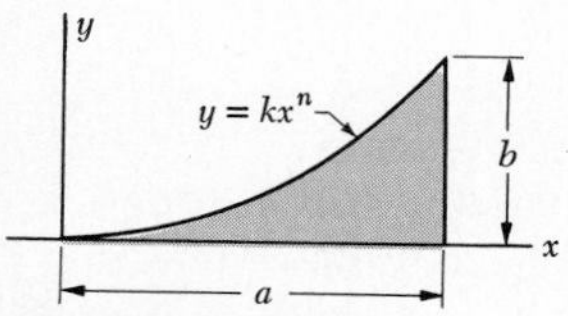

Fig. P9.1 and P9.5

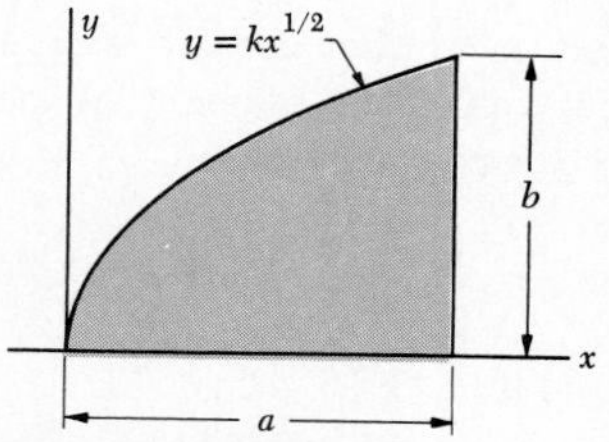

Fig. P9.2 and P9.6

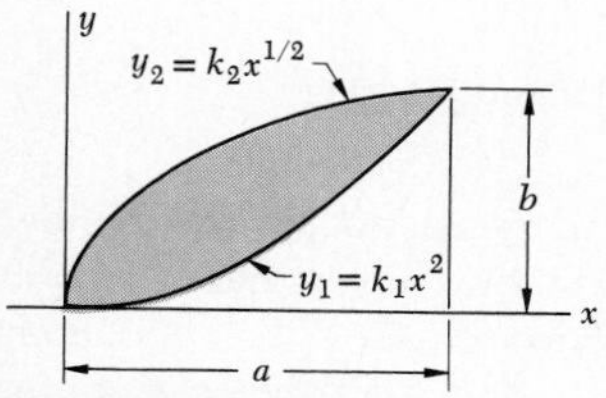

Fig. P9.3 and P9.7

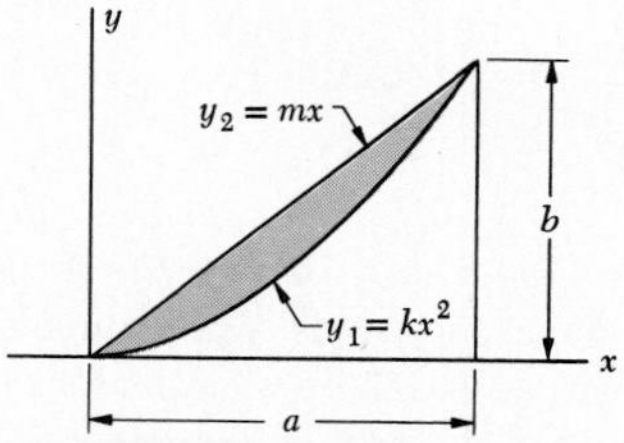

Fig. P9.4 and P9.8

9.9 and 9.10 Determine the moment of inertia and radius of gyration of the shaded area shown with respect to the x axis.

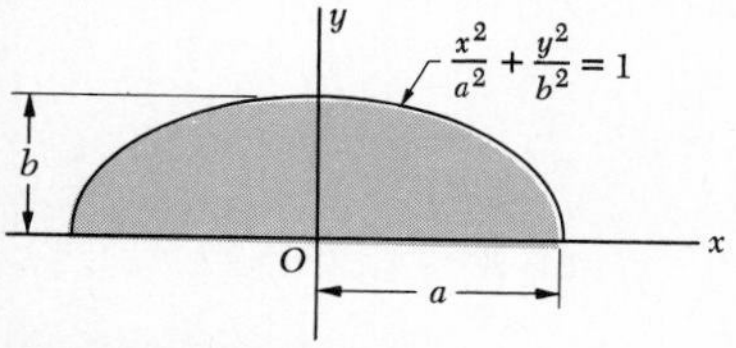

Fig. P9.9 and P9.11

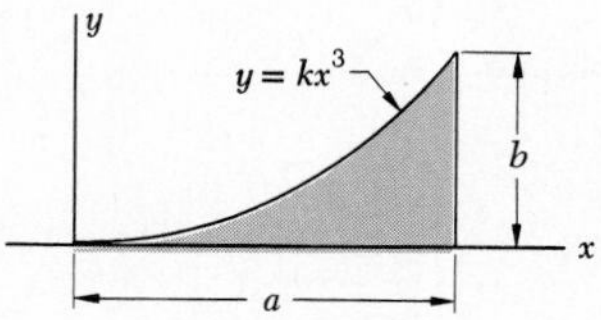

Fig. P9.10 and P9.12

9.11 and 9.12 Determine the moment of inertia and radius of gyration of the shaded area shown with respect to the y axis.

9.13 Determine the polar moment of inertia and polar radius of gyration of a square of side a with respect to the midpoint of one of its sides.

9.14 Determine the polar moment of inertia and the polar radius of gyration of an equilateral triangle of side a with respect to one of its vertices.

9.15 (*a*) Determine by direct integration the polar moment of inertia of the annular area shown. (*b*) Using the results of part *a*, determine the moment of inertia of the given area with respect to the x axis.

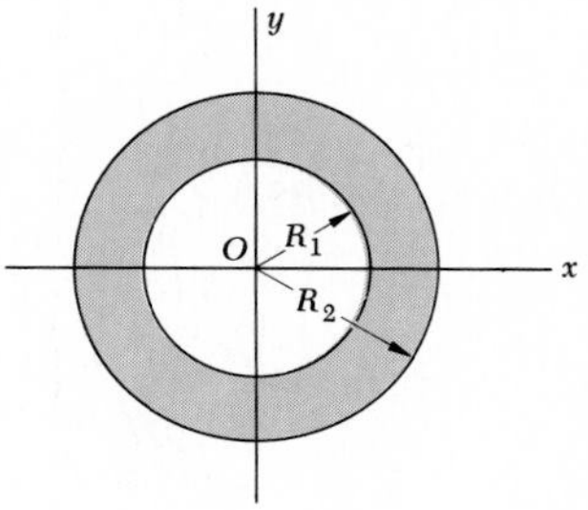

Fig. P9.15 and P9.16

9.16 (*a*) Show that the polar radius of gyration k_O of the annular area shown is approximately equal to the mean radius $R_m = (R_1 + R_2)/2$ for small values of the thickness $t = R_2 - R_1$. (*b*) Determine the percentage error introduced by using R_m in place of k_O for values of t/R_m respectively equal to 1, $\frac{1}{2}$, and $\frac{1}{10}$.

***9.17** Prove that the centroidal polar moment of inertia of a given area A cannot be smaller than $A^2/2\pi$. (*Hint.* Compare the moment of inertia of the given area with the moment of inertia of a circle of the same area and same centroid.)

***9.18** Determine the moment of inertia of the shaded area with respect to the x axis.

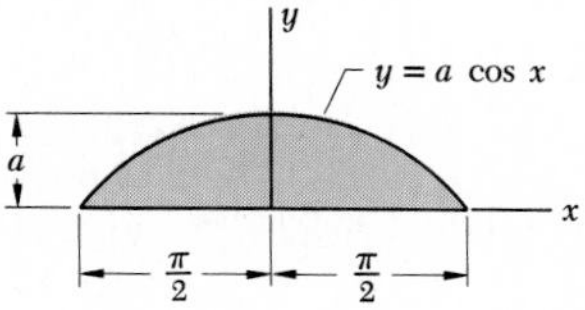

Fig. P9.18 and P9.19

***9.19** Determine the moment of inertia of the shaded area with respect to the y axis.

9.5. Parallel-Axis Theorem. Consider the moment of inertia I of an area A with respect to an axis AA' (Fig. 9.9). Denoting by y the distance from an element of area dA to AA', we write

$$I = \int y^2\, dA$$

Let us now draw an axis BB' parallel to AA' through the centroid C of the area; this axis is called a *centroidal axis.* Denoting by

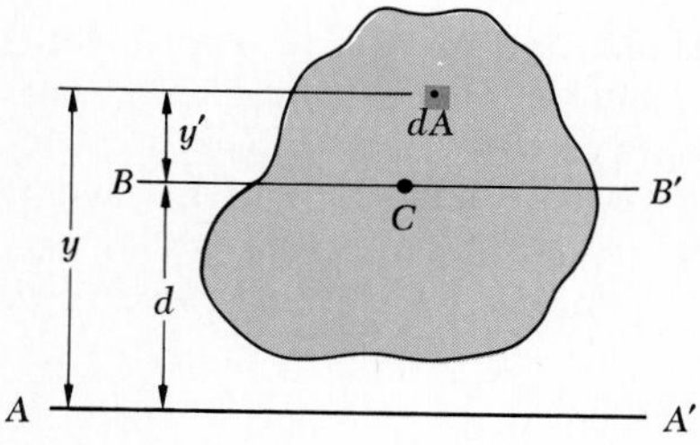

Fig. 9.9

y' the distance from the element dA to BB', we write $y = y' + d$, where d is the distance between the axes AA' and BB'. Substituting for y in the integral representing I, we write

$$\begin{aligned} I &= \int y^2\, dA = \int (y' + d)^2\, dA \\ &= \int y'^2\, dA + 2d\int y'\, dA + d^2 \int dA \end{aligned}$$

The first integral represents the moment of inertia $\bar{I}$ of the area with respect to the centroidal axis BB'. The second integral represents the first moment of the area with respect to BB'; since the centroid C of the area is located on that axis, the second integral must be zero. Finally, we observe that the last integral is equal to the total area A. We write therefore

$$I = \bar{I} + Ad^2 \tag{9.9}$$

This formula expresses that the moment of inertia I of an area with respect to any given axis AA' is equal to the moment of inertia $\bar{I}$ of the area with respect to a centroidal axis BB' parallel to AA' *plus* the product Ad^2 of the area A and of the square of the distance d between the two axes. This theorem is known as the *parallel-axis theorem.* Substituting k^2A for I and $\bar{k}^2A$ for $\bar{I}$, the theorem may also be expressed in the following way:

$$k^2 = \bar{k}^2 + d^2 \tag{9.10}$$

A similar theorem may be used to relate the polar moment of inertia J_O of an area about a point O and the polar moment of inertia $\bar{J}_C$ of the same area about its centroid C. Denoting by d the distance between O and C, we write

$$J_O = \bar{J}_C + Ad^2 \qquad \text{or} \qquad k_O^2 = \bar{k}_C^2 + d^2 \qquad (9.11)$$

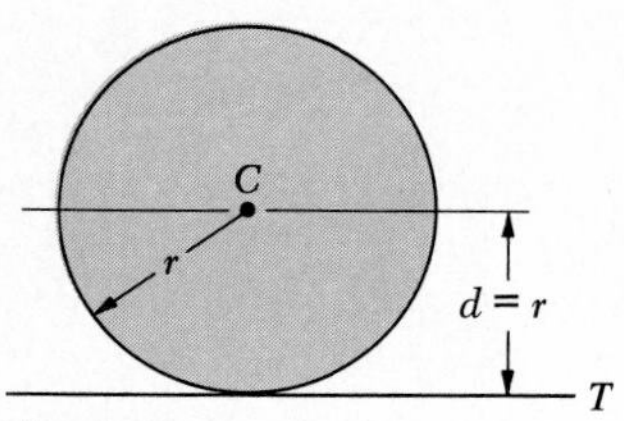

Fig. 9.10

Example 1. As an application of the parallel-axis theorem, we shall determine the moment of inertia I_T of a circular area with respect to a line tangent to the circle (Fig. 9.10). We found in Sample Prob. 9.2 that the moment of inertia of a circular area about a centroidal axis is $\bar{I} = \frac{1}{4}\pi r^4$. We may write, therefore,

$$I_T = \bar{I} + Ad^2 = \tfrac{1}{4}\pi r^4 + \pi r^2 r^2 = \tfrac{5}{4}\pi r^4$$

Example 2. The parallel-axis theorem may also be used to determine the centroidal moment of inertia of an area when the moment of inertia of this area with respect to some parallel axis is known. Consider, for instance, a triangular area (Fig. 9.11). We found in Sample Prob. 9.1 that the moment of inertia of a triangle with respect to its base AA' is equal to $\frac{1}{12}bh^3$. Using the parallel-axis theorem, we write

$$I_{AA'} = \bar{I}_{BB'} + Ad^2$$
$$\bar{I}_{BB'} = I_{AA'} - Ad^2 = \tfrac{1}{12}bh^3 - \tfrac{1}{2}bh(\tfrac{1}{3}h)^2 = \tfrac{1}{36}bh^3$$

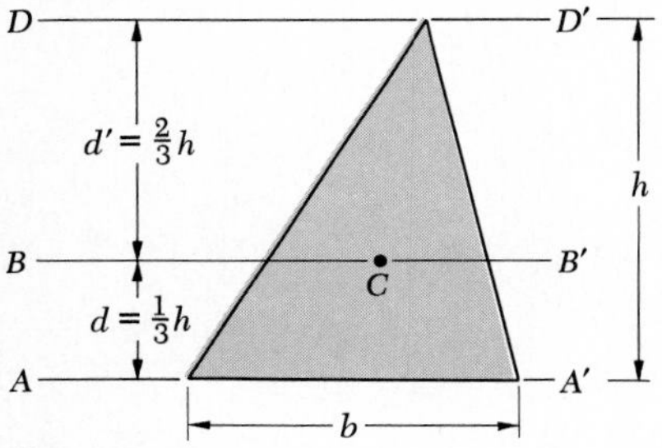

Fig. 9.11

It should be observed that the product Ad^2 was *subtracted* from the given moment of inertia in order to obtain the centroidal moment of inertia of the triangle. While this product is *added* in transferring *from* a centroidal axis to a parallel axis, it should be *subtracted* in transferring *to* a centroidal axis. In other words, the moment of inertia of an area is always smaller with respect to a centroidal axis than with respect to any other parallel axis.

Returning to Fig. 9.11, we observe that the moment of inertia of the triangle with respect to a line DD' drawn through a vertex may be obtained by writing

$$I_{DD'} = \bar{I}_{BB'} + Ad'^2 = \tfrac{1}{36}bh^3 + \tfrac{1}{2}bh(\tfrac{2}{3}h)^2 = \tfrac{1}{4}bh^3$$

Note that $I_{DD'}$ *could not* have been obtained directly from $I_{AA'}$. The parallel-axis theorem can be applied only if one of the two parallel axes passes through the centroid of the area.

9.6. Moments of Inertia of Composite Areas. Consider a composite area A made of several component areas A_1, A_2, etc. Since the integral representing the moment of inertia of A may be subdivided into integrals computed over A_1, A_2, etc., the moment of inertia of A with respect to a given axis will be obtained by adding the moments of inertia of the areas A_1, A_2, etc., with respect to the same axis. The moment of inertia of an area made of several of the common shapes shown in Fig. 9.12 may thus be obtained from the formulas given in that figure. Before adding the moments of inertia of the component areas, however, the parallel-axis theorem should be used to transfer each moment of inertia to the desired axis. This is shown in Sample Probs. 9.4 and 9.5.

Rectangle	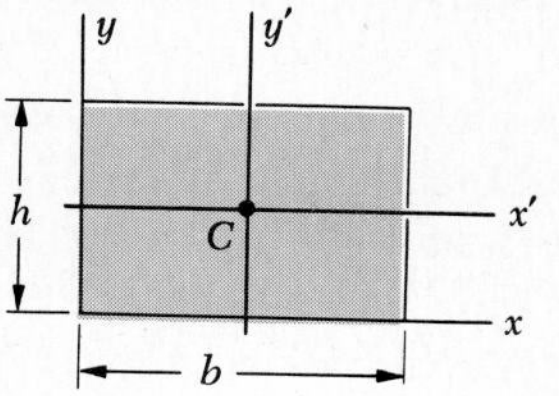	$\bar{I}_{x'} = \frac{1}{12}bh^3$ $\bar{I}_{y'} = \frac{1}{12}b^3h$ $I_x = \frac{1}{3}bh^3$ $I_y = \frac{1}{3}b^3h$ $J_C = \frac{1}{12}bh(b^2 + h^2)$
Triangle	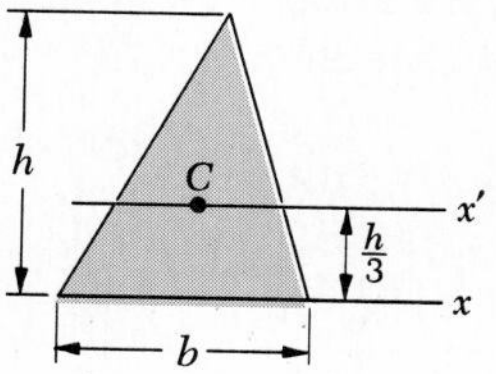	$\bar{I}_{x'} = \frac{1}{36}bh^3$ $I_x = \frac{1}{12}bh^3$
Circle	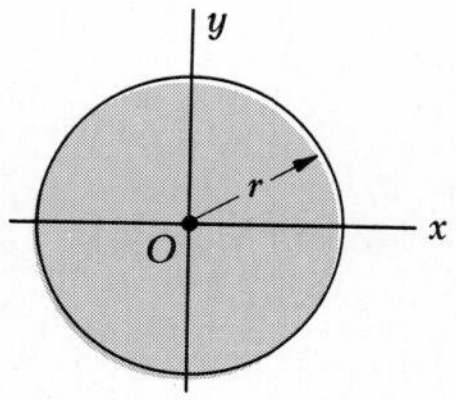	$\bar{I}_x = \bar{I}_y = \frac{1}{4}\pi r^4$ $J_O = \frac{1}{2}\pi r^4$
Semicircle	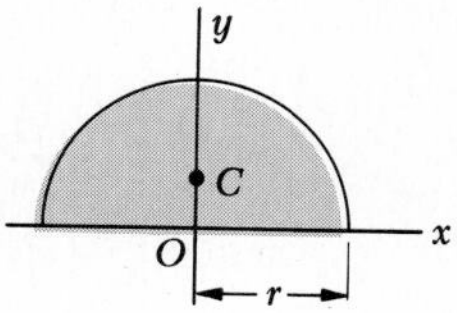	$I_x = I_y = \frac{1}{8}\pi r^4$ $J_O = \frac{1}{4}\pi r^4$
Quarter circle	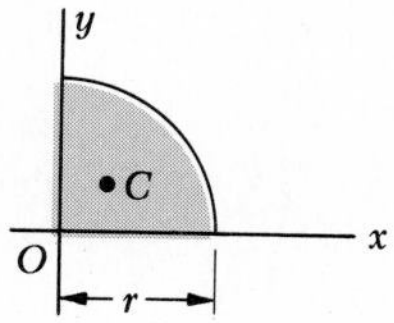	$I_x = I_y = \frac{1}{16}\pi r^4$ $J_O = \frac{1}{8}\pi r^4$
Ellipse	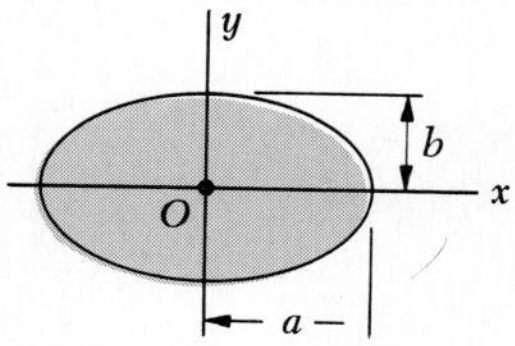	$\bar{I}_x = \frac{1}{4}\pi ab^3$ $\bar{I}_y = \frac{1}{4}\pi a^3b$ $J_O = \frac{1}{4}\pi ab(a^2 + b^2)$

Fig. 9.12 Moments of inertia of common geometric shapes.

The properties of the cross sections of various structural shapes are given in Fig. 9.13. As noted in Sec. 9.1, the moment of inertia of a beam section about its neutral axis is closely related to the value of the internal forces. The determination of moments of inertia is thus a prerequisite to the analysis and design of structural members.

It should be noted that the radius of gyration of a composite area is *not* equal to the sum of the radii of gyration of the component areas. In order to determine the radius of gyration of a composite area, it is necessary first to compute the moment of inertia of the area.

Fig. 9.13 Properties of rolled-steel structural shapes.

Shape		**Designation**	**Width, in.**
Wide-flange section	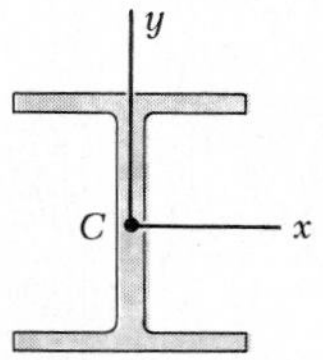	W16 × 64† W14 × 43 W8 × 31	$8\frac{1}{2}$ 8 8
American Standard beam	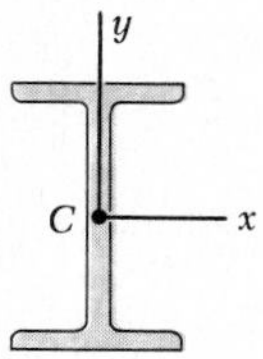	S18 × 70† S12 × 35 S6 × 12.5	$6\frac{1}{4}$ $5\frac{1}{8}$ $3\frac{3}{8}$
American Standard channel	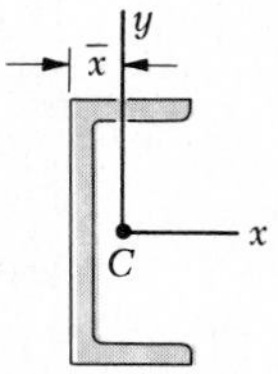	C10 × 25† C8 × 11.5 C6 × 8.2	$2\frac{7}{8}$ $2\frac{1}{4}$ $1\frac{7}{8}$
Angles	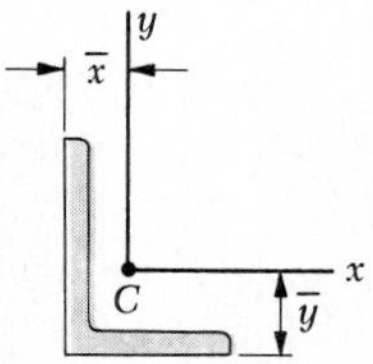	L6 × 6 × 1‡ L4 × 4 × $\frac{1}{2}$ L8 × 6 × 1 L5 × $3\frac{1}{2}$ × $\frac{1}{2}$	

† Depth in inches and weight per unit length in lb/ft.
‡ Depth, width, and thickness in inches.

Area, in^2	$\bar{I}_x$, in^4	$\bar{k}_x$, in.	$\bar{y}$, in.	$\bar{I}_y$, in^4	$\bar{k}_y$, in.	$\bar{x}$, in.
18.80	836	6.66	. . .	73.3	1.97	
12.60	429	5.82	. . .	45.1	1.89	
9.12	110	3.47	. . .	37.0	2.01	
20.6	926	6.71	. . .	24.1	1.08	
10.3	229	4.72	. . .	9.87	0.98	
3.67	21.1	2.45	. . .	1.8	0.71	
7.35	91.2	3.52	. . .	3.4	0.68	0.62
3.38	32.6	3.11	. . .	1.3	0.63	0.57
2.40	13.1	2.34	. . .	0.7	0.54	0.51
11.00	35.5	1.80	1.86	35.5	1.80	1.86
3.75	5.6	1.22	1.18	5.6	1.22	1.18
13.00	80.8	2.49	2.65	38.8	1.73	1.65
4.00	9.99	1.58	1.66	4.1	1.01	0.91

SAMPLE PROBLEM 9.4

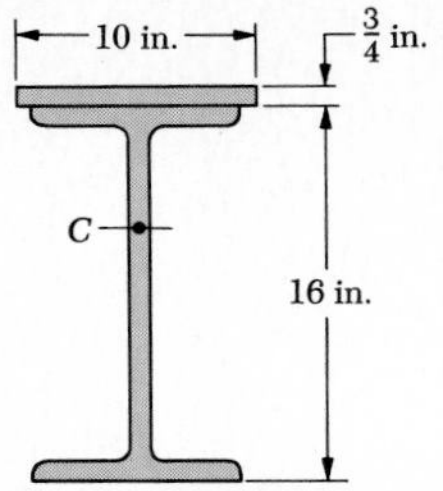

The strength of a 16-in., 64-lb wide-flange beam is increased by attaching a 10- by $\frac{3}{4}$-in. plate to its upper flange as shown. Determine the moment of inertia and the radius of gyration of the composite section with respect to an axis through its centroid C and parallel to the plate.

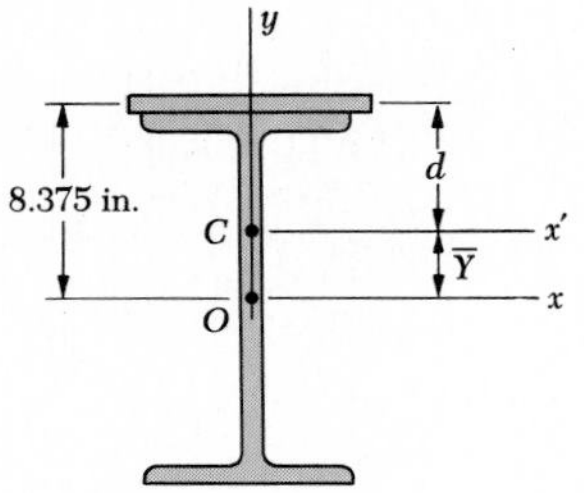

Solution. The origin of coordinates O is placed at the centroid of the wide-flange section, and the distance $\bar{Y}$ to the centroid of the composite section is computed by the methods of Chap. 5. The area of the wide-flange section is found by referring to Fig. 9.13.

Section	Area, in^2	$\bar{y}$, in.	$\bar{y}A$, in^3
Plate	7.5	8.375	62.8
Wide-flange section	18.8	0	0
	26.3	. . .	62.8

$$\bar{Y}\Sigma A = \Sigma\, \bar{y}A \qquad \bar{Y}(26.3) = 62.8 \qquad \bar{Y} = 2.39 \text{ in.}$$

Moment of Inertia. The parallel-axis theorem is used to determine the moments of inertia of the wide-flange section and of the plate with respect to the x' axis. This axis is a centroidal axis for the composite section, but *not* for either of the elements considered separately. The value of $\bar{I}_x$ for the wide-flange section is obtained from Fig. 9.13.

For the wide-flange section

$$I_{x'} = \bar{I}_x + A\bar{Y}^2 = 836 + (18.8)(2.39)^2 = 943 \text{ in}^4$$

For the plate

$$I_{x'} = \bar{I}_x + Ad^2 = (\tfrac{1}{12})(10)(\tfrac{3}{4})^3 + (7.5)(8.375 - 2.39)^2 = 269 \text{ in}^4$$

For the composite area

$$I_{x'} = 943 + 269 = 1212 \text{ in}^4$$

$$I_{x'} = 1212 \text{ in}^4 \quad ◀$$

Radius of Gyration

$$k_{x'}^2 = \frac{I_{x'}}{A} = \frac{1212 \text{ in}^4}{26.3 \text{ in}^2}$$

$$k_{x'} = 6.79 \text{ in.} \quad ◀$$

SAMPLE PROBLEM 9.5

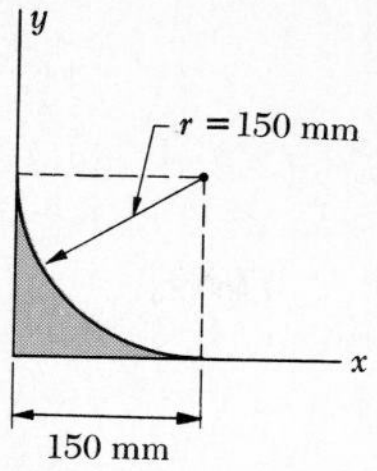

Determine the moment of inertia of the shaded area with respect to the y axis.

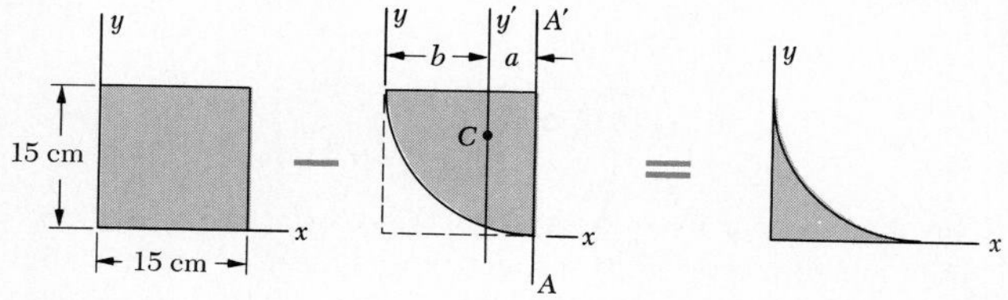

Solution. The given area may be obtained by subtracting a quarter circle from a square. The moments of inertia of the square and of the quarter circle are computed separately. To avoid excessively large numerical values, they will be expressed in cm^4 rather than in mm^4. We shall convert, therefore, the given data into centimeters.

I_y for Square. Referring to Fig. 9.12, we obtain

$$I_y = \tfrac{1}{3}b^3h = \tfrac{1}{3}(15\text{ cm})^3(15\text{ cm}) = 16\,875\text{ cm}^4$$

I_y for the Quarter Circle. Referring to Fig. 5.8, we locate the centroid C of the quarter circle with respect to side AA'.

$$a = \frac{4r}{3\pi} = \frac{4(15\text{ cm})}{3\pi} = 6.366\text{ cm}$$

The distance b from the centroid C to the y axis is

$$b = 15\text{ cm} - a = 15\text{ cm} - 6.366\text{ cm} = 8.634\text{ cm}$$

Referring now to Fig. 9.12, we compute the moment of inertia of the quarter circle with respect to side AA'; we also compute the area of the quarter circle.

$$I_{AA'} = \tfrac{1}{16}\pi r^4 = \tfrac{1}{16}\pi(15\text{ cm})^4 = 9940\text{ cm}^4$$
$$A = \tfrac{1}{4}\pi r^2 = \tfrac{1}{4}\pi(15\text{ cm})^2 = 176.7\text{ cm}^2$$

Using the parallel-axis theorem, we obtain the value of $\bar{I}_{y'}$,

$$I_{AA'} = \bar{I}_{y'} + Aa^2$$
$$\bar{I}_{y'} = I_{AA'} - Aa^2$$
$$= 9940\text{ cm}^4 - (176.7\text{ cm}^2)(6.366\text{ cm})^2 = 2779\text{ cm}^4$$

Again using the parallel-axis theorem, we obtain the value of I_y,

$$I_y = \bar{I}_{y'} + Ab^2 = 2779\text{ cm}^4 + (176.7\text{ cm}^2)(8.634\text{ cm})^2 = 15\,950\text{ cm}^4$$

I_y for Given Area. Subtracting the moment of inertia of the quarter circle from that of the square, we obtain

$$I_y = 16\,875\text{ cm}^4 - 15\,950\text{ cm}^4 = 925\text{ cm}^4$$

or, since $1\text{ cm}^4 = (10\text{ mm})^4 = 10^4\text{ mm}^4$,

$$I_y = 925 \times 10^4\text{ mm}^4 \qquad I_y = 9.25 \times 10^6\text{ mm}^4 ◄$$

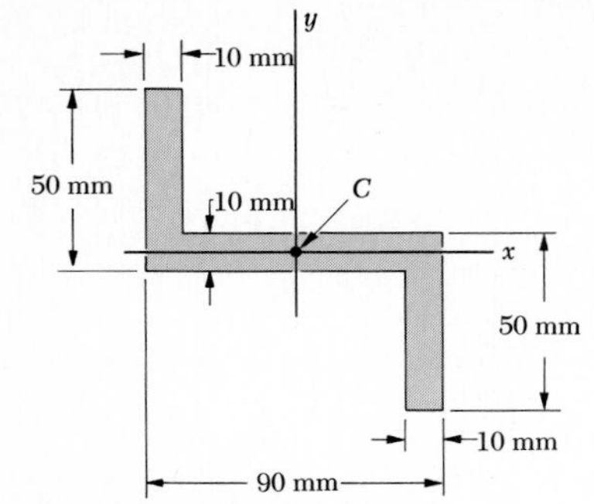

Fig. P9.20 and P9.21

PROBLEMS

9.20 and 9.22 Determine the moment of inertia and the radius of gyration of the shaded area with respect to the x axis.

9.21 and 9.23 Determine the moment of inertia and the radius of gyration of the shaded area with respect to the y axis.

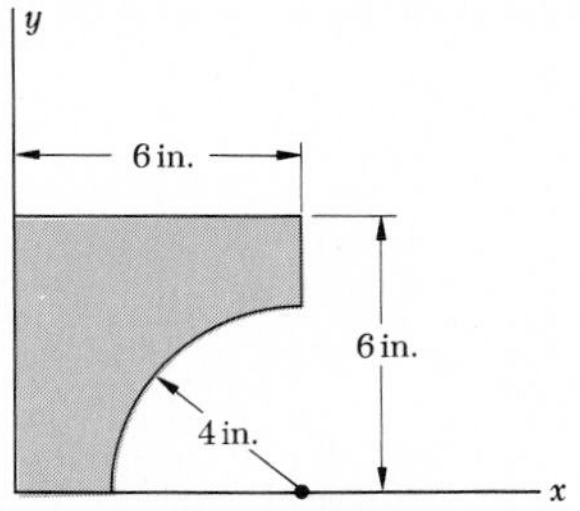

Fig. P9.22 and P9.23

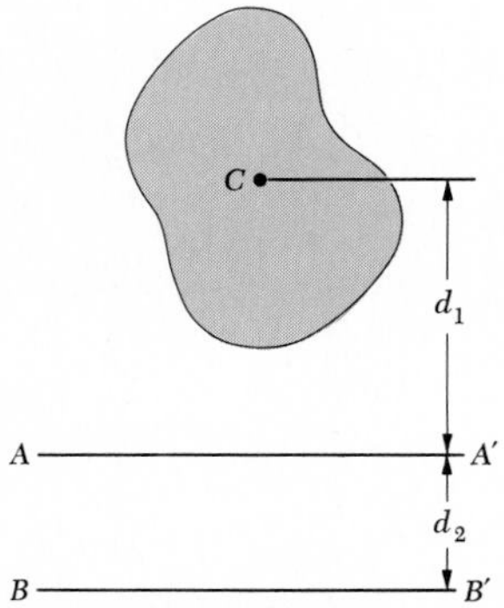

Fig. P9.24 and P9.25

9.24 The shaded area is equal to 20 in^2. Determine the distance d_1 and the moment of inertia with respect to a centroidal axis parallel to AA', knowing that $I_{AA'} = 900$ in^4, $I_{BB'} = 1800$ in^4, and $d_2 = 3$ in.

9.25 The shaded area is equal to 15 in^2 and its moment of inertia with respect to BB' is 400 in^4. Determine the moment of inertia with respect to AA' for $d_1 = 3$ in. and $d_2 = 2$ in.

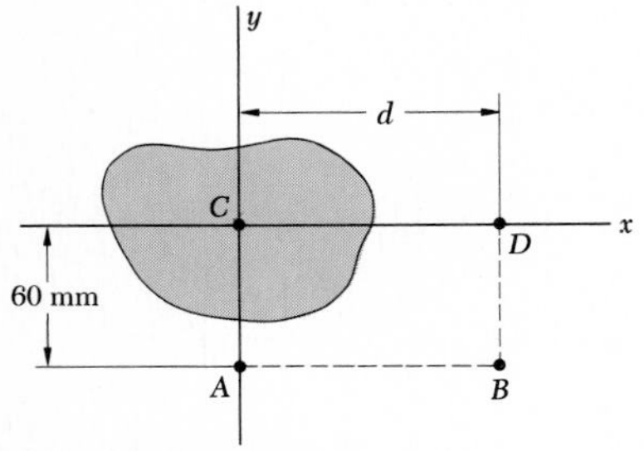

Fig. P9.26 and P9.27

9.26 The shaded area is equal to 5000 mm^2. Determine its centroidal moments of inertia $\bar{I}_x$ and $\bar{I}_y$, knowing that $2\bar{I}_x = \bar{I}_y$ and that the polar moment of inertia of the area about point A is $J_A = 22.5 \times 10^6$ mm^4.

9.27 The polar moments of inertia of the shaded area with respect to points A, B, and D are, respectively, $J_A = 28.8 \times 10^6$ mm^4, $J_B = 67.2 \times 10^6$ mm^4, and $J_D = 45.6 \times 10^6$ mm^4. Determine the shaded area, its centroidal moment of inertia $\bar{J}_C$, and the distance d from C to D.

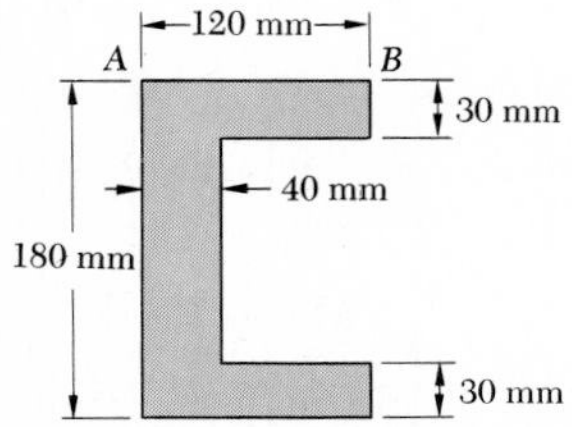

Fig. P9.28 and P9.29

9.28 Determine the centroidal polar moment of inertia of the area shown.

9.29 Determine the moments of inertia $\bar{I}_x$ and $\bar{I}_y$ of the area shown with respect to centroidal axes respectively parallel and perpendicular to the side AB.

9.30 Determine the moments of inertia $\bar{I}_x$ and $\bar{I}_y$ of the area shown with respect to centroidal axes respectively parallel and perpendicular to the side AB.

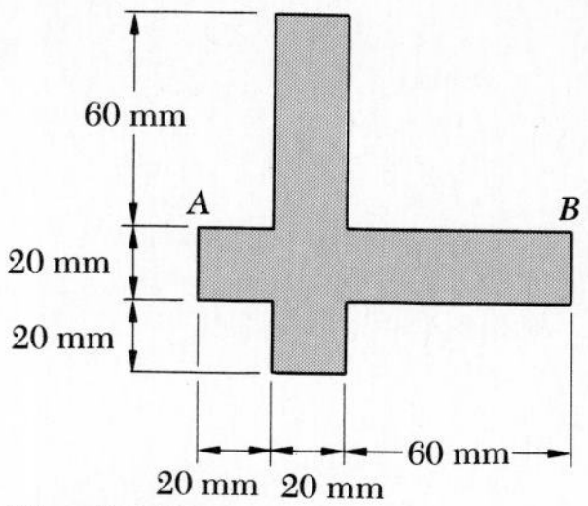

Fig. P9.30

9.31 Determine the polar moment of inertia of the area shown with respect to (*a*) point O, (*b*) the centroid of the area.

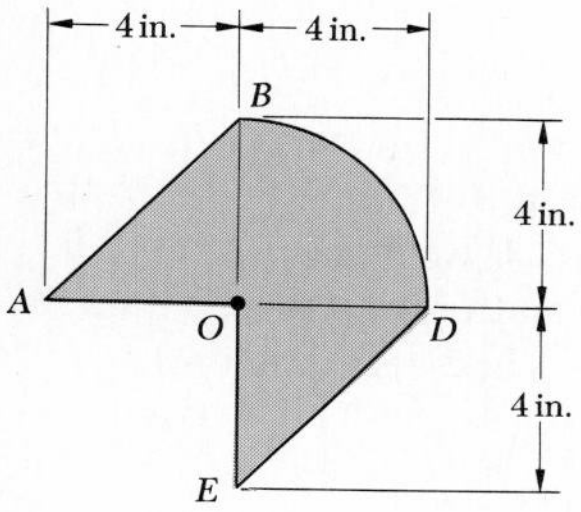

Fig. P9.31

9.32 Two 10-in., 25-lb American Standard channels and a 14- by $\frac{1}{2}$-in. plate are used to form the column section shown. For $b = 7$ in., determine the moments of inertia and the radii of gyration of the total section with respect to the centroidal axes shown.

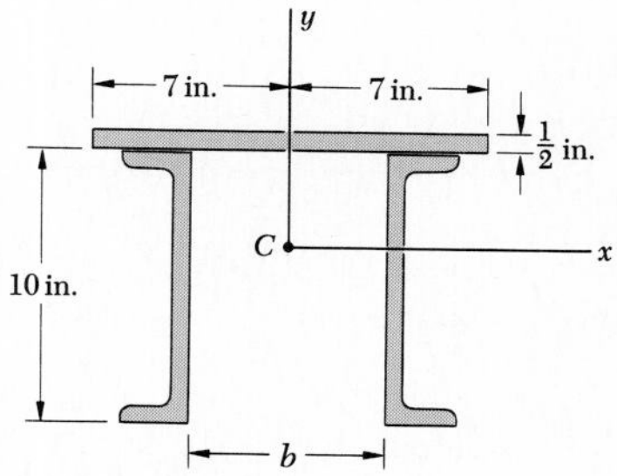

Fig. P9.32

9.33 In Prob. 9.32 determine the distance b for which the centroidal moments of inertia $\bar{I}_x$ and $\bar{I}_y$ of the column section are equal.

9.34 Three steel plates, 18 by 1 in., are riveted to four 6- by 6-in. angles, each 1 in. thick, to form the column whose cross section is shown. Determine the moments of inertia and the radii of gyration of the section with respect to centroidal axes respectively parallel and perpendicular to the flanges.

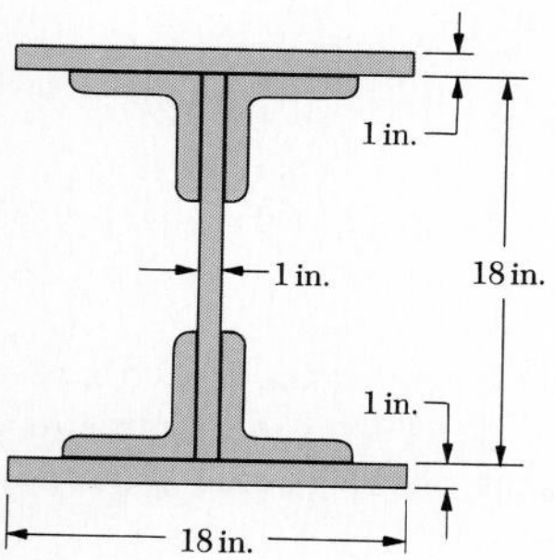

Fig. P9.34

9.35 Two 4- by 4- by $\frac{1}{2}$-in. angles are welded to a $\frac{1}{2}$- by 10-in. steel plate as shown. Determine the moments of inertia and the radii of gyration of the combined section with respect to centroidal axes respectively parallel and perpendicular to the plate.

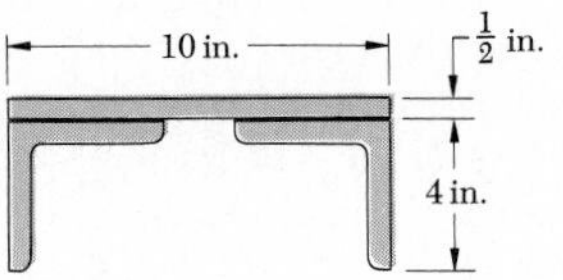

Fig. P9.35

9.36 through 9.38 The panel shown forms the end of a trough which is filled with water to the line AA'. Referring to Sec. 9.1, determine the depth of the point of application of the resultant of the hydrostatic forces acting on the panel (center of pressure).

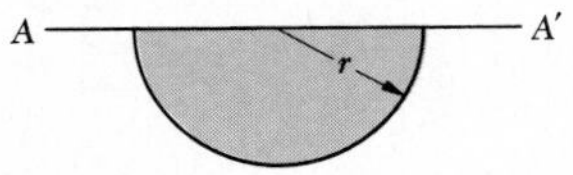

Fig. P9.36

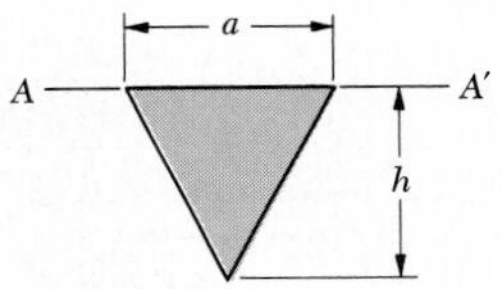

Fig. P9.37

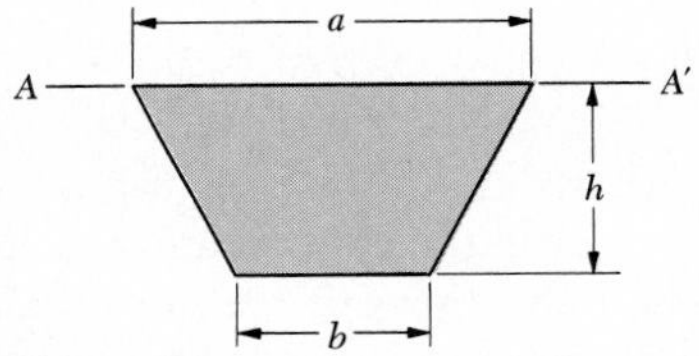

Fig. P9.38

***9.39** Determine the x coordinate of the centroid of the volume shown; this volume was obtained by intersecting a circular cylinder by the oblique plane. (*Hint.* The height of the volume is proportional to the x coordinate; consider an analogy between this height and the water pressure on a submerged surface.)

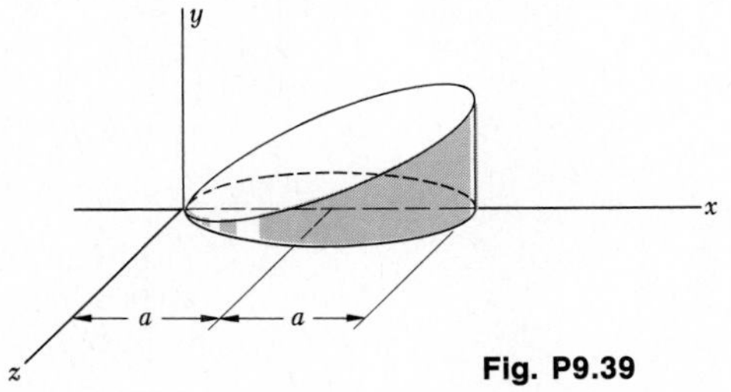

Fig. P9.39

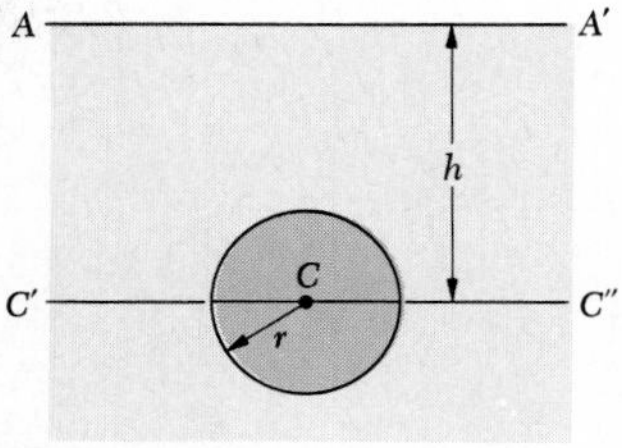

Fig. P9.40

9.40 A vertical circular gate of radius r is hinged about its diameter $C'C''$. Denoting the specific weight of water by γ, determine (*a*) the reaction at each hinge, (*b*) the moment of the couple required to keep the gate closed.

9.41 The center of a vertical circular gate, 2 m in diameter, is located 3 m below the water surface. The gate is held by three bolts equally spaced as shown. Determine the force on each bolt.

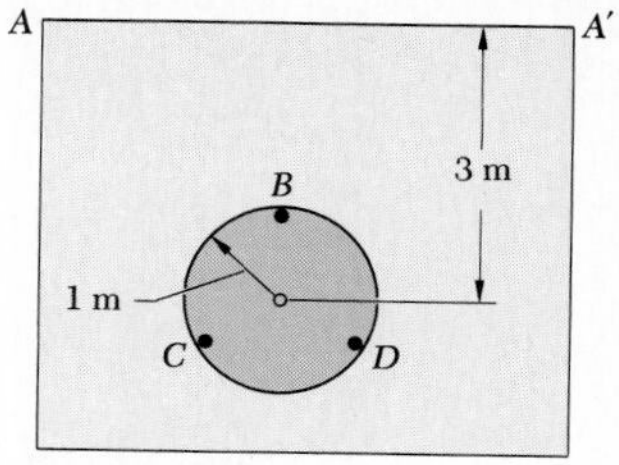

Fig. P9.41

***9.42** Show that the system of hydrostatic forces acting on a submerged plane area A may be reduced to a force $\mathbf{P}$ at the centroid C of the area and two couples. The force $\mathbf{P}$ is perpendicular to the area and of magnitude $P = \gamma A\bar{y} \sin\theta$, where γ is the specific weight of the liquid, and the couples are represented by vectors directed as shown and of magnitude $M_{x'} = \gamma \bar{I}_{x'} \sin\theta$ and $M_{y'} = \gamma \bar{P}_{x'y'} \sin\theta$, where $\bar{P}_{x'y'} = \int x'\, y'\, dA$ (see Sec. 9.7). Note that the couples are independent of the depth at which the area is submerged.

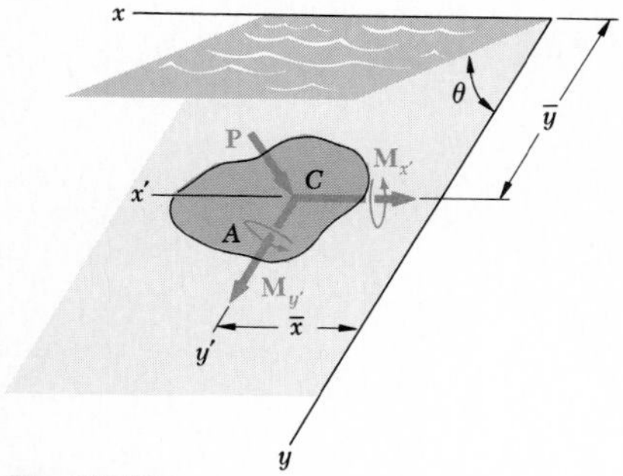

Fig. P9.42

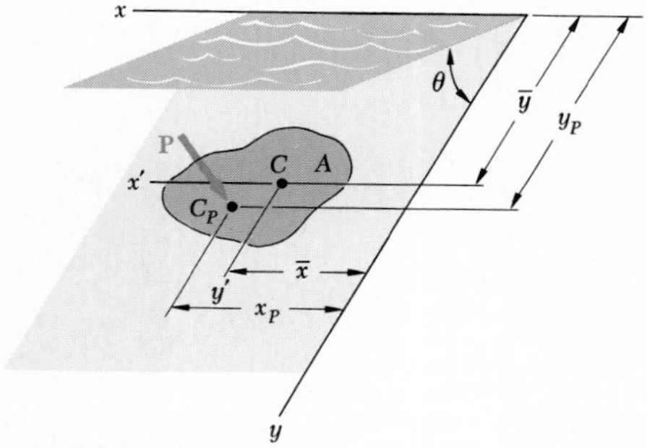

Fig. P9.43

***9.43** Show that the resultant of the hydrostatic forces acting on a submerged plane area A is a force $\mathbf{P}$ perpendicular to the area and of magnitude $P = \gamma A\bar{y} \sin\theta = \bar{p}A$, where γ is the specific weight of the liquid and $\bar{p}$ the pressure at the centroid C of the area. Show that $\mathbf{P}$ is applied at a point C_P, called the center of pressure, of coordinates $x_P = P_{xy}/A\bar{y}$ and $y_P = I_x/A\bar{y}$, where $P_{xy} = \int xy\, dA$ (see Sec. 9.7). Show also that the difference of ordinates $y_P - \bar{y}$ is equal to $\bar{k}^2_{x'}/\bar{y}$ and thus depends upon the depth at which the area is submerged.

*9.7. Product of Inertia.

The integral

$$P_{xy} = \int xy\, dA \tag{9.12}$$

obtained by multiplying each element dA of an area A by its coordinates x and y and integrating over the area (Fig. 9.14), is known as the *product of inertia* of the area A with respect to the x and y axes. Unlike the moments of inertia I_x and I_y, the product of inertia P_{xy} may be either positive or negative.

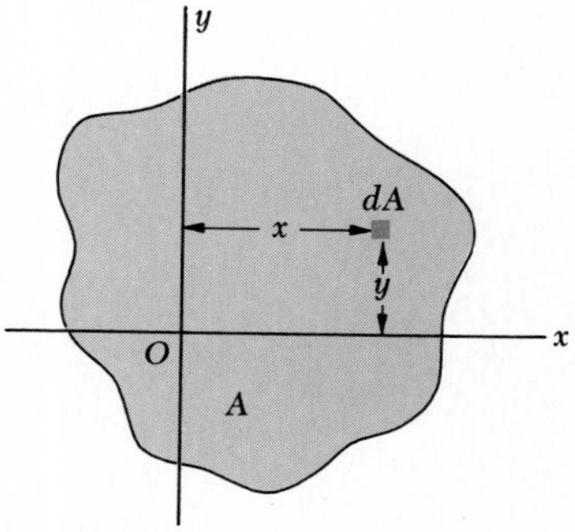

Fig. 9.14

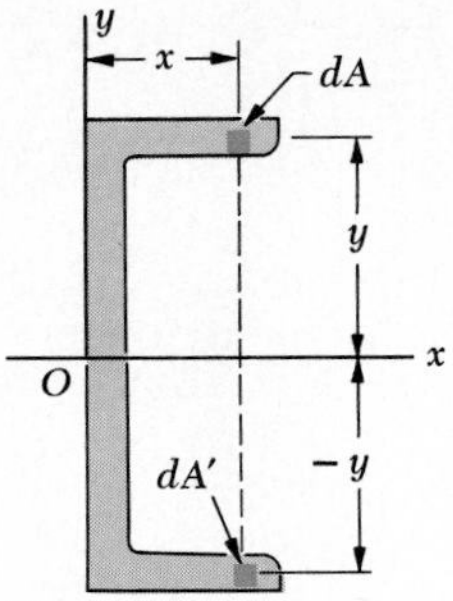

Fig. 9.15

When one or both of the x and y axes are axes of symmetry for the area A, the product of inertia P_{xy} is zero. Consider, for example, the channel section shown in Fig. 9.15. Since this section is symmetrical with respect to the x axis, we can associate to each element dA of coordinates x and y an element dA' of coordinates x and $-y$. Clearly, the contributions of any pair of elements chosen in this way cancel out, and the integral (9.12) reduces to zero.

A parallel-axis theorem similar to the one established in Sec. 9.5 for moments of inertia may be derived for products of inertia. Consider an area A and a system of rectangular coordinates x and y (Fig. 9.16). Through the centroid C of the area, of coordinates $\bar{x}$ and $\bar{y}$, we draw two *centroidal axes* x' and y'

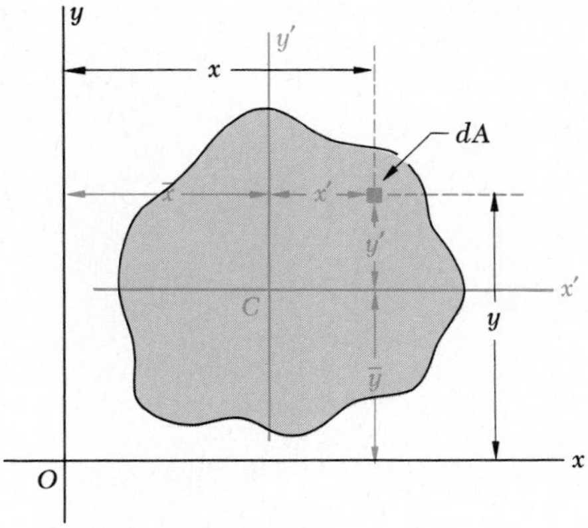

Fig. 9.16

parallel, respectively, to the x and y axes. Denoting by x and y the coordinates of an element of area dA with respect to the original axes, and by x' and y' the coordinates of the same element with respect to the centroidal axes, we write $x = x' + \bar{x}$ and $y = y' + \bar{y}$. Substituting into (9.12), we obtain the following expression for the product of inertia P_{xy}:

$$\begin{aligned} P_{xy} &= \int xy\, dA = \int (x' + \bar{x})(y' + \bar{y})\, dA \\ &= \int x'y'\, dA + \bar{y}\int x'\, dA + \bar{x}\int y'\, dA + \bar{x}\bar{y}\int dA \end{aligned}$$

The first integral represents the product of inertia $\bar{P}_{x'y'}$ of the area A with respect to the centroidal axes x' and y'. The next two integrals represent first moments of the area with respect to the centroidal axes; they reduce to zero, since the centroid C is located on these axes. Finally, we observe that the last integral is equal to the total area A. We write therefore

$$P_{xy} = \bar{P}_{x'y'} + \bar{x}\bar{y}A \tag{9.13}$$

***9.8. Principal Axes and Principal Moments of Inertia.** Consider the area A and the coordinate axes x and y (Fig. 9.17). We assume that the moments and product of inertia

$$I_x = \int y^2\,dA \qquad I_y = \int x^2\,dA \qquad P_{xy} = \int xy\,dA \tag{9.14}$$

of the area A are known, and we propose to determine the moments and product of inertia I_u, I_v, and P_{uv} of A with respect

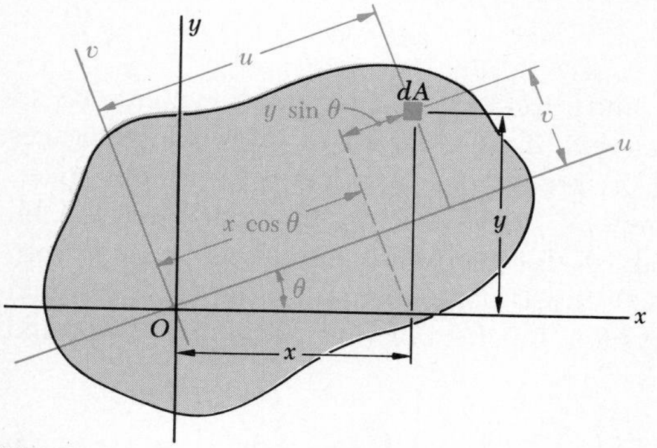

Fig. 9.17

to new axes u and v obtained by rotating the original axes about the origin through an angle θ.

We first note the following relations between the coordinates u, v and x, y of an element of area dA:

$$u = x\cos\theta + y\sin\theta \qquad v = y\cos\theta - x\sin\theta$$

Substituting for v in the expression for I_u, we write

$$\begin{aligned} I_u &= \int v^2\,dA = \int (y\cos\theta - x\sin\theta)^2\,dA \\ &= \cos^2\theta \int y^2\,dA - 2\sin\theta\cos\theta \int xy\,dA + \sin^2\theta \int x^2\,dA \end{aligned}$$

Taking the relations (9.14) into account, we write

$$I_u = I_x\cos^2\theta - 2P_{xy}\sin\theta\cos\theta + I_y\sin^2\theta \tag{9.15}$$

Similarly, we obtain for I_v and P_{uv} the expressions

$$I_v = I_x\sin^2\theta + 2P_{xy}\sin\theta\cos\theta + I_y\cos^2\theta \tag{9.16}$$

$$P_{uv} = I_x\sin\theta\cos\theta + P_{xy}(\cos^2\theta - \sin^2\theta) - I_y\sin\theta\cos\theta \tag{9.17}$$

We observe, by adding (9.15) and (9.16) member by member, that

$$I_u + I_v = I_x + I_y \tag{9.18}$$

This result could have been anticipated, since both members of (9.18) are equal to the polar moment of inertia J_O.

Using the trigonometric relations $\sin 2\theta = 2\sin\theta\cos\theta$ and

$\cos 2\theta = \cos^2\theta - \sin^2\theta$, we may write (9.15), (9.16), and (9.17) as follows:

$$I_u = \frac{I_x + I_y}{2} + \frac{I_x - I_y}{2}\cos 2\theta - P_{xy}\sin 2\theta \qquad (9.19)$$

$$I_v = \frac{I_x + I_y}{2} - \frac{I_x - I_y}{2}\cos 2\theta + P_{xy}\sin 2\theta \qquad (9.20)$$

$$P_{uv} = \frac{I_x - I_y}{2}\sin 2\theta + P_{xy}\cos 2\theta \qquad (9.21)$$

Equations (9.19) and (9.21) are the parametric equations of a circle. This means that, if we choose a set of rectangular axes and plot a point M of abscissa I_u and ordinate P_{uv} for any given value of the parameter θ, all the points thus obtained will lie on a circle. To establish this property we shall eliminate θ from Eqs. (9.19) and (9.21); this is done by transposing $(I_x + I_y)/2$ in Eq. (9.19), squaring both members of Eqs. (9.19) and (9.21), and adding. We write

$$\left(I_u - \frac{I_x + I_y}{2}\right)^2 + P_{uv}^2 = \left(\frac{I_x - I_y}{2}\right)^2 + P_{xy}^2 \qquad (9.22)$$

Setting

$$I_{\text{av}} = \frac{I_x + I_y}{2} \qquad \text{and} \qquad R = \sqrt{\left(\frac{I_x - I_y}{2}\right)^2 + P_{xy}^2} \qquad (9.23)$$

we write the identity (9.22) in the form

$$(I_u - I_{\text{av}})^2 + P_{uv}^2 = R^2 \qquad (9.24)$$

which is the equation of a circle of radius R centered at the point C of abscissa I_{av} and ordinate 0 (Fig. 9.18).

The two points A and B where the circle obtained intersects

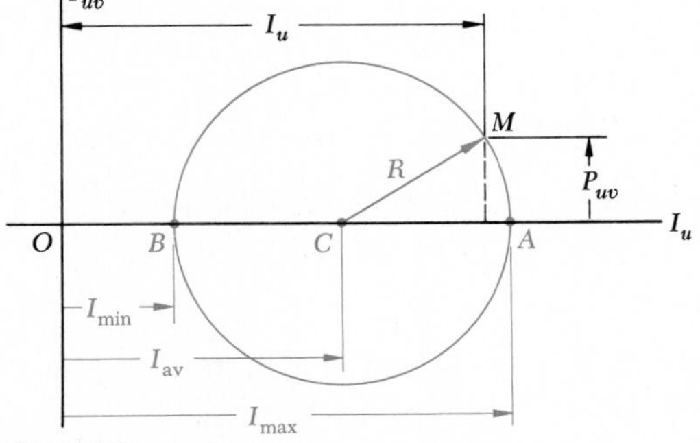

Fig. 9.18

the axis of abscissas are of special interest: Point A corresponds to the maximum value of the moment of inertia I_u, while point B corresponds to its minimum value. Besides, both points correspond to a zero value of the product of inertia P_{uv}. Thus, the values θ_m of the parameter θ which correspond to the points A and B may be obtained by setting $P_{uv} = 0$ in Eq. (9.21). We obtain†

$$\tan 2\theta_m = -\frac{2P_{xy}}{I_x - I_y} \tag{9.25}$$

This equation defines two values $2\theta_m$ which are 180° apart and thus two values θ_m which are 90° apart. One of them corresponds to point A in Fig. 9.18 and to an axis through O in Fig. 9.17 with respect to which the moment of inertia of the given area is maximum; the other value corresponds to point B and an axis through O with respect to which the moment of inertia of the area is minimum. The two axes thus defined, which are perpendicular to each other, are called the *principal axes of the area about O*, and the corresponding values I_{max} and I_{min} of the moment of inertia are called the *principal moments of inertia of the area about O*. We check from Fig. 9.18 that

$$I_{\text{max}} = I_{\text{av}} + R \qquad \text{and} \qquad I_{\text{min}} = I_{\text{av}} - R \tag{9.26}$$

Substituting for I_{av} and R from formulas (9.23), we write

$$I_{\text{max,min}} = \frac{I_x + I_y}{2} \pm \sqrt{\left(\frac{I_x - I_y}{2}\right)^2 + P_{xy}^2} \tag{9.27}$$

Since the two values θ_m defined by Eq. (9.25) were obtained by setting $P_{uv} = 0$ in Eq. (9.21), it is clear that the product of inertia of the given area with respect to its principal axes is zero. Referring to Sec. 9.7, we note that, if an area possesses an axis of symmetry through a point O, this axis must be a principal axis of the area about O. On the other hand, a principal axis does not need to be an axis of symmetry; whether or not an area possesses properties of symmetry, it will have two principal axes of inertia about any point O.

The properties established here hold for any point O located inside or outside the given area. If the point O is chosen to coincide with the centroid of the area, any axis through O is a centroidal axis; the two principal axes of the area about its centroid are referred to as the *principal centroidal axes of the area.*

† This relation may also be obtained by differentiating I_u in Eq. (9.19) and setting $dI_u/d\theta = 0$.

***9.9. Mohr's Circle for Moments and Products of Inertia.** The circle used in the preceding section to illustrate the relations existing between the moments and products of inertia of a given area with respect to axes passing through a fixed point O was first introduced by the German engineer Otto Mohr (1835–1918) and is known as *Mohr's circle.* We shall see that, if the moments and product of inertia of an area A are known with respect to two rectangular x and y axes through a point O, Mohr's circle may be used to determine graphically (a) the principal axes and principal moments of inertia of the area about O, or (b) the moments and product of inertia of the area with respect to any other pair of rectangular axes u and v through O.

Consider a given area A and two rectangular coordinate axes x and y (Fig. 9.19a). We shall assume that the moments of inertia I_x and I_y and the product of inertia P_{xy} are known, and we shall represent them on a diagram by plotting a point X of coordinates I_x and P_{xy} and a point Y of coordinates I_y and $-P_{xy}$ (Fig. 9.19b). Joining X and Y by a straight line, we define the point C of intersection of line XY with the I axis and draw the circle of center C and diameter XY. Noting that the abscissa of C and the radius of the circle are respectively equal to the quantities I_{av} and R defined by the formulas (9.23), we conclude that the circle obtained is Mohr's circle for the given area about point O. Thus the abscissas of the points A and B where the circle intersects the I axis represent respectively the principal moments of inertia I_{max} and I_{min} of the area.

We also note that, since $\tan(XCA) = 2P_{xy}/(I_x - I_y)$, the angle XCA is equal in magnitude to one of the angles $2\theta_m$ which satisfy Eq. (9.25); thus the angle θ_m which defines in Fig. 9.19a the principal axis Oa corresponding to point A in Fig. 9.19b may be obtained by dividing in half the angle XCA measured on Mohr's circle. We further observe that, if $I_x > I_y$ and $P_{xy} > 0$, as in the case considered here, the rotation which brings CX into CA is clockwise. But, in that case, the angle θ_m obtained from Eq. (9.25) and defining the principal axis Oa in Fig. 9.19a is negative; thus the rotation bringing Ox into Oa is also clockwise. We conclude that the senses of rotation in both parts of Fig. 9.19 are the same; if a clockwise rotation through $2\theta_m$ is required to bring CX into CA on Mohr's circle, a clockwise rotation through θ_m will bring Ox into the corresponding principal axis Oa in Fig. 9.19a.

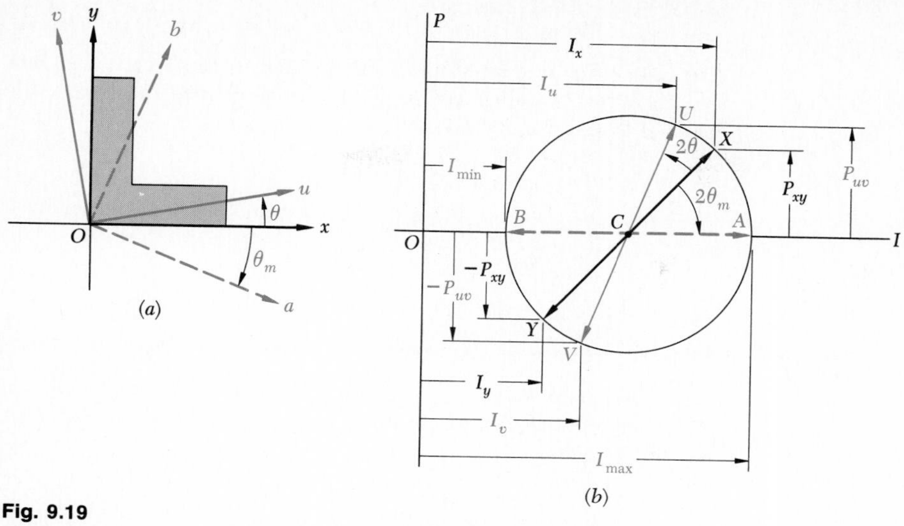

Fig. 9.19

Since Mohr's circle is uniquely defined, the same circle may be obtained by considering the moments and product of inertia of the area A with respect to rectangular axes u and v (Fig. 9.19*a*). The point U of coordinates I_u and P_{uv} and the point V of coordinates I_v and $-P_{uv}$ are therefore located on Mohr's circle, and the angle UCA in Fig. 9.19*b* must be equal to twice the angle uOa in Fig. 9.19*a*. Since, as noted above, the angle XCA is twice the angle xOa, it follows that the angle XCU in Fig. 9.19*b* is twice the angle xOu in Fig. 9.19*a*. Thus the diameter UV defining the moments and product of inertia I_u, I_v, and P_{uv} of the given area with respect to rectangular axes u and v forming an angle θ with the x and y axes may be obtained by rotating through an angle 2θ the diameter XY corresponding to the moments and product of inertia I_x, I_y, and P_{xy}. We note that the rotation which brings the diameter XY into the diameter UV in Fig. 9.19*b* has the same sense as the rotation which brings the xy axes into the uv axes in Fig. 9.19*a*.

It should be noted that the use of Mohr's circle is not limited to graphical solutions, i.e., to solutions based on the careful drawing and measuring of the various parameters involved. By merely sketching Mohr's circle and using trigonometry, one may easily derive the various relations required for a numerical solution of a given problem. Actual computations may then be carried out on a calculator or a slide rule (see Sample Prob. 9.8).

SAMPLE PROBLEM 9.6

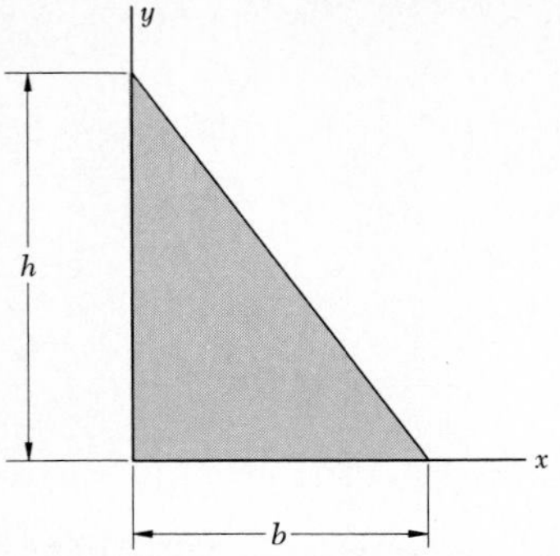

Determine the product of inertia of the right triangle shown (*a*) with respect to the x and y axes and (*b*) with respect to centroidal axes parallel to the x and y axes.

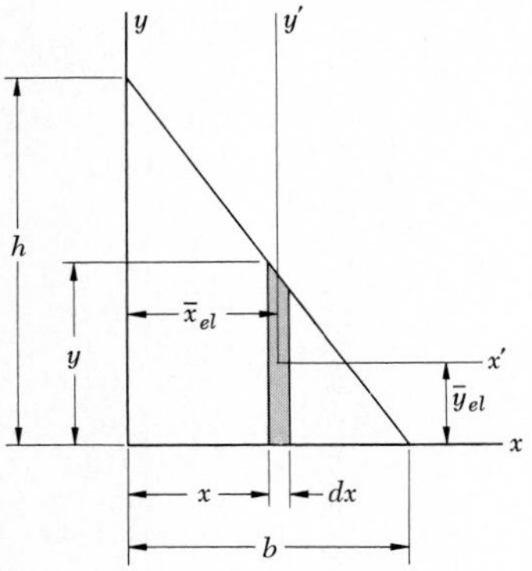

a. **Product of Inertia P_{xy}.** A vertical rectangular strip is chosen as the differential element of area. Using the parallel-axis theorem, we write

$$dP_{xy} = dP_{x'y'} + \bar{x}_{el}\,\bar{y}_{el}\,dA$$

Since the element is symmetrical with respect to the x' and y' axes, we note that $dP_{x'y'} = 0$. From the geometry of the triangle, we obtain

$$y = h\left(1 - \frac{x}{b}\right) \qquad dA = y\,dx = h\left(1 - \frac{x}{b}\right)dx$$

$$\bar{x}_{el} = x \qquad \bar{y}_{el} = \tfrac{1}{2}y = \tfrac{1}{2}h\left(1 - \frac{x}{b}\right)$$

Integrating dP_{xy} from $x = 0$ to $x = b$, we obtain

$$P_{xy} = \int dP_{xy} = \int \bar{x}_{el}\bar{y}_{el}\,dA = \int_0^b x(\tfrac{1}{2})h^2\left(1 - \frac{x}{b}\right)^2 dx$$

$$= h^2\int_0^b \left(\frac{x}{2} - \frac{x^2}{b} + \frac{x^3}{2b^2}\right)dx = h^2\left[\frac{x^2}{4} - \frac{x^3}{3b} + \frac{x^4}{8b^2}\right]_0^b$$

$$P_{xy} = \tfrac{1}{24}b^2h^2 \quad ◀$$

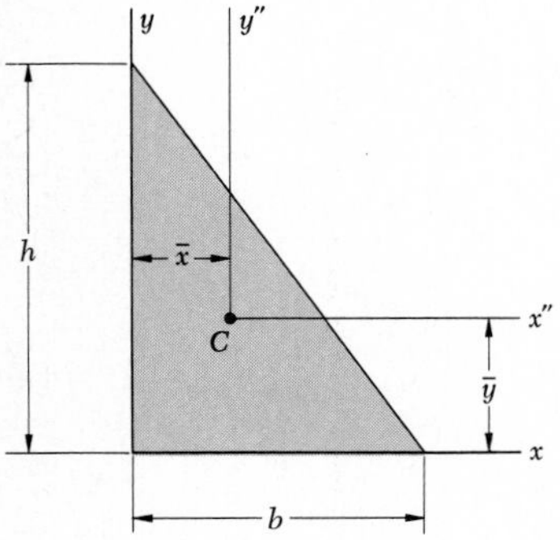

b. **Product of Inertia $\bar{P}_{x''y''}$.** The coordinates of the centroid of the triangle are

$$\bar{x} = \tfrac{1}{3}b \qquad \bar{y} = \tfrac{1}{3}h$$

Using the expression for P_{xy} obtained in part *a*, we apply the parallel-axis theorem and write

$$P_{xy} = \bar{P}_{x''y''} + \bar{x}\bar{y}A$$

$$\tfrac{1}{24}b^2h^2 = \bar{P}_{x''y''} + (\tfrac{1}{3}b)(\tfrac{1}{3}h)(\tfrac{1}{2}bh)$$

$$\bar{P}_{x''y''} = \tfrac{1}{24}b^2h^2 - \tfrac{1}{18}b^2h^2$$

$$\bar{P}_{x''y''} = -\tfrac{1}{72}b^2h^2 \quad ◀$$

SAMPLE PROBLEM 9.7

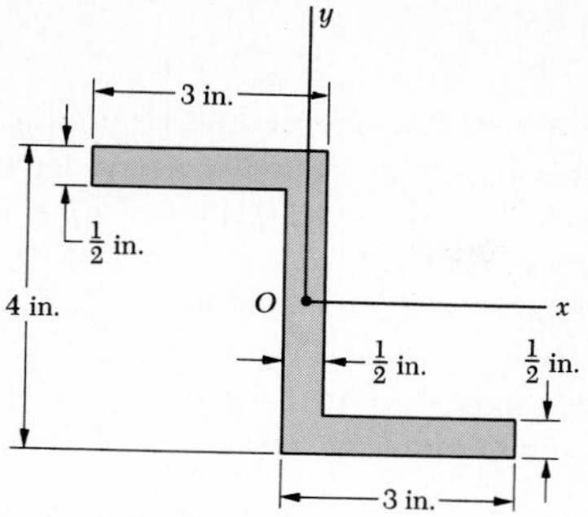

For the section shown, the moments of inertia with respect to the x and y axes have been computed and are known to be

$$I_x = 10.38 \text{ in}^4 \qquad I_y = 6.97 \text{ in}^4$$

Determine (*a*) the principal axes of the section about O, (*b*) the values of the principal moments of inertia of the section about O.

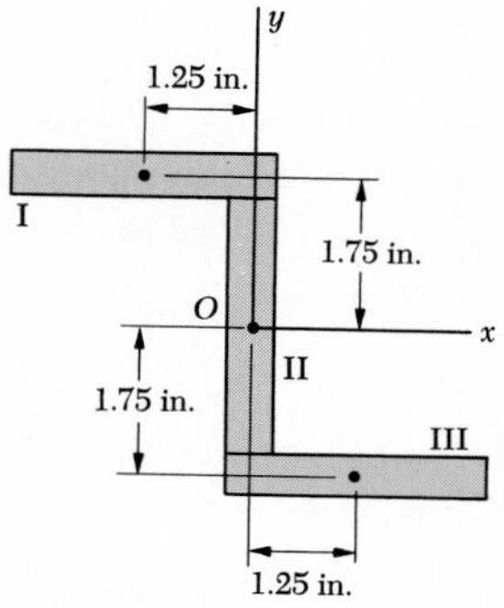

Solution. We first compute the product of inertia with respect to the x and y axes. The area is divided into three rectangles as shown. We note that the product of inertia $\bar{P}_{x'y'}$ with respect to centroidal axes parallel to the x and y axes is zero for each rectangle. Using the parallel-axis theorem $P_{xy} = \bar{P}_{x'y'} + \bar{x}\bar{y}A$, we thus find that, for each rectangle, P_{xy} reduces to $\bar{x}\bar{y}A$.

Rectangle	Area, in^2	$\bar{x}$, in.	$\bar{y}$, in.	$\bar{x}\bar{y}A$, in^4
I	1.5	−1.25	+1.75	−3.28
II	1.5	0	0	0
III	1.5	+1.25	−1.75	−3.28
				−6.56

$$P_{xy} = \Sigma \bar{x}\bar{y}A = -6.56 \text{ in}^4$$

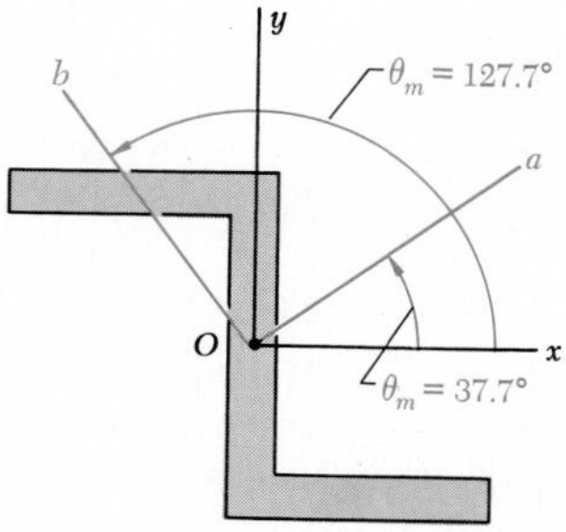

a. Principal Axes. Since the magnitudes of I_x, I_y, and P_{xy} are known, Eq. (9.25) is used to determine the values of θ_m,

$$\tan 2\theta_m = -\frac{2P_{xy}}{I_x - I_y} = -\frac{2(-6.56)}{10.38 - 6.97} = +3.85$$

$$2\theta_m = 75.4° \text{ and } 255.4°$$

$$\theta_m = 37.7° \quad \text{and} \quad \theta_m = 127.7° \blacktriangleleft$$

b. Principal Moments of Inertia. Using Eq. (9.27), we write

$$I_{\max,\min} = \frac{I_x + I_y}{2} \pm \sqrt{\left(\frac{I_x - I_y}{2}\right)^2 + P_{xy}^2}$$

$$= \frac{10.38 + 6.97}{2} \pm \sqrt{\left(\frac{10.38 - 6.97}{2}\right)^2 + (-6.56)^2}$$

$$I_{\max} = 15.45 \text{ in}^4 \qquad I_{\min} = 1.897 \text{ in}^4 \blacktriangleleft$$

Noting that the area of the section is farther away from the a axis than from the b axis, we conclude that $I_a = I_{\max} = 15.45 \text{ in}^4$ and $I_b = I_{\min} = 1.897 \text{ in}^4$. This conclusion may be verified by substituting $\theta = 37.7°$ into Eqs. (9.19) and (9.20).

SAMPLE PROBLEM 9.8

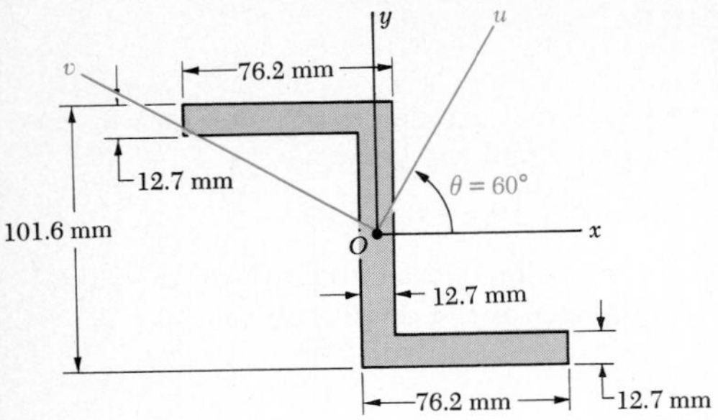

For the section shown, the moments and product of inertia with respect to the x and y axes have been computed and are known to be $I_x = 4.32 \times 10^6 \text{ mm}^4$, $I_y = 2.90 \times 10^6 \text{ mm}^4$, $P_{xy} = -2.73 \times 10^6 \text{ mm}^4$. Using Mohr's circle, determine (*a*) the principal axes of the section about O, (*b*) the values of the principal moments of inertia of the section about O, (*c*) the moments and product of inertia of the section with respect to the u and v axes forming an angle of 60° with the x and y axes. (*Note:* This section is the same as that of Sample Prob. 9.7.)

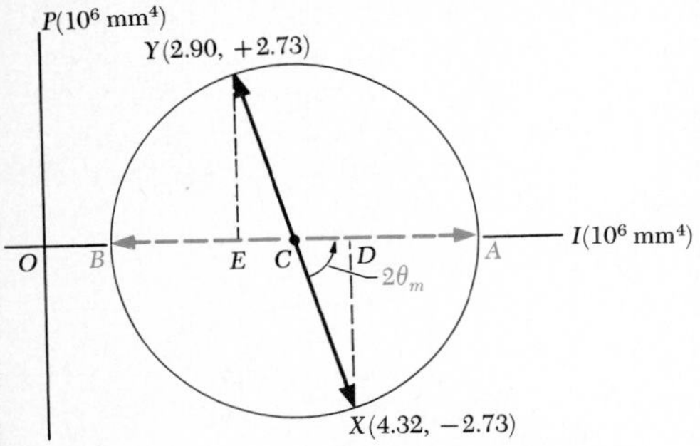

Solution. We first plot point X of coordinates $I_x = 4.32$, $P_{xy} = -2.73$, and point Y of coordinates $I_y = 2.90$, $-P_{xy} = +2.73$. Joining X and Y by a straight line, we define the center C of Mohr's circle. The abscissa of C, which represents I_{av}, and the radius R of the circle may be measured directly or calculated as follows:

$$I_{av} = OC = \tfrac{1}{2}(I_x + I_y) = \tfrac{1}{2}(4.32 \times 10^6 \text{ mm}^4 + 2.90 \times 10^6 \text{ mm}^4) = 3.61 \times 10^6 \text{ mm}^4$$

$$R = \sqrt{(CD)^2 + (DX)^2} = \sqrt{(0.71 \times 10^6 \text{ mm}^4)^2 + (2.73 \times 10^6 \text{ mm}^4)^2} = 2.82 \times 10^6 \text{ mm}^4$$

a. Principal Axes. The principal axes of the section correspond to points A and B on Mohr's circle and the angle through which we should rotate CX to bring it into CA defines $2\theta_m$. We have

$$\tan 2\theta_m = \frac{DX}{CD} = \frac{2.73}{0.71} = 3.85 \qquad 2\theta_m = 75.4° \circlearrowleft \qquad \theta_m = 37.7° \circlearrowleft \blacktriangleleft$$

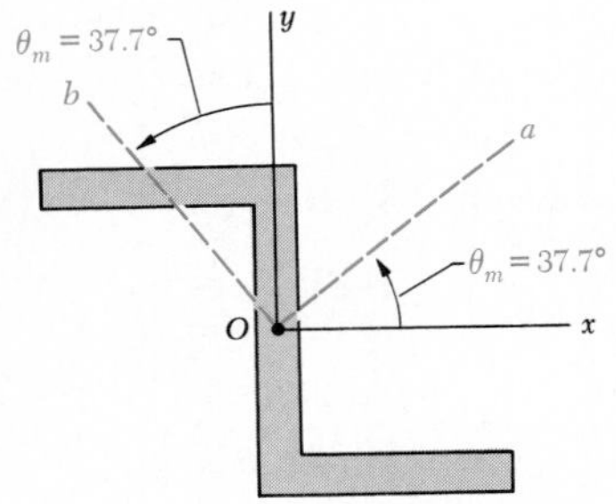

Thus the principal axis Oa corresponding to the maximum value of the moment of inertia is obtained by rotating the x axis through 37.7° counterclockwise; the principal axis corresponding to the minimum value of the moment of inertia may be obtained by rotating the y axis through the same angle.

b. Principal Moments of Inertia. The principal moments of inertia are represented by the abscissas of A and B. We have

$$I_{max} = OA = OC + CA = I_{av} + R = (3.61 + 2.82)10^6 \text{ mm}^4$$

$$I_{max} = 6.43 \times 10^6 \text{ mm}^4 \blacktriangleleft$$

$$I_{min} = OB = OC - BC = I_{av} - R = (3.61 - 2.82)10^6 \text{ mm}^4$$

$$I_{min} = 0.79 \times 10^6 \text{ mm}^4 \blacktriangleleft$$

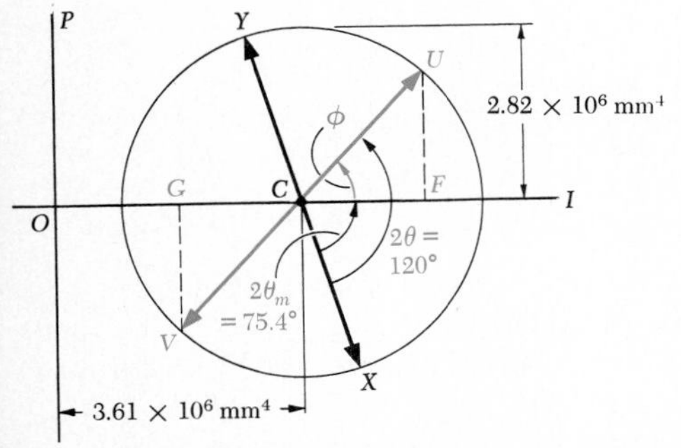

c. Moments and Product of Inertia with Respect to uv Axes. The points U and V on Mohr's circle which correspond to the u and v axes are obtained by rotating CX and CY through an angle $2\theta = 2(60°) = 120°$ counterclockwise. The coordinates of U and V yield the desired moments and product of inertia. Noting that the angle that CU forms with the I axis is $\phi = 120° - 75.4° = 44.6°$, we write

$$I_u = OF = OC + CF = 3.61 \times 10^6 \text{ mm}^4 + (2.82 \times 10^6 \text{ mm}^4) \cos 44.6°$$

$$I_u = 5.62 \times 10^6 \text{ mm}^4 \blacktriangleleft$$

$$I_v = OG = OC - GC = 3.61 \times 10^6 \text{ mm}^4 - (2.82 \times 10^6 \text{ mm}^4) \cos 44.6°$$

$$I_v = 1.60 \times 10^6 \text{ mm}^4 \blacktriangleleft$$

$$P_{uv} = FU = (2.82 \times 10^6 \text{ mm}^4) \sin 44.6°$$

$$P_{uv} = +1.98 \times 10^6 \text{ mm}^4 \blacktriangleleft$$

PROBLEMS

9.44 through 9.47 Determine by direct integration the product of inertia of the given area with respect to the x and y axes.

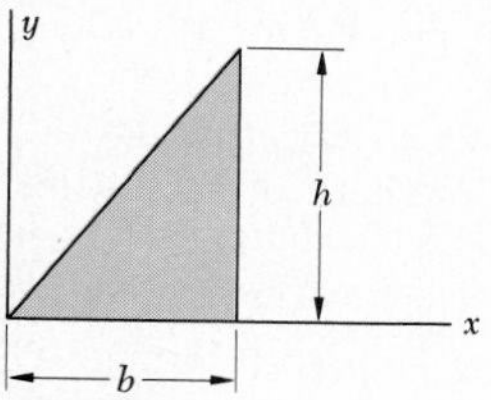

Fig. P9.44

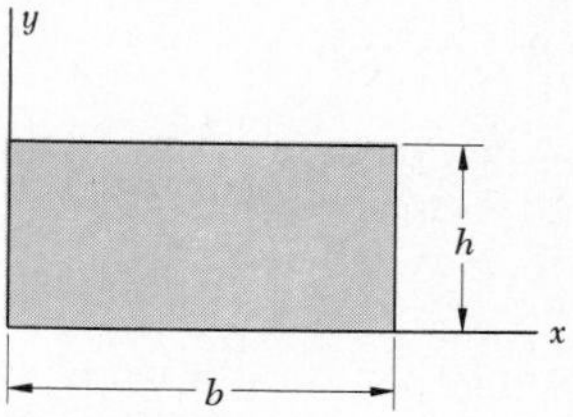

Fig. P9.45

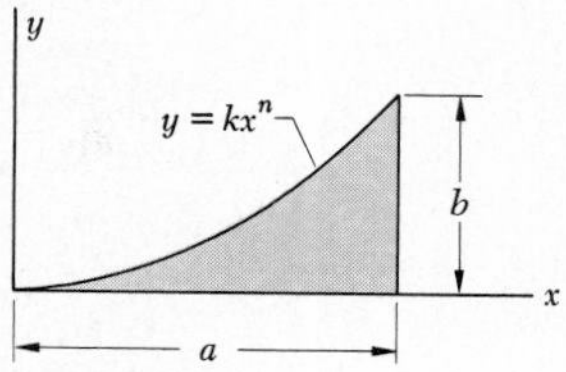

Fig. P9.46

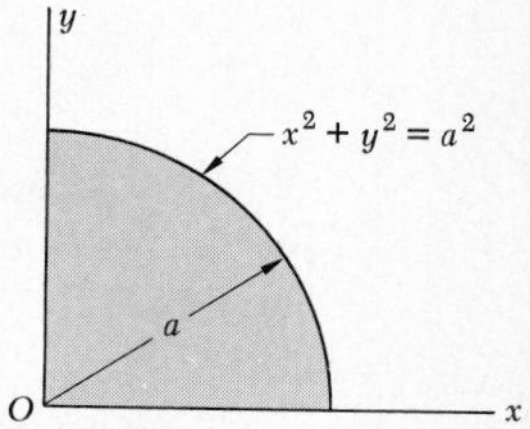

Fig. P9.47

9.48 through 9.51 Using the parallel-axis theorem, determine the product of inertia of the area shown with respect to the centroidal x and y axes.

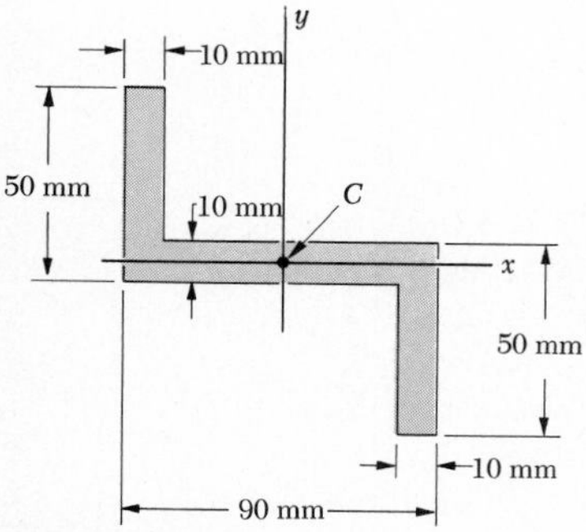

Fig. P9.48

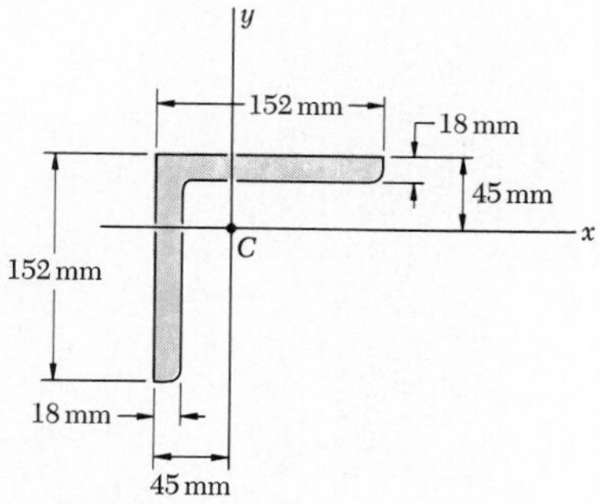

Fig. P9.49

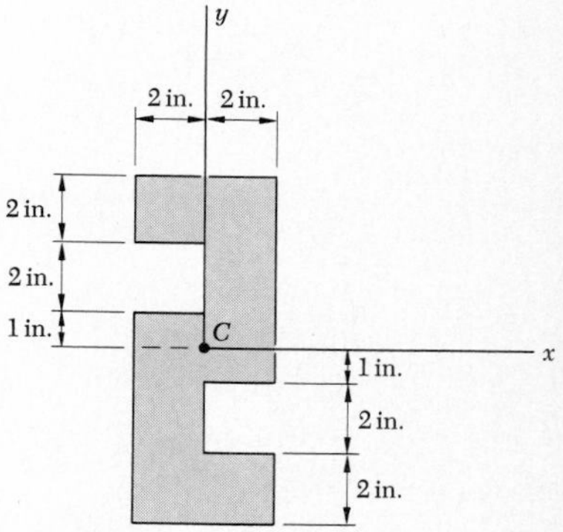

Fig. P9.50

9.52 Determine the moments of inertia and the product of inertia of the area of Prob. 9.48 with respect to new centroidal axes obtained by rotating the x and y axes through 30° counterclockwise.

9.53 Determine the moments of inertia and the product of inertia of the quarter circle of Prob. 9.47 with respect to new axes obtained by rotating the x and y axes about O (a) through 30° counterclockwise, (b) through 45° counterclockwise.

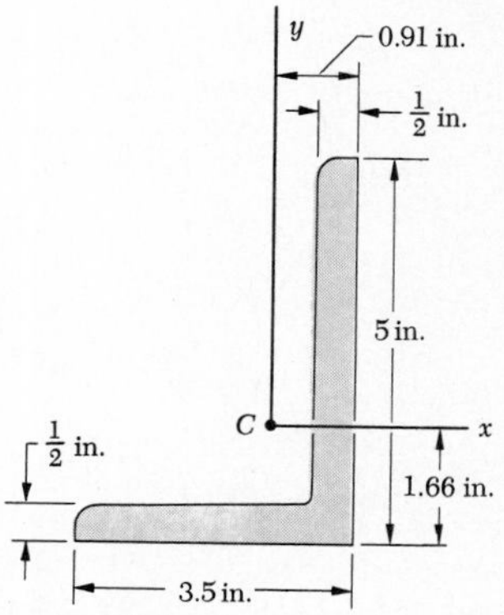

Fig. P9.51

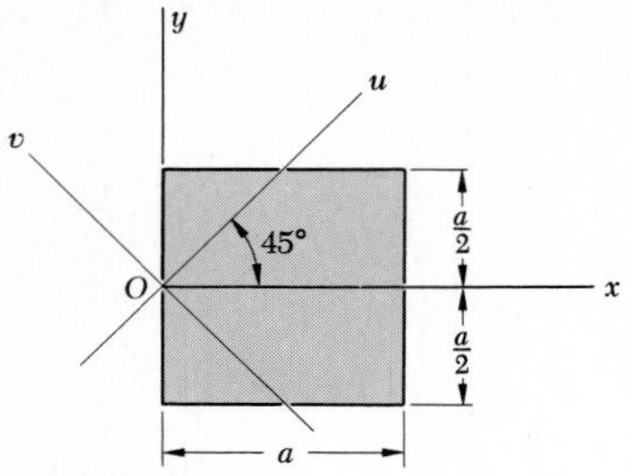

Fig. P9.54

9.54 Determine the moments of inertia and the product of inertia of the square shown with respect to the u and v axes.

9.55 Determine the moments of inertia and the product of inertia of the angle cross section of Prob. 9.51 with respect to new centroidal axes obtained by rotating the x and y axes through 45° clockwise. (The moments of inertia $\bar{I}_x$ and $\bar{I}_y$ are given in Fig. 9.13.)

9.56 Determine the orientation of the principal axes through the centroid and the corresponding values of the moment of inertia for the area of Prob. 9.48.

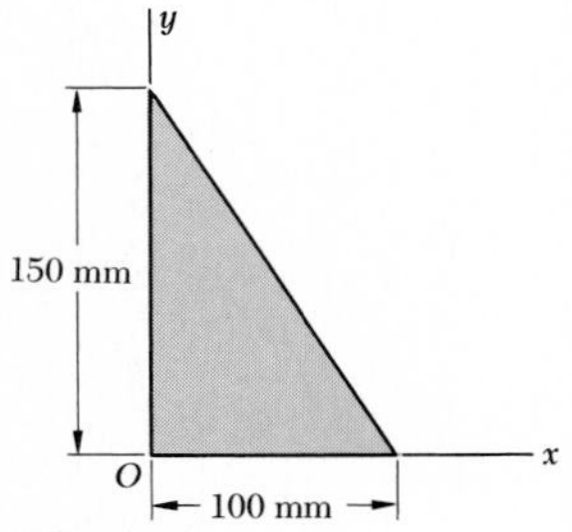

Fig. P9.57

9.57 Determine the orientation of the principal axes through O and the corresponding values of the moment of inertia for the triangle shown. (The product of inertia P_{xy} of the triangle is given in Sample Prob. 9.6.)

9.58 Determine the orientation of the principal axes through the centroid and the corresponding values of the moment of inertia for the angle cross section shown. Neglect the effect of fillets in computing $\bar{P}_{xy}$. (The moments of inertia $\bar{I}_x$ and $\bar{I}_y$ of the section are given in Fig. 9.13.)

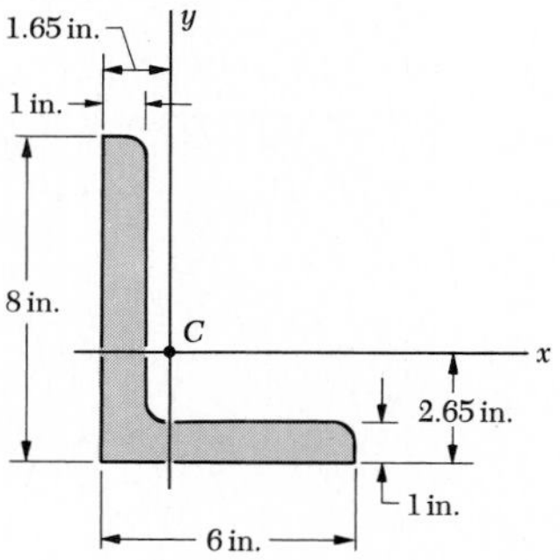

Fig. P9.58

9.59 Determine the orientation of the principal axes through the centroid and the corresponding values of the moment of inertia for the angle cross section of Prob. 9.51. (The moments of inertia $\bar{I}_x$ and $\bar{I}_y$ of the section are given in Fig. 9.13.)

9.60 Solve Prob. 9.54, using Mohr's circle.

9.61 Using Mohr's circle, determine the moments of inertia and the product of inertia of the angle cross section of Prob. 9.51 with respect to new centroidal axes obtained by rotating the x and y axes through 45° clockwise. (The moments of inertia $\bar{I}_x$ and $\bar{I}_y$ of the section are given in Fig. 9.13.)

9.62 Using Mohr's circle, determine the moments of inertia and the product of inertia of the area of Prob. 9.48 with respect to new centroidal axes obtained by rotating the x and y axes through 30° counterclockwise.

9.63 Using Mohr's circle, determine the moments of inertia and the product of inertia of the quarter circle of Prob. 9.47 with respect to new axes obtained by rotating the x and y axes about O (*a*) through 30° counterclockwise, (*b*) through 45° counterclockwise.

9.64 Solve Prob. 9.58, using Mohr's circle.

9.65 Using Mohr's circle, determine the orientation of the principal axes through the centroid and the corresponding values of the moment of inertia for the angle cross section of Prob. 9.51. (The moments of inertia $\bar{I}_x$ and $\bar{I}_y$ of the section are given in Fig. 9.13.)

9.66 Using Mohr's circle, determine the orientation of the principal axes through the centroid and the corresponding values of the moment of inertia for the area of Prob. 9.48.

9.67 Solve Prob. 9.57, using Mohr's circle.

9.68 Using Mohr's circle, show that, for any regular polygon (such as a pentagon), (*a*) the moment of inertia with respect to every axis through the centroid is the same, (*b*) the product of inertia with respect to any pair of rectangular axes through the centroid is zero.

9.69 The moments and product of inertia of a given area with respect to two rectangular axes x and y through O are, respectively, $I_x = 14.4 \times 10^6$ mm^4, $I_y = 8 \times 10^6$ mm^4, and $P_{xy} > 0$, while the minimum value of the moment of inertia of the area with respect to any axis through O is $I_{min} = 7.2 \times 10^6$ mm^4. Using Mohr's circle, determine (*a*) the product of inertia P_{xy} of the area, (*b*) the orientation of the principal axes, (*c*) the value of I_{max}.

***9.70** Prove that the expression $I_u I_v - P_{uv}^2$, where I_u, I_v, and P_{uv} represent, respectively, the moments and product of inertia of a given area with respect to two rectangular axes u and v through a given point O, is independent of the orientation of the u and v axes. Considering the particular case when the u and v axes correspond to the maximum value of P_{uv}, show that the given expression represents the square of the tangent drawn from the origin of the coordinates to Mohr's circle.

***9.71** Using the invariance property established in the preceding problem, express the product of inertia P_{xy} of an area A with respect to two rectangular axes through O in terms of the moments of inertia I_x and I_y of A and of the principal moments of inertia I_{min} and I_{max} of A about O. Apply the formula obtained to calculate the product of inertia $\bar{P}_{xy}$ of the 8- by 6- by 1-in.-angle cross section shown in Fig. 9.13, knowing that its minimum moment of inertia is 21.3 in^4.

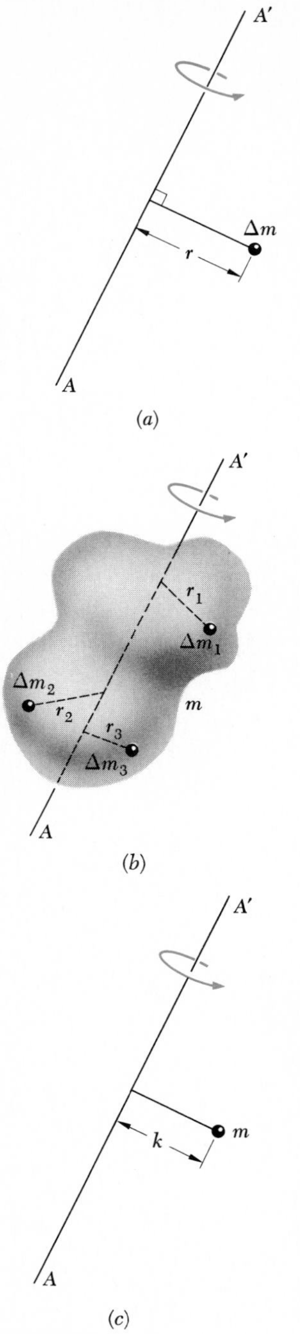

Fig. 9.20

MOMENTS OF INERTIA OF MASSES

9.10. Moment of Inertia of a Mass. Consider a small mass Δm mounted on a rod of negligible mass which may rotate freely about an axis AA' (Fig. 9.20*a*). If a couple is applied to the system, the rod and mass, assumed initially at rest, will start rotating about AA'. The details of this motion will be studied later in dynamics. At present, we wish only to indicate that the time required for the system to reach a given speed of rotation is proportional to the mass Δm and to the square of the distance r. The product $r^2\,\Delta m$ provides, therefore, a measure of the *inertia* of the system, i.e., of the resistance the system offers when we try to set it in motion. For this reason, the product $r^2\,\Delta m$ is called the *moment of inertia* of the mass Δm with respect to the axis AA'.

Consider now a body of mass m which is to be rotated about an axis AA' (Fig. 9.20*b*). Dividing the body into elements of mass Δm_1, Δm_2, etc., we find that the resistance offered by the body is measured by the sum $r_1^2\,\Delta m_1 + r_2^2\,\Delta m_2 + \cdots$. This sum defines, therefore, the moment of inertia of the body with respect to the axis AA'. Increasing the number of elements, we find that the moment of inertia is equal, at the limit, to the integral

$$I = \int r^2\,dm \tag{9.28}$$

The *radius of gyration* k of the body with respect to the axis AA' is defined by the relation

$$I = k^2 m \qquad \text{or} \qquad k = \sqrt{\frac{I}{m}} \tag{9.29}$$

The radius of gyration k represents, therefore, the distance at which the entire mass of the body should be concentrated if its moment of inertia with respect to AA' is to remain unchanged (Fig. 9.20*c*). Whether it is kept in its original shape (Fig. 9.20*b*) or whether it is concentrated as shown in Fig. 9.20*c*, the mass m will react in the same way to a rotation, or *gyration*, about AA'.

If SI units are used, the radius of gyration k is expressed in meters and the mass m in kilograms. The moment of inertia of a mass, therefore, will be expressed in kg · m^2. If U.S. customary units are used, the radius of gyration is expressed in feet and the mass in slugs, i.e., in lb · s²/ft. The moment of inertia of a mass, then, will be expressed in lb · ft · s².†

The moment of inertia of a body with respect to a coordinate axis may easily be expressed in terms of the coordinates x, y, z of the element of mass dm (Fig. 9.21). Noting, for example, that the square of the distance r from the element dm to the y axis is $z^2 + x^2$, we express the moment of inertia of the body with respect to the y axis as

$$I_y = \int r^2\, dm = \int (z^2 + x^2)\, dm$$

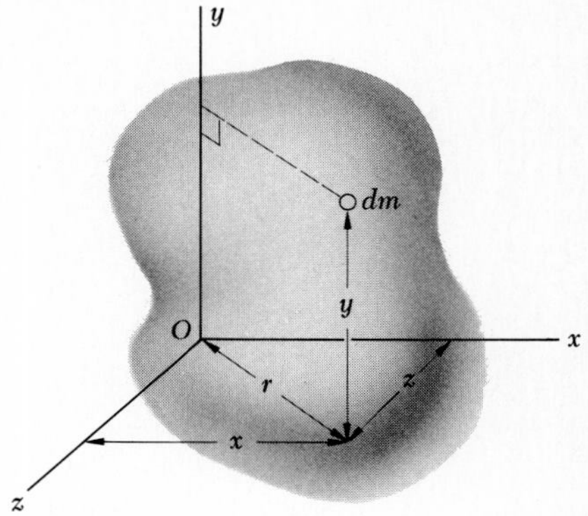

Fig. 9.21

Similar expressions may be obtained for the moments of inertia with respect to the x and z axes. We write

$$\begin{aligned} I_x &= \int (y^2 + z^2)\, dm \\ I_y &= \int (z^2 + x^2)\, dm \\ I_z &= \int (x^2 + y^2)\, dm \end{aligned} \qquad (9.30)$$

9.11. Parallel-Axis Theorem. Consider a body of mass m. Let $Oxyz$ be a system of rectangular coordinates with origin at an arbitrary point O, and $Gx'y'z'$ a system of parallel *centroidal axes*, i.e., a system with origin at the center of gravity G of the body‡ and with axes x', y', z', respectively parallel to x, y, z (Fig. 9.22). Denoting by $\bar{x}$, $\bar{y}$, $\bar{z}$ the coordinates of G with respect to $Oxyz$, we write the following relations between the coordinates x, y, z of the element dm with respect to $Oxyz$ and its coordinates x', y', z' with respect to the centroidal axes $Gx'y'z'$:

$$x = x' + \bar{x} \qquad y = y' + \bar{y} \qquad z = z' + \bar{z} \qquad (9.31)$$

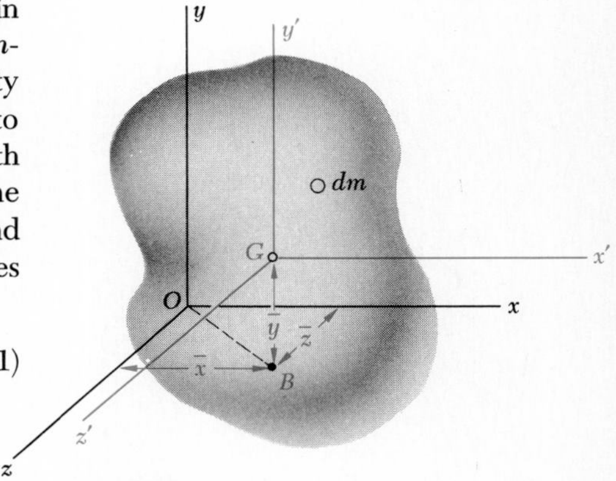

Fig. 9.22

†It should be kept in mind, when converting the moment of inertia of a mass from U.S. customary units to SI units, that the base unit pound used in the derived unit lb · ft · s² is a unit of force (*not* of mass) and, therefore, should be converted into newtons. We have

$$1 \text{ lb} \cdot \text{ft} \cdot \text{s}^2 = (4.45 \text{ N})(0.3048 \text{ m})(1 \text{ s})^2 = 1.356 \text{ N} \cdot \text{m} \cdot \text{s}^2$$

or, since N = kg · m/s²,

$$1 \text{ lb} \cdot \text{ft} \cdot \text{s}^2 = 1.356 \text{ kg} \cdot \text{m}^2$$

‡Note that the term centroidal is used to define an axis passing through the center of gravity G of the body, whether or not G coincides with the centroid of the volume of the body.

Referring to Eqs. (9.30), we may express the moment of inertia of the body with respect to the x axis as follows:

$$\begin{aligned} I_x &= \int (y^2 + z^2)\, dm = \int [(y' + \bar{y})^2 + (z' + \bar{z})^2]\, dm \\ &= \int (y'^2 + z'^2)\, dm + 2\bar{y}\int y'\, dm + 2\bar{z}\int z'\, dm + (\bar{y}^2 + \bar{z}^2)\int dm \end{aligned}$$

The first integral in the expression obtained represents the moment of inertia $\bar{I}_{x'}$ of the body with respect to the centroidal axis x'; the second and third integrals represent the first moment of the body with respect to the $z'x'$ and $x'y'$ planes, respectively, and, since both planes contain G, the two integrals are zero; the last integral is equal to the total mass m of the body. We write, therefore,

$$I_x = \bar{I}_{x'} + m(\bar{y}^2 + \bar{z}^2) \tag{9.32}$$

and, similarly,

$$I_y = \bar{I}_{y'} + m(\bar{z}^2 + \bar{x}^2) \qquad I_z = \bar{I}_{z'} + m(\bar{x}^2 + \bar{y}^2) \tag{9.32'}$$

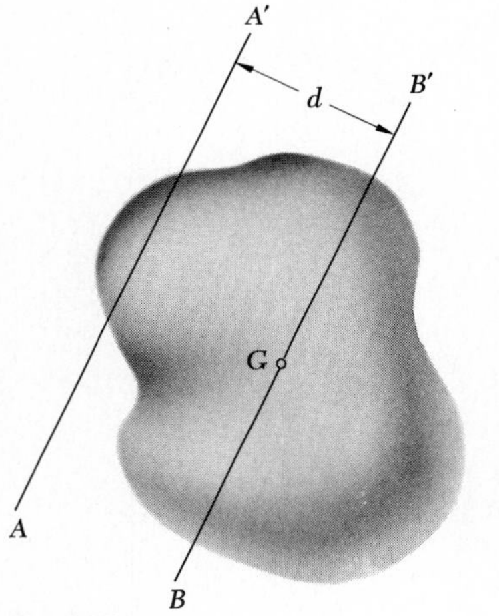

Fig. 9.23

We easily verify from Fig. 9.22 that the sum $\bar{z}^2 + \bar{x}^2$ represents the square of the distance OB between the y and y' axis. Similarly, $\bar{y}^2 + \bar{z}^2$ and $\bar{x}^2 + \bar{y}^2$ represent the squares of the distances between the x and x' axes, and the z and z' axes, respectively. Denoting by d the distance between an arbitrary axis AA' and a parallel centroidal axis BB' (Fig. 9.23), we may, therefore, write the following general relation between the moment of inertia I of the body with respect to AA' and its moment of inertia $\bar{I}$ with respect to BB':

$$I = \bar{I} + md^2 \tag{9.33}$$

Expressing the moments of inertia in terms of the corresponding radii of gyration, we may also write

$$k^2 = \bar{k}^2 + d^2 \tag{9.34}$$

where k and $\bar{k}$ represent the radii of gyration about AA' and BB', respectively.

9.12. Moments of Inertia of Thin Plates. Consider a thin plate of uniform thickness t, made of a homogeneous material of density ρ (density = mass per unit volume). The mass moment of inertia of the plate with respect to an axis AA' *contained in the plane* of the plate (Fig. 9.24*a*) is

$$I_{AA',\text{mass}} = \int r^2\, dm$$

Since $dm = \rho t\, dA$, we write

$$I_{AA',\text{mass}} = \rho t \int r^2\, dA$$

But r represents the distance of the element of area dA to the axis AA'; the integral is therefore equal to the moment of inertia of the area of the plate with respect to AA'. We have

$$I_{AA',\text{mass}} = \rho t I_{AA',\text{area}} \tag{9.35}$$

Similarly, we have with respect to an axis BB' perpendicular to AA' (Fig. 9.24*b*)

$$I_{BB',\text{mass}} = \rho t I_{BB',\text{area}} \tag{9.36}$$

Considering now the axis CC' *perpendicular* to the plate through the point of intersection C of AA' and BB' (Fig. 9.24*c*), we write

$$I_{CC',\text{ mass}} = \rho t J_{C,\text{ area}} \tag{9.37}$$

where J_C is the *polar* moment of inertia of the area of the plate with respect to point C.

Recalling the relation $J_C = I_{AA'} + I_{BB'}$ existing between polar and rectangular moments of inertia of an area, we write the following relation between the mass moments of inertia of a thin plate:

$$I_{CC'} = I_{AA'} + I_{BB'} \tag{9.38}$$

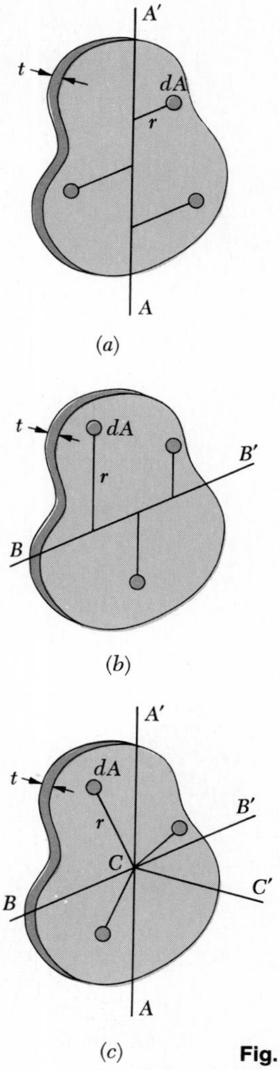

Fig. 9.24

Rectangular Plate. In the case of a rectangular plate of sides a and b (Fig. 9.25), we obtain the following mass moments of inertia with respect to axes through the center of gravity of the plate:

$$I_{AA',\text{mass}} = \rho t I_{AA',\text{area}} = \rho t(\tfrac{1}{12}a^3 b)$$
$$I_{BB',\text{mass}} = \rho t I_{BB',\text{area}} = \rho t(\tfrac{1}{12}ab^3)$$

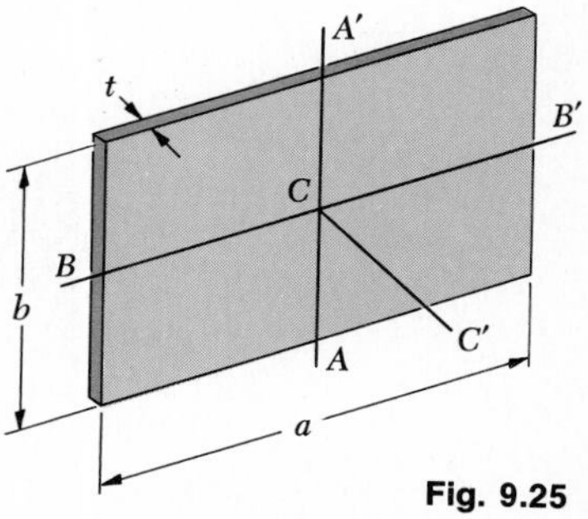

Fig. 9.25

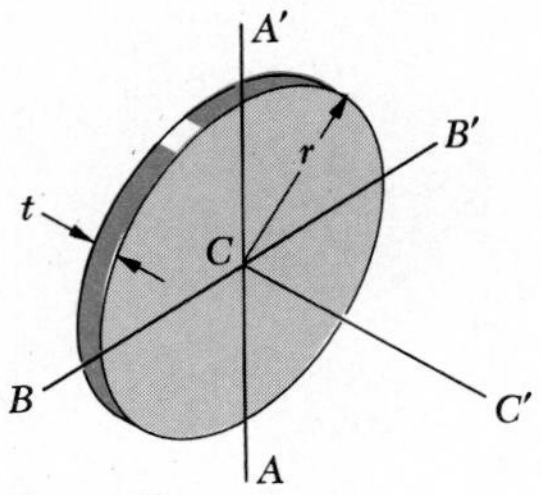

Fig. 9.26

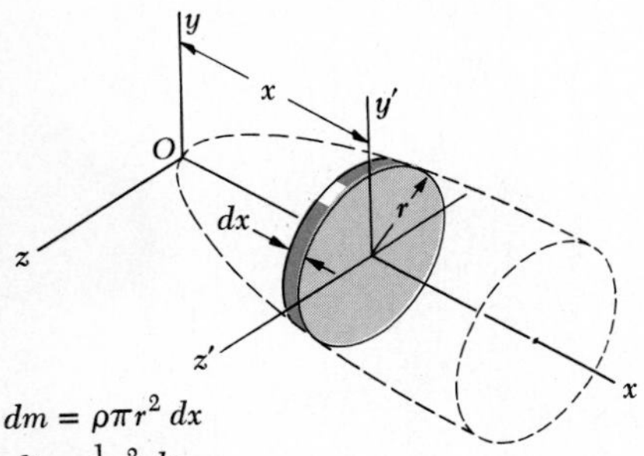

$dm = \rho\pi r^2\,dx$

$dI_x = \frac{1}{2}r^2\,dm$

$dI_y = dI_{y'} + x^2\,dm = (\frac{1}{4}r^2 + x^2)\,dm$

$dI_z = dI_{z'} + x^2\,dm = (\frac{1}{4}r^2 + x^2)\,dm$

Fig. 9.27 Determination of the moment of inertia of a body of revolution.

Observing that the product ρabt is equal to the mass m of the plate, we write the mass moments of inertia of a thin rectangular plate as follows:

$$I_{AA'} = \tfrac{1}{12}ma^2 \qquad I_{BB'} = \tfrac{1}{12}mb^2 \qquad (9.39)$$

$$I_{CC'} = I_{AA'} + I_{BB'} = \tfrac{1}{12}m(a^2 + b^2) \qquad (9.40)$$

Circular Plate. In the case of a circular plate, or disk, of radius r (Fig. 9.26), we write

$$I_{AA',\text{mass}} = \rho t I_{AA',\text{area}} = \rho t(\tfrac{1}{4}\pi r^4)$$

Observing that the product $\rho\pi r^2 t$ is equal to the mass m of the plate and that $I_{AA'} = I_{BB'}$, we write the mass moments of inertia of a circular plate as follows:

$$I_{AA'} = I_{BB'} = \tfrac{1}{4}mr^2 \qquad (9.41)$$

$$I_{CC'} = I_{AA'} + I_{BB'} = \tfrac{1}{2}mr^2 \qquad (9.42)$$

9.13. Determination of the Moment of Inertia of a Three-dimensional Body by Integration. The moment of inertia of a three-dimensional body is obtained by computing the integral $I = \int r^2\,dm$. If the body is made of a homogeneous material of density ρ, we have $dm = \rho\,dV$ and write $I = \rho\int r^2\,dV$. This integral depends only upon the shape of the body. In order to compute it, it will generally be necessary to perform a triple, or at least a double, integration.

However, if the body possesses two planes of symmetry, it is usually possible to determine its moment of inertia through a single integration by choosing as an element of mass dm the mass of a thin slab perpendicular to the planes of symmetry. In the case of bodies of revolution, for example, the element of mass should be a thin disk (Fig. 9.27). Using formula (9.42), the moment of inertia of the disk with respect to the axis of revolution may be readily expressed as indicated in Fig. 9.27. Its moment of inertia with respect to each of the other two axes of coordinates will be obtained by using formula (9.41) and the parallel-axis theorem. Integration of the expressions obtained will yield the desired moments of inertia of the body of revolution.

9.14. Moments of Inertia of Composite Bodies. The moments of inertia of a few common shapes are shown in Fig. 9.28. The moment of inertia with respect to a given axis of a body made of several of these simple shapes may be obtained by computing the moments of inertia of its component parts about the desired axis and adding them together. We should note, as we already have noted in the case of areas, that the radius of gyration of a composite body *cannot* be obtained by adding the radii of gyration of its component parts.

Slender rod	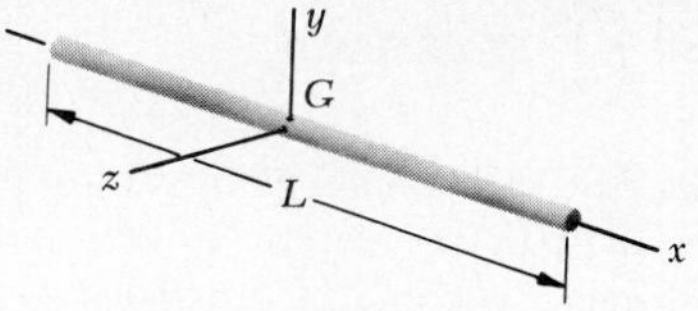	$I_y = I_z = \frac{1}{12}mL^2$
Thin rectangular plate	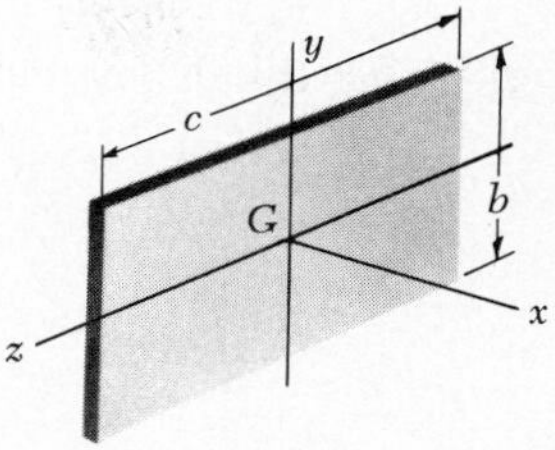	$I_x = \frac{1}{12}m(b^2 + c^2)$ $I_y = \frac{1}{12}mc^2$ $I_z = \frac{1}{12}mb^2$
Rectangular prism	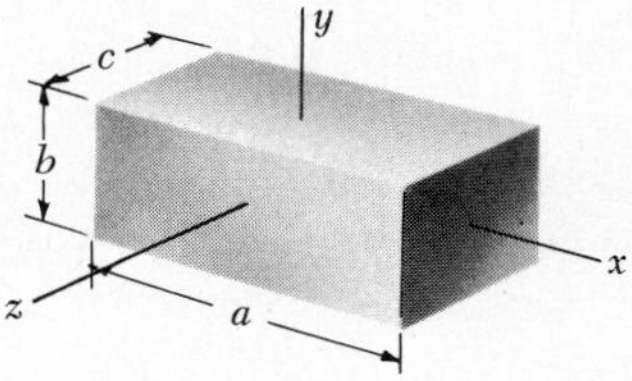	$I_x = \frac{1}{12}m(b^2 + c^2)$ $I_y = \frac{1}{12}m(c^2 + a^2)$ $I_z = \frac{1}{12}m(a^2 + b^2)$
Thin disk	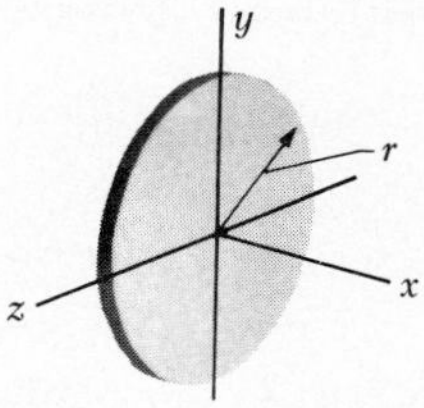	$I_x = \frac{1}{2}mr^2$ $I_y = I_z = \frac{1}{4}mr^2$
Circular cylinder	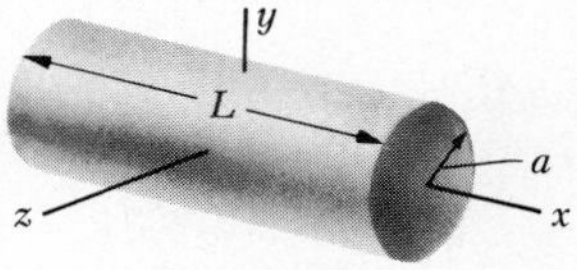	$I_x = \frac{1}{2}ma^2$ $I_y = I_z = \frac{1}{12}m(3a^2 + L^2)$
Circular cone	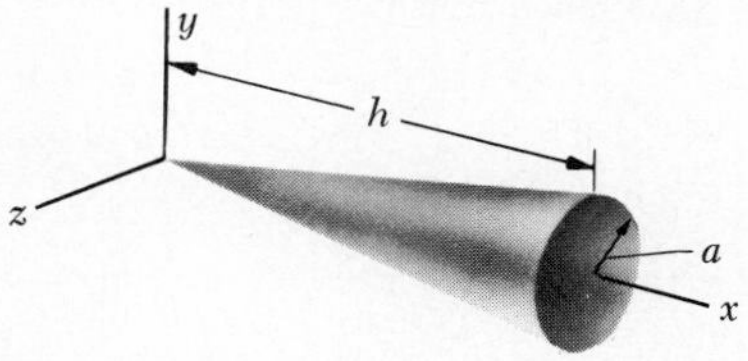	$I_x = \frac{3}{10}ma^2$ $I_y = I_z = \frac{3}{5}m(\frac{1}{4}a^2 + h^2)$
Sphere	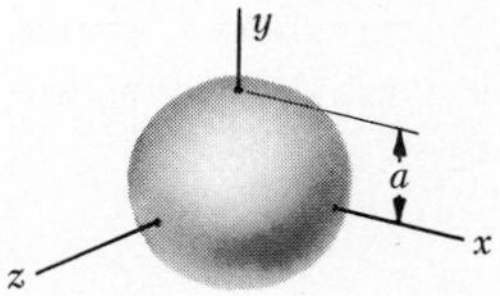	$I_x = I_y = I_z = \frac{2}{5}ma^2$

Fig. 9.28 Mass moments of inertia of common geometric shapes

SAMPLE PROBLEM 9.9

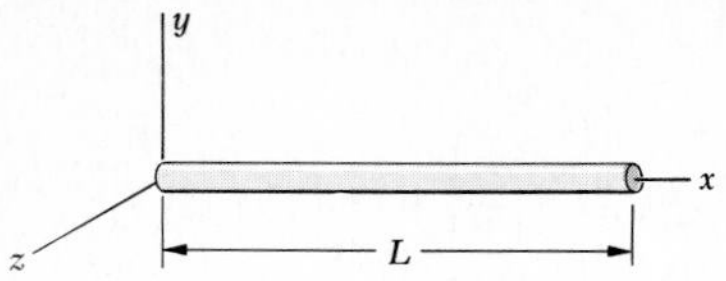

Determine the mass moment of inertia of a slender rod of length L and mass m with respect to an axis perpendicular to the rod and passing through one end of the rod.

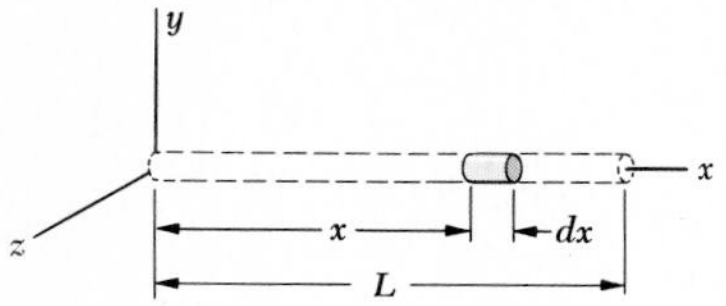

Solution. Choosing the differential element of mass shown, we write

$$dm = \frac{m}{L}\,dx$$

$$I_y = \int x^2\,dm = \int_0^L x^2 \frac{m}{L}\,dx = \left[\frac{m}{L}\frac{x^3}{3}\right]_0^L \qquad I_y = \frac{mL^2}{3} \blacktriangleleft$$

SAMPLE PROBLEM 9.10

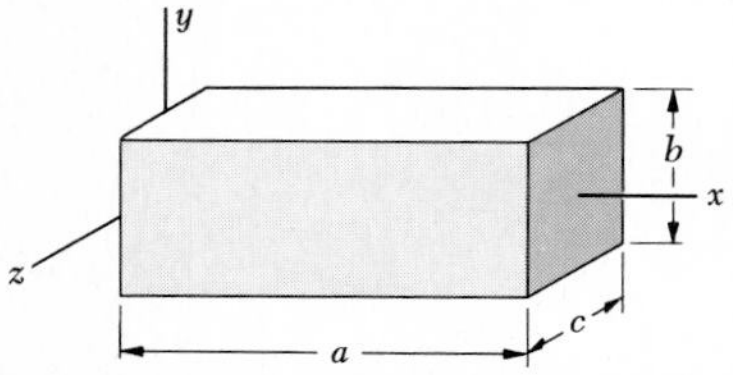

Determine the mass moment of inertia of the homogeneous rectangular prism shown with respect to the z axis.

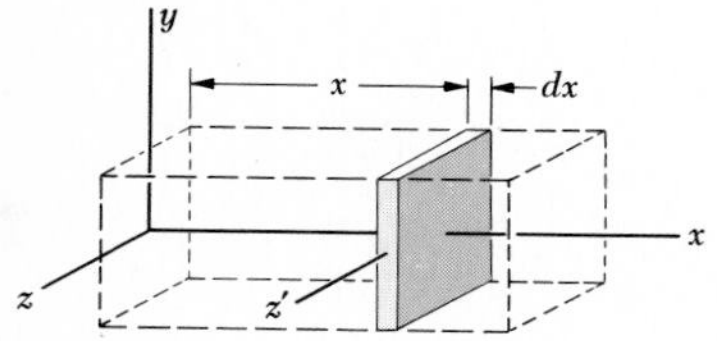

Solution. We choose as a differential element of mass the thin slab shown for which

$$dm = \rho bc\,dx$$

Referring to Sec. 9.12, we find that the moment of inertia of the element with respect to the z' axis is

$$dI_{z'} = \tfrac{1}{12}b^2\,dm$$

Applying the parallel-axis theorem, we obtain the mass moment of inertia of the slab with respect to the z axis.

$$dI_z = dI_{z'} + x^2\,dm = \tfrac{1}{12}b^2\,dm + x^2\,dm = (\tfrac{1}{12}b^2 + x^2)\,\rho bc\,dx$$

Integrating from $x = 0$ to $x = a$, we obtain

$$I_z = \int dI_z = \int_0^a (\tfrac{1}{12}b^2 + x^2)\,\rho bc\,dx = \rho abc\,(\tfrac{1}{12}b^2 + \tfrac{1}{3}a^2)$$

Since the total mass of the prism is $m = \rho abc$, we may write

$$I_z = m(\tfrac{1}{12}b^2 + \tfrac{1}{3}a^2) \qquad I_z = \tfrac{1}{12}m(4a^2 + b^2) \blacktriangleleft$$

We note that if the prism is slender, b is small compared to a and the expression for I_z reduces to $ma^2/3$, which is the result obtained in Sample Prob. 9.9 when $L = a$.

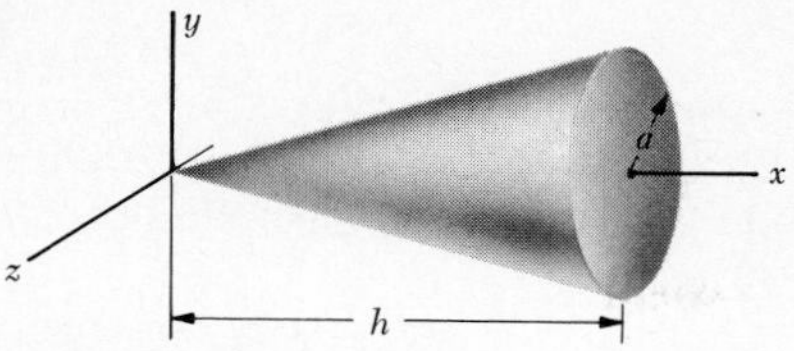

SAMPLE PROBLEM 9.11

Determine the mass moment of inertia of a right circular cone with respect to (*a*) its longitudinal axis, (*b*) an axis through the apex of the cone and perpendicular to its longitudinal axis, (*c*) an axis through the centroid of the cone and perpendicular to its longitudinal axis.

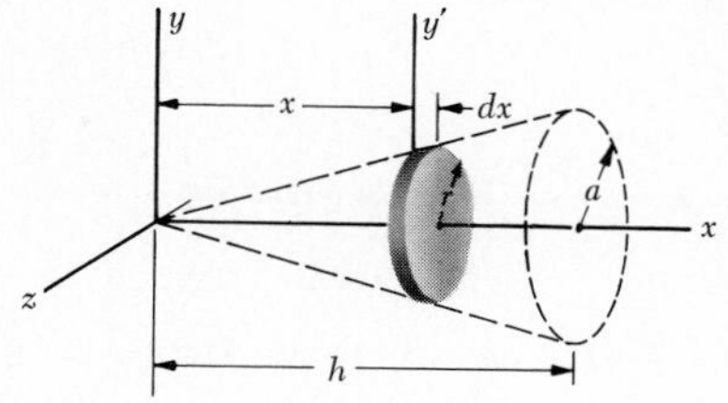

Solution. We choose the differential element of mass shown.

$$r = a\frac{x}{h} \qquad dm = \rho\pi r^2\,dx = \rho\pi\frac{a^2}{h^2}x^2\,dx$$

a. Moment of Inertia I_x. Using the expression derived in Sec. 9.12 for a thin disk, we compute the mass moment of inertia of the differential element with respect to the x axis.

$$dI_x = \tfrac{1}{2}r^2\,dm = \tfrac{1}{2}\left(a\frac{x}{h}\right)^2\left(\rho\pi\frac{a^2}{h^2}x^2\,dx\right) = \tfrac{1}{2}\rho\pi\frac{a^4}{h^4}x^4\,dx$$

Integrating from $x = 0$ to $x = h$, we obtain

$$I_x = \int dI_x = \int_0^h \tfrac{1}{2}\rho\pi\frac{a^4}{h^4}x^4\,dx = \tfrac{1}{2}\rho\pi\frac{a^4}{h^4}\frac{h^5}{5} = \tfrac{1}{10}\rho\pi a^4 h$$

Since the total mass of the cone is $m = \frac{1}{3}\rho\pi a^2 h$, we may write

$$I_x = \tfrac{1}{10}\rho\pi a^4 h = \tfrac{3}{10}a^2(\tfrac{1}{3}\rho\pi a^2 h) = \tfrac{3}{10}ma^2 \qquad I_x = \tfrac{3}{10}ma^2 \blacktriangleleft$$

b. Moment of Inertia I_y. The same differential element will be used. Applying the parallel-axis theorem and using the expression derived in Sec. 9.12 for a thin disk, we write

$$dI_y = dI_{y'} + x^2\,dm = \tfrac{1}{4}r^2\,dm + x^2\,dm = (\tfrac{1}{4}r^2 + x^2)\,dm$$

Substituting the expressions for r and dm, we obtain

$$dI_y = \left(\frac{1}{4}\frac{a^2}{h^2}x^2 + x^2\right)\left(\rho\pi\frac{a^2}{h^2}x^2\,dx\right) = \rho\pi\frac{a^2}{h^2}\left(\frac{a^2}{4h^2} + 1\right)x^4\,dx$$

$$I_y = \int dI_y = \int_0^h \rho\pi\frac{a^2}{h^2}\left(\frac{a^2}{4h^2} + 1\right)x^4\,dx = \rho\pi\frac{a^2}{h^2}\left(\frac{a^2}{4h^2} + 1\right)\frac{h^5}{5}$$

Introducing the total mass of the cone m, we rewrite I_y as follows:

$$I_y = \tfrac{3}{5}(\tfrac{1}{4}a^2 + h^2)\tfrac{1}{3}\rho\pi a^2 h \qquad I_y = \tfrac{3}{5}m(\tfrac{1}{4}a^2 + h^2) \blacktriangleleft$$

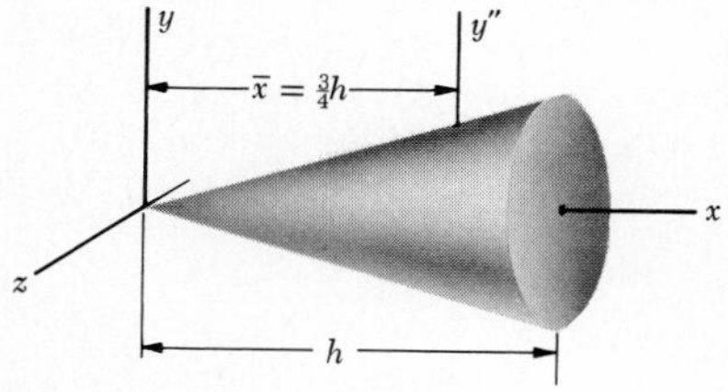

c. Moment of Inertia $\bar{I}_{y''}$. We apply the parallel-axis theorem and write

$$I_y = \bar{I}_{y''} + m\bar{x}^2$$

Solving for $\bar{I}_{y''}$ and recalling that $\bar{x} = \frac{3}{4}h$, we have

$$\bar{I}_{y''} = I_y - m\bar{x}^2 = \tfrac{3}{5}m(\tfrac{1}{4}a^2 + h^2) - m(\tfrac{3}{4}h)^2$$

$$\bar{I}_{y''} = \tfrac{3}{20}m(a^2 + \tfrac{1}{4}h^2) \blacktriangleleft$$

SAMPLE PROBLEM 9.12

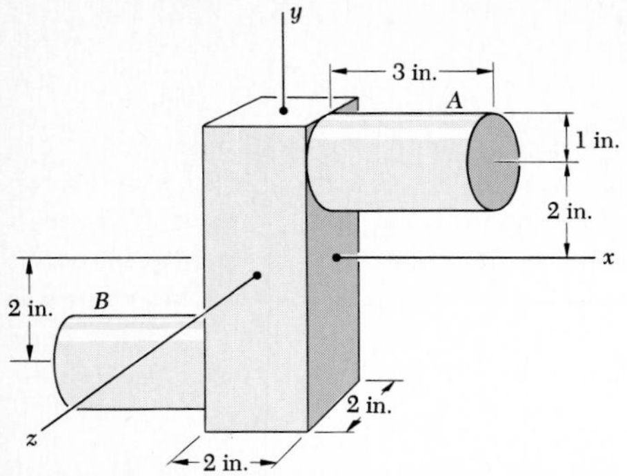

A steel forging consists of a rectangular prism 6 by 2 by 2 in. and of two cylinders of diameter 2 in. and length 3 in., as shown. Determine the mass moments of inertia with respect to the coordinate axes. (Specific weight of steel = 490 lb/ft³.)

Computation of Masses

Prism

$$V = 24\ \text{in}^3 \qquad W = \frac{(24\ \text{in}^3)(490\ \text{lb/ft}^3)}{1728\ \text{in}^3/\text{ft}^3} = 6.81\ \text{lb}$$

$$m = \frac{6.81\ \text{lb}}{32.2\ \text{ft/s}^2} = 0.211\ \text{lb}\cdot\text{s}^2/\text{ft}$$

Each Cylinder

$$V = \pi(1\ \text{in.})^2(3\ \text{in.}) = 9.42\ \text{in}^3$$

$$W = \frac{(9.42\ \text{in}^3)(490\ \text{lb/ft}^3)}{1728\ \text{in}^3/\text{ft}^3} = 2.67\ \text{lb}$$

$$m = \frac{2.67\ \text{lb}}{32.2\ \text{ft/s}^2} = 0.0829\ \text{lb}\cdot\text{s}^2/\text{ft}$$

Mass Moments of Inertia. The mass moments of inertia of each component are computed from Fig. 9.28, using the parallel-axis theorem when necessary. Note that all lengths should be expressed in feet.

Prism

$$I_x = I_z = \tfrac{1}{12}(0.211\ \text{lb}\cdot\text{s}^2/\text{ft})[(\tfrac{6}{12}\ \text{ft})^2 + (\tfrac{2}{12}\ \text{ft})^2] = 4.88 \times 10^{-3}\ \text{lb}\cdot\text{ft}\cdot\text{s}^2$$
$$I_y = \tfrac{1}{12}(0.211\ \text{lb}\cdot\text{s}^2/\text{ft})[(\tfrac{2}{12}\ \text{ft})^2 + (\tfrac{2}{12}\ \text{ft})^2] = 0.977 \times 10^{-3}\ \text{lb}\cdot\text{ft}\cdot\text{s}^2$$

Each cylinder

$I_{MASS} = \bar{I}_{MASS} + md^2$

$$I_x = \tfrac{1}{2}ma^2 + m\bar{y}^2 = \tfrac{1}{2}(0.0829\ \text{lb}\cdot\text{s}^2/\text{ft})(\tfrac{1}{12}\ \text{ft})^2 + (0.0829\ \text{lb}\cdot\text{s}^2/\text{ft})(\tfrac{2}{12}\ \text{ft})^2 = 2.59 \times 10^{-3}\ \text{lb}\cdot\text{ft}\cdot\text{s}^2$$
$$I_y = \tfrac{1}{12}m(3a^2 + L^2) + m\bar{x}^2 = \tfrac{1}{12}(0.0829\ \text{lb}\cdot\text{s}^2/\text{ft})[3(\tfrac{1}{12}\ \text{ft})^2 + (\tfrac{3}{12}\ \text{ft})^2] + (0.0829\ \text{lb}\cdot\text{s}^2/\text{ft})(\tfrac{2.5}{12}\ \text{ft})^2 = 4.17 \times 10^{-3}\ \text{lb}\cdot\text{ft}\cdot\text{s}^2$$
$$I_z = \tfrac{1}{12}m(3a^2 + L^2) + m(\bar{x}^2 + \bar{y}^2) = \tfrac{1}{12}(0.0829)[3(\tfrac{1}{12})^2 + (\tfrac{3}{12})^2] + (0.0829)[(\tfrac{2.5}{12})^2 + (\tfrac{2}{12})^2] = 6.48 \times 10^{-3}\ \text{lb}\cdot\text{ft}\cdot\text{s}^2$$

Entire Body. Adding the values obtained:

$$I_x = 4.88 \times 10^{-3} + 2(2.59 \times 10^{-3})$$
$$I_x = 10.06 \times 10^{-3}\ \text{lb}\cdot\text{ft}\cdot\text{s}^2 \ \blacktriangleleft$$
$$I_y = 0.977 \times 10^{-3} + 2(4.17 \times 10^{-3})$$
$$I_y = 9.32 \times 10^{-3}\ \text{lb}\cdot\text{ft}\cdot\text{s}^2 \ \blacktriangleleft$$
$$I_z = 4.88 \times 10^{-3} + 2(6.48 \times 10^{-3})$$
$$I_z = 17.84 \times 10^{-3}\ \text{lb}\cdot\text{ft}\cdot\text{s}^2 \ \blacktriangleleft$$

SAMPLE PROBLEM 9.13

Solve Sample Prob. 9.12 using SI units.

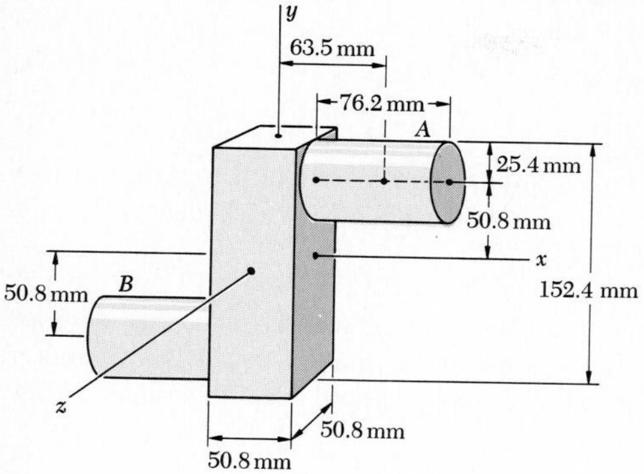

Solution. First, the dimensions are converted into millimeters (1 in. = 25.4 mm). Next, the density of steel ρ (mass per unit volume) is determined in SI units. Recalling that 1 ft = 0.3048 m and that the mass of a block weighing 1 lb is 0.454 kg, we have

$$\rho = (490\ \text{lb/ft}^3)\left(\frac{0.454\ \text{kg}}{1\ \text{lb}}\right)\left(\frac{1\ \text{ft}}{0.3048\ \text{m}}\right)^3 = 7850\ \text{kg/m}^3$$

Computation of Masses

Prism. $V = (50.8\ \text{mm})^2(152.4\ \text{mm}) = 0.393 \times 10^6\ \text{mm}^3$

or, since $1\ \text{mm}^3 = (10^{-3}\ \text{m})^3 = 10^{-9}\ \text{m}^3$,

$$V = 0.393 \times 10^6 \times 10^{-9}\ \text{m}^3 = 0.393 \times 10^{-3}\ \text{m}^3$$
$$m = \rho V = (7.85 \times 10^3\ \text{kg/m}^3)(0.393 \times 10^{-3}\ \text{m}^3) = 3.09\ \text{kg}$$

Each Cylinder

$$V = \pi r^2 h = \pi(25.4\ \text{mm})^2(76.2\ \text{mm}) = 0.1544 \times 10^6\ \text{mm}^3 = 0.1544 \times 10^{-3}\ \text{m}^3$$
$$m = \rho V = (7.85 \times 10^3\ \text{kg/m}^3)(0.1544 \times 10^{-3}\ \text{m}^3) = 1.212\ \text{kg}$$

Mass Moments of Inertia. The mass moments of inertia of each component are computed from Fig. 9.28, using the parallel-axis theorem when necessary. Note that all lengths should be expressed in millimeters.

Prism

$$I_x = I_z = \tfrac{1}{12}(3.09\ \text{kg})[(152.4\ \text{mm})^2 + (50.8\ \text{mm})^2] = 6640\ \text{kg}\cdot\text{mm}^2$$
$$I_y = \tfrac{1}{12}(3.09\ \text{kg})[(50.8\ \text{mm})^2 + (50.8\ \text{mm})^2] = 1329\ \text{kg}\cdot\text{mm}^2$$

Each Cylinder

$$I_x = \tfrac{1}{2}ma^2 + m\bar{y}^2 = \tfrac{1}{2}(1.212\ \text{kg})(25.4\ \text{mm})^2 + (1.212\ \text{kg})(50.8\ \text{mm})^2 = 3520\ \text{kg}\cdot\text{mm}^2$$
$$I_y = \tfrac{1}{12}m(3a^2 + L^2) + m\bar{x}^2 = \tfrac{1}{12}(1.212\ \text{kg})[3(25.4\ \text{mm})^2 + (76.2\ \text{mm})^2] + (1.212\ \text{kg})(63.5\ \text{mm})^2 = 5670\ \text{kg}\cdot\text{mm}^2$$
$$I_z = \tfrac{1}{12}m(3a^2 + L^2) + m(\bar{x}^2 + \bar{y}^2) = \tfrac{1}{12}(1.212\ \text{kg})[3(25.4\ \text{mm})^2 + (76.2\ \text{mm})^2] + (1.212\ \text{kg})[(63.5\ \text{mm})^2 + (50.8\ \text{mm})^2] = 8800\ \text{kg}\cdot\text{mm}^2$$

Entire Body. Adding the values obtained, and observing that $1\ \text{mm}^2 = (10^{-3}\ \text{m})^2 = 10^{-6}\ \text{m}^2$, we have

$$I_x = 6640\ \text{kg}\cdot\text{mm}^2 + 2(3520\ \text{kg}\cdot\text{mm}^2) = 13.68 \times 10^3\ \text{kg}\cdot\text{mm}^2$$
$$I_x = 13.68 \times 10^{-3}\ \text{kg}\cdot\text{m}^2 \quad ◀$$
$$I_y = 1329\ \text{kg}\cdot\text{mm}^2 + 2(5670\ \text{kg}\cdot\text{mm}^2) = 12.67 \times 10^3\ \text{kg}\cdot\text{mm}^2$$
$$I_y = 12.67 \times 10^{-3}\ \text{kg}\cdot\text{m}^2 \quad ◀$$
$$I_z = 6640\ \text{kg}\cdot\text{mm}^2 + 2(8800\ \text{kg}\cdot\text{mm}^2) = 24.2 \times 10^3\ \text{kg}\cdot\text{mm}^2$$
$$I_z = 24.2 \times 10^{-3}\ \text{kg}\cdot\text{m}^2 \quad ◀$$

Recalling that $1\ \text{lb}\cdot\text{ft}\cdot\text{s}^2 = 1.356\ \text{kg}\cdot\text{m}^2$ (see footnote, page 353) we may check these answers against the values obtained in Sample Prob. 9.12.

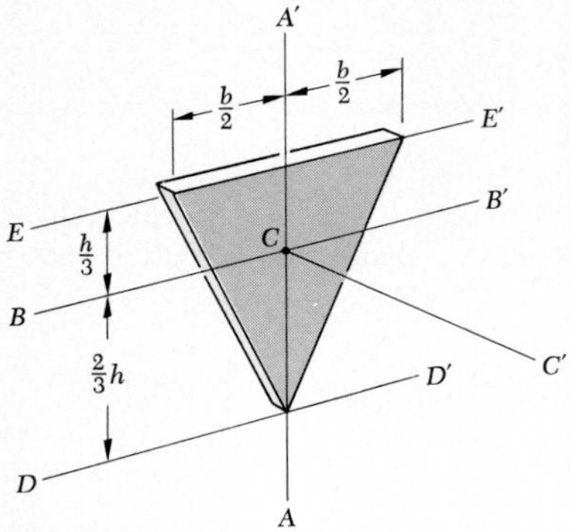

Fig. P9.72

PROBLEMS

9.72 A thin plate of mass m is cut in the shape of an isosceles triangle of width b and height h. Determine the mass moment of inertia of the plate with respect to (a) the centroidal axes AA' and BB' in the plane of the plate, (b) the centroidal axis CC' perpendicular to the plate.

9.73 Determine the mass moment of inertia of the plate of Prob. 9.72 with respect to (a) edge EE', (b) axis DD' which passes through a vertex and is parallel to edge EE'.

9.74 Determine the mass moment of inertia of a ring of mass m, cut from a thin uniform plate, with respect to (a) the diameter AA' of the ring, (b) the axis CC' perpendicular to the plane of the ring.

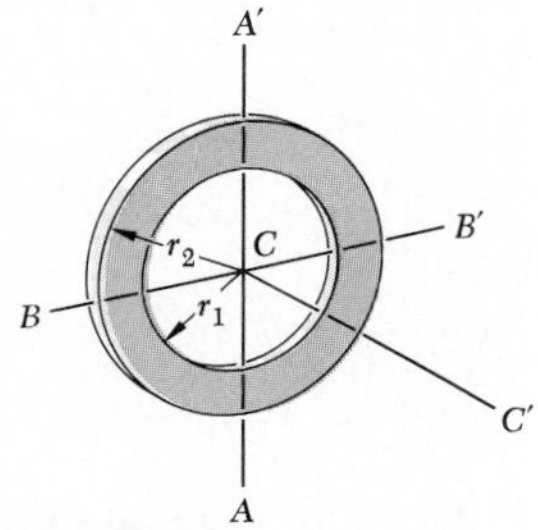

Fig. P9.74

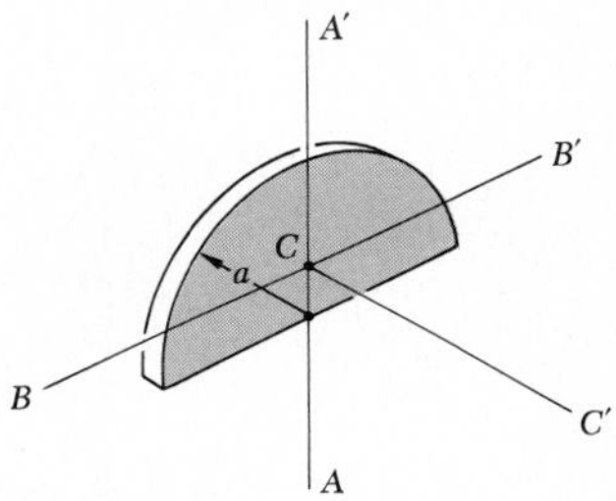

Fig. P9.75

9.75 A thin semicircular plate has a radius a and a mass m. Determine the mass moment of inertia of the plate with respect to (a) the centroidal axis BB', (b) the centroidal axis CC'.

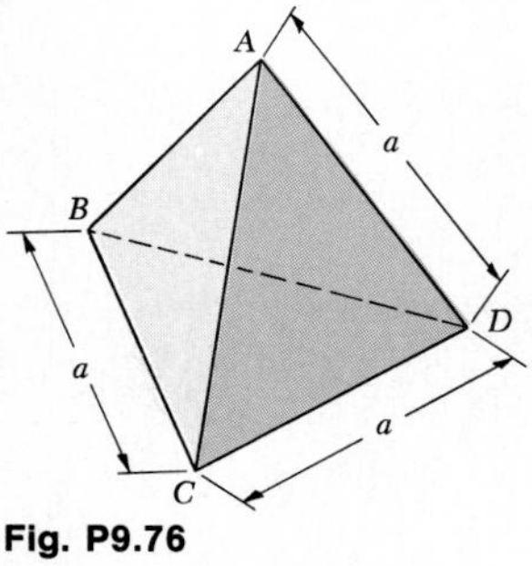

Fig. P9.76

9.76 Determine by direct integration the mass moment of inertia of the homogeneous regular tetrahedron of mass m and side a with respect to an axis through A perpendicular to the base BCD. (*Hint.* Use the result of part b of Prob. 9.72.)

9.77 The area shown is revolved about the x axis to form a homogeneous solid of revolution of mass m. Express the mass moment of inertia of the solid with respect to the x axis in terms of m, a, and n. The expression obtained may be used to verify (a) the value given in Fig. 9.28 for a cone (with $n = 1$) and (b) the answer to Prob. 9.78 (with $n = \frac{1}{2}$).

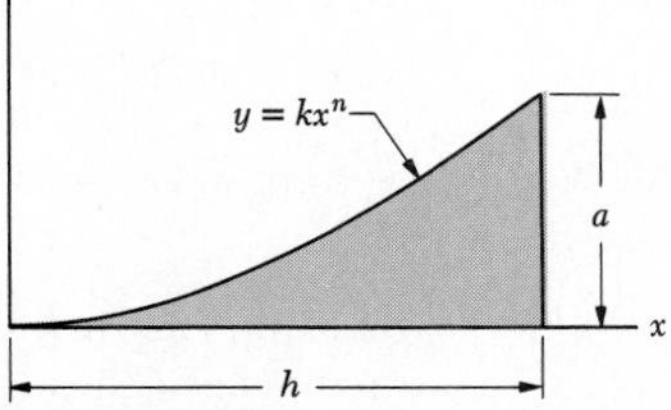

Fig. P9.77

9.78 Determine by direct integration the mass moment of inertia and the radius of gyration with respect to the x axis of the paraboloid shown, assuming a uniform density and a mass m.

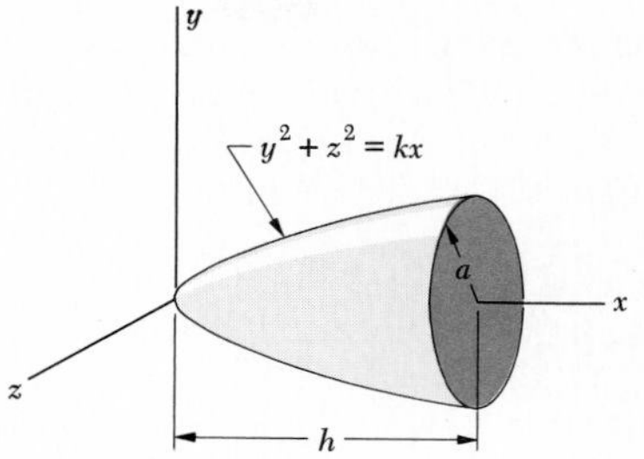

Fig. P9.78 and P9.80

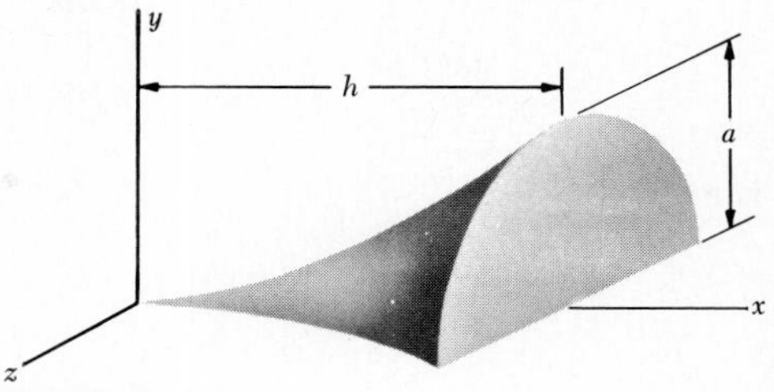

Fig. P9.79

9.79 The homogeneous solid shown was obtained by rotating the area of Prob. 9.77, with $n = 2$, through 180° about the x axis. Determine the mass moment of inertia $\bar{I}_x$ in terms of m and a.

9.80 Determine by direct integration the mass moment of inertia and the radius of gyration with respect to the y axis of the paraboloid shown, assuming a uniform density and a mass m.

9.81 Determine by direct integration the mass moment of inertia and the radius of gyration of a circular cylinder of radius a, length L, and uniform density ρ, with respect to a diameter of its base.

9.82 Determine by direct integration the mass moment of inertia with respect to the y axis of the pyramid shown, assuming a uniform density and a mass m.

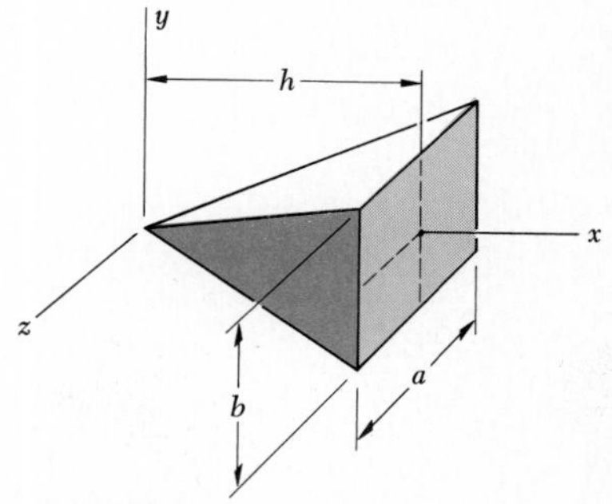

Fig. P9.82

9.83 Determine by direct integration the mass moment of inertia and the radius of gyration of the right circular cone with respect to the z axis, assuming a uniform density and a mass m.

9.84 Determine the mass moment of inertia of the frustum of a right circular cone of mass m with respect to its axis of symmetry.

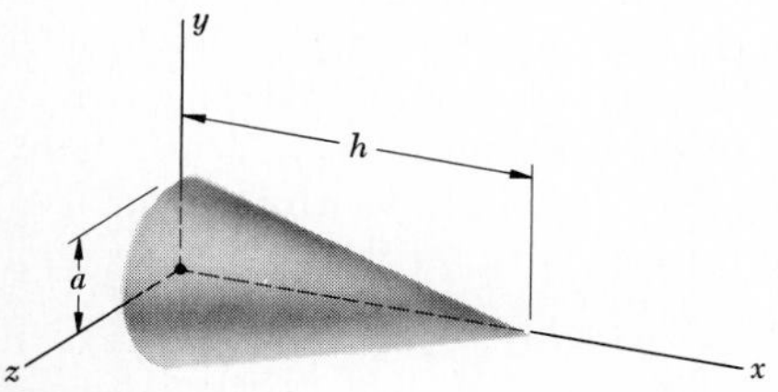

Fig. P9.83

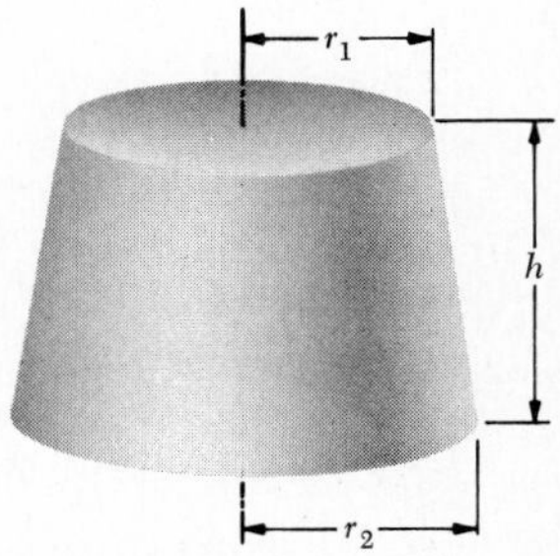

Fig. P9.84

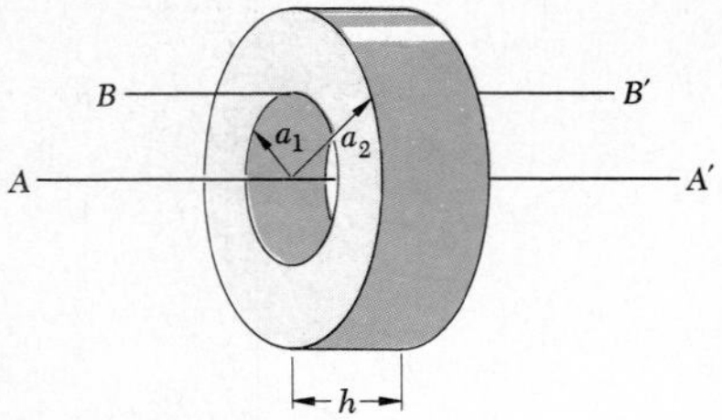

Fig. P9.85

9.85 Determine the mass moment of inertia of the homogeneous ring of mass m with respect to the axis AA'.

9.86 Determine the mass moment of inertia and the radius of gyration of the steel flywheel shown with respect to the axis of rotation. The web of the flywheel consists of a solid plate 25 mm thick. (Density of steel = 7850 kg/m^3.)

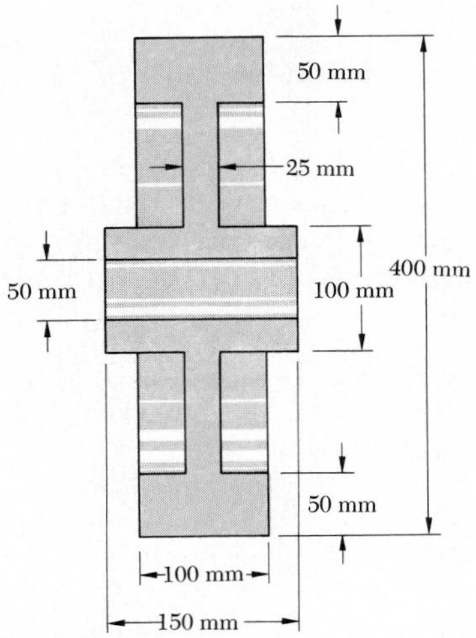

Fig. P9.86

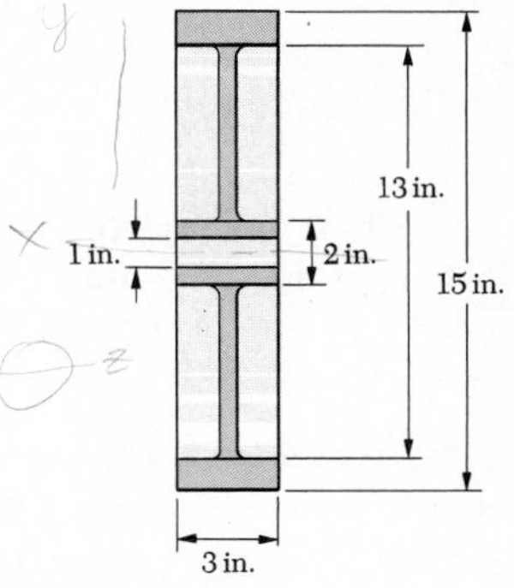

Fig. P9.87

9.87 The cross section of a small flywheel is shown. The rim and hub are connected by eight spokes (two of which are shown in the cross section). Each spoke has a cross-sectional area of 0.1875 in^2. Determine the mass moment of inertia and radius of gyration of the flywheel with respect to the axis of rotation. (Specific weight of steel = 490 lb/ft^3.)

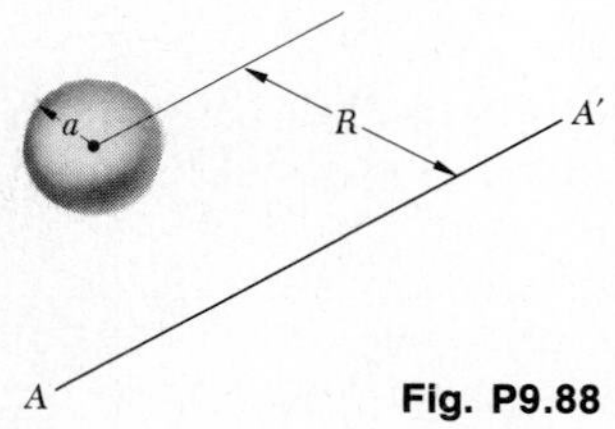

Fig. P9.88

9.88 In using the parallel-axis theorem, the error introduced by neglecting the centroidal moment of inertia is sometimes small. For a homogeneous sphere of radius a and mass m, (a) determine the mass moment of inertia with respect to an axis AA' at a distance R from the center of the sphere, (b) express as a function of a/R the relative error introduced by neglecting the centroidal moment of inertia, (c) determine the distance R in terms of a for which the relative error is 0.4 percent.

9.89 Two rods, *each* of length l and mass m, are welded together as shown. Knowing that rod CD is parallel to the x axis, determine the mass moment of inertia of the composite body with respect to (*a*) the x axis, (*b*) the y axis, (*c*) the z axis.

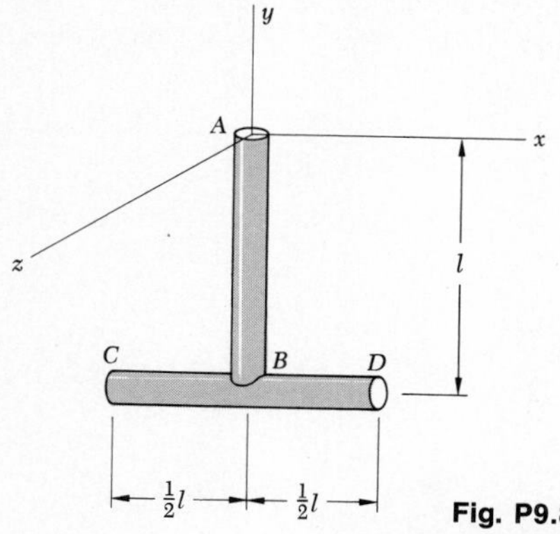

Fig. P9.89

9.90 In Prob. 9.89, determine the mass moments of inertia of the assembly with respect to centroidal axes parallel to the given x, y, and z axes.

9.91 For the homogeneous ring shown, which is of density ρ, determine (*a*) the mass moment of inertia with respect to the axis BB', (*b*) the value of a_1 for which, given a_2 and h, $I_{BB'}$ is maximum, (*c*) the corresponding value of $I_{BB'}$.

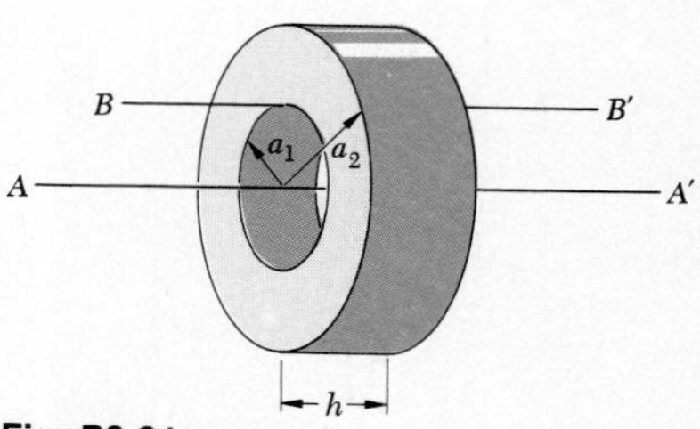

Fig. P9.91

9.92 and 9.93 Determine the mass moment of inertia and the radius of gyration of the steel link shown with respect to the x axis. (Specific weight of steel = 490 lb/ft³; density of steel = 7850 kg/m³.)

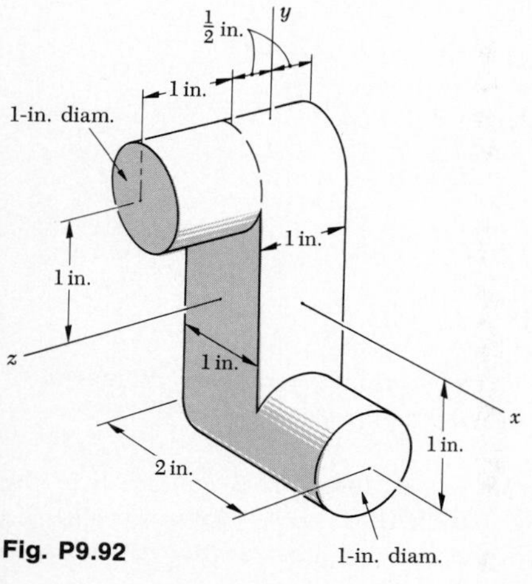

Fig. P9.92

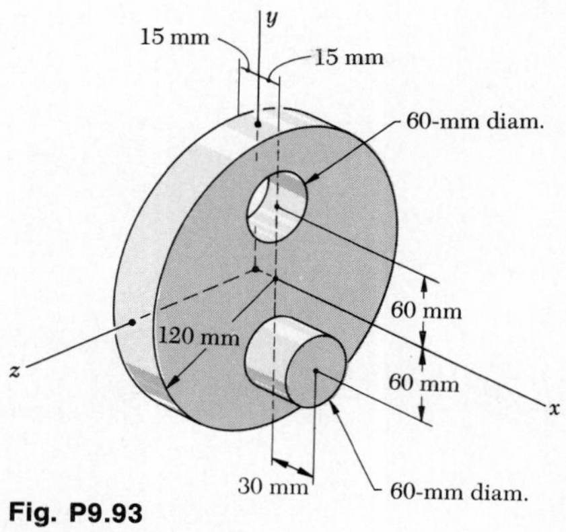

Fig. P9.93

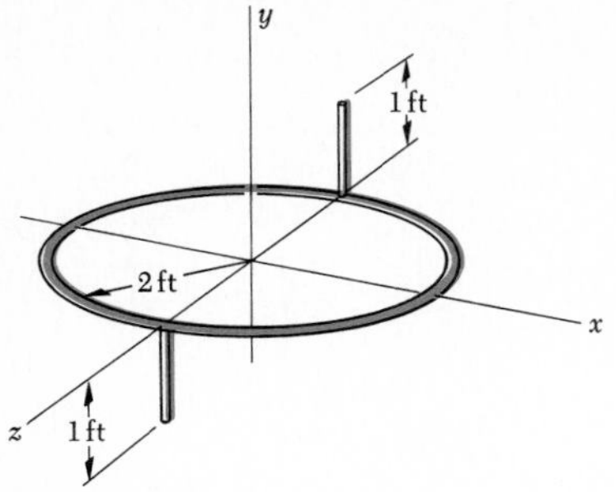

Fig. P9.94

9.94 A homogeneous wire, of weight 1.5 lb/ft, is used to form the figure shown. Determine the mass moment of inertia of the wire figure with respect to (*a*) the *x* axis, (*b*) the *y* axis, (*c*) the *z* axis.

9.95 Two holes, each of diameter 50 mm, are drilled through the steel block shown. Determine the mass moment of inertia of the body with respect to the axis of either of the holes. (Density of steel = 7850 kg/m^3.)

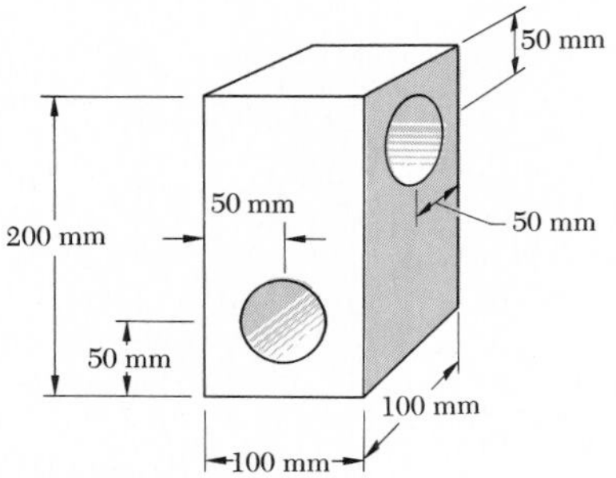

Fig. P9.95

REVIEW PROBLEMS

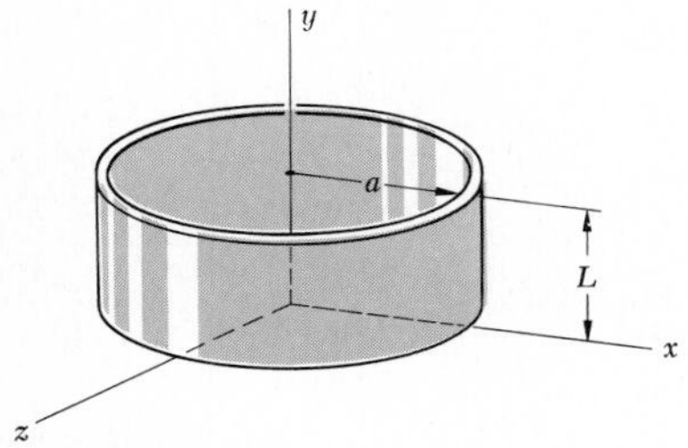

Fig. P9.96

9.96 Determine the mass moment of inertia I_x of the thin cylindrical shell of mass m which is shown. (*Hint.* Consider the solid bounded by cylindrical surfaces of radius a and $a + \varepsilon$, respectively, and let ε approach zero after the moment of inertia has been expressed in terms of a, ε, L, and m.)

9.97 Determine by direct integration the moment of inertia of the shaded area with respect to (*a*) the *x* axis, (*b*) the *y* axis.

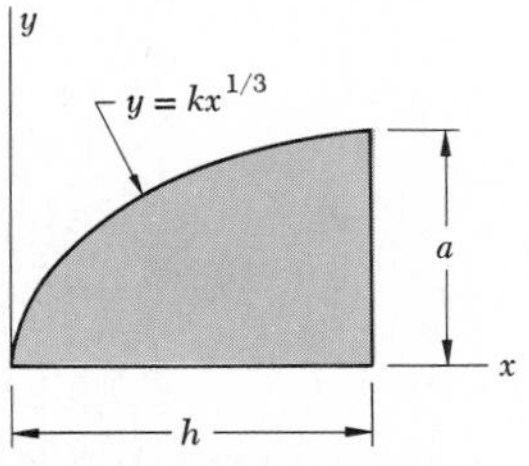

Fig. P9.97 and P9.98

9.98 Determine by direct integration the product of inertia of the shaded area with respect to the *x* and *y* axes.

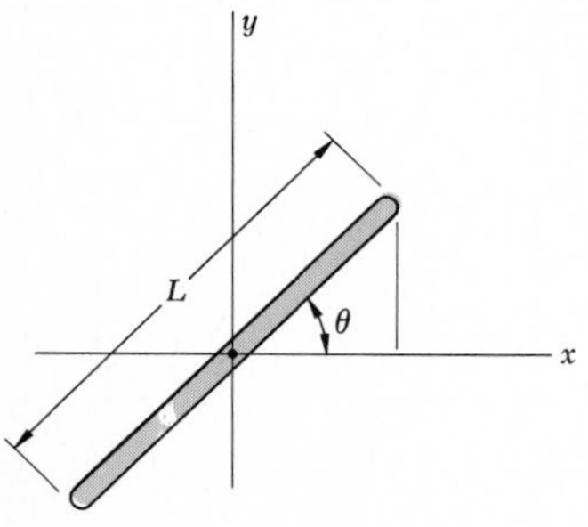

Fig. P9.99

9.99 A slender rod of mass m and length L is contained in the xy plane and forms an angle θ with the x axis. Determine (*a*) the mass moment of inertia $\bar{I}_x$, (*b*) the mass moment of inertia $\bar{I}_y$.

9.100 Determine the radius of gyration of an equilateral triangle of side a with respect to one of its sides.

9.101 Determine by direct integration the mass moment of inertia and the radius of gyration of a sphere of radius a and uniform density ρ with respect to a diameter.

9.102 Determine the centroidal moments of inertia $\bar{I}_x$ and $\bar{I}_y$ and the centroidal radii of gyration $\bar{k}_x$ and $\bar{k}_y$ for the structural shape shown. Neglect the effect of fillets.

9.103 Determine the orientation of the principal axes through the centroid C and the corresponding values of the moments of inertia for the structural shape shown. Neglect the effect of fillets and use the values of $\bar{I}_x$ and $\bar{I}_y$ obtained in Prob. 9.102.

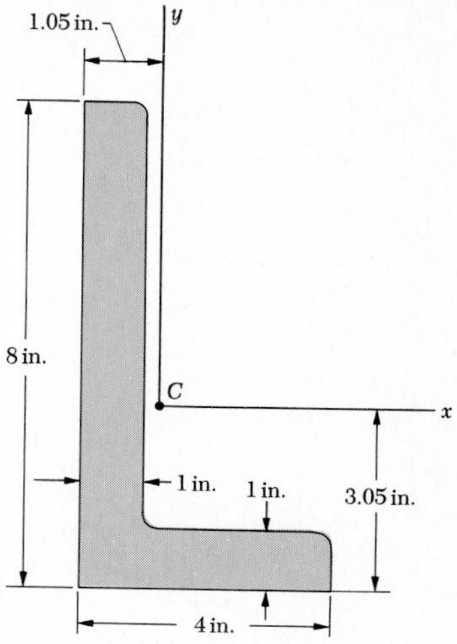

Fig. P9.102 and P9.103

9.104 Determine the mass moment of inertia of a thin elliptical plate of mass m with respect to (*a*) the axes AA' and BB' of the ellipse, (*b*) the axis CC' perpendicular to the plate.

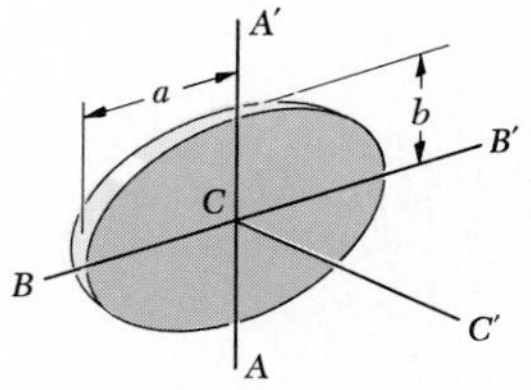

Fig. P9.104

9.105 A hemisphere of radius a and a cone of height h, of the same density ρ, are attached as shown. (*a*) Determine the ratio h/a for which the moment of inertia of the composite body with respect to its axis of symmetry is equal to the moment of inertia of a sphere, of radius a and density ρ, with respect to one of its diameters. (*b*) Repeat the solution considering now the moment of inertia of the composite body with respect to a diameter of the base common to the cone and the hemisphere. (*Hint.* Solve the resulting cubic equation by trial and error.)

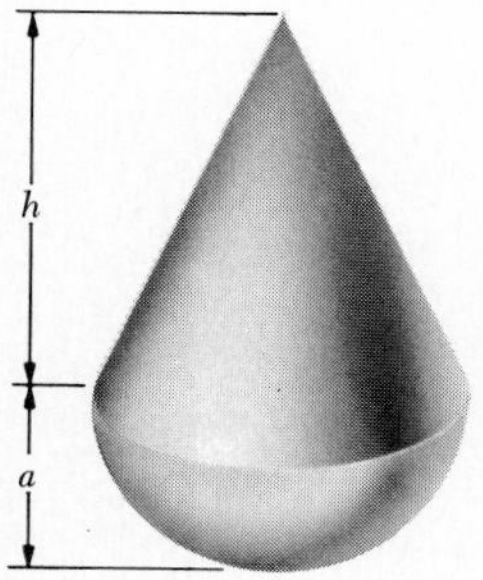

Fig. P9.105 and P9.106

9.106 A steel hemisphere of radius $a = 40$ mm and a steel cone of height $h = 100$ mm are attached as shown. Determine the mass moment of inertia of the composite body about its axis of symmetry. (Density of steel $= 7850\ \text{kg/m}^3$.)

9.107 (*a*) Determine I_x and I_y if $b = 250$ mm. (*b*) Determine the dimension b for which $I_x = I_y$.

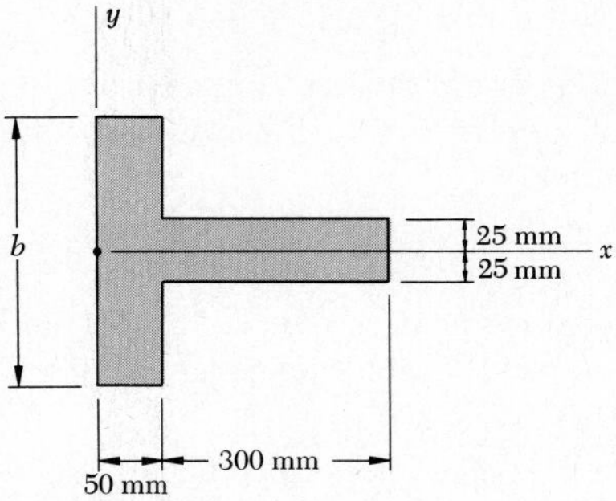

Fig. P9.107

Chapter

10

Method of Virtual Work

***10.1. Work of a Force.** In the preceding chapters, problems involving the equilibrium of rigid bodies were solved by expressing that the external forces acting on the bodies were balanced. The equations of equilibrium $\Sigma F_x = 0$, $\Sigma F_y = 0$, $\Sigma M_A = 0$ were written and solved for the desired unknowns. We shall now consider a different method, which will prove more effective for solving certain types of equilibrium problems. This method is based on the concept of the *work of a force* and was first formally used by the Swiss mathematician Jean Bernoulli in the eighteenth century.

We shall first define the terms *displacement* and *work* as they are used in mechanics. Consider a particle which moves from a point A to a neighboring point A' (Fig. 10.1*a*). The small vector $d\mathbf{s}$ of magnitude ds which joins A to A' is called the *displacement* of the particle. Now, let us assume that a force $\mathbf{F}$, of magnitude F and forming an angle α with the vector $d\mathbf{s}$, is acting on the particle. The *work of the force* $\mathbf{F}$ *during the displacement* $d\mathbf{s}$ is defined as the scalar quantity

$$dU = F\,ds\cos\alpha \tag{10.1}$$

Being a *scalar quantity,* work has a magnitude and a sign, but no direction. We also note that work should be expressed in

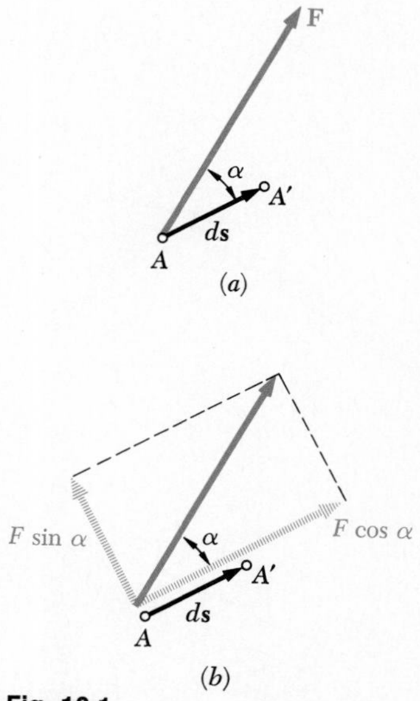

Fig. 10.1

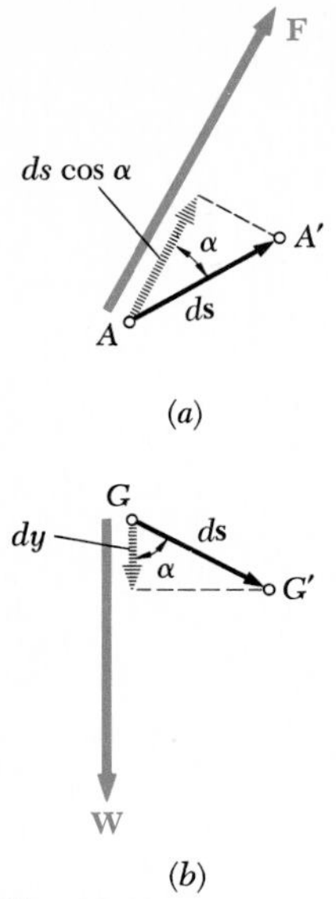

Fig. 10.2

units obtained by multiplying units of length by units of force. Thus, if U.S. customary units are used, work should be expressed in ft · lb or in · lb. If SI units are used, work should be expressed in N · m. The unit of work N · m is called a *joule* (J).†

It appears from (10.1) that the work dU may be considered as the product of the magnitude ds of the displacement and of the component $F \cos \alpha$ of the force **F** in the direction of ds (Fig. 10.1*b*). If this component and the displacement ds have the same sense, the work is positive; if they have opposite senses, the work is negative. Three particular cases are of special interest. If the force **F** has the same direction as ds, the work dU reduces to $F\,ds$. If **F** has a direction opposite to that of ds, the work is $dU = -F\,ds$. Finally, if **F** is perpendicular to ds, the work dU is zero.

The work dU of a force **F** during a displacement ds may also be considered as the product of F and of the component $ds \cos \alpha$ of the displacement ds along **F** (Fig. 10.2*a*). This view is particularly useful in the computation of the work done by the weight **W** of a body (Fig. 10.2*b*). The work of **W** is equal to the product of W and of the vertical displacement dy of the center of gravity G of the body. If the displacement is downward, the work is positive; if it is upward, the work is negative.

A number of forces frequently encountered in statics *do no work*. They are forces applied to fixed points ($ds = 0$) or acting in a direction perpendicular to the displacement ($\cos \alpha = 0$). Among the forces which do no work are the following: the reaction at a frictionless pin when the body supported rotates about the pin, the reaction at a frictionless surface when the body in contact moves along the surface, the reaction at a roller moving along its track, the weight of a body when its center of gravity moves horizontally, the friction force acting on a wheel rolling without slipping (since at any instant the point of contact does not move). Examples of forces which *do work* are the weight of a body (except in the case considered above), the friction force acting on a body sliding on a rough surface, and most forces applied on a moving body.

In certain cases, the sum of the work done by several forces is zero. Consider, for example, two rigid bodies AC and BC connected at C by a *frictionless pin* (Fig. 10.3*a*). Among the forces acting on AC is the force **F** exerted at C by BC. In general, the work of this force will not be zero, but it will be equal in

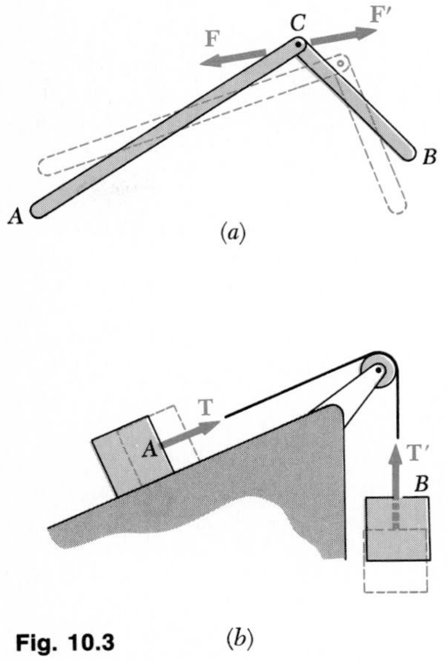

Fig. 10.3

† The joule is the SI unit of *energy*, whether in mechanical form (work, potential energy, kinetic energy) or in chemical, electrical, or thermal form. We should note that even though N · m = J, the moment of a force must be expressed in N · m, and not in joules, since the moment of a force is not a form of energy.

magnitude and opposite in sign to the work of the force **F**′ exerted by *AC* on *BC*, since these forces are equal and opposite and are applied to the same particle. Thus, when the total work done by all the forces acting on *AB* and *BC* is considered, the work of the two internal forces at *C* cancels out. A similar result is obtained if we consider a system consisting of two blocks connected by an *inextensible cord AB* (Fig. 10.3*b*). The work of the tension force **T** at *A* is equal in magnitude to the work of the tension force **T**′ at *B*, since these forces have the same magnitude and the points *A* and *B* move through the same distance; but in one case the work is positive, and in the other it is negative. Thus, the work of the internal forces again cancels out.

It may be shown that the total work of the internal forces holding together the particles of a rigid body is zero. Consider two particles *A* and *B* of a rigid body and the two equal and opposite forces **F** and **F**′ they exert on each other (Fig. 10.4). While, in general, small displacements *ds* and *ds*′ of the two particles are different, the components of these displacements along *AB* must be equal; otherwise, the particles would not remain at the same distance from each other, and the body would not be rigid. Therefore, the work of **F** is equal in magnitude and opposite in sign to the work of **F**′, and their sum is zero.

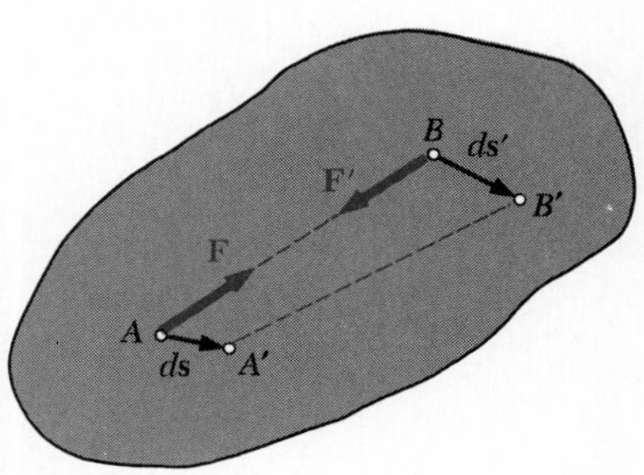

Fig. 10.4

In computing the work of the external forces acting on a rigid body, it is often convenient to determine the work of a couple without considering separately the work of each of the two forces forming the couple. Consider the two forces **F** and **F**′ forming a couple of moment $M = Fr$ and acting on a rigid body (Fig. 10.5). Any small displacement of the rigid body bringing *A* and *B*, respectively, into *A*′ and *B*″ may be divided into two parts, one in which points *A* and *B* undergo equal displacements ds_1, the other in which *A*′ remains fixed while *B*′ moves into *B*″ through a displacement ds_2 of magnitude $ds_2 = r\,d\theta$. In the first part of the motion, the work of **F** is equal in magnitude and opposite in sign to the work of **F**′, and their sum is zero. In the second part of the motion, only force **F** works, and its work is $dU = F\,ds_2 = Fr\,d\theta$. But the product Fr is equal to the moment M of the couple. Thus, the work of a couple of moment M acting on a rigid body is

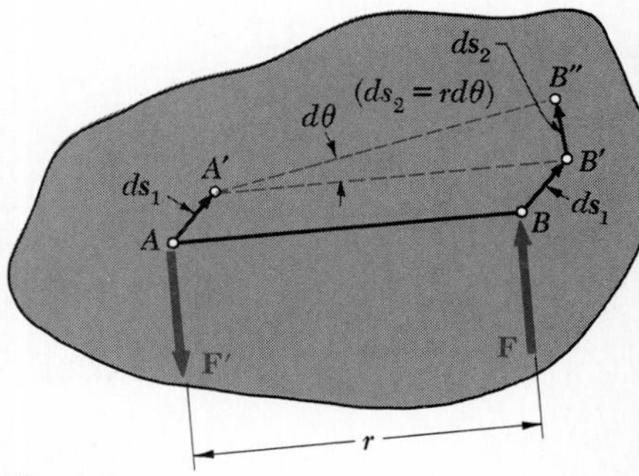

Fig. 10.5

$$dU = M\,d\theta \tag{10.2}$$

where $d\theta$ is the small angle expressed in radians through which the body rotates. We again note that work should be expressed in units obtained by multiplying units of force by units of length.

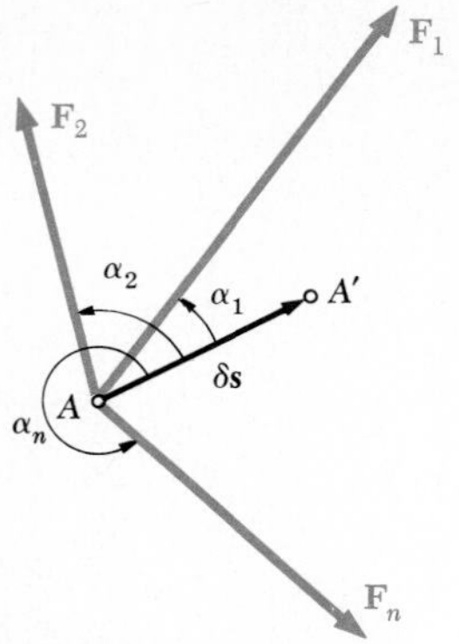

Fig. 10.6

***10.2. Principle of Virtual Work.** Consider a particle acted upon by several forces $\mathbf{F}_1, \mathbf{F}_2, \ldots, \mathbf{F}_n$ (Fig. 10.6). We shall assume that the particle undergoes a small displacement from A to A'. This displacement is possible, but it will not necessarily take place. The forces may be balanced and the particle at rest, or the particle may move under the action of the given forces in a direction different from that of AA'. The displacement considered is therefore an imaginary displacement; it is called a *virtual displacement* and is denoted by $\delta\mathbf{s}$. The symbol $\delta\mathbf{s}$ represents a differential of the first order; it is used to distinguish the virtual displacement from the displacement $d\mathbf{s}$ which would take place under actual motion. As we shall see, virtual displacements may be used to determine whether the conditions of equilibrium of a particle are satisfied.

The work of each of the forces $\mathbf{F}_1, \mathbf{F}_2, \ldots, \mathbf{F}_n$ during the virtual displacement $\delta\mathbf{s}$ is called *virtual work*. The virtual work of all the forces acting on the particle of Fig. 10.6 is

$$\begin{aligned}\delta U &= F_1 \cos\alpha_1\,\delta s + F_2\cos\alpha_2\,\delta s + \cdots + F_n\cos\alpha_n\,\delta s\\ &= (F_1\cos\alpha_1 + F_2\cos\alpha_2 + \cdots + F_n\cos\alpha_n)\,\delta s \qquad (10.3)\end{aligned}$$

Since the expression in parentheses represents the component of the resultant $\mathbf{R}$ of the forces $\mathbf{F}_1, \mathbf{F}_2, \ldots, \mathbf{F}_n$ along AA', it is seen that the total virtual work of the forces $\mathbf{F}_1, \mathbf{F}_2, \ldots, \mathbf{F}_n$ is equal to the virtual work of their resultant $\mathbf{R}$.

The principle of virtual work for a particle states that, *if a particle is in equilibrium, the total virtual work of the forces acting on the particle is zero for any virtual displacement of the particle.* This condition is necessary: if the particle is in equilibrium, the resultant $\mathbf{R}$ of the forces is zero, the expression in parentheses in (10.3) is zero, and the total virtual work is zero. The condition is also sufficient: choosing AA' successively along the x axis and the y axis, and making $\delta U = 0$ in (10.3), we find that the equilibrium equations $\Sigma F_x = 0$ and $\Sigma F_y = 0$ are satisfied.

In the case of a rigid body, the principle of virtual work states that, *if a rigid body is in equilibrium, the total virtual work of the external forces acting on the rigid body is zero for any virtual displacement of the body.* The condition is necessary: if the body is in equilibrium, all the particles forming the body are in equilibrium and the total virtual work of the forces acting on all the particles must be zero; but we have seen in the preceding section that the total work of the internal forces is zero; the total work of the external forces must therefore also be zero. The condition may also be proved to be sufficient.

The principle of virtual work may be extended to the case of a *system of connected rigid bodies.* If the system remains connected during the virtual displacement, *only the work of the forces external to the system need be considered*, since the total work of the internal forces at the various connections is zero.

***10.3. Applications of the Principle of Virtual Work.** The principle of virtual work is particularly effective when applied to the solution of problems involving machines or mechanisms consisting of several connected rigid bodies. Consider, for instance, the toggle vise ACB of Fig. 10.7*a*, used to compress a wooden block. We wish to determine the force exerted by the vise on the block when a given force **P** is applied at C, assuming that there is no friction. Denoting by **Q** the reaction of the block on the vise, we draw the free-body diagram of the vise and consider the virtual displacement obtained by giving to the angle θ a positive increment $\delta\theta$ (Fig. 10.7*b*).

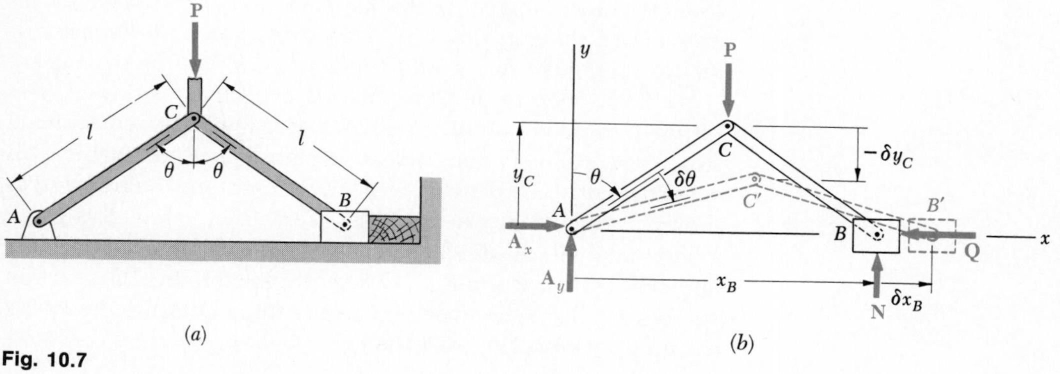

Fig. 10.7

Choosing a system of coordinate axes with origin at A, we note that x_B increases while y_C decreases. This is indicated in the figure by means of the positive increment δx_B and the negative increment $-\delta y_C$. The reactions $\mathbf{A}_x$, $\mathbf{A}_y$, and **N** will do no work during the virtual displacement considered, and we need only compute the work of **P** and **Q**. Since **Q** and δx_B have opposite senses, the virtual work of **Q** is $\delta U_Q = -Q\,\delta x_B$. Since **P** and the increment shown $(-\delta y_C)$ have the same sense, the virtual work of **P** is $\delta U_P = +P(-\delta y_C) = -P\,\delta y_C$. The minus signs obtained could have been predicted by simply noting that the forces **Q** and **P** are directed opposite to the positive x and y axes, respectively. Expressing the coordinates x_B and y_C in terms of the angle θ and differentiating, we obtain

$$x_B = 2l\sin\theta \qquad y_C = l\cos\theta$$
$$\delta x_B = 2l\cos\theta\,\delta\theta \qquad \delta y_C = -l\sin\theta\,\delta\theta \tag{10.4}$$

The total virtual work of the forces **Q** and **P** is thus

$$\delta U = \delta U_Q + \delta U_P = -Q\,\delta x_B - P\,\delta y_C$$
$$= -2Ql\cos\theta\,\delta\theta + Pl\sin\theta\,\delta\theta$$

Making $\delta U = 0$, we obtain

$$2Ql\cos\theta\,\delta\theta = Pl\sin\theta\,\delta\theta \tag{10.5}$$
$$Q = \tfrac{1}{2}P\tan\theta \tag{10.6}$$

The superiority of the method of virtual work over the conventional equilibrium equations in the problem considered here is clear: by using the method of virtual work, we were able to eliminate all unknown reactions, while the equation $\Sigma M_A = 0$ would have eliminated only two of the unknown reactions. We may take advantage of this characteristic of the method of virtual work to solve many problems involving machines and mechanisms. *If the virtual displacement considered is consistent with the constraints imposed by the supports and connections, all reactions and internal forces are eliminated and only the work of the loads, applied forces, and friction forces need be considered.*

We shall observe that the method of virtual work may also be used to solve problems involving completely constrained structures, although the virtual displacements considered will never actually take place. Consider, for example, the frame ACB shown in Fig. 10.8*a*. If point A is kept fixed, while B is given a horizontal virtual displacement (Fig. 10.8*b*), we need consider only the work of **P** and $\mathbf{B}_x$. We may thus determine the reaction component $\mathbf{B}_x$ in the same way as the force **Q** of the preceding example (Fig. 10.7*b*); we have

$$B_x = -\tfrac{1}{2}P\tan\theta$$

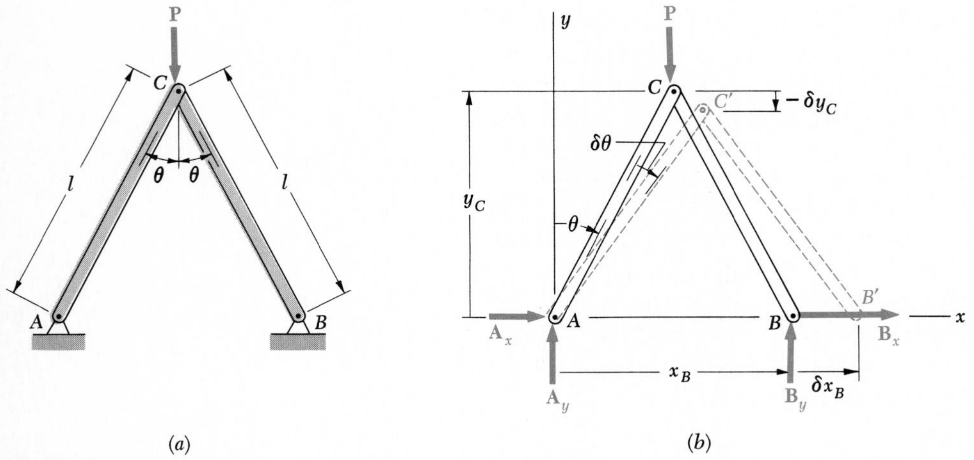

Fig. 10.8

Keeping B fixed and giving to A a horizontal virtual displacement, we may similarly determine the reaction component $\mathbf{A}_x$. The components $\mathbf{A}_y$ and $\mathbf{B}_y$ may be determined by rotating the frame ACB as a rigid body about B and A, respectively.

The method of virtual work may also be used to determine the configuration of a system in equilibrium under given forces. For example, the value of the angle θ for which the linkage of Fig. 10.7 is in equilibrium under two given forces **P** and **Q** may be obtained by solving Eq. (10.6) for $\tan\theta$.

It should be noted, however, that the attractiveness of the method of virtual work depends to a large extent upon the existence of simple geometric relations between the various virtual displacements involved in the solution of a given problem. When no such simple relations exist, it is usually advisable to revert to the conventional method of Chap. 6.

***10.4. Real Machines. Mechanical Efficiency.** In analyzing the toggle vise in the preceding section, we assumed that no friction forces were involved. Thus, the virtual work consisted only of the work of the applied force **P** and of the reaction **Q**. But the work of the reaction **Q** is equal in magnitude and opposite in sign to the work of the force exerted by the vise on the block. Equation (10.5), therefore, expresses that the *output work* $2Ql\cos\theta\,\delta\theta$ is equal to the *input work* $Pl\sin\theta\,\delta\theta$. A machine in which input and output work are equal is said to be an "ideal" machine. In a "real" machine, friction forces will always do some work, and the output work will be smaller than the input work.

Consider, for example, the toggle vise of Fig. 10.7*a*, and assume now that a friction force **F** develops between the sliding block B and the horizontal plane (Fig. 10.9). Using the conventional methods of statics and summing moments about A, we find $N = P/2$. Denoting by μ the coefficient of friction between block B and the horizontal plane, we have $F = \mu N = \mu P/2$. Recalling formulas (10.4), we find that the total virtual work of the forces **Q**, **P**, and **F** during the virtual displacement shown in Fig. 10.9 is

$$\begin{aligned}\delta U &= -Q\,\delta x_B - P\,\delta y_C - F\,\delta x_B \\ &= -2Ql\cos\theta\,\delta\theta + Pl\sin\theta\,\delta\theta - \mu Pl\cos\theta\,\delta\theta\end{aligned}$$

Making $\delta U = 0$, we obtain

$$2Ql\cos\theta\,\delta\theta = Pl\sin\theta\,\delta\theta - \mu Pl\cos\theta\,\delta\theta \qquad (10.7)$$

which expresses that the output work is equal to the input work minus the work of the friction force. Solving for Q, we have

$$Q = \tfrac{1}{2}P(\tan\theta - \mu) \qquad (10.8)$$

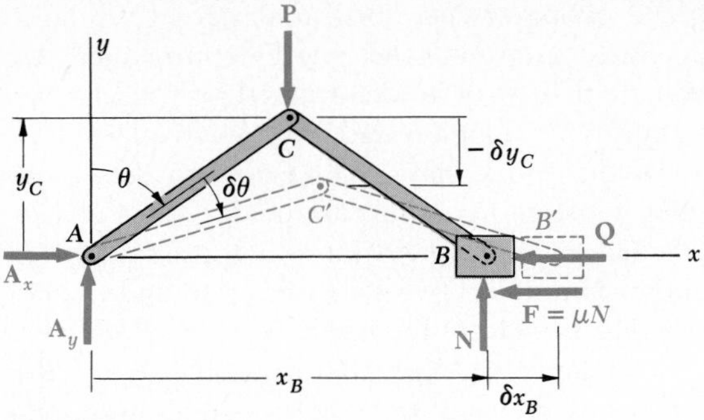

Fig. 10.9

We note that $Q = 0$ when $\tan\theta = \mu$, that is, when θ is equal to the angle of friction ϕ, and that $Q < 0$ when $\theta < \phi$. The toggle vise may thus be used only for values θ larger than the angle of friction.

The *mechanical efficiency* of a machine is defined as the ratio

$$\eta = \frac{\text{output work}}{\text{input work}} \tag{10.9}$$

Clearly, the mechanical efficiency of an ideal machine is $\eta = 1$, since input and output work are then equal, while the mechanical efficiency of a real machine will always be less than 1.

In the case of the toggle vise we have just analyzed, we write

$$\eta = \frac{\text{output work}}{\text{input work}} = \frac{2Ql\cos\theta\,\delta\theta}{Pl\sin\theta\,\delta\theta}$$

Substituting from (10.8) for Q, we obtain

$$\eta = \frac{P(\tan\theta - \mu)l\cos\theta\,\delta\theta}{Pl\sin\theta\,\delta\theta} = 1 - \mu\cot\theta \tag{10.10}$$

We check that, in the absence of friction forces, we would have $\mu = 0$ and $\eta = 1$. In the general case, when μ is different from zero, the efficiency η becomes zero for $\mu\cot\theta = 1$, that is, for $\tan\theta = \mu$, or $\theta = \tan^{-1}\mu = \phi$. We check again that the toggle vise may be used only for values of θ larger than the angle of friction ϕ.

SAMPLE PROBLEM 10.1

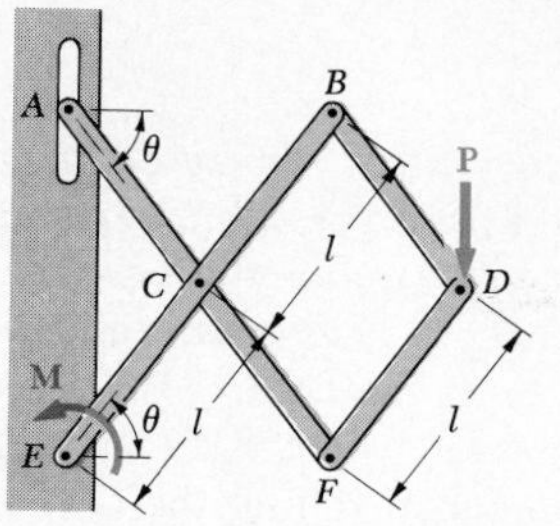

Using the method of virtual work, determine the moment of the couple **M** required to maintain the equilibrium of the mechanism shown.

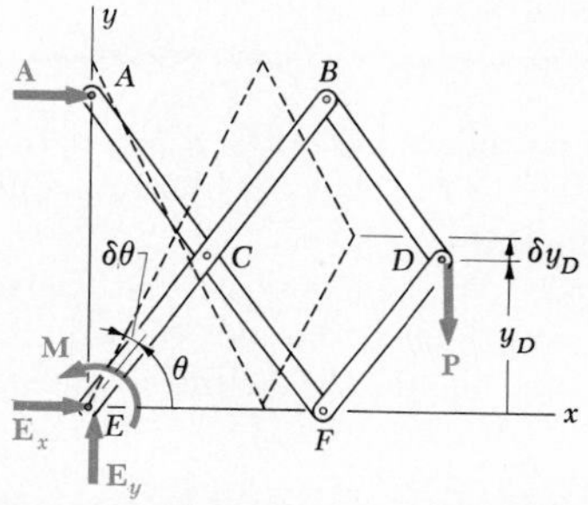

Solution. Choosing a coordinate system with origin at E, we write

$$y_D = l \sin \theta \qquad \delta y_D = l \cos \theta \, \delta\theta$$

Principle of Virtual Work. Since the reactions **A**, $\mathbf{E}_x$, and $\mathbf{E}_y$ will do no work during the virtual displacement, the total virtual work done by **M** and **P** must be zero.

$\delta U = 0$:

$$+M\,\delta\theta - P\,\delta y_D = 0$$

$$+M\,\delta\theta - P(l \cos\theta \, \delta\theta) = 0 \qquad M = Pl\cos\theta$$

SAMPLE PROBLEM 10.2

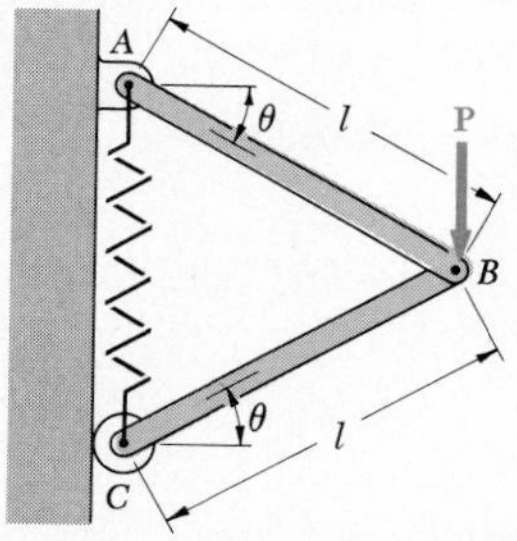

Determine the expressions for θ and for the tension in the spring which correspond to the equilibrium position of the mechanism. The unstretched length of the spring is h, and the constant of the spring is k. Neglect the weight of the mechanism.

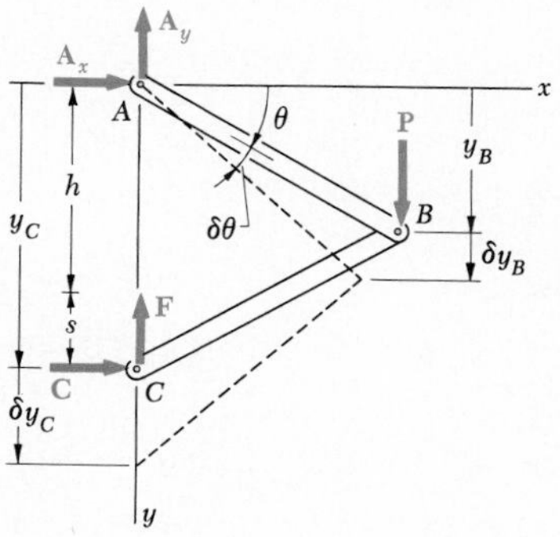

Solution. With the coordinate system shown

$$y_B = l\sin\theta \qquad y_C = 2l\sin\theta$$

$$\delta y_B = l\cos\theta\,\delta\theta \qquad \delta y_C = 2l\cos\theta\,\delta\theta$$

The elongation of the spring is

$$s = y_C - h = 2l\sin\theta - h$$

The magnitude of the force exerted at C by the spring is

$$F = ks = k(2l\sin\theta - h) \tag{1}$$

Principle of Virtual Work. Since the reactions $\mathbf{A}_x$, $\mathbf{A}_y$, and **C** do no work, the total virtual work done by **P** and **F** must be zero.

$\delta U = 0$:

$$P\,\delta y_B - F\,\delta y_C = 0$$

$$P(l\cos\theta\,\delta\theta) - k(2l\sin\theta - h)(2l\cos\theta\,\delta\theta) = 0$$

$$\sin\theta = \frac{P + 2kh}{4kl}$$

Substituting this expression into (1), we obtain

$$F = \tfrac{1}{2}P$$

SAMPLE PROBLEM 10.3

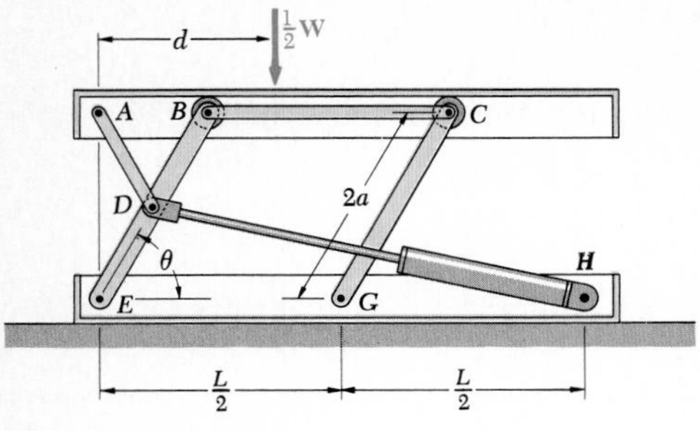

A hydraulic-lift table is used to raise a 1000-kg crate. It consists of a platform and of two identical linkages on which hydraulic cylinders exert equal forces. (Only one linkage and one cylinder are shown.) Members EDB and CG are each of length $2a$, and member AD is pinned to the midpoint of EDB. If the crate is placed on the table, so that half of its weight is supported by the system shown, determine the force exerted by each cylinder in raising the crate for $\theta = 60°$, $a = 0.70$ m, and $L = 3.20$ m. This mechanism has been previously considered in Sample Prob. 6.8.

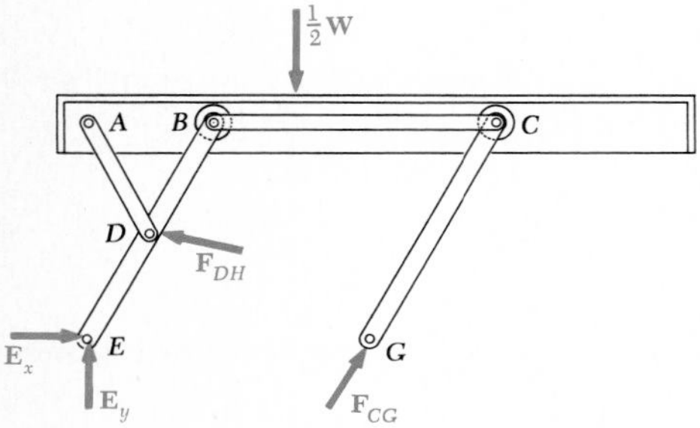

Solution. The machine considered consists of the platform and of the linkage, with an input force $\mathbf{F}_{DH}$ exerted by the cylinder and an output force equal and opposite to $\frac{1}{2}\mathbf{W}$.

Principle of Virtual Work. We first observe that the reactions at E and G do no work. Denoting by y the elevation of the platform about the base, and by s the length DH of the cylinder-and-piston assembly, we write

$\delta U = 0$: $$-\tfrac{1}{2}W\,\delta y + F_{DH}\,\delta s = 0 \qquad (1)$$

The vertical displacement δy of the platform is expressed in terms of the angular displacement $\delta\theta$ of EDB as follows:

$$y = (EB)\sin\theta = 2a\sin\theta$$
$$\delta y = 2a\cos\theta\,\delta\theta$$

To express δs similarly in terms of $\delta\theta$, we first note that, by the law of cosines,

$$s^2 = a^2 + L^2 - 2aL\cos\theta$$

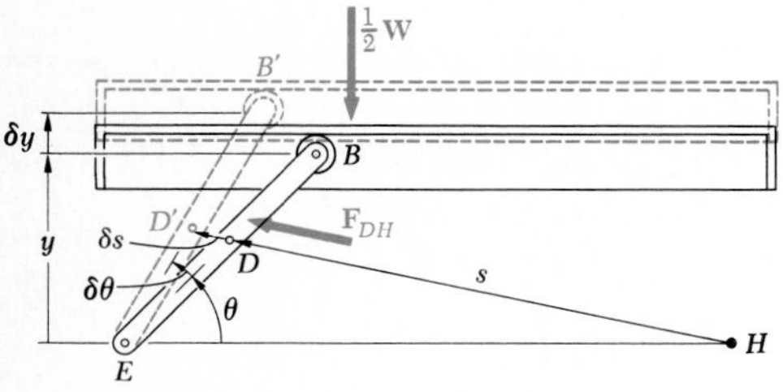

Differentiating:

$$2s\,\delta s = -2aL(-\sin\theta)\,\delta\theta$$
$$\delta s = \frac{aL\sin\theta}{s}\,\delta\theta$$

Substituting for δy and δs into (1), we write

$$(-\tfrac{1}{2}W)2a\cos\theta\,\delta\theta + F_{DH}\frac{aL\sin\theta}{s}\delta\theta = 0$$

$$F_{DH} = W\frac{s}{L}\cot\theta$$

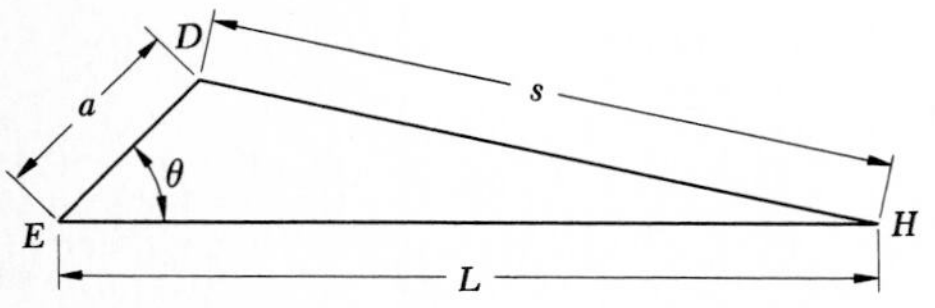

With the given numerical data, we have

$$W = mg = (1000\text{ kg})(9.81\text{ m/s}^2) = 9810\text{ N} = 9.81\text{ kN}$$
$$s^2 = a^2 + L^2 - 2aL\cos\theta$$
$$= (0.70)^2 + (3.20)^2 - 2(0.70)(3.20)\cos 60° = 8.49$$
$$s = 2.91\text{ m}$$

$$F_{DH} = W\frac{s}{L}\cot\theta = (9.81\text{ kN})\frac{2.91\text{ m}}{3.20\text{ m}}\cot 60°$$

$F_{DH} = 5.15$ kN ◄

PROBLEMS

10.1 Determine the magnitude of the force **P** required to maintain the equilibrium of the linkage shown.

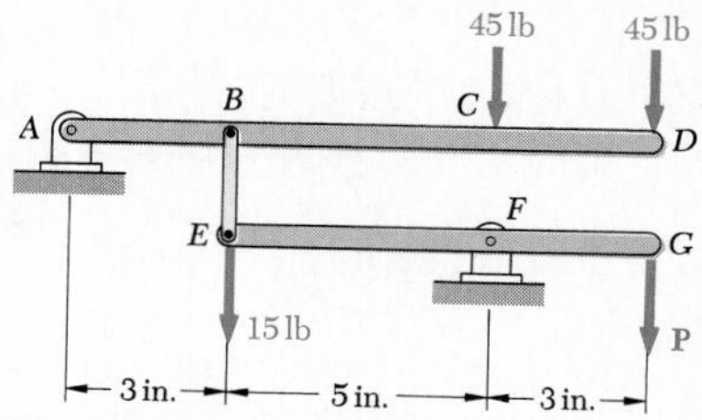

Fig. P10.1

10.2 Determine the weight W which balances the 45-lb load.

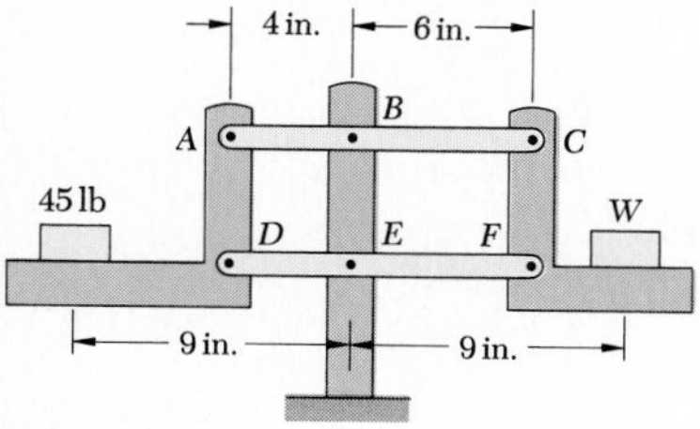

Fig. P10.2

10.3 Determine the force **P** required to maintain the equilibrium of the linkage shown. All members are of the same length and the wheels at A and B roll freely on the horizontal rod.

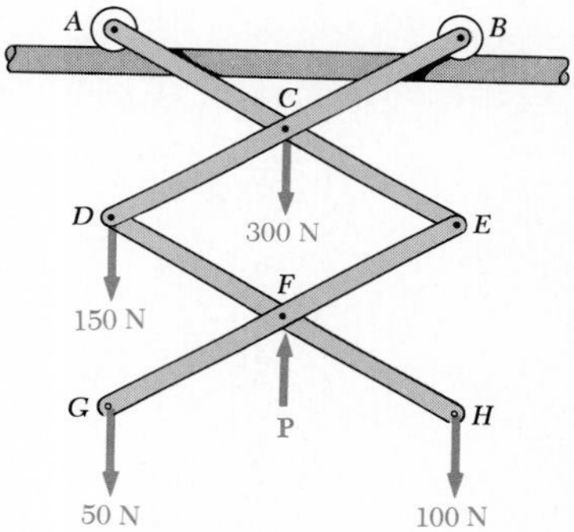

Fig. P10.3

10.4 Determine the horizontal force **P** which must be applied at A to maintain the equilibrium of the linkage.

10.5 Determine the couple **M** which must be applied to member AC to maintain the equilibrium of the linkage.

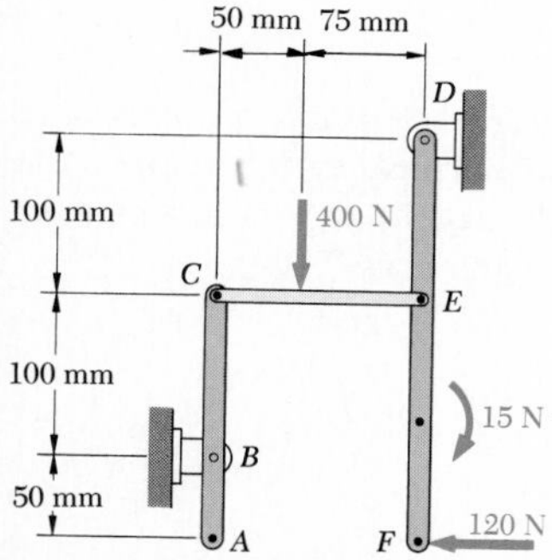

Fig. P10.4 and P10.5

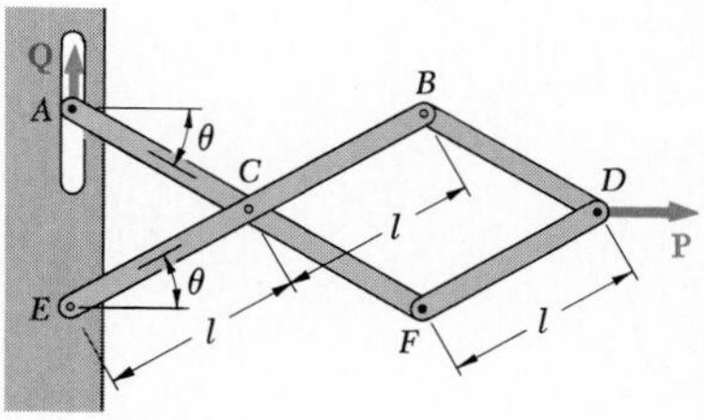

Fig. P10.6

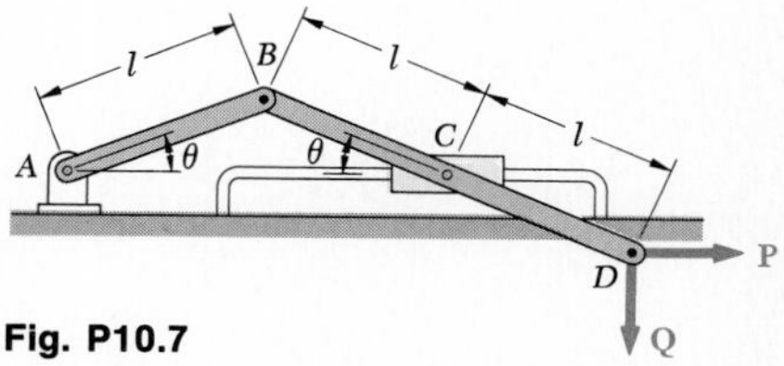

Fig. P10.7

10.6 and 10.7 The mechanism shown is acted upon by the force **P**; derive an expression for the magnitude of the force **Q** required to maintain equilibrium.

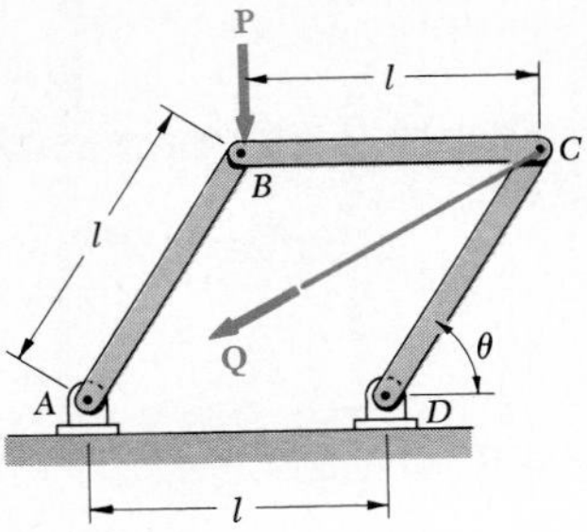

Fig. P10.8

10.8 Three links, each of length l, are connected as shown. Knowing that the line of action of the force **Q** passes through point A, derive an expression for the magnitude of **Q** required to maintain equilibrium.

10.9 The slender rod AB is attached to a collar at A and rests on a wheel at C. Neglecting the effect of friction, derive an expression for the magnitude of the force **Q** required to maintain equilibrium.

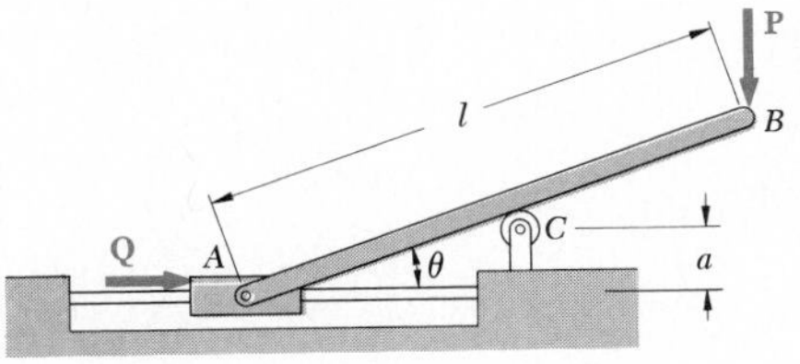

Fig. P10.9

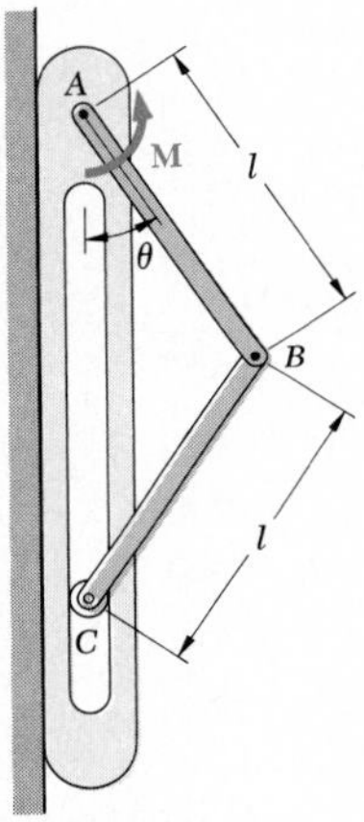

Fig. P10.10

10.10 Each of the uniform rods AB and BC is of weight W. Derive an expression for the moment of the couple **M** required to maintain equilibrium.

10.11 Derive an expression for the moment of the couple **M** required to maintain equilibrium.

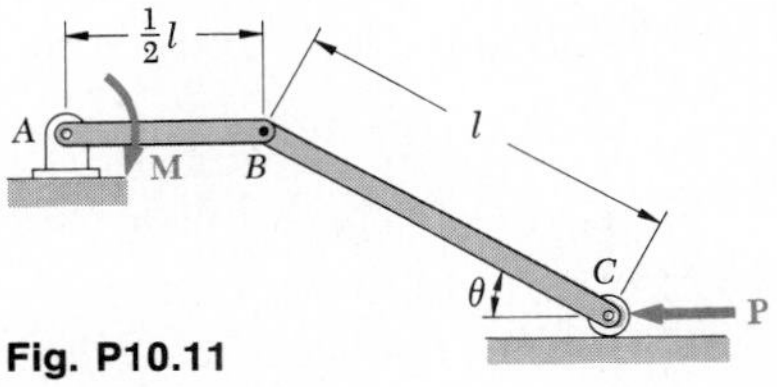

Fig. P10.11

10.12 If gripping forces of magnitude $Q = 450$ lb are desired, determine the magnitude P of the forces which must be applied to the pliers. Also show that the required magnitude P is independent of the position of the object gripped by the jaws.

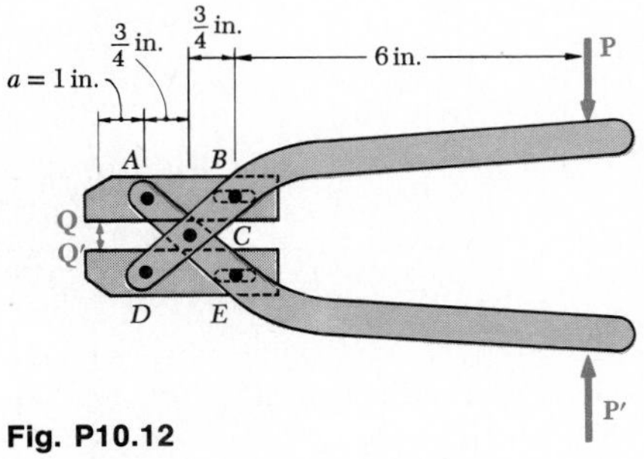

Fig. P10.12

10.13 Collar A weighs 5.60 lb and may slide freely on the smooth vertical rod. Knowing that the length of the wire AB is 18 in., determine the magnitude of **P** required for equilibrium when (*a*) $c = 2$ in., (*b*) $c = 8$ in.

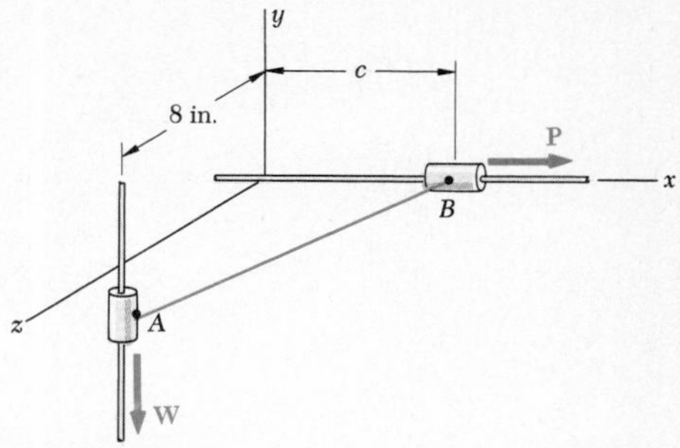

Fig. P10.13

10.14 In Prob. 10.9, determine the magnitude of the force **Q** required for equilibrium when $l = 500$ mm, $a = 100$ mm, $P = 500$ N, and $\theta = 30°$.

10.15 In Prob. 10.7, determine the magnitude of the force **Q** required for equilibrium when $l = 450$ mm, $\theta = 30°$, and $P = 600$ N.

10.16 Determine the value of θ corresponding to the equilibrium position of the mechanism of Prob. 10.6 when $P = 40$ lb and $Q = 160$ lb.

10.17 Determine the value of θ corresponding to the equilibrium position of the mechanism of Prob. 10.7 when $P = 100$ lb and $Q = 150$ lb.

10.18 Determine the value of θ corresponding to the equilibrium position of the mechanism of Prob. 10.8 when $P = 400$ N and $Q = 600$ N.

10.19 Determine the value of θ corresponding to the equilibrium position of the mechanism of Prob. 10.9 when $P = 500$ N, $Q = 400$ N, $a = 100$ mm, and $l = 500$ mm.

10.20 Two bars AB and BC, each of mass 4 kg, are connected by a pin at B and by a spring DE. When unstretched, the spring is 150 mm long; the constant of the spring is 800 N/m. Determine the value of x corresponding to equilibrium.

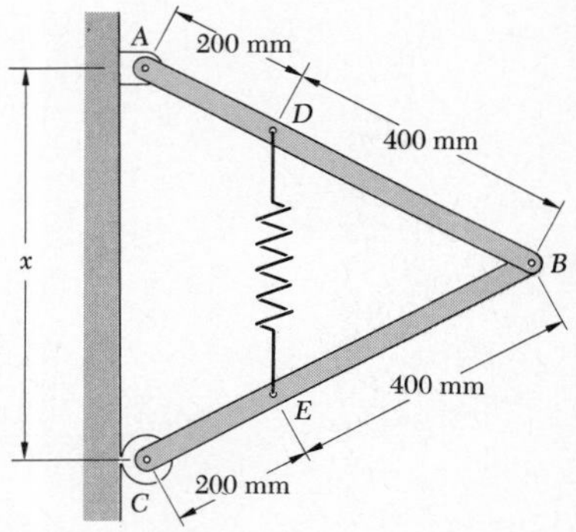

Fig. P10.20

10.21 A vertical load **W** is applied to the linkage at B. The constant of the spring is k, and the spring is unstretched when AB and BC are horizontal. Neglecting the weight of the linkage, derive an equation in θ, W, l, and k, which must be satisfied when the linkage is in equilibrium.

10.22 A load **W** of magnitude 150 lb is applied to the linkage at B. Neglecting the weight of the linkage and knowing that $l = 8$ in., determine the value of θ corresponding to equilibrium. The constant of the spring is $k = 10$ lb/in., and the spring is unstretched when AB and BC are horizontal. (*Hint.* Obtain the approximate value of θ by solving by trial and error the equation obtained.)

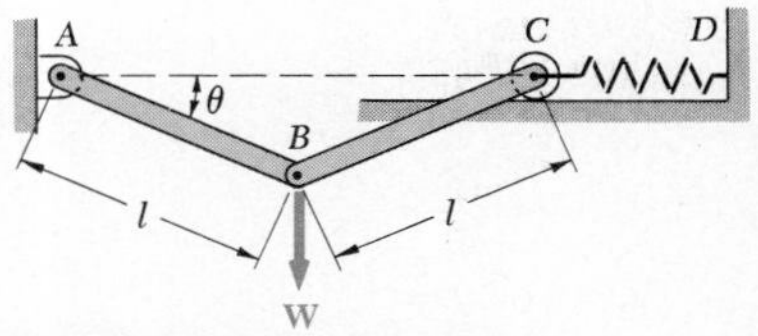

Fig. P10.21 and P10.22

10.23 The top surface of the work platform is maintained at a given elevation by the single hydraulic cylinder shown. Knowing that members AF and BD are each of length l, determine the magnitude of the force exerted by the hydraulic cylinder when $\theta = 30°$.

10.24 Solve Prob. 10.23, assuming that the hydraulic cylinder has been moved and attached to points E and H.

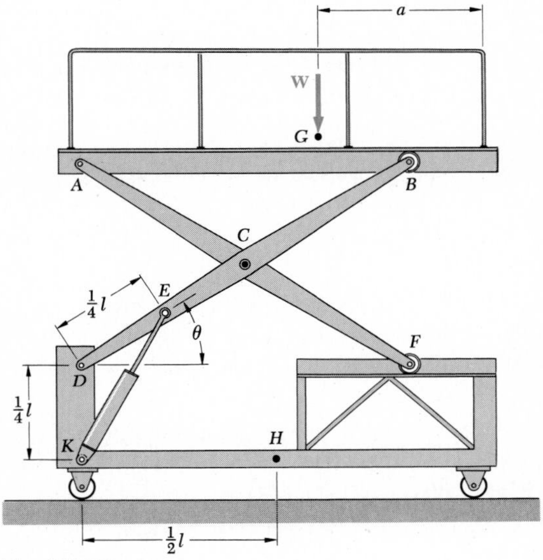

Fig. P10.23

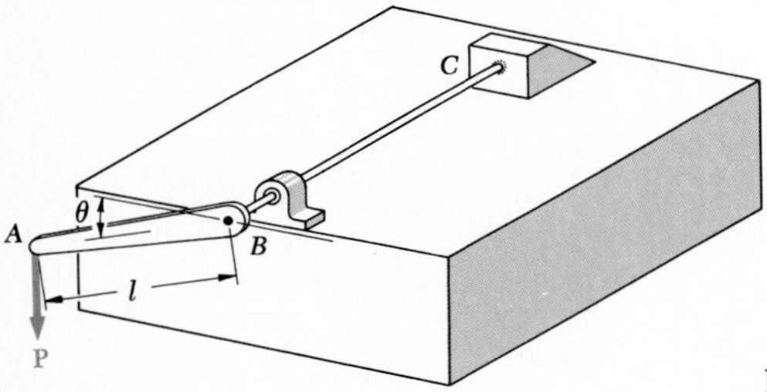

Fig. P10.25

10.25 The lever AB is attached to the horizontal shaft BC which passes through a bearing and is welded to a fixed support at C. The torsional spring constant of the shaft BC is K; that is, a couple of moment K is required to rotate end B through one radian. Knowing that the shaft is untwisted when AB is horizontal, determine the value of θ corresponding to the position of equilibrium, if $P = 20$ lb, $l = 10$ in., and $K = 100$ lb · in./rad.

10.26 Solve Prob. 10.25, if $P = 70$ lb, $l = 10$ in., and $K = 100$ lb · in./rad. Obtain an answer in each of the following quadrants: $0 < \theta < 90°$, $270° < \theta < 360°$, and $360° < \theta < 450°$. (It is assumed that K remains constant for the angles of twist considered.)

10.27 Derive an expression for the mechanical efficiency of the jack discussed in Sec. 8.6. Show that, if the jack is to be self-locking, the mechanical efficiency cannot exceed $\frac{1}{2}$.

10.28 A block of weight W is pulled up a plane forming an angle α with the horizontal by a force $\mathbf{P}$ directed along the plane. If μ is the coefficient of friction between the block and the plane, derive an expression for the mechanical efficiency of the system. Show that the mechanical efficiency cannot exceed $\frac{1}{2}$ if the block is to remain in place when the force $\mathbf{P}$ is removed.

10.29 In Prob. 10.6, assume that the coefficient of friction between the pin and the slot at A is $\mu = 0.30$, and determine the smallest and the largest magnitudes of the force $\mathbf{Q}$ for which equilibrium is maintained when $\theta = 30°$ and $P = 300$ lb.

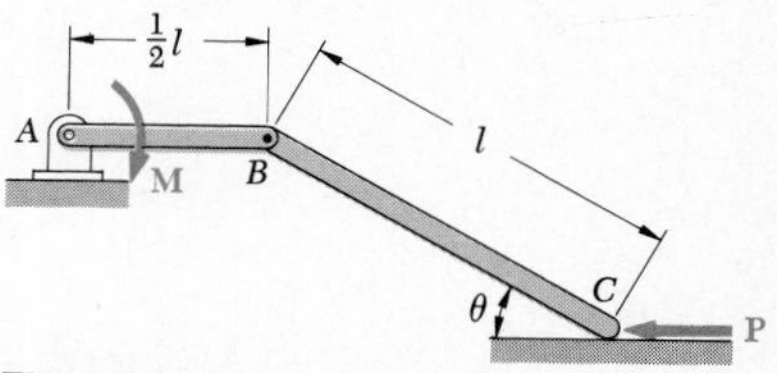

Fig. P10.30

10.30 Denoting by μ the coefficient of friction between point C and the horizontal surface, determine the largest moment of the couple $\mathbf{M}$ for which equilibrium is maintained.

10.31 Using the method of virtual work, determine the reaction at D.

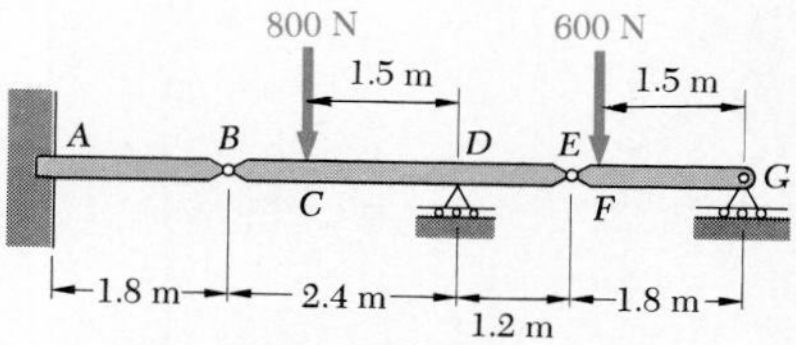

Fig. P10.31 and P10.32

10.32 Using the method of virtual work, determine separately the force and the couple representing the reaction at A.

10.33 Two uniform rods, each of weight W and length l, are connected as shown. Using the method of virtual work, determine θ_1 and θ_2 corresponding to equilibrium.

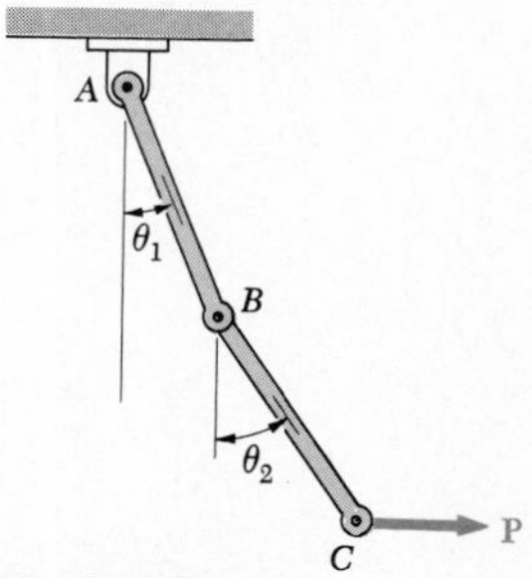

Fig. P10.33

10.34 In Prob. 10.3 the force $\mathbf{P}$ is removed and the linkage is maintained in equilibrium by a cord which is attached to pins E and H. Determine the tension in the cord.

10.35 Determine the vertical movement of joint G if member BD is shortened 1.5 in. (*Hint.* Apply a vertical load at joint G, and, using the methods of Chap. 6, compute the force exerted by member BD on joints B and D. Then apply the method of virtual work for a virtual displacement making member BD shorter. This method should be used only for small changes in the length of members.)

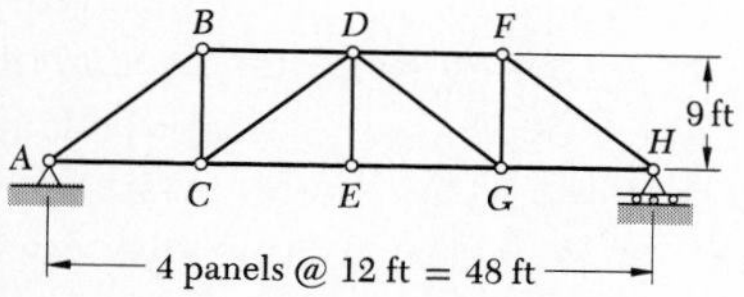

Fig. P10.35, P10.36, and P10.37

10.36 Determine the vertical movement of joint B if the length of member DG is changed to 15 ft $1\frac{1}{2}$ in. (See hint of Prob. 10.35.)

10.37 Determine the horizontal movement of joint B if the length of member DG is changed to 15 ft $1\frac{1}{2}$ in. (See hint of Prob. 10.35.)

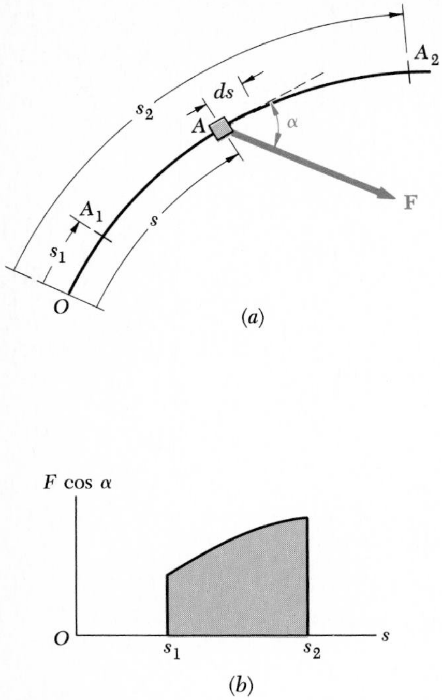

Fig. 10.10

*10.5. Work of a Force during a Finite Displacement.

Consider a force **F** acting on a particle. The work of **F** corresponding to an infinitesimal displacement ds of the particle was defined in Sec. 10.1 as

$$dU = F\,ds \cos\alpha \tag{10.1}$$

The work of **F** corresponding to a finite displacement of the particle from A_1 to A_2 (Fig. 10.10*a*) is denoted by $U_{1\to2}$ and is obtained by integrating (10.1) along the curve described by the particle:

$$U_{1\to2} = \int_{s_1}^{s_2} (F\cos\alpha)\,ds \tag{10.11}$$

where the variable of integration s measures the distance along the path traveled by the particle. The work $U_{1\to2}$ is represented by the area under the curve obtained by plotting $F\cos\alpha$ against s (Fig. 10.10*b*). In the case of a force **F** of constant magnitude acting in the direction of motion, formula (10.11) yields $U_{1\to2} = F(s_2 - s_1)$.

Recalling from Sec. 10.1 that the work of a couple of moment M during an infinitesimal rotation $d\theta$ of a rigid body is

$$dU = M\,d\theta \tag{10.2}$$

we express as follows the work of the couple during a finite rotation of the body:

$$U_{1\to2} = \int_{\theta_1}^{\theta_2} M\,d\theta \tag{10.12}$$

In the case of a constant couple, formula (10.12) yields

$$U_{1\to2} = M(\theta_2 - \theta_1)$$

Work of a Weight. It was stated in Sec. 10.1 that the work of the weight **W** of a body during an infinitesimal displacement of the body is equal to the product of W and of the vertical displacement of the center of gravity of the body. With the y axis pointing upward, the work of **W** during a finite displacement of the body (Fig. 10.11) is obtained by writing

$$dU = -W\,dy$$

$$U_{1\to2} = -\int_{y_1}^{y_2} W\,dy = Wy_1 - Wy_2 \tag{10.13}$$

or

$$U_{1\to2} = -W(y_2 - y_1) = -W\,\Delta y \tag{10.13'}$$

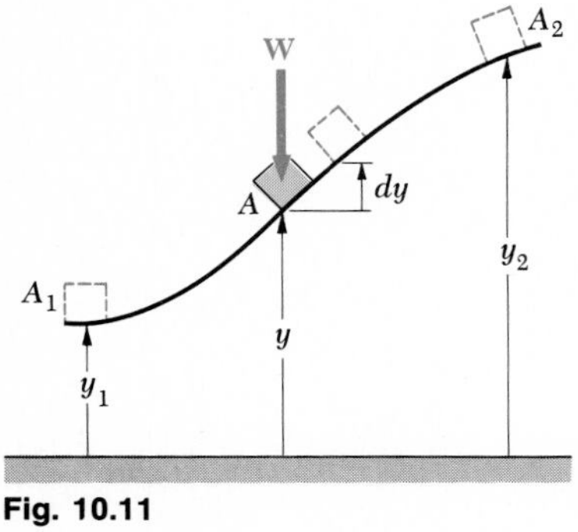

Fig. 10.11

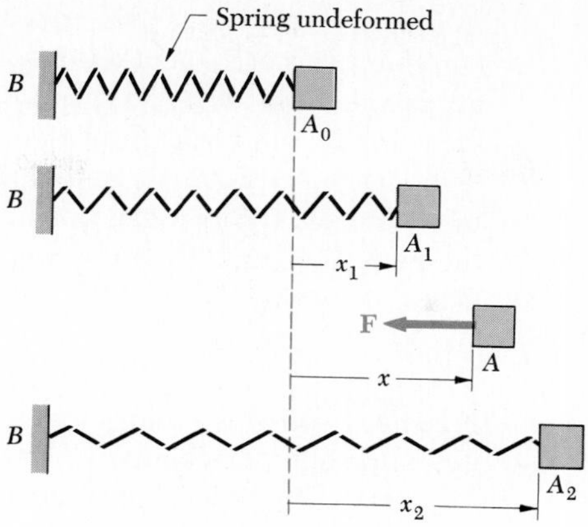

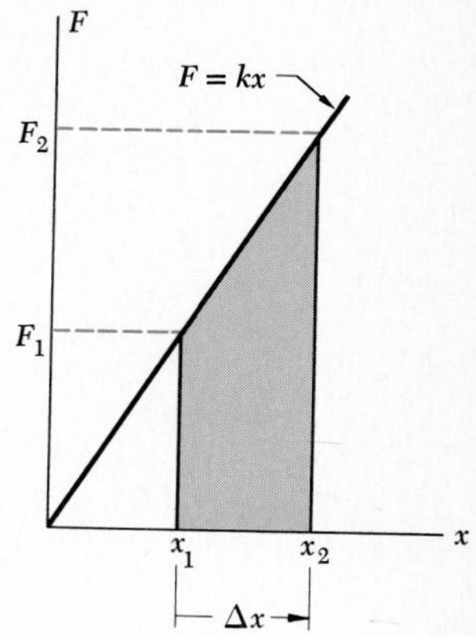

Fig. 10.12 (*a*) (*b*)

where Δy is the vertical displacement from A_1 to A_2. The work of the weight **W** is thus equal to *the product of W and of the vertical displacement of the center of gravity of the body.* The work is *positive* when $\Delta y < 0$, that is, *when the body moves down.*

Work of the Force Exerted by a Spring. Consider a body A attached to a fixed point B by a spring; it is assumed that the spring is undeformed when the body is at A_0 (Fig. 10.12*a*). Experimental evidence shows that the magnitude of the force **F** exerted by the spring on a body A is proportional to the deflection x of the spring measured from the position A_0. We have

$$F = kx \tag{10.14}$$

where k is the *spring constant,* expressed in N/m if SI units are used, and in lb/ft or lb/in. if U.S. customary units are used. The work of the force **F** exerted by the spring during a finite displacement of the body from $A_1(x = x_1)$ to $A_2(x = x_2)$ is obtained by writing

$$dU = -F\,dx = -kx\,dx$$

$$U_{1\to2} = -\int_{x_1}^{x_2} kx\,dx = \tfrac{1}{2}kx_1^2 - \tfrac{1}{2}kx_2^2 \tag{10.15}$$

Care should be taken to express k and x in consistent units. For example, if U.S. customary units are used, k should be expressed in lb/ft and x in feet, or k in lb/in. and x in inches; in the first case, the work is obtained in ft · lb; in the second case, in in · lb. We note that the work of the force **F** exerted by the spring on the body is *positive* when $x_2 < x_1$, that is, *when the spring is returning to its undeformed position.*

Since Eq. (10.14) is the equation of a straight line of slope k passing through the origin, the work $U_{1\to2}$ of **F** during the displacement from A_1 to A_2 may be obtained by evaluating the area of the trapezoid shown in Fig. 10.12*b*. This is done by computing the values F_1 and F_2 and multiplying the base Δx of the trapezoid by its mean height $\frac{1}{2}(F_1 + F_2)$. Since the work of the force **F** exerted by the spring is positive for a negative value of Δx, we write

$$U_{1\to2} = -\tfrac{1}{2}(F_1 + F_2)\,\Delta x \qquad (10.16)$$

Formula (10.16) is usually more convenient to use than (10.15) and affords fewer chances of confusing the units involved.

***10.6. Potential Energy.** Considering again the body of Fig. 10.11, we note from (10.13) that the work of the weight **W** during a finite displacement is obtained by subtracting the value of the function Wy corresponding to the second position of the body from its value corresponding to the first position. The work of **W** is thus independent of the actual path followed; it depends only upon the initial and final values of the function Wy. This function is called the *potential energy* of the body with respect to the *force of gravity* **W** and is denoted by V_g. We write

$$U_{1\to2} = (V_g)_1 - (V_g)_2 \qquad \text{with } V_g = Wy \qquad (10.17)$$

We note that if $(V_g)_2 > (V_g)_1$, that is, *if the potential energy increases* during the displacement (as in the case considered here), *the work $U_{1\to2}$ is negative.* If, on the other hand, the work of **W** is positive, the potential energy decreases. Therefore, the potential energy V_g of the body provides a measure of *the work which may be done* by its weight **W.** Since only the *change* in potential energy, and not the actual value of V_g, is involved in formula (10.17), an arbitrary constant may be added to the expression obtained for V_g. In other words, the level from which the elevation y is measured may be chosen arbitrarily. Note that potential energy is expressed in the same units as work, i.e., in joules (J) if SI units are used,† and in ft · lb or in · lb if U.S. customary units are used.

Considering now the body of Fig. 10.12*a*, we note from formula (10.15) that the work of the elastic force **F** is obtained by subtracting the value of the function $\frac{1}{2}kx^2$ corresponding to the second position of the body from its value corresponding to the first position. This function is denoted by V_e and is called the *potential energy* of the body with respect to the *elastic force* **F.** We write

† See footnote, page 370.

$$U_{1\to 2} = (V_e)_1 - (V_e)_2 \qquad \text{with } V_e = \tfrac{1}{2}kx^2 \tag{10.18}$$

and observe that, during the displacement considered, the work of the force **F** exerted by the spring on the body is negative and the potential energy V_e increases. We should note that the expression obtained for V_e is valid only if the deflection of the spring is measured from its undeformed position.

The concept of potential energy may be used when forces other than gravity forces and elastic forces are involved. It remains valid as long as the elementary work dU of the force considered is an *exact differential.* It is then possible to find a function V, called potential energy, such that

$$dU = -dV \tag{10.19}$$

Integrating (10.19) over a finite displacement, we obtain the general formula

$$U_{1\to 2} = V_1 - V_2 \tag{10.20}$$

which expresses that *the work of the force is independent of the path followed and is equal to minus the change in potential energy.* A force which satisfies Eq. (10.20) is said to be a *conservative force.*

***10.7. Potential Energy and Equilibrium.** The application of the principle of virtual work is considerably simplified when the potential energy of a system is known. In the case of a virtual displacement, formula (10.19) becomes $\delta U = -\delta V$. Besides, if the position of the system is defined by a single independent variable θ, we may write $\delta V = (dV/d\theta)\,\delta\theta$. Since $\delta\theta$ must be different from zero, the condition $\delta U = 0$ for the equilibrium of the system becomes

$$\frac{dV}{d\theta} = 0 \tag{10.21}$$

In terms of potential energy, the principle of virtual work states therefore that, *if a system is in equilibrium, the derivative of its total potential energy is zero.* If the position of the system depends upon several independent variables (the system is then said to possess *several degrees of freedom*), the partial derivatives of V with respect to each of the independent variables must be zero.

Consider, for example, a structure made of two members AC and CB and carrying a load W at C. The structure is supported by a pin at A and a roller at B, and a spring BD connects B

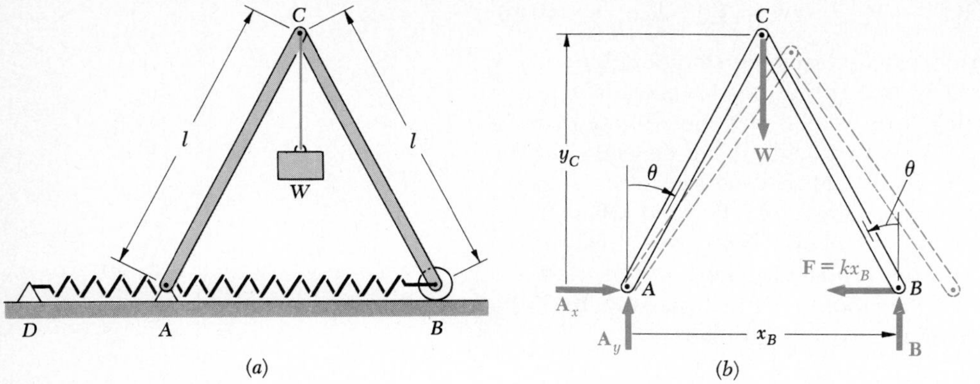

Fig. 10.13

to a fixed point D (Fig. 10.13*a*). The constant of the spring is k, and it is assumed that the natural length of the spring is equal to AD and thus that the spring is undeformed when B coincides with A. Neglecting the friction forces and the weight of the members, we find that the only forces which work during a displacement of the structure are the weight **W** and the force **F** exerted by the spring at point B (Fig. 10.13*b*). The total potential energy of the system will thus be obtained by adding the potential energy V_g corresponding to the gravity force **W** and the potential energy V_e corresponding to the elastic force **F**.

Choosing a coordinate system with origin at A and noting that the deflection of the spring, measured from its undeformed position, is $AB = x_B$, we write

$$V_e = \tfrac{1}{2}kx_B^2 \qquad V_g = Wy_C$$

Expressing the coordinates x_B and y_C in terms of the angle θ, we have

$$x_B = 2l \sin\theta \qquad y_C = l\cos\theta$$
$$V_e = \tfrac{1}{2}k(2l\sin\theta)^2 \qquad V_g = W(l\cos\theta)$$
$$V = V_e + V_g = 2kl^2\sin^2\theta + Wl\cos\theta \tag{10.22}$$

The positions of equilibrium of the system are obtained by equating to zero the derivative of the potential energy V,

$$\frac{dV}{d\theta} = 4kl^2\sin\theta\cos\theta - Wl\sin\theta = 0$$

$$\sin\theta = 0 \qquad 4kl\cos\theta - W = 0$$

There are therefore two positions of equilibrium, corresponding to the values $\theta = 0$ and $\theta = \cos^{-1}(W/4kl)$.†

† The second position does not exist if $W > 4kl$ (see Prob. 10.61 for a further discussion of the equilibrium of this system).

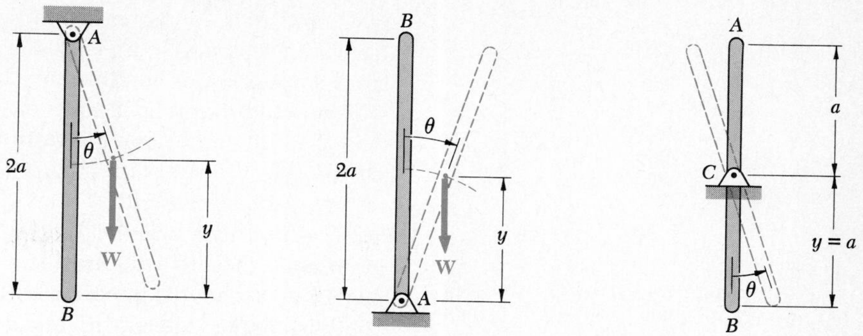

Fig. 10.14 (*a*) Stable equilibrium (*b*) Unstable equilibrium (*c*) Neutral equilibrium

***10.8. Stability of Equilibrium.** Consider the three uniform rods of length $2a$ and weight **W** shown in Fig. 10.14. While each rod is in equilibrium, there is an important difference between the three cases considered. Suppose that each rod is slightly disturbed from its position of equilibrium and then released: rod a will move back toward its original position, rod b will keep moving away from its original position, and rod c will remain in its new position. In case a, the equilibrium of the rod is said to be *stable;* in case b, the equilibrium is said to be *unstable;* and, in case c, to be *neutral.*

Recalling from Sec. 10.6 that the potential energy V_g with respect to gravity is equal to Wy, where y is the elevation of the point of application of **W** measured from an arbitrary level, we observe that the potential energy of rod a is minimum in the position of equilibrium considered, that the potential energy of rod b is maximum, and that the potential energy of rod c is constant. Equilibrium is thus *stable, unstable,* or *neutral* according to whether the potential energy is *minimum, maximum,* or *constant* (Fig. 10.15).

That the result obtained is quite general may be seen as follows: We first observe that a force always tends to do positive

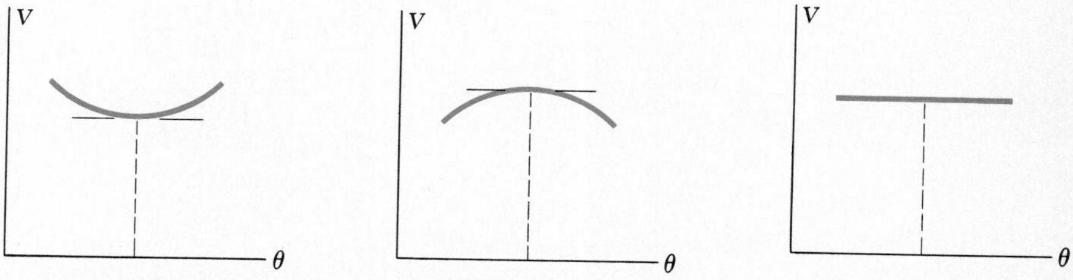

Fig. 10.15 (*a*) Stable equilibrium (*b*) Unstable equilibrium (*c*) Neutral equilibrium

work and thus to decrease the potential energy of the system on which it is applied. Therefore, when a system is disturbed from its position of equilibrium, the forces acting on the system will tend to bring it back to its original position if V is minimum (Fig. 10.15*a*) and to move it farther away if V is maximum (Fig. 10.15*b*). If V is constant (Fig. 10.15*c*), the forces will not tend to move the system either way.

Recalling from calculus that a function is minimum or maximum according to whether its second derivative is positive or negative, we may summarize as follows the conditions for the equilibrium of a system with one degree of freedom (i.e., a system the position of which is defined by a single independent variable θ):

$$\frac{dV}{d\theta} = 0 \qquad \frac{d^2V}{d\theta^2} > 0\text{: stable equilibrium}$$
$$\frac{dV}{d\theta} = 0 \qquad \frac{d^2V}{d\theta^2} < 0\text{: unstable equilibrium} \tag{10.23}$$

If both the first and the second derivatives of V are zero, it is necessary to examine derivatives of a higher order to determine whether the equilibrium is stable, unstable, or neutral. The equilibrium will be neutral if all derivatives are zero, since the potential energy V is then a constant. The equilibrium will be stable if the first derivative found to be different from zero is of even order, and if that derivative is positive. In all other cases the equilibrium will be unstable.

If the system considered possesses *several degrees of freedom*, the potential energy V depends upon several variables and it is thus necessary to apply the theory of functions of several variables to determine whether V is minimum. It may be verified that a system with two degrees of freedom will be stable, and the corresponding potential energy $V(\theta_1, \theta_2)$ will be minimum, if the following relations are satisfied simultaneously:

$$\frac{\partial V}{\partial \theta_1} = \frac{\partial V}{\partial \theta_2} = 0$$
$$\left(\frac{\partial^2 V}{\partial \theta_1\,\partial \theta_2}\right)^2 - \frac{\partial^2 V}{\partial \theta_1^2}\frac{\partial^2 V}{\partial \theta_2^2} < 0 \tag{10.24}$$
$$\frac{\partial^2 V}{\partial \theta_1^2} > 0 \qquad \text{or} \qquad \frac{\partial^2 V}{\partial \theta_2^2} > 0$$

SAMPLE PROBLEM 10.4

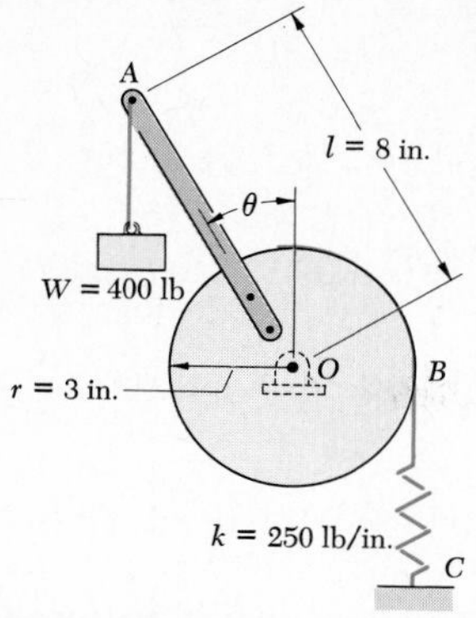

A 400-lb weight is attached to the lever AO as shown. The constant of the spring BC is $k = 250$ lb/in., and the spring is unstretched when $\theta = 0$. Determine the position or positions of equilibrium, and state in each case whether the equilibrium is stable, unstable, or neutral.

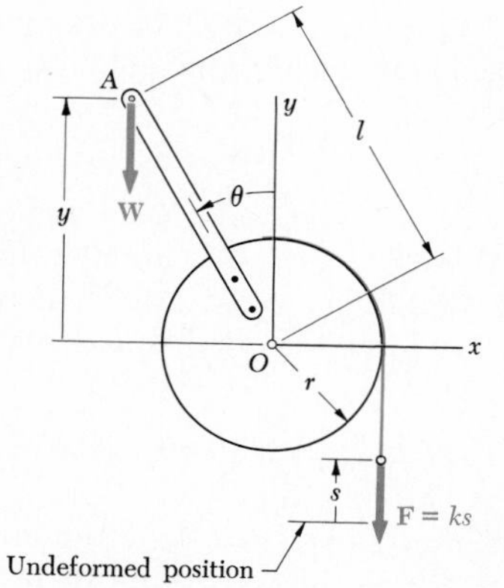

Potential Energy. Denoting by s the deflection of the spring from its undeformed position and placing the origin of coordinates at O, the potential energy of the system is

$$V_e = \tfrac{1}{2}ks^2 \qquad V_g = Wy$$

Measuring θ in radians, we have

$$s = r\theta \qquad y = l\cos\theta$$
$$V_e = \tfrac{1}{2}kr^2\theta^2 \qquad V_g = Wl\cos\theta$$
$$V = V_e + V_g = \tfrac{1}{2}kr^2\theta^2 + Wl\cos\theta$$

Positions of Equilibrium. Setting $dV/d\theta = 0$, we write

$$\frac{dV}{d\theta} = kr^2\theta - Wl\sin\theta = 0$$

$$\sin\theta = \frac{kr^2}{Wl}\theta$$

Substituting the given data, we obtain

$$\sin\theta = \frac{(250\text{ lb/in.})(3\text{ in.})^2}{(400\text{ lb})(8\text{ in.})}\theta = 0.703\,\theta$$

Solving by trial and error for θ, we find

$\theta = 0$ and $\theta = 80.4°$ ◄

Stability of Equilibrium. The second derivative of the potential energy V with respect to θ is

$$\frac{d^2V}{d\theta^2} = kr^2 - Wl\cos\theta = (250\text{ lb/in.})(3\text{ in.})^2 - (400\text{ lb})(8\text{ in.})\cos\theta$$
$$= 2250 - 3200\cos\theta$$

For $\theta = 0$: $\dfrac{d^2V}{d\theta^2} = 2250 - 3200\cos 0° = -950 < 0$

The equilibrium is unstable for $\theta = 0°$ ◄

For $\theta = 80.4°$: $\dfrac{d^2V}{d\theta^2} = 2250 - 3200\cos 80.4° = +1716 > 0$

The equilibrium is stable for $\theta = 80.4°$ ◄

PROBLEMS

10.38 Using the method of Sec. 10.7, solve Prob. 10.20.

10.39 Using the method of Sec. 10.7, solve Prob. 10.21.

10.40 Using the method of Sec. 10.7, solve Prob. 10.22.

10.41 Using the method of Sec. 10.7, solve Prob. 10.33.

10.42 Show that the equilibrium is neutral in Prob. 10.2.

10.43 Show that the equilibrium is neutral in Prob. 10.3.

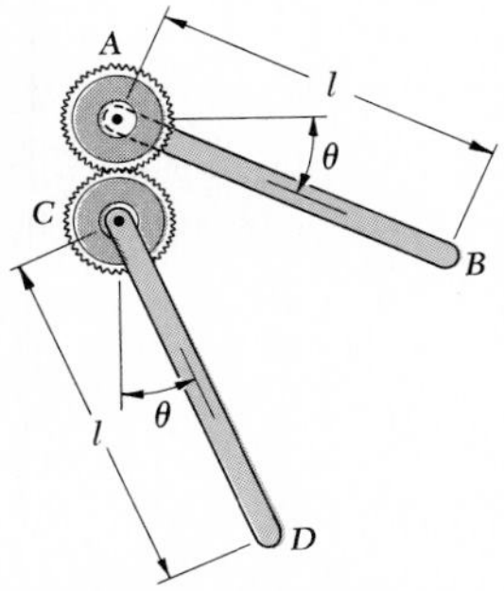

Fig. P10.44 and P10.45

10.44 Two uniform rods, each of weight W, are attached to gears of equal radii as shown. Determine the positions of equilibrium of the system and state in each case whether the equilibrium is stable, unstable, or neutral.

10.45 A vertical force **P** of magnitude 10 lb is applied to rod CD at D. Knowing that the uniform rods AB and CD weigh 5 lb each, determine the positions of equilibrium of the system and state in each case whether the equilibrium is stable, unstable, or neutral.

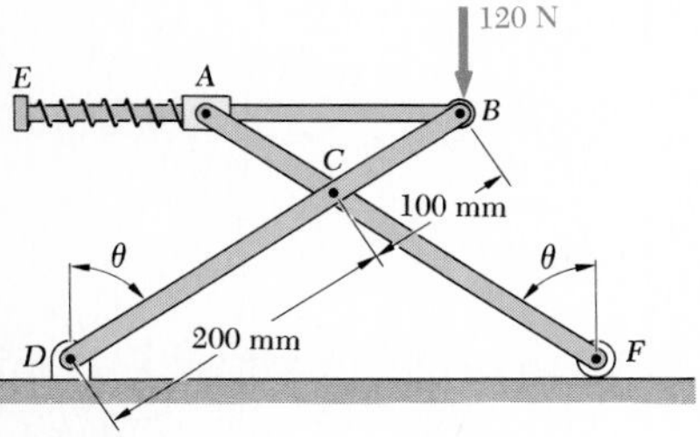

Fig. P10.46

10.46 Two rods AF and BD, each of length 300 mm, and a third rod BE are connected as shown. The spring EA is of constant $k = 1.8$ kN/m and is undeformed when $\theta = 0$. Determine the values of θ corresponding to equilibrium and state in each case whether the equilibrium is stable, unstable, or neutral.

10.47 Using the method of Sec. 10.7, solve Prob. 10.25. Determine whether the equilibrium is stable, unstable, or neutral. (*Hint.* The potential energy corresponding to the couple exerted by a torsion spring is $\frac{1}{2}K\theta^2$, where K is the torsional spring constant and θ is the angle of twist.)

10.48 In Prob. 10.26, determine whether each of the positions of equilibrium is stable, unstable, or neutral. (See hint of Prob. 10.47.)

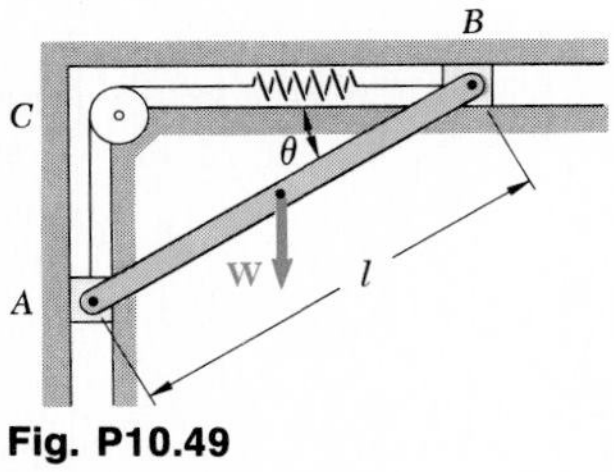

Fig. P10.49

10.49 A slender rod AB, of weight W, is attached to blocks A and B which may move freely in the guides shown. The constant of the spring is k and the spring is unstretched when AB is horizontal. Neglecting the weight of the blocks, derive an equation in θ, W, l, and k which must be satisfied when the rod is in equilibrium.

10.50 In Prob. 10.49, determine two values of θ corresponding to equilibrium when $W = 40$ lb, $k = 10$ lb/in., and $l = 15$ in. State in each case whether the equilibrium is stable, unstable, or neutral.

10.51 For the mechanism considered in Prob. 10.21, assuming $0° \leq \theta \leq 90°$, show that for any values of W and k there is only one equilibrium position. Further show that the equilibrium position is stable.

10.52 (*a*) Derive an equation defining the angle θ corresponding to the equilibrium position. (*b*) Determine the angle θ corresponding to the equilibrium position if $W = \frac{1}{2}P$.

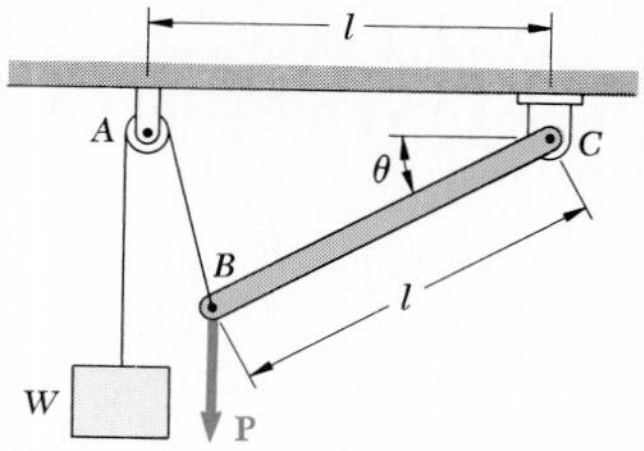

Fig. P10.52

10.53 The internal spring AC is of constant k and is undeformed when $\theta = 45°$. Derive an equation defining the values of θ corresponding to equilibrium positions.

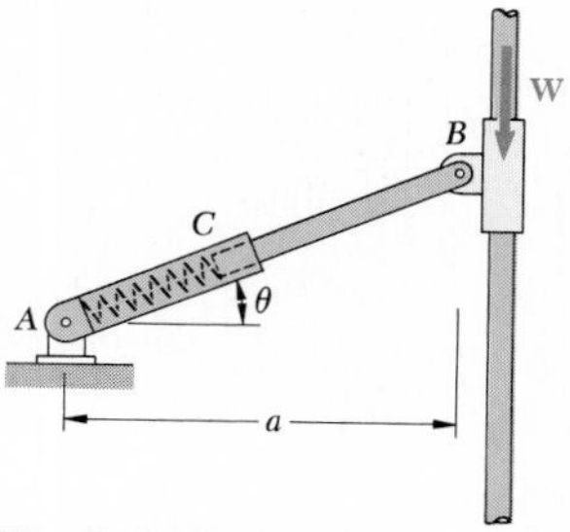

Fig. P10.53

10.54 In Prob. 10.53, determine the values of θ corresponding to equilibrium positions when $W = 75$ N, $k = 2$ kN/m, and $a = 500$ mm.

10.55 For the mechanism of Sample Prob. 10.4, determine the range of values of W for which stable equilibrium exists when lever AO is vertical. Express the result in terms of l, r, and k.

10.56 The rod AB is attached to a hinge at A and to two springs each of constant k. If $h = 450$ mm, $d = 300$ mm, and $m = 200$ kg, determine the range of values of k for which the equilibrium of rod AB is stable in the position shown. Each spring can act in either tension or compression.

10.57 If $m = 100$ kg, $h = 600$ mm, and the constant of each spring is $k = 3$ kN/m, determine the range of values of the distance d for which the equilibrium of the rod AB is stable in the position shown. Each spring can act in either tension or compression.

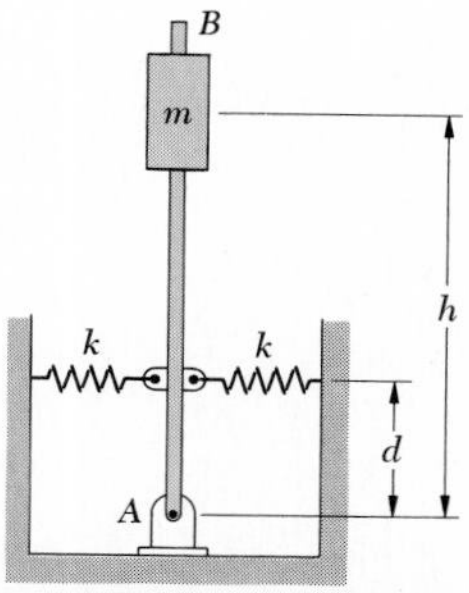

Fig. P10.56 and P10.57

10.58 and 10.59 Two bars AB and BC are attached to a single spring of constant k which is unstretched when the bars are horizontal. Determine the range of values of the magnitude P of the two equal and opposite forces **P** and **P′** for which the equilibrium of the system is stable in the position shown.

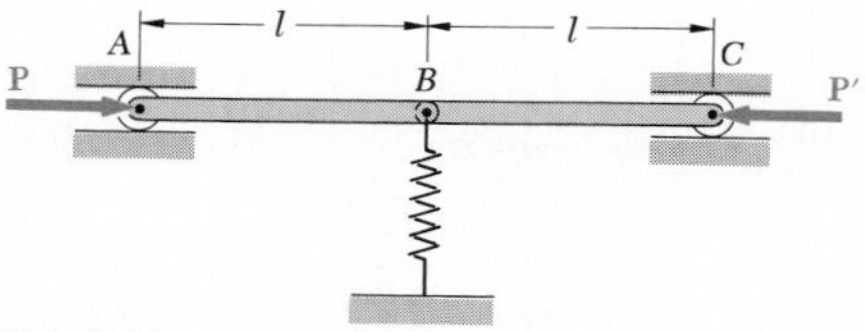

Fig. P10.58

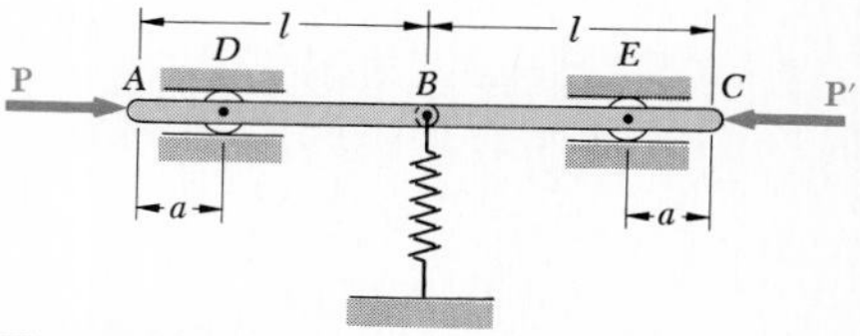

Fig. P10.59

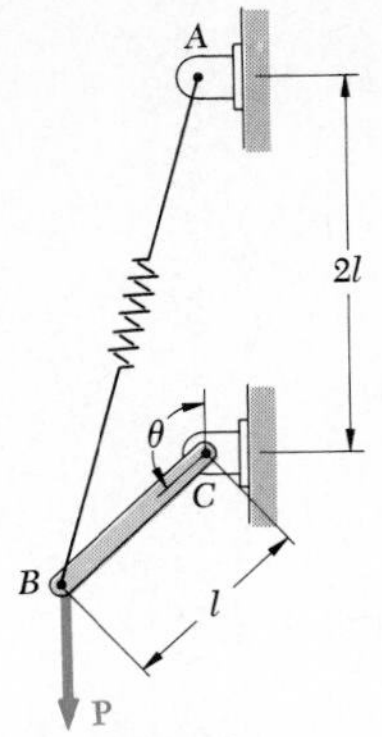

Fig. P10.60

10.60 The spring AB is of constant k and is unstretched when $\theta = 0$. Determine the range of values of k for which $\theta = 180°$ is a stable position of equilibrium.

10.61 In Sec. 10.7, two positions of equilibrium were obtained for the system shown in Fig. 10.13, namely, $\theta = 0$ and $\theta = \cos^{-1}(W/4kl)$. Show that (*a*) if $W < 4kl$, the equilibrium is stable in the first position ($\theta = 0$) and unstable in the second, (*b*) if $W = 4kl$, the two positions coincide and the equilibrium is unstable, (*c*) if $W > 4kl$, the equilibrium is unstable in the first position ($\theta = 0$) and the second position does not exist. (*Note.* It is assumed that the system must deform as shown and that the system cannot rotate as a single rigid body about A when A and B coincide.)

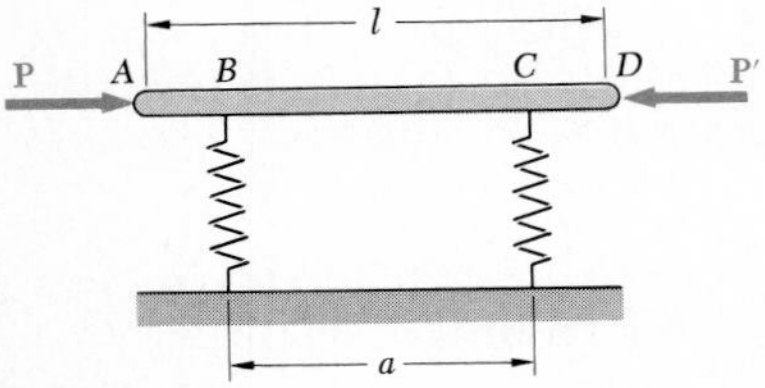

Fig. P10.62

10.62 The horizontal bar AD is attached to two springs of constant k and is in equilibrium in the position shown. Determine the range of values of the magnitude P of the two equal and opposite *horizontal* forces **P** and **P′** for which the equilibrium position is stable (*a*) if $AB = CD$, (*b*) if $AB = 2(CD)$.

***10.63 and 10.64** The bars shown, each of length l and of negligible weight, are attached to springs each of constant k. The springs are undeformed and the system is in equilibrium when $\theta_1 = \theta_2 = 0$. Determine the range of values of P for which the equilibrium position is stable.

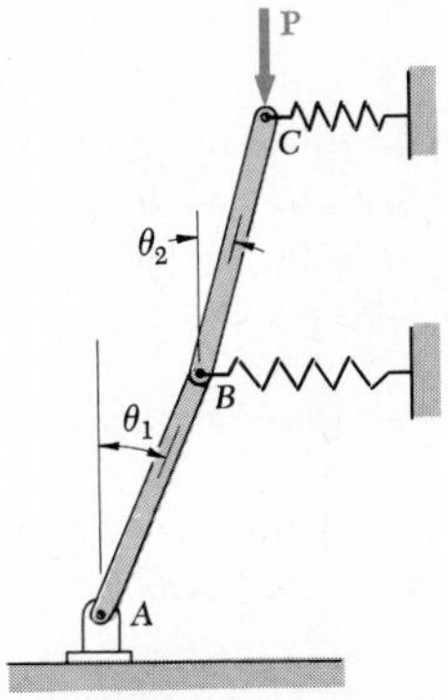

Fig. P10.63

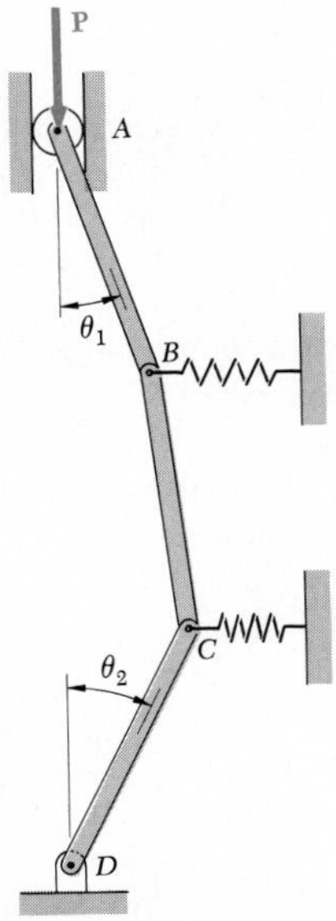

Fig. P10.64

REVIEW PROBLEMS

10.65 (*a*) Derive an expression for the moment of the couple **M** required to maintain equilibrium. (*b*) Determine the value of θ corresponding to the equilibrium position when $P = 20$ lb, $M = 80$ lb · in., $l = 4$ in., and $b = 10$ in., and state whether the equilibrium is stable, unstable, or neutral.

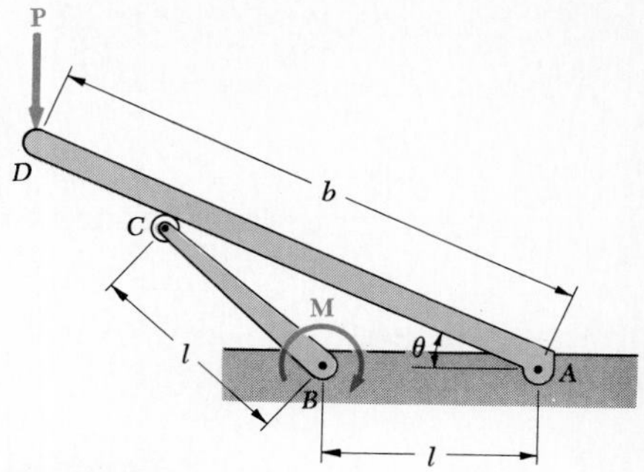

Fig. P10.65

10.66 The elevation of the overhead platform is controlled by two identical mechanisms, only one of which is shown. A load of 5 kN is applied to the mechanism shown. Neglecting the weight of the mechanism, determine the force exerted by the hydraulic cylinder on pin E when $\theta = 30°$.

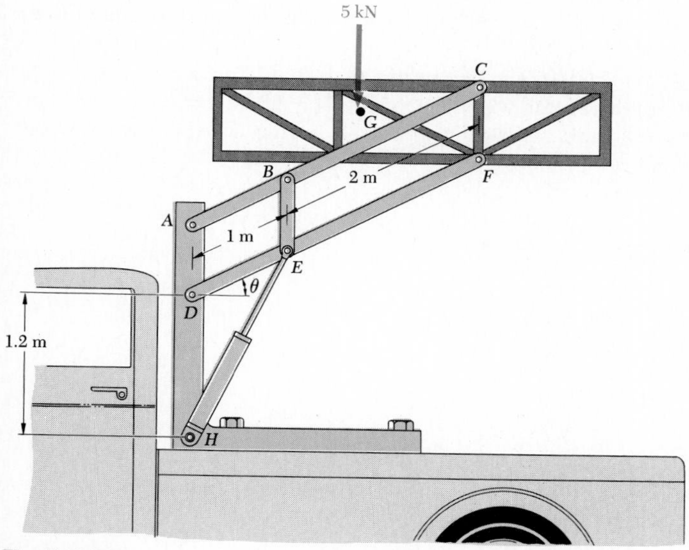

Fig. P10.66

10.67 For the mechanism of Prob. 10.66, (*a*) express the force exerted by the hydraulic cylinder on pin E as a function of the length HE, (*b*) determine the largest possible value of the angle θ if the maximum force that the cylinder can exert on pin E is 25 kN.

10.68 Knowing that rod AB may slide freely along the floor and the inclined plane, derive an expression for the magnitude of the force **Q** required to maintain equilibrium. Neglect the weight of the rod.

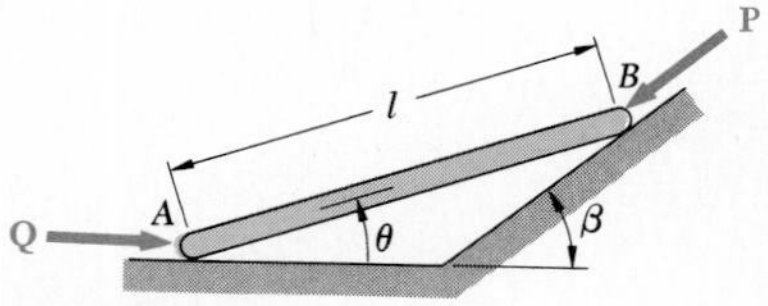

Fig. P10.68

10.69 Solve Prob. 10.68, taking into account the weight W of the uniform rod.

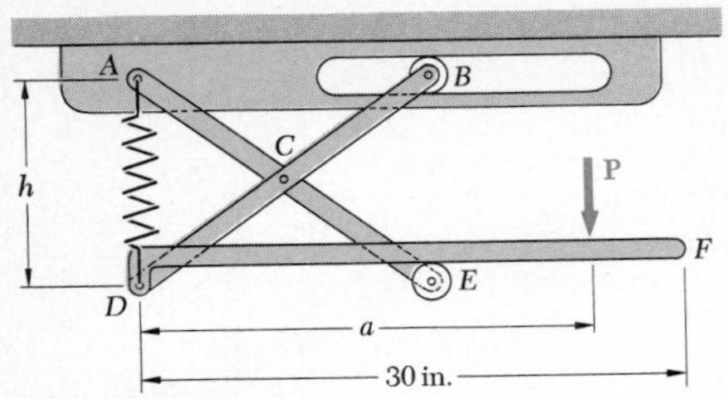

Fig. P10.70

10.70 Members *ACE* and *DCB* are each of length 25 in. and are connected by a pin at their midpoints *C*. A load **P** of magnitude 200 lb is applied to member *DF*. If $h = 15$ in. and $a = 30$ in., determine (*a*) the tension in the spring *AD*, (*b*) the unstretched length of the spring, knowing that the spring constant is 40 lb/in.

10.71 A slender rod of length l is attached to a collar at *B* and rests on a smooth circular cylinder of radius r. Knowing that the collar may slide freely along a vertical guide, derive an expression for the magnitude of the force **Q** required to maintain equilibrium.

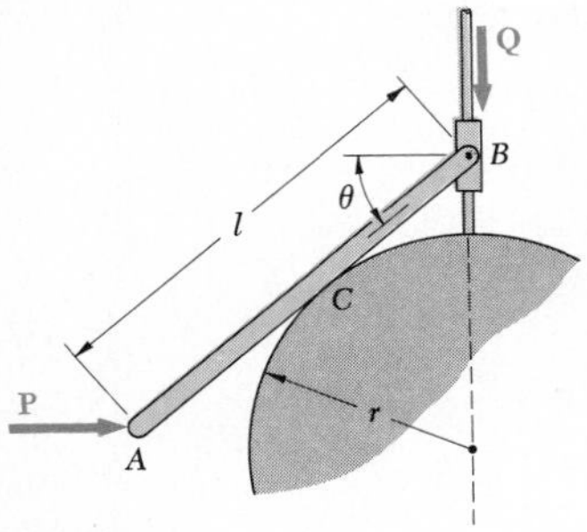

Fig. P10.71

10.72 Solve Prob. 10.71, assuming that the force **P** is removed and that a couple **M**, directed counterclockwise, is applied to rod *AB*.

10.73 The spring *AD* is of constant k and is undeformed when $x = \frac{1}{2}l$. Neglecting the weight of the smooth rod *AB*, derive an equation in θ which must be satisfied when the system is in equilibrium.

10.74 Determine the values of θ and x corresponding to the equilibrium position when $P = 35$ lb, $W = 20$ lb, $r = 3$ in., $l = 16$ in., and $k = 5$ lb/in., knowing that the spring is undeformed when $x = 8$ in.

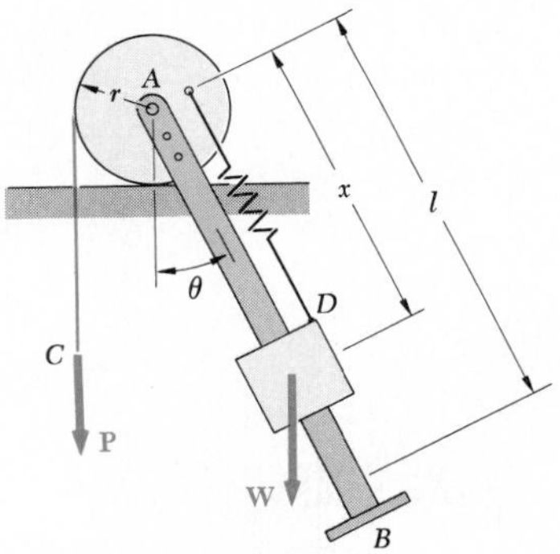

Fig. P10.73 and P10.74

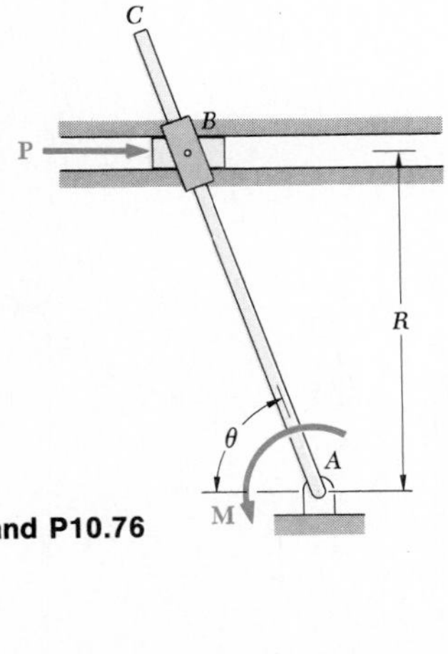

Fig. P10.75 and P10.76

10.75 Collar *B* may slide along rod *AC* and is attached by a pin to a block which may slide in the horizontal slot. Derive an expression for the moment of the couple **M** required to maintain equilibrium.

10.76 Determine the value of θ corresponding to the equilibrium position of rod *AC* when $R = 400$ mm, $P = 100$ N, and $M = 50$ N · m.

Mechanics for Engineers

DYNAMICS

Chapter
11

Kinematics of Particles

11.1. Introduction to Dynamics. Chapters 1 to 10 were devoted to *statics*, i.e., to the analysis of bodies at rest. We shall now begin the study of *dynamics*, which is the part of mechanics dealing with the analysis of bodies in motion.

While the study of statics goes back to the time of the Greek philosophers, the first significant contribution to dynamics was made by Galileo (1564–1642). His experiments on uniformly accelerated bodies led Newton (1642–1727) to formulate his fundamental laws of motion.

Dynamics is divided into two parts: (1) *Kinematics*, which is the study of the geometry of motion; kinematics is used to relate displacement, velocity, acceleration, and time, without reference to the cause of the motion. (2) *Kinetics*, which is the study of the relation existing between the forces acting on a body, the mass of the body, and the motion of the body; kinetics is used to predict the motion caused by given forces or to determine the forces required to produce a given motion.

Chapters 11 to 14 are devoted to the *dynamics of particles*, and Chap. 11 more particularly to the *kinematics of particles*. The use of the word particles does not imply that we shall restrict our study to that of small corpuscles; it rather indicates that in these first chapters we shall study the motion of bodies—possibly as large as cars, rockets, or airplanes—without regard to their size. By saying that the bodies are analyzed as particles, we mean that only their motion as an entire unit will be considered; any rotation about their own mass center will be neglected. There are cases, however, when such a rotation is not negligible; the bodies, then, may not be considered as particles. The analysis of such motions will be carried out in later chapters dealing with the *dynamics of rigid bodies*.

RECTILINEAR MOTION OF PARTICLES

11.2. Position, Velocity, and Acceleration. A particle moving along a straight line is said to be in *rectilinear motion*. At any given instant t, the particle will occupy a certain position on the straight line. To define the position P of the particle, we choose a fixed origin O on the straight line and a positive direction along the line. We measure the distance x from O to P and record it with a plus or minus sign, according to whether P is reached from O by moving along the line in the positive or the negative direction. The distance x, with the appropriate sign, completely defines the position of the particle; it is called the *position coordinate* of the particle considered. For example, the position coordinate corresponding to P in Fig. 11.1a is $x = +5$ m, while the coordinate corresponding to P' in Fig. 11.1b is $x' = -2$ m.

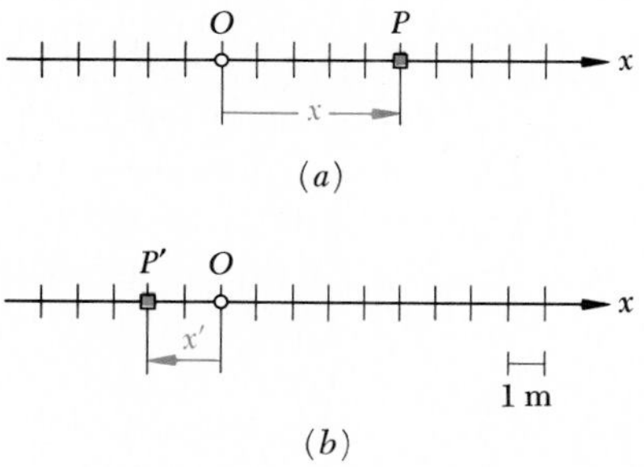

Fig. 11.1

When the position coordinate x of a particle is known for every value of time t, we say that the motion of the particle is known. The "timetable" of the motion may be given in the form of an equation in x and t, such as $x = 6t^2 - t^3$, or in the form of a graph of x vs. t as shown in Fig. 11.6. The units most generally used to measure the position coordinate x are the meter (m) in the SI system of units,† and the foot (ft) in the U.S. customary system of units. Time t will generally be measured in seconds (s).

† Cf. Sec. 1.3.

Consider the position P occupied by the particle at time t and the corresponding coordinate x (Fig. 11.2). Consider also the position P' occupied by the particle at a later time $t + \Delta t$; the position coordinate of P' may be obtained by adding to the coordinate x of P the small displacement Δx, which will be positive or negative according to whether P' is to the right or to the left of P. The *average velocity* of the particle over the time interval Δt is defined as the quotient of the displacement Δx and the time interval Δt,

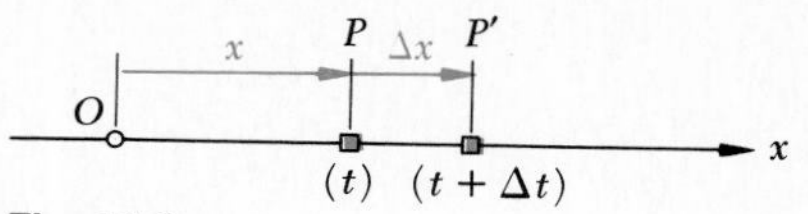

Fig. 11.2

$$\text{Average velocity} = \frac{\Delta x}{\Delta t}$$

If SI units are used, Δx is expressed in meters and Δt in seconds; the average velocity will thus be expressed in meters per second (m/s). If U.S. customary units are used, Δx is expressed in feet and Δt in seconds; the average velocity will then be expressed in feet per second (ft/s).

The *instantaneous velocity* v of the particle at the instant t is obtained from the average velocity by choosing shorter and shorter time intervals Δt and displacements Δx,

$$\text{Instantaneous velocity} = v = \lim_{\Delta t \to 0} \frac{\Delta x}{\Delta t}$$

The instantaneous velocity will also be expressed in m/s or ft/s. Observing that the limit of the quotient is equal, by definition, to the derivative of x with respect to t, we write

$$v = \frac{dx}{dt} \tag{11.1}$$

The velocity v is represented by an algebraic number which may be positive or negative.† A positive value of v indicates that x increases, i.e., that the particle moves in the positive direction (Fig. 11.3*a*); a negative value of v indicates that x decreases, i.e., that the particle moves in the negative direction (Fig. 11.3*b*). The magnitude of v is known as the *speed* of the particle.

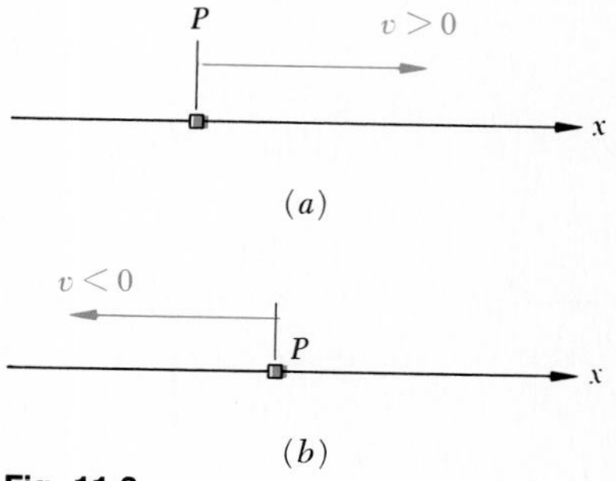

Fig. 11.3

† As we shall see in Sec. 11.9, the velocity is actually a vector quantity. However, since we are considering here the rectilinear motion of a particle, where the velocity of the particle has a known and fixed direction, we need only specify the sense and magnitude of the velocity; this may be conveniently done by using a scalar quantity with a plus or minus sign. The same remark will apply to the acceleration of a particle in rectilinear motion.

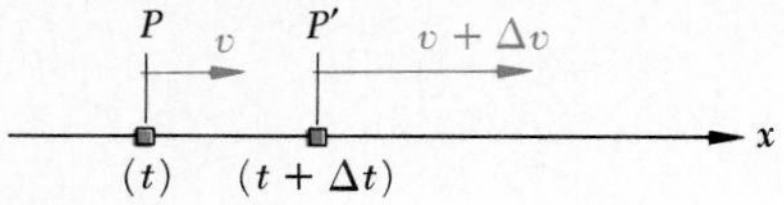

Fig. 11.4

Consider the velocity v of the particle at time t and also its velocity $v + \Delta v$ at a later time $t + \Delta t$ (Fig. 11.4). The *average acceleration* of the particle over the time interval Δt is defined as the quotient of Δv and Δt,

$$\text{Average acceleration} = \frac{\Delta v}{\Delta t}$$

If SI units are used, Δv is expressed in m/s and Δt in seconds; the average acceleration will thus be expressed in m/s^2. If U.S. customary units are used, Δv is expressed in ft/s and Δt in seconds; the average acceleration will then be expressed in ft/s^2.

The *instantaneous acceleration* a of the particle at the instant t is obtained from the average acceleration by choosing smaller and smaller values for Δt and Δv,

$$\text{Instantaneous acceleration} = a = \lim_{\Delta t \to 0} \frac{\Delta v}{\Delta t}$$

The instantaneous acceleration will also be expressed in m/s^2 or ft/s^2. The limit of the quotient is by definition the derivative of v with respect to t and measures the rate of change of the velocity. We write

$$a = \frac{dv}{dt} \tag{11.2}$$

or, substituting for v from (11.1),

$$a = \frac{d^2x}{dt^2} \tag{11.3}$$

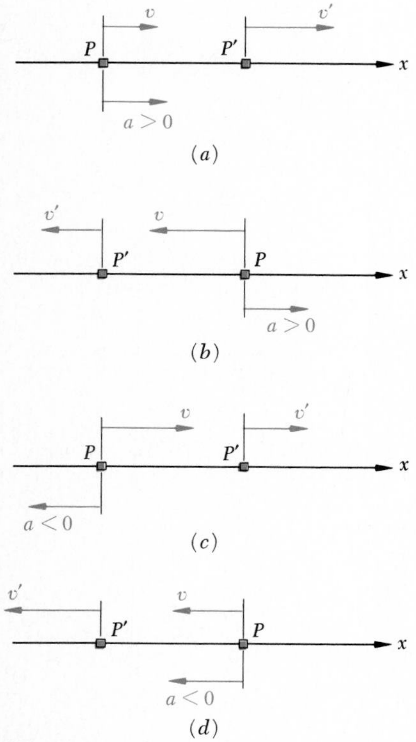

Fig. 11.5

The acceleration a is represented by an algebraic number which may be positive or negative.† A positive value of a indicates that the velocity (i.e., the algebraic number v) increases. This may mean that the particle is moving faster in the positive direction (Fig. 11.5*a*) or that it is moving more slowly in the negative direction (Fig. 11.5*b*); in both cases, Δv is positive. A negative value of a indicates that the velocity decreases; either the particle is moving more slowly in the positive direction (Fig. 11.5*c*), or it is moving faster in the negative direction (Fig. 11.5*d*).

The term *deceleration* is sometimes used to refer to a when the speed of the particle (i.e., the magnitude of v) decreases; the particle is then moving more slowly. For example, the particle of Fig. 11.5 is decelerated in parts b and c, while it is truly accelerated (i.e., moves faster) in parts a and d.

Another expression may be obtained for the acceleration by eliminating the differential dt in Eqs. (11.1) and (11.2). Solving

† See footnote, page 401.

(11.1) for dt, we obtain $dt = dx/v$; carrying into (11.2), we write

$$a = v\frac{dv}{dx} \tag{11.4}$$

Example. Consider a particle moving in a straight line, and assume that its position is defined by the equation

$$x = 6t^2 - t^3$$

where t is expressed in seconds and x in meters. The velocity v at any time t is obtained by differentiating x with respect to t,

$$v = \frac{dx}{dt} = 12t - 3t^2$$

The acceleration a is obtained by differentiating again with respect to t,

$$a = \frac{dv}{dt} = 12 - 6t$$

The position coordinate, the velocity, and the acceleration have been plotted against t in Fig. 11.6. The curves obtained are known as *motion curves*. It should be kept in mind, however, that the particle does not move along any of these curves; the particle moves in a straight line. Since the derivative of a function measures the slope of the corresponding curve, the slope of the x–t curve at any given time is equal to the value of v at that time and the slope of the v–t curve is equal to the value of a. Since $a = 0$ at $t = 2$ s, the slope of the v–t curve must be zero at $t = 2$ s; the velocity reaches a maximum at this instant. Also, since $v = 0$ at $t = 0$ and at $t = 4$ s, the tangent to the x–t curve must be horizontal for both of these values of t.

A study of the three motion curves of Fig. 11.6 shows that the motion of the particle from $t = 0$ to $t = \infty$ may be divided into four phases:

1. The particle starts from the origin, $x = 0$, with no velocity but with a positive acceleration. Under this acceleration, the particle gains a positive velocity and moves in the positive direction. From $t = 0$ to $t = 2$ s, x, v, and a are all positive.
2. At $t = 2$ s, the acceleration is zero; the velocity has reached its maximum value. From $t = 2$ s to $t = 4$ s, v is positive, but a is negative; the particle still moves in the positive direction but more and more slowly; the particle is decelerated.
3. At $t = 4$ s, the velocity is zero; the position coordinate x has reached its maximum value. From then on, both v and a are negative; the particle is accelerated and moves in the negative direction with increasing speed.
4. At $t = 6$ s, the particle passes through the origin; its coordinate x is then zero, while the total distance traveled since the beginning of the motion is 64 m. For values of t larger than 6 s, x, v, and a will all be negative. The particle keeps moving in the negative direction, away from O, faster and faster.

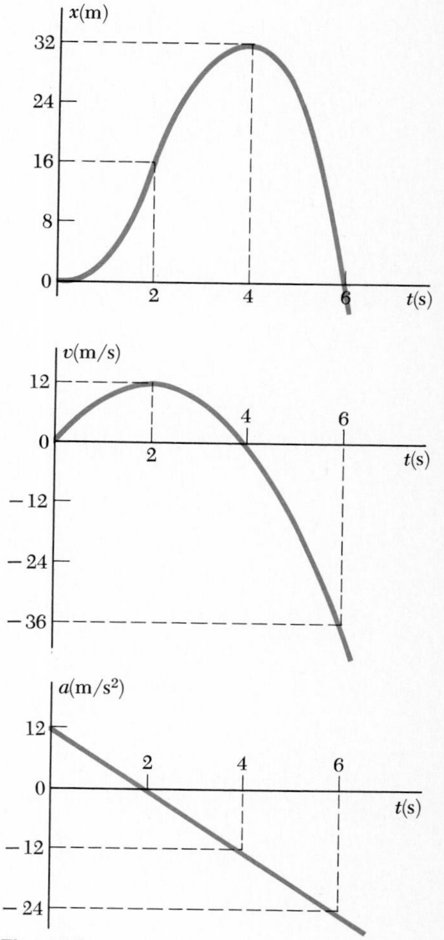

Fig. 11.6

11.3. Determination of the Motion of a Particle.

We saw in the preceding section that the motion of a particle is said to be known if the position of the particle is known for every value of the time t. In practice, however, a motion is seldom defined by a relation between x and t. More often, the conditions of the motion will be specified by the type of acceleration that the particle possesses. For example, a freely falling body will have a constant acceleration, directed downward and equal to 9.81 m/s^2 or 32.2 ft/s^2; a mass attached to a spring which has been stretched will have an acceleration proportional to the instantaneous elongation of the spring measured from the equilibrium position; etc. In general, the acceleration of the particle may be expressed as a function of one or more of the variables x, v, and t. In order to determine the position coordinate x in terms of t, it will thus be necessary to perform two successive integrations.

We shall consider three common classes of motion:

1. $a = f(t)$. *The Acceleration Is a Given Function of t.* Solving (11.2) for dv and substituting $f(t)$ for a, we write

$$dv = a\,dt$$
$$dv = f(t)\,dt$$

Integrating both members, we obtain the equation

$$\int dv = \int f(t)\,dt$$

which defines v in terms of t. It should be noted, however, that an arbitrary constant will be introduced as a result of the integration. This is due to the fact that there are many motions which correspond to the given acceleration $a = f(t)$. In order to uniquely define the motion of the particle, it is necessary to specify the *initial conditions* of the motion, i.e., the value v_0 of the velocity and the value x_0 of the position coordinate at $t = 0$. Replacing the indefinite integrals by *definite integrals* with lower limits corresponding to the initial conditions $t = 0$ and $v = v_0$ and upper limits corresponding to $t = t$ and $v = v$, we write

$$\int_{v_0}^{v} dv = \int_0^t f(t)\,dt$$

$$v - v_0 = \int_0^t f(t)\,dt$$

which yields v in terms of t.

We shall now solve (11.1) for dx,

$$dx = v\,dt$$

and substitute for v the expression just obtained. Both members are then integrated, the left-hand member with respect to x from $x = x_0$ to $x = x$, and the right-hand member with respect to t from $t = 0$ to $t = t$. The position coordinate x is thus obtained in terms of t; the motion is completely determined.

Two important particular cases will be studied in greater detail in Secs. 11.4 and 11.5: the case when $a = 0$, corresponding to a *uniform motion,* and the case when $a =$ constant, corresponding to a *uniformly accelerated motion.*

2. $a = f(x)$. *The Acceleration Is a Given Function of* x. Rearranging Eq. (11.4) and substituting $f(x)$ for a, we write

$$v\,dv = a\,dx$$
$$v\,dv = f(x)\,dx$$

Since each member contains only one variable, we may integrate the equation. Denoting again by v_0 and x_0, respectively, the initial values of the velocity and of the position coordinate, we obtain

$$\int_{v_0}^{v} v\,dv = \int_{x_0}^{x} f(x)\,dx$$
$$\tfrac{1}{2}v^2 - \tfrac{1}{2}v_0^2 = \int_{x_0}^{x} f(x)\,dx$$

which yields v in terms of x. We now solve (11.1) for dt,

$$dt = \frac{dx}{v}$$

and substitute for v the expression just obtained. Both members may be integrated, and the desired relation between x and t is obtained.

3. $a = f(v)$. *The Acceleration Is a Given Function of* v. We may then substitute $f(v)$ for a either in (11.2) or in (11.4) to obtain either of the following relations:

$$f(v) = \frac{dv}{dt} \qquad\qquad f(v) = v\frac{dv}{dx}$$
$$dt = \frac{dv}{f(v)} \qquad\qquad dx = \frac{v\,dv}{f(v)}$$

Integration of the first equation will yield a relation between v and t; integration of the second equation will yield a relation between v and x. Either of these relations may be used in conjunction with Eq. (11.1) to obtain the relation between x and t which characterizes the motion of the particle.

SAMPLE PROBLEM 11.1

The position of a particle which moves along a straight line is defined by the relation $x = t^3 - 6t^2 - 15t + 40$, where x is expressed in feet and t in seconds. Determine (a) the time at which the velocity will be zero, (b) the position and distance traveled by the particle at that time, (c) the acceleration of the particle at that time, (d) the distance traveled by the particle from $t = 4$ s to $t = 6$ s.

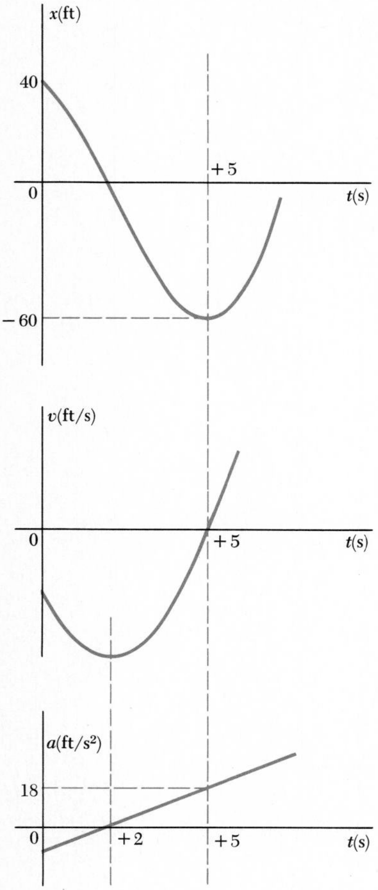

Solution. The equations of motion are

$$x = t^3 - 6t^2 - 15t + 40 \qquad (1)$$

$$v = \frac{dx}{dt} = 3t^2 - 12t - 15 \qquad (2)$$

$$a = \frac{dv}{dt} = 6t - 12 \qquad (3)$$

a. Time at Which $v = 0$. We make $v = 0$ in (2),

$$3t^2 - 12t - 15 = 0 \qquad t = -1\text{ s} \quad \text{and} \quad t = +5\text{ s} \;◄$$

Only the root $t = +5$ s corresponds to a time after the motion has begun: for $t < 5$ s, $v < 0$, the particle moves in the negative direction; for $t > 5$ s, $v > 0$, the particle moves in the positive direction.

b. Position and Distance Traveled When $v = 0$. Carrying $t = +5$ s into (1), we have

$$x_5 = (5)^3 - 6(5)^2 - 15(5) + 40 \qquad x_5 = -60\text{ ft} \;◄$$

The initial position at $t = 0$ was $x_0 = +40$ ft. Since $v \neq 0$ during the interval $t = 0$ to $t = 5$ s, we have

$$\text{Distance traveled} = x_5 - x_0 = -60\text{ ft} - 40\text{ ft} = -100\text{ ft}$$

$$\text{Distance traveled} = 100\text{ ft in the negative direction} \;◄$$

c. Acceleration When $v = 0$. We carry $t = +5$ s into (3):

$$a_5 = 6(5) - 12 \qquad a_5 = +18\text{ ft/s}^2 \;◄$$

d. Distance Traveled from $t = 4$ s to $t = 6$ s. Since the particle moves in the negative direction from $t = 4$ s to $t = 5$ s and in the positive direction from $t = 5$ s to $t = 6$ s, we shall compute separately the distance traveled during each of these time intervals.

From $t = 4$ s to $t = 5$ s: $\quad x_5 = -60$ ft

$$x_4 = (4)^3 - 6(4)^2 - 15(4) + 40 = -52\text{ ft}$$

$$\text{Distance traveled} = x_5 - x_4 = -60\text{ ft} - (-52\text{ ft}) = -8\text{ ft}$$
$$= 8\text{ ft in the negative direction}$$

From $t = 5$ s to $t = 6$ s: $\quad x_5 = -60$ ft

$$x_6 = (6)^3 - 6(6)^2 - 15(6) + 40 = -50\text{ ft}$$

$$\text{Distance traveled} = x_6 - x_5 = -50\text{ ft} - (-60\text{ ft}) = +10\text{ ft}$$
$$= 10\text{ ft in the positive direction}$$

Total distance traveled from $t = 4$ s to $t = 6$ s is

$$8\text{ ft} + 10\text{ ft} = 18\text{ ft} \;◄$$

SAMPLE PROBLEM 11.2

A ball is thrown from the top of a tower 18 m high, with a velocity of 12 m/s directed vertically upward. Knowing that the acceleration of the ball is constant and equal to 9.81 m/s² downward, determine (*a*) the velocity v and elevation y of the ball above the ground at any time t, (*b*) the highest elevation reached by the ball and the corresponding value of t, (*c*) the time when the ball will hit the ground and the corresponding velocity. Draw the v–t and y–t curves.

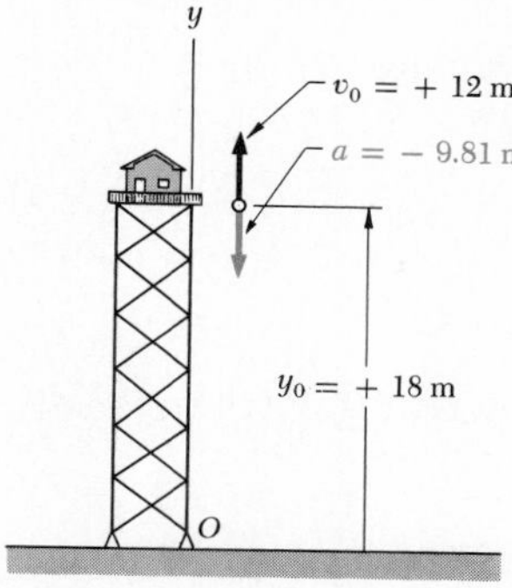

a. Velocity and Elevation. The y axis measuring the position coordinate (or elevation) is chosen with its origin O on the ground and its positive sense upward. The value of the acceleration and the initial values of v and y are as indicated. Substituting for a in $a = dv/dt$ and noting that, at $t = 0$, $v_0 = +12$ m/s, we have

$$\frac{dv}{dt} = a = -9.81 \text{ m/s}^2$$

$$\int_{v_0=12}^{v} dv = -\int_0^t 9.81\, dt$$

$$[v]_{12}^{v} = -[9.81t]_0^t$$

$$v - 12 = -9.81t$$

$$v = 12 - 9.81t \quad (1)$$

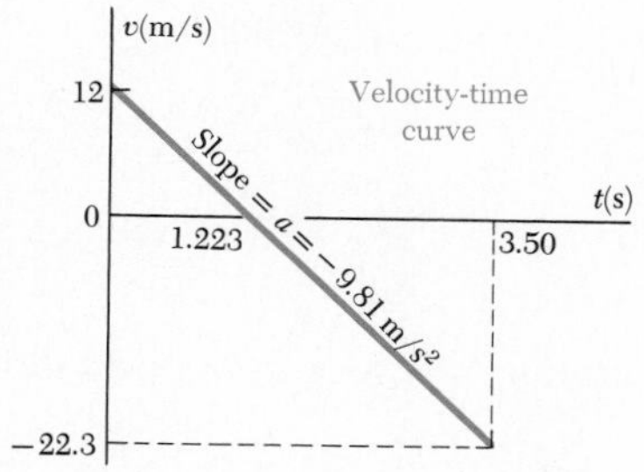

Substituting for v in $v = dy/dt$ and noting that, at $t = 0$, $y_0 = 18$ m, we have

$$\frac{dy}{dt} = v = 12 - 9.81t$$

$$\int_{y_0=18}^{y} dy = \int_0^t (12 - 9.81t)\, dt$$

$$[y]_{18}^{y} = [12t - 4.90t^2]_0^t$$

$$y - 18 = 12t - 4.90t^2$$

$$y = 18 + 12t - 4.90t^2 \quad (2)$$

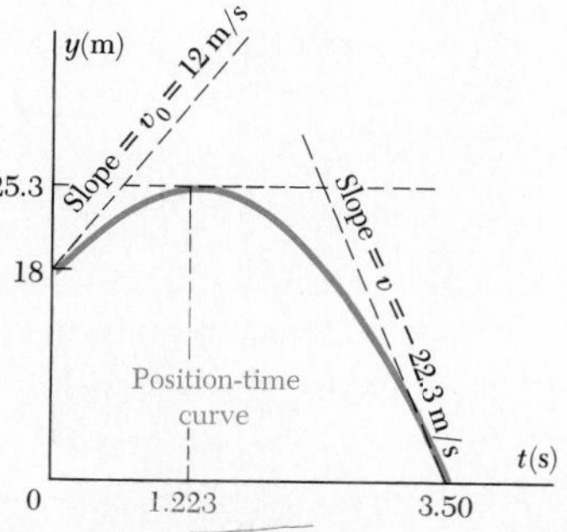

b. Highest Elevation. When the ball reaches its highest elevation, we have $v = 0$. Substituting into (1), we obtain

$$12 - 9.81t = 0 \qquad t = 1.223 \text{ s}$$

Carrying $t = 1.223$ s into (2), we have

$$y = 18 + 12(1.223) - 4.90(1.223)^2 \qquad y = 25.3 \text{ m}$$

c. Ball Hits the Ground. When the ball hits the ground, we have $y = 0$. Substituting into (2), we obtain

$$18 + 12t - 4.90t^2 = 0 \qquad t = -1.05 \text{ s} \quad \text{and} \quad t = +3.50 \text{ s}$$

Only the root $t = +3.50$ s corresponds to a time after the motion has begun. Carrying this value of t into (1), we have

$$v = 12 - 9.81(3.50) = -22.3 \text{ m/s} \qquad v = 22.3 \text{ m/s} \downarrow$$

$\frac{-b \pm \sqrt{b^2-4ac}}{2a}$ $\quad 3.67 + 2.45t - t^2 = 0$

$= \frac{-2.45 \pm \sqrt{2.45^2 - 4(-1)(3.67)}}{2(-1)}$

$= 3.50$ or -1.05

SAMPLE PROBLEM 11.3

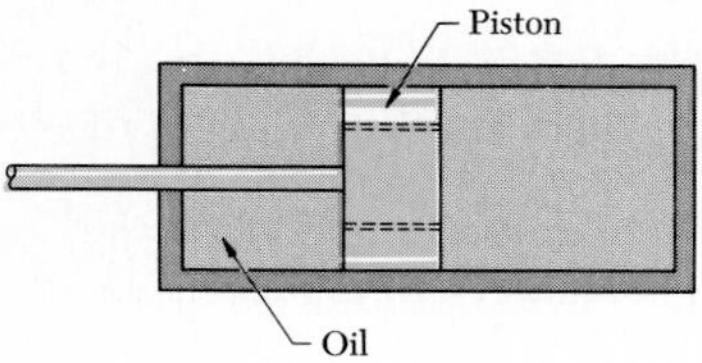

The brake mechanism used to reduce recoil in certain types of guns consists essentially of a piston which is attached to the barrel and may move in a fixed cylinder filled with oil. As the barrel recoils with an initial velocity v_0, the piston moves and oil is forced through orifices in the piston, causing the piston and the barrel to decelerate at a rate proportional to their velocity, i.e., $a = -kv$. Express (*a*) v in terms of t, (*b*) x in terms of t, (*c*) v in terms of x. Draw the corresponding motion curves.

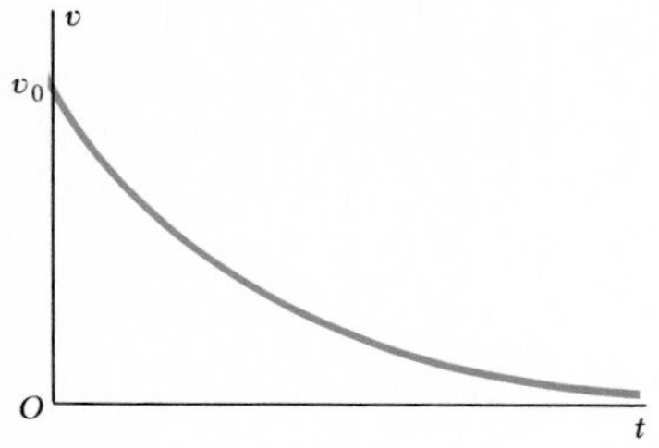

a. v in Terms of t. Substituting $-kv$ for a in the fundamental formula defining acceleration, $a = dv/dt$, we write

$$-kv = \frac{dv}{dt} \qquad \frac{dv}{v} = -k\,dt \qquad \int_{v_0}^{v} \frac{dv}{v} = -k \int_0^t dt$$

$$\ln \frac{v}{v_0} = -kt \qquad\qquad v = v_0 e^{-kt} \quad ◄$$

b. x in Terms of t. Substituting the expression just obtained for v into $v = dx/dt$, we write

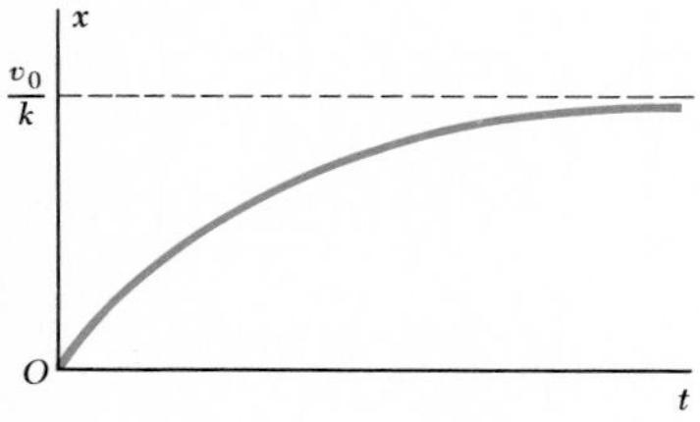

$$v_0 e^{-kt} = \frac{dx}{dt}$$

$$\int_0^x dx = v_0 \int_0^t e^{-kt}\,dt$$

$$x = -\frac{v_0}{k}[e^{-kt}]_0^t = -\frac{v_0}{k}(e^{-kt} - 1)$$

$$x = \frac{v_0}{k}(1 - e^{-kt}) \quad ◄$$

c. v in Terms of x. Substituting $-kv$ for a in $a = v\,dv/dx$, we write

$$-kv = v\frac{dv}{dx}$$

$$dv = -k\,dx$$

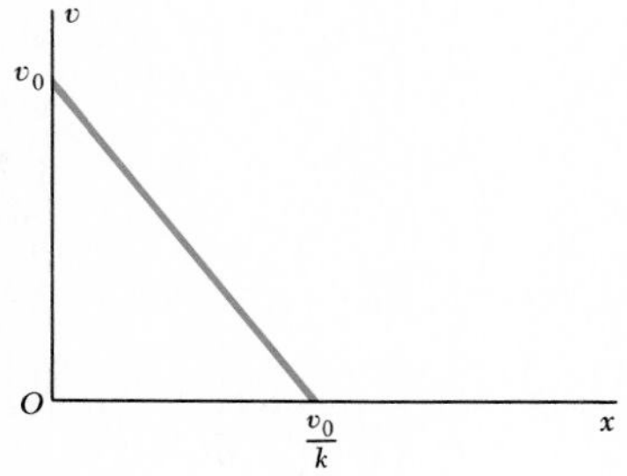

$$\int_{v_0}^{v} dv = -k \int_0^x dx$$

$$v - v_0 = -kx \qquad\qquad v = v_0 - kx \quad ◄$$

Check. Part *c* could have been solved by eliminating t from the answers obtained for parts *a* and *b*. This alternate method may be used as a check. From part *a* we obtain $e^{-kt} = v/v_0$; substituting in the answer of part *b*, we obtain

$$x = \frac{v_0}{k}(1 - e^{-kt}) = \frac{v_0}{k}\left(1 - \frac{v}{v_0}\right) \qquad v = v_0 - kx \qquad \text{(checks)}$$

PROBLEMS

11.1 The motion of a particle is defined by the relation $x = t^3 - 9t^2 + 24t - 2$, where x is expressed in inches and t in seconds. Determine the position, velocity, and acceleration when $t = 5$ s.

11.2 The motion of a particle is defined by the relation $x = 2t^3 - 6t^2 + 15$, where x is expressed in inches and t in seconds. Determine the time, position, and acceleration when $v = 0$.

11.3 The motion of a particle is defined by the relation $x = t^2 - 10t + 30$, where x is expressed in meters and t in seconds. Determine (*a*) when the velocity is zero, (*b*) the position and the total distance traveled when $t = 8$ s.

11.4 The motion of a particle is defined by the relation $x = \frac{1}{3}t^3 - 3t^2 + 8t + 2$, where x is expressed in meters and t in seconds. Determine (*a*) when the velocity is zero, (*b*) the position and the total distance traveled when the acceleration is zero.

11.5 The acceleration of a particle is directly proportional to the time t. At $t = 0$, the velocity of the particle is $v = -16$ m/s. Knowing that both the velocity and the position coordinate are zero when $t = 4$ s, write the equations of motion for the particle.

11.6 The acceleration of a particle is defined by the relation $a = -2$ m/s^2. If $v = +8$ m/s and $x = 0$ when $t = 0$, determine the velocity, position, and total distance traveled when $t = 6$ s.

11.7 The acceleration of a particle is defined by the relation $a = 64 - 12t^2$. The particle starts at $t = 0$ with $v = 0$ and $x = 25$ in. Determine (*a*) the time when the velocity is again zero, (*b*) the position and velocity when $t = 5$ s, (*c*) the total distance traveled by the particle from $t = 0$ to $t = 5$ s.

11.8 The acceleration of a particle is defined by the relation $a = kt^2$. (*a*) Knowing that $v = -32$ ft/s when $t = 0$ and that $v = +32$ ft/s when $t = 4$ s, determine the constant k. (*b*) Write the equations of motion knowing also that $x = 0$ when $t = 4$ s.

11.9 The acceleration of a particle is defined by the relation $a = 21 - 12x^2$, where a is expressed in m/s^2 and x in meters. The particle starts with no initial velocity at the position $x = 0$. Determine (*a*) the velocity when $x = 1.5$ m, (*b*) the position where the velocity is again zero, (*c*) the position where the velocity is maximum.

11.10 The acceleration of an oscillating particle is defined by the relation $a = -kx$. Find the value of k such that $v = 10$ m/s when $x = 0$ and $x = 2$ m when $v = 0$.

11.11 The acceleration of a particle is defined by the relation $a = -kx^{-2}$. The particle starts with no initial velocity at $x = 8$ in., and it is observed that its velocity is 5 in./s when $x = 4$ in. Determine (*a*) the value of k, (*b*) the velocity of the particle when $x = 2$ in.

11.12 The acceleration of a particle moving in a straight line is directed toward a fixed point O and is inversely proportional to the distance of the particle from O. At $t = 0$, the particle is 2 ft to the right of O, has a velocity of 4 ft/s to the right, and has an acceleration of 3 ft/s^2 to the left. Determine (*a*) the velocity of the particle when it is 3 ft away from O, (*b*) the position of the particle at which its velocity is zero.

11.13 The acceleration of a particle is defined by the relation $a = -10v$, where a is expressed in m/s^2 and v in m/s. Knowing that at $t = 0$ the velocity is 30 m/s, determine (*a*) the distance the particle will travel before coming to rest, (*b*) the time required for the particle to come to rest, (*c*) the time required for the velocity of the particle to be reduced to 1 percent of its initial value.

11.14 The acceleration of a particle is defined by the relation $a = -0.0125v^2$, where a is the acceleration in m/s^2 and v is the velocity in m/s. If the particle is given an initial velocity v_0, find the distance it will travel (*a*) before its velocity drops to half the initial value, (*b*) before it comes to rest.

11.15 The acceleration of a particle falling through the atmosphere is defined by the relation $a = g(1 - k^2v^2)$. Knowing that the particle starts at $t = 0$ and $x = 0$ with no initial velocity, (*a*) show that the velocity at any time t is $v = (1/k) \tanh kgt$, (*b*) write an equation defining the velocity for any value of x. (*c*) Why is $v_t = 1/k$ called the terminal velocity?

11.16 It has been determined experimentally that the magnitude in ft/s^2 of the deceleration due to air resistance of a projectile is $0.001v^2$, where v is expressed in ft/s. If the projectile is released from rest and keeps pointing downward, determine its velocity after it has fallen 500 ft. (*Hint.* The total acceleration is $g - 0.001v^2$, where $g = 32.2$ ft/s^2.)

11.17 The acceleration of a particle is defined by the relation $a = k \sin(\pi t/T)$. Knowing that both the velocity and the position coordinate of the particle are zero when $t = 0$, determine (*a*) the equations of motion, (*b*) the maximum velocity, (*c*) the position at $t = 2T$, (*d*) the average velocity during the interval $t = 0$ to $t = 2T$.

11.18 The position of an oscillating particle is defined by the relation $x = A \sin(pt + \phi)$. Denoting the velocity and position coordinate when $t = 0$ by v_0 and x_0, respectively, show (*a*) that $\tan \phi = x_0 p / v_0$, (*b*) that the maximum value of the position coordinate is

$$A = \sqrt{x_0^2 + \left(\frac{v_0}{p}\right)^2}$$

11.19 A more accurate expression for the acceleration due to gravity is

$$g = \frac{32.2}{\left(1 + \dfrac{y}{20.9 \times 10^6}\right)^2}$$

where g is expressed in ft/s^2 and y is the altitude in feet. Using this value of g, compute the height reached by a bullet fired vertically upward with the following initial velocities: (*a*) 1000 ft/s, (*b*) 10,000 ft/s, (*c*) 36,700 ft/s.

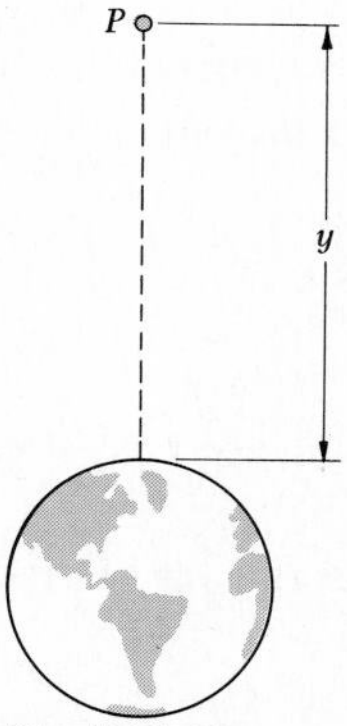

Fig. P11.19

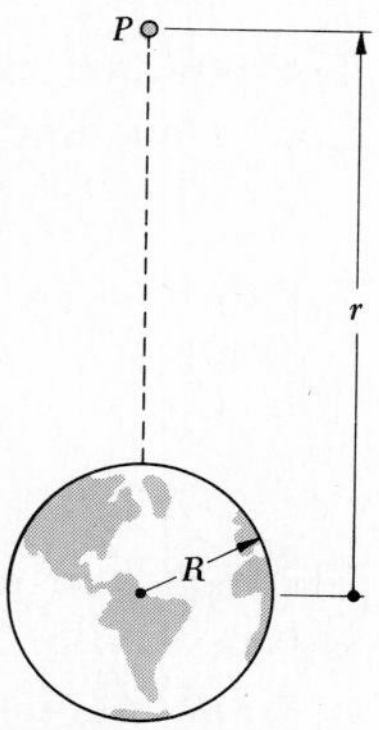

Fig. P11.20

11.20 The acceleration due to gravity of a particle falling toward the earth is $a = -gR^2/r^2$, where r is the distance from the *center* of the earth to the particle, R is the radius of the earth, and g is the acceleration due to gravity at the surface of the earth. Derive an expression for the *escape velocity*, i.e., for the minimum velocity with which a particle should be projected vertically upward from the surface of the earth if it is not to return to the earth. (*Hint.* $v = 0$ for $r = \infty$.)

$*$**11.21** A particle is released from rest at a distance R above the surface of the earth. Using the expression for the acceleration due to gravity given in Prob. 11.20, determine (*a*) the velocity of the particle as it strikes the earth, (*b*) the time required for it to fall to the surface of the earth. (Radius of earth $= R = 6370$ km, $g = 9.81\ \text{m/s}^2$.)

*11.22 When a package is dropped on a rigid surface, the acceleration of its cushioned contents may be defined by the relation $a = -k \tan(\pi x/2L)$, where L is the distance through which the cushioning material can be compressed. Denoting by v_0 the velocity when $x = 0$, show that $v^2 = v_0^2 + (4kL/\pi) \ln \cos(\pi x/2L)$. If $k = 300 \text{ m/s}^2$ and $L = 0.36$ m, compute the initial velocity v_0 for which the maximum value of the position coordinate x is (*a*) 0.18 m, (*b*) 0.36 m.

11.4. Uniform Rectilinear Motion. This is a type of straight-line motion which is frequently encountered in practical applications. In this motion, the acceleration a of the particle is zero for every value of t. The velocity v is therefore constant, and Eq. (11.1) becomes

$$\frac{dx}{dt} = v = \text{constant}$$

The position coordinate x is obtained by integrating this equation. Denoting by x_0 the initial value of x, we write

$$\int_{x_0}^{x} dx = v \int_0^t dt$$

$$x - x_0 = vt$$

$$x = x_0 + vt \qquad (11.5)$$

This equation may be used *only if the velocity of the particle is known to be constant.*

11.5. Uniformly Accelerated Rectilinear Motion. This is another common type of motion. In this motion, the acceleration a of the particle is constant, and Eq. (11.2) becomes

$$\frac{dv}{dt} = a = \text{constant}$$

The velocity v of the particle is obtained by integrating this equation,

$$\int_{v_0}^{v} dv = a \int_0^t dt$$

$$v - v_0 = at$$

$$v = v_0 + at \qquad (11.6)$$

where v_0 is the initial velocity. Substituting for v into (11.1), we write

$$\frac{dx}{dt} = v_0 + at$$

Denoting by x_0 the initial value of x and integrating, we have

$$\int_{x_0}^{x} dx = \int_0^t (v_0 + at)\, dt$$

$$x - x_0 = v_0 t + \tfrac{1}{2}at^2$$

$$x = x_0 + v_0 t + \tfrac{1}{2}at^2 \qquad (11.7)$$

We may also use Eq. (11.4) and write

$$v\frac{dv}{dx} = a = \text{constant}$$

$$v\, dv = a\, dx$$

Integrating both sides, we obtain

$$\int_{v_0}^{v} v\, dv = a\int_{x_0}^{x} dx$$

$$\tfrac{1}{2}(v^2 - v_0^2) = a(x - x_0)$$

$$v^2 = v_0^2 + 2a(x - x_0) \qquad (11.8)$$

The three equations we have derived provide useful relations among position coordinate, velocity, and time in the case of a uniformly accelerated motion, as soon as appropriate values have been substituted for a, v_0, and x_0. The origin O of the x axis should first be defined and a positive direction chosen along the axis; this direction will be used to determine the signs of a, v_0, and x_0. Equation (11.6) relates v and t and should be used when the value of v corresponding to a given value of t is desired, or inversely. Equation (11.7) relates x and t; Eq. (11.8) relates v and x. An important application of uniformly accelerated motion is the motion of a *freely falling body.* The acceleration of a freely falling body (usually denoted by g) is equal to 9.81 m/s^2 or 32.2 ft/s^2.

It is important to keep in mind that the three equations above may be used *only when the acceleration of the particle is known to be constant.* If the acceleration of the particle is variable, its motion should be determined from the fundamental equations (11.1) to (11.4), according to the methods outlined in Sec. 11.3.

11.6. Motion of Several Particles. When several particles move independently along the same line, independent equations of motion may be written for each particle. Whenever possible, time should be recorded from the same initial instant for all particles, and displacements should be measured from the same origin and in the same direction. In other words, a single clock and a single measuring tape should be used.

Relative Motion of Two Particles. Consider two particles A and B moving along the same straight line (Fig. 11.7). If the position coordinates x_A and x_B are measured from the same origin, the difference $x_B - x_A$ defines the *relative position coordinate of B with respect to A* and is denoted by $x_{B/A}$. We write

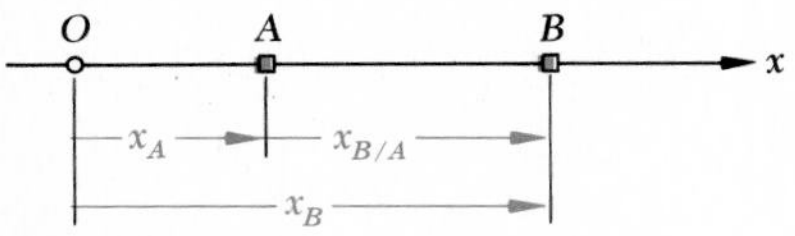

Fig. 11.7

$$x_{B/A} = x_B - x_A \qquad \text{or} \qquad x_B = x_A + x_{B/A} \tag{11.9}$$

A positive sign for $x_{B/A}$ means that B is to the right of A, a negative sign that B is to the left of A, regardless of the position of A and B with respect to the origin.

The rate of change of $x_{B/A}$ is known as the *relative velocity of B with respect to A* and is denoted by $v_{B/A}$. Differentiating (11.9), we write

$$v_{B/A} = v_B - v_A \qquad \text{or} \qquad v_B = v_A + v_{B/A} \tag{11.10}$$

A positive sign for $v_{B/A}$ means that B is *observed from A* to move in the positive direction; a negative sign, that it is observed to move in the negative direction.

The rate of change of $v_{B/A}$ is known as the *relative acceleration of B with respect to A* and is denoted by $a_{B/A}$. Differentiating (11.10), we obtain

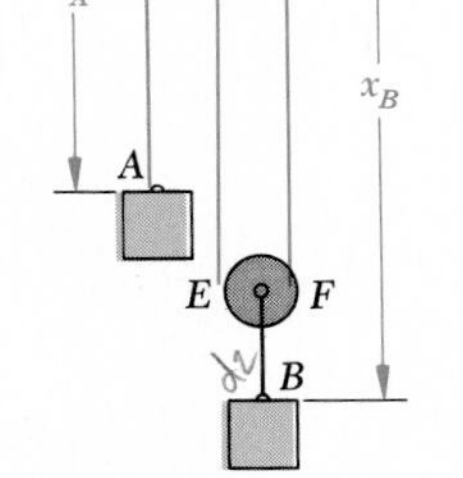

Fig. 11.8

$$a_{B/A} = a_B - a_A \qquad \text{or} \qquad a_B = a_A + a_{B/A} \tag{11.11}$$

Dependent Motions. Sometimes, the position of a particle will depend upon the position of another or of several other particles. The motions are then said to be dependent. For example, the position of block B in Fig. 11.8 depends upon the position of block A. Since the rope $ACDEFG$ is of constant length, and since the lengths of the portions of rope CD and EF wrapped around the pulleys remain constant, it follows that the sum of the lengths of the segments AC, DE, and FG is constant. Observing that the length of the segment AC differs from x_A only by a constant, and that, similarly, the lengths of the segments DE and FG differ from x_B only by a constant, we write

$$x_A + 2x_B = \text{constant}$$

Since only one of the two coordinates x_A and x_B may be chosen arbitrarily, we say that the system shown in Fig. 11.8 has *one degree of freedom*. From the relation between the position coordinates x_A and x_B, it follows that if x_A is given an increment Δx_A, i.e., if block A is lowered by an amount Δx_A, the coordinate x_B will receive an increment $\Delta x_B = -\frac{1}{2}\Delta x_A$, that is, block B will rise by half the same amount; this may easily be checked directly from Fig. 11.8.

In the case of the three blocks of Fig. 11.9, we may again observe that the length of the rope which passes over the pulleys is constant, and thus that the following relation must be satisfied by the position coordinates of the three blocks:

$$2x_A + 2x_B + x_C = \text{constant}$$

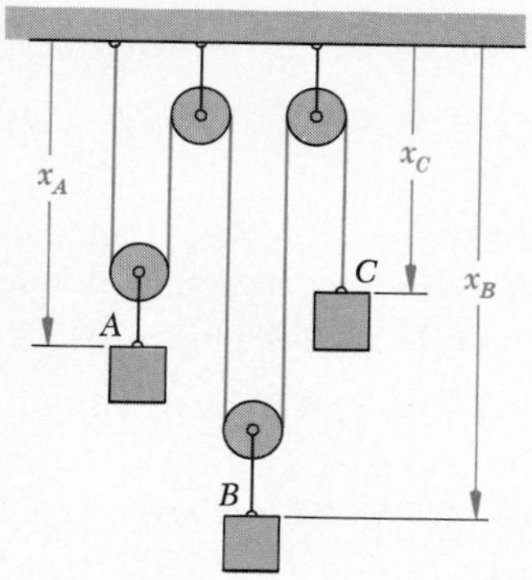

Fig. 11.9

Since two of the coordinates may be chosen arbitrarily, we say that the system shown in Fig. 11.9 has *two degrees of freedom*.

When the relation existing between the position coordinates of several particles is *linear*, a similar relation holds between the velocities and between the accelerations of the particles. In the case of the blocks of Fig. 11.9, for instance, we differentiate twice the equation obtained and write

$$2\frac{dx_A}{dt} + 2\frac{dx_B}{dt} + \frac{dx_C}{dt} = 0 \quad \text{or} \quad 2v_A + 2v_B + v_C = 0$$

$$2\frac{dv_A}{dt} + 2\frac{dv_B}{dt} + \frac{dv_C}{dt} = 0 \quad \text{or} \quad 2a_A + 2a_B + a_C = 0$$

SAMPLE PROBLEM 11.4

A ball is thrown vertically upward from the 40-ft level in an elevator shaft, with an initial velocity of 50 ft/s. At the same instant an open-platform elevator passes the 10-ft level, moving upward with a constant velocity of 5 ft/s. Determine (*a*) when and where the ball will hit the elevator, (*b*) the relative velocity of the ball with respect to the elevator when the ball hits the elevator.

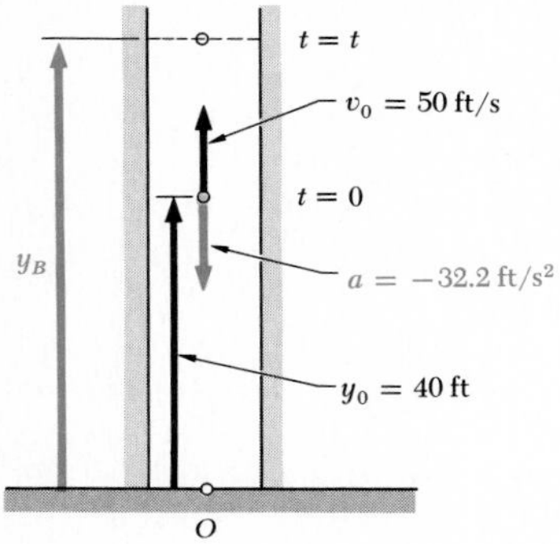

Motion of Ball. Since the ball has a constant acceleration, its motion is *uniformly accelerated.* Placing the origin O of the y axis at ground level and choosing its positive direction upward, we find that the initial position is $y_0 = +40$ ft, the initial velocity is $v_0 = +50$ ft/s, and the acceleration is $a = -32.2$ ft/s^2. Substituting these values in the equations for uniformly accelerated motion, we write

$$v_B = v_0 + at \qquad v_B = 50 - 32.2t \qquad (1)$$
$$y_B = y_0 + v_0 t + \tfrac{1}{2}at^2 \qquad y_B = 40 + 50t - 16.1t^2 \qquad (2)$$

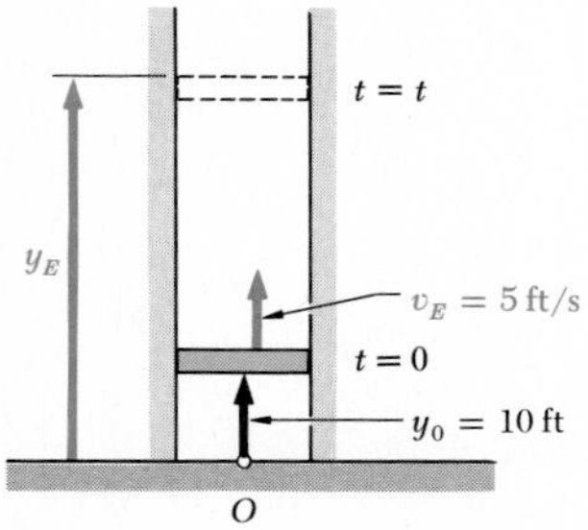

Motion of Elevator. Since the elevator has a constant velocity, its motion is *uniform.* Again placing the origin O at the ground level and choosing the positive direction upward, we note that $y_0 = +10$ ft and write

$$v_E = +5 \text{ ft/s} \qquad (3)$$
$$y_E = y_0 + v_E t \qquad y_E = 10 + 5t \qquad (4)$$

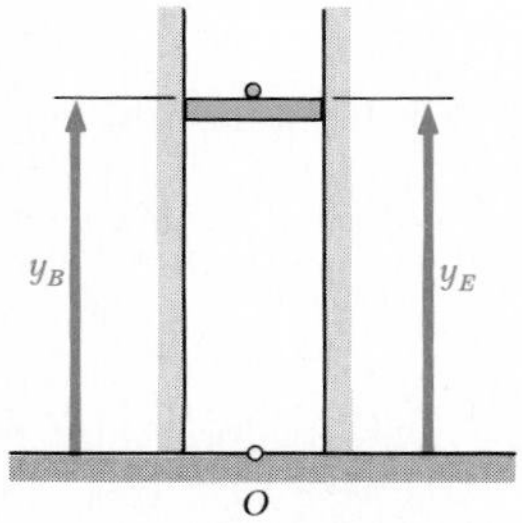

Ball Hits Elevator. We first note that the same time t and the same origin O were used in writing the equations of motion of both the ball and the elevator. We see from the figure that, when the ball hits the elevator,

$$y_E = y_B \qquad (5)$$

Substituting for y_E and y_B from (2) and (4) into (5), we have

$$10 + 5t = 40 + 50t - 16.1t^2$$
$$t = -0.56 \text{ s} \quad \text{and} \quad t = +3.35 \text{ s} \quad ◄$$

Only the root $t = 3.35$ s corresponds to a time after the motion has begun. Substituting this value into (4), we have

$$y_E = 10 + 5(3.35) = 26.7 \text{ ft}$$

Elevation from ground = 26.7 ft ◄

The relative velocity of the ball with respect to the elevator is

$$v_{B/E} = v_B - v_E = (50 - 32.2t) - 5 = 45 - 32.2t$$

When the ball hits the elevator at time $t = 3.35$ s, we have

$$v_{B/E} = 45 - 32.2(3.35) \qquad v_{B/E} = -62.9 \text{ ft/s} \quad ◄$$

The negative sign means that the ball is observed from the elevator to be moving in the negative sense (downward).

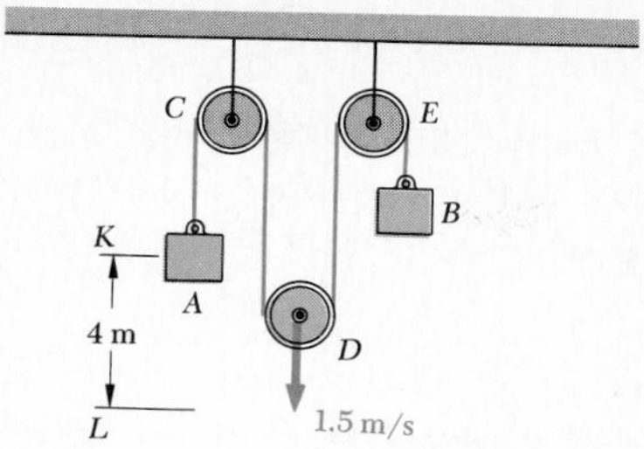

SAMPLE PROBLEM 11.5

Two blocks A and B are connected by a cord passing over three pulleys C, D, and E as shown. Pulleys C and E are fixed, while D is pulled downward with a constant velocity of 1.5 m/s. At $t = 0$, block A starts moving downward from the position K with a constant acceleration and no initial velocity. Knowing that the velocity of block A is 6 m/s as it passes through point L, determine the change in elevation, the velocity, and the acceleration of block B when A passes through L.

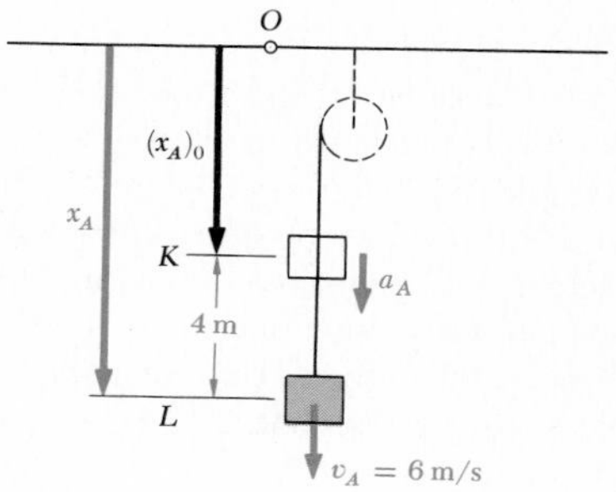

Motion of Block A. We place the origin O at the horizontal surface and choose the positive direction downward. We observe that when $t = 0$, block A is at position K and $(v_A)_0 = 0$. Since $v_A = 6$ m/s and $x_A - (x_A)_0 = 4$ m when the block passes through L, we write

$$v_A^2 = (v_A)_0^2 + 2a_A[x_A - (x_A)_0] \qquad (6)^2 = 0 + 2a_A(4)$$
$$a_A = 4.50 \text{ m/s}^2$$

The time at which block A reaches point L is obtained by writing

$$v_A = (v_A)_0 + a_A t \qquad 6 = 0 + 4.50t \qquad t = 1.333 \text{ s}$$

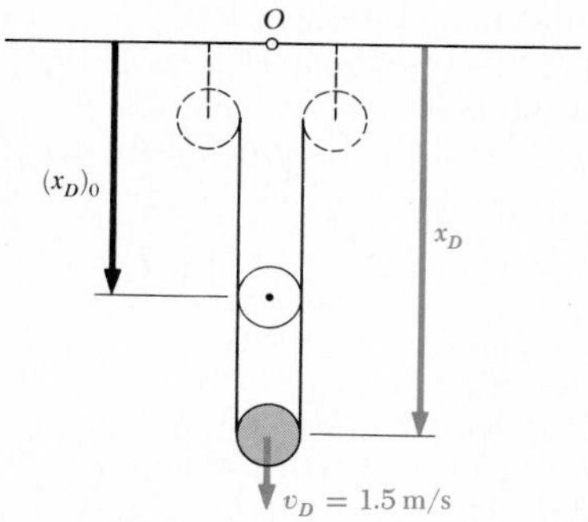

Motion of Pulley D. Recalling that the positive direction is downward, we write

$$a_D = 0 \qquad v_D = 1.5 \text{ m/s} \qquad x_D = (x_D)_0 + v_D t = (x_D)_0 + 1.5t$$

When block A reaches L, at $t = 1.333$ s, we have

$$x_D = (x_D)_0 + 1.5(1.333) = (x_D)_0 + 2$$

Thus,

$$x_D - (x_D)_0 = 2 \text{ m}$$

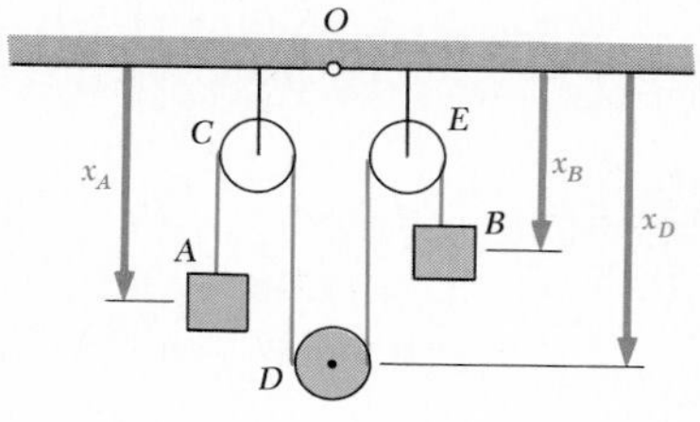

Motion of Block B. We note that the total length of cord $ACDEB$ differs from the quantity $(x_A + 2x_D + x_B)$ only by a constant. Since the cord length is constant during the motion, this quantity must also remain constant. Thus considering the times $t = 0$ and $t = 1.333$ s, we write

$$x_A + 2x_D + x_B = (x_A)_0 + 2(x_D)_0 + (x_B)_0 \qquad (1)$$
$$[x_A - (x_A)_0] + 2[x_D - (x_D)_0] + [x_B - (x_B)_0] = 0 \qquad (2)$$

But we know that $x_A - (x_A)_0 = 4$ m and $x_D - (x_D)_0 = 2$ m; substituting these values in (2), we find

$$4 + 2(2) + [x_B - (x_B)_0] = 0 \qquad x_B - (x_B)_0 = -8 \text{ m}$$

Thus: Change in elevation of $B = 8 \text{ m} \uparrow$ ◄

Differentiating (1) twice, we obtain equations relating the velocities and the accelerations of A, B, and D. Substituting for the velocities and accelerations of A and D at $t = 1.333$ s, we have

$$v_A + 2v_D + v_B = 0: \qquad 6 + 2(1.5) + v_B = 0$$
$$v_B = -9 \text{ m/s} \qquad v_B = 9 \text{ m/s} \uparrow \;◄$$

$$a_A + 2a_D + a_B = 0: \qquad 4.50 + 2(0) + a_B = 0$$
$$a_B = -4.50 \text{ m/s}^2 \qquad a_B = 4.50 \text{ m/s}^2 \uparrow \;◄$$

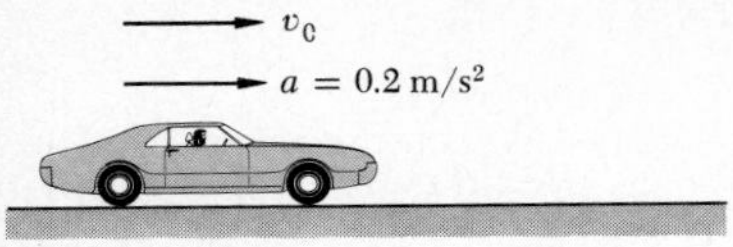

Fig. P11.23

PROBLEMS

11.23 An automobile travels 240 m in 30 s while being accelerated at a constant rate of $0.2\ \mathrm{m/s^2}$. Determine (*a*) its initial velocity, (*b*) its final velocity, (*c*) the distance traveled during the first 10 s.

11.24 A stone is released from an elevator moving up at a speed of 5 m/s and reaches the bottom of the shaft in 3 s. (*a*) How high was the elevator when the stone was released? (*b*) With what speed does the stone strike the bottom of the shaft?

11.25 A man jumps from a 20-ft cliff with no initial velocity. (*a*) How long does it take him to reach the ground, and with what velocity does he hit the ground? (*b*) If this takes place on the moon, where $g = 5.31\ \mathrm{ft/s^2}$, what are the values obtained for the time and velocity? (*c*) If a motion picture is taken on the earth, but if the scene is supposed to take place on the moon, how many frames per second should be used so that the scene would appear realistic when projected at the standard speed of 24 frames per second?

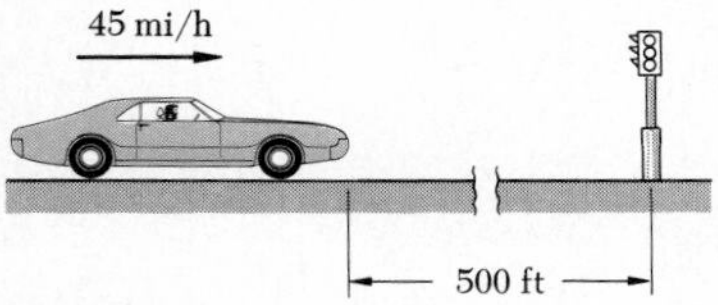

Fig. P11.26

11.26 A motorist is traveling at 45 mi/h when he observes that a traffic light 500 ft ahead of him turns red. The traffic light is timed to stay red for 10 s. If the motorist wishes to pass the light without stopping just as it turns green again, determine (*a*) the required uniform deceleration of the car, (*b*) the speed of the car as it passes the light.

11.27 Automobile *A* starts from *O* and accelerates at the constant rate of $4\ \mathrm{ft/s^2}$. A short time later it is passed by truck *B* which is traveling in the opposite direction at a constant speed of 45 ft/s. Knowing that truck *B* passes point *O*, 25 s after automobile *A* started from there, determine when and where the vehicles passed each other.

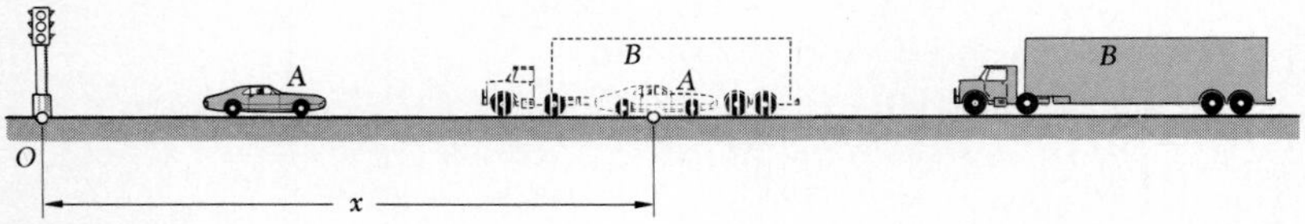

Fig. P11.27

11.28 An open-platform elevator is moving down a mine shaft at a constant velocity v_e when the elevator platform hits and dislodges a stone. Assuming that the stone starts falling with no initial velocity, (*a*) show that the stone will hit the platform with a relative velocity of magnitude v_e. (*b*) If $v_e = 20\ \mathrm{ft/s}$, determine when and where the stone will hit the elevator platform.

11.29 Two automobiles A and B are traveling in the same direction in adjacent highway lanes. Automobile B is stopped when it is passed by A, which travels at a constant speed of 36 km/h. Two seconds later automobile B starts and accelerates at a constant rate of 1.5 m/s². Determine (*a*) when and where B will overtake A, (*b*) the speed of B at that time.

11.30 Drops of water are observed to drip from a faucet at uniform intervals of time. As any drop B begins to fall freely, the preceding drop A has already fallen 0.3 m. Determine the distance drop A will have fallen by the time the distance between A and B will have increased to 0.9 m.

11.31 The elevator shown in the figure moves downward at the constant velocity of 15 ft/s. Determine (*a*) the velocity of the cable C, (*b*) the velocity of the counterweight W, (*c*) the relative velocity of the cable C with respect to the elevator, (*d*) the relative velocity of the counterweight W with respect to the elevator.

11.32 The elevator shown starts from rest and moves upward with a constant acceleration. If the counterweight W moves through 24 ft in 4 s, determine (*a*) the accelerations of the elevator and the cable C, (*b*) the velocity of the elevator after 4 s.

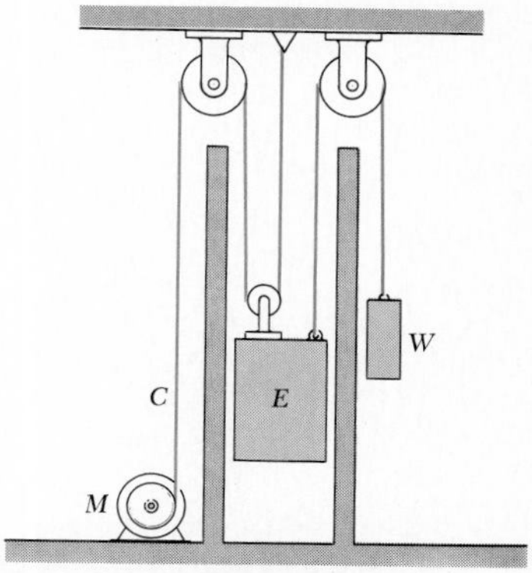

Fig. P11.31 and P11.32

11.33 The slider block A moves to the left at a constant velocity of 300 mm/s. Determine (*a*) the velocity of block B, (*b*) the velocities of portions C and D of the cable, (*c*) the relative velocity of A with respect to B, (*d*) the relative velocity of portion C of the cable with respect to portion D.

11.34 The slider block B starts from rest and moves to the right with a constant acceleration. After 4 s the relative velocity of A with respect to B is 60 mm/s. Determine (*a*) the accelerations of A and B, (*b*) the velocity and position of B after 3 s.

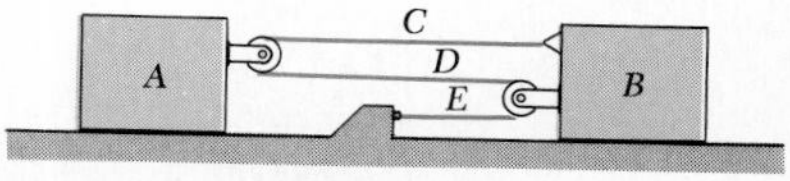

Fig. P11.33 and P11.34

11.35 Blocks A and C start from rest and move to the right with the following accelerations: $a_A = 12t$ ft/s² and $a_C = 3$ ft/s². Determine (*a*) the time at which the velocity of block B is zero, (*b*) the corresponding position of B.

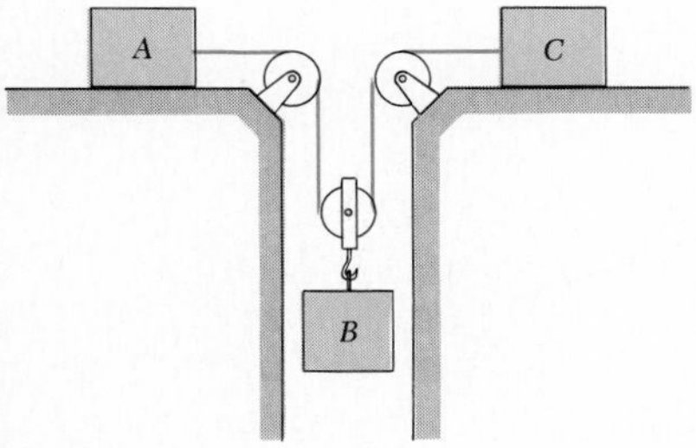

Fig. P11.35

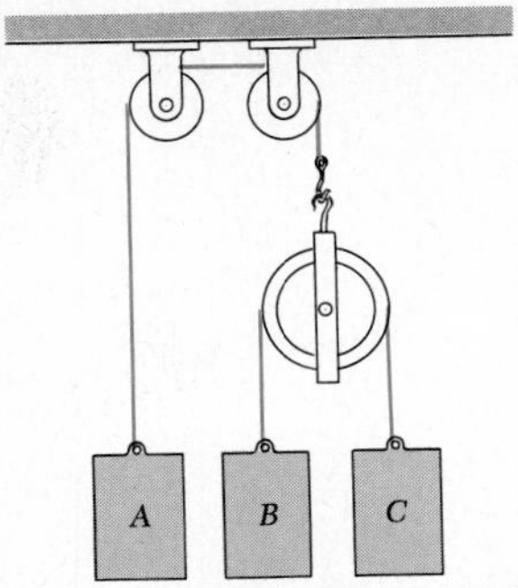

Fig. P11.36, P11.37, and P11.39

11.36 (*a*) Choosing the positive sense downward for each block, express the velocity of *A* in terms of the velocities of *B* and *C*. (*b*) Knowing that both blocks *A* and *C* start from rest and move downward with the respective accelerations $a_A = 2\text{ in./s}^2$ and $a_C = 5\text{ in./s}^2$, determine the position and velocity of *B* after 3 s.

11.37 The three blocks shown move with constant velocities. Find the velocity of each block, knowing that the relative velocity of *A* with respect to *C* is 200 mm/s upward and that the relative velocity of *B* with respect to *A* is 120 mm/s downward.

11.38 The three blocks of Fig. 11.9 move with constant velocities. Find the velocity of each block, knowing that *C* is observed from *B* to move downward with a relative velocity of 180 mm/s and *A* is observed from *B* to move downward with a relative velocity of 160 mm/s.

***11.39** The three blocks shown are equally spaced horizontally and move vertically with constant velocities. Knowing that initially they are at the same level and that the relative velocity of *C* with respect to *B* is 160 mm/s downward, determine the velocity of each block so that the three blocks will remain aligned during their motion.

***11.7. Graphical Solution of Rectilinear-Motion Problems.** It was observed in Sec. 11.2 that the fundamental formulas

$$v = \frac{dx}{dt} \qquad \text{and} \qquad a = \frac{dv}{dt}$$

have a geometrical significance. The first formula expresses that the velocity at any instant is equal to the slope of the x–t curve at the same instant (Fig. 11.10). The second formula expresses that the acceleration is equal to the slope of the v–t curve. These

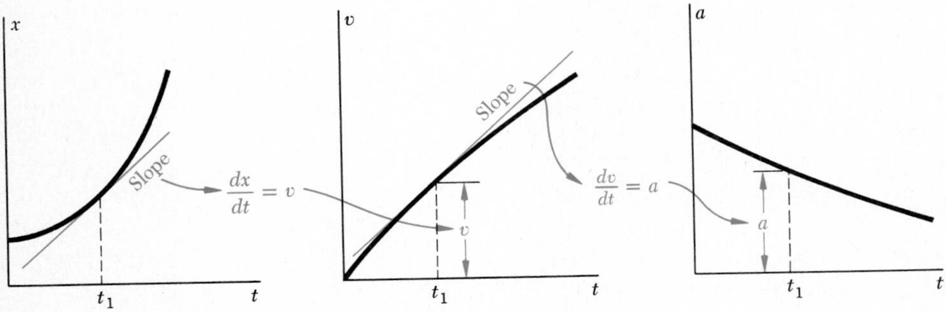

Fig. 11.10

two properties may be used to derive graphically the v–t and a–t curves of a motion when the x–t curve is known.

Integrating the two fundamental formulas from a time t_1 to a time t_2, we write

$$x_2 - x_1 = \int_{t_1}^{t_2} v\,dt \qquad \text{and} \qquad v_2 - v_1 = \int_{t_1}^{t_2} a\,dt \qquad (11.12)$$

The first formula expresses that the area measured under the v–t curve from t_1 to t_2 is equal to the change in x during that time interval (Fig. 11.11). The second formula expresses similarly that the area measured under the a–t curve from t_1 to t_2 is equal to the change in v during that time interval. These two properties may be used to determine graphically the x–t curve of a motion when its v–t curve or its a–t curve is known (see Sample Prob. 11.6).

Graphical solutions are particularly useful when the motion considered is defined from experimental data and when x, v, and a are not analytical functions of t. They may also be used to advantage when the motion consists of distinct parts and when its analysis requires writing a different equation for each of its parts. When using a graphical solution, however, one should be careful to note (1) that the area under the v–t curve measures the *change in* x, not x itself, and, similarly, that the area under the a–t curve measures the change in v; (2) that, while an area above the t axis corresponds to an increase in x or v, an area located below the t axis measures a decrease in x or v.

It will be useful to remember, in drawing motion curves, that, if the velocity is constant, it will be represented by a horizontal straight line; the position coordinate x will then be a linear function of t and will be represented by an oblique straight line. If the acceleration is constant and different from zero, it will be represented by a horizontal straight line; v will then be a linear function of t, represented by an oblique straight line; and x will be expressed as a second-degree polynomial in t, represented by a parabola. If the acceleration is a linear function of t, the velocity and the position coordinate will be equal, respectively, to second-degree and third-degree polynomials; a is then represented by an oblique straight line, v by a parabola, and x by a cubic. In general, if the acceleration is a polynomial of degree n in t, the velocity will be a polynomial of degree $n + 1$ and the position coordinate a polynomial of degree $n + 2$; these polynomials are represented by motion curves of a corresponding degree.

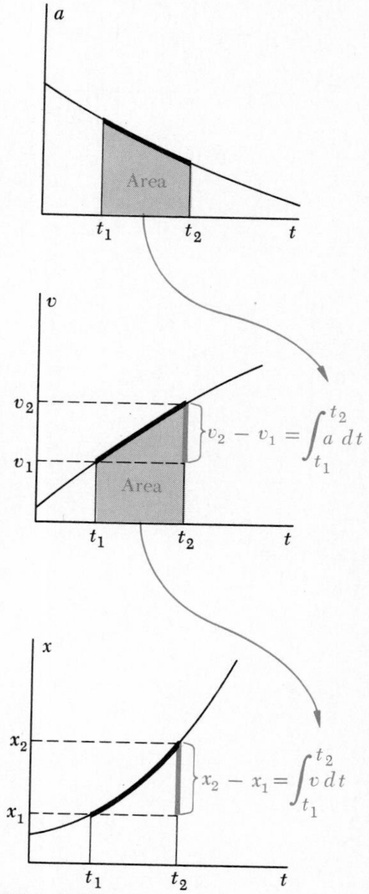

Fig. 11.11

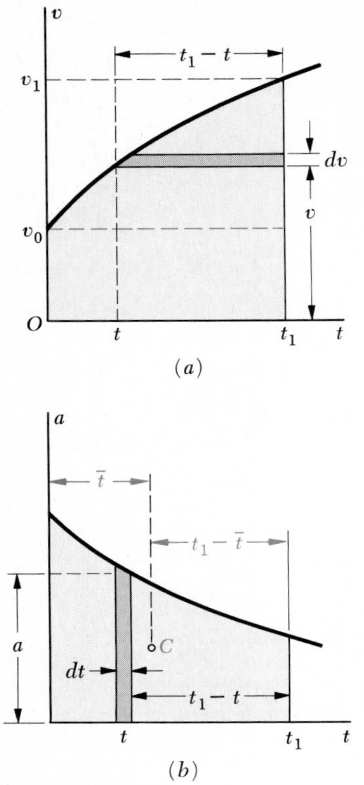

Fig. 11.12

*11.8. Other Graphical Methods.

An alternate graphical solution may be used to determine directly from the a–t curve the position of a particle at a given instant. Denoting respectively by x_0 and v_0 the values of x and v at $t = 0$, by x_1 and v_1 their values at $t = t_1$, and observing that the area under the v–t curve may be divided into a rectangle of area $v_0 t_1$ and horizontal differential elements of area $(t_1 - t)\,dv$ (Fig. 11.12a), we write

$$x_1 - x_0 = \text{area under } v\text{–}t \text{ curve} = v_0 t_1 + \int_{v_0}^{v_1} (t_1 - t)\,dv$$

Substituting $dv = a\,dt$ in the integral, we obtain

$$x_1 - x_0 = v_0 t_1 + \int_0^{t_1} (t_1 - t) a\,dt$$

Referring to Fig. 11.12b, we note that the integral represents the first moment of the area under the a–t curve with respect to the line $t = t_1$ bounding the area on the right. This method of solution is known, therefore, as the *moment-area method*. If the abscissa $\bar{t}$ of the centroid C of the area is known, the position coordinate x_1 may be obtained by writing

$$x_1 = x_0 + v_0 t_1 + (\text{area under } a\text{–}t \text{ curve})(t_1 - \bar{t}) \qquad (11.13)$$

If the area under the a–t curve is a composite area, the last term in (11.13) may be obtained by multiplying each component area by the distance from its centroid to the line $t = t_1$. Areas above the t axis should be considered as positive and areas below the t axis as negative.

Another type of motion curve, the v–x curve, is sometimes used. If such a curve has been plotted (Fig. 11.13), the acceleration a may be obtained at any time by drawing the normal to the curve and *measuring the subnormal BC*. Indeed, observing that the angle between AC and AB is equal to the angle θ between the horizontal and the tangent at A (the slope of which is $\tan\theta = dv/dx$), we write

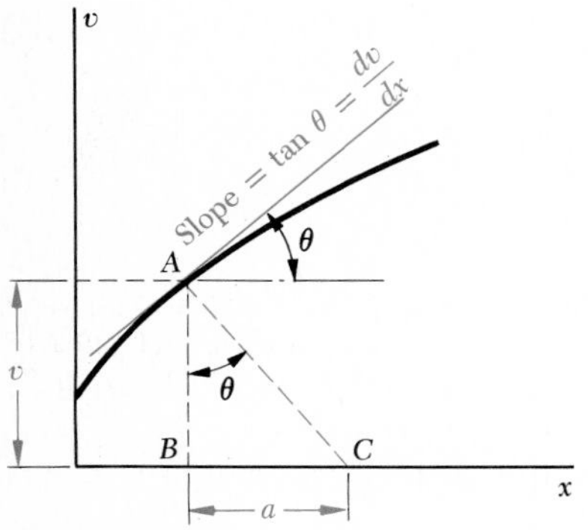

Fig. 11.13

$$BC = AB \tan\theta = v\,\frac{dv}{dx}$$

and thus, recalling formula (11.4),

$$BC = a$$

SAMPLE PROBLEM 11.6

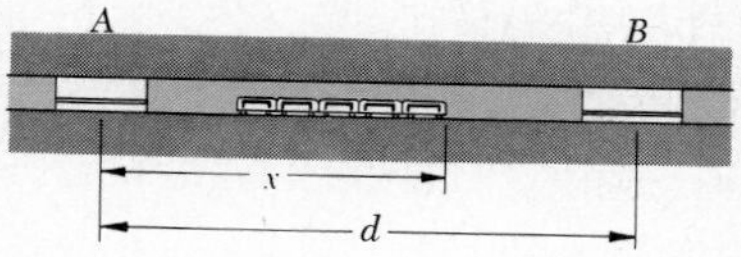

A subway train leaves station A; it gains speed at the rate of 4 ft/s^2 for 6 s, and then at the rate of 6 ft/s^2 until it has reached the speed of 48 ft/s. The train maintains the same speed until it approaches station B; brakes are then applied, giving the train a constant deceleration and bringing it to a stop in 6 s. The total running time from A to B is 40 s. Draw the a–t, v–t, and x–t curves, and determine the distance between stations A and B.

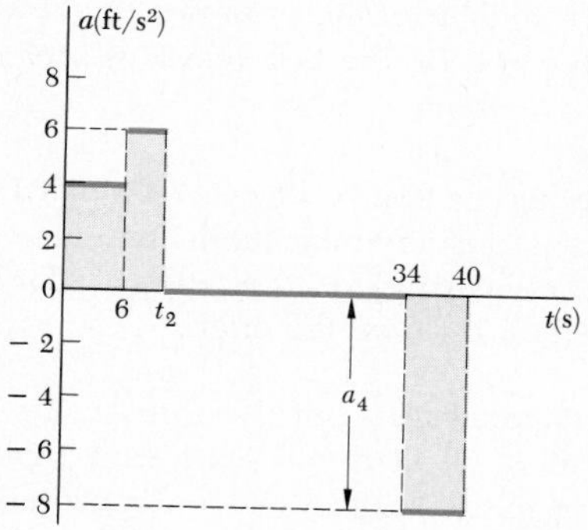

Acceleration-Time Curve. Since the acceleration is either constant or zero, the a–t curve is made of horizontal straight-line segments. The values of t_2 and a_4 are determined as follows:

$0 < t < 6$: Change in v = area under a–t curve

$$v_6 - 0 = (6\text{ s})(4\text{ ft/s}^2) = 24\text{ ft/s}$$

$6 < t < t_2$: Since the velocity increases from 24 to 48 ft/s,

Change in v = area under a–t curve

$$48 - 24 = (t_2 - 6)(6\text{ ft/s}^2) \qquad t_2 = 10\text{ s}$$

$t_2 < t < 34$: Since the velocity is constant, the acceleration is zero.

$34 < t < 40$: Change in v = area under a–t curve

$$0 - 48 = (6\text{ s})a_4 \qquad a_4 = -8\text{ ft/s}^2$$

The acceleration being negative, the corresponding area is below the t axis; this area represents a decrease in velocity.

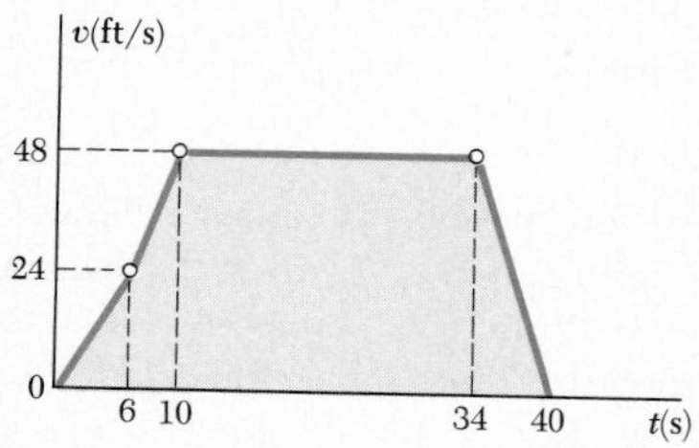

Velocity-Time Curve. Since the acceleration is either constant or zero, the v–t curve is made of segments of straight line connecting the points determined above.

Change in x = area under v–t curve

$0 < t < 6$: $\quad x_6 - 0 = \frac{1}{2}(6)(24) = 72\text{ ft}$

$6 < t < 10$: $\quad x_{10} - x_6 = \frac{1}{2}(4)(24 + 48) = 144\text{ ft}$

$10 < t < 34$: $\quad x_{34} - x_{10} = (24)(48) = 1152\text{ ft}$

$34 < t < 40$: $\quad x_{40} - x_{34} = \frac{1}{2}(6)(48) = 144\text{ ft}$

Adding the changes in x, we obtain the distance from A to B:

$$d = x_{40} - 0 = 1512\text{ ft}$$

$d = 1512\text{ ft}$ ◄

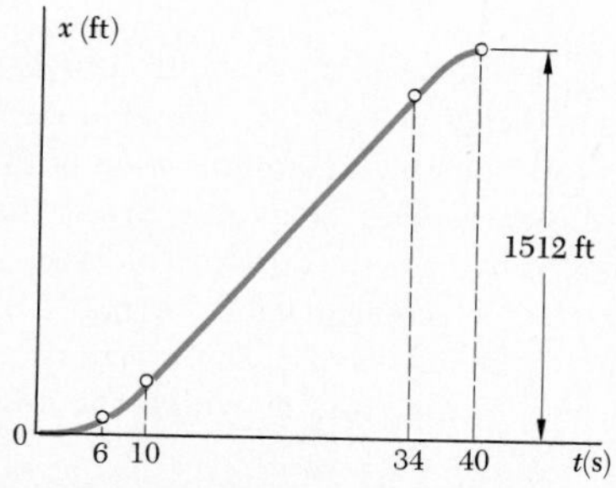

Position-Time Curve. The points determined above should be joined by three arcs of parabola and one segment of straight line. The construction of the x–t curve will be performed more easily and more accurately if we keep in mind that for any value of t the slope of the tangent to the x–t curve is equal to the value of v at that instant.

PROBLEMS

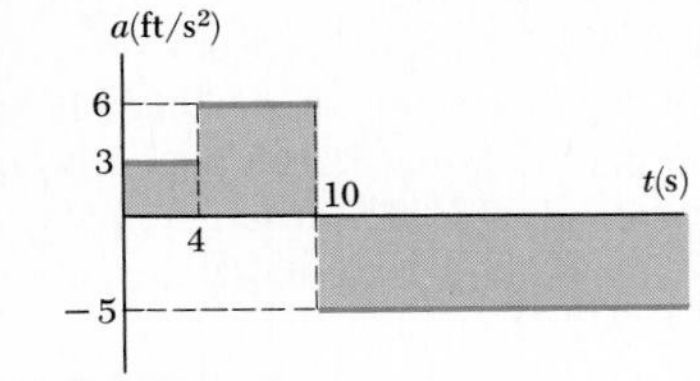

Fig. P11.40

11.40 A particle moves in a straight line with the acceleration shown in the figure. Knowing that it starts from the origin with $v_0 = -18$ ft/s, (*a*) plot the v–t and x–t curves for $0 < t < 20$ s, (*b*) determine its velocity, its position, and the total distance traveled after 12 s.

11.41 For the particle and motion of Prob. 11.40, plot the v–t and x–t curves for $0 < t < 20$ s and determine (*a*) the maximum value of the velocity of the particle, (*b*) the maximum value of its position coordinate.

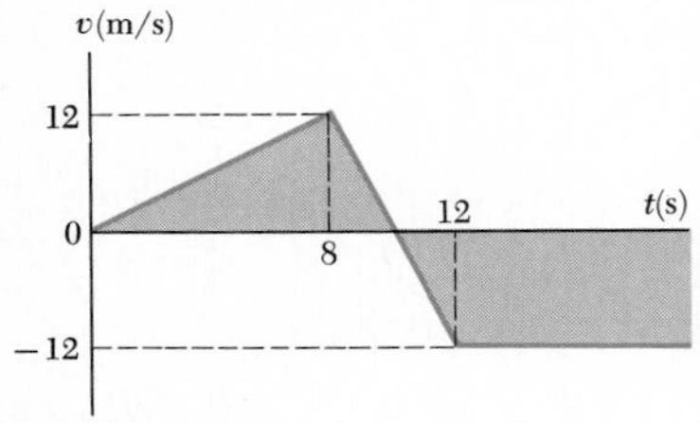

Fig. P11.42

11.42 A particle moves in a straight line with the velocity shown in the figure. Knowing that $x = -12$ m at $t = 0$, draw the a–t and x–t curves for $0 < t < 16$ s and determine (*a*) the total distance traveled by the particle after 12 s, (*b*) the two values of t for which the particle passes through the origin.

11.43 For the particle and motion of Prob. 11.42, plot the a–t and x–t curves for $0 < t < 16$ s and determine (*a*) the maximum value of the position coordinate of the particle, (*b*) the values of t for which the particle is at a distance of 15 m from the origin.

11.44 A series of city traffic signals is timed so that an automobile traveling at a constant speed of 30 mi/h will reach each signal just as it turns green. A motorist misses a signal and is stopped at signal *A*. Knowing that the next signal *B* is 800 ft ahead and that the maximum acceleration of his automobile is 5 ft/s², determine what the motorist should do to keep his maximum speed as small as possible, yet reach signal *B* just as it turns green. What is the maximum speed reached?

11.45 A bus starts from rest at point *A* and accelerates at the rate of 0.9 m/s² until it reaches a speed of 7.2 m/s. It then proceeds at 7.2 m/s until the brakes are applied; it comes to rest at point *B*, 18 m beyond the point where the brakes were applied. Assuming uniform deceleration and knowing that the distance between *A* and *B* is 90 m, determine the time required for the bus to travel from *A* to *B*.

11.46 The firing of a howitzer causes the barrel to recoil 800 mm before a braking mechanism brings it to rest. From a high-speed photographic record, it is found that the maximum value of the recoil velocity is 5.4 m/s and that this is reached 0.02 s after firing. Assuming that the recoil period consists of two phases during which the acceleration has, respectively, a constant positive value a_1 and a constant negative value a_2, determine (*a*) the values of a_1 and a_2, (*b*) the position of the barrel 0.02 s after firing, (*c*) the time at which the velocity of the barrel is zero.

11.47 A motorist is traveling at 60 mi/h when he observes that a traffic signal 1000 ft ahead of him turns red. He knows that the signal is timed to stay red for 20 s. What should he do to pass the signal at 60 mi/h just as it turns green again? Draw the v–t curve, selecting the solution which calls for the smallest possible deceleration and acceleration, and determine (*a*) the common value of the deceleration and acceleration in ft/s^2, (*b*) the minimum speed reached in mi/h.

11.48 A policeman on a motorcycle is escorting a motorcade which is traveling at 54 km/h. The policeman suddenly decides to take a new position in the motorcade, 70 m ahead. Assuming that he accelerates and decelerates at the rate of 2.5 m/s^2 and that he does not exceed at any time a speed of 72 km/h, draw the a–t and v–t curves for his motion and determine (*a*) the shortest time in which he can occupy his new position in the motorcade, (*b*) the distance he will travel in that time.

11.49 A freight elevator moving upward with a constant velocity of 5 m/s passes a passenger elevator which is stopped. Three seconds later, the passenger elevator starts upward with an acceleration of 1.25 m/s^2. When the passenger elevator has reached a velocity of 10 m/s, it proceeds at constant speed. Draw the v–t and y–t curves, and from them determine the time and distance required by the passenger elevator to overtake the freight elevator.

11.50 A car and a truck are both traveling at the constant speed of 45 mi/h; the car is 40 ft behind the truck. The driver of the car wants to pass the truck, i.e., he wishes to place his car at B, 40 ft in front of the truck, and then resume the speed of 45 mi/h. The maximum acceleration of the car is 5 ft/s^2 and the maximum deceleration obtained by applying the brakes is 20 ft/s^2. What is the shortest time in which the driver of the car can complete the passing operation if he does not at any time exceed a speed of 60 mi/h? Draw the v–t curve.

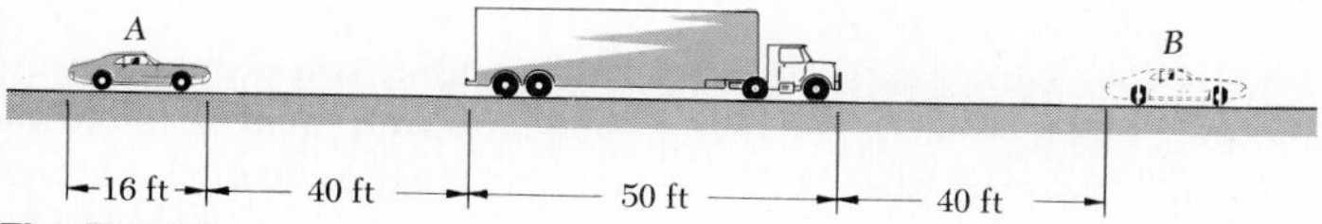

Fig. P11.50

11.51 Car A is traveling at the constant speed v_A. It approaches car B which is traveling in the same direction at the constant speed of 40 mi/h. The driver of car B notices car A when it is still 150 ft behind him. He then accelerates at the constant rate of 2 ft/s^2 to avoid being passed by A. Knowing that the closest that A comes to B is 50 ft, determine the speed of car A.

11.52 A car and a truck are both traveling at the constant speed of 60 mi/h; the car is 30 ft behind the truck. The truck driver suddenly applies his brakes, causing the truck to decelerate at the constant rate of 9 ft/s^2. Two seconds later the driver of the car applies his brakes and just manages to avoid a rear-end collision. Determine the constant rate at which the car decelerated.

11.53 Two cars are traveling toward each other on a single-lane road at 16 and 12 m/s, respectively. When 120 m apart, both drivers realize the situation and apply their brakes. They succeed in stopping simultaneously, and just short of colliding. Assuming a constant deceleration for each car, determine (*a*) the time required for the cars to stop, (*b*) the deceleration of each car, and (*c*) the distance traveled by each car while slowing down.

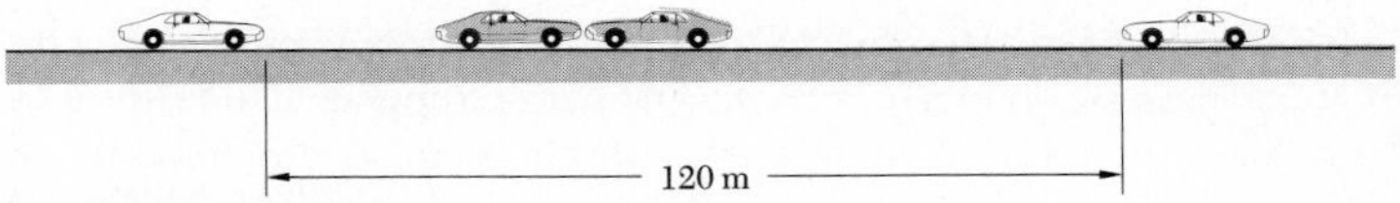

Fig. P11.53

11.54 An express subway train and a train making local stops run on parallel tracks between stations *A* and *E*, which are 1400 m apart. The local train makes stops of 30-s duration at each of the stations *B*, *C*, and *D*; the express train proceeds to station *E* without any intermediate stop. Each train accelerates at a rate of 1.25 m/s^2 until it reaches a speed of 12.5 m/s; it then proceeds at that constant speed. As the train approaches its next stop, the brakes are applied, providing a constant deceleration of 1.5 m/s^2. If the express train leaves station *A* 4 min after the local train has left *A*, determine (*a*) which of the two trains will arrive at station *E* first, (*b*) how much later the other train will arrive at station *E*.

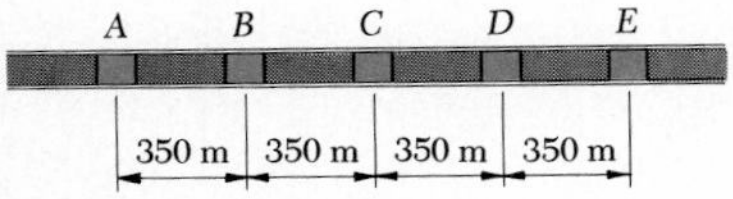

Fig. P11.54

11.55 An automobile at rest is passed by a truck traveling at a constant speed of 45 mi/h. The automobile starts and accelerates for 10 s at a constant rate until it reaches a speed of 60 mi/h. If the automobile then maintains a constant speed of 60 mi/h, determine when it will overtake the truck, assuming the automobile starts (*a*) just as the truck passes it, (*b*) 2 s after the truck has passed it.

11.56 The rate of change of acceleration is known as the *jerk;* large or abrupt rates of change of acceleration cause discomfort to elevator passengers. If the jerk, or rate of change of the acceleration, of an elevator is limited to ± 0.5 m/s^2 per second, determine the shortest time required for an elevator, starting from rest, to rise 8 m and stop.

11.57 The acceleration record shown was obtained for a truck traveling on a straight highway. Knowing that the initial velocity of the truck was 18 km/h, determine (*a*) the velocity at $t = 6$ s, (*b*) the distance the truck travels during the 6-s test.

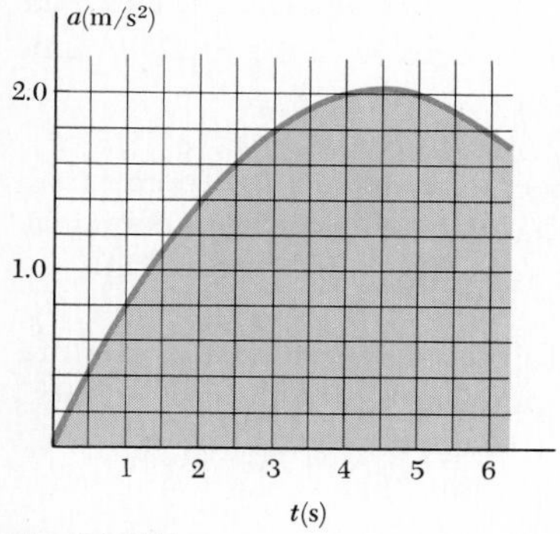

Fig. P11.57

11.58 A training airplane lands on an aircraft carrier and is brought to rest in 3.0 s by the arresting gear of the carrier. An accelerometer attached to the airplane provides the acceleration record shown. Determine by approximate means (*a*) the initial velocity of the airplane relative to the deck, (*b*) the distance the airplane travels along the deck before coming to rest.

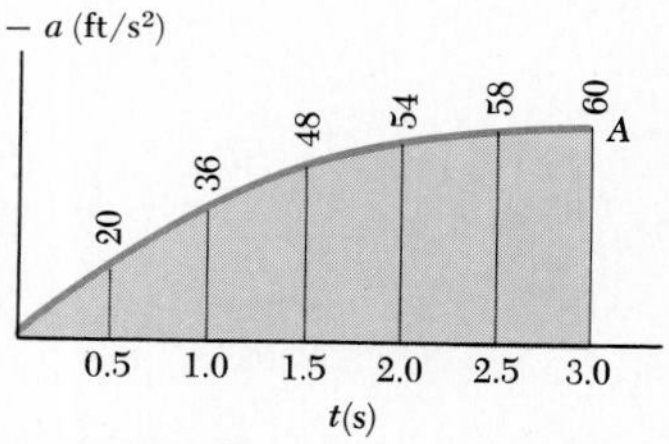

Fig. P11.58

11.59 The v–x curve shown was obtained experimentally during the motion of the bed of an industrial planer. Determine by approximate means the acceleration (*a*) when $x = 3$ in., (*b*) when $v = 40$ in./s.

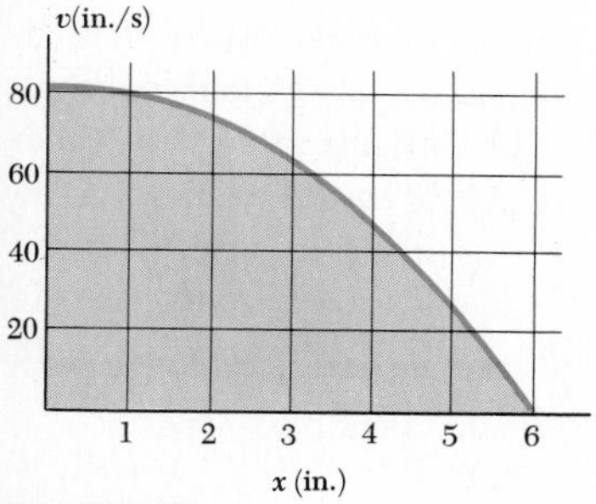

Fig. P11.59

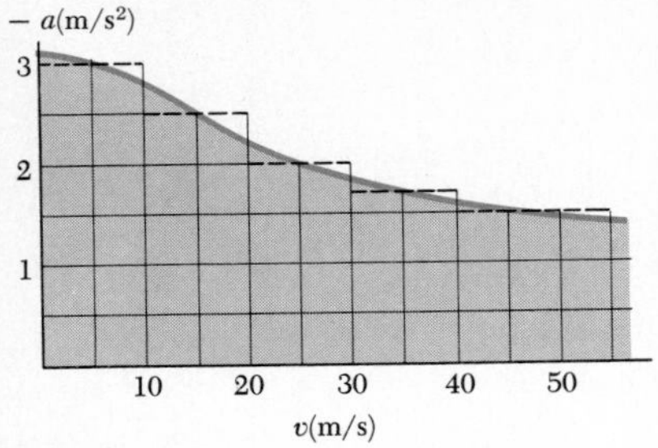

Fig. P11.60

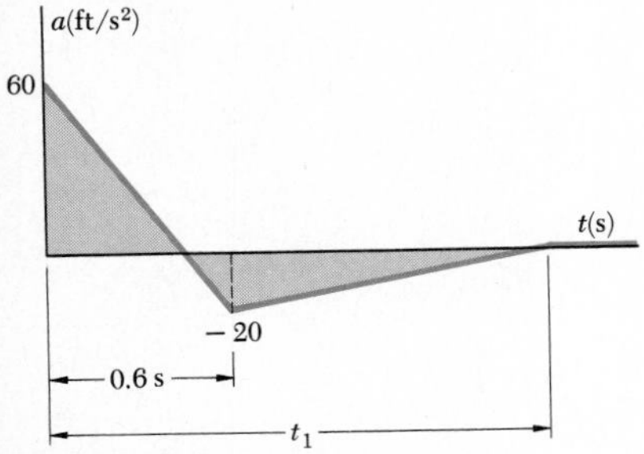

Fig. P11.63

11.60 The maximum possible acceleration of a passenger train under emergency conditions was determined experimentally; the results are shown (solid curve) in the figure. If the brakes are applied when the train is traveling at 90 km/h, determine by approximate means (*a*) the time required for the train to come to rest, (*b*) the distance traveled in that time.

11.61 Using the method of Sec. 11.8, derive the formula $x = x_0 + v_0t + \frac{1}{2}at^2$, for the position coordinate of a particle in uniformly accelerated rectilinear motion.

11.62 Using the method of Sec. 11.8, obtain an approximate solution for Prob. 11.58, assuming that the a–t curve is (*a*) a straight line from the origin to point *A*, (*b*) a parabola with vertex at *A*.

11.63 The acceleration of an object subjected to the pressure wave of a large explosion is defined approximately by the curve shown. The object is initially at rest and is again at rest at time t_1. Using the method of Sec. 11.8, determine (*a*) the time t_1, (*b*) the distance through which the object is moved by the pressure wave.

11.64 Using the method of Sec. 11.8, determine the position of the particle of Prob. 11.40 when $t = 14$ s.

11.65 For the particle of Prob. 11.42, draw the a–t curve and, using the method of Sec. 11.8, determine (*a*) the position of the particle when $t = 14$ s, (*b*) the maximum value of its position coordinate.

9-10-11

CURVILINEAR MOTION OF PARTICLES

11.9. Position Vector, Velocity, and Acceleration. When a particle moves along a curve other than a straight line, we say that the particle is in *curvilinear motion.* To define the position *P* occupied by the particle at a given time *t*, we select a fixed reference system, such as the *x* and *y* axes shown in Fig. 11.14*a*, and draw the vector **r** joining the origin *O* and point *P*. Since the vector **r** is characterized by its magnitude *r* and its direction with respect to the reference axes, it completely defines the position of the particle with respect to those axes; the vector **r** is referred to as the *position vector* of the particle at time *t*.

Consider now the vector $\mathbf{r}'$ defining the position P' occupied by the same particle at a later time $t + \Delta t$. The vector $\Delta\mathbf{r}$ joining P and P' represents the change in the position vector during the time interval Δt since, as we may easily check from Fig. 11.14*a*, the vector $\mathbf{r}'$ is obtained by adding the vectors $\mathbf{r}$ and $\Delta\mathbf{r}$ according to the triangle rule.† We note that $\Delta\mathbf{r}$ represents a change in *direction* as well as a change in *magnitude* of the position vector $\mathbf{r}$. The *average velocity* of the particle over the time interval Δt is defined as the quotient of $\Delta\mathbf{r}$ and Δt. Since $\Delta\mathbf{r}$ is a vector and Δt a scalar, the quotient $\Delta\mathbf{r}/\Delta t$ is a vector attached at P, of the same direction as $\Delta\mathbf{r}$, and of magnitude equal to the magnitude of $\Delta\mathbf{r}$ divided by Δt (Fig. 11.14*b*).

The *instantaneous velocity* of the particle at time t is obtained by choosing shorter and shorter time intervals Δt and, correspondingly, shorter and shorter vector increments $\Delta\mathbf{r}$. The instantaneous velocity is thus represented by the vector

$$\mathbf{v} = \lim_{\Delta t \to 0} \frac{\Delta\mathbf{r}}{\Delta t} \tag{11.14}$$

As Δt and $\Delta\mathbf{r}$ become shorter, the points P and P' get closer; the vector $\mathbf{v}$ obtained at the limit must therefore be tangent to the path of the particle (Fig. 11.14*c*). Its magnitude, denoted by v and called the *speed* of the particle, is obtained by substituting for the vector $\Delta\mathbf{r}$ in formula (11.14) its magnitude represented by the straight-line segment PP'. But the length of the segment PP' approaches the length Δs of the arc PP' as Δt decreases (Fig. 11.14*a*), and we may write

$$v = \lim_{\Delta t \to 0} \frac{PP'}{\Delta t} = \lim_{\Delta t \to 0} \frac{\Delta s}{\Delta t}$$

$$v = \frac{ds}{dt} \tag{11.15}$$

The speed v may thus be obtained by differentiating with respect to t the length s of the arc described by the particle.

It is recalled that in many problems of statics it was found convenient to resolve a force $\mathbf{F}$ into component forces $\mathbf{F}_x$ and $\mathbf{F}_y$. Similarly, we shall often find it convenient in kinematics to resolve the vector $\mathbf{v}$ into rectangular components $\mathbf{v}_x$ and $\mathbf{v}_y$. To obtain these components, we resolve the vector $\Delta\mathbf{r}$ into com-

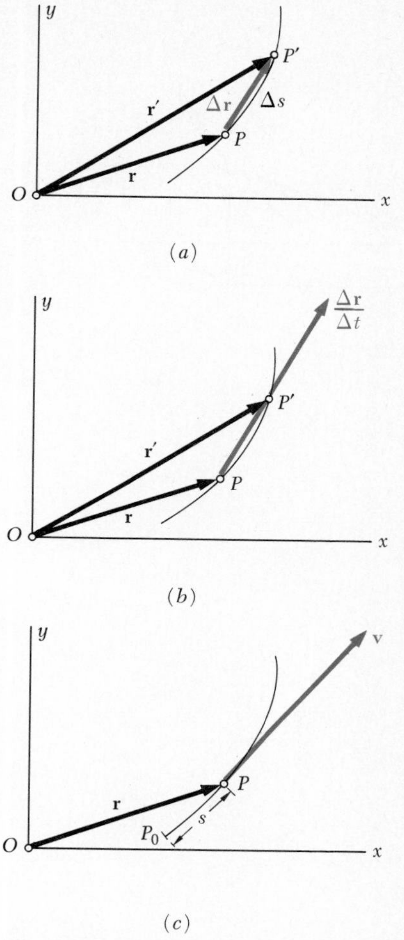

Fig. 11.14

† Since the displacement from O to P' represented by $\mathbf{r}'$ is clearly equivalent to the two successive displacements represented by $\mathbf{r}$ and $\Delta\mathbf{r}$, we verify that displacements satisfy the triangle rule and, therefore, the parallelogram law of addition. Thus, *displacements are truly vectors.*

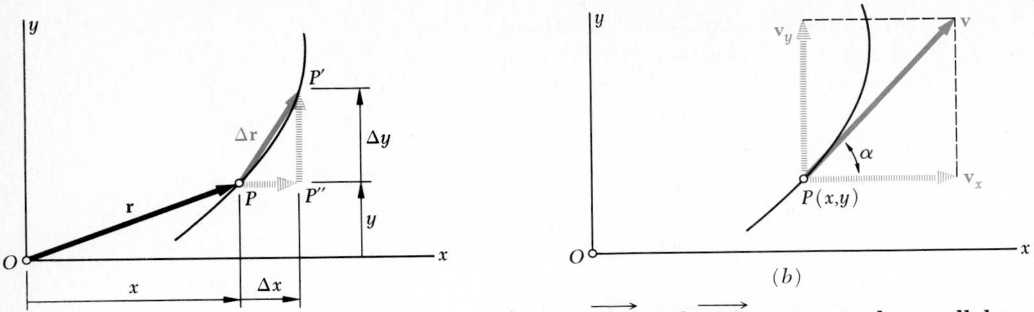

(a) (b)

Fig. 11.15

ponents $\overrightarrow{PP''}$ and $\overrightarrow{P''P'}$, respectively parallel to the x and y axes (Fig. 11.15*a*), and write

$$\Delta \mathbf{r} = \overrightarrow{PP''} + \overrightarrow{P''P'}$$

where the right-hand member represents a vector sum. Substituting in (11.14) the expression obtained for $\Delta\mathbf{r}$, we write

$$\mathbf{v} = \lim_{\Delta t \to 0} \frac{\overrightarrow{PP''}}{\Delta t} + \lim_{\Delta t \to 0} \frac{\overrightarrow{P''P'}}{\Delta t}$$

Since the first limit represents a vector parallel to the x axis, it must be equal to the component $\mathbf{v}_x$ of the velocity; similarly, the second limit must be equal to its component $\mathbf{v}_y$ (Fig. 11.15*b*). We write

$$\mathbf{v}_x = \lim_{\Delta t \to 0} \frac{\overrightarrow{PP''}}{\Delta t} \qquad \mathbf{v}_y = \lim_{\Delta t \to 0} \frac{\overrightarrow{P''P'}}{\Delta t}$$

The magnitudes v_x and v_y of the components of the velocity may be obtained by observing that the magnitudes of the vectors $\overrightarrow{PP''}$ and $\overrightarrow{P''P'}$ are respectively equal to the increments Δx and Δy of the coordinates x and y of the particle. We write

$$v_x = \lim_{\Delta t \to 0} \frac{\Delta x}{\Delta t} \qquad v_y = \lim_{\Delta t \to 0} \frac{\Delta y}{\Delta t}$$

$$v_x = \frac{dx}{dt} \qquad v_y = \frac{dy}{dt}$$

or, using dots to indicate time derivatives,

$$v_x = \dot{x} \qquad v_y = \dot{y} \tag{11.16}$$

The expressions obtained are known as the *scalar components* of the velocity. A positive value for v_x indicates that the vector component $\mathbf{v}_x$ is directed to the right, a negative value that it is directed to the left; similarly, a positive value for v_y indicates that $\mathbf{v}_y$ is directed upward, a negative value that it is directed downward. We verify from Fig. 11.15*b* that the magnitude and direction of the velocity **v** may be obtained from its scalar components v_x and v_y as follows:

$$v^2 = v_x^2 + v_y^2 \qquad \tan \alpha = \frac{v_y}{v_x} \tag{11.17}$$

Consider the velocity **v** of the particle at time t and also its velocity **v**′ at a later time $t + \Delta t$ (Fig. 11.16*a*). Let us draw both vectors **v** and **v**′ from the same origin O' (Fig. 11.16*b*). The vector $\Delta\mathbf{v}$ joining Q and Q' represents the change in the velocity of the particle during the time interval Δt, since the vector **v**′ may be obtained by adding the vectors **v** and $\Delta\mathbf{v}$. We should note that $\Delta\mathbf{v}$ represents a change in the *direction* of the velocity as well as a change in *speed.* The *average acceleration* of the particle over the time interval Δt is defined as the quotient of $\Delta\mathbf{v}$ and Δt. Since $\Delta\mathbf{v}$ is a vector and Δt a scalar, the quotient $\Delta\mathbf{v}/\Delta t$ is a vector of the same direction as $\Delta\mathbf{v}$.

The *instantaneous acceleration* of the particle at time t is obtained by choosing smaller and smaller values for Δt and $\Delta\mathbf{v}$. The instantaneous acceleration is thus represented by the vector

$$\mathbf{a} = \lim_{\Delta t \to 0} \frac{\Delta\mathbf{v}}{\Delta t} \tag{11.18}$$

It should be noted that, in general, *the acceleration is not tangent to the path of the particle;* also, in general, the magnitude of the acceleration *does not* represent the rate of change of the speed of the particle.

To obtain the rectangular components $\mathbf{a}_x$ and $\mathbf{a}_y$ of the acceleration, we resolve the vector $\Delta\mathbf{v}$ into components $\overrightarrow{QQ''}$ and $\overrightarrow{Q''Q'}$, respectively parallel to the x and y axes (Fig. 11.16*b*), and write

$$\Delta\mathbf{v} = \overrightarrow{QQ''} + \overrightarrow{Q''Q'}$$

Substituting in (11.18) the expression obtained for $\Delta\mathbf{v}$, we write

$$\mathbf{a} = \lim_{\Delta t \to 0} \frac{\overrightarrow{QQ''}}{\Delta t} + \lim_{\Delta t \to 0} \frac{\overrightarrow{Q''Q'}}{\Delta t}$$

Since the first limit represents a vector parallel to the x axis, it must be equal to the component $\mathbf{a}_x$ of the acceleration; similarly, the second limit must be equal to its component $\mathbf{a}_y$ (Fig. 11.17). We write

$$\mathbf{a}_x = \lim_{\Delta t \to 0} \frac{\overrightarrow{QQ''}}{\Delta t} \qquad \mathbf{a}_y = \lim_{\Delta t \to 0} \frac{\overrightarrow{Q''Q'}}{\Delta t}$$

The magnitudes a_x and a_y of the components of the acceleration, which are known as the scalar components of the acceleration,

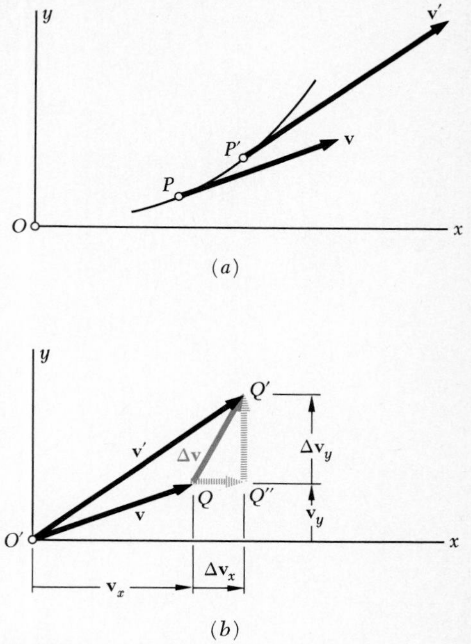

Fig. 11.16

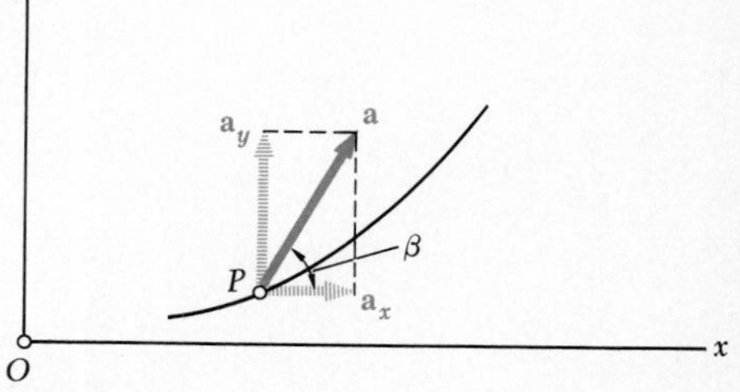

Fig. 11.17

may be obtained by observing that the magnitudes of the vectors $\overrightarrow{QQ''}$ and $\overrightarrow{Q''Q'}$ are respectively equal to the increments Δv_x and Δv_y of the scalar components v_x and v_y of the velocity of the particle. We write

$$a_x = \lim_{\Delta t \to 0} \frac{\Delta v_x}{\Delta t} \qquad a_y = \lim_{\Delta t \to 0} \frac{\Delta v_y}{\Delta t}$$

$$a_x = \frac{dv_x}{dt} \qquad a_y = \frac{dv_y}{dt}$$

$$a_x = \dot{v}_x \qquad a_y = \dot{v}_y \tag{11.19}$$

or, substituting for v_x and v_y from (11.16),

$$a_x = \ddot{x} \qquad a_y = \ddot{y} \tag{11.20}$$

The magnitude and direction of the acceleration **a** may be obtained from its scalar components a_x and a_y (Fig. 11.17) by formulas similar to formulas (11.17).

11.10. Component Motions. The use of rectangular components to describe the position, the velocity, and the acceleration of a particle is particularly effective when the component a_x of the acceleration is independent of the coordinate y and of the velocity component v_y and when the component a_y of the acceleration is independent of x and v_x. Equations (11.19) may then be integrated independently, and so may Eqs. (11.16). In other words, the motion of the particle in the x direction and its motion in the y direction may be considered separately.

In the case of the *motion of a projectile,* for example, it may be shown (Sec. 12.7), if the resistance of the air is neglected, that the components of the acceleration are

$$a_x = 0 \qquad a_y = -g$$

where g is 9.81 m/s² or 32.2 ft/s². Substituting into Eqs. (11.19), we write

$$\dot{v}_x = 0 \qquad \dot{v}_y = -g$$
$$v_x = (v_x)_0 \qquad v_y = (v_y)_0 - gt$$

where $(v_x)_0$ and $(v_y)_0$ are the components of the initial velocity. Substituting into Eqs. (11.16), and assuming that the projectile is fired from the origin O, we write

$$\dot{x} = (v_x)_0 \qquad \dot{y} = (v_y)_0 - gt$$
$$x = (v_x)_0 t \qquad y = (v_y)_0 t - \tfrac{1}{2}gt^2$$

These equations show that the motion of the projectile in the horizontal direction is uniform, while its motion in the vertical direction is uniformly accelerated. The motion of a projectile may thus be replaced by two independent rectilinear motions, which are easily visualized if we assume that the projectile is fired vertically with an initial velocity $(\mathbf{v}_y)_0$ from a platform moving with a constant horizontal velocity $(\mathbf{v}_x)_0$ (Fig. 11.18). The coordinate x of the projectile is equal at any instant to the distance traveled by the platform, while its coordinate y may be computed as if the projectile were moving along a vertical line.

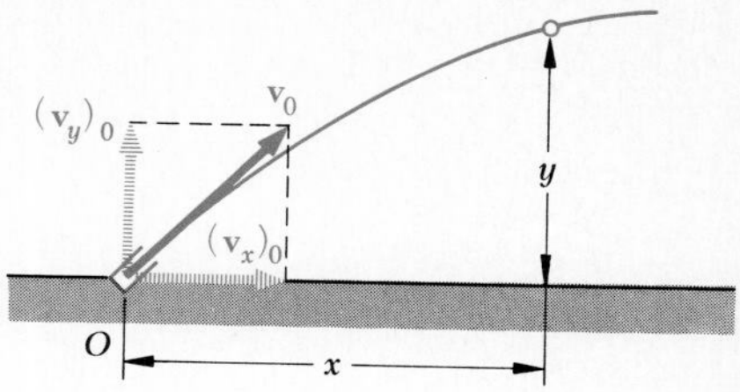

(a) Motion of a projectile

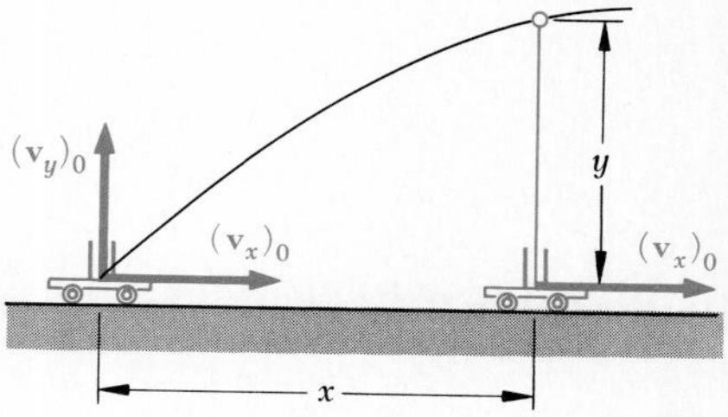

(b) Equivalent rectilinear motions

Fig. 11.18

It may be observed that the equations defining the coordinates x and y of a projectile at any instant are the parametric equations of a parabola. Thus, the trajectory of a projectile is *parabolic*. This result, however, ceases to be valid when the resistance of the air or the variation of g with altitude is taken into account.

11.11. Relative Motion. Consider two particles A and B moving in the same plane (Fig. 11.19); the vectors $\mathbf{r}_A$ and $\mathbf{r}_B$ define their position at any given instant with respect to a fixed system of axes xy centered at O. Consider now a system of axes $x'y'$ centered at A and parallel to the x and y axes. While the origin of these axes moves, their orientation remains the same; such a system of axes is said to be in *translation*. The vector $\mathbf{r}_{B/A}$ joining A and B defines the position of the particle B with respect to the moving reference system $x'y'$; it is therefore called the position vector of B relative to the reference system centered at A or, for short, the *position vector of B relative to A*.

We note from Fig. 11.19 that the position vector $\mathbf{r}_B$ of particle B may be obtained by adding by the triangle rule the position vector $\mathbf{r}_A$ of particle A and the position vector $\mathbf{r}_{B/A}$ of B relative to A; we write

$$\mathbf{r}_B = \mathbf{r}_A + \mathbf{r}_{B/A} \tag{11.21}$$

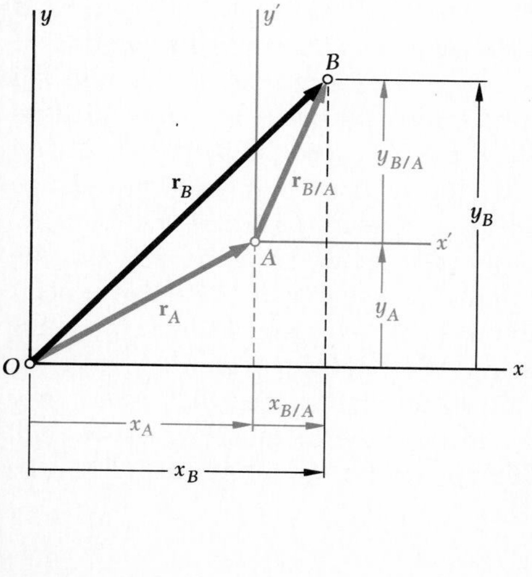

Fig. 11.19

In terms of x and y components, this relation reads

$$x_B = x_A + x_{B/A} \qquad y_B = y_A + y_{B/A} \tag{11.22}$$

where x_B and y_B denote the coordinates of particle B with respect to the fixed x and y axes, while $x_{B/A}$ and $y_{B/A}$ represent the coordinates of B with respect to the moving x' and y' axes. Differentiating Eqs. (11.22) with respect to t and using dots to indicate time derivatives, we write

$$\dot{x}_B = \dot{x}_A + \dot{x}_{B/A} \qquad \dot{y}_B = \dot{y}_A + \dot{y}_{B/A} \tag{11.23}$$

The derivatives $\dot{x}_A$ and $\dot{y}_A$ represent the x and y components of the velocity $\mathbf{v}_A$ of particle A, while the derivatives $\dot{x}_B$ and $\dot{y}_B$ represent the components of the velocity $\mathbf{v}_B$ of particle B. The vector $\mathbf{v}_{B/A}$ of components $\dot{x}_{B/A}$ and $\dot{y}_{B/A}$ is known as the *relative velocity of B with respect to A* or, more accurately, as the relative velocity of B with respect to a reference system in translation with A. Writing Eqs. (11.23) in vector form, we have

$$\mathbf{v}_B = \mathbf{v}_A + \mathbf{v}_{B/A} \tag{11.24}$$

Differentiating Eqs. (11.23), we write

$$\ddot{x}_B = \ddot{x}_A + \ddot{x}_{B/A} \qquad \ddot{y}_B = \ddot{y}_A + \ddot{y}_{B/A} \tag{11.25}$$

where the first two derivatives in each equation represent the components of the accelerations $\mathbf{a}_B$ and $\mathbf{a}_A$. The vector $\mathbf{a}_{B/A}$ of components $\ddot{x}_{B/A}$ and $\ddot{y}_{B/A}$ is known as the *relative acceleration of B with respect to A* or, more accurately, as the relative acceleration of B with respect to a reference system in translation with A. Writing Eqs. (11.25) in vector form, we obtain

$$\mathbf{a}_B = \mathbf{a}_A + \mathbf{a}_{B/A} \tag{11.26}$$

The motion of B with respect to the fixed x and y axes is referred to as the *absolute motion of B*, while the motion of B with respect to the x' and y' axes attached at A is referred to as the *relative motion of B with respect to A*. The equations derived in this section show that *the absolute motion of B may be obtained by combining the motion of A and the relative motion of B with respect to A*. Equation (11.24), for example, expresses that the absolute velocity $\mathbf{v}_B$ of particle B may be obtained by adding vectorially the velocity of A and the velocity of B relative to A. Equation (11.26) expresses a similar property in terms of the accelerations. We should keep in mind, however, that the x' and y' axes are in translation, i.e., that, while they move with A, they maintain the same orientation. As we shall see later (Sec. 15.11), different relations must be used when the moving reference axes rotate.

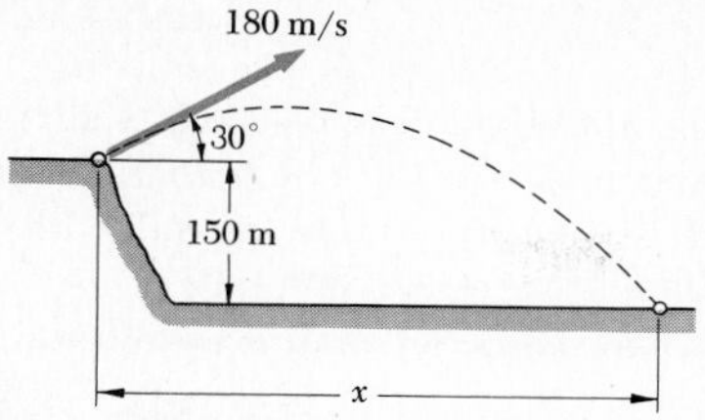

SAMPLE PROBLEM 11.7

A projectile is fired from the edge of a 150-m cliff with an initial velocity of 180 m/s, at an angle of 30° with the horizontal. Neglecting air resistance, find (*a*) the horizontal distance from the gun to the point where the projectile strikes the ground, (*b*) the greatest elevation above the ground reached by the projectile.

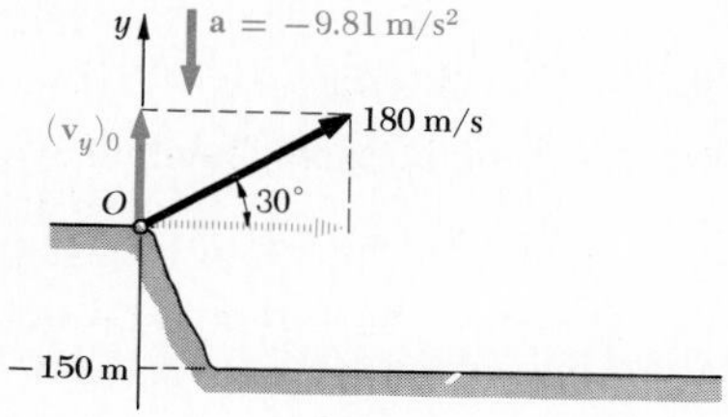

Solution. We shall consider separately the vertical and the horizontal motion.

Vertical Motion. Uniformly accelerated motion. Choosing the positive sense of the y axis upward and placing the origin O at the gun, we have

$$(v_y)_0 = (180 \text{ m/s}) \sin 30° = +90 \text{ m/s}$$
$$a = -9.81 \text{ m/s}^2$$

Substituting into the equations of uniformly accelerated motion, we have

$$v_y = (v_y)_0 + at \qquad v_y = 90 - 9.81t \tag{1}$$
$$y = (v_y)_0 t + \tfrac{1}{2}at^2 \qquad y = 90t - 4.90t^2 \tag{2}$$
$$v_y^2 = (v_y)_0^2 + 2ay \qquad v_y^2 = 8100 - 19.62y \tag{3}$$

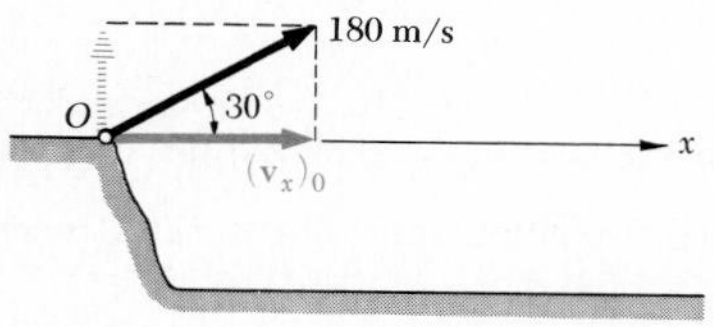

Horizontal Motion. Uniform motion. Choosing the positive sense of the x axis to the right, we have

$$(v_x)_0 = (180 \text{ m/s}) \cos 30° = +155.9 \text{ m/s}$$

Substituting into the equation of uniform motion, we obtain

$$x = (v_x)_0 t \qquad x = 155.9t \tag{4}$$

***a.* Horizontal Distance.** When the projectile strikes the ground, we have

$$y = -150 \text{ m}$$

Carrying this value into Eq. (2) for the vertical motion, we write

$$-150 = 90t - 4.90t^2 \qquad t^2 - 18.37t - 30.6 = 0 \qquad t = 19.91 \text{ s}$$

Carrying $t = 19.91$ s into Eq. (4) for the horizontal motion, we obtain

$$x = 155.9(19.91) \qquad x = 3100 \text{ m} \blacktriangleleft$$

***b.* Greatest Elevation.** When the projectile reaches its greatest elevation, we have $v_y = 0$; carrying this value into Eq. (3) for the vertical motion, we write

$$0 = 8100 - 19.62y \qquad y = 413 \text{ m}$$

Greatest elevation above ground $= 150 \text{ m} + 413 \text{ m}$
$= 563 \text{ m}$ ◄

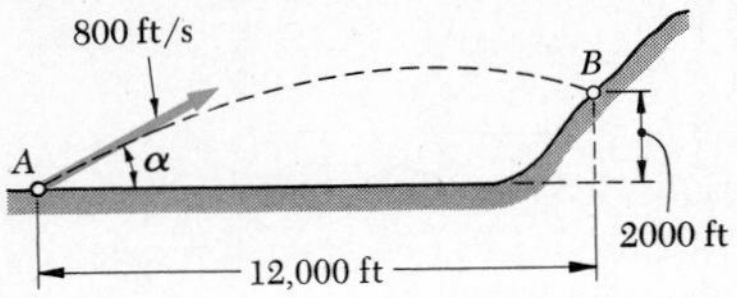

SAMPLE PROBLEM 11.8

A projectile is fired with an initial velocity of 800 ft/s at a target B located 2000 ft above the gun A and at a horizontal distance of 12,000 ft. Neglecting air resistance, determine the value of the firing angle α.

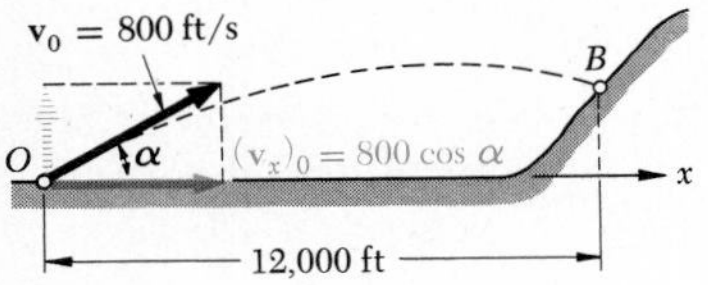

Solution. We shall consider separately the horizontal and the vertical motion.

Horizontal Motion. Placing the origin of coordinates at the gun, we have

$$(v_x)_0 = 800 \cos \alpha$$

Substituting into the equation of uniform horizontal motion, we obtain

$$x = (v_x)_0 t \qquad x = (800 \cos \alpha)t$$

The time required for the projectile to move through a horizontal distance of 12,000 ft is obtained by making x equal to 12,000 ft.

$$12{,}000 = (800 \cos \alpha)t$$

$$t = \frac{12{,}000}{800 \cos \alpha} = \frac{15}{\cos \alpha}$$

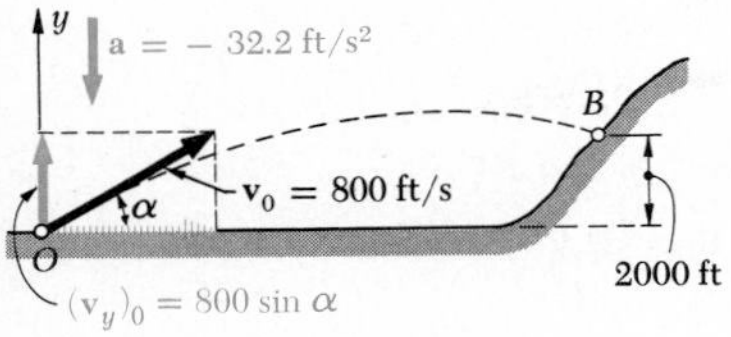

Vertical Motion

$$(v_y)_0 = 800 \sin \alpha \qquad a = -32.2 \text{ ft/s}^2$$

Substituting into the equation of uniformly accelerated vertical motion, we obtain

$$y = (v_y)_0 t + \tfrac{1}{2}at^2 \qquad y = (800 \sin \alpha)t - 16.1t^2$$

Projectile Hits Target. When $x = 12{,}000$ ft, we must have $y = 2000$ ft. Substituting for y and making t equal to the value found above, we write

$$2000 = 800 \sin \alpha \frac{15}{\cos \alpha} - 16.1\left(\frac{15}{\cos \alpha}\right)^2$$

Since $1/\cos^2 \alpha = \sec^2 \alpha = 1 + \tan^2 \alpha$, we have

$$2000 = 800(15) \tan \alpha - 16.1(15^2)(1 + \tan^2 \alpha)$$
$$3622 \tan^2 \alpha - 12{,}000 \tan \alpha + 5622 = 0$$

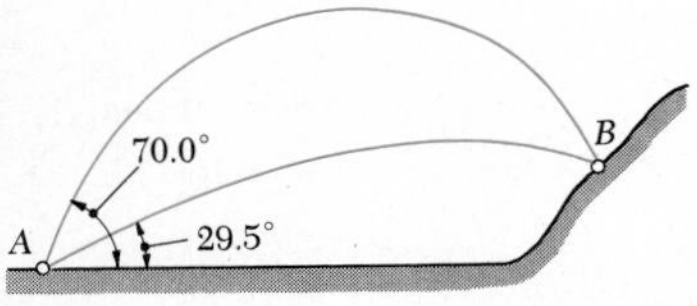

Solving this quadratic equation for $\tan \alpha$, we have

$$\tan \alpha = 0.565 \qquad \text{and} \qquad \tan \alpha = 2.75$$

$$\alpha = 29.5° \qquad \text{and} \qquad \alpha = 70.0° \quad ◀$$

The target will be hit if either of these two firing angles is used (see figure).

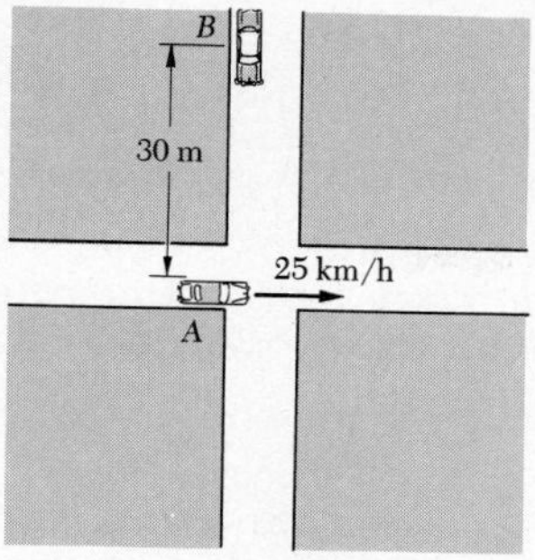

SAMPLE PROBLEM 11.9

Automobile A is traveling east at the constant speed of 25 km/h. As automobile A crosses the intersection shown, automobile B starts from rest 30 m north of the intersection and moves south with a constant acceleration of 1.2 m/s². Determine the position, velocity, and acceleration of B relative to A five seconds after A crosses the intersection.

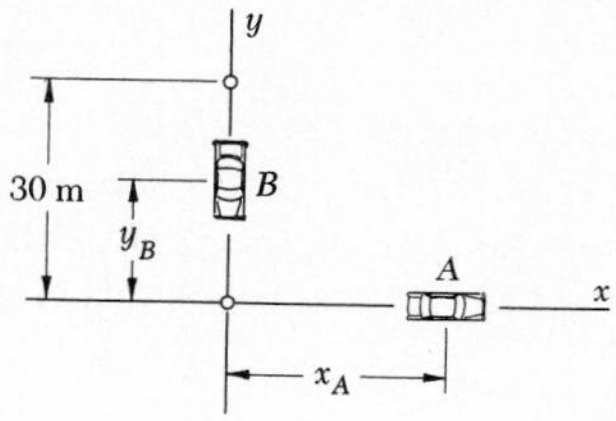

Solution. We choose x and y axes with origin at the intersection of the two streets and with positive senses directed respectively east and north.

Motion of Automobile A. First the speed is expressed in m/s:

$$25 \text{ km/h} = \frac{25 \text{ km}}{1 \text{ h}} = \frac{25\,000 \text{ m}}{3600 \text{ s}} = 6.94 \text{ m/s}$$

Noting that the motion of A is uniform, we write, for any time t,

$$a_A = 0$$
$$v_A = +6.94 \text{ m/s}$$
$$x_A = (x_A)_0 + v_A t = 0 + 6.94t$$

For $t = 5$ s, we have

$$a_A = 0 \qquad \mathbf{a}_A = 0$$
$$v_A = +6.94 \text{ m/s} \qquad \mathbf{v}_A = 6.94 \text{ m/s} \rightarrow$$
$$x_A = +(6.94 \text{ m/s})(5 \text{ s}) = +34.7 \text{ m} \qquad \mathbf{r}_A = 34.7 \text{ m} \rightarrow$$

Motion of Automobile B. We note that the motion of B is uniformly accelerated, and write

$$a_B = -1.2 \text{ m/s}^2$$
$$v_B = (v_B)_0 + at = 0 - 1.2t$$
$$y_B = (y_B)_0 + (v_B)_0 t + \tfrac{1}{2}a_B t^2 = 30 + 0 - \tfrac{1}{2}(1.2)t^2$$

For $t = 5$ s, we have

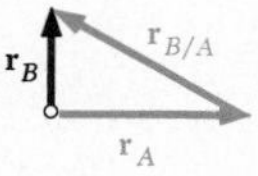

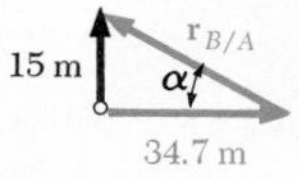

$$a_B = -1.2 \text{ m/s}^2 \qquad \mathbf{a}_B = 1.2 \text{ m/s}^2 \downarrow$$
$$v_B = -(1.2 \text{ m/s})(5 \text{ s}) = -6 \text{ m/s} \qquad \mathbf{v}_B = 6 \text{ m/s} \downarrow$$
$$y_B = 30 - \tfrac{1}{2}(1.2 \text{ m/s})(5 \text{ s})^2 = +15 \text{ m} \qquad \mathbf{r}_B = 15 \text{ m} \uparrow$$

Motion of B Relative to A. We draw the triangle corresponding to the vector equation $\mathbf{r}_B = \mathbf{r}_A + \mathbf{r}_{B/A}$ and obtain the magnitude and direction of the position vector of B relative to A.

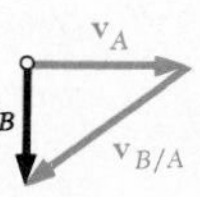

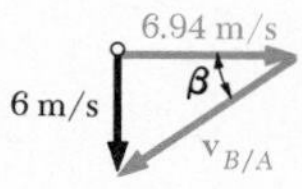

$$r_{B/A} = 37.8 \text{ m} \qquad \alpha = 23.4° \qquad \mathbf{r}_{B/A} = 37.8 \text{ m} \measuredangle 23.4°$$ ◀

Proceeding in a similar fashion, we find the velocity and acceleration of B relative to A.

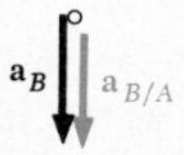

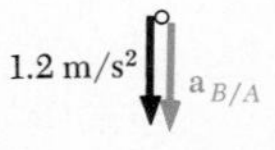

$$\mathbf{v}_B = \mathbf{v}_A + \mathbf{v}_{B/A}$$
$$v_{B/A} = 9.17 \text{ m/s} \qquad \beta = 40.8° \qquad \mathbf{v}_{B/A} = 9.17 \text{ m/s} \measuredangle 40.8°$$ ◀
$$\mathbf{a}_B = \mathbf{a}_A + \mathbf{a}_{B/A} \qquad \mathbf{a}_{B/A} = 1.2 \text{ m/s}^2 \downarrow$$ ◀

PROBLEMS

Note. Neglect air resistance in problems concerning projectiles.

11.66 The motion of a particle is defined by the equations $x = \frac{1}{2}t^3 - 2t^2$ and $y = \frac{1}{2}t^2 - 2t$, where x and y are expressed in meters and t in seconds. Determine the velocity and acceleration when (*a*) $t = 1$ s, (*b*) $t = 3$ s.

11.67 In Prob. 11.66, determine (*a*) the time at which the value of the y coordinate is minimum, (*b*) the corresponding velocity and acceleration of the particle.

11.68 The motion of a particle is defined by the equations $x = (t + 1)^2$ and $y = (t + 1)^{-2}$, where x and y are expressed in feet and t in seconds. Show that the path of the particle is a rectangular hyperbola and determine the velocity and acceleration when (*a*) $t = 0$, (*b*) $t = 1$ s.

11.69 The motion of a particle is defined by the equations $x = 25(1 - e^{-t})$ and $y = -3t^2 + 15$, where x and y are expressed in feet and t in seconds. Determine the velocity and acceleration when $t = 2$ s.

11.70 The motion of a vibrating particle is defined by the equations $x = 100 \sin \pi t$ and $y = 25 \cos 2\pi t$, where x and y are expressed in millimeters and t in seconds. (*a*) Determine the velocity and acceleration when $t = 1$ s. (*b*) Show that the path of the particle is parabolic.

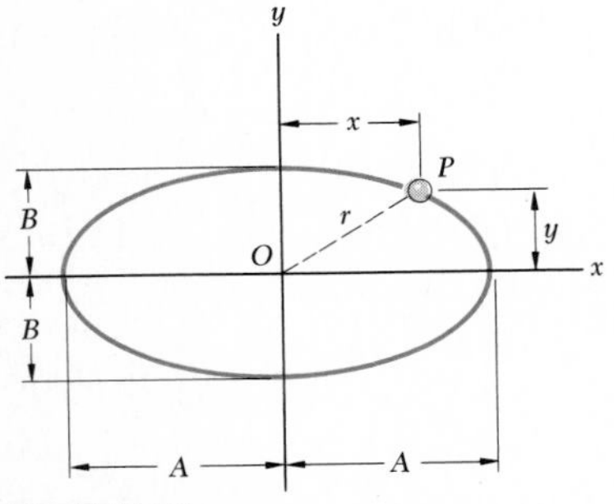

Fig. P11.71

11.71 A particle moves in an elliptic path according to the equations $x = A \cos pt$ and $y = B \sin pt$. Show that the acceleration (*a*) is directed toward the origin, (*b*) is proportional to the distance from the origin to the particle.

11.72 A particle moves in a circular path according to the equations $x = r \sin \omega t$ and $y = r \cos \omega t$. Show (*a*) that the magnitude v of the velocity is constant and equal to $r\omega$, (*b*) that the acceleration is of constant magnitude $r\omega^2$ (or v^2/r) and is directed toward the center of the circular path.

11.73 The motion of a vibrating particle is defined by the equations $x = A \sin pt$ and $y = B \sin (pt + \phi)$, where A, B, p, and ϕ are constants. If $A = B$, determine (*a*) the value of the constant ϕ for which the speed of the particle is constant, (*b*) the corresponding value of the speed, (*c*) the corresponding path of the particle.

11.74 A man standing on a bridge 20 m above the water throws a stone in a horizontal direction. Knowing that the stone hits the water 30 m from a point on the water directly below the man, determine (*a*) the initial velocity of the stone, (*b*) the distance at which the stone would hit the water if it were thrown with the same velocity from a bridge 5 m lower.

11.75 Water issues at *A* from a pressure tank with a horizontal velocity $\mathbf{v}_0$. For what range of values v_0 will the water enter the opening *BC*?

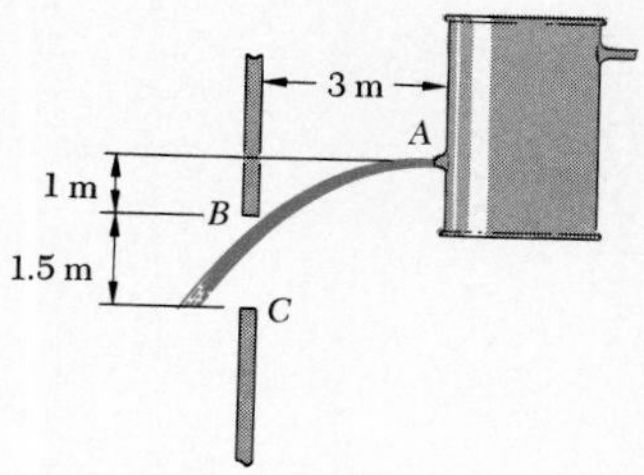

Fig. P11.75

11.76 A nozzle at *A* discharges water with an initial velocity of 40 ft/s at an angle of 60° with the horizontal. Determine where the stream of water strikes the roof. Check that the stream will clear the edge of the roof.

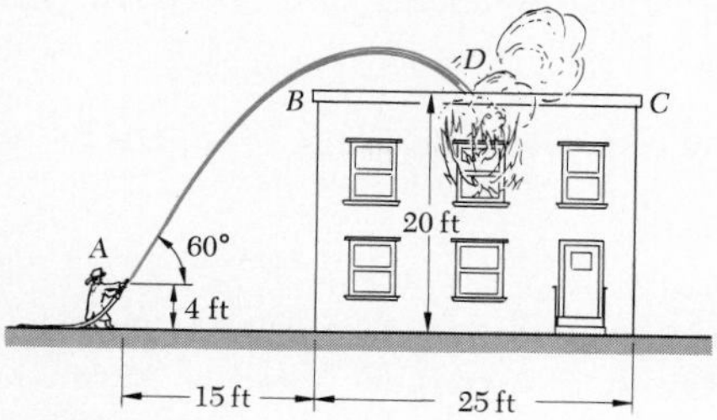

Fig. P11.76

11.77 In Prob. 11.76, determine the largest and smallest initial velocity for which the water will fall on the roof.

11.78 A ball is dropped vertically onto a 20° incline at *A*; the direction of rebound forms an angle of 40° with the vertical. Knowing that the ball next strikes the incline at *B*, determine (*a*) the velocity of rebound at *A*, (*b*) the time required for the ball to travel from *A* to *B*.

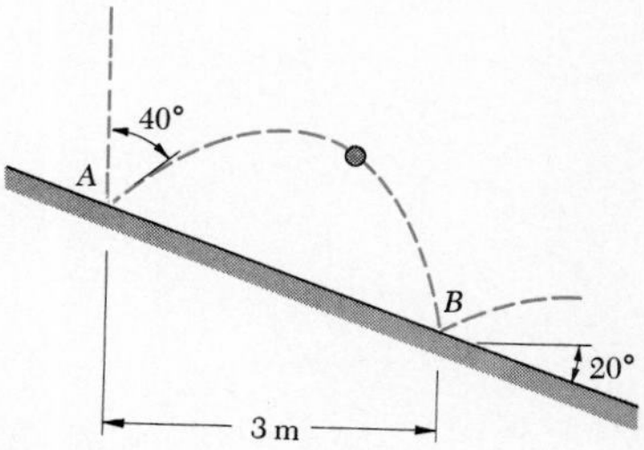

Fig. P11.78

11.79 Sand is discharged at *A* from a conveyor belt and falls onto a stockpile at *B*. Knowing that the conveyor belt forms an angle $\alpha = 20°$ with the horizontal and moves at a constant speed of 12 ft/s, determine the distance *d* to the top of the stockpile.

11.80 The conveyor belt moves at a constant speed of 12 ft/s. Knowing that $d = 12$ ft, determine the angle α for which the sand is deposited on the stockpile at *B*.

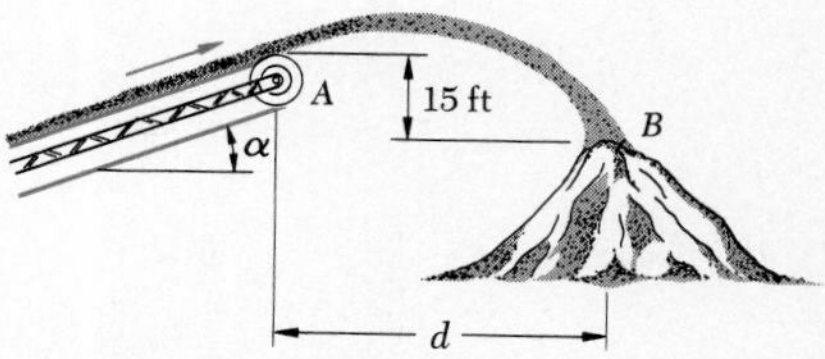

Fig. P11.79 and P11.80

11.81 A projectile is fired with an initial velocity of 210 m/s. Find the angle at which it should be fired if it is to hit a target located at a distance of 3600 m on the same level.

11.82 A boy can throw a baseball a maximum distance of 30 m in New York, where $g = 9.81 \text{ m/s}^2$. How far could he throw the baseball (*a*) in Singapore, where $g = 9.78 \text{ m/s}^2$? (*b*) On the moon, where $g = 1.618 \text{ m/s}^2$?

11.83 A projectile is fired with an initial velocity $\mathbf{v}_0$ at an angle α with the horizontal. Determine (*a*) the maximum height h reached by the projectile, (*b*) the horizontal range R of the projectile, (*c*) the maximum horizontal range R and the corresponding firing angle α.

11.84 If the maximum horizontal range of a given gun is R, determine the firing angle which should be used to hit a target located at a distance $\frac{1}{2}R$ on the same level.

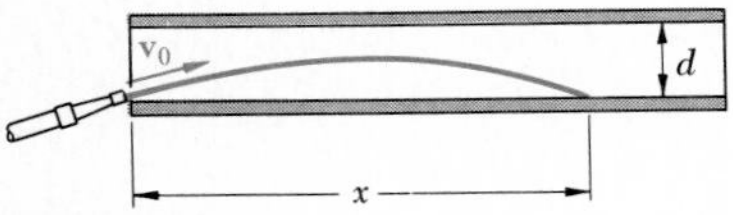

Fig. P11.85

11.85 A nozzle discharges a stream of water with an initial velocity $\mathbf{v}_0$ of 50 ft/s into the end of a horizontal pipe of inside diameter $d = 5$ ft. Determine the largest distance x that the stream can reach.

11.86 Show that the trajectory of a projectile is a parabola if air resistance is neglected.

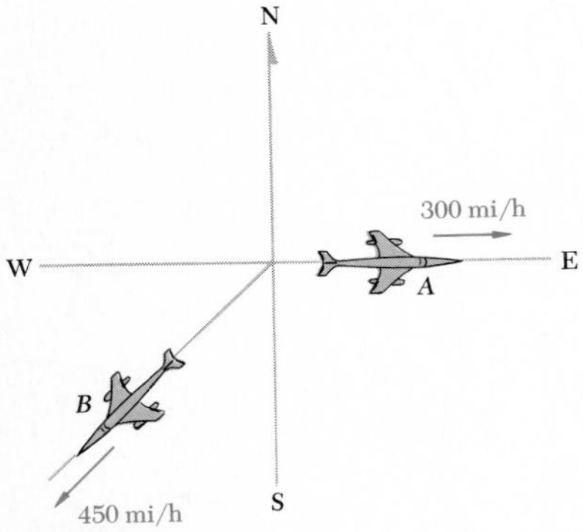

Fig. P11.87

11.87 Two airplanes A and B are each flying at a constant altitude of 3000 ft. Plane A is flying due east at a constant speed of 300 mi/h while plane B is flying southwest at a constant speed of 450 mi/h. Determine the change in position of plane B relative to plane A which takes place during a 2-min interval.

11.88 Instruments in an airplane indicate that, with respect to the air, the plane is moving west at a speed of 250 mi/h. At the same time ground-based radar indicates the plane to be moving at a speed of 300 mi/h in a direction 12° south of west. Determine the magnitude and direction of the velocity of the air.

11.89 As he passes a pole, a man riding in a truck tries to hit the pole by throwing a stone with a horizontal velocity of 20 m/s relative to the truck. Knowing that the speed of the truck is 40 km/h, determine (*a*) the direction in which he must throw the stone, (*b*) the horizontal velocity of the stone with respect to the ground.

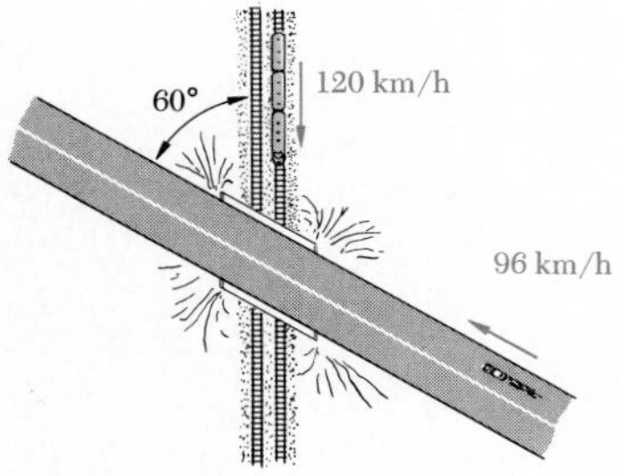

Fig. P11.90

11.90 An automobile and a train travel at the constant speeds shown. Three seconds after the train passes under the highway bridge the automobile crosses the bridge. Determine (*a*) the velocity of the train relative to the automobile, (*b*) the change in position of the train relative to the automobile during a 4-s interval, (*c*) the distance between the train and the automobile 5 s after the automobile has crossed the bridge.

11.91 During a rainstorm the paths of the raindrops appear to form an angle of 30° with the vertical when observed from a side window of a train moving at a speed of 15 km/h. A short time later, after the speed of the train has increased to 30 km/h, the angle between the vertical and the paths of the drops appears to be 45°. If the train were stopped, at what angle and with what velocity would the drops be observed to fall?

11.92 Ship *A* travels due south at 12 knots, while ship *B* travels toward the southeast at 10 knots. Determine the velocity (*a*) of ship *B* relative to ship *A*, (*b*) of ship *A* relative to ship *B*.

11.93 As observed from a ship moving due south at 10 mi/h, the wind appears to blow from the east. After the ship has changed course, and as it is moving due west at 10 mi/h, the wind appears to blow from the northeast. Assuming that the wind velocity is constant during the period of observation, determine the magnitude and direction of the true wind velocity.

11.94 An airplane is flying horizontally at an altitude of 2500 m and at a constant speed of 900 km/h on a path which passes directly over an antiaircraft gun. The gun fires a shell with a muzzle velocity of 500 m/s and hits the airplane. Knowing that the firing angle of the gun is 60°, determine the velocity and acceleration of the shell relative to the airplane at the time of impact.

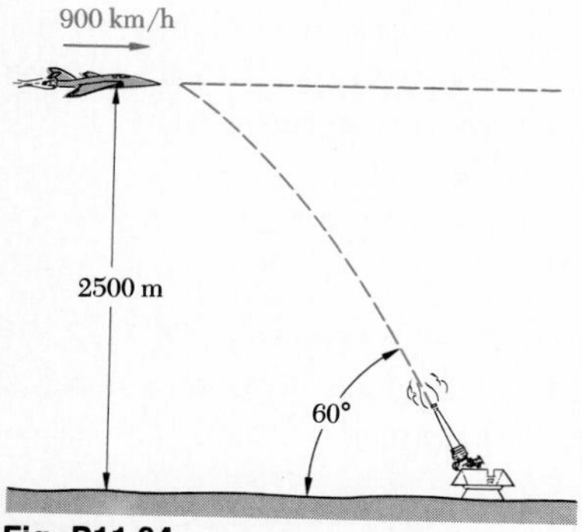

Fig. P11.94

11.95 Water is discharged at *A* with an initial velocity of 10 m/s and strikes a series of vanes at *B*. Knowing that the vanes move downward with a constant speed of 3 m/s, determine the velocity and acceleration of the water relative to the vane at *B*.

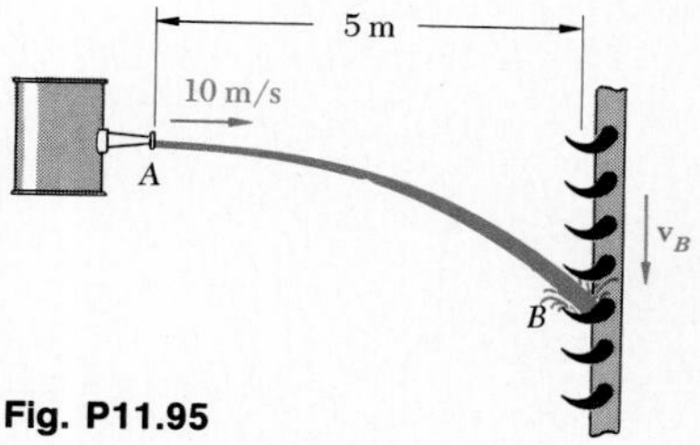

Fig. P11.95

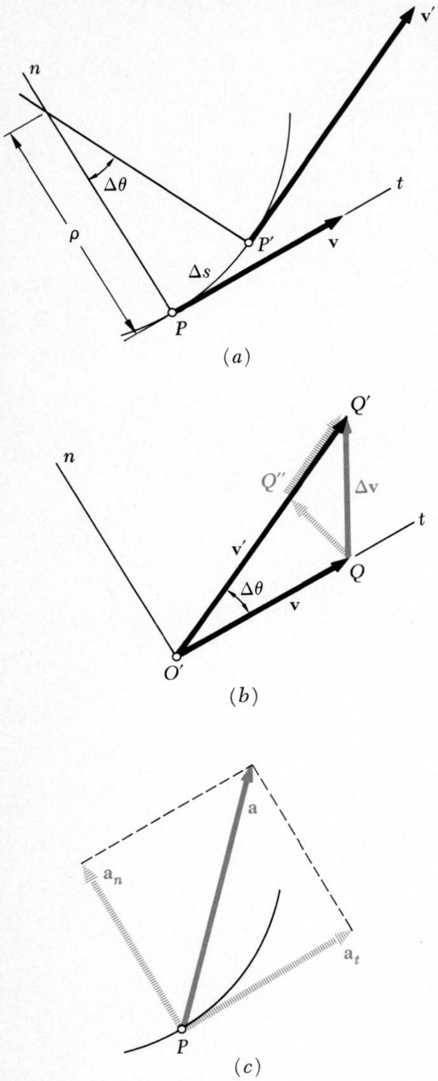

Fig. 11.20

11.12. Tangential and Normal Components. We saw in Sec. 11.9 that the velocity of a particle is a vector tangent to the path of the particle but that, in general, the acceleration is not tangent to the path. It is sometimes convenient to resolve the acceleration into components directed, respectively, along the tangent and the normal to the path of the particle. The positive sense along the tangent is chosen to coincide with the sense of motion of the particle, and the positive sense along the normal is chosen toward the inside of the path (Fig. 11.20).

We consider again, as we did in Sec. 11.9, the velocity $\mathbf{v}$ of the particle at a time t and its velocity $\mathbf{v}'$ at a time $t + \Delta t$ (Fig. 11.20*a*). Drawing both vectors from the same origin O', and joining the points Q and Q', we obtain the small vector $\Delta\mathbf{v}$ representing the difference between the vectors $\mathbf{v}$ and $\mathbf{v}'$ (Fig. 11.20*b*). As we saw earlier, the average acceleration is defined as the quotient of $\Delta\mathbf{v}$ and Δt, and the instantaneous acceleration $\mathbf{a}$ as the limit of this quotient. To obtain the tangential component $\mathbf{a}_t$ and the normal component $\mathbf{a}_n$ of the acceleration, we measure the distance $O'Q'' = O'Q$ along the line $O'Q'$ and resolve the vector $\Delta\mathbf{v}$ into its components $\overrightarrow{QQ''}$ and $\overrightarrow{Q''Q'}$ (Fig. 11.20*b*); we write

$$\Delta\mathbf{v} = \overrightarrow{QQ''} + \overrightarrow{Q''Q'}$$

The component $\overrightarrow{QQ''}$ represents a change in the *direction* of the velocity, while the component $\overrightarrow{Q''Q'}$ represents a change in the *magnitude* of the velocity, i.e., a change in *speed*. Substituting in (11.18) the expression obtained for $\Delta\mathbf{v}$, we write

$$\mathbf{a} = \lim_{\Delta t \to 0} \frac{\Delta\mathbf{v}}{\Delta t} = \lim_{\Delta t \to 0} \frac{\overrightarrow{QQ''}}{\Delta t} + \lim_{\Delta t \to 0} \frac{\overrightarrow{Q''Q'}}{\Delta t}$$

As Δt approaches zero, the vector $\overrightarrow{Q''Q'}$ becomes parallel to the tangent to the path at P and the vector $\overrightarrow{QQ''}$ becomes parallel to the normal. We must have, therefore,

$$\mathbf{a}_t = \lim_{\Delta t \to 0} \frac{\overrightarrow{Q''Q'}}{\Delta t} \qquad \mathbf{a}_n = \lim_{\Delta t \to 0} \frac{\overrightarrow{QQ''}}{\Delta t} \tag{11.27}$$

Considering first the tangential component, we observe that the length of the vector $\overrightarrow{Q''Q'}$ measures the change in speed $v' - v = \Delta v$. The scalar component a_t of the acceleration is therefore

$$a_t = \lim_{\Delta t \to 0} \frac{\Delta v}{\Delta t}$$

$$a_t = \frac{dv}{dt} \tag{11.28}$$

Formula (11.28) expresses that the *tangential component* a_t of the acceleration is equal to the *rate of change of the speed of the particle.*

Considering now the normal component of the acceleration, and denoting by $\Delta\theta$ the angle between the tangents to the path at P and P', we observe that, for small values of $\Delta\theta$, the length of the vector $\overrightarrow{QQ''}$ is equal to $v\,\Delta\theta$. Since the angle $\Delta\theta$ is equal to the angle between the normals at P and P' (Fig. 11.20*a*), we have $\Delta\theta = \Delta s/\rho$, where Δs is the length of the arc PP' and ρ the radius of curvature of the path at P. It thus follows from the second of the relations (11.27) that the scalar component a_n of the acceleration is

$$a_n = \lim_{\Delta t \to 0} \frac{v\,\Delta\theta}{\Delta t} = \lim_{\Delta t \to 0} \frac{v}{\rho}\frac{\Delta s}{\Delta t} = \frac{v}{\rho}\frac{ds}{dt}$$

Recalling from (11.15) that $ds/dt = v$, we have

$$a_n = \frac{v^2}{\rho} \qquad (11.29)$$

Formula (11.29) expresses that the *normal component* a_n of the acceleration at P is equal to the *square of the speed divided by the radius of curvature of the path at P.* It should be noted that the vector $\mathbf{a}_n$ is always directed *toward the center of curvature of the path* (Fig. 11.20*c*).

It appears from the above that the tangential component of the acceleration reflects a change in the speed of the particle, while its normal component reflects a change in the direction of motion of the particle. The acceleration of a particle will be zero only if both its components are zero. Thus, the acceleration of a particle moving with constant speed along a curve will not be zero, unless the particle happens to pass through a point of inflection of the curve (where the radius of curvature is infinite) or unless the curve is a straight line.

The fact that the normal component of the acceleration depends upon the radius of curvature of the path followed by the particle is taken into account in the design of structures or mechanisms as widely different as airplane wings, railroad tracks, and cams. In order to avoid sudden changes in the acceleration of the air particles flowing past a wing, wing profiles are designed without any sudden change in curvature. Similar care is taken in designing railroad curves, to avoid sudden changes in the acceleration of the cars (which would be hard on the equipment and unpleasant for the passengers). A straight section of track, for instance, is never directly followed by a circular section. Special transition sections are used, to help pass smoothly from the infinite radius of curvature of the straight section to the finite

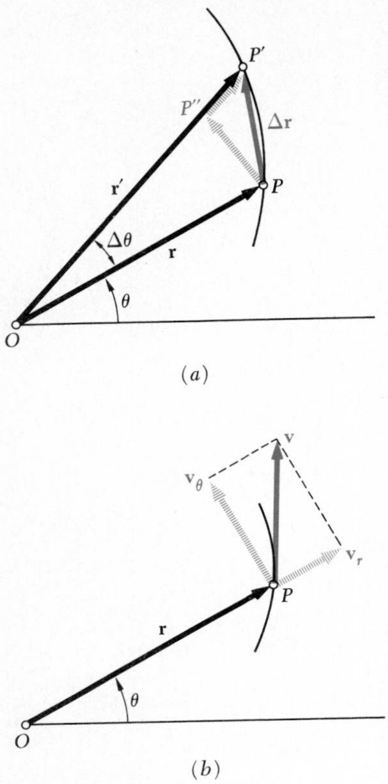

Fig. 11.21

radius of the circular track. Likewise, in the design of high-speed cams, abrupt changes in acceleration are avoided by using transition curves which produce a continuous change in acceleration.

11.13. Radial and Transverse Components. In certain problems, the position of a particle is defined by its polar coordinates r and θ. It is then convenient to resolve the velocity and acceleration of the particle into components parallel and perpendicular, respectively, to the position vector $\mathbf{r}$. These components are called *radial* and *transverse components*.

Consider the vectors $\mathbf{r}$ and $\mathbf{r}'$ defining the position of the particle respectively at times t and $t + \Delta t$ (Fig. 11.21a). We recall that the velocity of the particle at time t is defined as the limit of the quotient of $\Delta\mathbf{r}$ and Δt, where $\Delta\mathbf{r}$ is the vector joining P and P'. To obtain the radial component $\mathbf{v}_r$ and the transverse component $\mathbf{v}_\theta$ of the velocity, we measure a distance $OP'' = OP$ along the line OP' and resolve the vector $\Delta\mathbf{r}$ into its components $\overrightarrow{PP''}$ and $\overrightarrow{P''P'}$. The velocity $\mathbf{v}$ may thus be expressed as

$$\mathbf{v} = \lim_{\Delta t \to 0} \frac{\Delta\mathbf{r}}{\Delta t} = \lim_{\Delta t \to 0} \frac{\overrightarrow{PP''}}{\Delta t} + \lim_{\Delta t \to 0} \frac{\overrightarrow{P''P'}}{\Delta t}$$

As Δt approaches zero, the vectors $\overrightarrow{P''P'}$ and $\overrightarrow{PP''}$ become parallel and perpendicular, respectively, to $\mathbf{r}$. We must have, therefore,

$$\mathbf{v}_r = \lim_{\Delta t \to 0} \frac{\overrightarrow{P''P'}}{\Delta t} \qquad \mathbf{v}_\theta = \lim_{\Delta t \to 0} \frac{\overrightarrow{PP''}}{\Delta t}$$

We observe that the length of the vector $\overrightarrow{P''P'}$ measures the change in magnitude $r' - r = \Delta r$ of the radius vector. (Note that the change in magnitude Δr should not be confused with the magnitude of the vector $\Delta\mathbf{r}$, which represents a change in direction as well as in magnitude.) On the other hand, the length of the vector $\overrightarrow{PP''}$ is equal to $r\,\Delta\theta$ for small values of $\Delta\theta$. The scalar components of the velocity are therefore

$$v_r = \lim_{\Delta t \to 0} \frac{\Delta r}{\Delta t} \qquad v_\theta = \lim_{\Delta t \to 0} \frac{r\,\Delta\theta}{\Delta t}$$

$$v_r = \frac{dr}{dt} \qquad v_\theta = r\frac{d\theta}{dt}$$

or, using dots to indicate time derivatives,

$$v_r = \dot{r} \qquad v_\theta = r\dot{\theta} \tag{11.30}$$

Consider, now, the velocity $\mathbf{v}$ of P at a time t and its velocity $\mathbf{v}'$ at a time $t + \Delta t$ (Fig. 11.22a). Drawing the radial and transverse components of $\mathbf{v}$ and $\mathbf{v}'$ from the same origin O', we represent the change in velocity by the two vectors $\overrightarrow{RR'}$ and $\overrightarrow{TT'}$ (Fig. 11.22b). The vector $\overrightarrow{RR'}$ represents a change in magnitude and direction for the vector $\mathbf{v}_r$; similarly, the vector $\overrightarrow{TT'}$ represents a change in magnitude and direction for the vector $\mathbf{v}_\theta$. The sum of the vectors $\overrightarrow{RR'}$ and $\overrightarrow{TT'}$ represents the total change in velocity $\Delta\mathbf{v}$. As we saw earlier, the acceleration of the particle is defined as the limit of the quotient of $\Delta\mathbf{v}$ and Δt. To obtain the radial component $\mathbf{a}_r$ and the transverse component $\mathbf{a}_\theta$ of the acceleration, we resolve the vector $\overrightarrow{RR'}$ into its components $\overrightarrow{RR''}$ and $\overrightarrow{R''R'}$ (with $O'R'' = O'R$) and the vector $\overrightarrow{TT'}$ into its components $\overrightarrow{TT''}$ and $\overrightarrow{T''T'}$ (with $O'T'' = O'T$). The acceleration $\mathbf{a}$ of the particle may then be expressed as

$$\mathbf{a} = \lim_{\Delta t \to 0} \frac{\Delta \mathbf{v}}{\Delta t} = \lim_{\Delta t \to 0} \left(\frac{\overrightarrow{RR'}}{\Delta t} + \frac{\overrightarrow{TT'}}{\Delta t} \right)$$

$$= \lim_{\Delta t \to 0} \left(\frac{\overrightarrow{RR''}}{\Delta t} + \frac{\overrightarrow{R''R'}}{\Delta t} + \frac{\overrightarrow{TT''}}{\Delta t} + \frac{\overrightarrow{T''T'}}{\Delta t} \right)$$

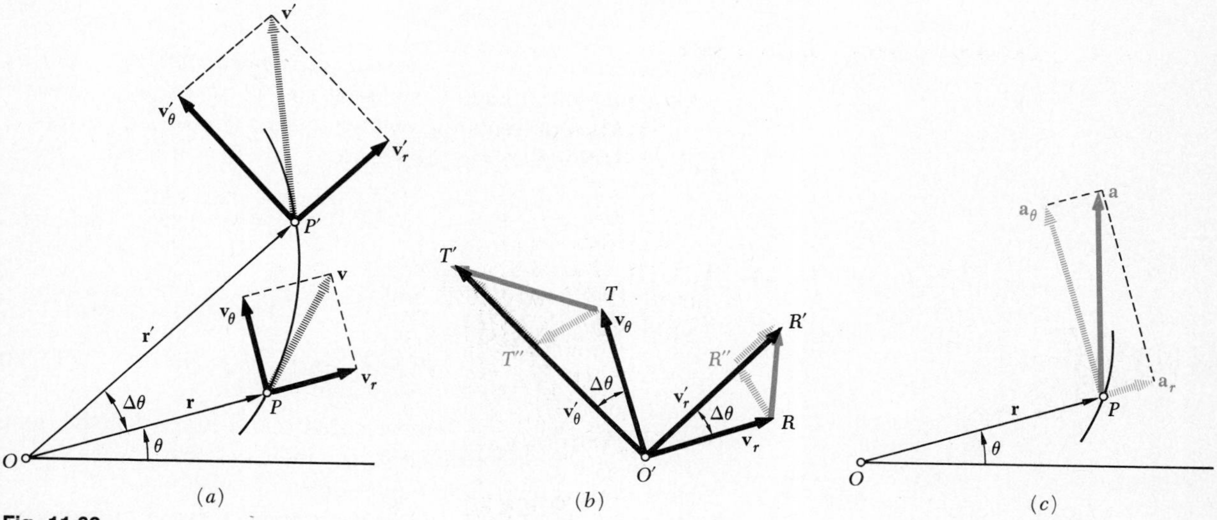

Fig. 11.22

As Δt approaches zero, the vectors $\overrightarrow{R''R'}$ and $\overrightarrow{TT''}$ become parallel to **r**, while the vectors $\overrightarrow{RR''}$ and $\overrightarrow{T''T'}$ become perpendicular to **r**. We must have, therefore,

$$\mathbf{a}_r = \lim_{\Delta t \to 0} \left(\frac{\overrightarrow{R''R'}}{\Delta t} + \frac{\overrightarrow{TT''}}{\Delta t} \right) \qquad \mathbf{a}_\theta = \lim_{\Delta t \to 0} \left(\frac{\overrightarrow{T''T'}}{\Delta t} + \frac{\overrightarrow{RR''}}{\Delta t} \right)$$

From Fig. 11.22*b* we observe that, for small values of Δt, the lengths of the various vectors involved are

$$R''R' = \Delta(v_r) \qquad T''T' = \Delta(v_\theta)$$
$$RR'' = v_r\,\Delta\theta \qquad TT'' = v_\theta\,\Delta\theta$$

where $\Delta(v_r)$ and $\Delta(v_\theta)$ represent the *change in magnitude* of the vectors $\mathbf{v}_r$ and $\mathbf{v}_\theta$, respectively. Noting also that the vector $\overrightarrow{TT''}$ has a sense opposite to that of **r**, we obtain the following expressions for the scalar components of the acceleration:

$$a_r = \lim_{\Delta t \to 0} \left[\frac{\Delta(v_r)}{\Delta t} - \frac{v_\theta\,\Delta\theta}{\Delta t} \right] \qquad a_\theta = \lim_{\Delta t \to 0} \left[\frac{\Delta(v_\theta)}{\Delta t} + \frac{v_r\,\Delta\theta}{\Delta t} \right]$$

$$a_r = \frac{d(v_r)}{dt} - v_\theta \frac{d\theta}{dt} \qquad a_\theta = \frac{d(v_\theta)}{dt} + v_r \frac{d\theta}{dt}$$

or, using dots to indicate time derivatives,

$$a_r = \dot{v}_r - v_\theta \dot{\theta} \qquad a_\theta = \dot{v}_\theta + v_r \dot{\theta} \tag{11.31}$$

We should note that, in general, a_r *is not equal to* $\dot{v}_r$ and a_θ *is not equal to* $\dot{v}_\theta$. Each of the relations (11.31) contains an extra term which represents the *change in direction* of the vectors $\mathbf{v}_r$ and $\mathbf{v}_\theta$. Differentiating the relations (11.30), we obtain

$$\dot{v}_r = \ddot{r} \qquad \dot{v}_\theta = \frac{d}{dt}(r\dot{\theta}) = r\ddot{\theta} + \dot{r}\dot{\theta} \tag{11.32}$$

Substituting from (11.30) and (11.32) into (11.31), we write

$$a_r = \ddot{r} - r\dot{\theta}^2 \qquad a_\theta = r\ddot{\theta} + 2\dot{r}\dot{\theta} \tag{11.33}$$

A positive value for a_r indicates that the vector $\mathbf{a}_r$ has the same sense as **r**; a positive value for a_θ indicates that the vector $\mathbf{a}_\theta$ points toward increasing values of θ.

In the case of a particle moving along a circle of center O, we have $r =$ constant, $\dot{r} = \ddot{r} = 0$, and the formulas (11.33) reduce to

$$a_r = -r\dot{\theta}^2 \qquad a_\theta = r\ddot{\theta} \tag{11.34}$$

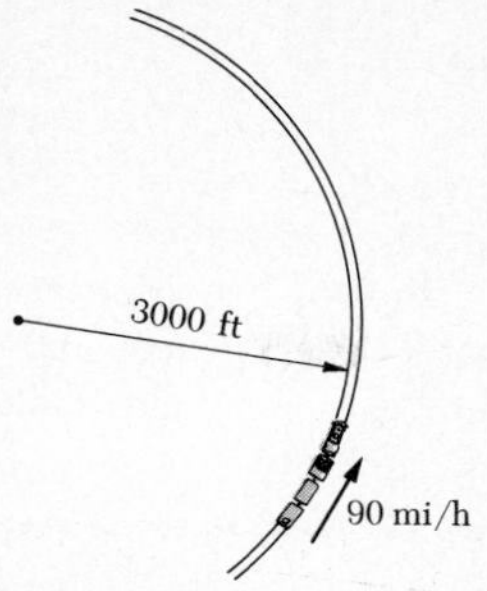

SAMPLE PROBLEM 11.10

A train is traveling on a curved section of track of radius 3000 ft at the speed of 90 mi/h. The brakes are suddenly applied, causing the train to slow down at a constant rate; after 6 s, the speed has been reduced to 60 mi/h. Determine the acceleration of a car immediately after the brakes have been applied.

Tangential Component of Acceleration. First the speeds are expressed in ft/s.

$$90\text{ mi/h} = \left(90\,\frac{\text{mi}}{\text{h}}\right)\left(\frac{5280\text{ ft}}{1\text{ mi}}\right)\left(\frac{1\text{ h}}{3600\text{ s}}\right) = 132\text{ ft/s}$$

$$60\text{ mi/h} = 88\text{ ft/s}$$

Since the train slows down at a constant rate, we have

$$a_t = \text{average } a_t = \frac{\Delta v}{\Delta t} = \frac{88\text{ ft/s} - 132\text{ ft/s}}{6\text{ s}} = -7.33\text{ ft/s}^2$$

Normal Component of Acceleration. Immediately after the brakes have been applied, the speed is still 132 ft/s, and we have

$$a_n = \frac{v^2}{\rho} = \frac{(132\text{ ft/s})^2}{3000\text{ ft}} = 5.81\text{ ft/s}^2$$

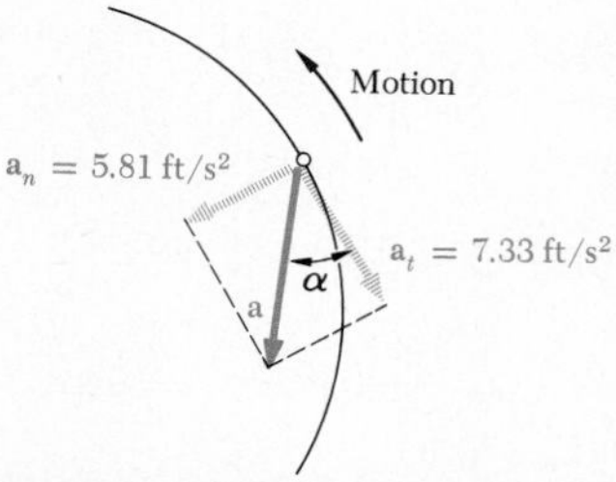

Magnitude and Direction of Acceleration. The magnitude and direction of the resultant **a** of the components $\mathbf{a}_n$ and $\mathbf{a}_t$ are

$$\tan\alpha = \frac{a_n}{a_t} = \frac{5.81\text{ ft/s}^2}{7.33\text{ ft/s}^2} \qquad \alpha = 38.4° ◀$$

$$a = \frac{a_n}{\sin\alpha} = \frac{5.81\text{ ft/s}^2}{\sin 38.4°} \qquad a = 9.35\text{ ft/s}^2 ◀$$

SAMPLE PROBLEM 11.11

Determine the minimum radius of curvature of the trajectory described by the projectile considered in Sample Prob. 11.7.

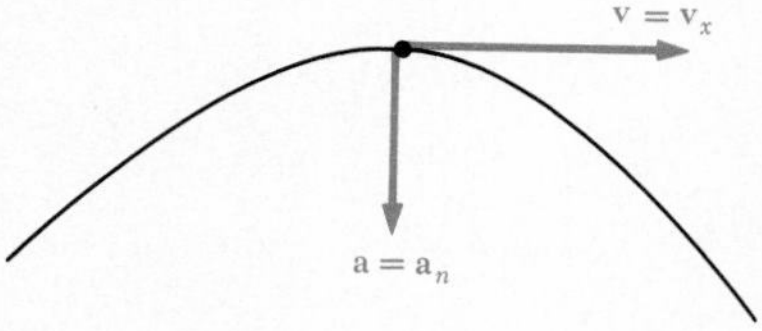

Solution. Since $a_n = v^2/\rho$, we have $\rho = v^2/a_n$. The radius will be small when v is small or when a_n is large. The speed v is minimum at the top of the trajectory since $v_y = 0$ at that point; a_n is maximum at that same point, since the direction of the vertical coincides with the direction of the normal. Therefore, the minimum radius of curvature occurs at the top of the trajectory. At this point, we have

$$v = v_x = 155.9\text{ m/s} \qquad a_n = a = 9.81\text{ m/s}^2$$

$$\rho = \frac{v^2}{a_n} = \frac{(155.9\text{ m/s})^2}{9.81\text{ m/s}^2} \qquad \rho = 2480\text{ m} ◀$$

SAMPLE PROBLEM 11.12

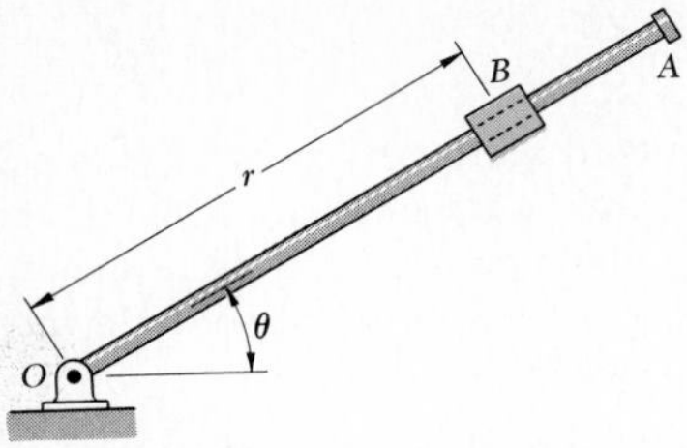

The rotation of the 3-ft arm OA about O is defined by the relation $\theta = 0.15t^2$, where θ is expressed in radians and t in seconds. Block B slides along the arm in such a way that its distance from O is $r = 3 - 0.40t^2$, where r is expressed in feet and t in seconds. Determine the total velocity and the total acceleration of block B after the arm OA has rotated through 30°.

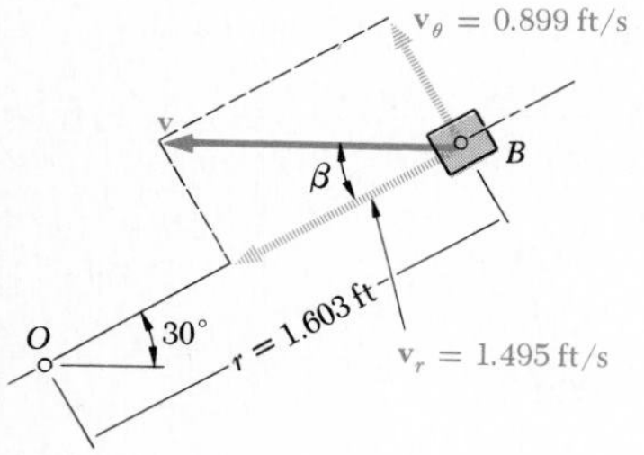

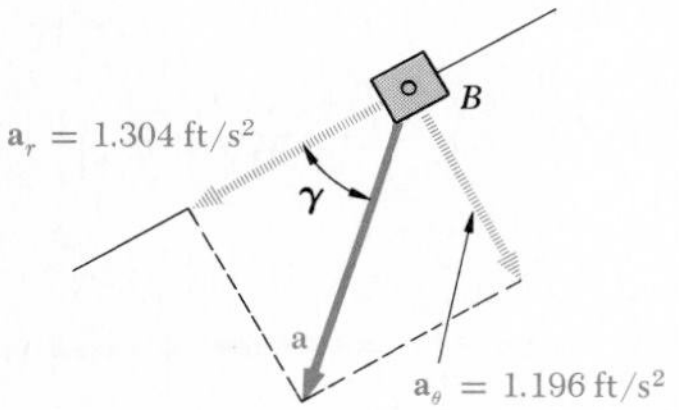

Solution. We first find the time t at which $\theta = 30°$. Substituting $\theta = 30° = 0.524$ rad into the expression for θ, we obtain

$$\theta = 0.15t^2 \qquad 0.524 = 0.15t^2 \qquad t = 1.869 \text{ s}$$

Equations of Motion. Substituting $t = 1.869$ s in the expressions for r, θ, and their first and second derivatives, we have

$$r = 3 - 0.40t^2 = 1.603 \text{ ft} \qquad \theta = 0.15t^2 = 0.524 \text{ rad}$$
$$\dot{r} = -0.80t = -1.495 \text{ ft/s} \qquad \dot{\theta} = 0.30t = 0.561 \text{ rad/s}$$
$$\ddot{r} = -0.80 = -0.800 \text{ ft/s}^2 \qquad \ddot{\theta} = 0.30 = 0.300 \text{ rad/s}^2$$

Velocity of B. Using Eqs. (11.30), we obtain the values of v_r and v_θ when $t = 1.869$ s.

$$v_r = \dot{r} = -1.495 \text{ ft/s}$$
$$v_\theta = r\dot{\theta} = 1.603(0.561) = 0.899 \text{ ft/s}$$

Solving the right triangle shown, we obtain the magnitude and direction of the velocity,

$$v = 1.744 \text{ ft/s} \qquad \beta = 31.0° \quad ◀$$

Acceleration of B. Using Eqs. (11.33), we obtain

$$a_r = \ddot{r} - r\dot{\theta}^2$$
$$= -0.800 - 1.603(0.561)^2 = -1.304 \text{ ft/s}^2$$
$$a_\theta = r\ddot{\theta} + 2\dot{r}\dot{\theta}$$
$$= 1.603(0.300) + 2(-1.495)(0.561) = -1.196 \text{ ft/s}^2$$
$$a = 1.770 \text{ ft/s} \qquad \gamma = 42.5° \quad ◀$$

96-104 108, 110

PROBLEMS

11.96 An automobile travels at a constant speed on a highway curve of 1000-m radius. If the normal component of the acceleration is not to exceed 1.2 m/s^2, determine the maximum allowable speed.

11.97 A car goes around a highway curve of 300-m radius at a speed of 90 km/h. (*a*) What is the normal component of its acceleration? (*b*) At what speed is the normal component of the acceleration one-half as large as that found in part *a*?

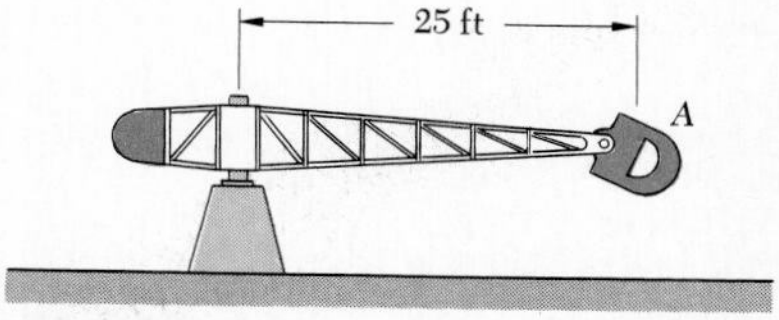

Fig. P11.98

11.98 Determine the peripheral speed of the centrifuge test cab *A* for which the normal component of the acceleration is 10*g*.

11.99 It is desired to have a normal component of acceleration of 8*g* during the testing of an airplane flying at 1200 mi/h. Determine the radius of the circular path along which the pilot should fly the airplane.

11.100 A motorist enters a curve of 500-ft radius at a speed of 30 mi/h. If he increases his speed at a constant rate of 3 ft/s^2, determine the magnitude of the total acceleration after he has traveled 400 ft along the curve.

11.101 A passenger train is traveling at 60 mi/h along a curve of 3000-ft radius. If the maximum total acceleration is not to exceed *g*/10, determine the maximum rate at which the speed may be decreased.

11.102 The speed of a racing car is increased at a constant rate from 90 km/h to 126 km/h over a distance of 150 m along a curve of 250-m radius. Determine the magnitude of the total acceleration of the car after it has traveled 100 m along the curve.

11.103 A monorail train is traveling at a speed of 144 km/h along a curve of 1000-m radius. Determine the maximum rate at which the speed may be decreased if the total acceleration of the train is not to exceed 2 m/s^2.

11.104 A nozzle discharges a stream of water in the direction shown with an initial velocity of 25 m/s. Determine the radius of curvature of the stream (*a*) as it leaves the nozzle, (*b*) at the maximum height of the stream.

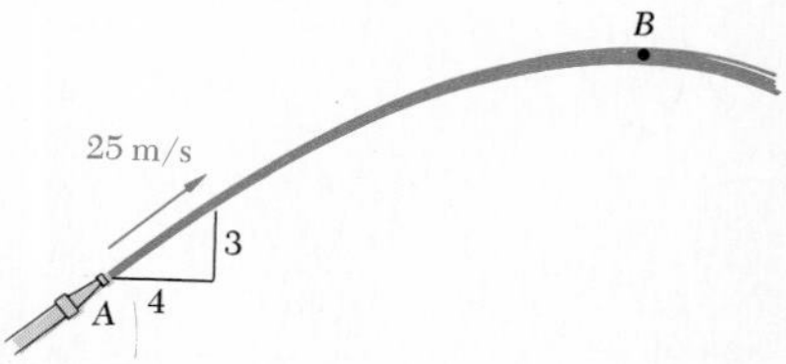

Fig. P11.104

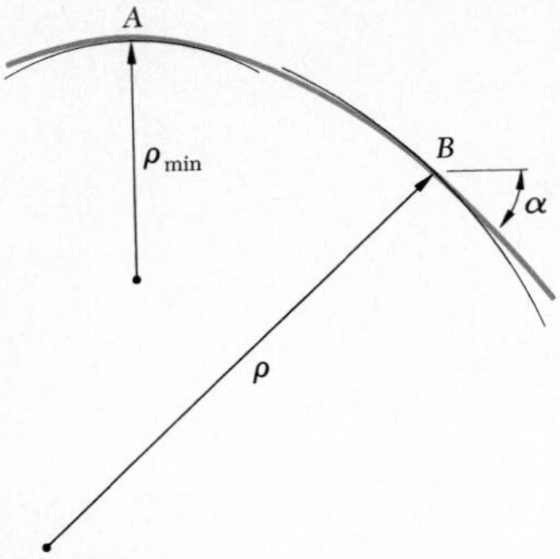

Fig. P11.105

11.105 (*a*) Show that the radius of curvature of the trajectory of a projectile reaches its minimum value at the highest point A of the trajectory. (*b*) Denoting by α the angle formed by the trajectory and the horizontal at a given point B, show that the radius of curvature of the trajectory at B is $\rho = \rho_{min}/\cos^3 \alpha$.

11.106 Determine the radius of curvature of the trajectory described by the projectile of Sample Prob. 11.7 as the projectile leaves the gun.

11.107 For each of the two firing angles obtained in Sample Prob. 11.8, determine the radius of curvature of the trajectory described by the projectile as it leaves the gun.

11.108 A satellite will travel indefinitely in a circular orbit around the earth if the normal component of its acceleration is equal to $g(R/r)^2$, where $g = 32.2\ \text{ft/s}^2$, R = radius of the earth = 3960 mi, and r = distance from the center of the earth to the satellite. Determine the height above the surface of the earth at which a satellite will travel indefinitely around the earth at a speed of 16,500 mi/h.

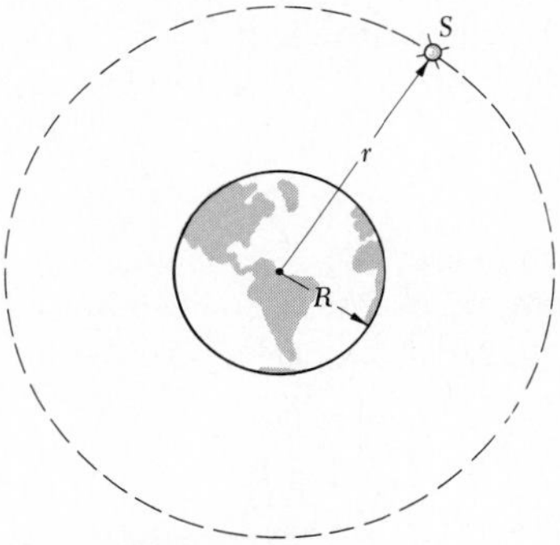

Fig. P11.108

11.109 Determine the speed of an earth satellite traveling in a circular orbit 1000 mi above the surface of the earth. (See information given in Prob. 11.108.)

11.110 Assuming the orbit of the moon to be a circle of radius 239,000 mi, determine the speed of the moon relative to the earth. (See information given in Prob. 11.108.)

11.111 Show that the speed of an earth satellite traveling in a circular orbit is inversely proportional to the square root of the radius of its orbit. Also, determine the minimum time in which a satellite can circle the earth. (See information given in Prob. 11.108.)

11.112 The two-dimensional motion of a particle is defined by the relations $r = 60t^2 - 20t^3$ and $\theta = 2t^2$, where r is expressed in millimeters, t in seconds, and θ in radians. Determine the velocity and acceleration of the particle when (*a*) $t = 0$, (*b*) $t = 1$ s.

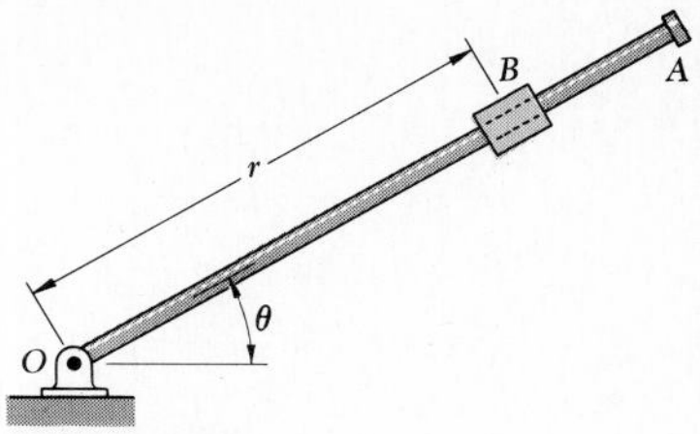

Fig. P11.112

11.113 The particle of Prob. 11.112 is at the origin at $t = 0$. Determine its velocity and acceleration as it returns to the origin.

11.114 The two-dimensional motion of a particle is defined by the relations $r = 2b \cos \omega t$ and $\theta = \omega t$, where b and ω are constants. Determine (*a*) the velocity and acceleration of the particle at any instant, (*b*) the radius of curvature of its path. What conclusion can you draw regarding the path of the particle?

11.115 The two-dimensional motion of a particle is defined by the relations $r = b \sin \pi t$ and $\theta = \frac{1}{2}\pi t$. Determine the velocity and acceleration of the particle when (*a*) $t = \frac{1}{2}$, (*b*) $t = 1$.

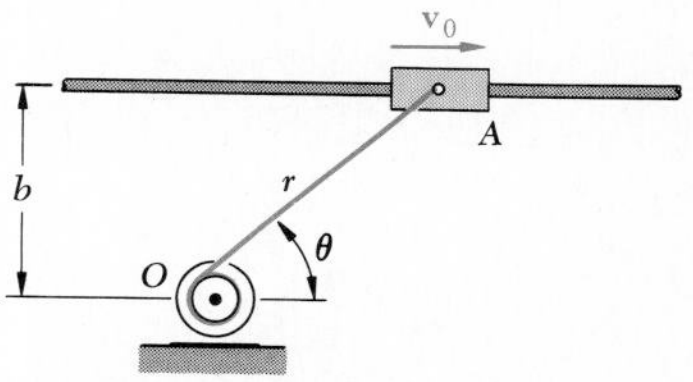

Fig. P11.116

11.116 A wire OA connects the collar A and a reel located at O. Knowing that the collar moves to the right with a constant speed v_0, determine $d\theta/dt$ in terms of v_0, b, and θ.

11.117 A rocket is fired vertically from a launching pad at B. Its flight is tracked by radar from point A. Determine the velocity of the rocket in terms of d, θ, and $\dot{\theta}$.

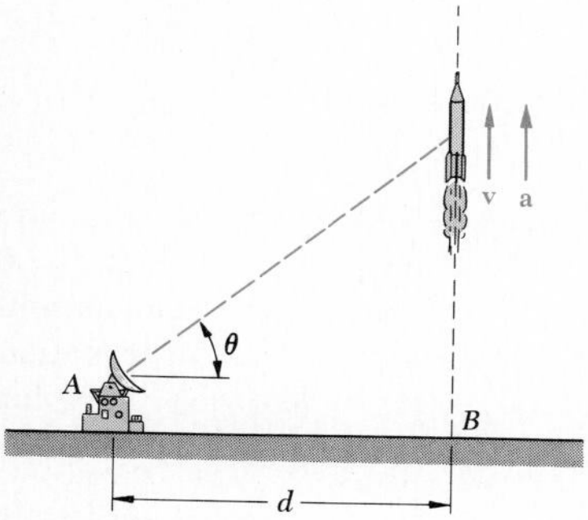

Fig. P11.117

11.118 Determine the acceleration of the rocket of Prob. 11.117 in terms of d, θ, $\dot{\theta}$, and $\ddot{\theta}$.

11.119 As the rod OA rotates, the pin P moves along the parabola BCD. Knowing that the equation of the parabola is $r = 2b/(1 + \cos \theta)$ and that $\theta = kt$, determine the velocity and acceleration of P when (*a*) $\theta = 0$, (*b*) $\theta = 90°$.

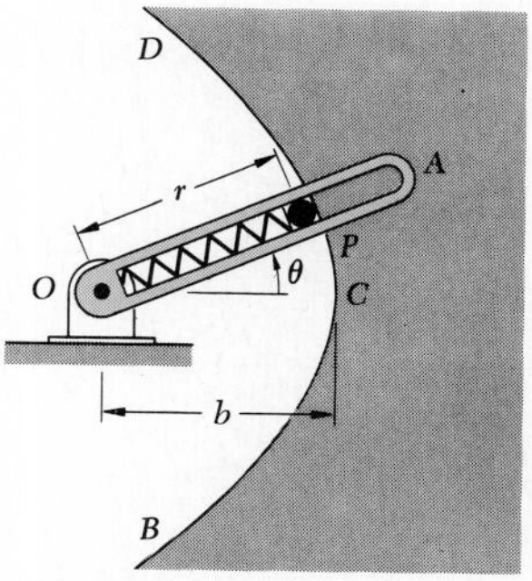

Fig. P11.119

REVIEW PROBLEMS

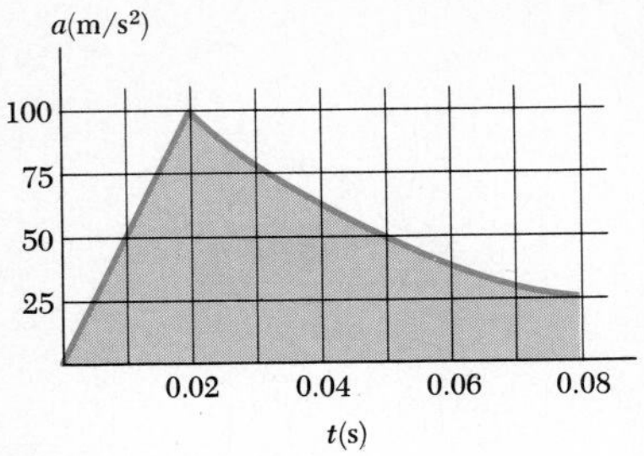

Fig. P11.120

11.120 The a–t curve shown was obtained during the motion of a test sled. Knowing that the sled started from rest at $t = 0$, determine the velocity and position of the sled at $t = 0.08$ s.

11.121 An experimental ion-propulsion engine is capable of giving a space vehicle a constant acceleration of 0.02 ft/s^2. If the engine is placed in operation when the speed of the vehicle is 18,000 mi/h, determine the time required to bring the speed of the vehicle to 19,000 mi/h. Assume that the vehicle is moving in a straight line, far from the sun or any planet.

11.122 As circle B rolls on the fixed circle A, point P describes a cardioid defined by the relations $r = 2b(1 + \cos \frac{1}{2}\pi t)$ and $\theta = \frac{1}{2}\pi t$. Determine the velocity and acceleration of P when (*a*) $t = 1$, (*b*) $t = 2$.

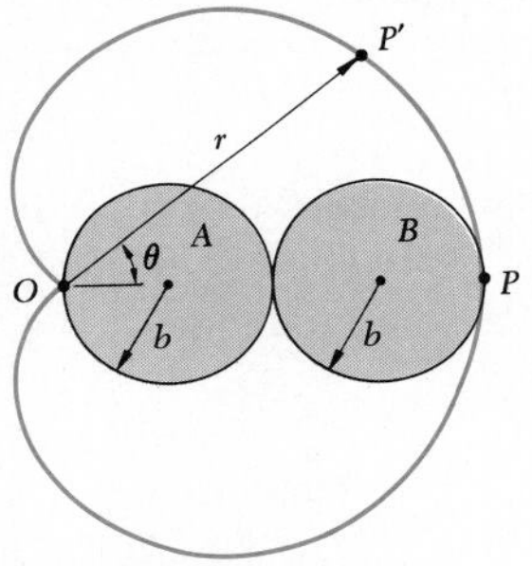

Fig. P11.122

11.123 The acceleration of a particle moving in a straight line is defined by the relation $a = -kv^{1.5}$. The particle starts at $t = 0$ and $x = 0$ with an initial velocity v_0. (*a*) Show that the velocity and position coordinate at any time t are related by the equation $x/t = \sqrt{v_0 v}$. (*b*) Determine the value of k, knowing that for $v_0 = 40$ m/s the particle comes to rest after traveling 2 m.

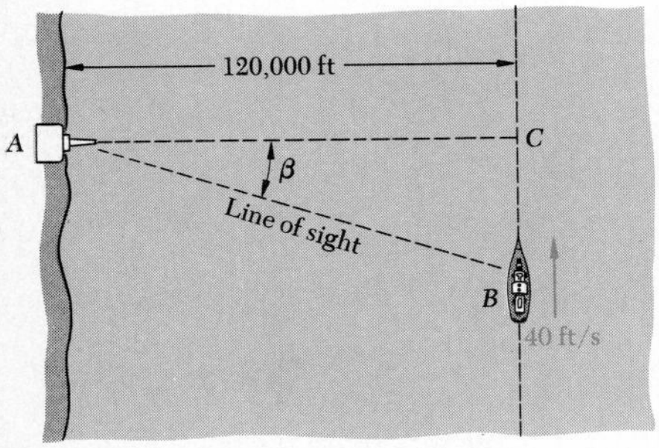

Fig. P11.124

11.124 Ship B is proceeding due north on the course shown. The shore gun at A is trained due east and will be fired in an attempt to hit the ship at point C. Knowing that the muzzle velocity of the gun is 2400 ft/s, determine (*a*) the required firing angle α, (*b*) the required angle β between AC and the line of sight at the time of firing.

11.125 The magnitude in m/s^2 of the deceleration due to air resistance of the nose cone of a small experimental rocket is known to be $6 \times 10^{-4}\, v^2$, where v is expressed in m/s. If the nose cone is projected vertically from the ground with an initial velocity of 100 m/s, determine the maximum height that it will reach.

11.126 Determine the velocity of the nose cone of Prob. 11.125 when it returns to the ground.

11.127 Standing on the side of a hill, an archer shoots an arrow with an initial velocity of 200 ft/s at an angle $\alpha = 15°$ with the horizontal. Determine the horizontal distance d traveled by the arrow before it strikes the ground at B.

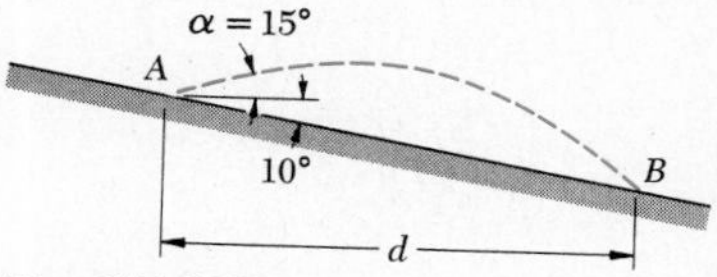

Fig. P11.127

11.128 In Prob. 11.127, determine the radius of curvature of the trajectory (*a*) immediately after the arrow has been shot, (*b*) as the arrow passes through its point of maximum elevation.

11.129 A train starts at a station and accelerates uniformly at a rate of 0.6 m/s^2 until it reaches a speed of 24 m/s; it then proceeds at the constant speed of 24 m/s. Determine the time and the distance traveled if its average velocity is (*a*) 16 m/s, (*b*) 22 m/s.

11.130 Knowing that block B moves downward with a constant velocity of 180 mm/s, determine (*a*) the velocity of block A, (*b*) the velocity of pulley D.

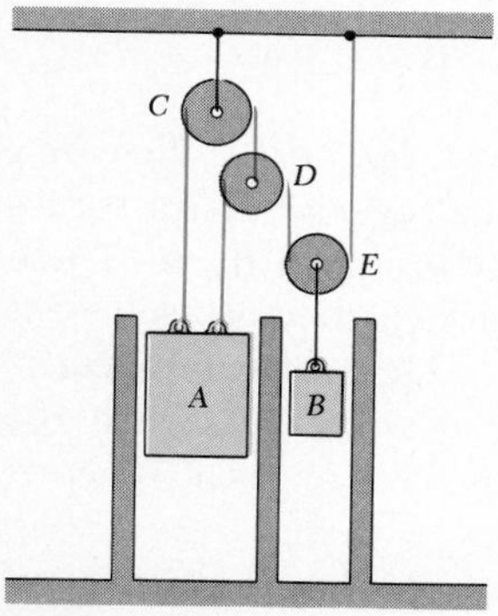

Fig. P11.130

11.131 Drops of water fall down a mine shaft at the uniform rate of one drop per second. A mine elevator moving up the shaft at 30 ft/s is struck by a drop of water when it is 300 ft below ground level. When and where will the next drop of water strike the elevator?

Chapter
12

Kinetics of Particles: Force, Mass, and Acceleration

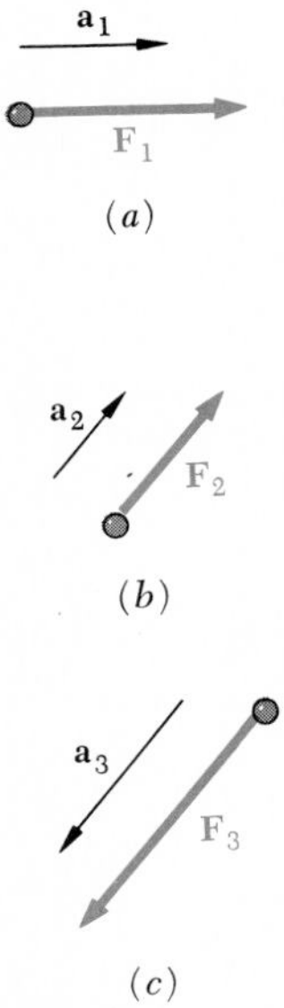

Fig. 12.1

12.1. Newton's Second Law of Motion. Newton's first and third laws of motion were used extensively in statics to study bodies at rest and the forces acting upon them. These two laws are also used in dynamics; in fact, they are sufficient for the study of the motion of bodies which have no acceleration. However, when bodies are accelerated, i.e., when the magnitude or the direction of their velocity changes, it is necessary to use the second law of motion in order to relate the motion of the body with the forces acting on it. This law may be stated as follows:

If the resultant force acting on a particle is not zero, the particle will have an acceleration proportional to the magnitude of the resultant and in the direction of this resultant force.

Newton's second law of motion may best be understood if we imagine the following experiment: A particle is subjected to a force $\mathbf{F}_1$ of constant direction and constant magnitude F_1. Under the action of that force, the particle will be observed to move in a straight line and *in the direction of the force* (Fig. 12.1*a*). By determining the position of the particle at various instants, we find that its acceleration has a constant magnitude a_1. If the experiment is repeated with forces $\mathbf{F}_2$, $\mathbf{F}_3$, etc., of different magnitude or direction (Fig. 12.1*b* and *c*), we find each time that the particle moves in the direction of the force acting on it and that the magnitudes a_1, a_2, a_3, etc., of the accelerations are proportional to the magnitudes F_1, F_2, F_3, etc., of the corresponding forces,

$$\frac{F_1}{a_1} = \frac{F_2}{a_2} = \frac{F_3}{a_3} = \cdots = \text{constant}$$

The constant value obtained for the ratio of the magnitudes of the forces and accelerations is a characteristic of the particle under consideration. It is called the *mass* of the particle and is denoted by m. When a particle of mass m is acted upon by a force **F**, the force **F** and the acceleration **a** of the particle must therefore satisfy the relation

$$\mathbf{F} = m\mathbf{a} \tag{12.1}$$

This relation provides a complete formulation of Newton's second law; it expresses not only that the magnitudes of **F** and **a** are proportional, but also (since m is a positive scalar) that the vectors **F** and **a** have the same direction (Fig. 12.2). We should note that Eq. (12.1) still holds when **F** is not constant but varies with t in magnitude or direction. The magnitudes of **F** and **a** remain proportional, and the two vectors have the same direction at any given instant. However, they will not, in general, be tangent to the path of the particle.

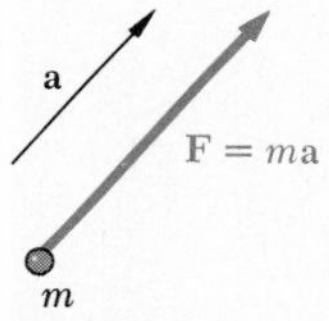

Fig. 12.2

When a particle is subjected simultaneously to several forces, Eq. (12.1) should be replaced by

$$\Sigma\mathbf{F} = m\mathbf{a} \tag{12.2}$$

where $\Sigma\mathbf{F}$ represents the sum, or resultant, of all the forces acting on the particle.

It should be noted that the system of axes with respect to which the acceleration **a** is determined is not arbitrary. These axes must have a constant orientation with respect to the stars, and their origin must either be attached to the sun† or move with a constant velocity with respect to the sun. Such a system of axes is called a *newtonian frame of reference.*‡ A system of axes attached to the earth does *not* constitute a newtonian frame of reference, since the earth rotates with respect to the stars and is accelerated with respect to the sun. However, in most

† More accurately, to the mass center of the solar system.

‡ Since the stars are not actually fixed, a more rigorous definition of a newtonian frame of reference (also called *inertial system*) is *one with respect to which Eq.* (12.2) *holds.*

engineering applications, the acceleration **a** may be determined with respect to axes attached to the earth and Eqs. (12.1) and (12.2) used without any appreciable error. On the other hand, these equations do not hold if **a** represents a relative acceleration measured with respect to moving axes, such as axes attached to an accelerated car or to a rotating piece of machinery.

We may observe that, if the resultant $\Sigma\mathbf{F}$ of the forces acting on the particle is zero, it follows from Eq. (12.2) that the acceleration **a** of the particle is also zero. If the particle is initially at rest ($\mathbf{v}_0 = 0$) with respect to the newtonian frame of reference used, it will thus remain at rest ($\mathbf{v} = 0$). If originally moving with a velocity $\mathbf{v}_0$, the particle will maintain a constant velocity $\mathbf{v} = \mathbf{v}_0$; that is, it will move with the constant speed v_0 in a straight line. This, we recall, is the statement of Newton's first law (Sec. 2.9). Thus, Newton's first law is a particular case of Newton's second law and may be omitted from the fundamental principles of mechanics.

12.2. Systems of Units. In using the fundamental equation $\mathbf{F} = m\mathbf{a}$, the units of force, mass, length, and time cannot be chosen arbitrarily. If they are, the magnitude of the force **F** required to give an acceleration **a** to the mass m will *not* be numerically equal to the product ma; it will only be proportional to this product. Thus, we may choose three of the four units arbitrarily but must choose the fourth unit so that the equation $\mathbf{F} = m\mathbf{a}$ is satisfied. The units are then said to form a consistent system of kinetic units.

Two systems of consistent kinetic units are currently used by American engineers, the International System of Units (SI units†), and the U.S. customary units. Since both systems have been discussed in detail in Sec. 1.3, we shall describe them only briefly in this section.

International System of Units (SI Units). In this system, the base units are the units of length, mass, and time, and are called, respectively, the *meter* (m), the *kilogram* (kg), and the *second* (s). All three are arbitrarily defined (Sec. 1.3). The unit of force is a derived unit. It is called the *newton* (N) and is defined as the force which gives an acceleration of 1 m/s^2 to a mass of 1 kg (Fig. 12.3). From Eq. (12.1) we write

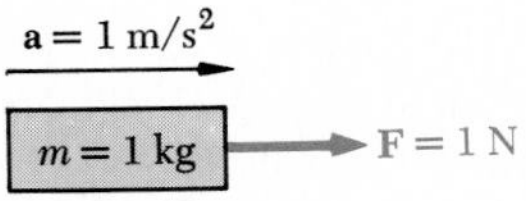

Fig. 12.3

$$1 \text{ N} = (1 \text{ kg})(1 \text{ m/s}^2) = 1 \text{ kg} \cdot \text{m/s}^2$$

The SI units are said to form an *absolute* system of units. This means that the three base units chosen are independent of the location where measurements are made. The meter, the kilogram, and the second may be used anywhere on the earth; they

† SI stands for *Système International d'Unités* (French).

may even be used on another planet. They will always have the same significance.

Like any other force, the *weight* **W** of a body should be expressed in newtons. Since a body subjected to its own weight acquires an acceleration equal to the acceleration of gravity g, it follows from Newton's second law that the magnitude W of the weight of a body of mass m is

$$W = mg \tag{12.3}$$

Recalling that $g = 9.81\ \text{m/s}^2$, we find that the weight of a body of mass 1 kg (Fig. 12.4) is

$$W = (1\ \text{kg})(9.81\ \text{m/s}^2) = 9.81\ \text{N}$$

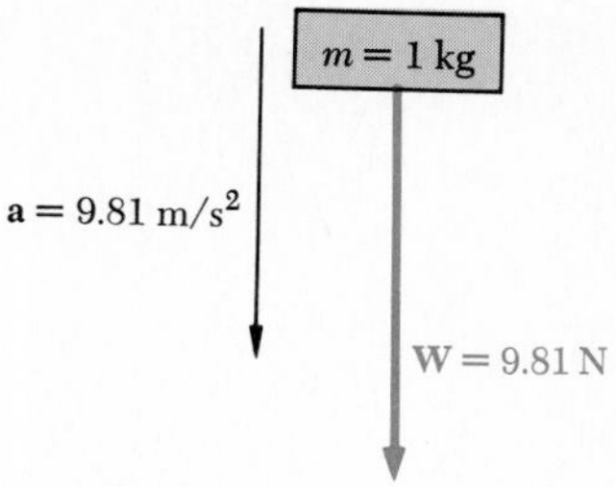

Fig. 12.4

Multiples and submultiples of the units of length, mass, and force are frequently used in engineering practice. They are, respectively, the *kilometer* (km) and the *millimeter* (mm); the *megagram*† (Mg) and the *gram* (g); and the *kilonewton* (kN). By definition

$$1\ \text{km} = 1000\ \text{m} \qquad 1\ \text{mm} = 0.001\ \text{m}$$
$$1\ \text{Mg} = 1000\ \text{kg} \qquad 1\ \text{g} = 0.001\ \text{kg}$$
$$1\ \text{kN} = 1000\ \text{N}$$

The conversion of these units to meters, kilograms, and newtons, respectively, can be effected by simply moving the decimal point three places to the right or to the left.

U.S. Customary Units. Most practicing American engineers still commonly use a system in which the base units are the units of length, force, and time. These units are, respectively, the *foot* (ft), the *pound* (lb), and the *second* (s). The second is the same as the corresponding SI unit. The foot is defined as 0.3048 m. The pound is defined as the *weight* of a platinum standard, called the *standard pound* and kept at the National Bureau of Standards in Washington, the mass of which is 0.453 592 43 kg. Since the weight of a body depends upon the gravitational attraction of the earth, which varies with location, it is specified that the standard pound should be placed at sea level and at the latitude of 45° to properly define a force of 1 lb. Clearly the U.S. customary units do not form an absolute system of units. Because of their dependence upon the gravitational attraction of the earth, they are said to form a *gravitational* system of units.

While the standard pound also serves as the unit of mass in commercial transactions in the United States, it cannot be so used in engineering computations since such a unit would not be consistent with the base units defined in the preceding para-

† Also known as a *metric ton*.

graph. Indeed, when acted upon by a force of 1 lb, that is, when subjected to its own weight, the standard pound receives the acceleration of gravity, $g = 32.2\ \text{ft/s}^2$ (Fig. 12.5), not the unit acceleration required by Eq. (12.1). The unit of mass consistent with the foot, the pound, and the second is the mass which receives an acceleration of $1\ \text{ft/s}^2$ when a force of 1 lb is applied to it (Fig. 12.6). This unit, sometimes called a *slug*, can be derived from the equation $F = ma$ after substituting 1 lb and $1\ \text{ft/s}^2$ for F and a, respectively. We write

$$F = ma \qquad 1\ \text{lb} = (1\ \text{slug})(1\ \text{ft/s}^2)$$

and obtain

$$1\ \text{slug} = \frac{1\ \text{lb}}{1\ \text{ft/s}^2} = 1\ \text{lb}\cdot\text{s}^2/\text{ft}$$

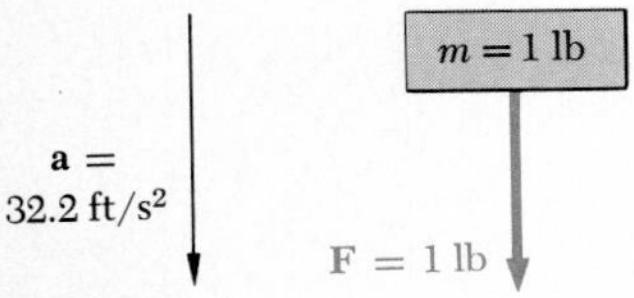

Fig. 12.5

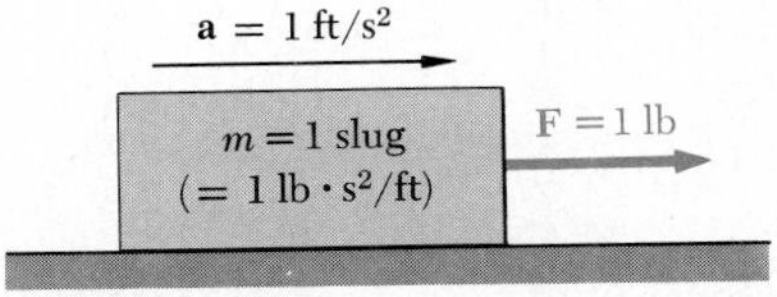

Fig. 12.6

Comparing Figs. 12.5 and 12.6, we conclude that the slug is a mass 32.2 times larger than the mass of the standard pound.

The fact that bodies are characterized in the U.S. customary system of units by their weight in pounds, rather than by their mass in slugs, was a convenience in the study of statics, where we were dealing constantly with weights and other forces and only seldom with masses. However, in the study of kinetics, where forces, masses, and accelerations are involved, we repeatedly shall have to express the mass m in slugs of a body, the weight W of which has been given in pounds. Recalling Eq. (12.3), we shall write

$$m = \frac{W}{g} \tag{12.4}$$

where g is the acceleration of gravity ($g = 32.2\ \text{ft/s}^2$).

The conversion from U.S. customary units to SI units, and vice versa, has been discussed in Sec. 1.4. We shall recall the conversion factors obtained respectively for the units of length, force, and mass:

Length:	$1\ \text{ft} = 0.3048\ \text{m}$
Force:	$1\ \text{lb} = 4.448\ \text{N}$
Mass:	$1\ \text{slug} = 1\ \text{lb}\cdot\text{s}^2/\text{ft} = 14.59\ \text{kg}$

Although it cannot be used as a consistent unit of mass, we also recall that the mass of the standard pound is, by definition,

$$1 \text{ pound-mass} = 0.4536 \text{ kg}$$

This constant may be used to determine the *mass* in SI units (kilograms) of a body which has been characterized by its *weight* in U.S. customary units (pounds).

12.3. Equations of Motion. Dynamic Equilibrium. Consider a particle of mass m acted upon by several forces. Newton's second law of motion states that the resultant of these forces must be equal to the vector $m\mathbf{a}$ obtained by multiplying the acceleration $\mathbf{a}$ of the particle by its mass m. The given system of forces must therefore be equivalent to the vector $m\mathbf{a}$ (Fig. 12.7). Using rectangular components, we write the equations of motion,

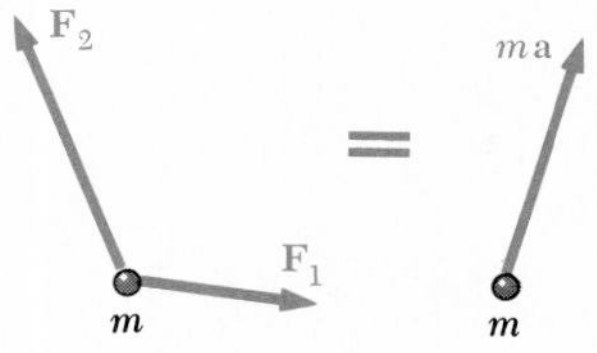

Fig. 12.7

$$\Sigma F_x = ma_x \qquad \Sigma F_y = ma_y \qquad \Sigma F_z = ma_z \qquad (12.5)$$

Newton's second law may also be expressed by considering a vector of magnitude ma, but of *sense opposite* to that of the acceleration. This vector is denoted by $(m\mathbf{a})_{\text{rev}}$, where the subscript indicates that the sense of the acceleration has been reversed, and is called an *inertia vector.* If the inertia vector $(m\mathbf{a})_{\text{rev}}$ is added to the forces acting on the particle, *we obtain a system of vectors equivalent to zero* (Fig. 12.8). The particle may thus be considered to be in equilibrium under the given forces and the inertia vector. The particle is said to be in *dynamic equilibrium,* and the problem under consideration may be solved by using the methods developed earlier in statics. We may, for instance, write that the sums of the components of the vectors shown in Fig. 12.8, *including the inertia vector,* are zero,

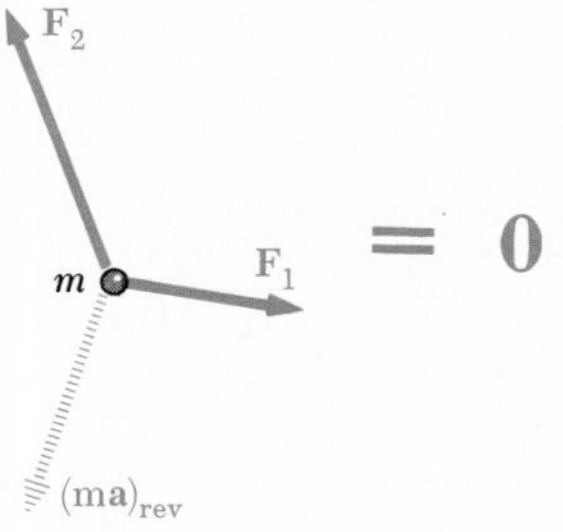

Fig. 12.8

$$\Sigma F_x = 0 \quad \Sigma F_y = 0 \quad \Sigma F_z = 0 \quad \textbf{\textit{including inertia vector}} \quad (12.6)$$

or if the forces are coplanar, we may draw all vectors tip to tail, including again the inertia vector, to form a closed vector polygon.

Because they measure the resistance that particles offer when we try to set them in motion or when we try to change the conditions of their motion, inertia vectors are often called *inertia forces.* The inertia forces, however, are not forces like the forces found in statics, which are either contact forces or gravitational forces (weights). Many people, therefore, object to the use of the word force when referring to the vector $(m\mathbf{a})_{\text{rev}}$ or even avoid altogether the concept of dynamic equilibrium. Others point out that inertia forces and actual forces such as gravitational forces affect our senses in the same way and cannot be distinguished by physical measurements. A man riding in an elevator which is accelerated upward will have the feeling that his weight has suddenly increased; and no measurement made within the elevator could establish whether the elevator is truly accelerated or whether the force of attraction exerted by the earth has suddenly increased.

Sample Problems have been solved in this text by the direct application of Newton's second law as illustrated in Fig. 12.7, rather than by the method of dynamic equilibrium.

12.4. Systems of Particles. When a problem involves the motion of several particles, the equations of motion, (12.5) or (12.6), may be written for each particle considered separately. However, in the case of a system involving a large number of particles, such as a rigid body, it will generally be more convenient to consider the system as a whole. In that case, we should note that the forces acting on a given particle of the system consist (1) of *external forces* exerted by bodies outside the system (such as the weight of the particle, which is exerted by the earth) and (2) of *internal forces* exerted by the other particles of the system. Figure 12.9 shows the particle P_1 of a system, the resultant $\mathbf{F}_1$ of the external forces exerted on P_1, and the internal forces $\mathbf{f}_{12}$, $\mathbf{f}_{13}$, etc., exerted by the other particles P_2, P_3, etc., of the system. According to Newton's second law, the resultant of the external and internal forces acting on P_1 is equal to the vector $m_1\mathbf{a}_1$ obtained by multiplying the acceleration $\mathbf{a}_1$ of the particle by its mass m_1. This vector is generally known as the *effective force* of the particle P_1.

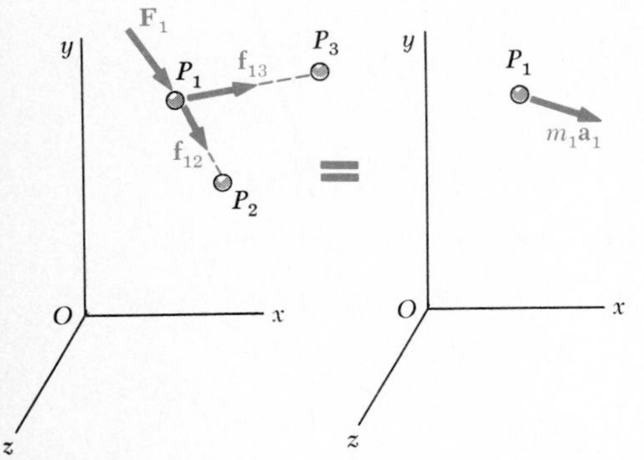

Fig. 12.9

We now consider simultaneously all the particles of the system (Fig. 12.10). Since the relation we have established holds for any particle, it follows that the system of all the external and internal forces acting on the various particles is equivalent to the system of the effective forces of the particles (i.e., one system may be replaced by the other). Thus the sum of the components

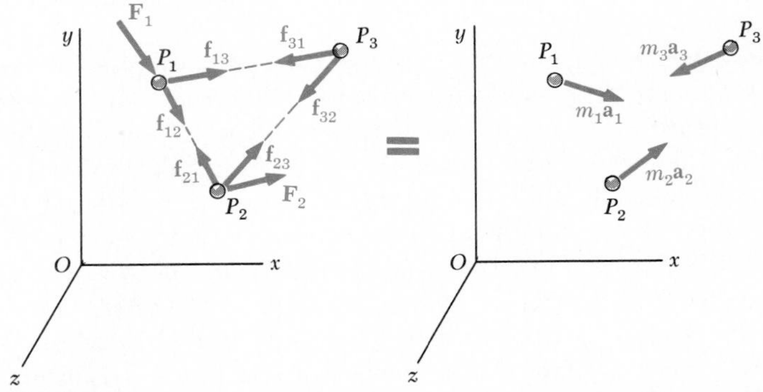

Fig. 12.10

in the x, y, or z direction of the external and internal forces must be equal to the sum of the components of the effective forces in the same direction. But the internal forces occur in pairs of *equal and opposite forces* (such as $\mathbf{f}_{12}$ and $\mathbf{f}_{21}$) and the sum of their components cancels out. We can therefore write the following relations between the components of the external and effective forces:

$$\Sigma(F_x)_{\text{ext}} = \Sigma m a_x \qquad \Sigma(F_y)_{\text{ext}} = \Sigma m a_y \qquad \Sigma(F_z)_{\text{ext}} = \Sigma m a_z \tag{12.7}$$

Consider now the moments about the x, y, or z axis of all the forces shown in Fig. 12.10. Since the internal forces occur in pairs of equal and opposite forces *directed along the same line of action,* the sum of their moments cancels out. Relations similar to Eqs. (12.7) therefore can be written between the moments of the external and effective forces.

Because the sums of their x, y, and z components and the sums of their moments about the x, y, and z axes are respectively equal, the system of the external forces and the system of the effective forces are said to be *equipollent*. This, however, does not mean that the two systems are *equivalent*. We note from Fig. 12.11 that the external force and the effective force acting on a given particle are *not* equal. The two systems of vectors, therefore, do not have the same effect on the individual particles; the two systems are *not* equivalent.

The distinction between equivalent and equipollent systems of vectors should be made in all cases involving a system of particles or a system of rigid bodies (as opposed to a single particle or a single rigid body). In all such cases we shall use a gray equals sign, as in Fig. 12.11, to indicate that the two systems of vectors are equipollent, i.e., have the same sums of

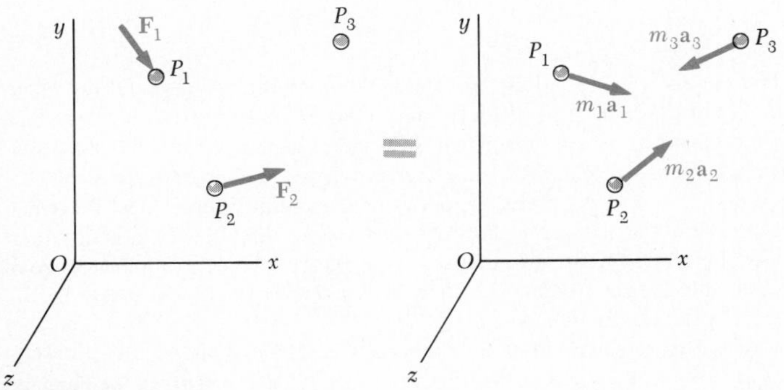

Fig. 12.11

components and the same sums of moments. We shall continue to use colored equals signs, as in Fig. 12.10, to indicate that two systems of vectors are truly equivalent.†

12.5. Motion of the Mass Center of a System of Particles. Equations (12.7) may be written in a modified form if the *mass center* of the system of particles is considered. The mass center of the system is the point G of coordinates $\bar{x}$, $\bar{y}$, $\bar{z}$ defined by the equations

$$(\Sigma m)\bar{x} = \Sigma mx \qquad (\Sigma m)\bar{y} = \Sigma my \qquad (\Sigma m)\bar{z} = \Sigma mz \tag{12.8}$$

Since $W = mg$, we note that G is also the center of gravity of the system. However, in order to avoid any confusion, we shall call G the *mass center* of the system of particles when discussing properties of the system associated with the *mass* of the particles, and we shall refer to it as the *center of gravity* of the system when considering properties associated with the *weight* of the particles. Particles located outside the gravitational field of the earth, for example, have a mass but no weight. We may then properly refer to their mass center, but obviously not to their center of gravity.‡

We differentiate twice with respect to t the relations (12.8):

$$(\Sigma m)\ddot{\bar{x}} = \Sigma m\ddot{x} \qquad (\Sigma m)\ddot{\bar{y}} = \Sigma m\ddot{y} \qquad (\Sigma m)\ddot{\bar{z}} = \Sigma m\ddot{z}$$

Recalling that the second derivatives of the coordinates of a point represent the components of the acceleration of that point, we have

$$(\Sigma m)\bar{a}_x = \Sigma ma_x \qquad (\Sigma m)\bar{a}_y = \Sigma ma_y \qquad (\Sigma m)\bar{a}_z = \Sigma ma_z \tag{12.9}$$

where $\bar{a}_x$, $\bar{a}_y$, $\bar{a}_z$ represent the components of the acceleration $\bar{\mathbf{a}}$ of the mass center G of the system. Combining the relations (12.7) and (12.9), we write the equations

† The relation established in this section between the external and the effective forces of a system of particles is often referred to as *D'Alembert's principle*, after the French mathematician Jean le Rond d'Alembert (1717–1783). However, d'Alembert's original statement refers to the motion of a system of connected bodies, with $\mathbf{f}_{12}$, $\mathbf{f}_{21}$, etc., representing constraint forces which, if applied by themselves, will not cause the system to move. Since this is in general not the case for the internal forces acting on a system of free particles, we shall postpone the consideration of D'Alembert's principle until the study of the motion of rigid bodies (Chap. 16).

‡ It may also be pointed out that the mass center and the center of gravity of a system of particles do not exactly coincide, since the weights of the particles are directed toward the center of the earth and thus do not truly form a system of parallel forces.

$$\Sigma(F_x)_{\text{ext}} = (\Sigma m)\bar{a}_x \qquad \Sigma(F_y)_{\text{ext}} = (\Sigma m)\bar{a}_y \qquad \Sigma(F_z)_{\text{ext}} = (\Sigma m)\bar{a}_z \tag{12.10}$$

which define the motion of the mass center of the system. We note that Eqs. (12.10) are identical with the equations we would obtain for a particle of mass Σm, acted upon by all the external forces. We state therefore: *The mass center of a system of particles moves as if the entire mass of the system and all the external forces were concentrated at that point.*

This principle is best illustrated by the motion of an exploding shell. We know that, if the resistance of the air is neglected, a shell may be assumed to travel along a parabolic path. After the shell has exploded, the mass center G of the fragments of shell will continue to travel along the same path. Indeed, point G must move as if the mass and the weight of all fragments were concentrated at G; it must move, therefore, as if the shell had not exploded.

The principle we have established shows that we may replace a rigid body by a particle of the same mass if we are interested in the motion of the mass center of the body, and not in the motion of the body about its mass center. We should keep in mind, however, that Eqs. (12.9) and (12.10) relate only the components of the forces and vectors involved. To study the motion of a rigid body about its mass center, it will be necessary to also consider the relations between the moments of the external and effective forces (Chap. 16).

12.6. Rectilinear Motion of a Particle. Consider a particle of mass m moving in a straight line under the action of coplanar forces $\mathbf{F}_1$, $\mathbf{F}_2$, $\mathbf{F}_3$, etc. The conditions of motion of the particle are expressed by Eq. (12.2). Since the particle moves in a straight line, its acceleration $\mathbf{a}$ must be directed along that line. Choosing the x axis in the same direction and the y axis in the plane of the forces (Fig. 12.12), we have $a_x = a$, $a_y = 0$ and write

$$\Sigma F_x = ma \qquad \Sigma F_y = 0 \tag{12.11}$$

The equations obtained may be solved for two unknowns.

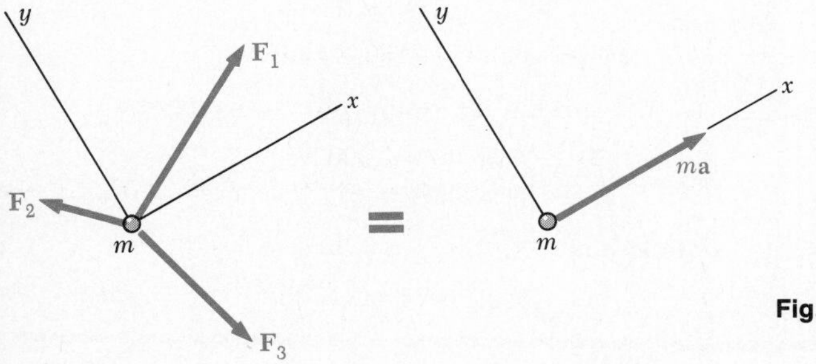

Fig. 12.12

SAMPLE PROBLEM 12.1

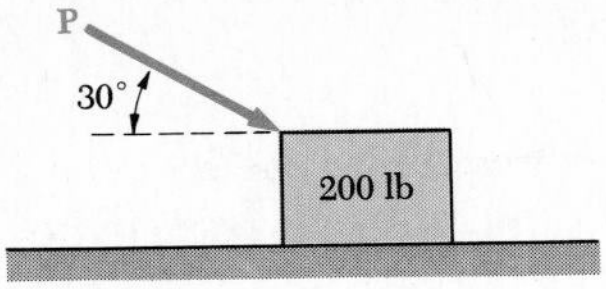

A 200-lb block rests on a horizontal plane. Find the magnitude of the force **P** required to give the block an acceleration of 10 ft/s² to the right. The coefficient of friction between the block and the plane is $\mu = 0.25$.

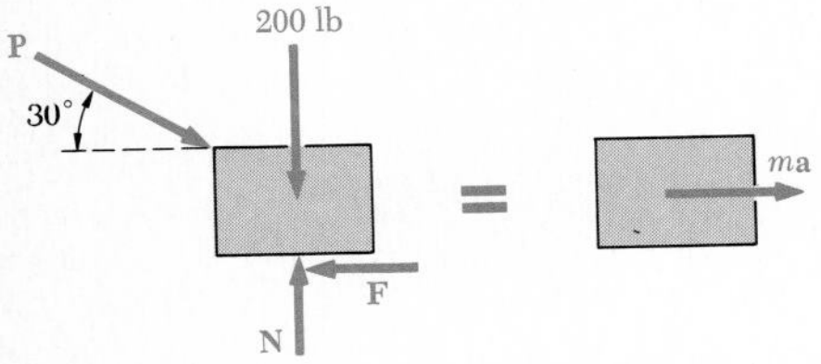

Solution. The mass of the block is

$$m = \frac{W}{g} = \frac{200\text{ lb}}{32.2\text{ ft/s}^2} = 6.21\text{ lb}\cdot\text{s}^2/\text{ft}$$

We note that $F = \mu N = 0.25N$ and that $a = 10\text{ ft/s}^2$. Expressing that the forces acting on the block are equivalent to the vector $m\mathbf{a}$, we write

$$\xrightarrow{+} \Sigma F_x = ma: \quad P\cos 30° - 0.25N = (6.21\text{ lb}\cdot\text{s}^2/\text{ft})(10\text{ ft/s}^2)$$
$$P\cos 30° - 0.25N = 62.1\text{ lb} \qquad (1)$$

$$+\uparrow\Sigma F_y = 0: \quad N - P\sin 30° - 200\text{ lb} = 0 \qquad (2)$$

Solving (2) for N and carrying the result into (1), we obtain

$$N = P\sin 30° + 200\text{ lb}$$
$$P\cos 30° - 0.25(P\sin 30° + 200\text{ lb}) = 62.1\text{ lb} \qquad P = 151\text{ lb} ◄$$

SAMPLE PROBLEM 12.2

Solve Sample Prob. 12.1 using SI units.

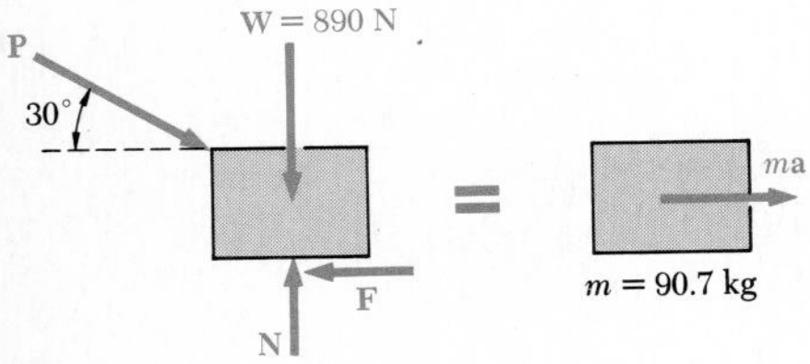

Solution. Using the conversion factors given in Sec. 12.2, we write

$$a = (10\text{ ft/s}^2)(0.3048\text{ m/ft}) = 3.05\text{ m/s}^2$$
$$W = (200\text{ lb})(4.448\text{ N/lb}) = 890\text{ N}$$

Recalling that, by definition, 1 lb is the weight of a mass of 0.4536 kg, we find that the mass of the 200-lb block is

$$m = 200(0.4536\text{ kg}) = 90.7\text{ kg}$$

Noting that $F = \mu N = 0.25N$ and expressing that the forces acting on the block are equivalent to the vector $m\mathbf{a}$, we write

$$\xrightarrow{+} \Sigma F_x = ma: \quad P\cos 30° - 0.25N = (90.7\text{ kg})(3.05\text{ m/s}^2)$$
$$P\cos 30° - 0.25N = 277\text{ N} \qquad (1)$$
$$+\uparrow\Sigma F_y = 0: \quad N - P\sin 30° - 890\text{ N} = 0 \qquad (2)$$

Solving (2) for N and carrying the result into (1), we obtain

$$N = P\sin 30° + 890\text{ N}$$
$$P\cos 30° - 0.25(P\sin 30° + 890\text{ N}) = 277\text{ N} \qquad P = 674\text{ N} ◄$$

or, in U.S. customary units,

$$P = (674\text{ N}) \div (4.448\text{ N/lb}) \qquad P = 151\text{ lb} ◄$$

SAMPLE PROBLEM 12.3

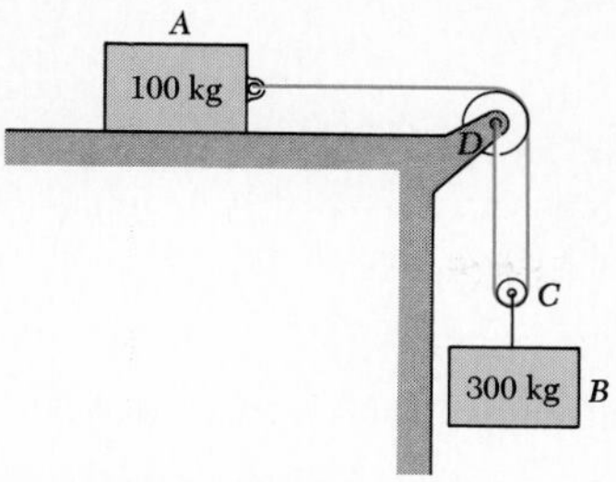

The two blocks shown start from rest. The horizontal plane and the pulley are frictionless, and the pulley is assumed to be of negligible mass. Determine the acceleration of each block and the tension in each cord.

Solution. We denote by T_1 the tension in cord ACD and by T_2 the tension in cord BC. We note that if block A moves through s_A, block B moves through

$$s_B = \tfrac{1}{2}s_A$$

Differentiating twice with respect to t, we have

$$a_B = \tfrac{1}{2}a_A \qquad (1)$$

We shall apply Newton's second law successively to block A, block B, and pulley C.

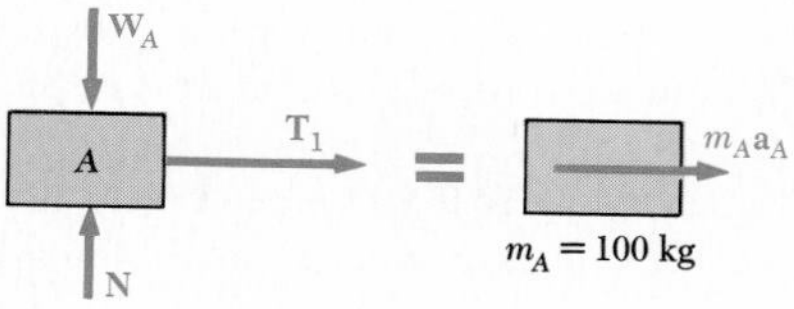

Block A

$$\overset{+}{\rightarrow} \Sigma F_x = m_A a_A: \qquad T_1 = 100a_A \qquad (2)$$

Block B. Observing that the weight of block B is

$$W_B = m_B g = (300 \text{ kg})(9.81 \text{ m/s}^2) = 2940 \text{ N}$$

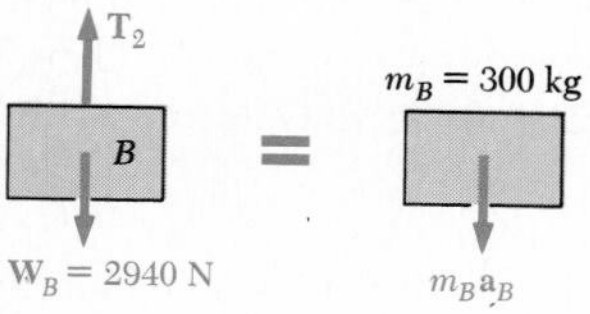

we write

$$+\!\downarrow \Sigma F_y = m_B a_B: \qquad 2940 - T_2 = 300a_B$$

or, substituting for a_B from (1),

$$2940 - T_2 = 300(\tfrac{1}{2}a_A)$$
$$T_2 = 2940 - 150a_A \qquad (3)$$

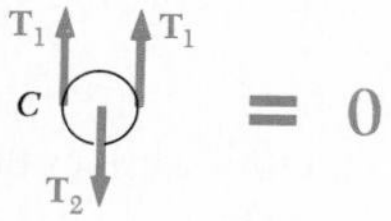

Pulley C. Since m_C is assumed to be zero, we have

$$+\!\downarrow \Sigma F_y = m_C a_C = 0: \qquad T_2 - 2T_1 = 0 \qquad (4)$$

Substituting for T_1 and T_2 from (2) and (3), respectively, into (4), we write

$$2940 - 150a_A - 2(100a_A) = 0$$
$$2940 - 350a_A = 0 \qquad a_A = 8.40 \text{ m/s}^2 \blacktriangleleft$$

Substituting the value obtained for a_A into (1) and (2), we have

$$a_B = \tfrac{1}{2}a_A = \tfrac{1}{2}(8.40 \text{ m/s}^2) \qquad a_B = 4.20 \text{ m/s}^2 \blacktriangleleft$$
$$T_1 = 100a_A = (100 \text{ kg})(8.40 \text{ m/s}^2) \qquad T_1 = 840 \text{ N} \blacktriangleleft$$

Recalling (4), we write

$$T_2 = 2T_1 \qquad T_2 = 2(840 \text{ N}) \qquad T_2 = 1680 \text{ N} \blacktriangleleft$$

We note that the value obtained for T_2 is *not* equal to the weight of block B.

PROBLEMS

12.1 The acceleration due to gravity on the moon is 5.31 ft/s^2. Determine the weight in pounds, the mass in pounds, and the mass in $lb \cdot s^2/ft$, on the moon, of a silver bar whose mass is officially defined as 100.00 lb.

12.2 The value of g at any latitude ϕ may be obtained from the formula

$$g = 9.7807(1 + 0.0053 \sin^2 \phi) \qquad m/s^2$$

Determine to four significant figures the weight in newtons and the mass in kilograms, at the latitudes of 0°, 45°, and 90°, of a silver bar whose mass is officially defined as 10 kg.

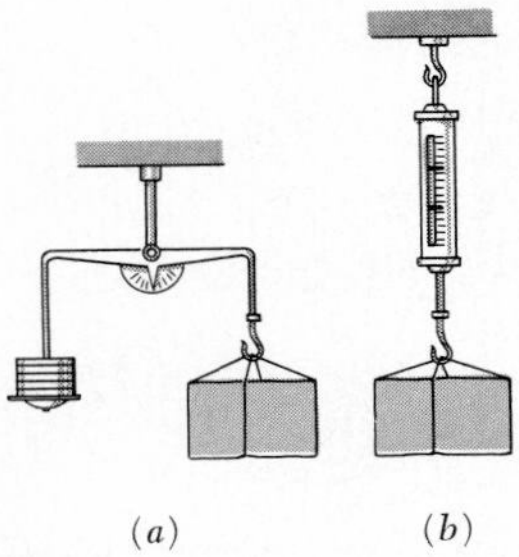

Fig. P12.3

12.3 Two boxes are weighed on the scales shown: scale a is a lever scale; scale b is a spring scale. The scales are attached to the roof of an elevator. When the elevator is at rest, each scale indicates a load of 20 lb. Determine the load which each scale will indicate when the elevator is accelerated upward at the rate of 5 ft/s^2.

12.4 If the spring scale of Prob. 12.3 indicates a load of 18 lb, determine the acceleration of the elevator and the load indicated by the lever scale.

12.5 (*a*) Knowing that the coefficient of friction between automobile tires and a wet concrete surface is 0.35, determine the shortest distance in which an automobile traveling at a speed of 75 mi/h can be stopped. (*b*) Solve part *a* assuming the surface is dry with a coefficient of friction of 0.80.

12.6 A motorist traveling at a speed of 45 mi/h suddenly applies his brakes and comes to a stop after skidding 150 ft. Determine (*a*) the time required for the car to stop, (*b*) the coefficient of friction between the tires and the pavement.

12.7 A 5-kg package is projected down the incline with an initial velocity of 4 m/s. Knowing that the coefficient of friction between the package and the incline is 0.35, determine (*a*) the velocity of the package after 3 m of motion, (*b*) the distance d at which the package comes to rest.

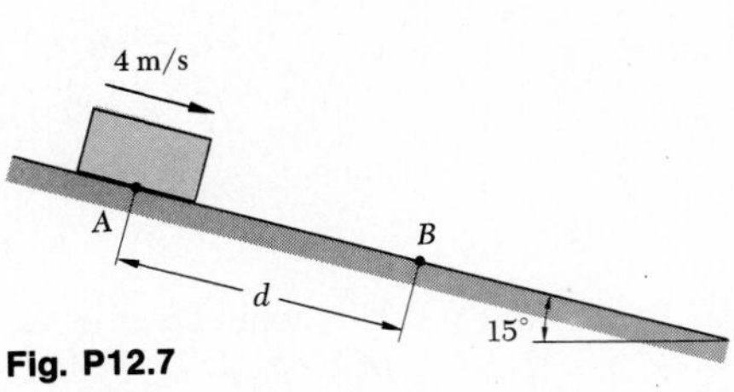

Fig. P12.7

12.8 A truck is proceeding up a long 3-percent grade at a constant speed of 60 km/h. If the driver does not change the setting of his throttle or shift gears, what will be the acceleration of the truck as it starts moving on the level section of the road?

Fig. P12.8

12.9 The 3-kg collar was moving down the rod with a velocity of 3 m/s when a force **P** was applied to the horizontal cable. Assuming negligible friction between the collar and the rod, determine the magnitude of the force **P** if the collar stopped after moving 1 m more down the rod.

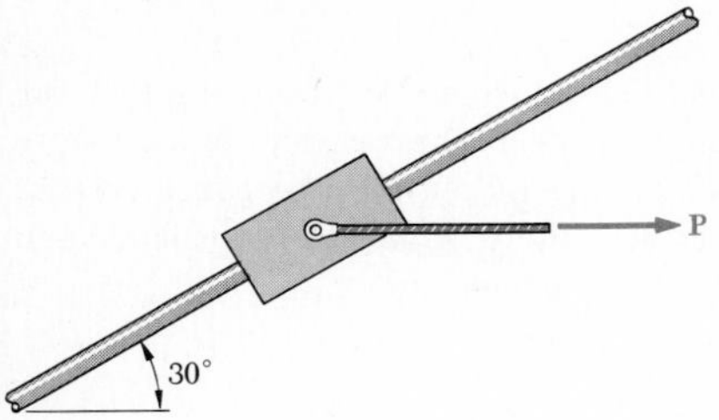

Fig. P12.9

12.10 Solve Prob. 12.9, assuming a coefficient of friction of 0.20 between the collar and the rod.

12.11 Two packages are placed on a conveyor belt which is at rest. The coefficient of friction is 0.20 between the belt and package *A*, and 0.10 between the belt and package *B*. If the belt is suddenly started to the right and slipping occurs between the belt and the packages, determine (*a*) the acceleration of the packages, (*b*) the force exerted by package *A* on package *B*.

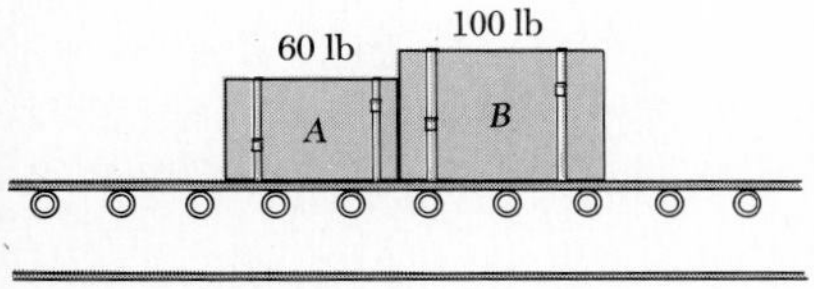

Fig. P12.11

12.12 The subway train shown travels at a speed of 30 mi/h. Determine the force in each coupling when the brakes are applied, knowing that the braking force is 5000 lb on each car.

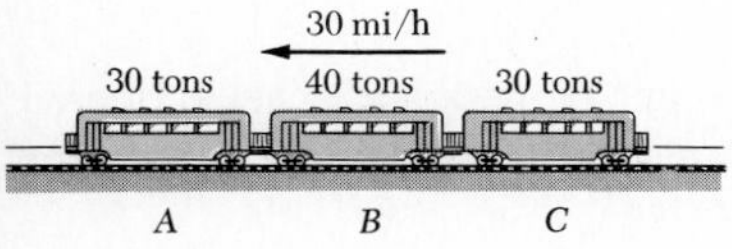

Fig. P12.12

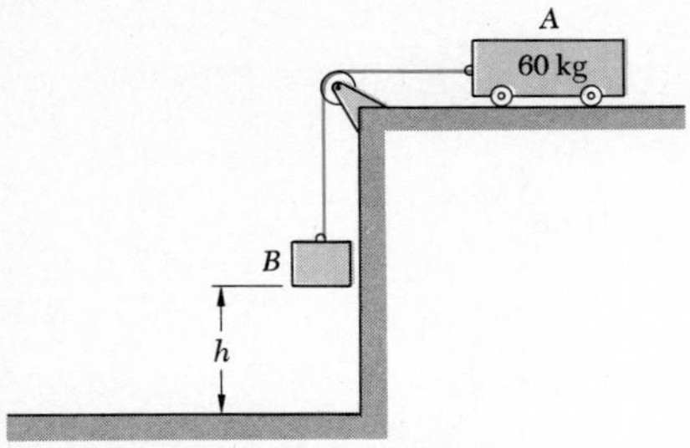

Fig. P12.13, P12.14, and P12.15

12.13 When the system shown is released from rest, the acceleration of block B is observed to be 3 m/s^2 downward. Neglecting the effect of friction, determine (*a*) the tension in the cable, (*b*) the mass of block B.

12.14 The system shown is released from rest when $h = 1.4$ m. (*a*) Determine the mass of block B, knowing that it strikes the ground with a speed of 3 m/s. (*b*) Attempt to solve part *a*, assuming the final speed to be 6 m/s; explain the difficulty encountered.

12.15 The system shown is released from rest. Knowing that the mass of block B is 30 kg, determine how far the cart will move before it reaches a speed of 2.5 m/s, (*a*) if the pulley may be considered as weightless and frictionless, (*b*) if the pulley "freezes" on its shaft and the cable must slip, with $\mu = 0.10$, over the pulley.

12.16 Each of the systems shown is initially at rest. Assuming the pulleys to be weightless and neglecting axle friction, determine for each system (*a*) the acceleration of block A, (*b*) the velocity of block A after 4 s, (*c*) the velocity of block A after it has moved 10 ft.

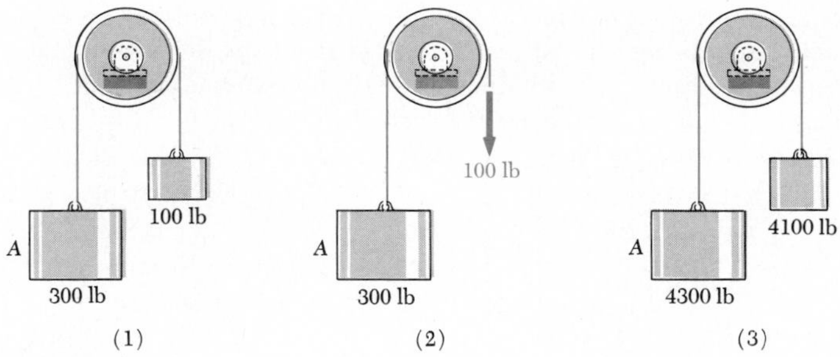

Fig. P12.16

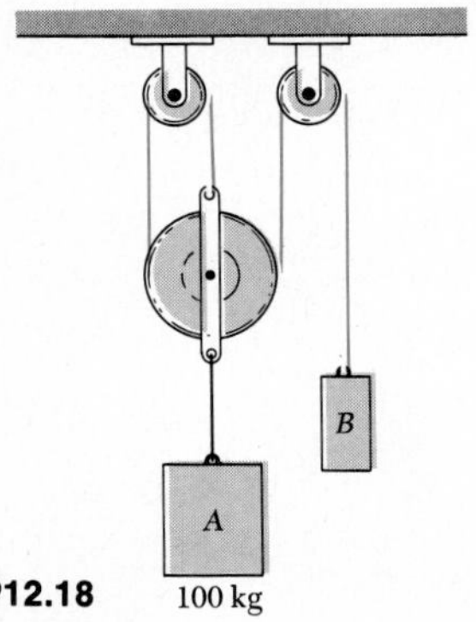

Fig. P12.17 and P12.18

12.17 The 100-kg block A is connected to a 25-kg counterweight B by the cable arrangement shown. If the system is released from rest, determine (*a*) the tension in the cable, (*b*) the velocity of B after 3 s, (*c*) the velocity of A after it has moved 1.2 m.

12.18 Block A is observed to move with an acceleration of 0.9 m/s^2 directed upward. Determine (*a*) the mass of block B, (*b*) the corresponding tension in the cable.

12.19 A small ore car and its load weigh 8000 lb. The car is partially counterweighted by a 500-lb weight. If the car is initially at rest, determine the force **P** required if the velocity of the car is to be 10 ft/s after it has moved 20 ft up the incline.

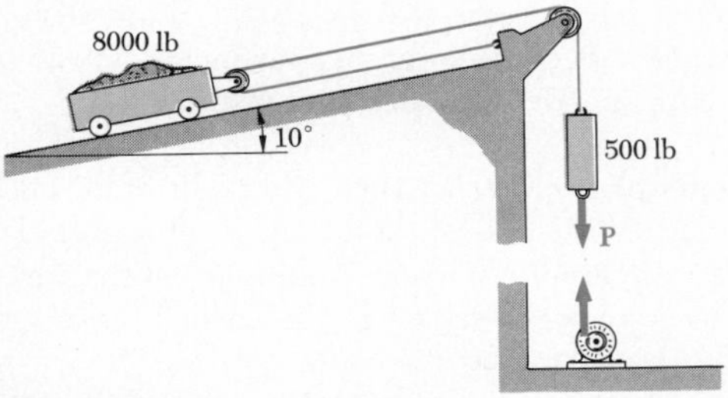

Fig. P12.19

12.20 Neglecting the effect of friction, determine (*a*) the acceleration of each block, (*b*) the tension in the cable.

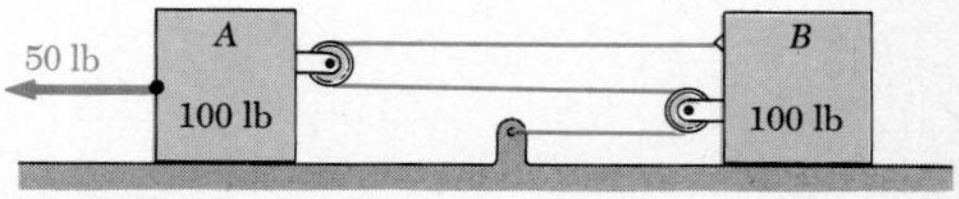

Fig. P12.20

12.21 It is assumed that friction exists between block *A* and the horizontal plane in Sample Prob. 12.3. If block *B* is observed to move 7.5 m in 2 s after starting from rest, determine the coefficient of friction between block *A* and the horizontal plane.

12.22 Solve Prob. 12.20, assuming the coefficient of friction between the blocks and the horizontal plane to be 0.20.

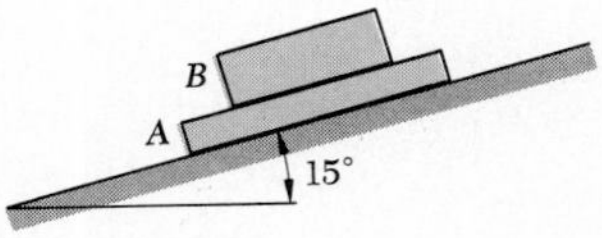

Fig. P12.23

12.23 Two plates *A* and *B*, each of mass 50 kg, are placed as shown on a 15° incline. The coefficient of friction between *A* and *B* is 0.10; the coefficient of friction between *A* and the incline is 0.20. (*a*) If the plates are released from rest, determine the acceleration of each plate. (*b*) Solve part *a* assuming that plates *A* and *B* are welded together and act as a single rigid body.

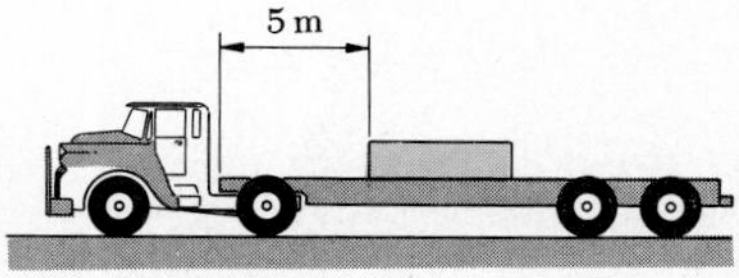

Fig. P12.24

12.24 The coefficient of friction between the load and the flat-bed trailer shown is 0.40. Knowing that the forward speed of the truck is 50 km/h, determine the shortest distance in which the truck can be brought to a stop if the load is not to shift.

12.25 The 400-lb plate *A* rests on small rollers which may be considered frictionless; the coefficient of friction between the two plates is $\mu = 0.20$. Determine the acceleration of plate *A* when a 160-lb force is suddenly applied to it, (*a*) for the system shown, (*b*) if the cable is cut.

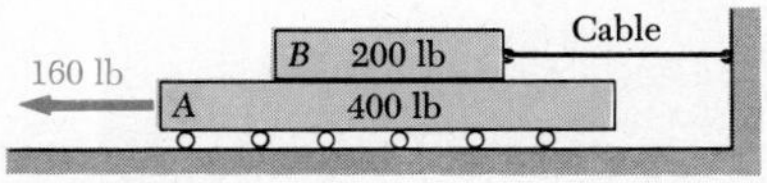

Fig. P12.25

*12.26 In Prob. 12.24, determine the shortest distance in which the truck can be stopped without having the load slide beyond the forward edge of the trailer. (*Note.* As the truck comes to a stop, the load is still moving relative to the trailer.)

12.27 A 60-lb crate rests on a 40-lb cart; the coefficient of static friction between the crate and the cart is 0.25. If the crate is not to slip with respect to the cart, determine (*a*) the maximum allowable magnitude of **P**, (*b*) the corresponding acceleration of the cart.

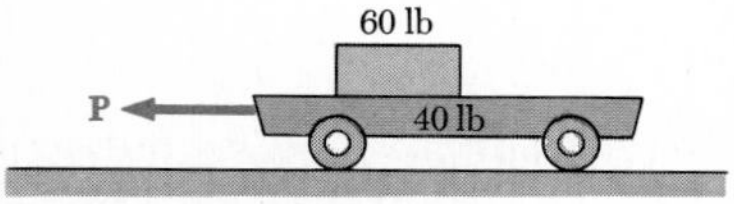

Fig. P12.27 and P12.28

12.28 The coefficients of friction between the 60-lb crate and the 40-lb cart are $\mu_s = 0.25$ and $\mu_k = 0.20$. If a force **P** of magnitude 30 lb is applied to the cart, determine the acceleration (*a*) of the cart, (*b*) of the crate, (*c*) of the crate with respect to the cart.

12.29 For the cart and crate of Prob. 12.28, determine the acceleration of the cart as a function of the magnitude of the force **P**.

12.30 A chain of length l and mass m per unit length rests over a small pulley of negligible mass. If the chain is released in the position shown ($h > \frac{1}{2}l$), determine the velocity of the chain when the end of the chain leaves the pulley. Also determine the maximum possible final velocity by letting h approach $\frac{1}{2}l$.

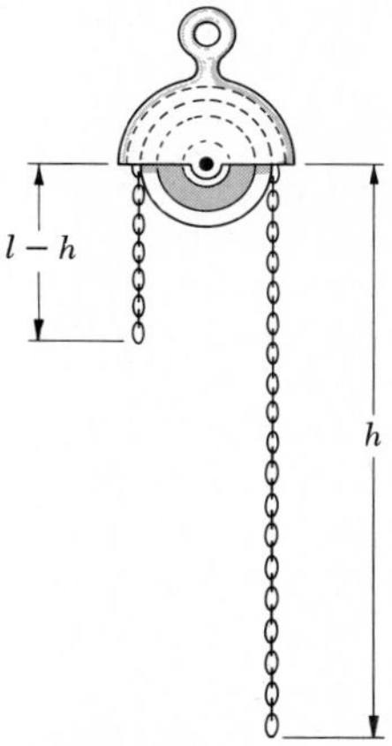

Fig. P12.30

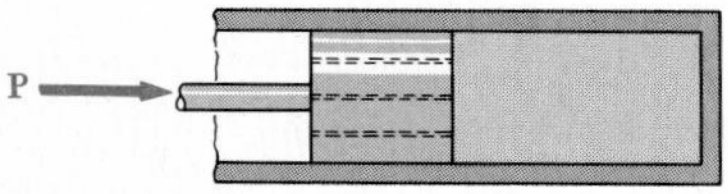

Fig. P12.31

12.31 A constant force **P** is applied to a piston and rod of total mass m in order to make them move in a cylinder filled with oil. As the piston moves, the oil is forced through orifices in the piston and exerts on the piston an additional force of magnitude kv, proportional to the speed v of the piston and in a direction opposite to its motion. Express the acceleration and velocity of the piston as a function of the time t, assuming that the piston starts from rest at time $t = 0$.

12.32 A ship of total mass m is anchored in the middle of a river which is flowing with a constant velocity $\mathbf{v}_0$. The horizontal component of the force exerted on the ship by the anchor chain is $\mathbf{T}_0$. If the anchor chain suddenly breaks, determine the time required for the ship to attain a velocity equal to $\frac{1}{2}\mathbf{v}_0$. Assume that the frictional resistance of the water is proportional to the velocity of the ship relative to the water.

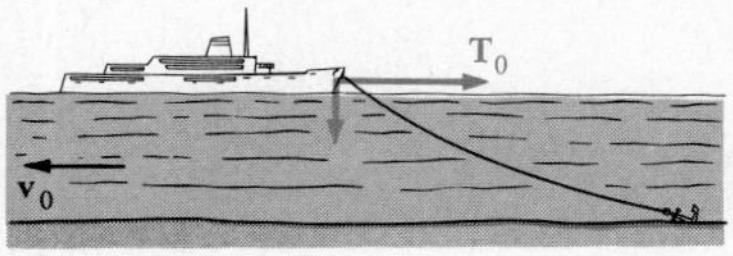

Fig. P12.32

12.33 A bomber flying horizontally in a straight line at a constant speed v_0 releases a powered missile which will fly ahead of the bomber in a parallel horizontal path. The missile engine, which is started at the time of release, provides a constant thrust T, while the atmosphere produces a drag equal to kv^2, where v is the speed of the missile. Neglecting any change in the mass of the missile, express the acceleration of the missile as a function of the distance traveled in free flight. It is assumed that $v_0 < \sqrt{T/k}$.

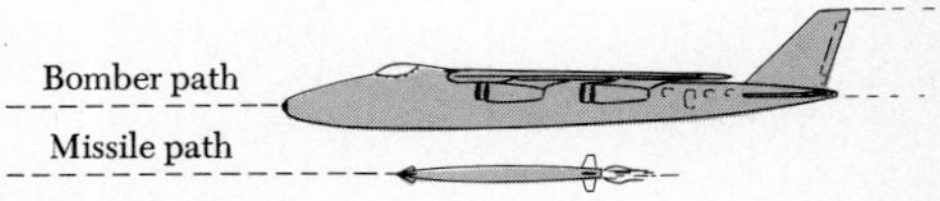

Fig. P12.33

12.34 Determine the acceleration of each block when $W_A = W_B = W_C$.

12.35 Determine the acceleration of each block when $W_A = 5$ lb, $W_B = 10$ lb, and $W_C = 10$ lb.

12.36 In the system shown, $W_A = 6$ lb and $W_C = 12$ lb. Determine the required weight W_B if block B is not to move when the system is released from rest.

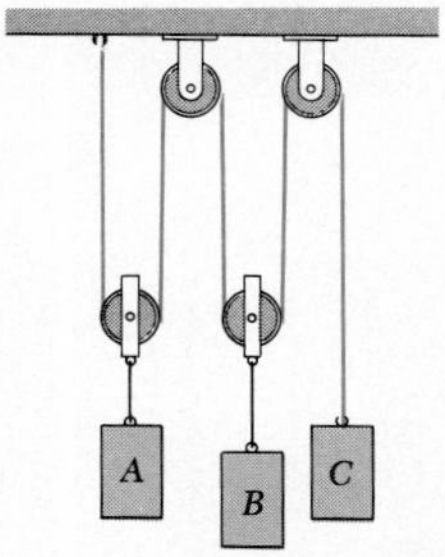

Fig. P12.34, P12.35, and P12.36

12.37 Determine the acceleration of each block when $m_A = 15$ kg, $m_B = 10$ kg, and $m_C = 5$ kg.

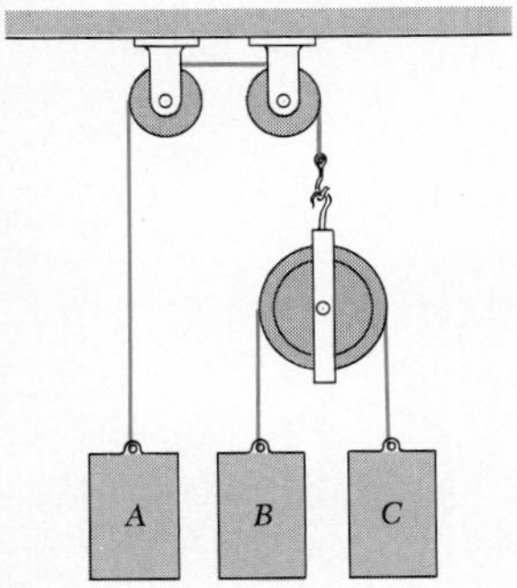

Fig. P12.37

12.38 Knowing that $\mu = 0.20$, determine the acceleration of each block when $m_A = 5$ kg, $m_B = 10$ kg, and $m_C = 10$ kg.

12.39 Knowing that $\mu = 0.20$, determine the acceleration of each block when $m_A = 5$ kg, $m_B = 10$ kg, and $m_C = 20$ kg.

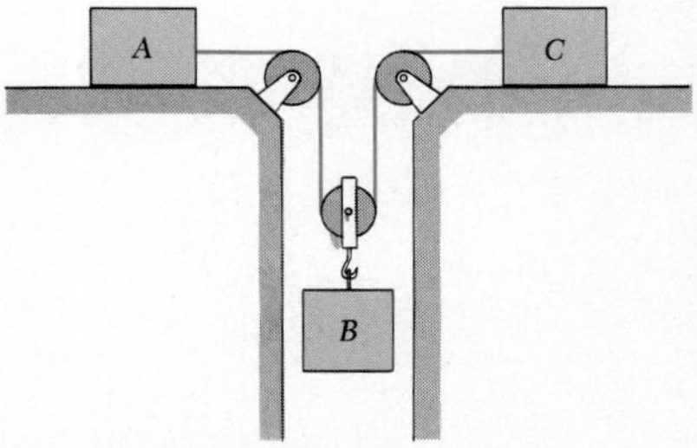

Fig. P12.38 and P12.39

12.7. Curvilinear Motion of a Particle. We saw in Chap. 11 that the acceleration of a particle moving in a plane along a curved path may be represented by its rectangular components $\mathbf{a}_x$ and $\mathbf{a}_y$, by its tangential and normal components $\mathbf{a}_t$ and $\mathbf{a}_n$, or by its radial and transverse components $\mathbf{a}_r$ and $\mathbf{a}_\theta$. We may therefore use any one of these methods of representation to express Newton's second law of motion.

Rectangular Components. Using the rectangular components $\mathbf{a}_x$ and $\mathbf{a}_y$ of the acceleration, and resolving the forces acting on the particle into their components $\mathbf{F}_x$ and $\mathbf{F}_y$, we write

$$\Sigma F_x = ma_x \qquad \Sigma F_y = ma_y \tag{12.12}$$

These equations may be solved for two unknowns.

Consider, as an example, the *motion of a projectile.* If the resistance of the air is neglected, the only force acting on the projectile after it has been fired is its weight **W**. Substituting $\Sigma F_x = 0$ and $\Sigma F_y = -W$ into Eqs. (12.12), we obtain

$$a_x = 0 \qquad a_y = -\frac{W}{m} = -g$$

where g is 9.81 m/s^2 or 32.2 ft/s^2. The equations obtained may be integrated independently, as was shown in Sec. 11.10, to obtain the velocity and displacement of the projectile at any instant.

Tangential and Normal Components. Using the tangential and normal components $\mathbf{a}_t$ and $\mathbf{a}_n$ of the acceleration, and resolving the forces acting on the particle into components $\mathbf{F}_t$ along the tangent to the path (in the direction of motion) and components $\mathbf{F}_n$ along the normal (toward the inside of the path), we write (Fig. 12.13)

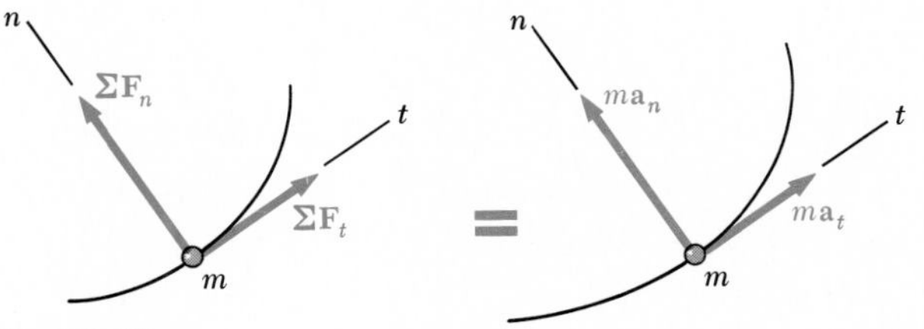

Fig. 12.13

$$\Sigma F_t = ma_t \qquad \Sigma F_n = ma_n \tag{12.13}$$

Substituting for a_t and a_n from (11.28) and (11.29), we have

$$\Sigma F_t = m\frac{dv}{dt} \qquad \Sigma F_n = m\frac{v^2}{\rho} \tag{12.14}$$

The equations obtained may be solved for two unknowns.

Radial and Transverse Components. Using the radial and transverse components $\mathbf{a}_r$ and $\mathbf{a}_\theta$ of the acceleration, and resolving the forces acting on the particle into components $\mathbf{F}_r$ along the line OP joining the origin O to the particle P, and components $\mathbf{F}_\theta$ in the direction perpendicular to OP (Fig. 12.14), we write

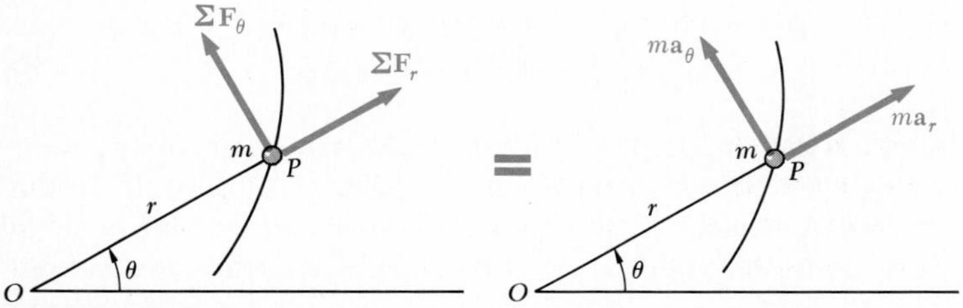

Fig. 12.14

$$\Sigma F_r = ma_r \qquad \Sigma F_\theta = ma_\theta \tag{12.15}$$

Substituting for a_r and a_θ from (11.33), we have

$$\Sigma F_r = m(\ddot{r} - r\dot{\theta}^2) \tag{12.16}$$
$$\Sigma F_\theta = m(r\ddot{\theta} + 2\dot{r}\dot{\theta}) \tag{12.17}$$

The equations obtained may be solved for two unknowns.

Dynamic Equilibrium. As an alternate method of solution, we may add the inertia vector $(m\mathbf{a})_{\text{rev}}$ to the forces acting on the particle and express that the system obtained is balanced. In practice, it is found convenient to resolve the inertia vector into two components. Resolving, for example, the inertia vector into its tangential and normal components, we obtain the system of vectors shown in Fig. 12.15. The tangential component of the inertia vector provides a measure of the resistance the particle offers to a change in speed, while its normal component (also called *centrifugal force*) represents the tendency of the particle to leave its curved path. We should note that either of these two components may be zero under special conditions: (1) if the particle starts from rest, its initial velocity is zero and the normal component of the inertia vector is zero at $t = 0$; (2) if the particle moves at constant speed along its path, the tangential component of the inertia vector is zero and only its normal component needs to be considered.

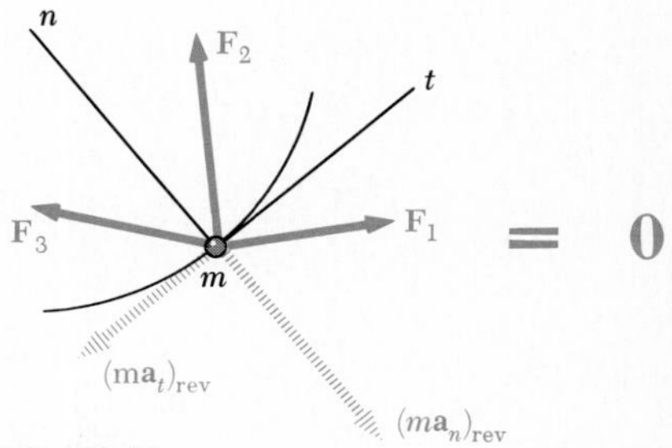

Fig. 12.15

12.8. Motion under a Central Force. When the only force acting on a particle P is a force $\mathbf{F}$ directed toward or away from a fixed point O, the particle is said to be moving *under*

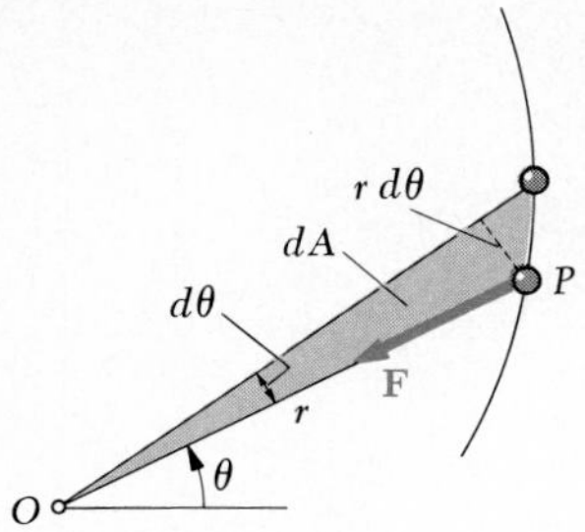

Fig. 12.16

a central force, and the point O is referred to as the *center of force* (Fig. 12.16). Selecting O as the origin of coordinates and using radial and transverse components, we note that the transverse component $\mathbf{F}_\theta$ of the force $\mathbf{F}$ is zero. Thus, Eq. (12.17) yields

$$r\ddot{\theta} + 2\dot{r}\dot{\theta} = 0 \qquad \text{or} \qquad \frac{1}{r}\frac{d}{dt}(r^2\dot{\theta}) = 0$$

It follows that, for any finite value of r, we must have

$$r^2\dot{\theta} = h \tag{12.18}$$

where h is a constant. Equation (12.18) may be given an interesting geometric interpretation. Observing from Fig. 12.16 that the radius vector OP sweeps an infinitesimal area $dA = \frac{1}{2}r^2\,d\theta$ as it rotates through an angle $d\theta$, and defining the *areal velocity* of the particle as the quotient dA/dt, we note that the left-hand member of Eq. (12.18) represents twice the areal velocity of the particle. We thus conclude that, *when a particle moves under a central force, its areal velocity is constant.*†

12.9. Newton's Law of Gravitation. The gravitational force exerted by the sun on a planet, or by the earth on an orbiting satellite, is an important example of a central force. In this section we shall learn how to determine the magnitude of a gravitational force.

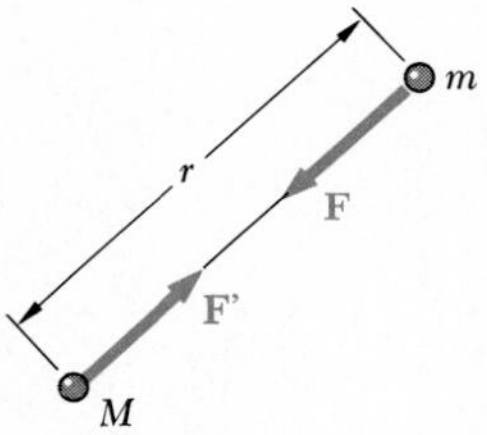

Fig. 12.17

In his *law of universal gravitation,* Newton states that two particles at a distance r from each other and, respectively, of mass M and m attract each other with equal and opposite forces $\mathbf{F}$ and $\mathbf{F}'$ directed along the line joining the particles (Fig. 12.17). The common magnitude F of the two forces is

$$F = G\frac{Mm}{r^2} \tag{12.19}$$

where G is a universal constant, called the *constant of gravitation.* Experiments show that the value of G is $(6.673 \pm 0.003) \times 10^{-11}\ \text{m}^3/\text{kg}\cdot\text{s}^2$ in SI units, or approximately $3.44 \times 10^{-8}\ \text{ft}^4/\text{lb}\cdot\text{s}^4$ in U.S. customary units. While gravitational forces exist between any pair of bodies, their effect is appreciable only when one of the bodies has a very large mass. The effect of gravitational forces is apparent in the case of the motion of a planet about the sun, of satellites orbiting about the earth, or of bodies falling on the surface of the earth.

† It will be seen in Sec. 14.10 that the left-hand member of Eq. (12.18) also represents the *angular momentum per unit mass* of the particle P. Thus, Eq. (12.18) may be considered as a statement of the *conservation of angular momentum* of a particle moving under a central force.

Since the force exerted by the earth on a body of mass m located on or near its surface is defined as the weight **W** of the body, we may substitute the magnitude $W = mg$ of the weight for F, and the radius R of the earth for r, in Eq. (12.19). We obtain

$$W = mg = \frac{GM}{R^2}m \qquad \text{or} \qquad g = \frac{GM}{R^2} \tag{12.20}$$

where M is the mass of the earth. Since the earth is not truly spherical, the distance R from the center of the earth depends upon the point selected on its surface, and the values of W and g will thus vary with the altitude and latitude of the point considered. Another reason for the variation of W and g with the latitude is that a system of axes attached to the earth does not constitute a newtonian frame of reference (see Sec. 12.1). A more accurate definition of the weight of a body should therefore include a component representing the centrifugal force due to the rotation of the earth. Values of g at sea level vary from 9.781 m/s^2 or 32.09 ft/s^2 at the equator to 9.833 m/s^2 or 32.26 ft/s^2 at the poles.†

The force exerted by the earth on a body of mass m located in space at a distance r from its center may be found from Eq. (12.19). The computations will be somewhat simplified if we note that, according to Eq. (12.20), the product of the constant of gravitation G and of the mass M of the earth may be expressed as

$$GM = gR^2 \tag{12.21}$$

where g and the radius R of the earth will be given their average values $g = 9.81 \text{ m/s}^2$ and $R = 6.37 \times 10^6 \text{ m}$ in SI units,‡ or $g = 32.2 \text{ ft/s}^2$ and $R = (3960 \text{ mi})(5280 \text{ ft/mi})$ in U.S. customary units.

The discovery of the law of universal gravitation has often been attributed to the fact that Newton, after observing an apple falling from a tree, had reflected that the earth must attract an apple and the moon in much the same way. While it is doubtful that this incident actually took place, it may be said that Newton would not have formulated his law if he had not first perceived that the acceleration of a falling body must have the same cause as the acceleration which keeps the moon in its orbit. This basic concept of continuity of the gravitational attraction is more easily understood now, when the gap between the apple and the moon is being filled with long-range ballistic missiles and artificial earth satellites.

† A formula expressing g in terms of the latitude ϕ was given in Prob. 12.2.

‡ The value of R is easily found if one recalls that the circumference of the earth is $2\pi R = 40 \times 10^6$ m.

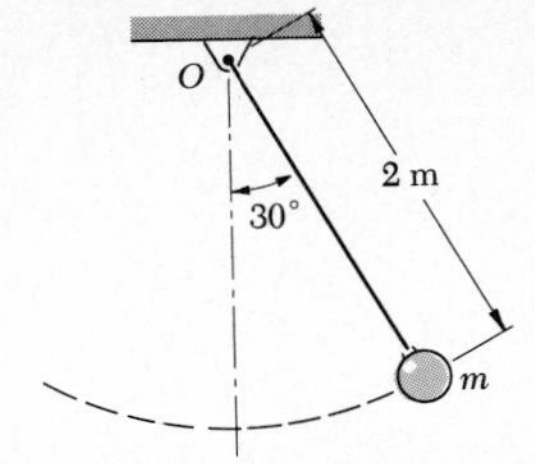

SAMPLE PROBLEM 12.4

The bob of a 2-m pendulum describes an arc of circle in a vertical plane. If the tension in the cord is 2.5 times the weight of the bob for the position shown, find the velocity and acceleration of the bob in that position.

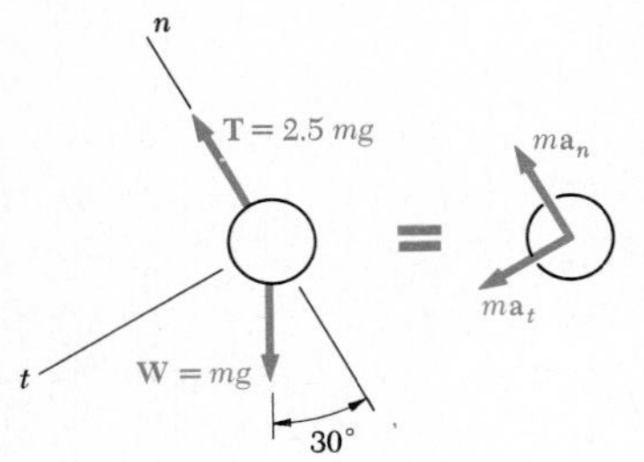

Solution. The weight of the bob is $W = mg$; the tension in the cord is thus 2.5 mg. Recalling that $\mathbf{a}_n$ is directed toward O and assuming $\mathbf{a}_t$ as shown, we apply Newton's second law and obtain

$+\swarrow \Sigma F_t = ma_t$: $\quad mg \sin 30° = ma_t$

$a_t = g \sin 30° = +4.90 \text{ m/s}^2 \qquad \mathbf{a}_t = 4.90 \text{ m/s}^2 \swarrow$ ◀

$+\nwarrow \Sigma F_n = ma_n$: $\quad 2.5\, mg - mg \cos 30° = ma_n$

$a_n = 1.634\, g = +16.03 \text{ m/s}^2 \qquad \mathbf{a}_n = 16.03 \text{ m/s}^2 \nwarrow$ ◀

Since $a_n = v^2/\rho$, we have $v^2 = \rho a_n = (2 \text{ m})(16.03 \text{ m/s}^2)$

$v = \pm 5.66 \text{ m/s} \qquad \mathbf{v} = 5.66 \text{ m/s}$ (up or down) ◀

SAMPLE PROBLEM 12.5

Determine the rated speed of a highway curve of radius $\rho = 400$ ft banked through an angle $\theta = 18°$. The rated speed of a banked curved road is the speed at which a car should travel if no lateral friction force is to be exerted on its wheels.

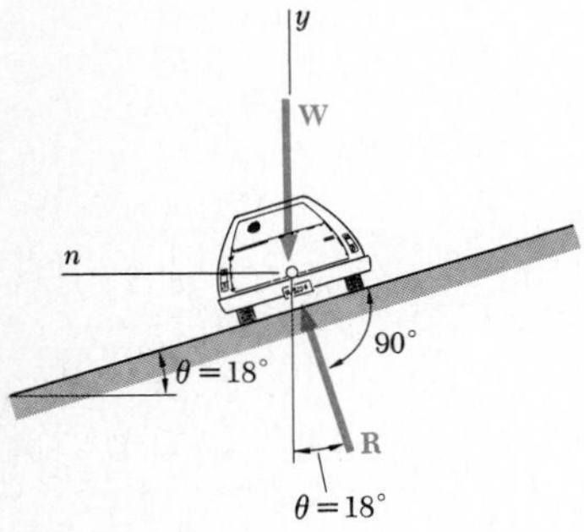

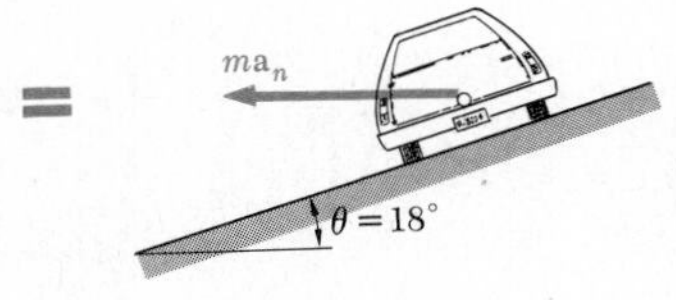

Solution. The car travels in a *horizontal* circular path of radius ρ. The normal component $\mathbf{a}_n$ of the acceleration is directed toward the center of the path; its magnitude is $a_n = v^2/\rho$, where v is the speed of the car in ft/s. The mass m of the car is W/g, where W is the weight of the car. Since no lateral friction force is to be exerted on the car, the reaction $\mathbf{R}$ of the road is shown perpendicular to the roadway. Applying Newton's second law, we write

$$+\uparrow \Sigma F_y = 0: \qquad R \cos \theta - W = 0 \qquad R = \frac{W}{\cos \theta} \tag{1}$$

$$\overset{+}{\leftarrow} \Sigma F_n = ma_n: \qquad R \sin \theta = \frac{W}{g} a_n \tag{2}$$

Substituting for R from (1) into (2), and recalling that $a_n = v^2/\rho$:

$$\frac{W}{\cos \theta} \sin \theta = \frac{W}{g} \frac{v^2}{\rho} \qquad v^2 = g\rho \tan \theta$$

Substituting the given data, $\rho = 400$ ft and $\theta = 18°$, into this equation, we obtain

$$v^2 = (32.2 \text{ ft/s}^2)(400 \text{ ft}) \tan 18°$$

$v = 64.7$ ft/s $\qquad v = 44.1$ mi/h ◀

SAMPLE PROBLEM 12.6

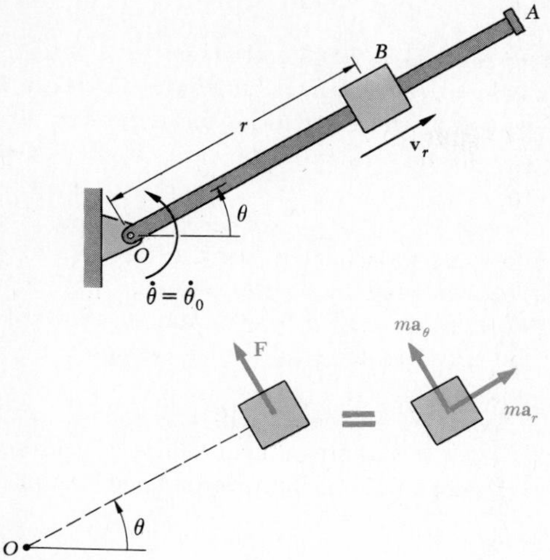

A block B of mass m may slide freely on a frictionless arm OA which rotates in a horizontal plane at a constant rate $\dot{\theta}_0$. Knowing that B is released at a distance r_0 from O, express as a function of r, (*a*) the component v_r of the velocity of B along OA, (*b*) the magnitude of the horizontal force $\mathbf{F}$ exerted on B by the arm OA.

Solution. Since all other forces are perpendicular to the plane of the figure, the only force shown acting on B is the force $\mathbf{F}$ perpendicular to OA.

Equations of Motion. Using radial and transverse components:

$$+\nearrow \Sigma F_r = ma_r: \qquad 0 = m(\ddot{r} - r\dot{\theta}^2) \qquad (1)$$

$$+\nwarrow \Sigma F_\theta = ma_\theta: \qquad F = m(r\ddot{\theta} + 2\dot{r}\dot{\theta}) \qquad (2)$$

a. Component $\mathbf{v}_r$ of Velocity. Since $v_r = \dot{r}$, we have

$$\ddot{r} = \dot{v}_r = \frac{dv_r}{dt} = \frac{dv_r}{dr}\frac{dr}{dt} = v_r \frac{dv_r}{dr}$$

Substituting for $\ddot{r}$ into (1), recalling that $\dot{\theta} = \dot{\theta}_0$, and separating the variables:

$$v_r\, dv_r = \dot{\theta}_0^2 r\, dr$$

Multiplying by 2, and integrating from 0 to v_r and from r_0 to r:

$$v_r^2 = \dot{\theta}_0^2(r^2 - r_0^2) \qquad v_r = \dot{\theta}_0(r^2 - r_0^2)^{1/2} \blacktriangleleft$$

b. Horizontal Force $\mathbf{F}$. Making $\dot{\theta} = \dot{\theta}_0$, $\ddot{\theta} = 0$, $\dot{r} = v_r$ in Eq. (2), and substituting for v_r the expression obtained in part *a*:

$$F = 2m\dot{\theta}_0(r^2 - r_0^2)^{1/2}\dot{\theta}_0 \qquad F = 2m\dot{\theta}_0^2(r^2 - r_0^2)^{1/2} \blacktriangleleft$$

SAMPLE PROBLEM 12.7

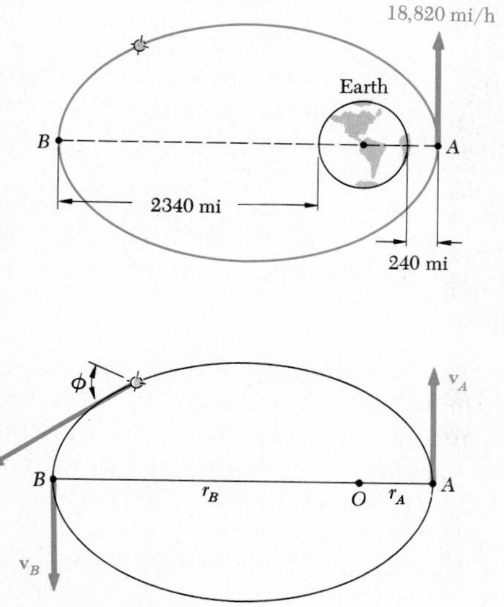

A satellite is launched in a direction parallel to the surface of the earth with a velocity of 18,820 mi/h from an altitude of 240 mi. Determine the velocity of the satellite as it reaches its maximum altitude of 2340 mi. It is recalled that the radius of the earth is 3960 mi.

Solution. Since the satellite is moving under a central force directed toward the center O of the earth, we may use Eq. (12.18) and write

$$r_A^2\dot{\theta}_A = r_B^2\dot{\theta}_B = h \qquad (1)$$

But at A and B the velocity of the satellite reduces to its transverse component; we have $v_A = r_A\dot{\theta}_A$ and $v_B = r_B\dot{\theta}_B$, and Eq. (1) yields

$$r_A v_A = r_B v_B$$

$$v_B = v_A\frac{r_A}{r_B} = (18{,}820 \text{ mi/h})\frac{3960 \text{ mi} + 240 \text{ mi}}{3960 \text{ mi} + 2340 \text{ mi}}$$

$$v_B = 12{,}550 \text{ mi/h} \blacktriangleleft$$

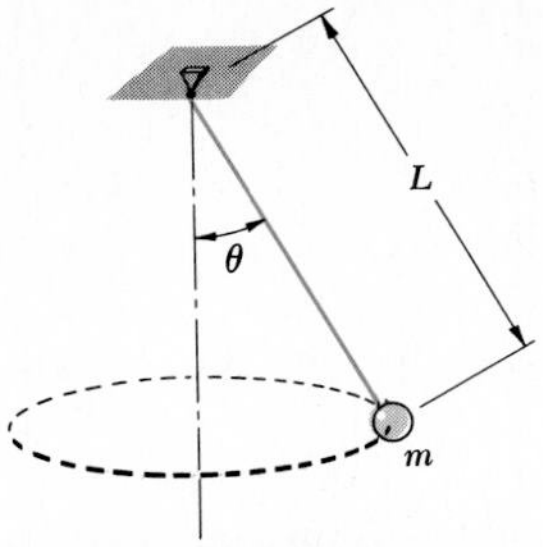

Fig. P12.40 and P12.41

PROBLEMS

12.40 A small ball of mass $m = 5$ kg is attached to a cord of length $L = 2$ m and is made to revolve in a horizontal circle at a constant speed v_0. Knowing that the cord forms an angle $\theta = 40°$ with the vertical, determine (*a*) the tension in the cord, (*b*) the speed v_0 of the ball.

12.41 A small ball of mass $m = 5$ kg is made to revolve in a horizontal circle as shown. Knowing that the maximum allowable tension in the cord is 100 N, determine (*a*) the maximum allowable velocity if $L = 2$ m, (*b*) the corresponding value of the angle θ.

12.42 Two wires *AC* and *BC* are each tied to a sphere at *C*. The sphere is made to revolve in a horizontal circle at a constant speed v. Determine the range of values of the speed v for which *both* wires are taut.

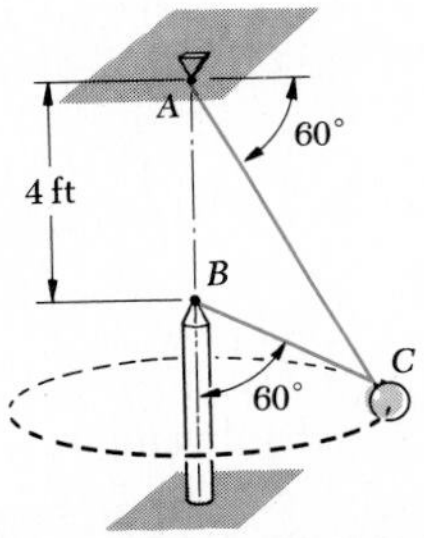

Fig. P12.42 and P12.43

12.43 Two wires *AC* and *BC* are each tied to a 10-lb sphere. The sphere is made to revolve in a horizontal circle at a constant speed of 12 ft/s. Determine the tension in each wire.

12.44 A 3-kg ball is swung in a vertical circle at the end of a cord of length $l = 0.8$ m. Knowing that when $\theta = 60°$ the tension in the cord is 25 N, determine the instantaneous velocity and acceleration of the ball.

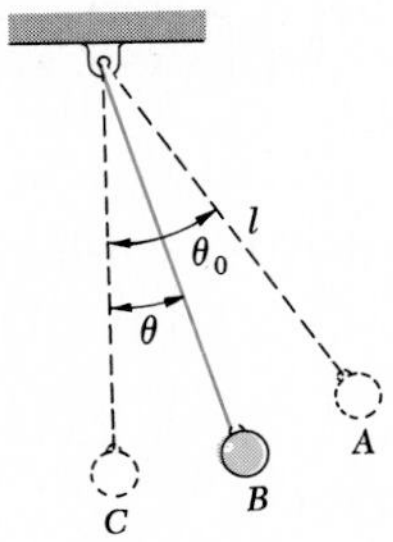

Fig. P12.44 and P12.45

12.45 A ball of weight *W* is released with no velocity from position *A* and oscillates in a vertical plane at the end of a cord of length *l*. Determine (*a*) the tangential component of the acceleration in position *B* in terms of the angle θ, (*b*) the velocity in position *B* in terms of θ, θ_0, and *l*, (*c*) the tension in the cord in terms of *W* and θ_0 when the ball passes through its lowest position *C*, (*d*) the value of θ_0 if the tension in the cord is $T = 2W$ when the ball passes through position *C*.

12.46 A small sphere of weight W is held as shown by two wires AB and CD. Wire AB is then cut. Determine (*a*) the tension in wire CD before AB was cut, (*b*) the tension in wire CD and the acceleration of the sphere just after AB has been cut.

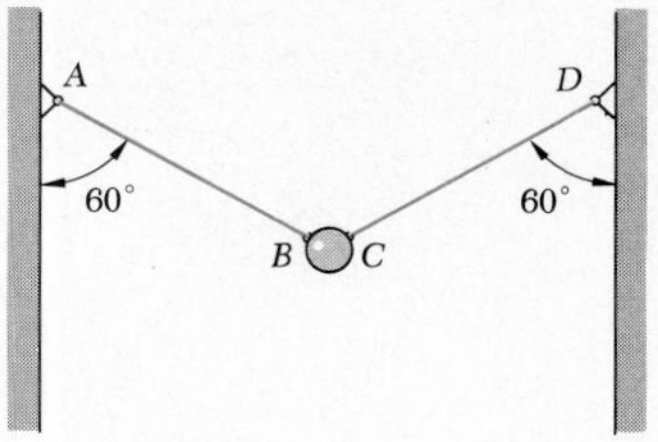

Fig. P12.46

12.47 A man swings a bucket full of water in a vertical plane in a circle of radius 0.75 m. What is the smallest velocity that the bucket should have at the top of the circle if no water is to be spilled?

12.48 A 175-lb pilot flies a small plane in a vertical loop of 300-ft radius. Determine the speed of the plane at points A and B, knowing that at point A the pilot experiences weightlessness and that at point B the pilot's apparent weight is 600 lb.

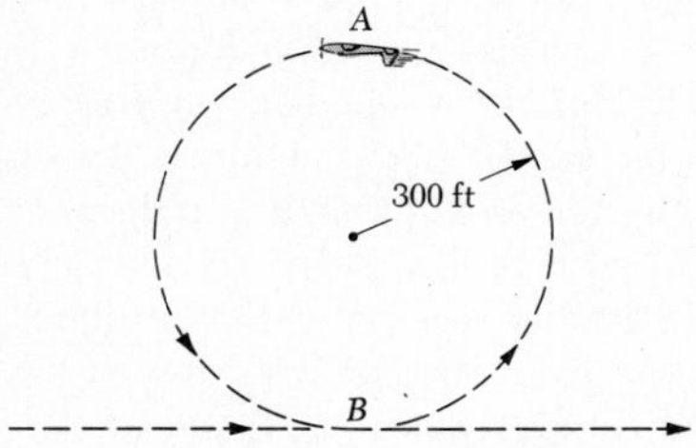

Fig. P12.48

12.49 Three automobiles are proceeding at a speed of 50 mi/h along the road shown. Knowing that the coefficient of friction between the tires and the road is 0.60, determine the tangential deceleration of each automobile if its brakes are suddenly applied and the wheels skid.

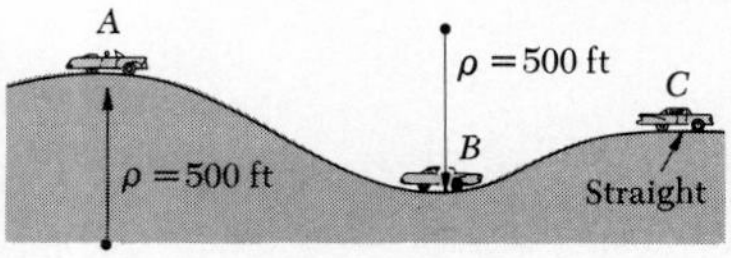

Fig. P12.49

12.50 A series of small packages, being moved by a conveyor belt at a constant speed v, passes over an idler roller as shown. Knowing that the coefficient of friction between the packages and the belt is 0.75, determine the maximum value of v for which the packages do not slip with respect to the belt.

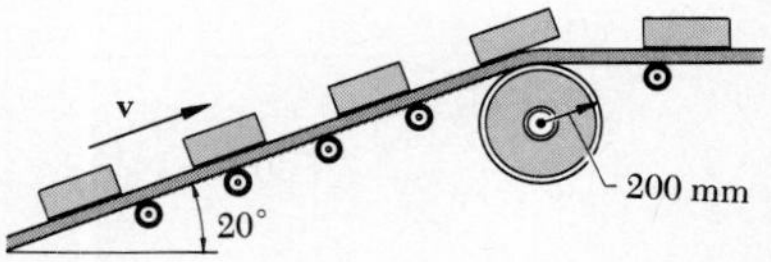

Fig. P12.50

12.51 Express the minimum and maximum safe speeds, with respect to skidding, of a car traveling on a banked road, in terms of the radius r of the curve, the banking angle θ, and the friction angle ϕ between the tires and the pavement.

12.52 A man on a motorcycle takes a turn on a flat unbanked road at 72 km/h. If the radius of the turn is 50 m, determine the minimum value of the coefficient of friction between the tires and the road which will ensure no skidding.

12.53 What angle of banking should be given to the road in Prob. 12.52 if the man on the motorcycle is to be able to take the turn at 72 km/h with a coefficient of friction $\mu = 0.30$?

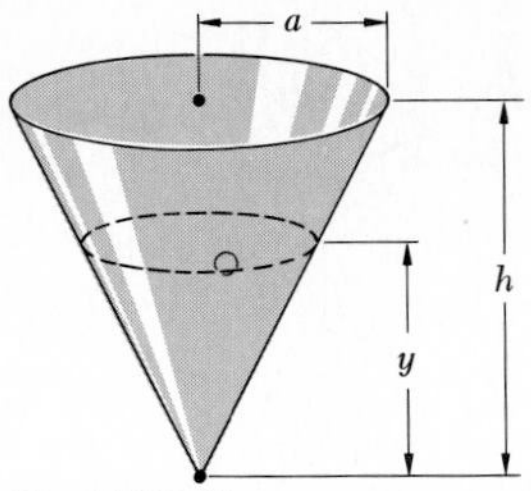

Fig. P12.55

12.54 A stunt driver proposes to drive a small automobile on the vertical wall of a circular pit of radius 50 ft. Knowing that the coefficient of friction between the tires and the wall is 0.75, determine the minimum speed at which the stunt can be performed.

12.55 A small ball rolls at a speed v_0 along a horizontal circle inside the circular cone shown. Express the speed v_0 in terms of the height y of the path above the apex of the cone.

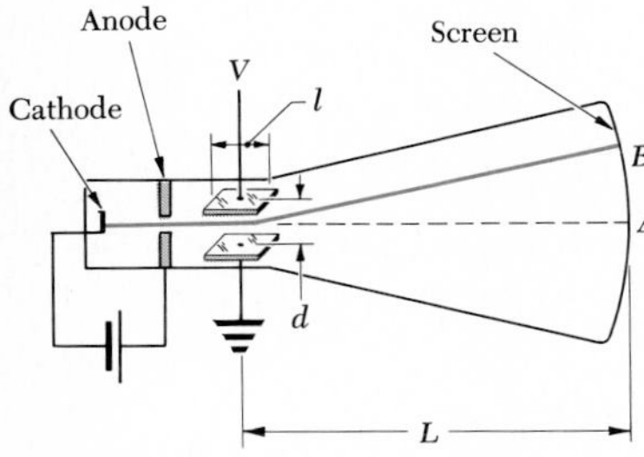

Fig. P12.56 and P12.58

12.56 In the cathode-ray tube shown, electrons emitted by the cathode and attracted by the anode pass through a small hole in the anode and keep traveling in a straight line with a speed v_0 until they strike the screen at A. However, if a difference of potential V is established between the two parallel plates, each electron will be subjected to a force $\mathbf{F}$ perpendicular to the plates while it travels between the plates and will strike the screen at point B at a distance δ from A. The magnitude of the force $\mathbf{F}$ is $F = eV/d$, where $-e$ is the charge of the electron and d is the distance between the plates. Derive an expression for the deflection δ in terms of V, v_0, the charge $-e$ of the electron, its mass m, and the dimensions d, l, and L.

12.57 In Prob. 12.56, determine the smallest allowable value of the ratio d/l in terms of e, m, v_0, and V if the electrons are not to strike the positive plate.

12.58 A manufacturer wishes to design a new cathode-ray tube which will be only half as long as his current model. If the size of the screen is to remain the same, how should the length l of the plates be modified if all the other characteristics of the circuit are to remain unchanged?

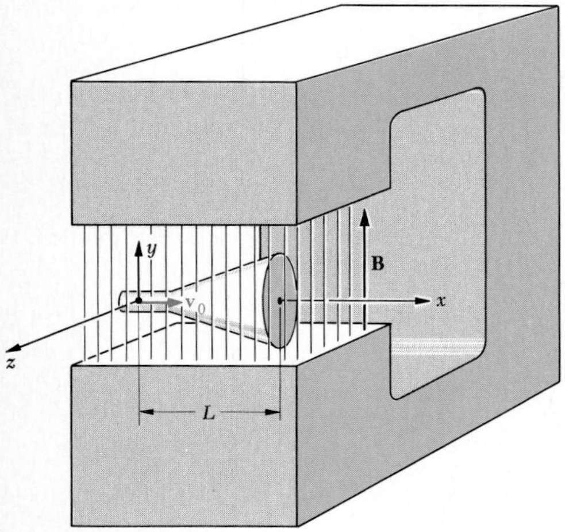

Fig. P12.59

12.59 A cathode-ray tube emitting electrons with a velocity $\mathbf{v}_0$ is placed as shown between the poles of a large electromagnet which creates a uniform magnetic field of strength $\mathbf{B}$. Determine the coordinates of the point where the electron beam strikes the tube screen when no difference of potential exists between the plates. It is known that an electron (mass m and charge $-e$) traveling with a velocity $\mathbf{v}$ at a right angle to the lines of force of a magnetic field of strength $\mathbf{B}$ is subjected to a force $\mathbf{F}$ perpendicular to $\mathbf{B}$ and $\mathbf{v}$, and of magnitude $F = eBv$; the sense of the force is such that $\mathbf{B}$, $\mathbf{v}$, and $\mathbf{F}$ form a right-handed triad.

12.60 The two-dimensional motion of particle B is defined by the relations $r = t^2 - \frac{1}{3}t^3$ and $\theta = 2t^2$, where r is expressed in meters, t in seconds, and θ in radians. If the particle has a mass of 2 kg and moves in a horizontal plane, determine the radial and transverse components of the force acting on the particle when (*a*) $t = 0$, (*b*) $t = 1$ s.

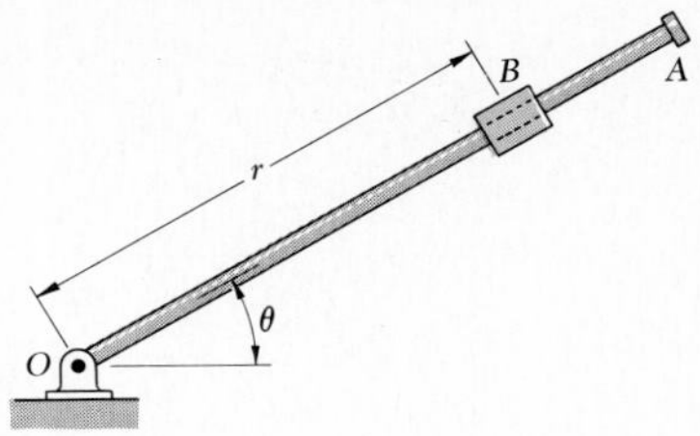

Fig. P12.60

12.61 For the motion defined in Prob. 12.60, determine the radial and transverse components of the force acting on the 2-kg particle as it returns to the origin at $t = 3$ s.

12.62 The two-dimensional motion of a particle is defined by the relations $r = 2 \sin \pi t$ and $\theta = \frac{1}{2}\pi t$, where r is expressed in feet, t in seconds, and θ in radians. If the particle weighs 3 lb and moves in a horizontal plane, determine the radial and transverse components of the force acting on the particle when (*a*) $t = \frac{1}{2}$ s, (*b*) $t = 1$ s.

12.63 A block B of mass m may slide on the frictionless arm OA which rotates in a horizontal plane at a constant rate $\dot{\theta}_0$. As the arm rotates, the cord wraps around a *fixed* drum of radius b and pulls the block toward O with a speed $b\dot{\theta}_0$. Express as a function of m, r, b, and $\dot{\theta}_0$, (*a*) the tension T in the cord, (*b*) the magnitude of the horizontal force $\mathbf{Q}$ exerted on B by the arm OA.

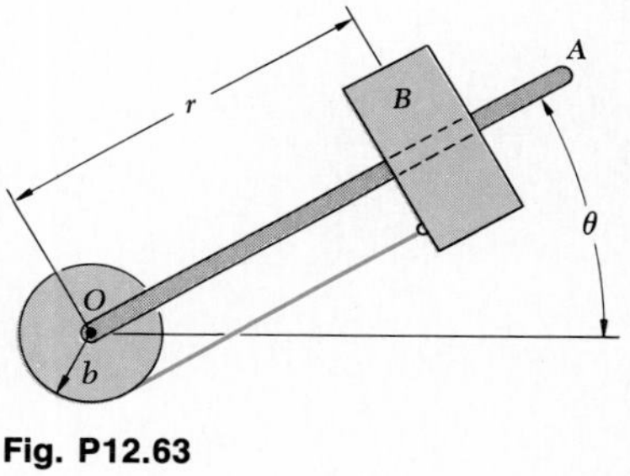

Fig. P12.63

12.64 Solve Prob. 12.63, knowing that the weight of the block is 3 lb and that $r = 30$ in., $b = 2$ in., and $\dot{\theta}_0 = 6$ rad/s.

12.65 In Prob. 12.63, determine the ratio r/b for which the tension T is equal to the magnitude of the horizontal force $\mathbf{Q}$.

12.66 While aiming at a moving target, a man rotates his rifle clockwise in a horizontal plane at the rate of 15° per second. Assuming that he can maintain the motion as the rifle is fired, determine the horizontal force exerted by the barrel on a 45-g bullet just before it leaves the barrel with a muzzle velocity of 550 m/s.

12.67 A particle moves under a central force in a path defined by the equation $r = r_0/\cos n\theta$, where n is a positive constant. Using Eq. (12.18), show that the radial and transverse components of the velocity are $v_r = nv_0 \sin n\theta$ and $v_\theta = v_0 \cos n\theta$, where v_0 is the velocity of the particle for $\theta = 0$. What is the motion of the particle when $n = 0$ and when $n = 1$?

12.68 For the particle and motion of Prob. 12.67, show that the radial and transverse components of the acceleration are $a_r = (n^2 - 1)(v_0^2/r_0) \cos^3 n\theta$ and $a_\theta = 0$.

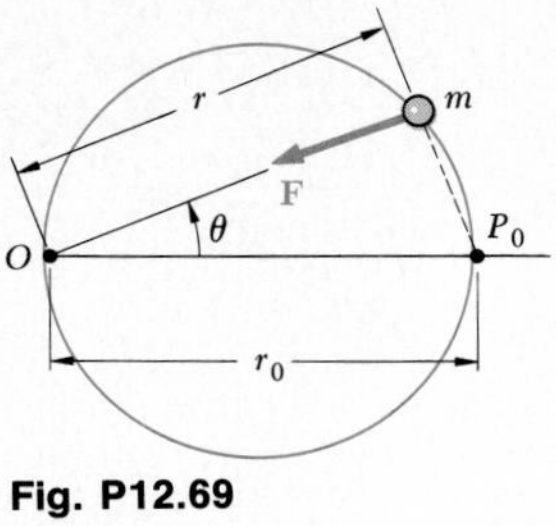

Fig. P12.69

12.69 A particle moves under a central force in a circular path of diameter r_0 which passes through the center of force O. Show that its speed is $v = v_0/\cos^2\theta$, where v_0 is the speed of the particle at point P_0 directly across the circle from O. [*Hint.* Use Eq. (12.18) with $r = r_0 \cos\theta$].

12.70 If a particle of mass m is attached to the end of a very light circular rod as shown in (*1*), the rod exerts on the mass a force **F** of magnitude $F = kr$ directed toward the origin O, as shown in (*2*). The path of the particle is observed to be an ellipse with axes $a = 6$ in. and $b = 2$ in. (*a*) Knowing that the speed of the particle at A is 8 in./s, determine the speed at B. (*b*) Further knowing that the constant k/m is equal to 16 s^{-2}, determine the radius of curvature of the path at A and at B.

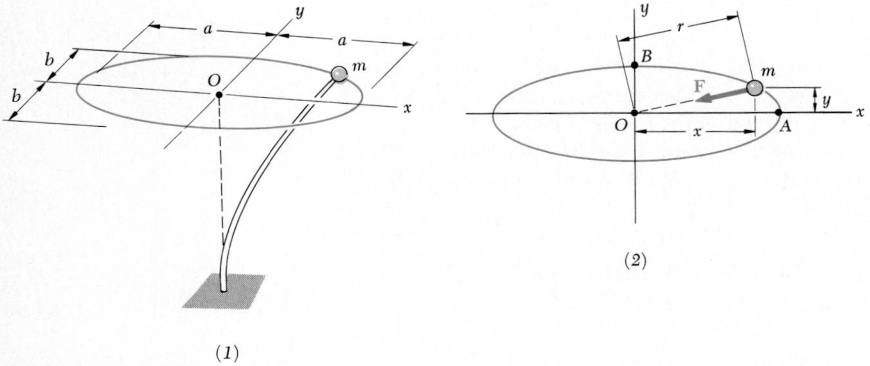

Fig. P12.70

12.71 Determine the mass of the earth from Newton's law of gravitation, knowing that it takes 94.14 min for a satellite to describe a circular orbit 300 mi above the surface of the earth.

12.72 Show that the radius r of the moon's orbit may be determined from the radius R of the earth, the acceleration of gravity g at the surface of the earth, and the time τ required by the moon to revolve once around the earth. Compute r knowing that $\tau = 27.3$ days.

12.73 Determine the time required for a spacecraft to describe a circular orbit 30 km above the moon's surface. It is known that the radius of the moon is 1740 km and that its mass is 81.3 times smaller than the mass of the earth.

12.74 Denoting by ρ the mean mass density of a planet, show that the periodic time of a satellite in a very low circular orbit is $(3\pi/G\rho)^{1/2}$, where G is the constant of gravitation.

12.75 A spacecraft describes a circular orbit of 6000-mi radius around the earth. In order to transfer it to a larger circular orbit of 24,000-mi radius, the spacecraft is first placed on an elliptic path AB by firing its engine as it passes through A, thus increasing its velocity by 3810 mi/h. By how much should the spacecraft's velocity be increased as it reaches B to insert it in the larger circular orbit?

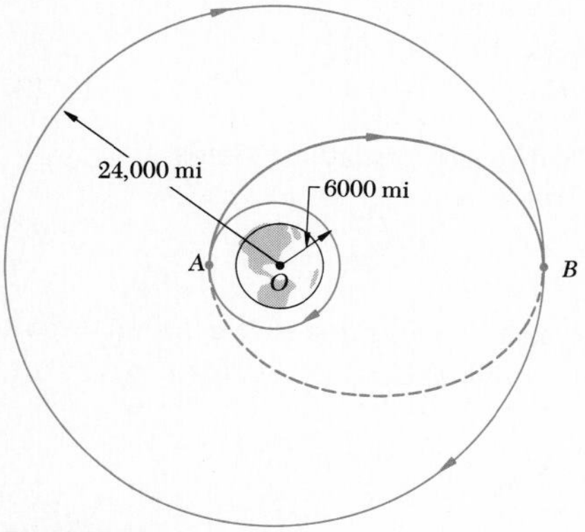

Fig. P12.75

12.76 An Apollo spacecraft describes a circular orbit of 2400-km radius around the moon with a velocity of 5140 km/h. In order to transfer it to a smaller circular orbit of 2000-km radius, the spacecraft is first placed on an elliptic path AB by reducing its velocity to 4900 km/h as it passes through A. Determine (*a*) the velocity of the spacecraft as it approaches B on the elliptic path, (*b*) the value to which its velocity must be reduced at B to insert it in the smaller circular orbit.

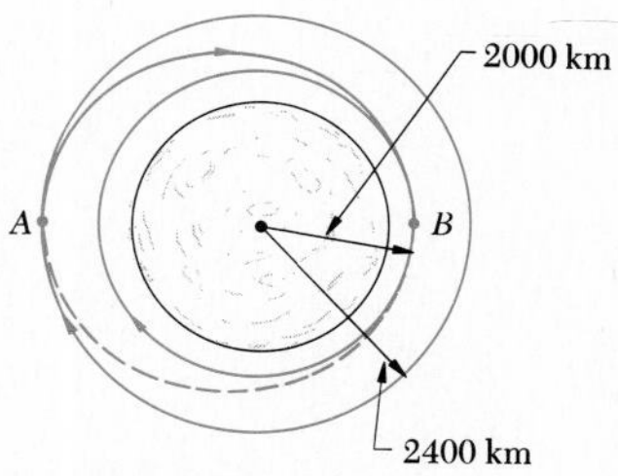

Fig. P12.76

12.77 Solve Prob. 12.76, assuming that the Apollo spacecraft is to be transferred from the orbit of 2400-km radius to a circular orbit of 1800-km radius and that its velocity is reduced to 4760 km/h as it passes through A.

12.78 Solve Prob. 12.75, assuming that the spacecraft is to be transferred from the orbit of 6000-mi radius to a circular orbit of 13,500-mi radius and that its velocity is increased by 2540 mi/h as it passes through A.

***12.10. Trajectory of a Particle under a Central Force.** Consider a particle P moving under a central force **F**. We propose to obtain the differential equation which defines its trajectory.

Assuming that the force **F** is directed toward the center of force O, we note that ΣF_r and ΣF_θ reduce respectively to $-F$ and zero in Eqs. (12.16) and (12.17). We therefore write

$$m(\ddot{r} - r\dot{\theta}^2) = -F \tag{12.22}$$
$$m(r\ddot{\theta} + 2\dot{r}\dot{\theta}) = 0 \tag{12.23}$$

But, as we saw in Sec. 12.8, the last equation yields

$$r^2\dot{\theta} = h \qquad \text{or} \qquad r^2\frac{d\theta}{dt} = h \tag{12.24}$$

Equation (12.24) may be used to eliminate the independent variable t from Eq. (12.22). Solving Eq. (12.24) for $\dot{\theta}$ or $d\theta/dt$, we have

$$\dot{\theta} = \frac{d\theta}{dt} = \frac{h}{r^2} \tag{12.25}$$

from which it follows that

$$\dot{r} = \frac{dr}{dt} = \frac{dr}{d\theta}\frac{d\theta}{dt} = \frac{h}{r^2}\frac{dr}{d\theta} = -h\frac{d}{d\theta}\left(\frac{1}{r}\right) \tag{12.26}$$

$$\ddot{r} = \frac{d\dot{r}}{dt} = \frac{d\dot{r}}{d\theta}\frac{d\theta}{dt} = \frac{h}{r^2}\frac{d\dot{r}}{d\theta}$$

or, substituting for $\dot{r}$ from (12.26),

$$\ddot{r} = \frac{h}{r^2}\frac{d}{d\theta}\left[-h\frac{d}{d\theta}\left(\frac{1}{r}\right)\right]$$

$$\ddot{r} = -\frac{h^2}{r^2}\frac{d^2}{d\theta^2}\left(\frac{1}{r}\right) \tag{12.27}$$

Substituting for $\dot{\theta}$ and $\ddot{r}$ from (12.25) and (12.27), respectively, into Eq. (12.22), and introducing the function $u = 1/r$, we obtain after reductions

$$\frac{d^2u}{d\theta^2} + u = \frac{F}{mh^2u^2} \tag{12.28}$$

In deriving Eq. (12.28), the force **F** was assumed directed toward O. The magnitude F should therefore be positive if **F** is actually directed toward O (attractive force) and negative if **F** is directed away from O (repulsive force). If F is a known function of r and thus of u, Eq. (12.28) is a differential equation in u and θ. This differential equation defines the trajectory followed by the particle under the central force **F**. The equation of the trajectory will be obtained by solving the differential equation (12.28) for u as a function of θ and determining the constants of integration from the initial conditions.

*12.11. Application to Space Mechanics.

After the last stage of their launching rockets has burned out, earth satellites and other space vehicles are subjected only to the gravitational pull of the earth. Their motion may therefore be determined from Eqs. (12.24) and (12.28), which govern the motion of a particle under a central force, after F has been replaced by the expression obtained for the force of gravitational attraction.† Setting in Eq. (12.28)

$$F = \frac{GMm}{r^2} = GMmu^2$$

where M = mass of earth
m = mass of space vehicle
r = distance from center of earth to vehicle
$u = 1/r$

we obtain the differential equation

$$\frac{d^2u}{d\theta^2} + u = \frac{GM}{h^2} \tag{12.29}$$

where the right-hand member is observed to be a constant.

The solution of the differential equation (12.29) is obtained by adding the particular solution $u = GM/h^2$ to the general solution $u = C \cos (\theta - \theta_0)$ of the corresponding homogeneous equation (i.e., the equation obtained by setting the right-hand member equal to zero). Choosing the polar axis so that $\theta_0 = 0$, we write

$$\frac{1}{r} = u = \frac{GM}{h^2} + C \cos \theta \tag{12.30}$$

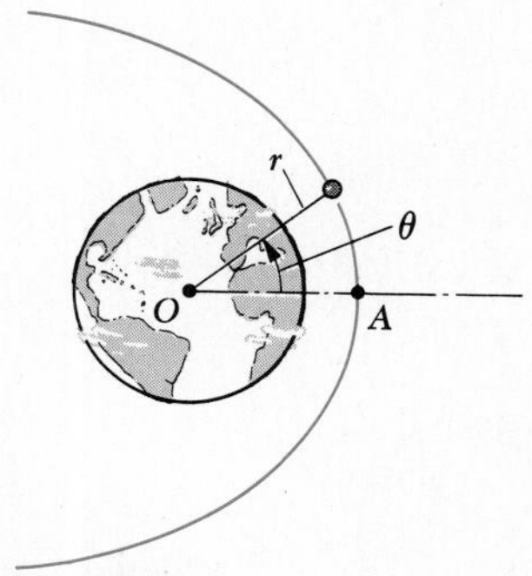

Fig. 12.18

Equation (12.30) is the equation of a *conic section* (ellipse, parabola, or hyperbola) in the polar coordinates r and θ. The origin O of the coordinates, which is located at the center of the earth, is a *focus* of this conic section, and the polar axis is one of its axes of symmetry (Fig. 12.18).

† It is assumed that the space vehicles considered here are attracted only by the earth and that their mass is negligible compared to the mass of the earth. If a vehicle moves very far from the earth, its path may be affected by the attraction of the sun, the moon, or another planet.

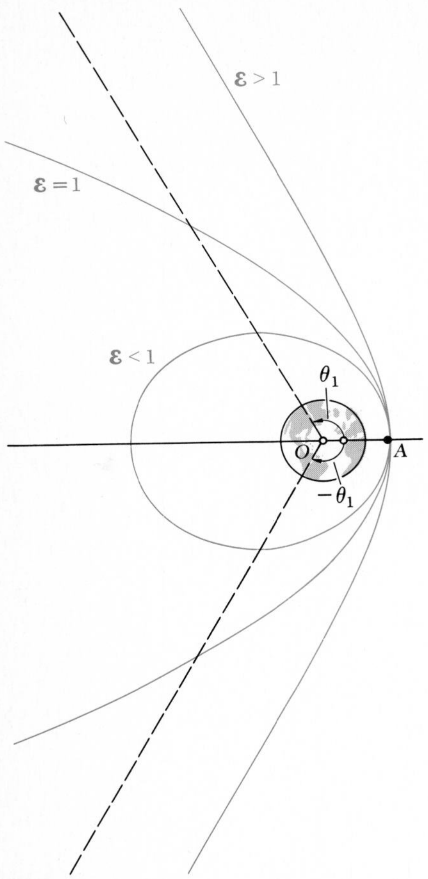

Fig. 12.19

The ratio of the constants C and GM/h^2 defines the *eccentricity* ε of the conic section; we write

$$\varepsilon = \frac{C}{GM/h^2} = \frac{Ch^2}{GM} \tag{12.31}$$

Three cases may be distinguished:

1. $\varepsilon > 1$, or $C > GM/h^2$: There are two values θ_1 and $-\theta_1$ of the polar angle, defined by $\cos\theta_1 = -GM/Ch^2$, for which the right-hand member of Eq. (12.30) becomes zero. For both of these values, the radius vector r becomes infinite; the conic section is a *hyperbola* (Fig. 12.19).
2. $\varepsilon = 1$, or $C = GM/h^2$: The radius vector becomes infinite for $\theta = 180°$; the conic section is a *parabola.*
3. $\varepsilon < 1$, or $C < GM/h^2$: The radius vector remains finite for every value of θ; the conic section is an *ellipse.* In the particular case when $\varepsilon = C = 0$, the length of the radius vector is constant; the conic section is a circle.

We shall see now how the constants C and GM/h^2 which characterize the trajectory of a space vehicle may be determined from the position and the velocity of the space vehicle at the beginning of its free flight. We shall assume, as it is generally the case, that the powered phase of its flight has been programmed in such a way that, as the last stage of the launching rocket burns out, the vehicle has a velocity parallel to the surface of the earth (Fig. 12.20). In other words, we shall assume that the space vehicle begins its free flight at the vertex A of its trajectory.†

Denoting respectively by r_0 and v_0 the radius vector and speed of the vehicle at the beginning of its free flight, we observe, since the velocity reduces to its transverse component, that $v_0 = r_0\dot{\theta}_0$. Recalling Eq. (12.24), we express the constant h as

$$h = r_0^2\dot{\theta}_0 = r_0 v_0 \tag{12.32}$$

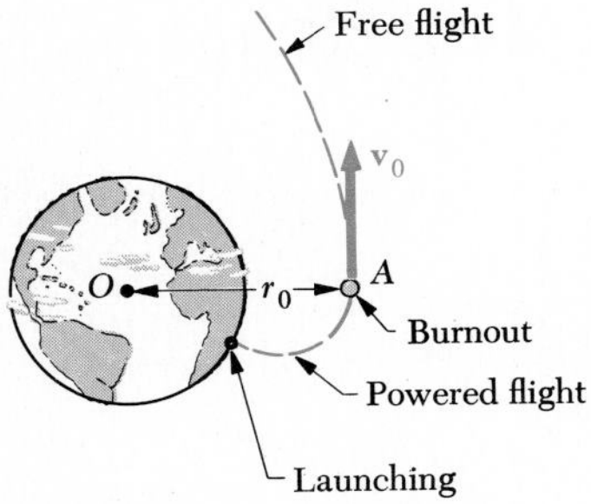

Fig. 12.20

The value obtained for h may be used to determine the constant GM/h^2. We also note that the computation of this constant will be simplified if we use the relation indicated in Sec. 12.9,

$$GM = gR^2 \tag{12.21}$$

where R is the radius of the earth ($R = 6.37 \times 10^6$ m or 3960 mi) and g the acceleration of gravity at the surface of the earth.

† Problems involving oblique launchings will be considered in Sec. 14.11.

The constant C will be determined by setting $\theta = 0$, $r = r_0$ in Eq. (12.30); we obtain

$$C = \frac{1}{r_0} - \frac{GM}{h^2} \tag{12.33}$$

Substituting for h from (12.32), we may then easily express C in terms of r_0 and v_0.

Let us now determine the initial conditions corresponding to each of the three fundamental trajectories indicated above. Considering first the parabolic trajectory, we set C equal to GM/h^2 in Eq. (12.33) and eliminate h between Eqs. (12.32) and (12.33). Solving for v_0, we obtain

$$v_0 = \sqrt{\frac{2GM}{r_0}}$$

We may easily check that a larger value of the initial velocity corresponds to a hyperbolic trajectory, and a smaller value to an elliptic orbit. Since the value of v_0 obtained for the parabolic trajectory is the smallest value for which the space vehicle does not return to its starting point, it is called the *escape velocity*. We write therefore

$$v_{esc} = \sqrt{\frac{2GM}{r_0}} \qquad \text{or} \qquad v_{esc} = \sqrt{\frac{2gR^2}{r_0}} \tag{12.34}$$

if we make use of Eq. (12.21). We note that the trajectory will be (1) hyperbolic if $v_0 > v_{esc}$; (2) parabolic if $v_0 = v_{esc}$; (3) elliptic if $v_0 < v_{esc}$.

Among the various possible elliptic orbits, one is of special interest, the *circular orbit*, which is obtained when $C = 0$. The value of the initial velocity corresponding to a circular orbit is easily found to be

$$v_{circ} = \sqrt{\frac{GM}{r_0}} \qquad \text{or} \qquad v_{circ} = \sqrt{\frac{gR^2}{r_0}} \tag{12.35}$$

if Eq. (12.21) is taken into account. We may note from Fig. 12.21 that, for values of v_0 comprised between v_{circ} and v_{esc}, point A where free flight begins is the point of the orbit closest to the earth; this point is called the *perigee*, while point A', which is farthest away from the earth, is known as the *apogee*. For values of v_0 smaller than v_{circ}, point A becomes the apogee, while point A'', on the other side of the orbit, becomes the perigee. For values of v_0 much smaller than v_{circ}, the trajectory of the space vehicle intersects the surface of the earth; in such a case, the vehicle does not go into orbit.

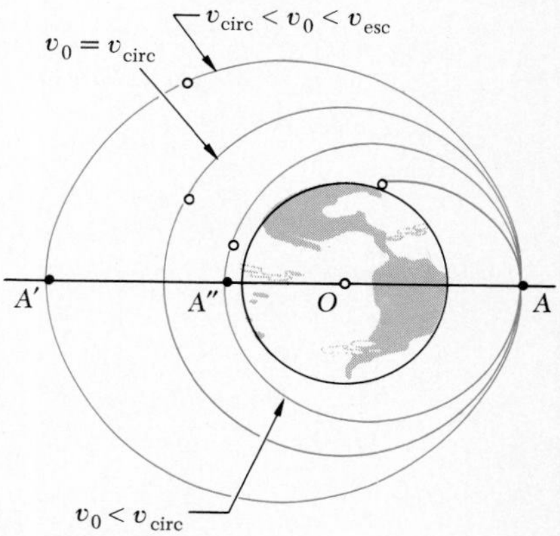

Fig. 12.21

Ballistic missiles, which are designed to hit the surface of the earth, also travel along elliptic trajectories. In fact, we should now realize that any object projected in vacuum with an initial velocity v_0 smaller than v_{esc} will move along an elliptic path. It is only when the distances involved are small that the gravitational field of the earth may be assumed uniform, and that the elliptic path may be approximated by a parabolic path, as was done earlier (Sec. 11.10) in the case of conventional projectiles.

Periodic Time. An important characteristic of the motion of an earth satellite is the time required by the satellite to describe its orbit. This time is known as the *periodic time* of the satellite and is denoted by τ. We first observe, in view of the definition of the areal velocity (Sec. 12.8), that τ may be obtained by dividing the area inside the orbit by the areal velocity. Since the area of an ellipse is equal to πab, where a and b denote, respectively, the semimajor and semiminor axes, and since the areal velocity is equal to $h/2$, we write

$$\tau = \frac{2\pi ab}{h} \tag{12.36}$$

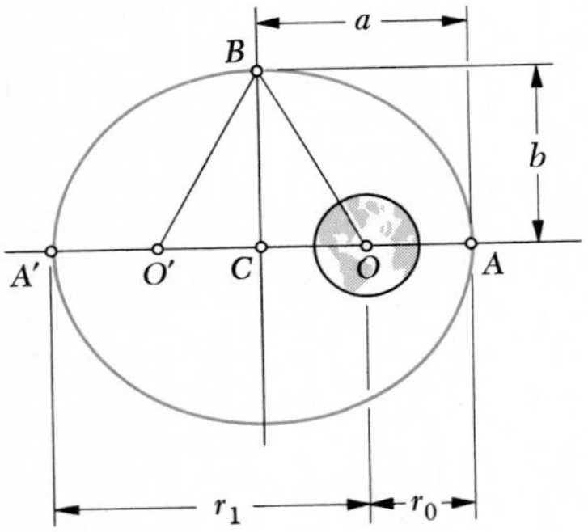

Fig. 12.22

While h may be readily determined from r_0 and v_0 in the case of a satellite launched in a direction parallel to the surface of the earth, the semiaxes a and b are not directly related to the initial conditions. Since, on the other hand, the values r_0 and r_1 of r corresponding to the perigee and apogee of the orbit may easily be determined from Eq. (12.30), we shall express the semiaxes a and b in terms of r_0 and r_1.

Consider the elliptic orbit shown in Fig. 12.22. The earth's center is located at O and coincides with one of the two foci of the ellipse, while the points A and A' represent, respectively, the perigee and apogee of the orbit. We easily check that

$$r_0 + r_1 = 2a$$

and thus

$$a = \tfrac{1}{2}(r_0 + r_1) \tag{12.37}$$

Recalling that the sum of the distances from each of the foci to any point of the ellipse is constant, we write

$$O'B + BO = O'A + OA = 2a \qquad \text{or} \qquad BO = a$$

On the other hand, we have $CO = a - r_0$. We may therefore write

$$b^2 = (BC)^2 = (BO)^2 - (CO)^2 = a^2 - (a - r_0)^2$$
$$b^2 = r_0(2a - r_0) = r_0 r_1$$

and thus

$$b = \sqrt{r_0 r_1} \tag{12.38}$$

Formulas (12.37) and (12.38) indicate that the semimajor and semiminor axes of the orbit are respectively equal to the arithmetic and geometric means of the maximum and minimum values of the radius vector. Once r_0 and r_1 have been determined, the lengths of the semiaxes may thus be easily computed and substituted for a and b in formula (12.36).

***12.12. Kepler's Laws of Planetary Motion.** The equations governing the motion of an earth satellite may be used to describe the motion of the moon around the earth. In that case, however, the mass of the moon is not negligible compared to the mass of the earth, and the results obtained are not entirely accurate.

The theory developed in the preceding sections may also be applied to the study of the motion of the planets around the sun. While another error is introduced by neglecting the forces exerted by the planets on each other, the approximation obtained is excellent. Indeed, the properties expressed by Eq. (12.30), where M now represents the mass of the sun, and by Eq. (12.24) had been discovered by the German astronomer Johann Kepler (1571–1630) from astronomical observations of the motion of the planets, even before Newton had formulated his fundamental theory.

Kepler's three *laws of planetary motion* may be stated as follows:

1. Each planet describes an ellipse, with the sun located at one of its foci.
2. The radius vector drawn from the sun to a planet sweeps equal areas in equal times.
3. The squares of the periodic times of the planets are proportional to the cubes of the semimajor axes of their orbits.

The first law states a particular case of the result established in Sec. 12.11, while the second law expresses that the areal velocity of each planet is constant (see Sec. 12.8). Kepler's third law may also be derived from the results obtained in Sec. 12.11.†

† See Prob. 12.95.

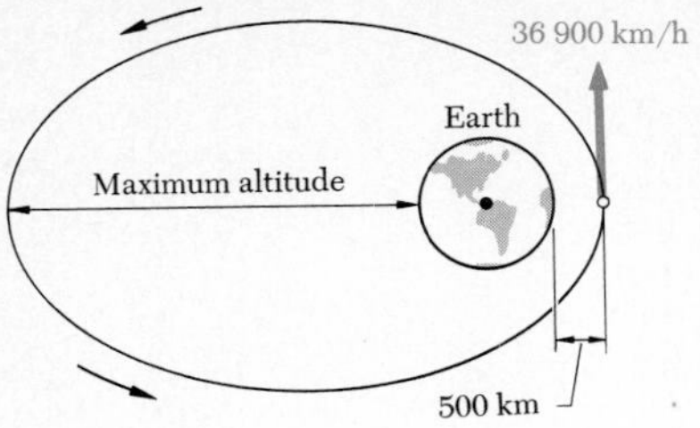

SAMPLE PROBLEM 12.8

A satellite is launched in a direction parallel to the surface of the earth with a velocity of 36 900 km/h from an altitude of 500 km. Determine (*a*) the maximum altitude reached by the satellite, (*b*) the periodic time of the satellite.

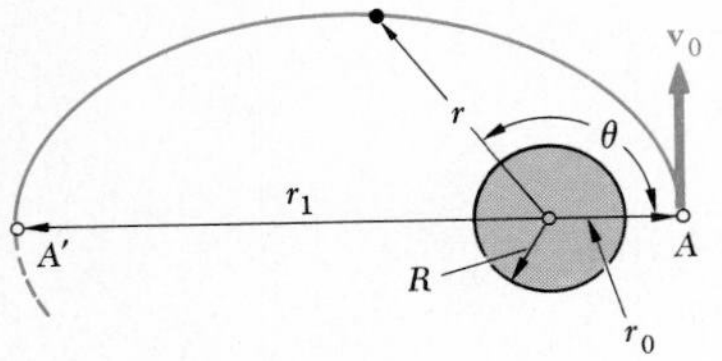

***a.* Maximum Altitude.** After launching, the satellite is subjected only to the gravitational attraction of the earth; its motion is thus governed by Eq. (12.30).

$$\frac{1}{r} = \frac{GM}{h^2} + C\cos\theta \qquad (1)$$

Since the radial component of the velocity is zero at the point of launching A, we have $h = r_0 v_0$. Recalling that the radius of the earth is $R = 6370$ km, we compute

$$r_0 = 6370 \text{ km} + 500 \text{ km} = 6870 \text{ km} = 6.87 \times 10^6 \text{ m}$$
$$v_0 = 36\,900 \text{ km/h} = \frac{3.69 \times 10^7 \text{ m}}{3.6 \times 10^3 \text{ s}} = 1.025 \times 10^4 \text{ m/s}$$
$$h = r_0 v_0 = (6.87 \times 10^6 \text{ m})(1.025 \times 10^4 \text{ m/s}) = 7.04 \times 10^{10} \text{ m}^2\text{/s}$$
$$h^2 = 4.96 \times 10^{21} \text{ m}^4\text{/s}^2$$

Since $GM = gR^2$, where R is the radius of the earth, we have

$$GM = gR^2 = (9.81 \text{ m/s}^2)(6.37 \times 10^6 \text{ m})^2 = 3.98 \times 10^{14} \text{ m}^3\text{/s}^2$$
$$\frac{GM}{h^2} = \frac{3.98 \times 10^{14} \text{ m}^3\text{/s}^2}{4.96 \times 10^{21} \text{ m}^4\text{/s}^2} = 8.03 \times 10^{-8} \text{ m}^{-1}$$

Substituting this value into (1), we obtain

$$\frac{1}{r} = 8.03 \times 10^{-8} + C\cos\theta \qquad (2)$$

Noting that at point A we have $\theta = 0$ and $r = r_0 = 6.87 \times 10^6$ m, we compute the constant C.

$$\frac{1}{6.87 \times 10^6 \text{ m}} = 8.03 \times 10^{-8} + C\cos 0° \qquad C = 6.53 \times 10^{-8} \text{ m}^{-1}$$

At A', the point on the orbit farthest from the earth, we have $\theta = 180°$. Using (2), we compute the corresponding distance r_1.

$$\frac{1}{r_1} = 8.03 \times 10^{-8} + (6.53 \times 10^{-8})\cos 180°$$
$$r_1 = 0.667 \times 10^8 \text{ m} = 66\,700 \text{ km}$$

Maximum altitude = 66 700 km − 6370 km = 60 300 km ◀

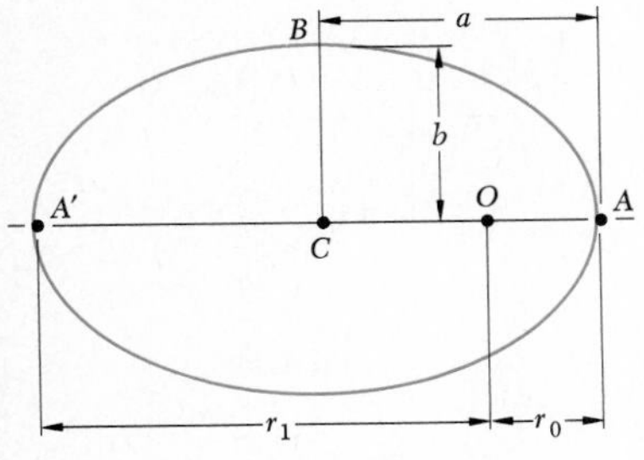

***b.* Periodic Time.** Since A and A' are the perigee and apogee, respectively, of the elliptic orbit, we use Eqs. (12.37) and (12.38) and compute the semimajor and semiminor axes of the orbit.

$$a = \tfrac{1}{2}(r_0 + r_1) = \tfrac{1}{2}(6.87 + 66.7)(10^6) \text{ m} = 36.8 \times 10^6 \text{ m}$$
$$b = \sqrt{r_0 r_1} = \sqrt{(6.87)(66.7)} \times 10^6 \text{ m} = 21.4 \times 10^6 \text{ m}$$
$$\tau = \frac{2\pi ab}{h} = \frac{2\pi(36.8 \times 10^6 \text{ m})(21.4 \times 10^6 \text{ m})}{7.04 \times 10^{10} \text{ m}^2\text{/s}}$$
$$\tau = 7.03 \times 10^4 \text{ s} = 1171 \text{ min} = 19 \text{ h } 31 \text{ min} \quad ◀$$

PROBLEMS

12.79 (*a*) Knowing that the radius of the moon is 1080 mi and that its mass is 0.01230 times the mass of the earth, determine the speed of an Apollo spacecraft describing a circular orbit 420 mi above the moon's surface. (*b*) To what value should the speed be increased if the spacecraft is to leave the moon's gravitation field? (Neglect the effect of the earth's gravity.)

12.80 A spacecraft is describing a circular orbit of 5000-mi radius around the earth when its engine is suddenly fired, increasing the speed of the spacecraft by 3000 mi/h. Determine the greatest distance from the center of the earth reached by the spacecraft.

12.81 A space vehicle approaches the planet Venus along a hyperbolic trajectory of eccentricity $\varepsilon = 2$. As the vehicle reaches point A, which is the point of the trajectory closest to the planet, retrorockets are fired to slow the vehicle and place it in a circular orbit. Knowing that the distance from the center O of the planet to A is 6000 mi, and that the mass of Venus is 0.82 times the mass of the earth, determine the velocity of the vehicle (*a*) as it approaches A, (*b*) after the retrorockets have been fired.

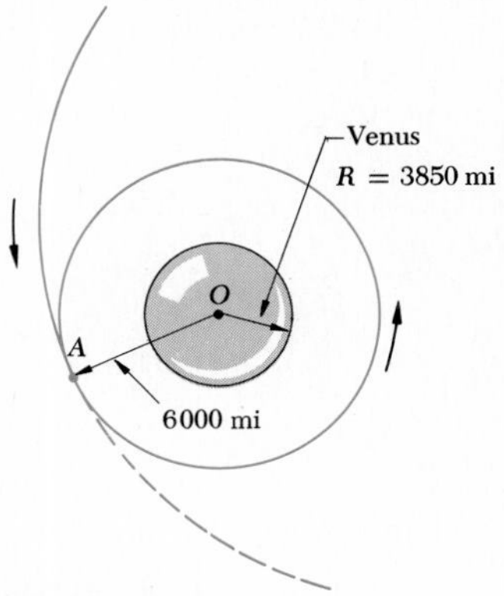

Fig. P12.81

12.82 The maximum altitude a_1 of an unmanned earth satellite is observed to be 8300 km, while its minimum altitude a_2 is known to be 225 km. Determine the maximum and minimum values of its velocity.

12.83 The speed of an earth satellite as it passes through its perigee A is 36×10^3 km/h. Knowing that the altitude of the perigee is $a_2 = 230$ km, determine (*a*) the maximum altitude reached by the satellite, (*b*) the corresponding minimum speed of the satellite.

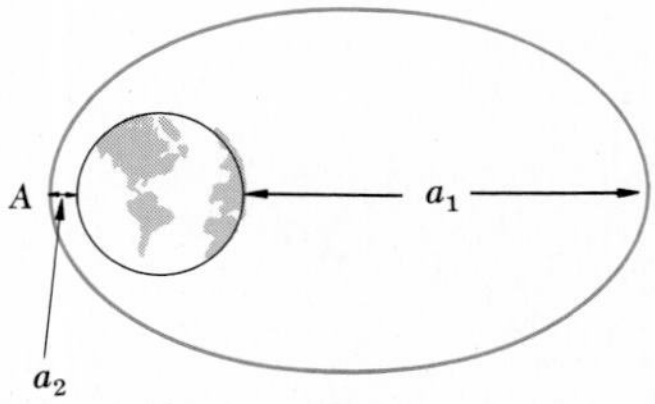

Fig. P12.82 and P12.83

12.84 After completing their moon-exploration mission, the two astronauts forming the crew of an Apollo lunar excursion module (LEM) prepare to rejoin the command module which is orbiting the moon at an altitude of 85 mi. They fire the LEM's engine, bring it along a curved path to a point A, 5 mi above the moon's surface, and shut off the engine. Knowing that the LEM is moving at that time in a direction parallel to the moon's surface and that it will coast along an elliptic path to a rendezvous at B with the command module, determine (*a*) the speed of the LEM at engine shutoff, (*b*) the relative velocity with which the command module will approach the LEM at B. (The radius of the moon is 1080 miles and its mass is 0.01230 times the mass of the earth.)

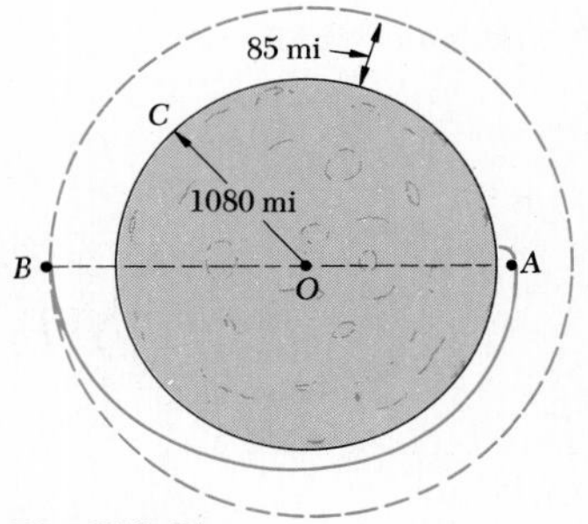

Fig. P12.84

12.85 Solve Prob. 12.84, assuming that the Apollo command module is orbiting the moon at an altitude of 55 mi.

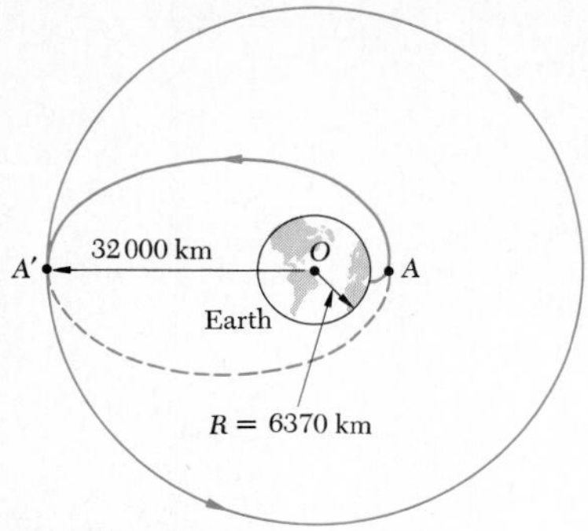

Fig. P12.86

12.86 In order to place a satellite in a circular orbit of radius 32×10^3 km around the earth, the satellite is first projected horizontally from A at an altitude of 500 km into an elliptic path whose apogee A' is at a distance of 32×10^3 km from the center of the earth. Auxiliary rockets are fired as the satellite reaches A' in order to place it in its final orbit. Determine (*a*) the initial velocity of the satellite at A, (*b*) the increase in velocity resulting from the firing of the rockets at A'.

12.87 Determine the time required for the LEM of Prob. 12.84 to travel from A to B.

12.88 Determine the periodic time of the spacecraft of Prob. 12.80 on its elliptic orbit.

12.89 Determine the periodic time of the satellite of Prob. 12.82.

12.90 Referring to Prob. 12.86, determine the time elapsed between the end of the launching of the satellite at A and the firing of the auxiliary rockets at A' which will place the satellite in its final orbit.

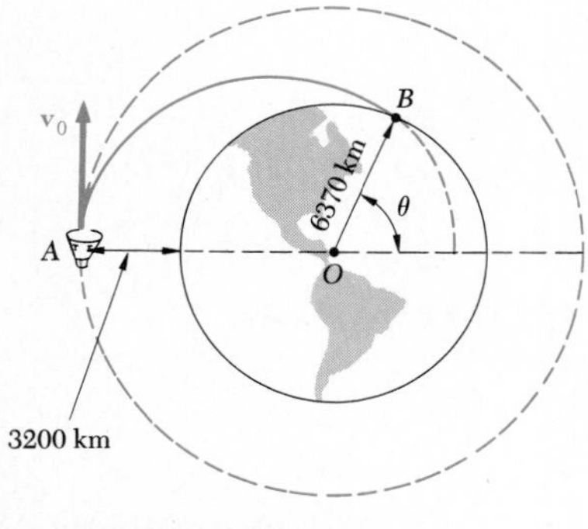

Fig. P12.91 and P12.92

12.91 A spacecraft describes a circular orbit at an altitude of 3200 km above the earth's surface. Preparatory to reentry it reduces its speed to a value $v_0 = 5400$ m/s, thus placing itself on an elliptic trajectory. Determine the value of θ defining the point B where splashdown will occur. (*Hint.* Point A is the apogee of the elliptic trajectory.)

12.92 A spacecraft describes a circular orbit at an altitude of 3200 km above the earth's surface. Preparatory to reentry it places itself on an elliptic trajectory by reducing its speed to a value v_0. Determine v_0 so that splashdown will occur at a point B corresponding to $\theta = 120°$. (See hint of Prob. 12.91.)

12.93 Upon the LEM's return to the command module, the Apollo spacecraft of Prob. 12.84 is turned around so that the LEM faces to the rear. After completing a full orbit, i.e., as the craft passes again through B, the LEM is cast adrift and crashes on the moon's surface at point C. Determine the velocity of the LEM relative to the command module as it is cast adrift, knowing that the angle BOC is 90°. (*Hint.* Point B is the apogee of the elliptic crash trajectory.)

12.94 Upon the LEM's return to the command module, the Apollo spacecraft of Prob. 12.84 is turned around so that the LEM faces to the rear. After completing a full orbit, i.e., as the craft passes again through B, the LEM is cast adrift with a velocity of 600 ft/s relative to the command module. Determine the point C where the LEM will crash on the moon's surface. (See hint of Prob. 12.93).

12.95 Derive Kepler's third law of planetary motion from Eqs. (12.30) and (12.36).

12.96 Two space stations S_1 and S_2 are describing coplanar circular counterclockwise orbits of radius r_0 and $8r_0$, respectively, around the earth. It is desired to send a vehicle from S_1 to S_2. The vehicle is to be launched in a direction tangent to the orbit of S_1 and is to reach S_2 with a velocity tangent to the orbit of S_2. After a short powered phase, the vehicle will travel in free flight from S_1 to S_2. (*a*) Determine the launching velocity (velocity of the vehicle relative to S_1) in terms of the velocity v_0 of S_1. (*b*) Briefly describe the docking operation with S_2. (*c*) Determine the angle θ defining the required position of S_2 relative to S_1 at the time of launching.

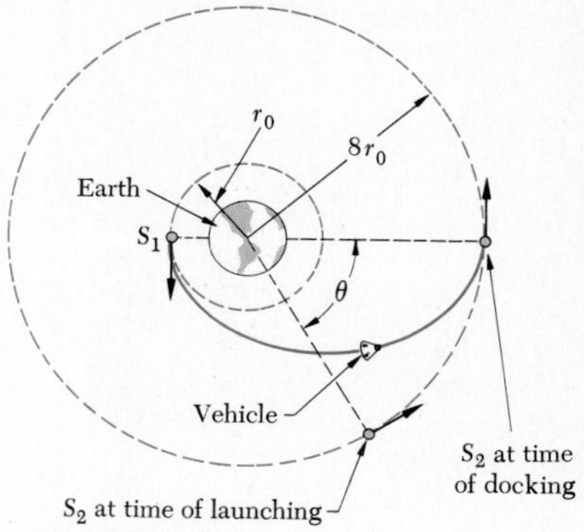

Fig. P12.96

***12.97** A space vehicle is launched in a direction parallel to the earth's surface from point *A* at an altitude of 340 mi. Knowing that the trajectory of the space vehicle is parabolic, determine (*a*) the launching velocity v_0 of the vehicle, (*b*) the angle θ described by the vehicle as it crosses the orbit of the moon at point *B*, 239,000 mi away from the center of the earth, (*c*) the velocity of the vehicle at *B* (magnitude and angle ϕ), (*d*) the time elapsed as the vehicle travels from *A* to *B*.

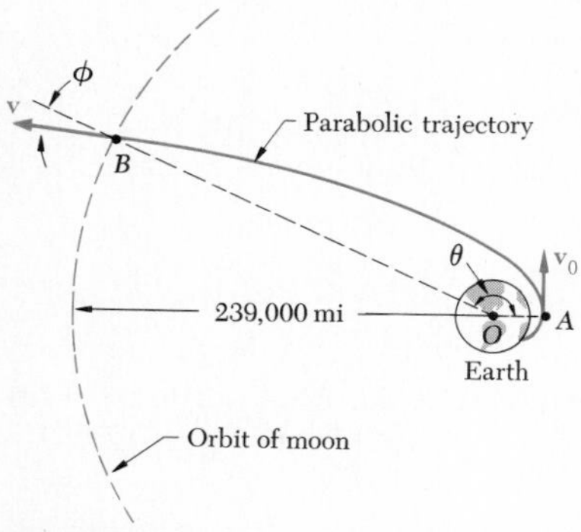

Fig. P12.97

REVIEW PROBLEMS

12.98 In a manufacturing process, disks are moved from one elevation to another by the lifting arm shown. Knowing that the coefficient of friction between the disks and the arm is 0.20, determine the magnitude of the downward acceleration **a** for which the disks slide on the arm.

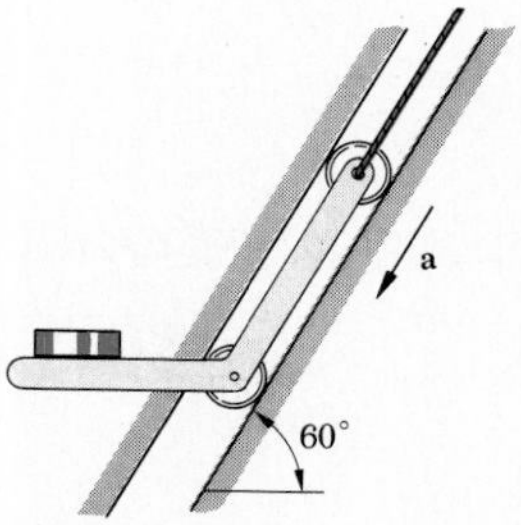

Fig. P12.98

12.99 Solve Prob. 12.98, assuming that the acceleration **a** is directed upward.

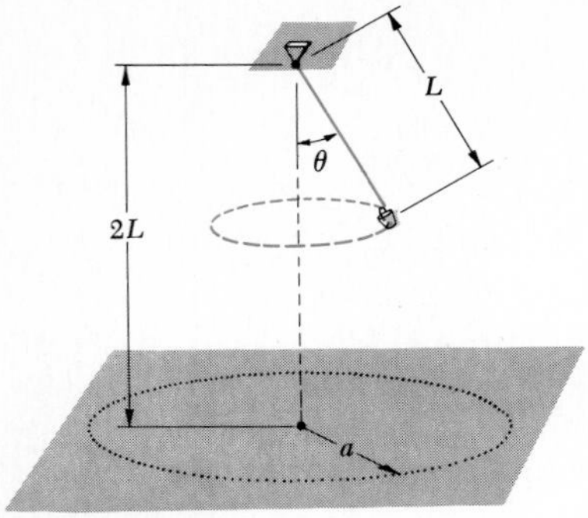

Fig. P12.100

12.100 A bucket is attached to a rope of length $L = 1.2$ m and is made to revolve in a horizontal circle. Drops of water leaking from the bucket fall and strike the floor along the perimeter of a circle of radius a. Determine the radius a when $\theta = 30°$.

12.101 Determine the radius a in Prob. 12.100, assuming that the speed of the bucket is 5 m/s. (The angle θ is not 30° in this case.)

12.102 Determine the required tension T if the acceleration of the 500-lb cylinder is to be (*a*) 6 ft/s^2 upward, (*b*) 6 ft/s^2 downward.

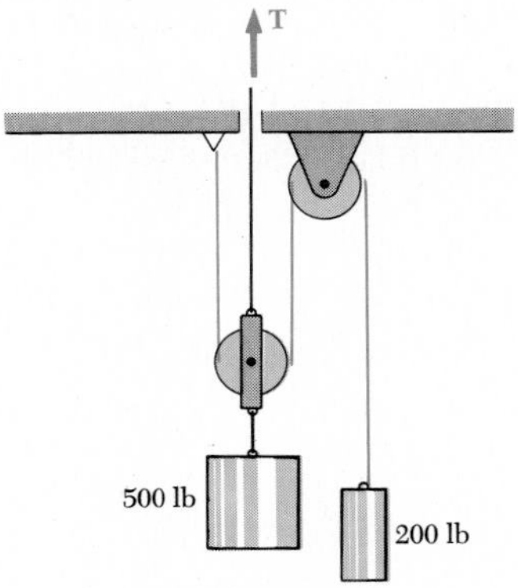

Fig. P12.102 and P12.103

12.103 Determine the acceleration of the 200-lb cylinder if (*a*) $T = 300$ lb, (*b*) $T = 800$ lb.

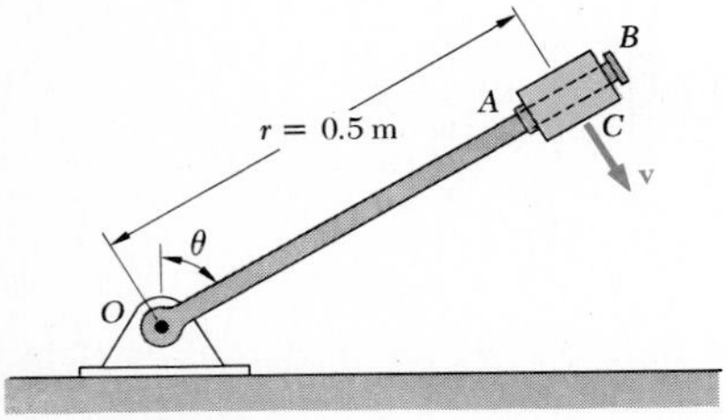

Fig. P12.104

12.104 The rod OAB rotates in a vertical plane at the constant rate $\dot{\theta} = 3$ rad/s, and collar C is free to slide on the rod between two stops A and B. Knowing that the distance between the stops is only slightly larger than the collar and neglecting the effect of friction, determine the range of values of θ for which the collar is in contact with stop A.

12.105 (*a*) Express the rated speed of a banked road in terms of the radius r of the curve and the banking angle θ. (*b*) What is the apparent weight of an automobile traveling at the rated speed? (See Sample Prob. 12.5 for the definition of rated speed.)

12.106 The assembly shown rotates about a vertical axis at a constant rate. Knowing that the coefficient of friction between the small block A and the cylindrical wall is 0.20, determine the lowest speed v for which the block will remain in contact with the wall.

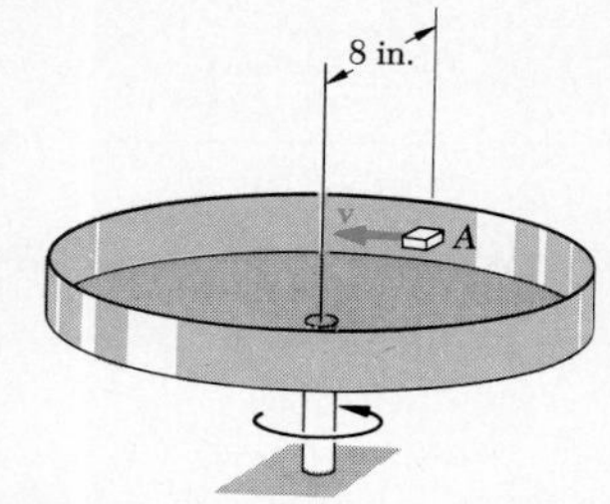

Fig. P12.106

12.107 Denoting by v_t the terminal speed of an object dropped from a great height, determine the distance the object will fall before its speed reaches the value $\frac{1}{2}v_t$. Assume that the frictional resistance of the air is proportional to the square of the speed of the object.

12.108 A spacecraft is describing a circular orbit of radius r_0 with a speed v_0 around an unspecified celestial body of center O, when its engine is suddenly fired, increasing the speed of the spacecraft from v_0 to αv_0, where $1 < \alpha^2 < 2$. Show that the maximum distance r_{max} from O reached by the spacecraft depends only upon r_0 and α, and express the ratio r_{max}/r_0 as a function of α.

12.109 A space vehicle approaches the planet Mars along a parabolic trajectory. As the vehicle reaches point A, which is the point of the trajectory closest to the planet, retrorockets are fired to slow the vehicle and place it in an elliptic path which will bring it to a tangential landing at B. Knowing that the distance from the center O of the planet to A is 12.5×10^3 km, that the radius of the planet is 3400 km, and that its mass is 0.108 times the mass of the earth, determine the velocity of the vehicle (a) as it approaches A, (b) after the retrorockets have been fired, (c) as it lands at B.

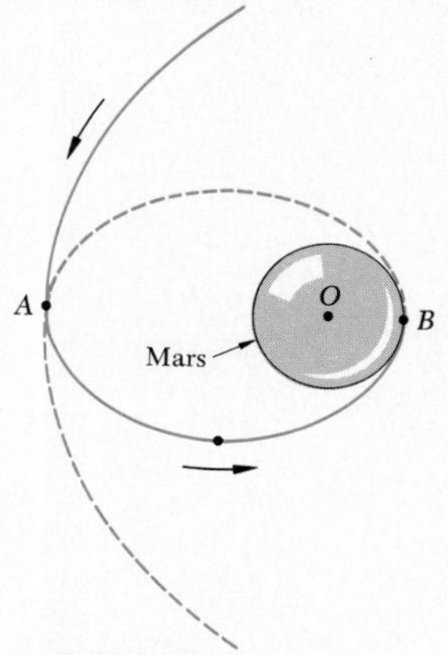

Fig. P12.109

Chapter 13

Kinetics of Particles: Work and Energy

13.1. Introduction. In the preceding chapter, problems dealing with the motion of particles were solved through the use of the fundamental equation of motion $\mathbf{F} = m\mathbf{a}$. Given a particle acted upon by a force $\mathbf{F}$, we could solve this equation for the acceleration $\mathbf{a}$; then, by applying the principles of kinematics, we could determine from $\mathbf{a}$ the velocity and position of the particle at any time.

If the equation $\mathbf{F} = m\mathbf{a}$ and the principles of kinematics are combined, two additional methods of analysis may be obtained, the *method of work and energy* and the *method of impulse and momentum*. The advantage of these methods lies in the fact that they make the determination of the acceleration unnecessary. Indeed, the method of work and energy relates directly force, mass, velocity, and displacement, while the method of impulse and momentum relates force, mass, velocity, and time.

The method of work and energy is treated in the present chapter. It is based on two important concepts, the concept of the *work of a force* and the concept of the *kinetic energy of a particle*. These concepts are defined in the following sections.

13.2. Work of a Force. We shall first define the terms *displacement* and *work* as they are used in mechanics.† Consider

† The definition of work was given in Sec. 10.1, and the basic properties of the work of a force were outlined in Secs. 10.1 and 10.5. For convenience, we repeat here the portions of this material which relate to the kinetics of particles.

a particle which moves from a point A to a neighboring point A' (Fig. 13.1*a*). The small vector which joins A to A' is called the *displacement* of the particle and is denoted by the differential $d\mathbf{r}$, where $\mathbf{r}$ represents the position vector of A. Now, let us assume that a force $\mathbf{F}$, of magnitude F and forming an angle α with $d\mathbf{r}$, is acting on the particle. The *work of the force* $\mathbf{F}$ *during the displacement* $d\mathbf{r}$, of magnitude ds, is defined as the scalar quantity

$$dU = F\,ds\cos\alpha \tag{13.1}$$

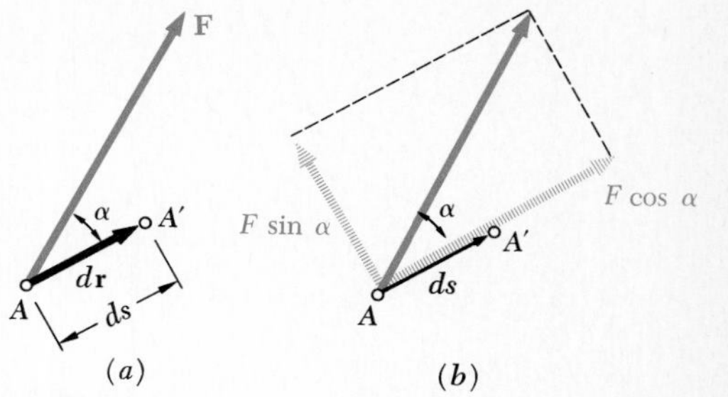

Fig. 13.1

Being a *scalar quantity*, work has a magnitude and a sign, but no direction. We also note that work should be expressed in units obtained by multiplying units of length by units of force. Thus, if U.S. customary units are used, work should be expressed in ft · lb or in · lb. If SI units are used, work should be expressed in N · m. The unit of work N · m is called a *joule* (J).† Recalling the conversion factors indicated in Sec. 12.2, we write

$$1\text{ ft}\cdot\text{lb} = (1\text{ ft})(1\text{ lb}) = (0.3048\text{ m})(4.448\text{ N}) = 1.356\text{ J}$$

It appears from (13.1) that the work dU may be considered as the product of the magnitude ds of the displacement and of the component $F\cos\alpha$ of the force $\mathbf{F}$ in the direction of $d\mathbf{r}$ (Fig. 13.1*b*). If this component and the displacement $d\mathbf{r}$ have the same sense, the work is positive; if they have opposite senses, the work is negative. Three particular cases are of special interest. If the force $\mathbf{F}$ has the same direction as $d\mathbf{r}$, the work dU reduces to $F\,ds$. If $\mathbf{F}$ has a direction opposite to that of $d\mathbf{r}$, the work is $dU = -F\,ds$. Finally, if $\mathbf{F}$ is perpendicular to $d\mathbf{r}$, the work dU is zero.

The work of $\mathbf{F}$ during a *finite* displacement of the particle from A_1 to A_2 (Fig. 13.2*a*) is obtained by integrating Eq. (13.1) from s_1 to s_2. This work, denoted by $U_{1\to 2}$, is

$$U_{1\to 2} = \int_{s_1}^{s_2} (F\cos\alpha)\,ds \tag{13.2}$$

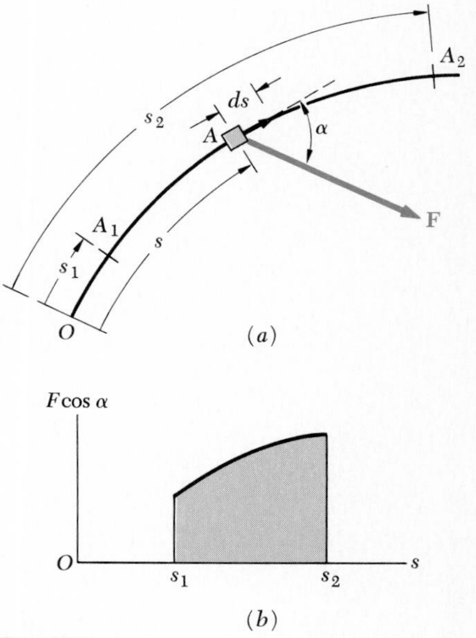

Fig. 13.2

where the variable of integration s measures the distance traveled by the particle along the path. The work $U_{1\to 2}$ is thus represented by the area under the curve obtained by plotting $F\cos\alpha$ against s (Fig. 13.2*b*).

† The joule (J) is the SI unit of *energy*, whether in mechanical form (work, potential energy, kinetic energy) or in chemical, electrical, or thermal form. We should note that, even though N · m = J, the moment of a force must be expressed in N · m, and not in joules, since the moment of a force is not a form of energy.

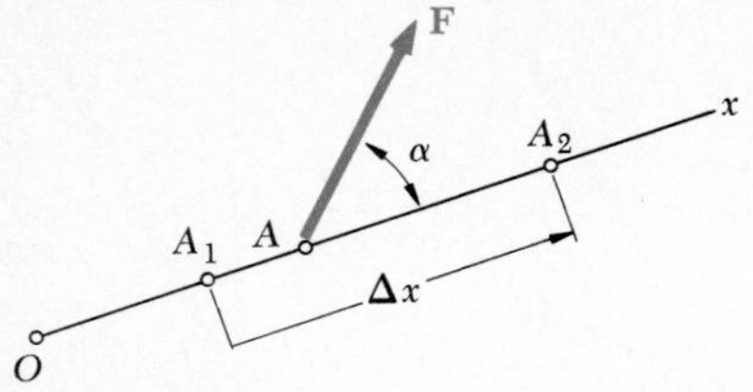

Fig. 13.3

Work of a Constant Force in Rectilinear Motion. When a particle moving in a straight line is acted upon by a force **F** of constant magnitude and of constant direction (Fig. 13.3), formula (13.2) yields

$$U_{1\to2} = (F\cos\alpha)\,\Delta x \tag{13.3}$$

where α = angle the force forms with direction of motion
Δx = displacement from A_1 to A_2

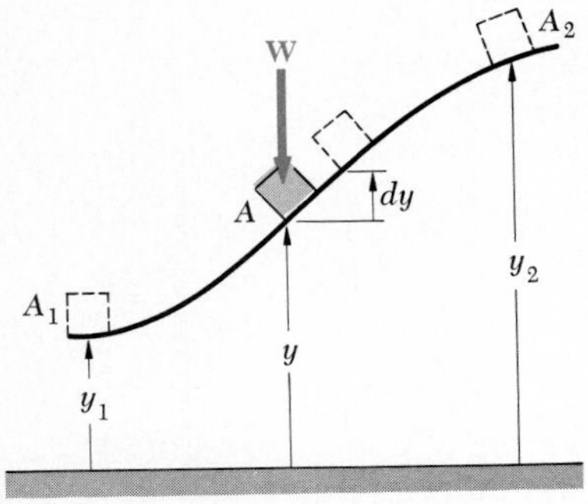

Fig. 13.4

Work of a Weight. The work dU of a force **F** during a displacement $d\mathbf{r}$ may also be considered as the product of the magnitude F of the force **F** and of the component $ds\cos\alpha$ of the displacement $d\mathbf{r}$ along **F**. This view is particularly useful in the computation of the work of the weight **W** of a body. The work dU of **W** is equal to the product of W and of the vertical displacement of the center of gravity G of the body. With the y axis chosen upward, the work of **W** during a finite displacement (Fig. 13.4) is obtained by writing

$$dU = -W\,dy$$

$$U_{1\to2} = -\int_{y_1}^{y_2} W\,dy = Wy_1 - Wy_2 \tag{13.4}$$

or

$$U_{1\to2} = -W(y_2 - y_1) = -W\,\Delta y \tag{13.4'}$$

where Δy is the vertical displacement from A_1 to A_2. The work of the weight **W** is thus equal to *the product of W and of the vertical displacement of the center of gravity of the body.* The work is *positive* when $\Delta y < 0$, that is, *when the body moves down.*

Work of the Force Exerted by a Spring. Consider a body A attached to a fixed point B by a spring; it is assumed that the spring is undeformed when the body is at A_0 (Fig. 13.5*a*). Experimental evidence shows that the magnitude of the force **F** exerted by the spring on body A is proportional to the deflection x of the spring measured from the position A_0. We have

$$F = kx \tag{13.5}$$

where k is the *spring constant,* expressed in N/m or kN/m if SI units are used and in lb/ft or lb/in. if U.S. customary units are used.†

† The relation $F = kx$ is correct under static conditions only. Under dynamic conditions, formula (13.5) should be modified to take the inertia of the spring into account. However, the error introduced by using the relation $F = kx$ in the solution of kinetics problems is small if the mass of the spring is small compared with the other masses in motion.

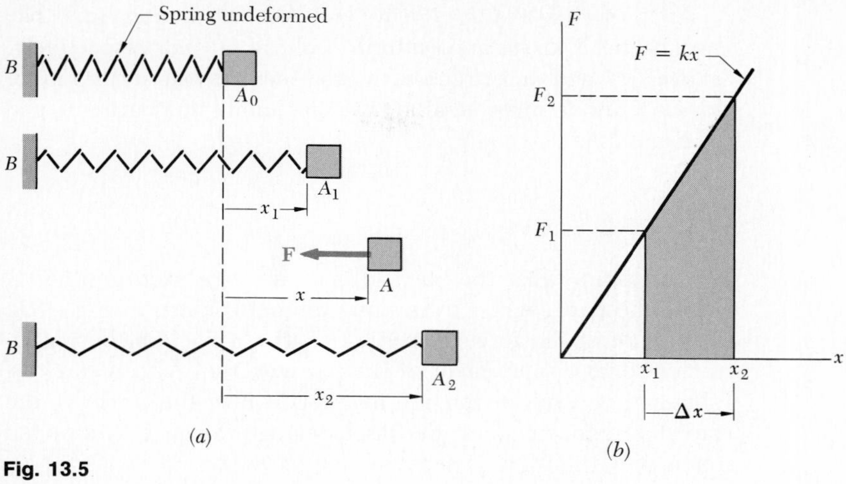

Fig. 13.5

The work of the force **F** exerted by the spring during a finite displacement of the body from $A_1(x = x_1)$ to $A_2(x = x_2)$ is obtained by writing

$$dU = -F\,dx = -kx\,dx$$

$$U_{1\to 2} = -\int_{x_1}^{x_2} kx\,dx = \tfrac{1}{2}kx_1^2 - \tfrac{1}{2}kx_2^2 \tag{13.6}$$

Care should be taken to express k and x in consistent units. For example, if U.S. customary units are used, k should be expressed in lb/ft and x in feet, or k in lb/in. and x in inches; in the first case, the work is obtained in ft · lb, in the second case, in in · lb. We note that the work of the force **F** exerted by the spring on the body is *positive* when $x_2 < x_1$, i.e., *when the spring is returning to its undeformed position.*

Since Eq. (13.5) is the equation of a straight line of slope k passing through the origin, the work $U_{1\to 2}$ of **F** during the displacement from A_1 to A_2 may be obtained by evaluating the area of the trapezoid shown in Fig. 13.5*b*. This is done by computing F_1 and F_2 and multiplying the base Δx of the trapezoid by its mean height $\frac{1}{2}(F_1 + F_2)$. Since the work of the force **F** exerted by the spring is positive for a negative value of Δx, we write

$$U_{1\to 2} = -\tfrac{1}{2}(F_1 + F_2)\,\Delta x \tag{13.6$'$}$$

Formula (13.6′) is usually more convenient to use than (13.6) and affords fewer chances of confusing the units involved.

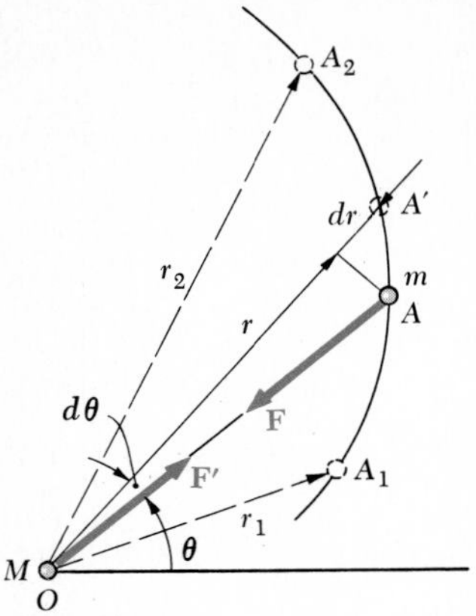

Fig. 13.6

Work of a Gravitational Force. We saw in Sec. 12.9 that two particles at distance r from each other and, respectively, of mass M and m, attract each other with equal and opposite forces **F** and **F**′ directed along the line joining the particles, and of magnitude

$$F = G\frac{Mm}{r^2}$$

Let us assume that the particle M occupies a fixed position O while the particle m moves along the path shown in Fig. 13.6. The work of the force **F** exerted on the particle m during an infinitesimal displacement of the particle from A to A' may be obtained by multiplying the magnitude F of the force by the radial component dr of the displacement. Since **F** is directed toward O, the work is negative and we write

$$dU = -F\,dr = -G\frac{Mm}{r^2}\,dr$$

The work of the gravitational force **F** during a finite displacement from $A_1(r = r_1)$ to $A_2(r = r_2)$ is therefore

$$U_{1\to2} = -\int_{r_1}^{r_2} \frac{GMm}{r^2}\,dr = \frac{GMm}{r_2} - \frac{GMm}{r_1} \tag{13.7}$$

The formula obtained may be used to determine the work of the force exerted by the earth on a body of mass m at a distance r from the center of the earth, when r is larger than the radius R of the earth. The letter M represents then the mass of the earth; recalling the first of the relations (12.20), we may thus replace the product GMm in Eq. (13.7) by WR^2, where R is the radius of the earth ($R = 6.37 \times 10^6$ m or 3960 mi) and W the value of the weight of the body at the surface of the earth.

A number of forces frequently encountered in problems of kinetics *do no work.* They are forces applied to fixed points ($ds = 0$) or acting in a direction perpendicular to the displacement ($\cos\alpha = 0$). Among the forces which do no work are the following: the reaction at a frictionless pin when the body supported rotates about the pin, the reaction at a frictionless surface when the body in contact moves along the surface, the reaction at a roller moving along its track, and the weight of a body when its center of gravity moves horizontally.

13.3. Kinetic Energy of a Particle. Principle of Work and Energy. Consider a particle of mass m acted upon by a force **F** and moving along a path which is either rectilinear or curved (Fig. 13.7). Expressing Newton's second law in terms

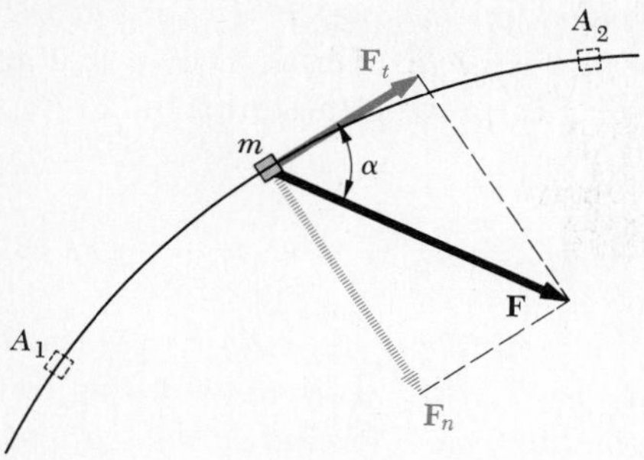

Fig. 13.7

of the tangential components of the force and of the acceleration (see Sec. 12.7), we write

$$F_t = ma_t \qquad \text{or} \qquad F \cos \alpha = m \frac{dv}{dt} \tag{13.8}$$

where α = angle the force forms with direction of motion
v = speed of particle

The speed was shown in Sec. 11.9 to be equal to the rate of change of the distance s traveled by the particle along its path; we write $v = ds/dt$. Solving for dt and substituting into (13.8), we obtain

$$F \cos \alpha = m \frac{dv}{ds} v \tag{13.9}$$

$$(F \cos \alpha)\, ds = mv\, dv$$

Integrating from A_1, where $s = s_1$ and $v = v_1$, to A_2, where $s = s_2$ and $v = v_2$, we write

$$\int_{s_1}^{s_2} (F \cos \alpha)\, ds = m \int_{v_1}^{v_2} v\, dv = \tfrac{1}{2} mv_2^2 - \tfrac{1}{2} mv_1^2 \tag{13.10}$$

The left-hand member of Eq. (13.10) represents the work $U_{1\to2}$ of the force **F** exerted on the particle during the displacement from A_1 to A_2; as indicated in Sec. 13.2, the work $U_{1\to2}$ is a scalar quantity. The expression $\frac{1}{2}mv^2$ is also a scalar quantity; it is defined as the kinetic energy of the particle and is denoted by T. We write

$$T = \tfrac{1}{2} mv^2 \tag{13.11}$$

Substituting into (13.10), we have

$$U_{1\to2} = T_2 - T_1 \tag{13.12}$$

which expresses that, when a particle moves from A_1 to A_2 under the action of a force **F**, *the work of the force* **F** *is equal to the change in kinetic energy of the particle.* This is known as the *principle of work and energy.* Rearranging the terms in (13.12), we write

$$T_1 + U_{1\to 2} = T_2 \tag{13.13}$$

Thus, *the kinetic energy of the particle at A_2 may be obtained by adding to its kinetic energy at A_1 the work done during the displacement from A_1 to A_2 by the force* **F** *exerted on the particle.* As Newton's second law from which it is derived, the principle of work and energy applies only with respect to a newtonian frame of reference (Sec. 12.1). The speed v used to determine the kinetic energy T should therefore be measured with respect to a newtonian frame of reference.

Since both work and kinetic energy are scalar quantities, their sum may be computed as an ordinary algebraic sum, the work $U_{1\to 2}$ being considered as positive or negative according to the direction of **F**. When several forces act on the particle, the expression $U_{1\to 2}$ represents the total work of the forces acting on the particle; it is obtained by adding algebraically the work of the various forces.

As noted above, the kinetic energy of a particle is a scalar quantity. It further appears from the definition $T = \frac{1}{2}mv^2$ that the kinetic energy is always positive, regardless of the direction of motion of the particle. Considering the particular case when $v_1 = 0$, $v_2 = v$, and substituting $T_1 = 0$, $T_2 = T$ into (13.12), we observe that the work done by the forces acting on the particle is equal to T. Thus, the kinetic energy of a particle moving with a speed v represents the work which must be done to bring the particle from rest to the speed v. Substituting $T_1 = T$ and $T_2 = 0$ into (13.12), we also note that, when a particle moving with a speed v is brought to rest, the work done by the forces acting on the particle is $-T$. Assuming that no energy is dissipated into heat, we conclude that the work done by the forces exerted *by the particle* on the bodies which cause it to come to rest is equal to T. Thus, the kinetic energy of a particle also represents *the capacity to do work associated with the speed of the particle.*

The kinetic energy is measured in the same units as work, i.e., in joules if SI units are used, and in ft · lb if U.S. customary units are used. We check that, in SI units,

$$T = \tfrac{1}{2}mv^2 = \text{kg(m/s)}^2 = (\text{kg}\cdot\text{m/s}^2)\text{m} = \text{N}\cdot\text{m} = \text{J}$$

while, in customary units,

$$T = \tfrac{1}{2}mv^2 = (\text{lb}\cdot\text{s}^2/\text{ft})(\text{ft/s})^2 = \text{lb}\cdot\text{ft}$$

13.4. Applications of the Principle of Work and Energy. The application of the principle of work and energy greatly simplifies the solution of many problems involving forces, displacements, and velocities. Consider, for example, the pendulum OA consisting of a bob A of weight W attached to a cord of length l (Fig. 13.8*a*). The pendulum is released with no initial velocity from a horizontal position OA_1 and allowed to swing in a vertical plane. We wish to determine the speed of the bob as it passes through A_2, directly under O.

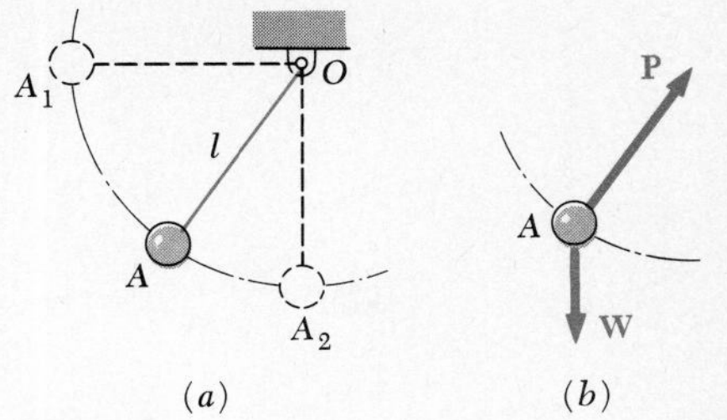

Fig. 13.8

We first determine the work done during the displacement from A_1 to A_2 by the forces acting on the bob. We draw a free-body diagram of the bob, showing all the *actual* forces acting on it, i.e., the weight **W** and the force **P** exerted by the cord (Fig. 13.8*b*). (An inertia vector is not an actual force and *should not* be included in the free-body diagram.) We note that the force **P** does no work, since it is normal to the path; the only force which does work is thus the weight **W**. The work of **W** is obtained by multiplying its magnitude W by the vertical displacement l (Sec. 13.2); since the displacement is downward, the work is positive. We therefore write $U_{1\rightarrow 2} = Wl$.

Considering, now, the kinetic energy of the bob, we find $T_1 = 0$ at A_1 and $T_2 = \frac{1}{2}(W/g)v_2^2$ at A_2. We may now apply the principle of work and energy; recalling formula (13.13), we write

$$T_1 + U_{1\rightarrow 2} = T_2 \qquad 0 + Wl = \frac{1}{2}\frac{W}{g}v_2^2$$

Solving for v_2, we find $v_2 = \sqrt{2gl}$. We note that the speed obtained is that of a body falling freely from a height l.

The example we have considered illustrates the following advantages of the method of work and energy:

1. In order to find the speed at A_2, there is no need to determine the acceleration in an intermediate position A and to integrate the expression obtained from A_1 to A_2.
2. All quantities involved are scalars and may be added directly, without using x and y components.
3. Forces which do no work are eliminated from the solution of the problem.

What is an advantage in one problem, however, may become a disadvantage in another. It is evident, for instance, that the

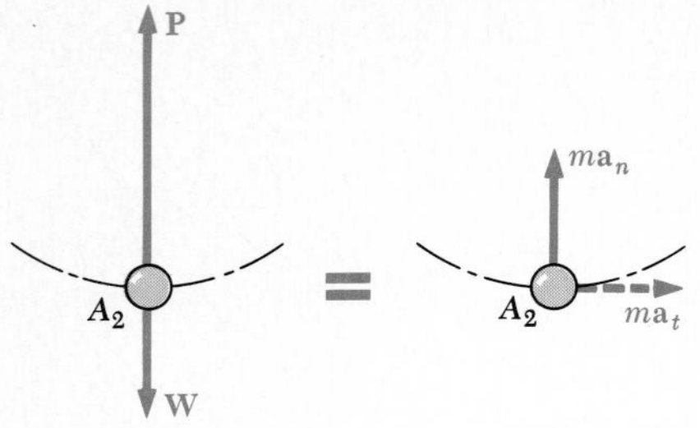

Fig. 13.9

method of work and energy cannot be used to directly determine an acceleration. We also note that it should be supplemented by the direct application of Newton's second law in order to determine a force which is normal to the path of the particle, since such a force does no work. Suppose, for example, that we wish to determine the tension in the cord of the pendulum of Fig. 13.8*a* as the bob passes through A_2. We draw a free-body diagram of the bob in that position (Fig. 13.9) and express Newton's second law in terms of tangential and normal components. The equations $\Sigma F_t = ma_t$ and $\Sigma F_n = ma_n$ yield, respectively, $a_t = 0$ and

$$P - W = ma_n = \frac{W}{g}\frac{v^2}{l}$$

But the speed at A_2 was determined earlier by the method of work and energy. Substituting $v^2 = 2gl$ and solving for P, we write

$$P = W + \frac{W}{g}\frac{2gl}{l} = 3W$$

13.5. Systems of Particles. When a problem involves several particles, each particle may be considered separately and the principle of work and energy may be applied to each particle. Adding the kinetic energies of all the particles, and considering the work of all the forces involved, we may also write the equation of work and energy for the entire system. We have

$$T_1 + U_{1\to2} = T_2 \tag{13.13}$$

where T represents the arithmetic sum of the kinetic energies of the particles forming the system (all terms are positive) and $U_{1\to2}$ the work of all the forces acting on the various particles, whether these forces are *internal* or *external* from the point of view of the system as a whole.

The method of work and energy is particularly useful in solving problems involving a system of bodies connected by *inextensible cords or links.* In this case, the internal forces occur by pairs of equal and opposite forces, and the points of application of the forces in each pair *move through equal distances.* As a result, the work of the internal forces is zero and $U_{1\to2}$ reduces to the work of the *external forces only* (see Sample Prob. 13.2).

SAMPLE PROBLEM 13.1

An automobile weighing 4000 lb is driven down a 5° incline at a speed of 60 mi/h when the brakes are applied, causing a constant total braking force (applied by the road on the tires) of 1500 lb. Determine the distance traveled by the automobile as it comes to a stop.

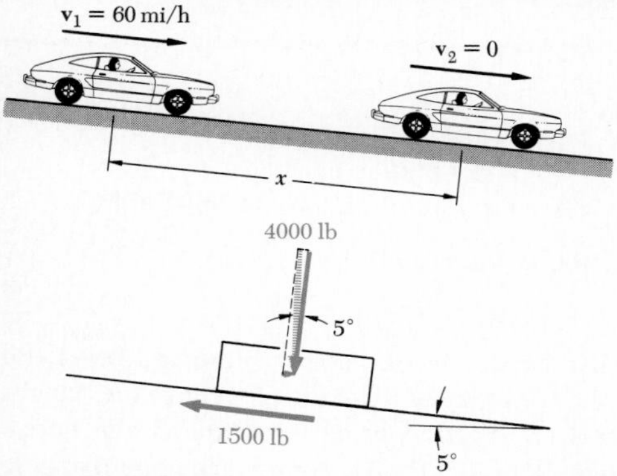

Solution. *Kinetic Energy*

Position *1*: $$v_1 = \left(60\frac{\text{mi}}{\text{h}}\right)\left(\frac{5280\text{ ft}}{1\text{ mi}}\right)\left(\frac{1\text{ h}}{3600\text{ s}}\right) = 88\text{ ft/s}$$

$$T_1 = \tfrac{1}{2}mv_1^2 = \tfrac{1}{2}(4000/32.2)(88)^2 = 481{,}000\text{ ft}\cdot\text{lb}$$

Position *2*: $$v_2 = 0 \qquad T_2 = 0$$

Work $$U_{1\to 2} = -1500x + (4000\sin 5°)x = -1151x$$

Principle of Work and Energy

$$T_1 + U_{1\to 2} = T_2$$
$$481{,}000 - 1151x = 0 \qquad x = 418\text{ ft} \blacktriangleleft$$

SAMPLE PROBLEM 13.2

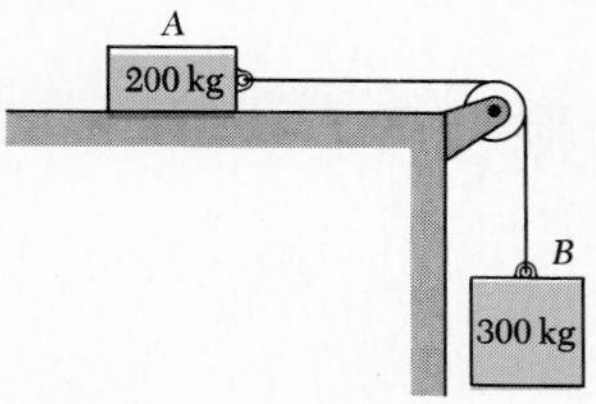

Two blocks are joined by an inextensible cable as shown. If the system is released from rest, determine the velocity of block *A* after it has moved 2 m. Assume that μ equals 0.25 between block *A* and the plane and that the pulley is weightless and frictionless.

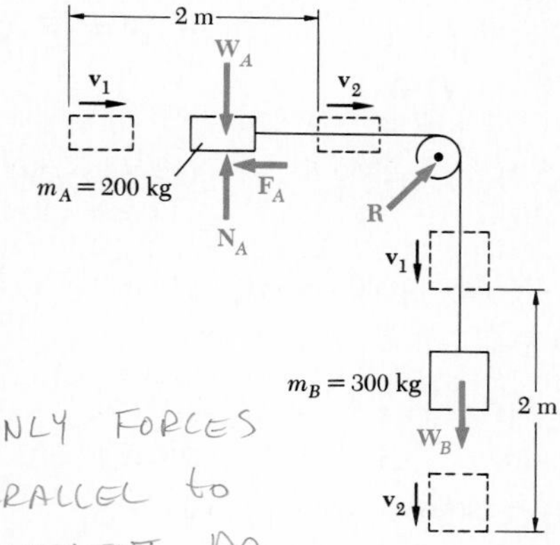

ONLY FORCES PARALLEL to MOVEMENT DO WORK.

Solution. Since the cable is inextensible, the work done by the forces exerted by the cable will cancel if we consider the two-block system.

Kinetic Energy. Position *1*: $v_1 = 0 \qquad T_1 = 0$

Position *2*: $v_2 = v \qquad T_2 = \tfrac{1}{2}m_A v^2 + \tfrac{1}{2}m_B v^2$

$$T_2 = \tfrac{1}{2}(200\text{ kg})v^2 + \tfrac{1}{2}(300\text{ kg})v^2 = \tfrac{1}{2}(500\text{ kg})v^2$$

Work. During the motion, only $\mathbf{W}_B$ and $\mathbf{F}_A$ do work. Since

$$W_B = (300\text{ kg})(9.81\text{ m/s}^2) = 2940\text{ N}$$
$$F_A = \mu N_A = \mu W_A = 0.25(200\text{ kg})(9.81\text{ m/s}^2) = 490\text{ N}$$

we have

$$U_{1\to 2} = W_B(2\text{ m}) - F_A(2\text{ m})$$
$$= (2940\text{ N})(2\text{ m}) - (490\text{ N})(2\text{ m}) = 4900\text{ J}$$

Principle of Work and Energy

$$T_1 + U_{1\to 2} = T_2: \qquad 0 + 4900\text{ J} = \tfrac{1}{2}(500\text{ kg})v^2 \qquad v = 4.43\text{ m/s}$$

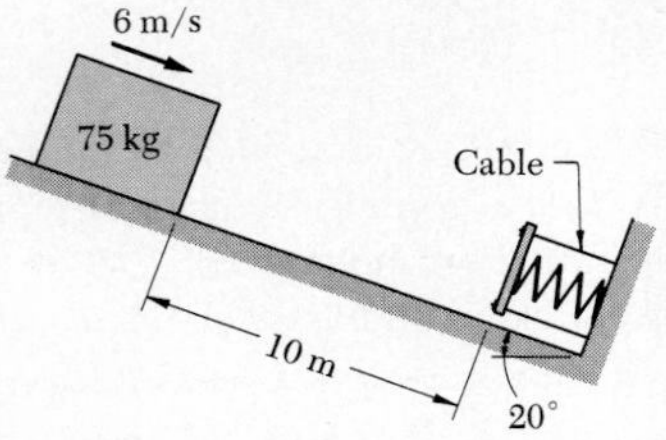

SAMPLE PROBLEM 13.3

A spring is used to stop a 75-kg package which is moving down a 20° incline. The spring has a constant $k = 25$ kN/m, and is held by cables so that it is initially compressed 100 mm. If the velocity of the package is 6 m/s when it is 10 m from the spring, determine the maximum additional deformation of the spring in bringing the package to rest. Assume $\mu = 0.20$.

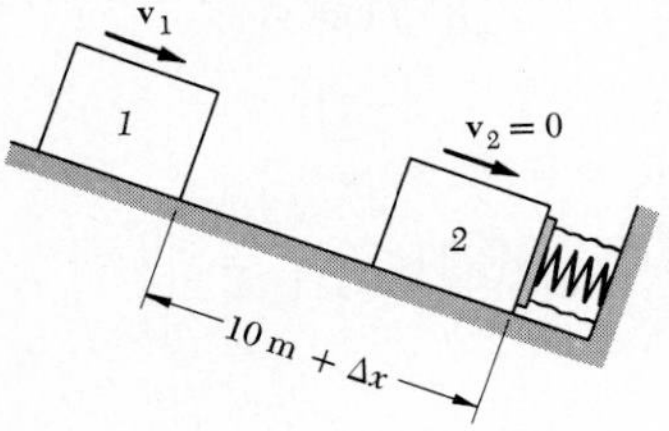

Kinetic Energy

Position *1*: $\qquad v_1 = 6$ m/s

$$T_1 = \tfrac{1}{2}mv_1^2 = \tfrac{1}{2}(75\text{ kg})(6\text{ m/s})^2 = 1350\text{ N}\cdot\text{m} = 1350\text{ J}$$

Position 2 (maximum spring deformation):

$$v_2 = 0 \qquad T_2 = 0$$

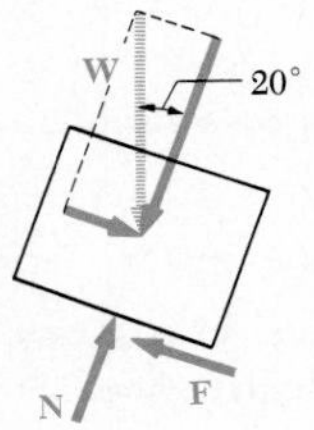

Work. We assume that when the package is brought to rest the additional deflection of the spring is Δx. The component of the weight parallel to the plane and the friction force act through the entire displacement, i.e., through 10 m + Δx. The total work done by these forces is

$$\begin{aligned} U_{1\to2} &= W_t(10\text{ m} + \Delta x) - F(10\text{ m} + \Delta x) \\ &= (W\sin 20°)(10\text{ m} + \Delta x) - 0.20(W\cos 20°)(10\text{ m} + \Delta x) \\ &= 0.1541W(10\text{ m} + \Delta x) \end{aligned}$$

or, since $W = mg = (75\text{ kg})(9.81\text{ m/s}^2) = 736$ N,

$$U_{1\to2} = 0.1541(736\text{ N})(10\text{ m} + \Delta x) = 1134\text{ J} + (113.4\text{ N})\,\Delta x$$

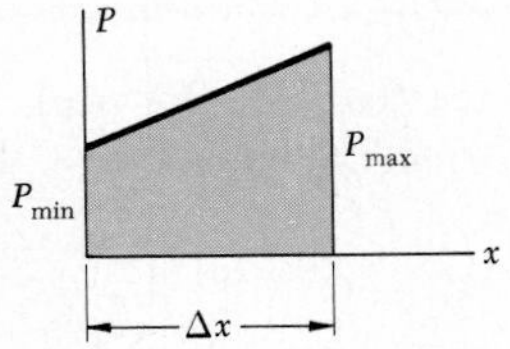

In addition, during the compression of the spring, the variable force **P** exerted by the spring does an amount of negative work equal to the area under the force-deflection curve of the spring force.

$$P_{min} = kx = (25\text{ kN/m})(100\text{ mm}) = (25\,000\text{ N/m})(0.100\text{ m}) = 2500\text{ N}$$

$$\begin{aligned} P_{max} &= P_{min} + k\,\Delta x \\ &= 2500\text{ N} + (25\,000\text{ N/m})\,\Delta x \end{aligned}$$

$$U_{1\to2} = -\tfrac{1}{2}(P_{min} + P_{max})\,\Delta x = -(2500\text{ N})\,\Delta x - (12\,500\text{ N/m})(\Delta x)^2$$

The total work is thus

$$U_{1\to2} = 1134\text{ J} + (113.4\text{ N})\,\Delta x - (2500\text{ N})\,\Delta x - (12\,500\text{ N/m})(\Delta x)^2$$

Principle of Work and Energy

$$T_1 + U_{1\to2} = T_2$$

$$1350 + 1134 + 113.4\,\Delta x - 2500\,\Delta x - 12\,500(\Delta x)^2 = 0$$

$$(\Delta x)^2 + 0.1909\,\Delta x - 0.1987 = 0$$

$$\Delta x = 0.360\text{ m} \qquad \Delta x = 360\text{ mm} \blacktriangleleft$$

SAMPLE PROBLEM 13.4

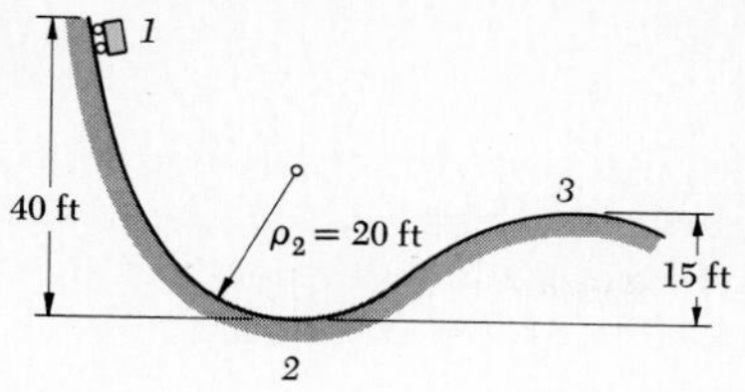

A 2000-lb car starts from rest at point *1* and moves without friction down the track shown. (*a*) Determine the force exerted by the track on the car at point *2*, where the radius of curvature of the track is 20 ft. (*b*) Determine the minimum safe value of the radius of curvature at point *3*.

a. Force Exerted by the Track at Point 2. The principle of work and energy is used to determine the velocity of the car as it passes through point *2*.

Kinetic Energy: $T_1 = 0 \qquad T_2 = \frac{1}{2} m v_2^2 = \frac{1}{2}\frac{W}{g} v_2^2$

Work. The only force which does work is the weight **W**. Since the vertical displacement from point *1* to point *2* is 40 ft downward, the work of the weight is

$$U_{1\to 2} = +W(40 \text{ ft})$$

Principle of Work and Energy

$$T_1 + U_{1\to 2} = T_2 \qquad 0 + W(40 \text{ ft}) = \frac{1}{2}\frac{W}{g} v_2^2$$

$$v_2^2 = 80g = 80(32.2) \qquad v_2 = 50.8 \text{ ft/s}$$

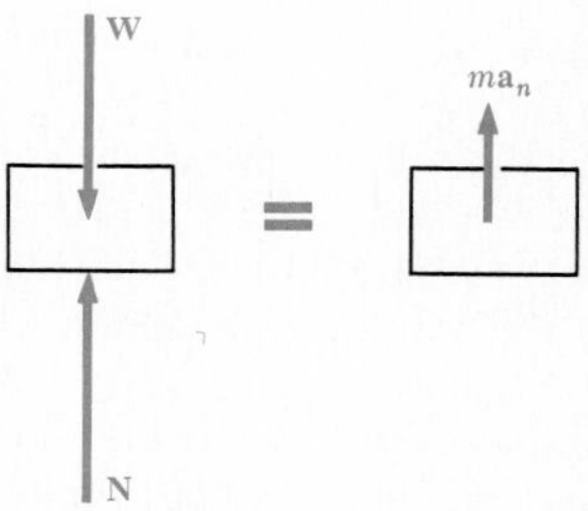

Newton's Second Law at Point 2. The acceleration $\mathbf{a}_n$ of the car at point *2* has a magnitude $a_n = v_2^2/\rho$ and is directed upward. Since the external forces acting on the car are **W** and **N**, we write

$$+\uparrow\Sigma F_n = ma_n: \qquad -W + N = ma_n$$

$$= \frac{W}{g}\frac{v_2^2}{\rho}$$

$$= \frac{W}{g}\frac{80g}{20}$$

$$N = 5W \qquad \mathrm{N} = 10{,}000 \text{ lb}\uparrow \blacktriangleleft$$

b. Minimum Value of ρ at Point 3. *Principle of Work and Energy.* Applying the principle of work and energy between point *1* and point *3*, we obtain

$$T_1 + U_{1\to 3} = T_3 \qquad 0 + W(25 \text{ ft}) = \frac{1}{2}\frac{W}{g} v_3^2$$

$$v_3^2 = 50g = 50(32.2) \qquad v_3 = 40.1 \text{ ft/s}$$

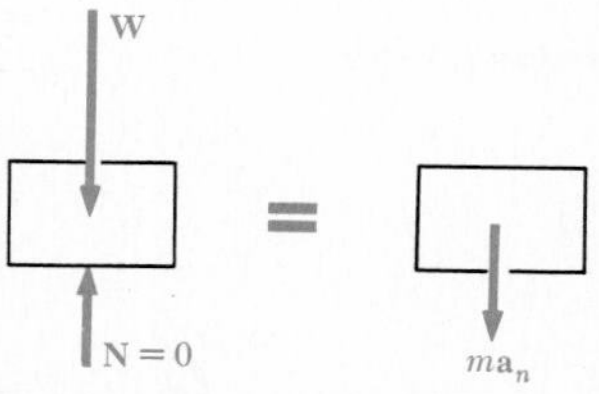

Newton's Second Law at Point 3. The minimum safe value of ρ occurs when $\mathbf{N} = 0$. In this case, the acceleration $\mathbf{a}_n$, of magnitude $a_n = v_3^2/\rho$, is directed downward, and we write

$$+\downarrow\Sigma F_n = ma_n: \qquad W = \frac{W}{g}\frac{v_3^2}{\rho}$$

$$= \frac{W}{g}\frac{50g}{\rho} \qquad \rho = 50 \text{ ft} \blacktriangleleft$$

PROBLEMS

13.1 A 100-kg satellite was placed in a circular orbit 2000 km above the surface of the earth. At this elevation the acceleration of gravity is 5.68 m/s^2. Determine the kinetic energy of the satellite, knowing that its orbital speed is 24.8×10^3 km/h.

13.2 A stone which weighs 8 lb is dropped from a height h and strikes the ground with a velocity of 75 ft/s. (*a*) Find the kinetic energy of the stone as it strikes the ground and the height h from which it was dropped. (*b*) Solve part *a*, assuming that the same stone is dropped on the moon. (Acceleration of gravity on the moon = 5.31 ft/s^2.)

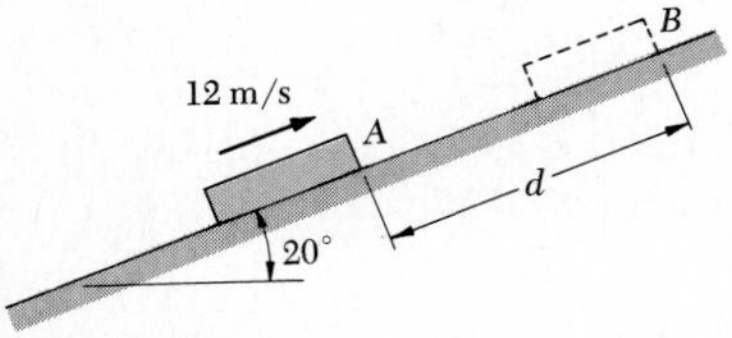

Fig. P13.4

13.3 Using the method of work and energy, solve Prob. 12.7.

13.4 A 20-kg package is projected up a 20° incline with an initial velocity of 12 m/s. The coefficient of friction between the incline and the package is 0.15. Determine (*a*) the maximum distance that the package will move up the incline, (*b*) the velocity of the package when it returns to its original position.

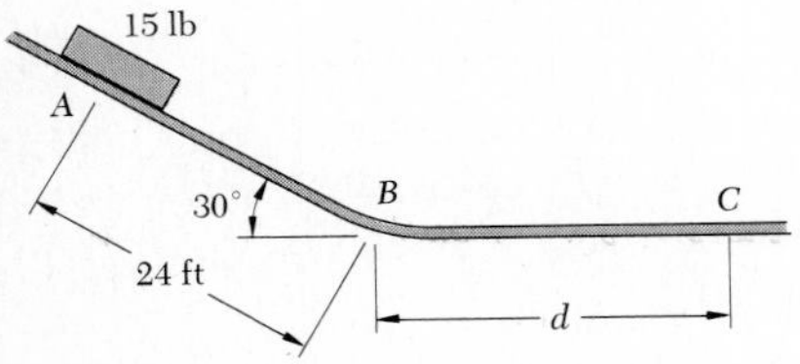

Fig. P13.5 and P13.6

13.5 A 15-lb package is placed on the chute shown at point *A*. The coefficient of friction between the package and the chute is 0.15. If the package is released from rest at *A*, determine the velocity of the package when it reaches the bottom of the chute *B*.

13.6 A 15-lb package slides down the chute and out along the floor until it comes to rest at point *C*. If $\mu = 0.15$ and if the initial velocity of the package at *A* is 3 ft/s directed down the chute, determine the distance d from the bottom of the chute to point *C*.

13.7 The 2-kg collar was moving down the rod with a velocity of 3 m/s when a force **P** was applied to the horizontal cable. Assuming negligible friction between the collar and the rod, determine the magnitude of the force **P** if the collar stopped after moving 1.2 m more down the rod.

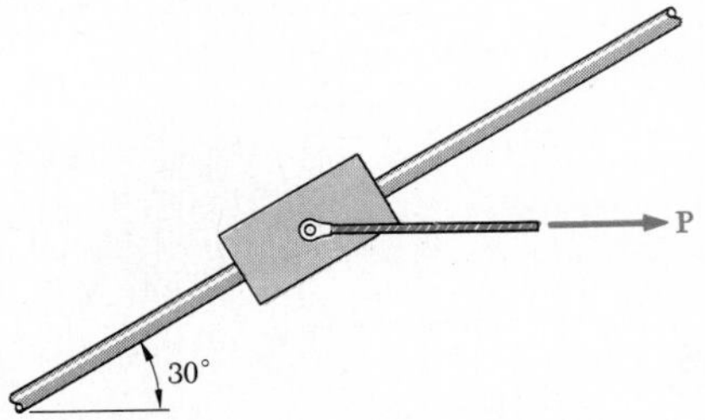

Fig. P13.7

13.8 Solve Prob. 13.7, assuming a coefficient of friction of 0.20 between the collar and the rod.

13.9 The total mass of loading car A and its load is 3500 kg. The car is connected to a 1000-kg counterweight and is at rest when a constant 22-kN force is applied as shown. (*a*) If the force acts through the entire motion, what is the speed of the car after it has traveled 30 m. (*b*) If after the car has moved a distance x the 22-kN force is removed, the car will coast to rest. After what distance x should the force be removed if the car is to come to rest after a total movement of 30 m?

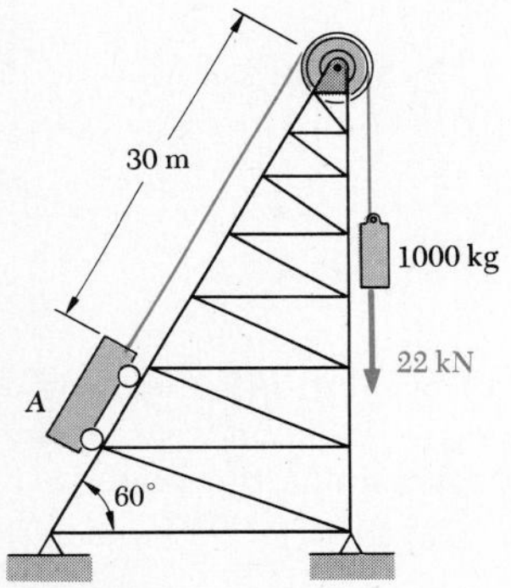

Fig. P13.9

13.10 The system shown is at rest when the 50-lb force is applied to block A. Neglecting the effect of friction, determine the velocity of block A after it has moved 9 ft.

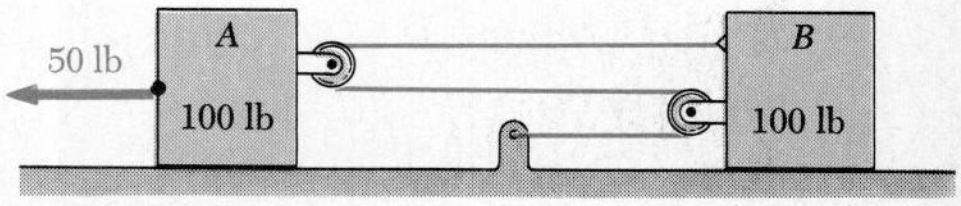

Fig. P13.10

13.11 Solve Prob. 13.10, assuming that the coefficient of friction between the blocks and the horizontal plane is 0.20.

13.12 Three 20-kg packages rest on a belt which passes over a pulley and is attached to a 40-kg block. Knowing that the coefficient of friction between the belt and the horizontal surface and also between the belt and the packages is 0.50, determine the speed of package B as it falls off the belt at E.

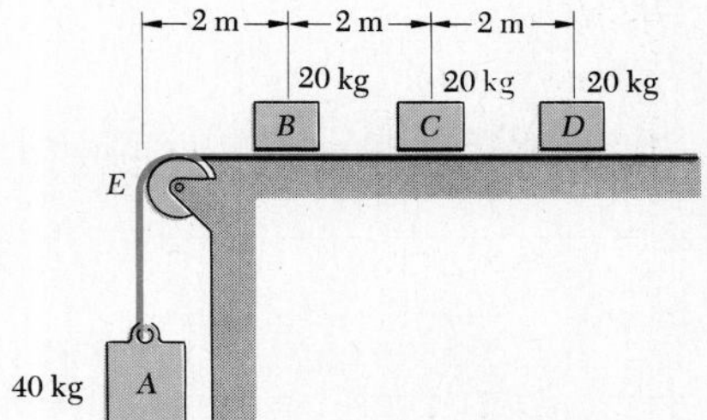

Fig. P13.12

13.13 In Prob. 13.12, determine the speed of package C as it falls off the belt at E.

13.14 Two blocks A and B, of mass 5 kg and 6 kg, respectively, are connected by a cord which passes over pulleys as shown. A collar C is placed on block A and the system is released from rest. After the blocks have moved 0.9 m, the collar C is removed and the blocks continue to move. Knowing that collar C has a mass of 4 kg, determine the speed of block A just before it strikes the ground.

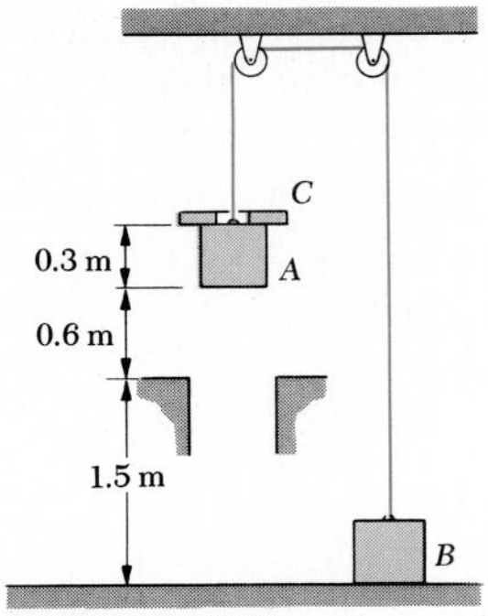

Fig. P13.14

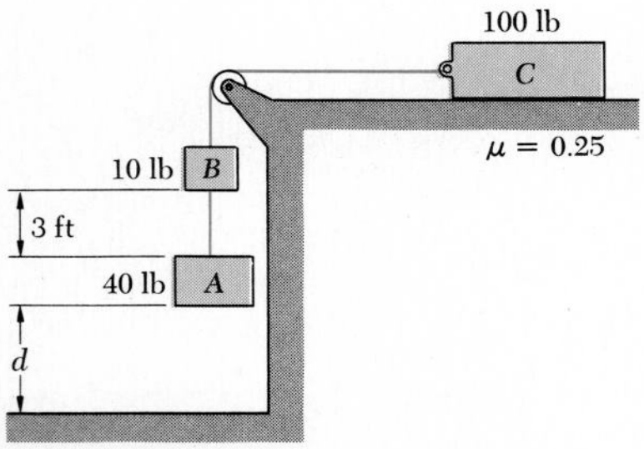

Fig. P13.15

13.15 The system shown is released from rest. After moving a distance d, block A strikes the ground and the cable between A and B becomes slack. Knowing that $d = 5$ ft, determine the velocity of block B as is strikes block A.

13.16 In Prob. 13.15, determine the smallest distance d for which the system will come to rest with block B touching block A.

13.17 Using the method of work and energy, solve Prob. 12.15*a*.

13.18 Using the method of work and energy, solve Prob. 12.14.

13.19 Using the method of work and energy, solve Prob. 12.17*c*.

13.20 Using the method of work and energy, solve Prob. 12.16*c*.

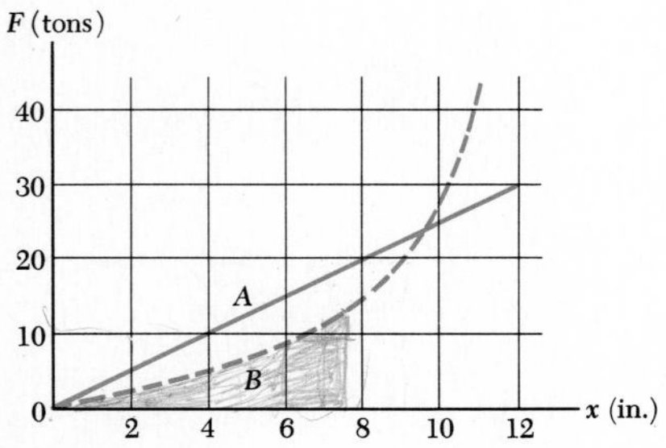

Fig. P13.21

13.21 Two types of energy-absorbing fenders designed to be used on a pier are statically loaded. The force-deflection curve for each type of fender is given in the graph. Determine the maximum deflection of each fender when a 100-ton ship moving at 1 mi/h strikes the fender and is brought to rest.

13.22 A 5000-kg airplane lands on an aircraft carrier and is caught by an arresting cable which is characterized by the force-deflection diagram shown. Knowing that the landing speed of the plane is 144 km/h, determine (*a*) the distance required for the plane to come to rest, (*b*) the maximum rate of deceleration of the plane.

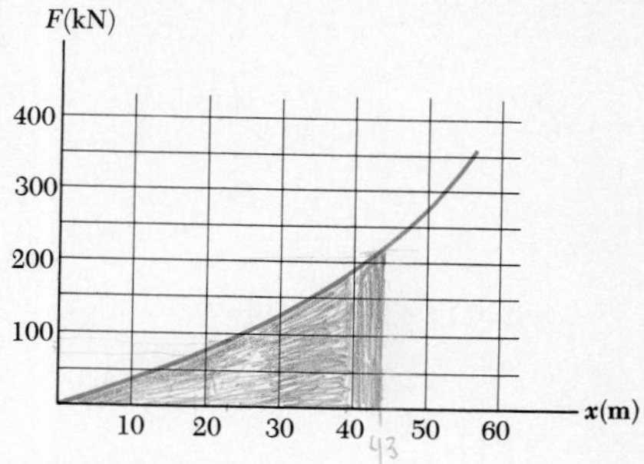

Fig. P13.22

13.23 A 2-kg block is at rest on a spring of constant 400 N/m. A 4-kg block is held above the 2-kg block so that it just touches it, and released. Determine (*a*) the maximum velocity attained by the blocks, (*b*) the maximum force exerted on the blocks.

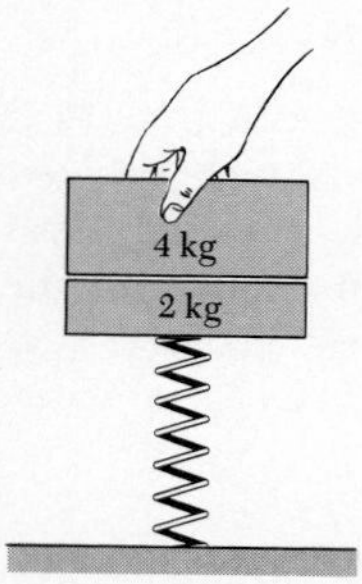

Fig. P13.23

13.24 As the bracket *ABC* is slowly rotated, the 6-kg block starts to slide toward the spring when $\theta = 15°$. The maximum deflection of the spring is observed to be 50 mm. Determine the values of the coefficients of static and kinetic friction.

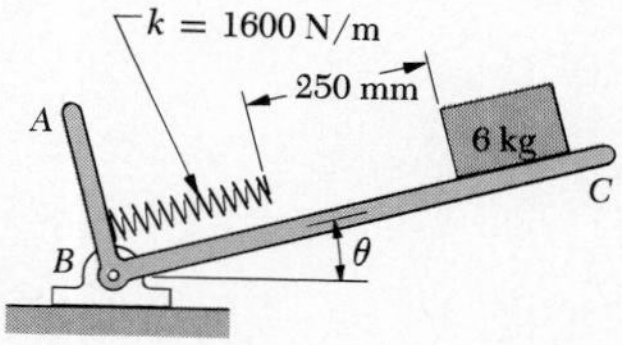

Fig. P13.24

13.25 A 30-ton railroad car was stopped by the bumper shown. Knowing that the maximum deflection of the bumper was 6 in., determine the speed of the car just before it struck the bumper.

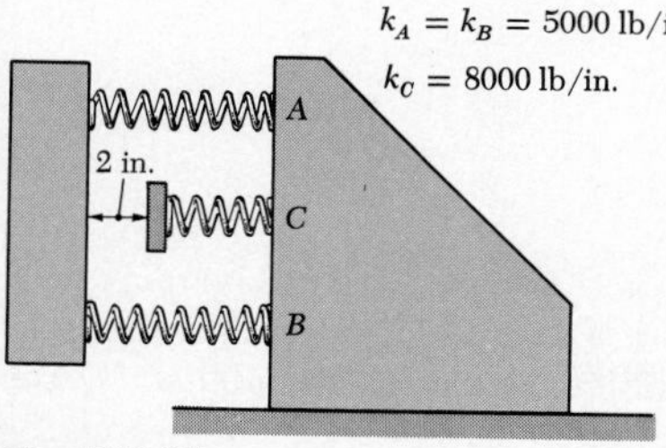

Fig. P13.25

13.26 A railroad car weighing 50,000 lb starts from rest and coasts down a 1-percent incline for a distance of 50 ft. It is stopped by a bumper having a spring constant of 8000 lb/in. (*a*) What is the speed of the car at the bottom of the incline? (*b*) How many inches will the spring be compressed?

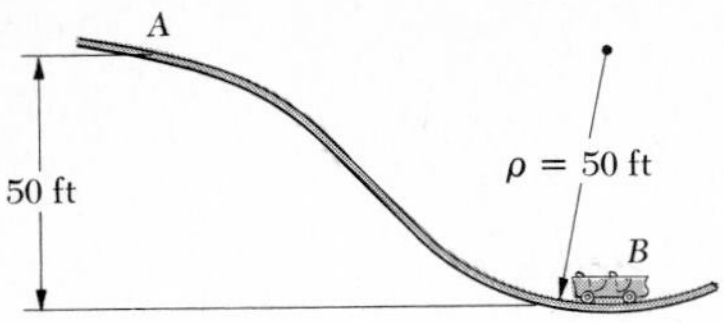

Fig. P13.27 and P13.28

13.27 A roller coaster starts from rest at A and rolls down the track shown. Assuming no energy loss and knowing that the radius of curvature of the track at B is 50 ft, determine the apparent weight of a 150-lb passenger at B.

13.28 A roller coaster is released with no velocity at A and rolls down the track shown. The brakes are suddenly applied as the car passes through point B, causing the wheels of the car to slide on the track ($\mu = 0.25$). Assuming no energy loss between A and B and knowing that the radius of curvature of the track at B is 50 ft, determine the normal and tangential components of the acceleration of the car just after the brakes have been applied.

13.29 A small package of mass m is projected into a vertical return loop at A with a velocity $\mathbf{v}_0$. The package travels without friction along a circle of radius r and is deposited on a horizontal surface at C. For each of the two loops shown, determine (a) the smallest velocity $\mathbf{v}_0$ for which the package will reach the horizontal surface at C, (b) the corresponding force exerted by the loop on the package as it passes point B.

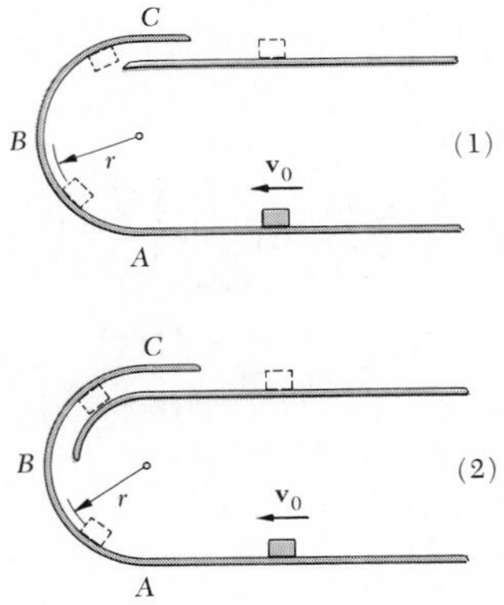

Fig. P13.29

13.30 In Prob. 13.29, it is desired to have the package deposited on the horizontal surface at C with a speed of 2 m/s. Knowing that $r = 0.6$ m, (a) show that this requirement cannot be fulfilled by the first loop, (b) determine the required initial velocity $\mathbf{v}_0$ when the second loop is used.

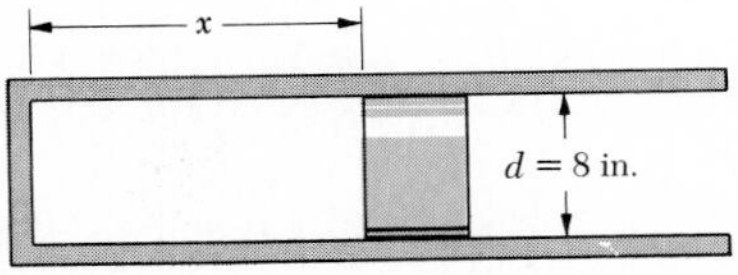

Fig. P13.31

13.31 An 8-in.-diameter piston weighing 10 lb slides without friction in a cylinder. When the piston is at a distance $x = 14$ in. from the end of the cylinder, the pressure in the cylinder is atmospheric ($p_a = 14.7\ \text{lb/in}^2$). If the pressure varies inversely as the volume, find the work done in moving the piston until $x = 6$ in.

13.32 The piston of Prob. 13.31 is moved to the left and released with no velocity when $x = 6$ in. Neglecting friction, determine (*a*) the maximum velocity attained by the piston, (*b*) the maximum value of the coordinate x.

13.33 An object is released with no velocity at an altitude equal to the radius of the earth. Neglecting air resistance, determine the velocity of the object as it strikes the earth. Give the answer in both SI and U.S. customary units.

13.34 A rocket is fired vertically from the ground. Knowing that at burnout the rocket is 80 km above the ground and has a velocity of 5000 m/s, determine the highest altitude it will reach.

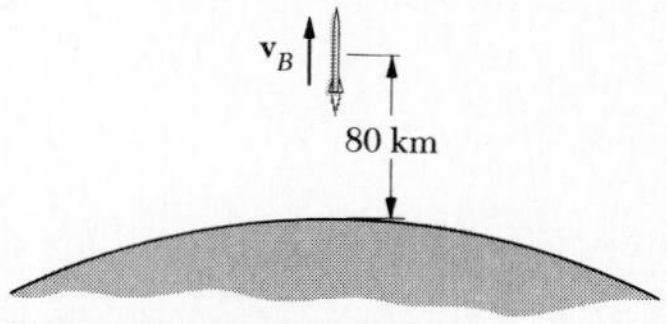

Fig. P13.34 and P13.35

13.35 A rocket is fired vertically from the ground. What should be its velocity $\mathbf{v}_B$ at burnout, 80 km above the ground, if it is to reach an altitude of 1000 km?

13.36 An object is released with no initial velocity at an altitude of 1000 mi. Neglecting air resistance, determine the velocity of the object (*a*) after it has fallen 500 mi, (*b*) as it strikes the ground.

13.6. Potential Energy.† Let us consider again a body of weight **W** which moves along a curved path from a point A_1 of elevation y_1 to a point A_2 of elevation y_2 (Fig. 13.4). We recall from Sec. 13.2 that the work of the weight **W** during this displacement is

$$U_{1\to2} = Wy_1 - Wy_2 \qquad (13.4)$$

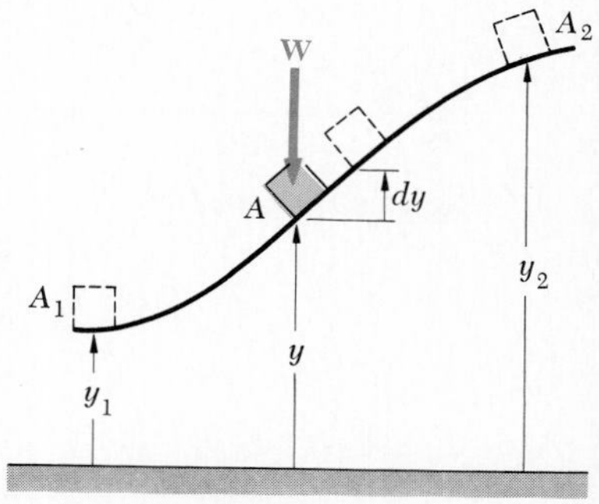

Fig. 13.4 (*repeated*)

The work of **W** may thus be obtained by subtracting the value of the function Wy corresponding to the second position of the body from its value corresponding to the first position. The work of **W** is independent of the actual path followed; it depends only upon the initial and final values of the function Wy. This function is called the *potential energy* of the body with respect to the *force of gravity* **W** and is denoted by V_g. We write

$$U_{1\to2} = (V_g)_1 - (V_g)_2 \qquad \text{with } V_g = Wy \qquad (13.14)$$

We note that if $(V_g)_2 > (V_g)_1$, i.e., *if the potential energy increases* during the displacement (as in the case considered here), *the work* $U_{1\to2}$ *is negative*. If, on the other hand, the work of **W** is positive, the potential energy decreases. Therefore, the potential energy V_g of the body provides a measure of the work which may be done by its weight **W**. Since only the *change* in potential energy,

†Most of the material in this section has already been considered in Sec. 10.6.

and not the actual value of V_g, is involved in formula (13.14), an arbitrary constant may be added to the expression obtained for V_g. In other words, the level, or datum, from which the elevation y is measured may be chosen arbitrarily. Note that potential energy is expressed in the same units as work, i.e., in joules if SI units are used, and in ft · lb or in · lb if U.S. customary units are used.

It should be noted that the expression just obtained for the potential energy of a body with respect to gravity is valid only as long as the weight **W** of the body may be assumed to remain constant, i.e., as long as the displacements of the body are small compared to the radius of the earth. In the case of a space vehicle, however, we should take into consideration the variation of the force of gravity with the distance r from the center of the earth. Using the expression obtained in Sec. 13.2 for the work of a gravitational force, we write (Fig. 13.6)

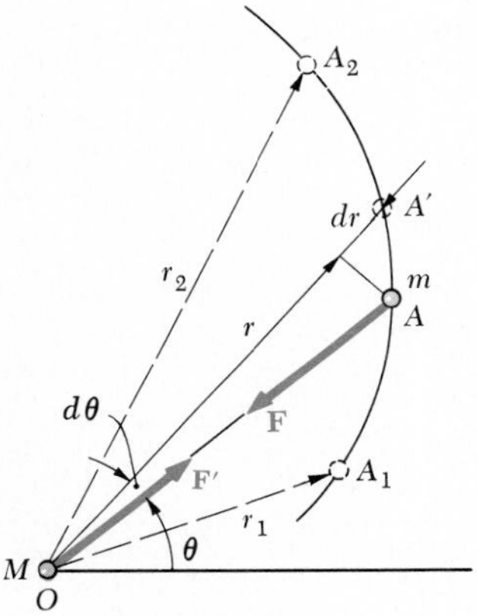

Fig. 13.6 (*repeated*)

$$U_{1\to 2} = \frac{GMm}{r_2} - \frac{GMm}{r_1} \tag{13.7}$$

The work of the force of gravity may therefore be obtained by subtracting the value of the function $-GMm/r$ corresponding to the second position of the body from its value corresponding to the first position. Thus, the expression which should be used for the potential energy V_g when the variation in the force of gravity cannot be neglected is

$$V_g = -\frac{GMm}{r} \tag{13.15}$$

Taking the first of the relations (12.20) into account, we write V_g in the alternate form

$$V_g = -\frac{WR^2}{r} \tag{13.15'}$$

where R is the radius of the earth and W the value of the weight of the body at the surface of the earth. When either of the relations (13.15) and (13.15′) is used to express V_g, the distance r should, of course, be measured from the center of the earth.† Note that V_g is always negative and that it approaches zero for very large values of r.

Consider, now, a body attached to a spring and moving from a position A_1, corresponding to a deflection x_1 of the spring, to a position A_2, corresponding to a deflection x_2 (Fig. 13.5). We

† The expressions given for V_g in (13.15) and (13.15′) are valid only when $r \geqq R$, that is, when the body considered is above the surface of the earth.

recall from Sec. 13.2 that the work of the force **F** exerted by the spring on the body is

$$U_{1\rightarrow 2} = \tfrac{1}{2}kx_1^2 - \tfrac{1}{2}kx_2^2 \qquad (13.6)$$

The work of the elastic force is thus obtained by subtracting the value of the function $\frac{1}{2}kx^2$ corresponding to the second position of the body from its value corresponding to the first position. This function is denoted by V_e and is called the *potential energy* of the body with respect to the *elastic force* **F**. We write

$$U_{1\rightarrow 2} = (V_e)_1 - (V_e)_2 \qquad \text{with } V_e = \tfrac{1}{2}kx^2 \qquad (13.16)$$

and observe that, during the displacement considered, the work of the force **F** exerted by the spring on the body is negative and the potential energy V_e increases. We should note that the expression obtained for V_e is valid only if the deflection of the spring is measured from its undeformed position. On the other hand, formula (13.16) may be used even when the spring is rotated about its fixed end (Fig. 13.10*a*). The work of the elastic force depends only upon the initial and final deflections of the spring (Fig. 13.10*b*).

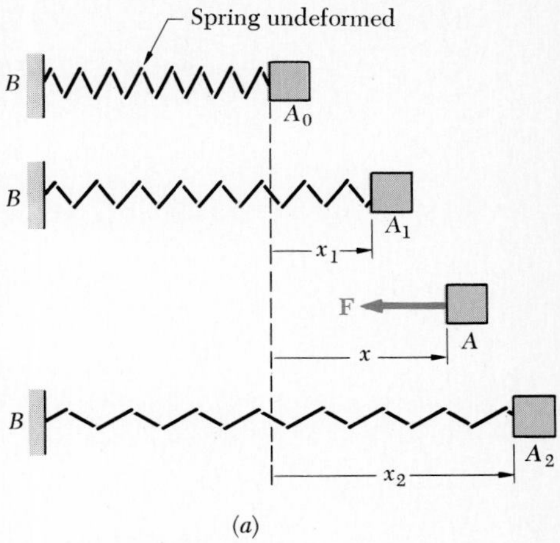

Fig. 13.5 (*repeated*)

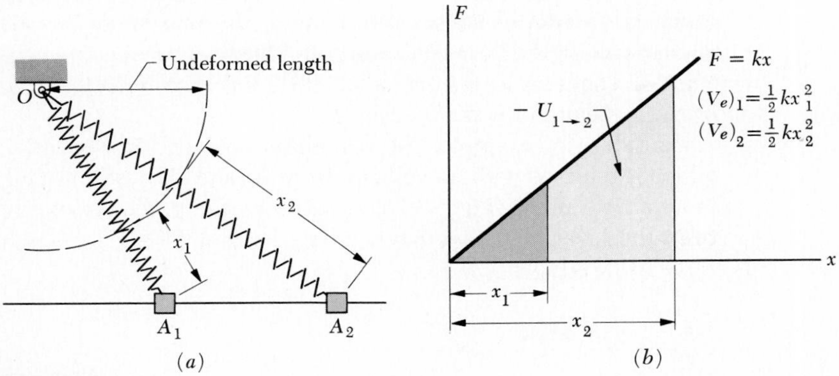

Fig. 13.10

The concept of potential energy may be used when forces other than gravity forces and elastic forces are involved. It remains valid as long as the elementary work dU of the force considered is an *exact differential.* It is then possible to find a function V, called potential energy, such that

$$dU = -dV$$

Integrating over a finite displacement, we obtain the general relationship

$$U_{1\to 2} = V_1 - V_2 \tag{13.17}$$

which expresses that *the work of the force is independent of the path followed and is equal to minus the change in potential energy.* A force which satisfies Eq. (13.17) is said to be a *conservative force.*

13.7. Conservation of Energy. We saw in the preceding section that the work of a conservative force, such as the weight of a particle or the force exerted by a spring, may be expressed as a change in potential energy. When a particle, or a system of particles, moves under the action of conservative forces, the principle of work and energy stated in Sec. 13.3 may be expressed in a modified form. Substituting for $U_{1\to 2}$ from (13.17) into (13.12), we write

$$V_1 - V_2 = T_2 - T_1$$

$$T_1 + V_1 = T_2 + V_2 \tag{13.18}$$

Formula (13.18) indicates that, when a system of particles moves under the action of conservative forces, *the sum of the kinetic energy and of the potential energy of the system remains constant.*† The sum $T + V$ is called the *total mechanical energy* of the system and is denoted by E.

Consider, for example, the pendulum analyzed in Sec. 13.4, which is released with no velocity from A_1 and allowed to swing in a vertical plane (Fig. 13.11). Measuring the potential energy from the level of A_2, we have, at A_1,

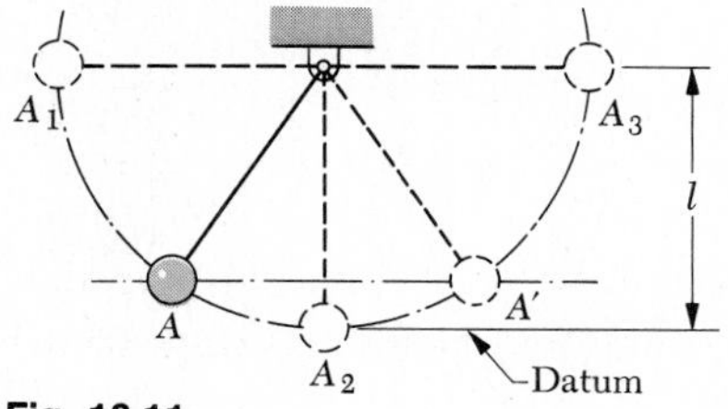

Fig. 13.11

† When the particles of a system move with respect to each other under the action of internal forces, the potential energy of the system must include the potential energy corresponding to the internal forces.

$$T_1 = 0 \qquad V_1 = Wl \qquad T_1 + V_1 = Wl$$

Recalling that, at A_2, the speed of the pendulum is $v_2 = \sqrt{2gl}$, we have

$$T_2 = \tfrac{1}{2} mv_2^2 = \frac{1}{2}\frac{W}{g}(2\,gl) = Wl \qquad V_2 = 0$$

$$T_2 + V_2 = Wl$$

We thus check that the total mechanical energy $E = T + V$ of the pendulum is the same at A_1 and A_2. While the energy is entirely potential at A_1, it becomes entirely kinetic at A_2 and, as the pendulum keeps swinging to the right, the kinetic energy is transformed back into potential energy. At A_3, we shall have $T_3 = 0$ and $V_3 = Wl$.

Since the total mechanical energy of the pendulum remains constant and since its potential energy depends only upon its elevation, the kinetic energy of the pendulum will have the same value at any two points located on the same level. Thus, the speed of the pendulum is the same at A and at A' (Fig. 13.11). This result may be extended to the case of a particle moving along any given path, regardless of the shape of the path, as long as the only forces acting on the particle are its weight and the normal reaction of the path. The particle of Fig. 13.12, for example, which slides in a vertical plane along a frictionless track, will have the same speed at A, A', and A''.

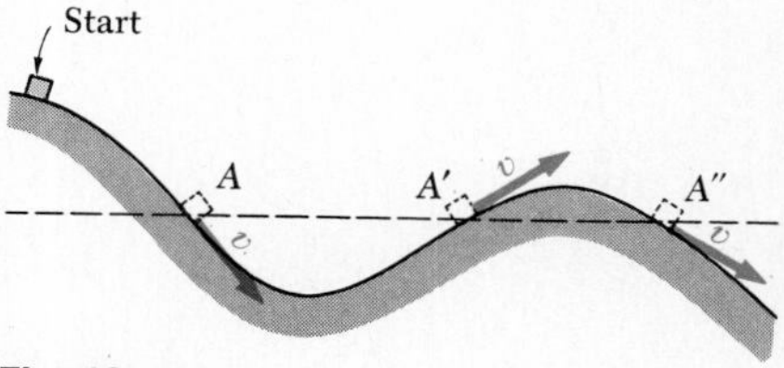

Fig. 13.12

While the weight of a particle and the force exerted by a spring are conservative forces, *friction forces are nonconservative forces.* In other words, *the work of a friction force cannot be expressed as a change in potential energy.* The work of a friction force depends upon the path followed by its point of application; and while the work $U_{1\to2}$ defined by (13.17) is positive or negative according to the sense of motion, *the work of a friction force is always negative.* It follows that, when a mechanical system involves friction, its total mechanical energy does not remain constant but decreases. The mechanical energy of the system, however, is not lost; it is transformed into heat, and the sum of the *mechanical energy* and of the *thermal energy* of the system remains constant.

Other forms of energy may also be involved in a system. For instance, a generator converts mechanical energy into *electric energy;* a gasoline engine converts *chemical energy* into mechanical energy; a nuclear reactor converts *mass* into thermal energy. If all forms of energy are considered, the energy of any system may be considered as constant and the principle of conservation of energy remains valid under all conditions.

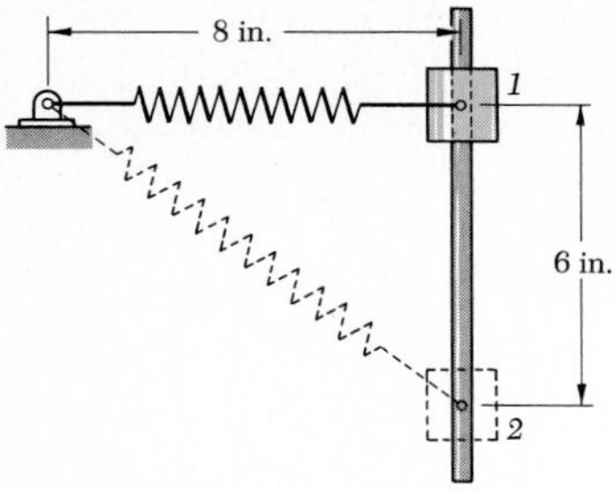

SAMPLE PROBLEM 13.5

A 20-lb collar slides without friction along a vertical rod as shown. The spring attached to the collar has an undeformed length of 4 in. and a constant of 3 lb/in. If the collar is released from rest in position *1*, determine its velocity after it has moved 6 in. to position *2*.

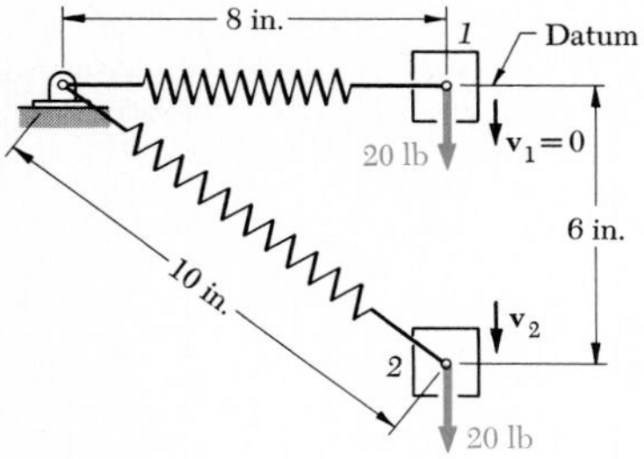

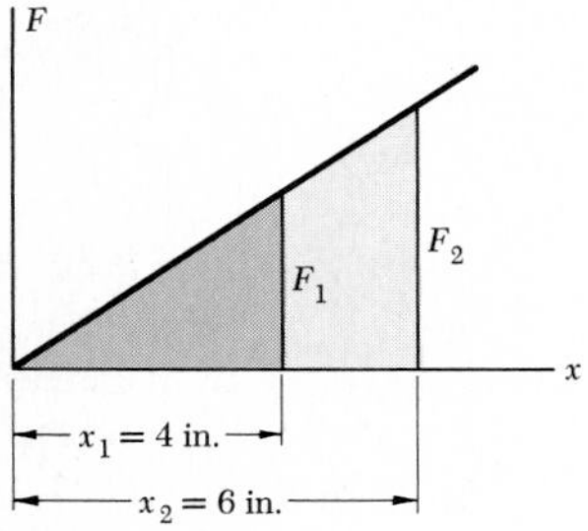

Position *1*. ***Potential Energy.*** The elongation of the spring is $x_1 = 8\text{ in.} - 4\text{ in.} = 4\text{ in.}$, and we have

$$V_e = \tfrac{1}{2}kx_1^2 = \tfrac{1}{2}(3\text{ lb/in.})(4\text{ in.})^2 = 24\text{ in}\cdot\text{lb}$$

Choosing the datum as shown, we have $V_g = 0$. Therefore,

$$V_1 = V_e + V_g = 24\text{ in}\cdot\text{lb} = 2\text{ ft}\cdot\text{lb}$$

Kinetic Energy. Since the velocity in position *1* is zero, $T_1 = 0$.

Position *2*. ***Potential Energy.*** The elongation of the spring is $x_2 = 10\text{ in.} - 4\text{ in.} = 6\text{ in.}$, and we have

$$V_e = \tfrac{1}{2}kx_2^2 = \tfrac{1}{2}(3\text{ lb/in.})(6\text{ in.})^2 = 54\text{ in}\cdot\text{lb}$$
$$V_g = Wy = (20\text{ lb})(-6\text{ in.}) = -120\text{ in}\cdot\text{lb}$$

Therefore,

$$V_2 = V_e + V_g = 54 - 120 = -66\text{ in}\cdot\text{lb} = -5.5\text{ ft}\cdot\text{lb}$$

Kinetic Energy

$$T_2 = \tfrac{1}{2}mv_2^2 = \frac{1}{2}\frac{20}{32.2}v_2^2 = 0.311v_2^2$$

Conservation of Energy. Applying the principle of conservation of energy between positions *1* and *2*, we write

$$T_1 + V_1 = T_2 + V_2$$
$$0 + 2\text{ ft}\cdot\text{lb} = 0.311v_2^2 - 5.5\text{ ft}\cdot\text{lb}$$
$$v_2 = \pm 4.91\text{ ft/s}$$

$\mathbf{v}_2 = 4.91\text{ ft/s}\downarrow$ ◄

SAMPLE PROBLEM 13.6

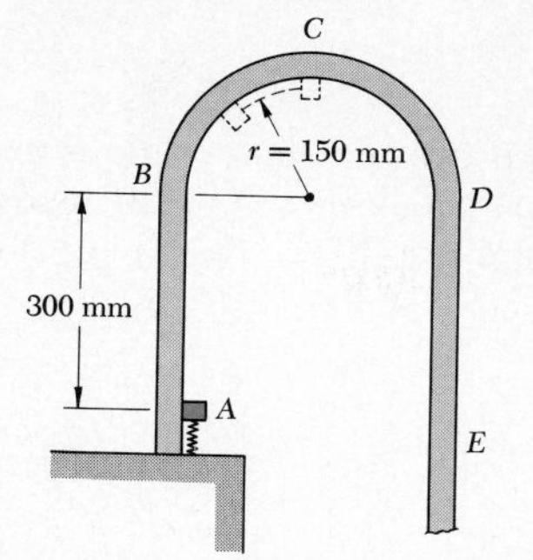

The 200-g pellet is released from rest at A when the spring is compressed 75 mm and travels around the loop $ABCDE$. Determine the smallest value of the spring constant for which the pellet will travel around the loop and will at all times remain in contact with the loop.

Required Speed at Point C. As the pellet passes through the highest point C, its potential energy with respect to gravity is maximum; thus, at the same point its kinetic energy and its speed are minimum. Since the pellet must remain in contact with the loop, the force **N** exerted on the pellet by the loop must be equal to, or greater than, zero. Setting $\mathbf{N} = 0$, we compute the smallest possible speed v_C.

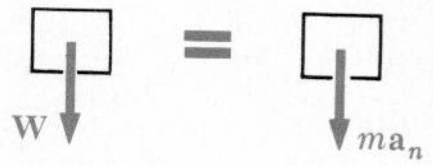

$$+\downarrow \Sigma F_n = ma_n: \qquad W = ma_n \qquad mg = ma_n \qquad a_n = g$$

$$a_n = \frac{v_C^2}{r}: \qquad v_C^2 = ra_n = rg = (0.150\text{ m})(9.81\text{ m/s}^2) = 1.472\text{ m}^2/\text{s}^2$$

Position 1. ***Potential Energy.*** Since the spring is compressed 0.075 m from its undeformed position, we have

$$V_e = \tfrac{1}{2}kx^2 = \tfrac{1}{2}k(0.075\text{ m})^2 = (0.00281\text{ m}^2)k$$

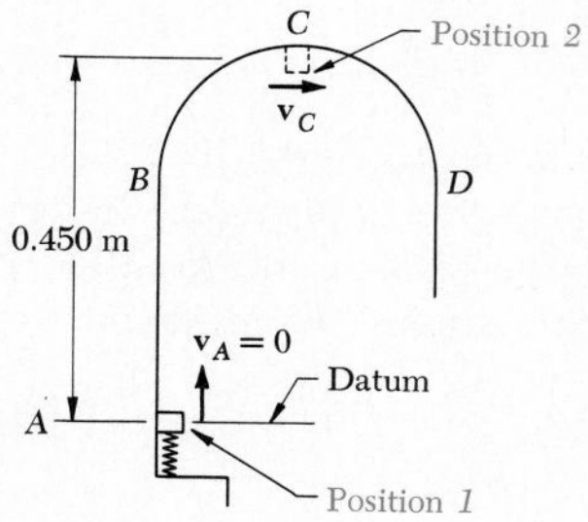

Choosing the datum at A, we have $V_g = 0$; therefore

$$V_1 = V_e + V_g = (0.00281\text{ m}^2)k$$

Kinetic Energy. Since the pellet is released from rest, $v_A = 0$ and we have $T_1 = 0$.

Position 2. ***Potential Energy.*** The spring is now undeformed; thus $V_e = 0$. Since the pellet is 0.450 m above the datum, and since $W = (0.200\text{ kg})(9.81\text{ m/s}^2) = 1.962\text{ N}$, we have

$$V_g = Wy = (1.962\text{ N})(0.450\text{ m}) = 0.883\text{ N}\cdot\text{m} = 0.883\text{ J}$$
$$V_2 = V_e + V_g = 0.883\text{ J}$$

Kinetic Energy. Using the value of v_C^2 obtained above, we write

$$T_2 = \tfrac{1}{2}mv_C^2 = \tfrac{1}{2}(0.200\text{ kg})(1.472\text{ m}^2/\text{s}^2) = 0.1472\text{ N}\cdot\text{m} = 0.1472\text{ J}$$

Conservation of Energy. Applying the principle of conservation of energy between positions *1* and *2*, we write

$$T_1 + V_1 = T_2 + V_2$$
$$0 + (0.00281\text{ m}^2)k = 0.1472\text{ J} + 0.883\text{ J}$$
$$k = 367\text{ J/m}^2 = 367\text{ N/m}$$

The required minimum value of k is therefore

$$k = 367\text{ N/m} \blacktriangleleft$$

PROBLEMS

13.37 The spring AB is of constant k and is unstretched when $\theta = 0$. Determine the potential energy of the system with respect (*a*) to the spring, (*b*) to gravity. (Place datum at C.)

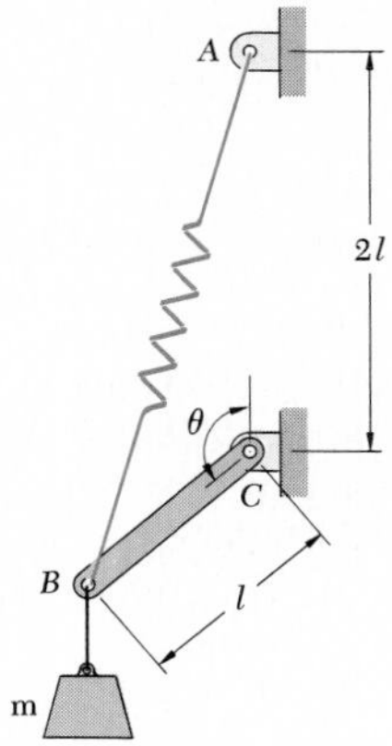

Fig. P13.37

13.38 A slender rod AB of negligible mass is attached to blocks A and B, each of mass m. The constant of the spring is k and the spring is undeformed when AB is horizontal. Determine the potential energy of the system with respect (*a*) to the spring, (*b*) to gravity. (Place datum at B.)

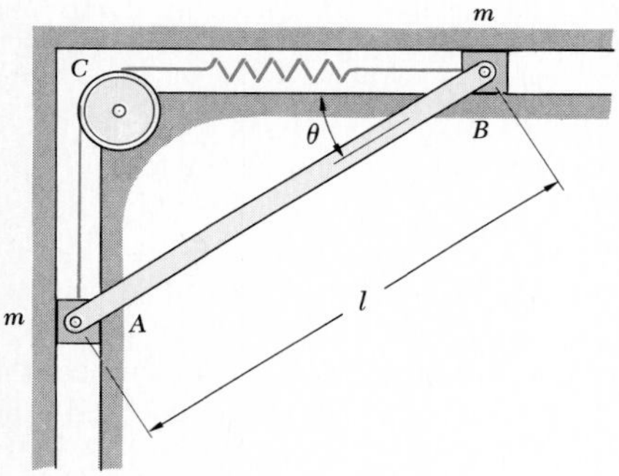

Fig. P13.38

13.39 The spring AB is of constant 3 lb/in. and is attached to the 2-lb collar B which moves freely along the horizontal rod. The unstretched length of the spring is 5 in. If the collar is released from rest in the position shown, determine the maximum velocity attained by the collar.

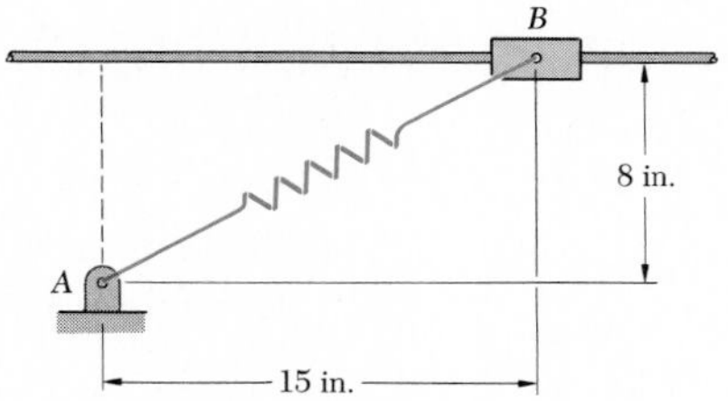

Fig. P13.39

13.40 In Prob. 13.39, determine the weight of the collar B for which the maximum velocity is 15 ft/s.

13.41 A collar of mass 1.5 kg is attached to a spring and slides without friction along a circular rod which lies in a *horizontal* plane. The spring is undeformed when the collar is at C and the constant of the spring is 400 N/m. If the collar is released from rest at B, determine the velocity of the collar as it passes through point C.

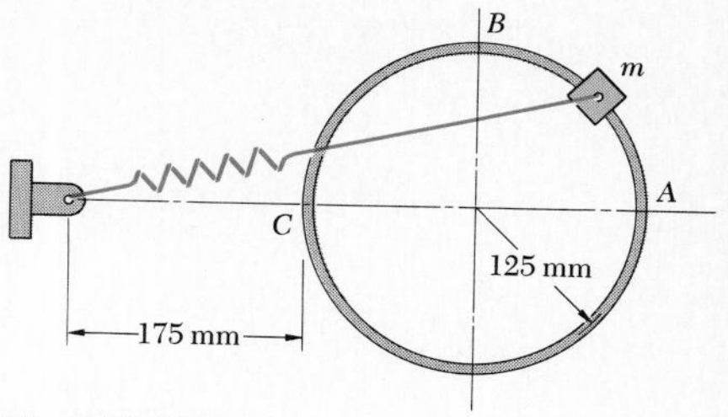

Fig. P13.41

13.42 A toy spring gun is used to shoot bullets of mass 30 g vertically upward. The undeformed length of the spring is 125 mm; it is compressed to a length of 25 mm when the gun is ready to be shot and expands to a length of 75 mm as the bullet leaves the gun. A force of 35 N is required to maintain the spring in firing position when the length of the spring is 25 mm. Determine (*a*) the velocity of the bullet as it leaves the gun, (*b*) the maximum height reached by the bullet.

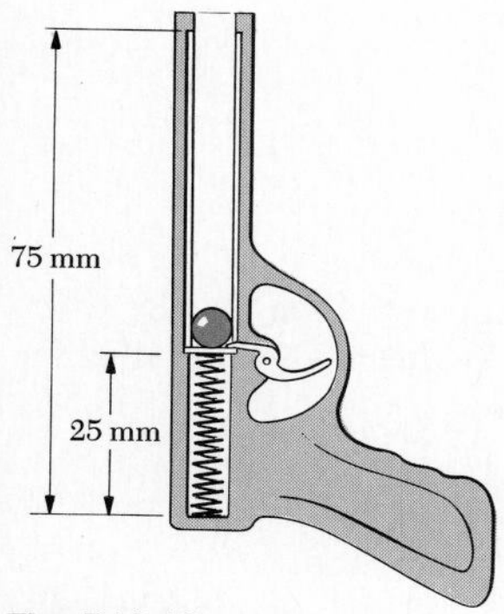

Fig. P13.42

13.43 The collar of Prob. 13.41 has a continuous, although nonuniform, motion along the rod. If the speed of the collar at A is to be half of its speed at C, determine (*a*) the required speed at C, (*b*) the corresponding speed at B.

13.44 An elastic cord is stretched between two points A and B, located 16 in. apart in the same horizontal plane. When stretched directly between A and B, the tension in the cord is 10 lb. The cord is then stretched as shown until its midpoint C has moved through 6 in. to C'; a force of 60 lb is required to hold the cord at C'. A 0.20-lb pellet is placed at C', and the cord is released. Determine the speed of the pellet as it passes through C.

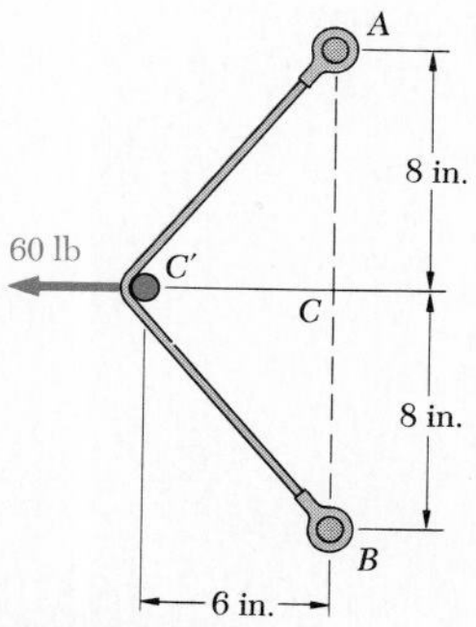

Fig. P13.44

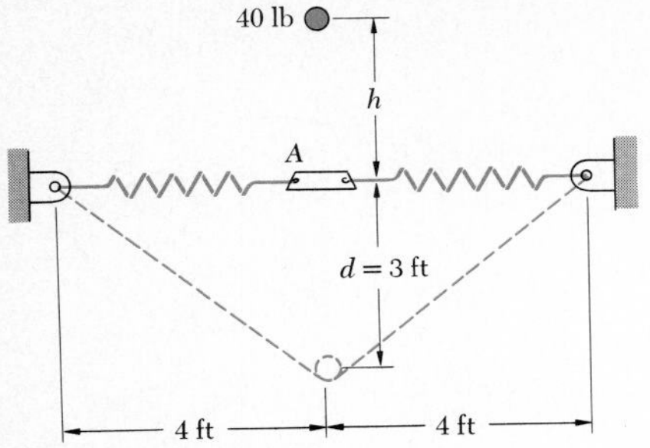

Fig. P13.45

13.45 Two springs are attached to a piece of cloth A of negligible weight, as shown. The initial tension in each spring is 120 lb, and the spring constant of each spring is $k = 10$ lb/in. A 40-lb ball is released from a height h above A; the ball hits the cloth, causing it to move through a maximum distance $d = 3$ ft, as shown by the dashed lines. Determine the height h.

13.46 The tension in the spring is zero when the arm ABC is horizontal ($\phi = 0$). If the 50-kg block is released when $\phi = 0$, determine the speed of the block (*a*) when $\phi = 90°$, (*b*) when $\phi = 180°$.

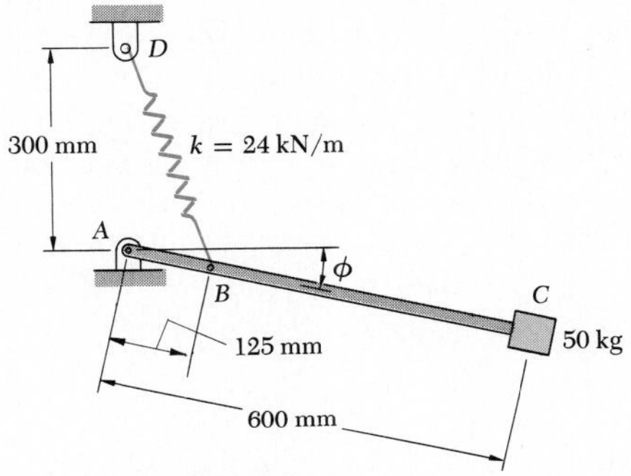

Fig. P13.46 and P13.47

13.47 The 50-kg block is released from rest when $\phi = 0$. If the speed of the block when $\phi = 90°$ is to be 2.5 m/s, determine the required value of the initial tension in the spring.

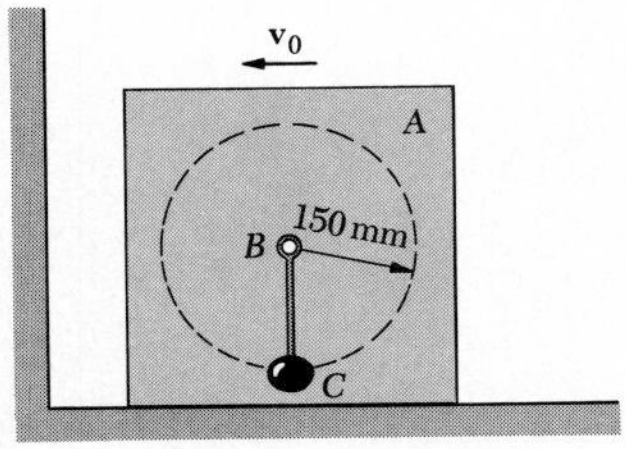

Fig. P13.48

13.48 The sphere C and the block A are both moving to the left with a velocity $\mathbf{v}_0$ when the block is suddenly stopped by the wall. Determine the smallest velocity $\mathbf{v}_0$ for which the sphere C will swing in a full circle about the pivot B (*a*) if BC is a slender rod of negligible weight, (*b*) if BC is a cord.

13.49 The collar of Prob. 13.41 is released from rest at point A. Determine the horizontal component of the force exerted by the rod on the collar as the collar passes through point B. Show that the force is independent of the mass of the collar.

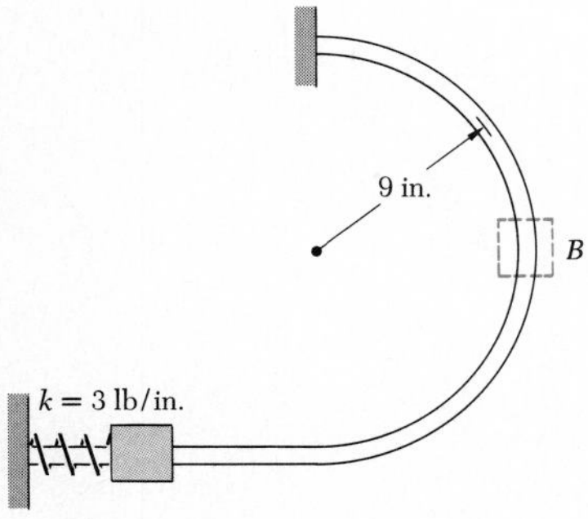

Fig. P13.50

13.50 A $\frac{1}{4}$-lb collar may slide without friction on a rod in a vertical plane. The collar is released from rest when the spring is compressed 1.5 in. As the collar passes through point B, determine (*a*) the speed of the collar, (*b*) the force exerted by the rod on the collar.

13.51 The pendulum shown is released from rest at A and swings through 90° before the cord touches the fixed peg B. Determine the smallest value of a for which the pendulum bob will describe a circle about the peg.

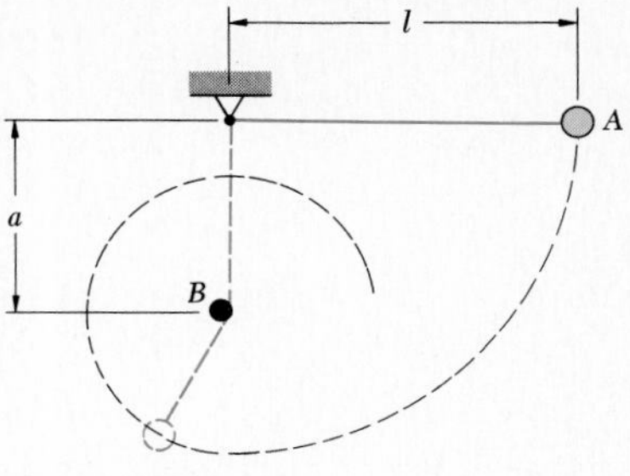

Fig. P13.51

13.52 A small block is released at A with zero velocity and moves along the frictionless guide to point B where it leaves the guide with a horizontal velocity. Knowing that $h = 8$ ft and $b = 3$ ft, determine (*a*) the speed of the block as it strikes the ground at C, (*b*) the corresponding distance c.

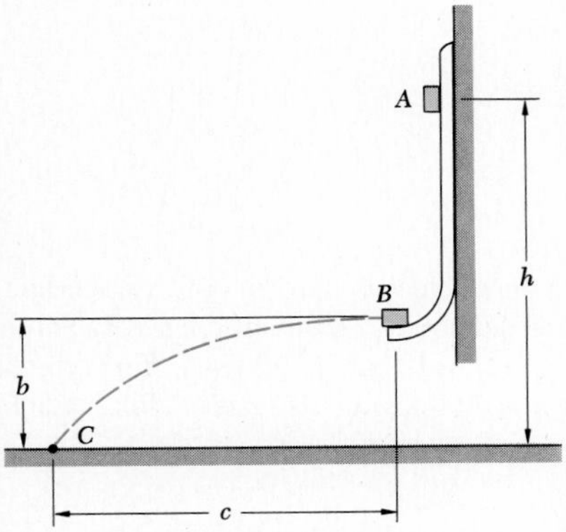

Fig. P13.52

13.53 Assuming a given height h in Prob. 13.52, (*a*) show that the speed at C is independent of the height b, (*b*) determine the height b for which the distance c is maximum and the corresponding value of c.

13.54 A bag is gently pushed off the top of a wall at A and swings in a vertical plane at the end of a 4-m rope which can withstand a maximum tension equal to twice the weight of the bag. (*a*) Determine the difference in elevation h between point A and point B where the rope will break. (*b*) How far from the vertical wall will the bag strike the floor?

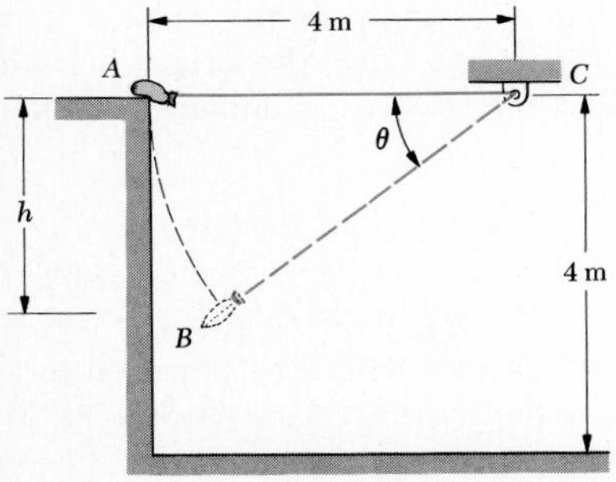

Fig. P13.54

13.55 A delicate instrument weighing 8 lb is placed on a spring of length l so that its base is just touching the undeformed spring. The instrument is then inadvertently released from that position. Determine the maximum deflection x of the spring and the maximum force exerted by the spring if the constant of the spring is $k = 10$ lb/in.

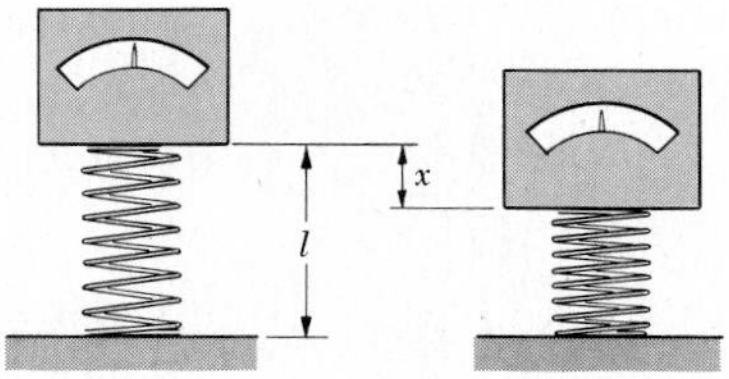

Fig. P13.55

13.56 Nonlinear springs are classified as hard or soft, depending upon the curvature of their force-deflection curves (see figure). Solve Prob. 13.55, assuming (*a*) that a hard spring is used, for which $F = 10x(1 + 0.1x^2)$, (*b*) that a soft spring is used, for which $F = 10x(1 - 0.1x^2)$.

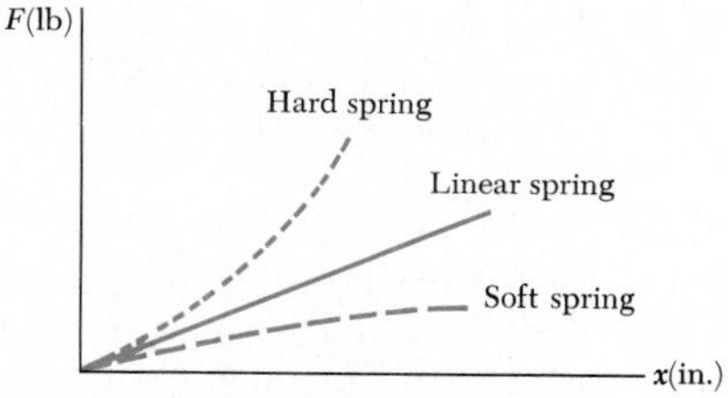

Fig. P13.56

13.57 Determine the escape velocity of a missile, i.e., the velocity with which it should be fired from the surface of the earth if it is to reach an infinite distance from the earth. Give the answer in both SI and U.S. customary units. Show that the result obtained is independent of the firing angle.

13.58 How much energy per kilogram should be imparted to a satellite in order to place it in a circular orbit at an altitude of (*a*) 500 km, (*b*) 5000 km?

13.59 A lunar excursion module (LEM) has been used in the Apollo moon-landing missions to save fuel by making it unnecessary to launch the entire Apollo spacecraft from the moon's surface on its return trip to the earth. Check the effectiveness of this approach by computing the energy per pound required for a spacecraft to escape the gravitational field of the moon if the spacecraft starts (*a*) from the moon's surface, (*b*) from a circular orbit 60 mi above the moon's surface. Neglect the effect of the earth's gravitational field. (The radius of the moon is 1080 mi and its mass is 0.01230 times the mass of the earth.)

13.60 A satellite of mass m describes a circular orbit around the earth. Express (*a*) its potential energy, (*b*) its kinetic energy, and (*c*) its total energy as a function of the radius r of the orbit. Denote by R the radius of the earth, by g the acceleration of gravity at the surface of the earth, and assume the potential energy of the satellite to be zero on the launching pad.

13.8. Power and Efficiency.

Power is defined as the time rate at which work is done. In the selection of a motor or engine, power is a much more important criterion than the actual amount of work to be performed. A small motor or a large power plant may both be used to do a given amount of work; but the small motor may require a month to do the work done by the power plant in a matter of minutes. If ΔU is the work done during the time interval Δt, then the average power during this time interval is

$$\text{Average power} = \frac{\Delta U}{\Delta t}$$

Letting Δt approach zero, we obtain at the limit

$$\text{Power} = \frac{dU}{dt} \qquad (13.19)$$

Since $dU = (F \cos \alpha)\, ds$ and $v = ds/dt$, the power may also be expressed as follows:

$$\text{Power} = \frac{dU}{dt} = \frac{(F \cos \alpha)\, ds}{dt}$$

$$\text{Power} = (F \cos \alpha) v \qquad (13.20)$$

where v = magnitude of velocity of point of application of force $\mathbf{F}$

α = angle between $\mathbf{F}$ and velocity $\mathbf{v}$

Since power was defined as the time rate at which work is done, it should be expressed in units obtained by dividing units of work by the unit of time. Thus, if SI units are used, power should be expressed in J/s; this unit is called a *watt* (W). We have

$$1 \text{ W} = 1 \text{ J/s} = 1 \text{ N} \cdot \text{m/s}$$

If U.S. customary units are used, power should be expressed in ft · lb/s or in *horsepower* (hp), with the latter defined as

$$1 \text{ hp} = 550 \text{ ft} \cdot \text{lb/s}$$

Recalling from Sec. 13.2 that 1 ft · lb = 1.356 J, we verify that

$$1 \text{ ft} \cdot \text{lb/s} = 1.356 \text{ J/s} = 1.356 \text{ W}$$
$$1 \text{ hp} = 550(1.356 \text{ W}) = 746 \text{ W} = 0.746 \text{ kW}$$

The *mechanical efficiency* of a machine was defined in Sec. 10.4 as the ratio of the output work to the input work:

$$\eta = \frac{\text{output work}}{\text{input work}} \tag{13.21}$$

This definition is based on the assumption that work is done at a constant rate. The ratio of the output to the input work is therefore equal to the ratio of the rates at which output and input work are done, and we have

$$\eta = \frac{\text{power output}}{\text{power input}} \tag{13.22}$$

Because of energy losses due to friction, the output work is always smaller than the input work, and, consequently, the power output is always smaller than the power input. The mechanical efficiency of a machine, therefore, is always less than 1.

When a machine is used to transform mechanical energy into electric energy, or thermal energy into mechanical energy, its *overall efficiency* may be obtained from formula (13.22). The overall efficiency of a machine is always less than 1; it provides a measure of all the various energy losses involved (losses of electric or thermal energy as well as frictional losses). We should note that it is necessary, before using formula (13.22), to express the power output and the power input in the same units.

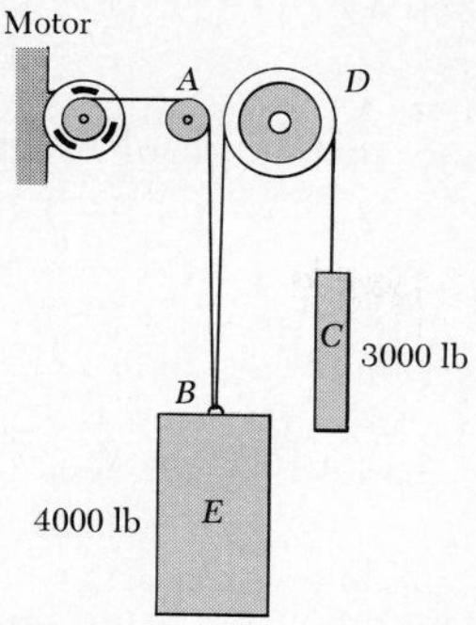

SAMPLE PROBLEM 13.7

The elevator shown weighs 4000 lb when fully loaded. It is connected to a 3000-lb counterweight C and is powered by an electric motor. Determine the power required when the elevator (a) is moving upward at a constant speed of 20 ft/s, (b) has an instantaneous velocity of 20 ft/s upward and an upward acceleration of 3 ft/s².

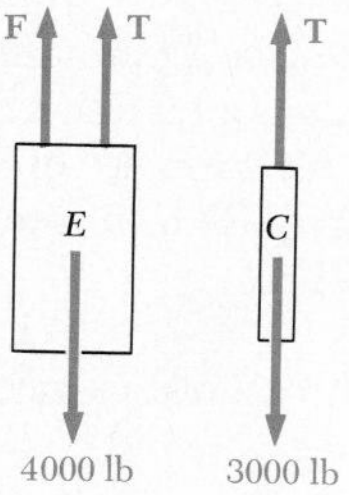

Solution. Since **F** and **v** have the same direction, the power is equal to Fv. We must first determine the force **F** exerted by cable AB on the elevator in each of the two given situations.

Force **F**. The forces acting on the elevator and on the counterweight are shown in the adjoining sketches.

a. Uniform Motion. We have $a = 0$; both bodies are in equilibrium.

Free Body C: $\quad +\uparrow\Sigma F_y = 0: \quad T - 3000 \text{ lb} = 0$

Free Body E: $\quad +\uparrow\Sigma F_y = 0: \quad F + T - 4000 \text{ lb} = 0$

Eliminating T, we find: $\quad F = 1000 \text{ lb}$

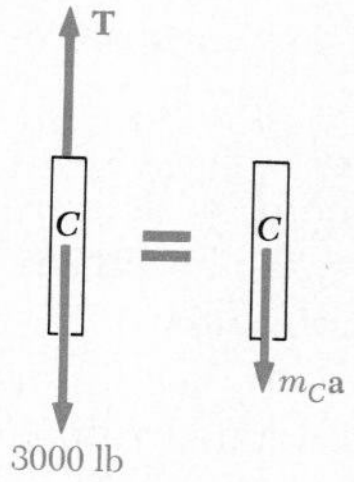

b. Accelerated Motion. We have $a = 3 \text{ ft/s}^2$. The equations of motion are

Free Body C: $\quad +\downarrow\Sigma F_y = m_C a: \quad 3000 - T = \dfrac{3000}{32.2}3$

Free Body E: $\quad +\uparrow\Sigma F_y = m_E a: \quad F + T - 4000 = \dfrac{4000}{32.2}3$

Eliminating T: $\quad F = 1000 + \dfrac{7000}{32.2}3 = 1652 \text{ lb}$

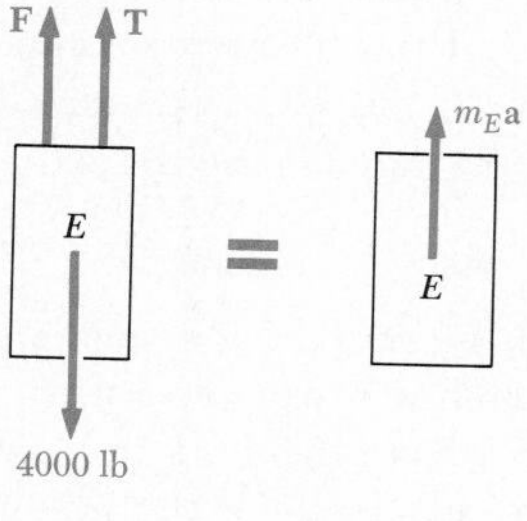

Power. Substituting the given values of v and the values found for F into the expression for the power, we have

a. $Fv = (1000 \text{ lb})(20 \text{ ft/s}) = 20{,}000 \text{ ft} \cdot \text{lb/s}$

$$\text{Power} = (20{,}000 \text{ ft} \cdot \text{lb/s})\frac{1 \text{ hp}}{550 \text{ ft} \cdot \text{lb/s}} = 36.4 \text{ hp}$$ ◄

b. $Fv = (1652 \text{ lb})(20 \text{ ft/s}) = 33{,}040 \text{ ft} \cdot \text{lb/s}$

$$\text{Power} = \frac{33{,}040}{550} = 60.1 \text{ hp}$$ ◄

PROBLEMS

13.61 A 70-kg man and an 80-kg man run up a flight of stairs in 5 s. If the flight of stairs is 4 m high, determine the average power required by each man.

13.62 A utility hoist can lift its maximum allowable load of 7000 lb at the rate of 65 ft/min. Knowing that the hoist is run by a 20-hp engine, determine the overall efficiency of the hoist.

13.63 A 3500-lb automobile starts from rest and accelerates uniformly to a speed of 60 mi/h in 30 s. Knowing that the rolling resistance of the automobile is 100 lb/ton and that the road is horizontal, determine the power required as a function of time.

13.64 A 1500-kg automobile travels 200 m while being accelerated at a uniform rate from 50 to 75 km/h. During the entire motion, the automobile is traveling on a horizontal road, and the rolling resistance is equal to 2 percent of the weight of the automobile. Determine (*a*) the maximum power required, (*b*) the power required to maintain a constant speed of 75 km/h.

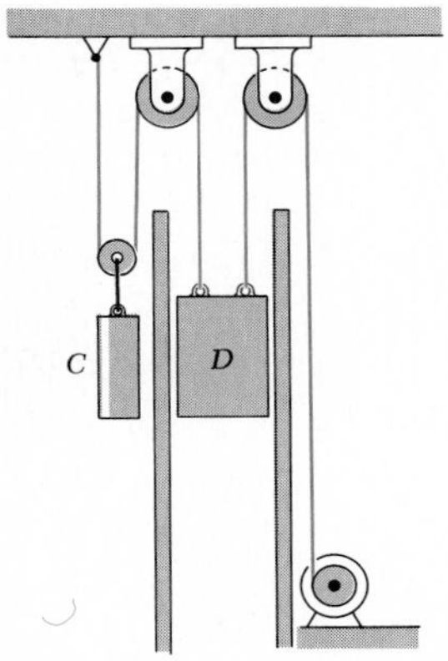

Fig. P13.65 and P13.66

13.65 The dumbwaiter *D* and its counterweight *C* weigh 600 lb each. Determine the power required when the dumbwaiter (*a*) is moving upward at a constant speed of 15 ft/s, (*b*) has an instantaneous velocity of 15 ft/s upward and an upward acceleration of 2 ft/s^2.

13.66 The dumbwaiter *D* and its counterweight *C* weigh 600 lb each. Knowing that the motor is delivering to the system 10 hp at the instant the speed of the dumbwaiter is 15 ft/s upward, determine the acceleration of the dumbwaiter.

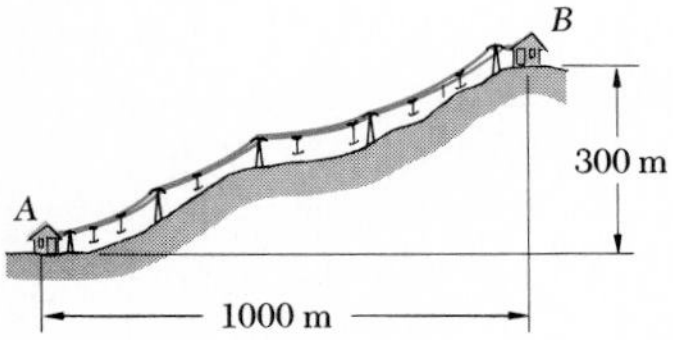

Fig. P13.67

13.67 A chair-lift is designed to transport 900 skiers per hour from the base *A* to the summit *B*. The average mass of a skier is 75 kg, and the average speed of the lift is 80 m/min. Determine (*a*) the average power required, (*b*) the required capacity of the motor if the mechanical efficiency is 85 percent and if a 300-percent overload is to be allowed.

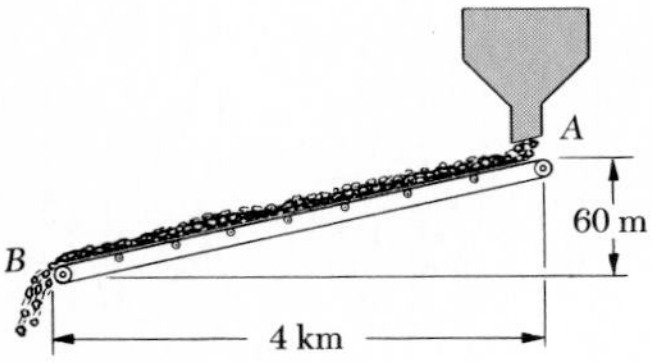

Fig. P13.68

13.68 Crushed stone is moved from a quarry at *A* to a construction site at *B* at the rate of 2000 Mg per 8-h period. An electric generator is attached to the system in order to maintain a constant belt speed. Knowing that the efficiency of the belt-generator system is 0.65, determine the average power developed by the generator (*a*) if the belt speed is 0.75 m/s, (*b*) if the belt speed is 2 m/s.

13.69 The fluid transmission of a truck of mass *m* permits the engine to deliver an essentially constant power *P* to the driving wheels. Determine the time elapsed and the distance traveled as the speed is increased from v_0 to v_1.

13.70 The fluid transmission of a 20-ton truck permits the engine to deliver an essentially constant power of 100 hp to the driving wheels. Determine the time required and the distance traveled as the speed of the truck is increased (*a*) from 30 to 45 mi/h, (*b*) from 45 to 60 mi/h.

13.71 The frictional resistance of a ship is known to vary directly as the 1.75 power of the speed v of the ship. A single tugboat at full power can tow the ship at a constant speed of 5 km/h by exerting a constant force of 200 kN. Determine (*a*) the power developed by the tugboat, (*b*) the maximum speed at which two tugboats, capable of delivering the same power, can tow the ship.

13.72 Determine the speed at which the single tugboat of Prob. 13.71 will tow the ship if the tugboat is developing half of its maximum power.

REVIEW PROBLEMS

13.73 Two cylinders are suspended from an inextensible cable as shown. If the system is released from rest, determine the maximum velocity attained by the 10-lb cylinder.

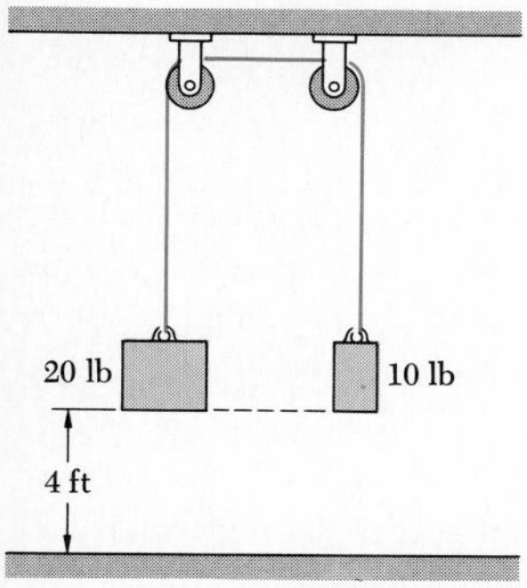

Fig. P13.73

13.74 In Prob. 13.73 determine the maximum height above the floor to which the 10-lb cylinder will rise.

13.75 Starting from rest, a 1500-kg automobile is accelerated at a constant rate of $1.25\ m/s^2$ until it reaches a speed of 60 km/h and then travels at a constant speed. During the entire motion, the automobile is traveling on a horizontal road, and the rolling resistance is equal to 1.8 percent of the weight of the automobile. Determine the power required as a function of time.

13.76 Two blocks *A* and *B* connected by a cord are released from rest in the position shown. Neglecting the effect of friction, determine the velocity of block *B* (*a*) as it leaves the table, (*b*) as it strikes the floor.

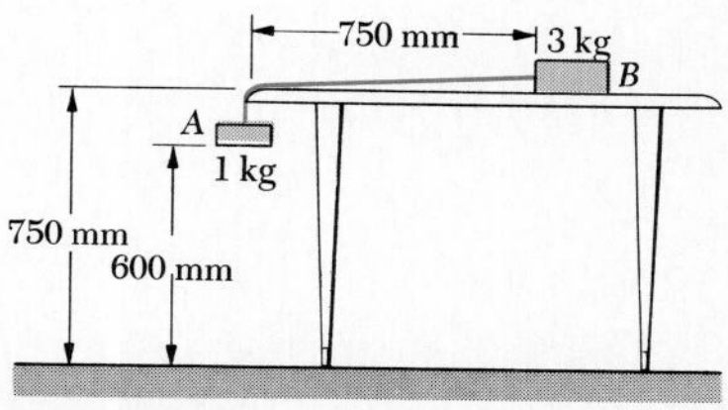

Fig. P13.76

13.77 An elevator travels upward at a constant speed of 2 m/s. A boy riding the elevator throws a 0.8-kg stone upward with a speed of 4 m/s *relative* to the elevator. Determine (*a*) the work done by the boy in throwing the stone, (*b*) the difference in the values of the kinetic energy of the stone before and after it was thrown. (*c*) Why are the values obtained in parts *a* and *b* not the same?

13.78 A ball of mass m attached to an inextensible cord rotates in a vertical circle of radius r. Show that the difference between the maximum value T_{max} of the tension in the cord and its minimum value T_{min} is independent of the speed at which the ball rotates, and determine $T_{max} - T_{min}$.

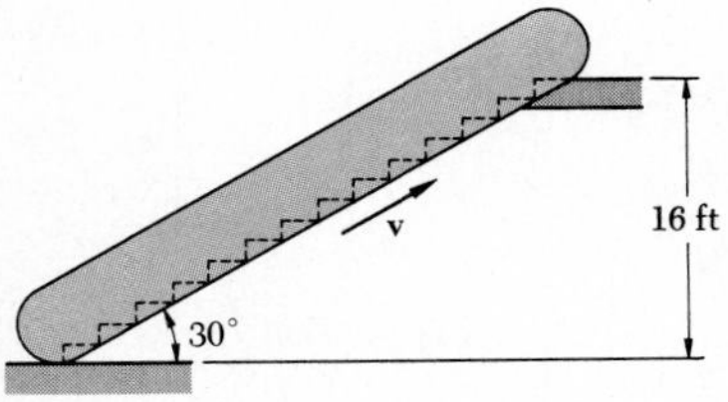

Fig. P13.79

13.79 The escalator shown is designed to transport 9000 persons per hour at a constant speed of 90 ft/min. Assuming an average weight of 150 lb per person, determine (*a*) the average power required, (*b*) the required capacity of the motor if the mechanical efficiency is 75 percent and if a 250-percent overload is to be allowed.

13.80 Two blocks are joined by an inextensible cable as shown. If the system is released from rest, determine the velocity of block A after it has moved 2 m. Assume that μ equals 0.25 between block A and the plane and neglect the mass and friction of the pulleys.

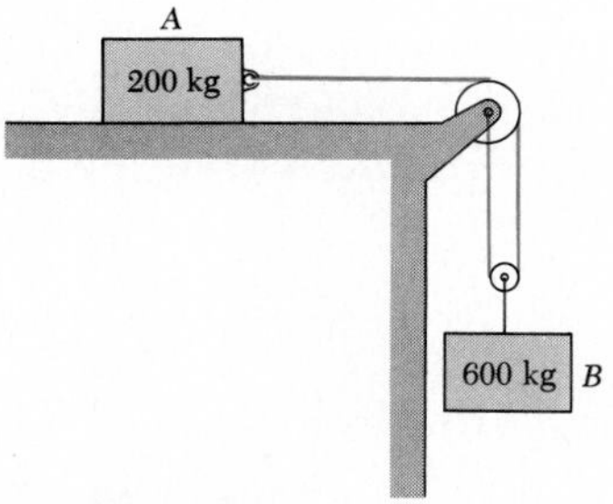

Fig. P13.80

13.81 The 0.5-lb pellet is released when the spring is compressed 6 in. and travels without friction around the vertical loop *ABCD*. Determine the force exerted by the loop on the pellet (*a*) at point A, (*b*) at point *B*, (*c*) at point *C*.

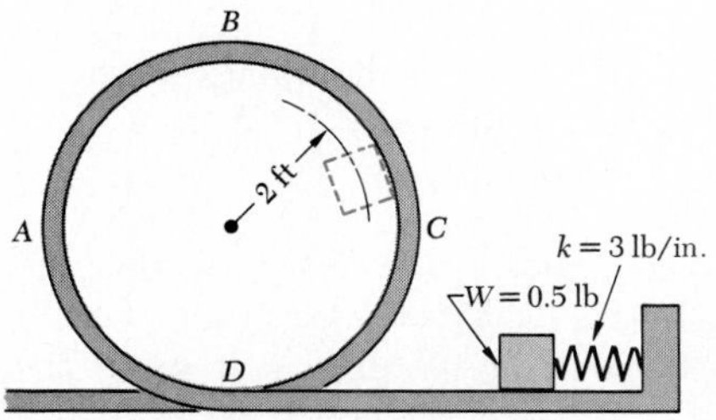

Fig. P13.81

13.82 In Prob. 13.81, determine the smallest allowable deflection of the spring if the pellet is to travel around the entire loop without leaving the track.

13.83 A 5-kg collar slides without friction along a rod which forms an angle of 30° with the vertical. The spring is unstretched when the collar is at *A*. If the collar is released from rest at *A*, determine the value of the spring constant *k* for which the collar has zero velocity at *B*.

13.84 In Prob. 13.83, determine the value of the spring constant *k* for which the velocity of the collar at *B* is 1.5 m/s.

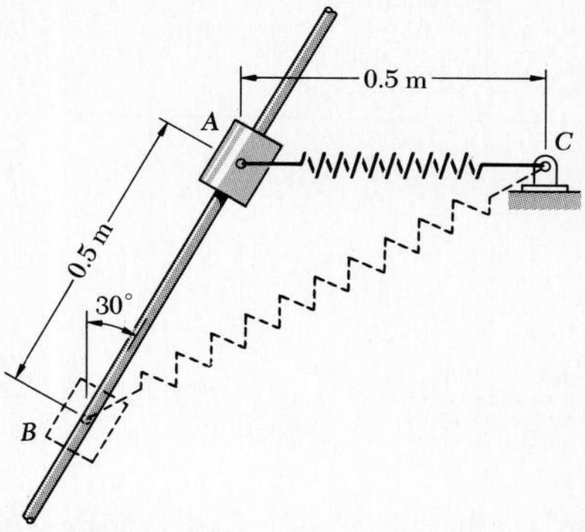

Fig. P13.83

Chapter
14

Kinetics of Particles: Impulse and Momentum

14.1. Principle of Impulse and Momentum. A third basic method for the solution of problems dealing with the motion of particles will be considered in this chapter. This method is based on the principle of impulse and momentum and is of particular interest in problems involving force, mass, velocity, and time.

Consider a particle of mass m acted upon by a force $\mathbf{F}$. Expressing Newton's second law $\mathbf{F} = m\mathbf{a}$ in terms of the x and y components of the force and of the acceleration, we write

$$F_x = ma_x \qquad F_y = ma_y$$
$$F_x = m\frac{dv_x}{dt} \qquad F_y = m\frac{dv_y}{dt}$$

Since the mass m of the particle is constant, we have

$$F_x = \frac{d}{dt}(mv_x) \qquad F_y = \frac{d}{dt}(mv_y) \tag{14.1}$$

or, in vector form,

$$\mathbf{F} = \frac{d}{dt}(m\mathbf{v}) \tag{14.2}$$

The vector $m\mathbf{v}$ is called the *linear momentum,* or simply the *momentum,* of the particle. It has the same direction as the velocity of the particle and its magnitude is expressed in N · s or in lb · s. We check that, with SI units,

$$mv = (\text{kg})(\text{m/s}) = (\text{kg} \cdot \text{m/s}^2)\text{s} = \text{N} \cdot \text{s}$$

while, with U.S. customary units,

$$mv = (\text{lb} \cdot \text{s}^2/\text{ft})(\text{ft/s}) = \text{lb} \cdot \text{s}$$

Equation (14.2) expresses that *the force* **F** *acting on the particle is equal to the rate of change of the momentum of the particle.* It is in this form that the second law of motion was originally stated by Newton.

Multiplying both sides of Eqs. (14.1) by dt and integrating from a time t_1 to a time t_2, we write

$$F_x\,dt = d(mv_x) \qquad F_y\,dt = d(mv_y)$$

$$\int_{t_1}^{t_2} F_x\,dt = (mv_x)_2 - (mv_x)_1 \qquad \int_{t_1}^{t_2} F_y\,dt = (mv_y)_2 - (mv_y)_1$$

$$(mv_x)_1 + \int_{t_1}^{t_2} F_x\,dt = (mv_x)_2 \qquad (mv_y)_1 + \int_{t_1}^{t_2} F_y\,dt = (mv_y)_2 \tag{14.3}$$

or, in vector form,

$$m\mathbf{v}_1 + \int_{t_1}^{t_2} \mathbf{F}\,dt = m\mathbf{v}_2 \tag{14.4}$$

The integral in Eq. (14.4) is a vector known as the *linear impulse,* or simply the *impulse,* of the force **F** during the interval of time considered. From Eqs. (14.3), we note that the components of the impulse of the force **F** are equal, respectively, to the areas under the curves obtained by plotting the components F_x and F_y against t (Fig. 14.1). In the case of a force **F** of constant magnitude and direction, the impulse is represented by the vector $\mathbf{F}(t_2 - t_1)$, which has the same direction as **F**. We easily check that the magnitude of the impulse of a force is expressed in N · s or in lb · s.

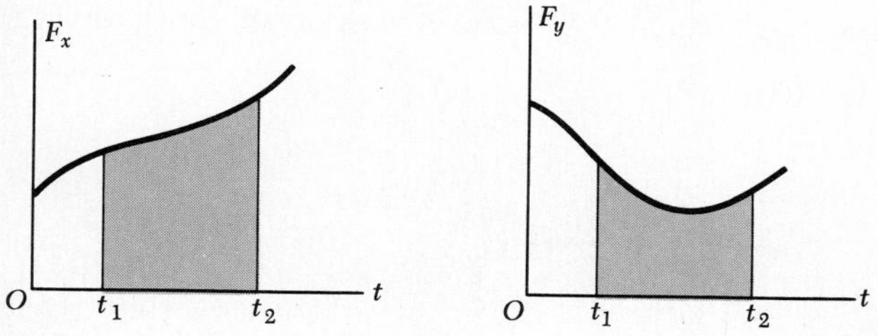

Fig. 14.1

Equation (14.4) expresses that, when a particle is acted upon by a force **F** during a given time interval, *the final momentum* $m\mathbf{v}_2$ *of the particle may be obtained by adding vectorially its initial momentum* $m\mathbf{v}_1$ *and the impulse of the force* **F** *during the time interval considered* (Fig. 14.2). We write

$$m\mathbf{v}_1 + \mathbf{Imp}_{1\to2} = m\mathbf{v}_2 \tag{14.5}$$

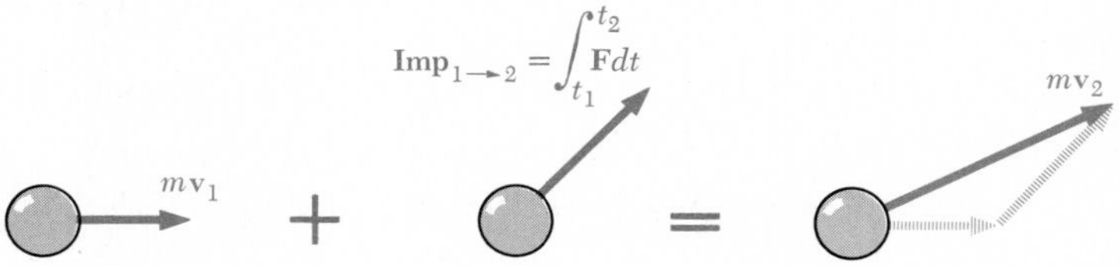

Fig. 14.2

We note that, while kinetic energy and work are scalar quantities, momentum and impulse are vector quantities. To obtain an analytic solution, it is thus necessary to replace Eq. (14.5) by the two corresponding component equations (14.3).

When several forces act on a particle, the impulse of each of the forces must be considered. We have

$$m\mathbf{v}_1 + \Sigma\,\mathbf{Imp}_{1\to2} = m\mathbf{v}_2 \tag{14.6}$$

Again, the equation obtained represents a relation between vector quantities; in the actual solution of a problem, it should be replaced by the two corresponding component equations.

14.2. Impulsive Motion. In some problems, a very large force may act during a very short time interval on a particle and produce a definite change in momentum. Such a force is called an *impulsive force* and the resulting motion an *impulsive motion*. For example, when a baseball is struck, the contact between bat and ball takes place during a very short time interval Δt. But the average value of the force **F** exerted by the bat on the ball is very large, and the resulting impulse $\mathbf{F}\,\Delta t$ is large enough to change the sense of motion of the ball (Fig. 14.3).

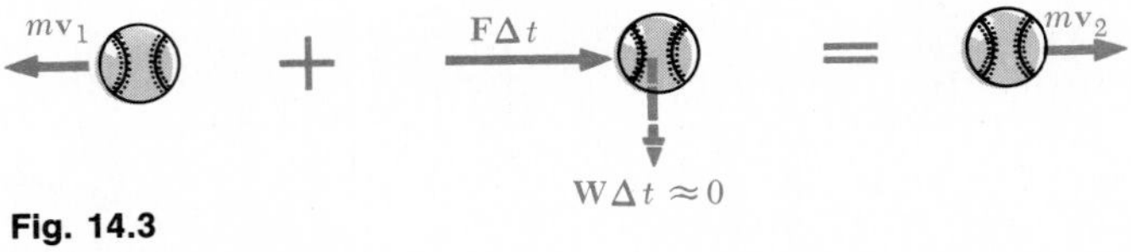

Fig. 14.3

When impulsive forces act on a particle, Eq. (14.6) becomes

$$m\mathbf{v}_1 + \Sigma\mathbf{F}\,\Delta t = m\mathbf{v}_2 \tag{14.7}$$

Any force which is not an impulsive force may be neglected, since the corresponding impulse $\mathbf{F}\,\Delta t$ is very small. *Nonimpulsive forces* include the weight of the body, the force exerted by a spring, or any other force which is *known* to be small compared with an impulsive force. Unknown reactions may or may not be impulsive; their impulse should therefore be included in Eq. (14.7) as long as it has not been proved negligible. The impulse of the weight of the baseball considered above, for example, may be neglected. If the motion of the bat is analyzed, the impulse of the weight of the bat may also be neglected. The impulses of the reactions of the player's hands on the bat, however, should be included; these impulses will not be negligible if the ball is incorrectly hit.

14.3. Systems of Particles. When a problem involves the motion of several particles, each particle may be considered separately and Eq. (14.6) may be written for each particle. We may also add vectorially the momenta of all the particles and the impulses of all the forces involved. We write then

$$\Sigma m\mathbf{v}_1 + \Sigma\,\mathbf{Imp}_{1\to 2} = \Sigma m\mathbf{v}_2$$

But since the internal forces occur by pairs of equal and opposite forces having the same line of action, their sum is zero and we need consider only the impulses of the external forces. We have

$$\Sigma m\mathbf{v}_1 + \Sigma\,\mathbf{Ext\ Imp}_{1\to 2} = \Sigma m\mathbf{v}_2 \tag{14.8}$$

Formula (14.8) may be written in a modified form if the mass center of the system of particles is considered. From Sec. 12.5, we recall that the mass center of the system is the point G of coordinates $\bar{x}$ and $\bar{y}$ defined by the equations

$$(\Sigma m)\bar{x} = \Sigma mx \qquad (\Sigma m)\bar{y} = \Sigma my \tag{14.9}$$

Differentiating (14.9) with respect to t, we have

$$(\Sigma m)\bar{v}_x = \Sigma mv_x \qquad (\Sigma m)\bar{v}_y = \Sigma mv_y \tag{14.10}$$

Writing (14.8) in terms of x and y components, and substituting for Σmv_x and Σmv_y from (14.10), we obtain

$$\begin{aligned}(\Sigma m)(\bar{v}_x)_1 + \Sigma\int_{t_1}^{t_2}(F_x)_{\text{ext}}\,dt &= (\Sigma m)(\bar{v}_x)_2\\ (\Sigma m)(\bar{v}_y)_1 + \Sigma\int_{t_1}^{t_2}(F_y)_{\text{ext}}\,dt &= (\Sigma m)(\bar{v}_y)_2\end{aligned} \tag{14.11}$$

We note that Eqs. (14.11) are identical with the equations we would obtain for a particle of mass Σm acted upon by all the external forces. We thus check that the mass center of a system of particles moves as if the entire mass of the system and all the external forces were concentrated at that point.

The equations derived in this section relate only the components of the impulses and momenta of a system of particles. By considering the *moments* as well as the *components* of the vectors involved, we shall obtain in Sec. 14.12 a more meaningful formulation of the principle of impulse and momentum for a system of particles.

14.4. Conservation of Momentum. We shall consider now the motion of a system of particles when the sum of the impulses of the external forces is zero. Equation (14.8) reduces then to

$$\Sigma m\mathbf{v}_1 = \Sigma m\mathbf{v}_2 \tag{14.12}$$

Thus, *when the sum of the impulses of the external forces acting on a system of particles is zero, the total momentum of the system remains constant.*

Introducing again the mass center G of the system, and using formulas (14.10), Eq. (14.12) reduces to

$$\overline{\mathbf{v}}_1 = \overline{\mathbf{v}}_2 \tag{14.13}$$

Thus, when the sum of the impulses of the external forces acting on a system of particles is zero, *the mass center of the system moves with a constant velocity.*

Two distinct cases of conservation of momentum are frequently encountered:

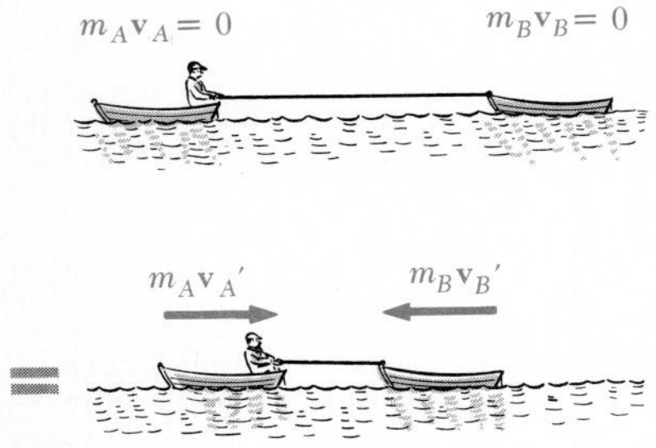

Fig. 14.4

1. *The external forces acting on the system during the interval of time considered are balanced.* No matter how long the time interval is, we have Σ **Ext Imp**$_{1\rightarrow 2} = 0$ and formula (14.12) applies. Consider, for example, two boats, of mass m_A and m_B, initially at rest, which are being pulled together (Fig. 14.4). If the resistance of the water is neglected, the only external forces acting on the boats are their weights and the buoyant forces exerted on them. Since these forces are balanced, we write

$$\Sigma m\mathbf{v}_1 = \Sigma m\mathbf{v}_2$$
$$0 = m_A\mathbf{v}_A' + m_B\mathbf{v}_B'$$

where $\mathbf{v}_A'$ and $\mathbf{v}_B'$ represent the velocities of the boats after a finite interval of time. The equation obtained indicates that the boats move in opposite directions (toward each other) with velocities inversely proportional to their masses. We also note that the mass center of the two boats, which was initially at rest, remains in the same position.

2. *The interval of time considered is very short, and all the external forces are nonimpulsive.* Again we have Σ **Ext Imp**$_{1\to2} = 0$, and formula (14.12) applies. Consider, for example, a bullet of mass m_A fired with a velocity $\mathbf{v}_A$ into a wooden sphere of mass m_B suspended from an inextensible wire and initially at rest (Fig. 14.5). The bullet penetrates the sphere and imparts to it a velocity $\mathbf{v}'$, which we propose to determine. After the sphere has been hit, the tension in the wire becomes zero and the combined weight of the sphere and bullet is unbalanced. But the impulse of the weight may be neglected since the time interval is very short. We write, therefore,

$$\Sigma m\mathbf{v}_1 = \Sigma m\mathbf{v}_2$$
$$m_A\mathbf{v}_A + 0 = (m_A + m_B)\mathbf{v}'$$

The equation obtained may be solved for $\mathbf{v}'$; we note that $\mathbf{v}'$ will have the same direction as $\mathbf{v}_A$.

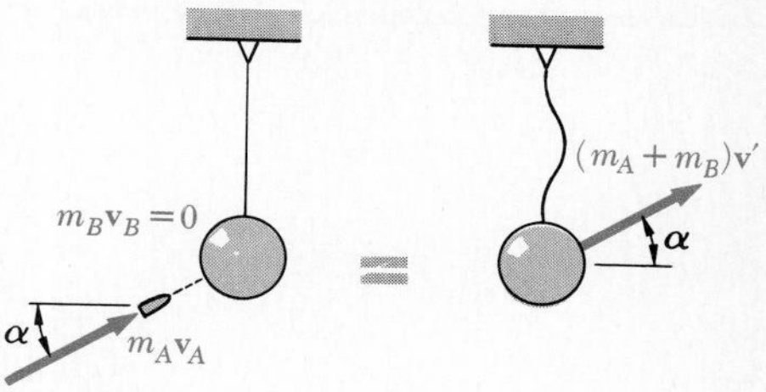

Fig. 14.5

Let us now consider the case when the bullet is fired downward into the sphere (Fig. 14.6). Since the wire is inextensible, it will prevent any downward motion of the sphere and exert on it a reaction **P.** This reaction is unknown and should therefore be assumed impulsive until proved to the contrary. Writing the general formula (14.8) in terms of horizontal and vertical components, and observing again that the impulse of the weight is negligible, we have

$$\Sigma m\mathbf{v}_1 + \Sigma\ \mathbf{Ext\ Imp}_{1\to2} = \Sigma m\mathbf{v}_2$$

$\xrightarrow{+}$ x components: $\quad m_Av_A\cos\alpha + 0 = (m_A + m_B)v'$

$+\uparrow y$ components: $\quad -m_Av_A\sin\alpha + P\,\Delta t = 0$

The first equation expresses that *the x component of the momentum is conserved;* it may be used to determine v'. The second equation indicates that the *y component of the linear momentum is not conserved;* this equation may be used to determine the magnitude $P\,\Delta t$ of the impulse of the force exerted by the wire. Thus, the momentum of the system considered is not conserved, and Eq. (14.12) does not hold, except in the x direction.

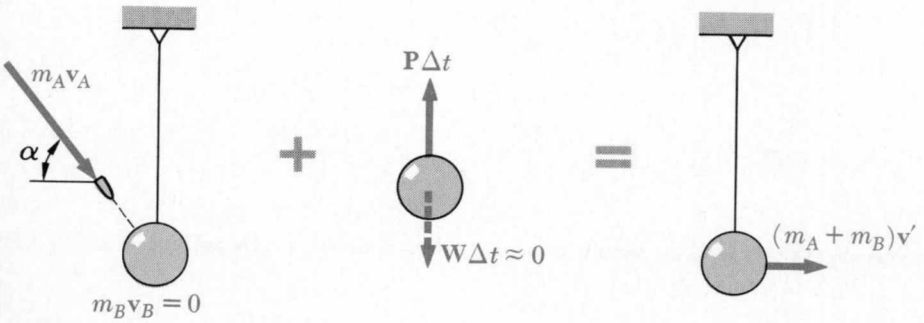

Fig. 14.6

SAMPLE PROBLEM 14.1

An automobile weighing 4000 lb is driven down a 5° incline at a speed of 60 mi/h when the brakes are applied, causing a constant total braking force (applied by the road on the tires) of 1500 lb. Determine the time required for the automobile to come to a stop.

Solution. We apply the principle of impulse and momentum. Since each force is constant in magnitude and direction, each corresponding impulse is equal to the product of the force and of the time interval t.

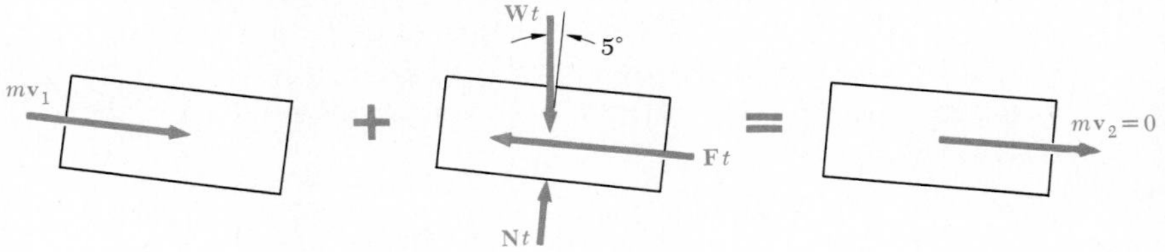

$$m\mathbf{v}_1 + \Sigma\,\mathbf{Imp}_{1\to 2} = m\mathbf{v}_2$$

$+\searrow x$ components: $\quad mv_1 + (W\sin 5°)t - Ft = 0$

$$(4000/32.2)(88\text{ ft/s}) + (4000\sin 5°)t - 1500t = 0$$

$$t = 9.49\text{ s} \quad ◄$$

SAMPLE PROBLEM 14.2

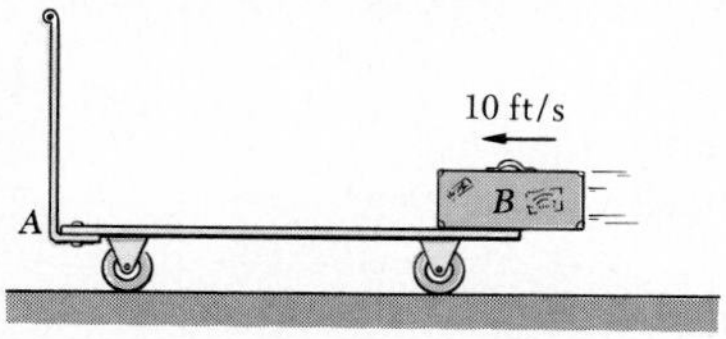

An airline employee tosses a 30-lb suitcase with a horizontal velocity of 10 ft/s onto a 70-lb baggage carrier. Knowing that the carrier can roll freely and is initially at rest, determine the velocity of the carrier after the suitcase has slid to a relative stop on the carrier.

Solution. We apply the principle of impulse and momentum to the carrier-suitcase system. Since the impulses of the internal forces cancel out, and since there are no horizontal external forces, the total momentum of the carrier and suitcase is conserved.

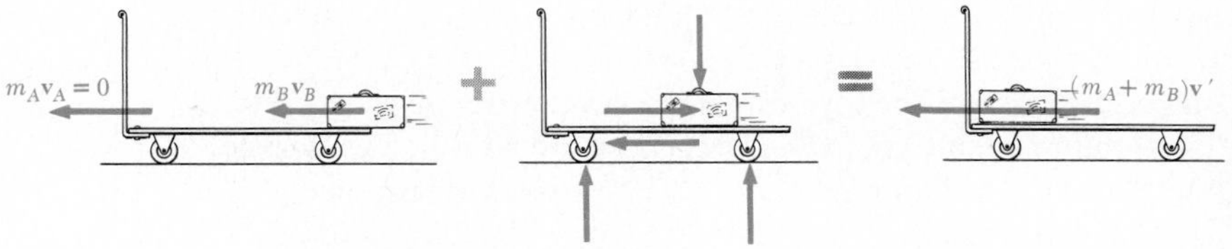

$$m_A\mathbf{v}_A + m_B\mathbf{v}_B = (m_A + m_B)\mathbf{v}'$$

$\overset{+}{\leftarrow} x$ components: $\quad 0 + \frac{30}{g}(10\text{ ft/s}) = \frac{70 + 30}{g}v'$

$$v' = 3\text{ ft/s} \quad ◄$$

SAMPLE PROBLEM 14.3

An old 2000-kg gun fires a 10-kg shell with an initial velocity of 600 m/s at an angle of 30°. The gun rests on a horizontal surface and is free to move horizontally. Assuming that the barrel of the gun is rigidly attached to the frame (no recoil mechanism) and that the shell leaves the barrel 6 ms after firing, determine the recoil velocity of the gun and the resultant **R** of the vertical impulsive forces exerted by the ground on the gun.

Solution. We apply the principle of impulse and momentum to the system consisting of the gun and the shell. Since the time interval $\Delta t = 6 \text{ ms} = 0.006 \text{ s}$ is very short, we neglect all nonimpulsive forces and consider only the impulse $\mathbf{R}\,\Delta t$.

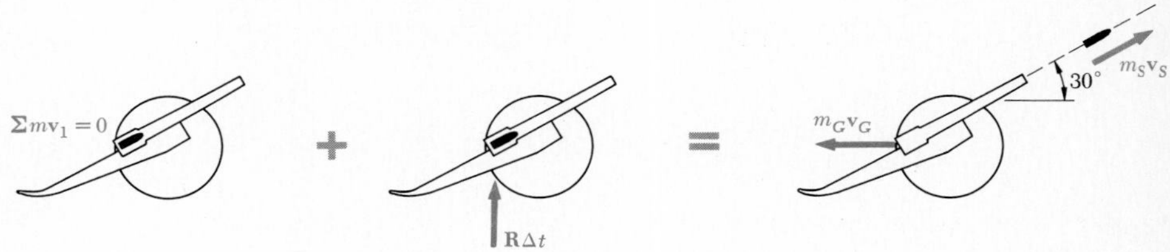

$$\Sigma m\mathbf{v}_1 + \Sigma \textbf{ Ext Imp}_{1\to 2} = \Sigma m\mathbf{v}_2$$

$\xrightarrow{+}$ x components:

$$0 + 0 = -m_G v_G + m_S v_S \cos 30^\circ$$
$$0 = -(2000 \text{ kg})v_G + (10 \text{ kg})(600 \text{ m/s}) \cos 30^\circ$$
$$v_G = +2.60 \text{ m/s}$$

$$\mathbf{v}_G = 2.60 \text{ m/s} \leftarrow \quad ◀$$

$+\uparrow y$ components:

$$0 + R\,\Delta t = m_S v_S \sin 30^\circ$$
$$R(0.006 \text{ s}) = (10 \text{ kg})(600 \text{ m/s}) \sin 30^\circ$$
$$R = +500\,000 \text{ N}$$

$$\mathbf{R} = 500 \text{ kN} \uparrow \quad ◀$$

The total force exerted by the ground on the gun during the time interval Δt should include the static reaction due to the weight of the gun as well as the impulsive reaction **R**. Since the weight of the gun is $(2000 \text{ kg})(9.81 \text{ m/s}^2) = 19.62 \text{ kN}$, the magnitude of this total force is $500 \text{ kN} + 19.62 \text{ kN} = 520 \text{ kN}$.

The high value obtained for the magnitude of **R** clearly indicates the need in modern guns for a recoil mechanism which will allow the barrel to move and will bring it to rest over a period of time substantially longer than Δt. Although the total vertical impulse must remain constant, the longer time interval will result in a smaller value for the magnitude of the reaction **R**.

PROBLEMS

14.1 A 3000-lb automobile is moving at a speed of 45 mi/h when the brakes are fully applied, causing all four wheels to skid. Determine the time required to stop the automobile (*a*) on concrete ($\mu = 0.75$), (*b*) on ice ($\mu = 0.08$).

14.2 A 60,000-ton ocean liner has an initial velocity of 2 mi/h. Neglecting the frictional resistance of the water, determine the time required to bring the liner to rest by using a single tugboat which exerts a constant force of 50,000 lb.

14.3 and 14.4 The initial velocity of the 50-kg car is 5 m/s to the left. Determine the time t at which the car has (*a*) no velocity, (*b*) a velocity of 5 m/s to the right.

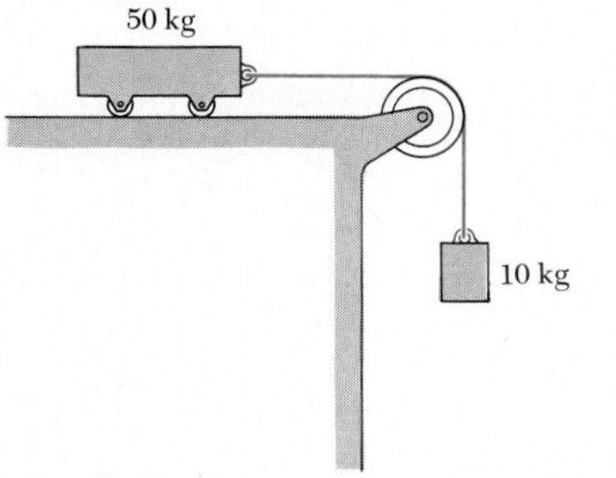

Fig. P14.3

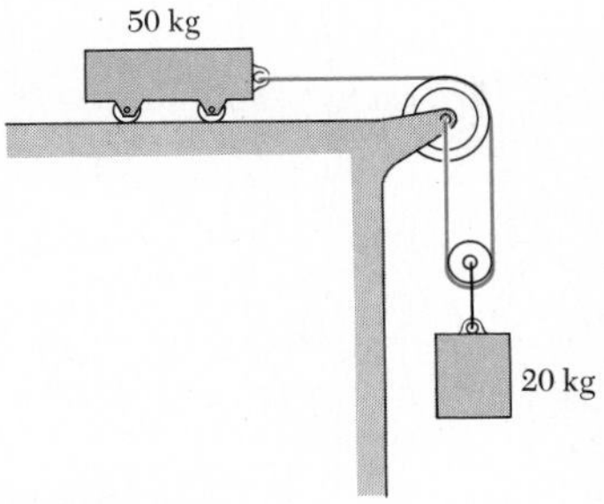

Fig. P14.4

14.5 Using the principle of impulse and momentum, solve Prob. 12.17*b*.

14.6 Using the principle of impulse and momentum, solve Prob. 12.16*b*.

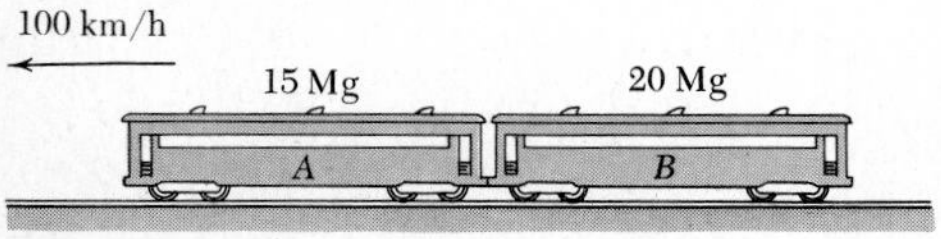

Fig. P14.7

14.7 A light train made of two cars travels at 100 km/h. The mass of car A is 15 Mg, and the mass of car B is 20 Mg. When the brakes are applied, a constant braking force of 25 kN is applied to each car. Determine (*a*) the time required for the train to stop after the brakes are applied, (*b*) the force in the coupling between the cars while the train is slowing down.

14.8 Solve Prob. 14.7, assuming that a constant braking force of 25 kN is applied to car B but that the brakes on car A are not applied.

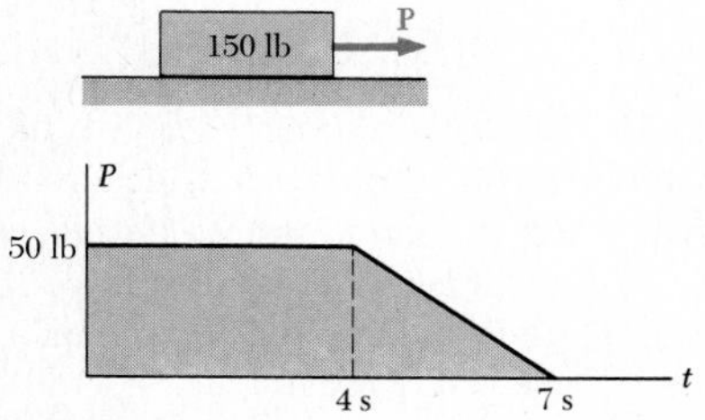

Fig. P14.9

14.9 A 150-lb block initially at rest is acted on by a force **P** which varies as shown. Knowing that $\mu = 0.20$, determine the velocity (*a*) at $t = 4$ s, (*b*) at $t = 7$ s.

14.10 In Prob. 14.9, determine (*a*) the maximum velocity reached by the block and the corresponding time, (*b*) the time at which the block comes to rest.

14.11 A 20-kg block is initially at rest and is subjected to a force **P** which varies as shown. Neglecting the effect of friction, determine (*a*) the maximum speed attained by the block, (*b*) the speed of the block at $t = 1.5$ s.

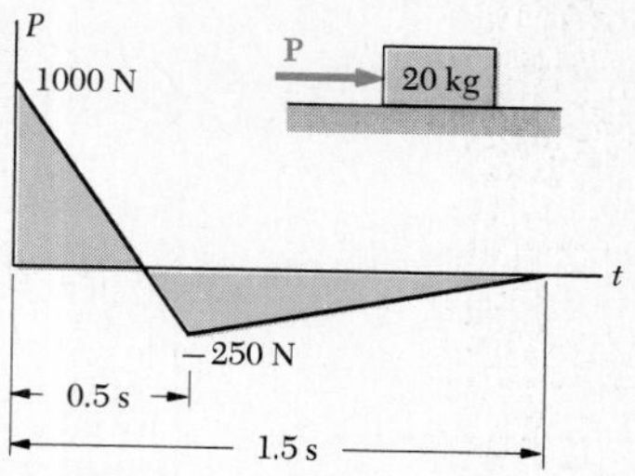

Fig. P14.11

14.12 Solve Prob. 14.11, assuming that $\mu = 0.25$ between the block and the surface.

14.13 A gun of mass 50 Mg is designed to fire a 250-kg shell with an initial velocity of 600 m/s. Determine the average force required to hold the gun motionless if the shell leaves the gun 0.02 s after being fired.

14.14 A 6000-kg plane lands on the deck of an aircraft carrier at a speed of 200 km/h relative to the carrier and is brought to a stop in 3.0 s. Determine the average horizontal force exerted by the carrier on the plane (*a*) if the carrier is at rest, (*b*) if the carrier is moving at a speed of 15 knots in the same direction as the airplane. (1 knot = 0.514 m/s.)

14.15 A 4-oz baseball is pitched with a velocity of 50 ft/s toward a batter. After the ball is hit by the bat *B*, it has a velocity of 150 ft/s in the direction shown. If the bat and ball are in contact 0.02 s, determine the average impulsive force exerted on the ball during the impact.

73.3 lb @ 34.2°

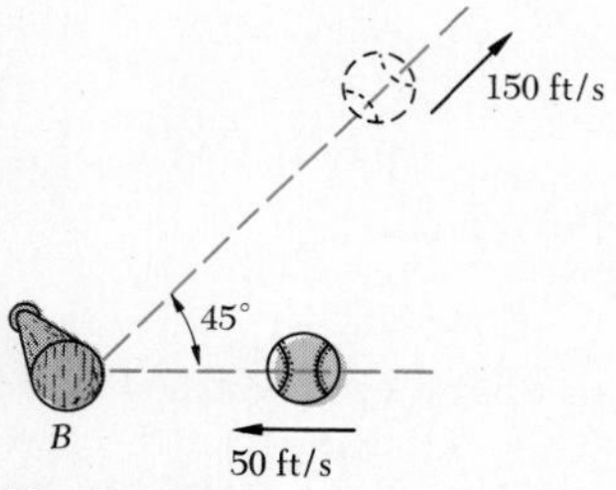

Fig. P14.15

14.16 A 175-lb man dives off the end of a pier with an initial velocity of 8 ft/s in the direction shown. Determine the horizontal and vertical components of the average force exerted on the pier during the 0.75 s that the man takes to leave the pier.

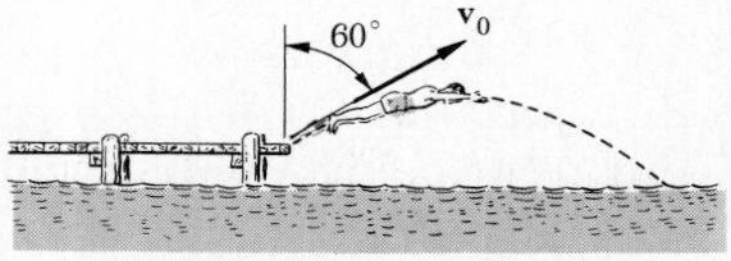

Fig. P14.16

14.17 Determine the initial recoil velocity of an 8-lb rifle which fires a $\frac{3}{4}$-oz bullet with a velocity of 1800 ft/s.

14.18 A 2-oz rifle bullet is fired horizontally with a velocity of 1200 ft/s into an 8-lb block of wood which can move freely in the horizontal direction. Determine the final velocity of the block.

14.19 Two men dive horizontally and to the right off the end of a 140-kg boat. The boat is initially at rest, and each man weighs 70 kg. If each man dives so that his relative horizontal velocity with respect to the boat is 4 m/s, determine (*a*) the velocity of the boat after the men dive simultaneously, (*b*) the velocity of the boat after one man dives and the velocity of the boat after the second man dives.

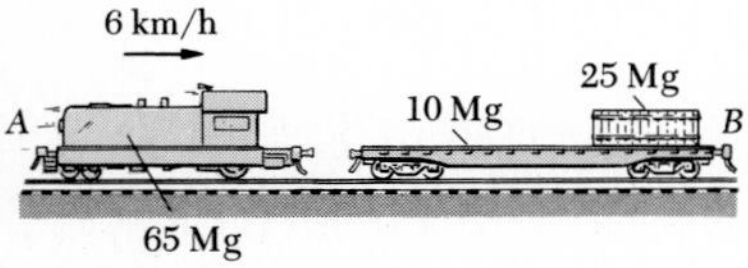

Fig. P14.20

14.20 A 65-Mg engine coasting at 6 km/h strikes, and is automatically coupled with, a 10-Mg flat car which carries a 25-Mg load. The load is *not* securely fastened to the car but may slide along the floor ($\mu = 0.20$). Knowing that the car was at rest with its brakes released and that the coupling takes place instantaneously, determine the velocity of the engine (*a*) immediately after the coupling, (*b*) after the load has slid to a stop relative to the car.

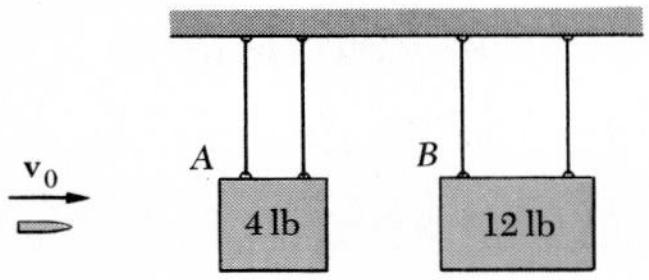

Fig. P14.21

14.21 A $\frac{3}{4}$-oz bullet is fired in a horizontal direction through block *A* and becomes embedded in block *B*. The bullet causes *A* and *B* to start moving with velocities of 7 and 5 ft/s, respectively. Determine (*a*) the initial velocity $\mathbf{v}_0$ of the bullet, (*b*) the velocity of the bullet as it travels from block *A* to block *B*.

Fig. P14.22

14.22 A 30-ton railroad car is to be coupled to a second car which weighs 50 tons. If initially the speed of the 30-ton car is 1 mi/h and the 50-ton car is at rest, determine (*a*) the final speed of the coupled cars, (*b*) the average impulsive force acting on each car if the coupling is completed in 0.40 s.

14.23 Solve Prob. 14.22, assuming that, initially, the 30-ton car is at rest and the 50-ton car has a speed of 1 mi/h.

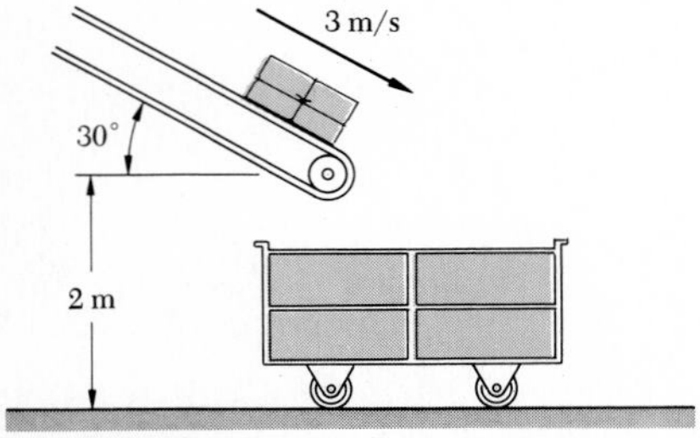

Fig. P14.24

14.24 A 10-kg package is discharged from a conveyor belt with a velocity of 3 m/s and lands in a 25-kg cart. Knowing that the cart is initially at rest and may roll freely, determine the final velocity of the cart.

14.25 Solve Prob. 14.24, assuming that the single 10-kg package is replaced by two 5-kg packages. The first 5-kg package comes to relative rest in the cart before the second package strikes the cart.

*****14.26** A machine part is forged in a small drop forge. The hammer weighs 400 lb and is dropped from a height of 5 ft. Determine the initial impulse exerted on the machine part, assuming that the 1000-lb anvil (*a*) is resting directly on hard ground, (*b*) is supported by springs.

14.5. Impact. A collision between two bodies which occurs in a very small interval of time, and during which the two bodies exert on each other relatively large forces, is called an *impact*. The common normal to the surfaces in contact during the impact is called the *line of impact*. If the mass centers of the two colliding bodies are located on this line, the impact is a *central impact*. Otherwise, the impact is said to be *eccentric*. We shall limit our present study to that of the central impact of two particles and postpone until later the analysis of the eccentric impact of two rigid bodies (Sec. 18.6).

If the velocities of the two particles are directed along the line of impact, the impact is said to be a *direct impact* (Fig. 14.7*a*). If, on the other hand, either or both particles move along a line other than the line of impact, the impact is said to be an *oblique impact* (Fig. 14.7*b*).

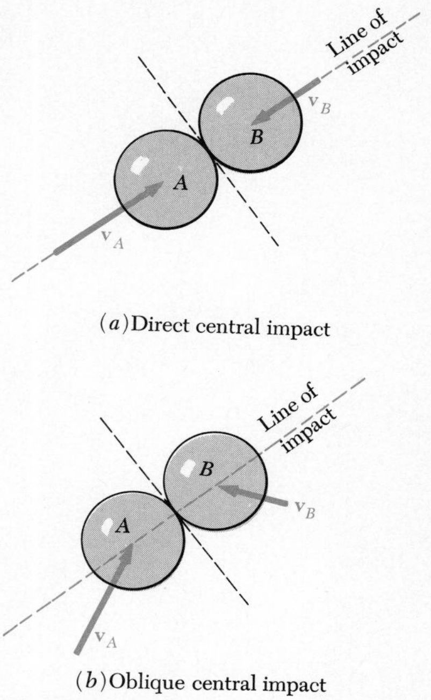

(*a*) Direct central impact

(*b*) Oblique central impact

Fig. 14.7

14.6. Direct Central Impact. Consider two particles A and B, of mass m_A and m_B, which are moving in the same straight line and to the right with known velocities $\mathbf{v}_A$ and $\mathbf{v}_B$ (Fig. 14.8*a*). If $\mathbf{v}_A$ is larger than $\mathbf{v}_B$, particle A will eventually strike particle B. Under the impact, the two particles will *deform* and, at the end of the period of deformation, they will have the same velocity $\mathbf{u}$ (Fig. 14.8*b*). A period of *restitution* will then take place, at the end of which, depending upon the magnitude of the impact forces and upon the materials involved, the two particles either will have regained their original shape or will stay permanently deformed. Our purpose here is to determine the velocities $\mathbf{v}'_A$ and $\mathbf{v}'_B$ of the particles at the end of the period of restitution (Fig. 14.8*c*).

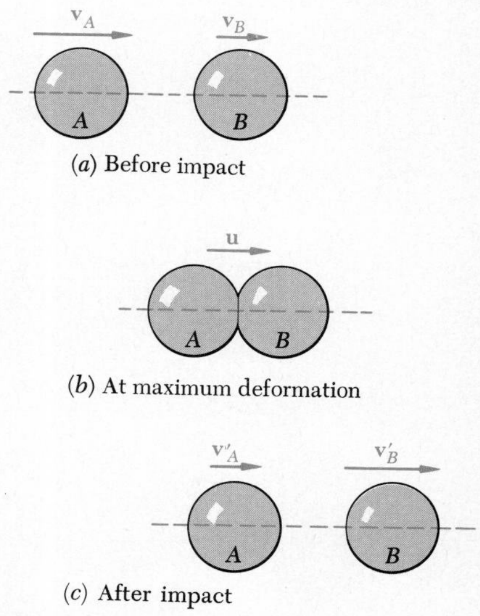

(*a*) Before impact

(*b*) At maximum deformation

(*c*) After impact

Fig. 14.8

Considering first the system of the two particles as a whole, we note that the only impulsive forces acting during the impact are internal forces. Thus, *the total momentum of the system is conserved*, and we write

$$m_A\mathbf{v}_A + m_B\mathbf{v}_B = m_A\mathbf{v}'_A + m_B\mathbf{v}'_B \tag{14.14}$$

Since all the velocities considered are directed along the same axis, we may replace the equation obtained by the following relation involving only scalar components:

$$m_A v_A + m_B v_B = m_A v'_A + m_B v'_B \tag{14.15}$$

A positive value for any of the scalar quantities v_A, v_B, v'_A, or v'_B means that the corresponding vector is directed to the right; a negative value indicates that the corresponding vector is directed to the left.

To obtain the velocities $\mathbf{v}'_A$ and $\mathbf{v}'_B$, it is necessary to establish a second relation between the scalars v'_A and v'_B. For this purpose, we shall consider now the motion of particle A during the period of deformation and apply the principle of impulse and momentum. Since the only impulsive force acting on A during this period is the force $\mathbf{P}$ exerted by B (Fig. 14.9*a*), we write, using again scalar components,

$$m_A v_A - \int P\,dt = m_A u \qquad (14.16)$$

where the integral extends over the period of deformation.

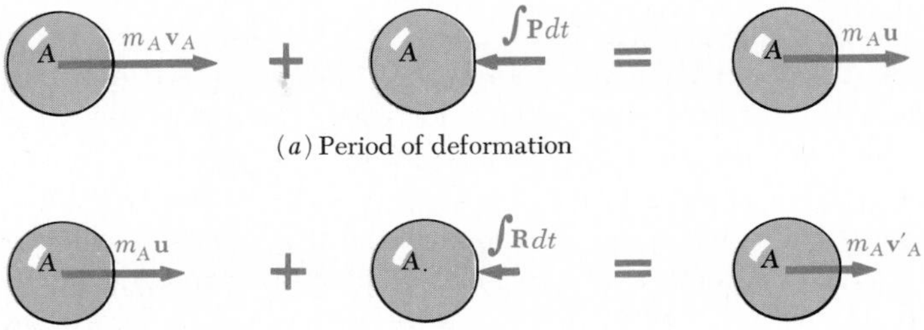

(*a*) Period of deformation

(*b*) Period of restitution

Fig. 14.9

Considering now the motion of A during the period of restitution, and denoting by $\mathbf{R}$ the force exerted by B on A during this period (Fig. 14.9*b*), we write

$$m_A u - \int R\,dt = m_A v'_A \qquad (14.17)$$

where the integral extends over the period of restitution.

In general, the force $\mathbf{R}$ exerted on A during the period of restitution differs from the force $\mathbf{P}$ exerted during the period of deformation, and the magnitude $\int R\,dt$ of its impulse is smaller than the magnitude $\int P\,dt$ of the impulse of $\mathbf{P}$. The ratio of the magnitudes of the impulses corresponding respectively to the period of restitution and to the period of deformation is called the *coefficient of restitution* and is denoted by e. We write

$$e = \frac{\int R\,dt}{\int P\,dt} \qquad (14.18)$$

The value of the coefficient e is always between 0 and 1 and depends to a large extent on the two materials involved. However, it also varies considerably with the impact velocity and the shape and size of the two colliding bodies.

Solving Eqs. (14.16) and (14.17) for the two impulses and substituting into (14.18), we write

$$e = \frac{u - v'_A}{v_A - u} \tag{14.19}$$

A similar analysis of particle B leads to the relation

$$e = \frac{v'_B - u}{u - v_B} \tag{14.20}$$

Since the quotients in (14.19) and (14.20) are equal, they are also equal to the quotient obtained by adding, respectively, their numerators and their denominators. We have, therefore,

$$e = \frac{(u - v'_A) + (v'_B - u)}{(v_A - u) + (u - v_B)} = \frac{v'_B - v'_A}{v_A - v_B}$$

and

$$v'_B - v'_A = e(v_A - v_B) \tag{14.21}$$

Since $v'_B - v'_A$ represents the relative velocity of the two particles after impact and $v_A - v_B$ their relative velocity before impact, formula (14.21) expresses that *the relative velocity of the two particles after impact may be obtained by multiplying their relative velocity before impact by the coefficient of restitution.* This property is used to determine experimentally the value of the coefficient of restitution of two given materials.

The velocities of the two particles after impact may now be obtained by solving Eqs. (14.15) and (14.21) simultaneously for v'_A and v'_B, i.e., by using the principle of conservation of momentum and the concept of coefficient of restitution. It is recalled that the derivation of Eqs. (14.15) and (14.21) was based on the assumption that particle B is located to the right of A, and that both particles are initially moving to the right. If particle B is initially moving to the left, the scalar v_B should be considered negative. The same sign convention holds for the velocities after impact: a positive sign for v'_A will indicate that particle A moves to the right after impact, and a negative sign that it moves to the left.

Two particular cases of impact are of special interest:

1. $e = 0$, *Perfectly Plastic Impact.* When $e = 0$, Eq. (14.21) yields $v'_B = v'_A$. There is no period of restitution, and both particles stay together after impact. Substituting $v'_B = v'_A = v'$ into Eq. (14.15), which expresses that the momentum of the system is conserved, we write

$$m_A v_A + m_B v_B = (m_A + m_B)v' \tag{14.22}$$

This equation may be solved for the common velocity v' of the two particles after impact.

2. $e = 1$, *Perfectly Elastic Impact.* When $e = 1$, Eq. (14.21) reduces to

$$v'_B - v'_A = v_A - v_B \tag{14.23}$$

which expresses that the relative velocities before and after impact are equal. The impulses received by each particle during the period of deformation and during the period of restitution are equal. The particles move away from each other after impact with the same velocity with which they approached each other before impact. The velocities v'_A and v'_B may be obtained by solving Eqs. (14.15) and (14.23) simultaneously.

It is worth noting that, *in the case of a perfectly elastic impact, the energy of the system,* as well as its momentum, *is conserved.* Equations (14.15) and (14.23) may be written as follows:

$$m_A(v_A - v'_A) = m_B(v'_B - v_B) \tag{14.15'}$$

$$v_A + v'_A = v_B + v'_B \tag{14.23'}$$

Multiplying (14.15′) and (14.23′) member by member, we have

$$m_A(v_A - v'_A)(v_A + v'_A) = m_B(v'_B - v_B)(v'_B + v_B)$$
$$m_A v_A^2 - m_A(v'_A)^2 = m_B(v'_B)^2 - m_B v_B^2$$

Rearranging the terms in the equation obtained, and multiplying by $\frac{1}{2}$, we write

$$\tfrac{1}{2}m_A v_A^2 + \tfrac{1}{2}m_B v_B^2 = \tfrac{1}{2}m_A(v'_A)^2 + \tfrac{1}{2}m_B(v'_B)^2$$

which expresses that the kinetic energy of the system is conserved. It should be noted, however, that *in the general case of impact,* i.e., when e is not equal to 1, *the energy of the system is not conserved.* This may be shown in any given case by comparing the kinetic energies before and after impact. The lost kinetic energy is in part transformed into heat and in part spent in generating elastic waves within the two colliding bodies.

14.7. Oblique Central Impact. Let us now consider the case when the velocities of the two colliding particles are *not* directed along the line of impact (Fig. 14.10). As indicated in Sec. 14.5, the impact is said to be *oblique.* Since the velocities $\mathbf{v}'_A$ and $\mathbf{v}'_B$ of the particles after impact are unknown in direction as well as in magnitude, their determination will require the use of four independent equations.

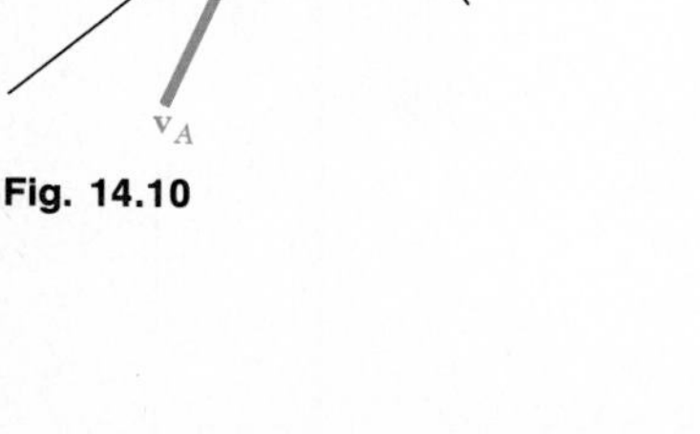

Fig. 14.10

We choose x and y axes, respectively, along the line of impact and along the common tangent to the surfaces in contact. Assuming that the particles are perfectly *smooth and frictionless,* we observe that the only impulsive forces acting on the particles during the impact are internal forces directed along the x axis. We may therefore express that:

1. The y component of the momentum of particle A is conserved.
2. The y component of the momentum of particle B is conserved.
3. The x component of the total momentum of the particles is conserved.
4. The x component of the relative velocity of the two particles after impact is obtained by multiplying the x component of their relative velocity before impact by the coefficient of restitution.

We thus obtain four independent equations which may be solved for the components of the velocities of A and B after impact. This method of solution is illustrated in Sample Prob. 14.6.

14.8. Problems Involving Energy and Momentum. We have now at our disposal three different methods for the solution of kinetics problems: the direct application of Newton's second law, $\Sigma\mathbf{F} = m\mathbf{a}$, the method of work and energy, and the method of impulse and momentum. To derive maximum benefit from these three methods, we should be able to choose the method best suited for the solution of a given problem. We should also be prepared to use different methods for solving the various parts of a problem when such a procedure seems advisable.

We have already seen that the method of work and energy is in many cases more expeditious than the direct application of Newton's second law. As indicated in Sec. 13.4, however, the method of work and energy has limitations, and it must sometimes be supplemented by the use of $\Sigma\mathbf{F} = m\mathbf{a}$. This is the case, for example, when we wish to determine an acceleration or a normal force.

There is generally no great advantage in using the method of impulse and momentum for the solution of problems involving no impulsive forces. It will usually be found that the equation $\Sigma\mathbf{F} = m\mathbf{a}$ yields a solution just as fast and that the method of work and energy, if it applies, is more rapid and more convenient. However, the method of impulse and momentum is the only practicable method in problems of impact. A solution based on the direct application of $\Sigma\mathbf{F} = m\mathbf{a}$ would be unwieldy, and the method of work and energy cannot be used since impact (unless perfectly elastic) involves a loss of mechanical energy.

Many problems involve only conservative forces, except for a short impact phase during which impulsive forces act. The solution of such problems may be divided into several parts. While the part corresponding to the impact phase calls for the use of the method of impulse and momentum and of the relation between relative velocities, the other parts may usually be solved by the method of work and energy. The use of the equation $\Sigma \mathbf{F} = m\mathbf{a}$ will be necessary, however, if the problem involves the determination of a normal force.

Consider, for example, a pendulum A, of weight $\mathbf{W}_A$ and length l, which is released with no velocity from a position A_1 (Fig. 14.11*a*). The pendulum swings freely in a vertical plane and hits a second pendulum B, of weight $\mathbf{W}_B$ and same length l, which is initially at rest. After the impact (with coefficient of restitution e), pendulum B swings through an angle θ that we wish to determine.

The solution of the problem may be divided into three parts:

1. *Pendulum A Swings from A_1 to A_2.* The principle of conservation of energy may be used to determine the velocity $(\mathbf{v}_A)_2$ of the pendulum at A_2 (Fig. 14.11*b*).
2. *Pendulum A Hits Pendulum B.* Using the principle of conservation of momentum and the relation between the relative velocities of the two pendulums, we determine the velocities $(\mathbf{v}_A)_3$ and $(\mathbf{v}_B)_3$ of the pendulums after impact (Fig. 14.11*c*).
3. *Pendulum B Swings from B_3 to B_4.* Applying the principle of conservation of energy, we determine the maximum elevation y_4 reached by pendulum B (Fig. 14.11*d*). The angle θ may then be determined by trigonometry.

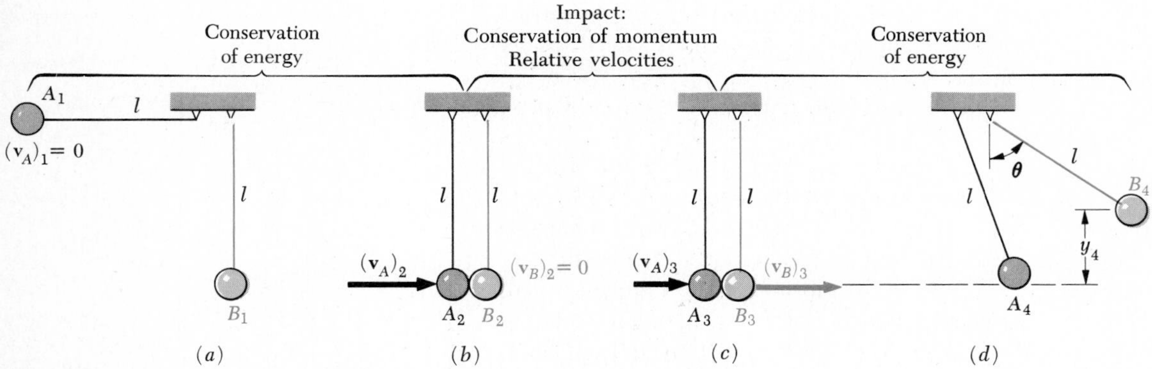

Fig. 14.11

We note that the method of solution just described should be supplemented by the use of $\Sigma \mathbf{F} = m\mathbf{a}$ if the tensions in the cords holding the pendulums are to be determined.

SAMPLE PROBLEM 14.4

A 20-Mg railroad car moving at a speed of 0.5 m/s to the right collides with a 35-Mg car which is at rest. If after the collision the 35-Mg car is observed to move to the right at a speed of 0.3 m/s, determine the coefficient of restitution between the two cars.

Solution. We consider the system consisting of the two cars and express that the total momentum is conserved.

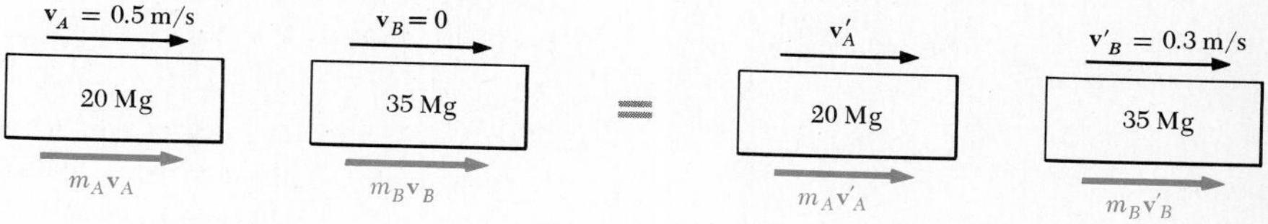

$$m_A \mathbf{v}_A + m_B \mathbf{v}_B = m_A \mathbf{v}'_A + m_B \mathbf{v}'_B$$

$$(20 \text{ Mg})(+0.5 \text{ m/s}) + (35 \text{ Mg})(0) = (20 \text{ Mg})v'_A + (35 \text{ Mg})(+0.3 \text{ m/s})$$

$$v'_A = -0.025 \text{ m/s} \qquad \mathbf{v}'_A = 0.025 \text{ m/s} \leftarrow$$

The coefficient of restitution is obtained by writing

$$e = \frac{v'_B - v'_A}{v_A - v_B} = \frac{+0.3 - (-0.025)}{+0.5 - 0} = \frac{0.325}{0.5} \qquad e = 0.65 \blacktriangleleft$$

SAMPLE PROBLEM 14.5

A ball is thrown against a frictionless vertical wall. Immediately before the ball strikes the wall, its velocity has a magnitude v and forms an angle of 30° with the horizontal. Knowing that $e = 0.90$, determine the magnitude and direction of the velocity of the ball as it rebounds from the wall.

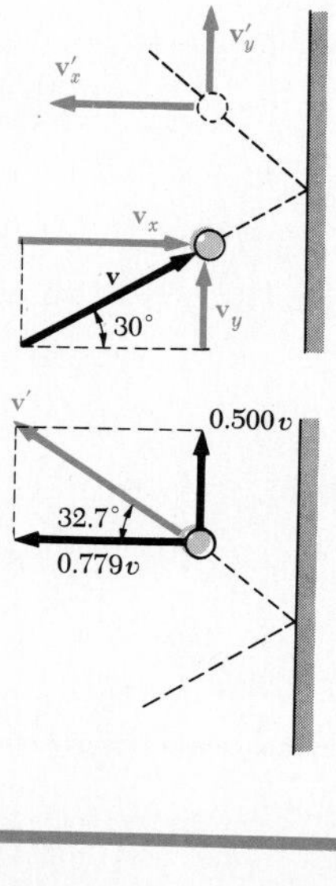

Solution. We resolve the initial velocity of the ball into components

$$v_x = v \cos 30° = 0.866v \qquad v_y = v \sin 30° = 0.500v$$

Vertical Motion. Since the wall is frictionless, no vertical impulsive force acts on the ball during the time it is in contact with the wall. The vertical component of the momentum, and hence the vertical component of the velocity, of the ball is thus unchanged:

$$\mathbf{v}'_y = \mathbf{v}_y = 0.500v \uparrow$$

Horizontal Motion. Since the mass of the wall (and earth) is essentially infinite, there is no point in expressing that the total momentum of the ball-wall system is conserved. Using the relation between relative velocities, we write

$$0 - v'_x = e(v_x - 0)$$

$$v'_x = -0.90(0.866v) = -0.779v \qquad \mathbf{v}'_x = 0.779v \leftarrow$$

Resultant Motion. Adding vectorially the components $\mathbf{v}'_x$ and $\mathbf{v}'_y$,

$$\mathbf{v}' = 0.926v \measuredangle 32.7° \blacktriangleleft$$

SAMPLE PROBLEM 14.6

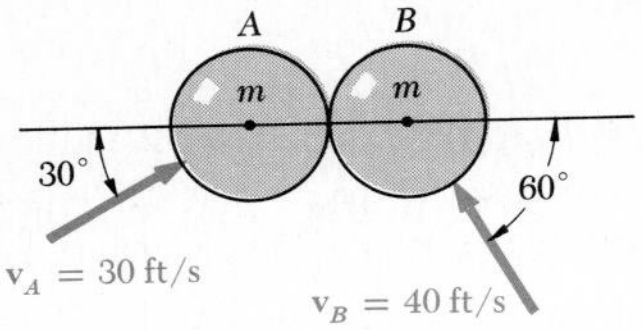

The magnitude and direction of the velocities of two identical frictionless balls before they strike each other are as shown. Assuming $e = 0.90$, determine the magnitude and direction of the velocity of each ball after the impact.

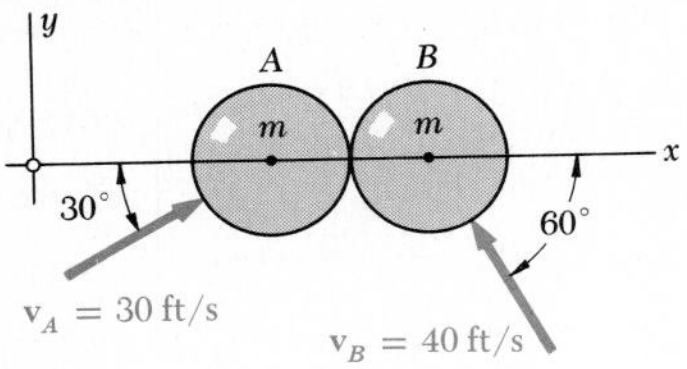

Solution. The impulsive forces acting between the balls during the impact are directed along a line joining the centers of the balls called the *line of impact*. Choosing x and y axes, respectively, parallel and perpendicular to the line of impact and directed as shown, we write

$$(v_A)_x = v_A \cos 30° = +26.0 \text{ ft/s}$$
$$(v_A)_y = v_A \sin 30° = +15.0 \text{ ft/s}$$
$$(v_B)_x = -v_B \cos 60° = -20.0 \text{ ft/s}$$
$$(v_B)_y = v_B \sin 60° = +34.6 \text{ ft/s}$$

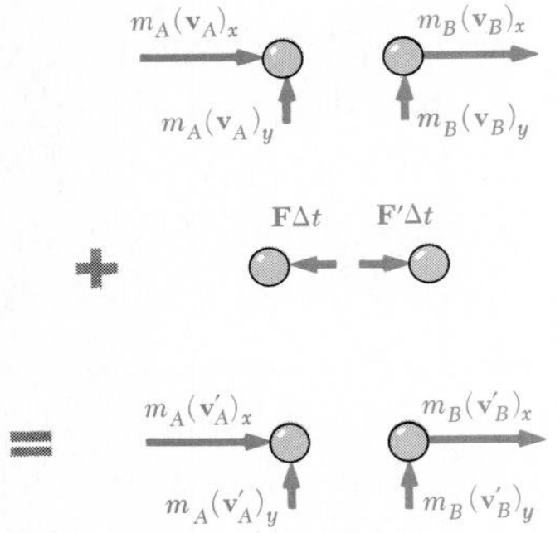

Principle of Impulse and Momentum. In the adjoining sketches we show in turn the initial momenta, the impulsive reactions, and the final momenta.

Motion Perpendicular to the Line of Impact. Considering only the y components, we apply the principle of impulse and momentum to each ball *separately*. Since no vertical impulsive force acts during the impact, the vertical component of the momentum, and hence the vertical component of the velocity, of each ball is unchanged.

$$(\mathbf{v}'_A)_y = 15.0 \text{ ft/s} \uparrow \qquad (\mathbf{v}'_B)_y = 34.6 \text{ ft/s} \uparrow$$

Motion Parallel to the Line of Impact. In the x direction, we consider the two balls as a *single* system and note that, by Newton's third law, the internal impulses $\mathbf{F}\,\Delta t$ and $\mathbf{F}'\,\Delta t$ cancel. Applying the principle of conservation of momentum, we write

$$m_A(v_A)_x + m_B(v_B)_x = m_A(v'_A)_x + m_B(v'_B)_x$$
$$m(26.0) + m(-20.0) = m(v'_A)_x + m(v'_B)_x$$
$$(v'_A)_x + (v'_B)_x = 6.0 \quad (1)$$

Using the relation between relative velocities, we write

$$(v'_B)_x - (v'_A)_x = e[(v_A)_x - (v_B)_x]$$
$$(v'_B)_x - (v'_A)_x = (0.90)[26.0 - (-20.0)]$$
$$(v'_B)_x - (v'_A)_x = 41.4 \quad (2)$$

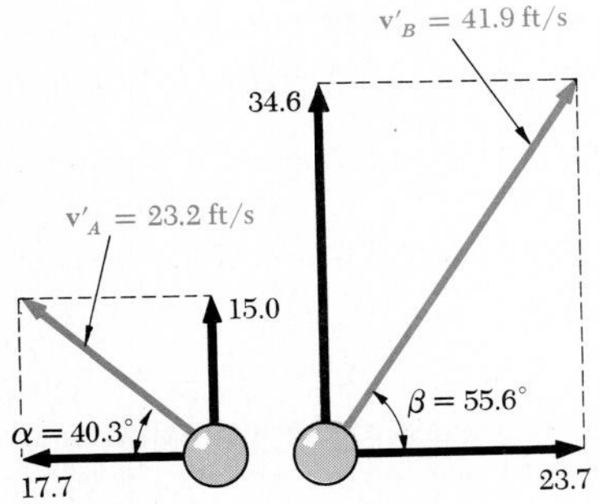

Solving Eqs. (1) and (2) simultaneously, we obtain

$$(v'_A)_x = -17.7 \qquad (v'_B)_x = +23.7$$
$$(\mathbf{v}'_A)_x = 17.7 \text{ ft/s} \leftarrow \qquad (\mathbf{v}'_B)_x = 23.7 \text{ ft/s} \rightarrow$$

Resultant Motion. Adding vectorially the velocity components of each ball, we obtain

$$\mathbf{v}'_A = 23.2 \text{ ft/s} \nwarrow 40.3° \qquad \mathbf{v}'_B = 41.9 \text{ ft/s} \nearrow 55.6° \blacktriangleleft$$

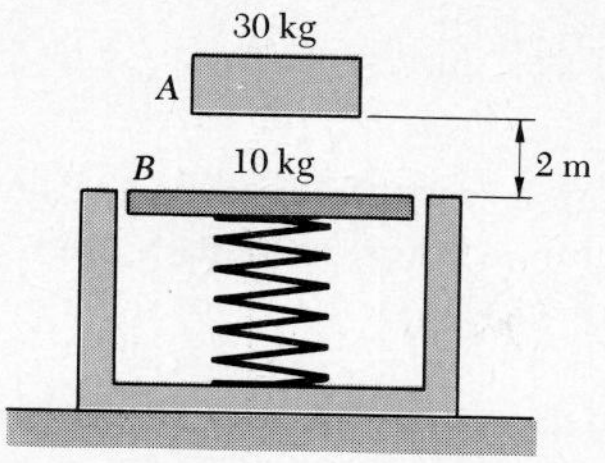

SAMPLE PROBLEM 14.7

A 30-kg block is dropped from a height of 2 m onto the 10-kg pan of a spring scale. Assuming the impact to be perfectly plastic, determine the maximum deflection of the pan. The constant of the spring is $k = 20$ kN/m.

Solution. The impact between the block and the pan *must* be treated separately; therefore we divide the solution into three parts.

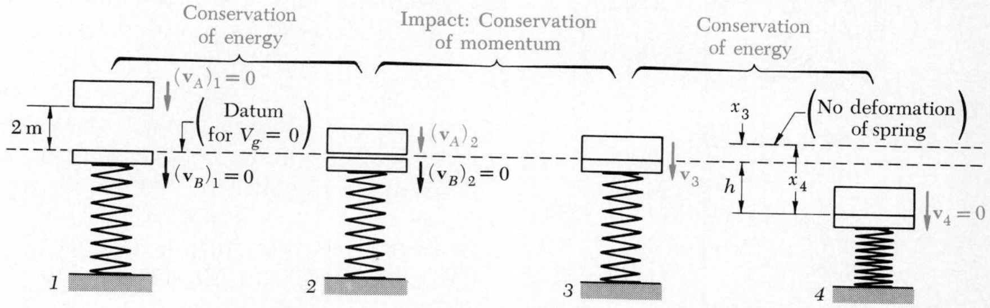

Conservation of Energy. Block: $W_A = (30 \text{ kg})(9.81 \text{ m/s}^2) = 294 \text{ N}$

$$T_1 = \tfrac{1}{2}m_A(v_A)_1^2 = 0 \qquad V_1 = W_A y = (294 \text{ N})(2 \text{ m}) = 588 \text{ J}$$

$$T_2 = \tfrac{1}{2}m_A(v_A)_2^2 = \tfrac{1}{2}(30 \text{ kg})(v_A)_2^2 \qquad V_2 = 0$$

$$T_1 + V_1 = T_2 + V_2: \qquad 0 + 588 \text{ J} = \tfrac{1}{2}(30 \text{ kg})(v_A)_2^2 + 0$$

$$(v_A)_2 = +6.26 \text{ m/s} \qquad (\mathbf{v}_A)_2 = 6.26 \text{ m/s} \downarrow$$

Impact: Conservation of Momentum. Since the impact is perfectly plastic, $e = 0$; the block and pan move together after the impact.

$$m_A(v_A)_2 + m_B(v_B)_2 = (m_A + m_B)v_3$$

$$(30 \text{ kg})(6.26 \text{ m/s}) + 0 = (30 \text{ kg} + 10 \text{ kg})v_3$$

$$v_3 = +4.70 \text{ m/s} \qquad \mathbf{v}_3 = 4.70 \text{ m/s} \downarrow$$

Conservation of Energy. Initially the spring supports the weight W_B of the pan; thus the initial deflection of the spring is

$$x_3 = \frac{W_B}{k} = \frac{(10 \text{ kg})(9.81 \text{ m/s}^2)}{20 \times 10^3 \text{ N/m}} = \frac{98.1 \text{ N}}{20 \times 10^3 \text{ N/m}} = 4.91 \times 10^{-3} \text{ m}$$

Denoting by x_4 the total maximum deflection of the spring, we write

$$T_3 = \tfrac{1}{2}(m_A + m_B)v_3^2 = \tfrac{1}{2}(30 \text{ kg} + 10 \text{ kg})(4.70 \text{ m/s})^2 = 442 \text{ J}$$

$$V_3 = V_g + V_e = 0 + \tfrac{1}{2}kx_3^2 = \tfrac{1}{2}(20 \times 10^3)(4.91 \times 10^{-3})^2 = 0.241 \text{ J}$$

$$T_4 = 0$$

$$V_4 = V_g + V_e = (W_A + W_B)(-h) + \tfrac{1}{2}kx_4^2 = -(392)h + \tfrac{1}{2}(20 \times 10^3)x_4^2$$

Noting that the displacement of the pan is $h = x_4 - x_3$, we write

$T_3 + V_3 = T_4 + V_4$:

$$442 + 0.241 = 0 - 392(x_4 - 4.91 \times 10^{-3}) + \tfrac{1}{2}(20 \times 10^3)x_4^2$$

$$x_4 = 0.230 \text{ m} \qquad h = x_4 - x_3 = 0.230 \text{ m} - 4.91 \times 10^{-3} \text{ m}$$

$$h = 0.225 \text{ m} \qquad h = 225 \text{ mm} \blacktriangleleft$$

PROBLEMS

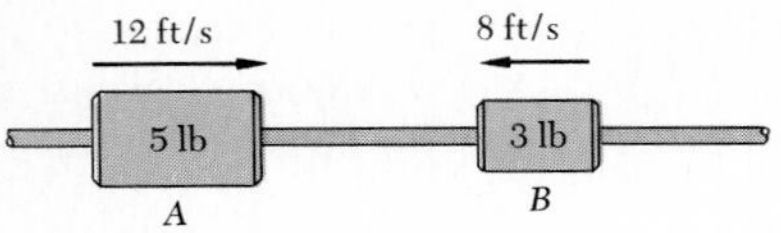

Fig. P14.27 and P14.28

14.27 The coefficient of restitution between the two collars is known to be 0.70; determine (*a*) their velocities after impact, (*b*) the energy loss during the impact.

14.28 Solve Prob. 14.27, assuming that the velocity of collar *B* is 8 ft/s to the right.

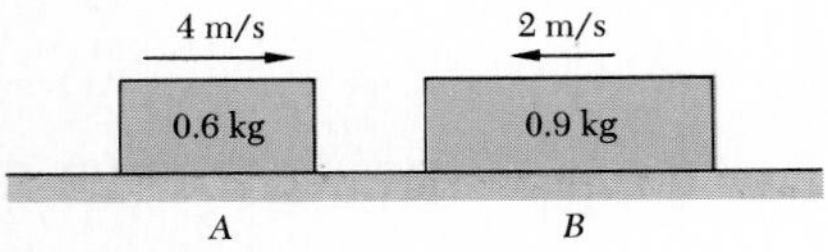

Fig. P14.29 and P14.30

14.29 Two steel blocks slide without friction on a horizontal surface; immediately before impact their velocities are as shown. Knowing that $e = 0.75$, determine (*a*) their velocities after impact, (*b*) the energy loss during impact.

14.30 The velocities of two steel blocks before impact are as shown. If after the impact the velocity of block *B* is observed to be 2.5 m/s to the right, determine the coefficient of restitution between the two blocks.

14.31 Two identical pucks *A* and *B*, of 80-mm diameter, may move freely on a hockey rink. Puck *B* is at rest and puck *A* has an initial velocity **v** as shown. (*a*) Knowing that $b = 40$ mm and $e = 0.80$, determine the velocity of each puck after the impact. (*b*) Show that if $e = 1$, the final velocities of the pucks form a right angle for all values of *b*.

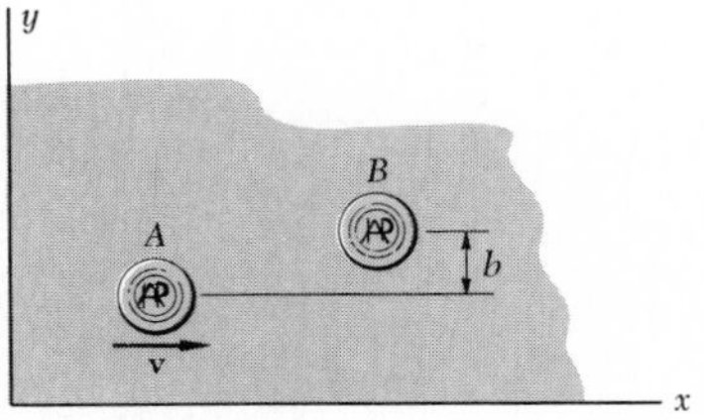

Fig. P14.31

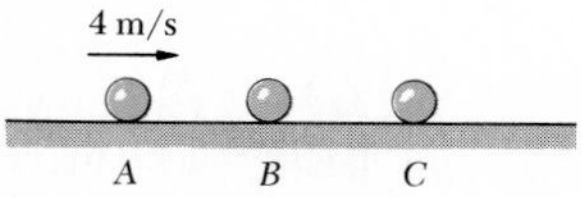

Fig. P14.32

14.32 Two identical balls *B* and *C* are at rest when ball *B* is struck by a ball *A* of the same mass, moving with a velocity of 4 m/s. This causes a series of collisions between the various balls. Knowing that $e = 0.40$, determine the velocity of each ball after *all* collisions have taken place.

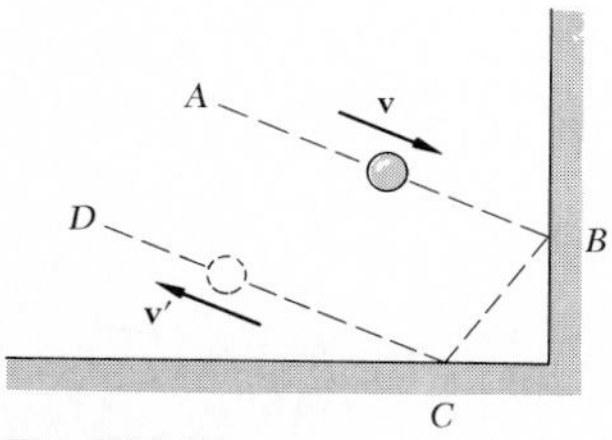

Fig. P14.33

14.33 A ball is thrown into a 90° corner with an initial velocity **v**. Denoting the coefficient of restitution by *e*, show that the final velocity **v**′ is of magnitude *ev* and that the initial and final paths *AB* and *CD* are parallel.

14.34 A steel ball falling vertically strikes a rigid plate A and rebounds horizontally as shown. Denoting by e the coefficient of restitution, determine (*a*) the required angle θ, (*b*) the magnitude of the velocity $\mathbf{v}_1$.

14.35 The ball of Prob. 14.34 strikes a second rigid plate B and rebounds vertically as shown. Determine (*a*) the required angle β, (*b*) the magnitude of the velocity $\mathbf{v}_2$ with which the ball leaves plate B.

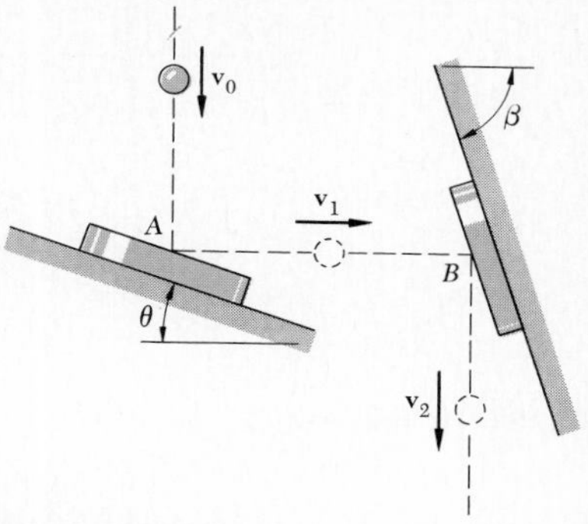

Fig. P14.34

14.36 A ball is dropped from a height $h_0 = 40$ in. onto a smooth floor. The coefficient of restitution is known to be 0.85 and the length of the first bounce is observed to be $d_1 = 12$ in. Determine (*a*) the height h_2 of the second bounce, (*b*) the length d_2 of the second bounce.

14.37 A ball is dropped onto a smooth floor and allowed to bounce several times as shown. Derive an expression for the coefficient of restitution in terms of (*a*) the heights of two successive bounces h_n and h_{n+1}, (*b*) the lengths of two successive bounces d_n and d_{n+1}, (*c*) the durations of two successive bounces t_n and t_{n+1}.

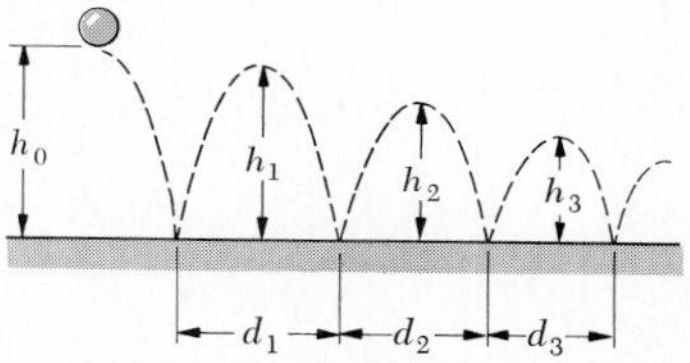

Fig. P14.36, P14.37, and P14.38

14.38 A ball is dropped from a height $h_0 = 48$ in. onto a smooth floor. Knowing that for the first bounce $h_1 = 40$ in. and $d_1 = 16$ in., determine (*a*) the coefficient of restitution, (*b*) the height and length of the second bounce.

∗**14.39** A 2-kg sphere moving to the left with a velocity of 10 m/s strikes the inclined surface of a 5-kg block which is at rest. The block rests on rollers and may move freely in the horizontal direction. Knowing that $e = 0.75$, determine the velocities of the block and of the sphere immediately after impact.

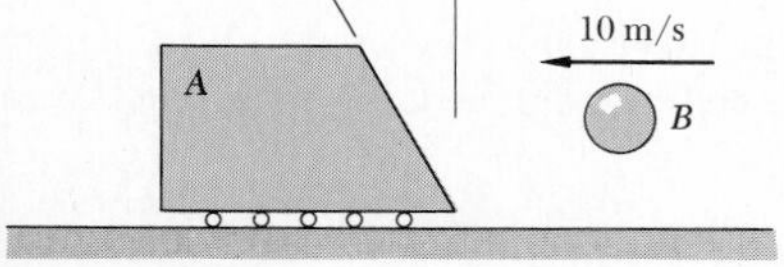

Fig. P14.39

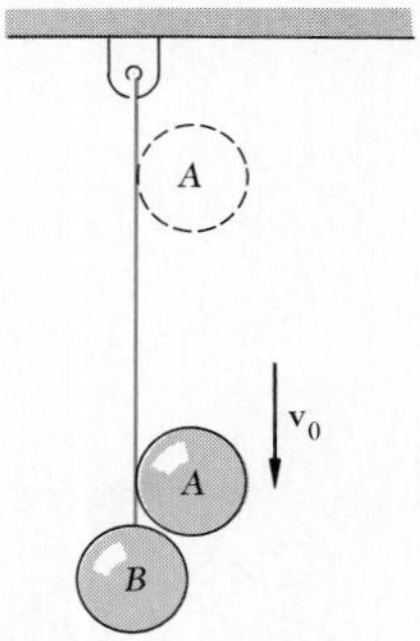

Fig. P14.40

∗**14.40** Ball B is suspended by an inextensible cord. An identical ball A is released from rest when it is just touching the cord and acquires a velocity $\mathbf{v}_0$ before striking ball B. Assuming $e = 1$, determine the velocity of each ball immediately after impact.

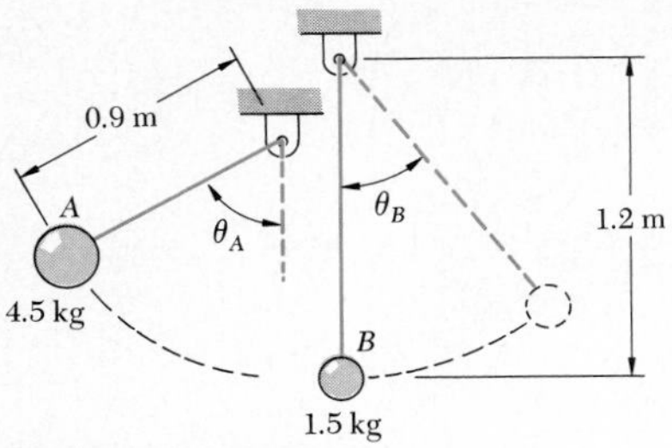

Fig. P14.41 and P14.42

14.41 The 4.5-kg sphere A strikes the 1.5-kg sphere B. Knowing that $e = 0.90$, determine the angle θ_A at which A must be released if the maximum angle θ_B reached by B is to be 90°.

14.42 The 4.5-kg sphere A is released from rest when $\theta_A = 60°$ and strikes the 1.5-kg sphere B. Knowing that $e = 0.90$, determine (*a*) the highest position to which sphere B will rise, (*b*) the maximum tension which will occur in the cord holding B.

14.43 The 4-lb sphere is released from rest when $\theta_A = 90°$. The coefficient of restitution between the sphere and the block is 0.70. Knowing that the coefficient of friction between the block and the horizontal surface is 0.30, determine how far the block will move after the impact.

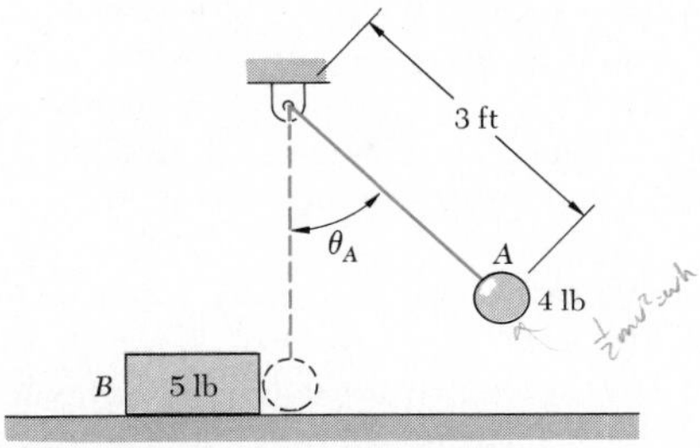

Fig. P14.43 and P14.44

14.44 The 4-lb sphere is released from rest when $\theta_A = 60°$. It is observed that the velocity of the sphere is zero after the impact and that the block moves 3 ft before coming to rest. Determine (*a*) the coefficient of restitution between the sphere and block, (*b*) the coefficient of friction between the block and the horizontal surface.

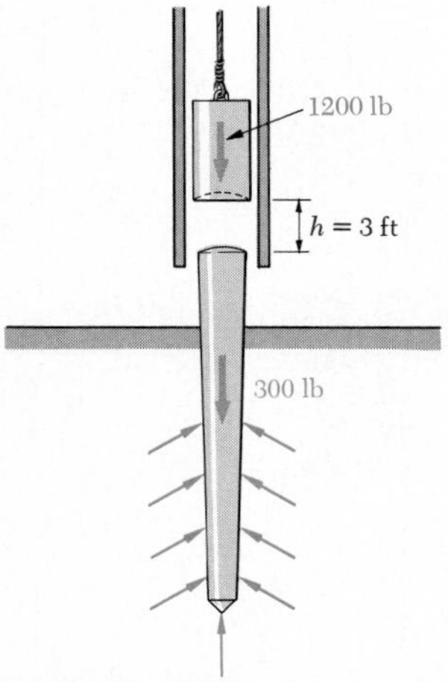

Fig. P14.45 and P14.46

14.45 It is desired to drive the 300-lb pile into the ground until the resistance to its penetration is 20,000 lb. Each blow of the 1200-lb hammer is the result of a 3-ft free fall onto the top of the pile. Determine how far the pile will be driven into the ground by a single blow when the 20,000-lb resistance is achieved. Assume that the impact is perfectly plastic.

14.46 The 1200-lb hammer of a drop-hammer pile driver falls from a height of 3 ft onto the top of a 300-lb pile. The pile is driven 4 in. into the ground. Assuming perfectly plastic impact, determine the average resistance of the ground to penetration.

14.47 Cylinder A is dropped 2 m onto cylinder B, which is resting on a spring of constant $k = 3$ kN/m. Assuming a perfectly plastic impact, determine (*a*) the maximum deflection of cylinder B, (*b*) the energy loss during the impact.

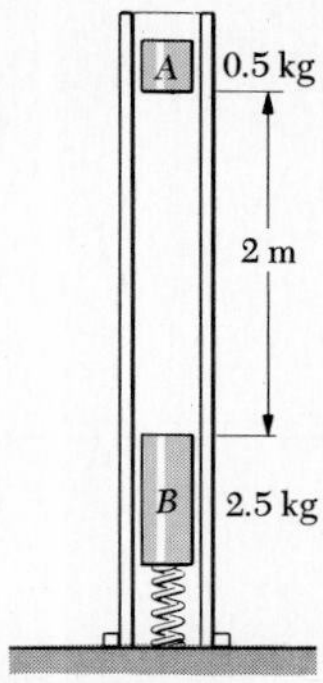

Fig. P14.47

14.48 The efficiency η of a drop-hammer pile driver may be defined as the ratio of the kinetic energy available after impact to the kinetic energy immediately before impact. Denoting by r the ratio of the pile mass m_p to the hammer mass m_h, and assuming perfectly plastic impact, show that $\eta = 1/(1 + r)$.

14.49 A bumper is designed to protect a 1600-kg automobile from damage when it hits a rigid wall at speeds up to 12 km/h. Assuming perfectly plastic impact, determine (*a*) the energy absorbed by the bumper during the impact, (*b*) the speed at which the automobile can hit another 1600-kg automobile without incurring any damage, if the other automobile is at rest and is similarly protected.

14.50 Solve Prob. 14.49, assuming a coefficient of restitution $e = 0.50$. Show that the answer to part b is independent of e.

14.51 A small rivet connecting two pieces of sheet metal is being clinched by hammering. Determine the energy absorbed by the rivet under each blow, knowing that the head of the hammer weighs 1.5 lb and that it strikes the rivet with a velocity of 20 ft/s. Assume that the anvil is supported by springs and (*a*) is infinite in weight (rigid support), (*b*) weighs 10 lb.

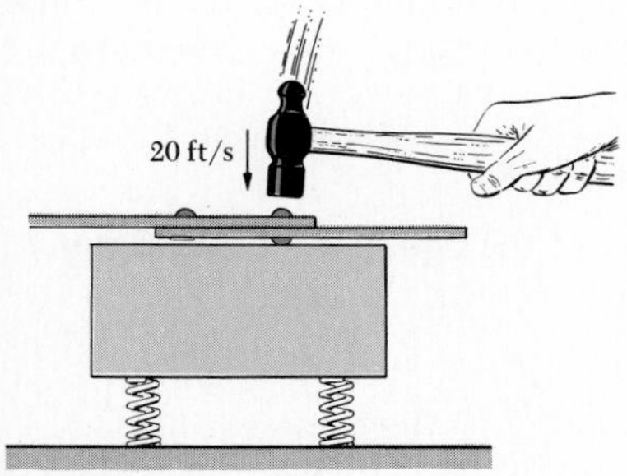

Fig. P14.51

14.52 In Prob. 14.51, determine the weight of the hammer required to deliver to the rivet an amount of energy equal to 8 ft · lb per blow if the hammer strikes the rivet with a velocity of 20 ft/s. Consider both types of support.

$*$**14.53** A mass m_A moving to the right with a velocity $\mathbf{v}_A$ strikes a second mass m_B which is at rest. Derive an expression for the kinetic-energy loss during the impact of the two masses. Assume that the balls strike each other squarely, and denote the coefficient of restitution by e.

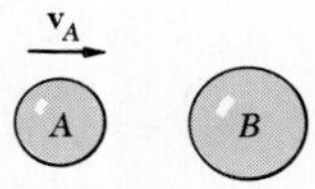

Fig. P14.53

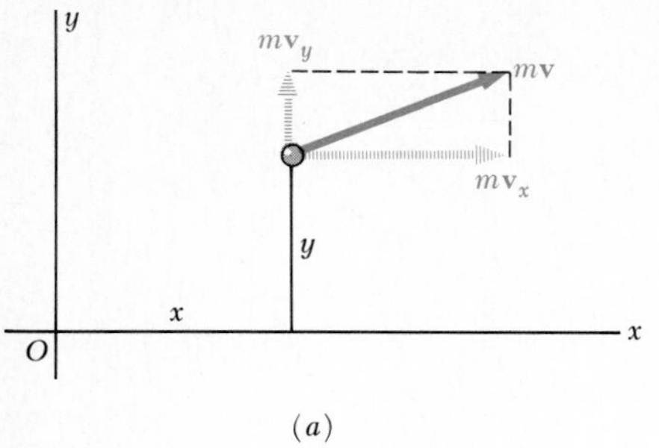

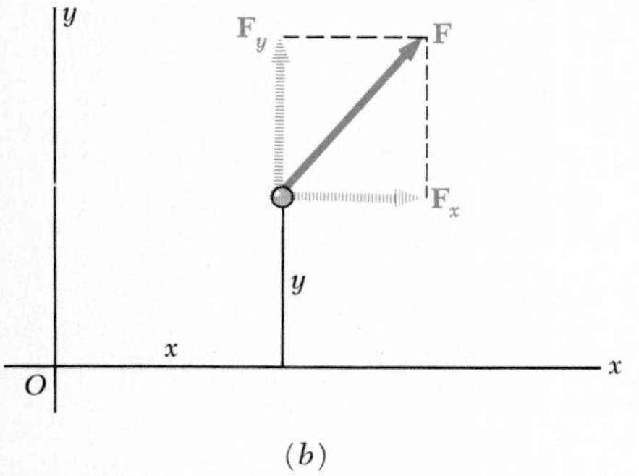

Fig. 14.12

14.9. Angular Momentum. Consider a particle of mass m moving in the xy plane. As we saw in Sec. 14.1, the momentum of the particle at a given instant is defined as the vector $m\mathbf{v}$ obtained by multiplying the velocity $\mathbf{v}$ of the particle by its mass m. The moment about O of the vector $m\mathbf{v}$ is called the *moment of momentum*, or the *angular momentum*, of the particle about O at that instant and is denoted by H_O. Resolving the vector $m\mathbf{v}$ into its x and y components (Fig. 14.12*a*), and recalling that counterclockwise is positive, we write

$$H_O = x(mv_y) - y(mv_x)$$

$$H_O = m(xv_y - yv_x) \qquad (14.24)$$

We shall now compute the derivative of the angular momentum H_O with respect to t; we have

$$\dot{H}_O = m\frac{d}{dt}(xv_y - yv_x)$$

$$= m(\dot{x}v_y + x\dot{v}_y - \dot{y}v_x - y\dot{v}_x)$$

or, since $\dot{x} = v_x$ and $\dot{y} = v_y$,

$$\dot{H}_O = m(x\dot{v}_y - y\dot{v}_x) \qquad (14.25)$$

But, according to Newton's second law, we have

$$m\dot{v}_x = ma_x = F_x \qquad m\dot{v}_y = ma_y = F_y$$

where F_x and F_y are the rectangular components of the force $\mathbf{F}$ acting on the particle. Equation (14.25) may therefore be written in the form

$$\dot{H}_O = xF_y - yF_x$$

or, since the right-hand member of the equation obtained represents the moment M_O of $\mathbf{F}$ about O (Fig. 14.12*b*),

$$M_O = \dot{H}_O \qquad (14.26)$$

14.10. Conservation of Angular Momentum. If the moment M_O about O of the force $\mathbf{F}$ acting on the particle of the preceding section is zero for every value of t, Eq. (14.26) yields

$$\dot{H}_O = 0$$

for any t; integrating with respect to t, we obtain

$$H_O = \text{constant} \qquad (14.27)$$

Thus, if the moment about O of the force $\mathbf{F}$ acting on a particle is zero for every value of t, *the angular momentum of the particle about O is conserved.*

Clearly, the angular momentum of a particle about a fixed point O will be conserved if the resultant of the forces acting on the particle is zero. But the angular momentum of the particle about O will also be conserved when the particle is acted upon by an unbalanced force **F** *if the force* **F** *is a central force whose line of action passes through O* (Fig. 14.13).

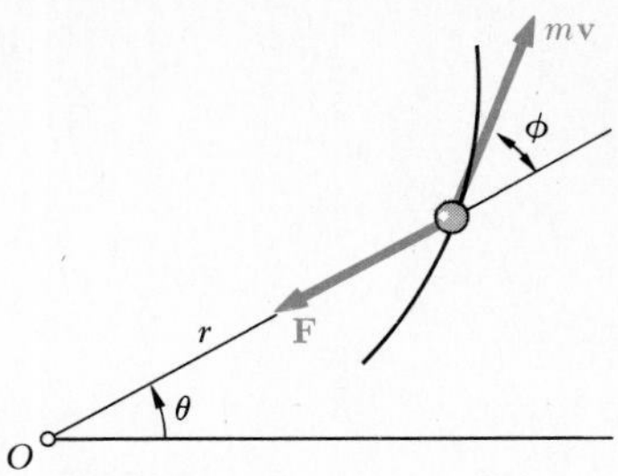

Fig. 14.13

The angular momentum H_O of a particle was expressed in Eq. (14.24) in terms of the retangular coordinates of the particle and of the rectangular components of its momentum. In the case of a particle moving under a central force **F**, it is found more convenient to use polar coordinates. Recalling that H_O is, by definition, the moment about O of the momentum $m\mathbf{v}$ of the particle, and denoting by ϕ the angle formed by the radius vector and the tangent to the path of the particle, we write Eq. (14.27) in the form

$$H_O = r(mv \sin \phi) = \text{constant} \qquad (14.28)$$

Observing that $v \sin \phi$ represents the transverse component v_θ of the velocity, which is equal to $r\dot{\theta}$ (Sec. 11.13), we may also write

$$H_O = mr^2\dot{\theta} = \text{constant} \qquad (14.29)$$

Comparing Eq. (14.29) and Eq. (12.18), we note that, when a particle moves under a central force, its angular momentum H_O about the center of force O is equal to the product of its mass and of the constant h introduced in Sec. 12.8. It follows that the angular momentum of the particle is equal to twice the product of its mass and of its areal velocity; thus, stating that the areal velocity of the particle is constant is equivalent to stating that its angular momentum is conserved.

14.11. Application to Space Mechanics. The principles of conservation of angular momentum and of conservation of energy may be used to solve many problems concerning the motion of earth satellites and other space vehicles.

Consider, for example, a space vehicle moving under the earth's gravitational force. We shall assume that it begins its free flight at point P_0 at a distance r_0 from the center of the earth, with a velocity $\mathbf{v}_0$ forming an angle ϕ_0 with the radius vector OP_0 (Fig. 14.14). Let P be a point of the trajectory described by the vehicle; we denote by r the distance from O to P, by **v** the velocity of the vehicle at P, and by ϕ the angle formed by **v** and the radius vector OP. Applying the principle of conservation of angular momentum about O between P_0 and P, we write

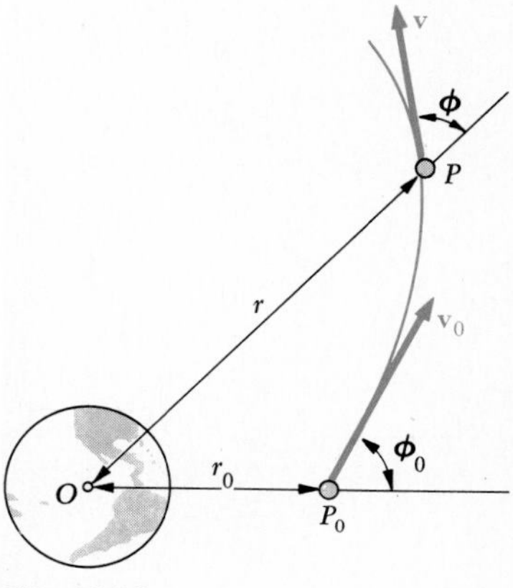

Fig. 14.14

$$r_0 m v_0 \sin \phi_0 = rmv \sin \phi \qquad (14.30)$$

Recalling expression (13.15) obtained for the potential energy due to a gravitational force, we apply the principle of conservation of energy between P_0 and P and write

$$T_0 + V_0 = T + V$$

$$\tfrac{1}{2}mv_0^2 - \frac{GMm}{r_0} = \tfrac{1}{2}mv^2 - \frac{GMm}{r} \tag{14.31}$$

where M is the mass of the earth.

Equation (14.31) may be solved for the magnitude v of the velocity of the vehicle at P when the distance r from O to P is known; Eq. (14.30) may then be used to determine the angle ϕ that the velocity forms with the radius vector OP.

Equations (14.30) and (14.31) may also be used to determine the maximum and minimum values of r in the case of a satellite launched from P_0 in a direction forming an angle ϕ_0 with the vertical OP_0 (Fig. 14.15). The desired values of r are obtained by making $\phi = 90°$ in (14.30) and eliminating v between Eqs. (14.30) and (14.31).

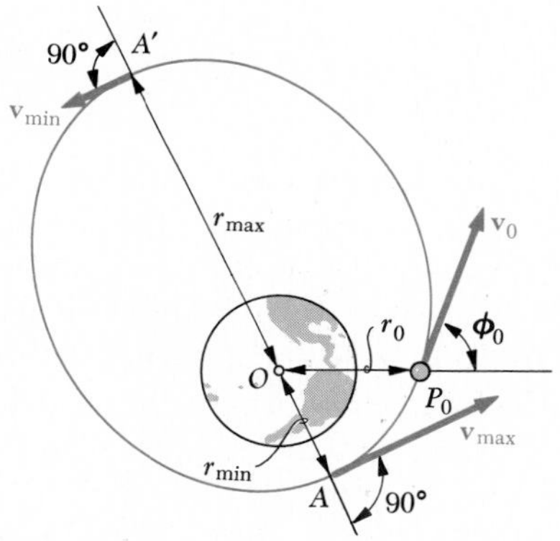

Fig. 14.15

It should be noted that the application of the principles of conservation of energy and of conservation of angular momentum leads to a more fundamental formulation of the problems of space mechanics than the method indicated in Sec. 12.11. In all cases involving oblique launchings, it will also result in much simpler computations. And while the method of Sec. 12.11 must be used when the actual trajectory or the periodic time of a space vehicle is to be determined, the calculations will be simplified if the conservation principles are first used to compute the maximum and minimum values of the radius vector r.

14.12. Principle of Impulse and Momentum for a System of Particles. The moment of momentum, or angular momentum, about O of a system of particles is defined as the sum of the moments about O of the momenta of the various particles of the system; we write

$$H_O = \Sigma m(xv_y - yv_x) \tag{14.32}$$

Since Eq. (14.26) is satisfied for each particle, we may write that the sum of the moments of all the forces acting on the various particles of the system is equal to the rate of change of the angular momentum of the system. However, since the internal forces occur by pairs of equal and opposite forces having the same line of action, the sum of their moments is zero and we need consider only the sum of the moments of the external forces. We thus write

$$\Sigma(M_O)_{\text{ext}} = \dot{H}_O \tag{14.33}$$

where H_O is the angular momentum of the system defined by relation (14.32).

We saw in Sec. 14.1 that the linear impulse of a force **F**, of components F_x and F_y, during a given time interval may be defined by the integrals of F_x and F_y over that time interval. We shall now, in a similar way, define the *angular impulse* of a couple **M** over a time interval from t_1 to t_2 as the integral

$$\int_{t_1}^{t_2} M\,dt$$

of the moment M of the couple over the given time interval. This new concept, as we shall see presently, will enable us to express in a more general form the principle of impulse and momentum for a system of particles.

Consider a system of particles and the *external forces* acting on the various particles. Each external force **F** may be replaced by an equal force **F** attached at the origin O, plus a couple $\mathbf{M}_O$ of moment equal to the moment about O of the original force (Chap. 3). Now we shall draw three separate sketches (Fig. 14.16). The first and third sketches will show the systems of vectors representing the momenta of the various particles at a

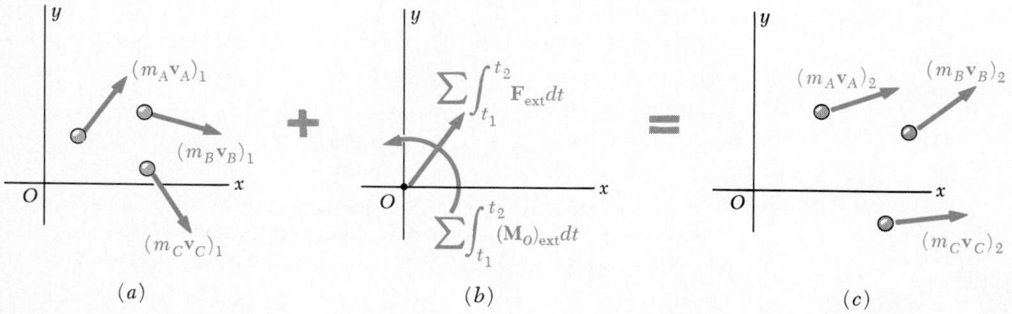

Fig. 14.16

time t_1 and a time t_2, respectively. The second sketch will show the resultant of the vectors representing the linear impulses of the external forces **F** during the given time interval, as well as the sum of the couples representing the angular impulses of the couples $\mathbf{M}_O$ during the same time interval. We recall from Sec. 14.3 that the sums of the x and y components of the vectors shown in parts a and b of Fig. 14.16 must be respectively equal to the sums of the x and y components of the vectors in part c of the same figure. Indeed, rewriting Eq. (14.8) in terms of x and y components, we have

$$\sum (mv_x)_1 + \sum \int_{t_1}^{t_2} (F_x)_{\text{ext}}\, dt = \sum (mv_x)_2$$
$$\sum (mv_y)_1 + \sum \int_{t_1}^{t_2} (F_y)_{\text{ext}}\, dt = \sum (mv_y)_2 \tag{14.34}$$

Returning now to Eq. (14.33) and integrating both members over the time interval from t_1 to t_2, we write

$$(H_O)_1 + \sum \int_{t_1}^{t_2} (M_O)_{\text{ext}}\, dt = (H_O)_2 \tag{14.35}$$

The equation obtained expresses that the sum of the moments about O of the vectors shown in parts a and b of Fig. 14.16 must be equal to the sum of the moments about O of the vectors shown in part c of the same figure. Together, Eqs. (14.34) and (14.35) thus express that *the momenta of the particles at time t_1 and the impulses of the external forces from t_1 to t_2 form a system of vectors equipollent to the system of the momenta of the particles at time t_2.*† We write

$$\textbf{Syst Momenta}_1 + \textbf{Syst Ext Imp}_{1\to2} = \textbf{Syst Momenta}_2 \tag{14.36}$$

This more general statement of the principle of impulse and momentum for a system of particles will prove particularly useful for the study of the motion of rigid bodies (Chap. 18).

When no external force acts on the particles of a system, the integrals in Eqs. (14.34) and (14.35) are zero, and part b of Fig. 14.16 may be omitted. The system of the momenta of the particles at time t_1 is equipollent to the system of the momenta at time t_2:

$$\textbf{Syst Momenta}_1 = \textbf{Syst Momenta}_2 \tag{14.37}$$

† Since we are dealing here with a system of particles, and not with a single rigid body, we cannot conclude that the two systems of vectors are actually equivalent. Hence we use gray plus and equals signs in Fig. 14.16 (cf. Sec. 12.4).

SAMPLE PROBLEM 14.8

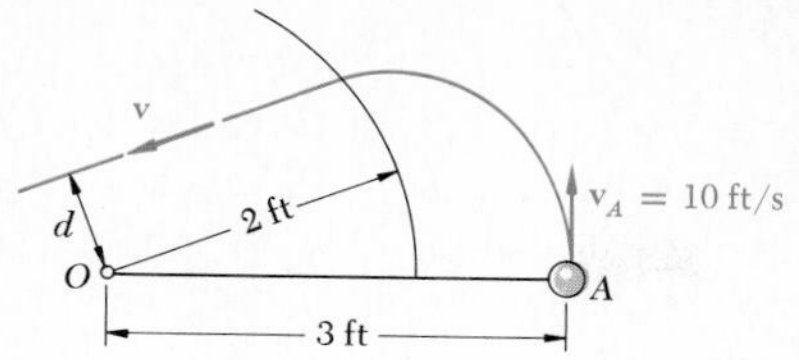

A ball weighing 0.5 lb is attached to a fixed point O by means of an elastic cord of constant $k = 10$ lb/ft and of undeformed length equal to 2 ft. The ball slides on a horizontal frictionless surface. If the ball is placed at point A, 3 ft from O, and is given an initial velocity of 10 ft/s in a direction perpendicular to OA, determine (*a*) the speed of the ball after the cord has become slack, (*b*) the closest distance d that the ball will come to O.

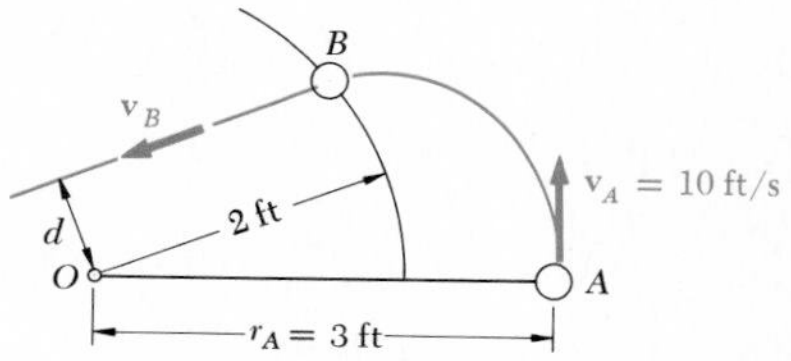

Solution. The force exerted by the cord on the ball passes through the fixed point O, and its work may be expressed as a change in potential energy. It is therefore a conservative central force, and both the total energy of the ball and its angular momentum about O are conserved between points A and B. After the cord has become slack at B, the resultant force acting on the ball is zero. The ball, therefore, will move in a straight line at a constant speed v. The straight line is the line of action of $\mathbf{v}_B$ and the speed v is equal to v_B.

a. Conservation of Energy.

At point A: $$T_A = \tfrac{1}{2}mv_A^2 = \frac{1}{2}\frac{0.5\text{ lb}}{32.2\text{ ft/s}^2}(10\text{ ft/s})^2 = 0.776\text{ ft}\cdot\text{lb}$$

$$V_A = \tfrac{1}{2}kx_A^2 = \tfrac{1}{2}(10\text{ lb/ft})(3\text{ ft} - 2\text{ ft})^2 = 5\text{ ft}\cdot\text{lb}$$

At point B: $$T_B = \tfrac{1}{2}mv_B^2 = \frac{1}{2}\frac{0.5}{32.2}v_B^2 = 0.00776v_B^2$$

$$V_B = 0$$

Applying the principle of conservation of energy between points A and B, we write

$$T_A + V_A = T_B + V_B$$
$$0.776 + 5 = 0.00776v_B^2$$
$$v_B^2 = 744 \qquad v = v_B = 27.3\text{ ft/s} \blacktriangleleft$$

b. Conservation of Angular Momentum About O. Since r_A and d represent the perpendicular distances to $\mathbf{v}_A$ and $\mathbf{v}_B$, respectively, we write

$$r_A(mv_A) = d(mv_B)$$
$$(3\text{ ft})\left(\frac{0.5}{g}\right)(10\text{ ft/s}) = d\left(\frac{0.5}{g}\right)(27.3\text{ ft/s})$$

$$d = 1.099\text{ ft} \blacktriangleleft$$

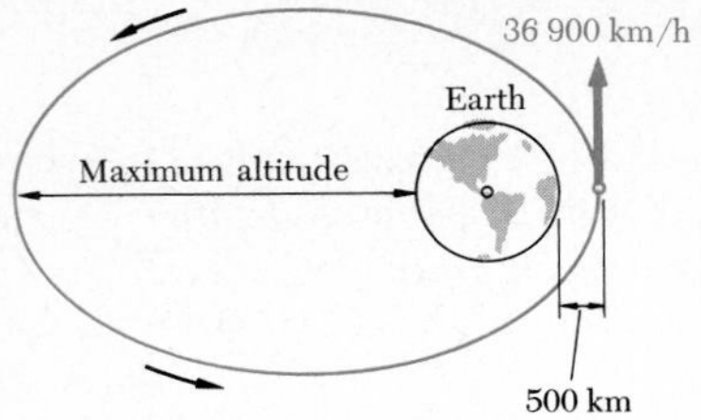

SAMPLE PROBLEM 14.9

A satellite is launched in a direction parallel to the surface of the earth with a velocity of 36 900 km/h from an altitude of 500 km. Determine (*a*) the maximum altitude reached by the satellite, (*b*) the maximum allowable error in the direction of launching if the satellite is to go into orbit and to come not closer than 200 km to the surface of the earth.

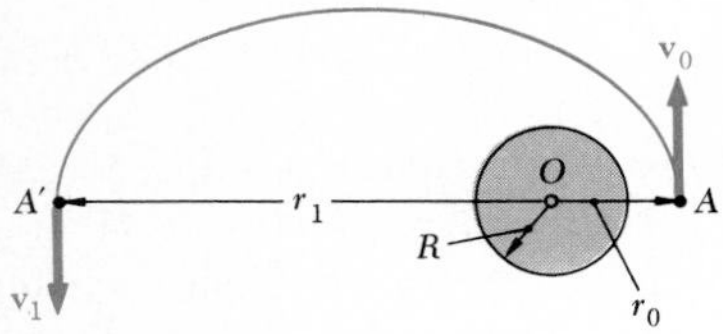

***a.* Maximum Altitude.** We denote by A' the point of the orbit farthest from the earth and by r_1 the corresponding distance from the center of the earth. Since the satellite is in free flight between A and A', we apply the principle of conservation of energy.

$$T_A + V_A = T_{A'} + V_{A'}$$

$$\tfrac{1}{2}mv_0^2 - \frac{GMm}{r_0} = \tfrac{1}{2}mv_1^2 - \frac{GMm}{r_1} \qquad (1)$$

Since the only force acting on the satellite is the force of gravity, which is a central force, the angular momentum of the satellite about O is conserved. Considering points A and A', we write

$$r_0 m v_0 = r_1 m v_1 \qquad v_1 = v_0 \frac{r_0}{r_1} \qquad (2)$$

Substituting this expression for v_1 into Eq. (1) and dividing each term by the mass m, we obtain after rearranging the terms,

$$\tfrac{1}{2}v_0^2\left(1 - \frac{r_0^2}{r_1^2}\right) = \frac{GM}{r_0}\left(1 - \frac{r_0}{r_1}\right) \qquad 1 + \frac{r_0}{r_1} = \frac{2GM}{r_0 v_0^2} \qquad (3)$$

Recalling that the radius of the earth is $R = 6370$ km, we compute

$$r_0 = 6370 \text{ km} + 500 \text{ km} = 6870 \text{ km} = 6.87 \times 10^6 \text{ m}$$
$$v_0 = 36\,900 \text{ km/h} = (3.69 \times 10^7 \text{ m})/(3.6 \times 10^3 \text{ s}) = 1.025 \times 10^4 \text{ m/s}$$
$$GM = gR^2 = (9.81 \text{ m/s}^2)(6.37 \times 10^6 \text{ m})^2 = 3.98 \times 10^{14} \text{ m}^3/\text{s}^2$$

Substituting these values into (3), we obtain $r_1 = 66.8 \times 10^6$ m

Maximum altitude $= 66.8 \times 10^6 \text{ m} - 6.37 \times 10^6 \text{ m} = 60.4 \times 10^6 \text{ m}$
$= 60\,400$ km ◄

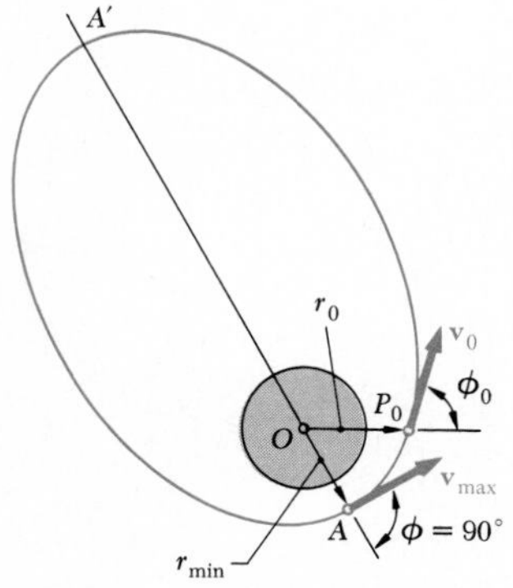

***b.* Allowable Error in Direction of Launching.** The satellite is launched from P_0 in a direction forming an angle ϕ_0 with the vertical OP_0. The value of ϕ_0 corresponding to $r_{min} = 6370 \text{ km} + 200 \text{ km} = 6570$ km is obtained by applying the principles of conservation of energy and of conservation of angular momentum between P_0 and A.

$$\tfrac{1}{2}mv_0^2 - \frac{GMm}{r_0} = \tfrac{1}{2}mv_{max}^2 - \frac{GMm}{r_{min}} \qquad (4)$$

$$r_0 m v_0 \sin\phi_0 = r_{min} m v_{max} \qquad (5)$$

Solving (5) for v_{max} and then substituting for v_{max} into (4), we may solve (4) for $\sin\phi_0$. Using the values of v_0 and GM computed in part *a* and noting that $r_0/r_{min} = 6870/6570 = 1.0457$, we find

$\sin\phi_0 = 0.9801 \qquad \phi_0 = 90° \pm 11.5° \qquad$ Allowable error $= \pm 11.5°$ ◄

PROBLEMS

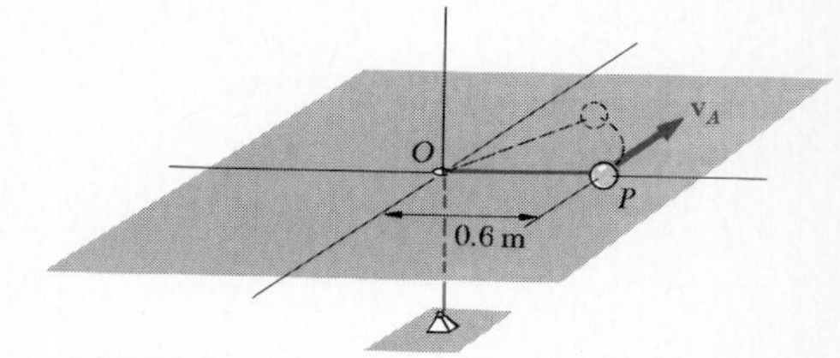

Fig. P14.54

14.54 A 2-kg sphere is attached to an elastic cord of constant 150 N/m which is undeformed when the sphere is located at the origin O. Knowing that in the position shown $\mathbf{v}_A$ is perpendicular to OP and has a magnitude of 10 m/s, determine (*a*) the maximum distance from the origin attained by the sphere, (*b*) the corresponding speed of the sphere.

14.55 In Prob. 14.54, determine the required initial speed v_A if the maximum distance from the origin attained by the sphere is to be 1.5 m.

14.56 In Sample Prob. 14.8, determine the required magnitude of $\mathbf{v}_A$ if the ball is to pass at a distance $d = 4$ in. from point O. Assume that the direction of $\mathbf{v}_A$ is not changed.

14.57 In Sample Prob. 14.8, determine the smallest magnitude of $\mathbf{v}_A$ for which the elastic cord will remain taut at all times.

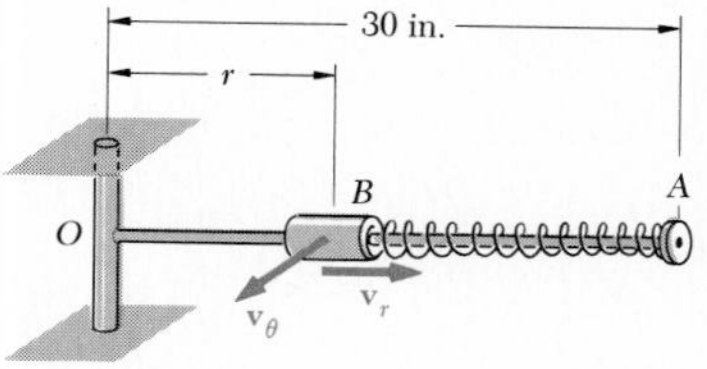

Fig. P14.58

14.58 Collar B weighs 10 lb and is attached to a spring of constant 50 lb/ft and of undeformed length equal to 18 in. The system is set in motion with $r = 12$ in., $v_\theta = 16$ ft/s, and $v_r = 0$. Neglecting the mass of the rod and the effect of friction, determine the radial and transverse components of the velocity of the collar when $r = 21$ in.

14.59 For the motion described in Prob. 14.58, determine (*a*) the maximum distance between the origin and the collar, (*b*) the corresponding velocity.

14.60 through 14.63 Using the principles of conservation of energy and of conservation of angular momentum, solve the following problems:

14.60 Prob. 12.82.
14.61 Prob. 12.84.
14.62 Prob. 12.86.
14.63 Prob. 12.85.

14.64 Upon the LEM's return to the command module, the Apollo spacecraft of Prob. 12.84 is turned around so that the LEM faces to the rear. The LEM is then cast adrift with a velocity of 600 ft/s relative to the command module. Determine the magnitude and direction (angle ϕ formed with the vertical OC) of the velocity $\mathbf{v}_C$ of the LEM just before it crashes at C on the moon's surface.

14.65 Determine the magnitude and direction (angle ϕ formed with the vertical OB) of the velocity $\mathbf{v}_B$ of the spacecraft of Prob. 12.91 just before splashdown at B. Neglect the effect of the atmosphere.

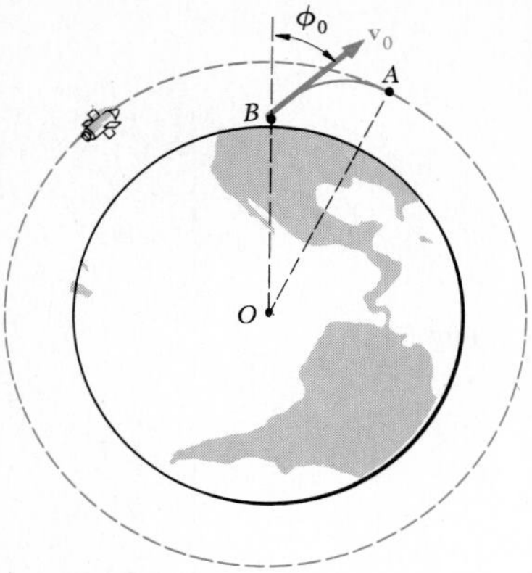

Fig. P14.67

14.66 To what value v_0 should the speed of the spacecraft of Prob. 12.92 be reduced preparatory to reentry if its velocity $\mathbf{v}_B$ just before splashdown at B is to form an angle $\phi = 30°$ with the vertical OB? Neglect the effect of the atmosphere.

14.67 A spacecraft is launched from the ground to rendezvous with a skylab orbiting the earth at a constant altitude of 1040 mi. The spacecraft has reached at burnout an altitude of 40 miles and a velocity $\mathbf{v}_0$ of magnitude 20,000 ft/s. What is the angle ϕ_0 that $\mathbf{v}_0$ should form with the vertical OB if the trajectory of the spacecraft is to be tangent at A to the orbit of the skylab?

14.68 In Prob. 14.67, determine the maximum and minimum values of v_0, and the corresponding values of ϕ_0, for which the trajectory of the spacecraft will be tangent to the orbit of the skylab.

14.69 At engine burnout a satellite has reached an altitude of 2400 km and has a velocity $\mathbf{v}_0$ of magnitude 8100 m/s forming an angle $\phi_0 = 76°$ with the vertical. Determine the maximum and minimum heights reached by the satellite.

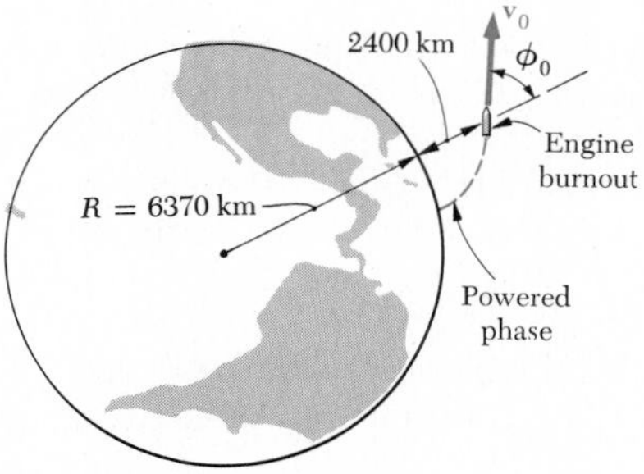

Fig. P14.69 and P14.70

14.70 At engine burnout a satellite has reached an altitude of 2400 km and has a velocity $\mathbf{v}_0$ of magnitude 8100 m/s. For what range of values of the angle ϕ_0, formed by $\mathbf{v}_0$ and the vertical, will the satellite go into a permanent orbit? (Assume that if the satellite gets closer than 300 km from the earth's surface, it will soon burn up.)

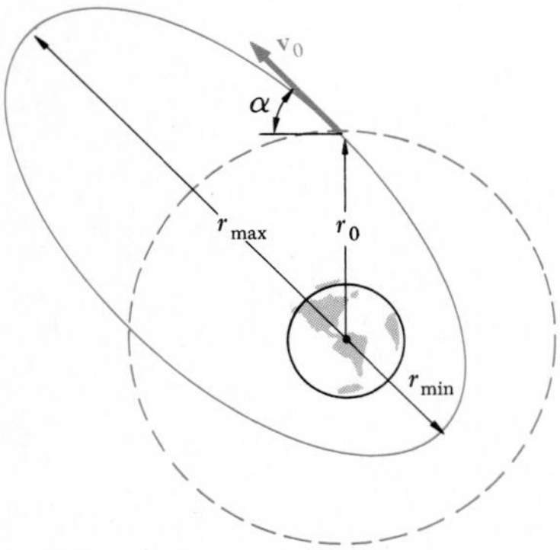

Fig. P14.71

14.71 A satellite is projected into space with a velocity $\mathbf{v}_0$ at a distance r_0 from the center of the earth by the last stage of its launching rocket. The velocity $\mathbf{v}_0$ was designed to send the satellite into a circular orbit of radius r_0. However, owing to a malfunction of control, the satellite is not projected horizontally but at an angle α with the horizontal and, as a result, is propelled into an elliptic orbit. Determine the maximum and minimum values of the distance from the center of the earth to the satellite.

***14.72** Using the answers obtained in Prob. 14.71, show that the intended circular orbit and the resulting elliptic orbit intersect at the ends of the minor axis of the elliptic orbit.

14.73 A spacecraft of mass m describes a circular orbit of radius r_1 around the earth. (*a*) Show that the additional energy ΔE which must be imparted to the spacecraft to transfer it to a circular orbit of larger radius r_2 is

$$\Delta E = \frac{GMm(r_2 - r_1)}{2r_1 r_2}$$

where M is the mass of the earth. (*b*) Further show that, if the transfer from one circular orbit to the other is executed by placing the spacecraft on a transitional semielliptic path AB, the amounts of energy ΔE_A and ΔE_B which must be imparted at A and B are respectively proportional to r_2 and r_1:

$$\Delta E_A = \frac{r_2}{r_1 + r_2}\Delta E \qquad \Delta E_B = \frac{r_1}{r_1 + r_2}\Delta E$$

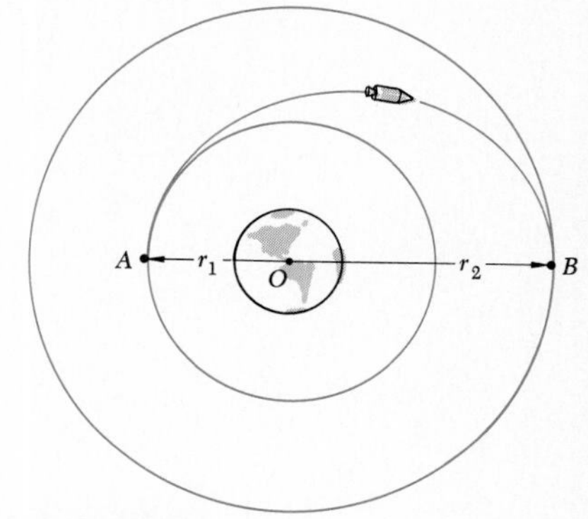

Fig. P14.73

14.74 Show that the total energy E of a satellite of mass m describing an elliptic orbit is $E = -GMm/(r_1 + r_2)$, where M is the mass of the earth, and r_1 and r_2 represent, respectively, the maximum and minimum distance of the orbit to the center of the earth. (It is recalled that the gravitational potential energy of a satellite was defined as being zero at an infinite distance from the earth.)

***14.75** A missile is fired from the ground with a velocity $\mathbf{v}_0$ of magnitude $v_0 = \sqrt{gR}$, forming an angle ϕ_0 with the vertical. (*a*) Express the maximum height d reached by the missile in terms of ϕ_0. (*b*) Show that the angle 2α subtending the trajectory BAC of the missile is equal to $2\phi_0$ and explain what happens when ϕ_0 approaches 90°. [*Hint.* Use Eq. (12.30) of Sec. 12.11 to solve part *b*, noting that $\theta = 180°$ for point A and $\theta = 180° - \alpha$ for point B.]

14.76 A missile is fired from the ground with an initial velocity $\mathbf{v}_0$ forming an angle ϕ_0 with the vertical. If the missile is to reach a maximum altitude equal to the radius of the earth, (*a*) show that the required angle ϕ_0 is defined by the relation

$$\sin\phi_0 = 2\sqrt{1 - \frac{1}{2}\left(\frac{v_{esc}}{v_0}\right)^2}$$

where v_{esc} is the escape velocity, (*b*) determine the maximum and minimum allowable values of v_0.

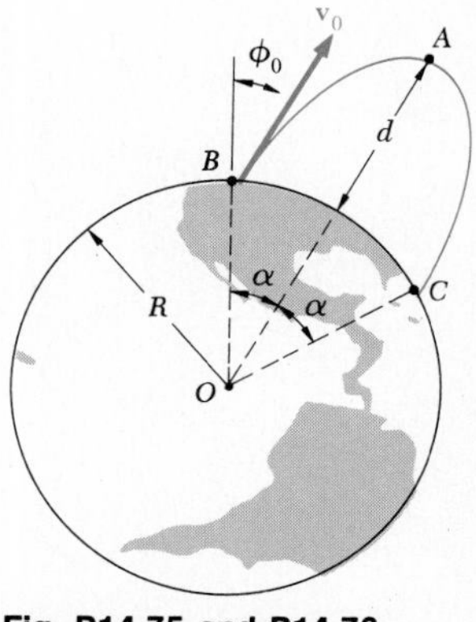

Fig. P14.75 and P14.76

***14.13. Variable Systems of Particles.** All the systems of particles considered so far consisted of well-defined particles. These systems did not gain or lose any particles during their motion. In a large number of engineering applications, however, it is necessary to consider *variable systems of particles*, i.e., systems which are continuously gaining or losing particles, or doing both at the same time. Consider, for example, a hydraulic turbine. Its analysis involves the determination of the forces exerted by a stream of water on rotating blades, and we note that the particles of water in contact with the blades form an everchanging system which continuously acquires and loses particles. Rockets furnish another example of variable systems, since their propulsion depends upon the continuous ejection of fuel particles.

We recall that all the kinetics principles established so far were derived for constant systems of particles, which neither gain nor lose particles. We must therefore find a way to reduce the analysis of a variable system of particles to that of an auxiliary constant system. The procedure to follow is indicated in Secs. 14.14 and 14.15 for two broad categories of applications.

***14.14. Steady Stream of Particles.** Consider a steady stream of particles, such as a stream of water diverted by a fixed vane or a flow of air through a duct or through a blower. In order to determine the resultant of the forces exerted on the particles in contact with the vane, duct, or blower, we isolate these particles and denote by S the system thus defined (Fig. 14.17). We observe that S is a variable system of particles, since it continuously gains particles flowing in and loses an equal number of particles flowing out. Therefore, the kinetics principles that have been established so far cannot be directly applied to S.

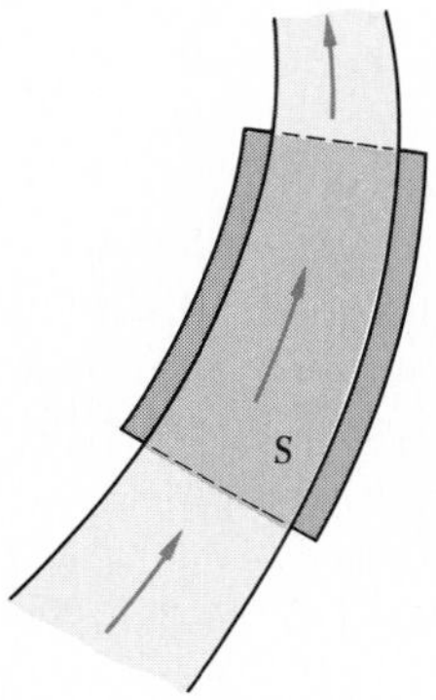

Fig. 14.17

However, we may easily define an auxiliary system of particles which does remain constant for a short interval of time Δt. Consider at time t the system S *plus* the particles which will enter S during the interval of time Δt (Fig. 14.18*a*). Next, consider at time $t + \Delta t$ the system S *plus* the particles which have left S during the interval Δt (Fig. 14.18*c*). Clearly, *the same particles are involved in both cases,* and we may apply to these particles the principle of impulse and momentum. Since the total mass m of the system S remains constant, the particles entering the system and those leaving the system in the time Δt must have the same mass Δm. Denoting by $\mathbf{v}_A$ and $\mathbf{v}_B$, respectively, the velocities of the particles entering S at A and leaving S at B, we represent the momentum of the particles entering S by $(\Delta m)\mathbf{v}_A$ (Fig. 14.18*a*) and the momentum of the particles leaving S by $(\Delta m)\mathbf{v}_B$ (Fig. 14.18*c*). We also represent the momenta of the particles forming S and the impulses of the forces exerted on S by the appropriate vectors, and indicate by gray plus and equals signs that the system of the momenta and impulses in parts a and b of Fig. 14.18 is equipollent to the system of the momenta in part c of the same figure.

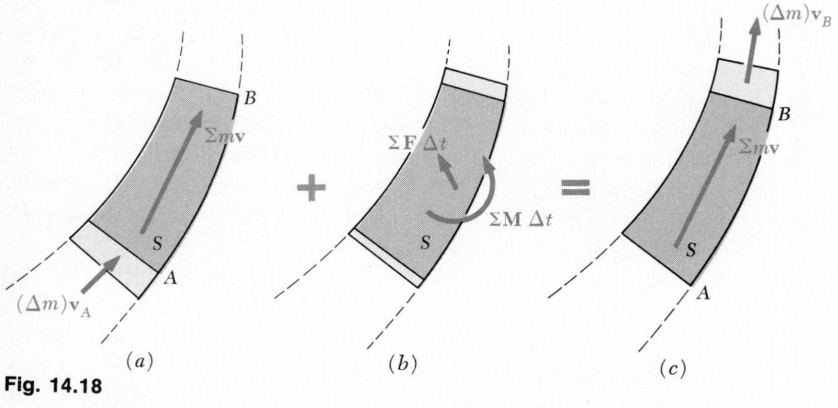

Fig. 14.18

Since the resultant $\Sigma m\mathbf{v}$ of the momenta of the particles of S is found on both sides of the equals sign, it may be omitted. We conclude that *the system formed by the momentum* $(\Delta m)\mathbf{v}_A$ *of the particles entering S in the time* Δt *and the impulses of the forces exerted on S during that time is equipollent to the momentum* $(\Delta m)\mathbf{v}_B$ *of the particles leaving S in the same time* Δt. We may therefore write

$$(\Delta m)\mathbf{v}_A + \Sigma \mathbf{F}\,\Delta t = (\Delta m)\mathbf{v}_B \qquad (14.38)$$

A similar equation may be obtained by taking the moments of the vectors involved (see Sample Prob. 14.10). Dividing all terms of Eq. (14.38) by Δt and letting Δt approach zero, we obtain at the limit

$$\Sigma \mathbf{F} = \frac{dm}{dt}(\mathbf{v}_B - \mathbf{v}_A) \tag{14.39}$$

where $\mathbf{v}_B - \mathbf{v}_A$ represents the difference between the *vectors* $\mathbf{v}_B$ and $\mathbf{v}_A$.

If SI units are used, dm/dt is expressed in kg/s and the velocities in m/s; we check that both members of Eq. (14.39) are expressed in the same units (newtons). If U.S. customary units are used, dm/dt must be expressed in slugs/s and the velocities in ft/s; we check again that both members of the equation are expressed in the same units (pounds).†

The principle we have established may be used to analyze a large number of engineering applications. Some of the most common are indicated below.

Fluid Stream Diverted by a Vane. If the vane is fixed, the method of analysis given above may be applied directly to find the force $\mathbf{F}$ exerted by the vane on the stream. We note that $\mathbf{F}$ is the only force which needs to be considered since the pressure in the stream is constant (atmospheric pressure). The force exerted by the stream on the vane will be equal and opposite to $\mathbf{F}$. If the vane moves with a constant velocity, the stream is not steady. However, it will appear steady to an observer moving with the vane. We should therefore choose a system of axes moving with the vane. Since this system of axes is not accelerated, Eq. (14.38) may still be used, but $\mathbf{v}_A$ and $\mathbf{v}_B$ must be replaced by the *relative velocities* of the stream with respect to the vane (see Sample Prob. 14.11).

Fluid Flowing through a Pipe. The force exerted by the fluid on a pipe transition such as a bend or a contraction may be determined by considering the system of particles S in contact with the transition. Since, in general, the pressure in the flow will vary, we should also consider the forces exerted on S by the adjoining portions of the fluid.

† It is often convenient to express the mass rate of flow dm/dt as the product ρQ, where ρ is the density of the stream (mass per unit volume) and Q its volume rate of flow (volume per unit time). If SI units are used, ρ is expressed in kg/m^3 (for instance, $\rho = 1000$ kg/m^3 for water) and Q in m^3/s. However, if U.S. customary units are used, ρ will generally have to be computed from the corresponding specific weight γ (weight per unit volume), $\rho = \gamma/g$. Since γ is expressed in lb/ft^3 (for instance, $\gamma = 62.4$ lb/ft^3 for water), ρ is obtained in slugs/ft^3. The volume rate of flow Q is expressed in ft^3/s.

Jet Engine. In a jet engine, air enters with no velocity through the front of the engine and leaves through the rear with a high velocity. The energy required to accelerate the air particles is obtained by burning fuel. While the exhaust gases contain burned fuel, the mass of the fuel is small compared with the mass of the air flowing through the engine and usually may be neglected. Thus, the analysis of a jet engine reduces to that of an air stream. This stream may be considered as a steady stream if all velocities are measured with respect to the airplane. The air stream shall be assumed, therefore, to enter the engine with a velocity **v** of magnitude equal to the speed of the airplane and to leave with a velocity **u** equal to the relative velocity of the exhaust gases (Fig. 14.19). Since the intake and exhaust

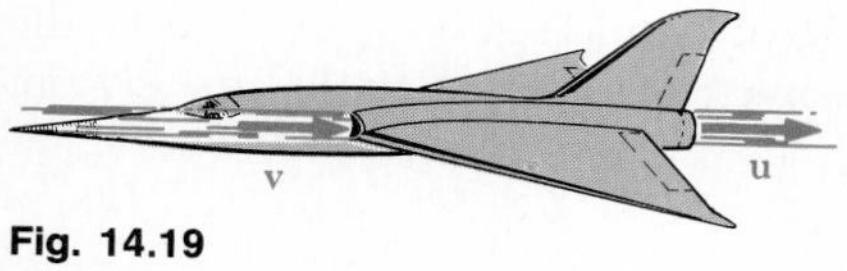

Fig. 14.19

pressures are nearly atmospheric, the only external force which needs to be considered is the force exerted by the engine on the air stream. This force is equal and opposite to the thrust.†

Fan. We consider the system of particles S shown in Fig. 14.20. The velocity $\mathbf{v}_A$ of the particles entering the system is assumed equal to zero, and the velocity $\mathbf{v}_B$ of the particles leaving the system is the velocity of the *slipstream*. The rate of flow may be obtained by multiplying v_B by the cross-sectional

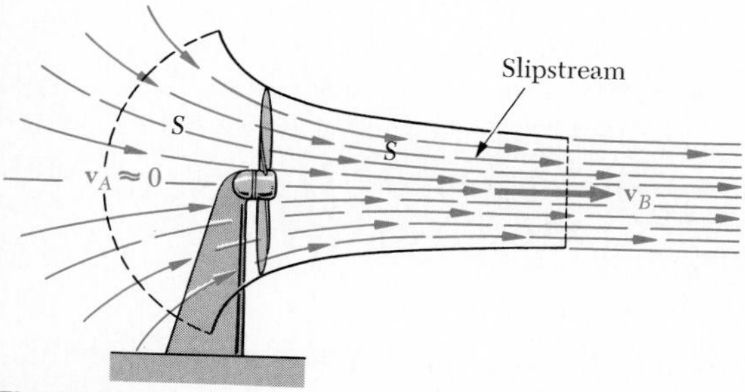

Fig. 14.20

† Note that, if the airplane is accelerated, it cannot be used as a newtonian frame of reference. The same result will be obtained for the thrust, however, by using a reference frame at rest with respect to the atmosphere, since the air particles will then be observed to enter the engine with no velocity and to leave it with a velocity of magnitude $u - v$.

area of the slipstream. Since the pressure all around S is atmospheric, the only external force acting on S is the thrust of the fan.

Airplane Propeller. In order to obtain a steady stream of air, velocities should be measured with respect to the airplane. Thus, the air particles will be assumed to enter the system with a velocity **v** of magnitude equal to the speed of the airplane and to leave with a velocity **u** equal to the relative velocity of the slipstream.

***14.15. Systems Gaining or Losing Mass.** We shall now analyze a different type of variable system of particles, namely, a system which gains mass by continuously absorbing particles or loses mass by continuously expelling particles. Consider the system S shown in Fig. 14.21. Its mass, equal to m at the instant t, increases by Δm in the interval of time Δt. In order to apply the principle of impulse and momentum to the analysis of this system, we must consider at time t the system

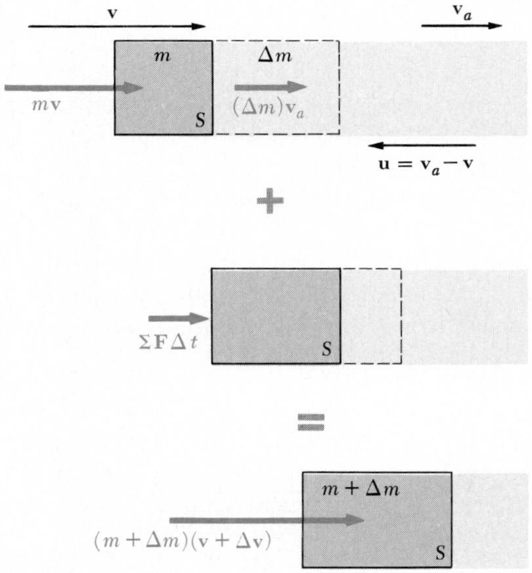

Fig. 14.21

S *plus* the particles of mass Δm which S absorbs during the time interval Δt. The velocity of S at time t is denoted by **v**, and its velocity at time $t + \Delta t$ is denoted by $\mathbf{v} + \Delta\mathbf{v}$, while the absolute velocity of the particles which are absorbed is denoted by $\mathbf{v}_a$. Applying the principle of impulse and momentum, we write

$$m\mathbf{v} + (\Delta m)\mathbf{v}_a + \Sigma\mathbf{F}\,\Delta t = (m + \Delta m)(\mathbf{v} + \Delta\mathbf{v})$$

Solving for the sum $\Sigma\mathbf{F}\,\Delta t$ of the impulses of the external forces acting on S (excluding the forces exerted by the particles being absorbed), we have

$$\Sigma\mathbf{F}\,\Delta t = m\,\Delta\mathbf{v} + \Delta m(\mathbf{v} - \mathbf{v}_a) + (\Delta m)(\Delta\mathbf{v}) \qquad (14.40)$$

Introducing the *relative velocity* $\mathbf{u}$ with respect to S of the particles which are absorbed, we write $\mathbf{u} = \mathbf{v}_a - \mathbf{v}$ and note, since $v_a < v$, that the relative velocity $\mathbf{u}$ is directed to the left, as shown in Fig. 14.21. Neglecting the last term in Eq. (14.40), which is of the second order, we write

$$\Sigma\mathbf{F}\,\Delta t = m\,\Delta\mathbf{v} - (\Delta m)\mathbf{u}$$

Dividing through by Δt and letting Δt approach zero, we have at the limit†

$$\Sigma\mathbf{F} = m\frac{d\mathbf{v}}{dt} - \frac{dm}{dt}\mathbf{u} \qquad (14.41)$$

Rearranging the terms, we obtain the equation

$$\Sigma\mathbf{F} + \frac{dm}{dt}\mathbf{u} = m\frac{d\mathbf{v}}{dt} \qquad (14.42)$$

which shows that the action on S of the particles being absorbed is equivalent to a thrust of magnitude $(dm/dt)u$ which tends to slow down the motion of S, since the relative velocity $\mathbf{u}$ of the particles is directed to the left. If SI units are used, dm/dt is expressed in kg/s, the relative velocity u in m/s, and the corresponding thrust in newtons. If U.S. customary units are used, dm/dt must be expressed in slugs/s and u in ft/s; the corresponding thrust will then be expressed in pounds.‡

The equations obtained may also be used to determine the motion of a system S losing mass. In this case, the rate of change of mass is negative, and the action on S of the particles being expelled is equivalent to a thrust in the direction of $-\mathbf{u}$, that is, in the direction opposite to that in which the particles are being expelled. A *rocket* represents a typical case of a system continuously losing mass (see Sample Prob. 14.12).

†When the absolute velocity $\mathbf{v}_a$ of the particles absorbed is zero, we have $\mathbf{u} = -\mathbf{v}$, and formula (14.41) becomes

$$\Sigma\mathbf{F} = \frac{d}{dt}(m\mathbf{v})$$

Comparing the formula obtained to Eq. (14.2) of Sec. 14.1, we observe that Newton's second law may be applied to a system gaining mass, *provided that the particles absorbed are initially at rest.* It may also be applied to a system losing mass, *provided that the velocity of the particles expelled is zero* with respect to the frame of reference selected.

‡See footnote on page 568.

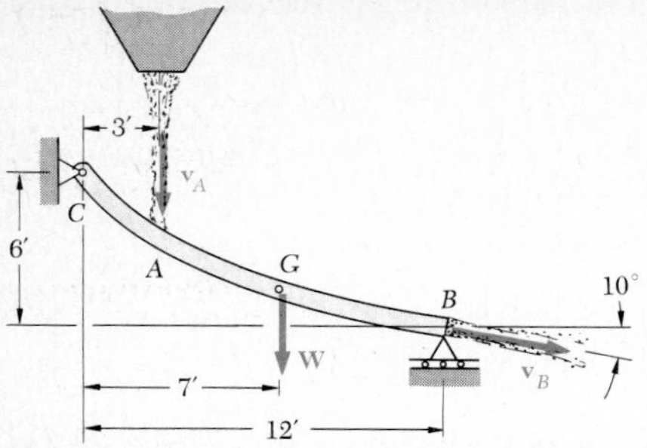

SAMPLE PROBLEM 14.10

Grain falls from a hopper onto a chute CB at the rate of 240 lb/s. It hits the chute at A with a velocity of 20 ft/s and leaves at B with a velocity of 15 ft/s, forming an angle of 10° with the horizontal. Knowing that the combined weight of the chute and of the grain it supports is a force $\mathbf{W}$ of magnitude 600 lb applied at G, determine the reaction at the roller-support B and the components of the reaction at the hinge C.

Solution. We apply the principle of impulse and momentum for the time interval Δt to the system consisting of the chute, the grain it supports, and the amount of grain which hits the chute in the interval Δt. Since the chute does not move, it has no momentum. We also note that the sum $\Sigma m\mathbf{v}$ of the momenta of the particles supported by the chute is the same at t and $t + \Delta t$ and thus may be omitted.

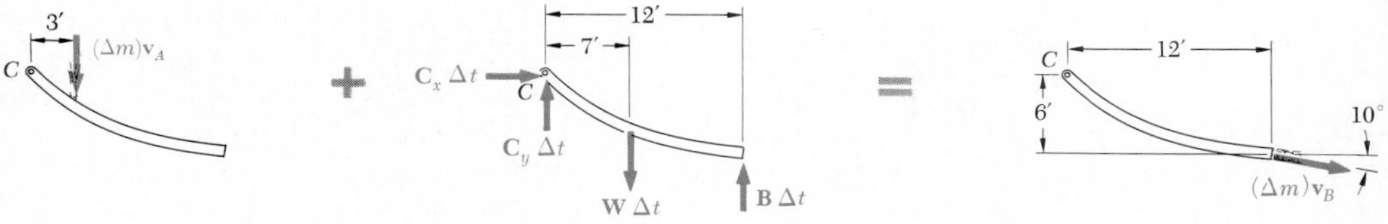

Since the system formed by the momentum $(\Delta m)\mathbf{v}_A$ and the impulses is equipollent to the momentum $(\Delta m)\mathbf{v}_B$, we write

$\xrightarrow{+}$ x components: $\quad C_x\,\Delta t = (\Delta m)v_B \cos 10° \quad (1)$

$+\uparrow y$ components: $\quad -(\Delta m)v_A + C_y\,\Delta t - W\,\Delta t + B\,\Delta t = -(\Delta m)v_B \sin 10° \quad (2)$

$+\circlearrowleft$ moments about C: $\quad -3(\Delta m)v_A - 7(W\,\Delta t) + 12(B\,\Delta t) = 6(\Delta m)v_B \cos 10° - 12(\Delta m)v_B \sin 10° \quad (3)$

Using the given data, $W = 600$ lb, $v_A = 20$ ft/s, $v_B = 15$ ft/s, $\Delta m/\Delta t = 240/32.2 = 7.45$ slugs/s, and solving Eq. (3) for B and Eq. (1) for C_x:

$$12B = 7(600) + 3(7.45)(20) + 6(7.45)(15)(\cos 10° - 2\sin 10°)$$

$$12B = 5075 \qquad B = 423 \text{ lb} \qquad \mathbf{B} = 423 \text{ lb}\uparrow \blacktriangleleft$$

$$C_x = (7.45)(15)\cos 10° = 110.1 \text{ lb} \qquad \mathbf{C}_x = 110.1 \text{ lb} \rightarrow \blacktriangleleft$$

Substituting for B and solving Eq. (2) for C_y:

$$C_y = 600 - 423 + (7.45)(20 - 15\sin 10°) = 307 \text{ lb}$$

$$\mathbf{C}_y = 307 \text{ lb}\uparrow \blacktriangleleft$$

SAMPLE PROBLEM 14.11

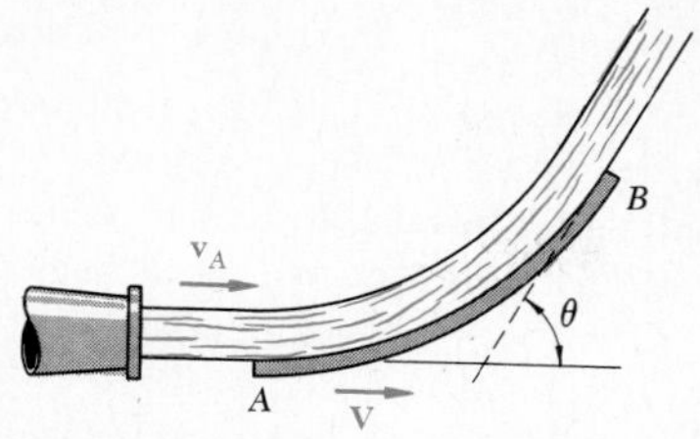

A nozzle discharges a stream of water of cross-sectional area A with a velocity $\mathbf{v}_A$. The stream is deflected by a *single* blade which moves to the right with a constant velocity $\mathbf{V}$. Assuming that the water moves along the blade at constant speed, determine (*a*) the components of the force $\mathbf{F}$ exerted by the blade on the stream, (*b*) the velocity $\mathbf{V}$ for which maximum power is developed.

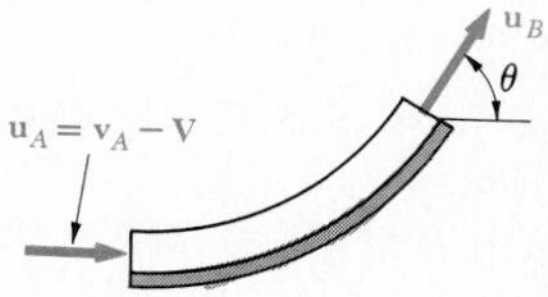

***a.* Components of Force Exerted on Stream.** We choose a coordinate system which moves with the blade at a constant velocity $\mathbf{V}$. The particles of water strike the blade with a relative velocity $\mathbf{u}_A = \mathbf{v}_A - \mathbf{V}$ and leave the blade with a relative velocity $\mathbf{u}_B$. Since the particles move along the blade at a constant speed, the relative velocities $\mathbf{u}_A$ and $\mathbf{u}_B$ have the same magnitude u. Denoting the density of water by ρ, the mass of the particles striking the blade during the time interval Δt is $\Delta m = A\rho(v_A - V)\,\Delta t$; an equal mass of particles leaves the blade during Δt. We apply the principle of impulse and momentum to the system formed by the particles in contact with the blade and by those striking the blade in the time Δt.

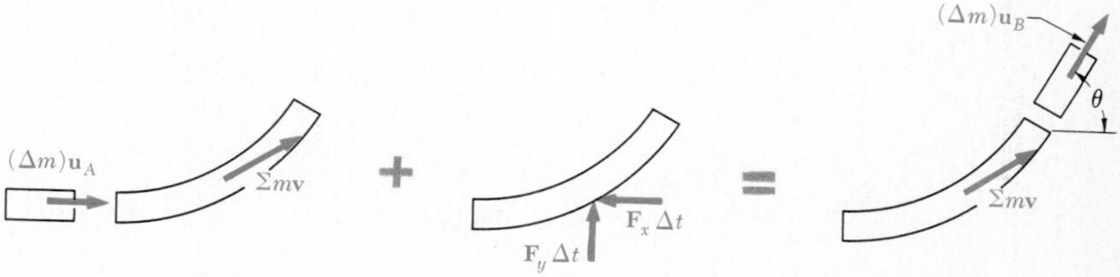

Recalling that $\mathbf{u}_A$ and $\mathbf{u}_B$ have the same magnitude u, and omitting the momentum $\Sigma m\mathbf{v}$ which appears on both sides, we write

$\overset{+}{\rightarrow} x$ components: $\quad (\Delta m)u - F_x\,\Delta t = (\Delta m)u\cos\theta$

$+\!\uparrow y$ components: $\quad +F_y\,\Delta t = (\Delta m)u\sin\theta$

Substituting $\Delta m = A\rho(v_A - V)\,\Delta t$ and $u = v_A - V$, we obtain

$$\mathbf{F}_x = A\rho(v_A - V)^2(1 - \cos\theta) \leftarrow \qquad \mathbf{F}_y = A\rho(v_A - V)^2 \sin\theta \uparrow \quad ◄$$

***b.* Velocity of Blade for Maximum Power.** The power is obtained by multiplying the velocity V of the blade by the component F_x of the force exerted by the stream on the blade.

$$\text{Power} = F_x V = A\rho(v_A - V)^2(1 - \cos\theta)V$$

Differentiating the power with respect to V and setting the derivative equal to zero, we obtain

$$\frac{d(\text{power})}{dV} = A\rho(v_A^2 - 4v_AV + 3V^2)(1 - \cos\theta) = 0$$

$$V = v_A \qquad V = \tfrac{1}{3}v_A \qquad \text{For maximum power } \mathbf{V} = \tfrac{1}{3}v_A \rightarrow \quad ◄$$

Note. These results are valid only when a *single* blade deflects the stream. Different results are obtained when a series of blades deflects the stream, as in a Pelton-wheel turbine. (See Prob. 14.97.)

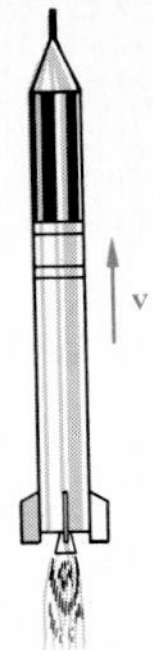

SAMPLE PROBLEM 14.12

A rocket of initial mass m_0 (including shell and fuel) is fired vertically at time $t = 0$. The fuel is consumed at a constant rate $q = dm/dt$ and is expelled at a constant speed u relative to the rocket. Derive an expression for the velocity of the rocket at time t, neglecting the resistance of the air.

Solution. At time t, the mass of the rocket shell and remaining fuel is $m = m_0 - qt$, and the velocity is $\mathbf{v}$. During the time interval Δt, a mass of fuel $\Delta m = q\,\Delta t$ is expelled with a speed u relative to the rocket. Denoting by $\mathbf{v}_e$ the absolute velocity of the expelled fuel, we apply the principle of impulse and momentum between time t and time $t + \Delta t$.

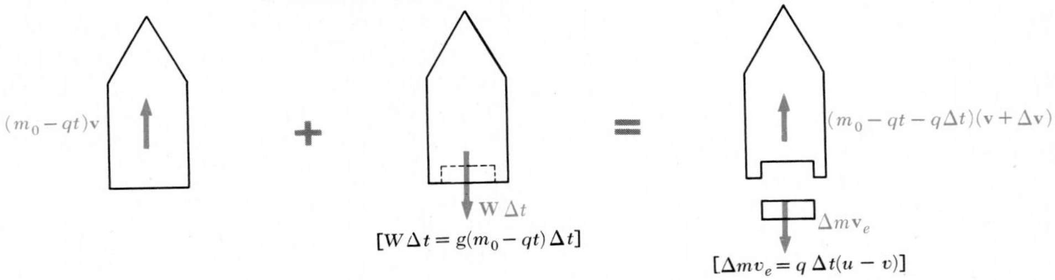

We write

$$(m_0 - qt)v - g(m_0 - qt)\,\Delta t = (m_0 - qt - q\,\Delta t)(v + \Delta v) - q\,\Delta t(u - v)$$

Dividing through by Δt, and letting Δt approach zero, we obtain

$$-g(m_0 - qt) = (m_0 - qt)\frac{dv}{dt} - qu$$

Separating variables and integrating from $t = 0$, $v = 0$ to $t = t$, $v = v$,

$$dv = \left(\frac{qu}{m_0 - qt} - g\right) dt \qquad \int_0^v dv = \int_0^t \left(\frac{qu}{m_0 - qt} - g\right) dt$$

$$v = [-u \ln (m_0 - qt) - gt]_0^t \qquad v = u \ln \frac{m_0}{m_0 - qt} - gt \quad ◄$$

Remark. The mass remaining at time t_f, after all the fuel has been expended, is equal to the mass of the rocket shell $m_s = m_0 - qt_f$, and the maximum velocity attained by the rocket is $v_m = u \ln (m_0/m_s) - gt_f$. Assuming that the fuel is expelled in a relatively short period of time, the term gt_f is small and we have $v_m \approx u \ln (m_0/m_s)$. In order to escape the gravitational field of the earth, a rocket must reach a velocity of 11 180 m/s. Assuming $u = 2200$ m/s and $v_m = 11\,180$ m/s, we obtain $m_0/m_s = 161$. Thus, to project each kilogram of the rocket shell into space, it is necessary to consume more than 161 kg of fuel if a propellant yielding $u = 2200$ m/s is used.

PROBLEMS

Note. In the following problems use $\rho = 1000\ \text{kg/m}^3$ for the density of water in SI units, and $\gamma = 62.4\ \text{lb/ft}^3$ for its specific weight in U.S. customary units. (See footnote on page 568.)

14.77 A stream of water of cross-sectional area A and velocity $\mathbf{v}_1$ strikes a plate which is held motionless by a force $\mathbf{P}$. Determine the magnitude of $\mathbf{P}$ when $A = 500\ \text{mm}^2$, $v_1 = 40\ \text{m/s}$, and $V = 0$.

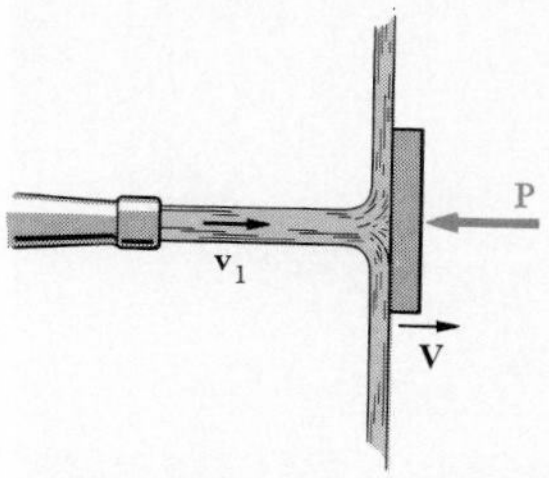

Fig. P14.77 and P14.78

14.78 A stream of water of cross-sectional area A and velocity $\mathbf{v}_1$ strikes a plate which moves to the right with a velocity $\mathbf{V}$. Determine the magnitude of the force $\mathbf{P}$ required to hold the plate when $A = 2\ \text{in.}^2$, $v_1 = 120\ \text{ft/s}$, and $V = 30\ \text{ft/s}$.

14.79 Water flows in a continuous sheet from between two plates A and B with a velocity $\mathbf{v}$. The stream is split into two equal streams *1* and *2* by a vane attached to plate C. Denoting the total rate of flow by Q, determine the force exerted by the stream on plate C.

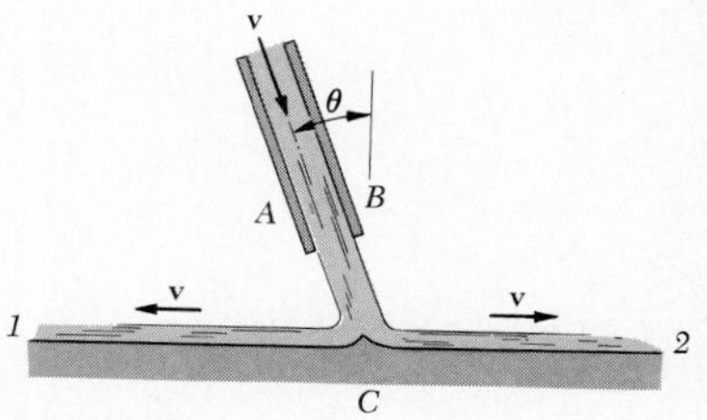

Fig. P14.79

14.80 Water flows in a continuous sheet from between two plates A and B with a velocity $\mathbf{v}$. The stream is split into two parts by a smooth horizontal plate C. Denoting the total rate of flow by Q, determine the rate of flow of each of the resulting streams. (*Hint.* The plate C can exert only a vertical force on the water.)

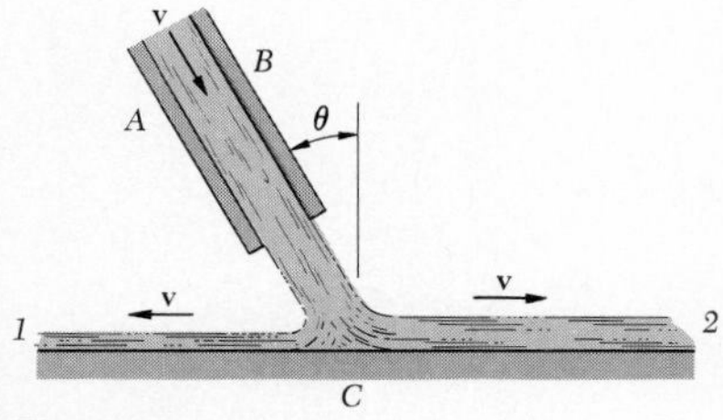

Fig. P14.80

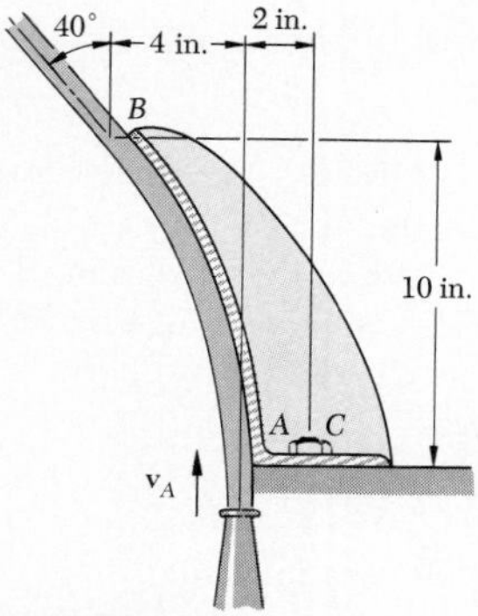

Fig. P14.81

14.81 The nozzle shown discharges 200 gal/min of water with a velocity $\mathbf{v}_A$ of 100 ft/s. The stream is deflected by the fixed vane AB. Determine the force-couple system which must be applied at C in order to hold the vane in place ($1\ \text{ft}^3 = 7.48\ \text{gal}$).

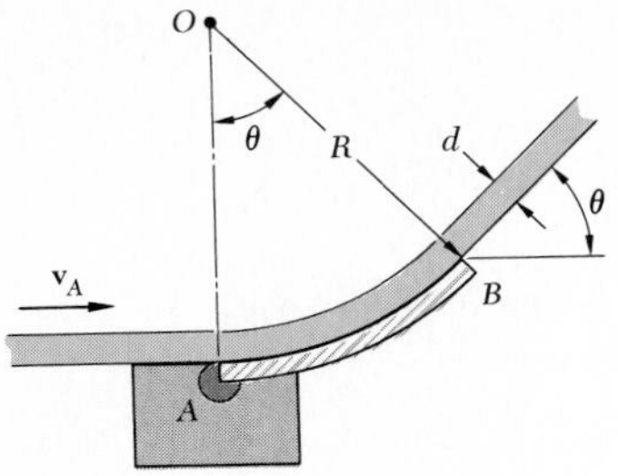

Fig. P14.82

14.82 A stream of water of cross-sectional area A and velocity $\mathbf{v}_A$ is deflected by a vane AB in the shape of an arc of circle of radius R. Knowing that the vane is welded to a fixed support at A, determine the components of the force-couple system exerted by the support on the vane.

14.83 Knowing that the blade AB of Sample Prob. 14.11 is in the shape of an arc of circle, show that the resultant force $\mathbf{F}$ exerted by the blade on the stream is applied at the midpoint C of the arc AB. (*Hint.* First show that the line of action of $\mathbf{F}$ must pass through the center O of the circle.)

14.84 The nozzle shown discharges water at the rate of $1.2\ \text{m}^3/\text{min}$. Knowing that at both A and B the stream of water moves with a velocity of magnitude 25 m/s and neglecting the weight of the vane, determine the components of the reactions at C and D.

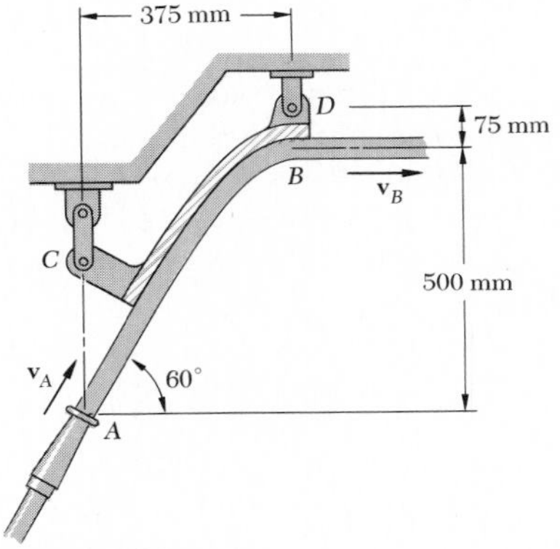

Fig. P14.84

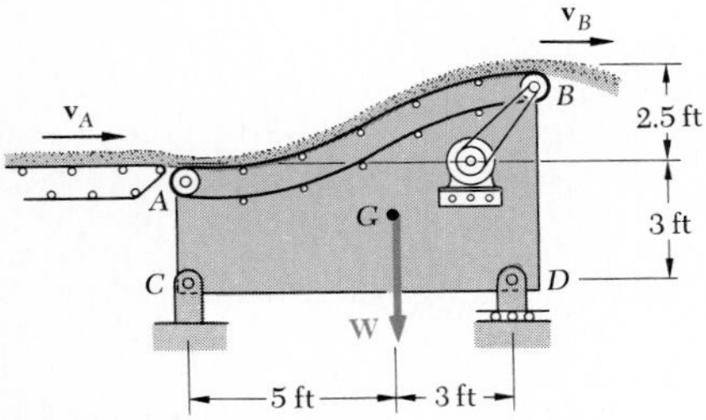

Fig. P14.85

14.85 The final component of a conveyor system receives sand at the rate of 180 lb/s at A and discharges it at B. The sand is moving horizontally at A and B with a velocity of magnitude $v_A = v_B =$ 12 ft/s. Knowing that the combined weight of the component and of the sand it supports is $W = 800$ lb, determine the reactions at C and D.

14.86 Solve Prob. 14.85, assuming the velocity of the belt of the final component of the conveyor system is increased in such a way that, while the sand is still received with a velocity $\mathbf{v}_A$ of 12 ft/s, it is discharged with a velocity $\mathbf{v}_B$ of 24 ft/s.

14.87 A jet airplane scoops in air at the rate of 200 lb/s and discharges it with a velocity of 2000 ft/s relative to the airplane. If the speed of the airplane is 600 mi/h, determine (*a*) the propulsive force developed, (*b*) the horsepower actually used to propel the airplane, (*c*) the horsepower developed by the engine.

14.88 The total drag due to air friction of a jet airplane traveling at 1000 km/h is 16 kN. Knowing that the exhaust velocity is 600 m/s relative to the airplane, determine the mass of air which must pass through the engine per second to maintain the speed of 1000 km/h in level flight.

14.89 While cruising in horizontal flight at a speed of 800 km/h, a 9000-kg jet airplane scoops in air at the rate of 70 kg/s and discharges it with a velocity of 600 m/s relative to the airplane. (*a*) Determine the total drag due to air friction. (*b*) Assuming that the drag is proportional to the square of the speed, determine the horizontal cruising speed if the flow of air through the jet is increased by 10 percent, i.e., to 77 kg/s.

14.90 The maximum speed of a jet airliner is known to be 600 mi/h. Each of the four engines discharges air with a velocity of 2000 ft/s relative to the plane. Assuming that the drag due to air resistance is proportional to the square of the speed, determine the maximum cruising speed of the airliner when only two of the engines are in operation.

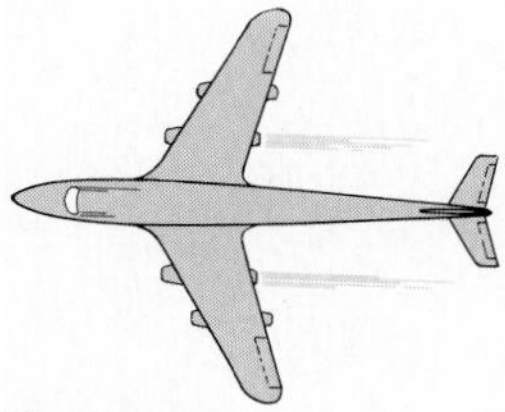

Fig. P14.90

14.91 A 9000-kg jet airplane maintains a constant speed of 900 km/h while climbing at an angle $\alpha = 5°$. The airplane scoops in air at the rate of 80 kg/s and discharges it with a velocity of 700 m/s relative to the airplane. If the pilot changes to a horizontal flight and the same engine conditions are maintained, determine (*a*) the initial acceleration of the plane, (*b*) the maximum horizontal speed attained. Assume that the drag due to air friction is proportional to the square of the speed.

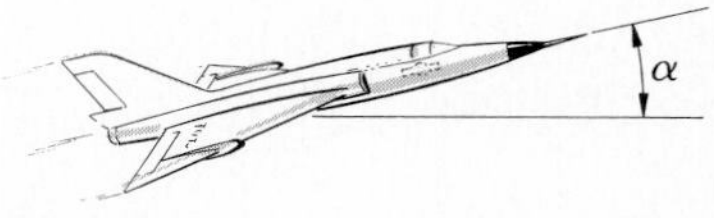

Fig. P14.91

14.92 In order to shorten the distance required for landing, a jet airplane is equipped with movable vanes which partially reverse the direction of the air discharged by each of its engines. Each engine scoops in air at the rate of 200 lb/s and discharges it with a velocity of 2000 ft/s relative to the engine. At an instant when the speed of the airplane is 120 mi/h, determine the reversed thrust provided by each of the engines.

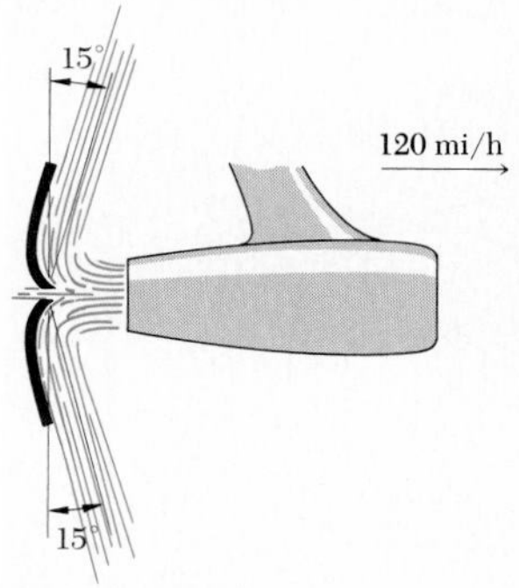

Fig. P14.92

14.93 An unloaded helicopter of weight 1800 lb produces a slipstream of 28-ft diameter. Assuming that air weighs 0.076 lb/ft^3, determine the vertical component of the velocity of the air in the slipstream when the helicopter is hovering in midair.

14.94 The helicopter shown weighs 1800 lb and can produce a maximum downward air speed of 45 ft/s in the 28-ft diameter slipstream. Determine the maximum load which the helicopter can carry while hovering in midair.

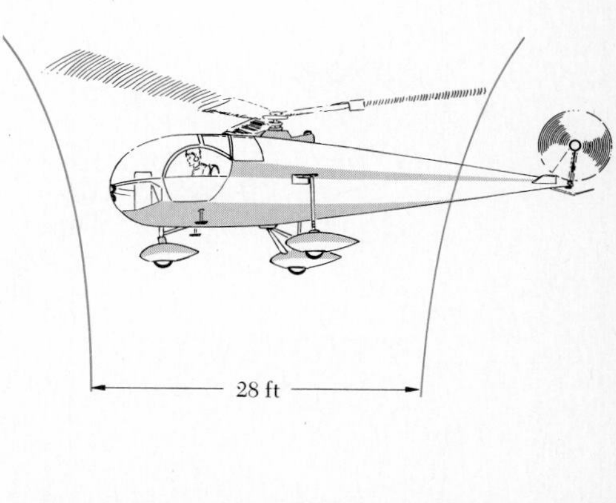

Fig. P14.93 and P14.94

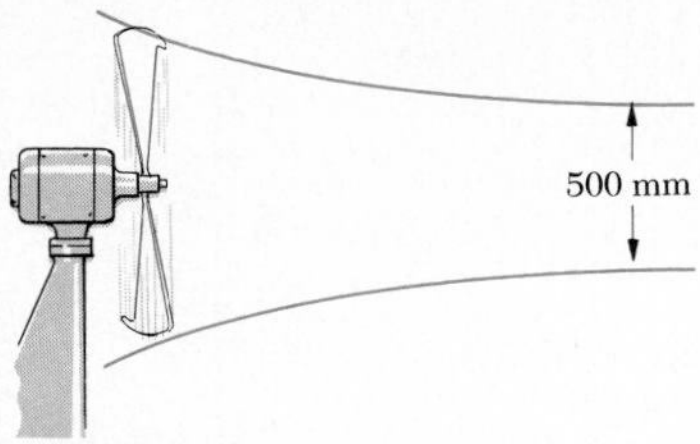

Fig. P14.95

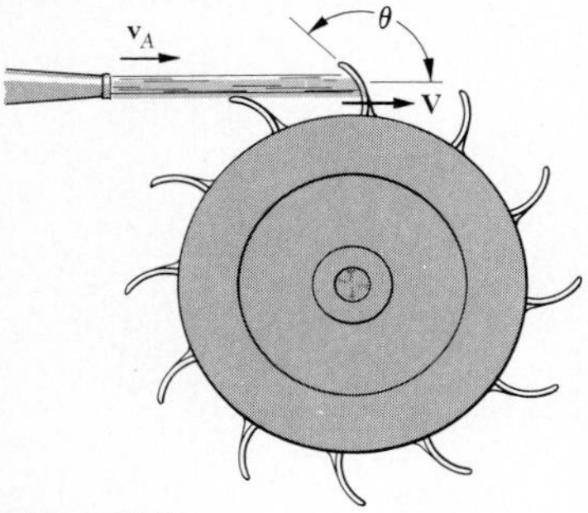

Fig. P14.97

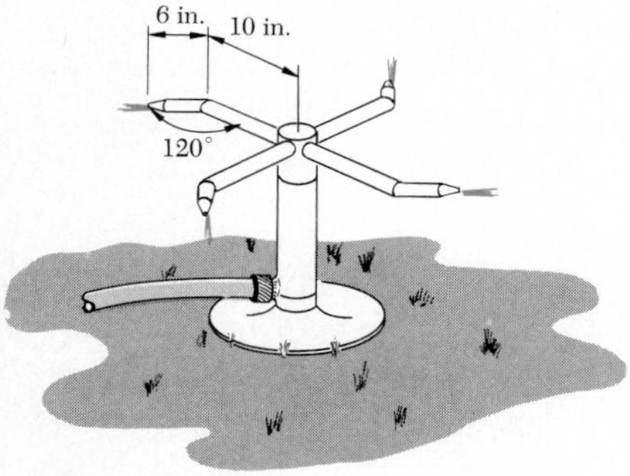

Fig. P14.98

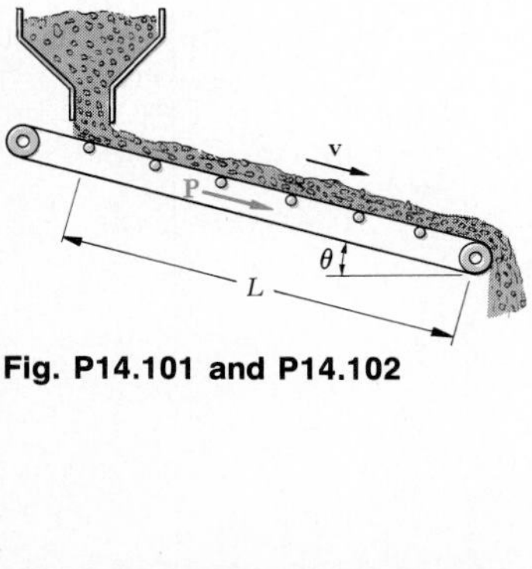

Fig. P14.101 and P14.102

14.95 The slipstream of a fan has a diameter of 500 mm and a velocity of 10 m/s relative to the fan. Assuming $\rho = 1.21\ \text{kg/m}^3$ for air and neglecting the velocity of approach of the air, determine the force required to hold the fan motionless.

14.96 The propeller of an airplane produces a thrust of 4000 N when the airplane is at rest on the ground and has a slipstream of 2-m diameter. Assuming $\rho = 1.21\ \text{kg/m}^3$ for air, determine (*a*) the speed of the air in the slipstream, (*b*) the volume of air passing through the propeller per second, (*c*) the kinetic energy imparted per second to the air of the slipstream.

14.97 In a Pelton-wheel turbine, a stream of water is deflected by a series of blades so that the rate at which water is deflected by the blades is equal to the rate at which water issues from the nozzle ($\Delta m/\Delta t = A\rho v_A$). Using the same notation as in Sample Prob. 14.11, (*a*) determine the velocity **V** of the blades for which maximum power is developed, (*b*) derive an expression for the maximum power, (*c*) derive an expression for the mechanical efficiency.

14.98 A garden sprinkler has four rotating arms, each of which consists of two horizontal straight sections of pipe forming an angle of 120°. Each arm discharges water at the rate of 2.50 gal/min with a velocity of 40 ft/s relative to the arm. Knowing that the friction between the moving and stationary parts of the sprinkler is equivalent to a couple of moment $M = 0.200$ lb · ft, determine the constant rate at which the sprinkler rotates (1 ft^3 = 7.48 gal).

14.99 Solve Prob. 14.98, assuming that the flow is increased and that each arm now discharges water at the rate of 3.00 gal/min with a velocity of 48 ft/s relative to the arm.

14.100 The kinetic energy available per second from the stream of Sample Prob. 14.11 is $\frac{1}{2}Q\rho v_A^2$, where Q is the flow of water per second, $Q = Av_A$. (*a*) Derive an expression for the mechanical efficiency of the single blade. (*b*) Determine the maximum possible value of the efficiency.

14.101 Gravel falls with practically zero velocity onto a conveyor belt at the constant rate $q = dm/dt$. A force **P** is applied to the belt to maintain a constant speed v. Derive an expression for the angle θ for which the force **P** is zero.

14.102 Gravel falls with practically zero velocity onto a conveyor belt at the constant rate $q = dm/dt$. (*a*) Determine the magnitude of the force **P** required to maintain a constant belt speed v, when $\theta = 0$. (*b*) Show that the kinetic energy acquired by the gravel in a given time interval is equal to half the work done in that interval by the force **P**. Explain what happens to the other half of the work done by **P**.

REVIEW PROBLEMS

14.118 Collar B has an initial velocity of 2 m/s. It strikes collar A causing a series of impacts involving the collars and the fixed support at C. Assuming $e = 1$ for all impacts and neglecting friction, determine (*a*) the number of impacts which will occur, (*b*) the final velocity of B, (*c*) the final position of A.

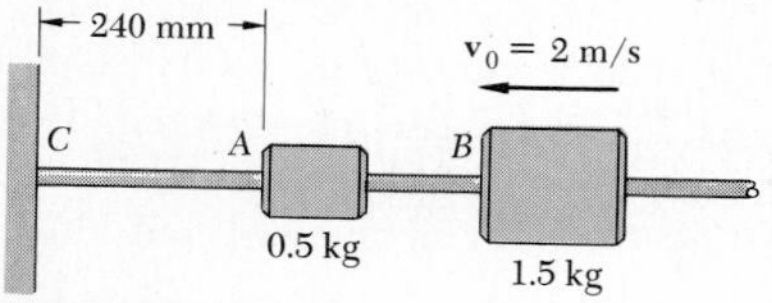

Fig. P14.118

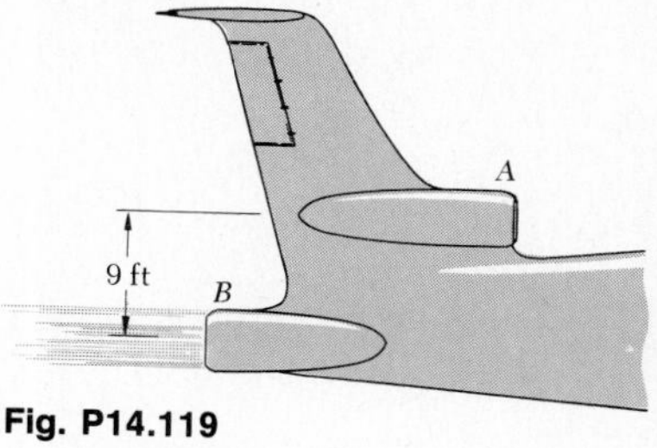

Fig. P14.119

14.119 The jet engine shown scoops in air at A at the rate of 145 lb/s and discharges it at B with a velocity of 2000 ft/s relative to the airplane. Determine the magnitude and line of action of the propulsive thrust developed by the engine when the speed of the airplane is (*a*) 300 mi/h, (*b*) 600 mi/h.

14.120 Solve Prob. 14.119, including the effect of the fuel which is consumed by the engine at the rate of 3 lb/s.

14.121 A 5-kg sphere is moving with a velocity of 60 m/s when it explodes into two fragments. Immediately after the explosion the fragments are observed to travel in the directions shown and the speed of fragment A is observed to be 90 m/s. Determine (*a*) the mass of fragment A, (*b*) the speed of fragment B.

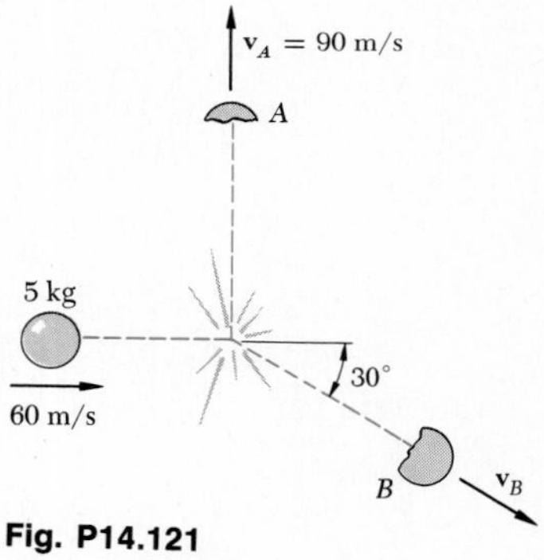

Fig. P14.121

14.122 Show that the values v_1 and v_2 of the speed of a satellite at the perigee A and the apogee A' of an elliptic orbit are defined by the relations

$$v_1^2 = \frac{2GM}{r_1 + r_2}\frac{r_2}{r_1} \qquad v_2^2 = \frac{2GM}{r_1 + r_2}\frac{r_1}{r_2}$$

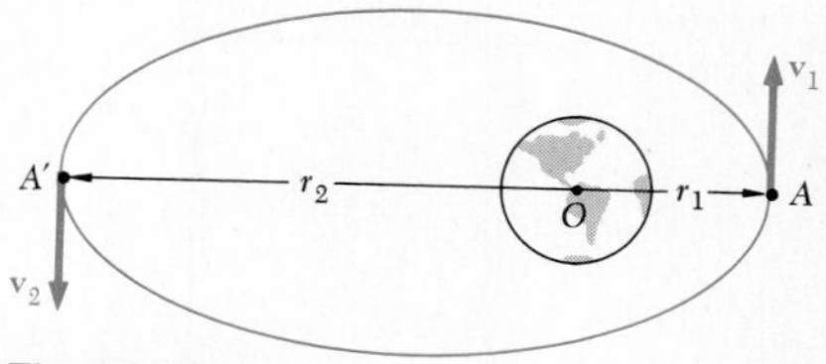

Fig. P14.122

14.123 A space vehicle equipped with a retrorocket, which may expel fuel with a relative velocity $\mathbf{u}$, is moving with a velocity $\mathbf{v}_0$. Denoting by m_s the net mass of the vehicle and by m_f the mass of the unexpended fuel, determine the minimum ratio m_f/m_s for which the velocity of the vehicle can be reduced to zero.

14.124 A chain of length l and mass m per unit length falls through a small hole in a plate. Initially, when y is very small, the chain is at rest. In each case shown, determine (*a*) the acceleration of the first link A as a function of y, (*b*) the velocity of the chain as the last link passes through the hole. In case *1* assume that the individual links are at rest until they fall through the hole; in case *2* assume that at any instant all links have the same speed. Ignore the effect of friction.

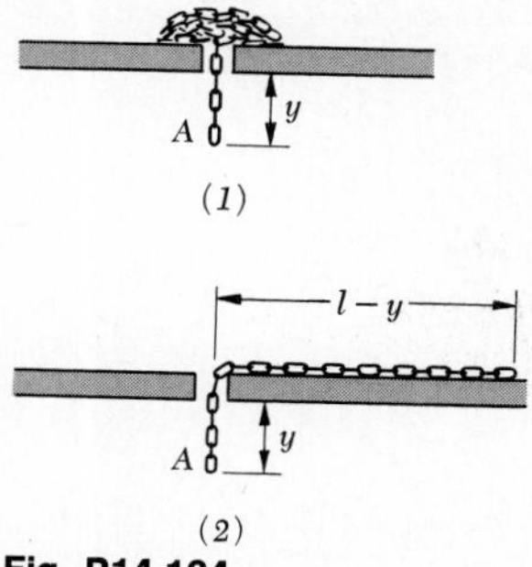

Fig. P14.124

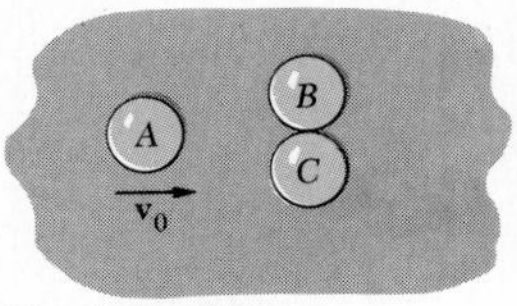

Fig. P14.125

14.125 Three identical balls A, B, and C may roll freely on a horizontal surface. Balls B and C are at rest and in contact when struck by ball A, which was moving to the right with a velocity $\mathbf{v}_0$. Assuming $e = 1$, determine the final velocity of ball A assuming (a) that the path of A is perfectly centered and that A strikes B and C simultaneously, (b) that the path of A is not perfectly centered and that A strikes B slightly before it strikes C.

14.126 A 1-oz bullet is fired with a velocity of 1600 ft/s into block A, which weighs 10 lb. The coefficient of friction between block A and the cart BC is 0.50. Knowing that the cart weighs 8 lb and can roll freely, determine (a) the final velocity of the cart and block, (b) the final position of the block on the cart.

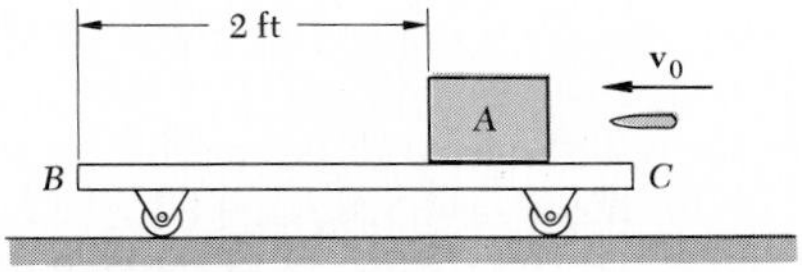

Fig. P14.126

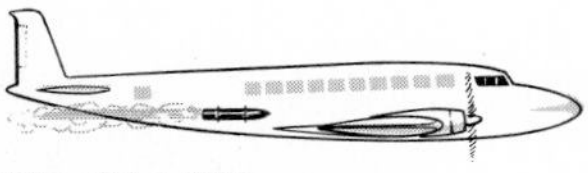

Fig. P14.127

14.127 A booster rocket is attached to the fuselage of an airplane weighing 40,000 lb. The rocket fuel is consumed at the rate of 10 lb/s and is ejected with a relative velocity of 5000 ft/s. Determine the additional propulsive thrust available while the rocket is being fired (a) if the speed of the airplane is 400 mi/h, (b) if the airplane is at rest on the ground.

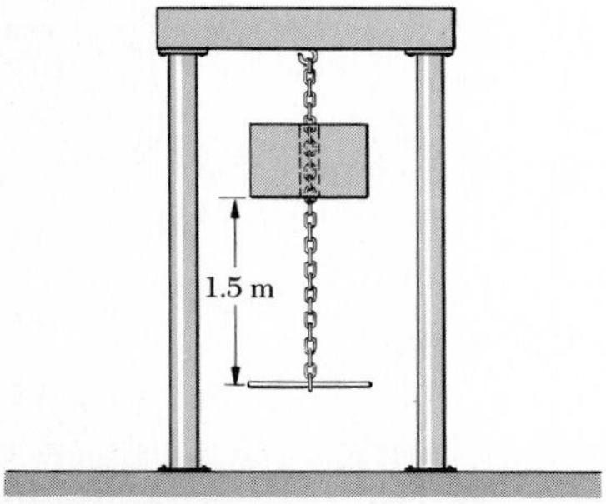

Fig. P14.128

14.128 In order to test the resistance of a chain to impact, the chain is suspended from a 100-kg block supported by two columns. A rod attached to the last link of the chain is then hit by a 25-kg cylinder dropped from a 1.5-m height. Determine the initial impulse exerted on the chain, assuming that the impact is perfectly plastic and that the columns supporting the dead weight (a) are perfectly rigid, (b) are equivalent to two perfectly elastic springs. (c) Determine the energy absorbed by the chain in parts a and b.

14.129 The 20-Mg truck and the 40-Mg railroad flatcar are both at rest with their brakes released. An engine bumps the flatcar and causes the flatcar alone to start moving with a velocity of 1 m/s to the right. Assuming $e = 1$ between the truck and the ends of the flatcar, determine the velocities of the truck and of the flatcar after end A strikes the truck. Describe the subsequent motion of the system. Neglect the effect of friction.

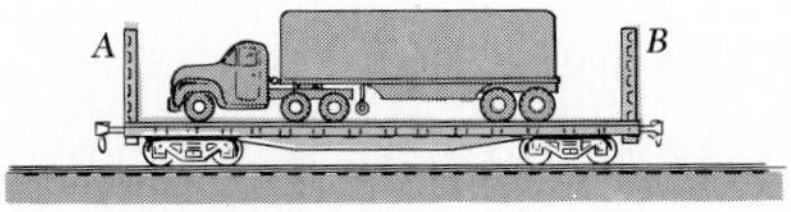

Fig. P14.129

Chapter 15

Kinematics of Rigid Bodies

15.1. Various Types of Plane Motion. In this chapter, we shall study the kinematics of *rigid bodies*. We shall investigate the relations existing between the time, the positions, the velocities, and the accelerations of the various particles forming a rigid body. Our study will be limited to that of the *plane motion* of a rigid body, which is characterized by the fact that each particle of the body remains at a constant distance from a fixed reference plane. Thus all particles of the body move in parallel planes, and the motion of the body may be represented by the motion of a representative slab in the reference plane.

The various types of plane motion may be grouped as follows:

1. *Translation.* A motion is said to be a translation if any straight line drawn on the body keeps the same direction during the motion. It may also be observed that in a translation all the particles forming the body move along parallel paths. If these paths are straight lines, the motion is said to be a *rectilinear translation* (Fig. 15.1*a*); if the paths are curved lines, the motion is a *curvilinear translation* (Fig. 15.1*b*).

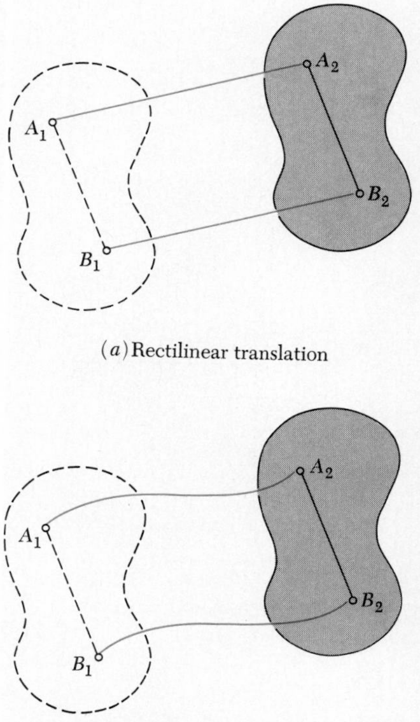

(*a*) Rectilinear translation

(*b*) Curvilinear translation

Fig. 15.1

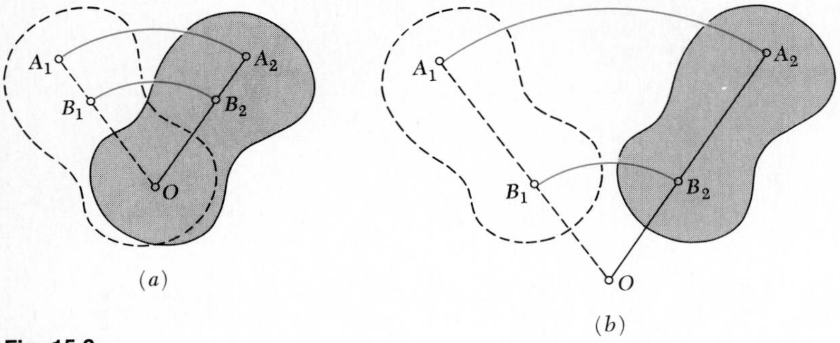

Fig. 15.2

2. *Rotation.* A motion is said to be a rotation when the particles of the representative slab move along concentric circles. If the common center O of the circles is located on the slab, the corresponding point remains fixed (Fig. 15.2*a*). In many cases of rotation, however, the common center O is located outside the slab, and none of the points of the slab remains fixed (Fig. 15.2*b*).

Rotation should not be confused with certain types of curvilinear translation. For example, the plate shown in Fig. 15.3*a* is in curvilinear translation, with all its particles moving along *parallel* circles, while the plate shown in Fig. 15.3*b* is in rotation, with all its particles moving along *concentric* circles.

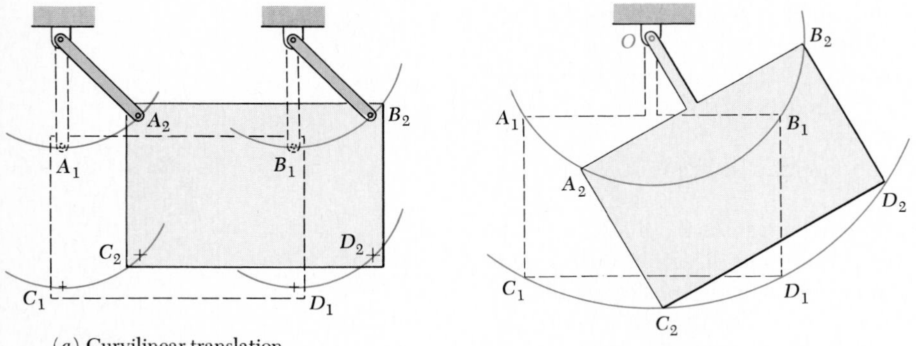

(*a*) Curvilinear translation

(*b*) Rotation

Fig. 15.3

In the first case, any given straight line drawn on the plate will maintain the same direction, while, in the second case, point O remains fixed.

3. *General Plane Motion.* Any plane motion which is neither a translation nor a rotation is referred to as a general plane motion. Two examples of general plane motion are given in Fig. 15.4.

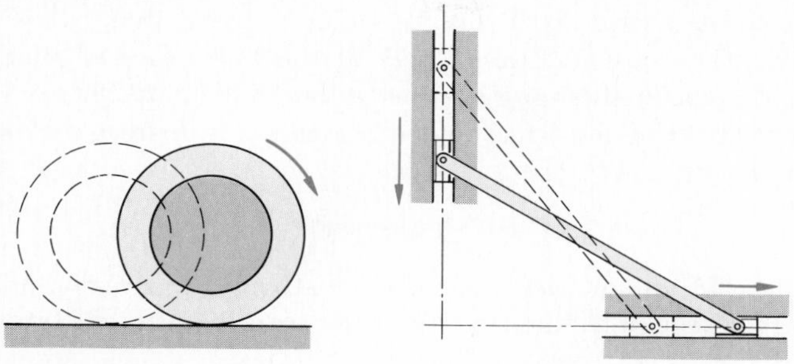

(*a*) Rolling wheel (*b*) Sliding rod

Fig. 15.4 Examples of general plane motion.

15.2. Translation. Consider a rigid slab in translation (either rectilinear or curvilinear translation), and denote by A and B the positions of two of its particles at time t. At time $t + \Delta t$, the two particles will occupy new positions A' and B' (Fig. 15.5*a*). Now, from the very definition of a translation, the line $A'B'$ must be parallel to AB. Besides, since the slab is rigid, we must have $A'B' = AB$. Thus, $ABB'A'$ is a parallelogram, and the two vectors $\Delta\mathbf{r}_A$ and $\Delta\mathbf{r}_B$ representing the displacements of the two particles during the time interval Δt must have the same magnitude and direction. We have, therefore, $\Delta\mathbf{r}_A/\Delta t = \Delta\mathbf{r}_B/\Delta t$ and, at the limit, as Δt approaches zero,

$$\mathbf{v}_A = \mathbf{v}_B \tag{15.1}$$

Since relation (15.1) holds at any time, the vectors $\Delta\mathbf{v}_A$ and $\Delta\mathbf{v}_B$ representing the change in velocity of the two particles during the time interval Δt must be equal. We have, therefore, $\Delta\mathbf{v}_A/\Delta t = \Delta\mathbf{v}_B/\Delta t$ and, at the limit, as Δt approaches zero,

$$\mathbf{a}_A = \mathbf{a}_B \tag{15.2}$$

Thus, *when a rigid slab is in translation, all the points of the slab have the same velocity and the same acceleration at any given instant* (Fig. 15.5*b* and *c*). In the case of curvilinear translation, the velocity and acceleration change in direction as well as in magnitude at every instant. In the case of rectilinear translation, all particles of the slab move along parallel straight lines, and their velocity and acceleration keep the same direction during the entire motion.

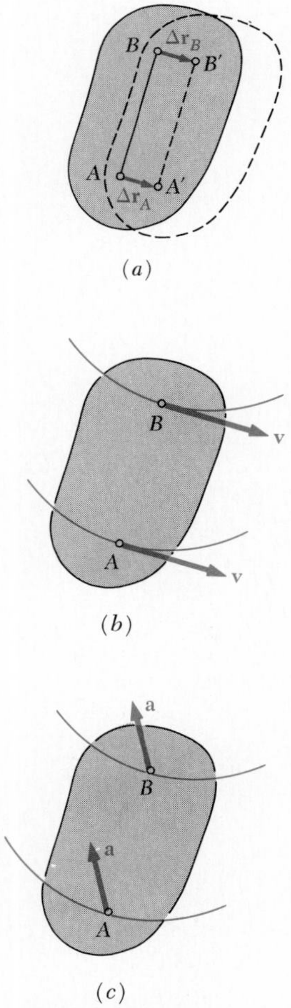

Fig. 15.5

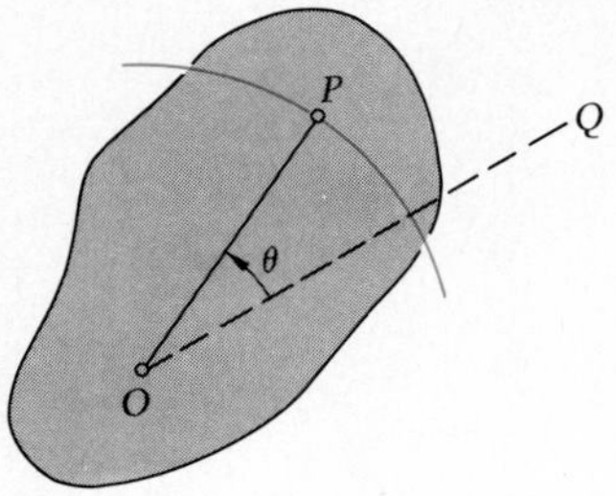

Fig. 15.6

15.3. Rotation. Consider a rigid slab which rotates about a fixed axis perpendicular to the plane of the slab and intersecting it at point O. Let P be a point of the slab (Fig. 15.6). The position of the slab will be entirely defined if the angle θ the line OP forms with a fixed direction OQ is given; the angle θ is known as the *angular coordinate* of the slab. The angular coordinate is defined as positive when counterclockwise and will be expressed in radians (rad) or, occasionally, in degrees (°) or revolutions (rev). We recall that

$$1 \text{ rev} = 2\pi \text{ rad} = 360°$$

In Chap. 11, the first and second derivatives, respectively, of the coordinate x of a particle were used to define the velocity and acceleration of a particle moving in a straight line. Similarly, the first derivative of the angular coordinate θ of a rotating slab may be used to define the *angular velocity* ω of the slab,

$$\omega = \frac{d\theta}{dt} \tag{15.3}$$

and its second derivative may be used to define the *angular acceleration* α of the slab,

$$\alpha = \frac{d\omega}{dt} = \frac{d^2\theta}{dt^2} \tag{15.4}$$

An alternate form for the angular acceleration is obtained by solving (15.3) for dt and substituting into (15.4). We write $dt = d\theta/\omega$ and

$$\alpha = \omega \frac{d\omega}{d\theta} \tag{15.5}$$

The angular velocity ω is expressed in rad/s or in rpm (revolutions per minute). The angular acceleration α is generally expressed in rad/s^2.

Since the angular velocity and the angular acceleration of a slab have a *sense,* as well as a *magnitude,* they should be considered as *vector quantities.*† We may, for instance, represent the angular velocity of the slab of Fig. 15.6 by a vector $\boldsymbol{\omega}$ of magnitude ω directed along the axis of rotation of the slab. This vector will be made to point out of the plane of the figure if the angular velocity is counterclockwise, and into the plane of the figure if the angular velocity is clockwise. The angular

† It may be shown, in the more general case of a rigid body rotating simultaneously about axes having different directions, that angular velocities and angular accelerations obey the parallelogram law of addition.

acceleration may similarly be represented by a vector **α** of magnitude α directed along the axis of rotation and pointing out of, or into, the plane of the figure, depending upon the sense of the angular acceleration. However, since in the cases considered in this chapter the axis of rotation will always be perpendicular to the plane of the figure, we need to specify only the sense and the magnitude of the angular velocity and of the angular acceleration. This may be done by using the scalar quantities ω and α, together with a plus or minus sign. As in the case of the angular coordinate θ, the plus sign corresponds to counterclockwise, and the minus sign to clockwise.

It should be observed that the values obtained for the angular velocity and the angular acceleration of a slab are independent of the choice of the reference point P. If a point P' were chosen instead of P (Fig. 15.7), the angular coordinate would be θ', which differs from θ only by the constant angle POP'. The successive derivatives of θ and θ' are thus equal, and the same values are obtained for ω and α.

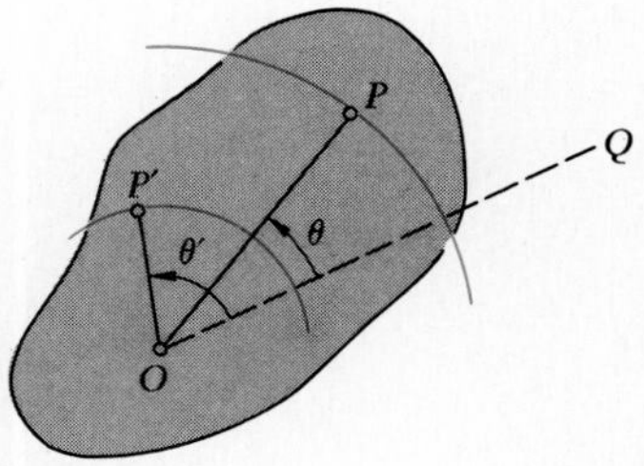

Fig. 15.7

The motion of a slab rotating about a fixed point O is said to be *known* when its coordinate θ may be expressed as a known function of t. In practice, however, the rotation of a slab is seldom defined by a relation between θ and t. More often, the conditions of motion will be specified by the type of angular acceleration that the slab possesses. For example, α may be given as a function of t, or as a function of θ, or as a function of ω. Equations (15.3) to (15.5) may then be used to determine the angular velocity ω and the angular coordinate θ, the procedure to follow being the same as that described in Sec. 11.3 for the rectilinear motion of a particle.

Two particular cases of rotation are frequently encountered:

1. *Uniform Rotation.* This case is characterized by the fact that the angular acceleration is zero. The angular velocity ω is thus constant, and the angular coordinate is given by the formula

$$\theta = \theta_0 + \omega t \tag{15.6}$$

2. *Uniformly Accelerated Rotation.* In this case, the angular acceleration α is constant. The following formulas relating angular velocity, angular coordinate, and time may then be derived in a manner similar to that described in Sec. 11.5. The similitude between the formulas derived here and those obtained for the rectilinear uniformly accelerated motion of a particle is easily noted.

$$\omega = \omega_0 + \alpha t \tag{15.7}$$

$$\theta = \theta_0 + \omega_0 t + \tfrac{1}{2}\alpha t^2 \tag{15.8}$$

$$\omega^2 = \omega_0^2 + 2\alpha(\theta - \theta_0) \tag{15.9}$$

It should be emphasized that formula (15.6) may be used only when $\alpha = 0$, and formulas (15.7) to (15.9) only when $\alpha =$ constant. In any other case, the general formulas (15.3) to (15.5) should be used.

15.4. Linear and Angular Velocity, Linear and Angular Acceleration in Rotation. Consider again a point P on a rotating slab (Fig. 15.8a). Point P describes a circle

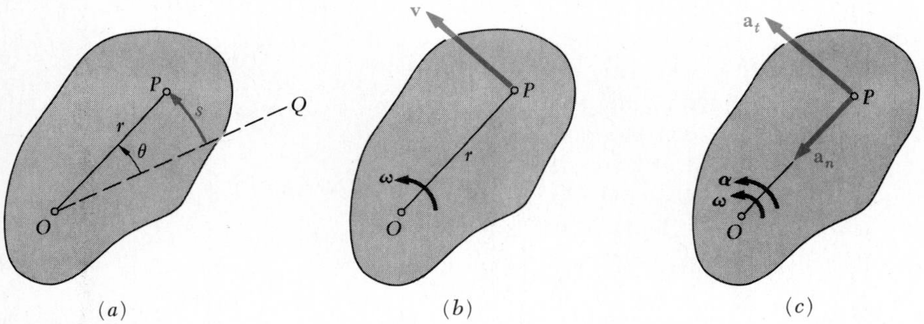

Fig. 15.8

of center O when the slab describes a full rotation. The arc s that P describes while the slab rotates through an angle θ depends not only upon θ but also upon the distance r from O to P. Expressing θ in radians and r and s in the same units of length, we write

$$s = r\theta \tag{15.10}$$

Differentiating, we obtain the following relation between the magnitude v of the *linear velocity* of P (Fig. 15.8b) and the *angular velocity* ω of the slab,

$$\frac{ds}{dt} = r\frac{d\theta}{dt} \qquad v = r\omega \tag{15.11}$$

While the value of the angular velocity ω is independent of the choice of P, the magnitude v of the linear velocity clearly depends upon the distance r. Recalling the definition of the tangential and normal components of the *linear acceleration* of a point P (Sec. 11.12), we write (Fig. 15.8c)

$$a_t = \frac{dv}{dt} = r\frac{d\omega}{dt} \qquad a_t = r\alpha \tag{15.12}$$

$$a_n = \frac{v^2}{r} = \frac{(r\omega)^2}{r} \qquad a_n = r\omega^2 \tag{15.13}$$

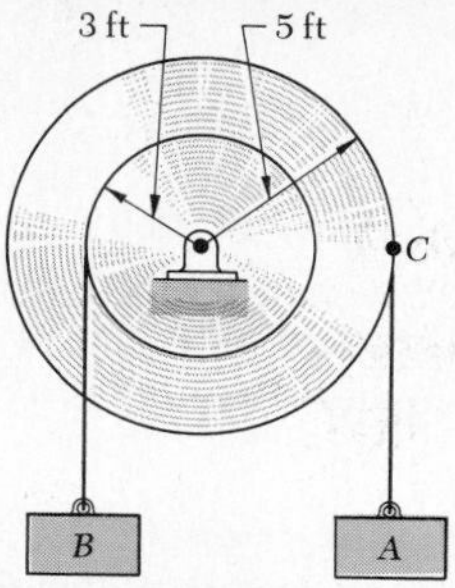

SAMPLE PROBLEM 15.1

A pulley and two loads are connected by inextensible cords as shown. Load A has a constant acceleration of 10 ft/s² and an initial velocity of 15 ft/s, both directed upward. Determine (*a*) the number of revolutions executed by the pulley in 3 s, (*b*) the velocity and position of load B after 3 s, (*c*) the acceleration of point C on the rim of the pulley at $t = 0$.

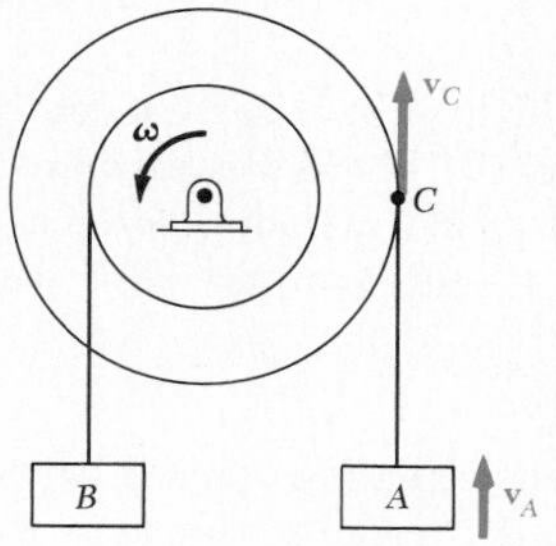

***a.* Motion of Pulley.** Since the cord connecting the pulley to load A is inextensible, the velocity of C is equal to the velocity of A and the tangential component of the acceleration of C is equal to the acceleration of A.

$$(\mathbf{v}_C)_0 = (\mathbf{v}_A)_0 = 15\text{ ft/s}\uparrow \qquad (\mathbf{a}_C)_t = \mathbf{a}_A = 10\text{ ft/s}^2\uparrow$$

Noting that the distance from C to the center of the pulley is 5 ft, we write

$$(v_C)_0 = r\omega_0 \qquad 15\text{ ft/s} = (5\text{ ft})\omega_0 \qquad \omega_0 = 3\text{ rad/s}\ \circlearrowleft$$
$$(a_C)_t = r\alpha \qquad 10\text{ ft/s}^2 = (5\text{ ft})\alpha \qquad \boldsymbol{\alpha} = 2\text{ rad/s}^2\ \circlearrowleft$$

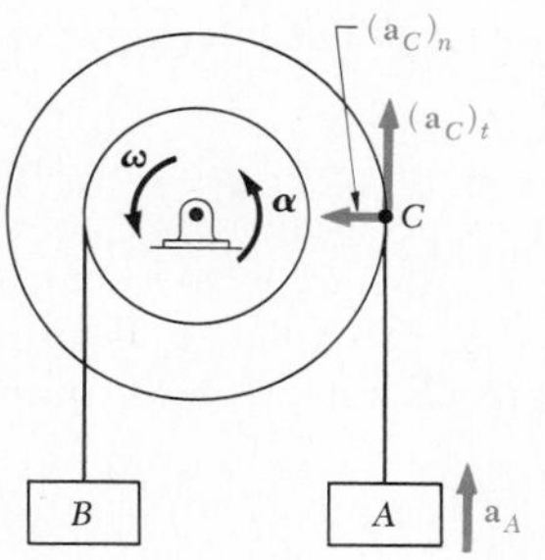

From the equations for uniformly accelerated motion, we obtain, for $t = 3$ s,

$$\omega = \omega_0 + \alpha t = 3\text{ rad/s} + (2\text{ rad/s}^2)(3\text{ s}) = 9\text{ rad/s}$$
$$\boldsymbol{\omega} = 9\text{ rad/s}\ \circlearrowleft$$
$$\theta = \omega_0 t + \tfrac{1}{2}\alpha t^2 = (3\text{ rad/s})(3\text{ s}) + \tfrac{1}{2}(2\text{ rad/s}^2)(3\text{ s})^2$$
$$\theta = 18\text{ rad}$$

$$\text{Number of revolutions} = (18\text{ rad})\left(\frac{1\text{ rev}}{2\pi\text{ rad}}\right) = 2.86\text{ rev} \quad ◄$$

***b.* Motion of Load *B*.** Using the following relations between the linear and angular motion, with $r = 3$ ft, we write

$$v_B = r\omega = (3\text{ ft})(9\text{ rad/s}) \qquad \mathbf{v}_B = 27\text{ ft/s}\downarrow \quad ◄$$
$$s_B = r\theta = (3\text{ ft})(18\text{ rad}) \qquad s_B = 54\text{ ft}\downarrow \quad ◄$$

***c.* Acceleration of Point *C* at $t = 0$.** The tangential component of the acceleration is

$$(\mathbf{a}_C)_t = \mathbf{a}_A = 10\text{ ft/s}^2\uparrow$$

Since, at $t = 0$, $\omega_0 = 3$ rad/s, the normal component of the acceleration is

$$(a_C)_n = r_C\omega_0^2 = (5\text{ ft})(3\text{ rad/s})^2 \qquad (\mathbf{a}_C)_n = 45\text{ ft/s}^2\leftarrow$$

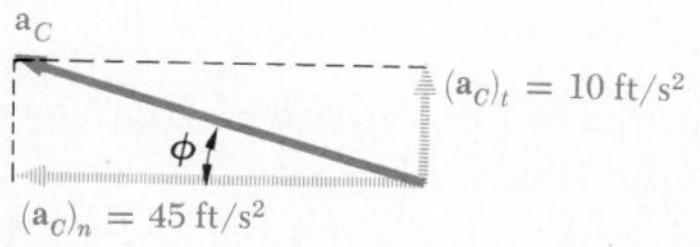

The magnitude and direction of the total acceleration are obtained by writing

$$\tan\phi = (10\text{ ft/s}^2)/(45\text{ ft/s}^2) \qquad \phi = 12.5°$$
$$a_C \sin 12.5° = 10\text{ ft/s}^2 \qquad a_C = 46.1\text{ ft/s}^2$$
$$\mathbf{a}_C = 46.1\text{ ft/s}^2 \ \measuredangle\ 12.5° \quad ◄$$

PROBLEMS

15.1 The motion of a cam is defined by the relation $\theta = t^3 - 2t^2 - 4t + 10$, where θ is expressed in radians and t in seconds. Determine the angular coordinate, the angular velocity, and the angular acceleration of the cam when (*a*) $t = 0$, (*b*) $t = 3$ s.

15.2 For the cam of Prob. 15.1 determine (*a*) the time at which the angular velocity is zero, (*b*) the corresponding angular coordinate and angular acceleration.

15.3 The motion of a disk rotating in an oil bath is defined by the relation $\theta = \theta_0(1 - e^{-t/3})$, where θ is expressed in radians and t in seconds. Knowing that $\theta_0 = 1.20$ rad, determine the angular coordinate, velocity, and acceleration of the disk when (*a*) $t = 0$, (*b*) $t = 3$ s, (*c*) $t = \infty$.

15.4 The motion of an oscillating crank is defined by the relation $\theta = \theta_0 \cos(2\pi t/T)$, where θ is expressed in radians and t in seconds. Knowing that $\theta_0 = 0.75$ rad and $T = 1.20$ s, determine the maximum angular velocity and the maximum angular acceleration of the crank.

15.5 The rotor of a steam turbine is rotating at a speed of 720 rpm when the steam supply is suddenly cut off. It is observed that 5 min are required for the rotor to come to rest. Assuming uniformly accelerated motion, determine (*a*) the angular acceleration, (*b*) the total number of revolutions that the rotor executes before coming to rest.

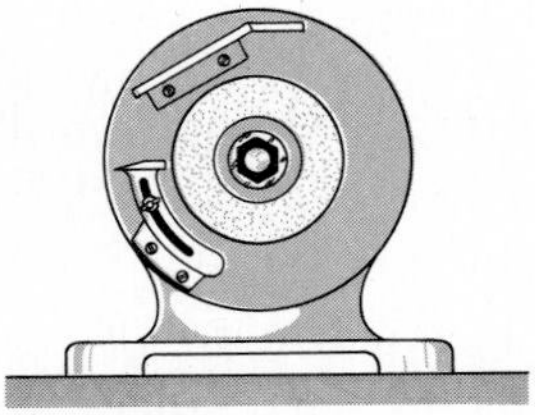

Fig. P15.6

15.6 A small grinding wheel is attached to the shaft of an electric motor which has a rated speed of 1800 rpm. When the power is turned on, the unit reaches its rated speed in 4 s, and when the power is turned off, the unit coasts to rest in 50 s. Assuming uniformly accelerated motion, determine the number of revolutions that the motor executes (*a*) in reaching its rated speed, (*b*) in coasting to rest.

15.7 The rotor of an electric motor has a speed of 1200 rpm when the power is cut off. The rotor is then observed to come to rest after executing 520 revolutions. Assuming uniformly accelerated motion, determine (*a*) the angular acceleration, (*b*) the time required for the rotor to come to rest.

15.8 The angular acceleration of the rotor of an electric motor is directly proportional to the time t. The rotor starts at $t = 0$ with no initial velocity; after 3 s, the rotor has completed 6 revolutions. Write the equations of motion for the rotor, and determine the angular velocity at $t = 2$ s.

15.9 The earth makes one complete revolution about the sun in 365.24 days. Assuming that the orbit of the earth is circular and has a radius of 93,000,000 mi, determine the velocity and acceleration of the earth.

15.10 The earth makes one complete revolution on its axis in 23.93 h. Knowing that the mean radius of the earth is 3960 mi, determine the linear velocity and acceleration of a point on the surface of the earth (*a*) at the equator, (*b*) at Philadelphia, latitude 40° north, (*c*) at the North Pole.

15.11 A small block *B* rests on a horizontal plate which rotates about a fixed vertical axis. If the plate is initially at rest at $t = 0$ and is accelerated at the constant rate α, derive an expression (*a*) for the total acceleration of the block at time *t*, (*b*) for the angle between the total acceleration and the radius *AB* at time *t*.

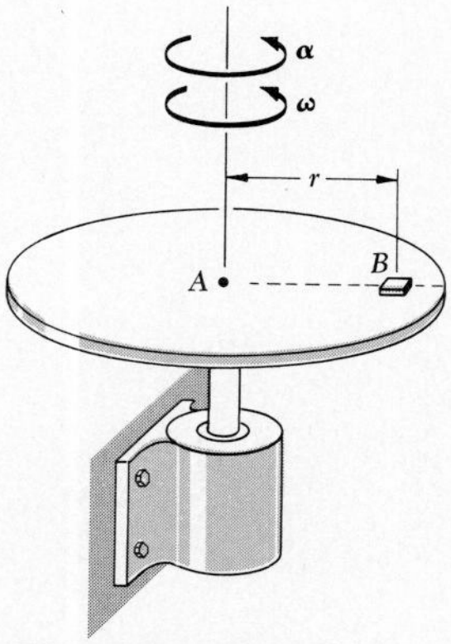

Fig. P15.11 and P15.12

15.12 It is known that the static-friction force between block *B* and the plate will be exceeded and that the block will start sliding on the plate when the total acceleration of the block reaches 5 m/s^2. If the plate starts from rest at $t = 0$ and is accelerated at the constant rate of 6 rad/s^2, determine the time *t* and the angular velocity of the plate when the block starts sliding, assuming $r = 100$ mm.

15.13 At a given instant, the acceleration of the rack is 15 in./s^2 directed downward, and the velocity of the rack is 6 in./s directed upward. Determine (*a*) the angular acceleration and the angular velocity of the gear, (*b*) the total acceleration of the gear tooth in contact with the rack.

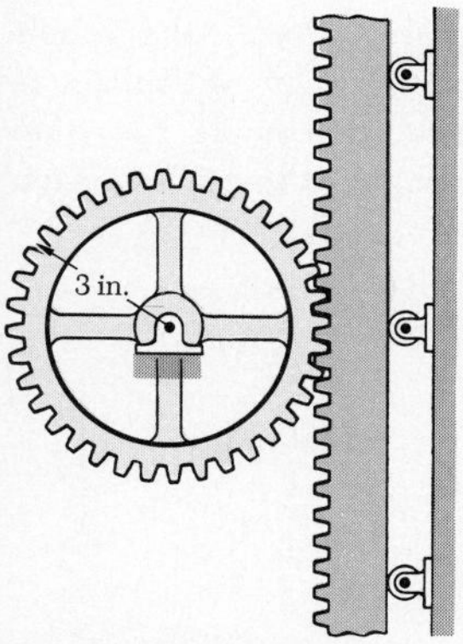

Fig. P15.13 and P15.14

15.14 The gear-and-rack system shown is initially at rest. If the acceleration of the rack is constant and equal to 0.5 ft/s^2 directed upward, determine (*a*) the angular velocity of the gear after the rack has moved 3 ft, (*b*) the time required for the gear to reach an angular velocity of 40 rpm.

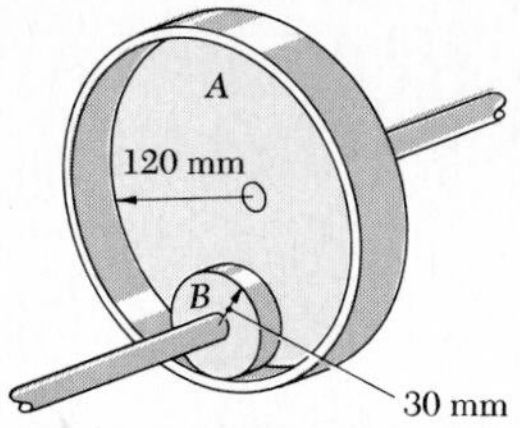

Fig. P15.15 and P15.16

15.15 The friction wheel B executes 100 revolutions about its fixed shaft during the time interval t, while its angular velocity is being increased uniformly from 200 to 600 rpm. Knowing that wheel B rolls without slipping on the inside rim of wheel A, determine (*a*) the angular acceleration of wheel A, (*b*) the time interval t.

15.16 The friction wheel B rotates at the constant angular velocity of 200 rpm about its fixed shaft. Knowing that wheel B rolls without slipping on the inside rim of wheel A, determine (*a*) the angular velocity of wheel A, (*b*) the linear accelerations of the points of A and B which are in contact.

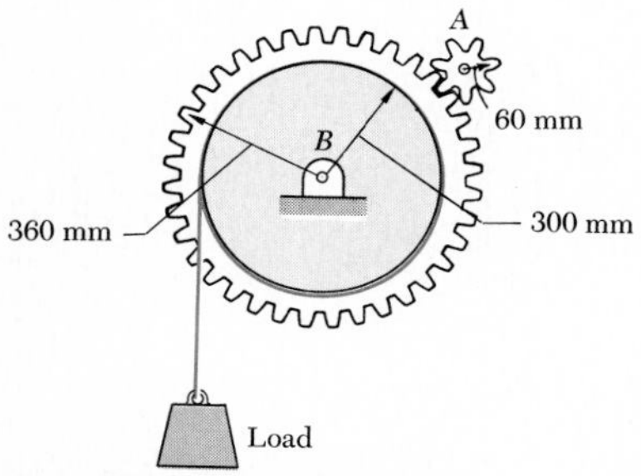

Fig. P15.17 and P15.18

15.17 A load is to be raised 6 m by the hoisting system shown. Assuming gear A is initially at rest, accelerates uniformly to a speed of 120 rpm in 5 s, and then maintains a constant speed of 120 rpm, determine (*a*) the number of revolutions executed by gear A in raising the load, (*b*) the time required to raise the load.

15.18 The system shown starts from rest at $t = 0$ and accelerates uniformly. Knowing that at $t = 4$ s the velocity of the load is 4.8 m/s downward, determine (*a*) the angular acceleration of gear A, (*b*) the number of revolutions executed by gear A during the 4-s interval.

15.19 The two pulleys shown may be operated with the V belt in any of three positions. If the angular acceleration of shaft A is 8 rad/s^2 and if the system is initially at rest, determine the time required for shaft B to reach a speed of 300 rpm with the belt in each of the three positions.

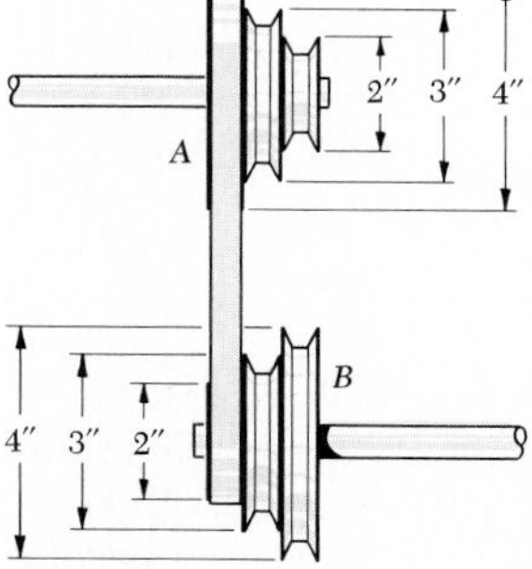

Fig. P15.19

15.20 A simple friction drive consists of two disks A and B. Initially, disk B has a clockwise angular velocity of 500 rpm, and disk A is at rest. It is known that disk B will coast to rest in 60 s. However, rather than waiting until both disks are at rest to bring them together, disk A is given a constant angular acceleration of 3 rad/s^2 counterclockwise. Determine (a) at what time the disks may be brought together if they are not to slip, (b) the angular velocity of each disk as contact is made.

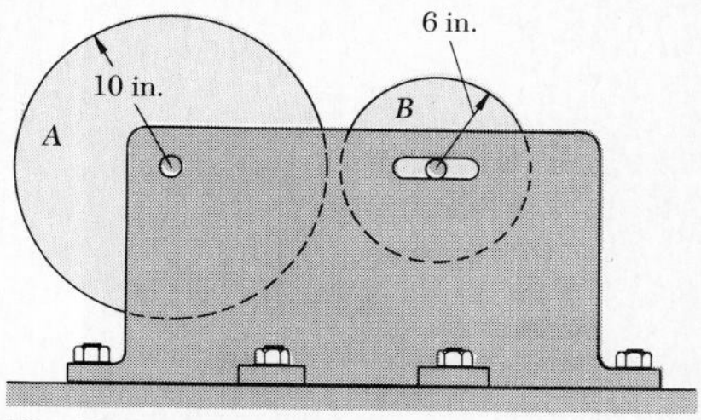

Fig. P15.20 and P15.21

15.21 Disk B is at rest when it is brought into contact with disk A which is rotating freely at 500 rpm clockwise. After 5 s of slippage, during which each disk has a constant angular acceleration, disk A reaches a final angular velocity of 300 rpm clockwise. Determine the angular acceleration of each disk during the period of slippage.

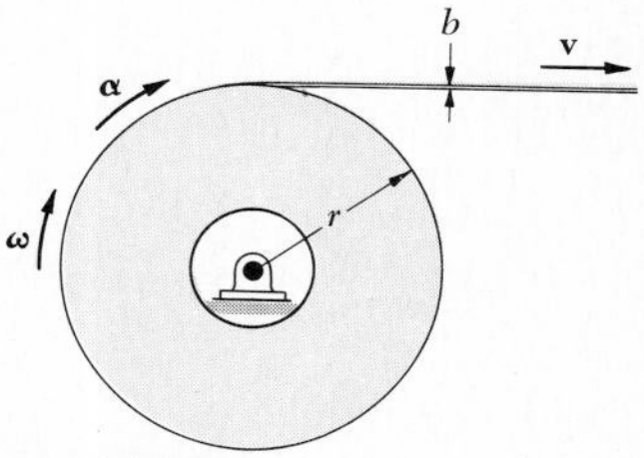

Fig. P15.22

***15.22** In a continuous printing process, paper is drawn into the presses at a constant speed v. Denoting by r the radius of paper on the roll at any given time and by b the thickness of the paper, derive an expression for the angular acceleration of the paper roll.

15.5. General Plane Motion. As indicated in Sec. 15.1, we understand by general plane motion a plane motion which is neither a translation nor a rotation. As we shall presently see, however, *a general plane motion may always be considered as the sum of a translation and a rotation.*

Consider, for example, a wheel rolling on a straight track (Fig. 15.9). Over a certain interval of time, two given points A and B will have moved, respectively, from A_1 to A_2 and from B_1 to B_2. The same result could be obtained through a translation which would bring A and B into A_2 and B_1' (the line AB remaining vertical), followed by a rotation about A bringing B into B_2. Although the original rolling motion differs from the combination of translation and rotation when these motions are taken in succession, the original motion may be completely duplicated by a combination of simultaneous translation and rotation.

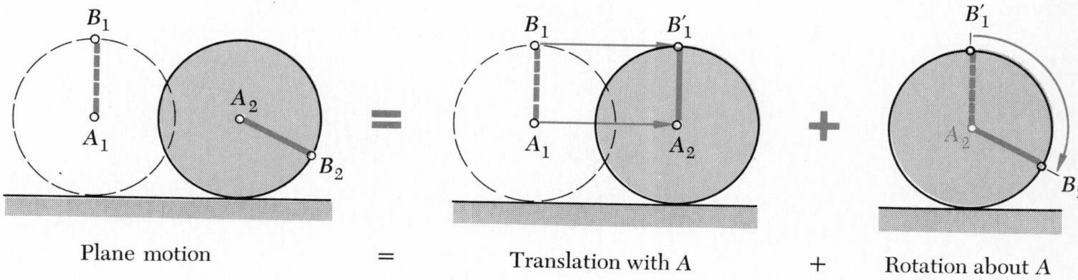

Fig. 15.9

Another example of plane motion is given in Fig. 15.10, which represents a rod whose extremities slide, respectively, along a horizontal and a vertical track. This motion may be replaced by a translation in a horizontal direction and a rotation about A (Fig. 15.10*a*) or by a translation in a vertical direction and a rotation about B (Fig. 15.10*b*).

In general, any plane motion of a rigid slab may be replaced by a translation in which all the particles of the slab move along paths parallel to the path actually followed by some arbitrary reference point A, and by a rotation about the reference point A. We saw in Sec. 11.11 that, when two particles A and B move in a plane, the absolute motion of B may be obtained by combining the motion of A and the relative motion of B with respect to A. We recall that by "absolute motion of B" we mean the motion of B with respect to fixed axes, while by "relative motion

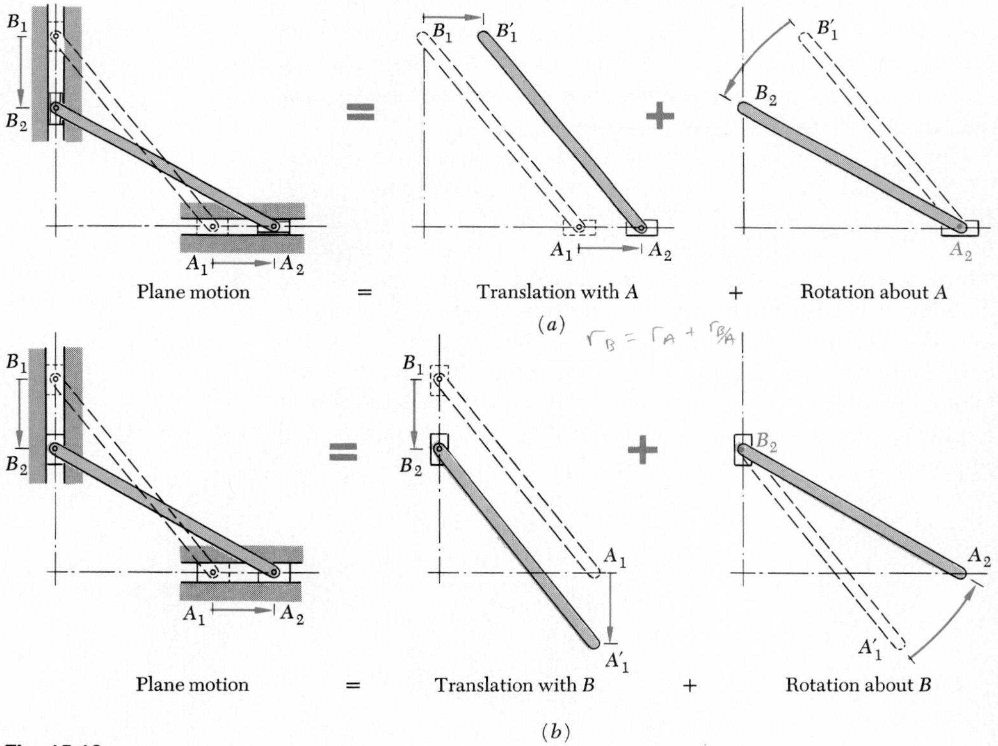

Fig. 15.10

of B with respect to A" we mean the motion of B with respect to a system of axes whose origin moves with A but which does not rotate. Now, in the present case, the two particles belong to the same rigid body. Particle B must thus remain at a constant distance from A and, *to an observer moving with A, but not rotating, particle B will appear to describe an arc of circle centered at A.* Considering successively each particle of the slab, and resolving its motion into the motion of A and its relative motion about A, we thus conclude that any plane motion of a rigid slab is the sum of a translation with A and a rotation about A.

15.6. Absolute and Relative Velocity in Plane Motion. We saw in the preceding section that any plane motion of a slab may be replaced by a translation defined by the motion of an arbitrary reference point A, and by a rotation about A. The absolute velocity $\mathbf{v}_B$ of a particle B of the slab is obtained from the relative-velocity formula derived in Sec. 11.11,

$$\mathbf{v}_B = \mathbf{v}_A + \mathbf{v}_{B/A} \tag{15.14}$$

where the right-hand member represents a vector sum. The velocity $\mathbf{v}_A$ corresponds to the translation of the slab with A, while the relative velocity $\mathbf{v}_{B/A}$ is associated with the rotation of the slab about A and is measured with respect to axes centered at A and of fixed orientation (Fig. 15.11). Denoting by r the

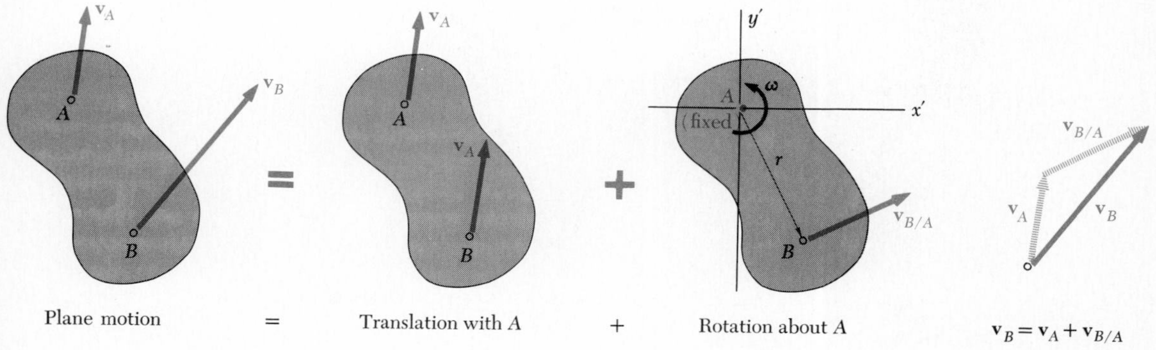

Fig. 15.11

distance from A to B, and by ω the magnitude of the angular velocity $\boldsymbol{\omega}$ of the slab with respect to axes of fixed orientation, we find that the magnitude of the relative velocity $\mathbf{v}_{B/A}$ is

$$v_{B/A} = r\omega \tag{15.15}$$

As an example, we shall consider again the rod AB of Fig. 15.10. Assuming that the velocity $\mathbf{v}_A$ of end A is known, we propose to find the velocity $\mathbf{v}_B$ of end B and the angular velocity $\boldsymbol{\omega}$ of the rod, in terms of the velocity $\mathbf{v}_A$, the length l, and the angle θ. Choosing A as reference point, we express that the given motion is equivalent to a translation with A and a rotation about A (Fig. 15.12). The absolute velocity of B must therefore be equal to the vector sum

$$\mathbf{v}_B = \mathbf{v}_A + \mathbf{v}_{B/A} \tag{15.14}$$

We note that, while the direction of $\mathbf{v}_{B/A}$ is known, its magnitude $l\omega$ is unknown. However, this is compensated by the fact that the direction of $\mathbf{v}_B$ is known. We may therefore complete the diagram of Fig. 15.12. Solving for the magnitudes v_B and ω, we write

$$v_B = v_A \tan\theta \qquad \omega = \frac{v_{B/A}}{l} = \frac{v_A}{l\cos\theta} \tag{15.16}$$

The same result may be obtained by using B as a point of reference. Resolving the given motion into a translation with

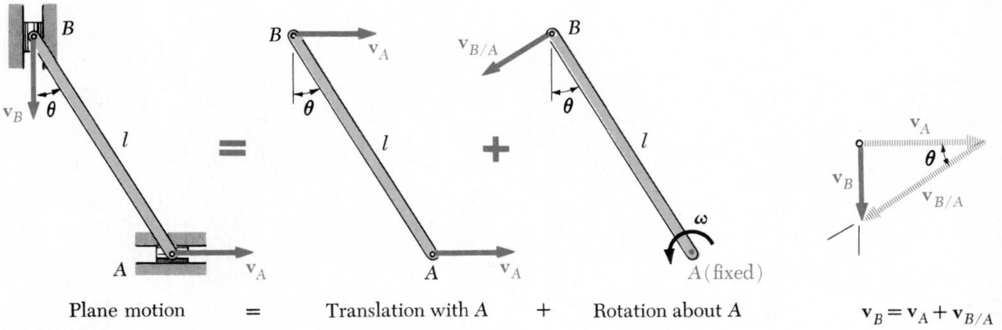

Fig. 15.12

B and a rotation about B (Fig. 15.13), we write the equation

$$\mathbf{v}_A = \mathbf{v}_B + \mathbf{v}_{A/B} \tag{15.17}$$

which is represented graphically in Fig. 15.13. We note that $\mathbf{v}_{A/B}$ and $\mathbf{v}_{B/A}$ have the same magnitude $l\omega$ but opposite sense. The sense of the relative velocity depends, therefore, upon the point of reference which has been selected and should be carefully ascertained from the appropriate diagram (Fig. 15.12 or 15.13).

Finally, we observe that the angular velocity $\boldsymbol{\omega}$ of the rod in its rotation about B is the same as in its rotation about A. It is measured in both cases by the rate of change of the angle θ. This result is quite general; we should therefore bear in mind that *the angular velocity* $\boldsymbol{\omega}$ *of a rigid body in plane motion is independent of the reference point.*

Most mechanisms consist, not of one, but of *several* moving parts. When the various parts of a mechanism are pin-connected, its analysis may be carried out by considering each part as a rigid body, while keeping in mind that the points where two parts are connected must have the same absolute velocity (see Sample Prob. 15.3). A similar analysis may be used when gears are involved, since the teeth in contact must also have the same absolute velocity. However, when a mechanism contains parts which slide on each other, the relative velocity of the parts in contact must be taken into account (see Secs. 15.10 and 15.11).

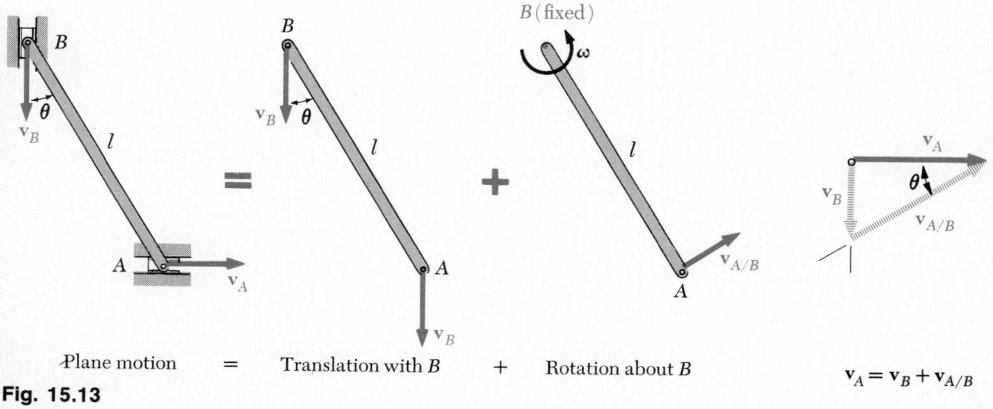

Fig. 15.13

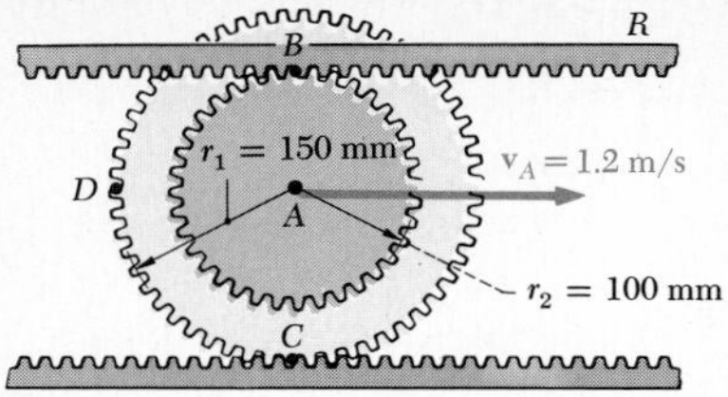

SAMPLE PROBLEM 15.2

The double gear shown rolls on the stationary lower rack; the velocity of its center A is 1.2 m/s directed to the right. Determine (*a*) the angular velocity of the gear, (*b*) the velocities of the upper rack R and of point D of the gear.

***a.* Angular Velocity of the Gear.** Since the gear rolls on the lower rack, its center A moves through a distance equal to the outer circumference $2\pi r_1$ for each full revolution of the gear. Noting that 1 rev = 2π rad, we obtain the coordinate x_A in terms of the corresponding angular coordinate θ (in radians) of the gear by a proportion,

$$\frac{x_A}{2\pi r_1} = \frac{\theta}{2\pi} \qquad x_A = r_1\theta$$

Differentiating with respect to the time t and substituting the known values $v_A = 1.2$ m/s and $r_1 = 150$ mm $= 0.150$ m, we obtain

$$v_A = r_1\omega \qquad 1.2\text{ m/s} = (0.150\text{ m})\omega \qquad \omega = 8\text{ rad/s} \downarrow \blacktriangleleft$$

***b.* Velocities.** The rolling motion is resolved into two component motions: a translation with the center A and a rotation about the center A. In the translation, all points of the gear move with the same velocity $\mathbf{v}_A$. In the rotation, each point P of the gear moves about A with a relative velocity of magnitude $r\omega$, where r is the distance from A.

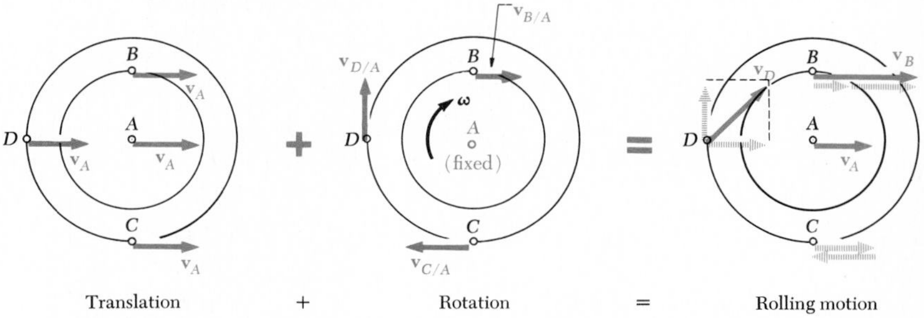

Velocity of Upper Rack. The velocity of the upper rack is equal to the velocity of point B; we write

$$\mathbf{v}_R = \mathbf{v}_B = \mathbf{v}_A + \mathbf{v}_{B/A}$$

where $\qquad \mathbf{v}_A = 1.2\text{ m/s} \rightarrow$

$$v_{B/A} = r_2\omega = (0.100\text{ m})(8\text{ rad/s}) = 0.8\text{ m/s} \qquad \mathbf{v}_{B/A} = 0.8\text{ m/s} \rightarrow$$

Since $\mathbf{v}_A$ and $\mathbf{v}_{B/A}$ are collinear, we add their magnitudes and obtain

$$\mathbf{v}_R = 2\text{ m/s} \rightarrow \blacktriangleleft$$

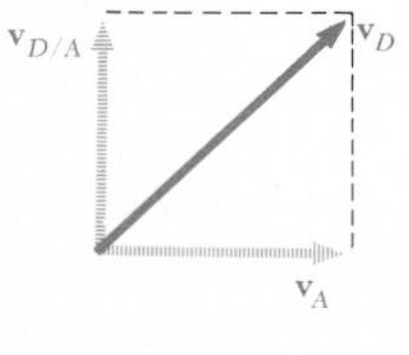

Velocity of Point D: $\qquad \mathbf{v}_D = \mathbf{v}_A + \mathbf{v}_{D/A}$

where $\qquad \mathbf{v}_A = 1.2\text{ m/s} \rightarrow$

$$v_{D/A} = r_1\omega = (0.150\text{ m})(8\text{ rad/s}) = 1.2\text{ m/s} \qquad \mathbf{v}_{D/A} = 1.2\text{ m/s} \uparrow$$

Adding these velocities vectorially, we obtain

$$\mathbf{v}_D = 1.697\text{ m/s} \angle 45° \blacktriangleleft$$

SAMPLE PROBLEM 15.3

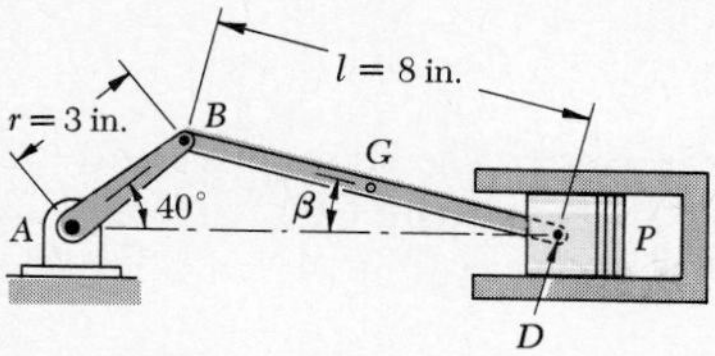

In the engine system shown, the crank AB has a constant clockwise angular velocity of 2000 rpm. For the crank position indicated, determine (*a*) the angular velocity of the connecting rod BD, (*b*) the velocity of the piston P.

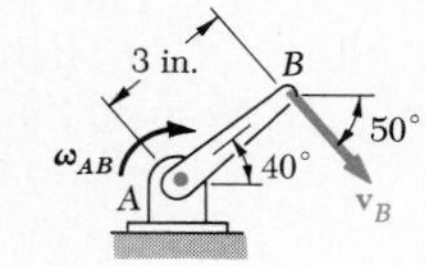

Motion of Crank AB. The crank AB rotates about point A. Expressing ω_{AB} in rad/s and writing $v_B = r\omega_{AB}$, we obtain

$$\omega_{AB} = \left(2000\,\frac{\text{rev}}{\text{min}}\right)\left(\frac{1\text{ min}}{60\text{ s}}\right)\left(\frac{2\pi\text{ rad}}{1\text{ rev}}\right) = 209\text{ rad/s}$$

$$v_B = (AB)\omega_{AB} = (3\text{ in.})(209\text{ rad/s}) = 627\text{ in./s}$$

$$\mathbf{v}_B = 627\text{ in./s} \searrow 50°$$

Motion of Connecting Rod BD. We consider this motion as a general plane motion. Using the law of sines, we compute the angle β between the connecting rod and the horizontal,

$$\frac{\sin 40°}{8\text{ in.}} = \frac{\sin\beta}{3\text{ in.}} \qquad \beta = 13.9°$$

The velocity $\mathbf{v}_D$ of the point D where the rod is attached to the piston must be horizontal, while the velocity of point B is equal to the velocity $\mathbf{v}_B$ obtained above. Resolving the motion of BD into a translation with B and a rotation about B, we obtain

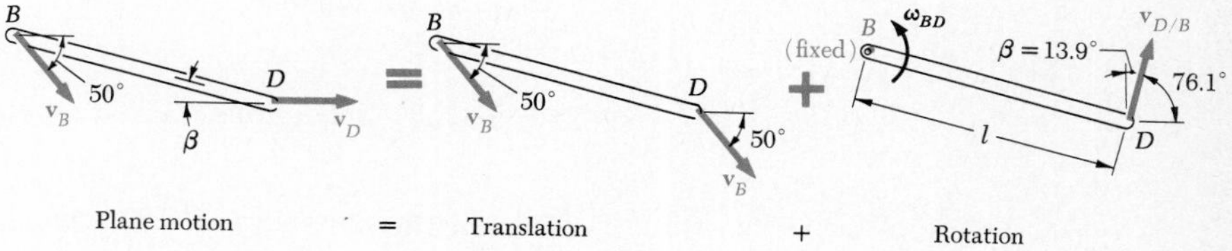

Expressing the relation between the velocities $\mathbf{v}_D$, $\mathbf{v}_B$, and $\mathbf{v}_{D/B}$, we write

$$\mathbf{v}_D = \mathbf{v}_B + \mathbf{v}_{D/B}$$

We draw the vector diagram corresponding to this equation. Recalling that $\beta = 13.9°$, we determine the angles of the triangle and write

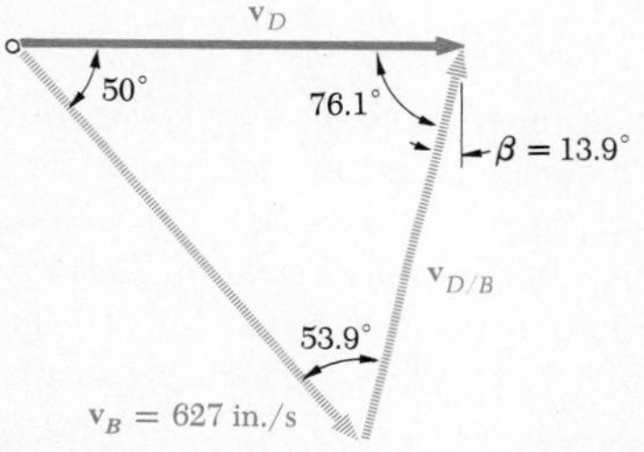

$$\frac{v_D}{\sin 53.9°} = \frac{v_{D/B}}{\sin 50°} = \frac{627\text{ in./s}}{\sin 76.1°}$$

$$v_{D/B} = 495\text{ in./s} \qquad \mathbf{v}_{D/B} = 495\text{ in./s} \measuredangle 76.1°$$

$$v_D = 522\text{ in./s} = 43.5\text{ ft/s} \qquad \mathbf{v}_D = 43.5\text{ ft/s} \rightarrow$$

$$\mathbf{v}_P = \mathbf{v}_D = 43.5\text{ ft/s} \rightarrow$$ ◄

Since $v_{D/B} = l\omega_{BD}$, we have

$$495\text{ in./s} = (8\text{ in.})\omega_{BD} \qquad \omega_{BD} = 61.9\text{ rad/s} \circlearrowleft$$ ◄

PROBLEMS

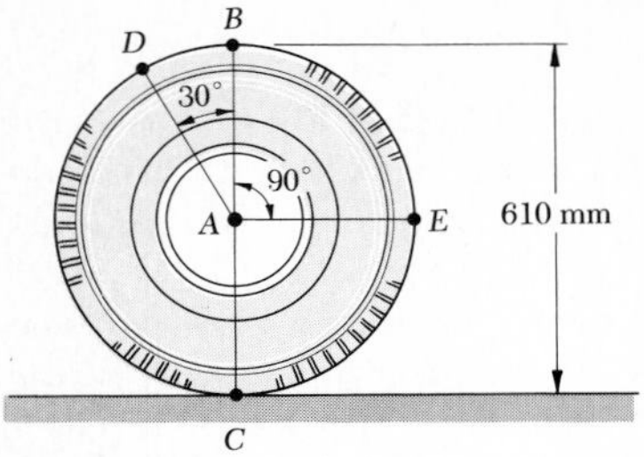

Fig. P15.23

15.23 An automobile travels to the right at a constant speed of 50 km/h. If the diameter of a wheel is 610 mm, determine the velocities of points B, C, D, and E on the rim of the wheel.

15.24 Rod AB is 6 ft long and slides with its ends in contact with the floor and the inclined plane. End A moves with a constant velocity of 20 ft/s to the right. At the instant when $\theta = 25°$, determine (*a*) the angular velocity of the rod, (*b*) the velocity of end B.

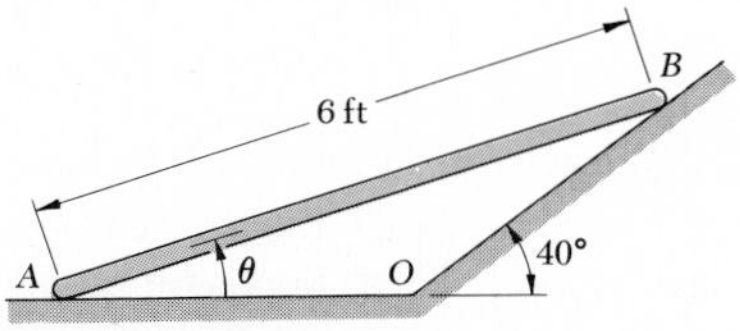

Fig. P15.24

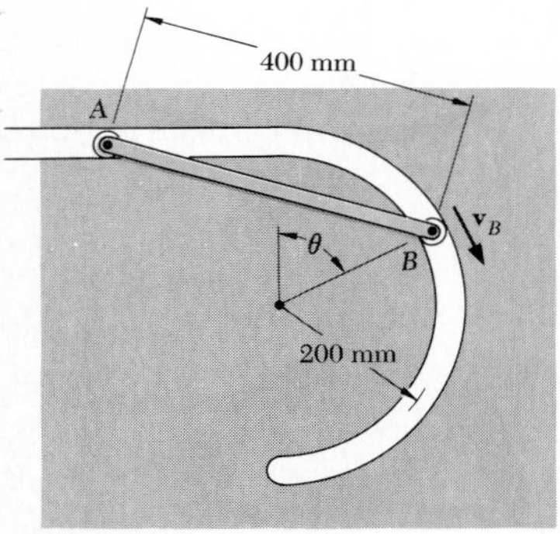

Fig. P15.26

15.25 Solve Prob. 15.24, assuming $\theta = 30°$.

15.26 A 400-mm rod AB is guided by wheels at A and B which roll in the track shown. Knowing that $\theta = 90°$ and that B moves at a constant speed $v_B = 1.2$ m/s, determine (*a*) the angular velocity of the rod, (*b*) the velocity of A.

15.27 Solve Prob. 15.26 assuming $\theta = 60°$.

15.28 Solve Prob. 15.26 assuming $\theta = 120°$.

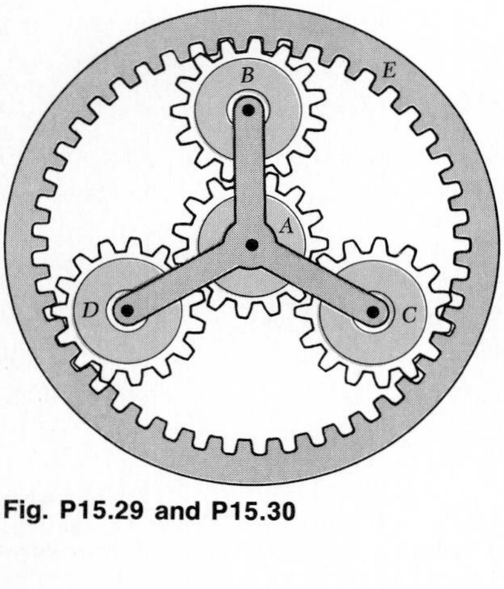

Fig. P15.29 and P15.30

15.29 In the planetary gear system shown, the radius of gears A, B, C, and D is a and the radius of the outer gear E is $3a$. Knowing that the angular velocity of gear A is ω_A clockwise and that the outer gear E is stationary, determine (*a*) the angular velocity of the spider connecting the planetary gears, (*b*) the angular velocity of each planetary gear.

15.30 In the planetary gear system shown, the radius of the central gear A is a, the radius of each of the planetary gears is b, and the radius of the outer gear E is $a + 2b$. The angular velocity of gear A is ω_A clockwise and the outer gear is stationary. If the angular velocity of the spider BCD is to be $\omega_A/5$, determine (*a*) the required value of the ratio b/a, (*b*) the corresponding angular velocity of each planetary gear.

15.31 Arm AB rotates with an angular velocity of 120 rpm clockwise. If the motion of gear B is to be a curvilinear translation, determine (*a*) the required angular velocity of gear A, (*b*) the corresponding velocity of the center of gear B.

15.32 Gear A rotates clockwise with a constant angular velocity of 60 rpm. Knowing that at the same time the arm AB rotates counterclockwise with a constant angular velocity of 30 rpm, determine the angular velocity of gear B.

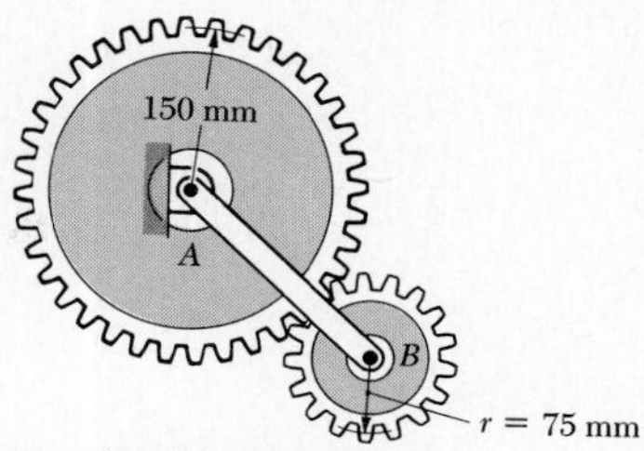

Fig. P15.31 and P15.32

15.33 In the engine system shown, $l = 160$ mm and $b = 60$ mm; the crank AB rotates with a constant angular velocity of 1000 rpm clockwise. Determine the velocity of the piston P and the angular velocity of the connecting rod for the position corresponding to (*a*) $\theta = 0$, (*b*) $\theta = 90°$, (*c*) $\theta = 180°$.

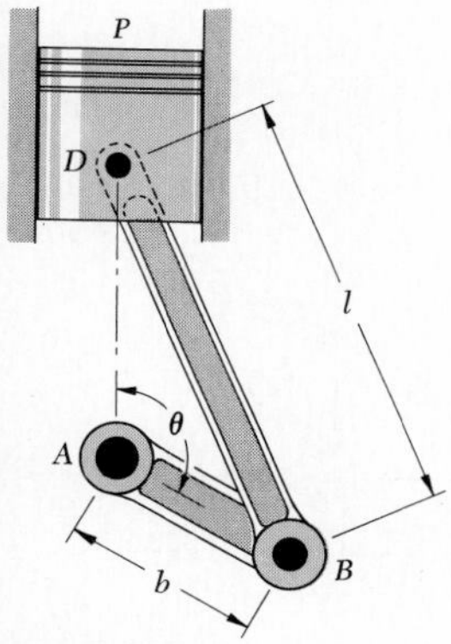

Fig. P15.33

15.34 Crank AB has a constant angular velocity of 12 rad/s clockwise. Determine the angular velocity of rod BD and the velocity of collar D when (*a*) $\theta = 0$, (*b*) $\theta = 90°$, (*c*) $\theta = 180°$.

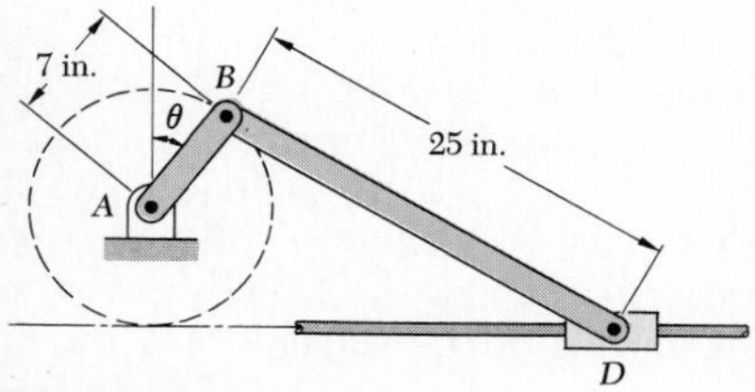

Fig. P15.34

15.35 Solve Prob. 15.34 for the position corresponding to $\theta = 120°$.

15.36 Solve Prob. 15.33 for the position corresponding to $\theta = 60°$.

15.37 through 15.40 In the position shown, bar AB has a constant angular velocity of 3 rad/s counterclockwise. Determine the angular velocity of bars BD and DE.

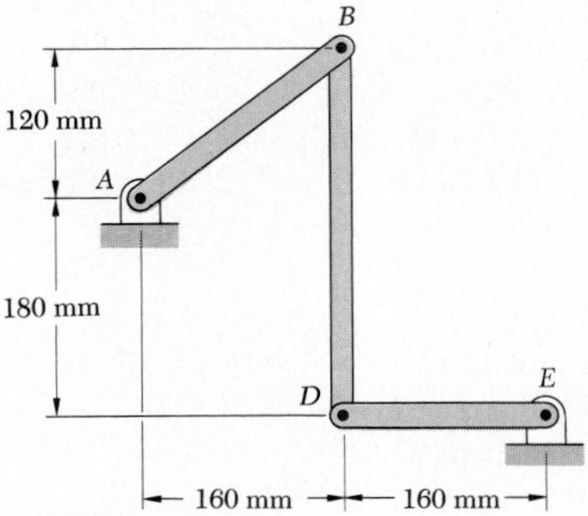

Fig. P15.37

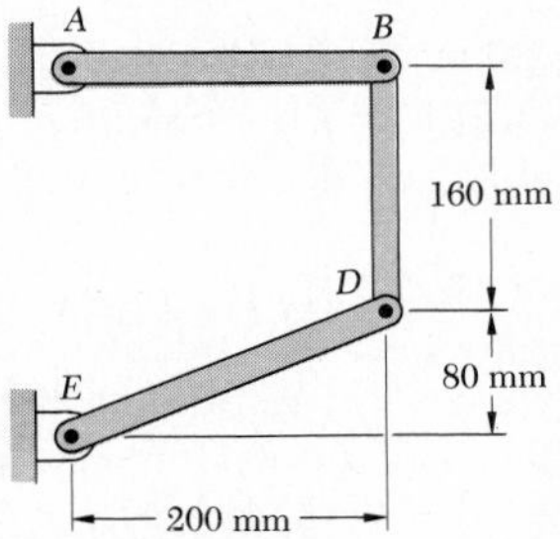

Fig. P15.38

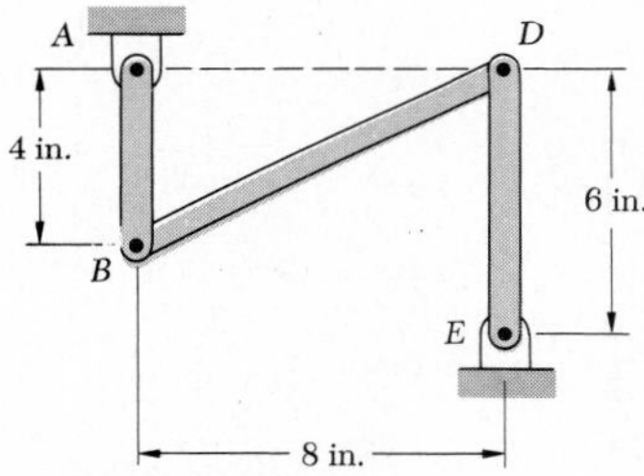

Fig. P15.39

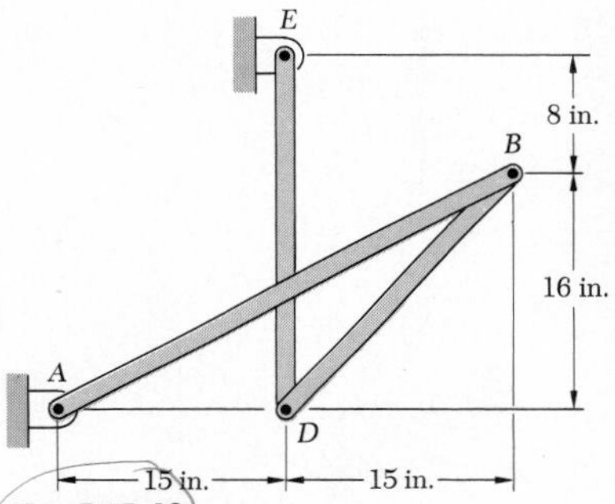

Fig. P15.40

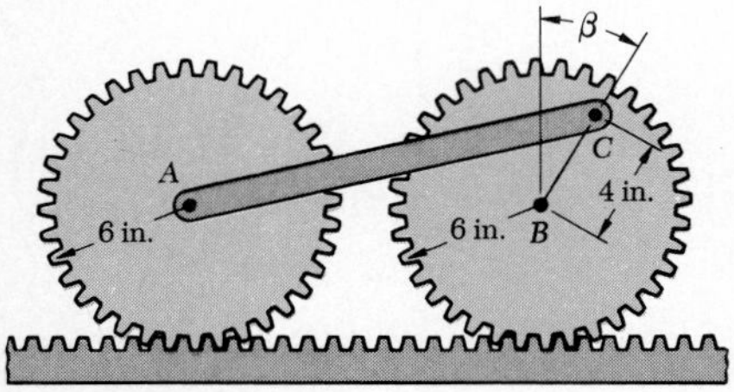

Fig. P15.41

15.41 Two gears, each of 12-in. diameter, are connected by an 18-in. rod AC. Knowing that the center of gear B has a constant velocity of 30 in./s to the right, determine the velocity of the center of gear A and the angular velocity of the connecting rod (*a*) when $\beta = 0$, (*b*) when $\beta = 90°$.

15.42 Solve Prob. 15.41, assuming (*a*) $\beta = 180°$, (*b*) $\beta = 60°$.

15.43 Rod AB of the mechanism shown rotates clockwise with an angular velocity ω_0. Determine in terms of l, ω_0, and θ, (*a*) the velocity of collar D, (*b*) the components of the velocity of point E.

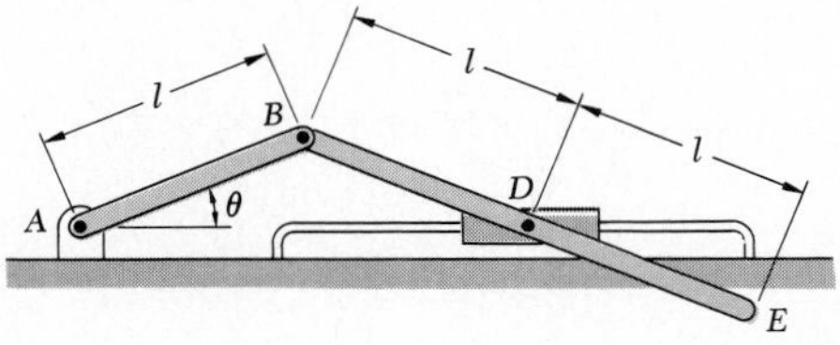

Fig. P15.43 and P15.44

15.44 Collar D of the mechanism shown moves to the right with a velocity of 300 mm/s. Knowing that $l = 200$ mm and $\theta = 30°$, determine (*a*) the angular velocity of rod AB, (*b*) the velocity of point E.

***15.45** Assuming that the crank AB of Prob. 15.33 rotates with a constant clockwise angular velocity ω and that $\theta = 0$ at $t = 0$, derive an expression for the velocity of the piston P in terms of the time t.

15.7. Instantaneous Center of Rotation in Plane Motion. Consider the general plane motion of a slab. We shall show that at any given instant the velocities of the various particles of the slab are the same as if the slab were rotating about a certain axis perpendicular to the plane of the slab, called the *instantaneous axis of rotation.* This axis intersects the plane of the slab at a point C, called the *instantaneous center of rotation* of the slab.

To prove our statement, we first recall that the plane motion of a slab may always be replaced by a translation defined by the motion of an arbitrary reference point A, and by a rotation about A. As far as the velocities are concerned, the translation is characterized by the velocity $\mathbf{v}_A$ of the reference point A and the rotation is characterized by the angular velocity ω of the slab (which is independent of the choice of A). Thus, the velocity $\mathbf{v}_A$ of point A and the angular velocity ω of the slab define completely the velocities of all the other particles of the slab (Fig. 15.14*a*). Now let us assume that $\mathbf{v}_A$ and ω are known and that they are both different from zero. (If $\mathbf{v}_A = 0$, point A is itself the instantaneous center of rotation, and if $\omega = 0$, the slab

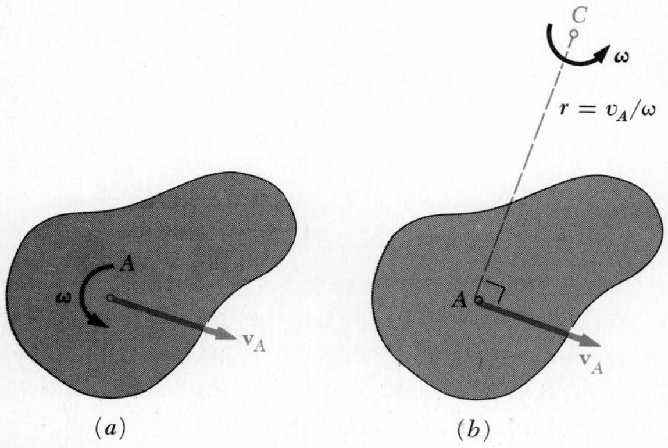

Fig. 15.14

is in translation.) These velocities could be obtained by letting the slab rotate with the angular velocity ω about a point C located on the perpendicular to $\mathbf{v}_A$ at a distance $r = v_A/\omega$ from A as shown in Fig. 15.14*b*. We check that the velocity of A would be perpendicular to AC and that its magnitude would be $r\omega = (v_A/\omega)\omega = v_A$. Thus the velocities of all the other particles of the slab would be the same as originally defined. Therefore, *as far as the velocities are concerned, the slab seems to rotate about the instantaneous center C* at the instant considered.

The position of the instantaneous center may be defined in two other ways. If the directions of the velocities of two particles A and B of the slab are known, and if they are different, the instantaneous center C is obtained by drawing the perpendicular to $\mathbf{v}_A$ through A and the perpendicular to $\mathbf{v}_B$ through B and determining the point in which these two lines intersect (Fig. 15.15*a*). If the velocities $\mathbf{v}_A$ and $\mathbf{v}_B$ of two particles A and B are perpendicular to the line AB, and if their magnitudes are known, the instantaneous center may be found by intersecting the line AB with the line joining the extremities of the vectors $\mathbf{v}_A$ and $\mathbf{v}_B$ (Fig. 15.15*b*). Note that, if $\mathbf{v}_A$ and $\mathbf{v}_B$ were parallel in Fig. 15.15*a*, or if $\mathbf{v}_A$ and $\mathbf{v}_B$ had the same magnitude in Fig. 15.15*b*, the instantaneous center C would be at an infinite distance and ω would be zero; the slab would be in translation.

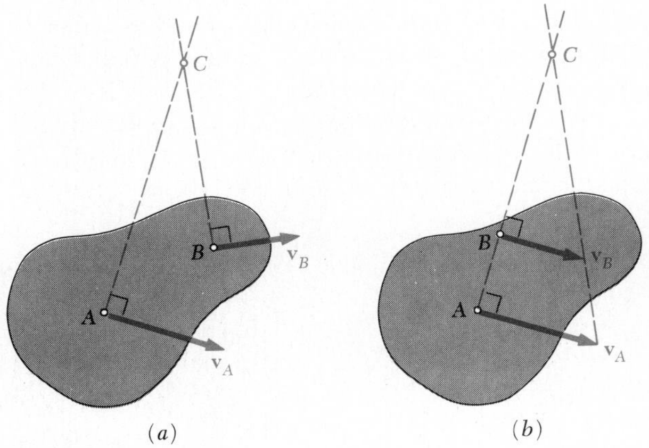

Fig. 15.15

To see how the concept of instantaneous center of rotation may be put to use, let us consider again the rod of Sec. 15.6. Drawing the perpendicular to $\mathbf{v}_A$ through A and the perpendicular to $\mathbf{v}_B$ through B (Fig. 15.16), we obtain the instantaneous center C. At the instant considered, the velocities of all the particles of the rod are thus the same as if the rod rotated about

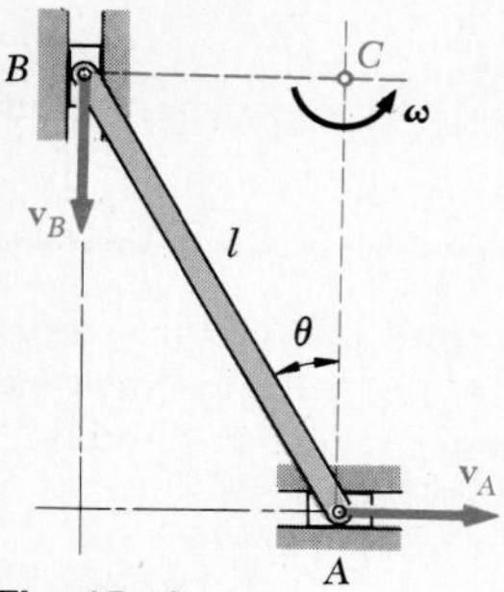

Fig. 15.16

C. Now, if the magnitude v_A of the velocity of A is known, the magnitude ω of the angular velocity of the rod may be obtained by writing

$$\omega = \frac{v_A}{AC} = \frac{v_A}{l \cos \theta}$$

The magnitude of the velocity of B may then be obtained by writing

$$v_B = (BC)\omega = l \sin \theta \frac{v_A}{l \cos \theta} = v_A \tan \theta$$

Note that only *absolute* velocities are involved in the computation.

The instantaneous center of a slab in plane motion may be located either on the slab or outside the slab. If it is located on the slab, the particle C coinciding with the instantaneous center at a given instant t must have zero velocity at that instant. The instantaneous-center method for the determination of velocities may thus be considered as a particular case of the method of relative velocities, where C is chosen as point of reference. Since $\mathbf{v}_C = 0$, the general formula $\mathbf{v}_B = \mathbf{v}_C + \mathbf{v}_{B/C}$ reduces to $\mathbf{v}_B = \mathbf{v}_{B/C}$, which means that the velocities of the various particles of the slab may be determined as if the slab were rotating about C.

It should be noted that the instantaneous center of rotation is valid only at a given instant. As the motion of the slab proceeds, other instantaneous centers should be used. The particle C of the slab which coincides with the instantaneous center at time t will generally not coincide with the instantaneous center at time $t + \Delta t$; while its velocity is zero at time t, it will probably be different from zero at time $t + \Delta t$. This means that, in general, the particle C *does not have zero acceleration,* and therefore that the *accelerations* of the various particles of the slab *cannot* be determined as if the slab were rotating about C.

SAMPLE PROBLEM 15.4

Solve Sample Prob. 15.2, using the method of the instantaneous center of rotation.

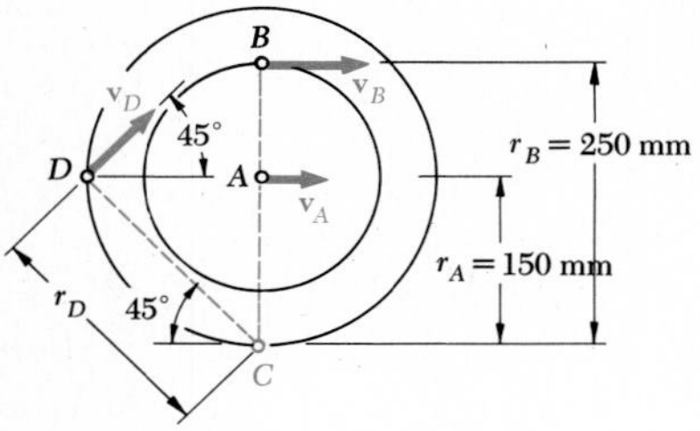

a. **Angular Velocity of the Gear.** Since the gear rolls on the stationary lower rack, the point of contact C of the gear with the rack has no velocity; point C is therefore the instantaneous center of rotation. We write

$$v_A = r_A\omega \qquad 1.2\text{ m/s} = (0.150\text{ m})\omega \qquad \omega = 8\text{ rad/s} \circlearrowright \blacktriangleleft$$

b. **Velocities.** All points of the gear seem to rotate about the instantaneous center as far as velocities are concerned.

Velocity of Upper Rack. Recalling that $v_R = v_B$, we write

$$v_R = v_B = r_B\omega \qquad v_R = (0.250\text{ m})(8\text{ rad/s}) = 2\text{ m/s}$$

$$\mathbf{v}_R = 2\text{ m/s} \rightarrow \blacktriangleleft$$

Velocity of Point D. Since $r_D = (0.150\text{ m})\sqrt{2} = 0.212$ m, we write

$$v_D = r_D\omega \qquad v_D = (0.212\text{ m})(8\text{ rad/s}) = 1.696\text{ m/s}$$

$$\mathbf{v}_D = 1.696\text{ m/s} \measuredangle 45^\circ \blacktriangleleft$$

SAMPLE PROBLEM 15.5

Solve Sample Prob. 15.3, using the method of the instantaneous center of rotation.

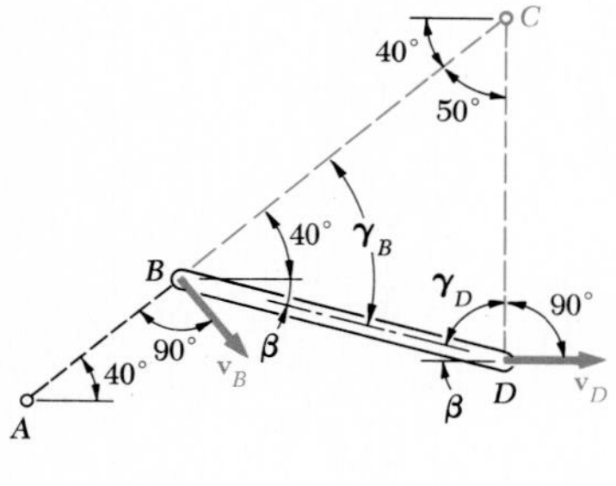

Motion of Crank *AB*. Referring to Sample Prob. 15.3, we obtain the velocity of point B; $\mathbf{v}_B = 627$ in./s $\measuredangle$ 50°.

Motion of the Connecting Rod *BD*. We first locate the instantaneous center C by drawing lines perpendicular to the absolute velocities $\mathbf{v}_B$ and $\mathbf{v}_D$. Recalling from Sample Prob. 15.3 that $\beta = 13.9^\circ$ and that $BD = 8$ in., we solve the triangle BCD.

$$\gamma_B = 40^\circ + \beta = 53.9^\circ \qquad \gamma_D = 90^\circ - \beta = 76.1^\circ$$

$$\frac{BC}{\sin 76.1^\circ} = \frac{CD}{\sin 53.9^\circ} = \frac{8\text{ in.}}{\sin 50^\circ}$$

$$BC = 10.14\text{ in.} \qquad CD = 8.44\text{ in.}$$

Since the connecting rod BD seems to rotate about point C, we write

$$v_B = (BC)\omega_{BD}$$

$$627\text{ in./s} = (10.14\text{ in.})\omega_{BD}$$

$$\omega_{BD} = 61.9\text{ rad/s} \circlearrowleft \blacktriangleleft$$

$$v_D = (CD)\omega_{BD} = (8.44\text{ in.})(61.9\text{ rad/s})$$

$$= 522\text{ in./s} = 43.5\text{ ft/s}$$

$$\mathbf{v}_P = \mathbf{v}_D = 43.5\text{ ft/s} \rightarrow \blacktriangleleft$$

PROBLEMS

15.46 A helicopter moves horizontally in the x direction at a speed of 45 mi/h. Knowing that the main blades rotate clockwise at an angular velocity of 120 rpm, determine the instantaneous axis of rotation of the main blades.

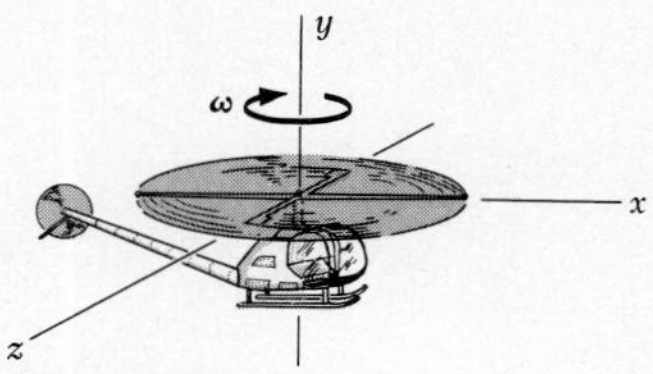

Fig. P15.46

15.47 A double pulley is attached to a slider block by a pin at A. The 50-mm-radius inner pulley is rigidly attached to the 100-mm-radius outer pulley. Knowing that each of the two cords is pulled at a constant speed of 300 mm/s as shown, determine (*a*) the instantaneous center of rotation of the double pulley, (*b*) the velocity of the slider block, (*c*) the length of cord wrapped or unwrapped on each pulley per second.

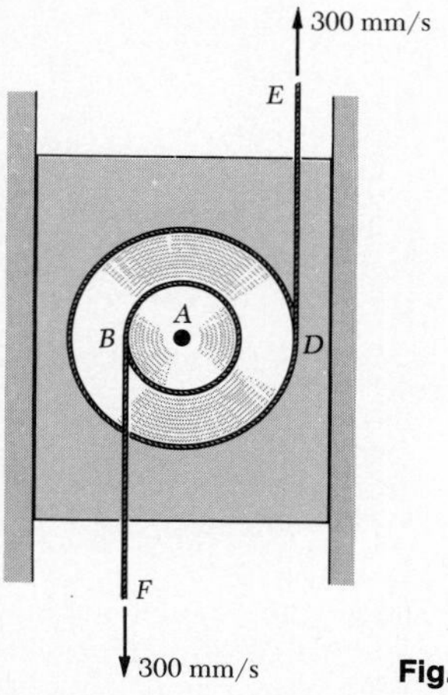

Fig. P15.47

15.48 Solve Sample Prob. 15.2, assuming that the lower rack is not stationary but moves to the left with a velocity of 0.6 m/s.

15.49 A drum, of radius 4.5 in., is mounted on a cylinder, of radius 6 in. A cord is wound around the drum, and its extremity D is pulled to the left at a constant velocity of 3 in./s, causing the cylinder to roll without sliding. Determine (*a*) the angular velocity of the cylinder, (*b*) the velocity of the center of the cylinder, (*c*) the length of cord which is wound or unwound per second.

a) $\omega = 2$ rad/s
b) $v_a = 12$ in/s ←
c) 9 in/s

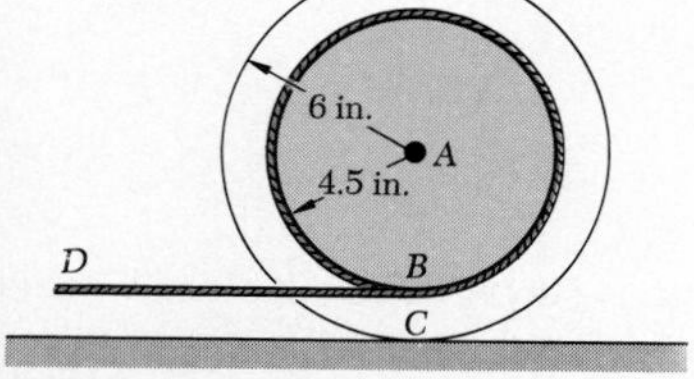

Fig. P15.49

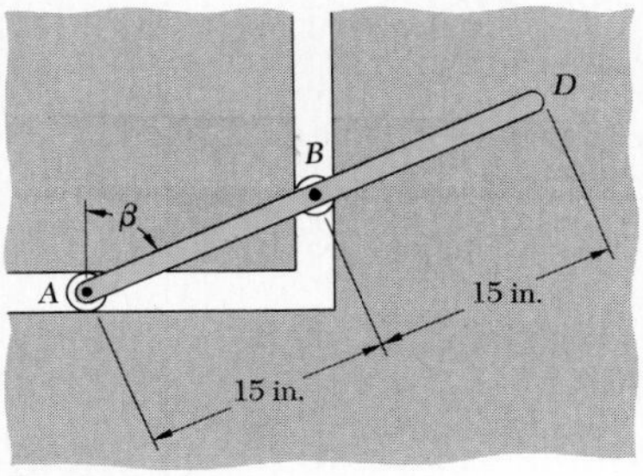

Fig. P15.50

15.50 The rod *ABD* is guided by wheels at *A* and *B* which roll in horizontal and vertical tracks as shown. Knowing that $\beta = 60°$ and that the velocity of *B* is 40 in./s downward, determine (*a*) the angular velocity of the rod, (*b*) the velocity of point *D*.

15.51 Knowing that at the instant shown the velocity of collar *D* is 20 in./s upward, determine (*a*) the angular velocity of rod *AD*, (*b*) the velocity of point *B*, (*c*) the velocity of point *A*.

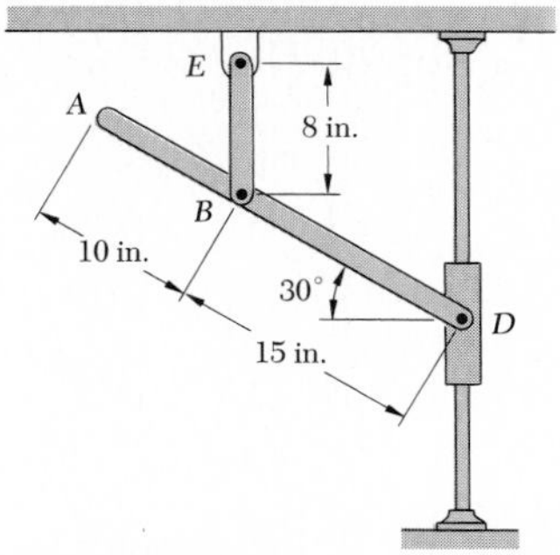

Fig. P15.51

15.52 Solve Prob. 15.50 assuming that $\beta = 30°$.

15.53 Knowing that at the instant shown the angular velocity of crank *AB* is 3 rad/s clockwise, determine (*a*) the angular velocity of link *BD*, (*b*) the velocity of collar *D*, (*c*) the velocity of the midpoint of link *BD*.

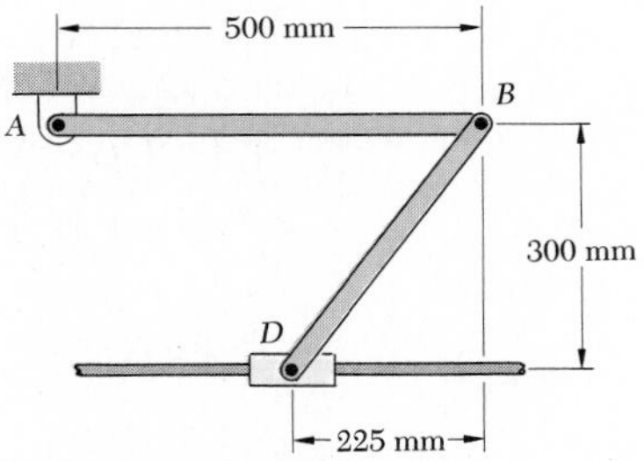

Fig. P15.53 and P15.54

15.54 Knowing that at the instant shown the velocity of collar *D* is 1.5 m/s to the right, determine (*a*) the angular velocities of crank *AB* and link *BD*, (*b*) the velocity of the midpoint of link *BD*.

15.55 Collar A slides downward with a constant velocity $\mathbf{v}_A$. Determine the angle θ corresponding to the position of rod AB for which the velocity of B is horizontal.

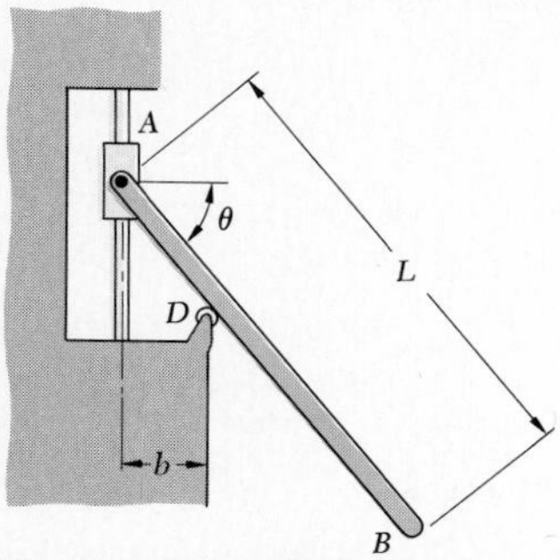

Fig. P15.55 and P15.57

15.56 Two rods AB and BD are connected to three collars as shown. Knowing that collar A moves downward with a constant velocity of 120 mm/s, determine (a) the angular velocity of each rod, (b) the velocity of collar D.

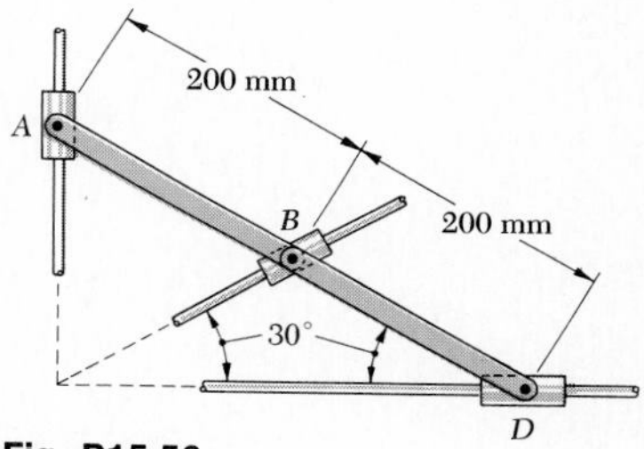

Fig. P15.56

15.57 Collar A slides downward with a constant speed of 16 in./s. Knowing that $b = 2$ in., $L = 10$ in., and $\theta = 60°$, determine (a) the angular velocity of rod AB, (b) the velocity of B.

15.58 The rectangular plate is supported by two 6-in. links as shown. Knowing that at the instant shown the angular velocity of link AB is 4 rad/s clockwise, determine (a) the angular velocity of the plate, (b) the velocity of the center of the plate, (c) the velocity of corner F.

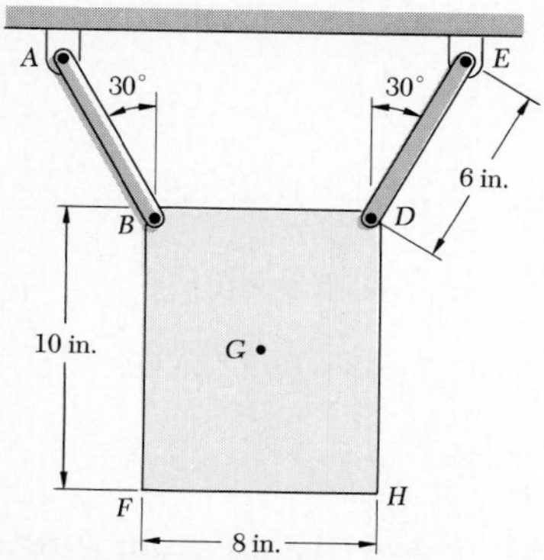

Fig. P15.58 and P15.59

15.59 Knowing that, at the instant shown, the angular velocity of link AB is 4 rad/s clockwise, determine (a) the angular velocity of the plate, (b) the points of the plate for which the magnitude of the velocity is equal to or less than 6 in./s.

15.60 At the instant shown, the velocity of the center of the gear is 200 mm/s to the right. Determine (*a*) the velocity of point *B*, (*b*) the velocity of collar *D*.

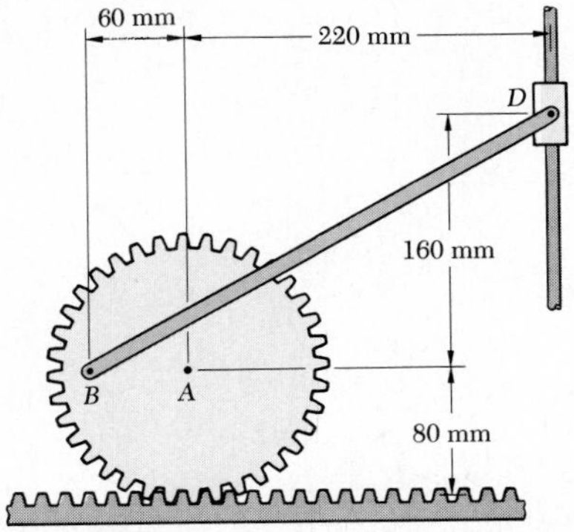

Fig. P15.60

15.61 Using the method of Sec. 15.7, solve Prob. 15.33.

15.62 Using the method of Sec. 15.7, solve Prob. 15.34.

15.63 Using the method of Sec. 15.7, solve Prob. 15.35.

15.64 Using the method of Sec. 15.7, solve Prob. 15.36.

15.65 Using the method of Sec. 15.7, solve Prob. 15.37.

15.66 Using the method of Sec. 15.7, solve Prob. 15.38.

15.67 Using the method of Sec. 15.7, solve Prob. 15.39.

15.68 Using the method of Sec. 15.7, solve Prob. 15.40.

15.69 Using the method of Sec. 15.7, solve Prob. 15.43.

15.70 Using the method of Sec. 15.7, solve Prob. 15.44.

15.71 Using the method of Sec. 15.7, solve Prob. 15.29.

15.72 Using the method of Sec. 15.7, solve Prob. 15.32.

15.8. Absolute and Relative Acceleration in Plane Motion. We saw in Sec. 15.5 that any plane motion may be replaced by a translation defined by the motion of an arbitrary reference point *A*, and by a rotation about *A*. This property was used in Sec. 15.6 to determine the velocity of the various points of a moving slab. We shall now use the same property to determine the acceleration of the points of the slab.

We first recall that the absolute acceleration $\mathbf{a}_B$ of a particle of the slab may be obtained from the relative-acceleration formula derived in Sec. 11.11,

$$\mathbf{a}_B = \mathbf{a}_A + \mathbf{a}_{B/A} \tag{15.18}$$

where the right-hand member represents a vector sum. The acceleration $\mathbf{a}_A$ corresponds to the translation of the slab with A, while the relative acceleration $\mathbf{a}_{B/A}$ is associated with the rotation of the slab about A and is measured with respect to axes of fixed orientation centered at A (Fig. 15.17). Denoting by r the distance from A to B and, respectively, by $\boldsymbol{\omega}$ and $\boldsymbol{\alpha}$ the angular velocity and angular acceleration of the slab with respect to axes of fixed orientation, we find that the relative acceleration $\mathbf{a}_{B/A}$ of B with respect to A consists of two components, a *normal component* $(\mathbf{a}_{B/A})_n$ of magnitude $r\omega^2$ directed toward A, and a *tangential component* $(\mathbf{a}_{B/A})_t$ of magnitude $r\alpha$ perpendicular to the line AB.

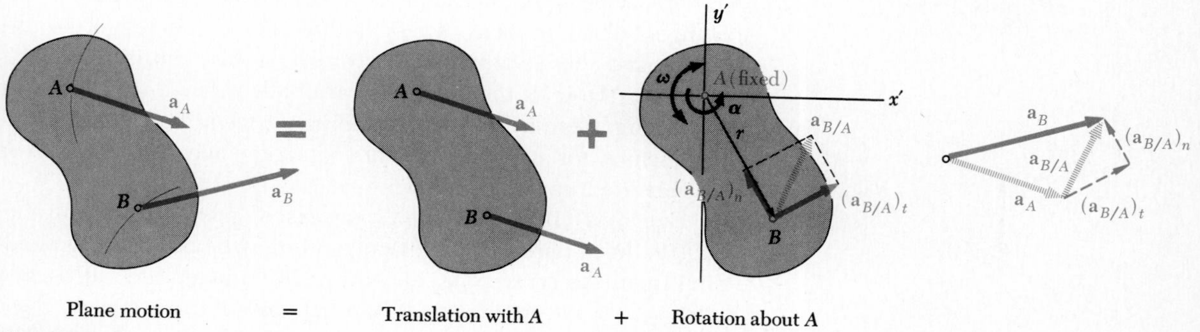

Fig. 15.17

As an example, we shall consider again the rod AB whose extremities slide, respectively, along a horizontal and a vertical track (Fig. 15.18). Assuming that the velocity $\mathbf{v}_A$ and the acceleration $\mathbf{a}_A$ of A are known, we propose to determine the acceleration $\mathbf{a}_B$ of B and the angular acceleration $\boldsymbol{\alpha}$ of the rod. Choosing A as a reference point, we express that the given motion is equivalent to a translation with A and a rotation about A. The

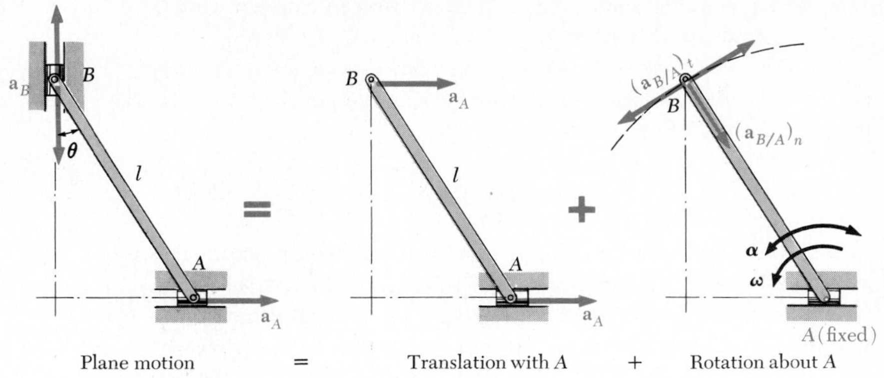

Fig. 15.18

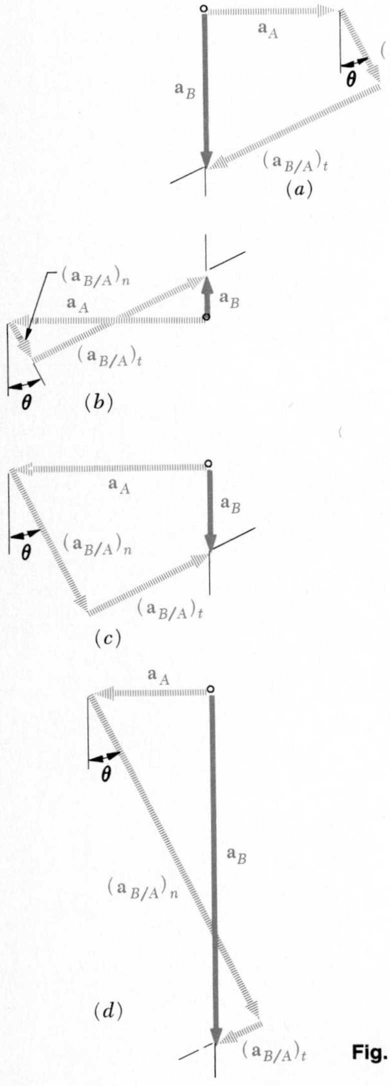

Fig. 15.19

absolute acceleration of B must be equal to the sum

$$\begin{aligned} \mathbf{a}_B &= \mathbf{a}_A + \mathbf{a}_{B/A} \\ &= \mathbf{a}_A + (\mathbf{a}_{B/A})_n + (\mathbf{a}_{B/A})_t \end{aligned} \tag{15.19}$$

where $(\mathbf{a}_{B/A})_n$ has the magnitude $l\omega^2$ and is *directed toward* A, while $(\mathbf{a}_{B/A})_t$ has the magnitude $l\alpha$ and is perpendicular to AB. There is no way of telling at the present time whether the tangential component $(\mathbf{a}_{B/A})_t$ is directed to the left or to the right, and the student should not rely on his "intuition" in this matter. We shall therefore indicate both possible directions for this component in Fig. 15.18. Similarly, we indicate both possible senses for $\mathbf{a}_B$, since we do not know whether point B is accelerated upward or downward.

Equation (15.19) has been expressed geometrically in Fig. 15.19. Four different vector polygons may be obtained, depending upon the sense of $\mathbf{a}_A$ and the relative magnitude of a_A and $(a_{B/A})_n$. If we are to determine a_B and α from one of these diagrams, we must know not only a_A and θ but also ω. The angular velocity of the rod, therefore, should be separately determined by one of the methods indicated in Secs. 15.6 and 15.7. The values of a_B and α may then be obtained by considering successively the x and y components of the vectors shown in Fig. 15.19. In the case of polygon a, for example, we write the equations

$\xrightarrow{+}$ x components: $\quad 0 = a_A + l\omega^2 \sin\theta - l\alpha\cos\theta$

$+\uparrow y$ components: $\quad -a_B = -l\omega^2\cos\theta - l\alpha\sin\theta$

which may be solved for a_B and α. The two unknowns may also be obtained by direct measurement on the vector polygon. In that case, care should be taken to draw first the known vectors $\mathbf{a}_A$ and $(\mathbf{a}_{B/A})_n$.

It is quite evident that the determination of accelerations is considerably more involved than the determination of velocities. Yet, in the example considered here, the extremities A and B of the rod were moving along straight tracks, and the diagrams drawn were relatively simple. If A and B had moved along curved tracks, the accelerations $\mathbf{a}_A$ and $\mathbf{a}_B$ should have been resolved into normal and tangential components and the solution of the problem would have involved six different vectors.

When a mechanism consists of several moving parts which are pin-connected, its analysis may be carried out by considering each part as a rigid body, while keeping in mind that the points where two parts are connected must have the same absolute acceleration (see Sample Prob. 15.7). In the case of meshed gears, the tangential components of the accelerations of the teeth in contact are equal, but their normal components are different.

***15.9. Analysis of Plane Motion in Terms of a Parameter.** In the case of certain mechanisms, it is possible to express the coordinates x and y of all the significant points of the mechanism by means of simple analytic expressions containing a single parameter. It may be advantageous in such a case to determine directly the absolute velocity and the absolute acceleration of the various points of the mechanism, since the components of the velocity and of the acceleration of a given point may be obtained by differentiating the coordinates x and y of that point.

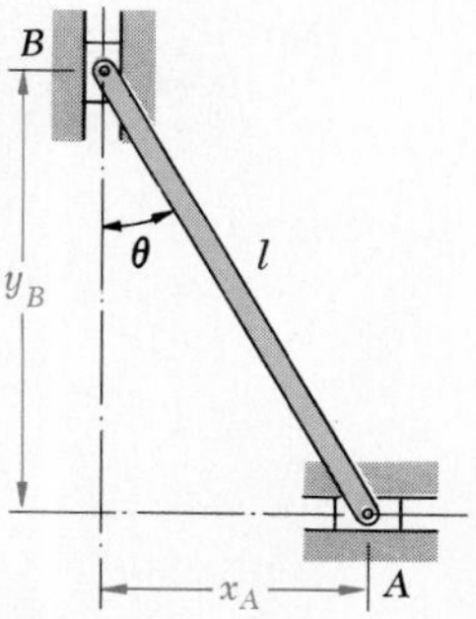

Fig. 15.20

Let us consider again the rod AB whose extremities slide, respectively, in a horizontal and a vertical track (Fig. 15.20). The coordinates x_A and y_B of the extremities of the rod may be expressed in terms of the angle θ the rod forms with the vertical,

$$x_A = l \sin\theta \qquad y_B = l\cos\theta \tag{15.20}$$

Differentiating Eqs. (15.20) twice with respect to t, we write

$$v_A = \dot{x}_A = l\dot{\theta}\cos\theta \qquad\qquad v_B = \dot{y}_B = -l\dot{\theta}\sin\theta$$
$$a_A = \ddot{x}_A = -l\dot{\theta}^2\sin\theta + l\ddot{\theta}\cos\theta$$
$$a_B = \ddot{y}_B = -l\dot{\theta}^2\cos\theta - l\ddot{\theta}\sin\theta$$

Recalling that $\dot{\theta} = \omega$ and $\ddot{\theta} = \alpha$, we obtain

$$v_A = l\omega\cos\theta \qquad v_B = -l\omega\sin\theta \tag{15.21}$$
$$a_A = -l\omega^2\sin\theta + l\alpha\cos\theta \qquad a_B = -l\omega^2\cos\theta - l\alpha\sin\theta \tag{15.22}$$

We note that a positive sign for v_A or a_A indicates that the velocity $\mathbf{v}_A$ or the acceleration $\mathbf{a}_A$ is directed to the right; a positive sign for v_B or a_B indicates that $\mathbf{v}_B$ or $\mathbf{a}_B$ is directed upward. Equations (15.21) may be used, for example, to determine v_B and ω when v_A and θ are known. Substituting for ω in (15.22), we may then determine a_B and α if a_A is known.

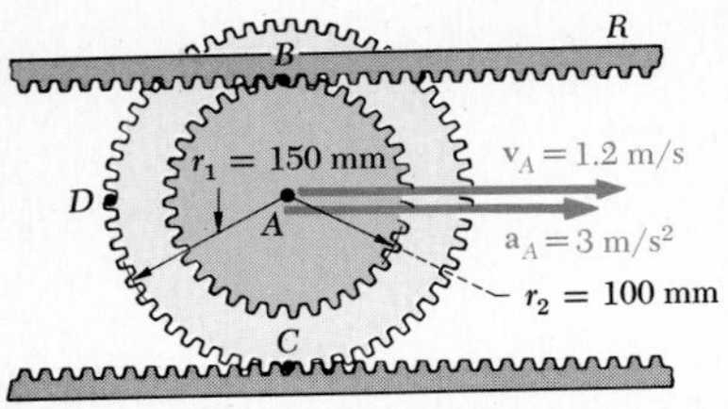

SAMPLE PROBLEM 15.6

The center of the double gear of Sample Prob. 15.2 has a velocity of 1.2 m/s to the right and an acceleration of 3 m/s² to the right. Determine (*a*) the angular acceleration of the gear, (*b*) the acceleration of points *B*, *C*, and *D* of the gear.

***a.* Angular Acceleration of the Gear.** In Sample Prob. 15.2, we found that $x_A = r_1\theta$ and $v_A = r_1\omega$. Differentiating the latter with respect to time, we obtain $a_A = r_1\alpha$.

$$v_A = r_1\omega \qquad 1.2 \text{ m/s} = (0.150 \text{ m})\omega \qquad \omega = 8 \text{ rad/s} \downarrow$$

$$a_A = r_1\alpha \qquad 3 \text{ m/s}^2 = (0.150 \text{ m})\alpha \qquad \alpha = 20 \text{ rad/s}^2 \downarrow \quad ◀$$

***b.* Accelerations.** The rolling motion of the gear is resolved into a translation with *A* and a rotation about *A*.

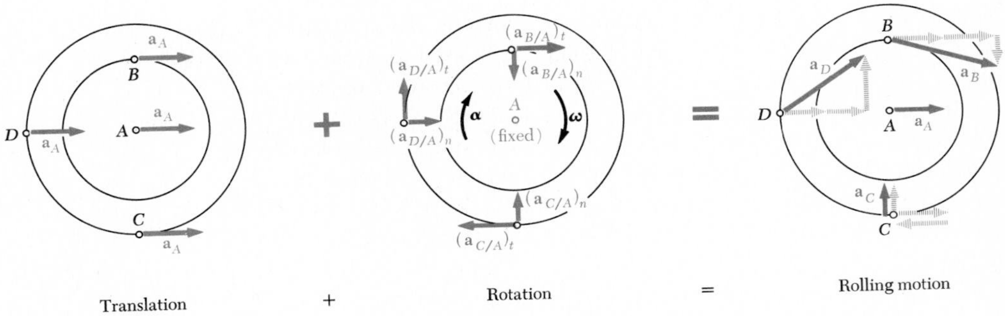

Acceleration of Point B. Adding vectorially the accelerations corresponding to the translation and to the rotation, we obtain

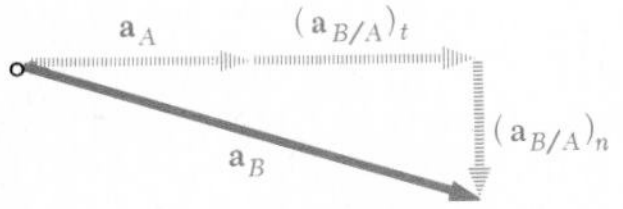

$$\mathbf{a}_B = \mathbf{a}_A + \mathbf{a}_{B/A} = \mathbf{a}_A + (\mathbf{a}_{B/A})_t + (\mathbf{a}_{B/A})_n$$

where $\qquad \mathbf{a}_A = 3 \text{ m/s}^2 \rightarrow$

$$(a_{B/A})_t = r_2\alpha = (0.100 \text{ m})(20 \text{ rad/s}^2) \qquad (\mathbf{a}_{B/A})_t = 2 \text{ m/s}^2 \rightarrow$$

$$(a_{B/A})_n = r_2\omega^2 = (0.100 \text{ m})(8 \text{ rad/s})^2 \qquad (\mathbf{a}_{B/A})_n = 6.4 \text{ m/s}^2 \downarrow$$

We have $\quad \mathbf{a}_B = [3 \text{ m/s}^2 \rightarrow] + [2 \text{ m/s}^2 \rightarrow] + [6.4 \text{ m/s}^2 \downarrow]$

$$\mathbf{a}_B = 8.12 \text{ m/s}^2 \searrow 52.0° \quad ◀$$

Acceleration of Point C

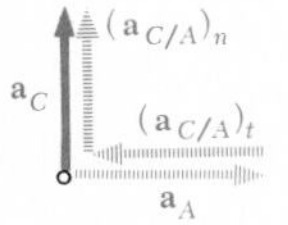

$$\mathbf{a}_C = \mathbf{a}_A + \mathbf{a}_{C/A} = \mathbf{a}_A + (\mathbf{a}_{C/A})_t + (\mathbf{a}_{C/A})_n$$

$$= [3 \text{ m/s}^2 \rightarrow] + [(0.150 \text{ m})(20 \text{ rad/s}^2) \leftarrow] + [(0.150 \text{ m})(8 \text{ rad/s})^2 \uparrow]$$

$$\mathbf{a}_C = 9.60 \text{ m/s}^2 \uparrow \quad ◀$$

Acceleration of Point D

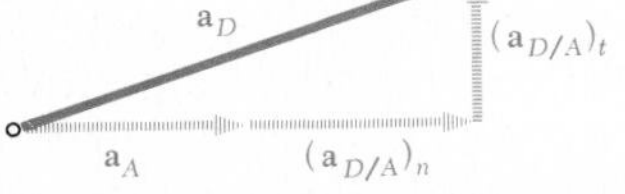

$$\mathbf{a}_D = \mathbf{a}_A + \mathbf{a}_{D/A} = \mathbf{a}_A + (\mathbf{a}_{D/A})_t + (\mathbf{a}_{D/A})_n$$

$$= [3 \text{ m/s}^2 \rightarrow] + [(0.150 \text{ m})(20 \text{ rad/s}^2) \uparrow] + [(0.150 \text{ m})(8 \text{ rad/s})^2 \rightarrow]$$

$$\mathbf{a}_D = 12.95 \text{ m/s}^2 \measuredangle 13.4° \quad ◀$$

SAMPLE PROBLEM 15.7

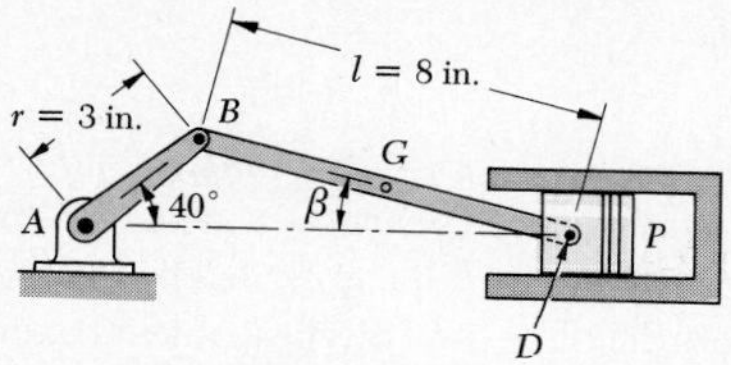

Crank AB of the engine system of Sample Prob. 15.3 has a constant clockwise angular velocity of 2000 rpm. For the crank position shown, determine the angular acceleration of the connecting rod BD and the acceleration of point D.

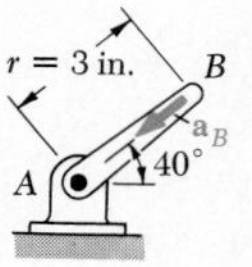

Motion of Crank AB. Since the crank rotates about A with constant $\omega_{AB} = 2000$ rpm $= 209$ rad/s, we have $\alpha_{AB} = 0$. The acceleration of B is therefore directed toward A and has a magnitude

$$a_B = r\omega_{AB}^2 = (\tfrac{3}{12}\text{ ft})(209\text{ rad/s})^2 = 10{,}920\text{ ft/s}^2$$

$$\mathbf{a}_B = 10{,}920\text{ ft/s}^2 \swarrow 40°$$

Motion of the Connecting Rod BD. The angular velocity $\boldsymbol{\omega}_{BD}$ and the value of β were obtained in Sample Prob. 15.3.

$$\omega_{BD} = 61.9\text{ rad/s} \circlearrowleft \qquad \beta = 13.9°$$

The motion of BD is resolved into a translation with B and a rotation about B. The relative acceleration $\mathbf{a}_{D/B}$ is resolved into normal and tangential components.

$$(a_{D/B})_n = (BD)\omega_{BD}^2 = (\tfrac{8}{12}\text{ ft})(61.9\text{ rad/s})^2 = 2550\text{ ft/s}^2$$

$$(\mathbf{a}_{D/B})_n = 2550\text{ ft/s}^2 \nwarrow 13.9°$$

$$(a_{D/B})_t = (BD)\alpha_{BD} = (\tfrac{8}{12})\alpha_{BD} = 0.667\alpha_{BD}$$

$$(\mathbf{a}_{D/B})_t = 0.667\alpha_{BD} \nearrow 76.1°$$

While $(\mathbf{a}_{B/D})_t$ must be perpendicular to BD, its sense is not known.

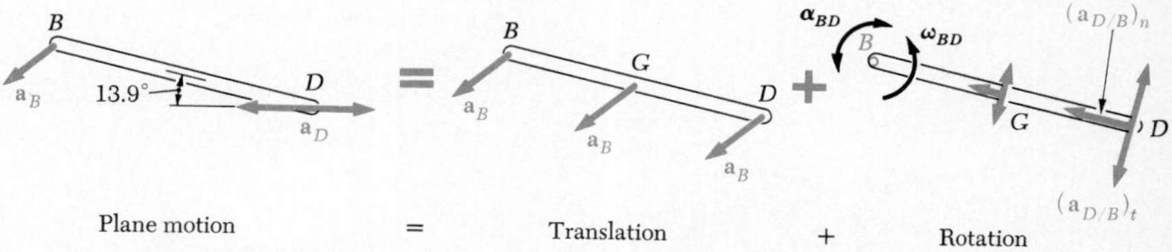

Noting that the acceleration $\mathbf{a}_D$ must be horizontal, we write

$$\mathbf{a}_D = \mathbf{a}_B + \mathbf{a}_{D/B} = \mathbf{a}_B + (\mathbf{a}_{D/B})_n + (\mathbf{a}_{D/B})_t$$

$$[a_D \leftrightarrow] = [10{,}920 \swarrow 40°] + [2560 \nwarrow 13.9°] + [0.667\alpha_{BD} \nearrow 76.1°]$$

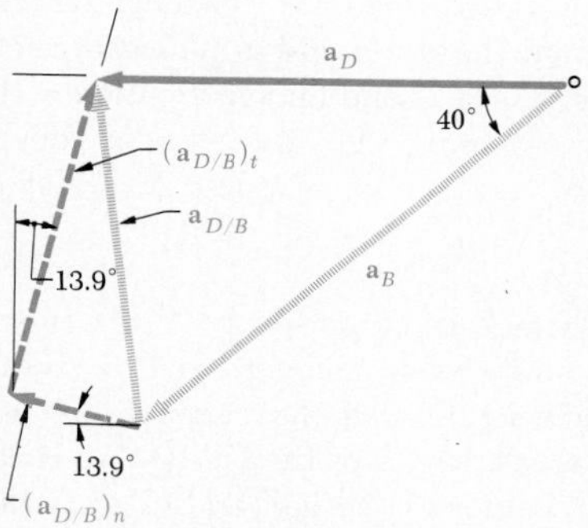

Equating x and y components, we obtain the following scalar equations:

$\xrightarrow{+}$ x components:

$$-a_D = -10{,}920\cos 40° - 2550\cos 13.9° + 0.667\alpha_{BD}\sin 13.9°$$

$+\uparrow y$ components:

$$0 = -10{,}920\sin 40° + 2550\sin 13.9° + 0.667\alpha_{BD}\cos 13.9°$$

Solving the equations simultaneously, we obtain $\alpha_{BD} = +9890$ rad/s² and $a_D = +9260$ ft/s². The positive signs indicate that the senses shown on the vector polygon are correct; we write

$$\alpha_{BD} = 9890\text{ rad/s}^2 \circlearrowleft \quad ◄$$

$$a_D = 9260\text{ ft/s}^2 \leftarrow \quad ◄$$

PROBLEMS

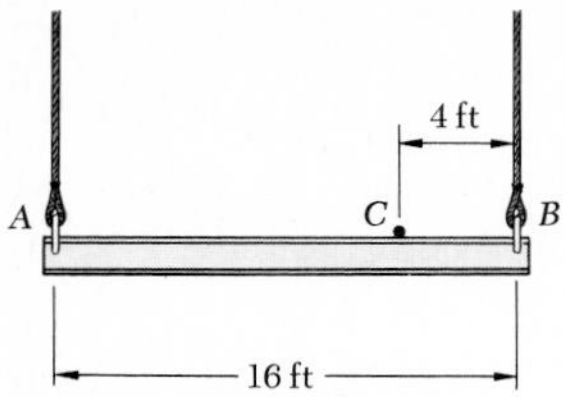

Fig. P15.73 and P15.74

15.73 A 16-ft steel beam is lowered by means of two cables unwinding at the same speed from overhead cranes. As the beam approaches the ground, the crane operators apply brakes to slow down the unwinding motion. The deceleration of the cable attached at A is 11 ft/s^2, while that of the cable attached at B is 3 ft/s^2. Determine the acceleration of point C and the angular acceleration of the beam at that instant.

15.74 The acceleration of point C is 2 ft/s^2 upward and the angular acceleration of the beam is 1.5 rad/s^2 clockwise. Knowing that the angular velocity of the beam is zero at the instant considered, determine the acceleration of each cable.

15.75 A 600-mm rod rests on a smooth horizontal table. A force **P** applied as shown produces the following accelerations: $\mathbf{a}_A = 0.8$ m/s^2 to the right, $\boldsymbol{\alpha} = 2$ rad/s^2 clockwise as viewed from above. Determine the acceleration (*a*) of point B, (*b*) of point G.

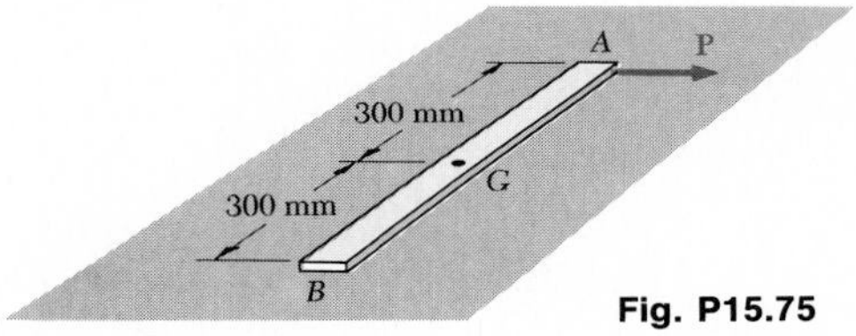

Fig. P15.75

15.76 In Prob. 15.75, determine the point of the rod which (*a*) has no acceleration, (*b*) has an acceleration of 0.350 m/s^2 to the right.

15.77 Determine the accelerations of points C and D of the wheel of Prob. 15.23, knowing that the automobile moves at a constant speed of 50 km/h.

15.78 Determine the accelerations of points B and E of the wheel of Prob. 15.23, knowing that the automobile moves at a constant speed of 50 km/h.

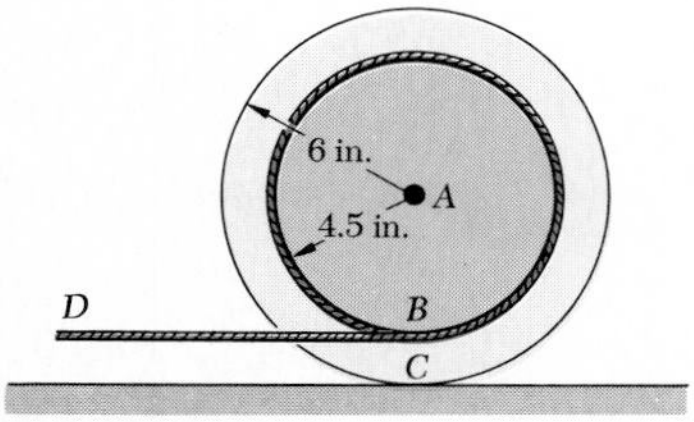

Fig. P15.79

15.79 A drum, of radius 4.5 in., is mounted on a cylinder, of radius 6 in. A cord is wound around the drum and is pulled in such a way that point D has a velocity of 3 in./s and an acceleration of 15 in./s^2, both directed to the left. Assuming that the cylinder rolls without slipping, determine the acceleration (*a*) of point A, (*b*) of point B, (*c*) of point C.

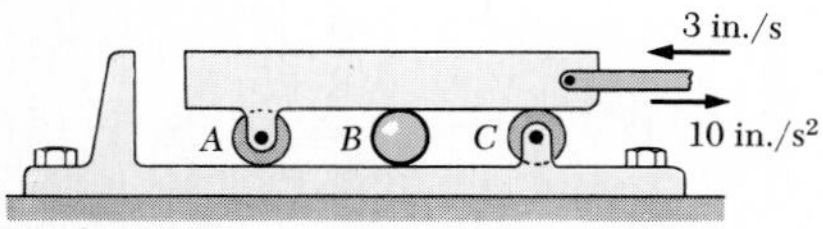

Fig. P15.80

15.80 The moving carriage is supported by two casters A and C, each of $\frac{1}{2}$-in. diameter, and by a $\frac{1}{2}$-in.-diameter ball B. If at a given instant the velocity and acceleration of the carriage are as shown, determine (*a*) the angular accelerations of the ball and of each caster, (*b*) the accelerations of the center of the ball and of each caster.

15.81 and 15.82 At the instant shown, the disk rotates with a constant angular velocity ω_0 clockwise. Determine the angular velocities and the angular accelerations of the rods AB and BC.

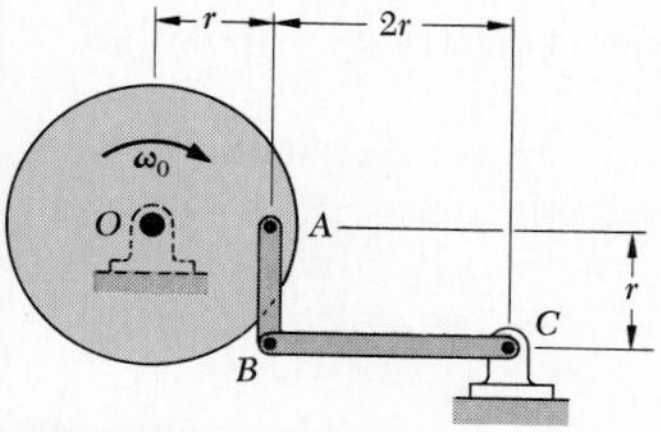

Fig. P15.81

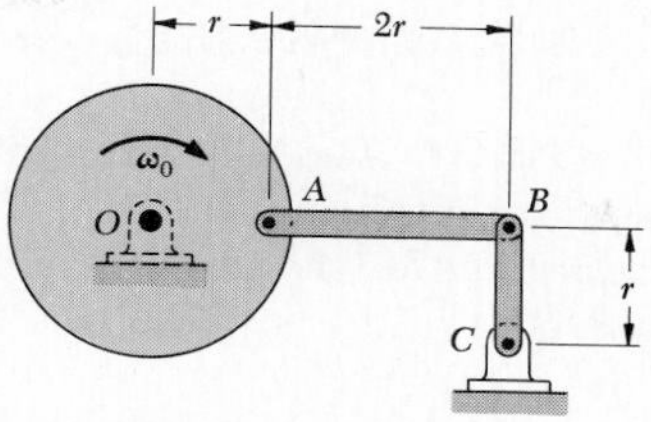

Fig. P15.82

15.83 Crank AB rotates about A with a constant angular velocity of 900 rpm clockwise. Determine the acceleration of the piston P when (a) $\theta = 90°$, (b) $\theta = 180°$.

15.84 Solve Prob. 15.83 when (a) $\theta = 0$, (b) $\theta = 270°$.

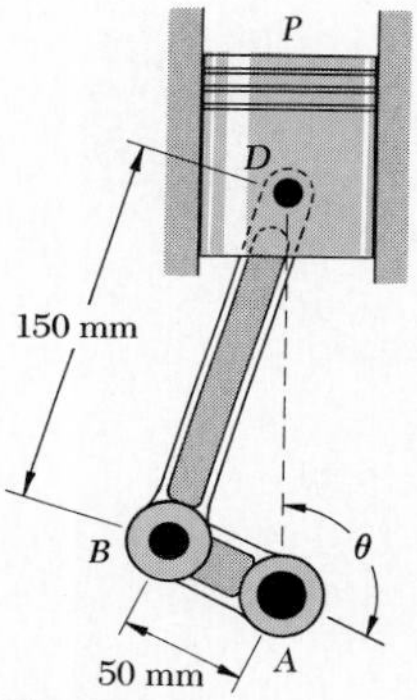

Fig. P15.83

15.85 Arm AB rotates with a constant angular velocity of 120 rpm clockwise. Knowing that gear A does not rotate, determine the acceleration of the tooth of gear B which is in contact with gear A.

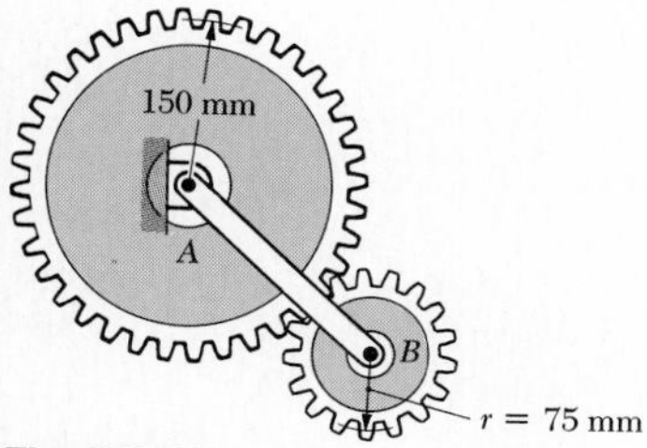

Fig. P15.85

15.86 and 15.87 For the linkage indicated, determine the angular acceleration (a) of bar BD, (b) of bar DE.

15.86 Linkage of Prob. 15.38.

15.87 Linkage of Prob. 15.39.

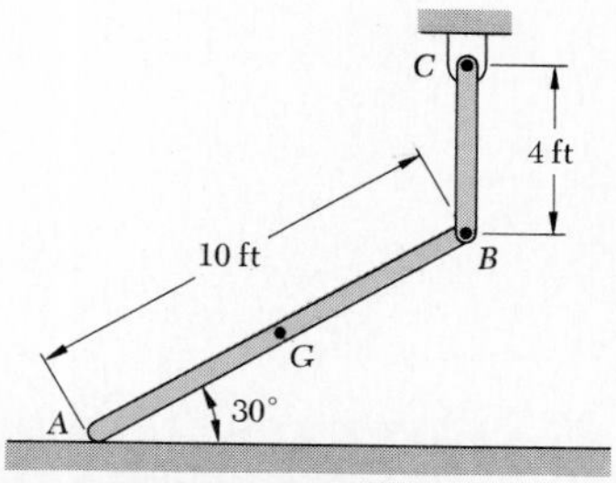

Fig. P15.88 and P15.90

15.88 and 15.89 The end A of the rod AB moves to the right with a constant velocity of 8 ft/s. For the position shown, determine (a) the angular acceleration of the rod, (b) the acceleration of the midpoint G of the rod.

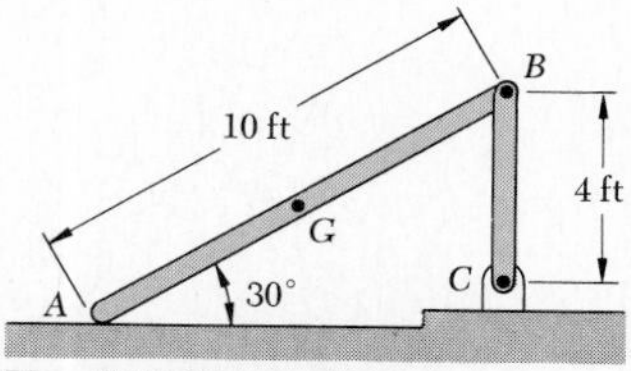

Fig. P15.89 and P15.91

15.90 and 15.91 In the position shown, end A of the rod AB has a velocity of 8 ft/s and an acceleration of 6 ft/s², both directed to the right. Determine (a) the angular acceleration of the rod, (b) the acceleration of the midpoint G of the rod.

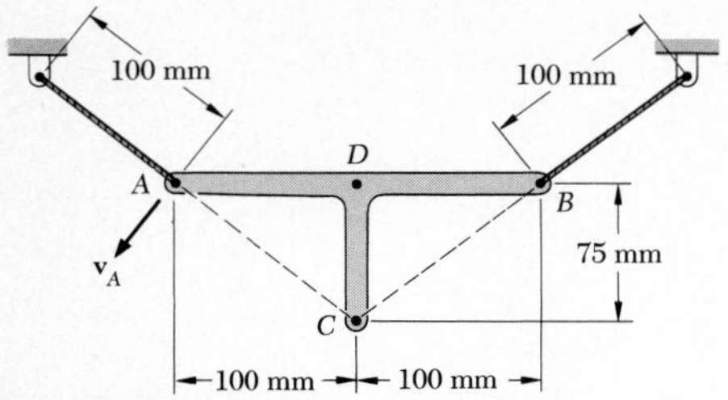

Fig. P15.92

15.92 Point A of the bracket $ABCD$ moves with a velocity of constant magnitude $v_A = 250$ mm/s. For the position shown, determine (*a*) the angular acceleration of the bracket, (*b*) the acceleration of point C.

15.93 In Prob. 15.92, determine the acceleration of point D.

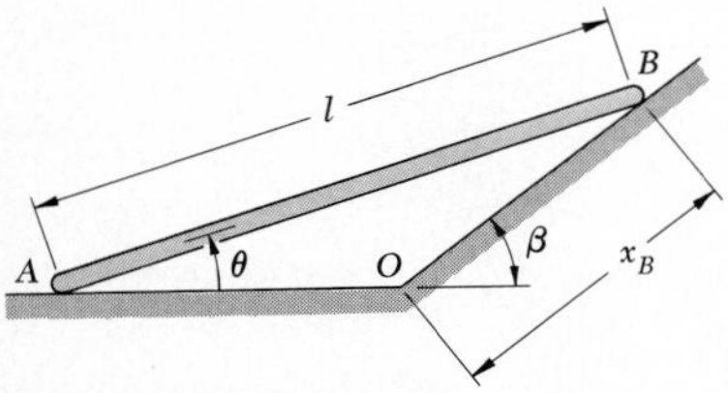

Fig. P15.94 and P15.95

***15.94** Rod AB slides with its ends in contact with the floor and the inclined plane. Using the method of Sec. 15.9, derive an expression for the angular velocity of the rod in terms of v_B, θ, l, and β.

***15.95** Derive an expression for the angular acceleration of the rod AB in terms of v_B, θ, l, and β, knowing that the acceleration of point B is zero.

***15.96** The drive disk of the Scotch crosshead mechanism shown has an angular velocity ω and an angular acceleration α, both directed clockwise. Using the method of Sec. 15.9, derive an expression (*a*) for the velocity of point B, (*b*) for the acceleration of point B.

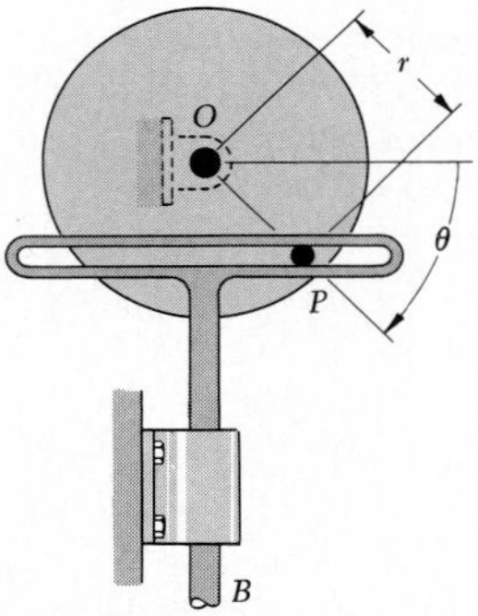

Fig. P15.96

***15.97** A disk of radius r rolls to the right with a constant velocity $\mathbf{v}$. Denoting by P the point of the rim in contact with the ground at $t = 0$, derive expressions for the horizontal and vertical components of the velocity of P at any time t. (The curve described by point P is called a *cycloid*.)

***15.98** For the disk of Prob. 15.97, derive expressions for the horizontal and vertical components of the acceleration of P at any time t.

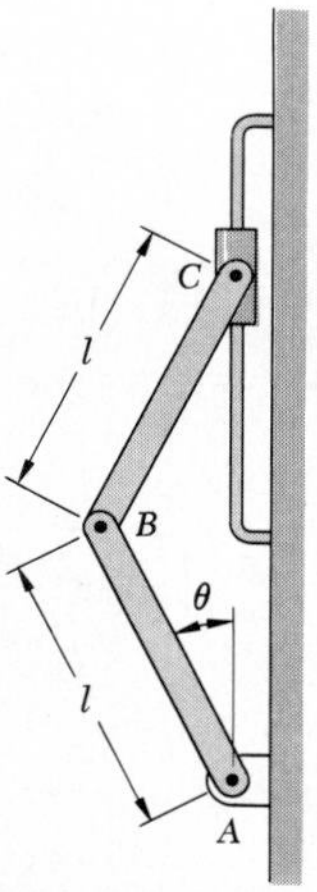

Fig. P15.99 and P15.100

***15.99** Knowing that rod AB rotates with an angular velocity ω and with an angular acceleration α, both counterclockwise, derive expressions for the velocity and acceleration of collar C.

***15.100** Knowing that collar C moves upward with a constant velocity $\mathbf{v}_0$, derive expressions for the angular velocity and angular acceleration of rod AB in the position shown.

***15.101** Collar B slides along rod OC and is attached to a sliding block which moves in a vertical slot. Knowing that rod OC rotates with an angular velocity ω and with an angular acceleration $\boldsymbol{\alpha}$, both counterclockwise, derive expressions for the velocity and acceleration of collar B.

***15.102** Collar B slides along rod OC and is attached to a sliding block which moves upward with a constant velocity $\mathbf{v}$ in a vertical slot. Using the method of Sec. 15.9, derive an expression (*a*) for the angular velocity of rod OC, (*b*) for the angular acceleration of rod OC.

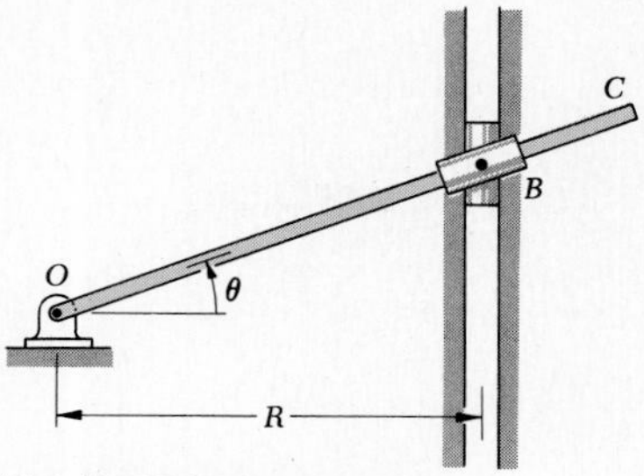

Fig. P15.101 and P15.102

***15.103** The crank AB of Prob. 15.33 rotates with a constant clockwise angular velocity ω, and $\theta = 0$ at $t = 0$. Using the method of Sec. 15.9, derive an expression for the velocity of the piston P in terms of the time t.

***15.104** A slender rod AB is attached to a collar at B and rests on a smooth circular cylinder of radius r. Knowing that the collar B moves upward with a constant velocity of magnitude v, derive an expression in terms of r, θ, and v (*a*) for the angular velocity of the rod, (*b*) for the angular acceleration of the rod.

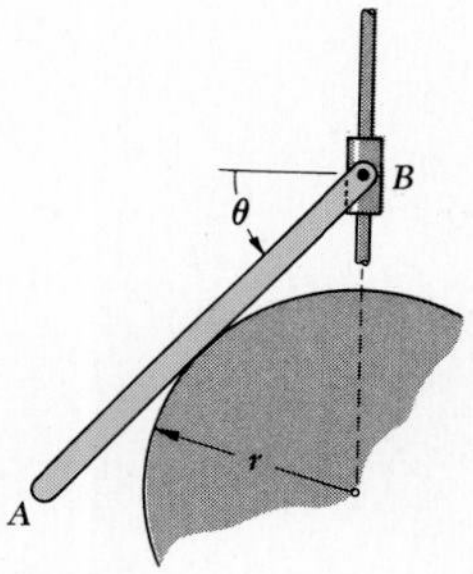

Fig. P15.104

***15.105** Collar A slides upward with a constant velocity $\mathbf{v}_A$. Using the method of Sec. 15.9, derive an expression for (*a*) the angular velocity of rod AB, (*b*) the components of the velocity of point B.

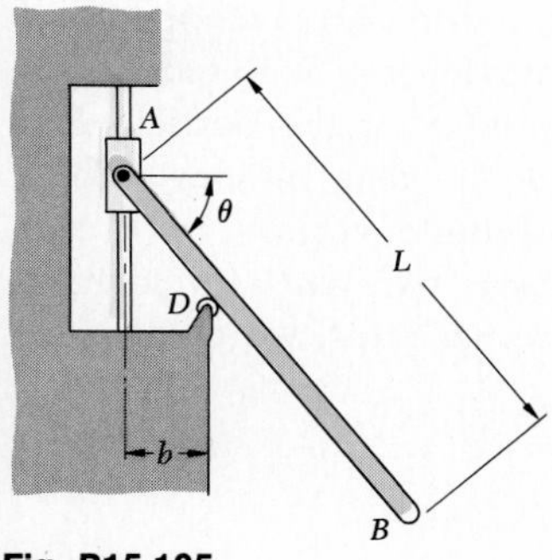

Fig. P15.105

***15.106** In Prob. 15.105, derive an expression for the angular acceleration of rod AB.

*15.10. Particle Moving on a Slab in Translation.

Let us consider the motion of a particle P which describes a path on a slab S which is itself in translation. The motion of P may be analyzed, either in terms of its coordinates x and y with respect to a fixed set of axes, or in terms of its coordinates x_1 and y_1 with respect to a set of axes attached to the slab S

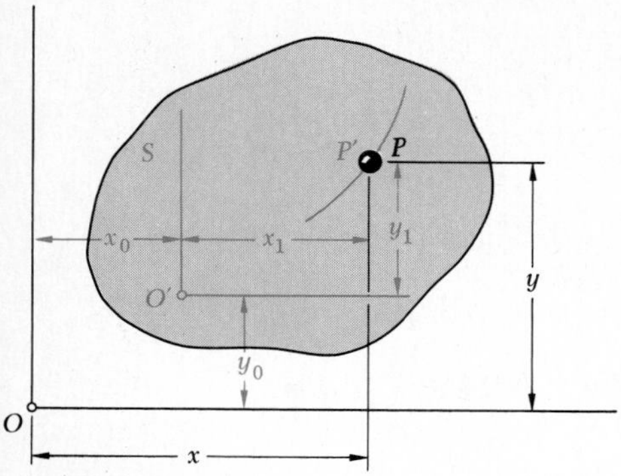

Fig. 15.21

and moving with it (Fig. 15.21). We propose to determine the relation existing between the absolute motion of P with respect to the fixed axes and its relative motion with respect to the axes moving with S.

Denoting by x_0 and y_0 the coordinates of O' with respect to the fixed axes, we write

$$x = x_0 + x_1 \qquad y = y_0 + y_1 \tag{15.23}$$

Differentiating with respect to t and using dots to indicate time derivatives, we have

$$\dot{x} = \dot{x}_0 + \dot{x}_1 \qquad \dot{y} = \dot{y}_0 + \dot{y}_1 \tag{15.24}$$

Now, $\dot{x}$ and $\dot{y}$ represent the components of the absolute velocity $\mathbf{v}_P$ of P, while $\dot{x}_1$ and $\dot{y}_1$ represent the components of the velocity $\mathbf{v}_{P/S}$ of P with respect to S. On the other hand, $\dot{x}_0$ and $\dot{y}_0$ represent the components of the velocity of O' or, since the slab S is in translation, the components of the velocity of any other point of S. We may, for example, consider that $\dot{x}_0$ and $\dot{y}_0$ represent the components of the velocity $\mathbf{v}_{P'}$ of the *point P' of the slab S which happens to coincide with the particle P at the instant considered.* We thus write

$$\mathbf{v}_P = \mathbf{v}_{P'} + \mathbf{v}_{P/S} \tag{15.25}$$

where the right-hand member represents a vector sum. We note that the velocity $\mathbf{v}_P$ reduces to $\mathbf{v}_{P/S}$ if the slab is stopped and if the particle is allowed to keep moving on S; it reduces to the velocity $\mathbf{v}_{P'}$ of the coinciding point P' if the particle P is immobilized on S while S is allowed to keep moving. Thus, formula (15.25) expresses that the absolute velocity of P may be obtained by adding vectorially these two partial velocities.

Differentiating Eqs. (15.24) with respect to t, we obtain

$$\ddot{x} = \ddot{x}_0 + \ddot{x}_1 \qquad \ddot{y} = \ddot{y}_0 + \ddot{y}_1 \tag{15.26}$$

or, in vector form,

$$\mathbf{a}_P = \mathbf{a}_{P'} + \mathbf{a}_{P/S} \tag{15.27}$$

where $\mathbf{a}_{P/S}$ = acceleration of P with respect to S
$\mathbf{a}_{P'}$ = acceleration of point P' of S coinciding with P at the instant considered

Formula (15.27) may be interpreted in the same way as formula (15.25).

Formulas (15.25) and (15.27) actually restate the results obtained in Sec. 11.11. Since S is in translation, the x_1 and y_1 axes, respectively, remain parallel to the fixed x and y axes, and the velocity $\mathbf{v}_{P/S}$ is equal to the velocity $\mathbf{v}_{P/P'}$ relative to point P'

as it was defined in Sec. 11.11; similarly, the acceleration $\mathbf{a}_{P/S}$ is equal to the acceleration $\mathbf{a}_{P/P'}$ relative to P'.

***15.11. Particle Moving on a Rotating Slab. Coriolis Acceleration.** We shall now consider the motion of a particle P which describes a path on a slab S which is itself rotating about a fixed point O. The motion of P may be analyzed either in terms of its polar coordinates r and θ with respect to fixed axes or in terms of its coordinates r and θ_1 with respect to axes attached to the slab S and rotating with it (Fig. 15.22). Again we propose to determine the relation existing between the absolute motion of P and its relative motion with respect to S.

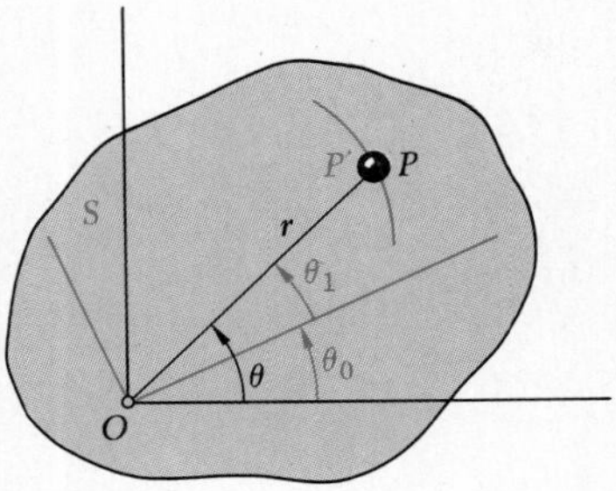

Fig. 15.22

We shall use formulas (11.30) to express the radial and transverse components of the absolute velocity $\mathbf{v}_P$ of P. Observing that $\theta = \theta_0 + \theta_1$, where θ_0 denotes the angular displacement of the slab at the instant considered, we write

$$(v_P)_r = \dot{r} \qquad (v_P)_\theta = r\dot{\theta} = r(\dot{\theta}_0 + \dot{\theta}_1) \tag{15.28}$$

Considering the particular case when P is immobilized on S and S is allowed to rotate, $\mathbf{v}_P$ reduces to the velocity $\mathbf{v}_{P'}$ of the point P' of the slab S which happens to coincide with P at the instant considered. Making $r =$ constant and $\theta_1 =$ constant in (15.28), we obtain

$$(v_{P'})_r = 0 \qquad (v_{P'})_\theta = r\,\dot{\theta}_0 \tag{15.29}$$

Considering now the particular case when the slab is maintained fixed and P is allowed to move, $\mathbf{v}_P$ reduces to the relative velocity $\mathbf{v}_{P/S}$. Making θ_0 constant in (15.28), we obtain, therefore,

$$(v_{P/S})_r = \dot{r} \qquad (v_{P/S})_\theta = r\dot{\theta}_1 \tag{15.30}$$

Comparing formulas (15.28) to (15.30), we find that

$$(v_P)_r = (v_{P'})_r + (v_{P/S})_r \qquad (v_P)_\theta = (v_{P'})_\theta + (v_{P/S})_\theta$$

or, in vector form,

$$\mathbf{v}_P = \mathbf{v}_{P'} + \mathbf{v}_{P/S} \tag{15.31}$$

Formula (15.31) expresses that $\mathbf{v}_P$ may be obtained by adding vectorially the velocities corresponding to the two particular cases considered above.

Using formulas (11.33), we now express the radial and transverse components of the absolute acceleration $\mathbf{a}_P$ of P,

$$\begin{aligned}(a_P)_r &= \ddot{r} - r\dot{\theta}^2 = \ddot{r} - r(\dot{\theta}_0 + \dot{\theta}_1)^2 \\ &= \ddot{r} - r(\dot{\theta}_0^2 + 2\dot{\theta}_0\dot{\theta}_1 + \dot{\theta}_1^2)\end{aligned} \tag{15.32}$$

$$\begin{aligned}(a_P)_\theta &= r\ddot{\theta} + 2\dot{r}\dot{\theta} \\ &= r(\ddot{\theta}_0 + \ddot{\theta}_1) + 2\dot{r}(\dot{\theta}_0 + \dot{\theta}_1)\end{aligned} \tag{15.33}$$

Considering again the particular case when r = constant and θ_1 = constant (P immobilized on S), Eqs. (15.32) and (15.33) yield

$$(a_{P'})_r = -r\dot{\theta}_0^2 \qquad (a_{P'})_\theta = r\ddot{\theta}_0 \tag{15.34}$$

Considering now the particular case when θ_0 = constant (slab fixed), we obtain

$$(a_{P/S})_r = \ddot{r} - r\dot{\theta}_1^2 \qquad (a_{P/S})_\theta = r\ddot{\theta}_1 + 2\dot{r}\dot{\theta}_1 \tag{15.35}$$

Comparing formulas (15.32) to (15.35), we find that the absolute acceleration $\mathbf{a}_P$ *cannot* be obtained by adding the accelerations $\mathbf{a}_{P'}$ and $\mathbf{a}_{P/S}$ corresponding to the two particular cases considered above. We have instead

$$\mathbf{a}_P = \mathbf{a}_{P'} + \mathbf{a}_{P/S} + \mathbf{a}_c \tag{15.36}$$

where $\mathbf{a}_c$ is a vector of components

$$(a_c)_r = -2r\dot{\theta}_0\dot{\theta}_1 \qquad (a_c)_\theta = 2\dot{r}\dot{\theta}_0$$

Noting that $\dot{\theta}_0$ represents the angular velocity ω of S, and recalling formulas (15.30), we have

$$(a_c)_r = -2\omega(v_{P/S})_\theta \qquad (a_c)_\theta = 2\omega(v_{P/S})_r \tag{15.37}$$

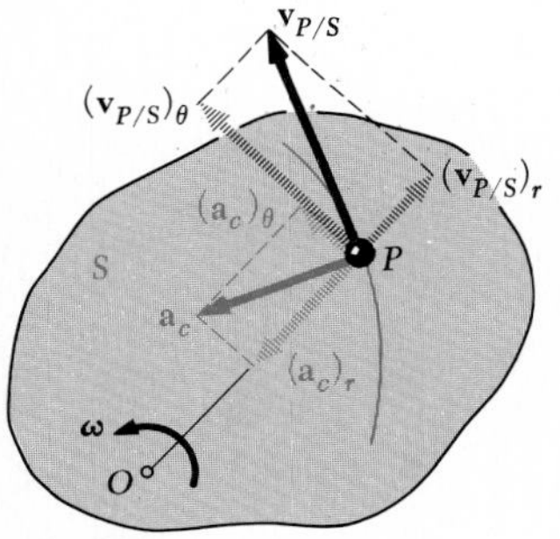

Fig. 15.23

The vector $\mathbf{a}_c$ is called the *complementary acceleration*, or *Coriolis acceleration*, after the French mathematician De Coriolis (1792–1843). The first of formulas (15.37) indicates that the vector $(\mathbf{a}_c)_r$ is obtained by multiplying the vector $(\mathbf{v}_{P/S})_\theta$ by 2ω and rotating it through 90° in the sense of rotation of the slab (Fig. 15.23); the second formula defines $(\mathbf{a}_c)_\theta$ from $(\mathbf{v}_{P/S})_r$ in a similar way. *The Coriolis acceleration* $\mathbf{a}_c$ *is thus a vector perpendicular to the relative velocity* $\mathbf{v}_{P/S}$, *and of magnitude equal to* $2\omega v_{P/S}$; *the sense of* $\mathbf{a}_c$ *is obtained by rotating the vector* $\mathbf{v}_{P/S}$ *through* 90° *in the sense of rotation of* S.

Comparing formulas (15.36) and (11.26), we note that the acceleration $\mathbf{a}_{P/S}$ of P relative to the slab S is *not* equal to the acceleration $\mathbf{a}_{P/P'}$ of P relative to the point P' of S; this is due to the fact that the first acceleration is defined with respect to *rotating axes,* while the second is defined with respect to *axes of fixed orientation.*

The following example will help in understanding the physical meaning of the Coriolis acceleration. Consider a collar P which is made to slide at a constant relative speed u along a rod OB rotating at a constant angular velocity $\boldsymbol{\omega}$ about O (Fig. 15.24a). According to formula (15.36), the absolute acceleration of P may be obtained by adding vectorially the acceleration $\mathbf{a}_A$ of the point A of the rod coinciding with P, the relative acceleration $\mathbf{a}_{P/OB}$ of P with respect to the rod, and the Coriolis acceleration $\mathbf{a}_c$. Since the angular velocity ω of the rod is constant, $\mathbf{a}_A$ reduces to its normal component $(\mathbf{a}_A)_n$ of magnitude $r\omega^2$; and since u

is constant, the relative acceleration $\mathbf{a}_{P/OB}$ is zero. According to the definition given above, the Coriolis acceleration is a vector perpendicular to OB, of magnitude $2\omega u$, and directed as shown in the figure. The acceleration of the collar P consists, therefore, of the two vectors shown in Fig. 15.24*a*. Note that the result obtained may be checked by applying formulas (11.33).

To understand better the significance of the Coriolis acceleration, we shall consider the absolute velocity of P at time t and at time $t + \Delta t$ (Fig. 15.24*b*). At time t, the velocity may be resolved into its components $\mathbf{u}$ and $\mathbf{v}_A$, and at time $t + \Delta t$ into its components $\mathbf{u}'$ and $\mathbf{v}_{A'}$. Drawing these components from the same origin (Fig. 15.24*c*), we note that the change in velocity during the time Δt may be represented by the sum of three vectors $\overrightarrow{RR'}$, $\overrightarrow{TT''}$, and $\overrightarrow{T''T'}$. The vector $\overrightarrow{TT''}$ measures the change in direction of the velocity $\mathbf{v}_A$, and the quotient $\overrightarrow{TT''}/\Delta t$ represents the acceleration $\mathbf{a}_A$ when Δt approaches zero. We check that the direction of $\overrightarrow{TT''}$ is that of $\mathbf{a}_A$ when Δt approaches zero and that

$$\lim_{\Delta t \to 0} \frac{TT''}{\Delta t} = \lim_{\Delta t \to 0} v_A \frac{\Delta\theta}{\Delta t} = r\omega\omega = r\omega^2 = a_A$$

The vector $\overrightarrow{RR'}$ measures the change in direction of $\mathbf{u}$ due to the rotation of the rod; the vector $\overrightarrow{T''T'}$ measures the change in magnitude of $\mathbf{v}_A$ due to the motion of P on the rod. The vectors $\overrightarrow{RR'}$ and $\overrightarrow{T''T'}$ result from the *combined effect* of the relative motion of P and of the rotation of the rod; they would vanish if *either* of these two motions stopped. We may easily verify that the sum of these two vectors defines the Coriolis acceleration. Their direction is that of $\mathbf{a}_c$ when Δt approaches zero and, since $RR' = u\,\Delta\theta$ and $T''T' = v_{A'} - v_A = (r + \Delta r)\omega - r\omega = \omega\,\Delta r$, we check that

$$\lim_{\Delta t \to 0}\left(\frac{RR'}{\Delta t} + \frac{T''T'}{\Delta t}\right) = \lim_{\Delta t \to 0}\left(u\frac{\Delta\theta}{\Delta t} + \omega\frac{\Delta r}{\Delta t}\right)$$

$$= u\omega + \omega u = 2\omega u = a_c$$

Formulas (15.31) and (15.36) may be used to analyze the motion of mechanisms which contain parts sliding on each other. They make it possible, for example, to relate the absolute and relative motions of sliding pins and collars (see Sample Probs. 15.8 and 15.9). The concept of Coriolis acceleration is also very useful in the study of long-range projectiles and of other bodies whose motions are appreciably affected by the rotation of the earth. As was pointed out in Sec. 12.1, a system of axes attached to the earth does not truly constitute a newtonian frame of reference; such a system of axes should actually be considered as rotating. The formulas derived in this section will therefore facilitate the study of the motion of bodies with respect to axes attached to the earth.

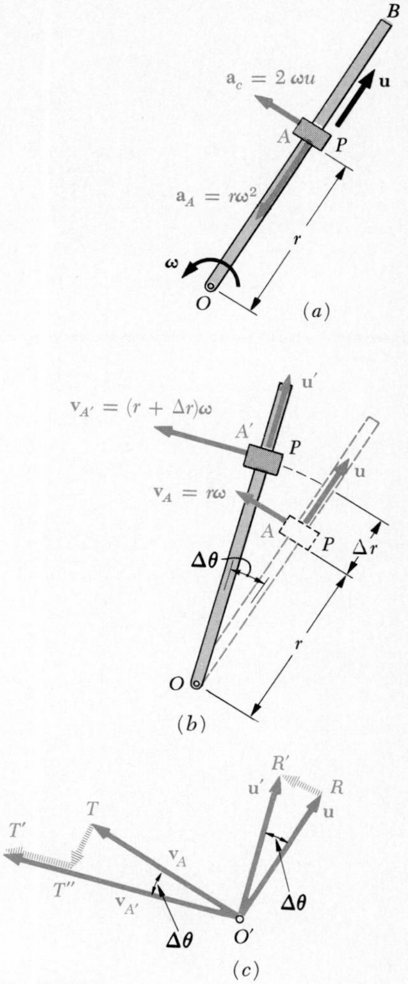

Fig. 15.24

SAMPLE PROBLEM 15.8

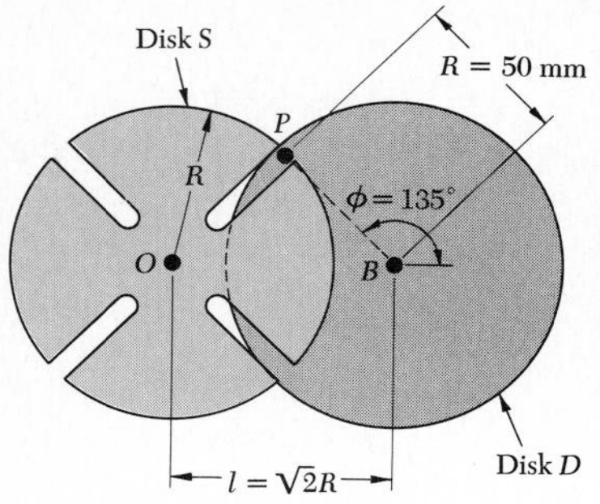

The Geneva mechanism shown is used in many counting instruments and in other applications where an intermittent rotary motion is required. Disk D rotates with a constant counterclockwise angular velocity ω_D of 10 rad/s. A pin P is attached to disk D and slides along one of several slots cut in disk S. It is desirable that the angular velocity of disk S be zero as the pin enters and leaves each slot; in the case of four slots, this will occur if the distance between the centers of the disks is $l = \sqrt{2}\,R$.

At the instant when $\phi = 150°$, determine (*a*) the angular velocity of disk S, (*b*) the velocity of pin P relative to disk S.

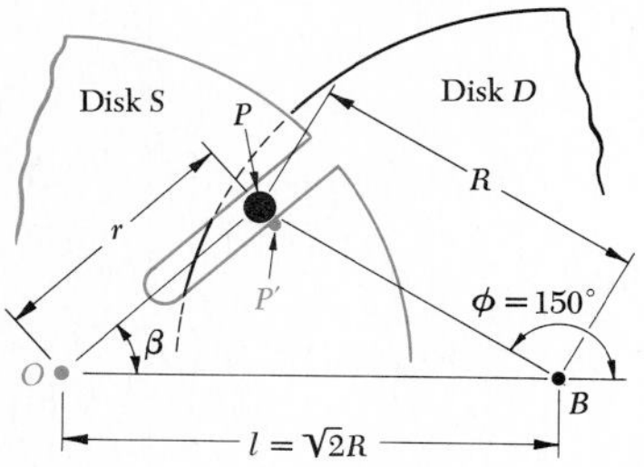

Solution. We solve triangle OPB, which corresponds to the position $\phi = 150°$. Using the law of cosines, we write

$$r^2 = R^2 + l^2 - 2Rl\cos 30° = 0.551R^2 \qquad r = 0.742R = 37.1\text{ mm}$$

From the law of sines

$$\frac{\sin\beta}{R} = \frac{\sin 30°}{r} \qquad \sin\beta = \frac{\sin 30°}{0.742} \qquad \beta = 42.4°$$

Since pin P is attached to disk D, and since disk D rotates about point B, the magnitude of the absolute velocity of P is

$$v_P = R\omega_D = (50\text{ mm})(10\text{ rad/s}) = 500\text{ mm/s}$$
$$\mathbf{v}_P = 500\text{ mm/s} \angle 60°$$

We consider now the motion of pin P along the slot in disk S. Denoting by P' the point of disk S which coincides with P at the instant considered, we write

$$\mathbf{v}_P = \mathbf{v}_{P'} + \mathbf{v}_{P/S}$$

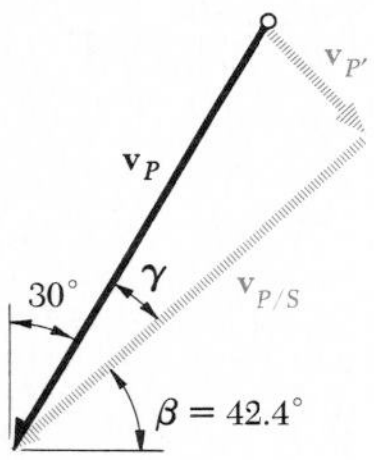

Noting that $\mathbf{v}_{P'}$ is perpendicular to the radius OP and that $\mathbf{v}_{P/S}$ is directed along the slot, we draw the velocity triangle corresponding to the above equation. From the triangle, we compute

$$\gamma = 90° - 42.4° - 30° = 17.6°$$
$$v_{P'} = v_P \sin\gamma = (500\text{ mm/s})\sin 17.6°$$
$$\mathbf{v}_{P'} = 151.2\text{ mm/s} \angle 42.4°$$
$$v_{P/S} = v_P\cos\gamma = (500\text{ mm/s})\cos 17.6°$$
$$\mathbf{v}_{P/S} = 477\text{ mm/s} \angle 42.4° \quad ◄$$

Since $\mathbf{v}_{P'}$ is perpendicular to the radius OP, we write

$$v_{P'} = r\omega_S \qquad 151.2\text{ mm/s} = (37.1\text{ mm})\omega_S$$
$$\omega_S = 4.08\text{ rad/s} \circlearrowright \quad ◄$$

SAMPLE PROBLEM 15.9

In the Geneva mechanism of Sample Prob. 15.8, disk D rotates with a constant counterclockwise angular velocity ω_D of 10 rad/s. At the instant when $\phi = 150°$, determine the angular acceleration of disk S.

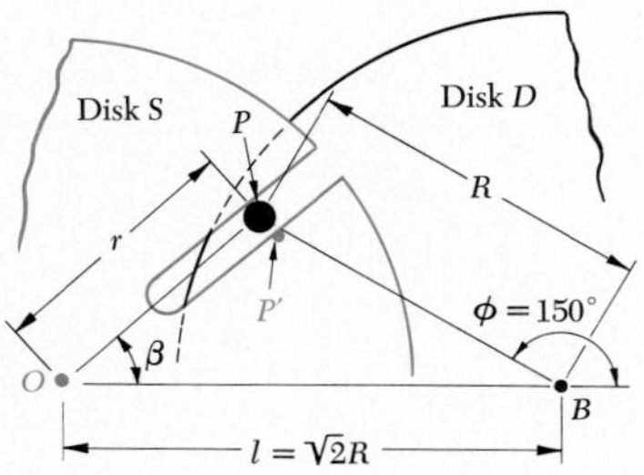

Solution. Referring to Sample Prob. 15.8, we obtain the angular velocity of disk S and the velocity of the pin relative to disk S.

$$\omega_S = 4.08 \text{ rad/s} \circlearrowright$$

$$\beta = 42.4° \qquad \mathbf{v}_{P/S} = 477 \text{ mm/s} \nearrow 42.4°$$

Since pin P moves with respect to the rotating disk S, we write

$$\mathbf{a}_P = \mathbf{a}_{P'} + \mathbf{a}_{P/S} + \mathbf{a}_c \qquad (1)$$

Each term of this vector equation is investigated separately.

Absolute Acceleration $\mathbf{a}_P$. Since disk D rotates with constant ω, the absolute acceleration $\mathbf{a}_P$ is directed toward B.

$$a_P = R\omega_D^2 = (50 \text{ mm})(10 \text{ rad/s})^2 = 5000 \text{ mm/s}^2$$

$$\mathbf{a}_P = 5000 \text{ mm/s}^2 \searrow 30°$$

Acceleration $\mathbf{a}_{P'}$ *of the Coinciding Point* P'. The acceleration $\mathbf{a}_{P'}$ of the point P' of disk S which coincides with P at the instant considered is resolved into normal and tangential components. (We recall from Sample Prob. 15.8 that $r = 37.1$ mm.)

$$(a_{P'})_n = r\omega_S^2 = (37.1 \text{ mm})(4.08 \text{ rad/s})^2 = 618 \text{ mm/s}^2$$

$$(\mathbf{a}_{P'})_n = 618 \text{ mm/s}^2 \swarrow 42.4°$$

$$(a_{P'})_t = r\alpha_S = 37.1\alpha_S \qquad (\mathbf{a}_{P'})_t = 37.1\alpha_S \nwarrow 42.4°$$

Relative Acceleration $\mathbf{a}_{P/S}$. Since the pin P moves in a straight slot cut in disk S, the relative acceleration $\mathbf{a}_{P/S}$ must be parallel to the slot; i.e., its direction must be $\swarrow 42.4°$.

Coriolis Acceleration $\mathbf{a}_c$. Rotating the relative velocity $\mathbf{v}_{P/S}$ through 90° in the sense of ω_S, we obtain the direction of the Coriolis component of the acceleration.

$$a_c = 2\omega_S v_{P/S} = 2(4.08 \text{ rad/s})(477 \text{ mm/s}) = 3890 \text{ mm/s}^2$$

$$\mathbf{a}_c = 3890 \text{ mm/s}^2 \nwarrow 42.4°$$

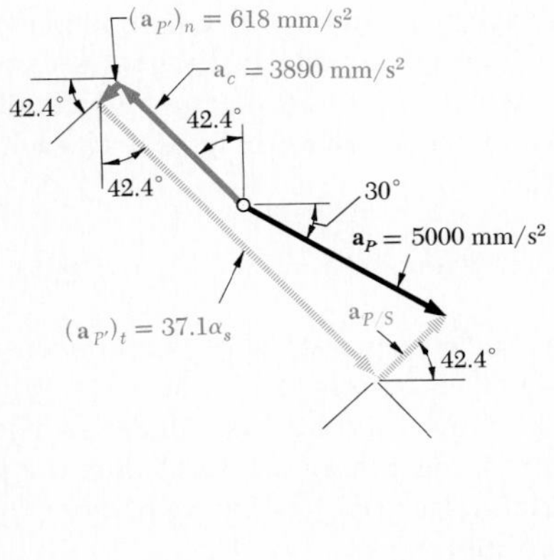

We rewrite Eq. (1) and substitute the accelerations found above.

$$\mathbf{a}_P = (\mathbf{a}_{P'})_n + (\mathbf{a}_{P'})_t + \mathbf{a}_{P/S} + \mathbf{a}_c$$

$$[5000 \searrow 30°] = [618 \swarrow 42.4°] + [37.1\alpha_S \nwarrow 42.4°]$$
$$+ [a_{P/S} \swarrow 42.4°] + [3890 \nwarrow 42.4°]$$

Equating components in a direction perpendicular to the slot:

$$5000 \cos 17.6° = 37.1\alpha_S - 3890$$

$$\alpha_S = 233 \text{ rad/s}^2 \circlearrowright \quad ◀$$

PROBLEMS

15.107 and 15.108 Two rotating rods are connected by a slider block *P*. The rod attached at *B* rotates with a constant clockwise angular velocity ω_B. For the given data, determine for the position shown (*a*) the angular velocity of the rod attached at *A*, (*b*) the relative velocity of the slider block *P* with respect to the rod on which it slides.

15.107 $b = 8$ in., $\omega_B = 6$ rad/s.

15.108 $b = 200$ mm, $\omega_B = 9$ rad/s.

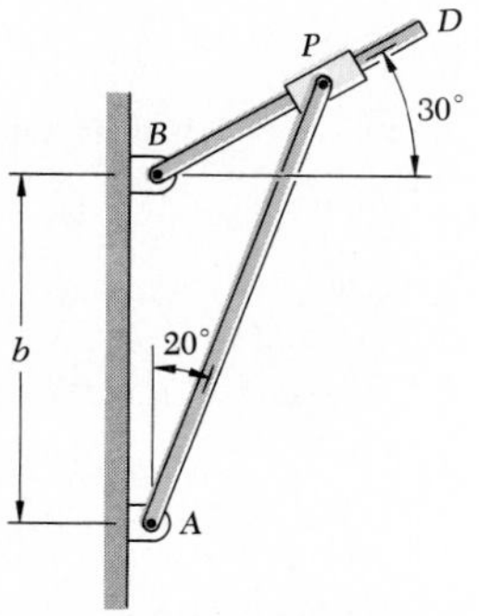

Fig. P15.107 and P15.109

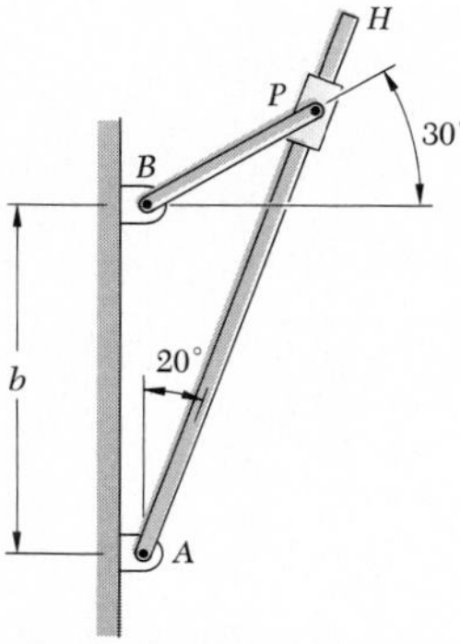

Fig. P15.108 and P15.110

15.109 and 15.110 Two rotating rods are connected by a slider block *P*. The velocity $\mathbf{v}_0$ of the slider block relative to the rod on which it slides is constant and is directed outward. For the given data, determine the angular velocity of each rod for the position shown.

15.109 $b = 200$ mm, $v_0 = 300$ mm/s.

15.110 $b = 8$ in., $v_0 = 12$ in./s.

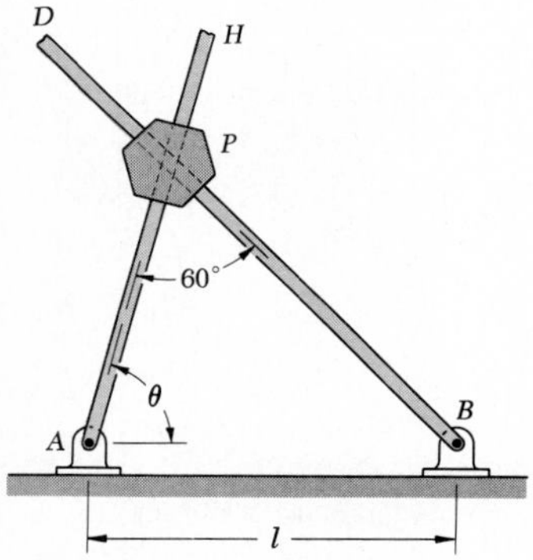

Fig. P15.111

15.111 Two rods *AH* and *BD* pass through smooth holes drilled in a hexagonal block. (The holes are drilled in different planes so that the rods will not hit each other.) Knowing that rod *AH* rotates counterclockwise at the rate ω, determine the angular velocity of rod *BD* and the relative velocity of the block with respect to each rod when (*a*) $\theta = 30°$, (*b*) $\theta = 15°$.

15.112 Solve Prob. 15.111 when (*a*) $\theta = 90°$, (*b*) $\theta = 60°$.

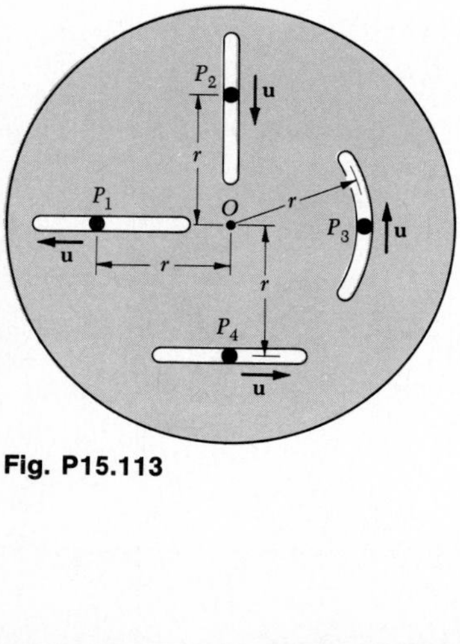

Fig. P15.113

15.113 Four pins slide in four separate slots cut in a circular plate as shown. When the plate is at rest, each pin has a velocity directed as shown and of the same constant magnitude *u*. If each pin maintains the same velocity in relation to the plate when the plate rotates about *O* with a constant *clockwise* angular velocity ω, determine the acceleration of each pin.

15.114 Solve Prob. 15.113, assuming that the plate rotates about O with a constant *counterclockwise* angular velocity $\boldsymbol{\omega}$.

15.115 Water flows through a straight pipe OB which rotates counterclockwise with an angular velocity of 120 rpm. If the velocity of the water relative to the pipe is 20 ft/s, determine the total acceleration (*a*) of the particle of water P_1, (*b*) of the particle of water P_2.

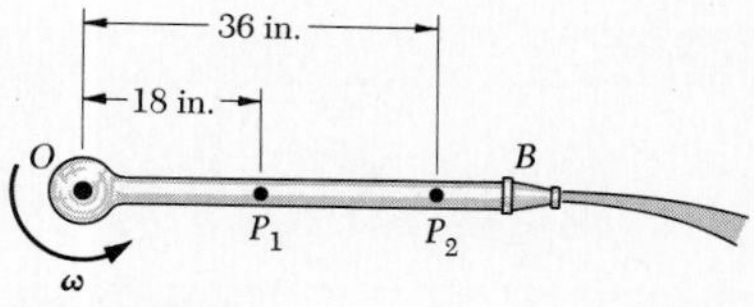

Fig. P15.115

15.116 A block P may slide on the arm OA which rotates at the constant angular velocity $\boldsymbol{\omega}$. As the arm rotates, the cord wraps around a *fixed* drum of radius b and pulls the block toward O with a speed $b\omega$. Determine the magnitude of the acceleration of the block in terms of r, b, and ω.

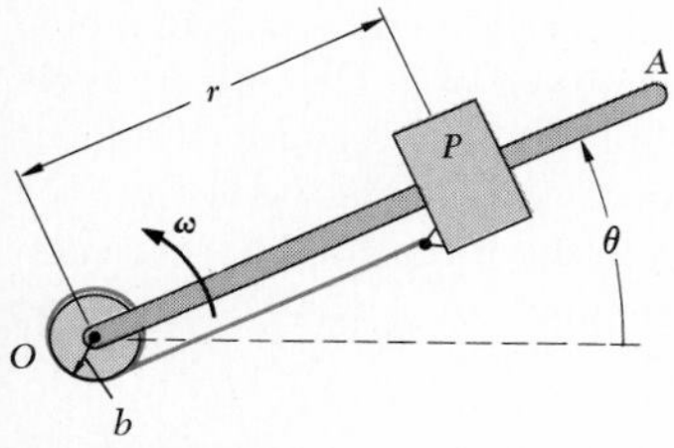

Fig. P15.116

15.117 A pin P slides in a circular slot of radius r_2 which is cut in the plate OE. The velocity of P relative to the plate is of constant magnitude u and is directed as shown. Knowing that the plate rotates counterclockwise with a constant angular velocity ω, derive an expression for (*a*) the magnitude u for which the acceleration of the pin is zero as it passes through point A, (*b*) the corresponding magnitude of the acceleration of the pin as it passes through point B.

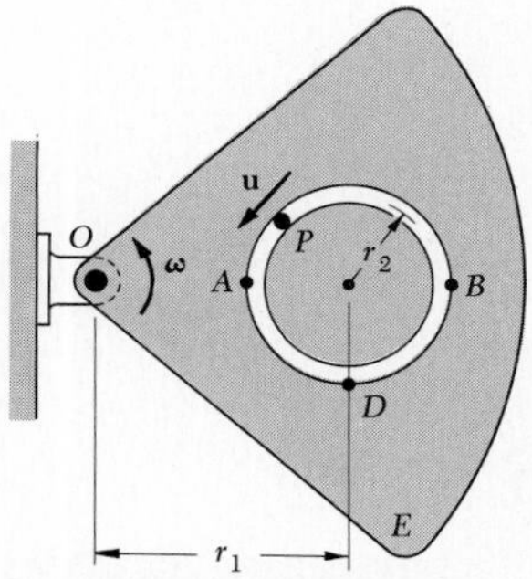

Fig. P15.117 and P15.118

15.118 Pin P slides in the circular slot cut in the plate OE at a constant relative speed $u = 0.5$ m/s as the plate rotates about O at the constant rate $\omega = 6$ rad/s. Knowing that $r_1 = 300$ mm and $r_2 = 100$ mm, determine the acceleration of the pin as it passes through (*a*) point A, (*b*) point B, (*c*) point D.

15.119 A rocket sled is tested on a straight track which is built along a meridian. Knowing that the track is located at a latitude of 40° north, determine the Coriolis acceleration of the sled when its velocity is 1000 km/h due north. (*Hint.* Consider separately the components of the motion parallel and perpendicular to the plane of the equator.)

15.120 A train crosses the parallel 50° north, traveling due north at a constant speed v. Determine the speed of the train if the Coriolis component of its acceleration is 0.01 ft/s². (See hint of Prob. 15.119.)

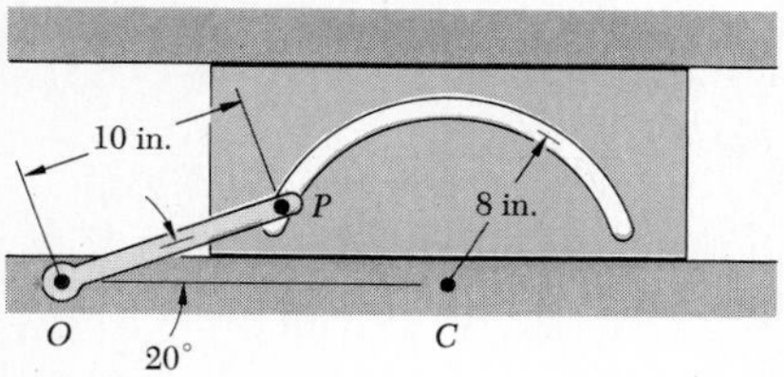

Fig. P15.123

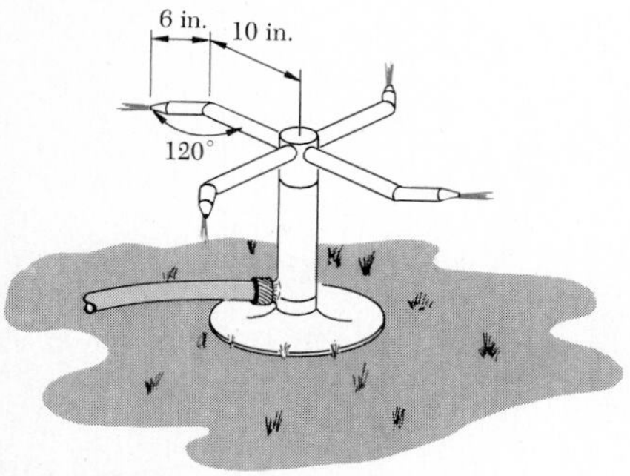

Fig. P15.125

15.121 In Prob. 15.107, determine the angular acceleration of the rod attached at A.

15.122 In Prob. 15.108, determine the angular acceleration of the rod attached at A.

15.123 At the instant shown, the slotted plate slides with a velocity of 10 in./s to the right and has an acceleration of 50 in./s^2 to the left. Determine the angular acceleration of rod OP.

15.124 At the instant when $\theta = 90°$ in Prob. 15.111, rod AH has a counterclockwise angular velocity ω and no angular acceleration. Determine at that instant (*a*) the angular acceleration of rod BD, (*b*) the total acceleration of the block P.

15.125 A garden sprinkler has four rotating arms, each of which consists of two horizontal straight sections of pipe forming an angle of 120°. The sprinkler when operating rotates with a constant angular velocity of 180 rpm. If the velocity of the water relative to the pipe sections is 12 ft/s, determine the magnitude of the total acceleration of a particle of water as it passes the midpoint of (*a*) the 10-in. section of pipe, (*b*) the 6-in. section of pipe.

15.126 Water flows through the curved pipe OB, which has a constant radius of 0.375 m and which rotates with a constant counterclockwise angular velocity of 120 rpm. If the velocity of the water relative to the pipe is 12 m/s, determine the total acceleration of the particle of water P.

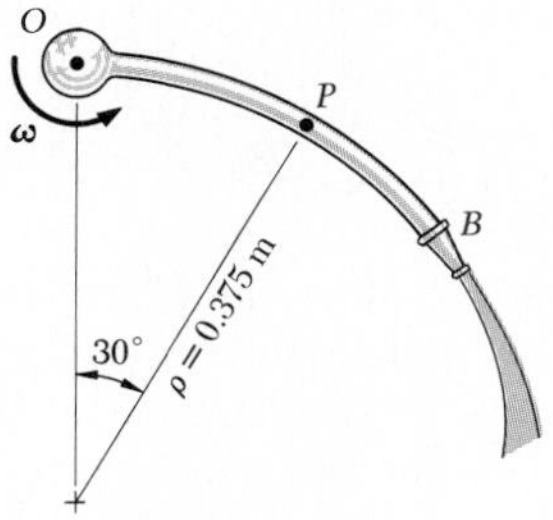

Fig. P15.126

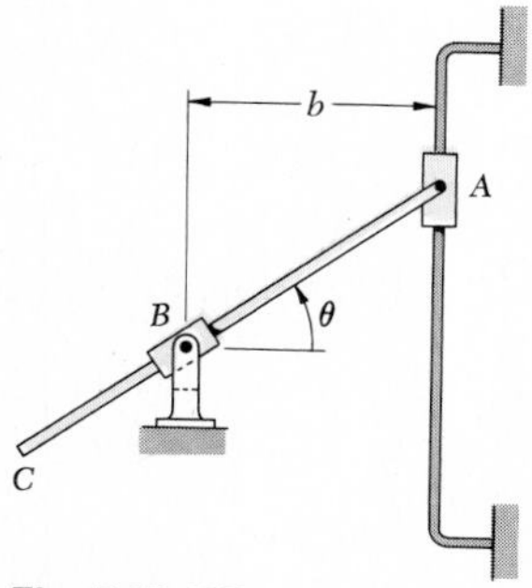

Fig. P15.127

REVIEW PROBLEMS

15.127 Rod AC of length $2b$ is attached to a collar at A and passes through a pivoted collar at B. Knowing that the collar A moves upward with a constant velocity $\mathbf{v}_A$, derive an expression for (*a*) the angular velocity of rod AC, (*b*) the velocity of the point of the rod in contact with the pivoted collar B, (*c*) the angular acceleration of rod AC.

15.128 In Prob. 15.26, determine (*a*) the angular acceleration of the rod, (*b*) the acceleration of *A*.

15.129 Three gears *A*, *B*, and *C* are pinned at their centers to rod *ABC*. Knowing that $r_A = 3r_B = 3r_C$ and that gear *A* does not rotate, determine the angular velocity of gears *B* and *C* when the rod *ABC* rotates clockwise with a constant angular velocity of 10 rpm.

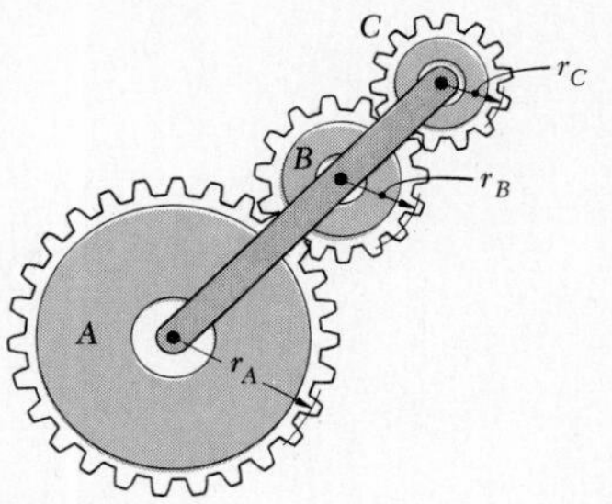

Fig. P15.129

15.130 In Prob. 15.129 it is known that $r_A = 12$ in., $r_B = r_C = 4$ in. Determine the acceleration of the tooth of gear *C* which is in contact with gear *B*.

15.131 Three links *AB*, *BC*, and *BD* are connected by a pin *B* as shown. Knowing that at the instant shown point *D* has a velocity of 200 mm/s to the right and no acceleration, determine (*a*) the angular acceleration of each link, (*b*) the accelerations of points *A* and *B*.

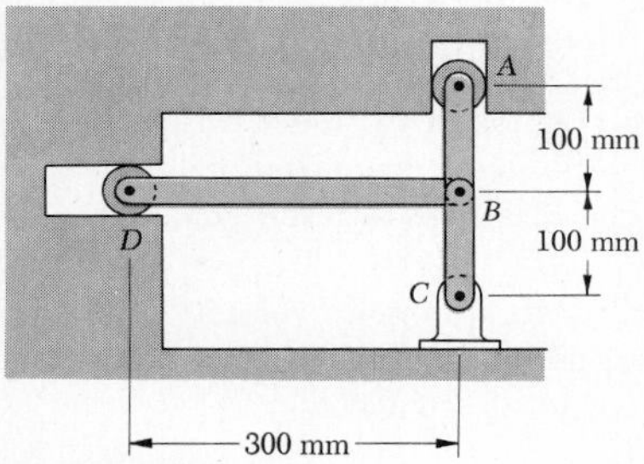

Fig. P15.131

15.132 The eccentric shown consists of a disk of 2-in. radius which revolves about a shaft *O* located $\frac{1}{2}$ in. from the center of the disk *A*. Assuming that the disk rotates about *O* with a constant angular velocity of 1800 rpm clockwise, determine the velocity and acceleration of block *B* when point *A* is directly below the shaft *O*.

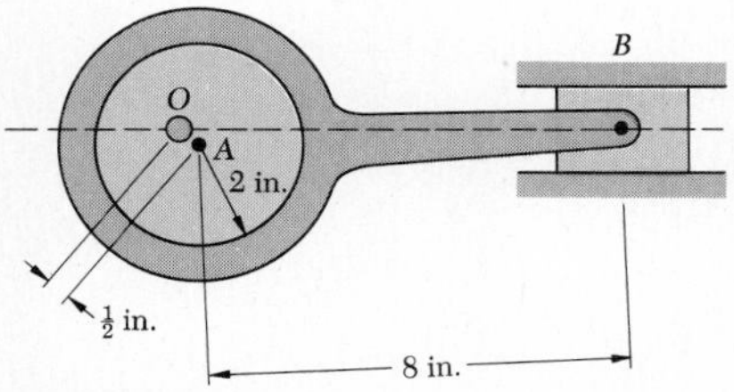

Fig. P15.132

15.133 It takes 0.8 s for the turntable of a 33-rpm record player to reach full speed after being started. Assuming uniformly accelerated motion, determine (*a*) the angular acceleration of the turntable, (*b*) the normal and tangential components of the acceleration of a point on the rim of the 12-in.-diameter turntable just before the speed of 33 rpm is reached, (*c*) the total acceleration of the same point at that time.

15.134 Two collars C and D move along the vertical rod shown. Knowing that the velocity of collar D is 0.210 m/s downward, determine (*a*) the velocity of collar C, (*b*) the angular velocity of member AB.

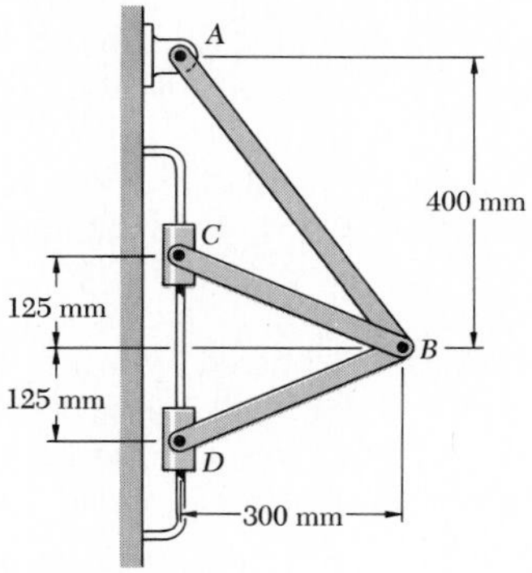

Fig. P15.134

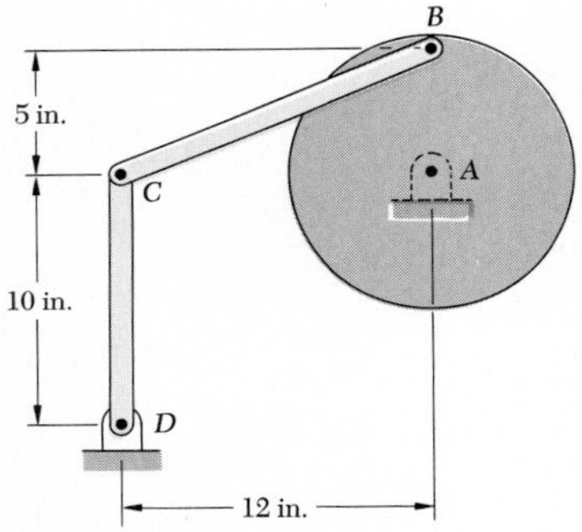

Fig. P15.136

∗**15.135** Prove for any given position of the mechanism of Prob. 15.134 that the ratio of the magnitudes of the velocities of collars C and D is equal to the ratio of the distances AC and AD.

15.136 The disk shown has a constant angular velocity of 6 rad/s clockwise. For the position shown, determine (*a*) the angular acceleration of each rod, (*b*) the acceleration of point C.

15.137 Solve Prob. 15.136 after the disk has rotated 180° and pin B is directly below A.

15.138 At the instant shown, the slotted plate slides with a velocity of 0.5 m/s upward and has an acceleration of 2 m/s^2 downward. Determine the angular velocity and the angular acceleration of rod OP.

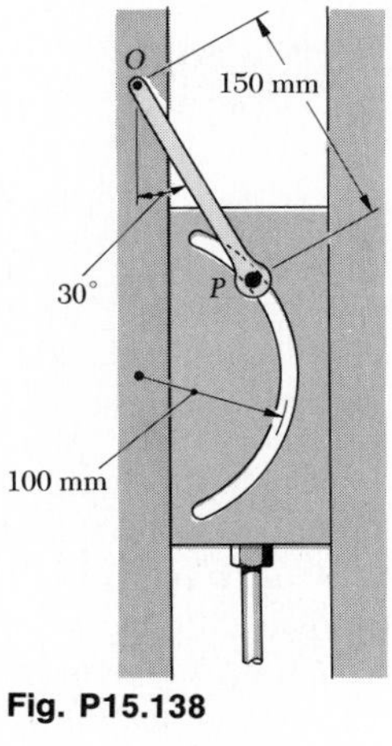

Fig. P15.138

Chapter
16

Kinetics of Rigid Bodies: Forces and Accelerations

16.1. Introduction. In this chapter and in Chaps. 17 and 18, we shall study the *kinetics of rigid bodies,* i.e., the relations existing between the forces acting on a rigid body, the shape and mass of the body, and the motion produced. In Chaps. 12 to 14, we studied similar relations, assuming then that the body could be considered as a particle, i.e., that its mass could be concentrated in one point and that all forces acted at that point. We shall now take the shape of the body into account, as well as the exact location of the points of application of the forces. Besides, we shall be concerned not only with the motion of the body as a whole but also with the rotation of the body about its mass center.

In this chapter, our study of the motion of rigid bodies will be based directly on the equation $\mathbf{F} = m\mathbf{a}$, while in the next two chapters we shall make use of the principles of work and energy and of impulse and momentum. Our study will be limited to that of the *plane motion* of rigid bodies. This motion was defined in Sec. 15.1 as a motion in which each particle of the body remains at a constant distance from a fixed reference plane. In the study of the kinetics of plane motion, the reference plane is chosen so that it contains the mass center of the body.

Throughout most of this chapter, our study will be further limited to that of plane slabs and of bodies which are symmet-

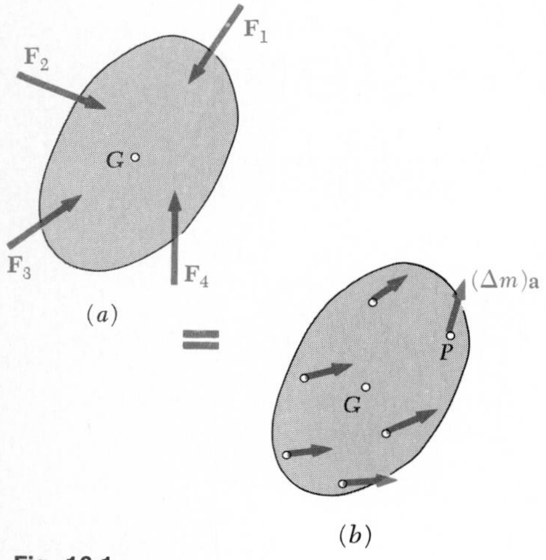

Fig. 16.1

rical with respect to the reference plane.† The study of the motion of three-dimensional bodies which are not symmetrical with respect to the reference plane will be postponed until Sec. 16.6, where problems such as the balancing of rotating shafts will be considered.

16.2. Plane Motion of a Rigid Body. D'Alembert's Principle. Consider a rigid slab of mass m moving under the action of several forces $\mathbf{F}_1$, $\mathbf{F}_2$, $\mathbf{F}_3$, etc., contained in the plane of the slab (Fig. 16.1*a*). We may regard the slab as being made of a large number of particles and use the results obtained in Sec. 12.4 for a system of particles. It was shown at that time that the system of the external forces acting on the particles is equipollent to the system of the effective forces of the particles. In other words, the sum of the components of the external forces in any given direction is equal to the sum of the components of the effective forces in that direction, and the sum of the moments of the external forces about any given axis is equal to the sum of the moments of the effective forces about that axis.

It follows that the external forces $\mathbf{F}_1$, $\mathbf{F}_2$, $\mathbf{F}_3$, etc., acting on the slab of Fig. 16.1 are equipollent to the system of the effective forces of the particles forming the slab. But since we are now dealing with a rigid body, the two systems of forces are not only equipollent, they are also *equivalent;* i.e., they both have the same effect on the slab (Sec. 3.11). This is shown in Fig. 16.1, where colored equals signs have been used to connect the system of the external forces and the system of the effective forces.

We may thus state that *the external forces acting on a rigid body are equivalent to the effective forces of the various particles forming the body.* This statement is referred to as *D'Alembert's principle,* after the French mathematician Jean le Rond d'Alembert (1717–1783), even though D'Alembert's original statement was written in a somewhat different form.

Again, because we are dealing with a rigid body, we may reduce the system of the effective forces to a force-couple system attached at a point of our choice (Sec. 3.10). By selecting the mass center G of the slab as the reference point, we shall be able to obtain a particularly simple result.

In order to determine the force-couple system at G which is equivalent to the system of the effective forces, we must evaluate the sums of the x and y components of the effective forces and the sum of their moments about G. We first write

$$\Sigma(F_x)_{\text{eff}} = \Sigma(\Delta m)a_x \qquad \Sigma(F_y)_{\text{eff}} = \Sigma(\Delta m)a_y \tag{16.1}$$

† Or, more generally, bodies which have a principal centroidal axis of inertia perpendicular to the reference plane.

But, according to Eqs. (12.9) of Sec. 12.5, we have

$$\Sigma(\Delta m)a_x = (\Sigma\, \Delta m)\bar{a}_x \qquad \Sigma(\Delta m)a_y = (\Sigma\, \Delta m)\bar{a}_y \tag{16.2}$$

where $\bar{a}_x$ and $\bar{a}_y$ are the components of the acceleration $\bar{\mathbf{a}}$ of the mass center G. Noting that $\Sigma\, \Delta m$ represents the total mass m of the slab, and substituting from (16.2) into (16.1), we have

$$\Sigma(F_x)_{\text{eff}} = m\bar{a}_x \qquad \Sigma(F_y)_{\text{eff}} = m\bar{a}_y \tag{16.3}$$

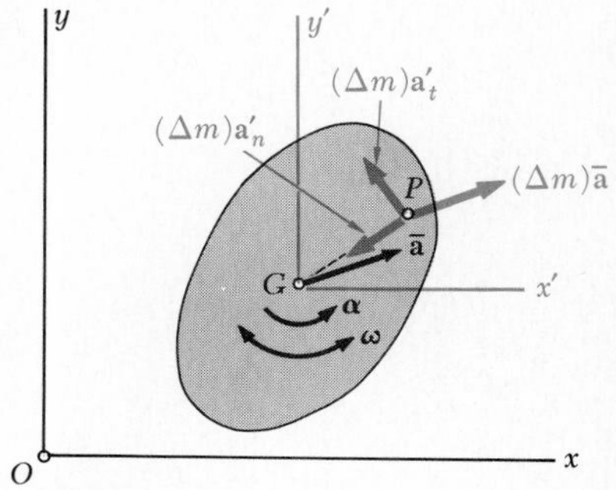

Fig. 16.2

Turning now our attention to the determination of the sum of the moments about G of the effective forces, we recall from Sec. 15.8 that the acceleration $\mathbf{a}$ of any given particle P of the slab (Fig. 16.2) may be expressed as the sum of the acceleration $\bar{\mathbf{a}}$ of the mass center G and of the acceleration $\mathbf{a}'$ of P relative to a frame $Gx'y'$ attached to G and of fixed orientation:

$$\mathbf{a} = \bar{\mathbf{a}} + \mathbf{a}'$$

Resolving $\mathbf{a}'$ into normal and tangential components, we write

$$\mathbf{a} = \bar{\mathbf{a}} + \mathbf{a}'_n + \mathbf{a}'_t \tag{16.4}$$

Each effective force $(\Delta m)\mathbf{a}$ may thus be resolved into the three component effective forces shown in Fig. 16.2. The effective forces $(\Delta m)\bar{\mathbf{a}}$ obtained in this fashion are associated with a translation of the slab with G, while the effective forces $(\Delta m)\mathbf{a}'_n$ and $(\Delta m)\mathbf{a}'_t$ are associated with a rotation of the slab about G.

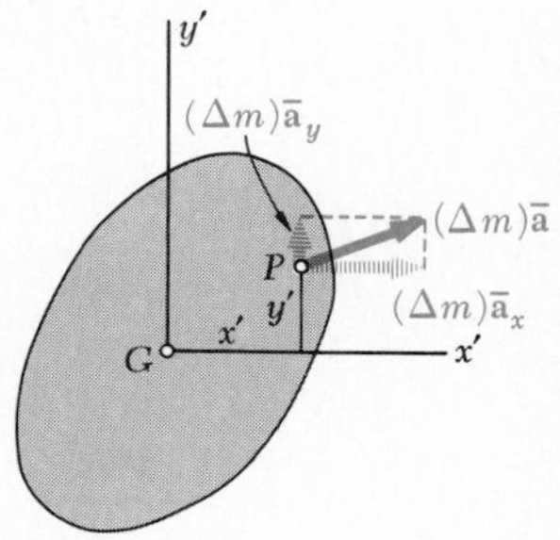

Fig. 16.3

Considering first the effective forces $(\Delta m)\bar{\mathbf{a}}$ associated with the translation of the slab, and resolving them into rectangular components (Fig. 16.3), we find that the sum of their moments about G is

$$\begin{aligned} +\circlearrowleft\ \Sigma(M_G)_{\text{eff, translation}} &= \Sigma x'(\Delta m)\bar{a}_y - \Sigma y'(\Delta m)\bar{a}_x \\ &= \bar{a}_y \Sigma x'\, \Delta m - \bar{a}_x \Sigma y'\, \Delta m \end{aligned} \tag{16.5}$$

Now, according to the definition of the mass center of a system of particles (Sec. 12.5), we have

$$\Sigma x'\, \Delta m = m\bar{x}' \qquad \Sigma y'\, \Delta m = m\bar{y}' \tag{16.6}$$

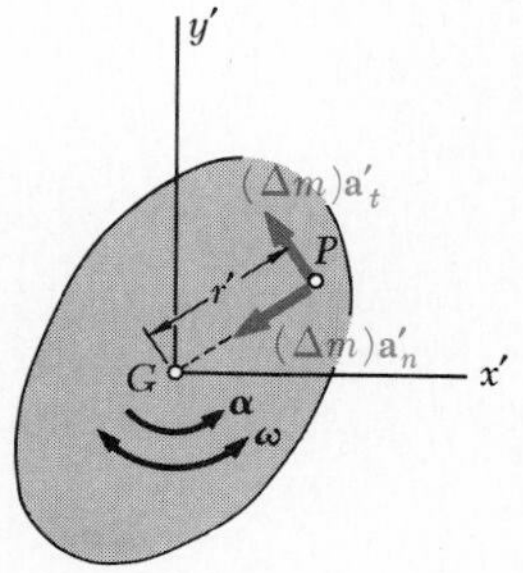

Fig. 16.4

where $\bar{x}'$ and $\bar{y}'$ are the coordinates of the mass center G and m the total mass. But $\bar{x}' = \bar{y}' = 0$, since the origin of the x' and y' coordinate axes was chosen at G, and Eq. (16.5) yields

$$\Sigma(M_G)_{\text{eff, translation}} = 0 \tag{16.7}$$

Considering now the effective forces $(\Delta m)\mathbf{a}'_n$ and $(\Delta m)\mathbf{a}'_t$ associated with the rotation of the slab about G (Fig. 16.4), we note that the moment of $(\Delta m)\mathbf{a}'_n$ about G is zero and that the moment of $(\Delta m)\mathbf{a}'_t$ about G is $r'(\Delta m)a'_t$. Recalling that $a'_t = r'\alpha$, and considering all the particles of the slab, we have

$$+\circlearrowleft\ \Sigma(M_G)_{\text{eff, rotation}} = \Sigma r'(\Delta m)r'\alpha = \alpha \Sigma r'^2\, \Delta m$$

Since the sum $\Sigma r'^2 \Delta m$ represents the moment of inertia $\bar{I}$ of the slab about an axis through G perpendicular to the slab, we write

$$\Sigma(M_G)_{\text{eff, rotation}} = \bar{I}\alpha \tag{16.8}$$

The sum of the moments about G of the effective forces of the slab is obtained by combining the information provided by Eqs. (16.7) and (16.8). We have

$$\Sigma(M_G)_{\text{eff}} = \bar{I}\alpha \tag{16.9}$$

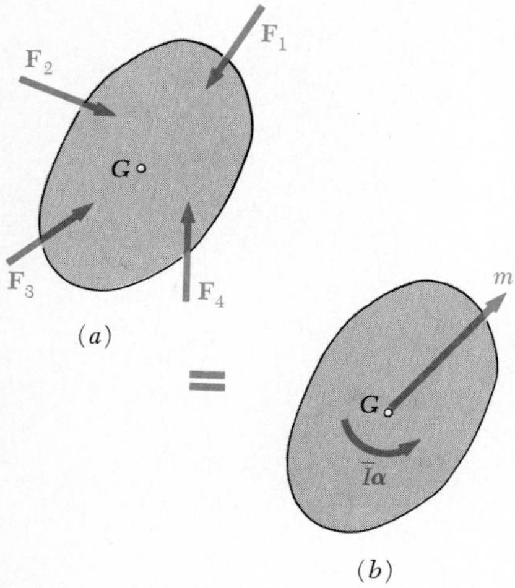

Fig. 16.5

Equations (16.3) and (16.9) indicate that the system of the effective forces shown in Fig. 16.1*b* may be replaced by an equivalent force-couple system consisting of a vector $m\bar{\mathbf{a}}$ attached at G and a couple of moment $\bar{I}\alpha$ and of the same sense as the angular acceleration $\boldsymbol{\alpha}$ (Fig. 16.5). Recalling from Sec. 15.3 that the angular acceleration of a rigid body rotating about a fixed axis may be represented by a vector $\boldsymbol{\alpha}$ of magnitude α directed along that axis, we note that the couple obtained may be represented by the vector $\bar{I}\boldsymbol{\alpha}$ of magnitude $\bar{I}\alpha$ directed along the axis of rotation.

In the case of the plane motion of a rigid slab,† D'Alembert's principle may thus be restated as follows: *The external forces acting on the body are equivalent to a force-couple system consisting of a vector* $m\bar{\mathbf{a}}$ *attached at the mass center G of the body and a couple* $\bar{I}\boldsymbol{\alpha}$.

The relation shown in Fig. 16.5 can be expressed algebraically by writing three equations relating respectively the x and y components and the moments about any given point A of the external and effective forces in Fig. 16.5. If the moments are computed about the mass center G of the rigid body, these equations of motion read

$$\Sigma F_x = m\bar{a}_x \qquad \Sigma F_y = m\bar{a}_y \qquad \Sigma M_G = \bar{I}\alpha \tag{16.10}$$

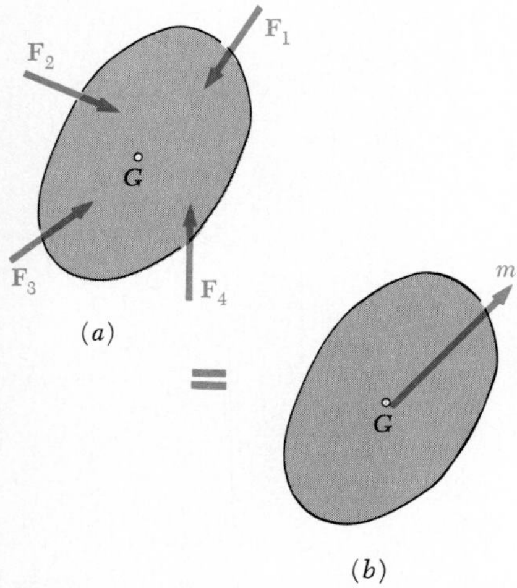

Fig. 16.6

Translation. When a rigid body is constrained to move in translation, its angular acceleration is identically equal to zero and its effective forces reduce to the vector $m\bar{\mathbf{a}}$ attached at G (Fig. 16.6). Thus, the resultant of the external forces acting on a rigid body in translation passes through the mass center of the body and is equal to $m\bar{\mathbf{a}}$.

† Or, more generally, in the case of the plane motion of a rigid body which has a principal centroidal axis of inertia perpendicular to the reference plane.

Centroidal Rotation. When a slab, or, more generally, a body symmetrical with respect to the reference plane, is constrained to rotate about a fixed axis perpendicular to the reference plane and passing through its mass center G, we say that the body is in *centroidal rotation*. Since the acceleration $\bar{\mathbf{a}}$ is identically equal to zero, the effective forces of the body reduce to the couple $\bar{I}\boldsymbol{\alpha}$ (Fig. 16.7). Thus, the external forces acting on a body in centroidal rotation are equivalent to a couple $\bar{I}\boldsymbol{\alpha}$.

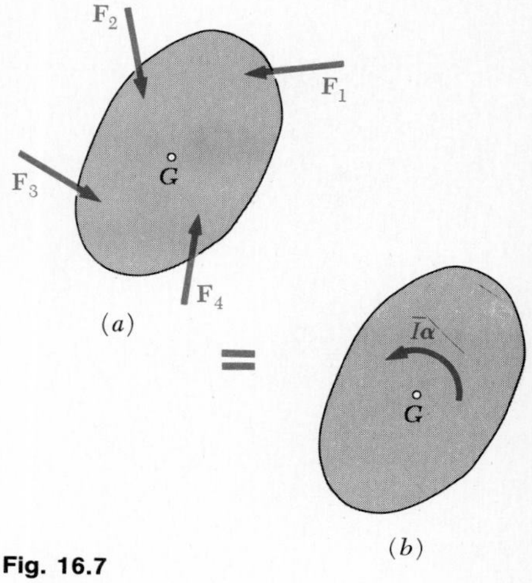

Fig. 16.7

General Plane Motion. Comparing Fig. 16.5 with Figs. 16.6 and 16.7, we observe that, from the point of view of *kinetics*, the most general plane motion of a rigid body symmetrical with respect to the reference plane may be replaced by the sum of a translation and a centroidal rotation. We should note that this statement is more restrictive than the similar statement made earlier from the point of view of *kinematics* (Sec. 15.5), since we now require that the mass center of the body be selected as the reference point.

Referring to Eqs. (16.10), we observe that the first two equations are identical with the equations of motion of a particle of mass m acted upon by the given forces $\mathbf{F}_1$, $\mathbf{F}_2$, $\mathbf{F}_3$, etc. We thus check that *the mass center G of a rigid body in plane motion moves as if the entire mass of the body were concentrated at that point, and as if all the external forces acted on it.* We recall that this result has already been obtained in Sec. 12.5 in the general case of a system of particles, the particles being not necessarily rigidly connected. We should note, however, that the system of the external forces does not, in general, reduce to a single vector $m\bar{\mathbf{a}}$ attached at G. Therefore, in the general case of the plane motion of a rigid body, *the resultant of the external forces acting on the body does not pass through the mass center of the body.*

Finally, we may observe that the last of Eqs. (16.10) would still be valid if the rigid body, while subjected to the same applied forces, were constrained to rotate about a fixed axis through G. Thus, *a rigid body in plane motion rotates about its mass center as if this point were fixed.*

16.3. Solution of Problems Involving the Plane Motion of Rigid Bodies. We saw in the preceding section that, when a rigid body is in plane motion, there exists a fundamental relation between the forces $\mathbf{F}_1$, $\mathbf{F}_2$, $\mathbf{F}_3$, etc., acting on the body, the acceleration $\bar{\mathbf{a}}$ of its mass center, and the angular acceleration $\boldsymbol{\alpha}$ of the body. This relation, which is represented in Fig. 16.5, may be used to determine the acceleration $\bar{\mathbf{a}}$ and the angular acceleration $\boldsymbol{\alpha}$ produced by a given system of forces acting on a rigid body or, conversely, to determine the forces which produce a given motion of the rigid body.

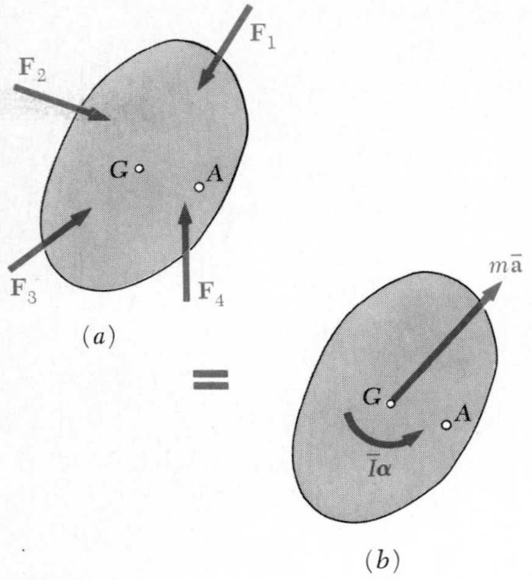

Fig. 16.8

While the three algebraic equations (16.10) may be used to solve problems of plane motion,† our experience in statics suggests that the solution of many problems involving rigid bodies could be simplified by an appropriate choice of the point about which the moments of the forces are computed. It is therefore preferable to remember the relation existing between the forces and the accelerations in the vectorial form shown in Fig. 16.8, and to derive from this fundamental relation the component or moment equations which fit best the solution of the problem under consideration. The equations most frequently used are

$$\Sigma F_x = \Sigma (F_x)_{\text{eff}} \qquad \Sigma F_y = \Sigma (F_y)_{\text{eff}} \qquad \Sigma M_A = \Sigma (M_A)_{\text{eff}} \quad (16.11)$$

We easily verify that the first two equations reduce to the corresponding equations in (16.10). The third equation, however, will generally be different from the corresponding equation in (16.10), since the selection of point A, about which the moments of the external and effective forces are to be computed, will be guided by our desire to simplify the solution of the problem as much as possible. In Sample Prob. 16.1, for example, the three unknown forces $\mathbf{F}_A$, $\mathbf{F}_B$, and $\mathbf{N}_A$ may be eliminated from this equation by equating moments about A. It should be noted, however, that, unless A happens to coincide with the mass center G, the determination of $\Sigma(M_A)_{\text{eff}}$ will involve the computation of the moments of the components of the vector $m\bar{\mathbf{a}}$, as well as the computation of the moment of the couple $\bar{I}\boldsymbol{\alpha}$.

The fundamental relation shown in Fig. 16.8 may be presented in an alternate form if we add to the external forces an inertia vector $(m\bar{\mathbf{a}})_{\text{rev}}$ of sense opposite to that of $\bar{\mathbf{a}}$, attached at G, and an inertia couple $(\bar{I}\boldsymbol{\alpha})_{\text{rev}}$ of moment equal in magnitude to $\bar{I}\alpha$ and of sense opposite to that of $\boldsymbol{\alpha}$ (Fig. 16.9). The system obtained is equivalent to zero, and the rigid body is said to be in dynamic equilibrium.

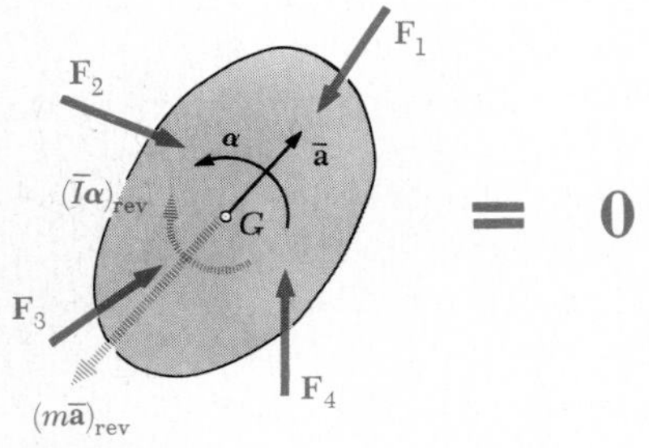

Fig. 16.9

Whether the principle of equivalence of external and effective forces is directly applied, as in Fig. 16.8, or whether the concept of dynamic equilibrium is introduced, as in Fig. 16.9, the use of free-body diagrams showing vectorially the relationship existing between the forces applied on the rigid body and the resulting linear and angular accelerations presents considerable advantages over the blind application of the formulas (16.10). These advantages may be summarized as follows:

† We recall that the last of Eqs. (16.10) is valid only in the case of the plane motion of a rigid body symmetrical with respect to the reference plane. The plane motion of bodies which are not symmetrical with respect to the reference plane is discussed in Sec. 16.6.

1. First of all, a much clearer understanding of the effect of the forces on the motion of the body will result from the use of a pictorial representation.
2. This approach makes it possible to divide the solution of a dynamics problem into two parts: In the first part, the analysis of the kinematic and kinetic characteristics of the problem leads to the free-body diagrams of Fig. 16.8 or 16.9; in the second part, the diagram obtained is used to analyze by the methods of Chap. 3 the various forces and vectors involved.
3. A unified approach is provided for the analysis of the plane motion of a rigid body, regardless of the particular type of motion involved. While the kinematics of the various motions considered may vary from one case to the other, the approach to the kinetics of the motion is consistently the same. In every case we shall draw a diagram showing the external forces, the vector $m\bar{a}$ associated with the motion of G, and the couple $\bar{I}\boldsymbol{\alpha}$ associated with the rotation of the body about G.
4. The resolution of the plane motion of a rigid body into a translation and a centroidal rotation, which is used here, is a basic concept which may be applied effectively throughout the study of mechanics. We shall use it again in Chap. 17 with the method of work and energy and in Chap. 18 with the method of impulse and momentum.

16.4. Systems of Rigid Bodies. The method described in the preceding section may also be used in problems involving the plane motion of several connected rigid bodies. A diagram similar to Fig. 16.8 or Fig. 16.9 may be drawn for each part of the system. The equations of motion obtained from these diagrams are solved simultaneously.

In some cases, as in Sample Prob. 16.3, a single diagram may be drawn for the entire system. This diagram should include all the external forces, as well as the vectors $m\bar{a}$ and the couples $\bar{I}\boldsymbol{\alpha}$ associated with the various parts of the system. However, internal forces, such as the forces exerted by connecting cables, may be omitted since they occur in pairs of equal and opposite forces and are thus equipollent to zero. The equations obtained by expressing that the system of the external forces is equipollent to the system of the effective forces may be solved for the remaining unknowns.†

This second approach may not be used in problems involving more than three unknowns, since only three equations of motion are available when a single diagram is used. We shall not elaborate upon this point, since the discussion involved would be completely similar to that given in Sec. 6.12 in the case of the equilibrium of a system of rigid bodies.

† Note that we cannot speak of *equivalent* systems since we are not dealing with a single rigid body.

SAMPLE PROBLEM 16.1

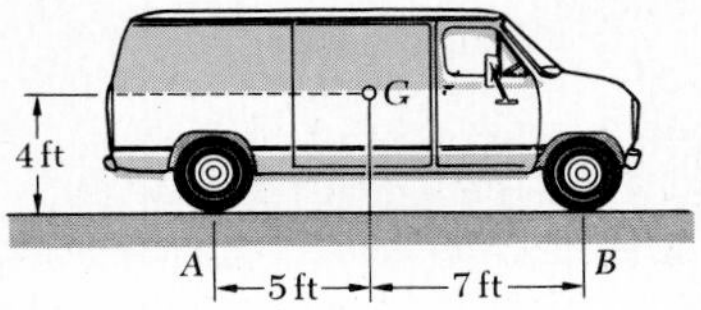

When the forward speed of the truck shown was 30 ft/s, the brakes were suddenly applied, causing all four wheels to stop rotating. It was observed that the truck skidded to rest in 20 ft. Determine the magnitude of the normal reaction and of the friction force at each wheel as the truck skidded to rest.

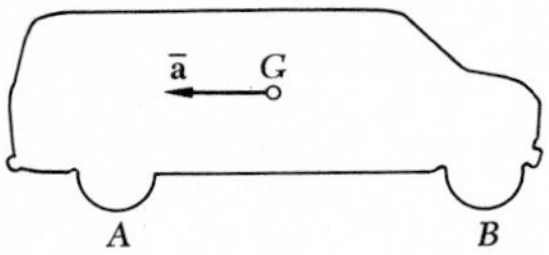

Kinematics of Motion. Choosing the positive sense to the right and using the equations of uniformly accelerated motion, we write

$$\bar{v}_0 = +30\text{ ft/s} \qquad \bar{v}^2 = \bar{v}_0^2 + 2\bar{a}\bar{x} \qquad 0 = (30)^2 + 2\bar{a}(20)$$
$$\bar{a} = -22.5\text{ ft/s}^2 \qquad \bar{\mathbf{a}} = 22.5\text{ ft/s}^2 \leftarrow$$

Equations of Motion. The external forces consist of the weight **W** of the truck and of the normal reactions and friction forces at the wheels. (The vectors $\mathbf{N}_A$ and $\mathbf{F}_A$ represent the sum of the reactions at the rear wheels, while $\mathbf{N}_B$ and $\mathbf{F}_B$ represent the sum of the reactions at the front wheels.) Since the truck is in translation, the effective forces reduce to the vector $m\bar{\mathbf{a}}$ attached at G. Three equations of motion are obtained by expressing that the system of the external forces is equivalent to the system of the effective forces.

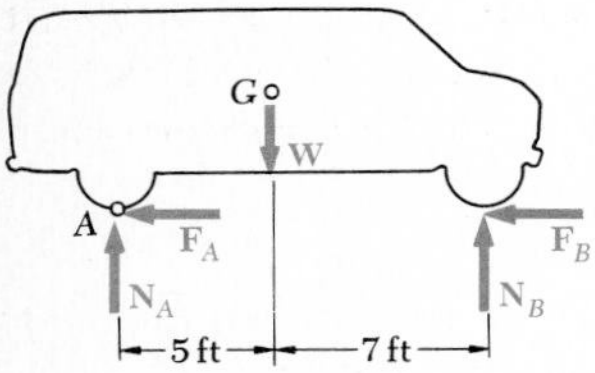

$$+\uparrow \Sigma F_y = \Sigma (F_y)_{\text{eff}}: \qquad N_A + N_B - W = 0$$

Since $F_A = \mu N_A$ and $F_B = \mu N_B$, we find

$$F_A + F_B = \mu(N_A + N_B) = \mu W$$

$$\overset{+}{\rightarrow} \Sigma F_x = \Sigma (F_x)_{\text{eff}}: \qquad -(F_A + F_B) = -m\bar{a}$$

$$-\mu W = -\frac{W}{32.2\text{ ft/s}^2}(22.5\text{ ft/s}^2)$$

$$\mu = 0.699$$

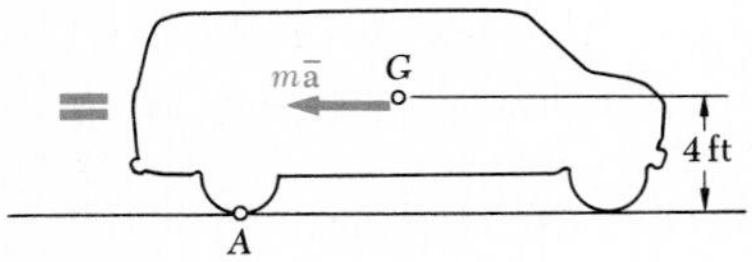

$$+\circlearrowleft \Sigma M_A = \Sigma (M_A)_{\text{eff}}: \qquad -W(5\text{ ft}) + N_B(12\text{ ft}) = m\bar{a}(4\text{ ft})$$

$$-W(5\text{ ft}) + N_B(12\text{ ft}) = \frac{W}{32.2\text{ ft/s}^2}(22.5\text{ ft/s}^2)(4\text{ ft})$$

$$N_B = 0.650W$$
$$F_B = \mu N_B = (0.699)(0.650W) \qquad F_B = 0.454W$$

$$+\uparrow \Sigma F_y = \Sigma (F_y)_{\text{eff}}: \qquad N_A + N_B - W = 0$$
$$N_A + 0.650W - W = 0$$
$$N_A = 0.350W$$
$$F_A = \mu N_A = (0.699)(0.350W) \qquad F_A = 0.245W$$

Reactions at Each Wheel. Recalling that the values computed above represent the sum of the reactions at the two front wheels or the two rear wheels, we obtain the magnitude of the reactions at each wheel by writing

$$N_{\text{front}} = \tfrac{1}{2}N_B = 0.325W \qquad N_{\text{rear}} = \tfrac{1}{2}N_A = 0.175W \quad \blacktriangleleft$$
$$F_{\text{front}} = \tfrac{1}{2}F_B = 0.227W \qquad F_{\text{rear}} = \tfrac{1}{2}F_A = 0.122W \quad \blacktriangleleft$$

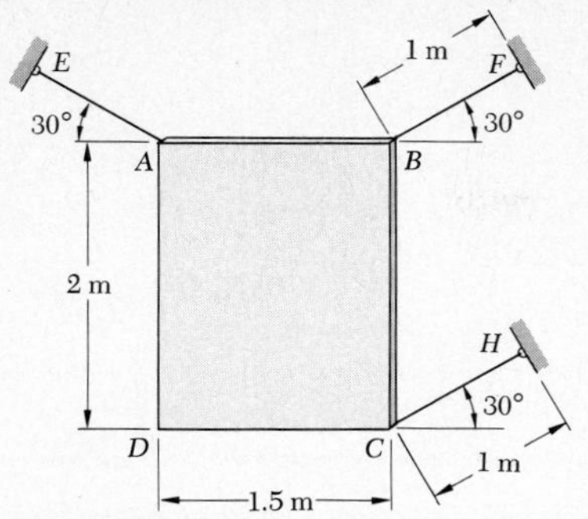

SAMPLE PROBLEM 16.2

The thin plate $ABCD$ has a mass of 50 kg and is held in position by the three inextensible wires AE, BF, and CH. Wire AE is then cut. Determine both (*a*) the acceleration of the plate, (*b*) the tension in wires BF and CH immediately after wire AE has been cut.

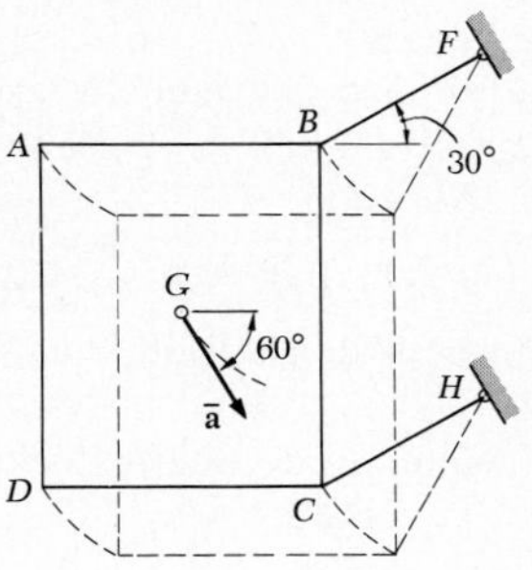

Motion of Plate. After wire AE has been cut, we observe that corners B and C move along parallel circles of radius 1 m centered, respectively, at F and H. The motion of the plate is thus a curvilinear translation; the particles forming the plate move along parallel circles of radius 1 m.

At the instant wire AE is cut, the velocity of the plate is zero; the acceleration $\bar{\mathbf{a}}$ of the mass center G is thus tangent to the circular path which will be described by G.

Equations of Motion. The external forces consist of the weight $\mathbf{W}$ and of the forces $\mathbf{T}_B$ and $\mathbf{T}_C$ exerted by the wires. Since the plate is in translation, the effective forces reduce to the vector $m\bar{\mathbf{a}}$ attached at G and directed along the t axis. Expressing that the system of the external forces is equivalent to the system of the effective forces, we write

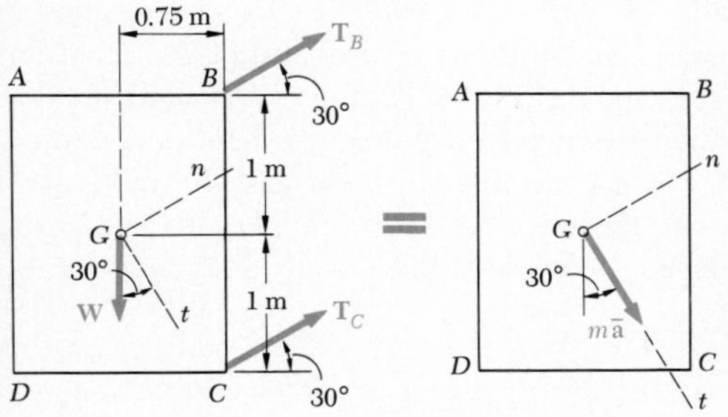

$$+\searrow \Sigma F_t = \Sigma(F_t)_{\text{eff}}: \qquad W\cos 30° = m\bar{a}$$

$$mg\cos 30° = m\bar{a} \qquad (1)$$

$$\bar{a} = g\cos 30° = (9.81\text{ m/s}^2)\cos 30°$$

$$\bar{\mathbf{a}} = 8.50\text{ m/s}^2 \nwarrow 60° \blacktriangleleft$$

$$+\nearrow \Sigma F_n = \Sigma(F_n)_{\text{eff}}: \qquad T_B + T_C - W\sin 30° = 0 \qquad (2)$$

$$+\circlearrowleft \Sigma M_G = \Sigma(M_G)_{\text{eff}}: \qquad (T_B\sin 30°)(0.75\text{ m}) - (T_B\cos 30°)(1\text{ m})$$

$$+ (T_C\sin 30°)(0.75\text{ m}) + (T_C\cos 30°)(1\text{ m}) = 0$$

$$-0.491T_B + 1.241T_C = 0$$

$$T_C = 0.396T_B \qquad (3)$$

Substituting for T_C from (3) into (2), we write

$$T_B + 0.396T_B - W\sin 30° = 0$$

$$T_B = 0.358W$$

$$T_C = 0.396(0.358W) = 0.1418W$$

Noting that $W = mg = (50\text{ kg})(9.81\text{ m/s}^2) = 491\text{ N}$, we have

$$T_B = 175.8\text{ N} \qquad T_C = 69.6\text{ N} \blacktriangleleft$$

SAMPLE PROBLEM 16.3

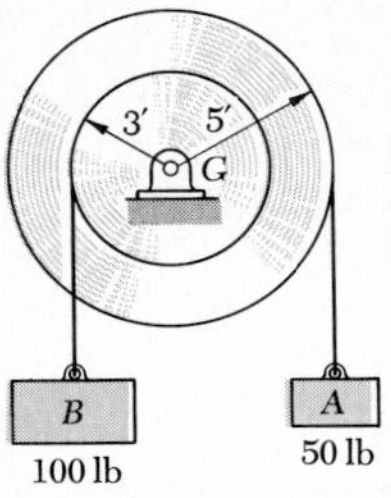

A pulley weighing 120 lb and having a radius of gyration of 4 ft is connected to two blocks as shown. Assuming no axle friction, determine the angular acceleration of the pulley.

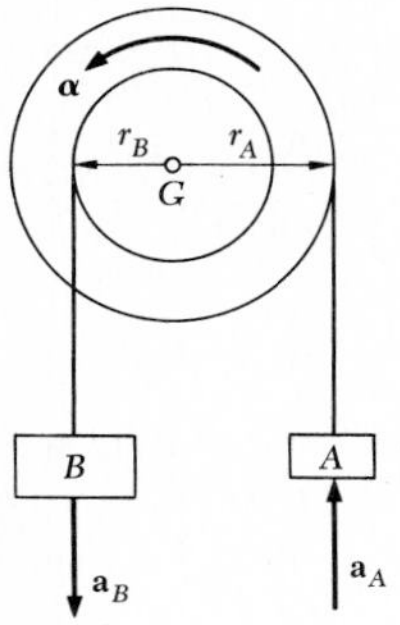

Sense of Motion. Although an arbitrary sense of motion may be assumed (since no friction forces are involved) and later checked by the sign of the answer, we may prefer first to determine the actual sense of rotation of the pulley. We first find the weight of block B required to maintain the equilibrium of the pulley when it is acted upon by the 50-lb block A. We write

$$+\circlearrowleft \Sigma M_G = 0: \qquad W_B(3\text{ ft}) - (50\text{ lb})(5\text{ ft}) = 0 \qquad W_B = 83.3\text{ lb}$$

Since block B actually weighs 100 lb, the pulley will rotate counterclockwise.

Kinematics of Motion. Assuming $\boldsymbol{\alpha}$ counterclockwise and noting that $a_A = r_A\alpha$ and $a_B = r_B\alpha$, we obtain

$$\mathbf{a}_A = 5\alpha \uparrow \qquad \mathbf{a}_B = 3\alpha \downarrow$$

Equations of Motion. A single system consisting of the pulley and the two blocks is considered. Forces external to this system consist of the weights of the pulley and the two blocks and of the reaction at G. (The forces exerted by the cables on the pulley and on the blocks are internal to the system considered and cancel out.) Since the motion of the pulley is a centroidal rotation and the motion of each block is a translation, the effective forces reduce to the couple $\bar{I}\boldsymbol{\alpha}$ and the two vectors $m\mathbf{a}_A$ and $m\mathbf{a}_B$. The centroidal moment of inertia of the pulley is

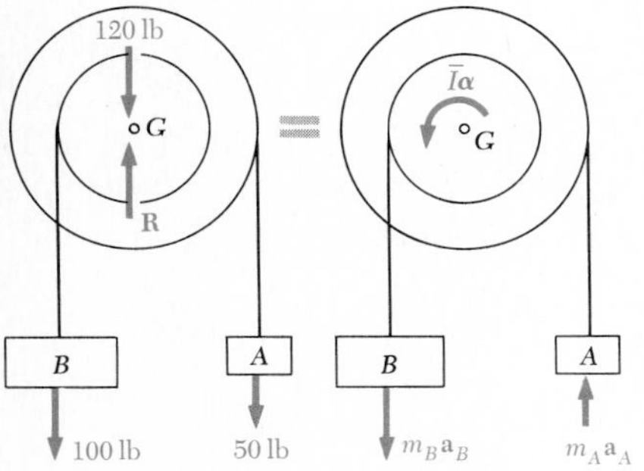

$$\bar{I} = mk^2 = \frac{W}{g}k^2 = \frac{120\text{ lb}}{32.2\text{ ft/s}^2}(4\text{ ft})^2 = 59.6\text{ lb}\cdot\text{ft}\cdot\text{s}^2$$

Since the system of the external forces is equipollent to the system of the effective forces, we write

$$+\circlearrowleft \Sigma M_G = \Sigma(M_G)_{\text{eff}}:$$

$$(100\text{ lb})(3\text{ ft}) - (50\text{ lb})(5\text{ ft}) = +\bar{I}\alpha + m_B a_B(3\text{ ft}) + m_A a_A(5\text{ ft})$$

$$(100)(3) - (50)(5) = +59.6\alpha + \frac{100}{32.2}(3\alpha)(3) + \frac{50}{32.2}(5\alpha)(5)$$

$$\alpha = +0.396\text{ rad/s}^2 \qquad \alpha = 0.396\text{ rad/s}^2 \circlearrowleft$$ ◄

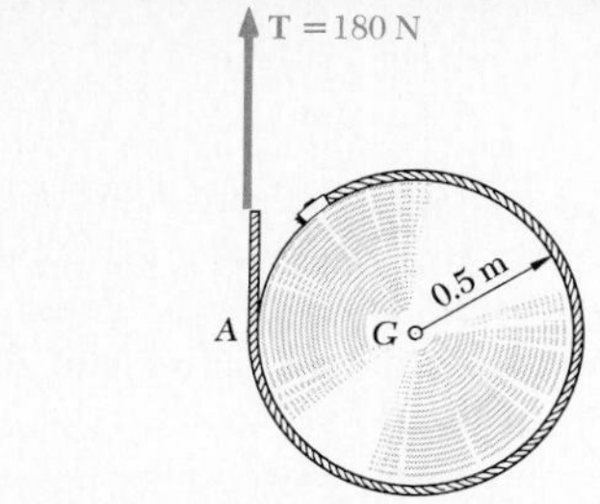

SAMPLE PROBLEM 16.4

A cord is wrapped around a homogeneous disk of radius $r = 0.5$ m and mass $m = 15$ kg. If the cord is pulled upward with a force **T** of magnitude 180 N, determine (*a*) the acceleration of the center of the disk, (*b*) the angular acceleration of the disk, (*c*) the acceleration of the cord.

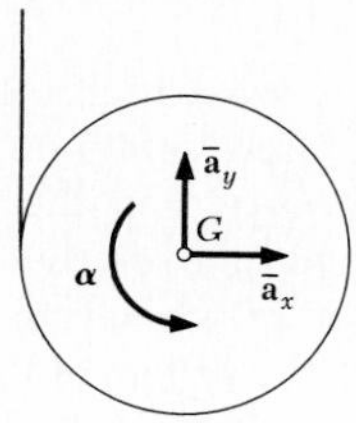

Equations of Motion. We assume that the components $\bar{\mathbf{a}}_x$ and $\bar{\mathbf{a}}_y$ of the acceleration of the center are directed, respectively, to the right and upward and that the angular acceleration of the disk is counterclockwise. The external forces acting on the disk consist of the weight **W** and the force **T** exerted by the cord. This system is equivalent to the system of the effective forces, which consists of a vector of components $m\bar{\mathbf{a}}_x$ and $m\bar{\mathbf{a}}_y$ attached at G and a couple $\bar{I}\boldsymbol{\alpha}$. We write

$$\xrightarrow{+} \Sigma F_x = \Sigma(F_x)_{\text{eff}}: \qquad 0 = m\bar{a}_x \qquad \bar{\mathbf{a}}_x = 0 \blacktriangleleft$$

$$+\uparrow \Sigma F_y = \Sigma(F_y)_{\text{eff}}: \qquad T - W = m\bar{a}_y$$

$$\bar{a}_y = \frac{T - W}{m}$$

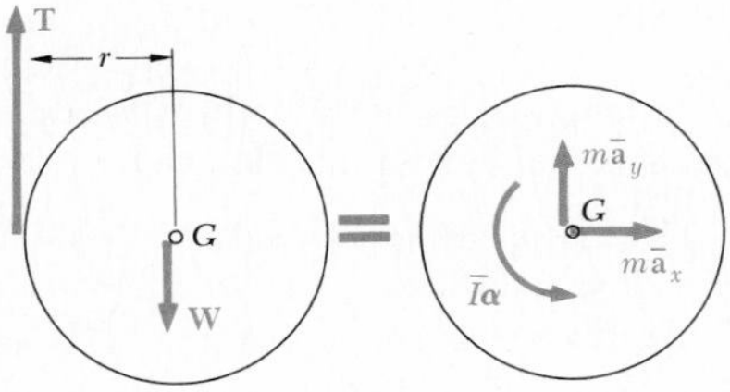

Since $T = 180$ N, $m = 15$ kg, and $W = (15\text{ kg})(9.81\text{ m/s}^2) = 147.1$ N, we have

$$\bar{a}_y = \frac{180\text{ N} - 147.1\text{ N}}{15\text{ kg}} = +2.19\text{ m/s}^2 \qquad \bar{\mathbf{a}}_y = 2.19\text{ m/s}^2 \uparrow \blacktriangleleft$$

$$+\circlearrowleft \Sigma M_G = \Sigma(M_G)_{\text{eff}}: \qquad -Tr = \bar{I}\alpha$$

$$-Tr = (\tfrac{1}{2}mr^2)\alpha$$

$$\alpha = -\frac{2T}{mr} = -\frac{2(180\text{ N})}{(15\text{ kg})(0.5\text{ m})} = -48.0\text{ rad/s}^2$$

$$\boldsymbol{\alpha} = 48.0\text{ rad/s}^2 \circlearrowright \blacktriangleleft$$

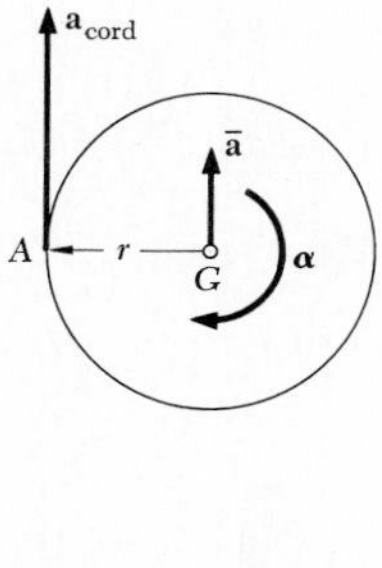

Acceleration of Cord. Since the acceleration of the cord is equal to the tangential component of the acceleration of point A on the disk, we write

$$\mathbf{a}_{\text{cord}} = (\mathbf{a}_A)_t = \bar{\mathbf{a}} + (\mathbf{a}_{A/G})_t$$
$$= [2.19\text{ m/s}^2 \uparrow] + [(0.5\text{ m})(48\text{ rad/s}^2) \uparrow]$$
$$\mathbf{a}_{\text{cord}} = 26.2\text{ m/s}^2 \uparrow \blacktriangleleft$$

SAMPLE PROBLEM 16.5

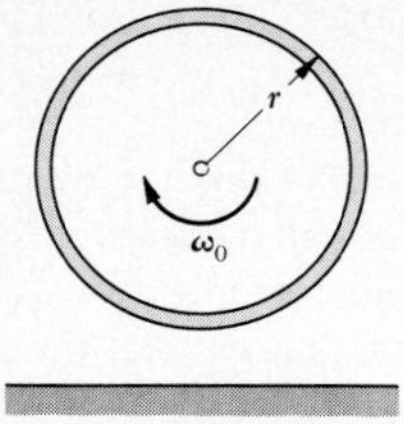

A hoop of radius r and mass m is placed on a horizontal surface with no linear velocity but with a clockwise angular velocity $\boldsymbol{\omega}_0$. Denoting by μ the coefficient of friction between the hoop and the floor, determine (*a*) the time t_1 at which the hoop will start rolling without sliding, (*b*) the linear and angular velocities of the hoop at time t_1.

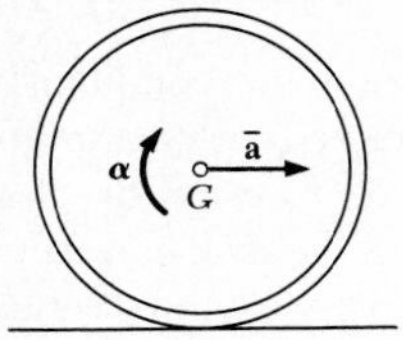

Solution. Since the entire mass is located at a distance r from the center of the hoop, we write $\bar{I} = mr^2$.

Equations of Motion. The positive sense is chosen to the right for $\bar{\mathbf{a}}$ and clockwise for $\boldsymbol{\alpha}$. The external forces acting on the hoop consist of the weight **W**, the normal reaction **N**, and the friction force **F**. While the hoop is sliding, the magnitude of the friction force is $F = \mu N$. The effective forces consist of the vector $m\bar{\mathbf{a}}$ attached at G and the couple $\bar{I}\boldsymbol{\alpha}$. Expressing that the system of the external forces is equivalent to the system of the effective forces, we write

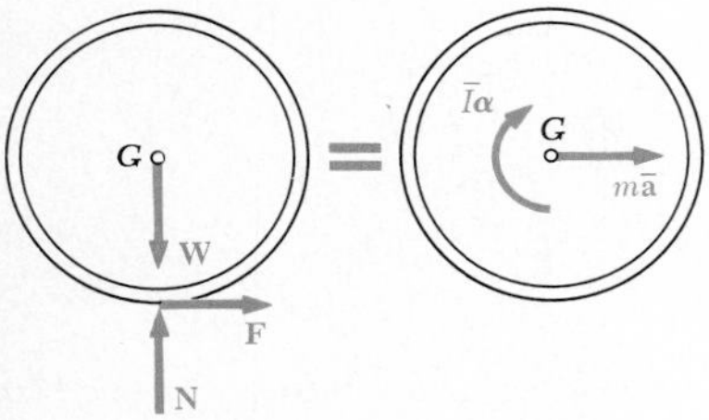

$$+\uparrow \Sigma F_y = \Sigma(F_y)_{\text{eff}}: \quad N - W = 0$$

$$N = W = mg \qquad F = \mu N = \mu mg$$

$$\xrightarrow{+} \Sigma F_x = \Sigma(F_x)_{\text{eff}}: \qquad F = m\bar{a} \qquad \mu mg = m\bar{a} \qquad \bar{a} = +\mu g$$

$$+\circlearrowright \Sigma M_G = \Sigma(M_G)_{\text{eff}}: \qquad -Fr = \bar{I}\alpha$$

$$-(\mu mg)r = (mr^2)\alpha \qquad \alpha = -\frac{\mu g}{r}$$

Kinematics of Motion. As long as the hoop both rolls and slides, its linear and angular motions are uniformly accelerated.

$$t = 0,\ \bar{v}_0 = 0 \qquad \bar{v} = \bar{v}_0 + \bar{a}t = 0 + \mu g t \tag{1}$$

$$t = 0,\ \omega = \omega_0 \qquad \omega = \omega_0 + \alpha t = \omega_0 + \left(-\frac{\mu g}{r}\right)t \tag{2}$$

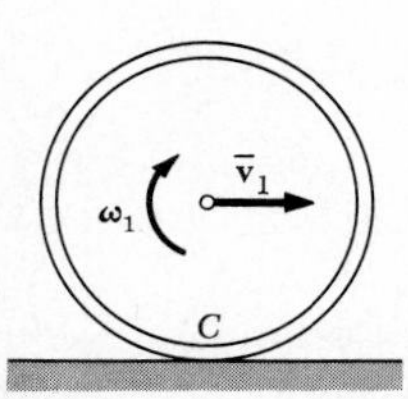

The hoop will start rolling without sliding when the velocity $\mathbf{v}_C$ of the point of contact is zero. At that time, $t = t_1$, point C becomes the instantaneous center of rotation, and we have

$$\bar{v}_1 = r\omega_1$$

$$\mu g t_1 = r\left(\omega_0 - \frac{\mu g}{r}t_1\right) \qquad t_1 = \frac{r\omega_0}{2\mu g} \blacktriangleleft$$

Substituting for t_1 into (1), we have

$$\bar{v}_1 = \mu g t_1 = \mu g \frac{r\omega_0}{2\mu g} \qquad \bar{v}_1 = \tfrac{1}{2}r\omega_0 \qquad \bar{\mathbf{v}}_1 = \tfrac{1}{2}r\omega_0 \rightarrow \blacktriangleleft$$

$$\bar{v}_1 = r\omega_1 \qquad \tfrac{1}{2}r\omega_0 = r\omega_1 \qquad \omega_1 = \tfrac{1}{2}\omega_0 \qquad \boldsymbol{\omega}_1 = \tfrac{1}{2}\omega_0 \circlearrowright \blacktriangleleft$$

PROBLEMS

16.1 The open tailgate of a truck is held by hinges at A and a chain as shown. Determine the maximum allowable acceleration of the truck if the tailgate is to remain in the position shown.

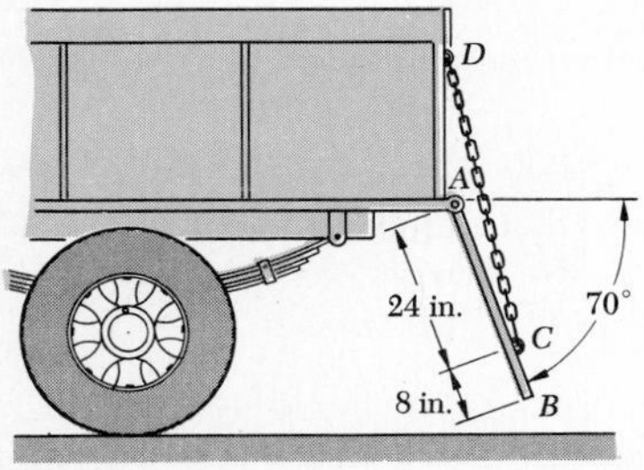

Fig. P16.1

16.2 A uniform rod ABC of mass 8 kg is connected to two collars of negligible mass which slide on smooth horizontal rods located in the same vertical plane. If a force **P** of magnitude 40 N is applied at C, determine (a) the acceleration of the rod, (b) the reactions at B and C.

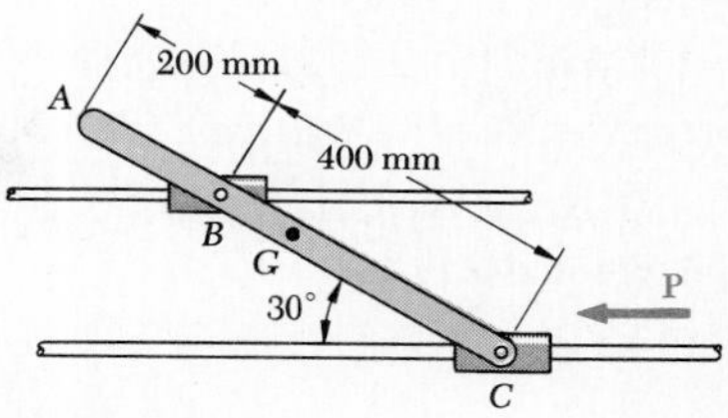

Fig. P16.2

16.3 In Prob. 16.2, determine (a) the required magnitude of **P** if the reaction at B is to be 45 N upward, (b) the corresponding acceleration of the rod.

16.4 A sphere of mass m and radius r is placed on a rough horizontal surface and pulled by a horizontal cord attached at A. The sphere is observed to slide to the right without rotating. Knowing that the acceleration of the sphere is $a = g/5$ directed to the right and that $\mu = 0.30$, determine (a) the magnitude of the force **P**, (b) the distance h from the surface to the cord.

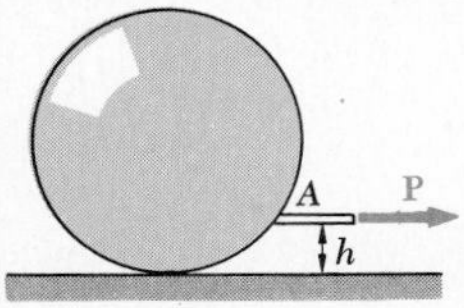

Fig. P16.4

16.5 Cylindrical cans are transported from one elevation to another by the moving horizontal arms shown. Assuming that $\mu = 0.20$ between the cans and the arms, determine (a) the magnitude of the downward acceleration **a** for which the cans slide on the horizontal arms, (b) the smallest ratio h/d for which the cans tip before they slide.

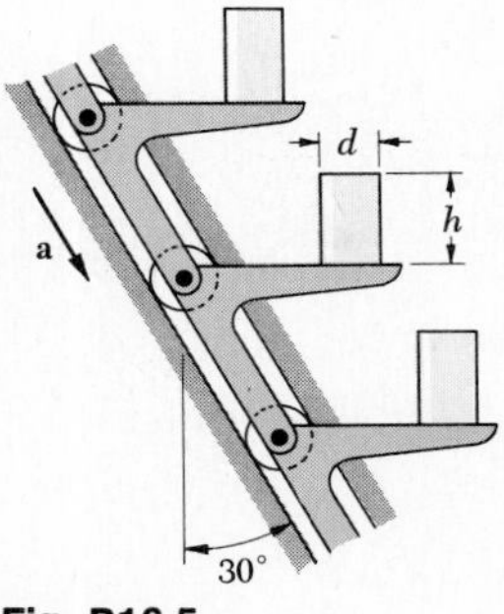

Fig. P16.5

16.6 Solve Prob. 16.5, assuming that the acceleration **a** of the horizontal arms is directed upward.

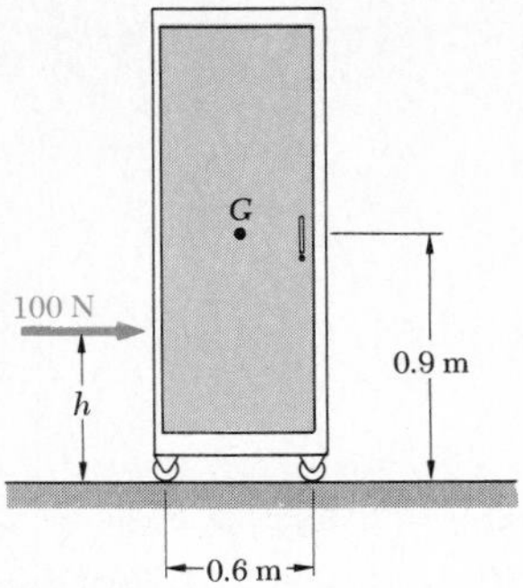

Fig. P16.7

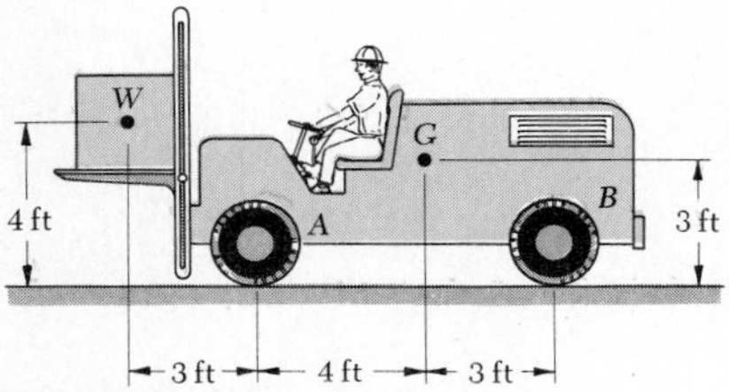

Fig. P16.9 and P16.10

16.7 A 20-kg cabinet is mounted on casters which allow it to move freely ($\mu = 0$) on the floor. If a 100-N force is applied as shown, determine (*a*) the acceleration of the cabinet, (*b*) the range of values of h for which the cabinet will not tip.

16.8 Solve Prob. 16.7, assuming that the casters are locked and slide along the rough floor ($\mu = 0.25$).

16.9 A fork-lift truck weighs 2250 lb and is used to lift a crate of weight $W = 2500$ lb. Determine (*a*) the upward acceleration of the crate for which the reactions at the rear wheels B are zero, (*b*) the corresponding reaction at each of the front wheels A.

16.10 The fork-lift truck shown weighs 2250 lb and carries a crate of weight $W = 2500$ lb. It is moving to the left with a speed of 10 ft/s when the brakes are applied on all four wheels. Determine the shortest distance in which the truck can be brought to a stop if the crate is not to slide and if the truck is not to tip forward. The coefficient of friction between the crate and the fork lift is 0.30.

16.11 Determine the distance through which the truck of Sample Prob. 16.1 will skid if the rear-wheel brakes fail to operate.

16.12 Determine the distance through which the truck of Sample Prob. 16.1 will skid if the front-wheel brakes fail to operate.

16.13 Knowing that the coefficient of friction between the tires and road is 0.80 for the car shown, determine the maximum possible acceleration on a level road, assuming (*a*) four-wheel drive, (*b*) conventional rear-wheel drive, (*c*) front-wheel drive.

a) 25.8 ft/s² →
b) 12.72 →
c) 13.32 →

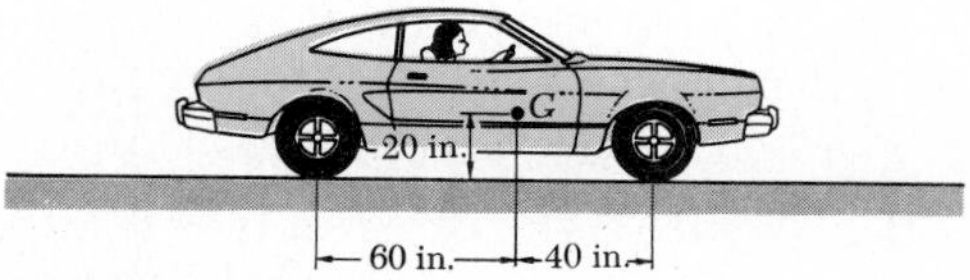

Fig. P16.13

16.14 A man rides a bicycle at a speed of 30 km/h. The distance between axles is 1050 mm, and the mass center of the man and bicycle is located 650 mm behind the front axle and 1000 mm above the ground. If the man applies the brakes on the front wheel only, determine the shortest distance in which he can stop without being thrown over the front wheel.

16.15 The total mass of the loading car and its load is 2500 kg. Neglecting the mass and friction of the wheels, determine (*a*) the minimum tension T in the cable for which the upper wheels are lifted from the track, (*b*) the corresponding acceleration of the car.

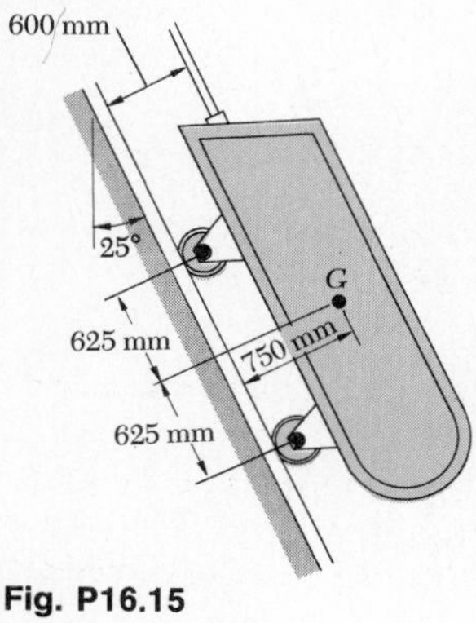

Fig. P16.15

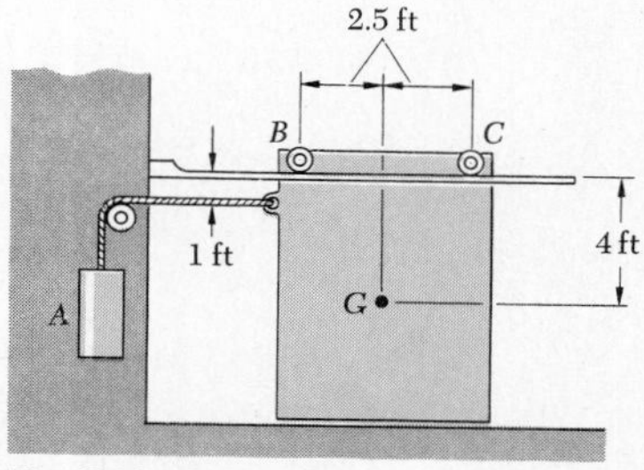

Fig. P16.16

16.16 The 500-lb fire door is supported by wheels B and C which may roll freely on the horizontal track. The counterweight A weighs 100 lb and is connected to the door by the cable shown. If the system is released from rest, determine (*a*) the acceleration of the door, (*b*) the reactions at B and C.

16.17 The cranks AB and CD rotate at a constant speed of 240 rpm. For the position $\phi = 30°$, determine the horizontal components of the forces exerted on the 5-kg uniform connecting rod BC by the pins B and C. Bx=205N ← Cx= 205N ←

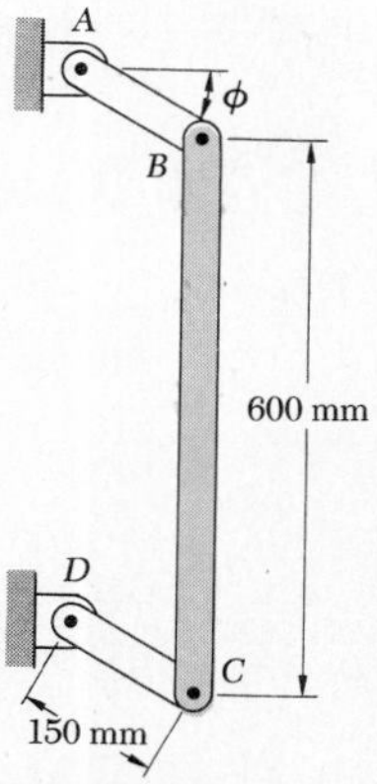

Fig. P16.17

16.18 Two uniform rods AB and BC, each of mass 3 kg, are welded together to form the rigid body ABC which is held in the position shown by the wire AE and the links AD and BE. Determine the force in each link immediately after wire AE has been cut. Neglect the mass of the links.

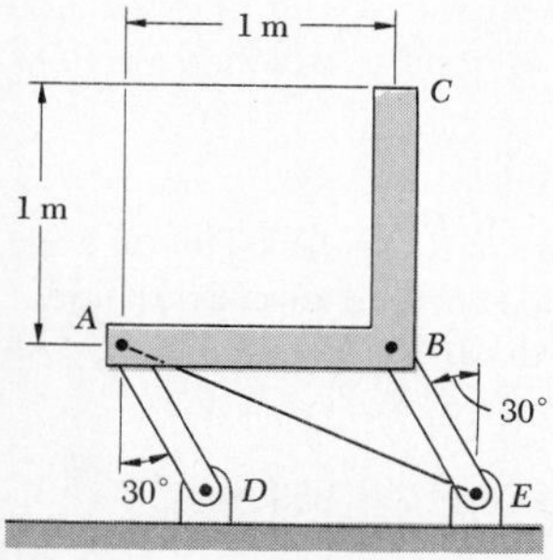

Fig. P16.18

16.19 The retractable shelf shown is supported by two identical linkage-and-spring systems; only one of the systems is shown. A 40-lb machine is placed on the shelf so that half of its weight is supported by the system shown. If the springs are removed and the system is released from rest, determine (*a*) the acceleration of the machine, (*b*) the tension in link *AB*. Neglect the weight of the shelf and links.

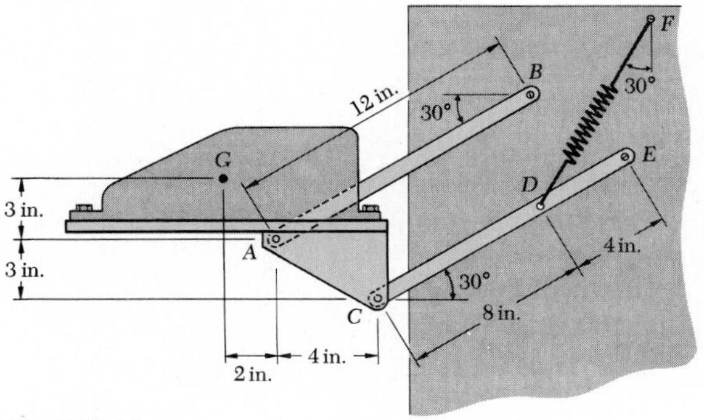

Fig. P16.19

16.20 Members *ACE* and *DCB* are each 20 in. long and are connected by a pin at *C*. The mass center of the 10-lb member *DE* is located at *G*. Determine (*a*) the acceleration of *DE* immediately after the system has been released from rest in the position shown, (*b*) the corresponding force exerted by roller *D* on member *DE*. Neglect the weight of members *ACE* and *DCB*.

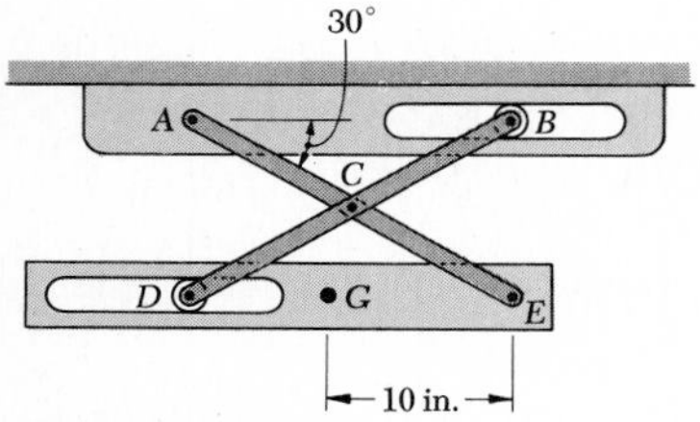

Fig. P16.20

16.21 The control rod *AC* is guided by two pins which slide freely in parallel curved slots of radius 200 mm. The rod has a mass of 10 kg, and its mass center is located at point *G*. Knowing that for the position shown the *vertical* component of the velocity of *C* is 1.25 m/s upward and the *vertical* component of the acceleration of *C* is 5 m/s² upward, determine the magnitude of the force **P**.

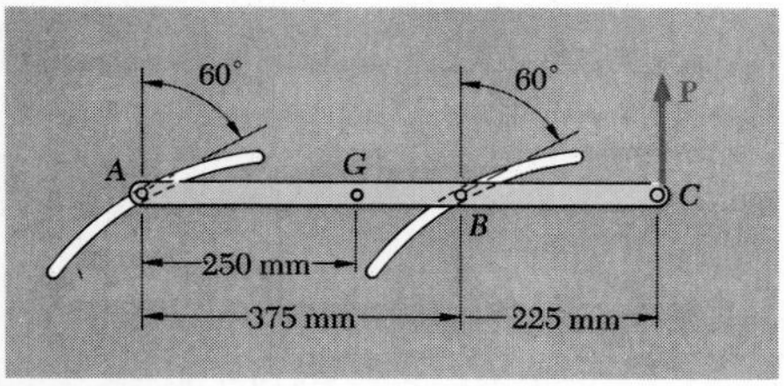

Fig. P16.21

∗16.22 A 12-kg block is placed on a 3-kg platform *BD* which is held in the position shown by three wires. Determine the accelerations of the block and of the platform immediately after wire *AB* has been cut. Assume (*a*) that the block is rigidly attached to *BD*, (*b*) that $\mu = 0$ between the block and *BD*.

∗16.23 The coefficient of friction between the 12-kg block and the 3-kg platform *BD* is 0.50. Determine the accelerations of the block and of the platform immediately after wire *AB* has been cut.

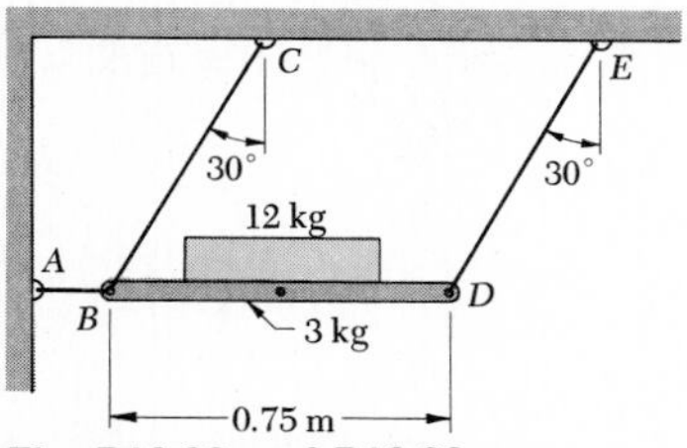

Fig. P16.22 and P16.23

∗16.24 Draw the shear and bending moment diagrams for the vertical rod *BC* of Prob. 16.18.

***16.25** Draw the shear and bending-moment diagrams for the rod BC of Prob. 16.17.

16.26 A turbine-generator unit is shut off when its rotor is rotating at 3600 rpm; it is observed that the rotor coasts to rest in 7.10 min. Knowing that the 1850-kg rotor has a radius of gyration of 234 mm, determine the average magnitude of the couple due to bearing friction.

16.27 An electric motor is rotating at 900 rpm when the load and power are cut off. The rotor weighs 200 lb and has a radius of gyration of 9 in. If the kinetic friction of the rotor produces a couple of moment 15 lb · in., how many revolutions will the rotor execute before stopping?

16.28 Disk A weighs 10 lb and is at rest when it is placed in contact with a conveyor belt moving at a constant speed. The link AB connecting the center of the disk to the support at B is of negligible weight. Knowing that $r = 5$ in. and $\mu = 0.40$, determine the angular acceleration of the disk while slipping occurs.

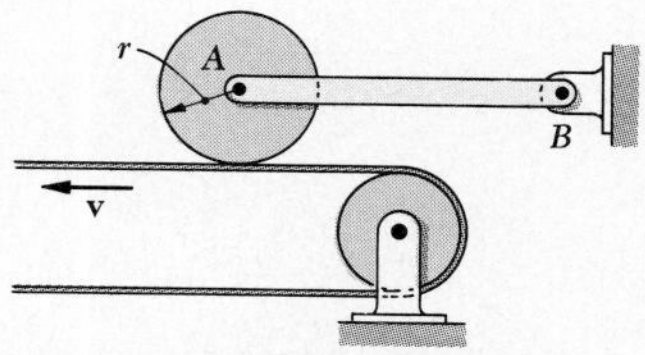

Fig. P16.28 and P16.29

16.29 The uniform disk A is at rest when it is placed in contact with a conveyor belt moving at a constant speed. Neglecting the weight of the link AB, derive an expression for the angular acceleration of the disk while slipping occurs.

16.30 Each of the double pulleys shown has a mass moment of inertia of $10\ \text{kg} \cdot \text{m}^2$ and is initially at rest. The outside radius is 400 mm, and the inner radius is 200 mm. Determine (*a*) the angular acceleration of each pulley, (*b*) the angular velocity of each pulley at $t = 2$ s, (*c*) the angular velocity of each pulley after point A on the cord has moved 2 m.

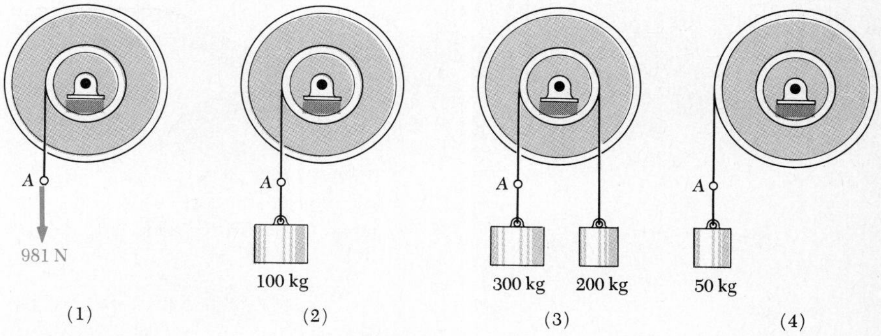

Fig. P16.30

16.31 Solve Prob. 12.16*a* assuming that each pulley is of 8-in. radius and has a centroidal mass moment of inertia $0.25\ \text{lb} \cdot \text{ft} \cdot \text{s}^2$.

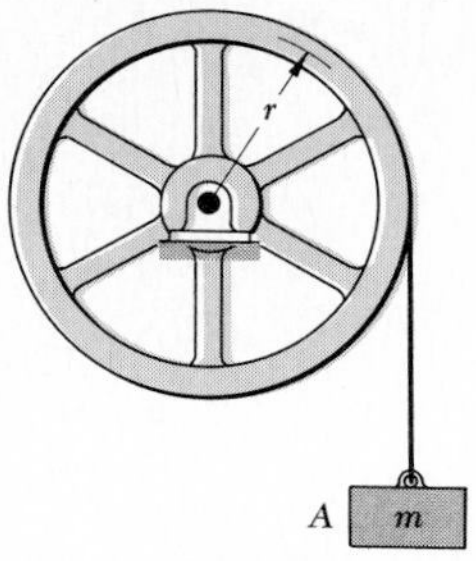

Fig. P16.32 and P16.34

16.32 The flywheel shown weighs 250 lb and has a radius of gyration of 15 in. A block A of weight 30 lb is attached to a wire wrapped around the rim of radius $r = 20$ in. The system is released from rest. Neglecting the effect of friction, determine (*a*) the acceleration of block A, (*b*) the speed of block A after it has moved 6 ft.

16.33 Solve Prob. 16.32 assuming that the weight of block A is 60 lb.

16.34 In order to determine the mass moment of inertia of a flywheel of radius $r = 600$ mm a block of mass 12 kg is attached to a cord which is wrapped around the rim of the flywheel. The block is released from rest and is observed to fall 3 m in 4.6 s. To eliminate bearing friction from the computation, a second block of mass 24 kg is used and is observed to fall 3 m in 3.1 s. Assuming that the moment of the couple due to friction is constant, determine the mass moment of inertia of the flywheel.

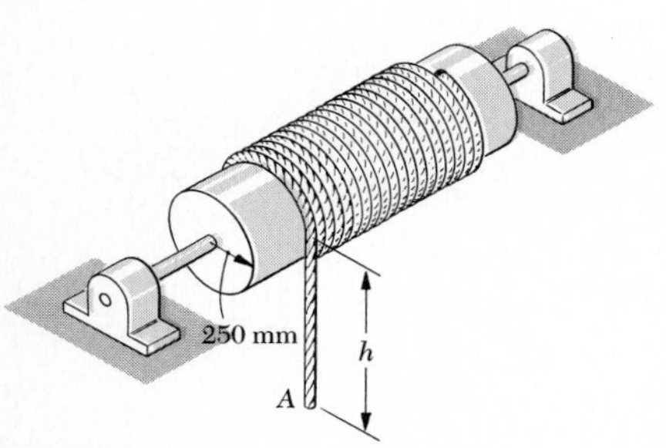

Fig. P16.35

16.35 A rope of total mass 10 kg and total length 20 m is wrapped around the drum of a hoist as shown. The mass of the drum and shaft is 18 kg, and they have a combined radius of gyration of 200 mm. Knowing that the system is released from rest when a length $h = 5$ m hangs from the drum, determine the initial angular acceleration of the drum.

16.36 The 15-in.-radius brake drum is attached to a larger flywheel which is not shown. The total mass moment of inertia of the flywheel and drum is 15 lb · ft · s^2. Knowing that the initial angular velocity is 150 rpm clockwise, determine the force which must be exerted by the hydraulic cylinder if the system is to stop in 10 revolutions.

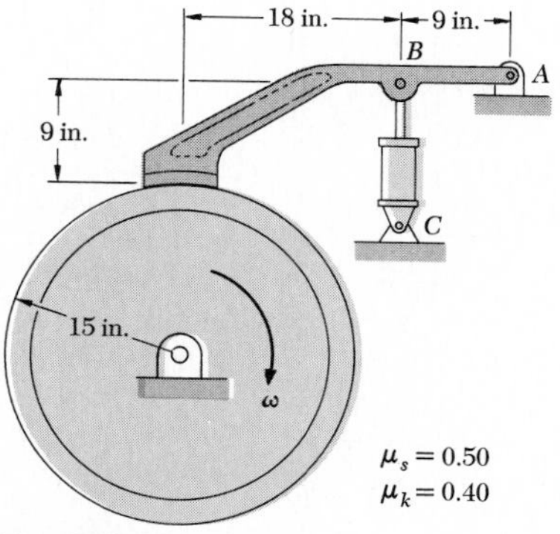

Fig. P16.36

16.37 Solve Prob. 16.36, assuming that the initial angular velocity of the flywheel is 150 rpm counterclockwise.

16.38 A cylinder of radius r and mass m is placed with no initial velocity on a belt as shown. Denoting by μ the coefficient of friction at A and at B and assuming that $\mu < 1$, determine the angular acceleration $\boldsymbol{\alpha}$ of the cylinder.

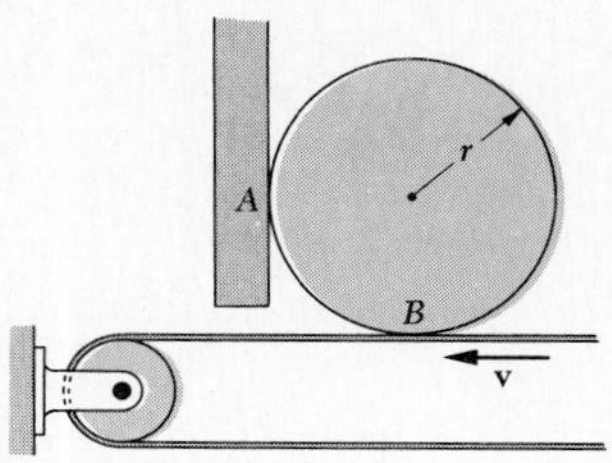

Fig. P16.38

16.39 Shaft A and friction disk B have a combined mass of 15 kg and a combined radius of gyration of 150 mm. Shaft D and friction wheel C rotate with a constant angular velocity of 1000 rpm. Disk B is at rest when it is brought into contact with the rotating wheel. Knowing that disk B accelerates uniformly for 12 s before acquiring its final angular velocity, determine the magnitude of the friction force between the disk and the wheel.

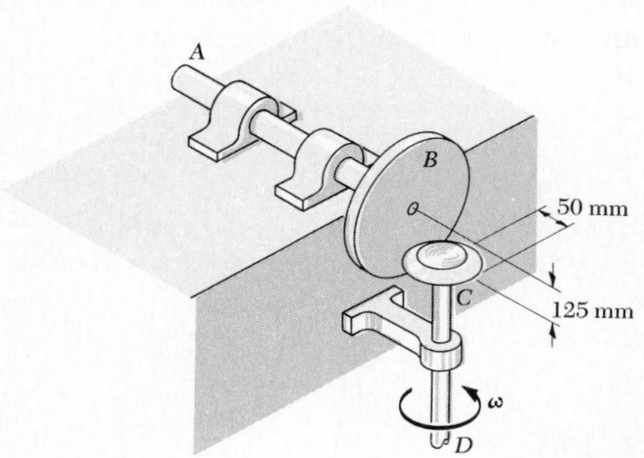

Fig. P16.39

16.40 Gear A weighs 30 lb and has a radius of gyration of 6 in.; gear B weighs 9 lb and has a radius of gyration of 2.5 in. A couple of moment $M = 15$ lb · in. is applied to gear B. Determine (a) the angular acceleration of gear A, (b) the time required for the angular velocity of gear A to increase from 100 to 500 rpm.

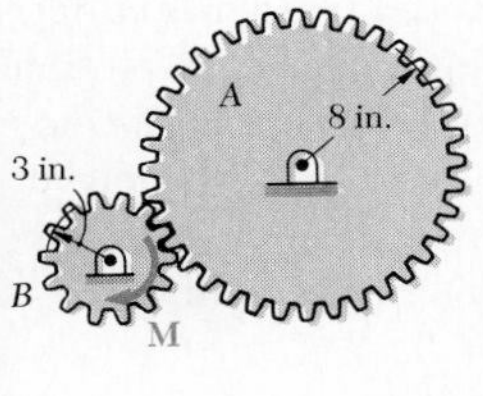

Fig. P16.40

16.41 Disk A is of mass 5 kg and has an initial angular velocity of 300 rpm clockwise. Disk B is of mass 1.8 kg and is at rest when it is placed in contact with disk A. Knowing that $\mu = 0.30$ between the disks and neglecting bearing friction, determine (a) the angular acceleration of each disk, (b) the reaction at the support C.

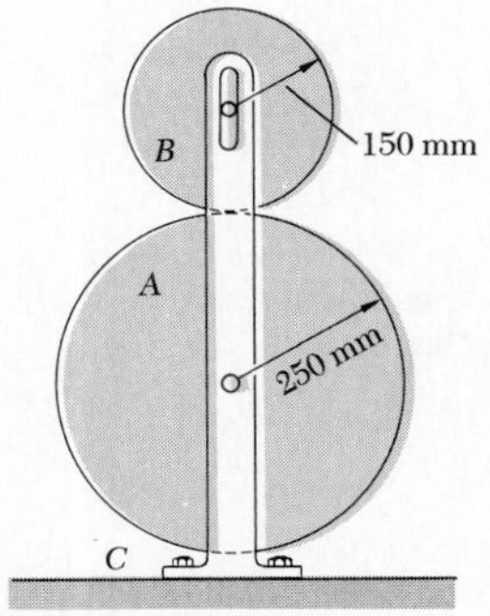

Fig. P16.41

16.42 In Prob. 16.41, (a) determine the final angular velocity of each disk, (b) show that the final angular velocities are independent of μ.

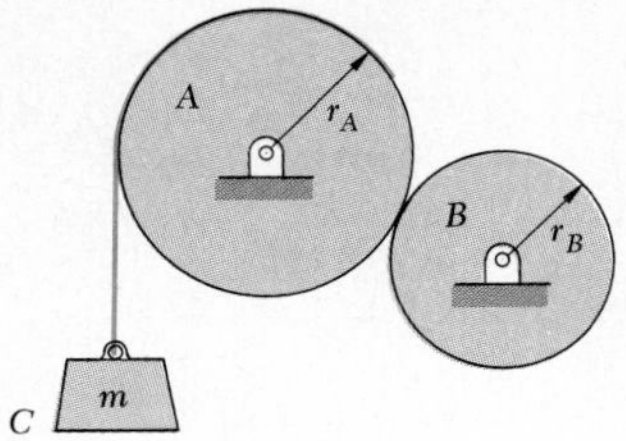

Fig. P16.43

16.43 Assuming that no slipping occurs, derive a formula for the acceleration of block C in terms of its mass m, the radii r_A and r_B, and the centroidal mass moments of inertia $\bar{I}_A$ and $\bar{I}_B$ of the two disks.

16.44 A coder C, used to record in digital form the rotation of a shaft S, is connected to the shaft by means of the gear train shown, which consists of four gears of the same thickness and of the same material. Two of the gears have a radius r and the other two a radius nr. Let I_R denote the ratio M/α of the moment M of the couple applied to the shaft S and of the resulting angular acceleration α of S. (I_R is sometimes called the "reflected moment of inertia" of the coder and gear train.) Determine I_R in terms of the gear ratio n, the moment of inertia I_0 of the first gear, and the moment of inertia I_C of the coder. Neglect the moments of inertia of the shafts.

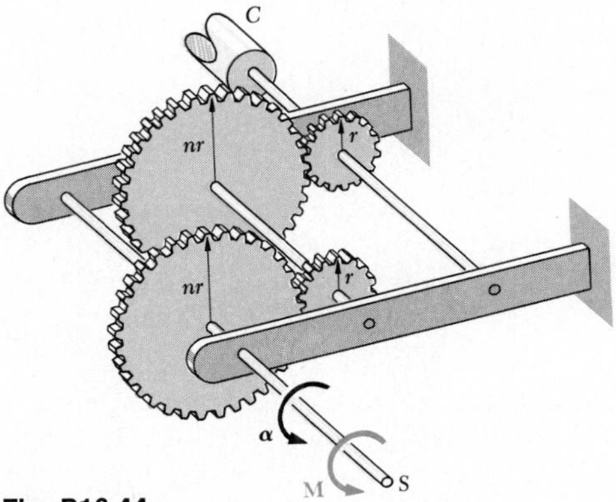

Fig. P16.44

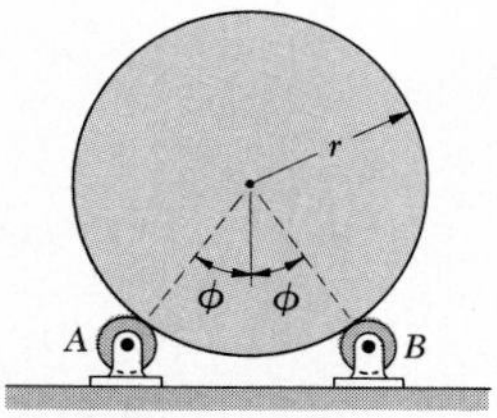

Fig. P16.45

16.45 A cylinder of radius r and mass m rests on two small casters A and B as shown. Initially, the cylinder is at rest and is set in motion by rotating caster B clockwise at high speed so that slipping occurs between the cylinder and caster B. Denoting by μ the coefficient of friction and neglecting the moment of inertia of the free caster A, derive an expression for the angular acceleration of the cylinder.

16.46 In Prob. 16.45, assume that no slipping can occur between caster B and the cylinder (such a case would exist if the cylinder and caster had gear teeth along their rims). Derive an expression for the maximum counterclockwise acceleration $\boldsymbol{\alpha}$ of the cylinder if it is not to lose contact with the caster at A.

16.47 Show that the system of the effective forces for a rigid slab in plane motion reduces to a single vector, and express the distance from the mass center G of the slab to the line of action of this vector in terms of the centroidal radius of gyration $\bar{k}$ of the slab, the magnitude $\bar{a}$ of the acceleration of G, and the angular acceleration α.

16.48 The uniform slender rod AB weighs 8 lb and is at rest on a frictionless horizontal surface. A force **P** of magnitude 2 lb is applied at A in a horizontal direction perpendicular to the rod. Determine (*a*) the angular acceleration of the rod, (*b*) the acceleration of the center of the rod, (*c*) the point of the rod which has no acceleration.

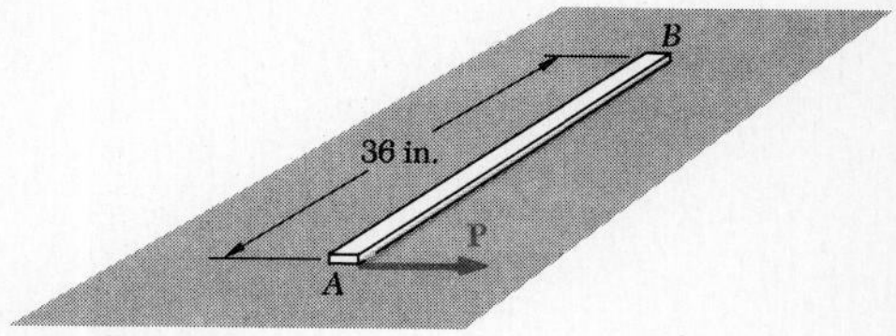

Fig. P16.48

16.49 In Prob. 16.48, determine the point of the rod AB at which the force **P** should be applied if the acceleration of point B is to be zero. Knowing that the magnitude of **P** is 2 lb, determine the corresponding angular acceleration of the rod and the acceleration of the center of the rod.

16.50 Shortly after being fired, the missile shown has a mass of 9000 kg and is moving upward with an acceleration of 16 m/s². If, at this instant, the rocket engine A fails, while rocket engine B continues to operate, determine (*a*) the acceleration of the mass center of the missile, (*b*) the angular acceleration of the missile. Assume the missile to be a uniform slender rod 15 m long.

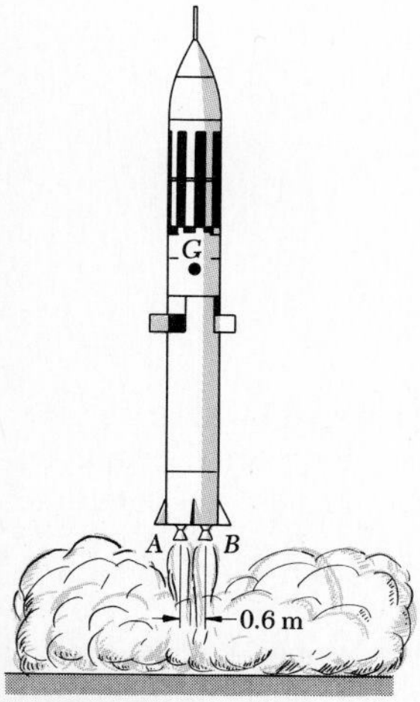

Fig. P16.50

16.51 A 50-kg space satellite has a radius of gyration of 450 mm with respect to the y axis, and is symmetrical with respect to the zx plane. The orientation of the satellite is changed by firing four small rockets A, B, C, and D which are equally spaced around the perimeter of the satellite. While being fired, each rocket produces a thrust **T** of magnitude 10 N directed as shown. Determine the angular acceleration of the satellite and the acceleration of its mass center G (*a*) when all four rockets are fired, (*b*) when all rockets except rocket D are fired.

16.52 Solve Prob. 16.51 assuming that only rocket A is fired.

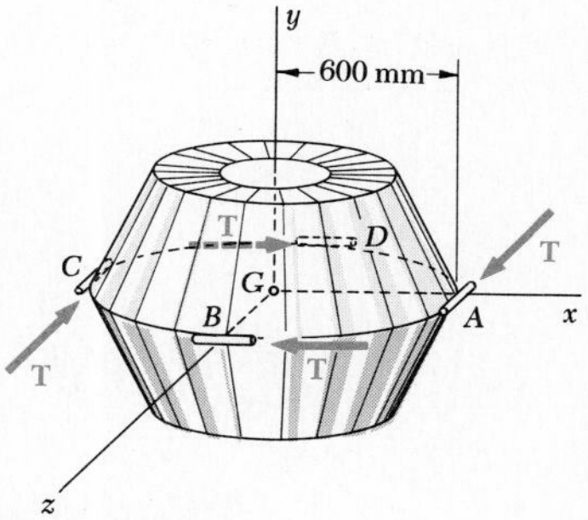

Fig. P16.51

16.53 An 18-ft beam weighing 400 lb is lowered from a considerable height by means of two cables unwinding from overhead cranes. As the beam approaches the ground, the crane operators apply brakes to slow the unwinding motion. The deceleration of cable A is 21 ft/s^2, while that of cable B is 3 ft/s^2. Determine the tension in each cable.

16.54 An 18-ft beam weighing 400 lb is lowered from a considerable height by means of two cables unwinding from overhead cranes. As the beam approaches the ground, the crane operators apply brakes to slow the unwinding motion. Determine the acceleration of each cable at that instant, knowing that $T_A = 325$ lb and $T_B = 275$ lb.

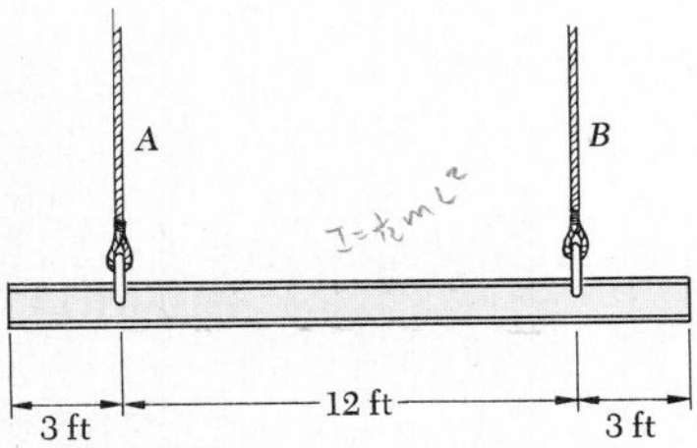

Fig. P16.53 and P16.54

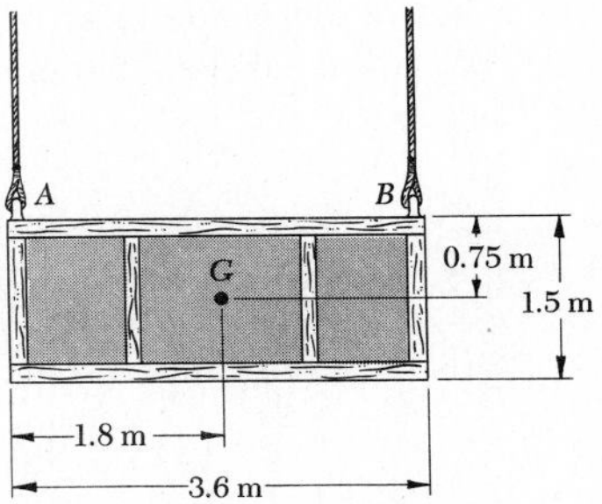

Fig. P16.55 and P16.56

16.55 The 180-kg crate shown is being lowered by means of two overhead cranes. Knowing that at the instant shown the deceleration of cable A is 7 m/s^2, while that of cable B is 1 m/s^2, determine the tension in each cable.

16.56 The 180-kg crate is being lowered by means of two overhead cranes. As the crate approaches the ground, the crane operators apply brakes to slow the motion. Determine the acceleration of each cable at that instant, knowing that $T_A = 1450$ N and $T_B = 1200$ N.

16.57 Solve Sample Prob. 16.4, assuming that the disk rests flat on a frictionless horizontal surface and that the cord is pulled horizontally with a force of magnitude 180 N.

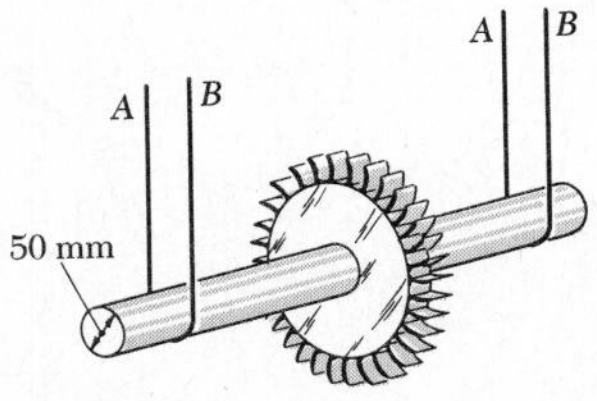

Fig. P16.58

16.58 A turbine disk and shaft have a combined mass of 100 kg and a centroidal radius of gyration of 50 mm. The unit is lifted by two ropes looped around the shaft as shown. Knowing that for each rope $T_A = 270$ N and $T_B = 320$ N, determine (*a*) the angular acceleration of the unit, (*b*) the acceleration of its mass center.

16.59 A uniform slender bar of mass m is suspended from two *springs* as shown. If spring BD breaks, determine at that instant (*a*) the angular acceleration of the bar, (*b*) the acceleration of point A, (*c*) the acceleration of point B.

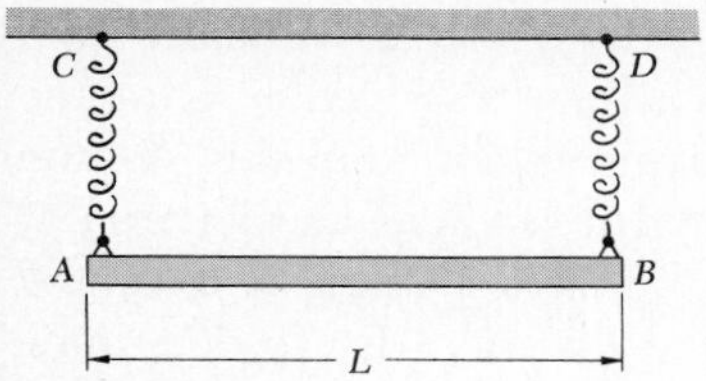

Fig. P16.59

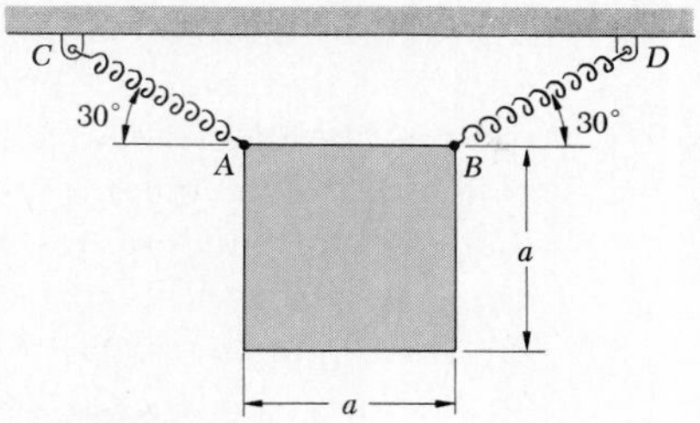

Fig. P16.60

16.60 Solve Prob. 16.59, assuming that the bar is replaced by the uniform square plate shown.

16.61 A sphere of mass m and radius r is projected along a rough horizontal surface with a linear velocity $\mathbf{v}_0$ and with $\boldsymbol{\omega}_0 = 0$. The sphere will decelerate and then reach a uniform motion. Denoting by μ the coefficient of friction, determine (*a*) the linear and angular acceleration of the sphere before it reaches a uniform motion, (*b*) the time required for the motion to become uniform, (*c*) the distance traveled before the motion becomes uniform, (*d*) the final linear and angular velocities of the sphere.

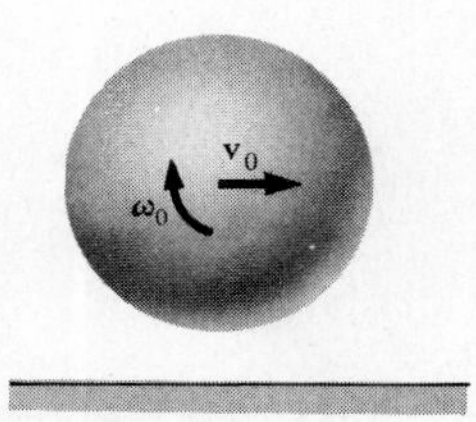

Fig. P16.61

$\alpha = \frac{5}{2}\frac{\mu g}{r}$ b) $t = \frac{2}{7}\frac{v_0}{\mu g}$ c) $x = \frac{12 v_0^2}{49 \mu g}$ d) $\omega_F = \frac{5}{7}\frac{v_0}{r}$ $v_F = \frac{5}{7} v_0$

16.62 Solve Prob. 16.61, assuming that the sphere is replaced by a uniform disk of radius r and mass m.

16.63 A sphere of mass m and radius r is dropped with no initial velocity on a belt which moves with a constant velocity $\mathbf{v}_0$. At first the sphere will both slide and roll on the belt. Denoting by μ the coefficient of friction between the sphere and the belt, determine the distance the sphere will move before it starts rolling without sliding.

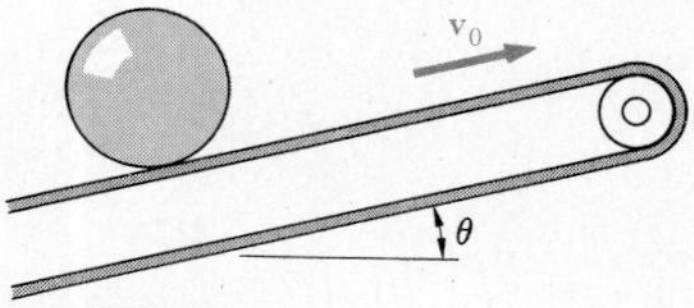

Fig. P16.63

16.64 Solve Prob. 16.63, assuming $\theta = 10°$, $\mu = 0.30$, and $v_0 = 3$ m/s.

16.5. Constrained Plane Motion. Most engineering applications deal with rigid bodies which are moving under given constraints. Cranks, for example, are constrained to rotate about a fixed axis, wheels roll without sliding, connecting rods must describe certain prescribed motions. In all such cases, definite relations exist between the components of the acceleration $\bar{\mathbf{a}}$ of the mass center G of the body considered and its angular acceleration $\boldsymbol{\alpha}$; the corresponding motion is said to be a *constrained motion*.

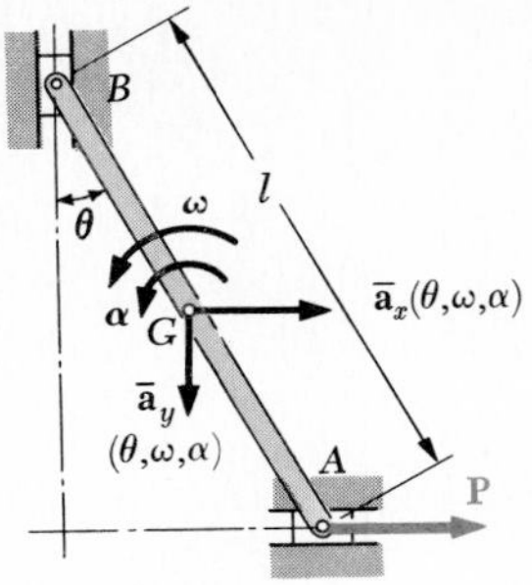

Fig. 16.10

The solution of a problem involving a constrained plane motion calls first for a *kinematic analysis* of the problem. Consider, for example, a slender rod AB of length l and mass m whose extremities are connected to blocks of negligible mass which slide along horizontal and vertical frictionless tracks. The rod is pulled by a force **P** applied at A (Fig. 16.10). We know from Sec. 15.8 that the acceleration $\bar{\mathbf{a}}$ of the mass center G of the rod may be determined at any given instant from the position of the rod, its angular velocity, and its angular acceleration at that instant. Suppose, for instance, that the values of θ, ω, and α are known at a given instant and that we wish to determine the corresponding value of the force **P**, as well as the reactions at A and B. We should first *determine the components* $\bar{a}_x$ *and* $\bar{a}_y$ *of the acceleration of the mass center* G by the method of Sec. 15.8. We then carry the values obtained into Fig. 16.11

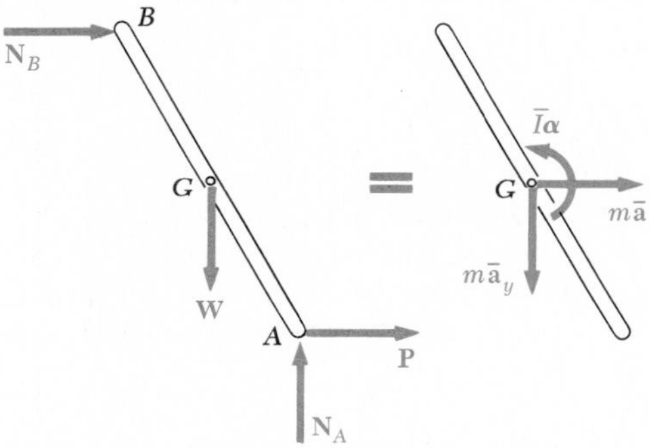

Fig. 16.11

or Fig. 16.12, depending upon whether D'Alembert's principle is applied directly or the method of dynamic equilibrium is used. The unknown forces **P**, $\mathbf{N}_A$, and $\mathbf{N}_B$ may then be determined by writing and solving the appropriate equations.

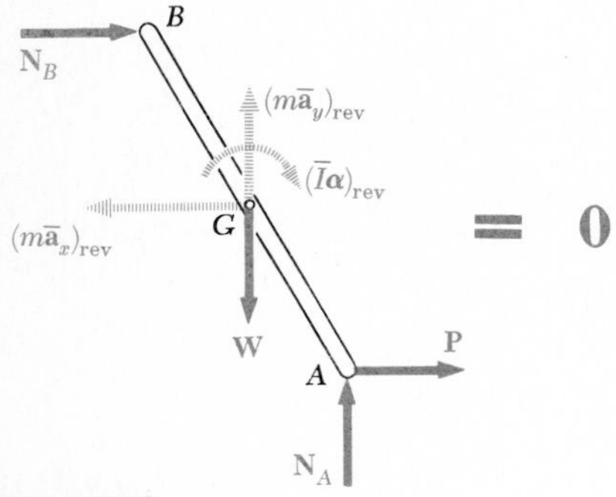

Fig. 16.12

Suppose now that the applied force **P**, the angle θ, and the angular velocity ω of the rod are known at a given instant and that we wish to find the angular acceleration α of the rod and the components $\bar{a}_x$ and $\bar{a}_y$ of the acceleration of its mass center at that instant, as well as the reactions at A and B. The preliminary kinematic study of the problem will have for its object *to express the components* $\bar{a}_x$ *and* $\bar{a}_y$ *of the acceleration of* G *in terms of the angular acceleration* α *of the rod.* This will be done by first expressing the acceleration of a suitable reference point such as A in terms of the angular acceleration α. The components $\bar{a}_x$ and $\bar{a}_y$ of the acceleration of G may then be determined in terms of α, and the expressions obtained carried into Fig. 16.11 or Fig. 16.12. Three equations may then be derived in terms of α, N_A, and N_B, and solved for the three unknowns (see Sample Prob. 16.10).

When a mechanism consists of *several moving parts,* the method just described may be applied to each part of the mechanism. The procedure required to determine the various unknowns is then similar to the procedure followed in the case of the equilibrium of a system of connected rigid bodies (Sec. 6.12).

We have analyzed earlier two particular cases of constrained plane motion, the translation of a rigid body, in which the angular acceleration of the body is constrained to be zero, and the centroidal rotation, in which the acceleration $\overline{\mathbf{a}}$ of the mass center of the body is constrained to be zero. Two other particular cases of constrained plane motion are of special interest, the *noncentroidal rotation* of a rigid body and the *rolling motion* of a disk or wheel. These two cases should be analyzed by one of the general methods described above. However, in view of the range of their applications, they deserve a few special comments.

Noncentroidal Rotation. This is the motion of a rigid body constrained to rotate about a fixed axis which does not pass through its mass center. Such a motion is called a *noncentroidal rotation*. The mass center G of the body moves along a circle of radius $\bar{r}$ centered at the point O, where the axis of rotation intersects the plane of reference (Fig. 16.13). Denoting, respectively, by $\boldsymbol{\omega}$ and $\boldsymbol{\alpha}$ the angular velocity and the angular acceleration of the line OG, we obtain the following expressions for the tangential and normal components of the acceleration of G:

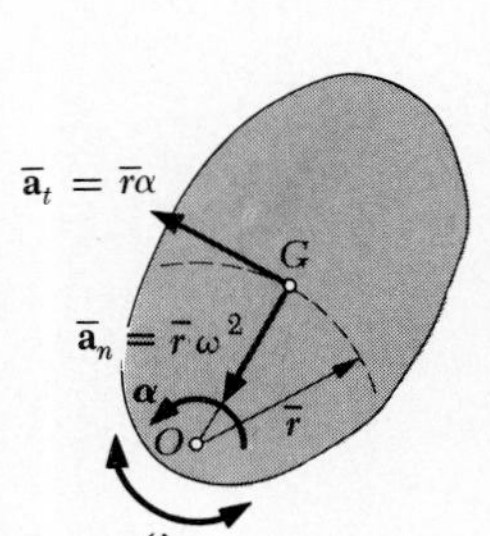

Fig. 16.13

$$\bar{a}_t = \bar{r}\alpha \qquad \bar{a}_n = \bar{r}\omega^2 \tag{16.12}$$

Since line OG belongs to the body, its angular velocity $\boldsymbol{\omega}$ and its angular acceleration $\boldsymbol{\alpha}$ also represent the angular velocity and the angular acceleration of the body in its motion relative to G. Equations (16.12) define, therefore, the kinematic relation existing between the motion of the mass center G and the motion of the body about G. They should be used to eliminate $\bar{a}_t$ and $\bar{a}_n$ from the equations obtained by applying D'Alembert's principle (Fig. 16.14) or the method of dynamic equilibrium (Fig. 16.15).

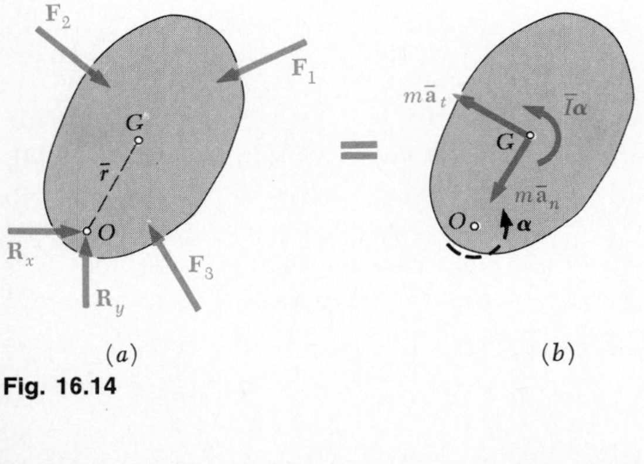

Fig. 16.14

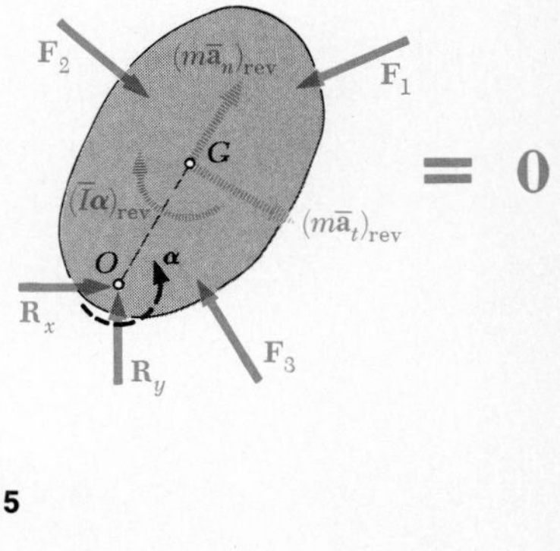

Fig. 16.15

An interesting relation may be obtained by equating the moments about the fixed point O of the forces and vectors shown respectively in parts a and b of Fig. 16.14. We write

$$+\circlearrowleft \Sigma M_O = \bar{I}\alpha + (m\bar{r}\alpha)\bar{r} = (\bar{I} + m\bar{r}^2)\alpha$$

But, according to the parallel-axis theorem, we have $\bar{I} + m\bar{r}^2 = I_O$, where I_O denotes the moment of inertia of the rigid body about the fixed axis. We write, therefore,

$$\Sigma M_O = I_O\alpha \tag{16.13}$$

While formula (16.13) expresses an important relation between the sum of the moments of the external forces about the fixed point O and the product $I_O\alpha$, it should be clearly understood that this formula *does not mean* that the system of the external forces is equivalent to a couple of moment $I_O\alpha$. The system of the effective forces, and thus the system of the external forces, reduces to a couple only when O coincides with G, that is, *only when the rotation is centroidal* (Sec. 16.2). In the more general case of noncentroidal rotation, the system of the external forces does not reduce to a couple.

A particular case of noncentroidal rotation is of special interest: the case of *uniform rotation*, in which the angular velocity $\boldsymbol{\omega}$ is constant. Since $\boldsymbol{\alpha}$ is zero, the inertia couple in Fig. 16.15 vanishes and the inertia vector reduces to its normal component. This component (also called *centrifugal force*) represents the tendency of the rigid body to break away from the axis of rotation.

Rolling Motion. Another important case of plane motion is the motion of a disk or wheel rolling on a plane surface. If the disk is constrained to roll without sliding, the acceleration $\bar{\mathbf{a}}$ of its mass center G and its angular acceleration $\boldsymbol{\alpha}$ are not independent. Assuming the disk to be balanced, so that its mass center and its geometric center coincide, we first write that the distance $\bar{x}$ traveled by G during a rotation θ of the disk is $\bar{x} = r\theta$, where r is the radius of the disk. Differentiating this relation twice, we write

$$\bar{a} = r\alpha \tag{16.14}$$

Recalling that the system of the effective forces in plane motion reduces to a vector $m\bar{\mathbf{a}}$ and a couple $\bar{I}\boldsymbol{\alpha}$, we find that, in the particular case of the rolling motion of a balanced disk, the effective forces reduce to a vector of magnitude $mr\alpha$ attached at G and to a couple of magnitude $\bar{I}\alpha$. We may thus express that the external forces are equivalent to the vector and couple shown in Fig. 16.16.

When a disk *rolls without sliding*, there is no relative motion between the point of the disk which is in contact with the ground and the ground itself. As far as the computation of the friction force **F** is concerned, a rolling disk may thus be compared with a block at rest on a surface. The magnitude F of the friction force may have any value, as long as it does not exceed the maximum value $F_m = \mu_s N$, where μ_s is the coefficient of static friction and N the magnitude of the normal force. In the case of a rolling disk, the magnitude F of the friction force should therefore be determined independently of N by solving the equation obtained from Fig. 16.16.

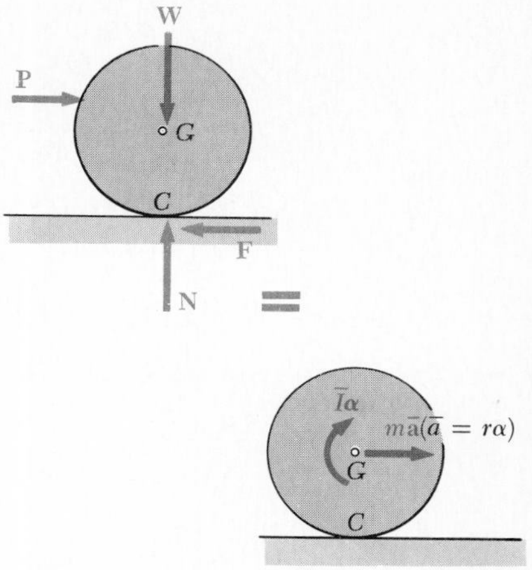

Fig. 16.16

When *sliding is impending*, the friction force reaches its maximum value $F_m = \mu_s N$ and may be obtained from N.

When the disk *rolls and slides* at the same time, a relative motion exists between the point of the disk which is in contact with the ground and the ground itself, and the force of friction has the magnitude $F_k = \mu_k N$, where μ_k is the coefficient of kinetic friction. In this case, however, the motion of the mass center G of the disk and the rotation of the disk about G are independent, and $\bar{a}$ is not equal to $r\alpha$.

These three different cases may be summarized as follows:

Rolling, no sliding:	$F \leq \mu_s N$	$\bar{a} = r\alpha$
Rolling, sliding impending:	$F = \mu_s N$	$\bar{a} = r\alpha$
Rolling and sliding:	$F = \mu_k N$	$\bar{a}$ *and* α *independent*

When it is not known whether a disk slides or not, it should first be assumed that the disk rolls without sliding. If F is found smaller than, or equal to, $\mu_s N$, the assumption is proved correct. If F is found larger than $\mu_s N$, the assumption is incorrect and the problem should be started again, assuming rolling and sliding.

When a disk is *unbalanced*, i.e., when its mass center G does not coincide with its geometric center O, the relation (16.14) does not hold between $\bar{a}$ and α. A similar relation will hold, however, between the magnitude a_O of the acceleration of the geometric center and the angular acceleration α,

$$a_O = r\alpha \qquad (16.15)$$

To determine $\bar{a}$ in terms of the angular acceleration α and the angular velocity ω of the disk, we may use the relative-acceleration formula,

$$\begin{aligned}\bar{\mathbf{a}} = \mathbf{a}_G &= \mathbf{a}_O + \mathbf{a}_{G/O} \\ &= \mathbf{a}_O + (\mathbf{a}_{G/O})_t + (\mathbf{a}_{G/O})_n \qquad (16.16)\end{aligned}$$

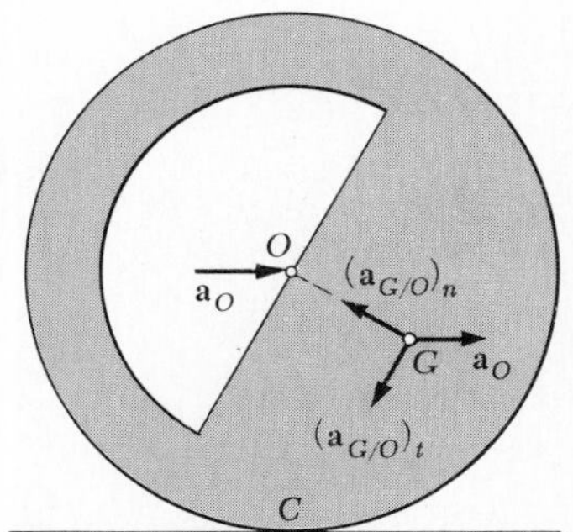

Fig. 16.17

where the three component accelerations obtained have the directions indicated in Fig. 16.17 and the magnitudes $a_O = r\alpha$, $(a_{G/O})_t = (OG)\alpha$, and $(a_{G/O})_n = (OG)\omega^2$.

SAMPLE PROBLEM 16.6

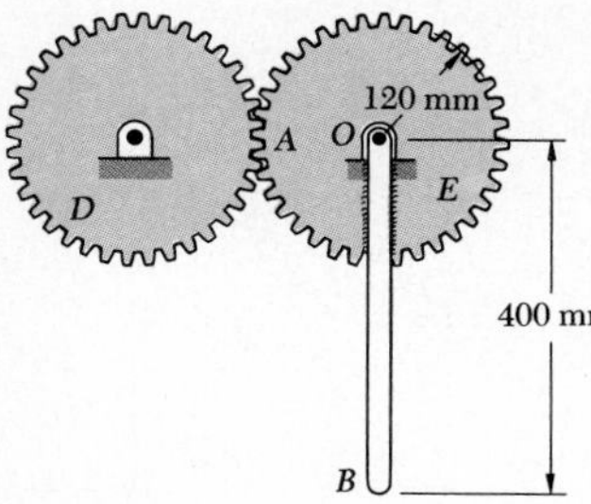

The portion AOB of a mechanism consists of a 400-mm steel rod OB welded to a gear E of radius 120 mm which may rotate about a horizontal shaft O. It is actuated by a gear D and, at the instant shown, has a clockwise angular velocity of 8 rad/s and a counterclockwise angular acceleration of 40 rad/s². Knowing that rod OB has a mass of 3 kg and gear E a mass of 4 kg and a radius of gyration of 85 mm, determine (*a*) the tangential force exerted by gear D on gear E, (*b*) the components of the reaction at shaft O.

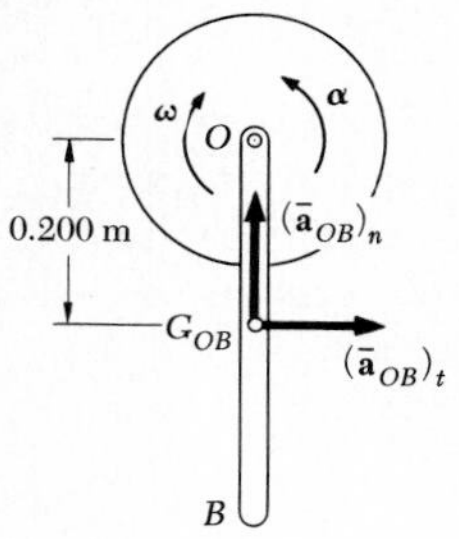

Solution. In determining the effective forces of the rigid body AOB we shall consider separately gear E and rod OB. Therefore, we shall first determine the components of the acceleration of the mass center G_{OB} of the rod:

$$(\bar{a}_{OB})_t = \bar{r}\alpha = (0.200\text{ m})(40\text{ rad/s}^2) = 8\text{ m/s}^2$$
$$(\bar{a}_{OB})_n = \bar{r}\omega^2 = (0.200\text{ m})(8\text{ rad/s})^2 = 12.8\text{ m/s}^2$$

Equations of Motion. Two sketches of the rigid body AOB have been drawn. The first shows the external forces consisting of the weight $\mathbf{W}_E$ of gear E, the weight $\mathbf{W}_{OB}$ of rod OB, the force $\mathbf{F}$ exerted by gear D, and the components $\mathbf{R}_x$ and $\mathbf{R}_y$ of the reaction at O. The magnitudes of the weights are, respectively,

$$W_E = m_E g = (4\text{ kg})(9.81\text{ m/s}^2) = 39.2\text{ N}$$
$$W_{OB} = m_{OB} g = (3\text{ kg})(9.81\text{ m/s}^2) = 29.4\text{ N}$$

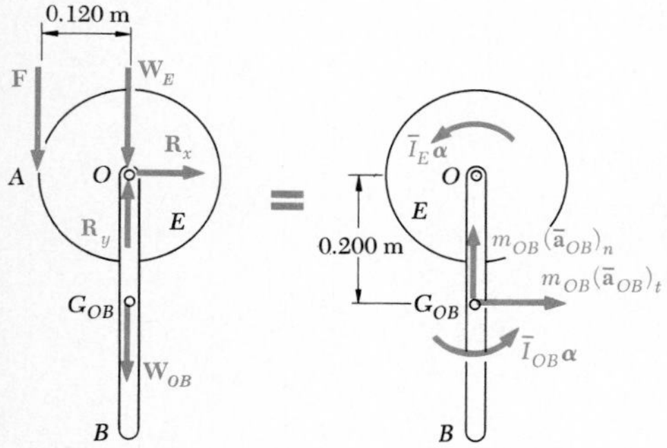

The second sketch shows the effective forces, which consist of a couple $\bar{I}_E\boldsymbol{\alpha}$ (since gear E is in centroidal rotation) and of a couple and two vector components at the mass center of OB. Since the accelerations are known, we compute the magnitudes of these components and couples:

$$\bar{I}_E\alpha = m_E\bar{k}_E^2\alpha = (4\text{ kg})(0.085\text{ m})^2(40\text{ rad/s}^2) = 1.156\text{ N}\cdot\text{m}$$
$$m_{OB}(\bar{a}_{OB})_t = (3\text{ kg})(8\text{ m/s}^2) = 24.0\text{ N}$$
$$m_{OB}(\bar{a}_{OB})_n = (3\text{ kg})(12.8\text{ m/s}^2) = 38.4\text{ N}$$
$$\bar{I}_{OB}\alpha = (\tfrac{1}{12}m_{OB}L^2)\alpha = \tfrac{1}{12}(3\text{ kg})(0.400\text{ m})^2(40\text{ rad/s}^2) = 1.600\text{ N}\cdot\text{m}$$

Expressing that the system of the external forces is equivalent to the system of the effective forces, we write the following equations:

$+\circlearrowleft \Sigma M_O = \Sigma(M_O)_{\text{eff}}$:

$$F(0.120\text{ m}) = \bar{I}_E\alpha + m_{OB}(\bar{a}_{OB})_t(0.200\text{ m}) + \bar{I}_{OB}\alpha$$
$$F(0.120\text{ m}) = 1.156\text{ N}\cdot\text{m} + (24.0\text{ N})(0.200\text{ m}) + 1.600\text{ N}\cdot\text{m}$$

$F = 63.0\text{ N}$ $\qquad$ $\mathbf{F} = 63.0\text{ N}\downarrow$ ◄

$\xrightarrow{+} \Sigma F_x = \Sigma(F_x)_{\text{eff}}$: $\qquad R_x = m_{OB}(\bar{a}_{OB})_t$

$R_x = 24.0\text{ N}$ $\qquad$ $\mathbf{R}_x = 24.0\text{ N}\rightarrow$ ◄

$+\uparrow \Sigma F_y = \Sigma(F_y)_{\text{eff}}$: $\qquad R_y - F - W_E - W_{OB} = m_{OB}(\bar{a}_{OB})_n$

$$R_y - 63.0\text{ N} - 39.2\text{ N} - 29.4\text{ N} = 38.4\text{ N}$$

$R_y = 170.0\text{ N}$ $\qquad$ $\mathbf{R}_y = 170.0\text{ N}\uparrow$ ◄

SAMPLE PROBLEM 16.7

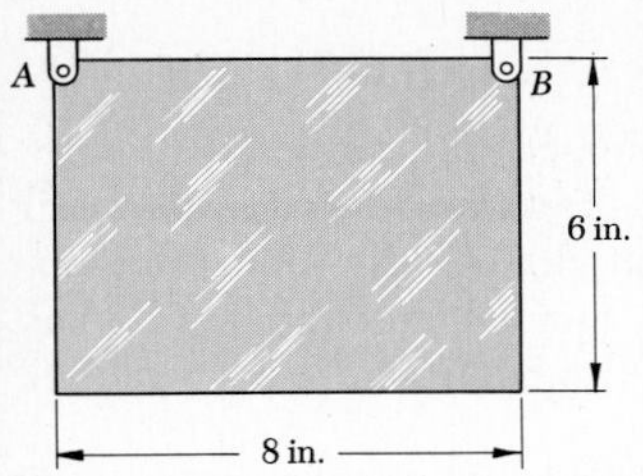

A rectangular plate, 6 by 8 in., weighs 60 lb and is suspended from two pins A and B. If pin B is suddenly removed, determine (*a*) the angular acceleration of the plate, (*b*) the components of the reactions at pin A, immediately after pin B has been removed.

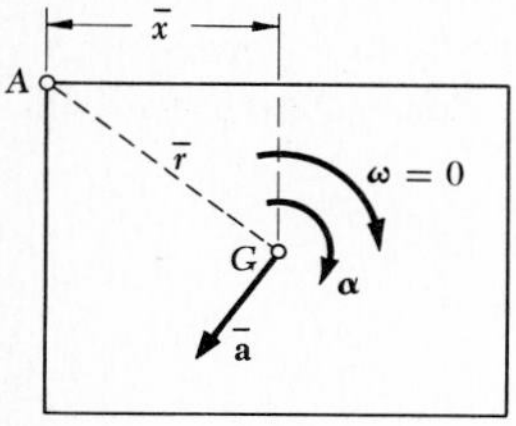

***a.* Angular Acceleration.** We observe that as the plate rotates about point A, its mass center G describes a circle of radius $\bar{r}$ with center at A.

Since the plate is released from rest ($\omega = 0$), the normal component of the acceleration of G is zero. The magnitude of the acceleration $\bar{\mathbf{a}}$ of the mass center G is thus $\bar{a} = \bar{r}\alpha$. We draw the diagram shown to express that the external forces are equivalent to the effective forces:

$$+\circlearrowright \Sigma M_A = \Sigma(M_A)_{\text{eff}}: \qquad W\bar{x} = (m\bar{a})\bar{r} + \bar{I}\alpha$$

Since $\bar{a} = \bar{r}\alpha$, we have

$$W\bar{x} = m(\bar{r}\alpha)\bar{r} + \bar{I}\alpha \qquad \alpha = \frac{W\bar{x}}{\frac{W}{g}\bar{r}^2 + \bar{I}} \tag{1}$$

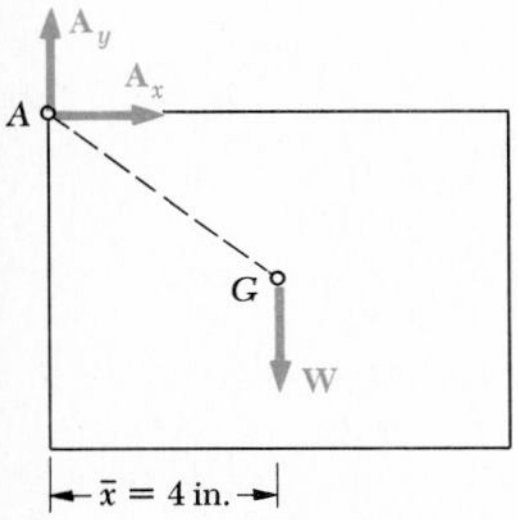

The centroidal moment of inertia of the plate is

$$\bar{I} = \frac{m}{12}(a^2 + b^2) = \frac{60 \text{ lb}}{12(32.2 \text{ ft/s}^2)}[(\tfrac{8}{12} \text{ ft})^2 + (\tfrac{6}{12} \text{ ft})^2]$$
$$= 0.1078 \text{ lb} \cdot \text{ft} \cdot \text{s}^2$$

Substituting this value of $\bar{I}$ together with $W = 60$ lb, $\bar{r} = \frac{5}{12}$ ft, and $\bar{x} = \frac{4}{12}$ ft into Eq. (1), we obtain

$$\alpha = +46.4 \text{ rad/s}^2 \qquad \alpha = 46.4 \text{ rad/s}^2 \circlearrowright \blacktriangleleft$$

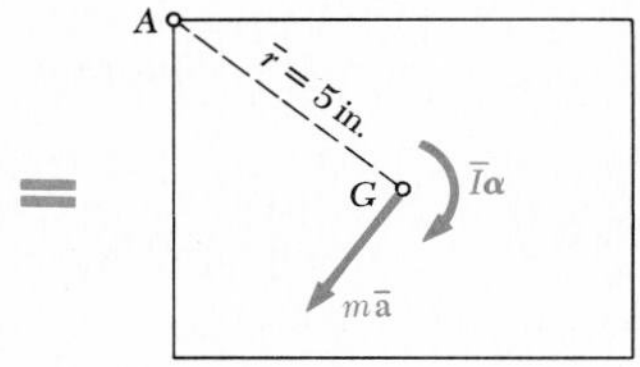

***b.* Reaction at *A*.** Using the computed value of α, we determine the magnitude of the vector $m\bar{\mathbf{a}}$ attached at G,

$$m\bar{a} = m\bar{r}\alpha = \frac{60 \text{ lb}}{32.2 \text{ ft/s}^2}(\tfrac{5}{12} \text{ ft})(46.4 \text{ rad/s}^2) = 36.0 \text{ lb}$$

Showing this result on the diagram, we write the equations of motion

$$\xrightarrow{+} \Sigma F_x = \Sigma(F_x)_{\text{eff}}: \qquad A_x = -\tfrac{3}{5}(36 \text{ lb})$$
$$A_x = -21.6 \text{ lb} \qquad A_x = 21.6 \text{ lb} \leftarrow \blacktriangleleft$$

$$+\uparrow \Sigma F_y = \Sigma(F_y)_{\text{eff}}: \qquad A_y - 60 \text{ lb} = -\tfrac{4}{5}(36 \text{ lb})$$
$$A_y = +31.2 \text{ lb} \qquad A_y = 31.2 \text{ lb} \uparrow \blacktriangleleft$$

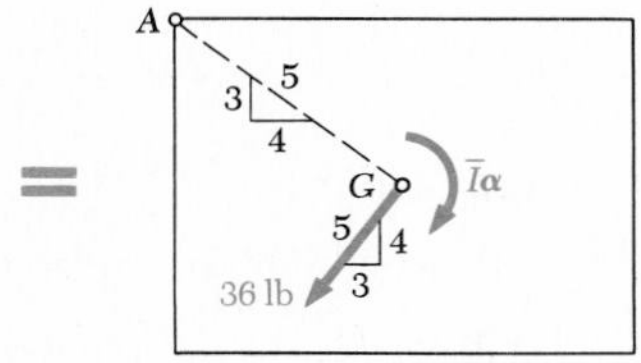

The couple $\bar{I}\boldsymbol{\alpha}$ is not involved in the last two equations; nevertheless, it should be indicated on the diagram.

SAMPLE PROBLEM 16.8

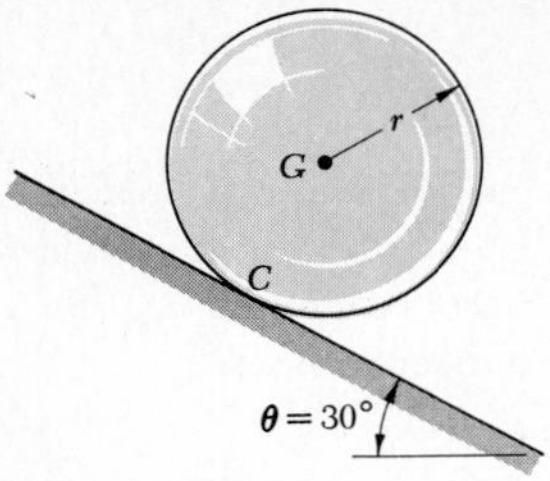

A sphere of radius r and weight W is released with no initial velocity on the incline and rolls without slipping. Determine (*a*) the minimum value of the coefficient of friction compatible with the rolling motion, (*b*) the velocity of the center G of the sphere after the sphere has rolled 10 ft, (*c*) the velocity of G if the sphere were to move 10 ft down a frictionless 30° incline.

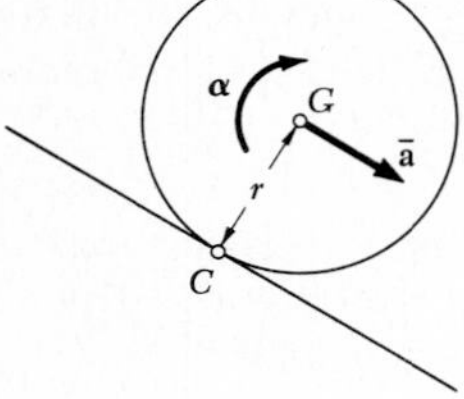

a. **Minimum μ for Rolling Motion.** The external forces **W**, **N**, and **F** form a system equivalent to the system of effective forces represented by the vector $m\bar{\mathbf{a}}$ and the couple $\bar{I}\boldsymbol{\alpha}$. Since the sphere rolls without sliding, we have $\bar{a} = r\alpha$.

$$+\circlearrowright \Sigma M_C = \Sigma (M_C)_{\text{eff}}: \qquad (W \sin\theta) r = (m\bar{a}) r + \bar{I}\alpha$$
$$(W \sin\theta) r = (mr\alpha) r + \bar{I}\alpha$$

Noting that $m = W/g$ and $\bar{I} = \frac{2}{5}mr^2$, we write

$$(W\sin\theta) r = \left(\frac{W}{g} r\alpha\right) r + \frac{2}{5}\frac{W}{g} r^2\alpha \qquad \alpha = +\frac{5g\sin\theta}{7r}$$

$$\bar{a} = r\alpha = \frac{5g\sin\theta}{7} = \frac{5(32.2 \text{ ft/s}^2)\sin 30°}{7} = 11.50 \text{ ft/s}^2$$

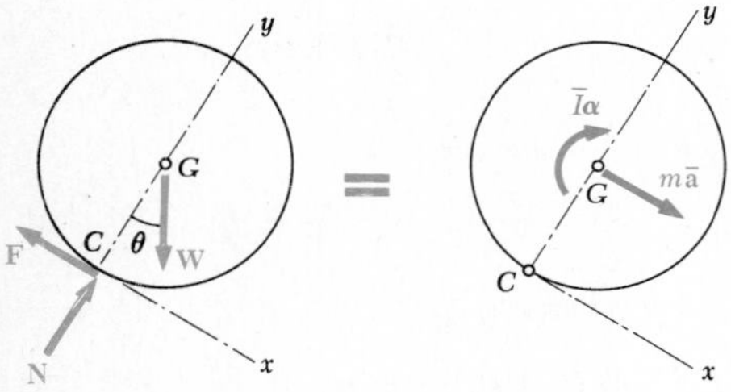

$$+\searrow \Sigma F_x = \Sigma (F_x)_{\text{eff}}: \qquad W\sin\theta - F = m\bar{a}$$
$$W\sin\theta - F = \frac{W}{g}\frac{5g\sin\theta}{7}$$
$$F = +\tfrac{2}{7}W\sin\theta = \tfrac{2}{7}W\sin 30° \qquad \mathbf{F} = 0.143W \angle 30°$$

$$+\nearrow \Sigma F_y = \Sigma (F_y)_{\text{eff}}: \qquad N - W\cos\theta = 0$$
$$N = W\cos\theta = 0.866W \qquad \mathbf{N} = 0.866W \angle 60°$$

$$\mu_{\min} = \frac{F}{N} = \frac{0.143W}{0.866W} \qquad \mu_{\min} = 0.165 \blacktriangleleft$$

b. **Velocity of Rolling Sphere.** We have uniformly accelerated motion,

$$\bar{v}_0 = 0 \qquad \bar{a} = 11.50 \text{ ft/s}^2 \qquad \bar{x} = 10 \text{ ft} \qquad \bar{x}_0 = 0$$
$$\bar{v}^2 = \bar{v}_0^2 + 2\bar{a}(\bar{x} - \bar{x}_0) \qquad \bar{v}^2 = 0 + 2(11.50 \text{ ft/s}^2)(10 \text{ ft})$$
$$\bar{v} = 15.17 \text{ ft/s} \qquad \bar{\mathbf{v}} = 15.17 \text{ ft/s} \searrow 30° \blacktriangleleft$$

c. **Velocity of Sliding Sphere.** Assuming now no friction, we have $F = 0$ and obtain

$$+\circlearrowright \Sigma M_G = \Sigma (M_G)_{\text{eff}}: \qquad 0 = \bar{I}\alpha \qquad \alpha = 0$$

$$+\searrow \Sigma F_x = \Sigma (F_x)_{\text{eff}}: \qquad W\sin 30° = m\bar{a} \qquad 0.50W = \frac{W}{g}\bar{a}$$
$$\bar{a} = +16.1 \text{ ft/s}^2 \qquad \bar{\mathbf{a}} = 16.1 \text{ ft/s}^2 \searrow 30°$$

Substituting $\bar{a} = 16.1 \text{ ft/s}^2$ into the equations for uniformly accelerated motion, we obtain

$$\bar{v}^2 = \bar{v}_0^2 + 2\bar{a}(\bar{x} - \bar{x}_0) \qquad \bar{v}^2 = 0 + 2(16.1 \text{ ft/s}^2)(10 \text{ ft})$$
$$\bar{v} = 17.94 \text{ ft/s} \qquad \bar{\mathbf{v}} = 17.94 \text{ ft/s} \searrow 30° \blacktriangleleft$$

SAMPLE PROBLEM 16.9

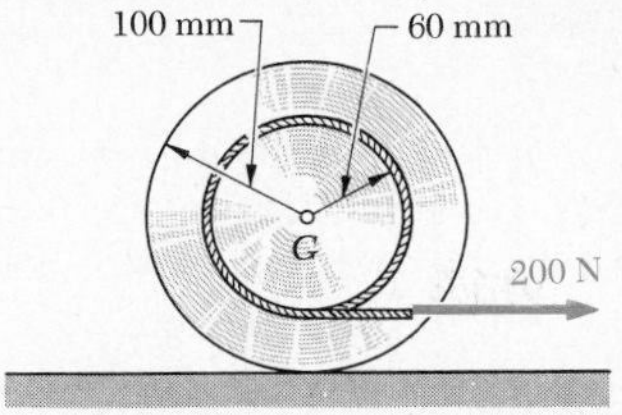

A cord is wrapped around the inner drum of a wheel and pulled horizontally with a force of 200 N. The wheel has a mass of 50 kg and a radius of gyration of 70 mm. Knowing that $\mu_s = 0.20$ and $\mu_k = 0.15$, determine the acceleration of G and the angular acceleration of the wheel.

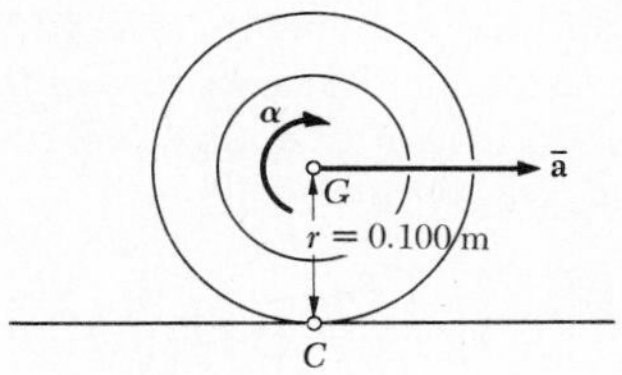

a. **Assume Rolling without Sliding.** In this case, we have

$$\bar{a} = r\alpha = (0.100 \text{ m})\alpha$$

By comparing the friction force obtained with the maximum available friction force, we shall determine whether this assumption is justified. The moment of inertia of the wheel is

$$\bar{I} = m\bar{k}^2 = (50 \text{ kg})(0.070 \text{ m})^2 = 0.245 \text{ kg} \cdot \text{m}^2$$

Equations of Motion

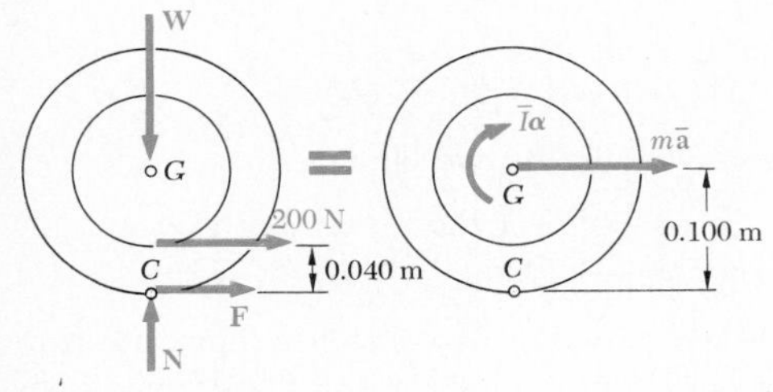

$$+\curvearrowright \Sigma M_C = \Sigma(M_C)_{\text{eff}}: \qquad (200 \text{ N})(0.040 \text{ m}) = m\bar{a}(0.100 \text{ m}) + \bar{I}\alpha$$
$$8.00 \text{ N} \cdot \text{m} = (50 \text{ kg})(0.100 \text{ m})\alpha(0.100 \text{ m}) + (0.245 \text{ kg} \cdot \text{m}^2)\alpha$$
$$\alpha = +10.74 \text{ rad/s}^2$$
$$\bar{a} = r\alpha = (0.100 \text{ m})(10.74 \text{ rad/s}^2) = 1.074 \text{ m/s}^2$$

$$\overset{+}{\rightarrow} \Sigma F_x = \Sigma(F_x)_{\text{eff}}: \qquad F + 200 \text{ N} = m\bar{a}$$
$$F + 200 \text{ N} = (50 \text{ kg})(1.074 \text{ m/s}^2)$$
$$F = -146.3 \text{ N} \qquad \mathbf{F} = 146.3 \text{ N} \leftarrow$$

$$+\uparrow \Sigma F_y = \Sigma(F_y)_{\text{eff}}:$$
$$N - W = 0 \qquad N = W = mg = (50 \text{ kg})(9.81 \text{ m/s}^2) = 490.5 \text{ N}$$
$$\mathbf{N} = 490.5 \text{ N} \uparrow$$

Maximum Available Friction Force

$$F_{\text{max}} = \mu_s N = 0.20\,(490.5 \text{ N}) = 98.1 \text{ N}$$

Since $F > F_{\text{max}}$, the assumed motion is impossible.

b. **Rolling and Sliding.** Since the wheel must roll and slide at the same time, we draw a new diagram, where $\bar{\mathbf{a}}$ and $\boldsymbol{\alpha}$ are independent and where

$$F = F_k = \mu_k N = 0.15\,(490.5 \text{ N}) = 73.6 \text{ N}$$

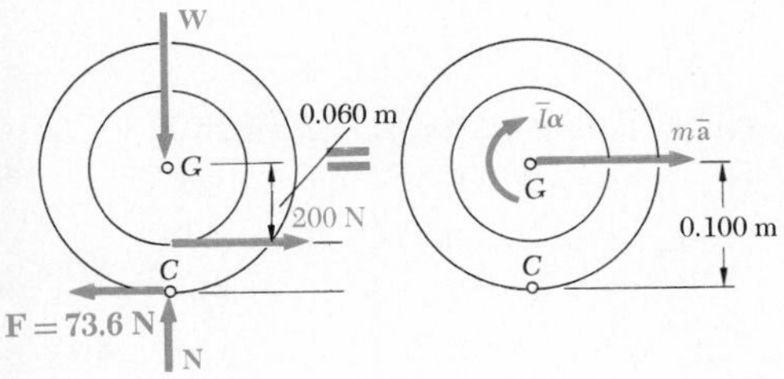

From the computation of part *a*, it appears that **F** should be directed to the left. We write the following equations of motion:

$$\overset{+}{\rightarrow} \Sigma F_x = \Sigma(F_x)_{\text{eff}}: \qquad 200 \text{ N} - 73.6 \text{ N} = (50 \text{ kg})\bar{a}$$
$$\bar{a} = +2.53 \text{ m/s}^2 \qquad \bar{\mathbf{a}} = 2.53 \text{ m/s}^2 \rightarrow \blacktriangleleft$$

$$+\curvearrowright \Sigma M_G = \Sigma(M_G)_{\text{eff}}:$$
$$(73.6 \text{ N})(0.100 \text{ m}) - (200 \text{ N})(0.060 \text{ m}) = (0.245 \text{ kg} \cdot \text{m}^2)\alpha$$
$$\alpha = -18.94 \text{ rad/s}^2 \qquad \boldsymbol{\alpha} = 18.94 \text{ rad/s}^2 \circlearrowleft \blacktriangleleft$$

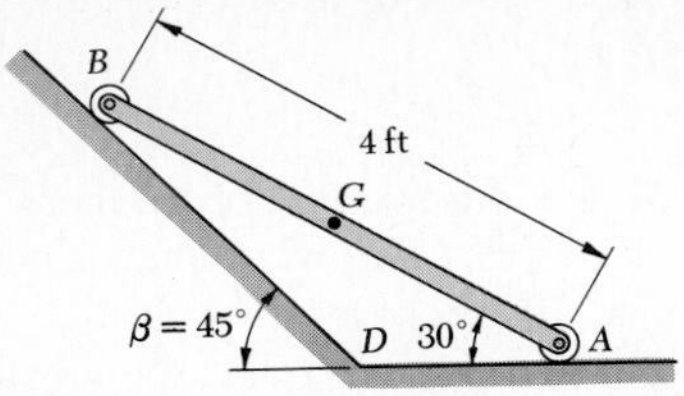

SAMPLE PROBLEM 16.10

The extremities of a 4-ft rod, weighing 50 lb, may move freely and with no friction along two straight tracks as shown. If the rod is released with no velocity from the position shown, determine (*a*) the angular acceleration of the rod, (*b*) the reactions at *A* and *B*.

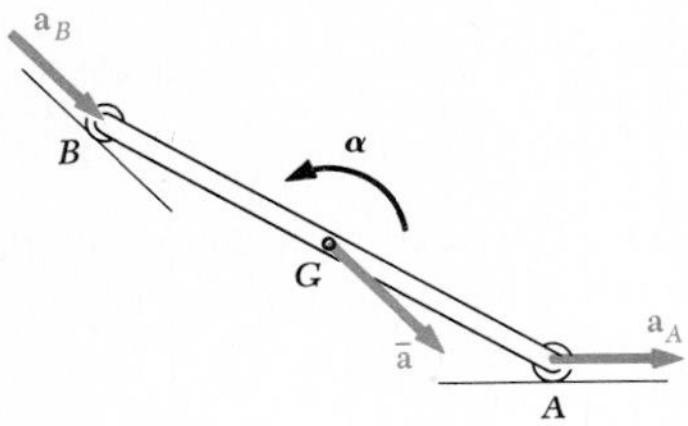

Kinematics of Motion. Since the motion is constrained, the acceleration of *G* must be related to the angular acceleration $\boldsymbol{\alpha}$. To obtain this relation, we shall first determine the magnitude of the acceleration $\mathbf{a}_A$ of point *A* in terms of α; assuming $\boldsymbol{\alpha}$ directed counterclockwise and noting that $a_{B/A} = 4\alpha$, we write

$$\mathbf{a}_B = \mathbf{a}_A + \mathbf{a}_{B/A}$$
$$[a_B \searrow 45°] = [a_A \rightarrow] + [4\alpha \nearrow 60°]$$

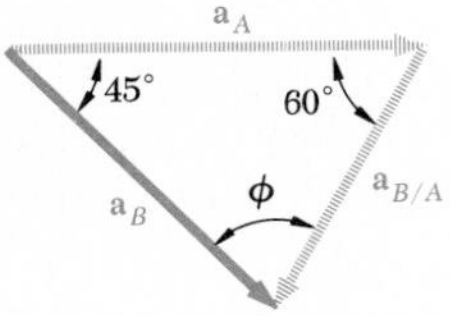

Noting that $\phi = 75°$ and using the law of sines, we obtain

$$a_A = 5.46\alpha \qquad a_B = 4.90\alpha$$

The acceleration of *G* is now obtained by writing

$$\bar{\mathbf{a}} = \mathbf{a}_G = \mathbf{a}_A + \mathbf{a}_{G/A}$$
$$\bar{\mathbf{a}} = [5.46\alpha \rightarrow] + [2\alpha \nearrow 60°]$$

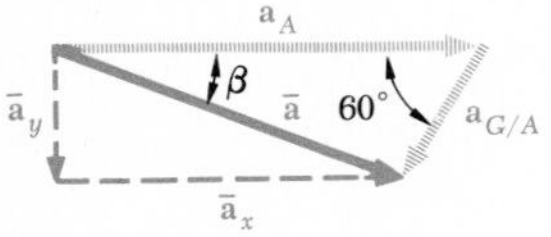

Resolving $\bar{\mathbf{a}}$ into *x* and *y* components, we obtain

$$\bar{a}_x = 5.46\alpha - 2\alpha \cos 60° = 4.46\alpha \qquad \bar{\mathbf{a}}_x = 4.46\alpha \rightarrow$$
$$\bar{a}_y = -2\alpha \sin 60° = -1.732\alpha \qquad \bar{\mathbf{a}}_y = 1.732\alpha \downarrow$$

Kinetics of Motion. We draw the two sketches shown to express that the system of external forces is equivalent to the system of effective forces represented by the vector of components $m\bar{\mathbf{a}}_x$ and $m\bar{\mathbf{a}}_y$ attached at *G* and the couple $\bar{I}\boldsymbol{\alpha}$. We compute the following magnitudes:

$$\bar{I} = \frac{1}{12}ml^2 = \frac{1}{12}\frac{50\text{ lb}}{32.2\text{ ft/s}^2}(4\text{ ft})^2 = 2.07\text{ lb}\cdot\text{ft}\cdot\text{s}^2 \qquad \bar{I}\alpha = 2.07\alpha$$

$$m\bar{a}_x = \frac{50}{32.2}(4.46\alpha) = 6.93\alpha \qquad m\bar{a}_y = -\frac{50}{32.2}(1.732\alpha) = -2.69\alpha$$

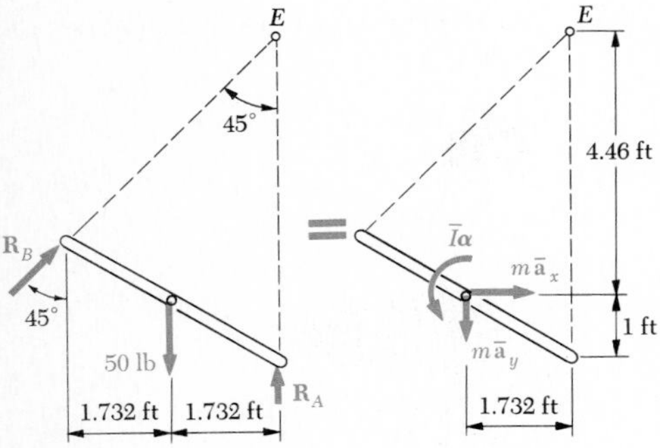

Equations of Motion

$+\circlearrowleft \Sigma M_E = \Sigma(M_E)_{\text{eff}}$:

$$(50)(1.732) = (6.93\alpha)(4.46) + (2.69\alpha)(1.732) + 2.07\alpha$$
$$\alpha = +2.30\text{ rad/s}^2 \qquad \boldsymbol{\alpha} = 2.30\text{ rad/s}^2 \circlearrowleft \blacktriangleleft$$

$\xrightarrow{+} \Sigma F_x = \Sigma(F_x)_{\text{eff}}$:

$$R_B \sin 45° = (6.93)(2.30) = 15.94$$
$$R_B = 22.5\text{ lb} \qquad \mathbf{R}_B = 22.5\text{ lb} \angle 45° \blacktriangleleft$$

$+\uparrow \Sigma F_y = \Sigma(F_y)_{\text{eff}}$:

$$R_A + R_B \cos 45° - 50 = -(2.69)(2.30)$$
$$R_A = -6.19 - 15.94 + 50 = 27.9\text{ lb} \qquad \mathbf{R}_A = 27.9\text{ lb} \uparrow \blacktriangleleft$$

PROBLEMS

16.65 Show that the couple $\bar{I}\boldsymbol{\alpha}$ of Fig. 16.14 may be eliminated by attaching the vectors $m\bar{\mathbf{a}}_t$ and $m\bar{\mathbf{a}}_n$ at a point P called the *center of percussion*, located on line OG at a distance $GP = \bar{k}^2/\bar{r}$ from the mass center of the body.

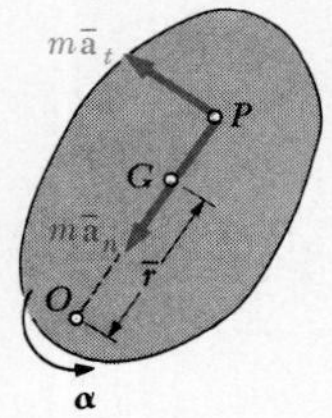

Fig. P16.65

16.66 A uniform slender rod, of length $L = 900$ mm and mass $m = 4$ kg, is supported as shown. A horizontal force **P** of magnitude 75 N is applied at end B. For $\bar{r} = \frac{1}{4}L = 225$ mm, determine (*a*) the angular acceleration of the rod, (*b*) the components of the reaction at C.

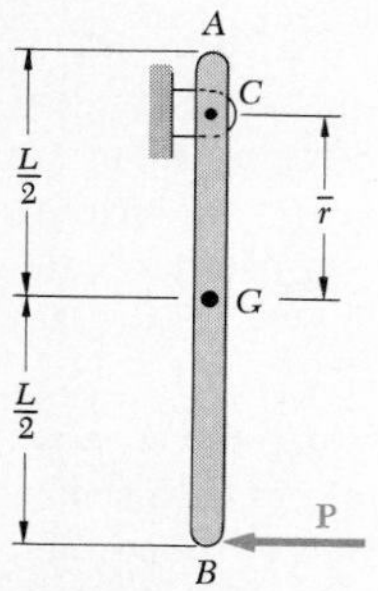

Fig. P16.66

16.67 In Prob. 16.66, determine (*a*) the distance $\bar{r}$ for which the horizontal component of the reaction at C is zero, (*b*) the corresponding angular acceleration of the rod.

16.68 A uniform slender rod, of length L and weight W, hangs freely from a hinge at A. A horizontal force **P** is applied as shown. For $h = L$, determine (*a*) the angular acceleration of the rod, (*b*) the components of the reaction at A.

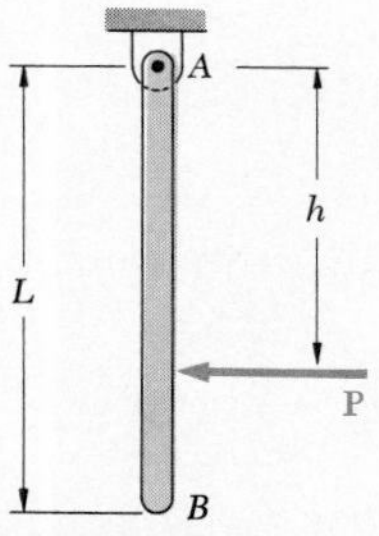

Fig. P16.68

16.69 In Prob. 16.68, determine (*a*) the distance h for which the horizontal component of the reaction at A is zero, (*b*) the corresponding angular acceleration of the rod.

16.70 A turbine disk of mass 75 kg rotates at a constant speed of 9600 rpm; the mass center of the disk coincides with the center of rotation O. Determine the reaction at O after a single vane at A, of mass 45 g, becomes loose and is thrown off.

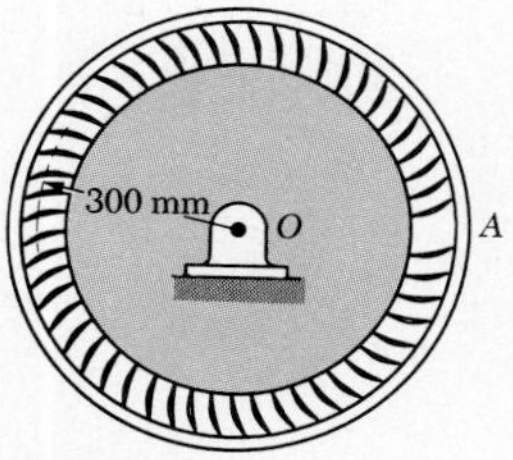

Fig. P16.70

16.71 A uniform slender rod of length l and mass m rotates about a vertical axis AA' at a constant angular velocity ω. Determine the tension in the rod at a distance x from the axis of rotation.

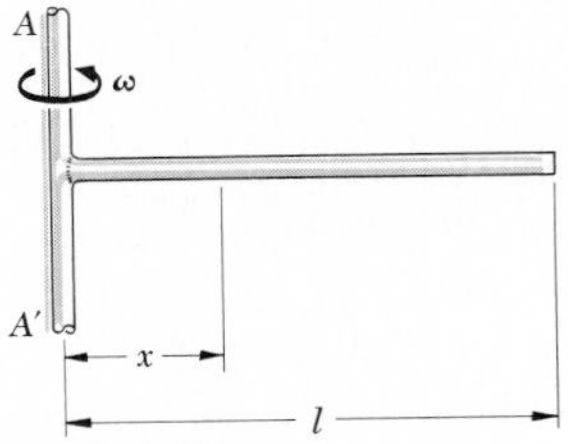

Fig. P16.71

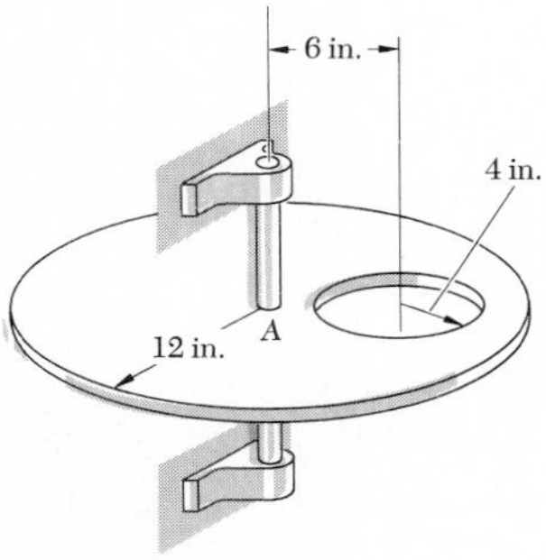

Fig. P16.72

16.72 An 8-in.-diameter hole is cut as shown in a thin disk of diameter 24 in. The disk rotates in a horizontal plane about its geometric center A at a constant angular velocity of 480 rpm. Knowing that the disk weighs 100 lb after the hole has been cut, determine the horizontal component of the force exerted by the shaft on the disk at A.

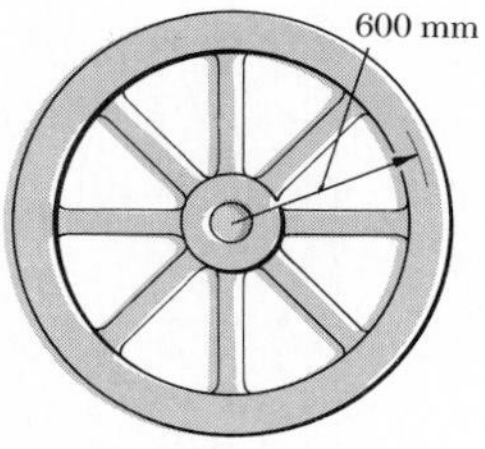

Fig. P16.73

16.73 The rim of a flywheel has a mass of 1800 kg and a mean radius of 600 mm. As the flywheel rotates at a constant angular velocity of 360 rpm, radial forces are exerted on the rim by the spokes and internal forces are developed within the rim. Neglecting the weight of the spokes, determine (*a*) the internal forces in the rim, assuming the radial forces exerted by the spokes to be zero, (*b*) the radial force exerted by each spoke, assuming the tangential forces in the rim to be zero.

16.74 and 16.75 A uniform beam of length L and weight W is supported as shown. If the cable attached at B suddenly breaks, determine (*a*) the reaction at the pin support, (*b*) the acceleration of point B.

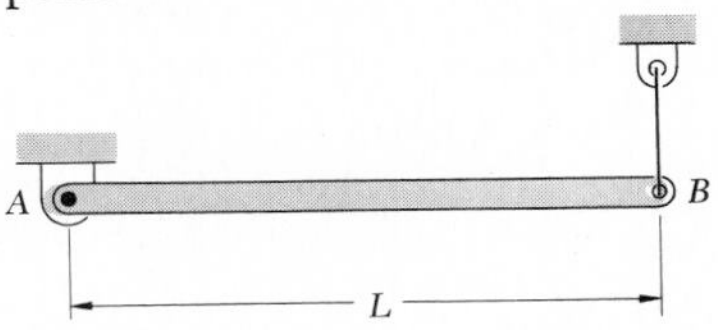

Fig. P16.74

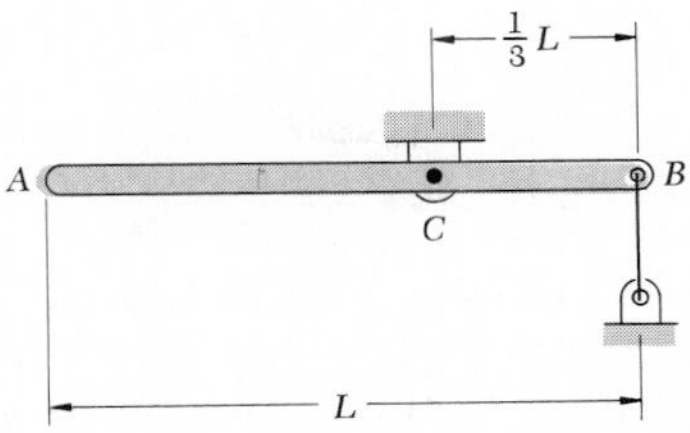

Fig. P16.75

16.76 A 12-ft plank, weighing 80 lb, rests on two horizontal pipes *AB* and *CD* of a scaffolding. The pipes are 10 ft apart, and the plank overhangs 1 ft at each end. A man weighing 160 lb is standing on the plank when pipe *CD* suddenly breaks. Determine the initial acceleration of the man, knowing that he was standing (*a*) in the middle of the plank, (*b*) just above pipe *CD*.

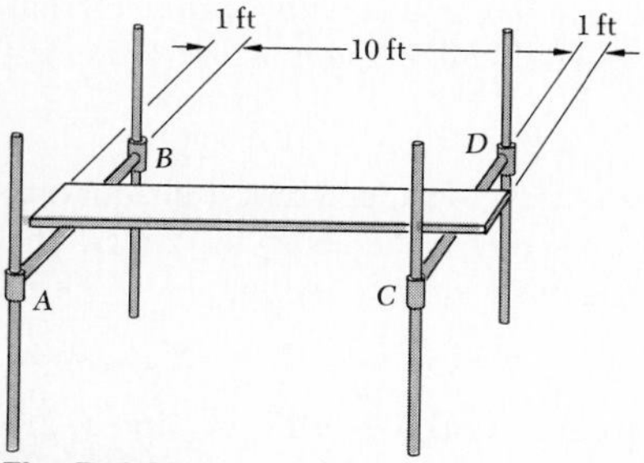

Fig. P16.76

16.77 A 2-kg slender rod is riveted to a 4-kg uniform disk as shown. The assembly swings freely in a vertical plane and, in the position shown, has an angular velocity of 4 rad/s clockwise. Determine (*a*) the angular acceleration of the assembly, (*b*) the components of the reaction at *A*.

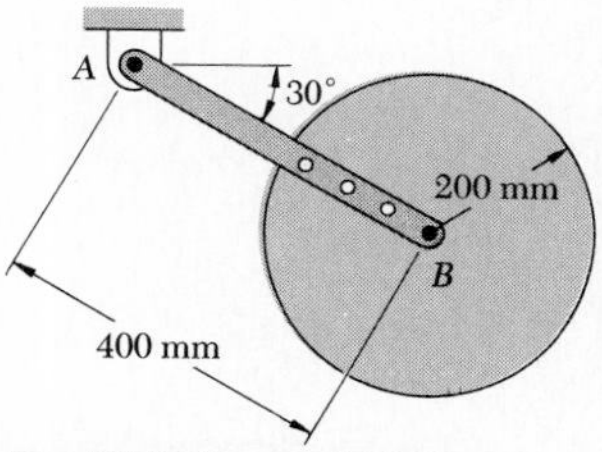

Fig. P16.77 and P16.78

16.78 A 2-kg slender rod is riveted to a 4-kg uniform disk as shown. The assembly rotates in a vertical plane under the combined effect of gravity and a couple **M** which is applied to rod *AB*. Knowing that at the instant shown the assembly has an angular velocity of 6 rad/s and an angular acceleration of 10 rad/s^2 both counterclockwise, determine (*a*) the moment of the couple **M**, (*b*) the components of the reaction at *A*.

16.79 After being released, the plate of Sample Prob. 16.7 is allowed to swing through 90°. Knowing that at that instant the angular velocity of the plate is 4.82 rad/s, determine (*a*) the angular acceleration of the plate, (*b*) the reaction at *A*.

16.80 Two uniform rods, *AB* of weight 6 lb and *CD* of weight 8 lb, are welded together to form the T-shaped assembly shown. The assembly rotates in a vertical plane about a horizontal shaft at *A*. Knowing that at the instant shown the assembly has a clockwise angular velocity of 8 rad/s and a counterclockwise angular acceleration of 24 rad/s^2, determine (*a*) the magnitude of the horizontal force **P**, (*b*) the components of the reaction at *A*.

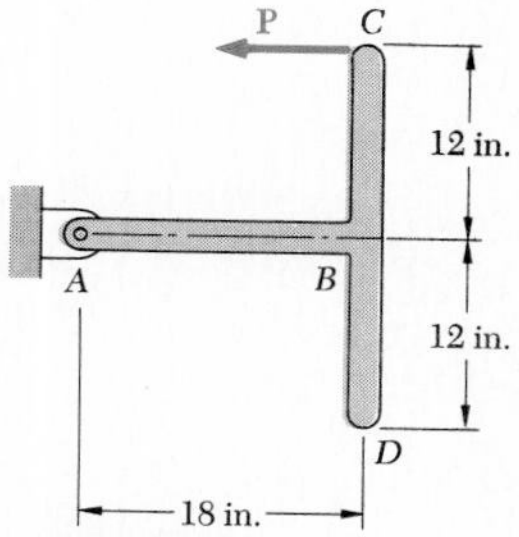

Fig. P16.80

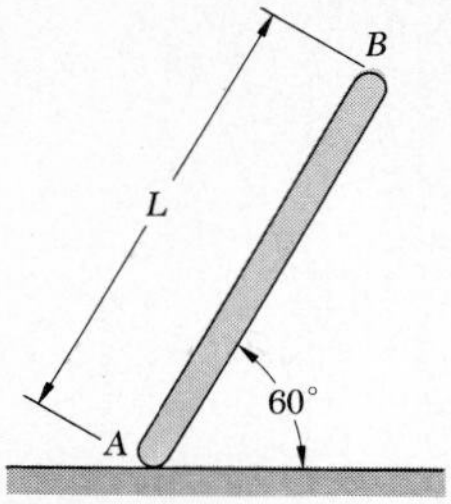

Fig. P16.81

16.81 A uniform rod of weight W is released from rest in the position shown. Assuming that the friction between end A and the surface is large enough to prevent sliding, determine (*a*) the angular acceleration of the rod just after release, (*b*) the normal reaction and the friction force at A, (*c*) the minimum value of μ compatible with the described motion.

16.82 Derive the equation $\Sigma M_C = I_C\alpha$ for the rolling disk of Fig. 16.16, where ΣM_C represents the sum of the moments of the external forces about the instantaneous center C and I_C the moment of inertia of the disk about C.

16.83 Show that, in the case of an unbalanced disk, the equation derived in Prob. 16.82 is valid only when the mass center G, the geometric center O, and the instantaneous center C happen to lie in a straight line.

16.84 A homogeneous cylinder C and a section of pipe P are in contact when they are released from rest. Knowing that both the cylinder and the pipe roll without slipping, determine the clear distance between them after 3 s.

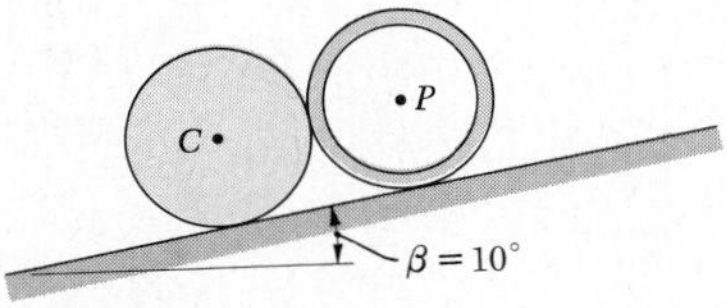

Fig. P16.84

16.85 A flywheel is rigidly attached to a shaft of 40-mm radius which may roll along parallel rails as shown. When released from rest, the system rolls a distance of 3 m in 30 s. Determine the centroidal radius of gyration of the system.

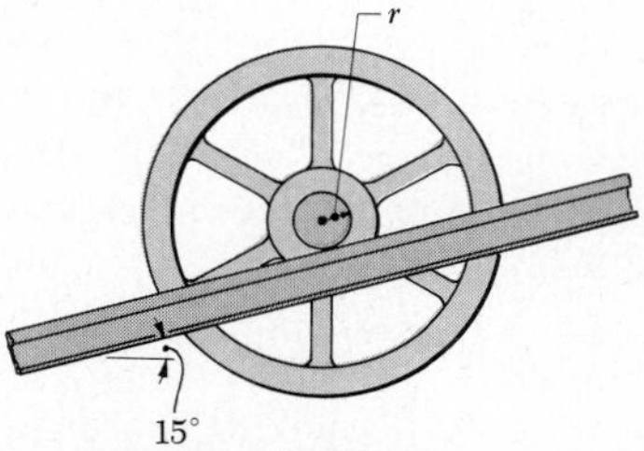

Fig. P16.85 and P16.86

16.86 A flywheel of centroidal radius of gyration $\bar{k} = 600$ mm is rigidly attached to a shaft of radius $r = 30$ mm which may roll along parallel rails. Knowing that the system is released from rest, determine the distance it will roll in 20 s.

16.87 through 16.90 A drum of 80-mm radius is attached to a disk of 160-mm radius. The disk and drum have a total mass of 5 kg and a radius of gyration of 120 mm. A cord is attached as shown and pulled with a force **P** of magnitude 20 N. Knowing that the disk rolls without sliding, determine (*a*) the angular acceleration of the disk and the acceleration of *G*, (*b*) the minimum value of the coefficient of friction compatible with this motion.

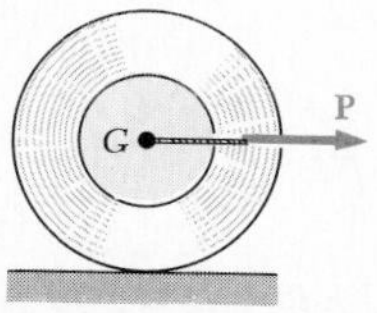

Fig. P16.87 and P16.91

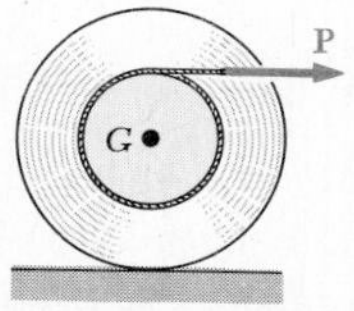

Fig. P16.88 and P16.92

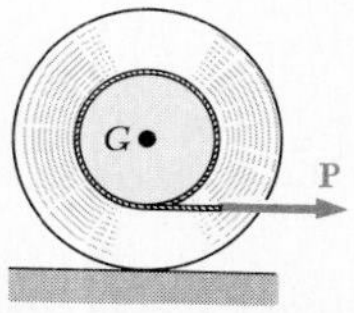

Fig. P16.89 and P16.93

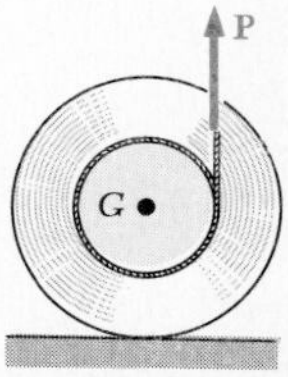

Fig. P16.90 and P16.94

16.91 through 16.94 A drum of 4-in. radius is attached to a disk of 8-in. radius. The disk and drum have a total weight of 10 lb and a radius of gyration of 6 in. A cord is attached as shown and pulled with a force **P** of magnitude 5 lb. Knowing that $\mu = 0.20$, determine (*a*) whether or not the disk slides, (*b*) the angular acceleration of the disk and the acceleration of *G*.

16.95 The cord attached to the drum of Prob. 16.90 is pulled to the right in a direction forming an angle β with the horizontal and with a force **P** of magnitude 20 N. Determine the range of values of β for which the disk (*a*) rolls to the right, (*b*) rolls to the left. Assume that μ is large enough to prevent sliding.

16.96 and 16.97 The 15-lb carriage is supported as shown by two uniform disks each of weight 10 lb and radius 3 in. Knowing that the disks roll without sliding, determine the acceleration of the carriage when a force of 5 lb is applied to it.

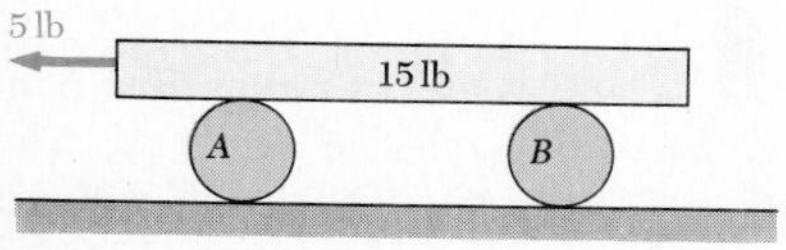

Fig. P16.96

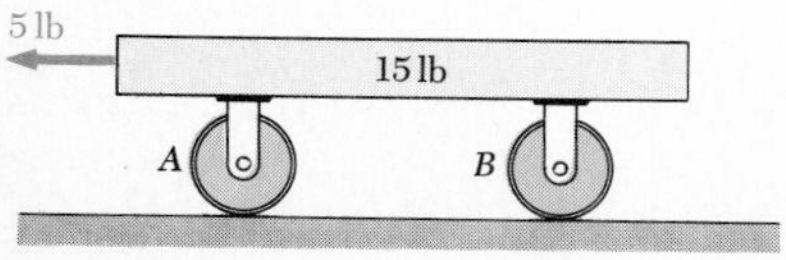

Fig. P16.97

16.98 A block B of mass m is attached to a cord wrapped around a cylinder of the same mass m and of radius r. The cylinder rolls without sliding on a horizontal surface. Determine the components of the accelerations of the center A of the cylinder and of the block B immediately after the system has been released from rest if (*a*) the block hangs freely, (*b*) the motion of the block is guided by a rigid member DAE, frictionless and of negligible mass, which is hinged to the cylinder at A.

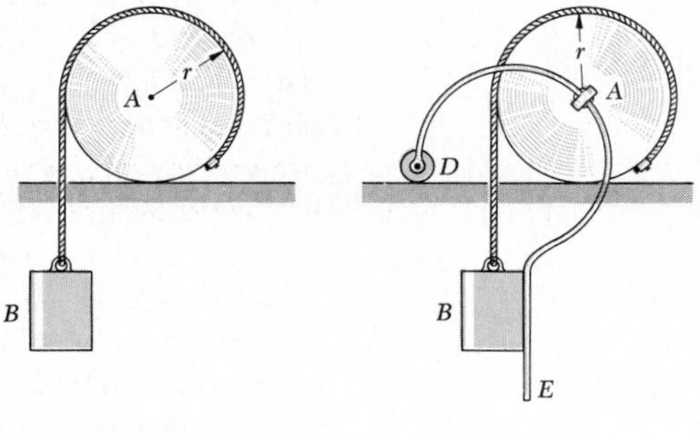

Fig. P16.98

16.99 A small block of mass m is attached at B to a hoop of mass m and radius r. Knowing that when the system is released from rest it starts to roll without sliding, determine (*a*) the angular acceleration of the hoop, (*b*) the acceleration of B.

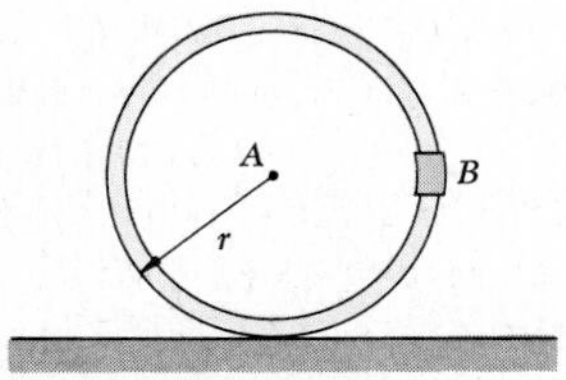

Fig. P16.99

16.100 A half section of pipe of mass m and radius r rests on a rough horizontal surface. A vertical force **P** is applied as shown. Assuming that the section rolls without sliding, derive an expression (*a*) for its angular acceleration, (*b*) for the minimum value of μ compatible with this motion. [*Hint.* Note that $OG = 2r/\pi$ and that, by the parallel-axis theorem, $\bar{I} = mr^2 - m(OG)^2$.]

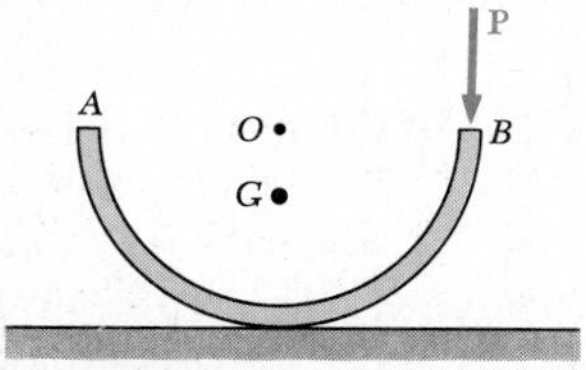

Fig. P16.100

16.101 Solve Prob. 16.100, assuming that the force **P** is applied at B and is directed horizontally to the right.

16.102 The mass center G of a 4-lb wheel of radius $R = 6$ in. is located at a distance $r = 2$ in. from its geometric center C. The centroidal radius of gyration is $\bar{k} = 3$ in. As the wheel rolls without sliding, its angular velocity varies and it is observed that $\omega = 12$ rad/s in the position shown. Determine the corresponding angular acceleration of the wheel.

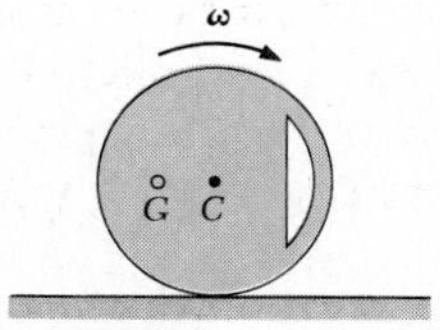

Fig. P16.102

B=10.8 ←

P=36.3 N ↑ A=21.7 ↑

16.103 The motion of a 4-kg slender rod is guided by pins at A and B which slide freely in the slots shown. A vertical force **P** is applied at B, causing the rod to start from rest with a counterclockwise angular acceleration of 12 rad/s². Determine (*a*) the magnitude and direction of **P**, (*b*) the reactions at A and B.

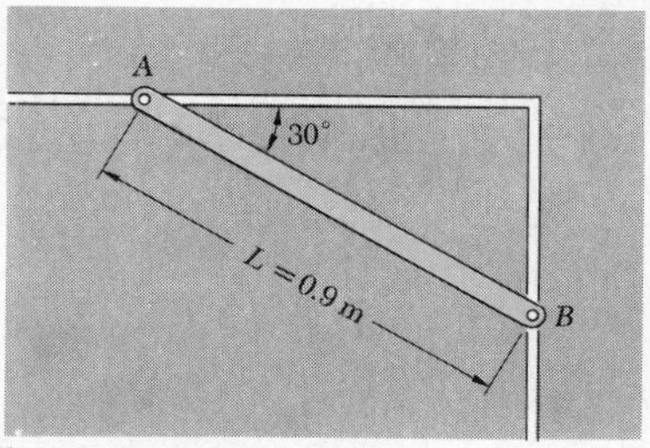

Fig. P16.103 and P16.104

16.104 The motion of a 4-kg slender rod is guided by pins at A and B which slide freely in the slots shown. A couple **M** is applied to the rod which is initially at rest. Knowing that the initial acceleration of point B is 6 m/s² upward, determine (*a*) the moment of the couple, (*b*) the reactions at A and B.

16.105 End A of the 100-lb beam AB moves along the frictionless floor, while end B is supported by a 4-ft cable. Knowing that at the instant shown end A is moving to the left with a constant velocity of 8 ft/s, determine (*a*) the magnitude of the force **P**, (*b*) the corresponding tension in the cable.

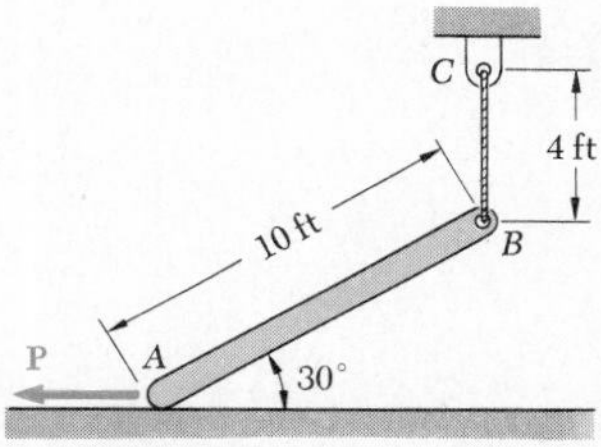

Fig. P16.105

16.106 Solve Prob. 16.105, assuming that at the instant shown end A moves with a velocity of 8 ft/s and an acceleration of 5 ft/s², both directed to the left.

16.107 The rod of Prob. 16.103 is released from rest in the position shown. Determine (*a*) the angular acceleration of the rod, (*b*) the reactions at A and B.

16.108 The 50-kg uniform rod AB is released from rest in the position shown. Knowing that end A may slide freely on the frictionless floor, determine (*a*) the angular acceleration of the rod, (*b*) the tension in wire BC, (*c*) the reaction at A.

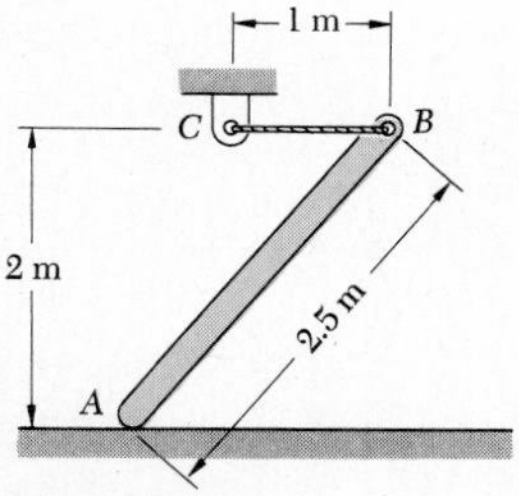

Fig. P16.108

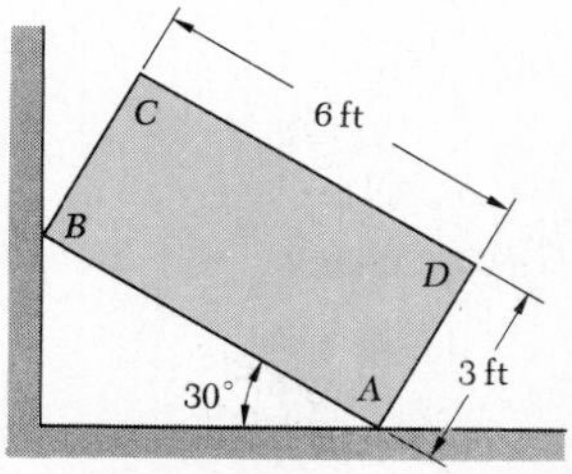

Fig. P16.109

16.109 A 3- by 6-ft crate weighing 200 lb is released from rest in the position shown. Assuming corners A and B slide without friction, determine (*a*) the angular acceleration of the crate, (*b*) the reactions at A and B.

16.110 Two uniform rods AB and BC, each of weight 10 lb, are welded together to form a single 20-lb rigid body which is held by the three wires shown. Determine the tension in wires BD and CF immediately after wire CE has been cut.

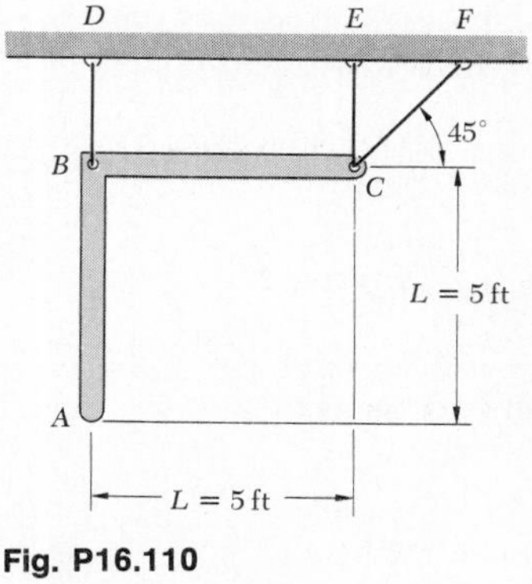

Fig. P16.110

16.111 Show that, for a rigid slab in plane motion, the equation $\Sigma M_A = I_A\alpha$, where ΣM_A represents the sum of the moments of the external forces about point A and I_A the moment of inertia of the slab about the same point A, is verified *if and only if* one of the following conditions is satisfied: (*a*) A is the mass center of the slab, (*b*) A has zero acceleration, (*c*) the acceleration of A is directed along a line joining point A and the mass center G.

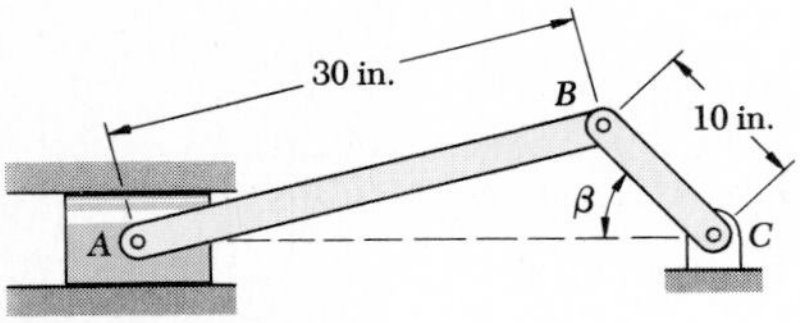

Fig. P16.112

16.112 The 10-lb sliding block is attached to the rotating crank BC by the uniform connecting rod AB which weighs 5 lb. Knowing that crank BC has a constant angular velocity of 240 rpm clockwise, determine the forces exerted on the connecting rod at A and B when $\beta = 0$.

16.113 Solve Prob. 16.112 when $\beta = 180°$.

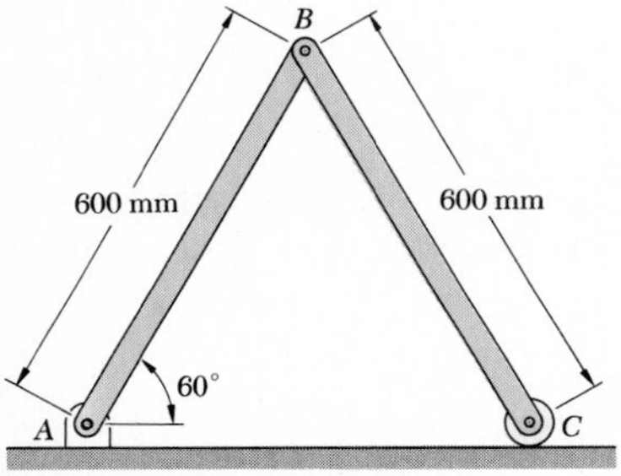

Fig. P16.114 and P16.115

16.114 Each of the bars shown is 600 mm long and has a mass of 4 kg. A horizontal and variable force **P** is applied at C, causing point C to move to the left with a constant speed of 10 m/s. Determine the force **P** for the position shown.

16.115 The two bars AB and BC are released from rest in the position shown. Each bar is 600 mm long and has a mass of 4 kg. Determine (*a*) the angular acceleration of each bar, (*b*) the reactions at A and C.

16.116 and 16.117 Two rods AB and BC, of mass m per unit length, are connected as shown to a disk which is made to rotate in a vertical plane at a constant angular velocity ω_0. For the position shown, determine the components of the forces exerted at A and B on rod AB.

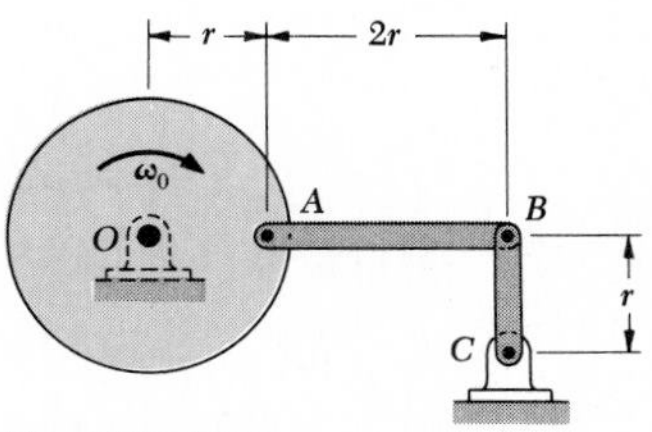

Fig. P16.116

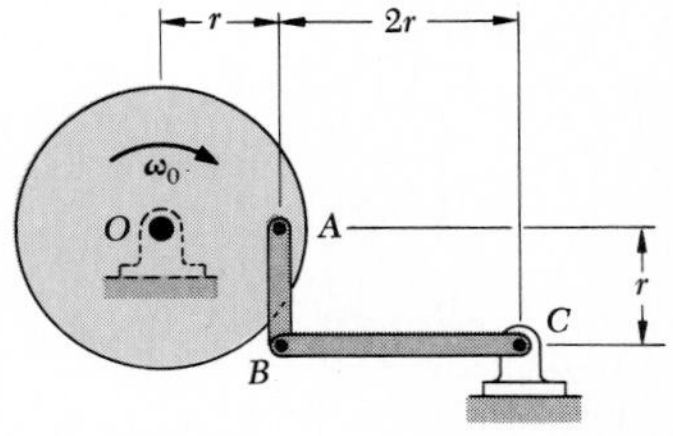

Fig. P16.117

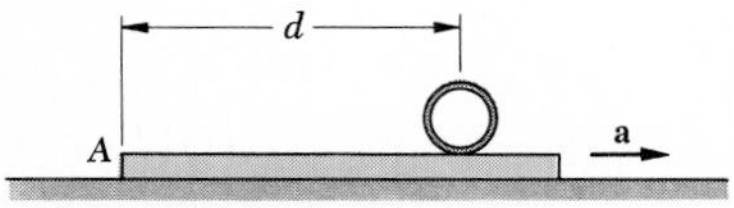

Fig. P16.118 and P16.119

16.118 A section of pipe rests on a plate. The plate is then given a constant acceleration **a** directed to the right. Assuming that the pipe rolls on the plate, determine (*a*) the acceleration of the pipe, (*b*) the distance through which the plate will move before the pipe reaches end A.

16.119 A section of pipe rests on a plate. Knowing that $\mu = 0.40$ between the plate and the pipe, determine the largest acceleration **a** of the plate for which the pipe rolls without sliding.

16.120 and 16.121 Gear C weighs 6 lb and has a centroidal radius of gyration of 3 in. The uniform bar AB weighs 5 lb and gear D is stationary. If the system is released from rest in the position shown, determine (*a*) the angular acceleration of gear C, (*b*) the acceleration of point B.

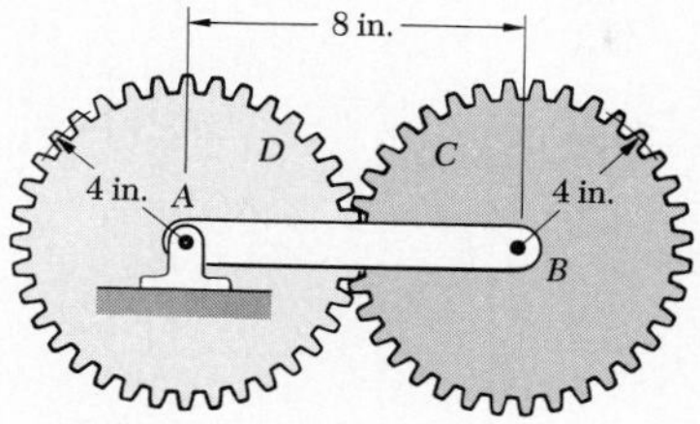

Fig. P16.120

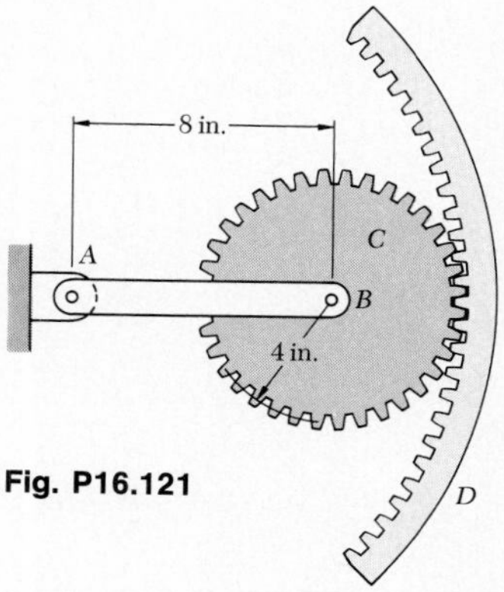

Fig. P16.121

16.122 The rectangular plate of Sample Prob. 16.7, which weighs 60 lb, is suspended from two inextensible wires as shown. If the wire at B is cut, determine at that instant (*a*) the angular acceleration of the plate, (*b*) the tension in the wire at A.

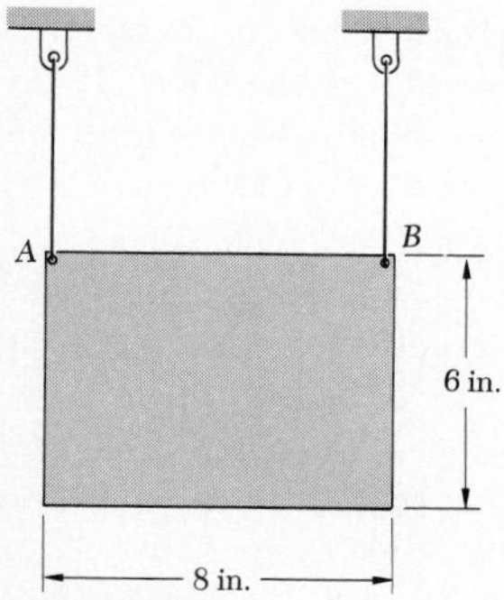

Fig. P16.122

***16.123** A uniform slender rod of length L and mass m is released from rest in the position shown. Derive an expression for (*a*) the angular acceleration of the rod, (*b*) the acceleration of end A, (*c*) the reaction at A, immediately after release. Neglect the mass and friction of the roller at A.

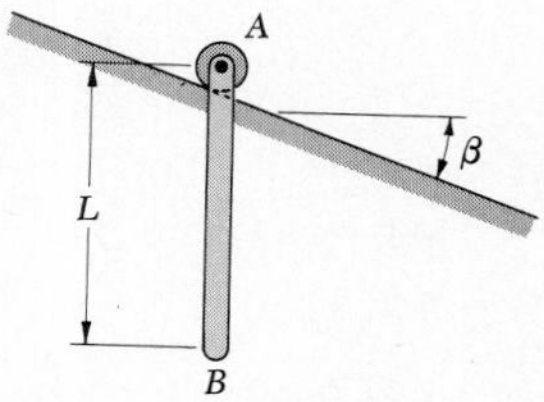

Fig. P16.123 and P16.124

***16.124** A uniform rod AB, of mass 3 kg and length $L = 1.2$ m, is released from rest in the position shown. Knowing that $\beta = 30°$, determine the values immediately after release of (*a*) the angular acceleration of the rod, (*b*) the acceleration of end A, (*c*) the reaction at A. Neglect the mass and friction of the roller at A.

∗16.125 Each of the bars AB and BC is of length $L = 2$ ft and weight 5 lb. A couple $\mathbf{M}_0$ of moment 6 lb · ft is applied to bar BC. Determine the angular acceleration of each bar.

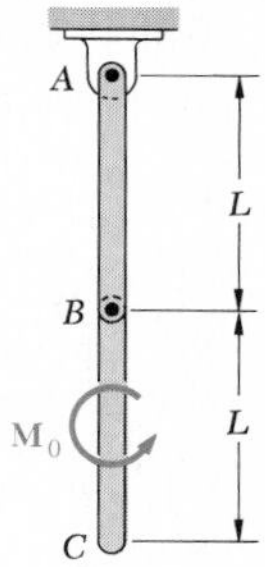

Fig. P16.125

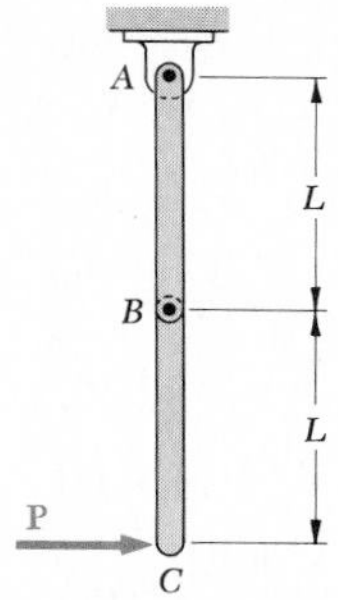

Fig. P16.126

∗16.126 Each of the bars AB and BC is of length $L = 2$ ft and weight 5 lb. A horizontal force $\mathbf{P}$ of magnitude 3 lb is applied at C. Determine the angular acceleration of each bar.

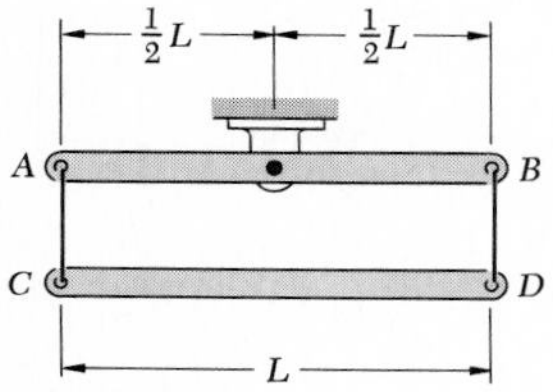

Fig. P16.127

∗16.127 Two uniform slender rods AB and CD, each of mass m, are connected by two wires as shown. If the wire BD is cut, determine at that instant the acceleration (*a*) of point D, (*b*) of point B.

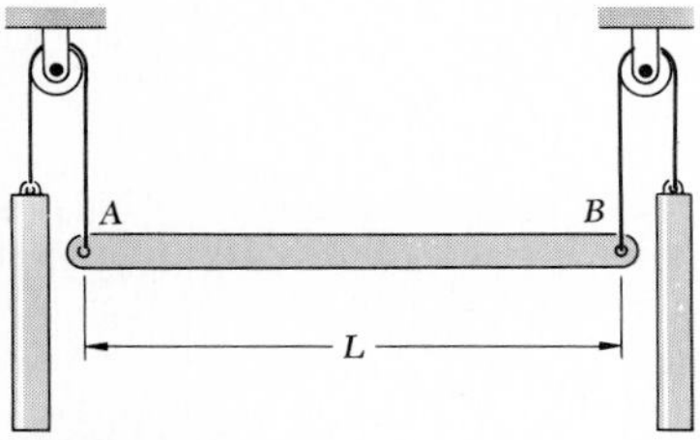

Fig. P16.128

∗16.128 The slender bar AB is of length L and mass m. It is held in equilibrium by two counterweights, each of mass $\frac{1}{2}m$. If the wire at B is cut, determine at that instant the acceleration of (*a*) point A, (*b*) point B.

∗16.129 Two uniform rods, each of weight W, are released from rest in the position shown. Determine the reaction at A immediately after release.

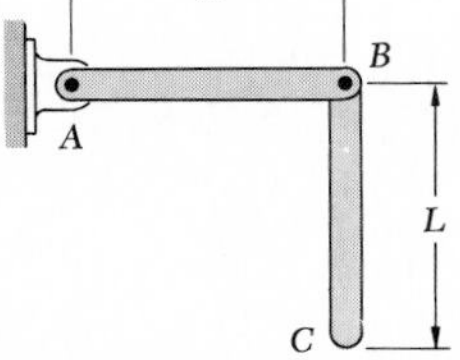

Fig. P16.129

∗16.130 Solve Prob. 16.129, assuming that the connection at B is welded and that the rods move as a single rigid body. Also, determine the corresponding bending moment at B.

∗16.131 In Prob. 16.125 the pin at B is severely rusted and the bars rotate as a single rigid body. Determine the bending moment which occurs at B.

∗16.132 (*a*) Determine the magnitude and the location of the maximum bending moment in the rod of Prob. 16.68. (*b*) Show that the answer to part *a* is independent of the weight W of the rod.

∗16.133 Draw the shear and bending-moment diagrams for the beam of Prob. 16.74 immediately after the cable at B breaks.

∗16.134 Draw the shear and bending-moment diagrams for the bar of Prob. 16.128 immediately after the wire at B has been cut.

16.6. Rotation of a Three-dimensional Body about a Fixed Axis. The rigid bodies considered in the preceding sections were assumed to be symmetrical with respect to the reference plane used to describe their motion. While such an assumption considerably simplifies the derivations involved, it also limits the range of application of the results obtained. One of the most important types of motion which *cannot* be analyzed directly by the methods considered so far is the rotation about a fixed axis of an unsymmetrical three-dimensional body.

We may, however, examine the particular case when the rotating body consists of several parallel slabs rigidly connected to a perpendicular shaft which is supported in bearings at A and B (Fig. 16.18). Since each slab rotates in its own plane, we may apply the method of Sec. 16.2 and reduce the effective forces of each slab to a force-couple system. By D'Alembert's principle the system of the external forces shown in part a of Fig. 16.18 must be equivalent to the system of the effective forces represented by the various force-couple systems shown in part b of the same figure. Equating components along, and moments about, each of the three coordinate axes, we write the following six equations:

$$\begin{aligned} \Sigma F_x &= \Sigma(F_x)_{\text{eff}} & \Sigma F_y &= \Sigma(F_y)_{\text{eff}} & \Sigma F_z &= \Sigma(F_z)_{\text{eff}} \\ \Sigma M_x &= \Sigma(M_x)_{\text{eff}} & \Sigma M_y &= \Sigma(M_y)_{\text{eff}} & \Sigma M_z &= \Sigma(M_z)_{\text{eff}} \end{aligned} \tag{16.17}$$

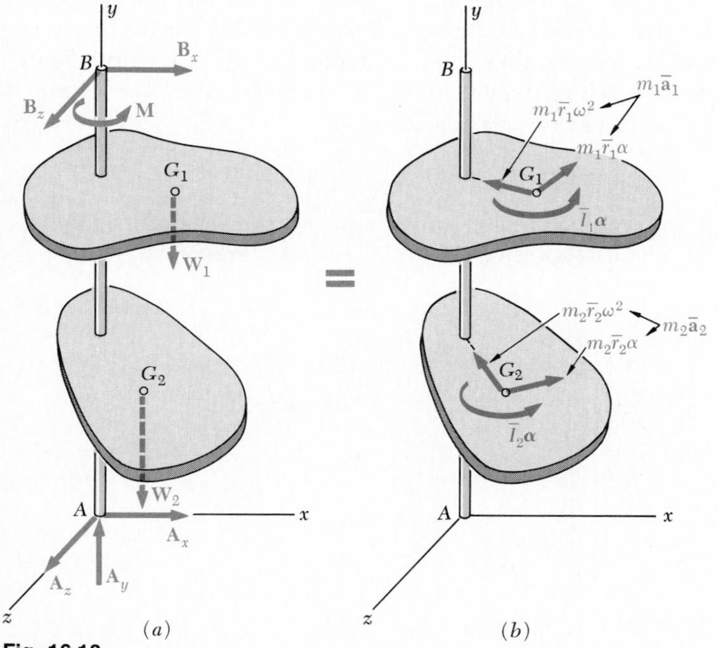

Fig. 16.18

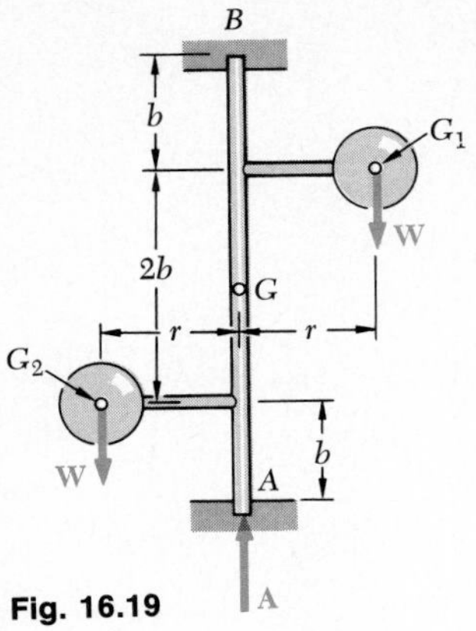

Fig. 16.19

It should be noted that this method may also be used when the bodies attached to the shaft are three-dimensional, *as long as each body is symmetrical with respect to a plane perpendicular to the shaft.*

As an example, we shall consider a shaft held in bearings at A and B and supporting two equal spheres of mass m and weight $W = mg$ (Fig. 16.19). We first observe that when the system is at rest, the shaft exerts no lateral thrust on its supports since the center of gravity G of the system is located directly above A. The system is said to be *statically balanced.* The reaction **A**, often referred to as a *static reaction,* is vertical, and its magnitude is $A = 2W = 2mg$. Let us now assume that the system rotates with a constant angular velocity $\boldsymbol{\omega}$. We shall apply D'Alembert's principle and express the fact that the system of the external forces, consisting of the weights and bearing reactions, is equivalent to the system of the effective forces. Since each sphere is symmetrical with respect to a plane perpendicular to the shaft and containing its own center, we may apply the method of Sec. 16.2 and replace the effective forces of each sphere by a force-couple system at its mass center. But in the present case the angular acceleration $\boldsymbol{\alpha}$ is zero and the force-couple systems reduce respectively to the vectors $m\bar{\mathbf{a}}_1$ and $m\bar{\mathbf{a}}_2$ directed toward the axis of rotation. Since these two vectors, shown in part b of Fig. 16.20, have the same magnitude $mr\omega^2$ and are located at a distance $2b$ from each other, we conclude that the system of the effective forces is equivalent to a counterclockwise couple of moment $2bmr\omega^2$. Additional bearing reactions $\mathbf{A}_x$ and $\mathbf{B}_x$, therefore, are required in part a of the figure. These reactions, called *dynamic reactions,* must also form a counterclockwise couple of moment $2bmr\omega^2$. They must be

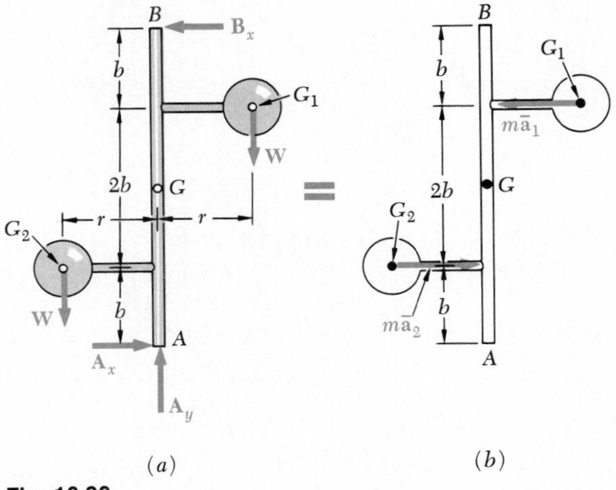

Fig. 16.20

directed as shown and have a magnitude $A_x = B_x = \frac{1}{2}mr\omega^2$. Although the system is statically balanced, it is clearly *not dynamically balanced.* A system which is not dynamically balanced will have a tendency to tear away from its bearings when rotating at high speed, as indicated by the values obtained for the bearing reactions. Moreover, since the bearing reactions are contained in the plane defined by the shaft and the centers of the spheres, they rotate with the shaft and cause the structure supporting it to vibrate. *Rotating systems should therefore be dynamically as well as statically balanced.* This may be done by rearranging the distribution of mass around the shaft, or by adding corrective masses, until the system of the effective forces becomes equivalent to zero.

The analysis of the rotating shaft of Figs. 16.19 and 16.20 could also be carried out by the method of dynamic equilibrium. The effective forces shown on the right-hand side of the equals sign in Fig. 16.20 would then be replaced by equal and opposite inertia vectors on the left-hand side of the equals sign (Fig. 16.21). The problem is thus reduced to a statics problem, in which the inertia vectors represent the "centrifugal forces" which tend to "pull" the spheres away from the shaft.

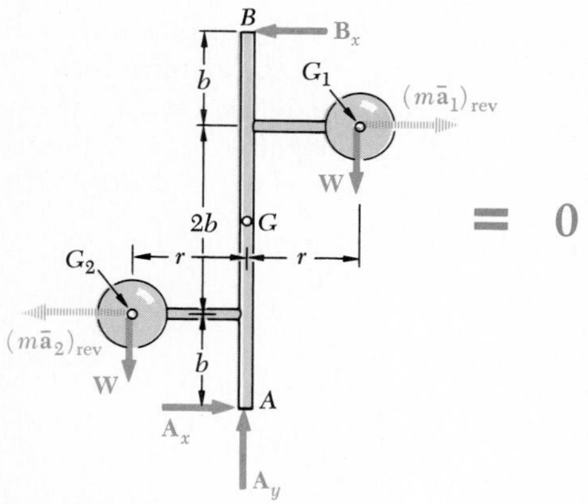

Fig. 16.21

In the general case of the rotation about a fixed axis of an unsymmetrical three-dimensional body, a typical element of mass should be isolated and its effective force should be determined. The sums of the components and moments of the effective forces may then be obtained by integration. While this approach is successfully used in Sample Prob. 16.12, it will lead, in general, to fairly involved computations. Other methods, beyond the scope of this chapter, may then be preferred. The approach indicated here shows clearly, however, that, even in as simple a motion as uniform rotation, the system of the effective forces *will not reduce to a single vector attached at the mass center of the rotating body* if the body does not possess a plane of symmetry perpendicular to the axis of rotation.

Consider, for example, an airplane making a turn of radius $\bar{r}$ in the banked position shown in Fig. 16.22. Since two symmetrical wing elements A and B are not at the same distance from

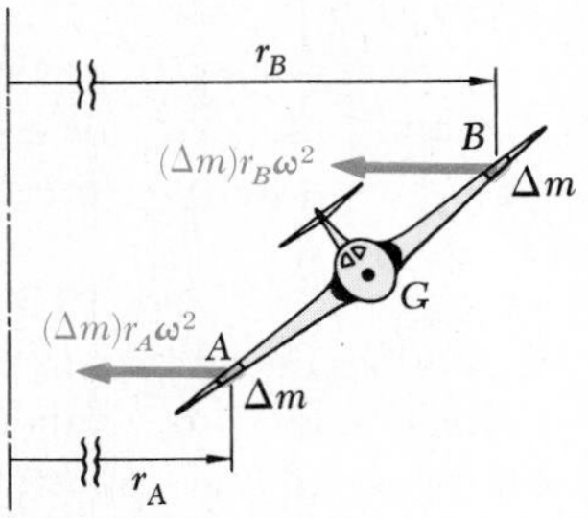

Fig. 16.22

the axis of rotation, the effective forces associated with A and B are not equal. The system of the effective forces corresponding to the various elements of the airplane, therefore, will *not* reduce to a single vector at the mass center G of the airplane. It will reduce to a force-couple system at G which, in turn, may be replaced by a single vector located above G (Fig. 16.23). However, since the difference between the magnitudes of the effective forces corresponding to A and B is relatively small, the

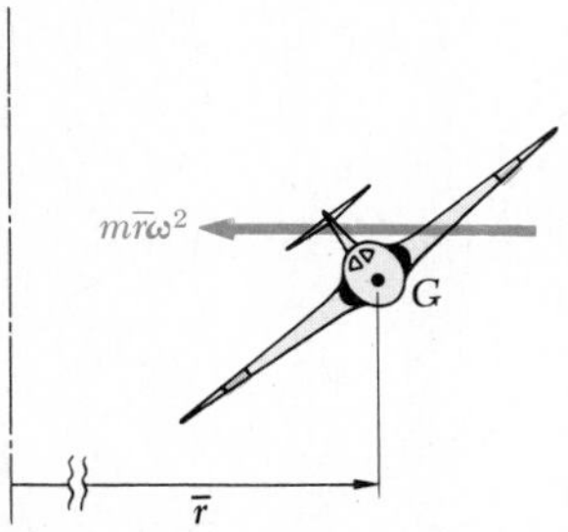

Fig. 16.23

distance from G to the resultant of the effective forces of the airplane is also relatively small. It is therefore customary in such problems, and in problems dealing with the motion of cars along banked highway curves, to neglect the effect of the nonsymmetry of the three-dimensional body on the position of the resultant of its effective forces.

SAMPLE PROBLEM 16.11

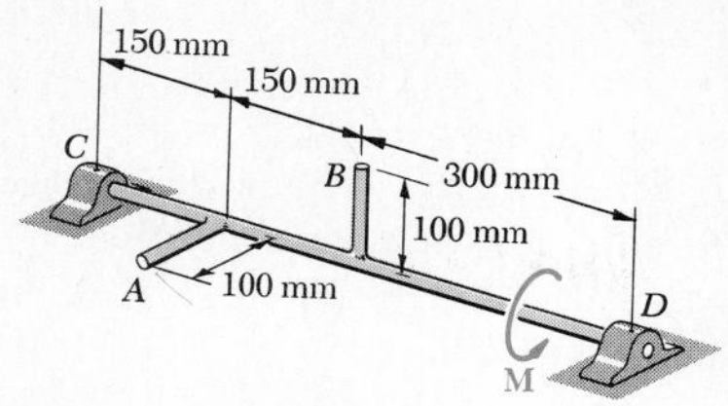

Two 100-mm rods A and B, each of mass 300 g, are welded to the shaft CD which is supported by bearings at C and D. If a couple $\mathbf{M}$ of moment equal to $6\ \text{N}\cdot\text{m}$ is applied to the shaft, determine the components of the dynamic reactions at C and D at the instant when the shaft has reached an angular velocity of 1200 rpm. Neglect the moment of inertia of the shaft itself.

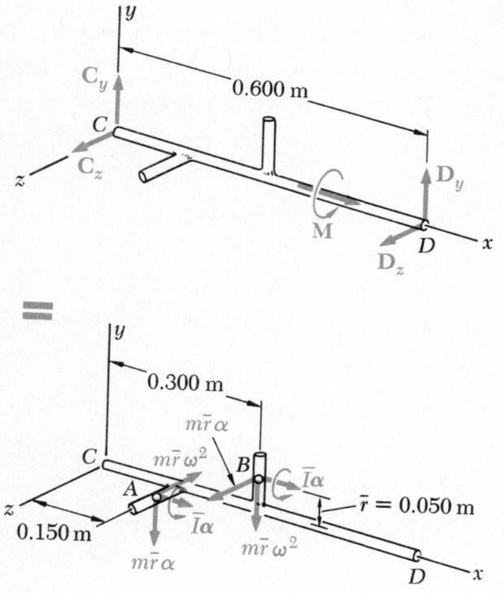

Solution. The external forces consist of the dynamic reactions at C and D. (The weights and the corresponding static reactions are ignored since they cancel each other.) The system of the effective forces of each rod reduces to a vector $m\bar{\mathbf{a}}$ at the mass center of the rod and a couple $\bar{I}\boldsymbol{\alpha}$ of axis parallel to the shaft. Each vector $m\bar{\mathbf{a}}$ is resolved into a tangential component of magnitude $m\bar{r}\alpha$ and a normal component of magnitude $m\bar{r}\omega^2$. The centroidal moment of inertia of each rod is

$$\bar{I} = \tfrac{1}{12}mL^2 = \tfrac{1}{12}(0.300\ \text{kg})(0.100\ \text{m})^2 = 2.5 \times 10^{-4}\ \text{kg}\cdot\text{m}^2$$

Angular Acceleration of Shaft. Equating the moments about the x axis of the external and effective forces, we first determine the angular acceleration of the shaft. The moments of the components of the vectors $m\bar{\mathbf{a}}$, as well as the moments of the couples $\bar{I}\boldsymbol{\alpha}$, must be taken into account.

$$+\circlearrowleft \Sigma M_x = \Sigma(M_x)_{\text{eff}}: \qquad M = 2(m\bar{r}\alpha)\bar{r} + 2\bar{I}\alpha$$
$$6\ \text{N}\cdot\text{m} = 2(0.300\ \text{kg})(0.050\ \text{m})^2\alpha + 2(2.5 \times 10^{-4}\ \text{kg}\cdot\text{m}^2)\alpha$$
$$\alpha = 3000\ \text{rad/s}^2$$

Dynamic Reactions. Using the value obtained for α and noting that $\omega = 1200\ \text{rpm} = 125.7\ \text{rad/s}$, we compute the components of the vectors $m\bar{\mathbf{a}}$:

$$m\bar{r}\alpha = (0.300\ \text{kg})(0.050\ \text{m})(3000\ \text{rad/s}^2) = 45\ \text{N}$$
$$m\bar{r}\omega^2 = (0.300\ \text{kg})(0.050\ \text{m})(125.7\ \text{rad/s})^2 = 237\ \text{N}$$

Equating successively the moments about and the components along the y and z axes of the external and effective forces, we write

$$+\circlearrowleft \Sigma M_y = \Sigma(M_y)_{\text{eff}}:$$
$$-(0.600\ \text{m})D_z = (0.150\ \text{m})(m\bar{r}\omega^2) - (0.300\ \text{m})(m\bar{r}\alpha)$$
$$-(0.600\ \text{m})D_z = (0.150\ \text{m})(237\ \text{N}) - (0.300\ \text{m})(45\ \text{N})$$
$$D_z = -36.8\ \text{N} \quad \blacktriangleleft$$

$$+\circlearrowleft \Sigma M_z = \Sigma(M_z)_{\text{eff}}:$$
$$(0.600\ \text{m})D_y = -(0.150\ \text{m})(m\bar{r}\alpha) - (0.300\ \text{m})(m\bar{r}\omega^2)$$
$$(0.600\ \text{m})D_y = -(0.150\ \text{m})(45\ \text{N}) - (0.300\ \text{m})(237\ \text{N})$$
$$D_y = -129.8\ \text{N} \quad \blacktriangleleft$$

$$\Sigma F_y = \Sigma(F_y)_{\text{eff}}: \qquad C_y + D_y = -m\bar{r}\alpha - m\bar{r}\omega^2$$
$$C_y - 129.8\ \text{N} = -45\ \text{N} - 237\ \text{N} \qquad C_y = -152.2\ \text{N} \quad \blacktriangleleft$$

$$\Sigma F_z = \Sigma(F_z)_{\text{eff}}: \qquad C_z + D_z = -m\bar{r}\omega^2 + m\bar{r}\alpha$$
$$C_z - 36.8\ \text{N} = -237\ \text{N} + 45\ \text{N} \qquad C_z = -155.2\ \text{N} \quad \blacktriangleleft$$

SAMPLE PROBLEM 16.12

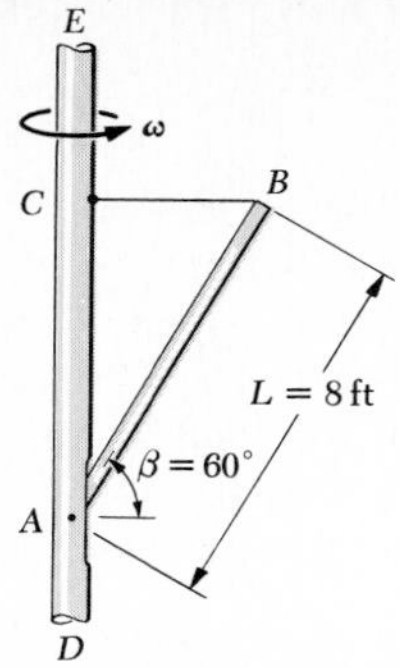

A slender rod AB of length $L = 8$ ft and weight $W = 40$ lb is pinned at A to a vertical axle DE which rotates with a constant angular velocity $\omega = 15$ rad/s. The rod is maintained in position by means of a wire BC attached to the axle and to the end B of the rod. Determine the tension in the wire BC.

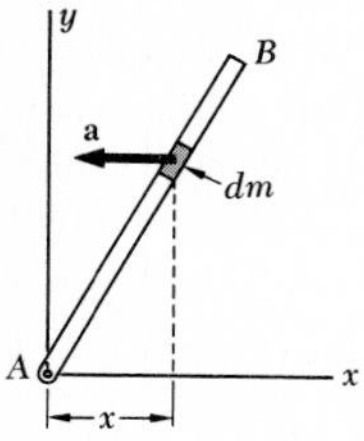

Kinematics of Motion. Choosing a coordinate system with origin at point A, we observe that each element dm of the rod moves in a horizontal circle of radius x. Each element is thus in plane motion and has an acceleration **a** of magnitude $a = x\omega^2$ directed horizontally toward the y axis.

Kinetics of Motion. By D'Alembert's principle, the system of the external forces, which consists of the weight **W**, the force **T** exerted by the wire at B and the components $\mathbf{A}_x$ and $\mathbf{A}_y$ of the reaction at A, must be equivalent to the system of the effective forces $\mathbf{a}\,dm$ of the various elements. We write

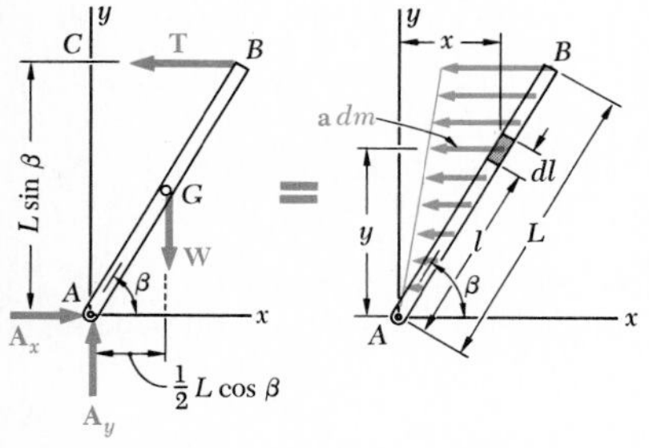

$$+\circlearrowleft \Sigma M_A = \Sigma (M_A)_{\text{eff}}: \qquad T(L \sin \beta) - W(\tfrac{1}{2}L \cos \beta) = \int y(a\,dm) \qquad (1)$$

Denoting by l the distance from A to the element dm, we have

$$y = l \sin \beta \qquad a = x\omega^2 = (l \cos \beta)\omega^2 \qquad dm = \frac{m}{L}\,dl$$

Substituting these expressions into the integral and integrating from $l = 0$ to $l = L$, we obtain

$$\int y(a\,dm) = \int_0^L (l \sin \beta)(l \cos \beta)\omega^2 \left(\frac{m}{L}\,dl\right) = \tfrac{1}{3}mL^2\omega^2 \sin \beta \cos \beta \qquad (2)$$

Substituting this expression into (1), we have

$$T(L \sin \beta) - W(\tfrac{1}{2}L \cos \beta) = \tfrac{1}{3}mL^2\omega^2 \sin \beta \cos \beta$$

$$T = \tfrac{1}{2}W \cot \beta + \tfrac{1}{3}mL\omega^2 \cos \beta \qquad (3)$$

Substituting $W = 40$ lb, $m = 1.242$ lb·s²/ft, $L = 8$ ft, $\beta = 60°$, and $\omega = 15$ rad/s into (3), we obtain

$$T = 384 \text{ lb} \quad ◄$$

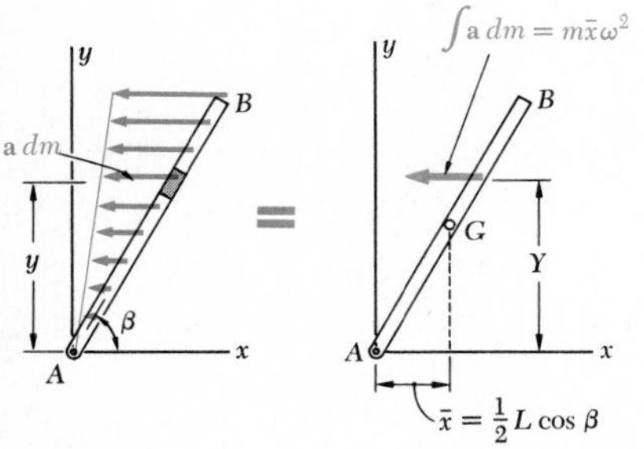

Note. The resultant of the effective forces $\mathbf{a}\,dm$ does not pass through the mass center of the rod. To show this, we first determine the magnitude of the resultant:

$$\int a\,dm = \int_0^L (l \cos \beta)\omega^2 \left(\frac{m}{L}\,dl\right) = \tfrac{1}{2}mL\omega^2 \cos \beta = m\bar{x}\omega^2 \qquad (4)$$

Since the resultant must be equivalent to the effective forces,

$$+\circlearrowright \Sigma M_A = \Sigma M_A: \qquad \int y(a\,dm) = Y \int a\,dm \qquad (5)$$

Substituting into (5) the expressions found in (2) and (4), we obtain

$$\tfrac{1}{3}mL^2\omega^2 \sin \beta \cos \beta = Y(\tfrac{1}{2}mL\omega^2 \cos \beta) \qquad Y = \tfrac{2}{3}L \sin \beta$$

PROBLEMS

16.135 The mass center of a loaded freight car is located 8 ft above the rails. Knowing that the gage is 4 ft $8\frac{1}{2}$ in. (distance between rails), determine the speed at which the car will overturn while rounding an unbanked curve of 2000-ft radius.

16.136 Solve Prob. 16.135 assuming that the rails are superelevated so that the outer rail is $2\frac{1}{2}$ in. higher than the inner rail.

16.137 The center of gravity of a 1200-kg automobile is 0.5 m above the road, and the transverse distance between the wheels is 1.4 m. If the automobile takes a turn of 70-m radius on an unbanked highway, determine (*a*) the maximum safe velocity with regard to overturning, (*b*) the value of μ between the tires and pavement if the maximum safe velocity with regard to skidding is equal to that found for overturning.

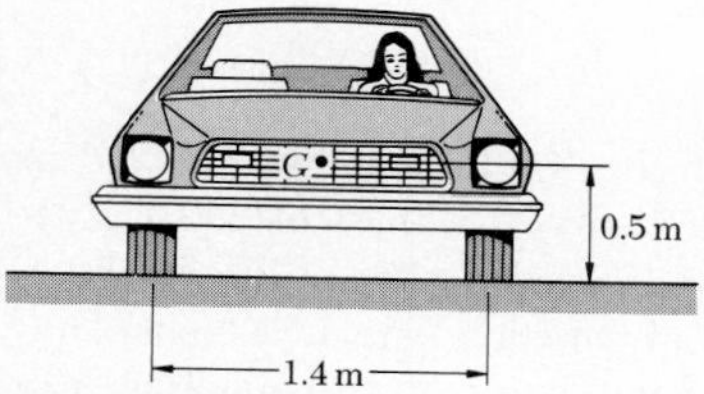

Fig. P16.137

16.138 The automobile described in Prob. 16.137 takes a turn of 70-m radius on a highway banked at 20°. Determine the maximum safe velocity with regard to overturning.

16.139 A stunt driver drives a small automobile of mass 600 kg (driver included) on the vertical wall of a cylindrical pit of radius 10 m at a speed of 50 km/h. Knowing that this is the slowest speed at which he can perform this stunt, determine (*a*) the coefficient of friction between tires and wall, (*b*) the normal component of the reaction of the wall at each wheel.

16.140 If the coefficient of friction between the wall and the tires of the automobile of Prob. 16.139 is 0.60, determine (*a*) the lowest safe speed at which the stunt can be performed, (*b*) the normal component of the reaction of the wall at each wheel.

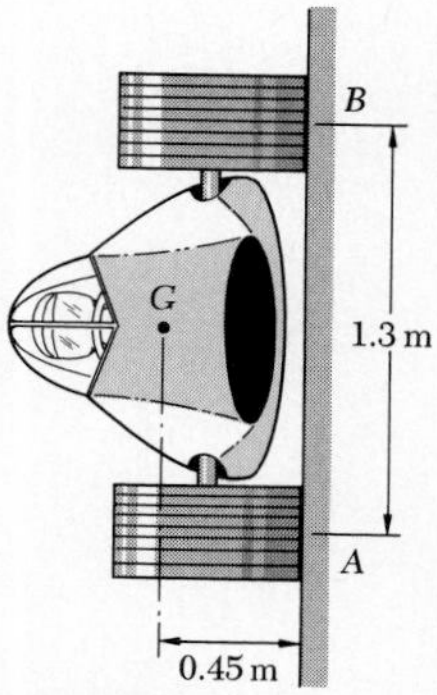

Fig. P16.139

16.141 A stunt driver rides a motorcycle at 45 mph inside a vertical cylindrical wall. Assuming that the mass center of the driver and motorcycle travels in a circle of radius 38 ft, determine (*a*) the angle β at which the motorcycle is inclined to the horizontal, (*b*) the minimum safe value of μ.

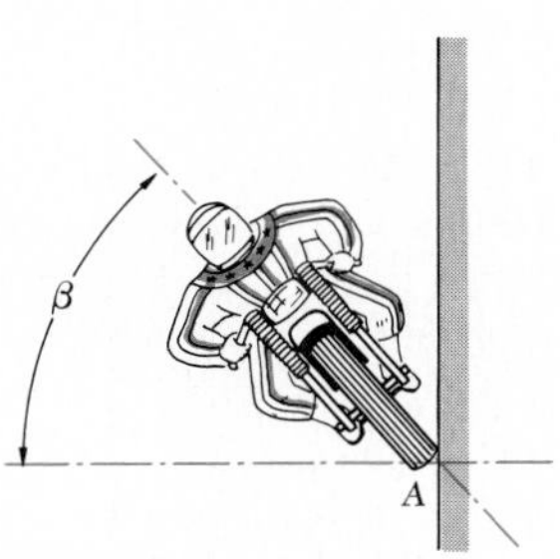

Fig. P16.141

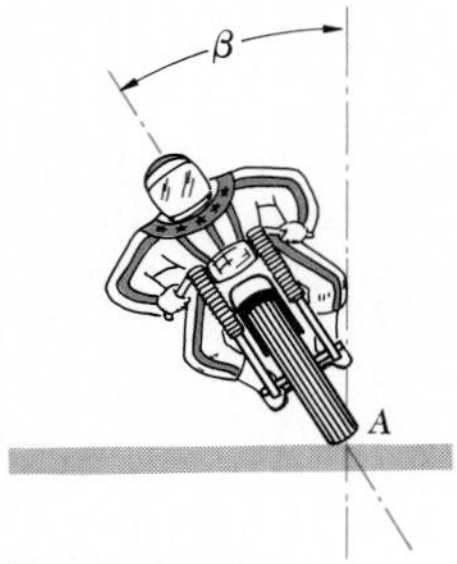

Fig. P16.142

16.142 A motorcyclist drives around an unbanked curve at a speed of 40 mi/h. Assuming that the mass center of the driver and motorcycle travels in a circle of radius 160 ft, determine (*a*) the angle between the plane of the motorcycle and the vertical, (*b*) the minimum safe value of μ.

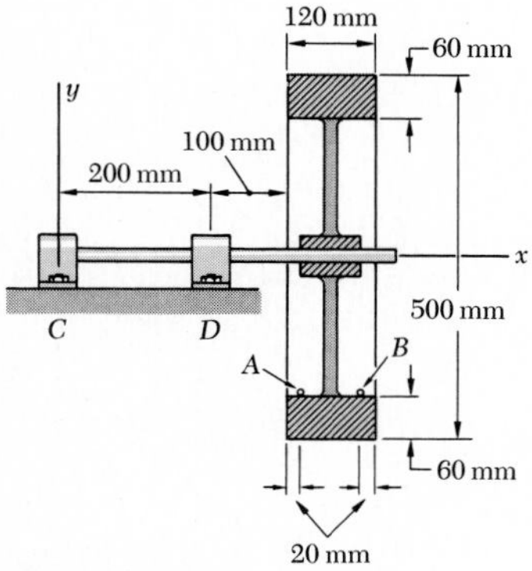

Fig. P16.143

16.143 It has been found that the flywheel shown will be in perfect balance (statically and dynamically) if a mass of 350 g is attached at point *A* located 20 mm from the inside edge of the flywheel. However, when the 350-g mass is permanently attached to the flywheel, it is mistakenly attached at point *B* located 20 mm from the outside edge. Determine the resulting dynamic bearing reactions at *C* and *D* for a speed of 1200 rpm.

16.144 In Prob. 16.143, determine the dynamic bearing reactions at *C* and *D* for a speed of 1200 rpm before the 350-g mass was attached to the flywheel.

16.145 Each element of the crankshaft shown is a homogeneous rod of weight w per unit length. Knowing that the crankshaft rotates with a constant angular velocity $\boldsymbol{\omega}$, determine the dynamic reactions at *A* and *B*.

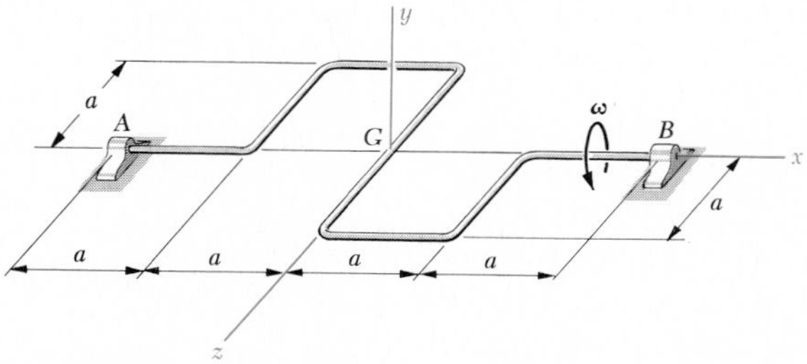

Fig. P16.145

16.146 Two L-shaped arms, each weighing 6 lb, are welded at the third points of the 3-ft shaft AB. Knowing that shaft AB rotates with a constant angular velocity ω of 300 rpm, determine the dynamic reactions at A and B.

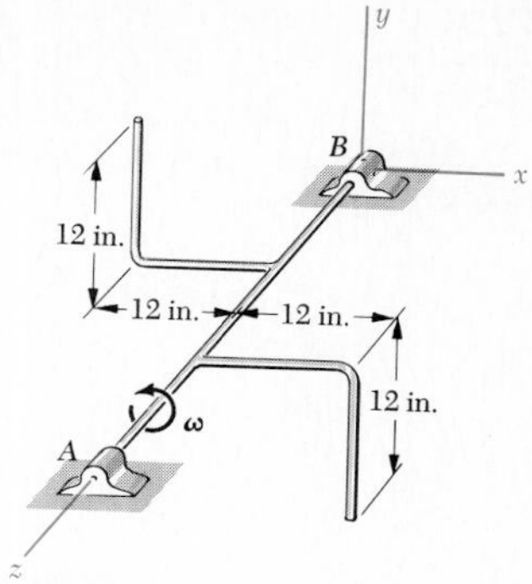

Fig. P16.146

16.147 The shaft of Prob. 16.145 is initially at rest ($\omega = 0$) and is accelerated at the rate $\alpha = 100\ \text{rad/s}^2$. Knowing that $w = 4\ \text{lb/ft}$ and $a = 3$ in., determine (*a*) the moment of the couple required to cause the acceleration, (*b*) the corresponding dynamic reactions at A and B.

16.148 The shaft of Prob. 16.146 is initially at rest ($\omega = 0$) when a couple $\mathbf{M}$ of moment 15 lb · ft is applied to the shaft AB. Determine (*a*) the angular acceleration of the shaft, (*b*) the corresponding dynamic reactions at A and B.

16.149 Two uniform rods CD and DE, each of mass 2 kg, are welded to the shaft AB, which is at rest. If a couple $\mathbf{M}$ of moment 10 N · m is applied to the shaft, determine the dynamic reactions at A and B.

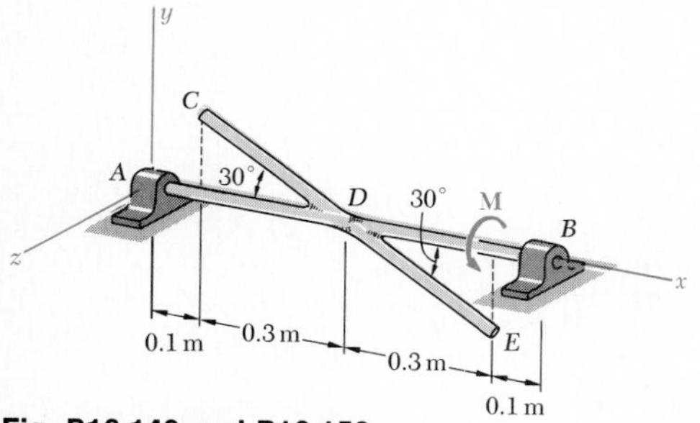

Fig. P16.149 and P16.150

16.150 Two uniform rods CD and DE, each of mass 2 kg, are welded to the shaft AB. At the instant shown the angular velocity of the shaft is 15 rad/s and the angular acceleration is $100\ \text{rad/s}^2$, both counterclockwise when viewed from the positive x axis. Determine (*a*) the moment of the couple $\mathbf{M}$ which must be applied to the shaft, (*b*) the corresponding dynamic reactions at A and B.

16.151 A door is made of a solid homogeneous panel 4 by 8 ft, weighing 75 lb. A man pushes on the knob D with a 10-lb force directed perpendicularly to the door. Determine (*a*) the angular acceleration of the door, (*b*) the horizontal components of the reactions at the hinges.

16.152 Locate the point on the face of the door of Prob. 16.151 at which a 10-lb force may be applied perpendicularly to the door without affecting the horizontal components of the reactions at the hinges.

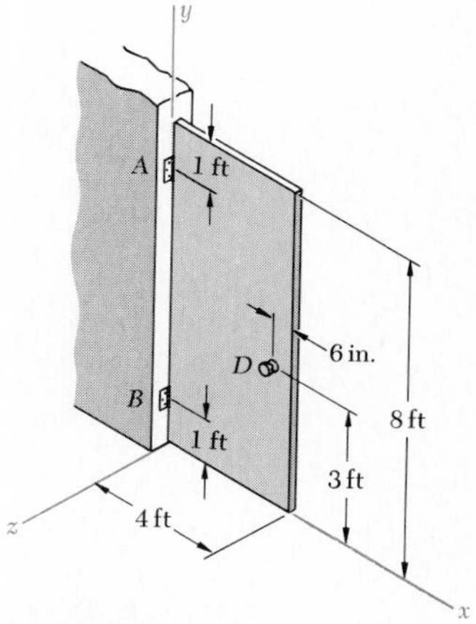

Fig. P16.151

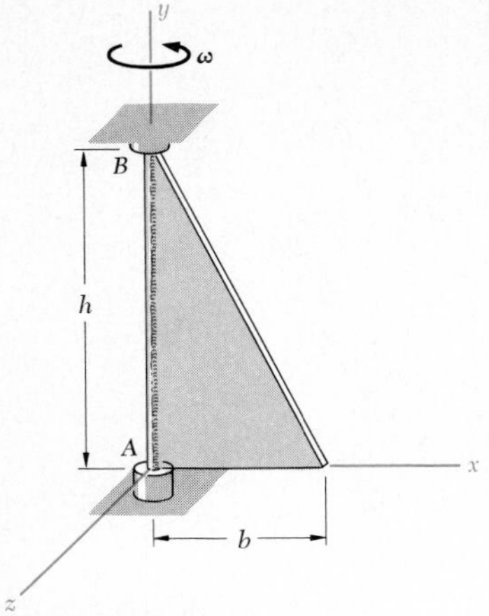

Fig. P16.153

16.153 A thin homogeneous triangular plate of mass m, base b, and height h is welded to a light axle which can rotate freely in bearings at A and B. Knowing that the plate rotates with a constant angular velocity $\boldsymbol{\omega}$, determine the dynamic reactions at A and B. (*Hint:* Consider the effective forces of elements consisting of horizontal strips.)

16.154 A thin homogeneous square plate of mass m and side a is welded to a vertical shaft AB with which it forms an angle of 45°. Knowing that the shaft rotates with a constant angular velocity $\boldsymbol{\omega}$, determine the force-couple system representing the dynamic reaction at A. (*Hint:* Consider the effective forces of elements consisting of horizontal strips.)

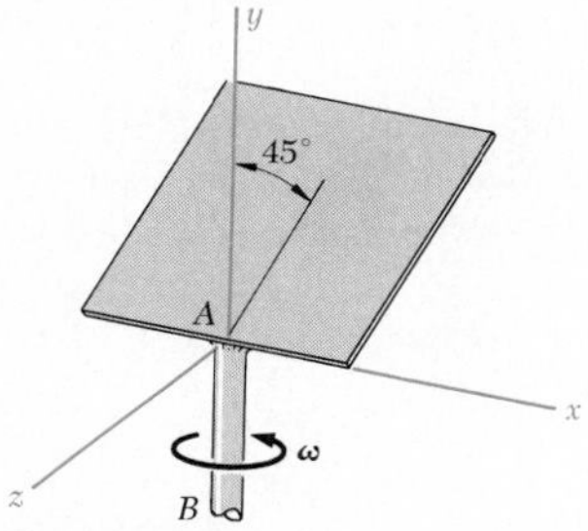

Fig. P16.154

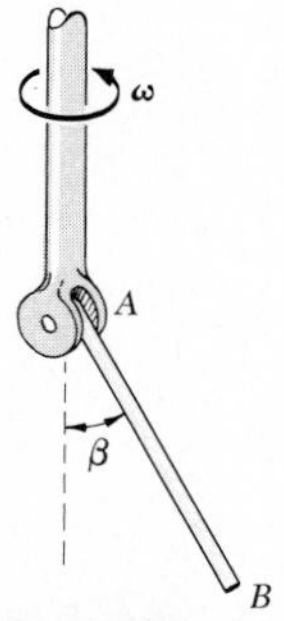

Fig. P16.155 and P16.156

16.155 A 2-ft uniform rod AB is attached at A to the pin of a clevis which rotates with a constant angular velocity $\boldsymbol{\omega}$. Determine (*a*) the angle β that the rod forms with the vertical when $\omega = 10$ rad/s, (*b*) the maximum value of ω for which the rod will remain vertical.

16.156 A uniform rod AB of length l and mass m is attached to the pin of a clevis which rotates with a constant angular velocity $\boldsymbol{\omega}$. Derive an expression (*a*) for the constant angle β that the rod forms with the vertical, (*b*) for the maximum value of ω for which the rod will remain vertical ($\beta = 0$).

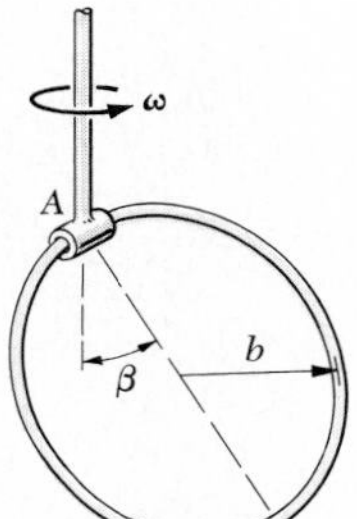

Fig. P16.157

16.157 A thin ring of radius b is attached by a collar at A to a vertical shaft which rotates with a constant angular velocity $\boldsymbol{\omega}$. Derive an expression (*a*) for the constant angle β that the plane of the ring forms with the vertical, (*b*) for the maximum value of ω for which the ring will remain vertical ($\beta = 0$).

***16.158** The square plate of Prob. 16.154 is at rest ($\boldsymbol{\omega} = 0$) when a couple of moment M_0 is applied to the shaft. Determine (*a*) the angular acceleration of the plate, (*b*) the force-couple system representing the dynamic reaction at A at that instant.

***16.159** The triangular plate of Prob. 16.153 is initially at rest ($\boldsymbol{\omega} = 0$) and has an angular acceleration $\boldsymbol{\alpha}$. Determine (*a*) the moment M of the couple required to cause the acceleration, (*b*) the corresponding dynamic reactions at A and B.

REVIEW PROBLEMS

16.160 A 15-lb uniform disk is suspended from a link AB of negligible weight. If a 10-lb force is applied at B, determine the acceleration of B (a) if the connection at B is a frictionless pin, (b) if the connection at B is "frozen" and the system rotates about A as a rigid body.

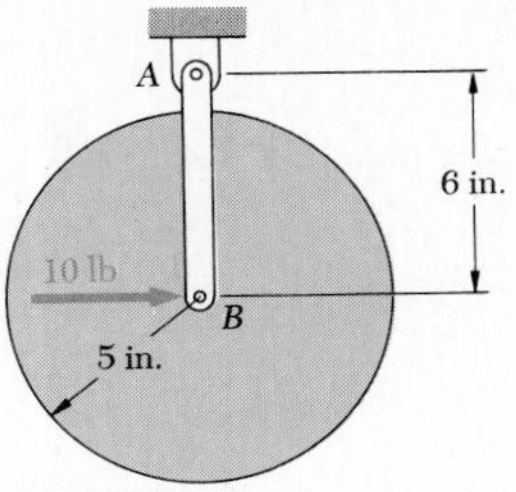

Fig. P16.160

16.161 and 16.162 A uniform plate of mass m is suspended in each of the ways shown. For each case determine the acceleration of the center of the plate immediately after the connection at B has been released.

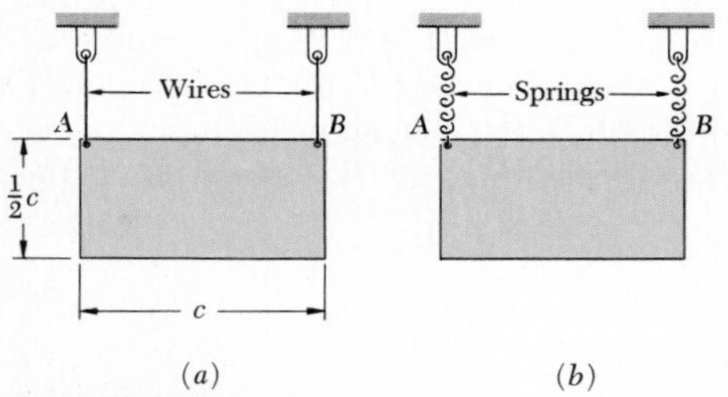

Fig. P16.161

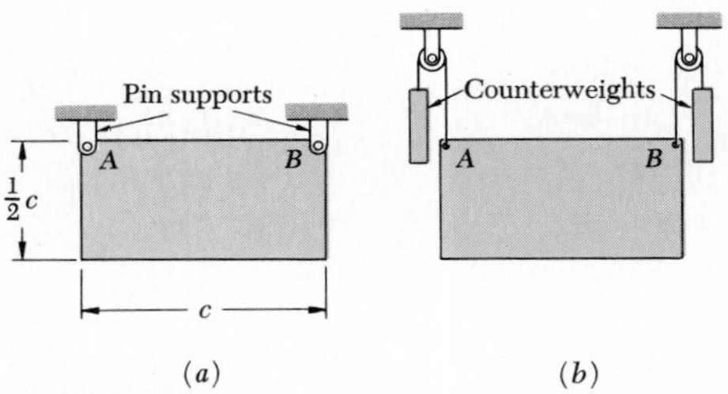

Fig. P16.162

16.163 A 100-lb uniform plate is suspended by two wires. Knowing that the plate is released from rest in the position shown, determine at that instant (a) the acceleration of the center of the plate, (b) the tension in each wire.

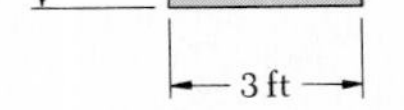

Fig. P16.163

16.164 The flanged wheel shown rolls to the right with a constant velocity of 1.5 m/s. The rod AB is 1.2 m long and has a mass of 5 kg. Knowing that point A slides without friction on the horizontal surface, determine the reaction at A (a) when $\beta = 0$, (b) when $\beta = 180°$.

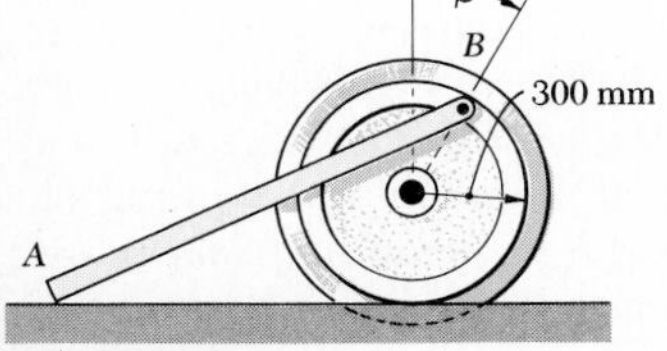

Fig. P16.164

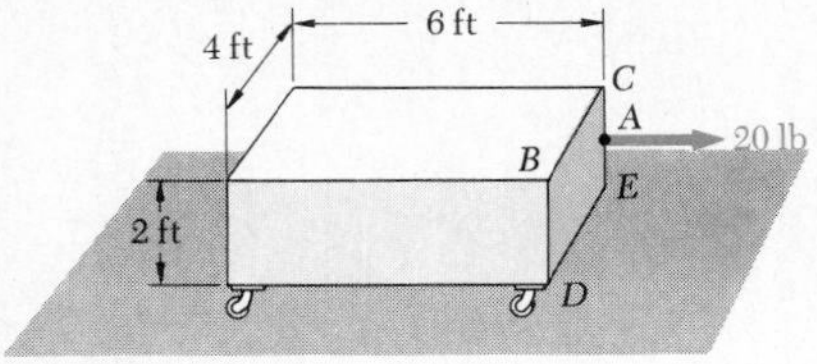

Fig. P16.165

16.165 The 80-lb crate shown rests on four casters which allow it to move without friction in any horizontal direction. A 20-lb horizontal force is applied at the midpoint A of edge CE. Knowing that the force is perpendicular to side $BCDE$, determine the angular acceleration of the crate and the acceleration of point A.

16.166 Solve Prob. 16.165, assuming that the 20-lb horizontal force applied at A is parallel to the edge BC.

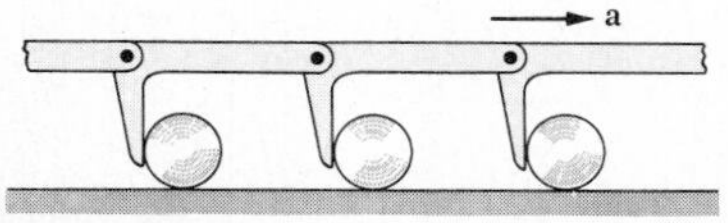

Fig. P16.167

16.167 Identical cylinders of mass m and radius r are pushed by a series of moving arms. Assuming the coefficient of friction between all surfaces to be $\mu < 1$, and denoting by a the magnitude of the acceleration of the arms, derive an expression for (*a*) the maximum allowable value of a if each cylinder is to roll without sliding, (*b*) the minimum allowable value of a if each cylinder is to move to the right without rotating.

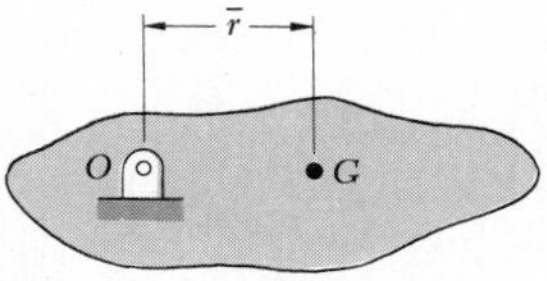

Fig. P16.168

16.168 A rigid slab of weight W and centroidal radius of gyration $\bar{k}$ may rotate freely about a shaft O located at a distance $\bar{r}$ from its mass center G. If the slab is released from rest in the position shown, determine (*a*) the distance $\bar{r}$ for which the angular acceleration of the slab is maximum, (*b*) the corresponding vertical component of the reaction at O.

16.169 A 6-kg bar is held between four disks as shown. Each disk has a mass of 3 kg and a diameter of 200 mm. The disks may rotate freely, and the normal reaction exerted by each disk on the bar is sufficient to prevent slipping. If the bar is released from rest, determine (*a*) its acceleration immediately after release, (*b*) its velocity after it has dropped 0.75 m

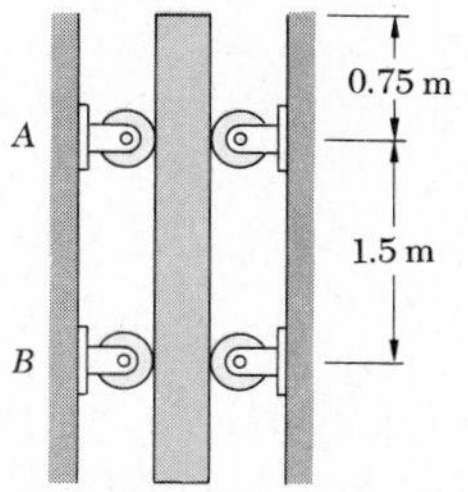

Fig. P16.169

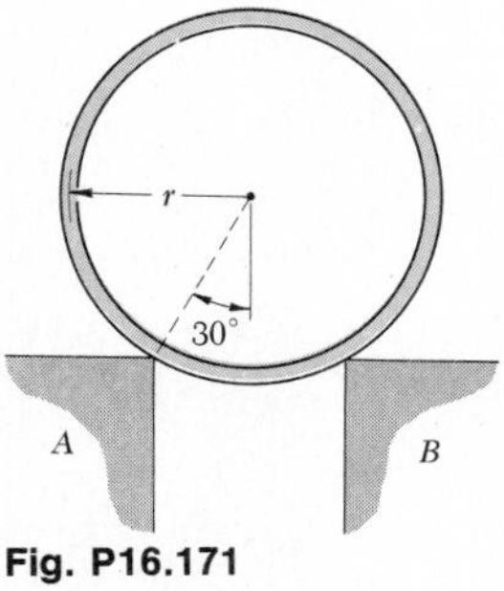

Fig. P16.171

16.170 In Prob. 16.169, determine (*a*) the acceleration of the bar after it has dropped 1.25 m, (*b*) the velocity of the bar after it has dropped 2.25 m.

16.171 A section of pipe, of mass 50 kg and radius 250 mm, rests on two corners as shown. Assuming that μ between the corners and the pipe is sufficient to prevent sliding, determine (*a*) the angular acceleration of the pipe just after corner B is removed, (*b*) the corresponding magnitude of the reaction at A.

Chapter

17

Kinetics of Rigid Bodies: Work and Energy

17.1. Principle of Work and Energy for a Rigid Body. In this chapter, the principle of work and energy will be used to analyze the motion of rigid bodies and of systems of rigid bodies. As was pointed out in Chap. 13, the method of work and energy is particularly well adapted to the solution of problems involving velocities and displacements. Its main advantage resides in the fact that the work of forces and the kinetic energy of particles are scalar quantities.

Consider a rigid body of mass m, and let P be a particle of the body, of mass Δm. The kinetic energy of this particle is $\Delta T = \frac{1}{2}(\Delta m)v^2$, where v is the speed of the particle. Consider now a displacement of the rigid body during which the particle P moves from a position P_1 to a position P_2, and denote by $\Delta U_{1\to2}$ the work of all the forces acting on P during the displacement. We recall from Sec. 13.3 that the principle of work and energy for a particle states that

$$\Delta T_1 + \Delta U_{1\to2} = \Delta T_2$$

where ΔT_1 = kinetic energy of the particle at P_1
ΔT_2 = kinetic energy of the particle at P_2

Similar relations may be written for all the particles forming the body. Adding the kinetic energy of all the particles, and the work of all the forces involved, we write

$$T_1 + U_{1\to2} = T_2 \tag{17.1}$$

where T_1, T_2 = initial and final values of total kinetic energy of the particles forming the rigid body

$U_{1\to2}$ = work of all forces acting on the various particles of the body

As indicated in Chap. 13, work and energy are expressed in joules (J) if SI units are used, and in ft · lb or in · lb if U.S. customary units are used.

The total kinetic energy

$$T = \Sigma\tfrac{1}{2}(\Delta m)v^2 \tag{17.2}$$

is obtained by adding positive scalar quantities and is itself a positive scalar quantity. We shall see later how T may be determined for various types of motion of a rigid body.

The expression $U_{1\to2}$ in (17.1) represents the work of all the forces acting on the various particles of the body, whether these forces are internal or external. However, as we shall see presently, the total work of the internal forces holding together the particles of a rigid body is zero. Consider two particles A and B of a rigid body and the two equal and opposite forces **F** and **F**′ they exert on each other (Fig. 17.1). While, in general, the

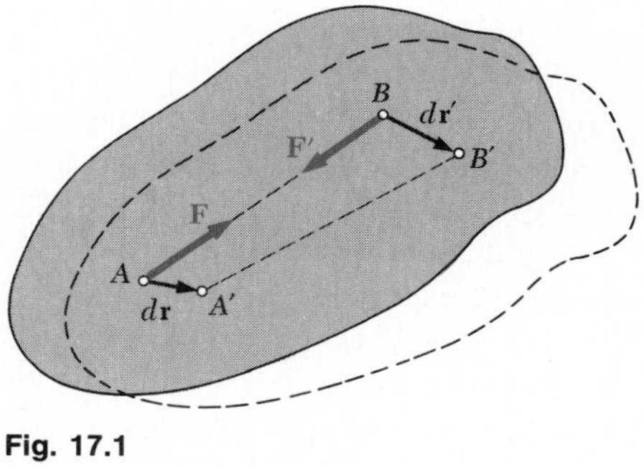

Fig. 17.1

displacements $d\mathbf{r}$ and $d\mathbf{r}'$ of the two particles are different, the components of these displacements along AB must be equal; otherwise, the particles would not remain at the same distance from each other, and the body would not be rigid. Therefore, the work of **F** is equal in magnitude and opposite in sign to the work of **F**′, and their sum is zero. Thus, the total work of the internal forces acting on the particles of a rigid body is zero, and *the expression* $U_{1\to2}$ *in Eq.* (17.1) *reduces to the work of the external forces* acting on the body during the displacement considered.

17.2. Work of Forces Acting on a Rigid Body. We saw in Sec. 13.2 that the work of a force **F** during a displacement of its point of application from A_1 $(s = s_1)$ to A_2 $(s = s_2)$ is

$$U_{1\rightarrow 2} = \int_{s_1}^{s_2} (F \cos \alpha)\, ds \tag{17.3}$$

where F is the magnitude of the force **F**, and α the angle that **F** forms at any given instant with the direction of motion of its point of application.

In computing the work of the external forces acting on a rigid body, it is often convenient to determine the work of a couple without considering separately the work of each of the two forces forming the couple. Consider the two forces **F** and **F′** forming a couple of moment $M = Fr$ and acting on a rigid body (Fig. 17.2). Any small displacement of the rigid body bringing A and B, respectively, into A' and B'' may be divided into two parts, one in which points A and B undergo equal displacements $d\mathbf{r}_1$, the other in which A' remains fixed while B' moves into B'' through a displacement $d\mathbf{r}_2$ of magnitude $ds_2 = r\, d\theta$. In the

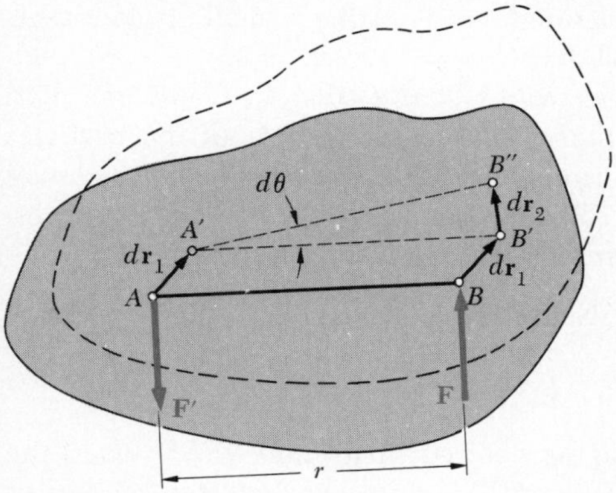

Fig. 17.2

first part of the motion, the work of **F** is equal in magnitude and opposite in sign to the work of **F′**, and their sum is zero. In the second part of the motion, only force **F** works, and its work is $dU = F\, ds_2 = Fr\, d\theta = M\, d\theta$. Thus, the work of a couple of moment M acting on a rigid body is

$$dU = M\, d\theta \tag{17.4}$$

where $d\theta$ is the small angle expressed in radians through which the body rotates. We again note that work should be expressed in units obtained by multiplying units of force by units of length. The work of the couple during a finite rotation of the rigid body is obtained by integrating both members of (17.4) from the initial value θ_1 of the angle θ to its final value θ_2. We write

$$U_{1\to 2} = \int_{\theta_1}^{\theta_2} M\,d\theta \tag{17.5}$$

When the moment M of the couple is constant, formula (17.5) reduces to

$$U_{1\to 2} = M(\theta_2 - \theta_1) \tag{17.6}$$

It was pointed out in Sec. 13.2 that a number of forces encountered in problems of kinetics *do no work.* They are forces applied to fixed points or acting in a direction perpendicular to the displacement of their point of application. Among the forces which do no work the following have been listed: the reaction at a frictionless pin when the body supported rotates about the pin, the reaction at a frictionless surface when the body in contact moves along the surface, the weight of a body when its center of gravity moves horizontally. We should also indicate now that, *when a rigid body rolls without sliding on a fixed surface, the friction force* **F** *at the point of contact C does no work.* The velocity $\mathbf{v}_C$ of the point of contact C is zero, and the work of the friction force **F** during a small displacement of the rigid body is $dU = F\,ds_C = F(v_C\,dt) = 0$.

17.3. Kinetic Energy in Translation. Consider a rigid body in translation. Since, at a given instant, all the particles of the body have the same velocity as the mass center, we may substitute $v = \bar{v}$ in (17.2). We thus obtain the following expression for the kinetic energy of a rigid body in translation:

$$T = \Sigma\tfrac{1}{2}(\Delta m)\bar{v}^2 = \tfrac{1}{2}(\Sigma\,\Delta m)\bar{v}^2$$

$$T = \tfrac{1}{2}m\bar{v}^2 \tag{17.7}$$

This expression should be entered in formula (17.1) when the method of work and energy is used to analyze the motion of a rigid body in translation (either rectilinear or curvilinear).

We note that the expression obtained in (17.7) represents the kinetic energy of a particle of mass m moving with the velocity $\bar{\mathbf{v}}$ of the mass center. The approach used in Chap. 13, where bodies as large as automobiles or trains were considered as single particles, is therefore correct as long as the bodies considered are in translation.

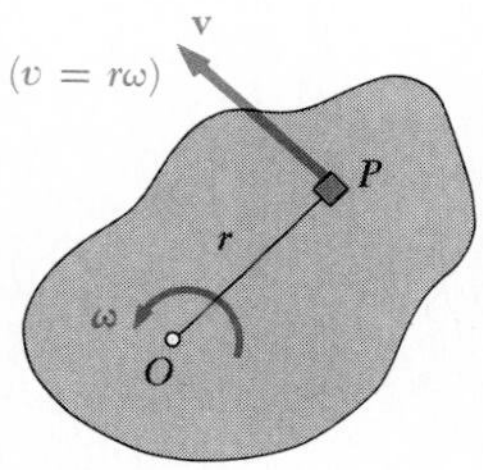

Fig. 17.3

17.4. Kinetic Energy in Rotation. Consider a rigid body rotating about a fixed axis intersecting the reference plane at O (Fig. 17.3). If the angular velocity of the body at a given instant is $\boldsymbol{\omega}$ (rad/s), the speed of a particle P located at a distance r from the axis of rotation is $v = r\omega$ at that instant. Substituting for v into (17.2), we obtain the following expression for the kinetic energy of the body:

$$T = \Sigma\tfrac{1}{2}(\Delta m)(r\omega)^2 = \tfrac{1}{2}\omega^2 \Sigma r^2\, \Delta m$$

Observing that the sum $\Sigma r^2\, \Delta m$ represents the moment of inertia I_O of the rigid body about the axis of rotation, we write

$$T = \tfrac{1}{2} I_O \omega^2 \tag{17.8}$$

This expression should be entered in formula (17.1) when the method of work and energy is used to analyze the motion of a rigid body rotating about a fixed axis.

We note that formula (17.8) is valid with respect to any fixed axis of rotation, whether or not this axis passes through the mass center of the body. Moreover, the result obtained is not limited to the rotation of plane slabs or to the rotation of bodies which are symmetrical with respect to the reference plane. Formula (17.8) may be applied to the rotation of any rigid body about a fixed axis, regardless of the shape of the body or of the location of the axis of rotation.

17.5. Systems of Rigid Bodies. When a problem involves several rigid bodies, each rigid body may be considered separately and the principle of work and energy may be applied to each body. Adding the kinetic energies of all the particles and considering the work of all the forces involved, we may also write the equation of work and energy for the entire system. We have

$$T_1 + U_{1\to 2} = T_2 \tag{17.9}$$

where T represents the arithmetic sum of the kinetic energies of the rigid bodies forming the system (all terms are positive) and $U_{1\to 2}$ the work of all the forces acting on the various bodies, whether these forces are *internal* or *external* from the point of view of the system as a whole.

The method of work and energy is particularly useful in solving problems involving pin-connected members, or blocks and pulleys connected by inextensible cords, or meshed gears. In all these cases, the internal forces occur by pairs of equal and opposite forces, and the points of application of the forces in each pair *move through equal distances* during a small displacement of the system. As a result, the work of the internal forces is zero, and $U_{1\to 2}$ reduces to the work of the *forces external to the system.*

SAMPLE PROBLEM 17.1

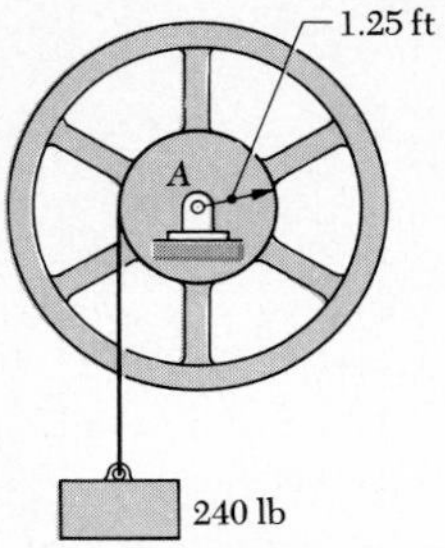

A 240-lb block is suspended from an inextensible cable which is wrapped around a drum of 1.25-ft radius rigidly attached to a flywheel. The drum and flywheel have a combined centroidal moment of inertia $\bar{I} = 10.5 \text{ lb} \cdot \text{ft} \cdot \text{s}^2$. At the instant shown, the velocity of the block is 6 ft/s directed downward. Knowing that the bearing at A is poorly lubricated and that the bearing friction is equivalent to a couple **M** of moment 60 lb · ft, determine the velocity of the block after it has moved 4 ft downward.

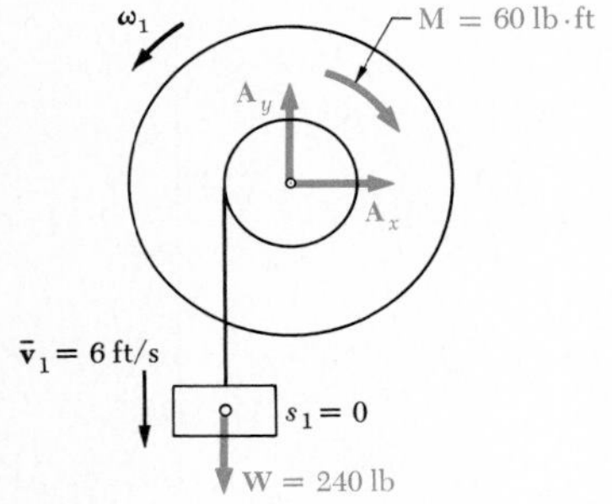

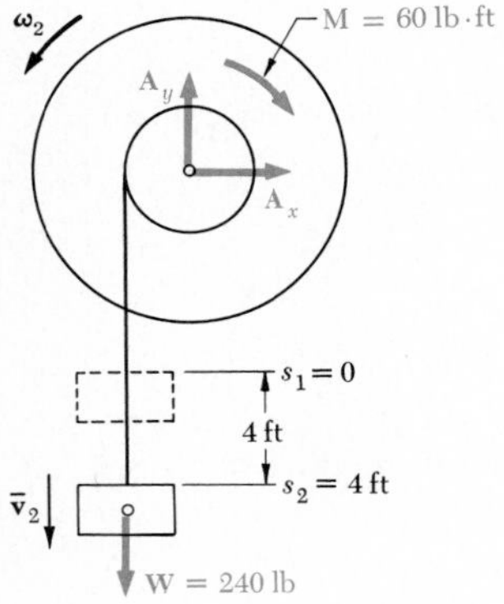

Solution. We consider the system formed by the flywheel and the block. Since the cable is inextensible, the work done by the internal forces exerted by the cable cancels. The initial and final positions of the system and the external forces acting on the system are as shown.

Kinetic Energy. Position 1. We have

$$\bar{v}_1 = 6 \text{ ft/s} \qquad \omega_1 = \frac{\bar{v}_1}{r} = \frac{6 \text{ ft/s}}{1.25 \text{ ft}} = 4.80 \text{ rad/s}$$

$$\begin{aligned} T_1 &= \tfrac{1}{2}m\bar{v}_1^2 + \tfrac{1}{2}\bar{I}\omega_1^2 \\ &= \frac{1}{2}\frac{240 \text{ lb}}{32.2 \text{ ft/s}^2}(6 \text{ ft/s})^2 + \tfrac{1}{2}(10.5 \text{ lb} \cdot \text{ft} \cdot \text{s}^2)(4.80 \text{ rad/s})^2 \\ &= 255 \text{ ft} \cdot \text{lb} \end{aligned}$$

Position 2. Noting that $\omega_2 = \bar{v}_2/1.25$, we write

$$\begin{aligned} T_2 &= \tfrac{1}{2}m\bar{v}_2^2 + \tfrac{1}{2}\bar{I}\omega_2^2 \\ &= \frac{1}{2}\frac{240}{32.2}(\bar{v}_2)^2 + (\tfrac{1}{2})(10.5)\left(\frac{\bar{v}_2}{1.25}\right)^2 = 7.09\bar{v}_2^2 \end{aligned}$$

Work. During the motion, only the weight **W** of the block and the friction couple **M** do work. Noting that **W** does positive work and that the friction couple **M** does negative work, we write

$$s_1 = 0 \qquad s_2 = 4 \text{ ft}$$

$$\theta_1 = 0 \qquad \theta_2 = \frac{s_2}{r} = \frac{4 \text{ ft}}{1.25 \text{ ft}} = 3.20 \text{ rad}$$

$$\begin{aligned} U_{1\to2} &= W(s_2 - s_1) - M(\theta_2 - \theta_1) \\ &= (240 \text{ lb})(4 \text{ ft}) - (60 \text{ lb} \cdot \text{ft})(3.20 \text{ rad}) \\ &= 768 \text{ ft} \cdot \text{lb} \end{aligned}$$

Principle of Work and Energy

$$T_1 + U_{1\to2} = T_2$$

$$255 \text{ ft} \cdot \text{lb} + 768 \text{ ft} \cdot \text{lb} = 7.09\bar{v}_2^2$$

$$\bar{v}_2 = 12.01 \text{ ft/s} \qquad \bar{\mathbf{v}}_2 = 12.01 \text{ ft/s} \downarrow \blacktriangleleft$$

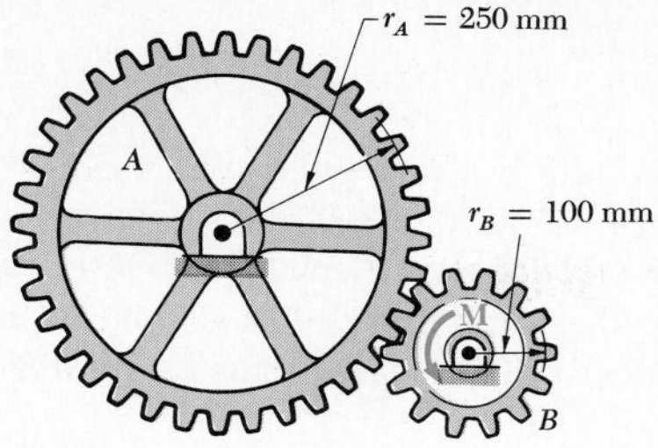

SAMPLE PROBLEM 17.2

Gear A has a mass of 10 kg and a radius of gyration of 200 mm, while gear B has a mass of 3 kg and a radius of gyration of 80 mm. The system is at rest when a couple $\mathbf{M}$ of moment 6 N · m is applied to gear B. Neglecting friction, determine (*a*) the number of revolutions executed by gear B before its angular velocity reaches 600 rpm, (*b*) the tangential force which gear B exerts on gear A.

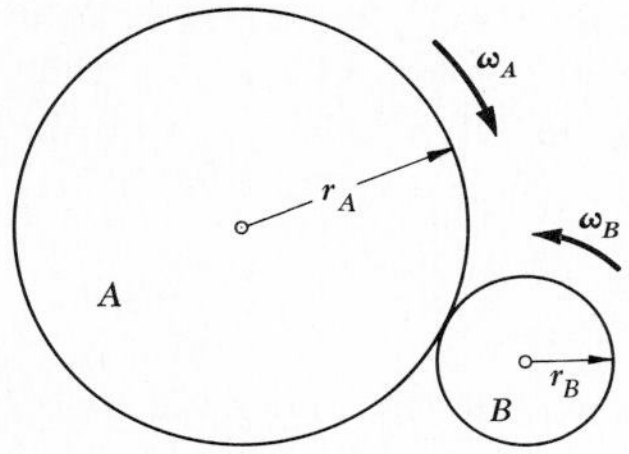

Motion of Entire System. Noting that the peripheral speeds of the gears are equal, we write

$$r_A\omega_A = r_B\omega_B \qquad \omega_A = \omega_B\frac{r_B}{r_A} = \omega_B\frac{100\text{ mm}}{250\text{ mm}} = 0.40\omega_B$$

For $\omega_B = 600$ rpm, we have

$$\omega_B = 62.8\text{ rad/s} \qquad \omega_A = 0.40\omega_B = 25.1\text{ rad/s}$$
$$\bar{I}_A = m_A\bar{k}_A^2 = (10\text{ kg})(0.200\text{ m})^2 = 0.400\text{ kg}\cdot\text{m}^2$$
$$\bar{I}_B = m_B\bar{k}_B^2 = (3\text{ kg})(0.080\text{ m})^2 = 0.0192\text{ kg}\cdot\text{m}^2$$

Kinetic Energy. Since the system is initially at rest, $T_1 = 0$. Adding the kinetic energies of the two gears when $\omega_B = 600$ rpm, we obtain

$$T_2 = \tfrac{1}{2}\bar{I}_A\omega_A^2 + \tfrac{1}{2}\bar{I}_B\omega_B^2$$
$$= \tfrac{1}{2}(0.400\text{ kg}\cdot\text{m}^2)(25.1\text{ rad/s})^2 + \tfrac{1}{2}(0.0192\text{ kg}\cdot\text{m}^2)(62.8\text{ rad/s})^2$$
$$= 163.9\text{ J}$$

Work. Denoting by θ_B the angular displacement of gear B, we have

$$U_{1\to2} = M\theta_B = (6\text{ N}\cdot\text{m})(\theta_B\text{ rad}) = (6\,\theta_B)\text{ J}$$

Principle of Work and Energy

$$T_1 + U_{1\to2} = T_2: \qquad 0 + (6\,\theta_B)\text{ J} = 163.9\text{ J}$$
$$\theta_B = 27.32\text{ rad} \qquad \theta_B = 4.35\text{ rev} \blacktriangleleft$$

Motion of Gear A. ***Kinetic Energy.*** Initially, gear A is at rest, $T_1 = 0$. When $\omega_B = 600$ rpm, the kinetic energy of gear A is

$$T_2 = \tfrac{1}{2}\bar{I}_A\omega_A^2 = \tfrac{1}{2}(0.400\text{ kg}\cdot\text{m}^2)(25.1\text{ rad/s})^2 = 126.0\text{ J}$$

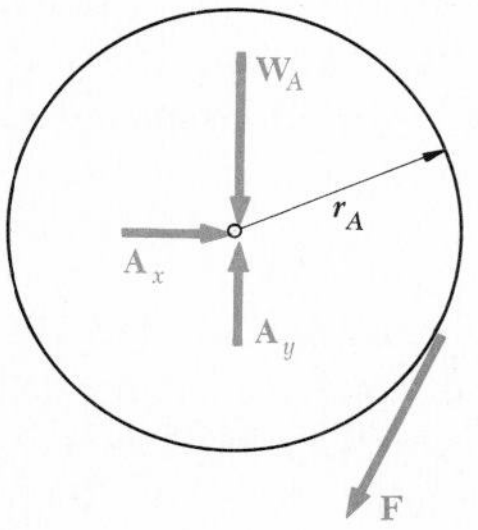

Work. The forces acting on gear A are as shown. The tangential force $\mathbf{F}$ does work equal to the product of its magnitude and of the length $\theta_A r_A$ of the arc described by the point of contact. Since $\theta_A r_A = \theta_B r_B$, we have

$$U_{1\to2} = F(\theta_B r_B) = F(27.3\text{ rad})(0.100\text{ m}) = F(2.73\text{ m})$$

Principle of Work and Energy

$$T_1 + U_{1\to2} = T_2: \qquad 0 + F(2.73\text{ m}) = 126.0\text{ J}$$
$$F = +46.1\text{ N} \qquad \mathbf{F} = 46.1\text{ N} \swarrow \blacktriangleleft$$

PROBLEMS

17.1 The rotor of a generator has an angular velocity of 3600 rpm when the generator is taken off line. The 150-kg rotor, which has a centroidal radius of gyration of 250 mm, then coasts to rest. Knowing that the kinetic friction of the rotor produces a couple of moment 2 N · m, determine the number of revolutions that the rotor executes before coming to rest.

17.2 A large flywheel of mass 1800 kg has a radius of gyration of 0.75 m. It is observed that 2500 revolutions are required for the flywheel to coast from an angular velocity of 450 rpm to rest. Determine the average moment of the couple due to kinetic friction in the bearings.

17.3 Three disks of the same thickness and same material are attached to a shaft as shown. Disks A and B each have a radius r; disk C has a radius nr. A couple $\mathbf{M}$ of constant moment is applied when the system is at rest and is removed after the system has executed one revolution. Determine the radius of disk C which results in the largest final speed of a point on the rim of disk C.

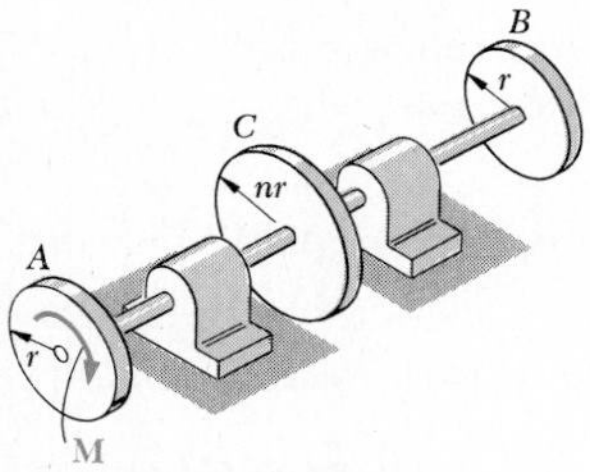

Fig. P17.3 and P17.4

17.4 Three disks of the same thickness and same material are attached to a shaft as shown. Disks A and B weigh 15 lb each and are of radius $r = 10$ in. A couple $\mathbf{M}$ of moment 30 lb · ft is applied to disk A when the system is at rest. Determine the radius nr of disk C if the angular velocity of the system is to be 600 rpm after 10 revolutions.

17.5 The flywheel of a punching machine weighs 900 lb and has a radius of gyration of 30 in. Each punching operation requires 2000 ft · lb of work. (*a*) Knowing that the speed of the flywheel is 200 rpm just before a punching, determine the speed immediately after the punching. (*b*) If a constant 25-lb · ft couple is applied to the shaft of the flywheel, determine the number of revolutions executed before the speed is again 200 rpm.

17.6 The flywheel of a small punch rotates at 300 rpm. It is known that 750 ft · lb of work must be done each time a hole is punched. It is desired that the speed of the flywheel after one punching be not less than 90 percent of the original speed of 300 rpm. (*a*) Determine the required moment of inertia of the flywheel. (*b*) If a constant 30-lb · ft couple is applied to the shaft of the flywheel, determine the number of revolutions which must occur between each punching, knowing that the initial velocity is to be 300 rpm at the start of each punching.

17.7 Two cylinders are attached by cords to an 18-kg double pulley which has a radius of gyration of 300 mm. When the system is at rest and in equilibrium, a 3-kg collar is added to the 12-kg cylinder. Neglecting friction, determine the velocity of each cylinder after the pulley has completed one revolution.

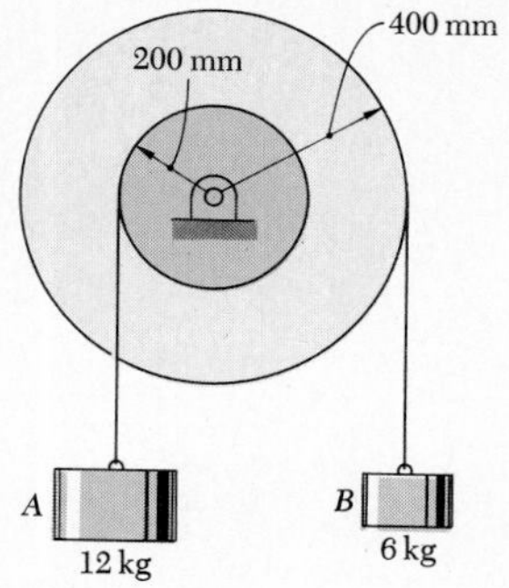

Fig. P17.7

17.8 Solve Prob. 17.7, assuming that the 3-kg collar is added to the 6-kg cylinder.

17.9 Using the principle of work and energy, solve Prob. 16.32*b*.

17.10 Using the principle of work and energy, solve Prob. 16.30*c*.

17.11 A disk of constant thickness and initially at rest is placed in contact with the belt, which moves with a constant velocity **v**. Denoting by μ the coefficient of friction between the disk and the belt, derive an expression for the number of revolutions executed by the disk before it reaches a constant angular velocity.

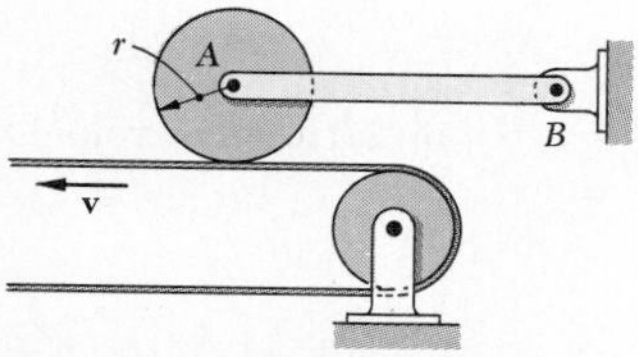

Fig. P17.11 and P17.12

17.12 Disk *A*, of weight 8 lb and radius $r = 5$ in., is at rest when it is placed in contact with the belt, which moves with a constant speed $v = 60$ ft/s. Knowing that $\mu = 0.15$ between the disk and the belt, determine the number of revolutions executed by the disk before it reaches a constant angular velocity.

17.13 Shafts *AB* and *CD* are connected by a friction disk and wheel as shown. The mass moment of inertia of the friction disk and its shaft *AB* is 0.360 kg · m^2, and the mass moment of inertia of the wheel and its shaft *CD* is 0.216 kg · m^2. The system is initially at rest when a couple **M** of constant moment 15 N · m is applied to shaft *AB* for 20 revolutions. Assuming no slipping occurs, determine the resulting angular velocity of each shaft when $x = 37.5$ mm.

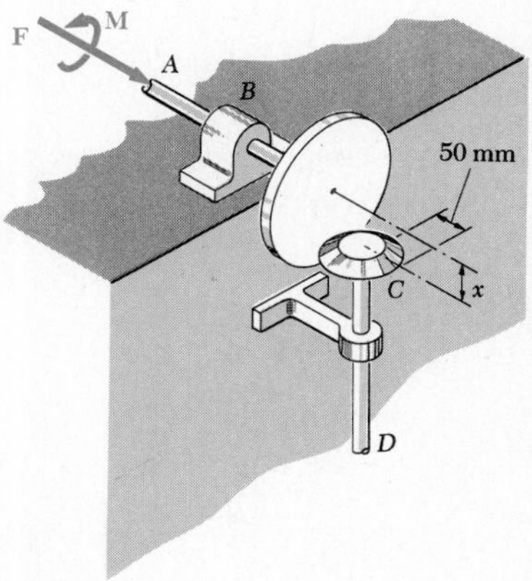

Fig. P17.13

17.14 The mechanism of Prob. 17.13 is set so that $x = 75$ mm; the initial angular velocity of shaft *CD* is 500 rpm. It is desired to increase its velocity to 600 rpm in 3 revolutions of shaft *CD*. Determine the moment of the couple **M** which must be applied to shaft *AB*.

17.15 The 15 in.-radius brake drum is attached to a larger flywheel which is not shown. The total mass moment of inertia of the flywheel and drum is 15 lb · ft · s^2. Knowing that the initial angular velocity is 150 rpm clockwise, determine the force which must be exerted by the hydraulic cylinder if the system is to come to rest in 10 revolutions.

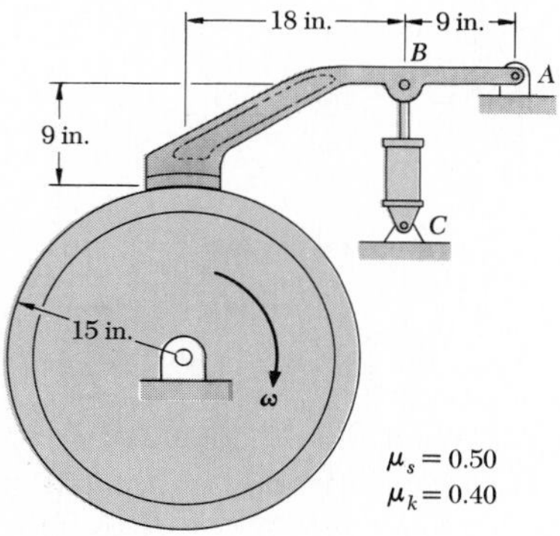

Fig. P17.15

17.16 Solve Prob. 17.15, assuming that the initial angular velocity of the flywheel is 150 rpm counterclockwise.

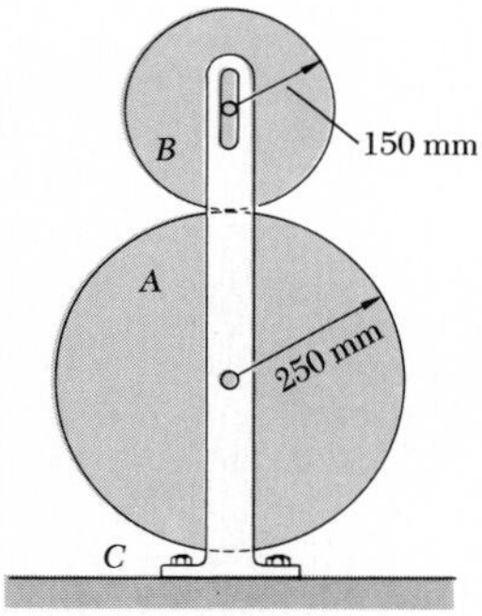

Fig. P17.17

17.17 Disk A is attached to a motor and is made to rotate with a constant angular velocity of 300 rpm clockwise. Disk B has a mass of 5 kg and is at rest when it is placed in contact with disk A. Knowing that $\mu = 0.25$ and neglecting bearing friction, determine the number of revolutions executed by disk B before it reaches a constant angular velocity.

17.18 Solve Prob. 17.17 assuming that disk B is replaced by a 6-kg wheel of radius $r = 150$ mm and radius of gyration $\bar{k} = 125$ mm.

17.19 The block C of mass m acquires a speed v after being released from rest and falling a distance h. Assuming that no slipping occurs, derive a formula for the speed v in terms of h, m, the radii r_A and r_B, and the centroidal mass moments of inertia $\bar{I}_A$ and $\bar{I}_B$ of the two disks.

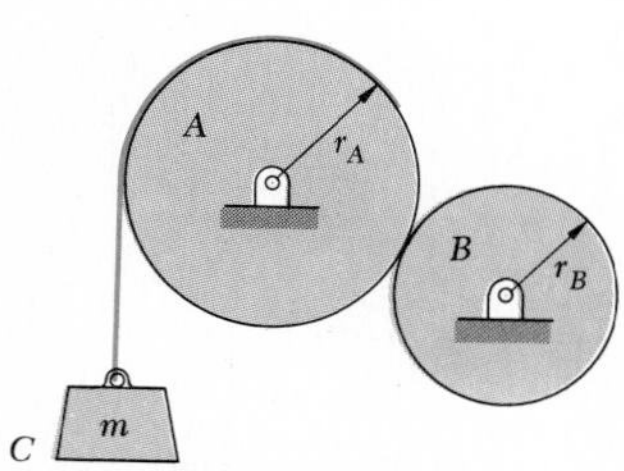

Fig. P17.19 and P17.20

17.20 Disk A weighs 8 lb, and its radius is $r_A = 10$ in.; disk B weighs 3 lb and its radius is $r_B = 6$ in. Knowing that the system is released from rest, determine the velocity of the 2-lb block C after it has fallen 5 ft.

17.21 Each of the gears A and B has a mass of 2 kg and a radius of gyration of 70 mm, while gear C has a mass of 10 kg and a radius of gyration of 175 mm. A couple **M** of constant moment 12 N · m is applied to gear C. Determine (*a*) the number of revolutions required for the angular velocity of gear C to increase from 100 to 450 rpm, (*b*) the corresponding tangential force acting on gear A.

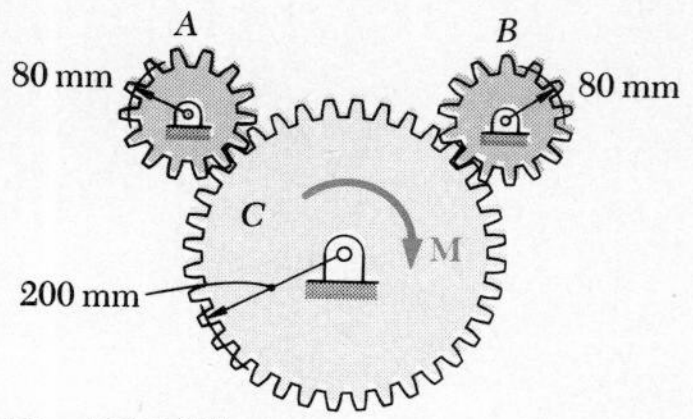

Fig. P17.21

17.22 Solve Prob. 17.21 assuming that the 12-N · m couple is applied to gear B.

17.23 Solve Sample Prob. 17.2 assuming that the 6-N · m couple is applied to gear A.

17.6. Kinetic Energy in Plane Motion. Consider a rigid body in plane motion. We saw in Sec. 15.7 that at any given instant the velocities of the various particles of the body are the same as if the body were rotating about an axis perpendicular to the reference plane, called the *instantaneous axis of rotation*. We may therefore use the results obtained in Sec. 17.4 to express at any instant the kinetic energy of a rigid body in plane motion. Recalling that the instantaneous axis of rotation intersects the reference plane at a point C called the *instantaneous center of rotation* (Fig. 17.4), we denote by I_C the moment of inertia of the body about the instantaneous axis. If the body has an angular velocity $\boldsymbol{\omega}$ at the instant considered, its kinetic energy is therefore

$$T = \tfrac{1}{2}I_C\omega^2 \qquad (17.10)$$

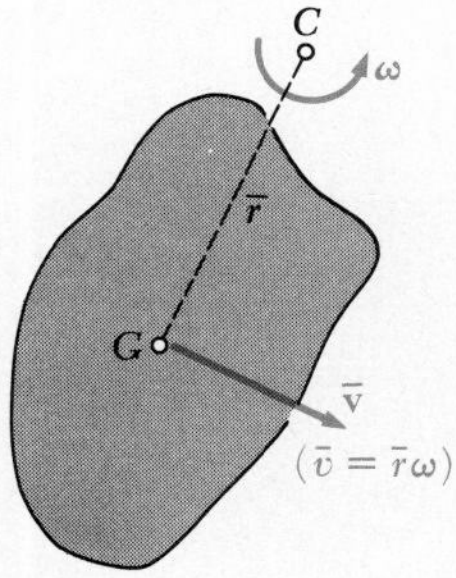

Fig. 17.4

But we know from the parallel-axis theorem (Sec. 9.11) that $I_C = m\bar{r}^2 + \bar{I}$, where $\bar{I}$ is the moment of inertia of the body about a centroidal axis perpendicular to the reference plane and $\bar{r}$ the distance from the instantaneous center C to the mass center G. Substituting for I_C into (17.10), we write

$$T = \tfrac{1}{2}(m\bar{r}^2 + \bar{I})\omega^2 = \tfrac{1}{2}m(\bar{r}\omega)^2 + \tfrac{1}{2}\bar{I}\omega^2$$

or, since $\bar{r}\omega$ represents the magnitude $\bar{v}$ of the velocity of the mass center,

$$T = \tfrac{1}{2}m\bar{v}^2 + \tfrac{1}{2}\bar{I}\omega^2 \qquad (17.11)$$

Either of the expressions (17.10) and (17.11) may be entered in formula (17.1) when the method of work and energy is used to analyze the plane motion of a rigid body.

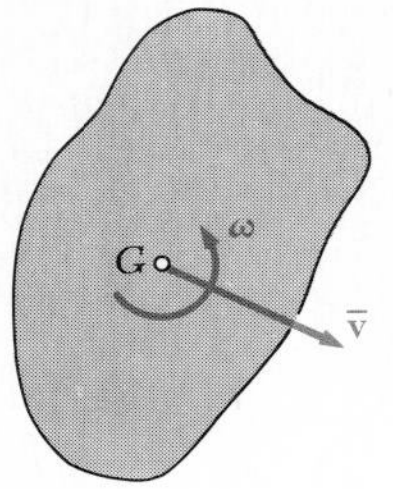

Fig. 17.5

We note that the expression given for the kinetic energy in (17.11) is separated into two distinct parts: (1) the kinetic energy $\frac{1}{2}m\bar{v}^2$ corresponding to the motion of the mass center G, and (2) the kinetic energy $\frac{1}{2}\bar{I}\omega^2$ corresponding to the rotation of the body about G (Fig. 17.5). This formula thus conforms to the unified approach we have used so far, and we shall apply it in preference to (17.10) in the Sample Problems of this chapter.

We also note that the results obtained in this section are not limited to the motion of plane slabs or to the motion of bodies which are symmetrical with respect to the reference plane. They may be applied to the study of the plane motion of any rigid body, regardless of its shape.

17.7. Conservation of Energy. We saw in Sec. 13.6 that the work of conservative forces, such as the weight of a body or the force exerted by a spring, may be expressed as a change in potential energy. When a rigid body, or a system of rigid bodies, moves under the action of conservative forces, the principle of work and energy stated in Sec. 17.1 may be expressed in a modified form. Substituting for $U_{1\to2}$ from (13.17) into (17.1), we write

$$T_1 + V_1 = T_2 + V_2 \tag{17.12}$$

Formula (17.12) indicates that, when a rigid body, or a system of rigid bodies, moves under the action of conservative forces, *the sum of the kinetic energy and of the potential energy of the system remains constant.* It should be noted that, in the case of the plane motion of a rigid body, the kinetic energy of the body should include both the *translational* term $\frac{1}{2}m\bar{v}^2$ and the *rotational* term $\frac{1}{2}\bar{I}\omega^2$.

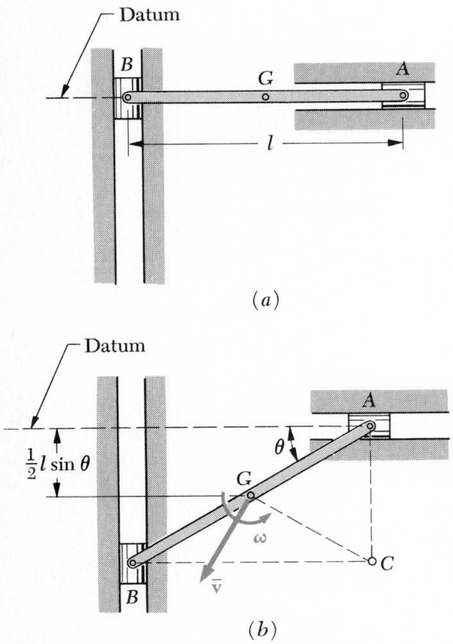

Fig. 17.6

As an example of application of the principle of conservation of energy, we shall consider a slender rod AB, of length l and mass m, whose extremities are connected to blocks of negligible mass sliding along horizontal and vertical tracks. We assume that the rod is released with no initial velocity from a horizontal position (Fig. 17.6*a*), and we wish to determine its angular velocity after it has rotated through an angle θ (Fig. 17.6*b*).

Since the initial velocity is zero, we have $T_1 = 0$. Measuring the potential energy from the level of the horizontal track, we write $V_1 = 0$. After the rod has rotated through θ, the center of gravity G of the rod is at a distance $\frac{1}{2}l\sin\theta$ below the reference level and we have

$$V_2 = -\tfrac{1}{2}Wl\sin\theta = -\tfrac{1}{2}mgl\sin\theta$$

Observing that, in this position, the instantaneous center of the rod is located at C, and that $CG = \frac{1}{2}l$, we write $\bar{v}_2 = \frac{1}{2}l\omega$ and obtain

$$T_2 = \tfrac{1}{2}m\bar{v}_2^2 + \tfrac{1}{2}\bar{I}\omega_2^2 = \tfrac{1}{2}m(\tfrac{1}{2}l\omega)^2 + \tfrac{1}{2}(\tfrac{1}{12}ml^2)\omega^2$$
$$= \frac{1}{2}\frac{ml^2}{3}\omega^2$$

Applying the principle of conservation of energy, we write

$$T_1 + V_1 = T_2 + V_2$$
$$0 = \frac{1}{2}\frac{ml^2}{3}\omega^2 - \tfrac{1}{2}mgl\sin\theta$$
$$\omega = \left(\frac{3g}{l}\sin\theta\right)^{1/2}$$

We recall that the advantages of the method of work and energy, as well as its shortcomings, were indicated in Sec. 13.4. In this connection, we wish to mention that the method of work and energy must be supplemented by the method of dynamic equilibrium when reactions at fixed axles, at rollers, or at sliding blocks are to be determined. For example, in order to compute the reactions at the extremities A and B of the rod of Fig. 17.6*b*, a diagram should be drawn to express that the system of the external forces applied to the rod is equivalent to the vector $m\bar{\mathbf{a}}$ and the couple $\bar{I}\boldsymbol{\alpha}$. The angular velocity $\boldsymbol{\omega}$ of the rod, however, is determined by the method of work and energy before the equations of motion are solved for the reactions. The complete analysis of the motion of the rod and of the forces exerted on the rod requires, therefore, the combined use of the method of work and energy and of the principle of equivalence of the external and effective forces.

17.8. Power. *Power* was defined in Sec. 13.8 as the time rate at which work is done. In the case of a body acted upon by a force **F**, and moving with a velocity **v**, the power was expressed as follows [Eq. (13.20)]:

$$\text{Power} = \frac{dU}{dt} = \frac{(F\cos\alpha)\,ds}{dt} = (F\cos\alpha)v$$

where α is the angle between **F** and **v**. In the case of a rigid body acted upon by a couple **M** and rotating at an angular velocity $\boldsymbol{\omega}$, the power may be expressed in a similar way in terms of M and ω,

$$\text{Power} = \frac{dU}{dt} = \frac{M\,d\theta}{dt} = M\omega \qquad (17.13)$$

The various units used to measure power, such as the watt and the horsepower, were defined in Sec. 13.8.

SAMPLE PROBLEM 17.3

A sphere, a cylinder, and a hoop, each having the same mass and the same radius, are released from rest on an incline. Determine the velocity of each body after it has rolled through a distance corresponding to a change in elevation h.

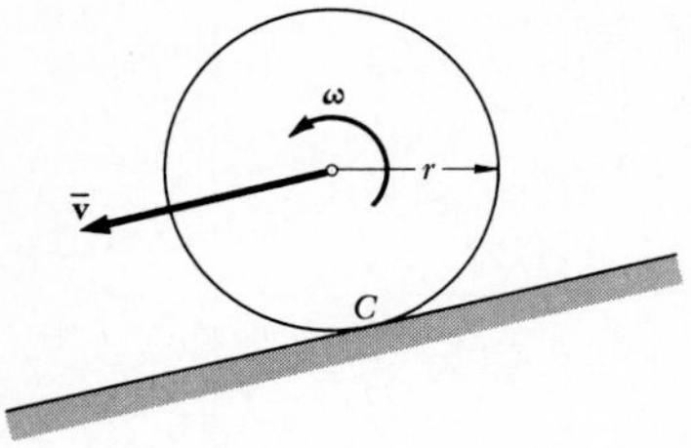

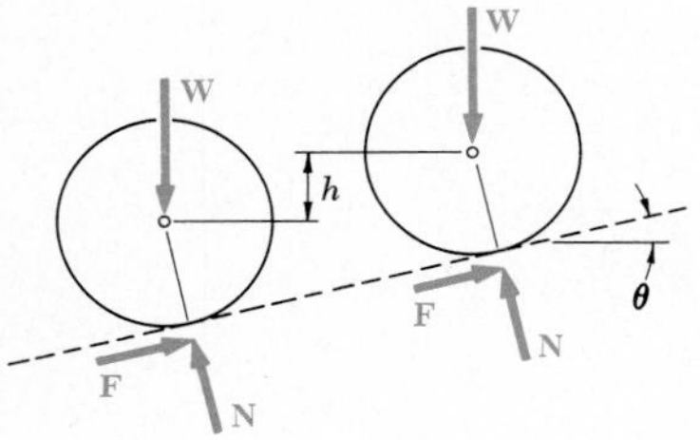

Solution. We shall first solve the problem in general terms and then find particular results for each body. We denote the mass by m, the centroidal moment of inertia by $\bar{I}$, the weight by W, and the radius by r.

Since each body rolls, the instantaneous center of rotation is located at C and we write

$$\omega = \frac{\bar{v}}{r}$$

Kinetic Energy

$$T_1 = 0$$
$$T_2 = \tfrac{1}{2}m\bar{v}^2 + \tfrac{1}{2}\bar{I}\omega^2$$
$$= \tfrac{1}{2}m\bar{v}^2 + \tfrac{1}{2}\bar{I}\left(\frac{\bar{v}}{r}\right)^2 = \tfrac{1}{2}\left(m + \frac{\bar{I}}{r^2}\right)\bar{v}^2$$

Work. Since the friction force **F** in rolling motion does no work,

$$U_{1\to2} = Wh$$

Principle of Work and Energy

$$T_1 + U_{1\to2} = T_2$$
$$0 + Wh = \tfrac{1}{2}\left(m + \frac{\bar{I}}{r^2}\right)\bar{v}^2 \qquad \bar{v}^2 = \frac{2Wh}{m + \bar{I}/r^2}$$

Noting that $W = mg$, we rearrange the result and obtain

$$\bar{v}^2 = \frac{2gh}{1 + \bar{I}/mr^2}$$

Velocities of Sphere, Cylinder, and Hoop. Introducing successively the particular expressions for $\bar{I}$, we obtain

Sphere:	$\bar{I} = \frac{2}{5}mr^2$	$\bar{v} = 0.845\sqrt{2gh}$ ◄
Cylinder:	$\bar{I} = \frac{1}{2}mr^2$	$\bar{v} = 0.816\sqrt{2gh}$ ◄
Hoop:	$\bar{I} = mr^2$	$\bar{v} = 0.707\sqrt{2gh}$ ◄

Remark. We may compare the results with the velocity attained by a frictionless block sliding through the same distance. The solution is identical to the above solution except that $\omega = 0$; we find $\bar{v} = \sqrt{2gh}$.

Comparing the results, we note that the velocity of the body is independent of both its mass and radius. However, the velocity does depend upon the quotient $\bar{I}/mr^2 = \bar{k}^2/r^2$, which measures the ratio of the rotational kinetic energy to the translational kinetic energy. Thus the hoop, which has the largest $\bar{k}$ for a given radius r, attains the smallest velocity, while the sliding block, which does not rotate, attains the largest velocity.

SAMPLE PROBLEM 17.4

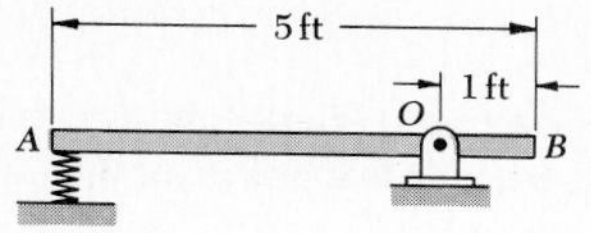

A 30-lb slender rod *AB* is 5 ft long and is pivoted about a point *O* which is 1 ft from end *B*. The other end is pressed against a spring of constant $k = 1800$ lb/in. until the spring is compressed 1 in. The rod is then in a horizontal position. If the rod is released from this position, determine its angular velocity and the reaction at the pivot *O* as the rod passes through a vertical position.

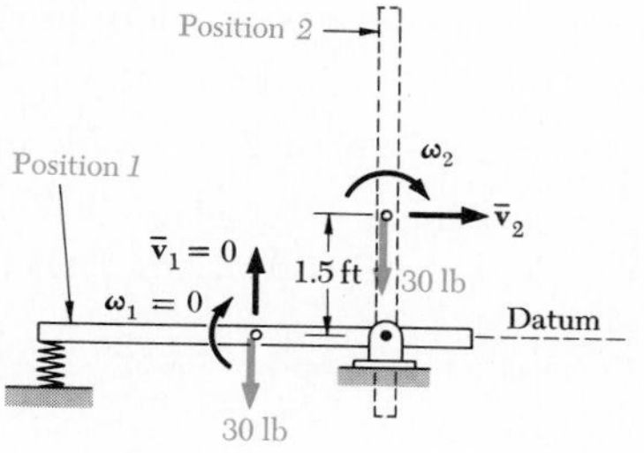

Position 1. *Potential Energy.* Since the spring is compressed 1 in., we have $x_1 = 1$ in.

$$V_e = \tfrac{1}{2}kx_1^2 = \tfrac{1}{2}(1800 \text{ lb/in.})(1 \text{ in.})^2 = 900 \text{ in}\cdot\text{lb}$$

Choosing the datum as shown, we have $V_g = 0$; therefore,

$$V_1 = V_e + V_g = 900 \text{ in}\cdot\text{lb} = 75 \text{ ft}\cdot\text{lb}$$

Kinetic Energy. Since the velocity in position *1* is zero, we have $T_1 = 0$.

Position 2. *Potential Energy.* The elongation of the spring is zero, and we have $V_e = 0$. Since the center of gravity of the rod is now 1.5 ft above the datum,

$$V_g = (30 \text{ lb})(+1.5 \text{ ft}) = 45 \text{ ft}\cdot\text{lb}$$
$$V_2 = V_e + V_g = 45 \text{ ft}\cdot\text{lb}$$

Kinetic Energy. Denoting by ω_2 the angular velocity of the rod in position *2*, we note that the rod rotates about *O* and write $\bar{v}_2 = \bar{r}\omega_2 = 1.5\omega_2$.

$$\bar{I} = \tfrac{1}{12}ml^2 = \frac{1}{12}\frac{30 \text{ lb}}{32.2 \text{ ft/s}^2}(5 \text{ ft})^2 = 1.941 \text{ lb}\cdot\text{ft}\cdot\text{s}^2$$

$$T_2 = \tfrac{1}{2}m\bar{v}_2^2 + \tfrac{1}{2}\bar{I}\omega_2^2 = \frac{1}{2}\frac{30}{32.2}(1.5\omega_2)^2 + \tfrac{1}{2}(1.941)\omega_2^2 = 2.019\omega_2^2$$

Conservation of Energy

$$T_1 + V_1 = T_2 + V_2: \qquad 0 + 75 \text{ ft}\cdot\text{lb} = 2.019\omega_2^2 + 45 \text{ ft}\cdot\text{lb}$$

$$\omega_2 = 3.86 \text{ rad/s} \circlearrowright \blacktriangleleft$$

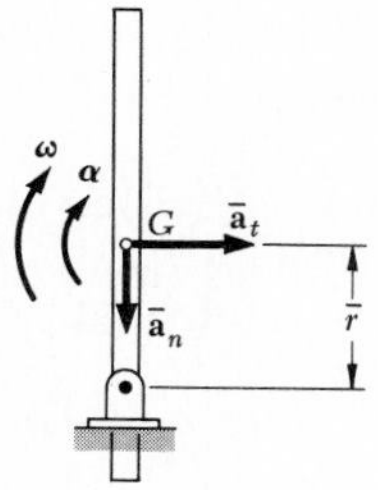

Reaction in Position 2. Since $\omega_2 = 3.86$ rad/s, the components of the acceleration of *G* as the rod passes through position *2* are

$$\bar{a}_n = \bar{r}\omega_2^2 = (1.5 \text{ ft})(3.86 \text{ rad/s})^2 = 22.3 \text{ ft/s}^2 \qquad \bar{\mathbf{a}}_n = 22.3 \text{ ft/s}^2 \downarrow$$
$$\bar{a}_t = \bar{r}\alpha \qquad \bar{\mathbf{a}}_t = \bar{r}\alpha \rightarrow$$

We express that the system of external forces is equivalent to the system of effective forces represented by the vector of components $m\bar{\mathbf{a}}_t$ and $m\bar{\mathbf{a}}_n$ attached at *G* and the couple $\bar{I}\boldsymbol{\alpha}$.

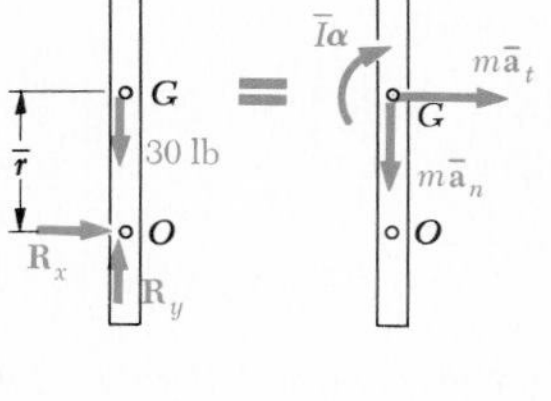

$$+\circlearrowright \Sigma M_O = \Sigma(M_O)_{\text{eff}}: \qquad 0 = \bar{I}\alpha + m(\bar{r}\alpha)\bar{r} \qquad \alpha = 0$$
$$\xrightarrow{+} \Sigma F_x = \Sigma(F_x)_{\text{eff}}: \qquad R_x = m(\bar{r}\alpha) \qquad R_x = 0$$
$$+\uparrow \Sigma F_y = \Sigma(F_y)_{\text{eff}}: \qquad R_y - 30 \text{ lb} = -m\bar{a}_n$$

$$R_y - 30 \text{ lb} = -\frac{30 \text{ lb}}{32.2 \text{ ft/s}^2}(22.3 \text{ ft/s}^2)$$

$$R_y = +9.22 \text{ lb} \qquad \mathbf{R} = 9.22 \text{ lb} \uparrow \blacktriangleleft$$

SAMPLE PROBLEM 17.5

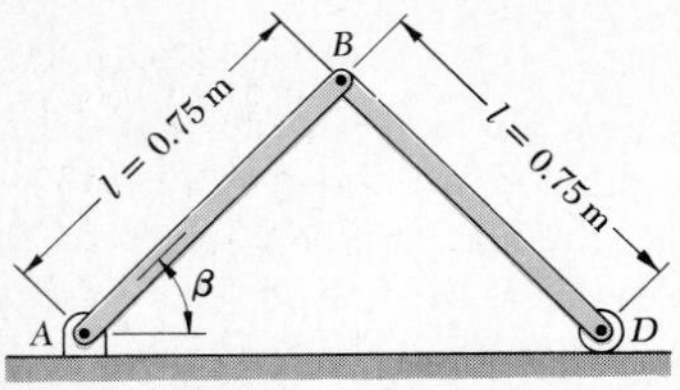

Each of the two slender rods shown is 0.75 m long and has a mass of 6 kg. If the system is released from rest when $\beta = 60°$, determine (*a*) the angular velocity of rod *AB* when $\beta = 20°$, (*b*) the velocity of point *D* at the same instant.

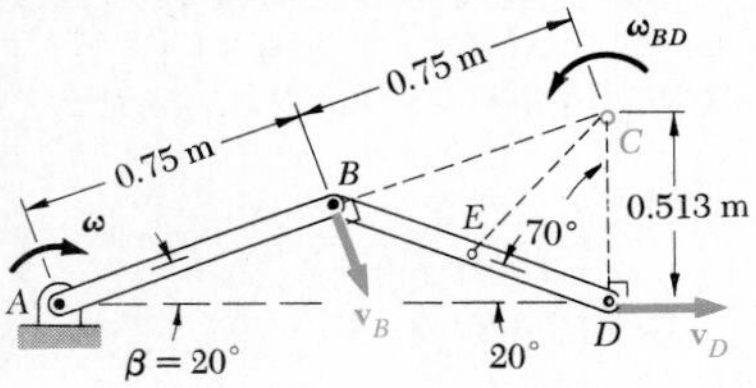

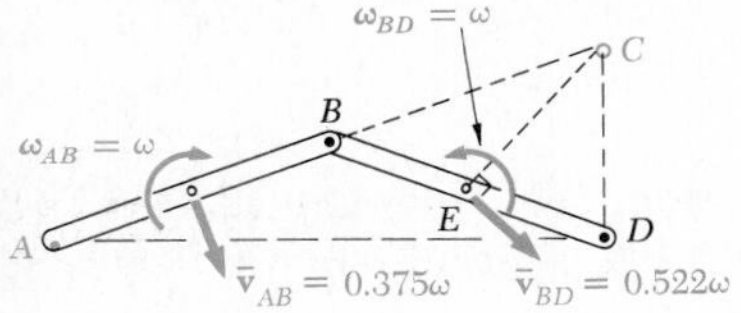

Kinematics of Motion When $\beta = 20°$. Since $\mathbf{v}_B$ is perpendicular to the rod *AB* and $\mathbf{v}_D$ is horizontal, the instantaneous center of rotation of rod *BD* is located at *C*. Considering the geometry of the figure, we obtain

$$BC = 0.75 \text{ m} \qquad CD = 2(0.75 \text{ m}) \sin 20° = 0.513 \text{ m}$$

Applying the law of cosines to triangle *CDE*, where *E* is located at the mass center of rod *BD*, we find $EC = 0.522$ m. Denoting by ω the angular velocity of rod *AB*, we have

$$\bar{v}_{AB} = (0.375 \text{ m})\omega \qquad \bar{\mathbf{v}}_{AB} = 0.375\omega \searrow$$
$$v_B = (0.75 \text{ m})\omega \qquad \mathbf{v}_B = 0.75\omega \searrow$$

Since rod *BD* seems to rotate about point *C*, we may write

$$v_B = (BC)\omega_{BD} \qquad (0.75 \text{ m})\omega = (0.75 \text{ m})\omega_{BD} \qquad \omega_{BD} = \omega \circlearrowleft$$
$$\bar{v}_{BD} = (EC)\omega_{BD} = (0.522 \text{ m})\omega \qquad \bar{\mathbf{v}}_{BD} = 0.522\omega \searrow$$

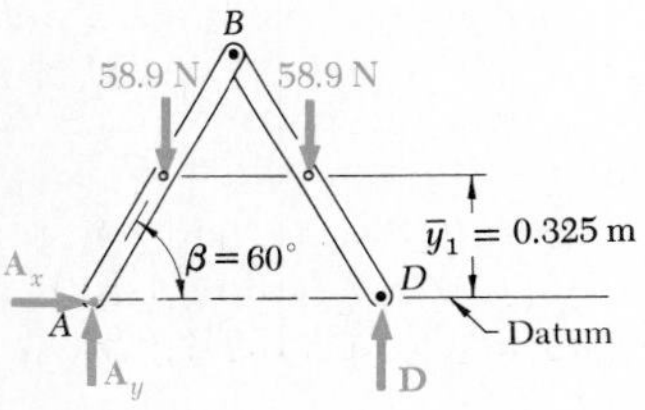

Position *1*

Position 1. *Potential Energy.* Choosing the datum as shown, and observing that $W = (6 \text{ kg})(9.81 \text{ m/s}^2) = 58.9$ N, we have

$$V_1 = 2W\bar{y}_1 = 2(58.9 \text{ N})(0.325 \text{ m}) = 38.3 \text{ J}$$

Kinetic Energy. Since the system is at rest, $T_1 = 0$.

Position 2. *Potential Energy*

$$V_2 = 2W\bar{y}_2 = 2(58.9 \text{ N})(0.128 \text{ m}) = 15.1 \text{ J}$$

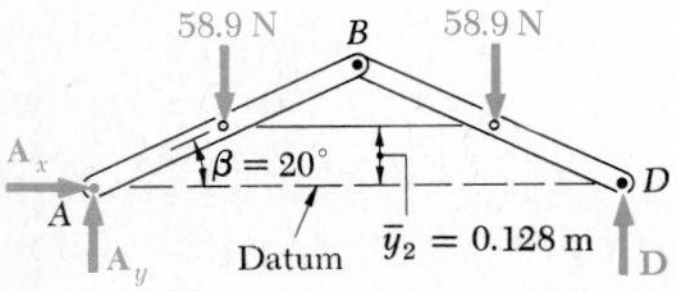

Position 2

Kinetic Energy

$$\bar{I}_{AB} = \bar{I}_{BD} = \tfrac{1}{12}ml^2 = \tfrac{1}{12}(6 \text{ kg})(0.75 \text{ m})^2 = 0.281 \text{ kg} \cdot \text{m}^2$$
$$\begin{aligned} T_2 &= \tfrac{1}{2}m\bar{v}_{AB}^2 + \tfrac{1}{2}\bar{I}_{AB}\omega_{AB}^2 + \tfrac{1}{2}m\bar{v}_{BD}^2 + \tfrac{1}{2}\bar{I}_{BD}\omega_{BD}^2 \\ &= \tfrac{1}{2}(6)(0.375\omega)^2 + \tfrac{1}{2}(0.281)\omega^2 + \tfrac{1}{2}(6)(0.522\omega)^2 + \tfrac{1}{2}(0.281)\omega^2 \\ &= 1.520\omega^2 \end{aligned}$$

Conservation of Energy

$$T_1 + V_1 = T_2 + V_2: \qquad 0 + 38.3 \text{ J} = 1.520\omega^2 + 15.1 \text{ J}$$
$$\omega = 3.91 \text{ rad/s} \qquad \omega_{AB} = 3.91 \text{ rad/s} \circlearrowright \blacktriangleleft$$

Velocity of Point *D*

$$v_D = (CD)\omega = (0.513 \text{ m})(3.91 \text{ rad/s}) = 2.01 \text{ m/s}$$
$$\mathbf{v}_D = 2.01 \text{ m/s} \rightarrow \blacktriangleleft$$

PROBLEMS

17.24 A cord is wrapped around a cylinder of radius r and mass m as shown. If the cylinder is released from rest, determine the velocity of the center of the cylinder after it has moved through a distance h.

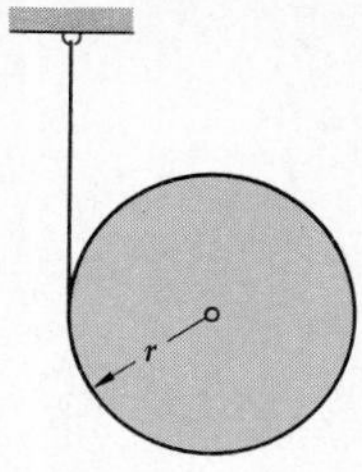

Fig. P17.24

17.25 Two 10-kg disks, each of radius $r = 0.3$ m, are connected by a cord. At the instant shown, the angular velocity of disk B is 20 rad/s clockwise. Determine how far disk A will rise before the angular velocity of disk B is reduced to 4 rad/s.

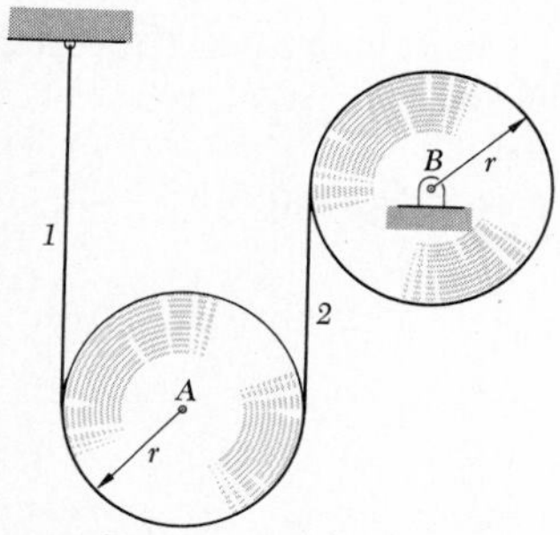

Fig. P17.25 and P17.26

17.26 Two identical disks, each of weight W and radius r, are connected by a cord as shown. If the system is released from rest, determine (*a*) the velocity of the center of disk A after it has moved through a distance h, (*b*) the tension in portion 2 of the cord.

17.27 A half cylinder of weight W and radius r is released from rest in the position shown. Assuming that the half cylinder rolls without sliding, determine (*a*) its angular velocity after it has rolled through 90°, (*b*) the normal reaction at the surface at the same instant. [*Hint.* Note that $GO = 4r/3\pi$ and that, by the parallel-axis theorem, $\bar{I} = \frac{1}{2}mr^2 - m(GO)^2$.]

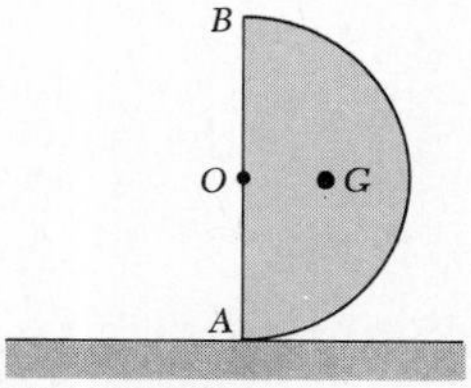

Fig. P17.27

17.28 A sphere of weight W and radius r rolls without slipping inside a curved surface of radius R. Knowing that the sphere is released from rest in the position shown, derive an expression (*a*) for the linear velocity of the sphere as it passes through B, (*b*) for the magnitude of the vertical reaction at that instant.

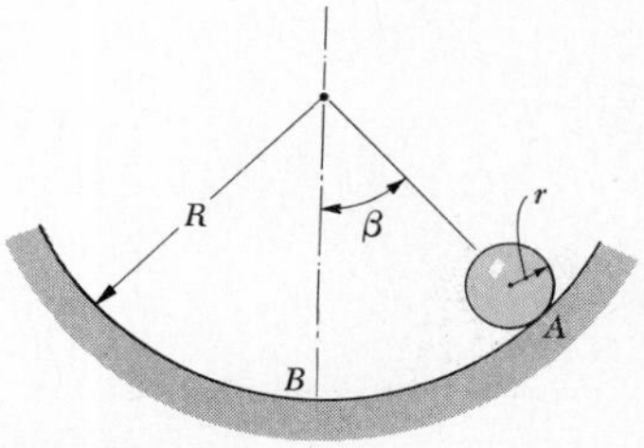

Fig. P17.28

17.29 Solve Prob. 17.28, assuming that the sphere is replaced by a uniform cylinder of weight W and radius r.

17.30 A slender rod of length l and mass m is pivoted at one end as shown. It is released from rest in a horizontal position and swings freely. (*a*) Determine the angular velocity of the rod as it passes through a vertical position and the corresponding reaction at the pivot. (*b*) Solve part *a* for $m = 1.5$ kg and $l = 0.9$ m.

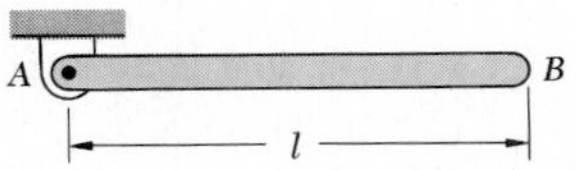

Fig. P17.30

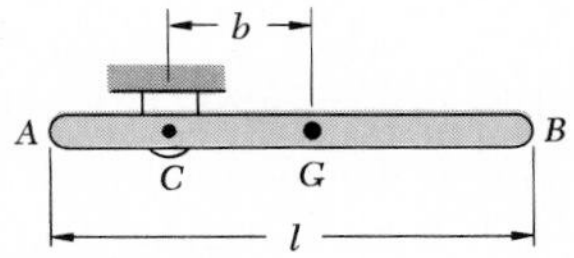

Fig. P17.31

17.31 A uniform rod of length l is pivoted about a point C located at a distance b from its center G. The rod is released from rest in a horizontal position. Determine (*a*) the distance b so that the angular velocity of the rod as it passes through a vertical position is maximum, (*b*) the value of the maximum angular velocity.

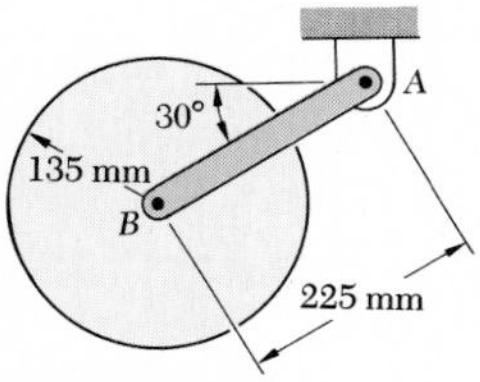

Fig. P17.32

17.32 A 15-kg disk is attached to a 6-kg rod AB as shown. The system is released from rest in the position shown. Determine the velocity of point B as it passes through its lowest position, (*a*) assuming that the disk is welded to the rod, (*b*) assuming that the disk is attached to the rod by a frictionless pin at B.

17.33 and 17.34 Gear C weighs 6 lb and has a centroidal radius of gyration of 3 in. The uniform bar AB weighs 5 lb, and gear D is stationary. If the system is released from rest in the position shown, determine the velocity of point B after bar AB has rotated through 90°.

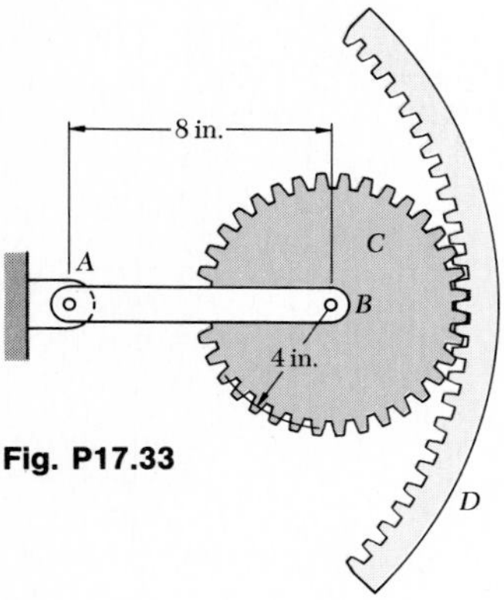

Fig. P17.33

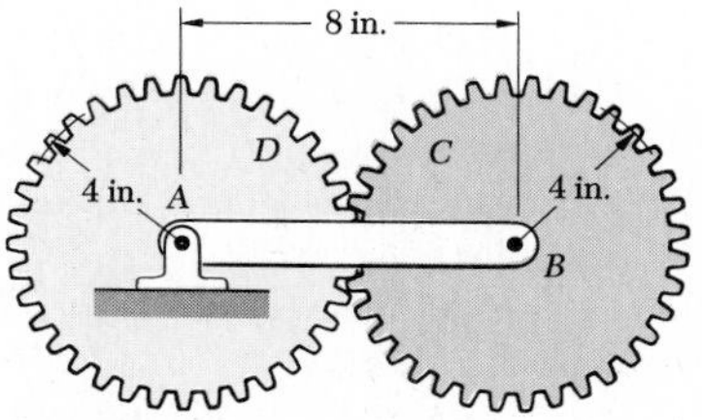

Fig. P17.34

17.35 The mass center G of a 1.5-kg wheel of radius $R = 150$ mm is located at a distance $r = 50$ mm from its geometric center C. The centroidal radius of gyration of the wheel is $\bar{k} = 75$ mm. As the wheel rolls without sliding, its angular velocity is observed to vary. Knowing that in position *1* the angular velocity is 10 rad/s, determine the angular velocity of the wheel (*a*) in position *2*, (*b*) in position *3*.

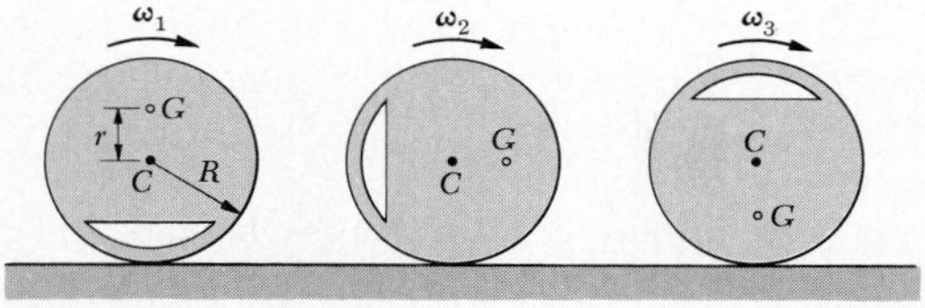

Fig. P17.35 and P17.36

17.36 The mass center G of a wheel of radius R is located at a distance r from its geometric center C. The centroidal radius of gyration of the wheel is denoted by $\bar{k}$. As the wheel rolls freely and without sliding on a horizontal plane, its angular velocity is observed to vary. Denoting by ω_1, ω_2, and ω_3, respectively, the angular velocity of the wheel when G is directly above C, level with C, and directly below C, show that ω_1, ω_2, and ω_3 satisfy the relation

$$\frac{\omega_2^2 - \omega_1^2}{\omega_3^2 - \omega_2^2} = \frac{g/R + \omega_1^2}{g/R + \omega_3^2}$$

17.37 The motion of a 0.6-m slender rod is guided by pins at A and B which slide freely in the slots shown. Knowing that the rod is released from rest when $\theta = 0$ and that end A is given a slight push to the right, determine (*a*) the angle θ for which the speed of end A is maximum, (*b*) the corresponding maximum speed of A.

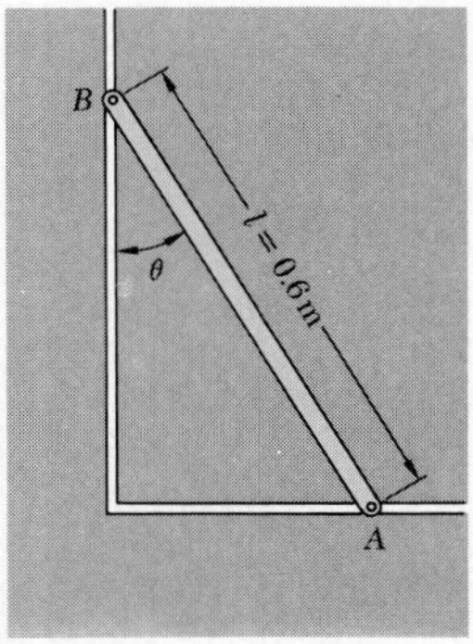

Fig. P17.37

17.38 In Prob. 17.37, determine the velocity of ends A and B (*a*) when $\theta = 30°$, (*b*) when $\theta = 90°$.

17.39 and 17.40 The 15-lb carriage is supported as shown by two uniform disks, each of weight 10 lb and radius 3 in. Knowing that the system is initially at rest, determine the velocity of the carriage after it has moved 2 ft. Assume that the disks roll without sliding.

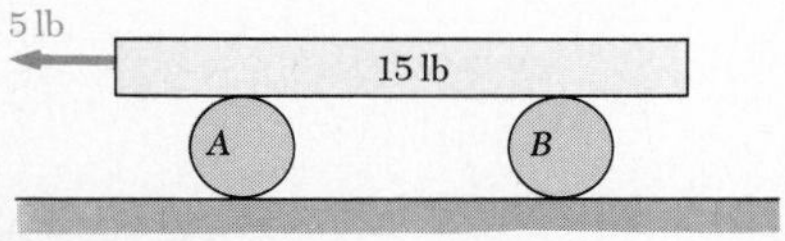

Fig. P17.39

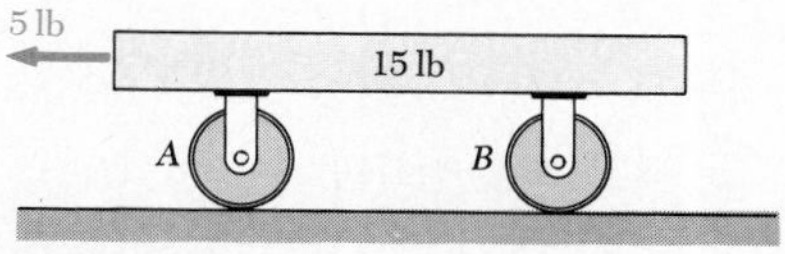

Fig. P17.40

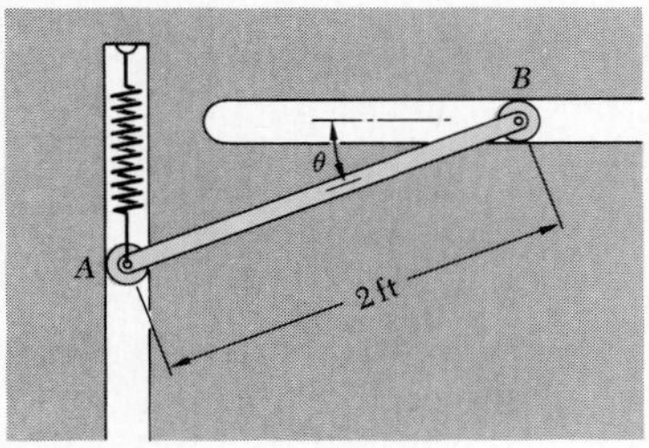

Fig. P17.41 and P17.42

17.41 The ends of a 25-lb rod AB are constrained to move along the slots shown. A spring of constant 3 lb/in. is attached to end A. Knowing that the rod is released from rest when $\theta = 0$ and that the initial tension in the spring is zero, determine the maximum distance through which end A will move.

17.42 The ends of a 25-lb rod AB are constrained to move along the slots shown. A spring of constant 10 lb/in. is attached to end A in such a way that its tension is zero when $\theta = 0$. If the rod is released from rest when $\theta = 60°$, determine the angular velocity of the rod when $\theta = 30°$.

17.43 The uniform rods AB and BC are of mass 4.5 and 1.5 kg respectively. If the system is released from rest in the position shown, determine the angular velocity of rod BC as it passes through a vertical position.

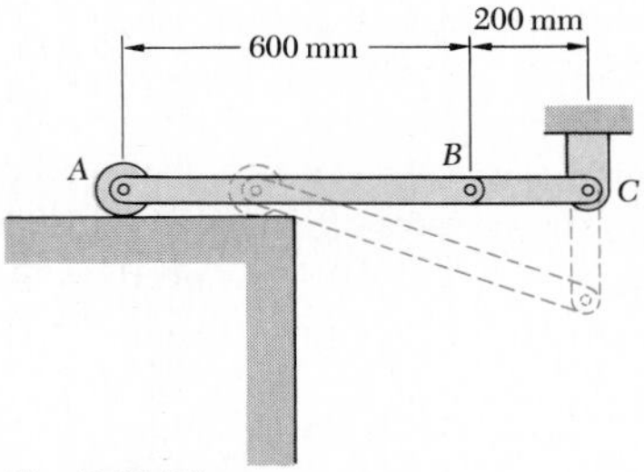

Fig. P17.43

17.44 Each of the two rods shown is of length $L = 3$ ft and weight $W = 5$ lb. Point D is constrained to move along a vertical line. If the system is released from rest when rod AB is horizontal, determine the velocity of points B and D as rod BD passes through a horizontal position.

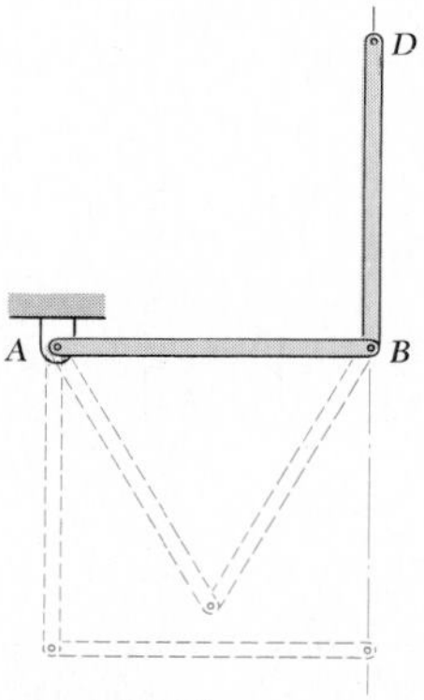

Fig. P17.44

17.45 In Prob. 17.44, determine the velocity of points B and D when points A and D are at the same elevation.

17.46 Determine the velocity of pin B as the rods of Sample Prob. 17.5 strike the horizontal surface.

17.47 A uniform rod of length L and weight W is attached to two wires, each of length b. The rod is released from rest when $\theta = 0$ and swings to the position $\theta = 90°$, at which time wire BD suddenly breaks. Determine the tension in wire AC (*a*) immediately before wire BD breaks, (*b*) immediately after wire BD breaks.

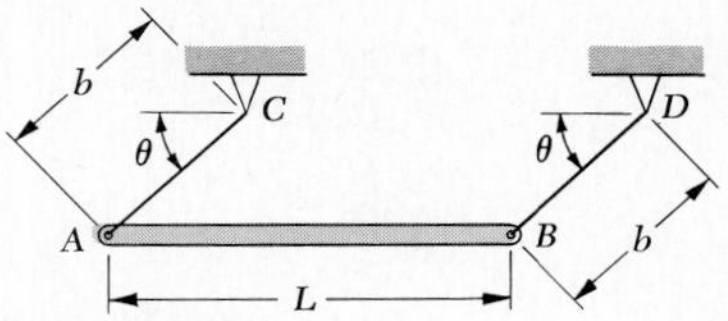

Fig. P17.47

***17.48** A small matchbox is placed on top of the rod AB. End B of the rod is given a slight horizontal push, causing it to slide on the horizontal floor. Assuming no friction and neglecting the weight of the matchbox, determine the angle θ through which the rod will have rotated when the matchbox loses contact with the rod.

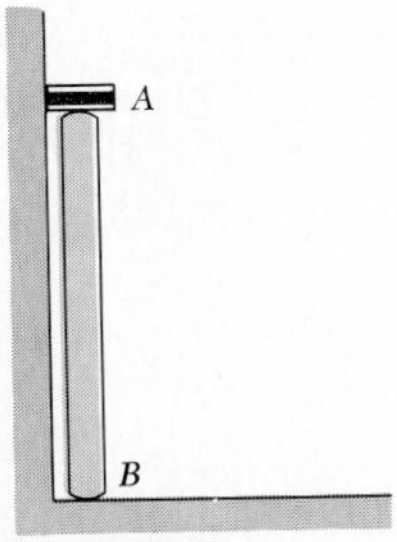

Fig. P17.48

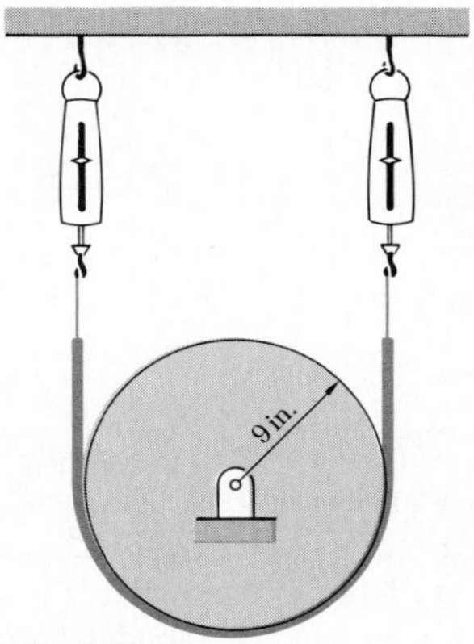

Fig. P17.49

17.49 The experimental setup shown is used to measure the power output of a small turbine. When the turbine is operating at 200 rpm, the readings of the two spring scales are 10 and 22 lb, respectively. Determine the power being developed by the turbine.

17.50 In Sample Prob. 17.2 determine the power being delivered to gear B at the instant when (*a*) the gear starts rotating, (*b*) the gear attains an angular velocity $\omega_B = 300$ rpm.

17.51 Knowing that the maximum allowable couple which can be applied to a shaft is 12 kN · m, determine the maximum power which can be transmitted by the shaft (*a*) at 100 rpm, (*b*) at 1000 rpm.

17.52 Determine the moment of the couple which must be exerted by a motor to develop $\frac{1}{4}$ hp at a speed of (*a*) 3600 rpm, (*b*) 720 rpm.

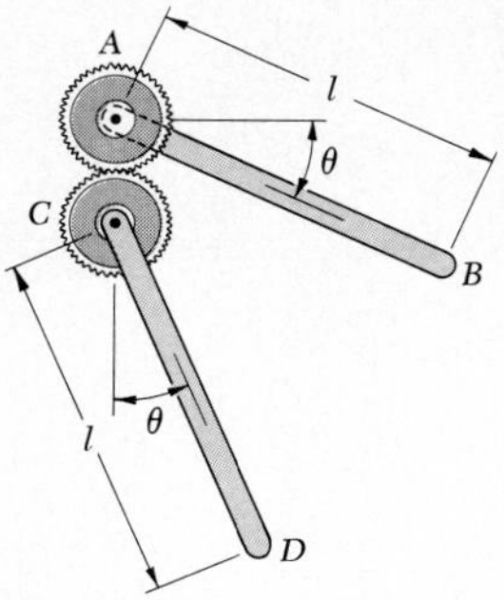

Fig. P17.53

REVIEW PROBLEMS

17.53 Two uniform rods, each of mass m, are attached to two gears of the same radius as shown. The rods are released from rest in the position $\theta = 0$. Neglecting the mass of the gears, determine the angular velocity of the rods when (*a*) $\theta = 30°$, (*b*) $\theta = 45°$, (*c*) $\theta = 60°$.

17.54 The motion of a 16-kg sliding panel is guided by rollers at *B* and *C*. The counterweight *A* has a mass of 12 kg and is attached to a cable as shown. If the system is released from rest, determine for each case shown the velocity of the counterweight as it strikes the ground. Neglect the effect of friction.

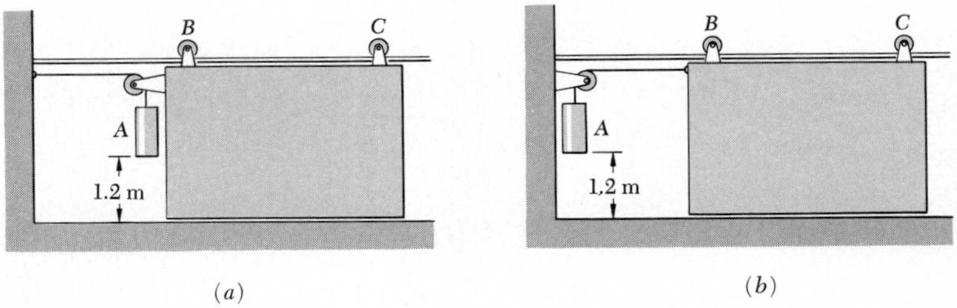

Fig. P17.54

17.55 The gear train shown consists of four gears of the same thickness and of the same material; two gears are of radius r, and the other two are of radius nr. The system is at rest when the couple $\mathbf{M}_0$ is applied to shaft *C*. Denoting by I_0 the moment of inertia of a gear of radius r, determine the angular velocity of shaft *A* if the couple $\mathbf{M}_0$ is applied for one revolution of shaft *C*.

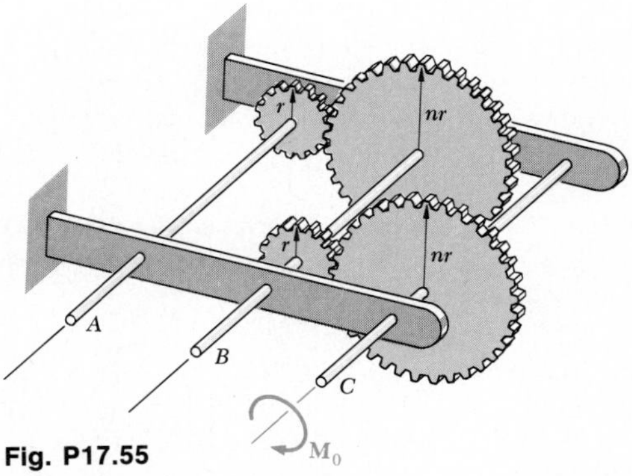

Fig. P17.55

17.56 Solve Prob. 17.55, assuming that the couple $\mathbf{M}_0$ is applied to shaft *A* for one revolution of shaft *A*.

17.57 A 200-lb bin is suspended from two rods AB and CD of negligible weight. The bin is released from rest in the position $\theta = 0$. As the bin passes through its lowest position, determine (*a*) the velocity of the bin, (*b*) the reactions at A and C. (*c*) Show that the answer to part *b* is independent of the length l of the rods.

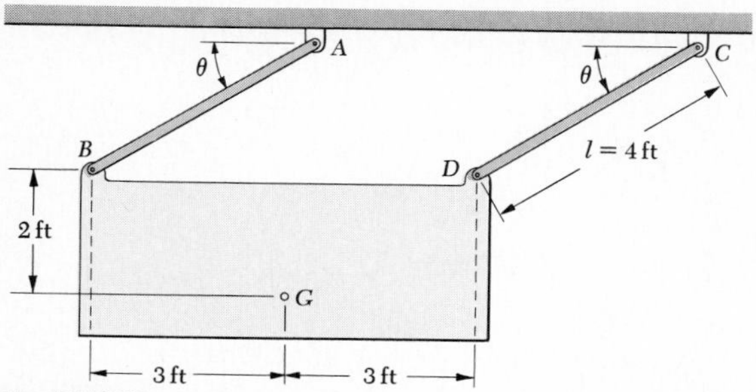

Fig. P17.57

17.58 Solve Prob. 17.57, assuming that each rod weighs 50 lb.

17.59 A flywheel of centroidal radius of gyration $\bar{k} = 600$ mm is rigidly attached to a shaft of radius $r = 30$ mm which may roll along parallel rails. Knowing that the system is released from rest, determine the velocity of the center of the shaft after it has moved 4 m.

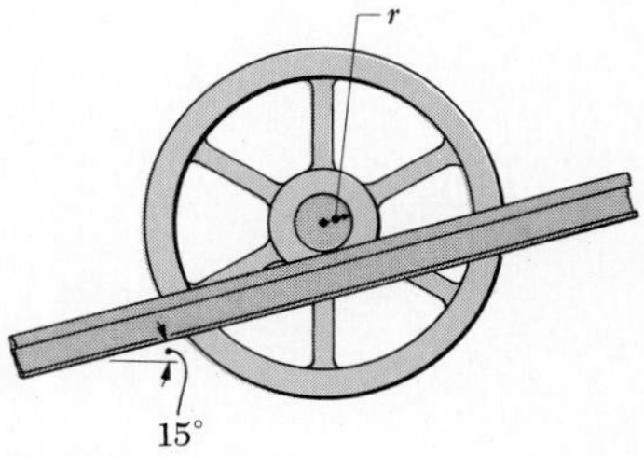

Fig. P17.59

17.60 A small collar of mass m is attached at B to the rim of a hoop of mass m and radius r. The hoop rolls without sliding on a horizontal plane. Find the angular velocity ω_1 of the hoop when B is directly above the center A in terms of g and r, knowing that the angular velocity of the hoop is $3\omega_1$ when B is directly below A.

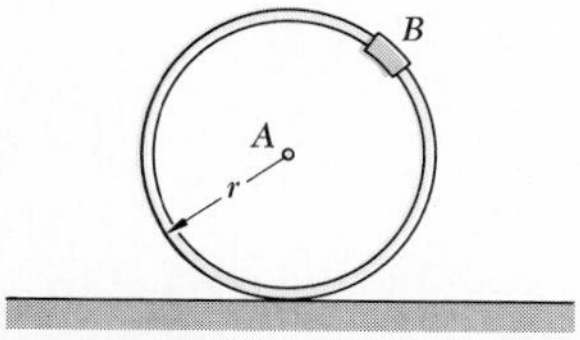

Fig. P17.60

17.61 A small disk A is driven at a constant angular velocity of 1200 rpm and is pressed against disk B, which is initially at rest. The normal force between disks is 10 lb, and $\mu_k = 0.20$. Knowing that disk B weighs 50 lb, determine the number of revolutions executed by disk B before its speed reaches 120 rpm.

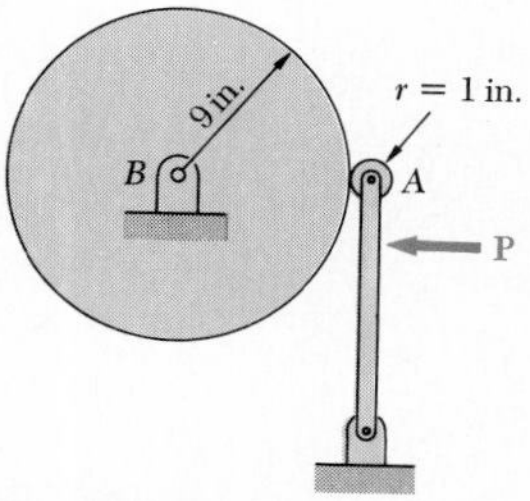

Fig. P17.61

17.62 Two uniform rods, AB of weight 4 lb and BC of weight 2 lb, are released from rest in the position shown. Neglecting both friction and the weight of the collar, determine the velocity of point B when $\theta = 90°$.

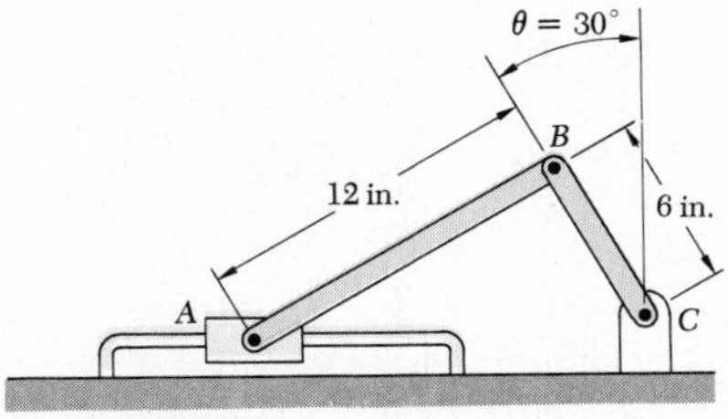

Fig. P17.62

17.63 Using the methods of Chap. 17, solve Prob. 16.169*b*.

17.64 Using the methods of Chap. 17, solve Prob. 16.170*b*.

Chapter

18

Kinetics of Rigid Bodies: Impulse and Momentum

18.1. Principle of Impulse and Momentum for a Rigid Body. In this chapter, the principle of impulse and momentum will be used to analyze the motion of rigid bodies and of systems of rigid bodies. As was pointed out in Chap. 14, the method of impulse and momentum is particularly well adapted to the solution of problems involving time and velocities. Moreover, the principle of impulse and momentum provides the only practicable method for the solution of problems of impact.

Since a rigid body of mass m may be considered as made of a large number of particles of mass Δm, Eq. (14.36), which was derived in Sec. 14.12 for a system of particles, may now be used to analyze the motion of a rigid body. We write

$$\textbf{Syst Momenta}_1 + \textbf{Syst Ext Imp}_{1\rightarrow 2} = \textbf{Syst Momenta}_2 \quad (18.1)$$

This equation relates the three systems of vectors shown in Fig. 18.1. It expresses that the system formed by the momenta of the particles of the rigid body at time t_1 and the system of the impulses of the external forces applied to the body from t_1 to t_2 are together equivalent to the system formed by the momenta of the particles at time t_2.

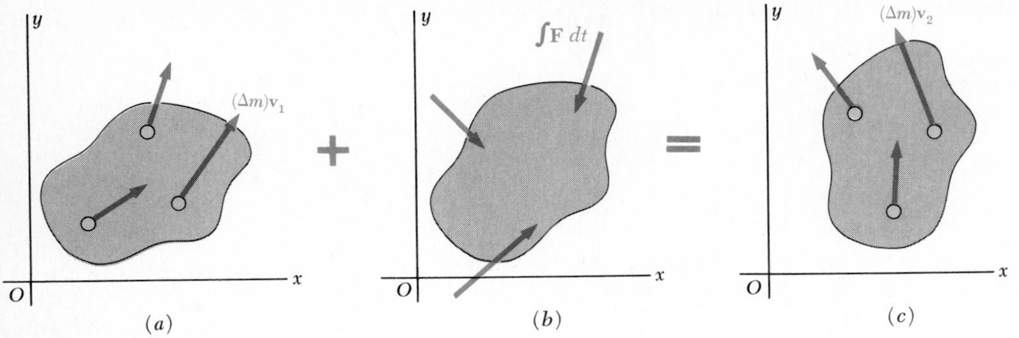

Fig. 18.1

We recall from Sec. 14.12 that, in the case of particles moving in the xy plane, three scalar equations are required to express the same relationship. The first two equations may be obtained by considering respectively the x and y components of the vectors shown in Fig. 18.1; they will relate the linear momenta of the particles and the linear impulses of the given forces in the x and y directions respectively. The third equation, which is obtained by computing the moments about O of the same vectors, will relate the angular momenta of the particles and the angular impulses of the forces.

Before we may conveniently apply the principle of impulse and momentum to the study of the plane motion of a rigid body, we must learn to express more simply the systems of vectors representing the momenta of the particles forming the rigid body. This is the object of the next section.

18.2. Momentum of a Rigid Body in Plane Motion. We shall see in this section that the system of the momenta of the particles forming a rigid body may be replaced at any given instant by an equivalent system consisting of a *momentum vector* attached at the mass center of the body and a *momentum couple*. This reduction of the system of momenta to a vector and a couple is similar to the reduction of the system of effective forces described in Chap. 16. While the method involved is quite general, our present analysis will be limited to the plane motion of rigid slabs and of rigid bodies which are symmetrical with respect to the reference plane.†

Translation. Consider first the case of a rigid body in translation. Since each particle, at any given instant, has the same velocity $\bar{\mathbf{v}}$ as the mass center G of the body, the momentum of a particle of mass Δm is represented by the vector $(\Delta m)\bar{\mathbf{v}}$.

† Or, more generally, to the motion of rigid bodies which have a principal axis of inertia perpendicular to the reference plane.

The resultant of the momenta of the various particles of the body is thus

$$\Sigma(\Delta m)\bar{\mathbf{v}} = (\Sigma\,\Delta m)\bar{\mathbf{v}} = m\bar{\mathbf{v}}$$

where m is the mass of the body. On the other hand, a reasoning similar to that used in Sec. 16.2 for the effective forces of a rigid body in translation shows that the sum of the moments about G of the momenta $(\Delta m)\bar{\mathbf{v}}$ is zero. We thus conclude that, in the case of a rigid body in translation, *the system of the momenta of the particles forming the body is equivalent to a single momentum vector* $m\bar{\mathbf{v}}$ *attached at the mass center* G (Fig. 18.2). This vector is the *linear momentum* of the body.

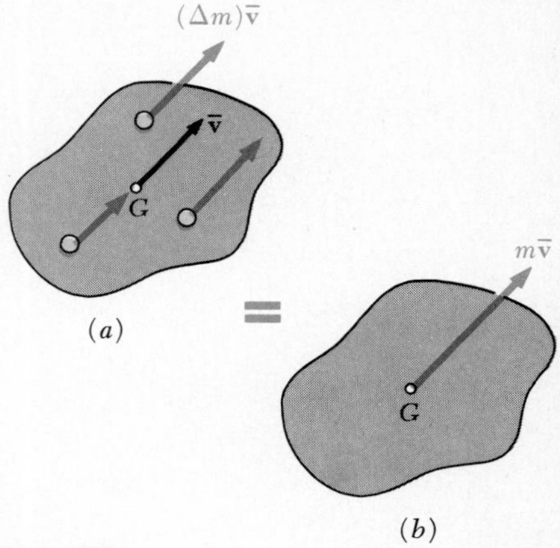

Fig. 18.2

Centroidal Rotation. When a rigid body rotates about a fixed centroidal axis perpendicular to the reference plane, the velocity $\bar{\mathbf{v}}$ of its mass center G is zero, since G is fixed, and it follows from Eqs. (14.10) of Sec. 14.3 that the sums of the x and y components of the momenta of the various particles of the body are zero:

$$\Sigma(\Delta m)v_x = (\Sigma\,\Delta m)\bar{v}_x = 0 \qquad \Sigma(\Delta m)v_y = (\Sigma\,\Delta m)\bar{v}_y = 0$$

Therefore, in the case of a rigid body in centroidal rotation, *the system of the momenta* of the particles forming the body *is equivalent to a single momentum couple.* If, as we have assumed in this section, the body is symmetrical with respect to the reference plane, the axis of the momentum couple coincides with the axis of rotation.

The moment of the momentum couple is obtained by summing the moments about G of the momenta $(\Delta m)\mathbf{v}$ of the various particles. This sum is denoted by H_G and referred to as the *angular momentum* of the body about the axis through G. Observing that the momentum $(\Delta m)\mathbf{v}$ of a particle P is perpendicular to the line GP (Fig. 18.3*a*) and that $v = r\omega$, where r is the distance from G to P and ω the angular velocity of the body at the instant considered, we write

$$+\circlearrowleft\, H_G = \Sigma r(\Delta m)v = \Sigma r(\Delta m)r\omega = \omega\Sigma r^2\,\Delta m$$

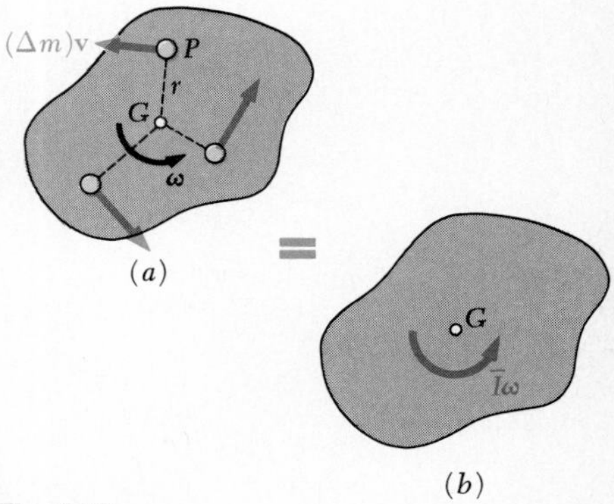

Fig. 18.3

Since $\Sigma r^2\,\Delta m$ represents the centroidal moment of inertia $\bar{I}$ of the body, we have

$$H_G = \bar{I}\omega \tag{18.2}$$

Recalling from Sec. 15.3 that the angular velocity of a rigid body rotating about a fixed axis may be represented by a vector $\boldsymbol{\omega}$ of magnitude ω directed along that axis, we note that the momentum couple may be represented by a couple vector $\bar{I}\boldsymbol{\omega}$ of magnitude $\bar{I}\omega$ directed along the axis of rotation. We therefore conclude that, in the case of the centroidal rotation of a rigid body symmetrical with respect to the reference plane, *the system of the momenta* of the particles of the body *is equivalent to a momentum couple* $\bar{I}\boldsymbol{\omega}$, where $\boldsymbol{\omega}$ denotes the angular velocity of the body (Fig. 18.3*b*).

General Plane Motion. In the general case of the plane motion of a rigid body, the velocity of any given particle P of the body may be resolved into the velocity $\bar{\mathbf{v}}$ of the mass center G and the relative velocity $\mathbf{v}'$ of P with respect to G. The

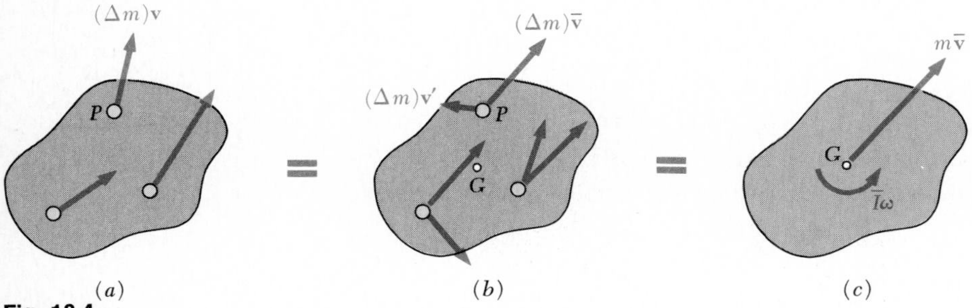

Fig. 18.4

momentum $(\Delta m)\mathbf{v}$ of each particle (Fig. 18.4*a*) may thus be resolved into the vectors $(\Delta m)\bar{\mathbf{v}}$ and $(\Delta m)\mathbf{v}'$ shown in Fig. 18.4*b*. Recalling the results obtained above in the particular cases of translation and centroidal rotation, we find that the vectors $(\Delta m)\bar{\mathbf{v}}$ may be replaced by their resultant $m\bar{\mathbf{v}}$ attached at G and that the vectors $(\Delta m)\mathbf{v}'$ reduce to the couple $\bar{I}\boldsymbol{\omega}$. While this vector and this couple may in turn be replaced by a single vector,† it is generally found more convenient to represent the system of momenta by the momentum vector $m\bar{\mathbf{v}}$ attached at G and the momentum couple $\bar{I}\boldsymbol{\omega}$ (Fig. 18.4*c*). The momentum vector is associated with the translation of the body with G and represents the *linear momentum* of the body. The momentum couple corresponds to the rotation of the body about G and its moment $\bar{I}\omega$ represents the *angular momentum* H_G *of the body about an axis through* G. Recalling the definition given in Sec. 14.12 for the angular momentum of a system of particles, we note that the angular momentum H_A of the body about an axis through an arbitrary point A may be obtained by adding algebraically the moment about A of the momentum vector $m\bar{\mathbf{v}}$ and the moment $\bar{I}\omega$ of the momentum couple.‡

† See Probs. 18.13 and 18.14.

‡ Note that, in general, H_A is *not* equal to $I_A\omega$ (see Prob. 18.15).

18.3. Application of the Principle of Impulse and Momentum to the Analysis of the Plane Motion of a Rigid Body. We saw in the preceding section that, in the most general case of the plane motion of a rigid body, the system of the momenta of the particles of the body reduces to a vector $m\bar{\mathbf{v}}$ attached at G and a couple $\bar{I}\boldsymbol{\omega}$. Replacing the system of momenta in parts a and c of Fig. 18.1 by the equivalent momentum vector and momentum couple, we obtain the three diagrams shown in Fig. 18.5. This figure expresses graphically, in the case

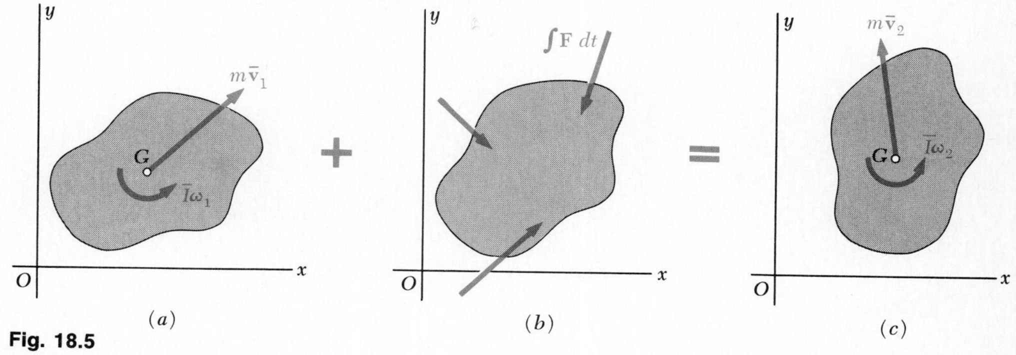

Fig. 18.5

of the plane motion of a rigid body symmetrical with respect to the reference plane, the fundamental relation

$$\textbf{Syst Momenta}_1 + \textbf{Syst Ext Imp}_{1\to2} = \textbf{Syst Momenta}_2 \quad (18.1)$$

Three equations of motion may be derived from Fig. 18.5. Two equations are obtained by summing and equating the x and y *components* of the momenta and impulses, and the third by summing and equating the *moments* of these vectors *about any given point.* The coordinate axes may be chosen fixed in space, or they may be allowed to move with the mass center of the body while maintaining a fixed direction. In either case, the point about which moments are taken should keep the same position relative to the coordinate axes during the interval of time considered.

In deriving the three equations of motion for a rigid body, care should be taken not to add indiscriminately linear and angular momenta. Confusion will be avoided if it is kept in mind that $m\bar{v}_x$ and $m\bar{v}_y$ represent the *components of a vector,* namely, the linear momentum vector $m\bar{\mathbf{v}}$, while $\bar{I}\omega$ represents the *moment of a couple,* namely, the momentum couple $\bar{I}\boldsymbol{\omega}$. Thus the angular momentum $\bar{I}\omega$ should be added only to the *moment* of the linear momentum $m\bar{\mathbf{v}}$, never to this vector itself nor to its components. All quantities involved will then be expressed in the same units, namely, N · m · s or lb · ft · s.

As an example, we shall consider the motion of a wheel of radius r, mass m, and centroidal moment of inertia $\bar{I}$, which rolls down an incline without sliding. Assuming that the initial angular velocity ω_1 of the wheel is known, we propose to determine its angular velocity ω_2, after a time interval t, and the values of the components F and N of the reaction of the incline.

The system of momenta at times t_1 and t_2 are shown, respectively, in parts a and c of Fig. 18.6. They are represented in both cases by a vector $m\bar{\mathbf{v}}$ attached at the center G of the wheel and a couple $\bar{I}\boldsymbol{\omega}$. Choosing coordinate axes centered at G with

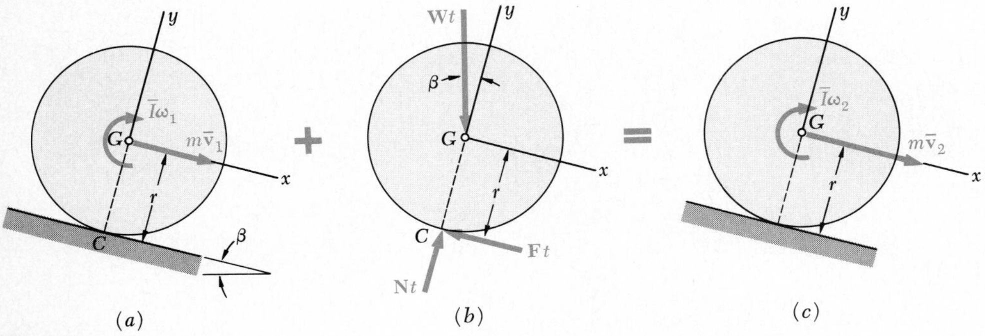

Fig. 18.6

the x and y axes parallel and perpendicular, respectively, to the incline, we note that the lines of action of the forces **W**, **F**, and **N** remain fixed with respect to the coordinate axes. Since these forces are also constant in magnitude, their impulses will be represented by the vectors **W**t, **F**t, and **N**t shown in part b of Fig. 18.6. Denoting by β the angle the incline forms with the horizontal, we add and equate successively the x components, y components, and moments about C (positive clockwise) of the momenta and impulses shown in Fig. 18.6. We write

$+\searrow$ x components:

$$m\bar{v}_1 - Ft + Wt \sin \beta = m\bar{v}_2 \tag{18.3}$$

$+\nearrow$ y components:

$$Nt - Wt \cos \beta = 0 \tag{18.4}$$

$+\circlearrowright$ moments about C:

$$\bar{I}\omega_1 + m\bar{v}_1 r + (Wt \sin \beta)r = \bar{I}\omega_2 + m\bar{v}_2 r \tag{18.5}$$

We note that all quantities involved in Eqs. (18.3) and (18.4) are expressed in the same units, namely, N · s or lb · s. Since, in Eq. (18.5), the magnitudes of the linear impulse **W**$t \sin \beta$ and of the linear momenta $m\bar{\mathbf{v}}_1$ and $m\bar{\mathbf{v}}_2$ are multiplied by the radius

r, all quantities involved in that equation are also expressed in the same units, namely, N · m · s or lb · ft · s. Observing that $\bar{v} = r\omega$, we write

$$mr\omega_1 - Ft + Wt \sin \beta = mr\omega_2 \qquad (18.3')$$
$$Nt - Wt \cos \beta = 0 \qquad (18.4')$$
$$(\bar{I} + mr^2)\omega_1 + (Wt \sin \beta)r = (\bar{I} + mr^2)\omega_2 \qquad (18.5')$$

Equation (18.5′) may be solved for ω_2 and Eq. (18.4′) for N; Eq. (18.3′) may be solved for F after substituting for ω_2.

Noncentroidal Rotation. In this particular case of plane motion, the magnitude of the velocity of the mass center of the body is $\bar{v} = \bar{r}\omega$, where $\bar{r}$ represents the distance from the mass center to the fixed axis of rotation and ω the angular velocity of the body at the instant considered; the magnitude of the momentum vector attached at G is thus $m\bar{v} = m\bar{r}\omega$. Summing the moments about O of the momentum vector and momentum couple (Fig. 18.7) and using the parallel-axis theorem for moments of inertia, we find that the angular momentum of the body about O is

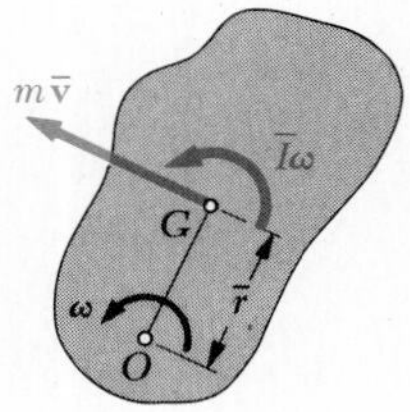

Fig. 18.7

$$\bar{I}\omega + (m\bar{r}\omega)\bar{r} = (\bar{I} + m\bar{r}^2)\omega = I_O\omega \qquad (18.6)$$

Equating the moments about O of the momenta and impulses in (18.1), we write

$$I_O\omega_1 + \sum \int_{t_1}^{t_2} M_O\,dt = I_O\omega_2 \qquad (18.7)$$

In the general case of plane motion of a rigid body symmetrical with respect to the reference plane, Eq. (18.7) may be used with respect to the instantaneous axis of rotation under certain conditions. It is recommended, however, that all problems of plane motion be solved by the general method described earlier in this section.

18.4. Systems of Rigid Bodies. The motion of several rigid bodies may be analyzed by applying the principle of impulse and momentum to each body separately (Sample Prob. 18.1).

However, in solving problems involving no more than three unknowns (including the impulses of unknown reactions), it is often found convenient to apply the principle of impulse and momentum to the system as a whole. The momentum and impulse diagrams are drawn for the entire system of bodies. The diagrams of momenta should include a momentum vector, a momentum couple, or both, for each moving part of the system. Impulses of forces internal to the system may be omitted from the impulse diagram since they occur in pairs of equal and opposite vectors. Summing and equating successively the x components, y components, and moments of all vectors involved, one obtains three relations which express that the momenta at time t_1 and the impulses of the external forces form a system equipollent to the system of the momenta at time t_2.† Again, care should be taken not to add indiscriminately linear and angular momenta; each equation should be checked to make sure that consistent units have been used. This approach has been used in Sample Prob. 18.3 and, further on, in Sample Probs. 18.4 and 18.5.

18.5. Conservation of Angular Momentum. When no external force acts on a rigid body or a system of rigid bodies, the impulses of the external forces are zero and the system of the momenta at time t_1 is equipollent to the system of the momenta at time t_2. Summing and equating successively the x components, y components, and moments of the momenta at times t_1 and t_2, we conclude that the total linear momentum of the system is conserved in any direction and that its total angular momentum is conserved about any point.

There are many engineering applications, however, in which *the linear momentum is not conserved,* yet in which *the angular momentum* H_O of the system about a given point O is *conserved:*

$$(H_O)_1 = (H_O)_2 \tag{18.8}$$

Such cases occur when the lines of action of all external forces pass through O or, more generally, when the sum of the angular impulses of the external forces about O is zero.

Problems involving *conservation of angular momentum* about a point O may be solved by the general method of impulse and momentum, i.e., by drawing momentum and impulse diagrams as described in Secs. 18.3 and 18.4. Equation (18.8) is then obtained by summing and equating moments about O (Sample Prob. 18.3). As we shall see later in Sample Prob. 18.4, two additional equations may be written by summing and equating x and y components; these equations may be used to determine two unknown linear impulses, such as the impulses of the reaction components at a fixed point.

† Note that, as in Sec. 16.4, we cannot speak of *equivalent* systems since we are not dealing with a single rigid body.

SAMPLE PROBLEM 18.1

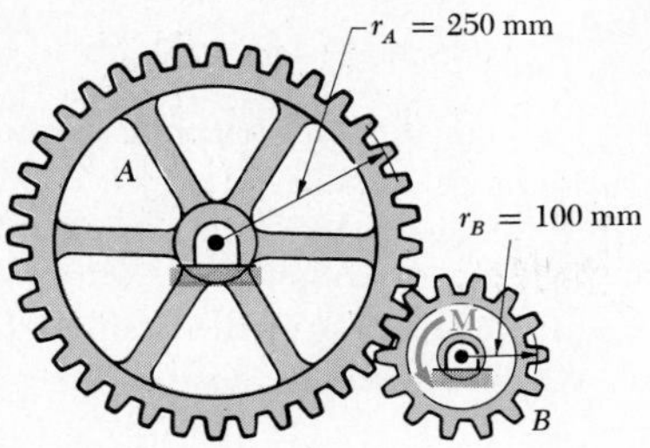

Gear A has a mass of 10 kg and a radius of gyration of 200 mm, while gear B has a mass of 3 kg and a radius of gyration of 80 mm. The system is at rest when a couple **M** of moment 6 N · m is applied to gear B. Neglecting friction, determine (a) the time required for the angular velocity of gear B to reach 600 rpm, (b) the tangential force which gear B exerts on gear A. These gears have been previously considered in Sample Prob. 17.2.

Solution. We apply the principle of impulse and momentum to each gear separately. Since all forces and the couple are constant, their impulses are obtained by multiplying them by the unknown time t. We recall from Sample Prob. 17.2 that the centroidal moments of inertia and the final angular velocities are

$$\bar{I}_A = 0.400 \text{ kg} \cdot \text{m}^2 \qquad \bar{I}_B = 0.0192 \text{ kg} \cdot \text{m}^2$$
$$(\omega_A)_2 = 25.1 \text{ rad/s} \qquad (\omega_B)_2 = 62.8 \text{ rad/s}$$

Principle of Impulse and Momentum for Gear A. The systems of initial momenta, impulses, and final momenta are shown in three separate sketches.

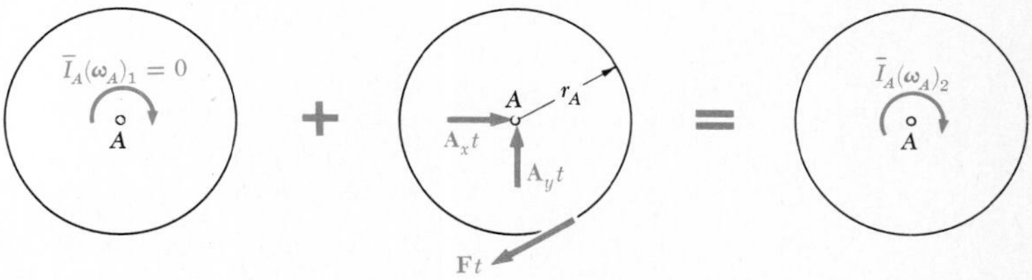

Syst Momenta$_1$ + Syst Ext Imp$_{1\to2}$ = Syst Momenta$_2$

$+\circlearrowleft$ moments about A: $\qquad 0 - Ftr_A = -\bar{I}_A(\omega_A)_2$

$$Ft(0.250 \text{ m}) = (0.400 \text{ kg} \cdot \text{m}^2)(25.1 \text{ rad/s})$$
$$Ft = 40.2 \text{ N} \cdot \text{s}$$

Principle of Impulse and Momentum for Gear B.

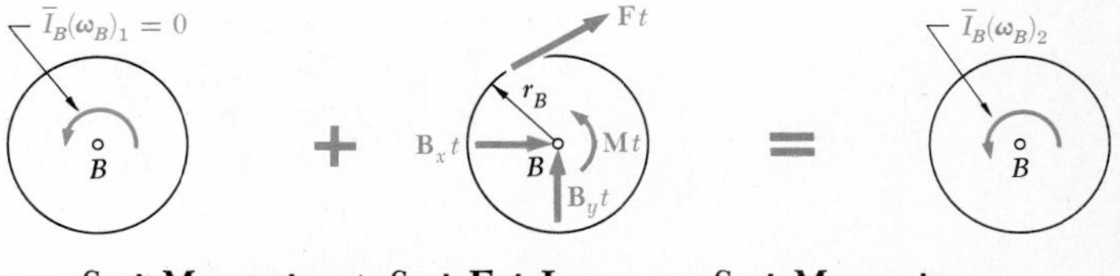

Syst Momenta$_1$ + Syst Ext Imp$_{1\to2}$ = Syst Momenta$_2$

$+\circlearrowleft$ moments about B: $\qquad 0 + Mt - Ftr_B = \bar{I}_B(\omega_B)_2$

$$+(6 \text{ N} \cdot \text{m})t - (40.2 \text{ N} \cdot \text{s})(0.100 \text{ m}) = (0.0192 \text{ kg} \cdot \text{m}^2)(62.8 \text{ rad/s})$$

$t = 0.871$ s ◄

Recalling that $Ft = 40.2$ N · s, we write

$$F(0.871 \text{ s}) = 40.2 \text{ N} \cdot \text{s} \qquad F = +46.2 \text{ N}$$

Thus, the force exerted by gear B on gear A is $\quad$ **F** = 46.2 N ↙ ◄

SAMPLE PROBLEM 18.2

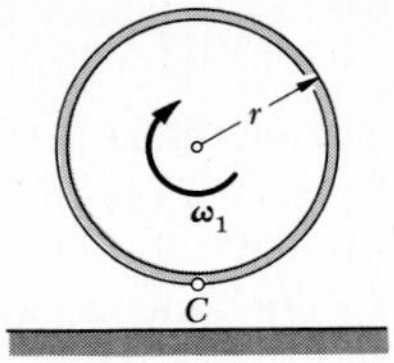

A hoop of radius r and mass m is placed on a horizontal surface with no linear velocity but with a clockwise angular velocity $\boldsymbol{\omega}_1$. Denoting by μ the coefficient of friction between the hoop and the surface, determine (*a*) the time t_2 at which the hoop will start rolling without sliding, (*b*) the linear and angular velocities of the hoop at time t_2.

Solution. Since the entire mass m is located at a distance r from the center of the hoop, we have $\bar{I} = mr^2$. While the hoop is sliding relative to the surface, it is acted upon by the normal force **N**, the friction force **F**, and its weight **W** of magnitude $W = mg$.

Principle of Impulse and Momentum. We apply the principle of impulse and momentum to the hoop from the time t_1 when it is placed on the surface until the time t_2 when it starts rolling without sliding.

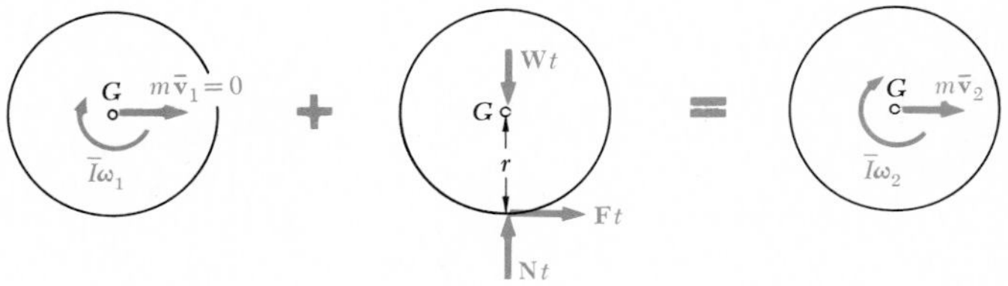

Syst Momenta$_1$ + Syst Ext Imp$_{1\to2}$ = Syst Momenta$_2$

$+\uparrow y$ components: $\quad Nt - Wt = 0 \qquad (1)$

$\xrightarrow{+} x$ components: $\quad Ft = m\bar{v}_2 \qquad (2)$

$+\circlearrowright$ moments about G: $\quad -\bar{I}\omega_1 + Ftr = -\bar{I}\omega_2 \qquad (3)$

From (1) we obtain $N = W = mg$. For $t < t_2$, sliding occurs at point C and we have $F = \mu N = \mu mg$. Substituting for F into (2), we write

$$\mu mgt = m\bar{v}_2 \qquad \bar{v}_2 = \mu gt \qquad (4)$$

Substituting $F = \mu mg$ and $\bar{I} = mr^2$ into (3),

$$-mr^2\omega_1 + \mu mgtr = -mr^2\omega_2 \qquad \omega_2 = \omega_1 - \frac{\mu g}{r}t \qquad (5)$$

The hoop will start rolling without sliding when the velocity $\mathbf{v}_C$ of the point of contact is zero. At that time, $t = t_2$, point C becomes the instantaneous center of rotation, and we have $\bar{v}_2 = r\omega_2$. Substituting from (4) and (5), we write

$$\bar{v}_2 = r\omega_2 \qquad \mu g t_2 = r\left(\omega_1 - \frac{\mu g}{r}t_2\right) \qquad t_2 = \frac{r\omega_1}{2\mu g} \quad ◄$$

Substituting this expression for t_2 into (4),

$$\bar{v}_2 = \mu g t_2 = \mu g\frac{r\omega_1}{2\mu g} \qquad \bar{v}_2 = \tfrac{1}{2}r\omega_1 \qquad \bar{\mathbf{v}}_2 = \tfrac{1}{2}r\omega_1 \rightarrow \quad ◄$$

$$\omega_2 = \frac{\bar{v}_2}{r} \qquad \omega_2 = \tfrac{1}{2}\omega_1 \qquad \boldsymbol{\omega}_2 = \tfrac{1}{2}\omega_1 \circlearrowright \quad ◄$$

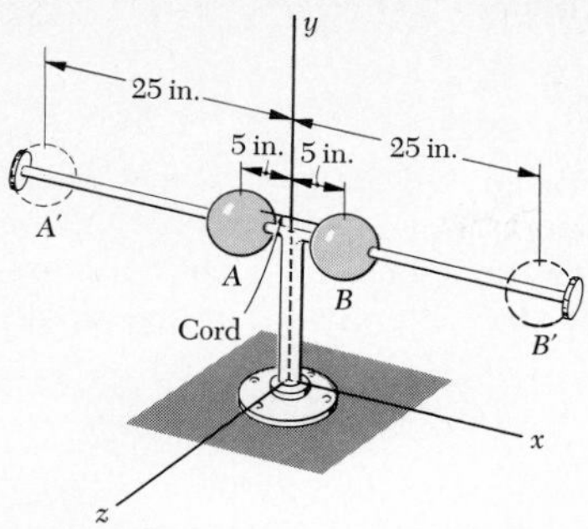

SAMPLE PROBLEM 18.3

Two solid spheres of radius 3 in., weighing 2 lb each, are mounted at A and B on the horizontal rod $A'B'$, which rotates freely about the vertical with a counterclockwise angular velocity of 6 rad/s. The spheres are held in position by a cord which is suddenly cut. Knowing that the centroidal moment of inertia of the rod and pivot is $\bar{I}_R = 0.25 \text{ lb}\cdot\text{ft}\cdot\text{s}^2$, determine (*a*) the angular velocity of the rod after the spheres have moved to positions A' and B', (*b*) the energy lost due to the plastic impact of the spheres and the stops at A' and B'.

a. **Principle of Impulse and Momentum.** In order to determine the final angular velocity of the rod, we shall express that the initial momenta of the various parts of the system and the impulses of the external forces are together equipollent to the final momenta of the system.

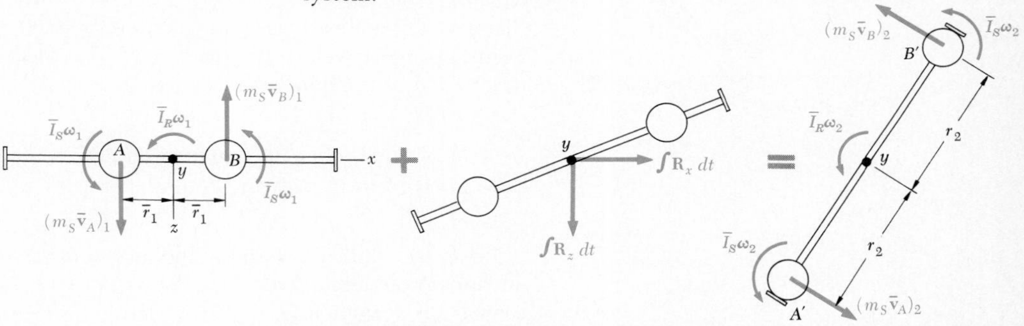

Observing that the external forces consist of the weights and the reaction at the pivot, which have no moment about the y axis, and noting that $\bar{v}_A = \bar{v}_B = \bar{r}\omega$, we write

$+\circlearrowleft$ moments about y axis:

$$2(m_S\bar{r}_1\omega_1)\bar{r}_1 + 2\bar{I}_S\omega_1 + \bar{I}_R\omega_1 = 2(m_S\bar{r}_2\omega_2)\bar{r}_2 + 2\bar{I}_S\omega_2 + \bar{I}_R\omega_2$$

$$(2m_S\bar{r}_1^2 + 2\bar{I}_S + \bar{I}_R)\omega_1 = (2m_S\bar{r}_2^2 + 2\bar{I}_S + \bar{I}_R)\omega_2 \qquad (1)$$

which expresses that *the angular momentum of the system about the y axis is conserved.* We now compute

$$\bar{I}_S = \tfrac{2}{5}m_S a^2 = \frac{2}{5}\left(\frac{2 \text{ lb}}{32.2 \text{ ft/s}^2}\right)(\tfrac{3}{12}\text{ ft})^2 = 0.00155 \text{ lb}\cdot\text{ft}\cdot\text{s}^2$$

$$m_S\bar{r}_1^2 = \frac{2}{32.2}\left(\frac{5}{12}\right)^2 = 0.0108 \qquad m_S\bar{r}_2^2 = \frac{2}{32.2}\left(\frac{25}{12}\right)^2 = 0.2696$$

Substituting these values and $\bar{I}_R = 0.25$, $\omega_1 = 6$ rad/s into (1):

$$0.275(6 \text{ rad/s}) = 0.792\omega_2 \qquad \omega_2 = 2.08 \text{ rad/s} \circlearrowleft \blacktriangleleft$$

b. **Energy Lost.** The kinetic energy of the system at any instant is

$$T = 2(\tfrac{1}{2}m_S\bar{v}^2 + \tfrac{1}{2}\bar{I}_S\omega^2) + \tfrac{1}{2}\bar{I}_R\omega^2 = \tfrac{1}{2}(2m_S\bar{r}^2 + 2\bar{I}_S + \bar{I}_R)\omega^2$$

Recalling the numerical values found above, we have

$$T_1 = \tfrac{1}{2}(0.275)(6)^2 = 4.95 \text{ ft}\cdot\text{lb} \qquad T_2 = \tfrac{1}{2}(0.792)(2.08)^2 = 1.713 \text{ ft}\cdot\text{lb}$$

$$\Delta T = T_2 - T_1 = 1.71 - 4.95 \qquad \Delta T = -3.24 \text{ ft}\cdot\text{lb} \blacktriangleleft$$

PROBLEMS

18.1 The rotor of a steam turbine has an angular velocity of 5400 rpm when the steam supply is suddenly cut off. The rotor weighs 200 lb and has a radius of gyration of 6 in. If kinetic friction produces a couple of magnitude 24 lb · in., determine the time required for the rotor to coast to rest.

18.2 A large flywheel weighs 3 tons and has a radius of gyration of 42 in. It is observed that 9 min 18 s is required for the flywheel to coast to rest from an angular velocity of 300 rpm. Determine the average moment of the couple due to kinetic friction in the bearings of the flywheel.

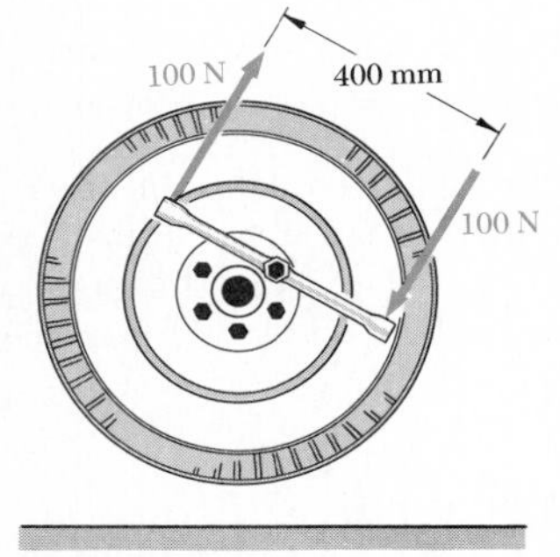

Fig. P18.3

18.3 A bolt located 50 mm from the center of an automobile wheel is tightened by applying the couple shown for 0.1 s. Assuming that the wheel is free to rotate and is initially at rest, determine the resulting angular velocity of the wheel. The 20-kg wheel has a radius of gyration of 250 mm.

18.4 Solve Prob. 17.3, assuming that the couple **M** is applied for a time t_0 and then removed.

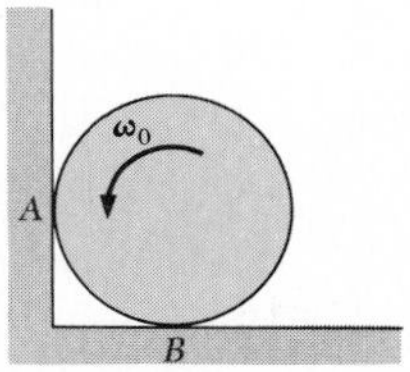

Fig. P18.5

18.5 A cylinder of radius r and mass m is placed in a corner with an initial counterclockwise angular velocity $\boldsymbol{\omega}_0$. Denoting by μ the coefficient of friction at A and B, derive an expression for the time required for the cylinder to come to rest.

18.6 Solve Prob. 18.5, assuming that the surface at A is frictionless.

18.7 Using the principle of impulse and momentum, solve Prob. 16.39.

18.8 Using the principle of impulse and momentum, solve Prob. 16.30*b*.

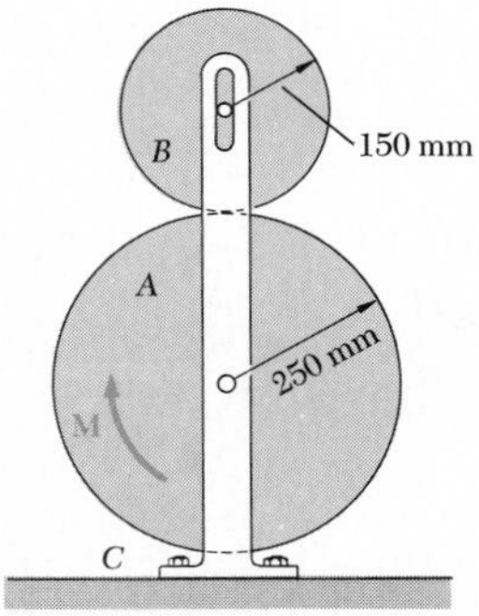

Fig. P18.9

18.9 Disks A and B are of mass 5 and 1.8 kg, respectively. The disks are initially at rest and the coefficient of friction between them is 0.20. A couple **M** of moment 4 N · m is applied to disk A for 1.50 s and then removed. Determine (*a*) whether slipping occurs between the disks, (*b*) the final angular velocity of each disk.

18.10 In Prob. 18.9, determine (*a*) the largest couple **M** for which no slipping occurs, (*b*) the corresponding final angular velocity of each disk.

18.11 The flywheel of a small hoisting engine has a radius of gyration of 2 ft and weighs 750 lb. If the power is cut off when the angular velocity of the flywheel is 100 rpm clockwise, determine the time required for the system to come to rest.

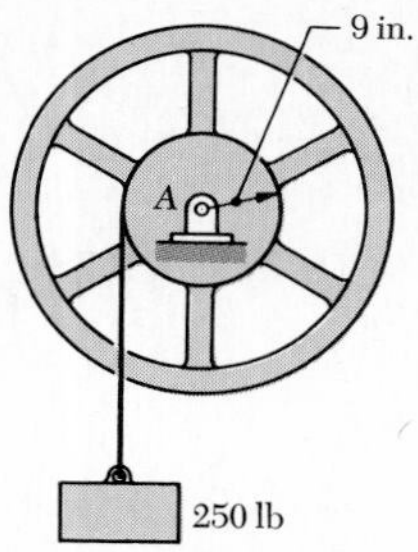

Fig. P18.11

18.12 In Prob. 18.11, determine the time required for the angular velocity of the flywheel to be reduced to 50 rpm.

18.13 Show that the system of momenta for a rigid slab in plane motion reduces to a single vector, and express the distance from the mass center G to the line of action of this vector in terms of the centroidal radius of gyration $\bar{k}$ of the slab, the magnitude $\bar{v}$ of the velocity of G, and the angular velocity ω.

18.14 Show that, when a rigid slab rotates about a fixed axis through O perpendicular to the slab, the system of momenta of its particles is equivalent to a single vector of magnitude $m\bar{r}\omega$, perpendicular to the line OG, and applied to a point P on this line, called the *center of percussion*, at a distance $GP = \bar{k}^2/\bar{r}$ from the mass center of the slab.

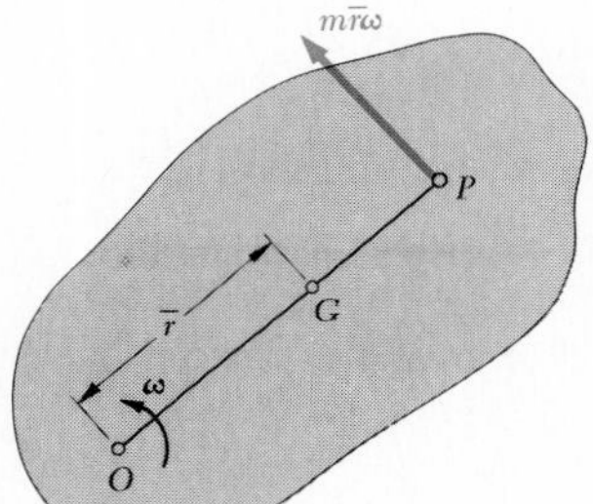

Fig. P18.14

18.15 Show that the sum H_A of the moments about a point A of the momenta of the particles of a rigid slab in plane motion is equal to $I_A\omega$, where ω is the angular velocity of the slab at the instant considered and I_A the moment of inertia of the slab about A, *if and only if* one of the following conditions is satisfied: (*a*) A is the mass center of the slab, (*b*) A is the instantaneous center of rotation, (*c*) the velocity of A is directed along a line joining point A and the mass center G.

18.16 Consider a rigid slab initially at rest and subjected to an impulsive force **F** contained in the plane of the slab. We define the *center of percussion P* as the point of intersection of the line of action of **F** with the perpendicular drawn from G. (*a*) Show that the instantaneous center of rotation C of the slab is located on line GP at a distance $GC = \bar{k}^2/GP$ on the opposite side of G. (*b*) Show that, if the center of percussion were located at C, the instantaneous center of rotation would be located at P.

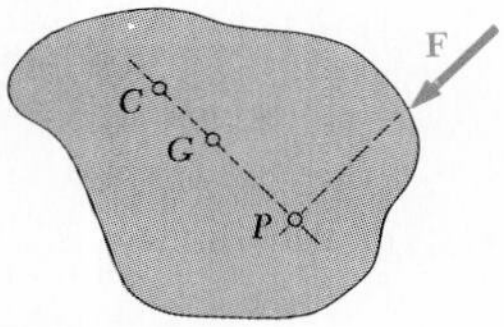

Fig. P18.16

18.17 A cord is wrapped around a solid cylinder of radius r and mass m as shown. If the cylinder is released from rest at time $t = 0$, determine the velocity of the center of the cylinder at a time t.

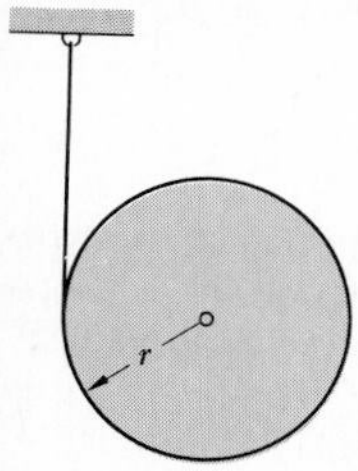

Fig. P18.17

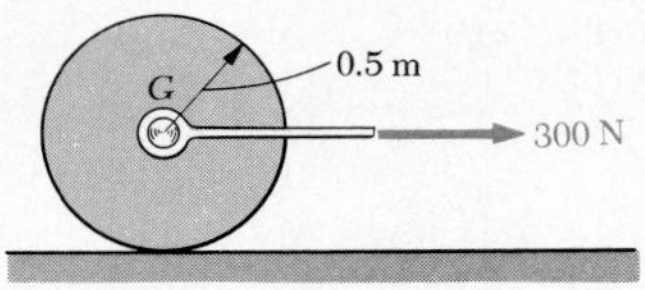

Fig. P18.18

18.18 A 100-kg cylindrical roller is initially at rest and is acted upon by a 300-N force as shown. Assuming that the body rolls without slipping, determine (*a*) the velocity of the center G after 6 s, (*b*) the friction force required to prevent slipping.

18.19 A section of thin-walled pipe of radius r is released from rest at time $t = 0$. Assuming that the pipe rolls without slipping, determine (*a*) the velocity of the center at time t, (*b*) the coefficient of friction required to prevent slipping.

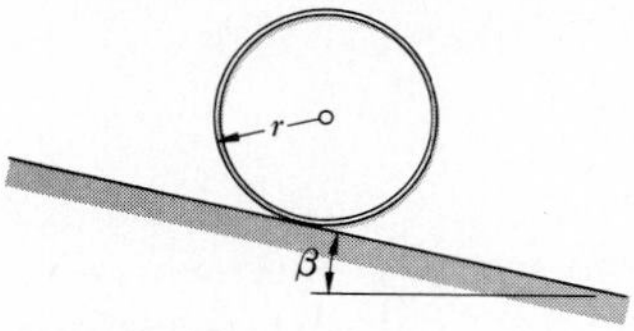

Fig. P18.19

18.20 and 18.21 The 15-lb carriage is supported as shown by two uniform disks, each of weight 10 lb and radius 3 in. Knowing that the carriage is initially at rest, determine the velocity of the carriage 3 s after the 5-lb force is applied. Assume that the disks roll without sliding.

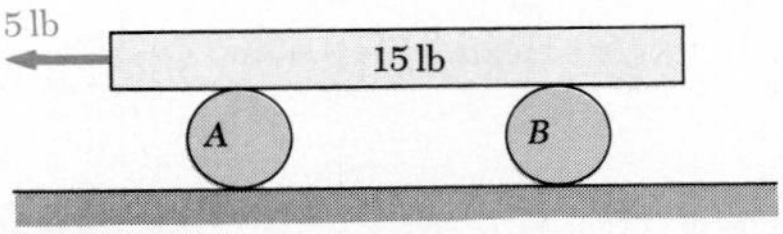

Fig. P18.20

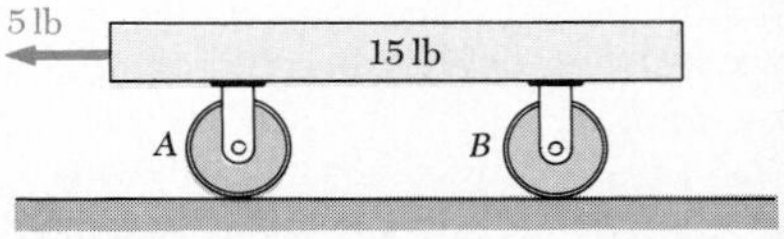

Fig. P18.21

18.22 A sphere of mass m and radius r is projected along a rough horizontal surface with a linear velocity $\bar{\mathbf{v}}_0$ but with no angular velocity ($\boldsymbol{\omega}_0 = 0$). Determine (*a*) the final velocity of the sphere, (*b*) the time at which the velocity of the sphere becomes constant in terms of $\bar{v}_0$ and μ.

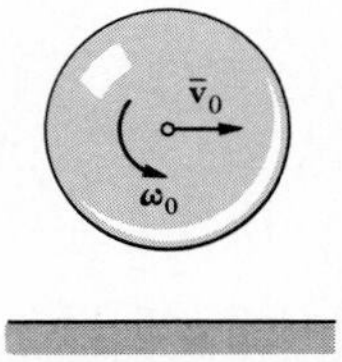

Fig. P18.22 and P18.23

18.23 A sphere of mass m and radius r is projected along a rough horizontal surface with the initial velocities indicated. If the final velocity of the sphere is to be zero, express (*a*) the required ω_0 in terms of $\bar{v}_0$ and r, (*b*) the time required for the sphere to come to rest in terms of $\bar{v}_0$ and μ.

18.24 Solve Sample Prob. 18.3, assuming that, after the cord is cut, sphere B moves to position B' but that an obstruction prevents sphere A from moving.

18.25 A 4-kg bar AB is attached by a pin at D to a 5-kg circular plate which may rotate freely about a vertical axis. Knowing that when the bar is vertical the angular velocity of the plate is 90 rpm, determine (a) the angular velocity of the plate after the bar has swung into a horizontal position and has come to rest against pin C, (b) the energy lost due to the plastic impact at C.

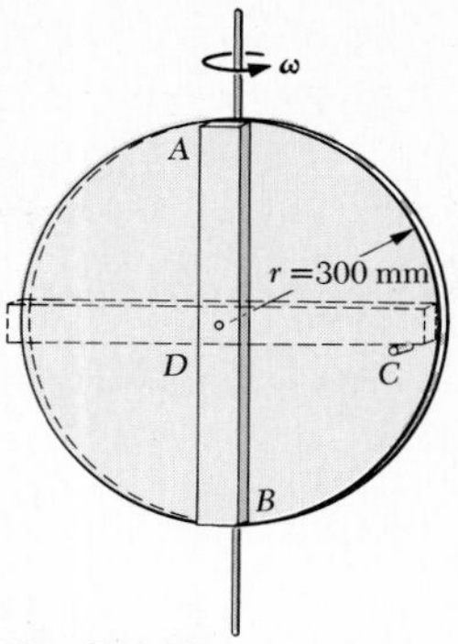

Fig. P18.25

18.26 Disks A and B are made of the same material and are of the same thickness; they may rotate freely about the vertical shaft. Disk B is at rest when it is dropped onto disk A which is rotating with an angular velocity of 400 rpm. Knowing that the mass of disk A is 4 kg, determine (a) the final angular velocity of the disks, (b) the change in kinetic energy of the system.

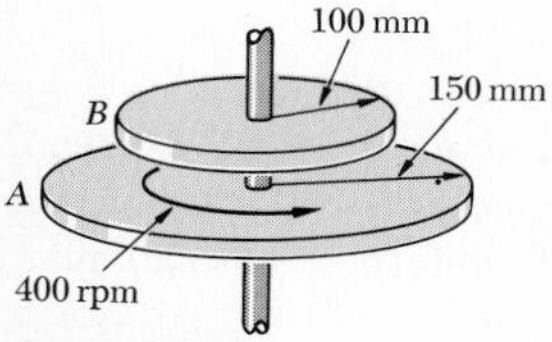

Fig. P18.26

18.27 Solve Prob. 18.26 assuming that disk B has an initial angular velocity of 800 rpm counterclockwise when viewed from above.

18.28 A small 250-g ball may slide in a slender tube of length 1 m and of mass 1 kg which rotates freely about a vertical axis passing through its center C. If the angular velocity of the tube is 10 rad/s as the ball passes through C, determine the angular velocity of the tube (a) just before the ball leaves the tube, (b) just after the ball has left the tube.

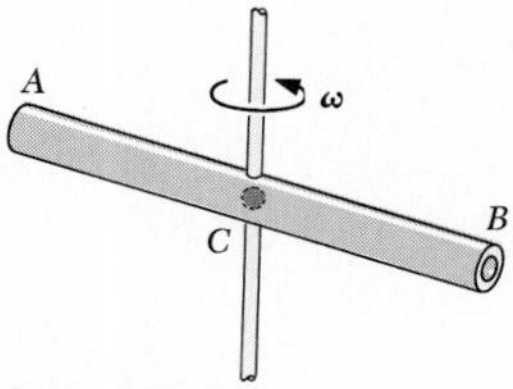

Fig. P18.28

18.29 The rod AB is of mass m and slides freely inside the tube CD which is also of mass m. The angular velocity of the assembly was ω_1 when the rod was entirely inside the tube ($x = 0$). Neglecting the effect of friction, determine the angular velocity of the assembly when $x = \frac{1}{2}L$.

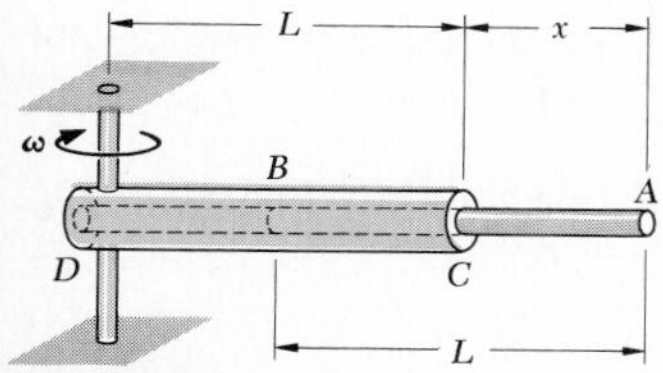

Fig. P18.29

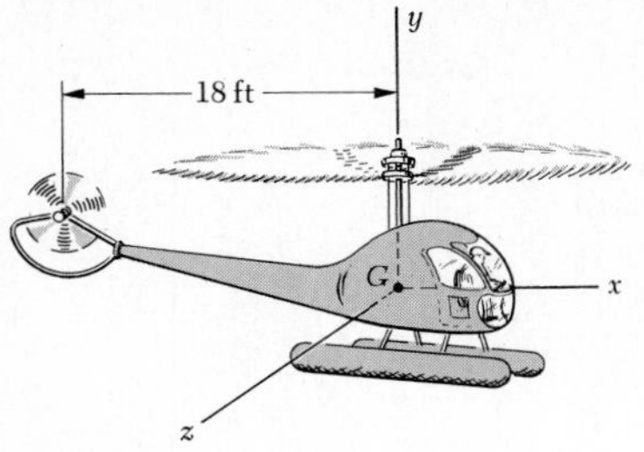

Fig. P18.30

18.30 In the helicopter shown, a vertical tail propeller is used to prevent rotation of the cab as the speed of the main blades is changed. Assuming that the tail propeller is not operating, determine the final angular velocity of the cab after the speed of the main blades has been changed from 200 to 300 rpm. The speed of the main blades is measured relative to the cab, which has a centroidal moment of inertia of 800 lb · ft · s^2. Each of the four main blades is assumed to be a 15-ft slender rod weighing 60 lb.

18.31 Assuming that the tail propeller in Prob. 18.30 is operating and that the angular velocity of the cab remains zero, determine the final horizontal velocity of the cab when the speed of the main blades is changed from 200 to 300 rpm. The cab weighs 1500 lb and is initially at rest. Also determine the force exerted by the tail propeller if this change in speed takes places uniformly in 12 s.

18.32 Two 4-kg disks and a small motor are mounted on a 6-kg rectangular platform which is free to rotate about a central vertical spindle. The normal operating speed of the motor is 180 rpm. If the motor is started when the system is at rest, determine the angular velocity of all elements of the system after the motor has attained its normal operating speed. Neglect the weight of the motor, and assume that the system is perfectly lubricated.

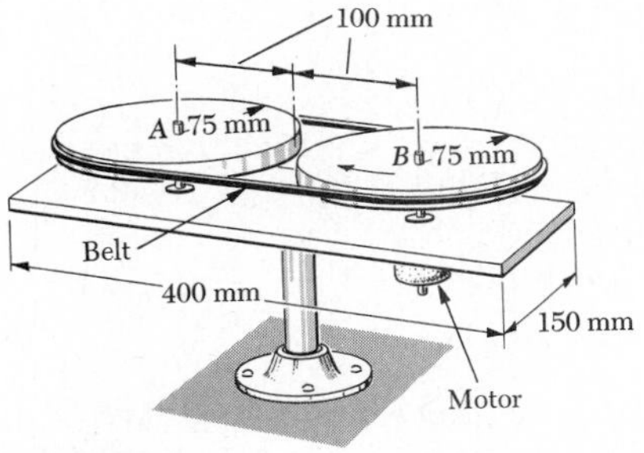

Fig. P18.32

18.33 Solve Prob. 18.32, assuming (*a*) that the belt is removed, (*b*) that the belt is looped around the disks in a figure 8.

18.34 Knowing that in Prob. 18.28 the speed of the ball is 1.2 m/s as it passes through *C*, determine the radial and transverse components of the velocity of the ball as it leaves the tube at *B*.

18.35 In Prob. 18.29, determine the velocity of the rod relative to the tube when $x = \frac{1}{2}L$.

18.6. Impulsive Motion. We saw in Chap. 14 that the method of impulse and momentum is the only practicable method for the solution of problems involving impulsive motion. Now we shall also find that, compared with the various problems considered in the preceding sections, problems involving impulsive motion are particularly well adapted to a solution by the method of impulse and momentum. The computation of linear impulses and angular impulses is quite simple, since, the time interval considered being very short, the bodies involved may be assumed to occupy the same position during that time interval.

18.7. Eccentric Impact. In Secs. 14.5 to 14.7, we learned to solve problems of *central impact,* i.e., problems in which the mass centers of the two colliding bodies are located on the line of impact. We shall now analyze the *eccentric impact* of two rigid bodies. Consider two bodies which collide, and denote by $\mathbf{v}_A$ and $\mathbf{v}_B$ the velocities before impact *of the two points of contact* A and B (Fig. 18.8*a*). Under the impact, the two bodies will *deform* and, at the end of the period of deformation, the velocities $\mathbf{u}_A$ and $\mathbf{u}_B$ of A and B will have equal components along the line of impact nn (Fig. 18.8*b*). A period of *restitution* will then take place, at the end of which A and B will have velocities $\mathbf{v}'_A$ and $\mathbf{v}'_B$ (Fig. 18.8*c*). Assuming the bodies frictionless, we find that the forces they exert on each other are directed along the line of impact. Denoting, respectively, by $\int P\,dt$ and $\int R\,dt$ the magnitude of the impulse of one of these forces during the period of deformation and during the period of restitution, we recall that the coefficient of restitution e is defined as the ratio

$$e = \frac{\int R\,dt}{\int P\,dt} \tag{18.9}$$

We propose to show that the relation established in Sec. 14.6 between the relative velocities of two particles before and after impact also holds between the components along the line of impact of the relative velocities of the two points of contact A and B. We propose to show, therefore, that

$$(v'_B)_n - (v'_A)_n = e[(v_A)_n - (v_B)_n] \tag{18.10}$$

We shall first assume that the motion of each of the two colliding bodies of Fig. 18.8 is unconstrained. Thus the only impulsive forces exerted on the bodies during the impact are applied at A and B respectively. Consider the body to which point A belongs and draw the three momentum and impulse

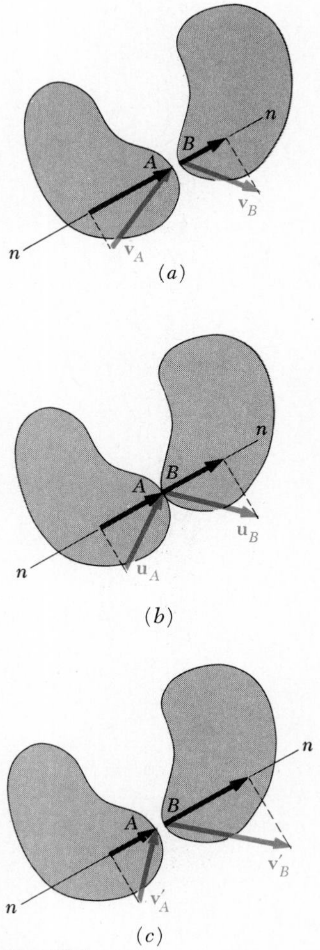

Fig. 18.8

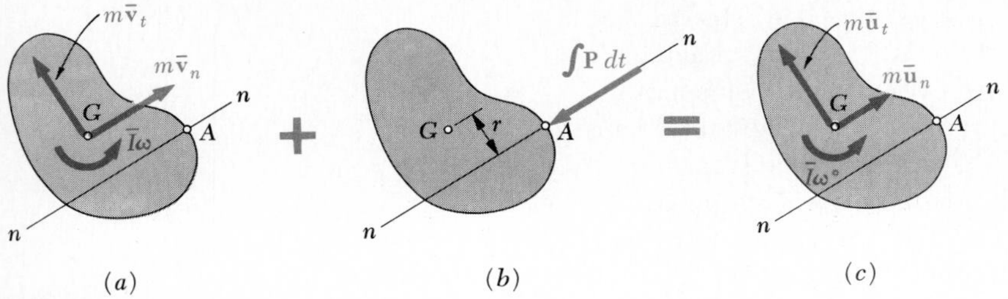

Fig. 18.9

diagrams corresponding to the period of deformation (Fig. 18.9). We denote by $\overline{\mathbf{v}}$ and $\overline{\mathbf{u}}$, respectively, the velocity of the mass center at the beginning and at the end of the period of deformation, and by $\boldsymbol{\omega}$ and $\boldsymbol{\omega}^*$ the angular velocity of the body at the same instants. Summing and equating the components of the momenta and impulses along the line of impact nn, we write

$$m\bar{v}_n - \int P\,dt = m\bar{u}_n \tag{18.11}$$

Summing and equating the moments about G of the momenta and impulses, we also write

$$\bar{I}\omega - r\int P\,dt = \bar{I}\omega^* \tag{18.12}$$

where r represents the perpendicular distance from G to the line of impact. Considering now the period of restitution, we obtain in a similar way

$$m\bar{u}_n - \int R\,dt = m\bar{v}'_n \tag{18.13}$$
$$\bar{I}\omega^* - r\int R\,dt = \bar{I}\omega' \tag{18.14}$$

where $\overline{\mathbf{v}}'$ and $\boldsymbol{\omega}'$ represent, respectively, the velocity of the mass center and the angular velocity of the body after impact. Solving (18.11) and (18.13) for the two impulses and substituting into (18.9), and then solving (18.12) and (18.14) for the same two impulses and substituting again into (18.9), we obtain the following two alternate expressions for the coefficient of restitution:

$$e = \frac{\bar{u}_n - \bar{v}'_n}{\bar{v}_n - \bar{u}_n} \qquad e = \frac{\omega^* - \omega'}{\omega - \omega^*} \tag{18.15}$$

Multiplying by r the numerator and denominator of the second expression obtained for e, and adding respectively to the numerator and denominator of the first expression, we have

$$e = \frac{\bar{u}_n + r\omega^* - (\bar{v}'_n + r\omega')}{\bar{v}_n + r\omega - (\bar{u}_n + r\omega^*)} \tag{18.16}$$

Observing that $\bar{v}_n + r\omega$ represents the component $(v_A)_n$ along nn of the velocity of the point of contact A and that, similarly,

$\bar{u}_n + r\omega^*$ and $\bar{v}'_n + r\omega'$ represent, respectively, the components $(u_A)_n$ and $(v'_A)_n$, we write

$$e = \frac{(u_A)_n - (v'_A)_n}{(v_A)_n - (u_A)_n} \tag{18.17}$$

The analysis of the motion of the second body leads to a similar expression for e in terms of the components along nn of the successive velocities of point B. Recalling that $(u_A)_n = (u_B)_n$, and eliminating these two velocity components by a manipulation similar to the one used in Sec. 14.6, we obtain relation (18.10).

If one or both of the colliding bodies is constrained to rotate about a fixed point O, as in the case of a compound pendulum (Fig. 18.10*a*), an impulsive reaction will be exerted at O (Fig. 18.10*b*). We shall verify that, while their derivation must be

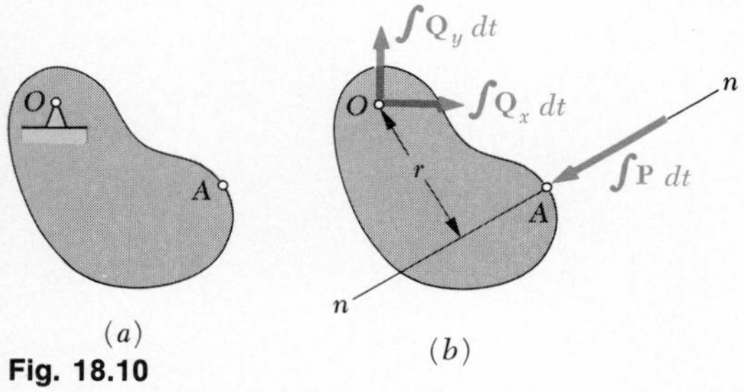

Fig. 18.10

modified, Eqs. (18.17) and (18.10) remain valid. Applying formula (18.7) to the period of deformation and to the period of restitution, we write

$$I_O\omega - r\int P\,dt = I_O\omega^* \tag{18.18}$$
$$I_O\omega^* - r\int R\,dt = I_O\omega' \tag{18.19}$$

where r represents the perpendicular distance from the fixed point O to the line of impact. Solving (18.18) and (18.19) for the two impulses and substituting into (18.9), and then observing that $r\omega$, $r\omega^*$, and $r\omega'$ represent the components along nn of the successive velocities of point A, we write

$$e = \frac{\omega^* - \omega'}{\omega - \omega^*} = \frac{r\omega^* - r\omega'}{r\omega - r\omega^*} = \frac{(u_A)_n - (v'_A)_n}{(v_A)_n - (u_A)_n}$$

and check that Eq. (18.17) still holds. Thus Eq. (18.10) remains valid when one or both of the colliding bodies is constrained to rotate about a fixed point O.

In order to determine the velocities of the two colliding bodies after impact, relation (18.10) should be used in conjunction with one or several other equations obtained by applying the principle of impulse and momentum (Sample Prob. 18.5).

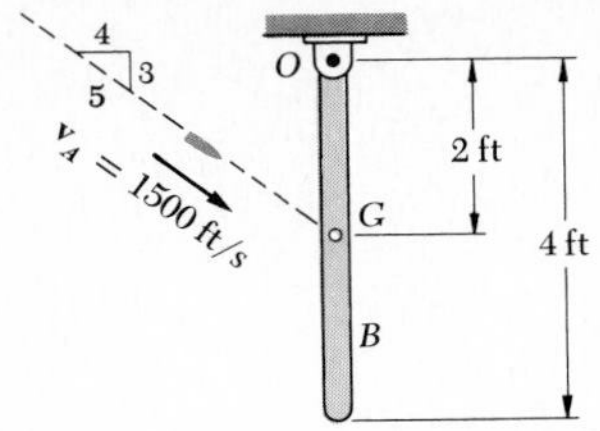

SAMPLE PROBLEM 18.4

A 0.125-lb bullet A is fired with an initial velocity of 1500 ft/s into a 50-lb wooden beam B which is suspended from a hinge at O. Knowing that the beam is initially at rest, determine (*a*) the angular velocity of the beam immediately after the bullet becomes embedded in the beam, (*b*) the impulsive reaction at the hinge, assuming that the bullet becomes embedded in 0.0002 s.

Solution. *Principle of Impulse and Momentum.* We consider the bullet and the beam as a single system and express that the initial momenta of the bullet and beam and the impulses of the external forces are together equipollent to the final momenta of the system. Since the time interval $\Delta t = 0.0002$ s is very short, we neglect all nonimpulsive forces and consider only the external impulses $\mathbf{R}_x \Delta t$ and $\mathbf{R}_y \Delta t$.

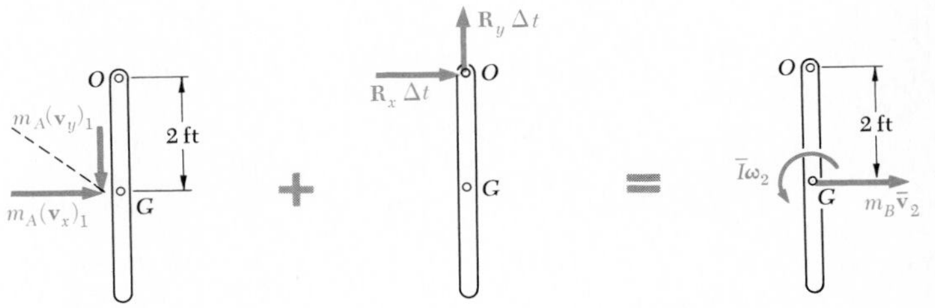

$+\circlearrowright$ moments about O: $\quad m_A(v_x)_1(2\text{ ft}) + 0 = \bar{I}\omega_2 + m_B\bar{v}_2(2\text{ ft}) \quad (1)$

$\xrightarrow{+}$ x components: $\quad m_A(v_x)_1 + R_x \Delta t = m_B\bar{v}_2 \quad (2)$

$+\uparrow$ y components: $\quad -m_A(v_y)_1 + R_y \Delta t = 0 \quad (3)$

The components of the velocity of the bullet and the centroidal moment of inertia of the beam are

$$(v_x)_1 = \tfrac{4}{5}(1500\text{ ft/s}) = 1200\text{ ft/s} \qquad (v_y)_1 = \tfrac{3}{5}(1500\text{ ft/s}) = 900\text{ ft/s}$$

$$\bar{I} = \tfrac{1}{12}ml^2 = \frac{1}{12}\frac{50\text{ lb}}{32.2\text{ ft/s}^2}(4\text{ ft})^2 = 2.07\text{ lb}\cdot\text{ft}\cdot\text{s}^2$$

Substituting these values into (1) and noting that $\bar{v}_2 = (2\text{ ft})\omega_2$:

$$(0.125/32.2)(1200)(2) = 2.07\omega_2 + (50/32.2)(2\omega_2)(2)$$

$$\omega_2 = 1.125\text{ rad/s} \qquad \omega_2 = 1.125\text{ rad/s}\circlearrowright \blacktriangleleft$$

Substituting $\bar{v}_2 = (2\text{ ft})(1.125\text{ rad/s}) = 2.25\text{ ft/s}$ into (2), we solve Eqs. (2) and (3) for R_x and R_y, respectively.

$$(0.125/32.2)(1200) + R_x(0.0002) = (50/32.2)(2.25)$$

$$R_x = -5820\text{ lb} \qquad \mathbf{R}_x = 5820\text{ lb} \leftarrow \blacktriangleleft$$

$$-(0.125/32.2)(900) + R_y(0.0002) = 0$$

$$R_y = +17{,}470\text{ lb} \qquad \mathbf{R}_y = 17{,}470\text{ lb} \uparrow \blacktriangleleft$$

SAMPLE PROBLEM 18.5

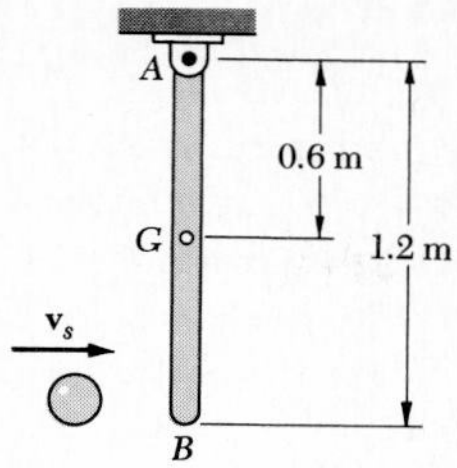

A 2-kg sphere moving horizontally to the right with an initial velocity of 5 m/s strikes the lower end of an 8-kg rigid rod AB. The rod is suspended from a hinge at A and is initially at rest. Knowing that the coefficient of restitution between the rod and sphere is 0.80, determine the angular velocity of the rod and the velocity of the sphere immediately after the impact.

Principle of Impulse and Momentum. We consider the rod and sphere as a single system and express that the initial momenta of the rod and sphere and the impulses of the external forces are together equipollent to the final momenta of the system. We note that the only impulsive force external to the system is the impulsive reaction at A.

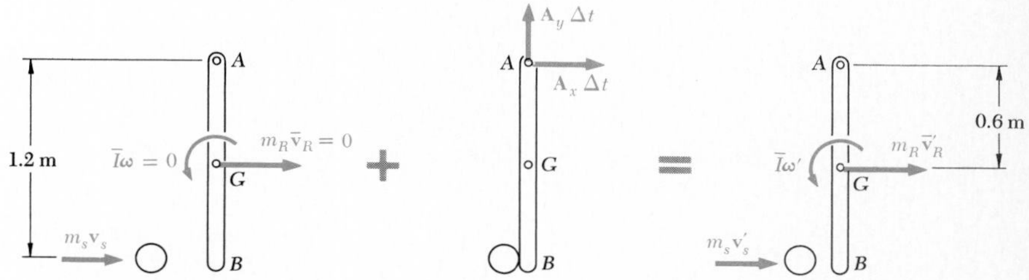

+↻ moments about A:

$$m_s v_s(1.2\text{ m}) = m_s v'_s(1.2\text{ m}) + m_R\bar{v}'_R(0.6\text{ m}) + \bar{I}\omega' \quad (1)$$

Since the rod rotates about A, we have $\bar{v}'_R = \bar{r}\omega' = (0.6\text{ m})\omega'$. Also,

$$\bar{I} = \tfrac{1}{12}mL^2 = \tfrac{1}{12}(8\text{ kg})(1.2\text{ m})^2 = 0.96\text{ kg}\cdot\text{m}^2$$

Substituting these values and the given data into Eq. (1), we have

$$(2\text{ kg})(5\text{ m/s})(1.2\text{ m})$$
$$= (2\text{ kg})v'_s(1.2\text{ m}) + (8\text{ kg})(0.6\text{ m})\omega'(0.6\text{ m}) + (0.96\text{ kg}\cdot\text{m}^2)\omega'$$
$$12 = 2.4v'_s + 3.84\omega' \quad (2)$$

Relative Velocities. Choosing positive to the right, we write

$$v'_B - v'_s = e(v_s - v_B)$$

Substituting $v_s = 5\text{ m/s}$, $v_B = 0$, and $e = 0.80$, we obtain

$$v'_B - v'_s = 0.80(5\text{ m/s}) \quad (3)$$

Again noting that the rod rotates about A, we write

$$v'_B = (1.2\text{ m})\omega' \quad (4)$$

Solving Eqs. (2) to (4) simultaneously, we obtain

$\omega' = +3.21\text{ rad/s}$ $\qquad$ $\omega' = 3.21\text{ rad/s}$ ↻ ◀

$v'_s = -0.143\text{ m/s}$ $\qquad$ $\mathbf{v}'_s = 0.143\text{ m/s} \leftarrow$ ◀

SAMPLE PROBLEM 18.6

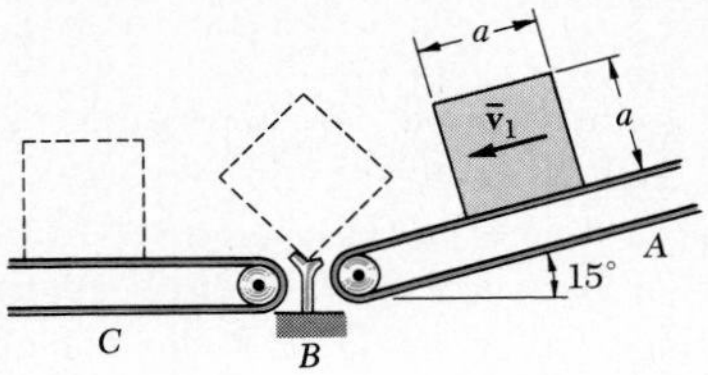

A square package of side a and mass m moves down a conveyor belt A with a constant velocity $\bar{\mathbf{v}}_1$. At the end of the conveyor belt, the corner of the package strikes a rigid support at B. Assuming that the impact at B is perfectly plastic, derive an expression for the smallest magnitude of the velocity $\bar{\mathbf{v}}_1$ for which the package will rotate about B and reach conveyor belt C.

Principle of Impulse and Momentum. Since the impact between the package and the support is perfectly plastic, the package rotates about B during the impact. We apply the principle of impulse and momentum to the package and note that the only impulsive force external to the package is the impulsive reaction at B.

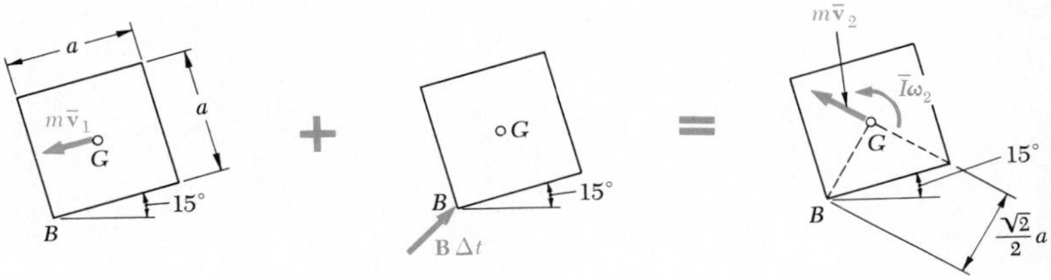

$$\textbf{Syst Momenta}_1 + \textbf{Syst Ext Imp}_{1\to2} = \textbf{Syst Momenta}_2$$

$+\circlearrowleft$ moments about B: $\quad (m\bar{v}_1)(\frac{1}{2}a) + 0 = (m\bar{v}_2)(\frac{1}{2}\sqrt{2}a) + \bar{I}\omega_2 \qquad (1)$

Since the package rotates about B, we have $\bar{v}_2 = (GB)\omega_2 = \frac{1}{2}\sqrt{2}a\omega_2$. We substitute this expression, together with $\bar{I} = \frac{1}{6}ma^2$, into Eq. (1):

$$(m\bar{v}_1)(\tfrac{1}{2}a) = m(\tfrac{1}{2}\sqrt{2}a\omega_2)(\tfrac{1}{2}\sqrt{2}a) + \tfrac{1}{6}ma^2\omega_2 \qquad \bar{v}_1 = \tfrac{4}{3}a\omega_2 \qquad (2)$$

Principle of Conservation of Energy. We apply the principle of conservation of energy between position 2 and position 3.

Position 2. $V_2 = Wh_2$. Recalling that $\bar{v}_2 = \frac{1}{2}\sqrt{2}a\omega_2$, we write

$$T_2 = \tfrac{1}{2}m\bar{v}_2^2 + \tfrac{1}{2}\bar{I}\omega_2^2 = \tfrac{1}{2}m(\tfrac{1}{2}\sqrt{2}a\omega_2)^2 + \tfrac{1}{2}(\tfrac{1}{6}ma^2)\omega_2^2 = \tfrac{1}{3}ma^2\omega_2^2$$

Position 2

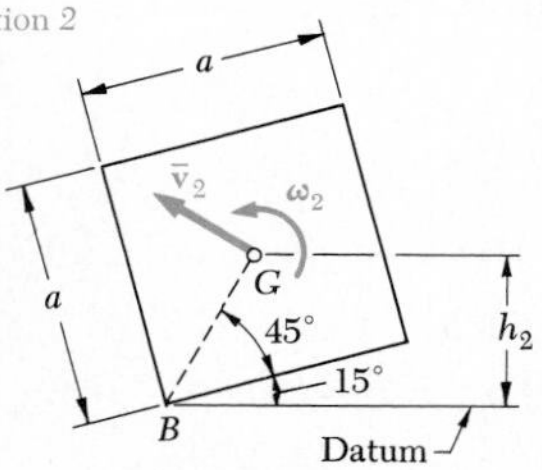

$GB = \frac{1}{2}\sqrt{2}a = 0.707a$
$h_2 = GB\sin(45° + 15°)$
$= 0.612a$

Position 3. Since the package must reach conveyor belt B, it must pass through position 3 where G is directly above B. Also, since we wish to determine the smallest velocity for which the package will reach this position, we choose $\bar{v}_3 = \omega_3 = 0$. Therefore $T_3 = 0$ and $V_3 = Wh_3$.

Position 3

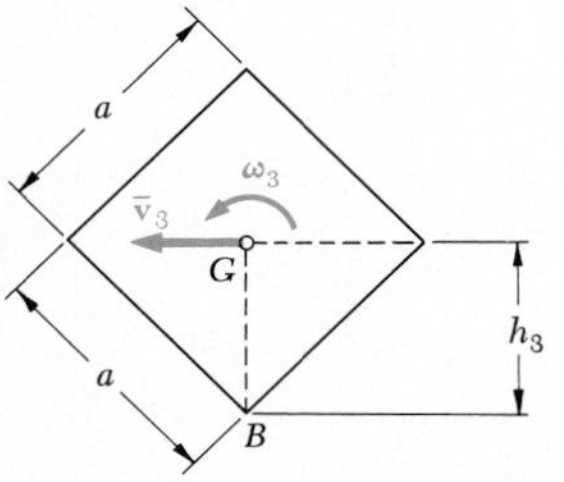

$h_3 = GB = 0.707a$

Conservation of Energy

$$T_2 + V_2 = T_3 + V_3$$
$$\tfrac{1}{3}ma^2\omega_2^2 + Wh_2 = 0 + Wh_3$$
$$\omega_2^2 = \frac{3W}{ma^2}(h_3 - h_2) = \frac{3g}{a^2}(h_3 - h_2) \qquad (3)$$

Substituting the computed values of h_2 and h_3 into Eq. (3), we obtain

$$\omega_2^2 = \frac{3g}{a^2}(0.707a - 0.612a) = \frac{3g}{a^2}(0.095a) \qquad \omega_2 = \sqrt{0.285g/a}$$

$$\bar{v}_1 = \tfrac{4}{3}a\omega_2 = \tfrac{4}{3}a\sqrt{0.285g/a} \qquad \bar{v}_1 = 0.712\sqrt{ga} \blacktriangleleft$$

PROBLEMS

18.36 A 45-g bullet is fired with a horizontal velocity of 400 m/s into a 9-kg wooden disk of radius $R = 200$ mm. Knowing that $h = 200$ mm and that the disk is initially at rest, determine (*a*) the velocity of the center of the disk immediately after the bullet becomes embedded, (*b*) the impulsive reaction at A, assuming that the bullet becomes embedded in 0.001 s.

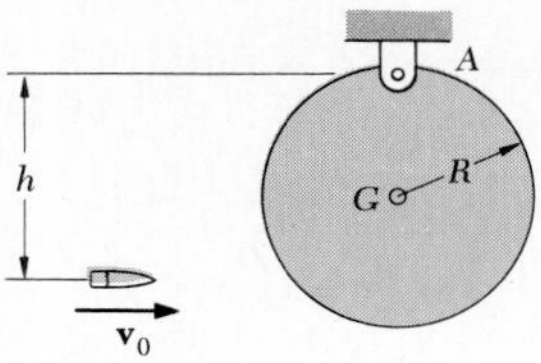

Fig. P18.36

18.37 In Prob. 18.36, determine (*a*) the required distance h if the impulsive reaction at A is to be zero, (*b*) the corresponding velocity of the center of the disk after the bullet becomes embedded.

18.38 A bullet weighing 0.08 lb is fired with a horizontal velocity of 1500 ft/s into the 15-lb wooden rod AB of length $L = 24$ in. The rod, which is initially at rest, is suspended by a cord of length $L = 24$ in. Knowing that $h = 6$ in., determine the velocity of each end of the rod immediately after the bullet becomes embedded.

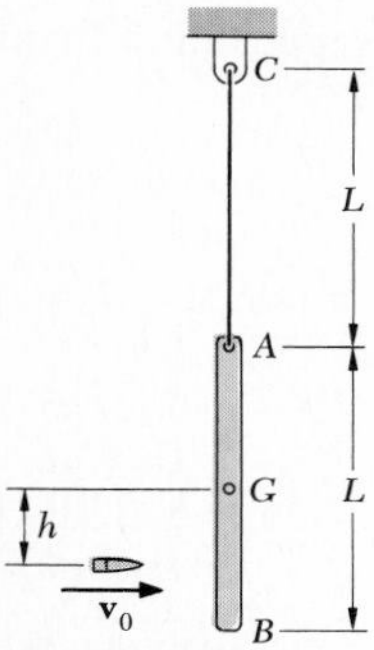

Fig. P18.38

18.39 In Prob. 18.38, determine the distance h for which, immediately after the bullet becomes embedded, the instantaneous center of rotation of the rod is point C.

18.40 At what height h above its center G should a billiard ball of radius r be struck horizontally by a cue if the ball is to start rolling without sliding?

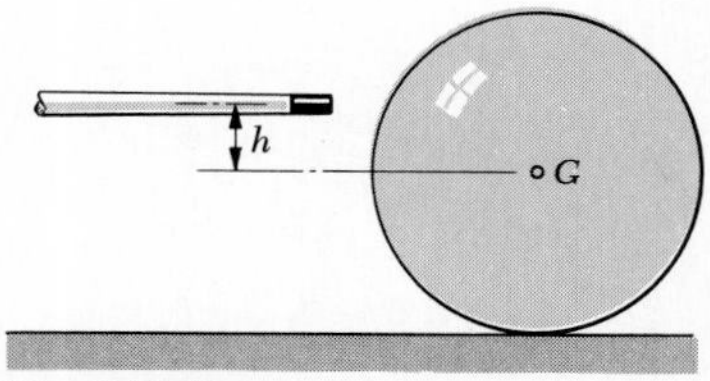

Fig. P18.40

18.41 A uniform slender rod AB is equipped at both ends with the hooks shown and is supported by a frictionless horizontal table. Initially the rod is hooked at A to a fixed pin C about which it rotates with the constant angular velocity ω_1. Suddenly end B of the rod hits and gets hooked to the pin D, causing end A to be released. Determine the magnitude of the angular velocity ω_2 of the rod in its subsequent rotation about D.

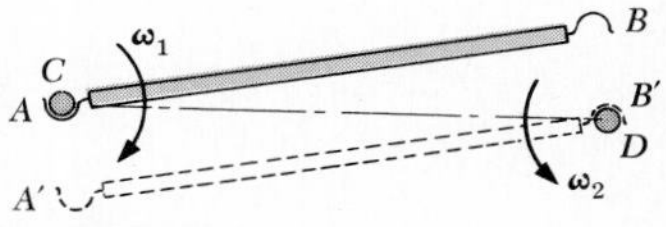

Fig. P18.41

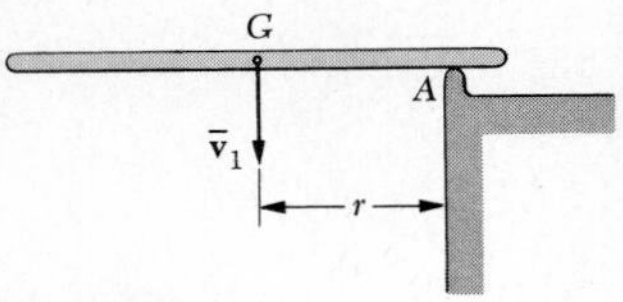

Fig. P18.42

18.42 A slender rigid rod of mass m and with a centroidal radius of gyration $\bar{k}$ strikes a rigid knob at A with a vertical velocity of magnitude $\bar{v}_1$ and no angular velocity. Assuming that the impact is perfectly plastic, show that the magnitude of the velocity of G after impact is

$$\bar{v}_2 = \bar{v}_1 \frac{r^2}{r^2 + \bar{k}^2}$$

18.43 Assuming that in Prob. 18.42 the impact with knob A is perfectly elastic ($e = 1$), (a) show that the magnitude of the velocity of G after impact is

$$\bar{v}_2 = \bar{v}_1 \frac{r^2 - \bar{k}^2}{r^2 + \bar{k}^2}$$

(b) derive an expression for the angular velocity of the rod after impact.

18.44 A slender rod of mass m and length L is released from rest in the position shown. The coefficient of friction is sufficient to prevent sliding at A. Assuming perfectly elastic impact ($e = 1$) at D, determine the distance d for which the rod will rebound with no angular velocity.

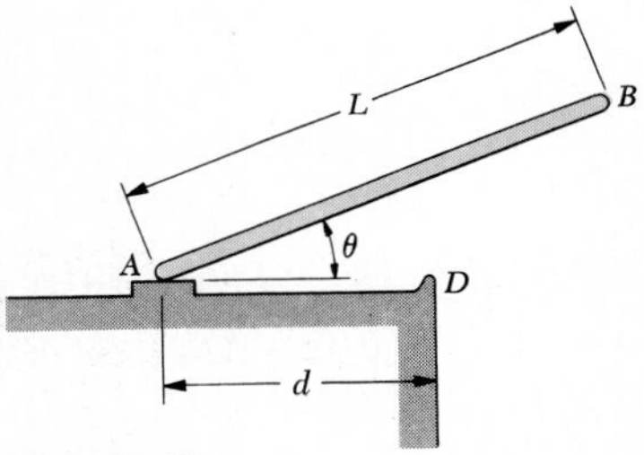

Fig. P18.44

18.45 Solve Prob. 18.44, assuming $e = \frac{1}{2}$.

18.46 A slender rod of length L is falling with a velocity $\bar{v}_1$ at the instant when the cords simultaneously become taut. Assuming that the impacts are perfectly plastic, determine the angular velocity of the rod and the velocity of its mass center immediately after the cords become taut.

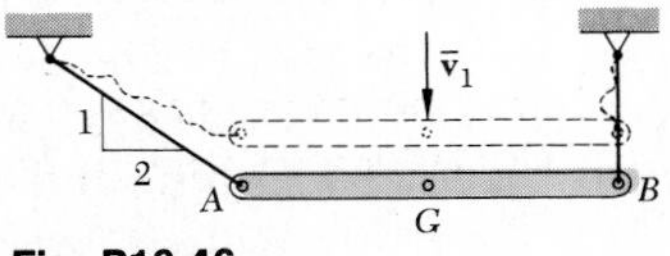

Fig. P18.46

18.47 A square block of mass m is falling with a velocity $\bar{\mathbf{v}}_1$ when it strikes a small obstruction at B. Assuming that the impact between corner A and the obstruction B is perfectly plastic, determine the angular velocity of the block and the velocity of its mass center G immediately after impact.

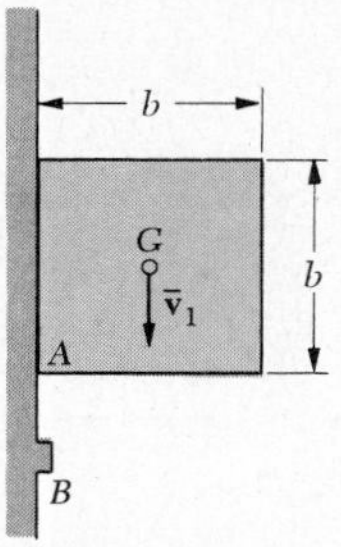

Fig. P18.47

18.48 Solve Prob. 18.47, assuming that the impact between corner A and the obstruction B is perfectly elastic.

18.49 A uniformly loaded rectangular crate is released from rest in the position shown. Assuming that the floor is sufficiently rough to prevent slipping and that the impact at B is perfectly plastic, determine the largest value of the ratio b/a for which corner A will remain in contact with the floor.

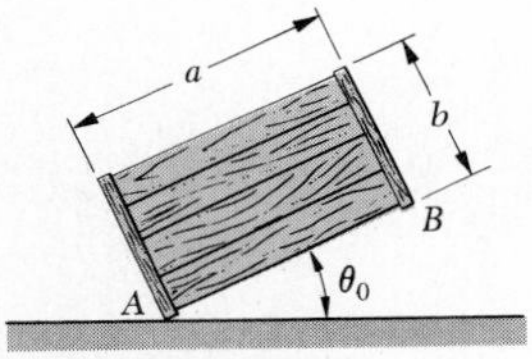

Fig. P18.49

18.50 A uniform sphere of radius r rolls without slipping down the incline shown. It hits the horizontal surface and, after slipping for a while, starts rolling again. Assuming that the sphere does not bounce as it hits the horizontal surface, determine its angular velocity and the velocity of its mass center after it has resumed rolling.

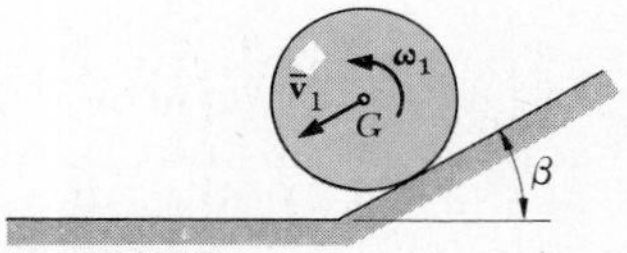

Fig. P18.50

18.51 Solve Prob. 18.50, assuming $r = 4$ in., $\beta = 30°$, and $v_1 = 5$ ft/s.

18.52 A sphere A of mass m and radius r rolls without slipping with a velocity $\mathbf{v}_0$ on a horizontal plane. It hits squarely an identical sphere B which is at rest. Denoting by μ the coefficient of friction between the spheres and the plane, neglecting the friction between the spheres, and assuming perfectly elastic impact ($e = 1$), determine (*a*) the linear and angular velocity of each sphere immediately after impact, (*b*) the velocity of each sphere after it has started rolling uniformly. (*c*) Discuss the special case when $\mu = 0$.

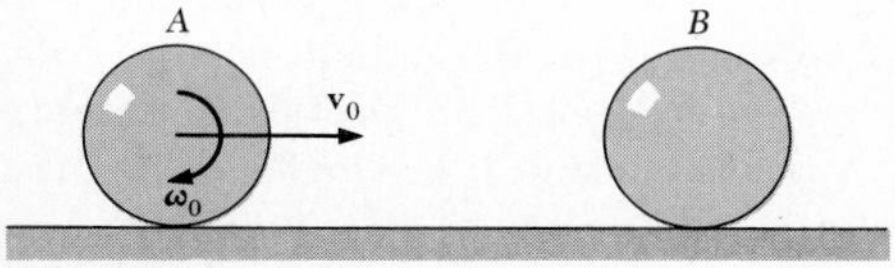

Fig. P18.52

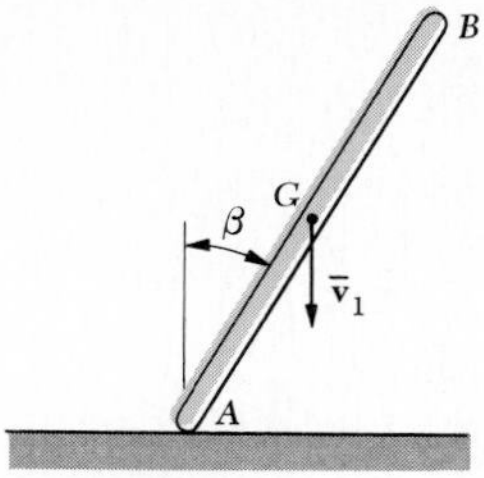

Fig. P18.53

18.53 A slender rod of length l strikes a frictionless floor at A with a vertical velocity $\bar{\mathbf{v}}_1$ and no angular velocity. Assuming that the impact at A is perfectly elastic, derive an expression for the angular velocity of the rod immediately after impact.

18.54 Solve Prob. 18.53, assuming that the impact at A is perfectly plastic.

18.55 A slender rod of mass m and length l is held in the position shown. Roller B is given a slight push to the right and moves along the horizontal plane, while roller A is constrained to move vertically. Determine the magnitudes of the impulses exerted on the rollers A and B as roller A strikes the ground. Assume perfectly plastic impact.

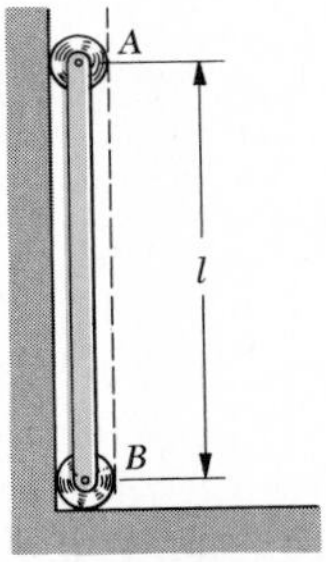

Fig. P18.55

18.56 In Prob. 18.47, determine the line of action of the impulsive force exerted on the block by the obstruction at B.

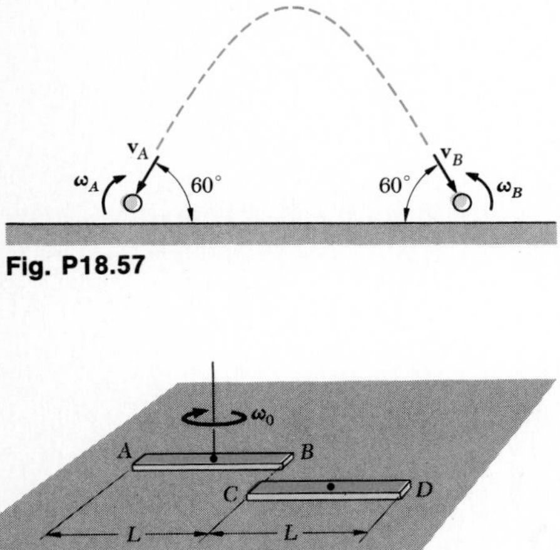

Fig. P18.57

18.57 A small rubber ball of radius r is thrown against a rough floor with a velocity $\mathbf{v}_A$ of magnitude v_0 and a "backspin" $\boldsymbol{\omega}_A$ of magnitude ω_0. It is observed that the ball bounces from A to B, then from B to A, then from A to B, etc. Assuming perfectly elastic impact, determine (*a*) the required magnitude ω_0 of the "backspin" in terms of v_0 and r, (*b*) the minimum required value of the coefficient of friction.

Fig. P18.58

18.58 Two identical rods AB and CD, each of length L, may move freely on a frictionless horizontal surface. Rod AB is rotating about its mass center with an angular velocity $\boldsymbol{\omega}_0$ when end B strikes end C of rod CD, which is at rest. Knowing that at the instant of impact the rods are parallel and assuming perfectly elastic impact ($e = 1$), determine the angular velocity of each rod and the velocity of its mass center immediately after impact.

18.59 Solve Prob. 18.58, assuming that the impact is perfectly plastic ($e = 0$).

***18.8. Gyroscopes.** Our study of kinetics is limited to the plane motion of rigid bodies. The analysis of the general motion of a rigid body in space is indeed beyond the scope of this text. Among the problems which we shall *not* attempt to solve with the means at our disposal are those involving the general motion of a top or of a gyroscope. However, we shall consider in this section a particular case of gyroscopic motion which requires only a very limited knowledge of the motion of a rigid body in space and yet has many useful engineering applications.

A *gyroscope* consists of a body of revolution mounted on an axle which coincides with its geometric axis. Consider a gyroscope which spins at a very large rate ω about an axle AB directed along the z axis (Fig. 18.11). This motion may be represented by a vector $\boldsymbol{\omega}$ of magnitude equal to the *rate of spin* ω of the gyroscope and directed along the *spin axis* AB in such a way that an observer located at the tip of the vector $\boldsymbol{\omega}$ will see the gyroscope rotate counterclockwise.

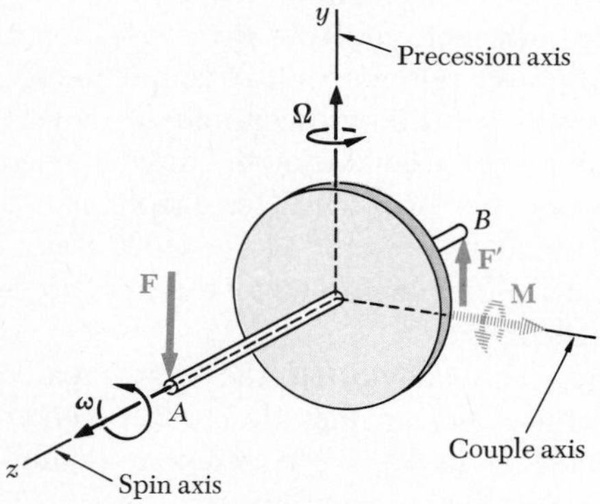

Fig. 18.11

If we attempt to rotate the axle of the gyroscope about the x axis by applying two vertical forces **F** and **F′** at A and B, forming a couple of moment M, we find that the axle will resist the motion we try to impose and that *it will rotate about the y axis* with an angular velocity Ω. This motion of the gyroscope is called *precession;* it may be represented by a vector $\boldsymbol{\Omega}$ of magnitude equal to the *rate of precession* Ω of the gyroscope and directed along the y axis, or *precession axis,* in such a way that a man located at the tip of the vector will observe this motion as counterclockwise.

In order to explain this apparently odd behavior, we shall consider the gyroscope in two different positions and apply the principle of impulse and momentum. We first consider the gyro-

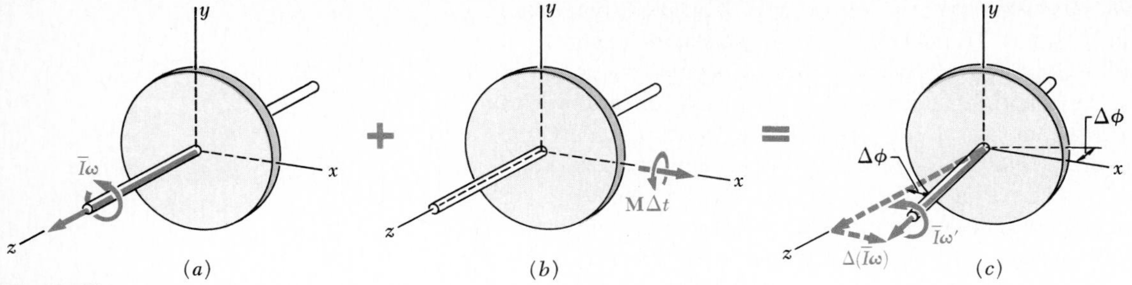

Fig. 18.12

scope in its initial position at $t_1 = 0$. Denoting by $\bar{I}$ its centroidal moment of inertia, we find that the system of the momenta of its particles reduces to a couple of moment $\bar{I}\omega$ (Sec. 18.2). This momentum couple is contained in the xy plane and may therefore be represented by the couple vector $\bar{I}\boldsymbol{\omega}$ directed along the z axis (Fig. 18.12a). Next we consider the gyroscope at time $t_2 = \Delta t$, after its axle has rotated about the y axis through an angle $\Delta\phi$ as a result of our miscalculated effort. We may neglect the relatively small angular momentum of the gyroscope about the y axis and consider only the large momentum couple vector $\bar{I}\boldsymbol{\omega}'$ associated with the rotation of the gyroscope about its axle (Fig. 18.12c). The change in momentum during the time interval Δt is represented by the couple vector $\Delta(\bar{I}\boldsymbol{\omega})$. This couple vector has the same direction as the couple vector $\mathbf{M}\,\Delta t$, which represents the angular impulse of the couple formed by the two forces applied to the axle (Fig. 18.12b).

Far from being strange, the behavior of the gyroscope is therefore entirely predictable. Noting that the couple vector $\Delta(\bar{I}\boldsymbol{\omega})$ has a magnitude equal to $\bar{I}\omega\,\Delta\phi$, we equate the angular impulse and the change in angular momentum and write

$$M\,\Delta t = \bar{I}\omega\,\Delta\phi$$

Dividing both members by Δt and letting Δt approach zero, we find that the moment M of the *gyroscopic couple* is equal to

$$M = \bar{I}\omega\frac{d\phi}{dt}$$

$$M = I\omega\Omega \tag{18.20}$$

where ω = rate of spin

$\Omega = d\phi/dt$ = rate of precession of the axis of the gyroscope

We note that the spin axis, the couple axis (about which the force couple is applied), and the precession axis are at right angles to each other. The relative position of these axes may easily be found if we remember that the extremity of the vector couple $\bar{I}\boldsymbol{\omega}$ representing the momentum of the gyroscope tends to move in the direction defined by the vector **M** representing the gyroscopic couple.

If the motion just described is maintained for an appreciable length of time, the gyroscope is said to be in *steady precession*. Steady precession may be obtained only if the axis of the applied couple is maintained perpendicular to the spin axis and to the precession axis.

We shall now consider another case of steady precession, in which the axis of the gyroscope forms with the precession axis an angle β different from 90° (Fig. 18.13). Neglecting again the angular momentum associated with the motion of precession, we observe that the change in momentum during a time interval

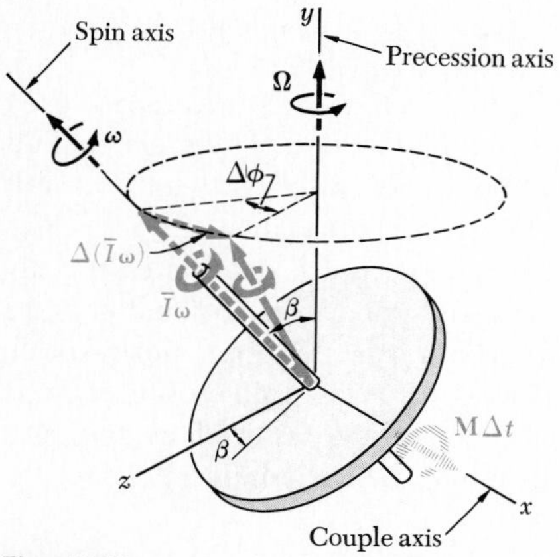

Fig. 18.13

Δt is represented by the vector $\Delta(\bar{I}\boldsymbol{\omega})$ of magnitude $\bar{I}\omega \sin\beta\,\Delta\phi$. Equating angular impulse and change in angular momentum, we write

$$M\,\Delta t = \bar{I}\omega \sin\beta\,\Delta\phi$$

and obtain

$$M = \bar{I}\omega\Omega \sin\beta \qquad (18.21)$$

where Ω represents the rate of precession $d\phi/dt$. Thus, the motion described will be obtained if a couple of moment M satisfying Eq. (18.21) is applied about an axis which is maintained perpendicular to the spin axis and to the precession axis.

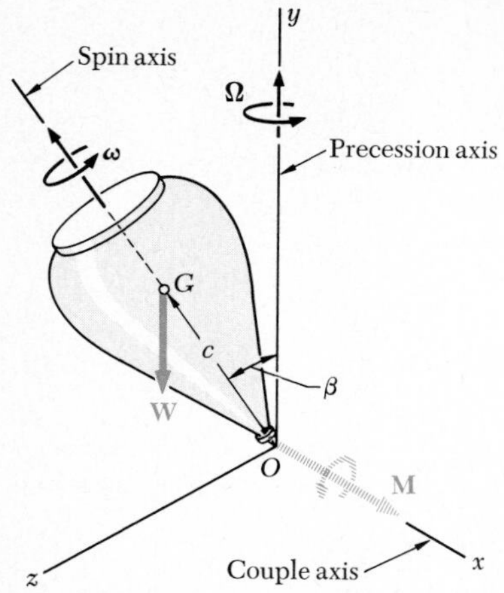

Fig. 18.14

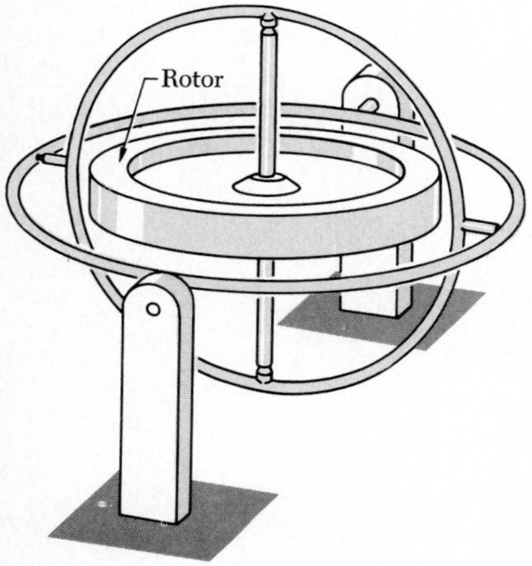

Fig. 18.15

An interesting example of gyroscopic motion is provided by a top supported at a fixed point O (Fig. 18.14). The weight $\mathbf{W}$ of the top and the normal component of the reaction at O form a couple of moment $M = Wc \sin \beta$. Assuming that the angle β the axis of the top forms with the vertical remains constant, we substitute for M in formula (18.21) and write

$$Wc = \bar{I}\omega\Omega \tag{18.22}$$

We should note that the result obtained is valid only for a very large rate of spin and that the top will just fall if it does not spin. Besides, the top will move in *steady* precession, at a constant rate Ω and with its axis forming a constant angle β with the vertical, only if it is given initially a rate of spin ω and a rate of precession Ω which satisfy Eq. (18.22). Otherwise, *the precession will be unsteady* and will be accompanied by variations in β, referred to as *nutation*. The study of the general case of motion of the top is a difficult one which requires a full knowledge of the motion of rigid bodies in space.†

A gyroscope mounted in a Cardan's suspension (Fig. 18.15) is free from any applied couple and thus will maintain a fixed direction with respect to a newtonian frame of reference. This was one of the facts used by the French physicist Jean Foucault (1819–1868) to prove that the earth rotates about its axis. The *gyrocompass* used to indicate the true north consists essentially of a gyroscope whose axis is maintained in a horizontal plane and oscillates symmetrically about the direction north-south. Other sensitive instruments based on gyroscopic motion include the inertial guidance systems of rockets, as well as the turn indicator, the artificial horizon, and the automatic pilot used on airplanes.

Because of the relatively large couples required to change the orientation of their axles, gyroscopes are used as stabilizers in torpedoes and ships. Spinning bullets and shells remain tangent to their trajectory because of gyroscopic action. And a bicycle is easier to keep balanced at high speeds because of the stabilizing effect of its spinning wheels. However, gyroscopic action is not always welcome and must be taken into account in the design of bearings supporting rotating shafts subjected to forced precession. The reactions exerted by its propellers on an airplane which changes its direction of flight must also be taken into consideration and compensated for whenever possible.

† Cf. Synge and Griffith, "Principles of Mechanics," McGraw-Hill Book Company, New York, sec. 14.2, or Becker, "Introduction to Theoretical Mechanics," McGraw-Hill Book Company, New York, sec. 12.14.

SAMPLE PROBLEM 18.7

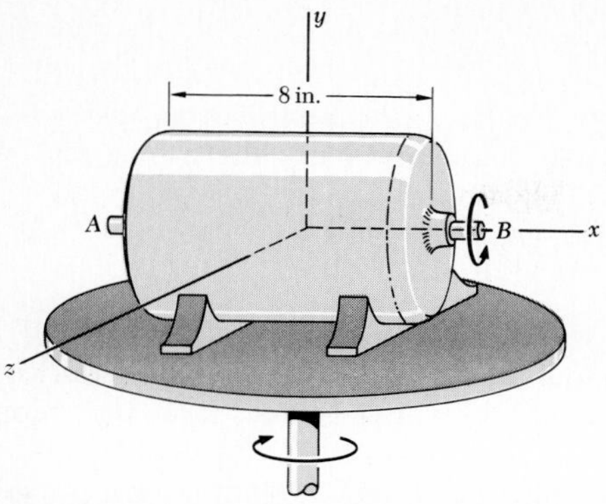

The rotor of an electric motor weighs 6 lb and has a radius of gyration of 2 in. The angular velocity of the rotor is 3600 rpm counterclockwise as viewed from the positive x axis. Determine the reactions exerted by the bearings on the axle AB when the motor is rotated about the y axis clockwise as viewed from above and at a rate of 6 rpm.

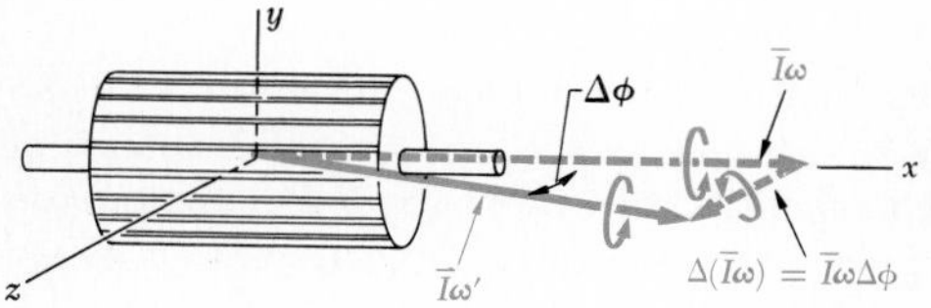

Solution. During the time interval Δt, the momentum couple vector $\bar{I}\boldsymbol{\omega}$ maintains a constant magnitude $\bar{I}\omega$ but is rotated in a horizontal plane through an angle $\Delta\phi$ into a new position $\bar{I}\boldsymbol{\omega}'$. Thus, the change in angular momentum is represented by the vector $\Delta(\bar{I}\boldsymbol{\omega})$ parallel to the z axis and of magnitude $\bar{I}\omega\,\Delta\phi$. Denoting by $\mathbf{M}_z$ the couple applied about the z axis, we write that the angular impulse $\mathbf{M}_z\,\Delta t$ must equal the change in angular momentum.

$$M_z\,\Delta t = \bar{I}\omega\,\Delta\phi \qquad M_z = \bar{I}\omega\,\frac{\Delta\phi}{\Delta t}$$

Denoting by $\boldsymbol{\Omega}$ the angular velocity about the vertical y axis, we write $\Omega = \Delta\phi/\Delta t$ and obtain

$$M_z = \bar{I}\omega\Omega \tag{1}$$

Using the given data, we have

$$\omega = 3600\text{ rpm} = 377\text{ rad/s}$$
$$\Omega = 6\text{ rpm} = 0.628\text{ rad/s}$$
$$\bar{I} = m\bar{k}^2 = \frac{6\text{ lb}}{32.2\text{ ft/s}^2}(\tfrac{2}{12}\text{ ft})^2 = 0.00518\text{ lb}\cdot\text{ft}\cdot\text{s}^2$$

Substituting these values into (1), we obtain the moment of the gyroscopic couple.

$$M_z = (0.00518\text{ lb}\cdot\text{ft}\cdot\text{s}^2)(377\text{ rad/s})(0.628\text{ rad/s})$$
$$= 1.226\text{ lb}\cdot\text{ft}$$

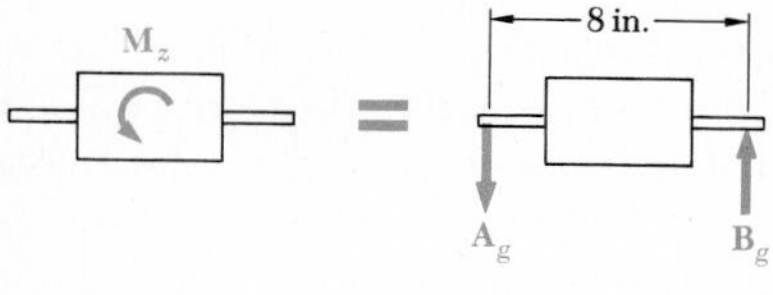

Reactions at A and B. The gyroscopic reactions $\mathbf{A}_g$ and $\mathbf{B}_g$ must be *equivalent* to the gyroscopic couple $\mathbf{M}_z$.

$$M_z = A_g(\tfrac{8}{12}\text{ ft})$$
$$1.226\text{ lb}\cdot\text{ft} = A_g(\tfrac{8}{12}\text{ ft})$$
$$A_g = 1.84\text{ lb} \qquad \mathbf{A}_g = 1.84\text{ lb}\downarrow$$
$$B_g = 1.84\text{ lb} \qquad \mathbf{B}_g = 1.84\text{ lb}\uparrow$$

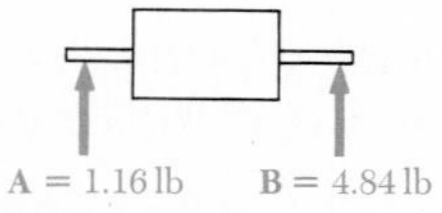

Since the static reactions are $\mathbf{A}_s = 3\text{ lb}\uparrow$ and $\mathbf{B}_s = 3\text{ lb}\uparrow$, the total reactions are

$$\mathbf{A} = 1.16\text{ lb}\uparrow \qquad \mathbf{B} = 4.84\text{ lb}\uparrow$$ ◄

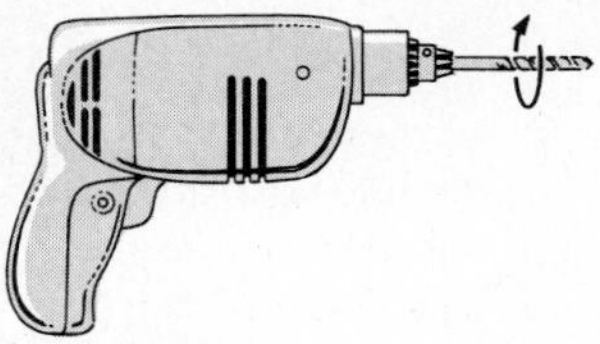

Fig. P18.60

PROBLEMS

18.60 The main rotating parts of a hand drill rotate clockwise when viewed from behind. In what direction will the front end of the drill tend to move if the man holding the drill (*a*) rotates it counterclockwise in a horizontal plane, (*b*) tips it downward while maintaining it in the same vertical plane?

18.61 The major rotating elements of a given propeller-driven airplane rotate clockwise when viewed from behind. If the pilot turns the airplane to his right, determine in what direction the nose of the airplane will tend to move under the action of the gyroscopic couple exerted by the rotating elements.

18.62 A three-bladed airplane propeller has a mass of 150 kg and a radius of gyration of 0.9 m. Knowing that the propeller rotates at 1500 rpm, determine the moment of the couple applied by the propeller to its shaft when the airplane travels in a circular path of 400-m radius at 600 km/h.

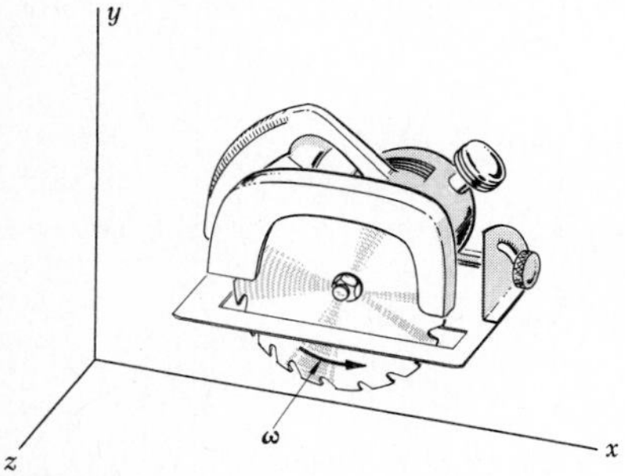

Fig. P18.63

18.63 The blade of a portable saw and the rotor of its motor have a combined mass of 1.2 kg and a radius of gyration of 35 mm. Determine the couple that a man must exert on the handle to rotate the saw about the y axis with a constant angular velocity of 3 rad/s clockwise, as viewed from above, when the blade rotates at the rate $\omega = 1800$ rpm as shown.

18.64 The flywheel of an automobile engine, which is mounted on the crankshaft, is equivalent to a 16-in.-diameter steel plate of $\frac{15}{16}$-in. thickness. At a time when the flywheel is rotating at 4000 rpm the automobile is traveling around a curve of 600-ft radius at a speed of 60 mi/h. Determine, at that time, the magnitude of the gyroscopic couple exerted by the flywheel on the horizontal crankshaft. (Specific weight of steel $= 490$ lb/ft^3.)

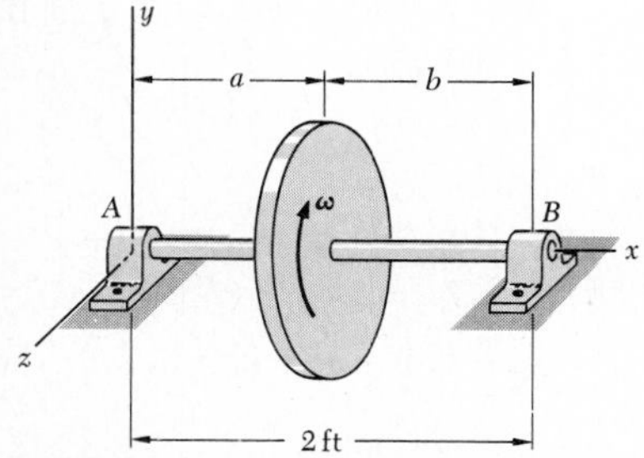

Fig. P18.65

18.65 The rotor of a given turbine may be approximated by a 50-lb disk of 12-in. radius. Knowing that the turbine rotates clockwise at 10,000 rpm as viewed from the positive x axis, determine the components due to gyroscopic action of the forces exerted by the bearings on axle AB if the instantaneous angular velocity of the turbine housing is 2 rad/s clockwise as viewed from (*a*) the positive y axis, (*b*) the positive x axis.

18.66 Solve Prob. 18.65 at an instant when the angular velocity of the turbine housing is 2 rad/s clockwise as viewed from the positive z axis.

18.67 The essential structure of a certain type of aircraft turn indicator is shown. Springs AC and BD are initially stretched and exert equal vertical forces at A and B when the airplane is traveling in a straight path. Knowing that the disk weighs $\frac{1}{2}$ lb and spins at the rate of 10,000 rpm, determine the angle through which the yoke will rotate when the airplane executes a horizontal turn of radius 2000 ft at a speed of 400 mi/h. The constant of each spring is 3 lb/in.

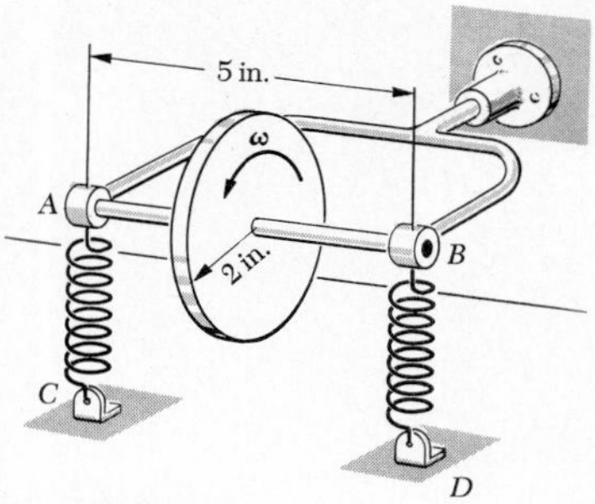

Fig. P18.67

18.68 The rotor shown spins at the rate of 12 000 rpm about the 150-mm shaft CD. At the same time the gimbal rotates about the vertical axis AB at the rate of 20 rpm. Knowing that the mass of the rotor is 30 g and that it has a centroidal radius of gyration of 20 mm, determine the magnitude of the reactions at C and D due to gyroscopic action.

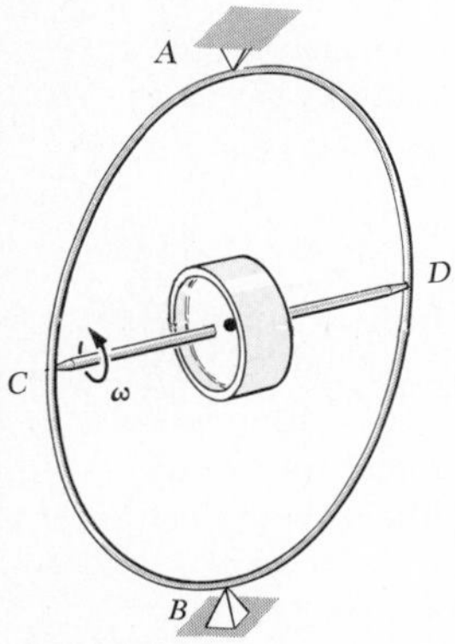

Fig. P18.68

18.69 A toy top has a radius of gyration of 20 mm about its geometric axis and is set spinning at the rate of 1500 rpm. Assuming steady precession, determine the rate of precession (*a*) for $\beta = 10°$, (*b*) for $\beta = 20°$.

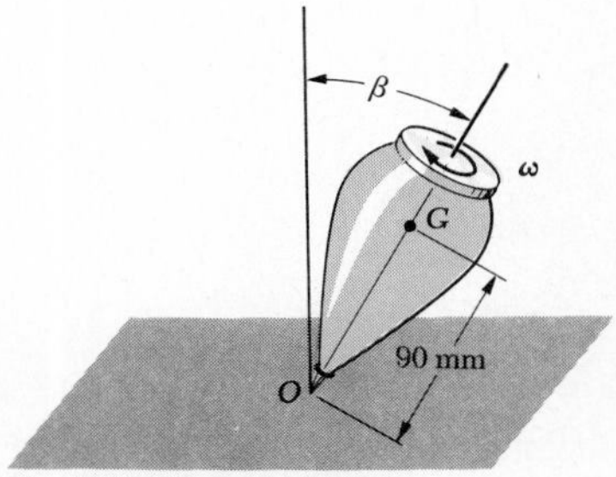

Fig. P18.69 and P18.70

18.70 A toy top having a radius of gyration of 20 mm with respect to its geometric axis is set in motion as shown. It is observed that the top precesses through two complete turns in 1 s. Assuming steady precession, determine the rate at which the top spins about its geometric axis.

18.71 A 5-lb disk of 9-in. diameter is attached to the end of a rod AB of negligible weight which is supported by a ball-and-socket joint at A. If the rate of steady precession Ω of the disk about the vertical is observed to be 24 rpm, determine the rate of spin ω of the disk about AB when $\beta = 30°$.

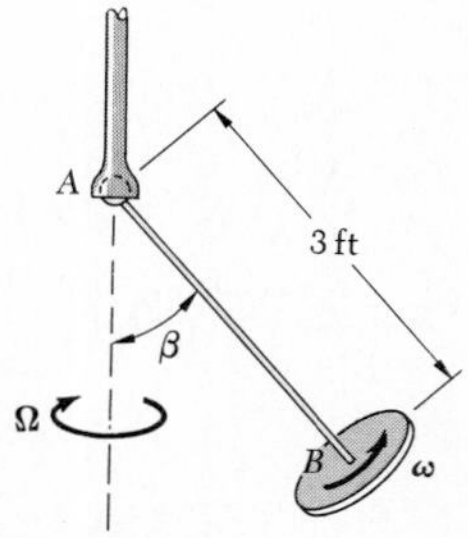

Fig. P18.71

18.72 Solve Prob. 18.71 assuming that the rod AB weighs 2 lb.

18.73 A homogeneous sphere of radius a and mass m is attached to a light rod of length $4a$. The rod forms an angle of 30° with the vertical and rotates about AC with a constant angular velocity $\mathbf{\Omega}$. (*a*) Assuming that the sphere does not spin about the rod ($\omega = 0$), determine the tension in the cord BC. (*b*) Determine the spin ω (magnitude and sense) which should be given to the sphere if the tension in the cord BC is to be zero.

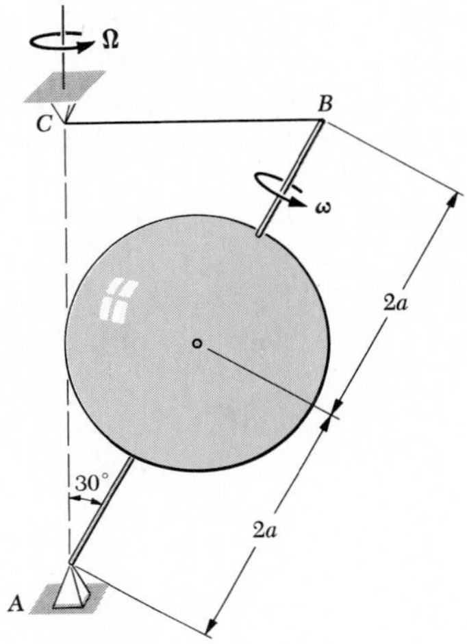

Fig. P18.73

REVIEW PROBLEMS

Fig. P18.74

18.74 A 300-mm-diameter pipe of mass 20 kg rests on a 5-kg plate. The pipe and plate are initially at rest when a force $\mathbf{P}$ of magnitude 150 N is applied for 0.50 s. Knowing that $\mu = 0.20$ between the plate and *both* the pipe and the floor, determine (*a*) whether the pipe slides with respect to the plate, (*b*) the resulting velocities of the pipe and of the plate.

18.75 Solve Prob. 18.74, assuming that the magnitude of the force $\mathbf{P}$ is 100 N.

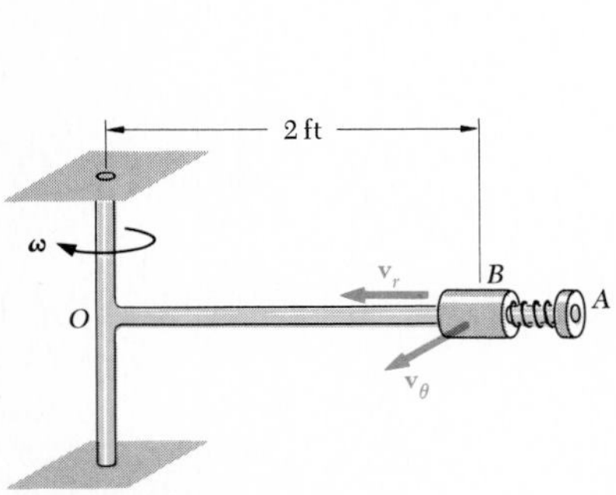

Fig. P18.76

18.76 Collar B weighs 3 lb and may slide freely on rod OA which in turn may rotate freely in the horizontal plane. The assembly is rotating with an angular velocity $\omega = 1.5$ rad/s when a spring located between A and B is released, projecting the collar along the rod with an initial relative speed $v_r = 5$ ft/s. Knowing that the moment of inertia about O of the rod and spring is $0.15 \text{ lb} \cdot \text{ft} \cdot \text{s}^2$, determine (*a*) the minimum distance between the collar and point O in the ensuing motion, (*b*) the corresponding angular velocity of the assembly.

18.77 In Prob. 18.76, determine the required magnitude of the initial relative velocity $\mathbf{v}_r$ if during the ensuing motion the minimum distance between collar *B* and point *O* is to be 1 ft.

18.78 A uniform slender rod of length *L* is dropped onto rigid supports at *A* and *B*. Immediately before striking *A* the velocity of the rod is $\bar{\mathbf{v}}_1$. Since support *B* is slightly lower than support *A*, the rod strikes *A* before it strikes *B*. Assuming perfectly elastic impact at both *A* and *B*, determine the angular velocity of the rod and the velocity of its mass center immediately after the rod (*a*) strikes support *A*, (*b*) strikes support *B*, (*c*) again strikes support *A*.

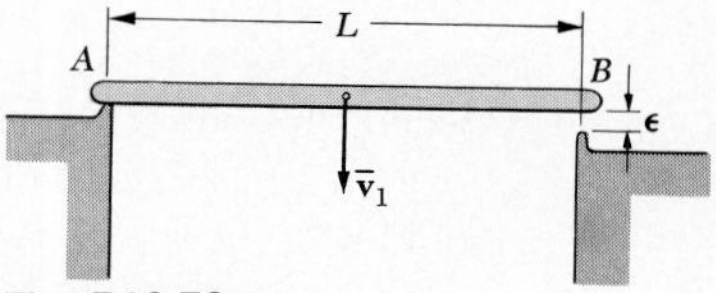

Fig. P18.78

18.79 A rigid rod of length *l* and mass *m* is released in the position shown. It is observed that the rod rebounds to a horizontal position after striking the inclined surface. (*a*) Determine the coefficient of restitution between the knob *K* and the surface. (*b*) Show that the same rebound may be expected for any position of the knob *K*.

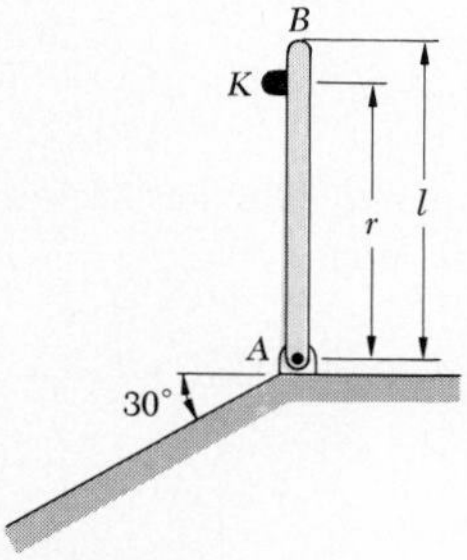

Fig. P18.79

18.80 The plank *CDE* of mass m_P rests on top of a small pivot at *D*. A gymnast *A* of mass *m* stands on the plank at end *C*; a second gymnast *B* of the same mass *m* jumps from a height *h* and strikes the plank at *E*. Assuming perfectly plastic impact, determine the height to which gymnast *A* will rise. (Assume that gymnast *A* stands completely rigid.)

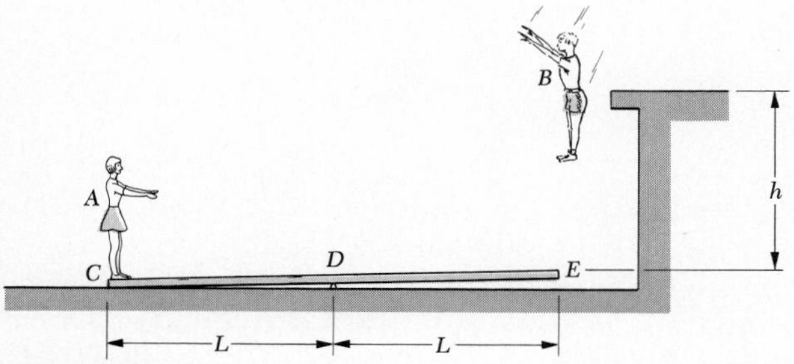

Fig. P18.80

18.81 Solve Prob. 18.80, assuming that the impact between gymnast *B* and the plank is perfectly elastic.

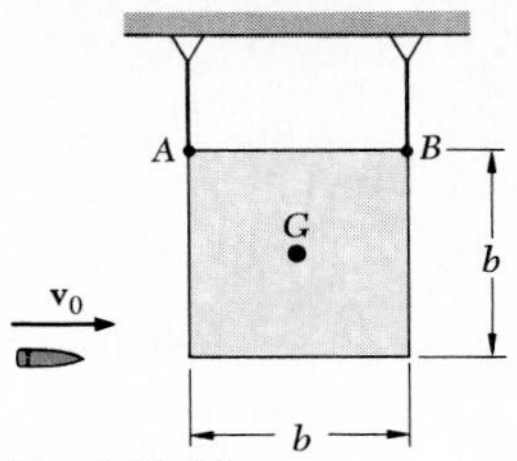

Fig. P18.82

18.82 A bullet of mass m is fired with a horizontal velocity $\mathbf{v}_0$ into the lower corner of a square panel of much larger mass M. The panel is held by two vertical wires as shown. Determine the velocity of the center G of the panel immediately after the bullet becomes embedded.

18.83 A sphere of mass m_S is dropped from a height h and strikes at A the uniform slender plank AB of mass m which is held by two inextensible cords. Knowing that the impact is perfectly plastic and that the sphere remains attached to the plank, determine the velocity of the sphere immediately after impact.

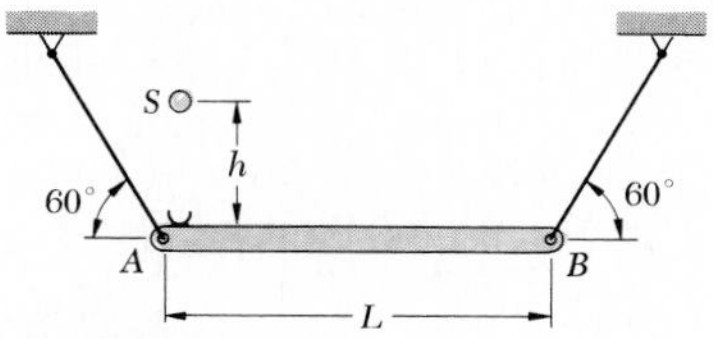

Fig. P18.83

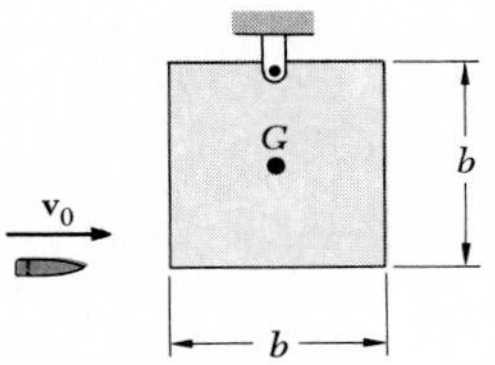

Fig. P18.84

18.84 Solve Prob. 18.82, assuming that the panel is held as shown by a single hinge located at the center of its top edge.

18.85 Solve Prob. 18.83 when $m = 3$ kg, $m_S = 1$ kg, $L = 1.2$ m, and $h = 1.5$ m.

Chapter
19

Mechanical Vibrations

19.1. Introduction. A *mechanical vibration* is the motion of a particle or a body which oscillates about a position of equilibrium. Most vibrations in machines and structures are undesirable because of the increased stresses and energy losses which accompany them. They should therefore be eliminated or reduced as much as possible by appropriate design. The analysis of vibrations has become increasingly important in recent years owing to the current trend toward higher-speed machines and lighter structures. There is every reason to expect that this trend will continue and that an even greater need for vibration analysis will develop in the future.

The analysis of vibrations is a very extensive subject to which entire texts have been devoted. We shall therefore limit our present study to the simpler types of vibrations, namely, the vibrations of a body or a system of bodies with one degree of freedom.

A mechanical vibration generally results when a system is displaced from a position of stable equilibrium. The system tends to return to this position under the action of restoring forces (either elastic forces, as in the case of a mass attached to a spring, or gravitational forces, as in the case of a pendulum). But the system generally reaches its original position with a certain acquired velocity which carries it beyond that position. Since the process can be repeated indefinitely, the system keeps moving back and forth across its position of equilibrium. The time interval required for the system to complete a full cycle of motion is called the *period* of the vibration. The number of cycles per unit time defines the *frequency,* and the maximum displacement of the system from its position of equilibrium is called the *amplitude* of the vibration.

When the motion is maintained by the restoring forces only, the vibration is said to be a *free vibration* (Secs. 19.2 to 19.6). When a periodic force is applied to the system, the resulting motion is described as a *forced vibration* (Sec. 19.7). When the effects of friction may be neglected, the vibrations are said to be *undamped.* However, all vibrations are actually *damped* to some degree. If a free vibration is only slightly damped, its amplitude slowly decreases until, after a certain time, the motion comes to a stop. But damping may be large enough to prevent any true vibration; the system then slowly regains its original position (Sec. 19.8). A damped forced vibration is maintained as long as the periodic force which produces the vibration is applied. The amplitude of the vibration, however, is affected by the magnitude of the damping forces (Sec. 19.9).

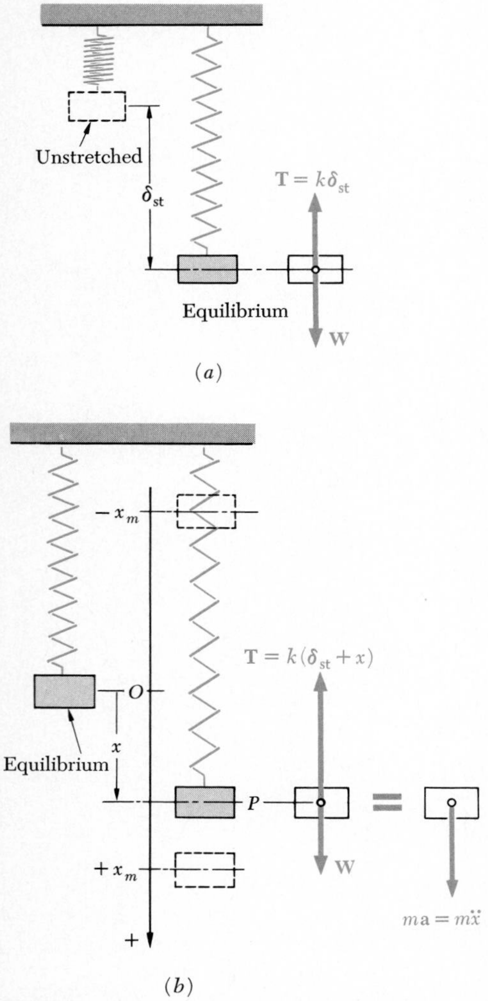

Fig. 19.1

VIBRATIONS WITHOUT DAMPING

19.2. Free Vibrations of Particles. Simple Harmonic Motion. Consider a body of mass m attached to a spring of constant k (Fig. 19.1*a*). Since, at the present time, we are concerned only with the motion of its mass center, we shall refer to this body as a particle. When the particle is in static equilibrium, the forces acting on it are its weight **W** and the force **T** exerted by the spring, of magnitude $T = k\delta_{st}$, where δ_{st} denotes the elongation of the spring. We have, therefore,

$$W = k\delta_{st} \tag{19.1}$$

Suppose now that the particle is displaced through a distance x_m from its equilibrium position and released with no initial velocity. If x_m has been chosen smaller than δ_{st}, the particle will move back and forth through its equilibrium position; a vibration of amplitude x_m has been generated. Note that the vibration may also be produced by imparting a certain initial velocity to the particle when it is in its equilibrium position $x = 0$ or, more generally, by starting the particle from any given position $x = x_0$ with a given initial velocity $\mathbf{v}_0$.

To analyze the vibration, we shall consider the particle in a position P at some arbitrary time t (Fig. 19.1*b*). Denoting by x the displacement OP measured from the equilibrium position

O (positive downward), we note that the forces acting on the particle are its weight $\mathbf{W}$ and the force $\mathbf{T}$ exerted by the spring which, in this position, has a magnitude $T = k(\delta_{st} + x)$. Recalling (19.1), we find that the magnitude of the resultant $\mathbf{F}$ of the two forces (positive downward) is

$$F = W - k(\delta_{st} + x) = -kx \tag{19.2}$$

Thus the *resultant* of the forces exerted on the particle is proportional to the displacement *OP measured from the equilibrium position.* Recalling the sign convention, we note that $\mathbf{F}$ is always directed *toward* the equilibrium position O. Substituting for F into the fundamental equation $F = ma$ and recalling that a is the second derivative $\ddot{x}$ of x with respect to t, we write

$$m\ddot{x} + kx = 0 \tag{19.3}$$

Note that the same sign convention should be used for the acceleration $\ddot{x}$ and for the displacement x, namely, positive downward.

Equation (19.3) is a linear differential equation of the second order. Setting

$$p^2 = \frac{k}{m} \tag{19.4}$$

we write (19.3) in the form

$$\ddot{x} + p^2x = 0 \tag{19.5}$$

The motion defined by Eq. (19.5) is called *simple harmonic motion.* It is characterized by the fact that *the acceleration is proportional to the displacement and of opposite direction.* We note that each of the functions $x_1 = \sin pt$ and $x_2 = \cos pt$ satisfies (19.5). These functions, therefore, constitute two *particular solutions* of the differential equation (19.5). As we shall see presently, the *general solution* of (19.5) may be obtained by multiplying the two particular solutions by arbitrary constants A and B and adding. We write

$$x = Ax_1 + Bx_2 = A \sin pt + B \cos pt \tag{19.6}$$

Differentiating, we obtain successively the velocity and acceleration at time t,

$$v = \dot{x} = Ap \cos pt - Bp \sin pt \tag{19.7}$$

$$a = \ddot{x} = -Ap^2 \sin pt - Bp^2 \cos pt \tag{19.8}$$

Substituting from (19.6) and (19.8) into (19.5), we verify that the expression (19.6) provides a solution of the differential equation

(19.5). Since this expression contains two arbitrary constants A and B, the solution obtained is the general solution of the differential equation. The values of the constants A and B depend upon the *initial conditions* of the motion. For example, we have $A = 0$ if the particle is displaced from its equilibrium position and released at $t = 0$ with no initial velocity, and we have $B = 0$ if P is started from O at $t = 0$ with a certain initial velocity. In general, substituting $t = 0$ and the initial values x_0 and v_0 of the displacement and velocity into (19.6) and (19.7), we find $A = v_0/p$ and $B = x_0$.

The expressions obtained for the displacement, velocity, and acceleration of a particle may be written in a more compact form if we observe that (19.6) expresses that the displacement $x = OP$ is the sum of the x components of two vectors **A** and **B**, respectively of magnitude A and B, directed as shown in Fig. 19.2*a*. As t varies, both vectors rotate clockwise; we also note that the magnitude of their resultant $\overrightarrow{OQ}$ is equal to the maximum displacement x_m. The simple harmonic motion of P along the x axis may thus be obtained by projecting on this axis the motion of a point Q describing an *auxiliary circle* of radius x_m *with a constant angular velocity* p. Denoting by ϕ the angle formed by the vectors $\overrightarrow{OQ}$ and **A**, we write

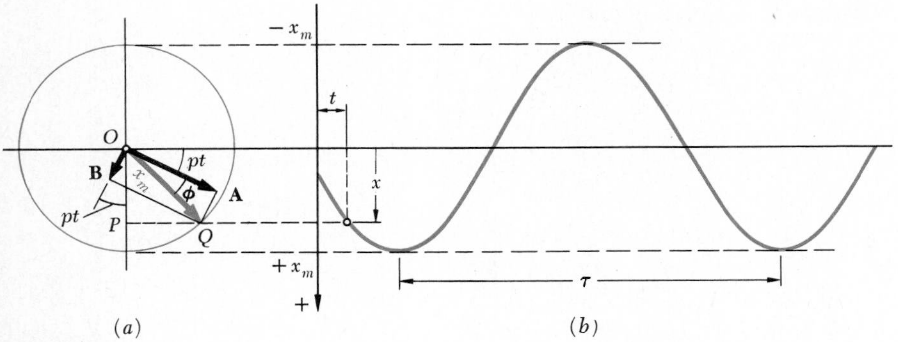

Fig. 19.2

$$OP = OQ \sin(pt + \phi) \tag{19.9}$$

which leads to new expressions for the displacement, velocity, and acceleration of P,

$$x = x_m \sin(pt + \phi) \tag{19.10}$$

$$v = \dot{x} = x_m p \cos(pt + \phi) \tag{19.11}$$

$$a = \ddot{x} = -x_m p^2 \sin(pt + \phi) \tag{19.12}$$

The displacement-time curve is represented by a sine curve (Fig. 19.2*b*), and the maximum value x_m of the displacement is called the *amplitude* of the vibration. The angular velocity p of the point Q which describes the auxiliary circle is known as the *circular frequency* of the vibration and is measured in rad/s, while the angle ϕ which defines the initial position of Q on the circle is called the *phase angle*. We note from Fig. 19.2 that a full *cycle* has been described after the angle pt has increased by 2π rad. The corresponding value of t, denoted by τ, is called the *period* of the vibration and is measured in seconds. We have

$$\text{Period} = \tau = \frac{2\pi}{p} \tag{19.13}$$

The number of cycles described per unit of time is denoted by f and is known as the *frequency* of the vibration. We write

$$\text{Frequency} = f = \frac{1}{\tau} = \frac{p}{2\pi} \tag{19.14}$$

The unit of frequency is a frequency of 1 cycle per second, corresponding to a period of 1 s. In terms of base units the unit of frequency is thus 1/s or s^{-1}. It is called a *hertz* (Hz) in the SI system of units. It also follows from Eq. (19.14) that a frequency of $1\ \mathrm{s}^{-1}$ or 1 Hz corresponds to a circular frequency of 2π rad/s. In problems involving angular velocities expressed in revolutions per minute (rpm), we have $1\ \text{rpm} = \frac{1}{60}\ \mathrm{s}^{-1} = \frac{1}{60}\ \mathrm{Hz}$, or $1\ \text{rpm} = (2\pi/60)\ \text{rad/s}$.

Recalling that p was defined in (19.4) in terms of the constant k of the spring and the mass m of the particle, we observe that the period and the frequency are independent of the initial conditions and of the amplitude of the vibration. Note that τ and f depend on the *mass* rather than on the *weight* of the particle and thus are independent of the value of g.

The velocity-time and acceleration-time curves may be represented by sine curves of the same period as the displacement-time curve, but with different phase angles. From (19.11) and (19.12), we note that the maximum values of the magnitudes of the velocity and acceleration are

$$v_m = x_m p \qquad a_m = x_m p^2 \tag{19.15}$$

Since the point Q describes the auxiliary circle, of radius x_m, at the constant angular velocity p, its velocity and acceleration are equal, respectively, to the expressions (19.15). Recalling Eqs. (19.11) and (19.12), we find, therefore, that the velocity and acceleration of P may be obtained at any instant by projecting on the x axis vectors of magnitudes $v_m = x_m p$ and $a_m = x_m p^2$ representing respectively the velocity and acceleration of Q at the same instant (Fig. 19.3).

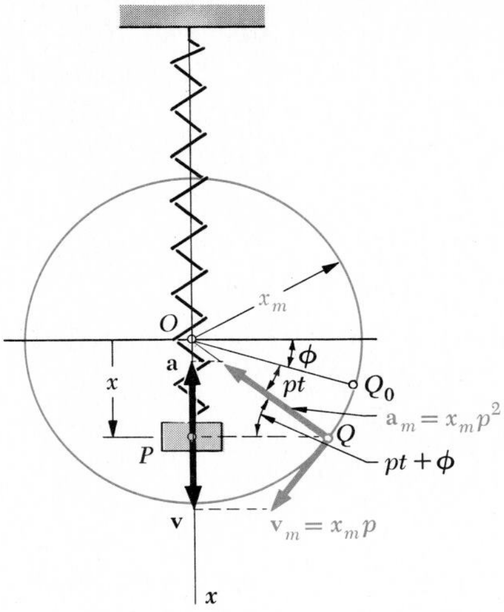

Fig. 19.3

The results obtained are not limited to the solution of the problem of a mass attached to a spring. They may be used to analyze the rectilinear motion of a particle *whenever the resultant* **F** *of the forces acting on the particle is proportional to the displacement x and directed toward O.* The fundamental equation of motion $F = ma$ may then be written in the form (19.5), which characterizes simple harmonic motion. Observing that the coefficient of x in (19.5) represents the square of the circular frequency p of the vibration, we easily obtain p and, after substitution into (19.13) and (19.14), the period τ and the frequency f of the vibration.

19.3. Simple Pendulum (Approximate Solution). Most of the vibrations encountered in engineering applications may be represented by a simple harmonic motion. Many others, although of a different type, may be *approximated* by a simple harmonic motion, provided that their amplitude remains small. Consider for example a *simple pendulum,* consisting of a bob of mass m attached to a cord of length l, which may oscillate in a vertical plane (Fig. 19.4a). At a given time t, the cord forms an angle θ with the vertical. The forces acting on the bob are its weight **W** and the force **T** exerted by the cord (Fig. 19.4b). Resolving the vector $m\mathbf{a}$ into tangential and normal components,

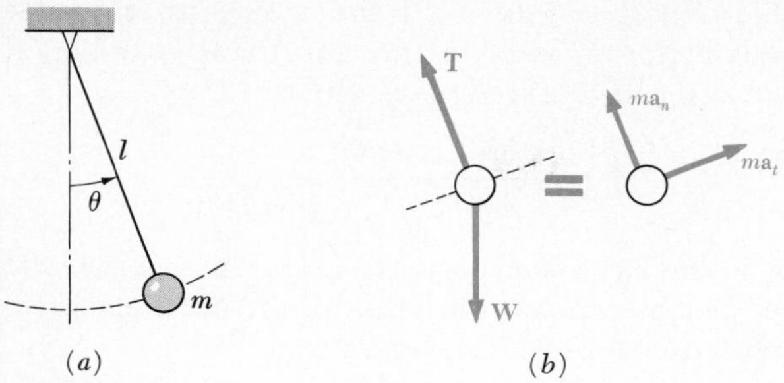

Fig. 19.4

with ma_t directed to the right, i.e., in the direction corresponding to increasing values of θ, and observing that $a_t = l\alpha = l\ddot{\theta}$, we write

$$\Sigma F_t = ma_t: \qquad -W \sin\theta = ml\ddot{\theta}$$

Noting that $W = mg$ and dividing through by ml, we obtain

$$\ddot{\theta} + \frac{g}{l}\sin\theta = 0 \tag{19.16}$$

For oscillations of small amplitude, we may replace $\sin\theta$ by θ, expressed in radians, and write

$$\ddot{\theta} + \frac{g}{l}\theta = 0 \tag{19.17}$$

Comparison with (19.5) shows that the equation obtained is that of a simple harmonic motion and that the circular frequency p of the oscillations is equal to $(g/l)^{1/2}$. Substitution into (19.13) yields the period of the small oscillations of a pendulum of length l,

$$\tau = \frac{2\pi}{p} = 2\pi\sqrt{\frac{l}{g}} \tag{19.18}$$

***19.4. Simple Pendulum (Exact Solution).** Formula (19.18) is only approximate. To obtain an exact expression for the period of the oscillations of a simple pendulum, we must return to (19.16). Multiplying both terms by $2\dot{\theta}$ and integrating from an initial position corresponding to the maximum deflection, that is, $\theta = \theta_m$ and $\dot{\theta} = 0$, we write

$$\dot{\theta}^2 = \frac{2g}{l}(\cos\theta - \cos\theta_m)$$

or

$$\left(\frac{d\theta}{dt}\right)^2 = \frac{2g}{l}(\cos\theta - \cos\theta_m)$$

Replacing $\cos\theta$ by $1 - 2\sin^2(\theta/2)$ and $\cos\theta_m$ by a similar expression, solving for dt, and integrating over a quarter period from $t = 0$, $\theta = 0$ to $t = \tau/4$, $\theta = \theta_m$, we have

$$\tau = 2\sqrt{\frac{l}{g}}\int_0^{\theta_m}\frac{d\theta}{\sqrt{\sin^2(\theta_m/2) - \sin^2(\theta/2)}}$$

The integral in the right-hand member is known as an *elliptic integral;* it cannot be expressed in terms of the usual algebraic or trigonometric functions. However, setting

$$\sin(\theta/2) = \sin(\theta_m/2)\sin\phi$$

we may write

$$\tau = 4\sqrt{\frac{l}{g}}\int_0^{\pi/2}\frac{d\phi}{\sqrt{1 - \sin^2(\theta_m/2)\sin^2\phi}} \qquad (19.19)$$

where the integral obtained, commonly denoted by K, may be found in *tables of elliptic integrals* for various values of $\theta_m/2$.† In order to compare the result just obtained with that of the preceding section, we write (19.19) in the form

$$\tau = \frac{2K}{\pi}\left(2\pi\sqrt{\frac{l}{g}}\right) \qquad (19.20)$$

Formula (19.20) shows that the actual value of the period of a simple pendulum may be obtained by multiplying the approximate value (19.18) by the correction factor $2K/\pi$. Values of the correction factor are given in Table 19.1 for various values of

Table 19.1 Correction Factor for the Period of a Simple Pendulum

θ_m	0°	10°	20°	30°	60°	90°	120°	150°	180°
K	1.571	1.574	1.583	1.598	1.686	1.854	2.157	2.768	∞
$2K/\pi$	1.000	1.002	1.008	1.017	1.073	1.180	1.373	1.762	∞

the amplitude θ_m. We note that for ordinary engineering computations the correction factor may be omitted as long as the amplitude does not exceed 10°.

† See, for example, Dwight, "Table of Integrals and Other Mathematical Data," The Macmillan Company, or Peirce, "A Short Table of Integrals," Ginn and Company.

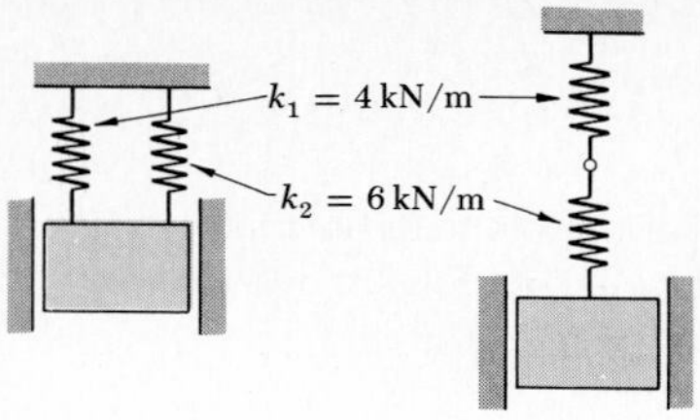

(a) (b)

SAMPLE PROBLEM 19.1

A 50-kg block moves between vertical guides as shown. The block is pulled 40 mm down from its equilibrium position and released. For each spring arrangement, determine the period of the vibration, the maximum velocity of the block, and the maximum acceleration of the block.

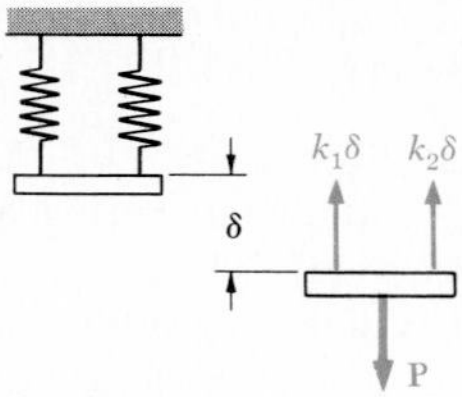

a. Springs Attached in Parallel. We first determine the constant k of a single spring equivalent to the two springs *by finding the magnitude of the force* **P** required to cause a given deflection δ. Since for a deflection δ the magnitudes of the forces exerted by the springs are, respectively, $k_1\delta$ and $k_2\delta$, we have

$$P = k_1\delta + k_2\delta = (k_1 + k_2)\delta$$

The constant k of the single equivalent spring is

$$k = \frac{P}{\delta} = k_1 + k_2 = 4\text{ kN/m} + 6\text{ kN/m} = 10\text{ kN/m} = 10^4\text{ N/m}$$

Period of Vibration: Since $m = 50$ kg, Eq. (19.4) yields

$$p^2 = \frac{k}{m} = \frac{10^4\text{ N/m}}{50\text{ kg}} \qquad p = 14.14\text{ rad/s}$$

$$\tau = 2\pi/p \qquad \tau = 0.444\text{ s} \blacktriangleleft$$

Maximum Velocity: $v_m = x_m p = (0.040\text{ m})(14.14\text{ rad/s})$

$$v_m = 0.566\text{ m/s} \qquad \mathbf{v}_m = 0.566\text{ m/s} \updownarrow \blacktriangleleft$$

Maximum Acceleration: $a_m = x_m p^2 = (0.040\text{ m})(14.14\text{ rad/s})^2$

$$a_m = 8.00\text{ m/s}^2 \qquad \mathbf{a}_m = 8.00\text{ m/s}^2 \updownarrow \blacktriangleleft$$

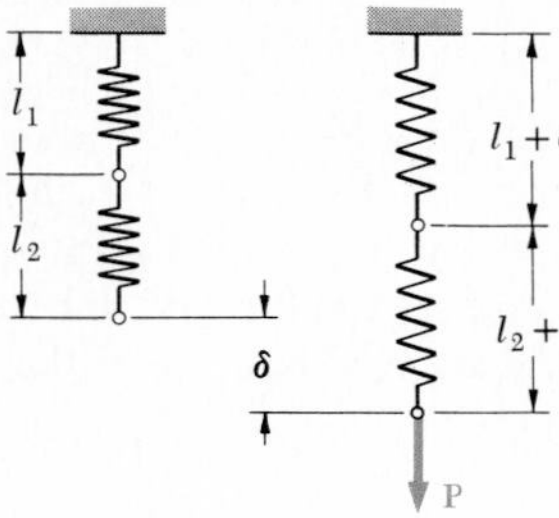

b. Springs Attached in Series. We first determine the constant k of a single spring equivalent to the two springs *by finding the total elongation* δ of the springs under a given static load **P**. To facilitate the computation, a static load of magnitude $P = 12$ kN is used.

$$\delta = \delta_1 + \delta_2 = \frac{P}{k_1} + \frac{P}{k_2} = \frac{12\text{ kN}}{4\text{ kN/m}} + \frac{12\text{ kN}}{6\text{ kN/m}} = 5\text{ m}$$

$$k = \frac{P}{\delta} = \frac{12\text{ kN}}{5\text{ m}} = 2.4\text{ kN/m} = 2400\text{ N/m}$$

Period of Vibration: $p^2 = \dfrac{k}{m} = \dfrac{2400\text{ N/m}}{50\text{ kg}} \qquad p = 6.93\text{ rad/s}$

$$\tau = \frac{2\pi}{p} \qquad \tau = 0.907\text{ s} \blacktriangleleft$$

Maximum Velocity: $v_m = x_m p = (0.040\text{ m})(6.93\text{ rad/s})$

$$v_m = 0.277\text{ m/s} \qquad \mathbf{v}_m = 0.277\text{ m/s} \updownarrow \blacktriangleleft$$

Maximum Acceleration: $a_m = x_m p^2 = (0.040\text{ m})(6.93\text{ rad/s})^2$

$$a_m = 1.920\text{ m/s}^2 \qquad \mathbf{a}_m = 1.920\text{ m/s}^2 \updownarrow \blacktriangleleft$$

PROBLEMS

19.1 A particle moves in simple harmonic motion with an amplitude of 3 in. and a period of 0.60 s. Find the maximum velocity and the maximum acceleration.

19.2 The analysis of the motion of a particle shows a maximum acceleration of 30 m/s^2 and a frequency of 120 cycles per minute. Assuming that the motion is simple harmonic, determine (*a*) the amplitude, (*b*) the maximum velocity.

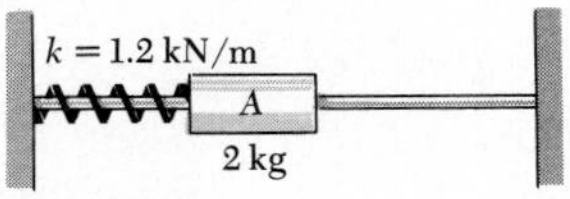

Fig. P19.3

19.3 Collar *A* is attached to the spring shown and may slide without friction on the horizontal rod. If the collar is moved 75 mm from its equilibrium position and released, determine the period, the maximum velocity, and the maximum acceleration of the resulting motion.

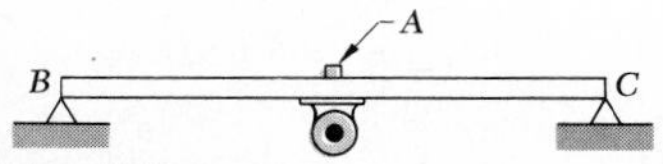

Fig. P19.4

19.4 A variable-speed motor is rigidly attached to the beam *BC*. The rotor is slightly unbalanced and causes the beam to vibrate with a circular frequency equal to the motor speed. When the speed of the motor is less than 600 rpm or more than 1200 rpm, a small object placed at *A* is observed to remain in contact with the beam. For speeds between 600 and 1200 rpm the object is observed to "dance" and actually to lose contact with the beam. Determine the amplitude of the motion of *A* when the speed of the motor is (*a*) 600 rpm, (*b*) 1200 rpm. Give answers in both SI and U.S. customary units.

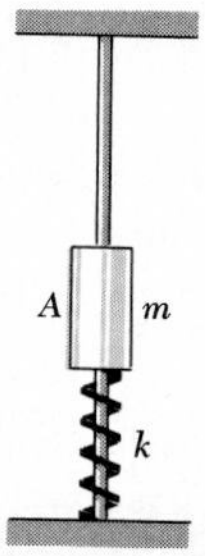

Fig. P19.5, P19.6, and P19.8

19.5 The 10-lb collar rests on, but is not attached to, the spring shown. The collar is depressed 1.5 in. and released. If the ensuing motion is to be simple harmonic, determine (*a*) the largest permissible value of the spring constant *k*, (*b*) the corresponding frequency of the motion.

19.6 The 5-kg collar is attached to a spring of constant $k = 800$ N/m as shown. If the collar is given a displacement of 50 mm from its equilibrium position and released, determine for the ensuing motion (*a*) the period, (*b*) the maximum velocity of the collar, (*c*) the maximum acceleration of the collar.

19.7 In Prob. 19.6, determine the position, velocity, and acceleration of the collar 0.20 s after it has been released.

19.8 An 8-lb collar is attached to a spring of constant $k = 5$ lb/in. as shown. If the collar is given a displacement of 2 in. downward from its equilibrium position and released, determine (*a*) the time required for the collar to move 3 in. upward, (*b*) the corresponding velocity and acceleration of the collar.

19.9 and 19.10 A 35-kg block is supported by the spring arrangement shown. If the block is moved vertically downward from its equilibrium position and released, determine (*a*) the period and frequency of the resulting motion, (*b*) the maximum velocity and acceleration of the block if the amplitude of the motion is 20 mm.

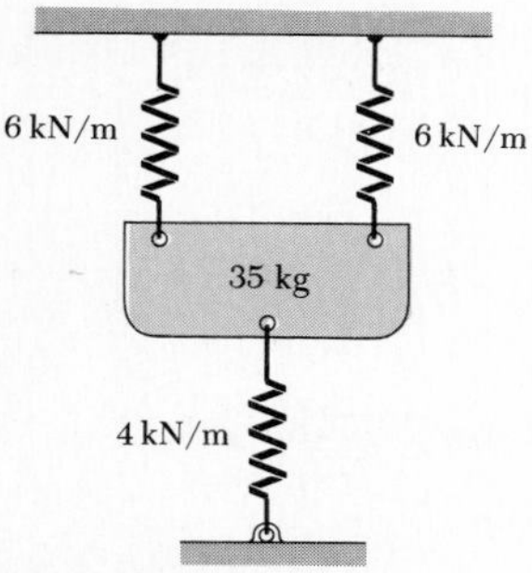

Fig. P19.9

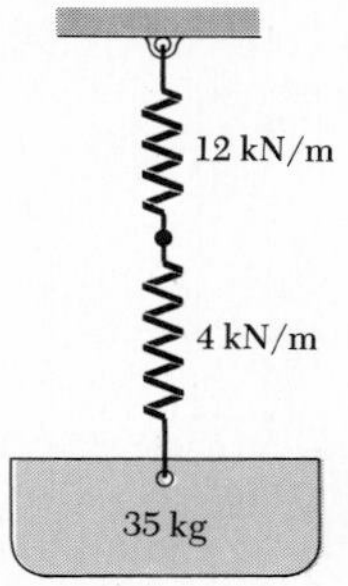

Fig. P19.10

19.11 Denoting by δ_{st} the static deflection of a beam under a given load, show that the frequency of vibration of the load is

$$f = \frac{1}{2\pi}\sqrt{\frac{g}{\delta_{st}}}$$

Neglect the mass of the beam, and assume that the load remains in contact with the beam.

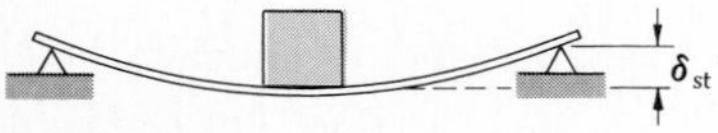

Fig. P19.11

19.12 The frequency of vibration of the system shown is observed to be 2.5 cycles per second. After cylinder *B* is removed, the frequency is observed to be 3 cycles per second. Knowing that cylinder *B* weighs 2 lb, determine the weight of cylinder *A*.

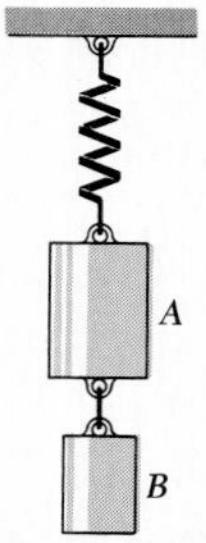

Fig. P19.12

19.13 A simple pendulum of length l is suspended in an elevator. A mass m is attached to a spring of constant k and is carried in the same elevator. Determine the period of vibration of both the pendulum and the mass if the elevator has an upward acceleration $\mathbf{a}$.

19.14 Determine (*a*) the required length l of a simple pendulum if the period of small oscillations is to be 2 s, (*b*) the required amplitude of this pendulum if the maximum velocity of the bob is to be 200 mm/s.

19.15 and 19.16 An 80-lb block is attached to the rod AB and to the two springs shown. Knowing that $h = 30$ in., determine the period of small oscillations of the block. Neglect the weight of the rod and assume that each spring can act in either tension or compression.

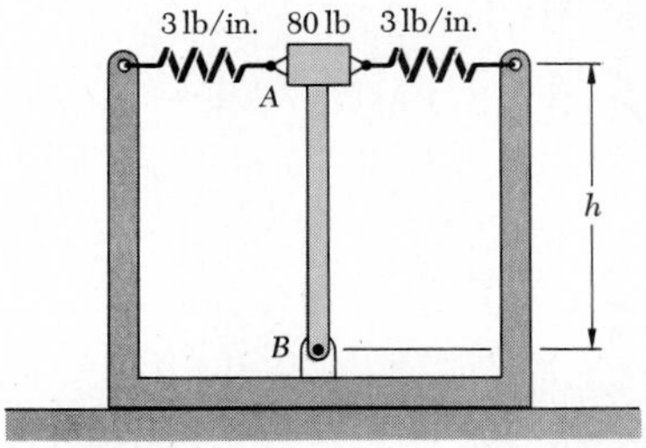

Fig. P19.15

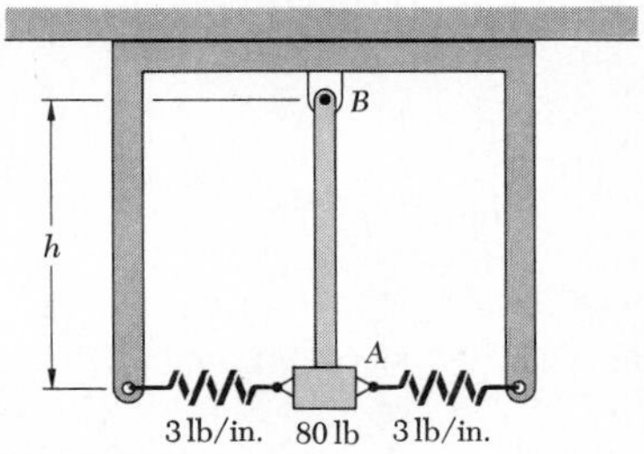

Fig. P19.16

19.17 In Prob. 19.15, determine the length h of rod AB for which the period of small oscillations is (*a*) 2 s, (*b*) infinite.

***19.18** A particle is placed with no initial velocity on a frictionless *plane* tangent to the surface of the earth. (*a*) Show that the particle will theoretically execute simple harmonic motion with a period of oscillation equal to that of a simple pendulum of length equal to the radius of the earth. (*b*) Compute the theoretical period of oscillation and show that it is equal to the periodic time of an earth satellite describing a low-altitude circular orbit. [*Hint.* See Eq. (12.35).]

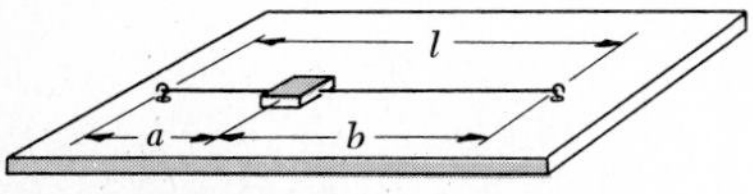

Fig. P19.19 and P19.20

19.19 A small block of mass m rests on a frictionless horizontal surface and is attached to a taut string. Denoting by T the tension in the string, determine the period and frequency of small oscillations of the block in a direction perpendicular to the string. Show that the longest period occurs when $a = b = \frac{1}{2}l$.

19.20 A 3-lb block rests on a frictionless horizontal surface and is attached to a taut string. Knowing that the tension in the string is 8 lb, determine the frequency of small oscillations of the block when $a = 12$ in. and $b = 18$ in.

***19.21** Expanding the integrand in (19.19) into a series of even powers of $\sin\phi$ and integrating, show that the period of a simple pendulum of length l may be approximated by the formula

$$\tau = 2\pi\sqrt{\frac{l}{g}}\left(1 + \tfrac{1}{4}\sin^2\frac{\theta_m}{2}\right)$$

where θ_m is the amplitude of the oscillations.

***19.22** Using the data of Table 19.1, determine the period of a simple pendulum of length 800 mm (*a*) for small oscillations, (*b*) for oscillations of amplitude $\theta_m = 30°$, (*c*) for oscillations of amplitude $\theta_m = 90°$.

***19.23** Using the formula given in Prob. 19.21, determine the amplitude θ_m for which the period of a simple pendulum is $\frac{1}{2}$ percent longer than the period of the same pendulum for small oscillations.

***19.24** Using a table of elliptic integrals, determine the period of a simple pendulum of length $l = 800$ mm if the amplitude of the oscillations is $\theta_m = 40°$.

19.5. Free Vibrations of Rigid Bodies. The analysis of the vibrations of a rigid body or of a system of rigid bodies possessing a single degree of freedom is similar to the analysis of the vibrations of a particle. An appropriate variable, such as a distance x or an angle θ, is chosen to define the position of the body or system of bodies, and an equation relating this variable and its second derivative with respect to t is written. If the equation obtained is of the same form as (19.5), i.e., if we have

$$\ddot{x} + p^2x = 0 \qquad \text{or} \qquad \ddot{\theta} + p^2\theta = 0 \tag{19.21}$$

the vibration considered is a simple harmonic motion. The period and frequency of the vibration may then be obtained by identifying p and substituting into (19.13) and (19.14).

In general, a simple way to obtain one of Eqs. (19.21) is to express that the system of the external forces is equivalent to the system of the effective forces by drawing a diagram of the body for an arbitrary value of the variable and writing the appropriate equation of motion. We recall that our goal should be *the determination of the coefficient* of the variable x or θ, *not* the determination of the variable itself or of the derivatives $\ddot{x}$ or $\ddot{\theta}$. Setting this coefficient equal to p^2, we obtain the circular frequency p, from which τ and f may be determined.

The method we have outlined may be used to analyze vibrations which are truly represented by a simple harmonic motion, or vibrations of small amplitude which can be *approximated* by a simple harmonic motion. As an example, we shall determine the period of the small oscillations of a square plate of side $2b$

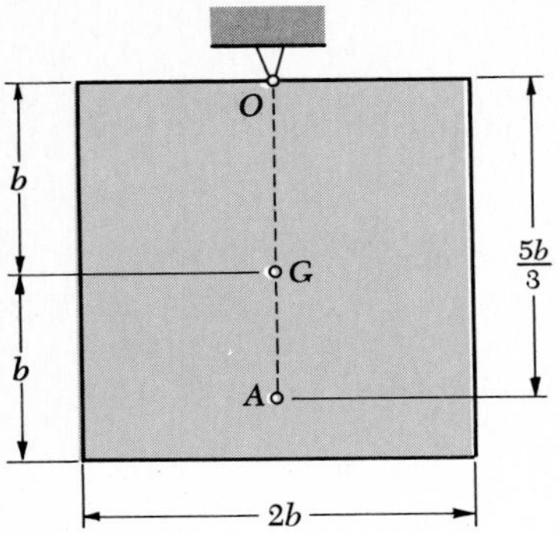

(a)

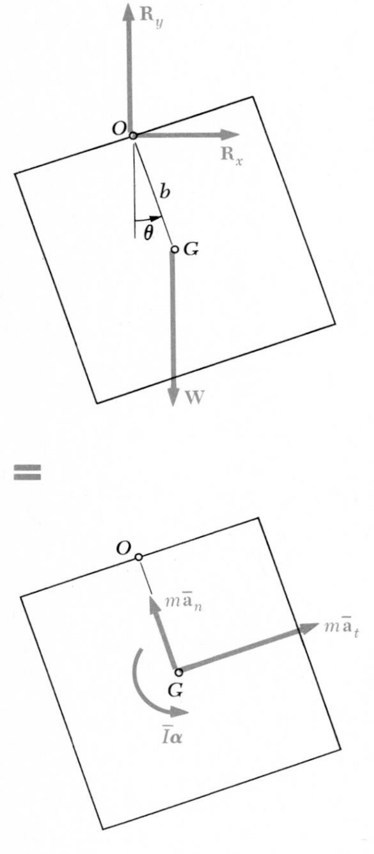

(b)

Fig. 19.5

which is suspended from the midpoint O of one side (Fig. 19.5*a*). We consider the plate in an arbitrary position defined by the angle θ that the line OG forms with the vertical and draw a diagram to express that the weight $\mathbf{W}$ of the plate and the components $\mathbf{R}_x$ and $\mathbf{R}_y$ of the reaction at O are equivalent to the vectors $m\bar{\mathbf{a}}_t$ and $m\bar{\mathbf{a}}_n$ and to the couple $\bar{I}\boldsymbol{\alpha}$ (Fig. 19.5*b*). Since the angular velocity and angular acceleration of the plate are equal, respectively, to $\dot{\theta}$ and $\ddot{\theta}$, the magnitudes of the two vectors are, respectively, $mb\ddot{\theta}$ and $mb\dot{\theta}^2$, while the moment of the couple is $\bar{I}\ddot{\theta}$. In previous applications of this method (Chap. 16), we tried whenever possible to assume the correct sense for the acceleration. Here, however, we must assume the same positive sense for θ and $\ddot{\theta}$ in order to obtain an equation of the form (19.21). Consequently, the angular acceleration $\ddot{\theta}$ will be assumed positive counterclockwise, even though this assumption is obviously unrealistic. Equating moments about O, we write

$$+\circlearrowleft \qquad -W(b\sin\theta) = (mb\ddot{\theta})b + \bar{I}\ddot{\theta}$$

Noting that $\bar{I} = \frac{1}{12}m[(2b)^2 + (2b)^2] = \frac{2}{3}mb^2$ and $W = mg$, we obtain

$$\ddot{\theta} + \frac{3}{5}\frac{g}{b}\sin\theta = 0 \tag{19.22}$$

For oscillations of small amplitude, we may replace $\sin\theta$ by θ, expressed in radians, and write

$$\ddot{\theta} + \frac{3}{5}\frac{g}{b}\theta = 0 \tag{19.23}$$

Comparison with (19.21) shows that the equation obtained is that of a simple harmonic motion and that the circular frequency p of the oscillations is equal to $(3g/5b)^{1/2}$. Substituting into (19.13), we find that the period of the oscillations is

$$\tau = \frac{2\pi}{p} = 2\pi\sqrt{\frac{5b}{3g}} \tag{19.24}$$

The result obtained is valid only for oscillations of small amplitude. A more accurate description of the motion of the plate is obtained by comparing Eqs. (19.16) and (19.22). We note that the two equations are identical if we choose l equal to $5b/3$. This means that the plate will oscillate as a simple pendulum of length $l = 5b/3$, and the results of Sec. 19.4 may be used to correct the value of the period given in (19.24). The point A of the plate located on line OG at a distance $l = 5b/3$ from O is defined as the *center of oscillation* corresponding to O (Fig. 19.5*a*).

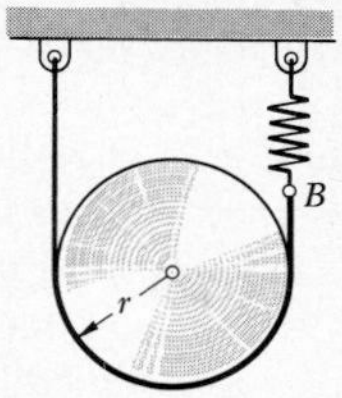

SAMPLE PROBLEM 19.2

A cylinder of weight W and radius r is suspended from a looped cord as shown. One end of the cord is attached directly to a rigid support, while the other end is attached to a spring of constant k. Determine the period and frequency of vibration of the cylinder.

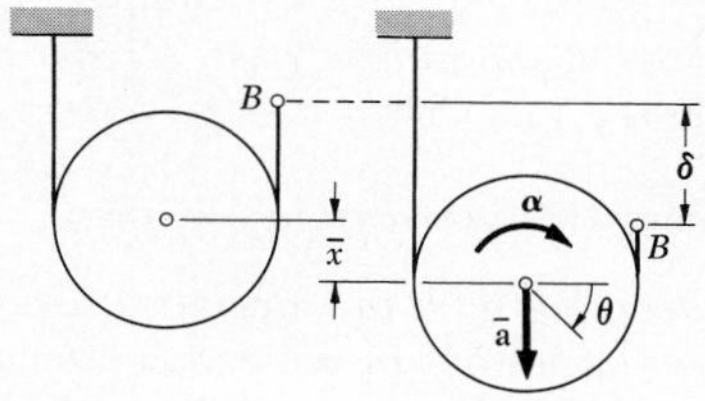

Kinematics of Motion. We express the linear displacement and the acceleration of the cylinder in terms of the angular displacement θ. Choosing the positive sense clockwise and measuring the displacements from the equilibrium position, we write

$$\bar{x} = r\theta \qquad \delta = 2\bar{x} = 2r\theta$$

$$\boldsymbol{\alpha} = \ddot{\theta} \downarrow \qquad \bar{a} = r\alpha = r\ddot{\theta} \qquad \bar{\mathbf{a}} = r\ddot{\theta} \downarrow \tag{1}$$

Equations of Motion. The system of external forces acting on the cylinder consists of the weight $\mathbf{W}$ and of the forces $\mathbf{T}_1$ and $\mathbf{T}_2$ exerted by the cord. We express that this system is equivalent to the system of effective forces represented by the vector $m\bar{\mathbf{a}}$ attached at G and the couple $\bar{I}\boldsymbol{\alpha}$.

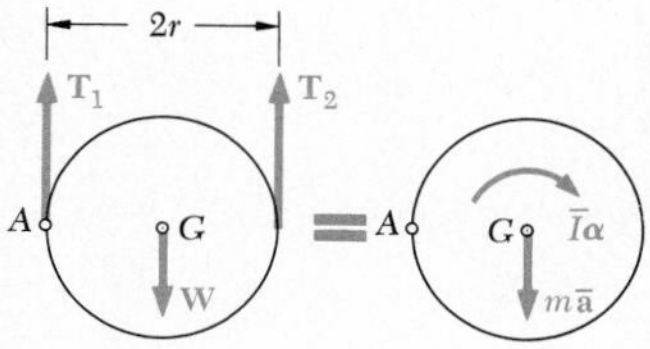

$$+\downarrow \Sigma M_A = \Sigma(M_A)_{\text{eff}}: \qquad Wr - T_2(2r) = m\bar{a}r + \bar{I}\alpha \tag{2}$$

When the cylinder is in its position of equilibrium, the tension in the cord is $T_0 = \frac{1}{2}W$. We note that, for an angular displacement θ, the magnitude of $\mathbf{T}_2$ is

$$T_2 = T_0 + k\delta = \tfrac{1}{2}W + k\delta = \tfrac{1}{2}W + k(2r\theta) \tag{3}$$

Substituting from (1) and (3) into (2), and recalling that $\bar{I} = \frac{1}{2}mr^2$, we write

$$Wr - (\tfrac{1}{2}W + 2kr\theta)(2r) = m(r\ddot{\theta})r + \tfrac{1}{2}mr^2\ddot{\theta}$$

$$\ddot{\theta} + \frac{8}{3}\frac{k}{m}\theta = 0$$

The motion is seen to be simple harmonic, and we have

$$p^2 = \frac{8}{3}\frac{k}{m} \qquad p = \sqrt{\frac{8}{3}\frac{k}{m}}$$

$$\tau = \frac{2\pi}{p} \qquad\qquad \tau = 2\pi\sqrt{\frac{3}{8}\frac{m}{k}} \blacktriangleleft$$

$$f = \frac{p}{2\pi} \qquad\qquad f = \frac{1}{2\pi}\sqrt{\frac{8}{3}\frac{k}{m}} \blacktriangleleft$$

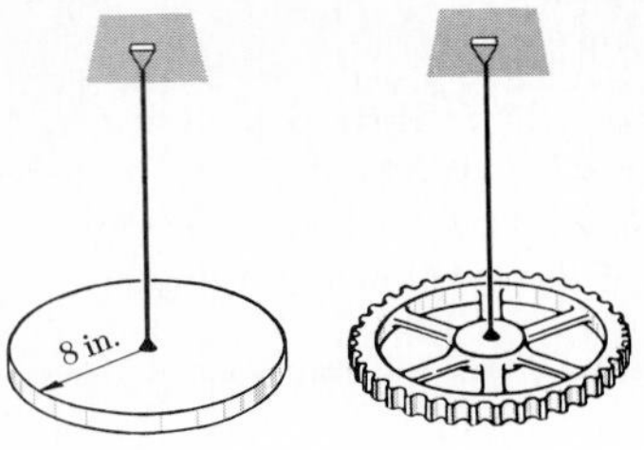

SAMPLE PROBLEM 19.3

A circular disk, weighing 20 lb and of radius 8 in., is suspended from a wire as shown. The disk is rotated (thus twisting the wire) and then released; the period of the torsional vibration is observed to be 1.13 s. A gear is then suspended from the same wire, and the period of torsional vibration for the gear is observed to be 1.93 s. Assuming that the moment of the couple exerted by the wire is proportional to the angle of twist, determine (*a*) the torsional spring constant of the wire, (*b*) the centroidal moment of inertia of the gear, (*c*) the maximum angular velocity reached by the gear if it is rotated through 90° and released.

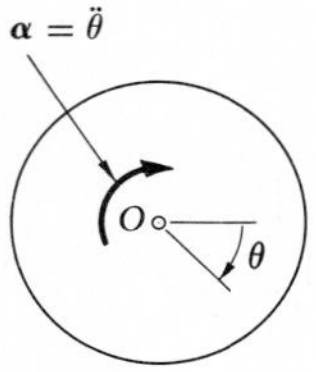

a. **Vibration of Disk.** Denoting by θ the angular displacement of the disk, we express that the magnitude of the couple exerted by the wire is $M = K\theta$, where K is the torsional spring constant of the wire. Since this couple must be equivalent to the couple $\bar{I}\boldsymbol{\alpha}$ representing the effective forces of the disk, we write

$$+\circlearrowleft \Sigma M_O = \Sigma(M_O)_{\text{eff}}: \qquad +K\theta = -\bar{I}\ddot{\theta}$$

$$\ddot{\theta} + \frac{K}{\bar{I}}\theta = 0$$

The motion is seen to be simple harmonic, and we have

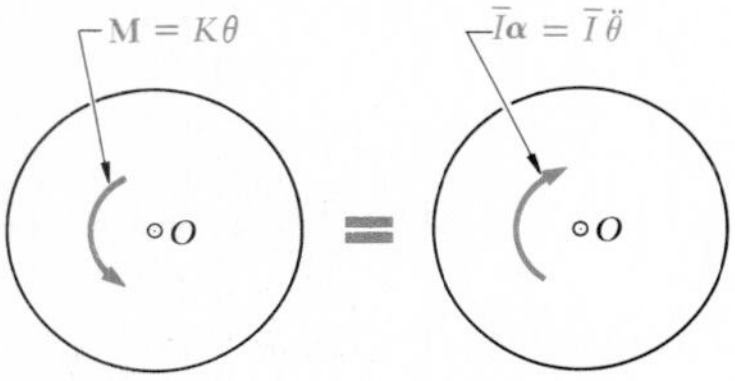

$$p^2 = \frac{K}{\bar{I}} \qquad \tau = \frac{2\pi}{p} \qquad \tau = 2\pi\sqrt{\frac{\bar{I}}{K}} \tag{1}$$

For the disk, we have

$$\tau = 1.13 \text{ s} \qquad \bar{I} = \tfrac{1}{2}mr^2 = \frac{1}{2}\left(\frac{20 \text{ lb}}{32.2 \text{ ft/s}^2}\right)(\tfrac{8}{12} \text{ ft})^2 = 0.138 \text{ lb} \cdot \text{ft} \cdot \text{s}^2$$

Substituting into (1), we obtain

$$1.13 = 2\pi\sqrt{\frac{0.138}{K}} \qquad K = 4.27 \text{ lb} \cdot \text{ft/rad} \blacktriangleleft$$

b. **Vibration of Gear.** Since the period of vibration of the gear is 1.93 s and $K = 4.27$ lb · ft/rad, Eq. (1) yields

$$1.93 = 2\pi\sqrt{\frac{\bar{I}}{4.27}} \qquad \bar{I}_{\text{gear}} = 0.403 \text{ lb} \cdot \text{ft} \cdot \text{s}^2 \blacktriangleleft$$

c. **Maximum Angular Velocity of the Gear.** Since the motion is simple harmonic, we have

$$\theta = \theta_m \sin pt \qquad \omega = \theta_m p \cos pt \qquad \omega_m = \theta_m p$$

Recalling that $\theta_m = 90° = 1.571$ rad and $\tau = 1.93$ s, we write

$$\omega_m = \theta_m p = \theta_m(2\pi/\tau) = (1.571 \text{ rad})(2\pi/1.93 \text{ s})$$

$$\omega_m = 5.11 \text{ rad/s} \blacktriangleleft$$

PROBLEMS

19.25 and 19.26 The uniform rod shown weighs 10 lb and is attached to a spring of constant $k = 3$ lb/in. If end A of the rod is depressed 1.5 in. and released, determine (*a*) the period of vibration, (*b*) the maximum velocity of end A.

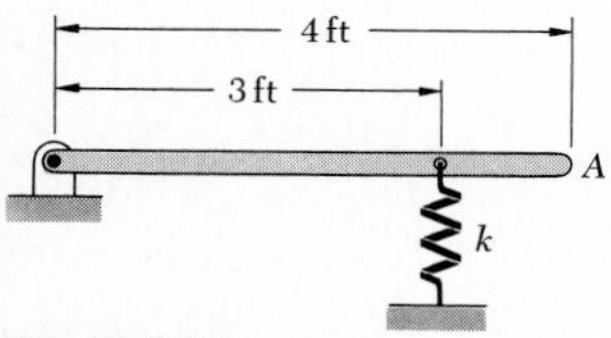

Fig. P19.25

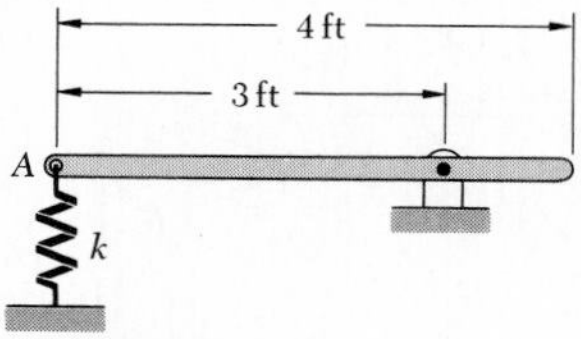

Fig. P19.26

19.27 A belt is placed over the rim of a 15-kg disk as shown and then attached to a 5-kg cylinder and to a spring of constant $k = 600$ N/m. If the cylinder is moved 50 mm down from its equilibrium position and released, determine (*a*) the period of vibration, (*b*) the maximum velocity of the cylinder. Assume friction is sufficient to prevent the belt from slipping on the rim.

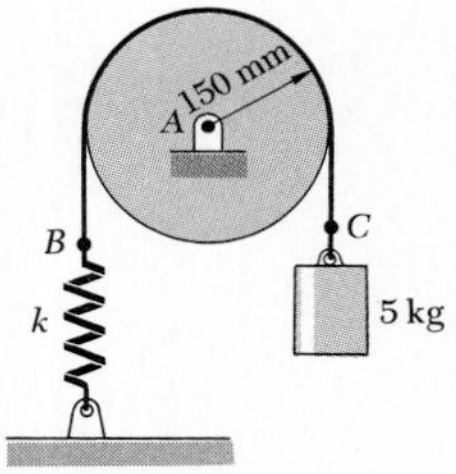

Fig. P19.27

19.28 In Prob. 19.27, determine (*a*) the frequency of vibration, (*b*) the maximum tension which occurs in the belt at B and at C.

19.29 A disk is rigidly attached to a gear of radius r which may roll on the gear rack shown. The disk and gear have a total mass m and a combined radius of gyration $\bar{k}$. A spring of constant K is attached to the center of the gear. If the system is displaced slightly to the right and released, determine the period of the resulting motion.

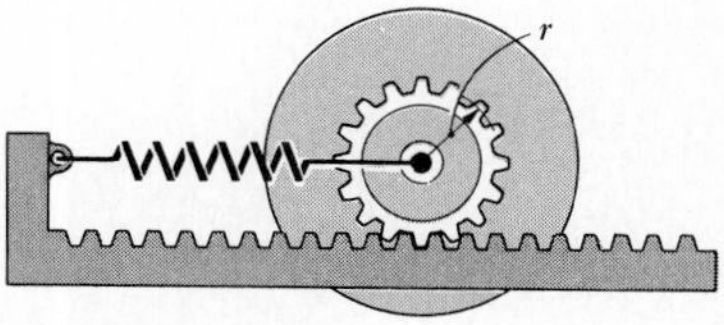

Fig. P19.29

19.30 A uniform square plate of mass m is supported in a horizontal plane by a vertical pin at B and is attached at A to a spring of constant k. If corner A is given a small displacement and released, determine the period of the resulting motion.

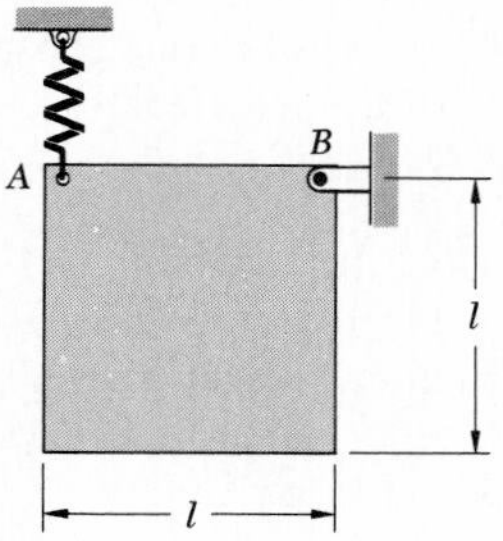

Fig. P19.30

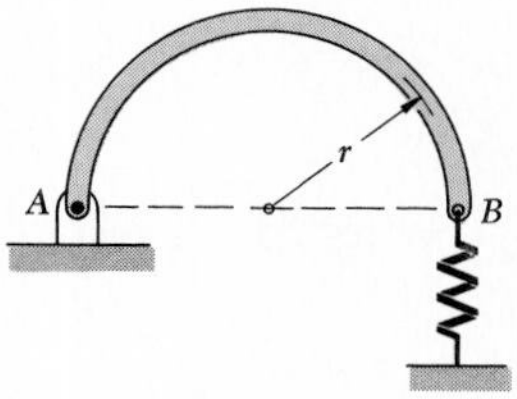

Fig. P19.31

19.31 A semicircular rod of mass m is supported in a horizontal plane by a vertical pin at A and is attached at B to a spring of constant k. If end B is given a small displacement and released, determine the frequency of the resulting motion.

19.32 A square plate of mass m is held by eight springs, each of constant k. Knowing that each spring can act in either tension or compression, determine the frequency of the resulting vibration (*a*) if the plate is given a small vertical displacement and released, (*b*) if the plate is rotated through a small angle about G and released.

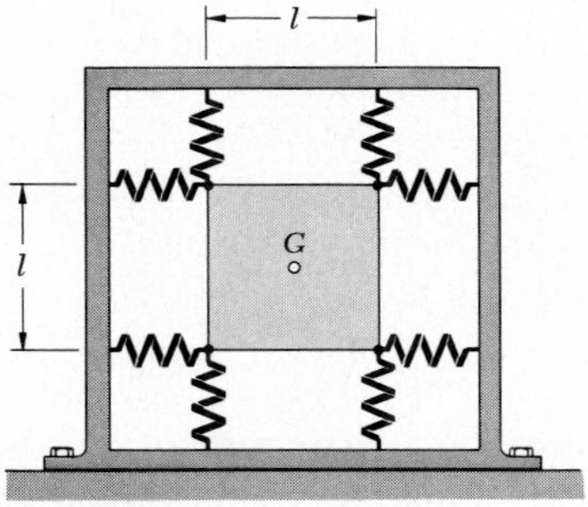

Fig. P19.32

19.33 A *compound pendulum* is defined as a rigid slab which oscillates about a fixed point O, called the center of suspension. Show that the period of oscillation of a compound pendulum is equal to the period of a simple pendulum of length OA, where the distance from A to the mass center G is $GA = \bar{k}^2/\bar{r}$. Point A is defined as the center of oscillation and coincides with the center of percussion defined in Prob. 18.14.

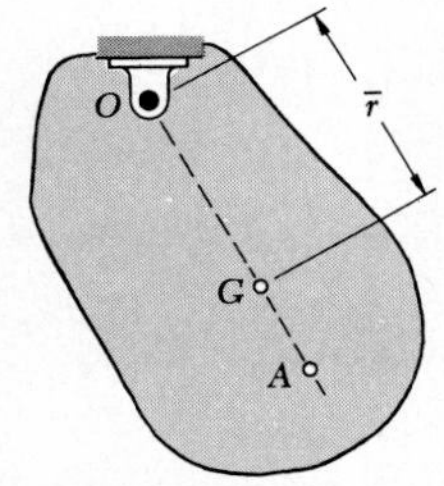

Fig. P19.33 and P19.35

19.34 Show that, if the compound pendulum of Prob. 19.33 is suspended from A instead of O, the period of oscillation is the same as before and the new center of oscillation is located at O.

19.35 A rigid slab oscillates about a fixed point O. Show that the smallest period of oscillation occurs when the distance $\bar{r}$ from point O to the mass center G is equal to $\bar{k}$.

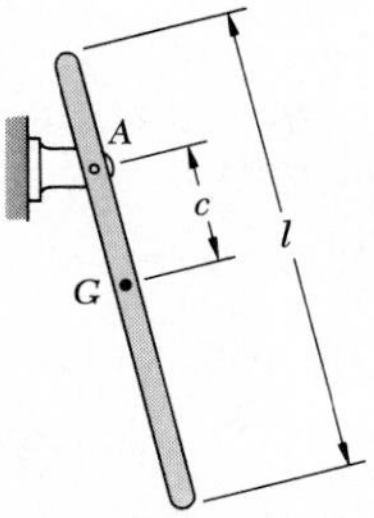

Fig. P19.36

19.36 A uniform bar of length l may oscillate about a hinge at A located a distance c from its mass center G. (*a*) Determine the frequency of small oscillations if $c = \frac{1}{2}l$. (*b*) Determine a second value of c for which the frequency of small oscillations is the same as that found in part *a*.

19.37 For the rod of Prob. 19.36, determine (*a*) the distance c for which the frequency of oscillation is maximum, (*b*) the corresponding minimum period.

19.38 A homogeneous wire is bent to form a square of side l which is supported by a ball-and-socket joint at A. Determine the period of small oscillations of the square (a) in the plane of the square, (b) in a direction perpendicular to the square.

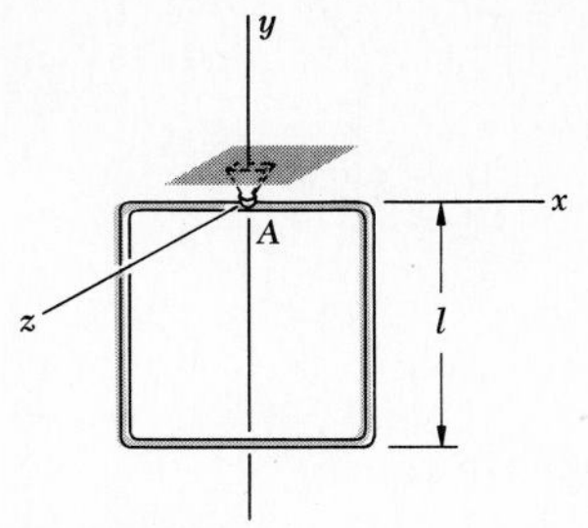

Fig. P19.38

19.39 A thin disk of radius r may oscillate about the axis AB located in the plane of the disk at a distance c from the mass center G. (a) Determine the period of small oscillations if $c = r$. (b) Determine a second value of c for which the period of oscillation is the same as that found in part a.

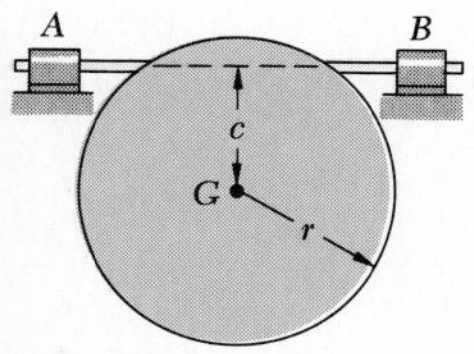

Fig. P19.39

19.40 Solve Prob. 19.39 for $r = 600$ mm.

19.41 A 75-mm-radius hole is cut in a 200-mm-radius uniform disk which is attached to a frictionless pin at its geometric center O. Determine (a) the period of small oscillations of the disk, (b) the length of a simple pendulum which has the same period.

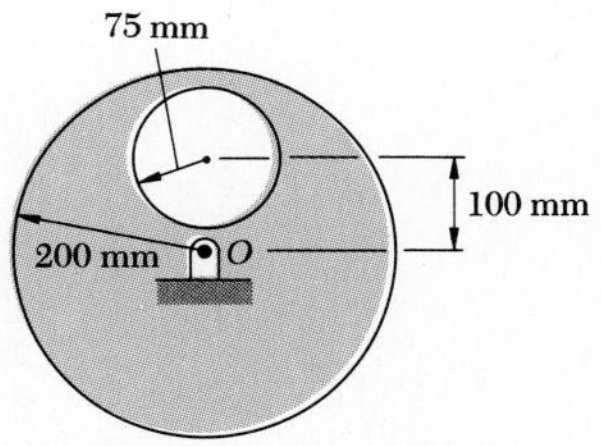

Fig. P19.41

19.42 The centroidal radii of gyration $\bar{k}_x$ and $\bar{k}_z$ of an airplane are determined by letting the airplane oscillate as a compound pendulum. In the arrangement shown, the distance from the mass center G to the point of suspension O is known to be 12 ft. Knowing that the observed periods of oscillation about axes through O parallel to the x and z axes are, respectively, 4.3 and 4.4 s, determine the centroidal radii of gyration $\bar{k}_x$ and $\bar{k}_z$.

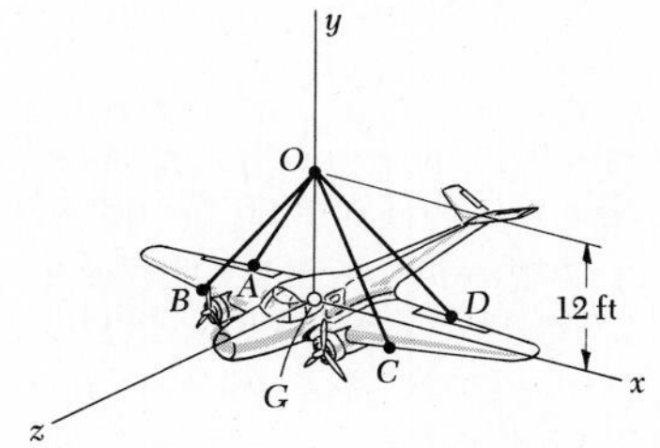

Fig. P19.42

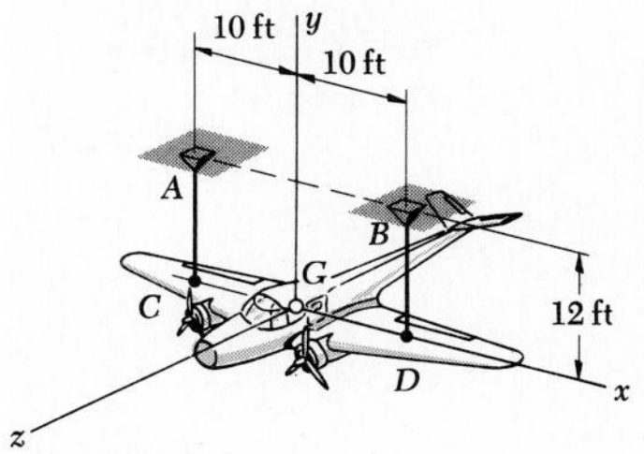

Fig. P19.43

19.43 The centroidal radius of gyration $\bar{k}_y$ of an airplane is determined by suspending the airplane by two 12-ft-long cables as shown. The airplane is rotated through a small angle about the vertical through G and then released. Knowing that the observed period of oscillation is 3.3 s, determine the centroidal radius of gyration $\bar{k}_y$.

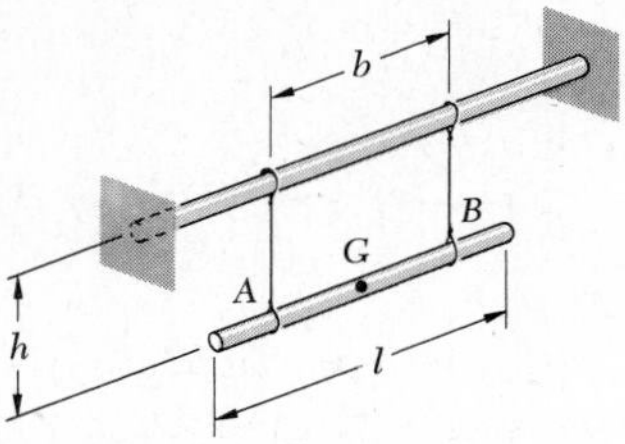

Fig. P19.44

19.44 A slender rod of length l is suspended from two vertical wires of length h, each located a distance $\frac{1}{2}b$ from the mass center G. Determine the period of oscillation when (*a*) the rod is rotated through a small angle about a vertical axis passing through G and released, (*b*) the rod is given a small horizontal translation along AB and released.

19.45 A slender rod of weight 6 lb is suspended from a steel wire which is known to have a torsional spring constant $K = 0.25\ \text{lb}\cdot\text{in./rad}$. If the rod is rotated through 180° about the vertical and then released, determine (*a*) the period of oscillation, (*b*) the maximum velocity of end A of the rod.

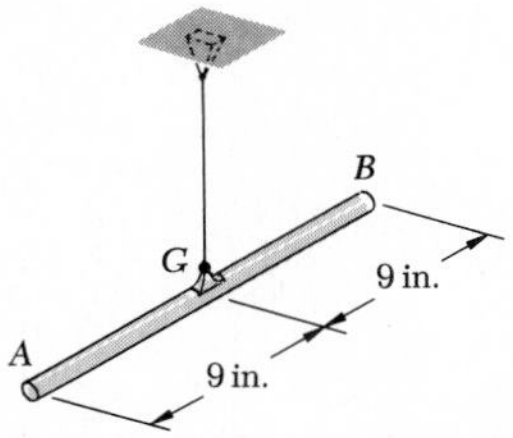

Fig. P19.45

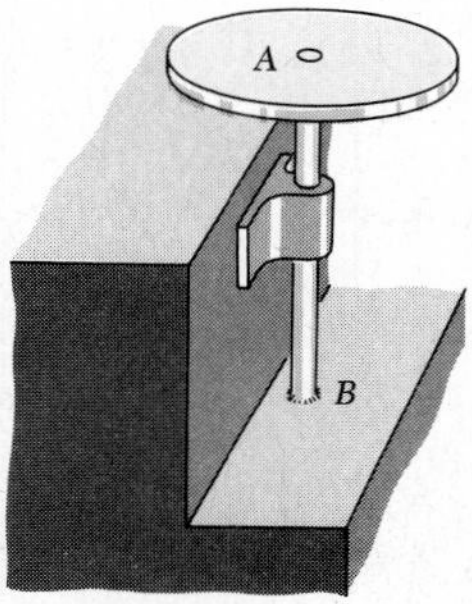

Fig. P19.46

19.46 A uniform disk of radius 200 mm and of mass 8 kg is attached to a vertical shaft which is rigidly held at B. It is known that the disk rotates through 3° when a 4 N · m static couple is applied to the disk. If the disk is rotated through 6° and then released, determine (*a*) the period of the resulting vibration, (*b*) the maximum velocity of a point on the rim of the disk.

19.47 A steel casting is rigidly bolted to the disk of Prob. 19.46. Knowing that the period of torsional vibration of the disk and casting is 0.650 s, determine the moment of inertia of the casting with respect to the shaft AB.

19.48 A period of 4.10 s is observed for the angular oscillations of a 1-lb gyroscope rotor suspended from a wire as shown. Knowing that a period of 6.20 s is obtained when a 2-in.-diameter steel sphere is suspended in the same fashion, determine the centroidal radius of gyration of the rotor. (Specific weight of steel = 490 lb/ft^3.)

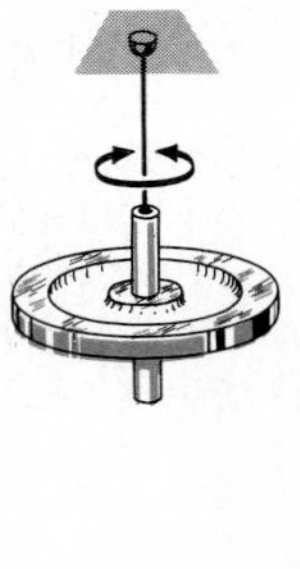

Fig. P19.48

19.6. Application of the Principle of Conservation of Energy. We saw in Sec. 19.2 that, when a particle of mass m is in simple harmonic motion, the resultant $\mathbf{F}$ of the forces exerted on the particle has a magnitude proportional to the displacement x measured from the position of equilibrium O and is directed toward O; we write $F = -kx$. Referring to Sec. 13.6, we note that $\mathbf{F}$ is a *conservative force* and that the corresponding potential energy is $V = \frac{1}{2}kx^2$, where V is assumed equal to zero in the equilibrium position $x = 0$. Since the velocity of the particle is equal to $\dot{x}$, its kinetic energy is $T = \frac{1}{2}m\dot{x}^2$ and we may express that the total energy of the particle is conserved by writing

$$T + V = \text{constant} \qquad \tfrac{1}{2}m\dot{x}^2 + \tfrac{1}{2}kx^2 = \text{constant}$$

Setting $p^2 = k/m$ as in (19.4), where p is the circular frequency of the vibration, we have

$$\dot{x}^2 + p^2x^2 = \text{constant} \tag{19.25}$$

Equation (19.25) is characteristic of simple harmonic motion; it may be obtained directly from (19.5) by multiplying both terms by $2\dot{x}$ and integrating.

The principle of conservation of energy provides a convenient way for determining the period of vibration of a rigid body or of a system of rigid bodies possessing a single degree of freedom, once it has been established that the motion of the system is a simple harmonic motion, or that it may be approximated by a simple harmonic motion. Choosing an appropriate variable, such as a distance x or an angle θ, we consider two particular positions of the system:

1. *The displacement of the system is maximum;* we have $T_1 = 0$, and V_1 may be expressed in terms of the amplitude x_m or θ_m (choosing $V = 0$ in the equilibrium position).
2. *The system passes through its equilibrium position;* we have $V_2 = 0$, and T_2 may be expressed in terms of the maximum velocity $\dot{x}_m$ or $\dot{\theta}_m$.

We then express that the total energy of the system is conserved and write $T_1 + V_1 = T_2 + V_2$. Recalling from (19.15) that for simple harmonic motion the maximum velocity is equal to the product of the amplitude and of the circular frequency p, we find that the equation obtained may be solved for p.

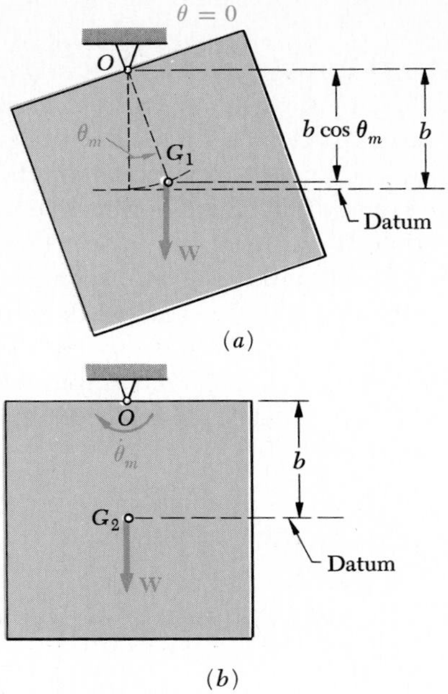

Fig. 19.6

As an example, we shall consider again the square plate of Sec. 19.5. In the position of maximum displacement (Fig. 19.6*a*), we have

$$T_1 = 0 \qquad V_1 = W(b - b\cos\theta_m) = Wb(1 - \cos\theta_m)$$

or, since $1 - \cos\theta_m = 2\sin^2(\theta_m/2) \approx 2(\theta_m/2)^2 = \theta_m^2/2$ for oscillations of small amplitude,

$$T_1 = 0 \qquad V_1 = \tfrac{1}{2}Wb\theta_m^2 \tag{19.26}$$

As the plate passes through its position of equilibrium (Fig. 19.6*b*), its velocity is maximum and we have

$$T_2 = \tfrac{1}{2}m\bar{v}_m^2 + \tfrac{1}{2}\bar{I}\omega_m^2 = \tfrac{1}{2}mb^2\dot{\theta}_m^2 + \tfrac{1}{2}\bar{I}\dot{\theta}_m^2 \qquad V_2 = 0$$

or, recalling from Sec. 19.5 that $\bar{I} = \frac{2}{3}mb^2$,

$$T_2 = \tfrac{1}{2}(\tfrac{5}{3}mb^2)\dot{\theta}_m^2 \qquad V_2 = 0 \tag{19.27}$$

Substituting from (19.26) and (19.27) into $T_1 + V_1 = T_2 + V_2$, and noting that the maximum velocity $\dot{\theta}_m$ is equal to the product $\theta_m p$, we write

$$\tfrac{1}{2}Wb\theta_m^2 = \tfrac{1}{2}(\tfrac{5}{3}mb^2)\theta_m^2 p^2 \tag{19.28}$$

which yields $p^2 = 3g/5b$ and

$$\tau = \frac{2\pi}{p} = 2\pi\sqrt{\frac{5b}{3g}} \tag{19.29}$$

as previously obtained.

SAMPLE PROBLEM 19.4

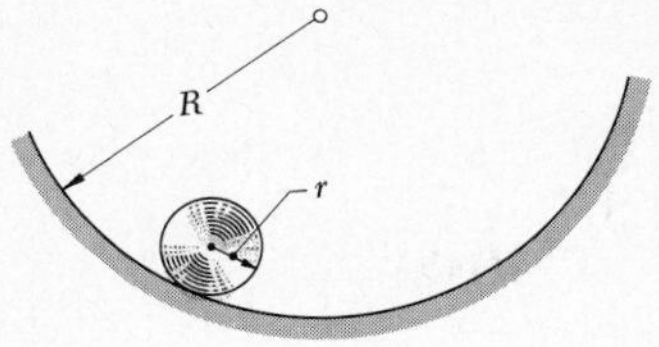

Determine the period of small oscillations of a cylinder of radius r which rolls without slipping inside a curved surface of radius R.

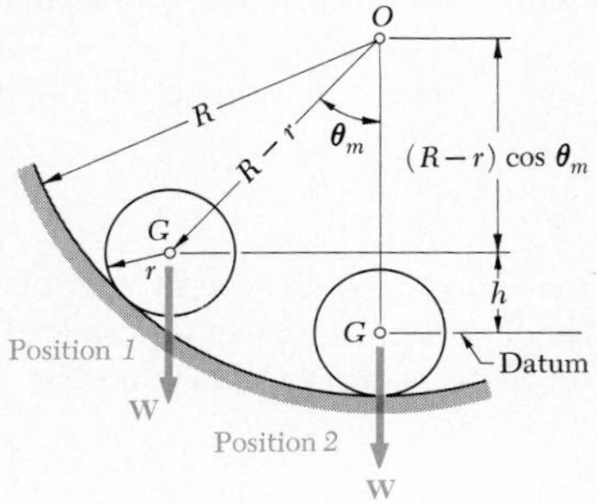

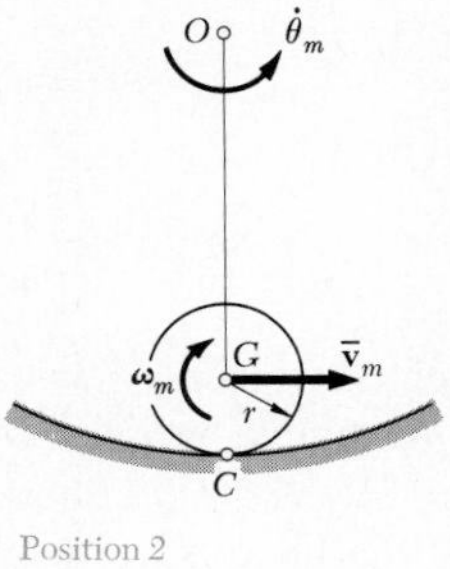

Solution. We denote by θ the angle which line OG forms with the vertical. Since the cylinder rolls without slipping, we may apply the principle of conservation of energy between position *1*, where $\theta = \theta_m$, and position *2*, where $\theta = 0$.

Position *1*. ***Kinetic Energy.*** Since the velocity of the cylinder is zero, we have $T_1 = 0$.

Potential Energy. Choosing a datum as shown and denoting by W the weight of the cylinder, we have

$$V_1 = Wh = W(R - r)(1 - \cos\theta)$$

Noting that for small oscillations $(1 - \cos\theta) = 2\sin^2(\theta/2) \approx \theta^2/2$, we have

$$V_1 = W(R - r)\frac{\theta_m^2}{2}$$

Position *2*. Denoting by $\dot{\theta}_m$ the angular velocity of line OG as the cylinder passes through position *2*, and observing that point C is the instantaneous center of rotation of the cylinder, we write

$$\bar{v}_m = (R - r)\dot{\theta}_m \qquad \omega_m = \frac{\bar{v}_m}{r} = \frac{R - r}{r}\dot{\theta}_m$$

Kinetic Energy:

$$\begin{aligned} T_2 &= \tfrac{1}{2}m\bar{v}_m^2 + \tfrac{1}{2}\bar{I}\omega_m^2 \\ &= \tfrac{1}{2}m(R - r)^2\dot{\theta}_m^2 + \tfrac{1}{2}(\tfrac{1}{2}mr^2)\left(\frac{R - r}{r}\right)^2\dot{\theta}_m^2 \\ &= \tfrac{3}{4}m(R - r)^2\dot{\theta}_m^2 \end{aligned}$$

Potential Energy: $V_2 = 0$

Conservation of Energy

$$T_1 + V_1 = T_2 + V_2$$

$$0 + W(R - r)\frac{\theta_m^2}{2} = \tfrac{3}{4}m(R - r)^2\dot{\theta}_m^2 + 0$$

Since $\dot{\theta}_m = p\theta_m$ and $W = mg$, we write

$$mg(R - r)\frac{\theta_m^2}{2} = \tfrac{3}{4}m(R - r)^2(p\theta_m)^2 \qquad p^2 = \frac{2}{3}\frac{g}{R - r}$$

$$\tau = \frac{2\pi}{p} \qquad \tau = 2\pi\sqrt{\frac{3}{2}\frac{R - r}{g}} \quad ◄$$

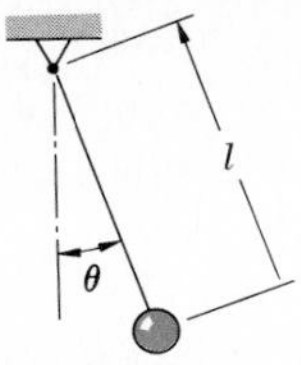

Fig. P19.49

PROBLEMS

19.49 Using the method of Sec. 19.6, determine the period of a simple pendulum of length l.

19.50 Using the method of Sec. 19.6, solve Prob. 19.6.

19.51 Using the method of Sec. 19.6, solve Prob. 19.9.

19.52 Using the method of Sec. 19.6, solve Prob. 19.10.

19.53 Using the method of Sec. 19.6, solve Prob. 19.15.

19.54 It is observed that when a 15-lb weight is attached to the rim of a 6-ft-diameter flywheel, the period of small oscillations of the flywheel is 21 s. Neglecting axle friction, determine the centroidal moment of inertia of the flywheel.

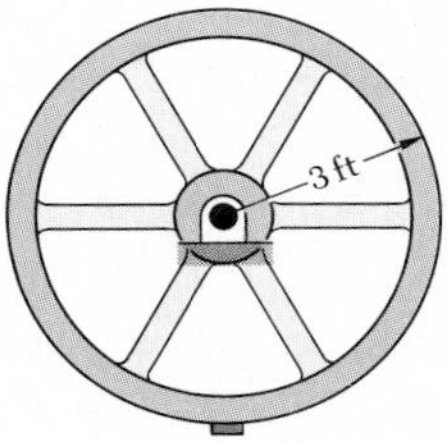

Fig. P19.54

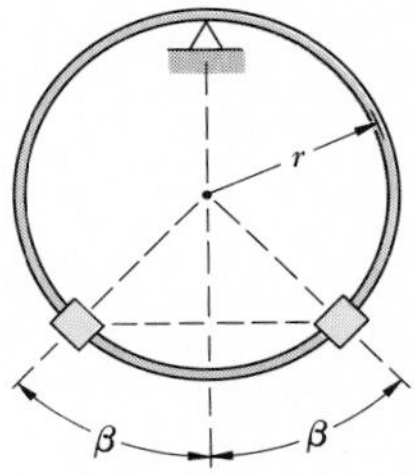

Fig. P19.55

19.55 Two collars, each of weight W, are attached as shown to a hoop of radius r and of negligible weight. (*a*) Show that for any value of β the period is $\tau = 2\pi\sqrt{2r/g}$. (*b*) Show that the same result is obtained if the weight of the hoop is not neglected.

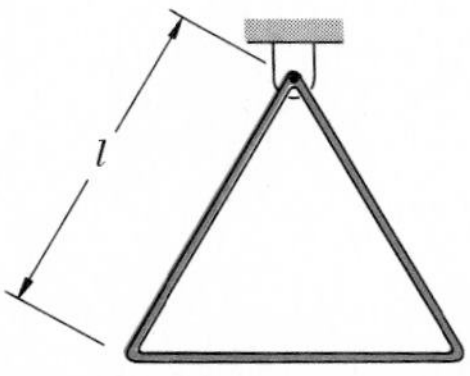

Fig. P19.56

19.56 A thin homogeneous wire is bent into the shape of an equilateral triangle of side l. Determine the period of small oscillations if the wire figure is suspended as shown.

19.57 Solve Prob. 19.56, assuming that the wire triangle is suspended from a pin located at the midpoint of one side.

19.58 Using the method of Sec. 19.6, solve Prob. 19.44.

19.59 Using the method of Sec. 19.6, solve Prob. 19.41.

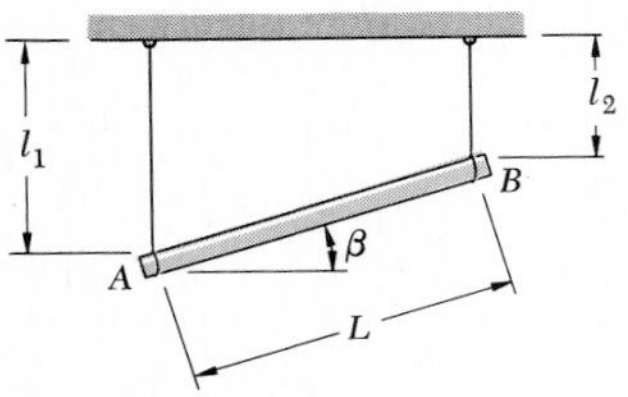

Fig. P19.60

19.60 A section of uniform pipe is suspended from two vertical cables attached at A and B. Determine the period of oscillation in terms of l_1 and l_2 when point B is given a small horizontal displacement to the right and released.

19.61 The motion of the uniform rod AB is guided by the cord BC and by the small roller at A. Determine the frequency of oscillation when the end B of the rod is given a small horizontal displacement and released.

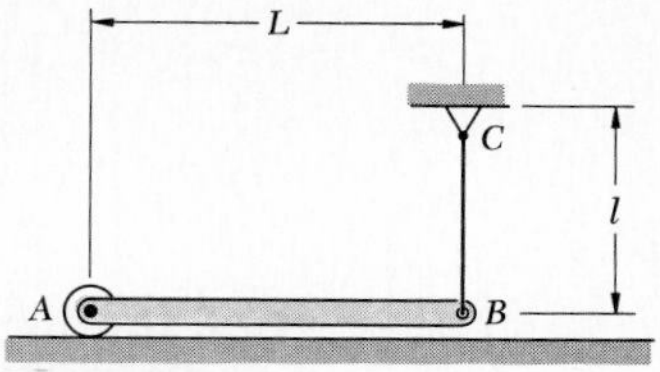

Fig. P19.61

19.62 A uniform disk of weight W and radius r may roll on the horizontal surface shown. In case a, a weight W of negligible dimensions is bolted to the disk at a distance $\frac{1}{2}r$ from the geometric center of the disk. In case b, a vertical force of constant magnitude W is applied to a cord which is attached to the disk at a distance $\frac{1}{2}r$ from the center of the disk. Determine the ratio τ_b/τ_a of the two periods of oscillation.

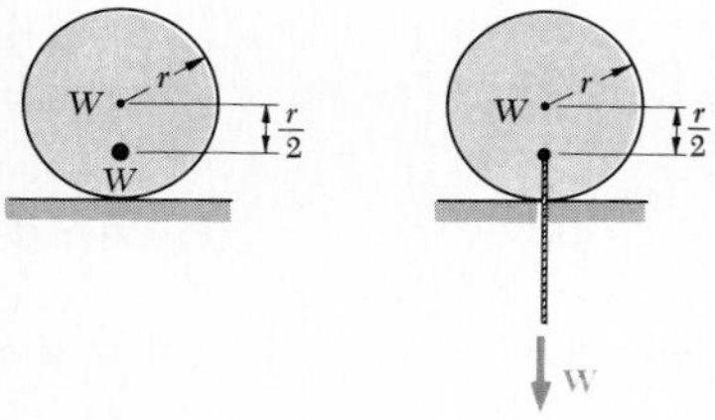

(a) (b)

Fig. P19.62

19.63 The 8-kg rod AB is bolted to the 12-kg disk. Knowing that the disk rolls without sliding, determine the period of small oscillations of the system.

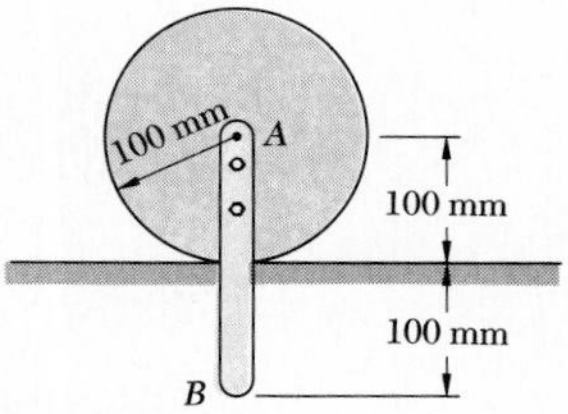

Fig. P19.63

19.64 Using the method of Sec. 19.6, solve Prob. 19.30.

19.65 Using the method of Sec. 19.6, solve Prob. 19.31.

19.66 The slender rod AB of mass m is attached to two collars of negligible mass. Knowing that the system lies in a horizontal plane and is in equilibrium in the position shown, determine the period of vibration if the collar A is given a small displacement and released.

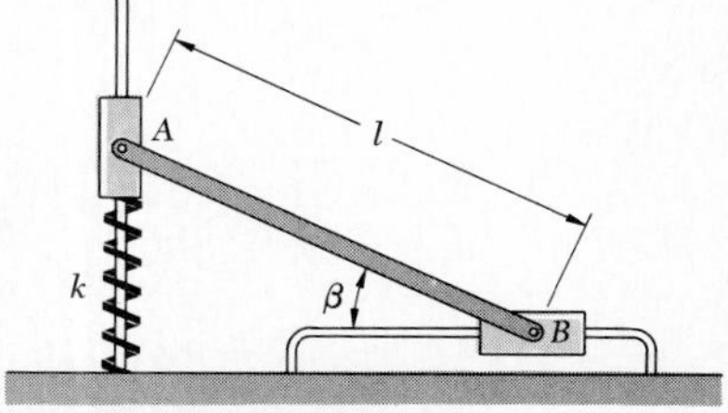

Fig. P19.66 and P19.68

19.67 Solve Prob. 19.66, assuming that rod AB is of mass m and that each collar is of mass m_c.

19.68 A slender rod AB of negligible mass connects two collars, each of mass m_c. Knowing that the system lies in a horizontal plane and is in equilibrium in the position shown, determine the period of vibration if the collar at A is given a small displacement and released.

19.69 Two uniform rods AB and CD, each of length l and weight W, are attached to two gears as shown. Neglecting the weight of the gears, determine for each of the positions shown the period of small oscillations of the system.

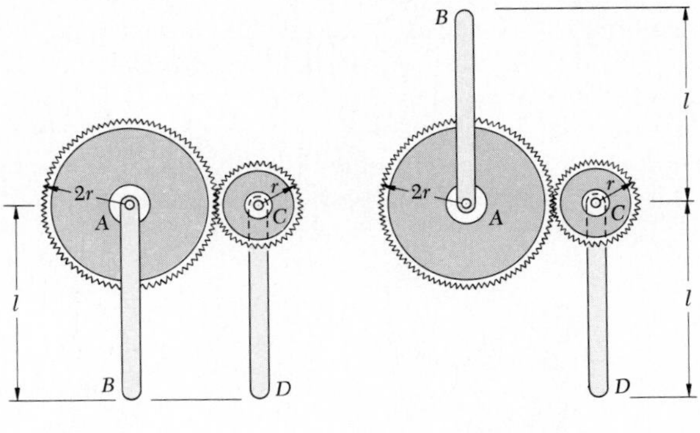

Fig. P19.69

19.70 Solve Prob. 16.69, assuming that $l = 30$ in., $W = 5$ lb, and $r = 4$ in.

19.71 A uniform rod of length L is supported by a ball-and-socket joint at A and by the vertical wire CD. Derive an expression for the period of oscillation of the rod if end B is given a small horizontal displacement and then released.

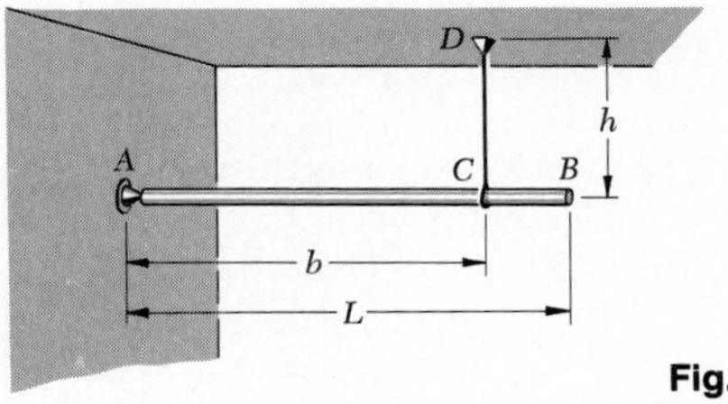

Fig. P19.71

19.72 Solve Prob. 19.71, assuming that $L = 10$ ft, $b = 8$ ft, and $h = 6$ ft.

Fig. P19.73

***19.73** As a submerged body moves through a fluid, the particles of the fluid flow around the body and thus acquire kinetic energy. In the case of a sphere moving in an ideal fluid, the total kinetic energy acquired by the fluid is $\frac{1}{4}\rho V v^2$, where ρ is the mass density of the fluid, V the volume of the sphere, and v the velocity of the sphere. Consider a 1-lb hollow spherical shell of radius 3 in. which is held submerged in a tank of water by a spring of constant 3 lb/in. (*a*) Neglecting fluid friction, determine the period of vibration of the shell when it is displaced vertically and then released. (*b*) Solve part *a*, assuming that the tank is accelerated upward at the constant rate of 10 ft/s^2.

19.7. Forced Vibrations. The most important vibrations from the point of view of engineering applications are the *forced vibrations* of a system. These vibrations occur when a system is subjected to a periodic force or when it is elastically connected to a support which has an alternating motion.

Consider first the case of a body of mass m suspended from a spring and subjected to a periodic force $\mathbf{P}$ of magnitude $P = P_m \sin \omega t$ (Fig. 19.7). This force may be an actual external force applied to the body, or it may be a centrifugal force produced by the rotation of some unbalanced part of the body. (See Sample Prob. 19.5.) Denoting by x the displacement of the body measured from its equilibrium position, we write the equation of motion,

$$+\downarrow \Sigma F = ma: \qquad P_m \sin \omega t + W - k(\delta_{st} + x) = m\ddot{x}$$

Recalling that $W = k\delta_{st}$, we have

$$m\ddot{x} + kx = P_m \sin \omega t \tag{19.30}$$

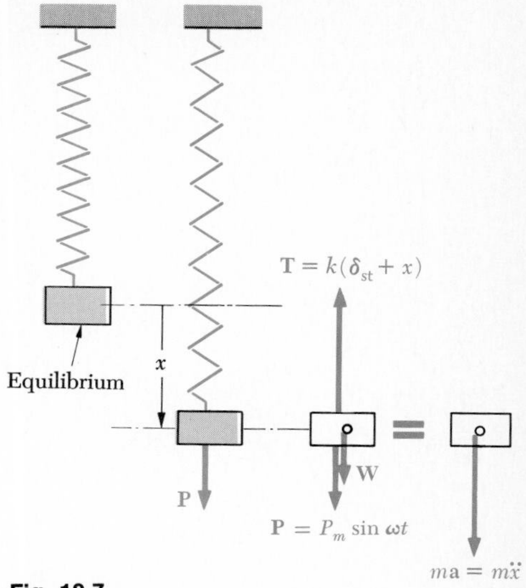

Fig. 19.7

Next we consider the case of a body of mass m suspended from a spring attached to a moving support whose displacement δ is equal to $\delta_m \sin \omega t$ (Fig. 19.8). Measuring the displacement x of the body from the position of static equilibrium corresponding to $\omega t = 0$, we find that the total elongation of the spring at time t is $\delta_{st} + x - \delta_m \sin \omega t$. The equation of motion is thus

$$+\downarrow \Sigma F = ma: \qquad W - k(\delta_{st} + x - \delta_m \sin \omega t) = m\ddot{x}$$

Recalling that $W = k\delta_{st}$, we have

$$m\ddot{x} + kx = k\delta_m \sin \omega t \tag{19.31}$$

We note that Eqs. (19.30) and (19.31) are of the same form and that a solution of the first equation will satisfy the second if we set $P_m = k\delta_m$.

A differential equation like (19.30) or (19.31), possessing a right-hand member different from zero, is said to be *nonhomogeneous*. Its general solution is obtained by adding a particular solution of the given equation to the general solution of the corresponding *homogeneous* equation (with right-hand member equal to zero). A *particular solution* of (19.30) or (19.31) may be obtained by trying a solution of the form

$$x_{part} = x_m \sin \omega t \tag{19.32}$$

Substituting x_{part} for x into (19.30), we find

$$-m\omega^2 x_m \sin \omega t + kx_m \sin \omega t = P_m \sin \omega t$$

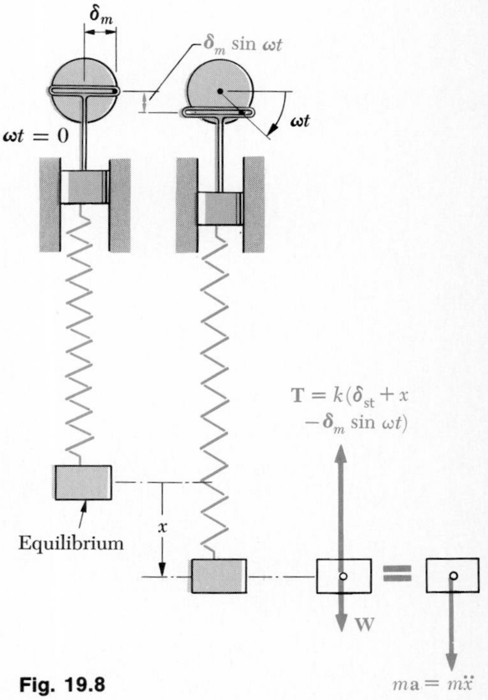

Fig. 19.8

which may be solved for the amplitude,

$$x_m = \frac{P_m}{k - m\omega^2}$$

Recalling from (19.4) that $k/m = p^2$, where p is the circular frequency of the free vibration of the body, we write

$$x_m = \frac{P_m/k}{1 - (\omega/p)^2} \tag{19.33}$$

Substituting from (19.32) into (19.31), we obtain in a similar way

$$x_m = \frac{\delta_m}{1 - (\omega/p)^2} \tag{19.33$'$}$$

The homogeneous equation corresponding to (19.30) or (19.31) is Eq. (19.3), defining the free vibration of the body. Its general solution, called the *complementary function,* was found in Sec. 19.2,

$$x_{\text{comp}} = A \sin pt + B \cos pt \tag{19.34}$$

Adding the particular solution (19.32) and the complementary function (19.34), we obtain the *general solution* of Eqs. (19.30) and (19.31),

$$x = A \sin pt + B \cos pt + x_m \sin \omega t \tag{19.35}$$

We note that the vibration obtained consists of two superposed vibrations. The first two terms in (19.35) represent a free vibration of the system. The frequency of this vibration, called the *natural frequency* of the system, depends only upon the constant k of the spring and the mass m of the body, and the constants A and B may be determined from the initial conditions. This free vibration is also called a *transient* vibration since, in actual practice, it will soon be damped out by friction forces (Sec. 19.9).

The last term in (19.35) represents the *steady-state* vibration produced and maintained by the impressed force or impressed support movement. Its frequency is the *forced frequency* imposed by this force or movement, and its amplitude x_m, defined by (19.33) or (19.33′), depends upon the *frequency ratio* ω/p. The ratio of the amplitude x_m of the steady-state vibration to the static deflection P_m/k caused by a force P_m, or to the amplitude δ_m of the support movement, is called the *magnification factor*. From (19.33) and (19.33′), we obtain

$$\text{Magnification factor} = \frac{x_m}{P_m/k} = \frac{x_m}{\delta_m} = \frac{1}{1 - (\omega/p)^2} \qquad (19.36)$$

The magnification factor has been plotted in Fig. 19.9 against the frequency ratio ω/p. We note that, when $\omega = p$, the amplitude of the forced vibration becomes infinite. The impressed force or impressed support movement is said to be in *resonance* with the given system. Actually, the amplitude of the vibration remains finite because of damping forces (Sec. 19.9); nevertheless, such a situation should be avoided, and the forced frequency should not be chosen too close to the natural frequency of the system. We also note that for $\omega < p$ the coefficient of $\sin \omega t$ in (19.35) is positive, while for $\omega > p$ this coefficient is negative. In the first case the forced vibration is *in phase* with the impressed force or impressed support movement, while in the second case it is 180° *out of phase*.

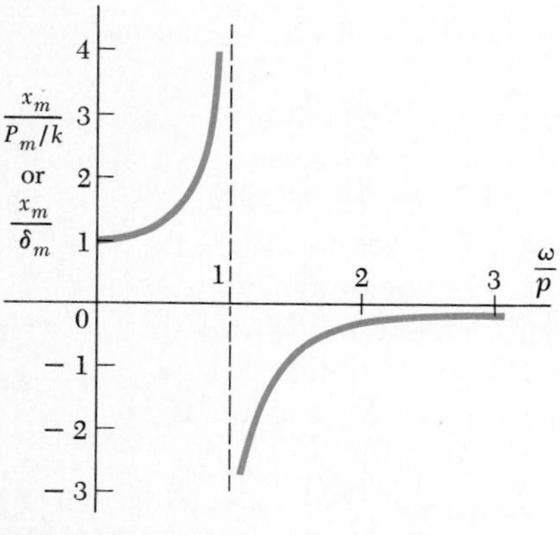

Fig. 19.9

SAMPLE PROBLEM 19.5

A motor weighing 350 lb is supported by four springs, each having a constant of 750 lb/in. The unbalance of the rotor is equivalent to a weight of 1 oz located 6 in. from the axis of rotation. Knowing that the motor is constrained to move vertically, determine (*a*) the speed in rpm at which resonance will occur, (*b*) the amplitude of the vibration of the motor at a speed of 1200 rpm.

a. **Resonance Speed.** The resonance speed is equal to the circular frequency (in rpm) of the free vibration of the motor. The mass of the motor and the equivalent constant of the supporting springs are

$$m = \frac{350 \text{ lb}}{32.2 \text{ ft/s}^2} = 10.87 \text{ lb} \cdot \text{s}^2/\text{ft}$$

$$k = 4(750 \text{ lb/in.}) = 3000 \text{ lb/in.} = 36{,}000 \text{ lb/ft}$$

$$p = \sqrt{\frac{k}{m}} = \sqrt{\frac{36{,}000}{10.87}} = 57.5 \text{ rad/s} = 549 \text{ rpm}$$

Resonance speed = 549 rpm ◄

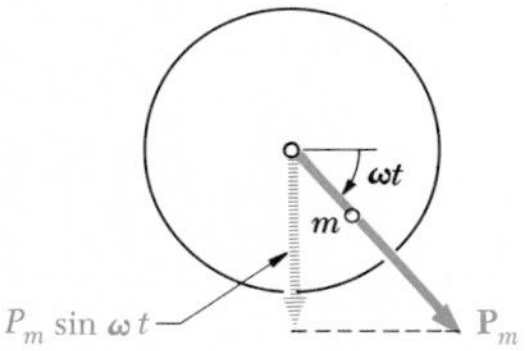

b. **Amplitude of Vibration at 1200 rpm.** The angular velocity of the motor and the mass of the equivalent 1-oz weight are

$$\omega = 1200 \text{ rpm} = 125.7 \text{ rad/s}$$

$$m = (1 \text{ oz})\frac{1 \text{ lb}}{16 \text{ oz}}\frac{1}{32.2 \text{ ft/s}^2} = 0.00194 \text{ lb} \cdot \text{s}^2/\text{ft}$$

The magnitude of the centrifugal force due to the unbalance of the rotor is

$$P_m = ma_n = mr\omega^2 = (0.00194 \text{ lb} \cdot \text{s}^2/\text{ft})(\tfrac{6}{12} \text{ ft})(125.7 \text{ rad/s})^2 = 15.3 \text{ lb}$$

The static deflection that would be caused by a constant load P_m is

$$\frac{P_m}{k} = \frac{15.3 \text{ lb}}{3000 \text{ lb/in.}} = 0.00510 \text{ in.}$$

Substituting the value of P_m/k together with the known values of ω and p into Eq. (19.33), we obtain

$$x_m = \frac{P_m/k}{1 - (\omega/p)^2} = \frac{0.00510 \text{ in.}}{1 - (125.7/57.5)^2}$$

$x_m = 0.00135$ in. ◄

Note. Since $\omega > p$, the vibration is 180° out of phase with the centrifugal force due to the unbalance of the rotor. For example, when the unbalanced mass is directly below the axis of rotation, the position of the motor is $x_m = 0.00135$ in. above the position of equilibrium.

PROBLEMS

19.74 A block of mass m is suspended from a spring of constant k and is acted upon by a vertical periodic force of magnitude $P = P_m \sin \omega t$. Determine the range of values of ω for which the amplitude of the vibration exceeds twice the static deflection caused by a constant force of magnitude P_m.

19.75 In Prob. 19.74, determine the range of values of ω for which the amplitude of the vibration is less than the static deflection caused by a constant force of magnitude P_m.

19.76 A simple pendulum of length l is suspended from a collar C which is forced to move horizontally according to the relation $x_C = \delta_m \sin \omega t$. Determine the range of values of ω for which the amplitude of the motion of the bob exceeds $2\delta_m$. (Assume δ_m is small compared to the length l of the pendulum.)

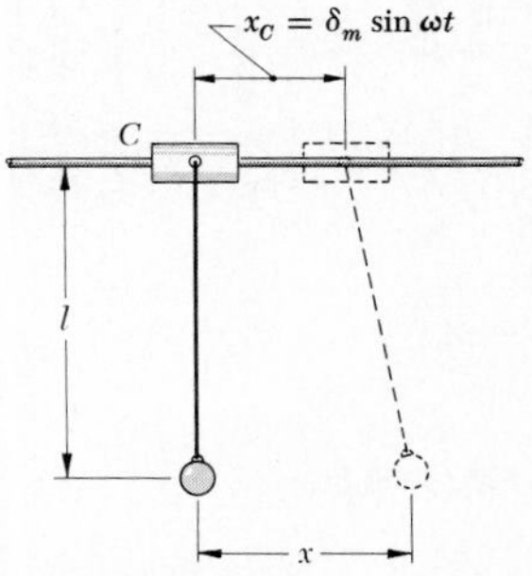

Fig. P19.76

19.77 In Prob. 19.76, determine the range of values of ω for which the amplitude of the motion of the bob is less than δ_m.

19.78 A 500-lb motor is supported by a light horizontal beam. The unbalance of the rotor is equivalent to a weight of 1 oz located 10 in. from the axis of rotation. Knowing that the static deflection of the beam due to the weight of the motor is 0.220 in., determine (*a*) the speed (in rpm) at which resonance will occur, (*b*) the amplitude of the steady-state vibration of the motor at a speed of 800 rpm.

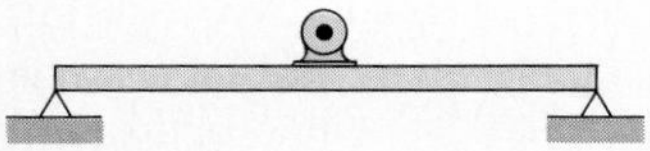
Fig. P19.78

19.79 Solve Prob. 19.78, assuming that the 500-lb motor is supported by a nest of springs having a total constant of 400 lb/in.

19.80 A motor of mass 45 kg is supported by four springs, each of constant 100 kN/m. The motor is constrained to move vertically, and the amplitude of its movement is observed to be 0.5 mm at a speed of 1200 rpm. Knowing that the mass of the rotor is 14 kg, determine the distance between the mass center of the rotor and the axis of the shaft.

19.81 In Prob. 19.80, determine the amplitude of the vertical movement of the motor at a speed of (*a*) 200 rpm, (*b*) 1600 rpm, (*c*) 900 rpm.

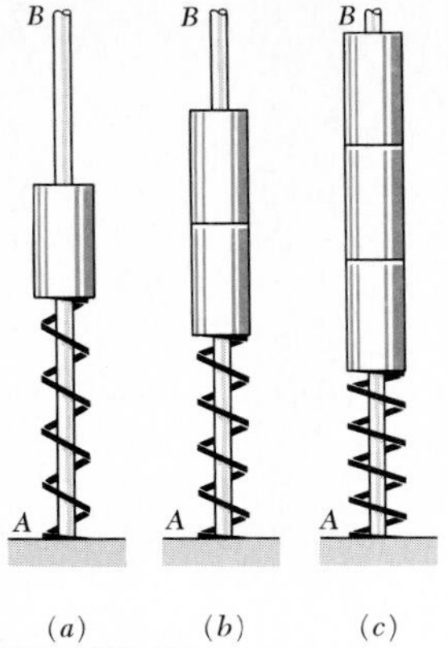

Fig. P19.82

19.82 Rod AB is rigidly attached to the frame of a motor running at a constant speed. When a collar of mass m is placed on the spring, it is observed to vibrate with an amplitude of 0.5 in. When two collars, each of mass m, are placed on the spring, the amplitude is observed to be 0.6 in. What amplitude of vibration should be expected when three collars, each of mass m, are placed on the spring? (Obtain two answers.)

19.83 Solve Prob. 19.82, assuming that the speed of the motor is changed and that one collar has an amplitude of 0.60 in. and two collars have an amplitude of 0.20 in.

19.84 A disk of mass m is attached to the midpoint of a vertical shaft which revolves at an angular velocity ω. Denoting by k the spring constant of the system for a horizontal movement of the disk and by e the eccentricity of the disk with respect to the shaft, show that the deflection of the center of the shaft may be written in the form

$$r = \frac{e(\omega/p)^2}{1 - (\omega/p)^2}$$

where $p = \sqrt{k/m}$.

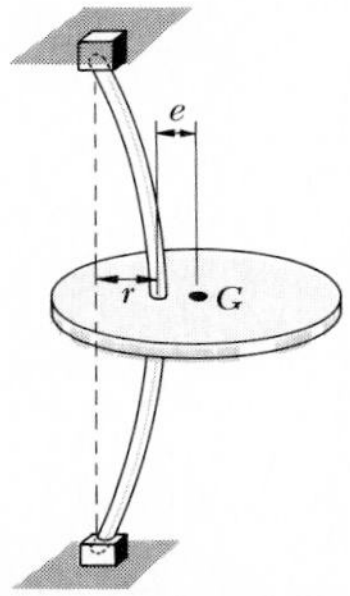

Fig. P19.84 and P19.85

19.85 A disk of mass 30 kg is attached to the midpoint of a shaft. Knowing that a static force of 200 N will deflect the shaft 0.6 mm, determine the speed of the shaft in rpm at which resonance will occur.

19.86 Knowing that the disk of Prob. 19.85 is attached to the shaft with an eccentricity $e = 0.2$ mm, determine the deflection r of the shaft at a speed of 900 rpm.

19.87 A variable-speed motor is rigidly attached to the beam BC. When the speed of the motor is less than 600 rpm or more than 1200 rpm, a small object placed at A is observed to remain in contact with the beam. For speeds between 600 and 1200 rpm the object is observed to "dance" and actually to lose contact with the beam. Determine the speed at which resonance will occur.

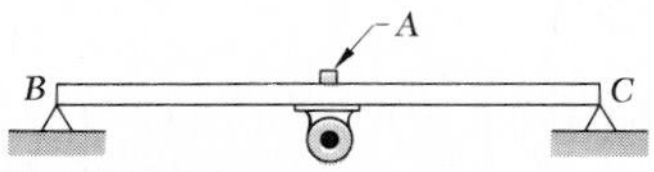

Fig. P19.87

19.88 As the speed of a spring-supported motor is slowly increased from 300 to 400 rpm, the amplitude of the vibration due to the unbalance of the rotor is observed to decrease continuously from 0.075 to 0.040 in. Determine the speed at which resonance will occur.

19.89 In Prob. 19.88, determine the speed for which the amplitude of the vibration is 0.100 in.

19.90 The amplitude of the motion of the pendulum bob shown is observed to be 60 mm when the amplitude of the motion of collar C is 15 mm. Knowing that the length of the pendulum is $l = 720$ mm, determine the two possible values of the frequency of the horizontal movement of the collar C.

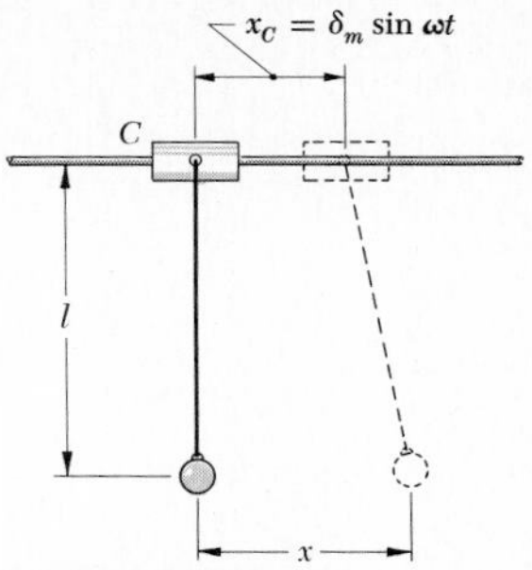

Fig. P19.90

19.91 A small trailer of mass 200 kg with its load is supported by two springs, each of constant 20 kN/m. The trailer is pulled over a road, the surface of which may be approximated by a sine curve of amplitude 30 mm and of period 5 m (i.e., the distance between two successive crests is 5 m, and the vertical distance from a crest to a trough is 60 mm). Determine (*a*) the speed at which resonance will occur, (*b*) the amplitude of the vibration of the trailer at a speed of 60 km/h.

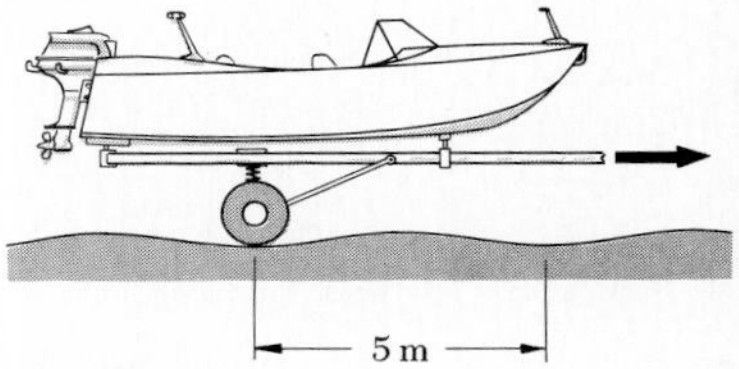

Fig. P19.91

19.92 Knowing that the amplitude of the vibration of the trailer of Prob. 19.91 is not to exceed 15 mm, determine the smallest speed at which the trailer can be pulled over the road.

DAMPED VIBRATIONS

***19.8. Damped Free Vibrations.** The vibrating systems considered in the first part of this chapter were assumed free of damping. Actually all vibrations are damped to some degree by friction forces. These forces may be caused by *dry friction*, or *Coulomb friction*, between rigid bodies, by *fluid friction* when a rigid body moves in a fluid, or by *internal friction* between the molecules of a seemingly elastic body.

A type of damping of special interest is the *viscous damping* caused by fluid friction at low and moderate speeds. Viscous damping is characterized by the fact that the friction force is *directly proportional to the speed* of the moving body. As an example, we shall consider again a body of mass m suspended from a spring of constant k, and we shall assume that the body is attached to the plunger of a dashpot (Fig. 19.10). The magnitude of the friction force exerted on the plunger by the surrounding fluid is equal to $c\dot{x}$, where the constant c, expressed in N · s/m or lb · s/ft and known as the *coefficient of viscous damping*, depends upon the physical properties of the fluid and the construction of the dashpot. The equation of motion is

$$+\downarrow \Sigma F = ma: \qquad W - k(\delta_{st} + x) - c\dot{x} = m\ddot{x}$$

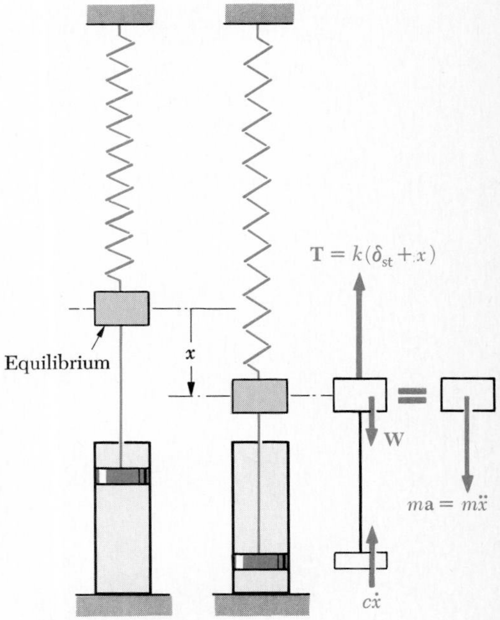

Fig. 19.10

Recalling that $W = k\delta_{st}$, we write

$$m\ddot{x} + c\dot{x} + kx = 0 \tag{19.37}$$

Substituting $x = e^{\lambda t}$ into (19.37) and dividing through by $e^{\lambda t}$, we write the *characteristic equation*

$$m\lambda^2 + c\lambda + k = 0 \tag{19.38}$$

and obtain the roots

$$\lambda = -\frac{c}{2m} \pm \sqrt{\left(\frac{c}{2m}\right)^2 - \frac{k}{m}} \tag{19.39}$$

Defining the *critical damping coefficient* c_c as the value of c which makes the radical in (19.39) equal to zero, we write

$$\left(\frac{c_c}{2m}\right)^2 - \frac{k}{m} = 0 \qquad c_c = 2m\sqrt{\frac{k}{m}} = 2mp \tag{19.40}$$

where p is the circular frequency of the system in the absence of damping. We may distinguish three different cases of damping, depending upon the value of the coefficient c.

1. *Heavy damping:* $c > c_c$. The roots λ_1 and λ_2 of the characteristic equation (19.38) are real and distinct, and the general solution of the differential equation (19.37) is

$$x = Ae^{\lambda_1 t} + Be^{\lambda_2 t} \tag{19.41}$$

   This solution corresponds to a nonvibratory motion. Since λ_1 and λ_2 are both negative, x approaches zero as t increases indefinitely. However, the system actually regains its equilibrium position after a finite time.
2. *Critical damping:* $c = c_c$. The characteristic equation has a double root $\lambda = -c_c/2m = -p$, and the general solution of (19.37) is

$$x = (A + Bt)e^{-pt} \tag{19.42}$$

   The motion obtained is again nonvibratory. Critically damped systems are of special interest in engineering applications since they regain their equilibrium position in the shortest possible time without oscillation.
3. *Light damping:* $c < c_c$. The roots of (19.38) are complex and conjugate, and the general solution of (19.37) is of the form

$$x = e^{-(c/2m)t}(A \sin qt + B \cos qt) \tag{19.43}$$

where q is defined by the relation

$$q^2 = \frac{k}{m} - \left(\frac{c}{2m}\right)^2$$

Substituting $k/m = p^2$ and recalling (19.40), we write

$$q = p\sqrt{1 - \left(\frac{c}{c_c}\right)^2} \tag{19.44}$$

where the constant c/c_c is known as the *damping factor*. A substitution similar to the one used in Sec. 19.2 enables us to write the general solution of (19.37) in the form

$$x = x_m e^{-(c/2m)t} \sin(qt + \phi) \tag{19.45}$$

The motion defined by (19.45) is vibratory with diminishing amplitude (Fig. 19.11). Although this motion does not actually repeat itself, the time interval $\tau = 2\pi/q$, corresponding to two successive points where the curve (19.45) touches one of the limiting curves shown in Fig. 19.11, is commonly referred to as the period of the damped vibration. Recalling (19.44), we observe that τ is larger than the period of vibration of the corresponding undamped system.

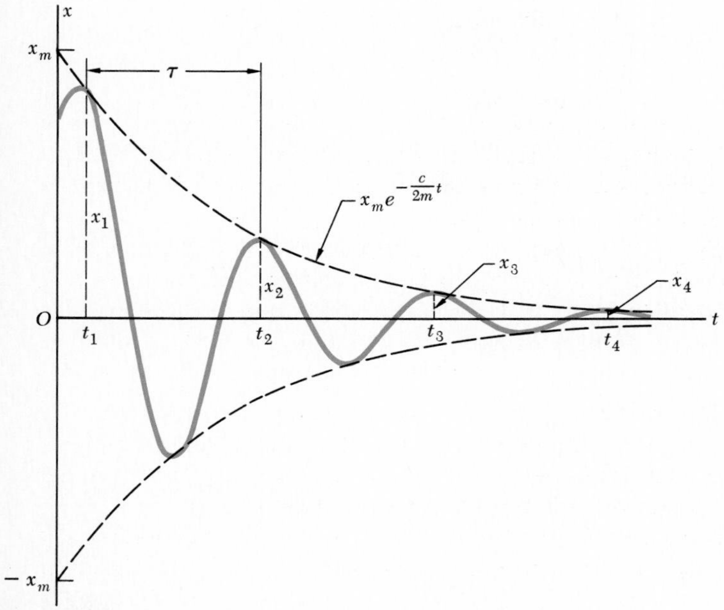

Fig. 19.11

***19.9. Damped Forced Vibrations.** If the system considered in the preceding section is subjected to a periodic force **P** of magnitude $P = P_m \sin \omega t$, the equation of motion becomes

$$m\ddot{x} + c\dot{x} + kx = P_m \sin \omega t \tag{19.46}$$

The general solution of (19.46) is obtained by adding a particular solution of (19.46) to the complementary function or general solution of the homogeneous equation (19.37). The complementary function is given by (19.41), (19.42), or (19.43), depending upon the type of damping considered. It represents a *transient* motion which is eventually damped out.

Our interest in this section is centered on the steady-state vibration represented by a particular solution of (19.46) of the form

$$x_{\text{part}} = x_m \sin (\omega t - \varphi) \tag{19.47}$$

Substituting x_{part} for x into (19.46), we obtain

$$-m\omega^2 x_m \sin (\omega t - \varphi) + c\omega x_m \cos (\omega t - \varphi) + kx_m \sin (\omega t - \varphi) = P_m \sin \omega t$$

Making $\omega t - \varphi$ successively equal to 0 and to $\pi/2$, we write

$$c\omega x_m = P_m \sin \varphi \tag{19.48}$$

$$(k - m\omega^2)x_m = P_m \cos \varphi \tag{19.49}$$

Squaring both members of (19.48) and (19.49) and adding, we have

$$[(k - m\omega^2)^2 + (c\omega)^2]x_m^2 = P_m^2 \tag{19.50}$$

Solving (19.50) for x_m and dividing (19.48) and (19.49) member by member, we obtain, respectively,

$$x_m = \frac{P_m}{\sqrt{(k - m\omega^2)^2 + (c\omega)^2}} \qquad \tan \varphi = \frac{c\omega}{k - m\omega^2} \tag{19.51}$$

Recalling from (19.4) that $k/m = p^2$, where p is the circular frequency of the undamped free vibration, and from (19.40) that $2mp = c_c$, where c_c is the critical damping coefficient of the system, we write

$$\frac{x_m}{P_m/k} = \frac{x_m}{\delta_m} = \frac{1}{\sqrt{[1 - (\omega/p)^2]^2 + [2(c/c_c)(\omega/p)]^2}} \tag{19.52}$$

$$\tan \varphi = \frac{2(c/c_c)(\omega/p)}{1 - (\omega/p)^2} \tag{19.53}$$

Formula (19.52) expresses the magnification factor in terms of the frequency ratio ω/p and damping factor c/c_c. It may be used to determine the amplitude of the steady-state vibration produced by an impressed force of magnitude $P = P_m \sin \omega t$ or by an impressed support movement $\delta = \delta_m \sin \omega t$. Formula (19.53) defines in terms of the same parameters the *phase difference* φ between the impressed force or impressed support movement and the resulting steady-state vibration of the damped system. The magnification factor has been plotted against the frequency ratio in Fig. 19.12 for various values of the damping factor. We observe that the amplitude of a forced vibration may be kept small by choosing a large coefficient of viscous damping c or by keeping the natural and forced frequencies far apart.

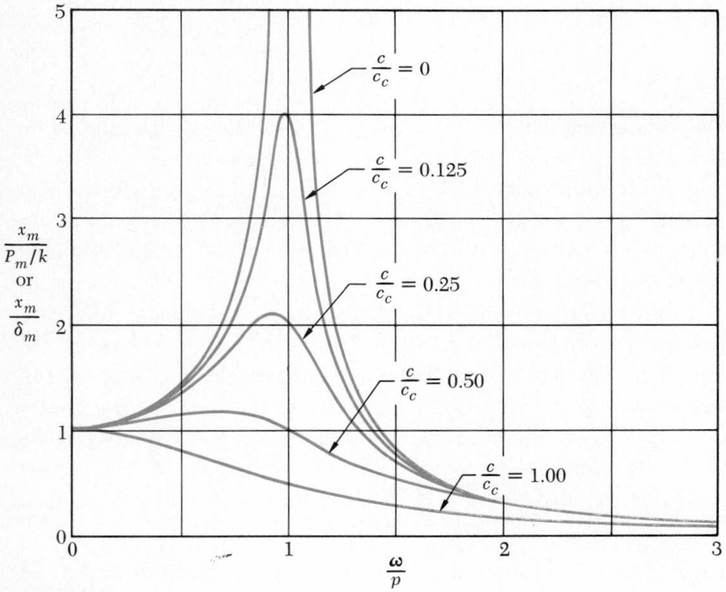

Fig. 19.12

*19.10. Electrical Analogues.

Oscillating electrical circuits are characterized by differential equations of the same type as those obtained in the preceding sections. Their analysis is therefore similar to that of a mechanical system, and the results obtained for a given vibrating system may be readily extended to the equivalent circuit. Conversely, any result obtained for an electrical circuit will also apply to the corresponding mechanical system.

Consider an electrical circuit consisting of an inductor of inductance L, a resistor of resistance R, and a capacitor of

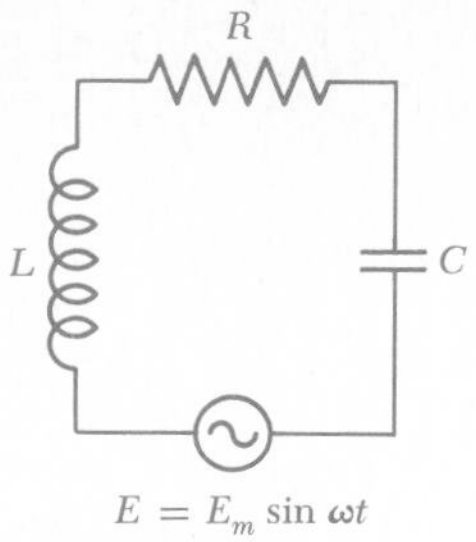

Fig. 19.13

capacitance C, connected in series with a source of alternating voltage $E = E_m \sin \omega t$ (Fig. 19.13). It is recalled from elementary electromagnetic theory† that, if i denotes the current in the circuit and q the electric charge on the capacitor, the drop in potential is $L(di/dt)$ across the inductor, Ri across the resistor, and q/C across the capacitor. Expressing that the algebraic sum of the applied voltage and of the drops in potential around the circuit loop is zero, we write

$$E_m \sin \omega t - L\frac{di}{dt} - Ri - \frac{q}{C} = 0 \qquad (19.54)$$

Rearranging the terms and recalling that, at any instant, the current i is equal to the rate of change $\dot{q}$ of the charge q, we have

$$L\ddot{q} + R\dot{q} + \frac{1}{C}q = E_m \sin \omega t \qquad (19.55)$$

We verify that Eq. (19.55), which defines the oscillations of the electrical circuit of Fig. 19.13, is of the same type as Eq. (19.46), which characterizes the damped forced vibrations of the mechanical system of Fig. 19.10. By comparing the two equations, we may construct a table of the analogous mechanical and electrical expressions.

Table 19.2 may be used to extend to their electrical analogues the results obtained in the preceding sections for various mechanical systems. For instance, the amplitude i_m of the current in the circuit of Fig. 19.13 may be obtained by noting that it corresponds to the maximum value v_m of the velocity in the analogous mechanical system. Recalling that $v_m = \omega x_m$, substituting for x_m from Eq. (19.51), and replacing the constants of

Table 19.2 Characteristics of a Mechanical System and of Its Electrical Analogue

Mechanical System		Electrical Circuit	
m	mass	L	Inductance
c	Coefficient of viscous damping	R	Resistance
k	Spring constant	$1/C$	Reciprocal of capacitance
x	Displacement	q	Charge
v	Velocity	i	Current
P	Applied force	E	Applied voltage

† See Hammond, "Electrical Engineering," McGraw-Hill Book Company, or Smith, "Circuits, Devices, and Systems," John Wiley & Sons.

the mechanical system by the corresponding electrical expressions, we have

$$i_m = \frac{\omega E_m}{\sqrt{\left(\frac{1}{C} - L\omega^2\right)^2 + (R\omega)^2}}$$

$$i_m = \frac{E_m}{\sqrt{R^2 + \left(L\omega - \frac{1}{C\omega}\right)^2}} \tag{19.56}$$

The radical in the expression obtained is known as the *impedance* of the electrical circuit.

The analogy between mechanical systems and electrical circuits holds for transient as well as steady-state oscillations. The oscillations of the circuit shown in Fig. 19.14, for instance, are analogous to the damped free vibrations of the system of Fig. 19.10. As far as the initial conditions are concerned, we may note that closing the switch S when the charge on the capacitor is $q = q_0$ is equivalent to releasing the mass of the mechanical system with no initial velocity from the position $x = x_0$. We should also observe that, if a battery of constant voltage E is introduced in the electrical circuit of Fig. 19.14, closing the switch S will be equivalent to suddenly applying a force of constant magnitude P to the mass of the mechanical system of Fig. 19.10.

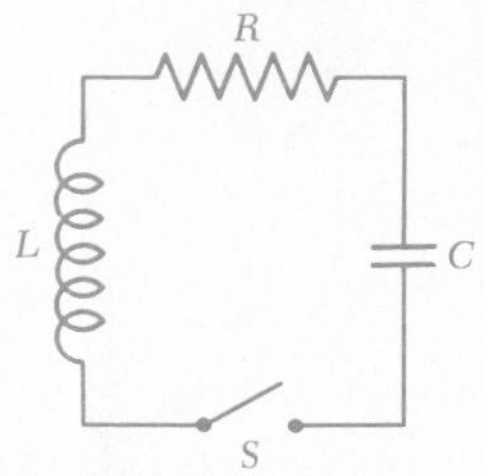

Fig. 19.14

The above discussion would be of questionable value if its only result were to make it possible for mechanics students to analyze electrical circuits without learning the elements of electromagnetism. It is hoped, rather, that this discussion will encourage the students to apply to the solution of problems in mechanical vibrations the mathematical techniques they may learn in later courses in electrical circuits theory. The chief value of the concept of electrical analogue, however, resides in its application to *experimental methods* for the determination of the characteristics of a given mechanical system. Indeed, an electrical circuit is much more easily constructed than a mechanical model, and the fact that its characteristics may be modified by varying the inductance, resistance, or capacitance of its various components makes the use of the electrical analogue particularly convenient.

To determine the electrical analogue of a given mechanical system, we shall focus our attention on each moving mass in the system and observe which springs, dashpots, or external forces are applied directly to it. An equivalent electrical loop may then be constructed to match each of the mechanical units

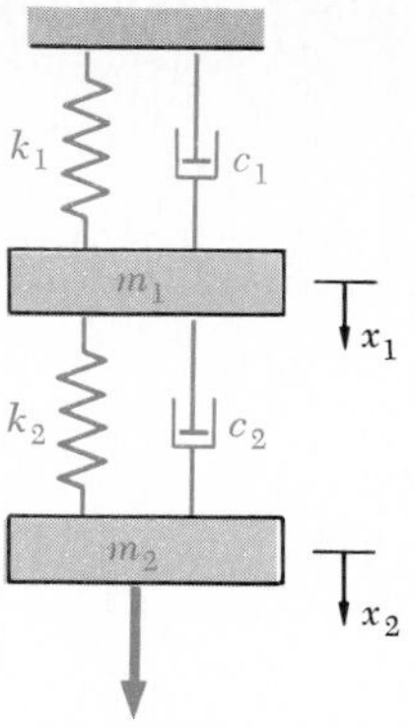

$\mathbf{P} = P_m \sin \omega t$

Fig. 19.15

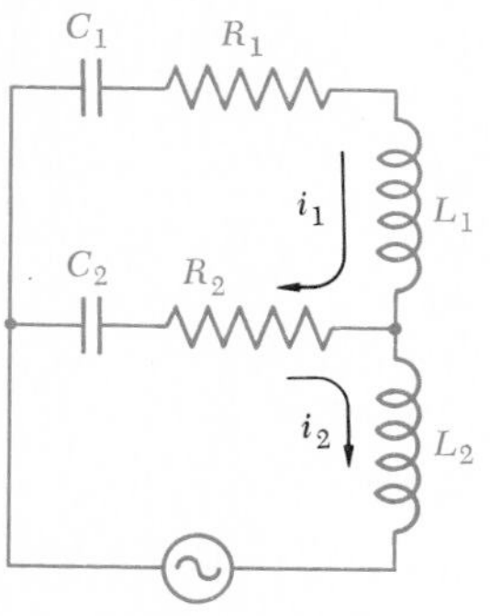

$E = E_m \sin \omega t$

Fig. 19.16

thus defined; the various loops obtained in that way will form together the desired circuit. Consider, for instance, the mechanical system of Fig. 19.15. We observe that the mass m_1 is acted upon by two springs of constants k_1 and k_2 and by two dashpots characterized by the coefficients of viscous damping c_1 and c_2. The electrical circuit should therefore include a loop consisting of an inductor of inductance L_1 proportional to m_1, of two capacitors of capacitance C_1 and C_2 inversely proportional to k_1 and k_2, respectively, and of two resistors of resistance R_1 and R_2, proportional to c_1 and c_2, respectively. Since the mass m_2 is acted upon by the spring k_2 and the dashpot c_2, as well as by the force $P = P_m \sin \omega t$, the circuit should also include a loop containing the capacitor C_2, the resistor R_2, the new inductor L_2, and the voltage source $E = E_m \sin \omega t$ (Fig. 19.16).

To check that the mechanical system of Fig. 19.15 and the electrical circuit of Fig. 19.16 actually satisfy the same differential equations, we shall first derive the equations of motion for m_1 and m_2. Denoting respectively by x_1 and x_2 the displacements of m_1 and m_2 from their equilibrium positions, we observe that the elongation of the spring k_1 (measured from the equilibrium position) is equal to x_1, while the elongation of the spring k_2 is equal to the relative displacement $x_2 - x_1$ of m_2 with respect to m_1. The equations of motion for m_1 and m_2 are therefore

$$m_1\ddot{x}_1 + c_1\dot{x}_1 + c_2(\dot{x}_1 - \dot{x}_2) + k_1x_1 + k_2(x_1 - x_2) = 0 \qquad (19.57)$$

$$m_2\ddot{x}_2 + c_2(\dot{x}_2 - \dot{x}_1) + k_2(x_2 - x_1) = P_m \sin \omega t \qquad (19.58)$$

Consider now the electrical circuit of Fig. 19.16; we denote respectively by i_1 and i_2 the current in the first and second loops, and by q_1 and q_2 the integrals $\int i_1\,dt$ and $\int i_2\,dt$. Noting that the charge on the capacitor C_1 is q_1, while the charge on C_2 is $q_1 - q_2$, we express that the sum of the potential differences in each loop is zero:

$$L_1\ddot{q}_1 + R_1\dot{q}_1 + R_2(\dot{q}_1 - \dot{q}_2) + \frac{q_1}{C_1} + \frac{q_1 - q_2}{C_2} = 0 \qquad (19.59)$$

$$L_2\ddot{q}_2 + R_2(\dot{q}_2 - \dot{q}_1) + \frac{q_2 - q_1}{C_2} = E_m \sin \omega t \qquad (19.60)$$

We easily check that Eqs. (19.59) and (19.60) reduce to (19.57) and (19.58), respectively, when the substitutions indicated in Table 19.2 are performed.

PROBLEMS

19.93 Show that, in the case of heavy damping ($c > c_c$), a body never passes through its position of equilibrium O (*a*) if it is released with no initial velocity from an arbitrary position or (*b*) if it is started from O with an arbitrary initial velocity.

19.94 Show that, in the case of heavy damping ($c > c_c$), a body released from an arbitrary position with an arbitrary initial velocity cannot pass more than once through its equilibrium position.

19.95 In the case of light damping, the displacements x_1, x_2, x_3, etc., shown in Fig. 19.11 may be assumed equal to the maximum displacements. Show that the ratio of any two successive maximum displacements x_n and x_{n+1} is a constant and that the natural logarithm of this ratio, called the *logarithmic decrement*, is

$$\ln \frac{x_n}{x_{n+1}} = \frac{2\pi(c/c_c)}{\sqrt{1 - (c/c_c)^2}}$$

19.96 In practice, it is often difficult to determine the logarithmic decrement defined in Prob. 19.95 by measuring two successive maximum displacements. Show that the logarithmic decrement may also be expressed as $(1/n) \ln (x_1/x_{n+1})$, where n is the number of cycles between readings of the maximum displacement.

19.97 Successive maximum displacements of a spring-mass-dashpot system similar to that shown in Fig. 19.10 are 75, 60, 48, and 38.4 mm. Knowing that $m = 20$ kg and $k = 800$ N/m, determine (*a*) the damping factor c/c_c, (*b*) the value of the coefficient of viscous damping c. (*Hint.* See Probs. 19.95 and 19.96.)

19.98 In a system with light damping ($c < c_c$), the period of vibration is commonly defined as the time interval $\tau = 2\pi/q$ corresponding to two successive points where the displacement-time curve touches one of the limiting curves shown in Fig. 19.11. Show that the interval of time (*a*) between a maximum positive displacement and the following maximum negative displacement is $\frac{1}{2}\tau$, (*b*) between two successive zero displacements is $\frac{1}{2}\tau$, (*c*) between a maximum positive displacement and the following zero displacement is greater than $\frac{1}{4}\tau$.

19.99 The barrel of a field gun weighs 1400 lb and is returned into firing position after recoil by a recuperator of constant $k = 10{,}000$ lb/ft. Determine the value of the coefficient of damping of the recoil mechanism which causes the barrel to return into firing position in the shortest possible time without oscillation.

19.100 A critically damped system is released from rest at an arbitrary position x_0 when $t = 0$. (*a*) Determine the position of the system at any time t. (*b*) Apply the result obtained in part *a* to the barrel of the gun of Prob. 19.99, and determine the time at which the barrel is halfway back to its firing position.

19.101 Assuming that the barrel of the gun of Prob. 19.99 is modified, with a resulting increase in weight of 400 lb, determine the constant k of the recuperator which should be used if the recoil mechanism is to remain critically damped.

19.102 In the case of the forced vibration of a system with a given damping factor c/c_c, determine the frequency ratio ω/p for which the amplitude of the vibration is maximum.

19.103 Show that for a small value of the damping factor c/c_c, (*a*) the maximum amplitude of a forced vibration occurs when $\omega \approx p$, (*b*) the corresponding value of the magnification factor is approximately $\frac{1}{2}(c_c/c)$.

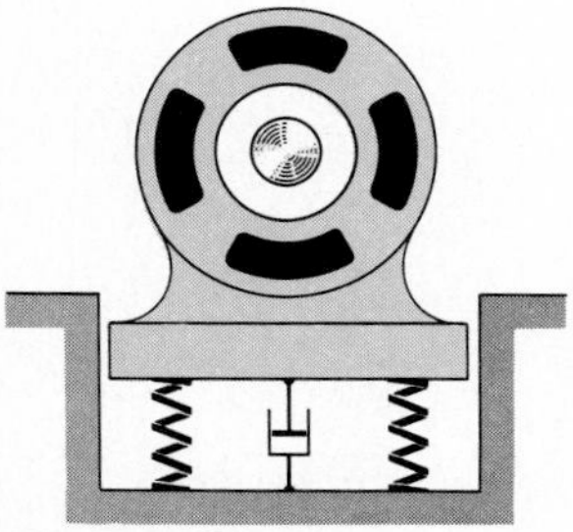

Fig. P19.104

19.104 A motor of mass 25 kg is supported by four springs, each having a constant of 200 kN/m. The unbalance of the rotor is equivalent to a mass of 30 g located 125 mm from the axis of rotation. Knowing that the motor is constrained to move vertically, determine the amplitude of the steady-state vibration of the motor at a speed of 1800 rpm, assuming (*a*) that no damping is present, (*b*) that the damping factor c/c_c is equal to 0.125.

19.105 Assume that the 25-kg motor of Prob. 19.104 is directly supported by a light horizontal beam. The static deflection of the beam due to the weight of the motor is observed to be 5.75 mm, and the amplitude of the vibration of the motor is 0.5 mm at a speed of 400 rpm. Determine (*a*) the damping factor c/c_c, (*b*) the coefficient of damping c.

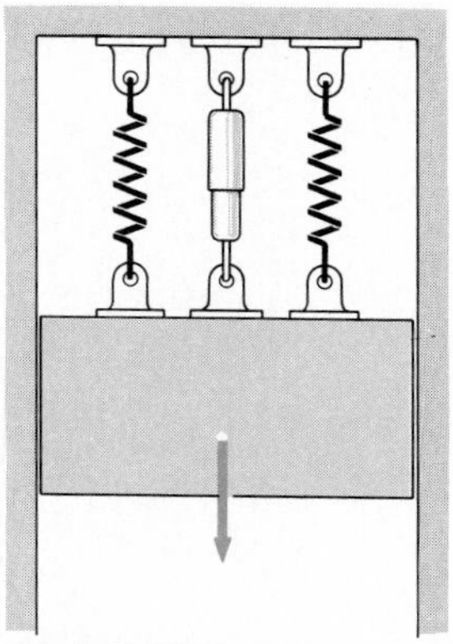

Fig. P19.106

19.106 A machine element weighing 800 lb is supported by two springs, each having a constant of 200 lb/in. A periodic force of maximum value 20 lb is applied to the element with a frequency of 2.5 cycles per second. Knowing that the coefficient of damping is 8 lb · s/in., determine the amplitude of the steady-state vibration of the element.

19.107 In Prob. 19.106, determine the required value of the coefficient of damping if the amplitude of the steady-state vibration of the element is to be 0.15 in.

19.108 A platform of mass 100 kg, supported by a set of springs equivalent to a single spring of constant $k = 80$ kN/m, is subjected to a periodic force of maximum magnitude 500 N. Knowing that the coefficient of damping is 2 kN · s/m, determine (*a*) the natural frequency in rpm of the platform *if* there were no damping, (*b*) the frequency in rpm of the periodic force corresponding to the maximum value of the magnification factor, assuming damping, (*c*) the amplitude of the *actual* motion of the platform for each of the frequencies found in parts *a* and *b*.

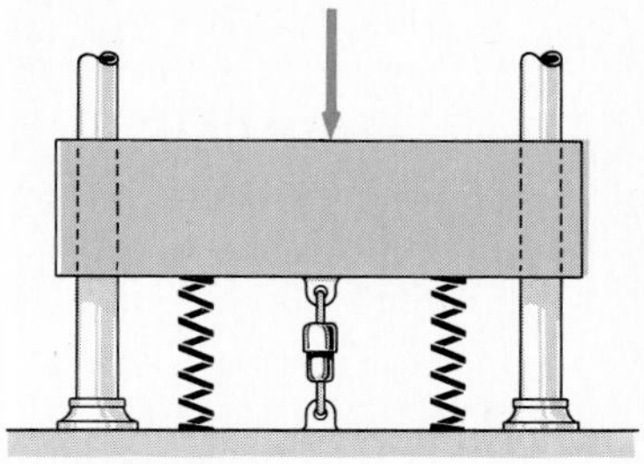

Fig. P19.108

***19.109** The suspension of an automobile may be approximated by the simplified spring-and-dashpot system shown. (*a*) Write the differential equation defining the absolute motion of the mass m when the system moves at a speed v over a road of sinusoidal cross section as shown. (*b*) Derive an expression for the amplitude of the absolute motion of m.

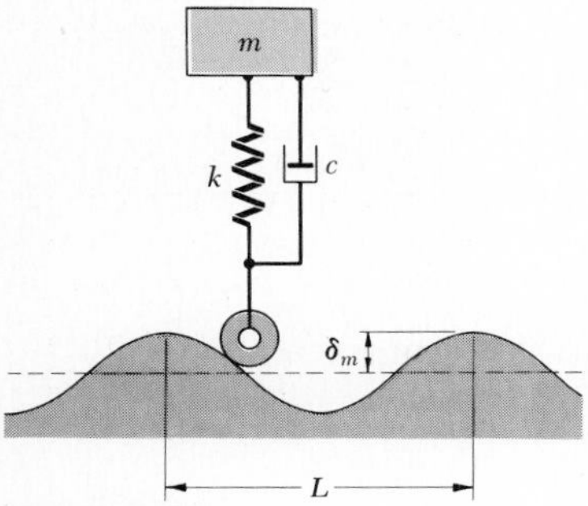

Fig. P19.109

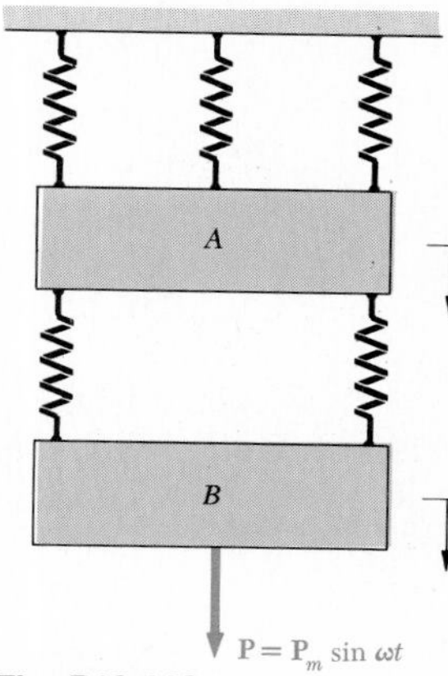

Fig. P19.110

***19.110** Two loads A and B, each of mass m, are suspended as shown by means of five springs of the same constant k. Load B is subjected to a force of magnitude $P = P_m \sin \omega t$. Write the differential equations defining the displacements x_A and x_B of the two loads from their equilibrium positions.

19.111 Determine the range of values of the resistance R for which oscillations will take place in the circuit shown when the switch S is closed.

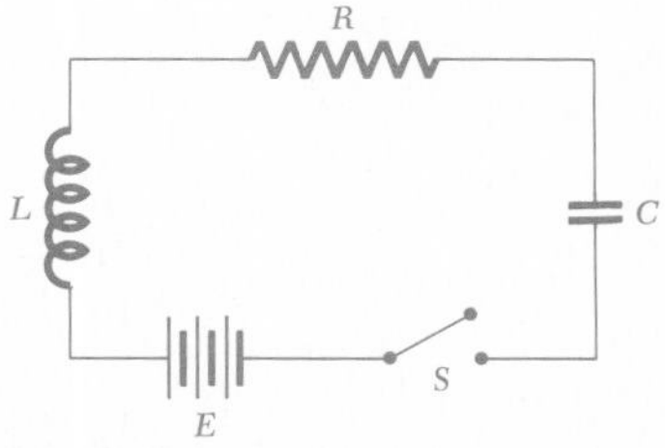

Fig. P19.111

19.112 Consider the circuit of Prob. 19.111 when the capacitance C is equal to zero. If the switch S is closed at time $t = 0$, determine (*a*) the final value of the current in the circuit, (*b*) the time t at which the current will have reached $(1 - 1/e)$ times its final value. (The desired value of t is known as the *time constant* of the circuit.)

19.113 through 19.116 Draw the electrical analogue of the mechanical system shown. (*Hint.* In Probs. 19.113 and 19.114, draw the loops corresponding to the free bodies m and A.)

19.117 and 19.118 Write the differential equations defining (a) the displacements of mass m and point A, (b) the currents in the corresponding loops of the electrical analogue.

19.119 and 19.120 Write the differential equations defining (a) the displacements of the masses m_1 and m_2, (b) the currents in the corresponding loops of the electrical analogue.

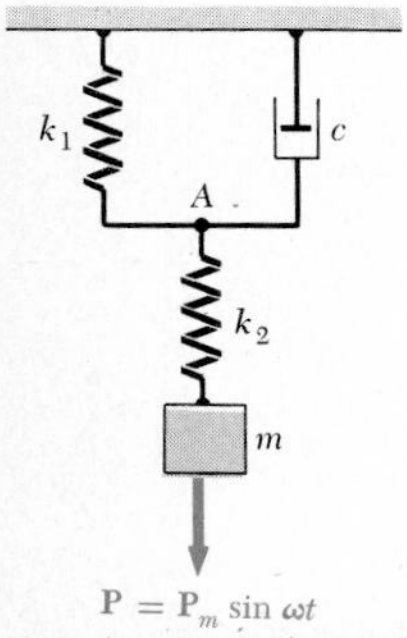

Fig. P19.113 and P19.117

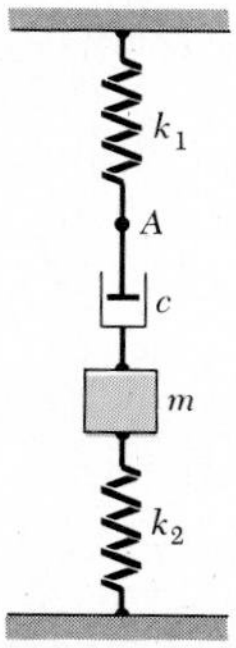

Fig. P19.114 and P19.118

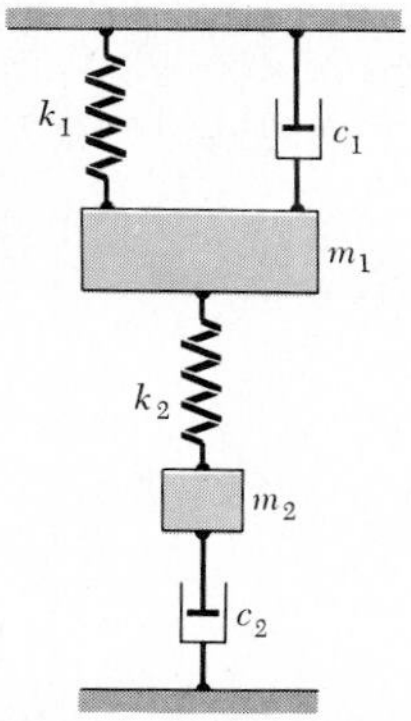

Fig. P19.115 and P19.119

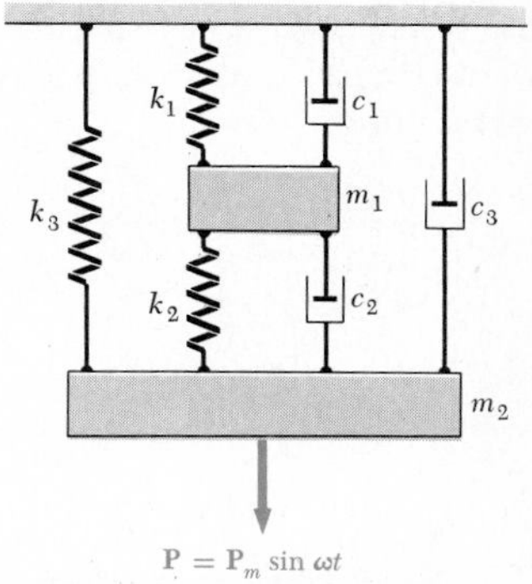

Fig. P19.116 and P19.120

REVIEW PROBLEMS

19.121 A homogeneous wire of length $2l$ is bent as shown and allowed to oscillate about a smooth pin at B. Denoting by τ_0 the period of small oscillations when $\beta = 0$, determine the angle β for which the period of small oscillations is $2\tau_0$.

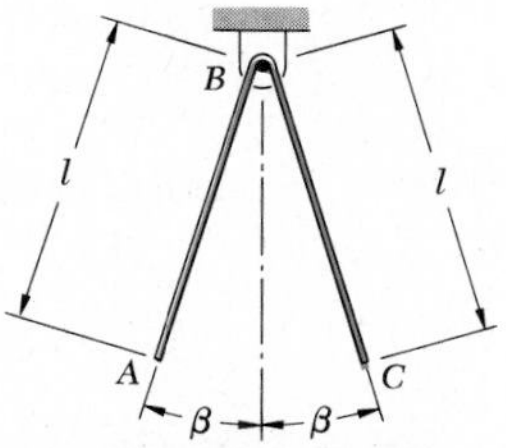

Fig. P19.121 and P19.122

19.122 Knowing that $l = 0.6$ m and $\beta = 30°$, determine the period of oscillation of the bent homogeneous wire shown.

19.123 An automobile wheel-and-tire assembly of total weight 47 lb is attached to a mounting plate of negligible weight which is suspended from a steel wire. The torsional spring constant of the wire is known to be $K = 0.40$ lb · in./rad. The wheel is rotated through 90° about the vertical and then released. Knowing that the period of oscillation is observed to be 30 s, determine the centroidal mass moment of inertia and the centroidal radius of gyration of the wheel-and-tire assembly.

Fig. P19.123

19.124 The period of small oscillations about A of a connecting rod is observed to be 1.12 s. Knowing that the distance r_a is 7.50 in., determine the centroidal radius of gyration of the connecting rod.

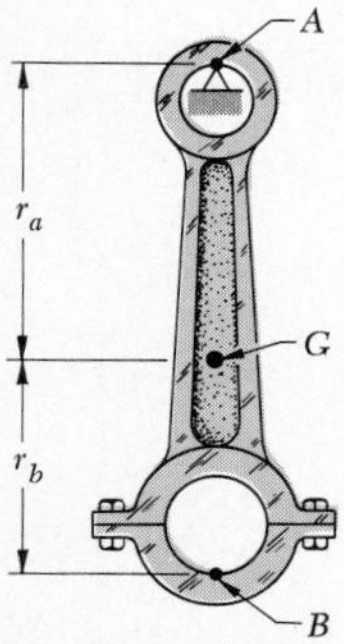

Fig. P19.124 and P19.125

19.125 A connecting rod is supported by a knife-edge at point A; the period of small oscillations is observed to be 0.945 s. The rod is then inverted and supported by a knife-edge at point B and the period of small oscillations is observed to be 0.850 s. Knowing that $r_a + r_b = 11.50$ in., determine (*a*) the location of the mass center G, (*b*) the centroidal radius of gyration $\bar{k}$.

19.126 A 150-kg electromagnet is at rest and is holding 100 kg of scrap steel when the current is turned off and the steel is dropped. Knowing that the cable and the supporting crane have a total stiffness equivalent to a spring of constant 200 kN/m, determine (*a*) the frequency, the amplitude, and the maximum velocity of the resulting motion, (*b*) the minimum tension which will occur in the cable during the motion, (*c*) the velocity of the magnet 0.03 s after the current is turned off.

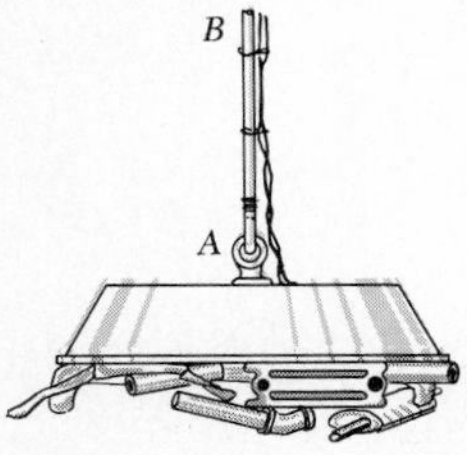

Fig. P19.126

Fig. P19.127

19.127 A certain vibrometer used to measure vibration amplitudes consists essentially of a box containing a slender rod to which a mass m is attached; the natural frequency of the mass-rod system is known to be 5 Hz. When the box is rigidly attached to the casing of a motor rotating at 600 rpm, the mass is observed to vibrate with an amplitude of 1.6 mm relative to the box. Determine the amplitude of the vertical motion of the motor.

19.128 Solve Prob. 19.127, assuming that the speed of the motor is 150 rpm and that the other data are unchanged.

19.129 A rod of mass m and length L rests on the two pulleys A and B which rotate in the directions shown. Denoting by μ the coefficient of friction between the rod and the pulleys, determine the frequency of vibration if the rod is given a small displacement to the right and released.

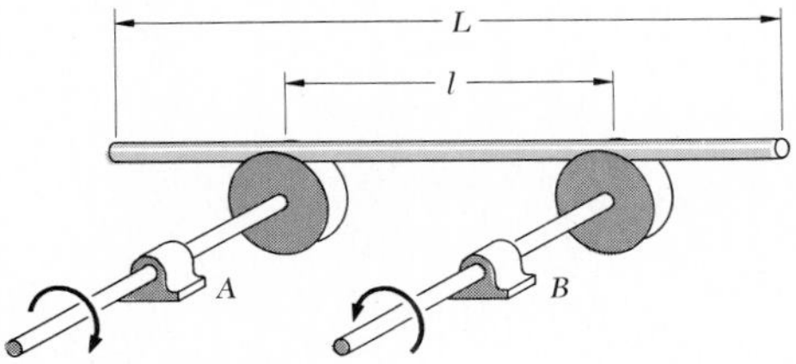

Fig. P19.129

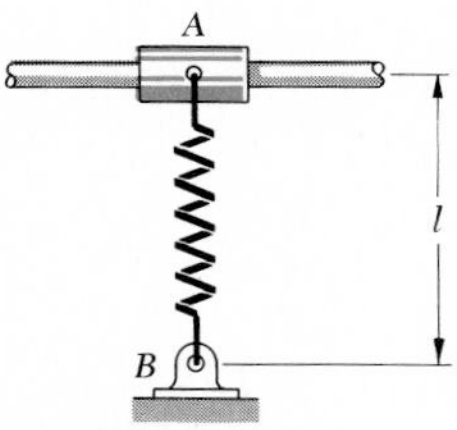

Fig. P19.130

19.130 A collar of mass m slides without friction on a horizontal rod and is attached to a spring AB of constant k. (a) If the unstretched length of the spring is just equal to l, show that the collar does not execute simple harmonic motion even when the amplitude of the oscillations is small. (b) If the unstretched length of the spring is less than l, show that the motion may be approximated by a simple harmonic motion for small oscillations.

19.131 The 30-lb beam is supported as shown by two uniform disks, each of weight 20 lb and radius 5 in. Knowing that the disks roll without sliding, determine the period of vibration of the system if the beam is given a small displacement to the right and released.

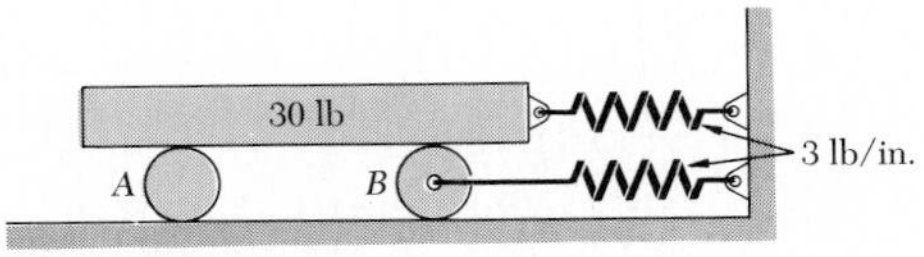

Fig. P19.131

19.132 Solve Prob. 19.131, assuming that the spring attached to the beam is removed.

INDEX

INDEX

Answers to Even-Numbered Problems

Answers to Even-Numbered Problems

CHAPTER 2

2.2 3240 N $\angle$40.9°.
2.4 (*a*) 14.21 kips $\angle$5.1°. (*b*) 14.85 kips.
SI: (*a*) 63.2 kN $\angle$5.1°. (*b*) 66.1 kN.
2.6 123.4°.
2.8 14.73 lb; 30.2 lb. SI: 65.5 N; 134.3 N.
2.10 43.6 lb $\angle$78.4°. SI: 193.9 N $\angle$78.4°.
2.12 2990 N $\angle$72.8°.
2.14 0.855 kN $\rightarrow$, 2.35 kN $\uparrow$.
2.16 (40 lb) 20 lb $\rightarrow$, 34.6 lb $\uparrow$; (50 lb) 47.0 lb $\rightarrow$, 17.10 lb $\uparrow$; (60 lb) 38.6 lb $\rightarrow$, 46.0 lb $\downarrow$,
SI: (40 lb) 89.0 N $\rightarrow$, 153.9 N $\uparrow$; (50 lb) 209 N $\rightarrow$, 76.1 N $\uparrow$; (60 lb) 171.7 N $\rightarrow$, 205 N $\downarrow$.
2.18 250 N $\rightarrow$, 600 N $\downarrow$.
2.20 200 lb $\angle$4.3°. SI: 890 N $\angle$4.3°.
2.22 105.6 lb $\angle$3.1°. SI: 470 N $\angle$3.1°.
2.26 14.04 kips. SI: 62.5 kN.
2.28 6.3° and 133.7°.
2.30 $T_{AC} = 120.1$ lb; $T_{BC} = 156.3$ lb.
SI: $T_{AC} = 534$ N; $T_{BC} = 695$ N.
2.32 $T_{AC} = 981$ N; $T_{BC} = 2350$ N.
2.34 (*a*) 60°. (*b*) $T_{AC} = T_{BC} = 173.2$ lb.
SI: (*a*) 60°. (*b*) $T_{AC} = T_{BC} = 770$ N.
2.36 2.87 kN $\angle$75°.
2.38 $C = 2170$ N; $D = 3750$ N.
2.40 $T_{AC} = 46.2$ lb; $T_{BC} = 36.9$ lb.
SI: $T_{AC} = 206$ N; $T_{BC} = 164.1$ N.
2.42 39 in. SI: 0.991 m.
2.44 (*b*) 654 N. (*d*) 491 N.
2.46 1.252 in. SI: 31.8 mm.
2.48 (*a*) 18 lb. (*b*) 24 lb.
SI: (*a*) 80.1 N. (*b*) 106.8 N.
2.50 (*a*) +113.3 N, +217 N, −52.8 N.
(*b*) 63.1°, 30.0°, 102.2°.
2.52 (*a*) −65.5 lb, +235 lb, −55.0 lb.
(*b*) 105.2°, 20.0°, 102.7°.
SI: (*a*) −291 N, +1045 N, −245 N.
2.54 48.8°; +51.8 lb, 141.4 lb, 131.6 lb.
SI: 48.8°; 230 N, 629 N, 585 N.
2.56 700 N; 64.6°, 120.9°, 42.0°.
2.58 −800 lb, +800 lb, −1400 lb.
SI: −3560 N, +3560 N, −6230 N.
2.60 116.4°, 63.6°, 141.1°.
2.62 (*a*) 54.7°, 125.3°. (*b*) 60°, 120°.
2.64 $T_{AC} = 21$ kN; $T_{AD} = 64.3$ kN.
2.66 1110 lb. SI: 4940 N.
2.68 $T_{AD} = 105$ lb; $T_{BD} = 78$ lb; $T_{CD} = 75$ lb.
SI: $T_{AD} = 467$ N; $T_{BD} = 347$ N; $T_{CD} = 334$ N.

2.70 $T_{AD} = T_{BD} = 1.75$ kN; $T_{CD} = 3$ kN.
2.72 $T_{AD} = 139.9$ N; $T_{BD} = T_{CD} = 115.7$ N.
2.74 (*a*) 6.3 lb. (*b*) 7.2 lb.
SI: (*a*) 28.0 N. (*b*) 32.0 N.
2.76 $T_{AD} = 103.5$ N; $T_{BD} = 51.8$ N;
$T_{CD} = 89.7$ N.
2.78 $T_{AC} = 5950$ N; $T_{BC} = 4490$ N.
2.80 $1223 \text{ N} \leqslant P \leqslant 3173$ N.
2.82 (*a*) 500 lb. (*b*) 375 lb.
SI: (*a*) 2220 N. (*b*) 1668 N.
2.84 1200 lb, down. SI: 5340 N, down.
2.86 $T_{AC} = 547$ N; $T_{BC} = 800$ N.
2.88 418 N ∠32.3° or 767 N ∠32.3°.
2.90 $P = 27$ lb; $T_{AB} = 51$ lb.
SI: $P = 120.1$ N; $T_{AB} = 227$ N.

CHAPTER 3

3.2 -37.5 N · m; $\alpha = 20°$.
3.4 (*a*) -88.8 N · m. (*b*) 395 N ←.
(*c*) 280 N ∠45°.
3.6 167.0 lb · in. SI: 18.87 N · m.
3.8 -547 lb · in. SI: -61.8 N · m.
3.12 $x = aQ/(Q - P)$.
3.14 (*a*) 271 N. (*b*) 390 N. (*c*) 250 N.
3.16 280 lb · in. ↻. SI: 31.6 N · m ↻.
3.18 (*a*) 400 N ∠60°, 77.8 N · m ↻.
(*b*) 400 N ∠60°, 55.1 N · m ↻.
3.20 16 kips ↓; 3 in. to left of *AB*.
SI: 71.2 kN ↓; 76.2 mm to left of *AB*.
3.22 (*a*) 960 N ∠60°; 28.9 mm to right of *O*.
(*b*) 960 N ∠60°; 50 mm below *O*.
3.24 $\mathbf{F}_A = Pb/a$ ↑; $\mathbf{F}_C = P(a + b)/a$ ↓.
3.26 (*a*) 600 N ∠36.9°; 28.8 N · m ↺.
(*b*) $\mathbf{F}_B = 200$ N ∠36.9°;
$\mathbf{F}_D = 400$ N ∠36.9°.
3.28 *e*.
3.30 (*a*) 300 lb →; 10.67 in. above *A*.
(*b*) 150 lb ←; 50.7 in. above *A*.
(*c*) System reduces to 4000-lb · in. ↺ couple
SI: (*a*) 1334 N →; 271 mm above *A*.
(*b*) 667 N ←; 1288 mm above *A*.
(*c*) System reduces to a 452-N · m ↺ couple.
3.32 (*a*) 0.5 m. (*b*) 0.875 m. (*c*) 1.625 m.
3.34 224 N ∠26.6°; 265 mm to the left of *C*, 132.5 mm below *C*.
3.36 1050 lb; 9.33 ft to the right of *A*.
SI: 4670 N; 2.84 m to the right of *A*.

3.38 (*a*) 758 lb →; 15,190 lb · ft ↺.
(*b*) 758 lb →, 20 ft below *DE*.
SI: (*a*) 3370 N →; 20.6 kN · m ↺.
(*b*) 3370 N →, 6.10 m below *DE*.
3.40 (*a*) 184.7 N ←. (*b*) 92.4 N ∠60°.
(*c*) 184.7 N ∠60°.
3.42 150.6° or $-52.5°$.
3.44 (*a*) 183.5 N. (*b*) 527 N ∠73.4°.
3.46 (*a*) $T = (W \cos\theta)/(2 \cos \frac{1}{2}\theta)$.
(*b*) 11.74 lb. SI: (*b*) 52.2 N.
3.48 $\pm 60°$.
3.50 2.4 kN.
3.52 $\mathbf{A}_x = 57.8$ kN →, $\mathbf{A}_y = 44.2$ kN ↑,
$\mathbf{B} = 35.6$ kN ∠60°.
3.54 $\mathbf{A}_x = 26.2$ kN →, $\mathbf{A}_y = 29.7$ kN ↑,
$\mathbf{B} = 52.3$ kN ∠60°.
3.56 $T = 120$ lb; $\mathbf{A} = 750$ lb ←;
$\mathbf{B} = 750$ lb →. SI: $T = 534$ N;
$\mathbf{A} = 3340$ N ←; $\mathbf{B} = 3340$ N →.
3.58 (*a*) $\mathbf{A} = 600$ N ↑; $\mathbf{B} = 2400$ N →;
$\mathbf{C} = 2400$ N ←. (*b*) $\mathbf{A} = 520$ N ↑;
$\mathbf{B} = 1780$ N →; $\mathbf{C} = 1480$ N ←.
3.60 $\mathbf{A} = 166.3$ lb ∠30°; $\mathbf{B} = 166.3$ lb ∠30°.
SI: $\mathbf{A} = 740$ N ∠30°; $\mathbf{B} = 740$ N ∠30°.
3.62 (*a*) $\mathbf{A} = M/2L$ ↑; $\mathbf{B} = M/2L$ ↓.
(*b*) $\mathbf{A} = M/L$ ←; $\mathbf{B} = M/L$ →.
(*c*) $\mathbf{B} = 0$; $\mathbf{M}_B = M$ ↻. (*d*) $\mathbf{A} = M/L$ ↑;
$\mathbf{B} = M/L$ ←; $\mathbf{D} = \sqrt{2}M/L$ ∠45°.
3.64 (*a*) $\mathbf{A} = 30$ kN ↑; $\mathbf{B} = 105$ kN ↑.
(*b*) $\mathbf{A} = 80$ kN ↑; $\mathbf{B} = 130$ kN ↑.
3.66 $22.5 \text{ lb} \leqslant P \leqslant 627$ lb.
SI: $100.1 \text{ N} \leqslant P \leqslant 2790$ N.
3.68 (*a*) $\mathbf{A} = 236$ N ∠28.3°; $\mathbf{B} = 232$ N ↓.
(*b*) $\mathbf{A} = 133.3$ N ←;
$\mathbf{B} = 141.2$ N ∠58.2°.
(*c*) $\mathbf{A} = 240$ N ∠30°; $\mathbf{M}_A = 46.4$ N · m ↻.
3.70 (*a*) $\mathbf{A}_x = 78.8$ lb →; $\mathbf{A}_y = 314$ lb ↑;
$\mathbf{M}_A = 1260$ lb · ft ↻. (*b*) 123.7 lb; 43.9 lb.
SI: (*a*) $\mathbf{A}_x = 351$ N →;
$\mathbf{A}_y = 1397$ N ↑; $\mathbf{M}_A = 1708$ N · m ↻.
(*b*) 550 N; 195.3 N.
3.72 (*a*) 377 N. (*b*) 849 N ∠39.7°.
3.74 18.4°.
3.76 (*a*) $T = \frac{1}{2}W/(1 - \tan\theta)$. (*b*) 36.9°.
3.78 (*a*) $\cos \frac{1}{2}\theta = \frac{1}{4}\left(\frac{W}{P} + \sqrt{\frac{W^2}{P^2} + 8}\right)$.
(*b*) 65.0°.
3.80 (*a*) 0° or 180°. (*b*) 73.3°. (*c*) 90°.
(*d*) 26.6°. (*e*) 8 in. SI: (*e*) 203 mm.

3.82 (1) Completely constrained; det.; **A** = **C** = 50 lb ↑. (2) Completely constrained; det.; **B** = 0, **C** = **D** = 50 lb ↑. (3) Completely constrained; indet.; $\mathbf{A}_x$ = 75 lb →, $\mathbf{D}_x$ = 75 lb ←. (4) Partially constrained; det.; equil.; **C** = **D** = 50 lb ↑. (5) Completely constrained; det.; **B** = 75 lb →, **D** = 125 lb ⦪53.1°. (6) Partially constrained; no equil. (7) Improperly constrained; indet.; no equil. (8) Improperly constrained; indet.; no equil. (9) Completely constrained; indet.; **B** = 50 lb ↑, $\mathbf{D}_y$ = 50 lb ↑. SI: Substitute as follows: 50 lb → 222 N; 75 lb → 334 N; 125 lb → 556 N.

3.86 (*a*) 76.0°. (*b*) **A** = 145.2 N ↑; **C** = 582 N →.

3.88 10 in. SI: 254 mm.

3.90 **A** = 418 N ⦣44.2°; **B** = **C** = 758 N ⦩22.6°.

3.92 (*a*) 49.1°. (*b*) 1.323 *W*.

3.94 2.65 m.

3.96 23.2°.

3.98 **R** = 0; **M** = 504 lb · ft ⟲. SI: **R** = 0; **M** = 683 N · m ⟲.

3.100 (*a*) **B** = 230 lb ↙; **C** = 20 lb ↙; **D** = 150 lb ↑. (*b*) Rollers 1 and 3. SI: (*a*) **B** = 1023 N ↙; **C** = 89.0 N ↙; **D** = 667 N ↑. (*b*) Rollers 1 and 3.

3.102 (*a*) **A** = 534 N ⦣69.4°; **E** = 187.5 N ←. (*b*) 36.9° and 69.4°.

3.104 $-45\ \text{N}\cdot\text{m} \leqslant M \leqslant 3\ \text{N}\cdot\text{m}$.

3.106 $-49.1° \leqslant \alpha \leqslant 49.1°$.

3.108 $\mathbf{A}_x$ = 68.0 lb →, $\mathbf{A}_y$ = 119.2 lb ↑; *T* = 45.3 lb. SI: $\mathbf{A}_x$ = 302 N ←, $\mathbf{A}_y$ = 530 N ↑; *T* = 201 N.

CHAPTER 4

4.2 $M_x = 0$, $M_y = -1200$ lb · ft, $M_z = 2400$ lb · ft. SI: $M_x = 0$, $M_y = -1627$ N · m, $M_z = 3250$ N · m.

4.4 $M_x = 40$ N · m, $M_y = -90$ N · m, $M_z = -40$ N · m.

4.6 (*a*) $M_x = -480 \sin\theta$; $M_y = 0$; $M_z = -360$ lb · in. (*b*) $M_x = -416$ lb · in.; $M_y = 0$; $M_z = -360$ lb · in. (*c*) $M_x = +416$ lb · in.; $M_y = 0$; $M_z = -360$ lb · in. SI: (*a*) $M_x = -54.2$ N · m; $M_y = 0$; $M_z = -40.7$ N · m. (*b*) $M_x = -47.0$ N · m; $M_y = 0$; $M_z = -40.7$ N · m. (*c*) $M_x = +47.0$ N · m; $M_y = 0$; $M_z = -40.7$ N · m.

4.8 4.88 kN.

4.10 −280 lb · in. SI: −31.6 N · m.

4.14 $M_x = -180$ lb · in., $M_y = 80$ lb · in., $M_z = 60$ lb · in. SI: $M_x = -20.3$ N · m, $M_y = 9.04$ N · m, $M_z = 6.78$ N · m.

4.16 $M_x = 3.6$ kN · m, $M_y = 7.72$ kN · m, $M_z = 0$.

4.18 $R_x = 0$, $R_y = -100$ kN, $R_z = 0$; $M_x = -5$ kN · m, $M_y = 0$, $M_z = -12.5$ kN · m.

4.20 (*a*) $R_x = -600$ lb, $R_y = 600$ lb, $R_z = 300$ lb; $M_x = 0$, $M_y = -1200$ lb · ft, $M_z = 2400$ lb · ft. (*b*) **R** = same as *a*; $M_x = 5400$ lb · ft, $M_y = 4200$ lb · ft, $M_z = 2400$ lb · ft. SI: (*a*) $R_x = -2670$ N, $R_y = 2670$ N, $R_z = 1335$ N; $M_x = -1627$ N · m, $M_y = 0$, $M_z = 3250$ N · m. (*b*) **R** = same as *a*; $M_x = 7320$ N · m, $M_y = 5690$ N · m, $M_z = 3250$ N · m.

4.22 (*a*) $C_y = 15$ lb, $C_z = 0$. (*b*) $R_x = -10$ lb, $R_y = -15$ lb, $R_z = 0$; $M_x = -180$ lb · in., $M_y = M_z = 0$. SI: (*a*) $C_y = 66.7$ N, $C_z = 0$. (*b*) $R_x = -44.5$ N, $R_y = -66.7$ N, $R_z = 0$; $M_x = -20.3$ N · m, $M_y = M_z = 0$.

4.24 $P = 2.24$ kN; $\theta_x = 116.6°$, $\theta_y = 90°$, $\theta_z = 26.6°$.

4.26 72.2 N.

4.28 **A** = 4 kips ↓, **B** = 28 kips ↓. SI: **A** = 17.79 kN ↓, **B** = 124.5 kN ↓.

4.30 $\cos(AOB) = (P_3^2 - P_1^2 - P_2^2)/2P_1P_2$; $\cos(AOC) = (P_2^2 - P_1^2 - P_3^2)/2P_1P_3$.

4.32 (*a*) $R = 50$ N, $\theta_x = \theta_y = 90°$, $\theta_z = 0$; $M = 28$ N · m, $\phi_x = 90°$, $\phi_y = 153.4°$, $\phi_z = 116.6°$. (*b*) $R = 50$ N, $M = -12.5$ N · m; at $x = 500$ mm, $y = z = 0$; direction of axis, $\theta_x = \theta_y = 90°$, $\theta_z = 0$; pitch = −250 mm.

4.34 $R = 10$ lb, $M = 200$ lb · in.; at $x = 0$, $y = -14$ in., $z = 16$ in.; direction of axis, $\theta_x = 180°$, $\theta_y = \theta_z = 90°$; pitch = 20 in. SI: $R = 44.5$ N, $M = 22.6$ N · m; at $x = 0$, $y = -356$ mm, $z = 406$ mm; direction of axis, $\theta_x = 180°$, $\theta_y = \theta_z = 90°$; pitch = 508 mm.
4.36 $R = \sqrt{2}P$, $M = -Pa/\sqrt{2}$; at $x = 0$, $y = a$, $z = \frac{1}{2}a$; direction of axis, $\theta_x = \theta_y = 45°$, $\theta_z = 90°$.
4.38 $F_3 = -\frac{1}{2}P$; $R = \frac{3}{2}P$, at $x = 0$, $y = a$, $z = \frac{1}{2}a$; direction of line of action, $\theta_x = \theta_y = 48.2°$, $\theta_z = 109.5°$.
4.42 $A_x = 1040$ N, $A_y = -200$ N, $A_z = 0$, $T_{BD} = T_{BE} = 849$ N.
4.44 $T_{BC} = 780$ lb, $T_{BD} = 560$ lb; $A_x = 1200$ lb, $A_y = A_z = 0$. SI: $T_{BC} = 3470$ N, $T_{BD} = 2490$ N; $A_x = 5340$ N, $A_y = A_z = 0$.
4.46 $T_{AD} = 2.29$ kN, $T_{CD} = 2.07$ kN, $T_{ED} = 2.19$ kN.
4.48 $T_{BG} = 101.3$ lb, $T_{EF} = 315$ lb; $A_x = 338$ lb, $A_y = 101.2$ lb, $A_z = -22.5$ lb. SI: $T_{BG} = 451$ N, $T_{EF} = 1401$ N; $A_x = 1503$ N, $A_y = 450$ N, $A_z = -100.1$ N.
4.50 $T_A = 5$ lb; $T_B = 15$ lb; $T_C = 20$ lb. SI: $T_A = 22.2$ N; $T_B = 66.7$ N; $T_C = 89.0$ N.
4.52 50 lb; $x = 16$ in., $z = 13$ in. SI: 222 N; $x = 406$ mm, $z = 330$ mm.
4.54 $\frac{1}{3}L$.
4.56 $P = 462$ N; $A_x = A_z = 0$, $A_y = 123$ N; $B_y = 1139$ N, $B_z = 0$.
4.58 $A_x = -9600$ lb, $A_z = -2400$ lb; $B_x = 9600$ lb, $B_z = 2400$ lb; A_y and B_y indeterminate ($A_y + B_y = 12{,}000$ lb). SI: $A_x = -42.7$ kN, $A_z = -10.68$ kN; $B_x = 42.7$ kN, $B_z = 10.68$ kN; A_y and B_y indeterminate ($A_y + B_y = 53.4$ kN).
4.60 $F_{CD} = 184.2$ N; $A_x = 178.3$ N, $A_y = 322$ N; $B_x = 0$, $B_y = 368$ N, $B_z = 0$.
4.62 $F_{CE} = 168.3$ lb; $A_x = -46.9$ lb, $A_y = 0$; $B_x = -46.9$ lb, $B_y = 125$ lb, $B_z = -62.5$ lb. SI: $F_{CE} = 749$ N; $A_x = -209$ N, $A_y = 0$; $B_x = -209$ N, $B_y = 556$ N, $B_z = -278$ N.
4.64 $F_{CE} = 168.3$ lb; $B_x = -93.8$ lb, $B_y = 125$ lb, $B_z = -62.5$ lb; $(M_B)_y = -1875$ lb · in., $(M_B)_z = 0$. SI: $F_{CE} = 749$ N; $B_x = -417$ N, $B_y = 556$ N, $B_z = -278$ N; $(M_B)_y = -212$ N · m, $(M_B)_z = 0$.
4.66 $T_{BD} = 7.8$ kN, $T_{CF} = T_{BE} = 6.5$ kN; $A_x = 19.2$ kN, $A_y = 0$, $A_z = -3$ kN.
4.68 $A_x = 0$, $A_y = 400$ N, $A_z = -500$ N; $B_x = 600$ N, $B_y = 0$, $B_z = 500$ N; $C_x = -600$ N, $C_y = 400$ N, $C_z = 0$.
4.70 $M_x = 25$ N · m, $M_y = 360$ N · m, $M_z = -190$ N · m.
4.72 (*a*) 0.70 lb. (*b*) 3.20 lb. SI: (*a*) 3.11 N. (*b*) 14.23 N.
4.74 $(M_0h)/(Wr^2)$.
4.76 (*a*) $R = P$, $M = Qr\sin\theta$; parallel to the x axis at $x = 0$, $y = (rQ/P)\cos\theta$, $z = r\sin\theta$. (*b*) $(P/Q)^2y^2 + z^2 = r^2$.
4.78 $10W$; $x = 0.6a$, $z = 0.3a$.
4.80 (*a*) $R_x = 400$ N, $R_y = R_z = 0$; $M_x = 15$ N · m, $M_y = 66$ N · m, $M_z = -60$ N · m. (*b*) $R = 400$ N, $M = 15$ N · m; at $x = 0$, $y = 150$ mm, $z = 165$ mm, direction of axis $\theta_x = 0$, $\theta_y = \theta_z = 90°$; pitch = 37.5 mm.

CHAPTER 5

5.2 $\bar{x} = 55.4$ mm, $\bar{y} = 66.2$ mm.
5.4 $\bar{x} = 1.333$ in., $\bar{y} = 2.83$ in. SI: $x = 33.9$ mm, $\bar{y} = 71.9$ mm.
5.6 $\bar{x} = 0$, $\bar{y} = 3.23$ in. SI: $\bar{x} = 0$, $\bar{y} = 82.0$ mm.
5.8 $\bar{x} = 0$, $\bar{y} = 9.26$ mm.
5.10 $\bar{x} = 0$, $\bar{y} = 4.57$ ft. SI: $\bar{x} = 0$, $\bar{y} = 1.393$ m.
5.12 $\bar{x} = 152.5$ mm, $\bar{y} = 17.2$ mm.
5.14 $\dfrac{2}{3}\dfrac{\sin\alpha}{\alpha}\dfrac{(a^3 - b^3)}{(a^2 - b^2)}$.
5.16 $\bar{x} = 53.0$ mm, $\bar{y} = 68.6$ mm.
5.18 $\bar{x} = 0$, $\bar{y} = 2.94$ in. SI: $\bar{x} = 0$, $\bar{y} = 74.7$ mm.
5.20 0.520.
5.22 $a = 2.54r$; $\alpha = 51.8°$.
5.24 27.3 in. SI: 693 mm.
5.26 $\cos\alpha = \frac{1}{2}\csc\theta$.
5.28 $\bar{x} = 22.6$ mm, $\bar{y} = 0$.
5.30 $\bar{x} = 0.7424a$; -1.01%.
5.32 $\bar{x} = \frac{2}{3}b$; $\bar{y} = \frac{1}{3}h$.

5.34 $\bar{x} = \frac{1}{2}a$; $\bar{y} = \frac{2}{5}b$.
5.42 $\bar{x} = 0.611L$.
5.44 $\bar{x} = 0.549R$, $\bar{y} = 0$.
5.46 $\bar{x} = \bar{y} = 1.028$ in. SI: $\bar{x} = \bar{y} = 26.1$ mm.
5.48 $\bar{x} = \bar{y} = \dfrac{(n+1)^2}{(n+2)(2n+1)}$.
5.50 (a) $\frac{1}{2}\pi a^2 h$. (b) $8\pi a h^2/15$.
5.52 $4\pi R^2 \sin^2 (\phi/2)$.
5.54 0.655 m^3.
5.56 (a) 867 ft^3. (b) 1093 ft.^{3}.
SI: (a) 24.6 m^3. (b) 31.0 m^3.
5.58 $\bar{y} = 10.80$ in. SI: $\bar{y} = 274$ mm.
5.60 $V = 120.4 \times 10^3$ mm^3;
$A = 19.86 \times 10^3$ mm^2.
5.62 (a) $\pi R^2 h$. (b) $\frac{2}{3}\pi R^2 h$. (c) $\frac{1}{2}\pi R^2 h$.
(d) $\frac{1}{3}\pi R^2 h$. (e) $\frac{1}{6}\pi R^2 h$.
5.66 $\mathbf{R} = 6$ kN ↓, 3 m to the right of A;
$\mathbf{A} = 1.5$ kN ↑; $\mathbf{B} = 4.5$ kN ↑.
5.68 $\mathbf{A} = 825$ lb ↑; $\mathbf{M}_A = 6900$ lb · in. ↺.
SI: $\mathbf{A} = 3670$ N ↑; $\mathbf{M}_A = 780$ N · m ↺.
5.70 $\mathbf{B} = 1200$ N ↑; $\mathbf{M}_B = 800$ N · m ↺.
5.72 (a) 0.500; $\mathbf{M}_A = \frac{1}{4}w_A L^2$ ↻. (b) 1.000;
$\mathbf{A} = \frac{1}{3}w_A L$ ↑.
5.74 (a) $w_2 = w_1(a/L)[4 - 3(a/L)]$;
$w_3 = w_1(a/L)[3(a/L) - 2]$. (b) $a/L \geq \frac{2}{3}$.
5.76 (a) $\mathbf{H} = 17{,}970$ lb →, $\mathbf{V} = 60{,}100$ lb ↑;
12.63 ft to the right of A.
(b) $\mathbf{R} = 19{,}190$ lb ⦣ 20.6°.
SI: (a) $\mathbf{H} = 79.9$ kN →, $\mathbf{V} = 267$ kN ↑;
3.85 m to the right of A.
(b) $\mathbf{R} = 85.4$ kN ⦣ 20.6°.
5.78 8.12 ft. SI: 2.47 m.
5.80 90 mm.
5.82 7.83 ft. SI: 2.39 m.
5.84 $\mathbf{A}_x = 1.635$ kN →, $\mathbf{A}_y = 9.81$ kN ↓;
$\mathbf{D} = 8.18$ kN →.
5.86 1.491 m.
5.88 2.58 m.
5.90 407 lb ⦪ 23.2°. SI: 1810 N ⦪ 23.2°.
5.92 $\sqrt{3}$.
5.94 $\sqrt{3}$.
5.96 69.5 mm above the base.
5.98 $\bar{x} = \bar{z} = 0.1803$ in., $\bar{y} = 0$.
SI: $\bar{x} = \bar{z} = 4.58$ mm, $\bar{y} = 0$.
5.100 82.2 mm above the base.
5.102 $\bar{x} = 1.823$ mm, $\bar{y} = -39.3$ mm, $\bar{z} = 0$.
5.104 $\bar{x} = 5.45$ in., $\bar{y} = -1.505$ in., $\bar{z} = 0$.
SI: $\bar{x} = 138.4$ mm, $\bar{y} = -38.2$ mm,
$\bar{z} = 0$.
5.106 $\bar{x} = 0$, $\bar{y} = -h/3$, $\bar{z} = -4a/3\pi$.
5.110 $\bar{x} = 2h/3$, $\bar{y} = 16a/15\pi$, $\bar{z} = 0$.
5.112 $\bar{x} = \frac{3}{8}a$, $\bar{y} = \frac{1}{4}h$, $\bar{z} = \frac{3}{8}b$.
5.116 3.
5.118 $\bar{x} = a/2$, $\bar{y} = \pi^2 h/32$, $\bar{z} = b/2$.
5.120 $\bar{x} = 0$, $\bar{y} = 32.9$ mm.
5.122 105.7×10^3 mm^2.
5.124 $\bar{x} = 4.50$ in., $\bar{y} = 0.75$ in., $\bar{z} = 3.00$ in.
SI: $\bar{x} = 114.3$ mm, $\bar{y} = 19.05$ mm,
$\bar{z} = 76.2$ mm.
5.126 $\frac{1}{3}$
5.128 $W_2 = W_1/\cos\theta$ at point B.
5.130 60°.

CHAPTER 6

6.2 and 6.22 *1-2* = 3000 lb *T*; *1-3* = 3200 lb *T*;
2-3 = 4000 lb *C*. SI: *1-2* = 13.34 kN *T*;
1-3 = 14.23 kN *T*; *2-3* = 17.79 kN *C*.
6.4 and 6.24 *1-2* = 3600 N *C*; *1-3* = 3900 N *T*;
2-3 = 4500 N *C*.
6.6 and 6.26 *1-2* = *5-6* = *9-10* = 400 N *C*;
1-3 = *3-4* = *7-8* = *7-9* = 0;
2-3 = *7-10* = 849 N *C*;
2-4 = *4-6* = *6-8* = *8-10* = 600 N *T*;
3-5 = *5-7* = 800 N *C*;
3-6 = *6-7* = 283 N *T*.
6.8 and 6.28 *1-2* = 800 lb *C*; *1-3* = 600 lb *C*;
1-4 = 1000 lb *T*; *2-4* = *5-6* = 0;
3-4 = 1600 lb *C*; *3-5* = 1800 lb *C*;
3-6 = 2000 lb *T*; *4-6* = 600 lb *T*.
SI: *1-2* = 3.56 kN *C*; *1-3* = 2.67 kN *C*;
1-4 = 4.45 kN *T*; *2-4* = *5-6* = 0;
3-4 = 7.12 kN *C*; *3-5* = 8.01 kN *C*;
3-6 = 8.90 kN *T*; *4-6* = 2.67 kN *T*.
6.10 and 6.30 *1-2* = *6-8* = 8 kN *C*;
1-3 = *3-5* = *5-7* = *7-8* = 6.93 kN *T*;
2-3 = *4-5* = *6-7* = 4 kN *T*;
2-4 = *2-5* = *4-6* = *5-6* = 4 kN *C*.
6.12 and 6.32 *1-2* = *2-4* = *4-6* = *6-8* = 30 kips *T*; *1-3* = *3-5* = *5-7* = 34 kips *C*;
2-3 = *3-4* = *4-5* = *5-6* = 0;
6-7 = 24 kips *C*; *7-8* = 50 kips *C*.
SI: *1-2* = *2-4* = *4-6* = *6-8* = 133.4 kN *T*;
1-3 = *3-5* = *5-7* = 151.2 kN *C*;
2-3 = *3-4* = *4-5* = *5-6* = 0;
6-7 = 106.8 kN *C*; *7-8* = 222 kN *C*.
6.14 *2-3; 6-7; 6-9; 8-9; 8-11; 10-11.*
6.16 *1-6; 3-8; 4-5; 5-10; 7-12; 9-14.*
6.18 $F_{BE} = F_{BC} = F_{DC} = F_{DE} = P\sqrt{2}/4$ *comp.*; in all other members: $P\sqrt{2}/4$ *ten.*

6.20 (*a*) $A_x = +1600$ lb, $A_y = -533$ lb, $A_z = -800$ lb; $B_x = 0$; $C_x = -1600$ lb, $C_y = +533$ lb. (*b*) $F_{AC} = 400$ lb C; $F_{AE} = 1733$ lb C; $F_{CE} = 1733$ lb T; for all other members: $F = 0$.
SI: (*a*) $A_x = +7.12$ kN, $A_y = -2.37$ kN, $A_z = -3.56$ kN; $B_x = 0$; $C_x = -7.12$ kN, $C_y = +2.37$ kN. (*b*) $F_{AC} = 1.779$ kN T; $F_{AE} = 7.71$ kN C; $F_{CE} = 7.71$ kN T; for all other members: $F = 0$.

6.34 *1-2* = 1118 lb C; *1-5* = 1000 lb T; *2-3* = *3-4* = 894 lb C; *2-5* = 224 lb C; *3-5* = 200 lb C; *4-5* = 361 lb T; *5-6* = 600 lb T; symmetrical about center. SI: *1-2* = 4970 N C; *1-5* = 4450 N T; *2-3* = *3-4* = 3980 N C; *2-5* = 996 N C; *3-5* = 890 N C; *4-5* = 1606 N T; *5-6* = 2670 N T; symmetrical about center.

6.36 *1-2* = *1-8* = 268 N C; *1-3* = *1-4* = *1-5* = *1-6* = *1-7* = 536 N C; *2-3* = *3-4* = *4-5* = *5-6* = *6-7* = *7-8* = 1035 N T.

6.38 *1-2* = *2-4* = 0; *1-3* = 4040 lb T; *1-4* = 4040 lb C; *3-4* = 1500 lb T; *3-5* = 5390 lb T; *3-6* = 1601 lb C; *4-6* = 3750 lb C; *5-6* = 1000 lb T.
SI: *1-2* = *2-4* = 0; *1-3* = 17.97 kN T; *1-4* = 17.97 kN C; *3-4* = 6.67 kN T; *3-5* = 24.0 kN T; *3-6* = 7.12 kN C; *4-6* = 16.68 kN C; *5-6* = 4.45 kN T.

6.40 $F_{CE} = 48$ kips C; $F_{CF} = 10$ kips T. SI: $F_{CE} = 214$ kN C; $F_{CF} = 44.5$ kN T.

6.42 $F_{DF} = 91.4$ kN T; $F_{DE} = 38.6$ kN C.

6.44 60 kN T.

6.46 $F_{EG} = 45$ kN T; $F_{EF} = 55$ kN C; $F_{DF} = 45$ kN C.

6.48 $F_{EC} = 4690$ lb T; $F_{CD} = 3600$ lb C; $F_{CB} = 0$. SI: $F_{EC} = 20.9$ kN T; $F_{CD} = 16.01$ kN C; $F_{CB} = 0$.

6.50 $F_{DF} = 26$ kips C; $F_{DG} = 15$ kips T. SI: $F_{DF} = 115.6$ kN C; $F_{DG} = 66.7$ kN T.

6.52 $F_{AB} = 5P/6$ *ten.*; $F_{KL} = 7P/6$ *ten.*

6.54 22.5 kN C.

6.56 $F_{AB} = 0$; $F_{EJ} = \frac{2}{3}P$ *comp.*

6.58 $F_{BD} = 21.3$ kN C; $F_{CE} = 10.67$ kN T. $F_{CD} = 13.33$ kN T.

6.60 $F_{BE} = 16$ kips C; $F_{CG} = 10$ kips C; $F_{BG} = 10$ kips T. SI: $F_{BE} = 71.2$ kN C; $F_{CG} = 44.5$ kN C; $F_{BG} = 44.5$ kN T.

6.62 (*a*) Completely constrained; det. (*b*) Completely constrained; indet. (*c*) Completely constrained; det. (*d*) Improperly constrained.

6.64 (*a*) Completely constrained; det. (*b*) Improperly constrained. (*c*) Completely constrained; indet. (*d*) Completely constrained; det.

6.66 $F_{BD} = 450$ lb T; $\mathbf{C}_x = 270$ lb ←, $\mathbf{C}_y = 240$ lb ↓. SI: $F_{BD} = 2000$ N T; $\mathbf{C}_x = 1201$ N ←, $\mathbf{C}_y = 1068$ N ↓.

6.68 $\mathbf{A}_x = 560$ N →, $\mathbf{A}_y = 160$ N ↑; $\mathbf{B}_x = 560$ N ←, $\mathbf{B}_y = 280$ N ↓; $\mathbf{C} = 160$ N ↓; $\mathbf{D} = 280$ N ↑.

6.70 $\mathbf{E}_x = 25$ lb →, $\mathbf{E}_y = 55$ lb ↓. SI: $\mathbf{E}_x = 111.2$ N →, $\mathbf{E}_y = 245$ N ↓.

6.72 (*a*) $\mathbf{B}_x = 500$ N →, $\mathbf{B}_y = 200$ N ↓. (*b*) $\mathbf{B} = 200$ N ↑.

6.74 $\mathbf{D}_x = 450$ lb →, $\mathbf{D}_y = 600$ lb ↓; $\mathbf{E}_x = 0$, $\mathbf{E}_y = 1200$ lb ↑; $\mathbf{F}_x = 450$ lb ←, $\mathbf{F}_y = 600$ lb ↓. SI: $\mathbf{D}_x = 2.00$ kN →, $\mathbf{D}_y = 2.67$ kN ↓; $\mathbf{E}_x = 0$, $\mathbf{E}_y = 5.34$ kN ↑; $\mathbf{F}_x = 2.00$ kN ←, $\mathbf{F}_y = 2.67$ kN ↓.

6.76 $\mathbf{E} = 150$ lb ↑; $\mathbf{F} = 450$ lb ↑; $\mathbf{C} = 900$ lb ↓. SI: $\mathbf{E} = 0.667$ kN ↑; $\mathbf{F} = 2.00$ kN ↑; $\mathbf{C} = 4.00$ kN ↓.

6.78 $\mathbf{A} = 0$, $\mathbf{M}_A = 1500$ lb · ft ↻; $\mathbf{D} = 500$ lb ↑. SI: $\mathbf{A} = 0$, $\mathbf{M}_A = 2.03$ kN · m ↻; $\mathbf{D} = 2.22$ kN ↑.

6.80 (*a*) 79.4°. (*b*) $\mathbf{A} = 573$ lb ←; $\mathbf{B} = 508$ lb ⦣79.4°. SI: (*a*) 79.4°. (*b*) $\mathbf{A} = 2.55$ kN ←; $\mathbf{B} = 2.26$ kN ⦣79.4°.

6.82 (*a*) *At each wheel:* $\mathbf{A} = 20{,}200$ lb ↑; $\mathbf{B} = 29{,}800$ lb ↑. (*b*) $\mathbf{C} = 2920$ lb →; $\mathbf{D}_x = 2920$ lb ←, $\mathbf{D}_y = 43{,}600$ lb ↓. SI: (*a*) *At each wheel:* $\mathbf{A} = 89.9$ kN ↑; $\mathbf{B} = 132.6$ kN ↑. (*b*) $\mathbf{C} = 12.99$ kN →; $\mathbf{D}_x = 12.99$ kN ←, $\mathbf{D}_y = 193.9$ kN ↓.

6.84 *At each wheel:* (*a*) $\mathbf{A} = 1268$ lb ↑; $\mathbf{B} = 713$ lb ↑; $\mathbf{C} = 994$ lb ↑. (*b*) $\mathbf{B} = 73$ lb ↑; $\mathbf{C} = 34$ lb ↑. SI: *At each wheel:* (*a*) $\mathbf{A} = 5640$ N ↑; $\mathbf{B} = 3170$ N ↑; $\mathbf{C} = 4420$ N ↑. (*b*) $\mathbf{B} = 324$ N ↑; $\mathbf{C} = 151$ N ↑.

6.86 $\mathbf{A} = 960$ N →; $\mathbf{B} = 160$ N ←; $\mathbf{C} = 1000$ N ⦣36.9°.

6.88 $F_{AF} = \frac{1}{4}P$ *comp.*;

$F_{BG} = F_{GD} = P/\sqrt{2}$ *ten.*;
$F_{EH} = \frac{1}{4}P$ *ten.*

6.90 $F_{AF} = M_0/4a$ *ten.*;
$F_{BG} = F_{GD} = M_0/a\sqrt{2}$ *comp.*;
$F_{EH} = 3M_0/4a$ *ten.*

6.92 (*a*) Rigid; $\mathbf{A} = 20\ \text{kN}\uparrow$; $\mathbf{B} = 20\ \text{kN}\downarrow$;
$\mathbf{C} = 40\ \text{kN}\uparrow$. (*b*) Not rigid.

6.94 (*a*) Rigid; $\mathbf{A} = 30\ \text{kN}\uparrow$; $\mathbf{B} = 40\ \text{kN}\downarrow$;
$\mathbf{C} = 50\ \text{kN}\uparrow$. (*b*) Not rigid.

6.96 (*a*) $\mathbf{A} = 550\ \text{lb}\rightarrow$; $\mathbf{B} = 1050\ \text{lb}\leftarrow$;
$\mathbf{D} = 788\ \text{lb}\uparrow$; $\mathbf{E} = 412\ \text{lb}\uparrow$.
(*b*) $\mathbf{C}_x = 550\ \text{lb}\leftarrow$, $\mathbf{C}_y = 412\ \text{lb}\downarrow$.
SI: (*a*) $\mathbf{A} = 2.45\ \text{kN}\rightarrow$; $\mathbf{B} = 4.67\ \text{kN}\leftarrow$;
$\mathbf{D} = 3.51\ \text{kN}\uparrow$; $\mathbf{E} = 1.833\ \text{kN}\uparrow$.
(*b*) $\mathbf{C}_x = 2.45\ \text{kN}\leftarrow$, $\mathbf{C}_y = 1.833\ \text{kN}\downarrow$.

6.98 $\mathbf{A} = 2.5\ \text{kips}\uparrow$; $\mathbf{B} = 1.5\ \text{kips}\uparrow$;
$\mathbf{C} = 5\ \text{kips}\uparrow$; $\mathbf{D} = 3\ \text{kips}\uparrow$.
SI: $\mathbf{A} = 11.12\ \text{kN}\uparrow$; $\mathbf{B} = 6.67\ \text{kN}\uparrow$;
$\mathbf{C} = 22.2\ \text{kN}\uparrow$; $\mathbf{D} = 13.34\ \text{kN}\uparrow$.

6.100 $\mathbf{A} = P/15\uparrow$; $\mathbf{D} = 2P/15\uparrow$; $\mathbf{E} = 8P/15\uparrow$;
$\mathbf{H} = 4P/15\uparrow$.

6.102 $\mathbf{D} = 400\ \text{N}\downarrow$; $\mathbf{E} = 800\ \text{N}\uparrow$; $\mathbf{F} = 200\ \text{N}\uparrow$.

6.104 $\mathbf{D} = 1179\ \text{lb}\ \angle 58.0°$; $\mathbf{C} = 625\ \text{lb}\rightarrow$.
SI: $\mathbf{D} = 5.24\ \text{kN}\ \angle 58.0°$;
$\mathbf{C} = 2.78\ \text{kN}\rightarrow$.

6.106 (*a*) 2500 N. (*b*) 2760 N $\angle 63.1°$.

6.108 (*a*) 1680 lb. (*b*) 4200 lb.
SI: (*a*) 7.47 kN. (*b*) 18.68 kN.

6.110 (*a*) $36.8\ \text{N}\cdot\text{m}$ ↺. (*b*) $\mathbf{C}_x = 120\ \text{N}\leftarrow$,
$\mathbf{C}_y = 160\ \text{N}\uparrow$.

6.112 1689 lb. SI: 7.51 kN.

6.114 $\mathbf{A}_x = 1100\ \text{lb}\leftarrow$, $\mathbf{A}_y = 4950\ \text{lb}\downarrow$;
$\mathbf{C}_x = 1225\ \text{lb}\rightarrow$, $\mathbf{C}_y = 4950\ \text{lb}\uparrow$.
SI: $\mathbf{A}_x = 4.89\ \text{kN}\leftarrow$, $\mathbf{A}_y = 22.0\ \text{kN}\downarrow$;
$\mathbf{C}_x = 5.45\ \text{kN}\rightarrow$, $\mathbf{C}_y = 22.0\ \text{kN}\uparrow$.

6.116 1540 N.

6.118 $F_{DE} = 1897$ lb *C*; $F_{EF} = 4200$ lb *T*.
SI: $F_{DE} = 8.44$ kN *C*; $F_{EF} = 18.68$ kN *T*.

6.120 (*a*) 18.75 kN *T*. (*b*) $\mathbf{H}_x = 10\ \text{kN}\leftarrow$,
$\mathbf{H}_y = 20\ \text{kN}\downarrow$.

6.122 50 lb. SI: 222 N.

6.124 (*a*) $-21.0\ \text{N}\cdot\text{m}$. (*b*) $G = 0$,
$M_G = 2.75\ \text{N}\cdot\text{m}$; $H = 0$,
$M_H = -11.70\ \text{N}\cdot\text{m}$; couples parallel to x axis.

6.126 (*a*) 1.5. (*b*) $4\,M_A$ ↻.

6.128 (*a*) $M_A = 43.3\ \text{N}\cdot\text{m}$.
(*b*) $B_x = B_y = D_x = D_y = E_x = E_y = 0$,
$B_z = -250$ N, $D_z = 350$ N,
$E_z = -100$ N.

6.130 5.29 oz. SI: 150.0 g (mass).

6.132 18.4°.

6.134 $F_{DF} = 40$ kN *C*; $F_{EF} = 90$ kN *C*;
$F_{EG} = 247$ kN *T*.

6.136 *On the frame:* $\mathbf{F} = 1611\ \text{lb}\uparrow$;
$\mathbf{A} = 500\ \text{lb}\leftarrow$; $\mathbf{D}_x = 500\ \text{lb}\rightarrow$,
$\mathbf{D}_y = 861\ \text{lb}\downarrow$. SI: *On the frame:*
$\mathbf{F} = 7.17\ \text{kN}\uparrow$; $\mathbf{A} = 2.22\ \text{kN}\leftarrow$;
$\mathbf{D}_x = 2.22\ \text{kN}\rightarrow$, $\mathbf{D}_y = 3.83\ \text{kN}\downarrow$.

6.138 $F_{AB} = F_{BC} = 1600$ lb *T*; $F_{AE} = F_{BF} = 0$;
$F_{CF} = 1200$ lb *C*; $F_{CG} = 2000$ lb *T*;
$F_{DE} = 1600$ lb *C*.
SI: $F_{AB} = F_{BC} = 7.12$ kN *T*;
$F_{AE} = F_{BF} = 0$; $F_{CF} = 5.34$ kN *C*;
$F_{CG} = 8.90$ kN *T*; $F_{DE} = 7.12$ kN *C*.

6.140 $45.4\ \text{N}\cdot\text{m}$ ↺.

6.142 $\mathbf{D} = 1209\ \text{lb}\ \angle 7.1°$; $F_{BF} = 1000$ lb *T*;
$F_{CE} = 500$ lb *C*. SI: $\mathbf{D} = 5.38\ \text{kN}\ \angle 7.1°$;
$F_{BF} = 4.45$ kN *T*; $F_{CE} = 2.22$ kN *C*.

CHAPTER 7

7.2 (On *AJ*) $\mathbf{F} = 560\ \text{N}\rightarrow$; $\mathbf{V} = 160\ \text{N}\uparrow$;
$\mathbf{M} = 16\ \text{N}\cdot\text{m}$ ↻.

7.4 (On *DJ*) $\mathbf{F} = 450\ \text{lb}\leftarrow$; $\mathbf{V} = 600\ \text{lb}\uparrow$;
$\mathbf{M} = 1200\ \text{lb}\cdot\text{ft}$ ↻.
SI: (On *DJ*) $\mathbf{F} = 2.00\ \text{kN}\leftarrow$;
$\mathbf{V} = 2.67\ \text{kN}\uparrow$; $\mathbf{M} = 1.627\ \text{kN}\cdot\text{m}$ ↻.

7.6 (On *CA*) $\mathbf{F} = 300\ \text{N}\leftarrow$; $\mathbf{V} = 50\ \text{N}\downarrow$;
$\mathbf{M} = 40\ \text{N}\cdot\text{m}$ ↺.

7.8 (On *CB*) (*a*) $\mathbf{F} = \frac{1}{2}wL\csc\theta \nearrow$; $\mathbf{V} = 0$;
$\mathbf{M} = \frac{1}{8}wL^2\cos\theta$ ↻. (*b*) $\mathbf{F} = 250\ \text{lb}\nearrow$;
$\mathbf{V} = 0$; $\mathbf{M} = 450\ \text{lb}\cdot\text{ft}$ ↻.
SI: (*b*) $\mathbf{F} = 1112\ \text{N}\nearrow$; $\mathbf{V} = 0$;
$\mathbf{M} = 610\ \text{N}\cdot\text{m}$ ↻.

7.10 (On left section) $\mathbf{F} = 1028\ \text{N}\swarrow$;
$\mathbf{V} = 79.1\ \text{N}\nwarrow$; $\mathbf{M} = 30\ \text{N}\cdot\text{m}$ ↻.

7.12 (On *JC*) $95.5\ \text{lb}\cdot\text{in.}$ ↺.
SI: (On *JC*) $10.79\ \text{N}\cdot\text{m}$ ↺.

7.14 (On *JC*) $\mathbf{F} = (W\theta/\pi)\cos\theta \nearrow$;
$\mathbf{V} = (W\theta/\pi)\sin\theta \nwarrow$;
$\mathbf{M} = (Wr/\pi)(\sin\theta - \theta\cos\theta)$ ↺.

7.16 $V = 0.1786W$ at $\theta = 40.7°$.

7.18 $M_{\max} = \frac{1}{8}wL^2$.

7.20 $V = 0$ everywhere; $M = +T$ between B and C; $M = 0$ elsewhere.

7.22 $M_B = -PL/2$; $M_C = -3PL/2$.

7.24 $M_C = -3500\ \text{lb}\cdot\text{ft}$.
SI: $M_C = -4.75\ \text{kN}\cdot\text{m}$.

7.26 $M_D = +28$ N · m.
7.28 24 kN.
7.30 $M = -5000$ lb · ft at center.
SI: $M = -6.78$ kN · m at center.
7.32 Just to the left of D: $M = +20$ N · m.
7.34 $M_C = +1500$ lb · ft.
SI: $M_C = +2.03$ kN · m.
7.36 (*a*) $\frac{1}{3}P$. (*b*) $\frac{1}{4}P$.
7.46 $M_D = +48$ kN · m.
7.48 $M_D = +530$ lb · in.
SI: $M_D = +59.9$ N · m.
7.50 $M = +2000$ lb · ft, 10 ft to the right of A. SI: $M = +2.71$ kN · m, 3.05 m to the right of A.
7.52 $M = -2$ kN · m at C;
$M = +0.667$ kN · m, 1.333 m to the left of B.
7.54 $V = w(\frac{1}{2}L - x)$; $M = \frac{1}{2}w(Lx - x^2)$.
7.56 $V = \frac{1}{2}w_0L[\frac{1}{3} - (x/L)^2]$;
$M = \frac{1}{6}w_0L^2[x/L) - (x/L)^3]$;
$M_{\text{max}} = +0.0641w_0L^2$, at $x = 0.577L$.
7.58 (*a*) 0.500.
(*b*) $V = \frac{1}{4}w_1L[1 - 4(x/L) + 3(x/L)^2]$;
$M = \frac{1}{4}w_1L^2[(x/L) - 2(x/L)^2 + (x/L)^3]$.
(*c*) $M_{\text{max}} = w_1L^2/27$, at $x = L/3$.
7.60 $P = 100$ lb; $Q = 50$ lb; $V_A = +230$ lb.
SI: $P = 445$ N; $Q = 222$ N;
$V_A = +1023$ N.
7.62 (*a*) $\mathbf{E}_x = 750$ lb →, $\mathbf{E}_y = 500$ lb ↑.
(*b*) 901 lb. SI: (*a*) $\mathbf{E}_x = 3.34$ kN →,
$\mathbf{E}_y = 2.22$ kN ↑. (*b*) 4.01 kN.
7.64 $\mathbf{E}_x = 1600$ N →. $\mathbf{E}_y = 1100$ N ↑.
7.66 1 m.
7.68 3575 ft. SI: 1090 m.
7.70 0.831 m.
7.72 (*a*) $\sqrt{3L\Delta/8}$. (*b*) 12.25 ft. SI: (*b*) 3.73 m.
7.74 8 m to left of B; 14.72 kN.
7.76 $x > 0$, $y = 8h(x/L)^3$;
$x < 0$, $y = -8h(x/L)^3$; $T_0 = w_0L^2/24h$.
7.80 $y = h(1 - \cos \pi x/L)$;
$T_{\text{max}} = (w_0L/\pi)[1 + (L/h\pi)^2]^{1/2}$;
$T_0 = w_0L^2/h\pi^2$.
7.84 89.0 ft; 69.0 lb. SI: 27.1 m; 307 N.
7.86 (*a*) 102.8 m. (*b*) 10.4 m. (*c*) 977 N.
7.88 (*a*) 11.27 m. (*b*) 29.5 kg/m.
7.90 50.3 ft; 55.5 lb. SI: 15.33 m; 247 N.
7.92 (*a*) 8.28 lb. (*b*) 6.69 ft. (*c*) 21.64 lb.
SI: (*a*) 36.8 N. (*b*) 2.04 m. (*c*) 96.3 N.
7.94 (*a*) $L = 1.326T_m/w$. (*b*) 7.53 mi.
SI: (*b*) 12.12 km.
7.96 0.1408.
7.98 (*a*) $1.029(2\pi r)$. (*b*) $0.657wr$.
7.100 $M_C = +750$ lb · ft.
SI: $M_C = +1017$ N · m.
7.102 $M_C = M_D = M_E = M_F = -12.5$ kip · ft.
SI: $M_C = M_D = M_E =$
$M_F = -16.95$ kN · m.
7.104 (*a*) 395 N. (*b*) 1456 N.
7.106 $a = 0.293L$.
7.108 (On AJ) $\mathbf{F} = 10.40$ kN ↖;
$\mathbf{V} = 0.928$ kN ↙; $\mathbf{M} = 0.5$ kN · m ↻.
7.110 (On JB) (*a*) $\mathbf{M}_J = \frac{1}{2}Wr$ ↻.
(*b*) $\mathbf{M}_{\text{max}} = 0.579Wr$ ↻, $\theta = 116.3°$.

CHAPTER 8

8.2 (*a*) 930 N ←. (*b*) 151.2 N ←.
8.4 $P = W \sin (\alpha + \phi_s)$; $\theta = \alpha + \phi_s$.
8.6 A and B move; $\mathbf{F}_A = 9.48$ N ↗;
$\mathbf{F}_B = 3.79$ N ↗; $\mathbf{F}_C = 12.71$ N ↗.
8.8 $\theta \geqslant 29.4°$.
8.10 (*a*) 36 lb. (*b*) 40 in.
SI: (*a*) 160.1 N. (*b*) 1.016 m.
8.12 (*a*) 66.4°. (*b*) 512 N.
8.14 40 lb →. SI: 177.9 N →.
8.16 (*a*) 54.5 N · m. (*b*) 46.2 N · m.
8.18 36.6°.
8.20 (*b*) until $b = 2.26$ ft.
SI: until $b = 0.689$ m.
8.22 371 N (motion impends at C).
8.24 600 lb. SI: 2.67 kN.
8.26 92.4 mm.
8.28 0.80.
8.30 $\mu_A = 0.222$; $\mu_C = 0.247$.
8.32 (*a*) 10.8 lb (slips at A). (*b*) 13.5 lb (slips at C). SI: (*a*) 48.0 N. (*b*) 60.1 N.
8.34 (*a*) $\frac{2}{3}L$.
8.36 (*a*) $P = W \tan \phi/\cos \theta$.
(*b*) $M = \frac{1}{2}Wr \sin 2\phi/\cos \theta$.
8.38 60.1 N · m ↺.
8.40 0.520 lb ↗; A rolls. SI: 2.31 N ↗; A rolls.
8.42 $\theta \leqslant 45°$.
8.44 54.4°.
8.46 1454 N.
8.48 474 lb. SI: 2.11 kN.
8.50 190.4 N.
8.52 80.9°.
8.54 (*a*) $\mathbf{P} = 41.7$ kN ←. (*b*) 30 kN →.

8.56 (a) 58.5 lb ←. (b) Machine will not move. SI: (a) 260 N ←.
8.58 832 lb · ft. SI: 1128 N · m.
8.60 16.88 N · m.
8.62 (a) 9440 lb · ft. (b) 2280 lb · ft. SI: (a) 12.80 kN · m. (b) 3.09 kN · m.
8.64 3.60 N · m.
8.66 14.60 lb · in. SI: 1.650 N · m.
8.68 (a) 0.24. (b) 55.6 lb. SI: (b) 247 N.
8.70 0.10.
8.72 $T_{AB} = 310$ N; $T_{CD} = 290$ N; $T_{EF} = 272$ N.
8.74 5.0 in. SI: 127 mm.
8.76 3.47 lb. SI: 15.44 N.
8.78 514 N; 1200 N.
8.82 2.60 lb. SI: 11.57 N.
8.84 58.9 N.
8.86 12 in. SI: 305 mm.
8.88 (a) 0.33. (b) 2.66 turns.
8.90 0.116
8.92 89.9 lb · ft. SI: 121.9 N · m.
8.94 88.8 N · m ↺.
8.96 109.5 kg.
8.98 13.17 ft. SI: 4.01 m.
8.100 0.253.
8.102 531 N · m.
8.106 0.27.
8.108 (a) 160.2 lb. (b) 202.8 lb. SI: (a) 713 N. (b) 902 N.
8.110 (a) 235 N. (b) 207 N.
8.112 (a) 26.0 N. (b) Motion impends at B.
8.114 412 lb. SI: 1833 N.

CHAPTER 9

9.2 $2a^3b/7$.
9.4 $a^3b/20$.
9.6 $2ab^3/15$.
9.8 $ab^3/28$.
9.10 $ab^3/30$; $b\sqrt{2/15}$.
9.12 $a^3b/6$; $a\sqrt{2/3}$.
9.14 $(5\sqrt{3}/48)a^4$; $a\sqrt{5/12}$.
9.16 (b) 10.6%; 2.99%; 0.125%.
9.18 $4a^3/9$.
9.20 614×10^3 mm^4; 19.01 mm.
9.22 382 in^4; 4.04 in. SI: 159.0×10^6 mm^4; 102.6 mm.
9.24 6 in.; 180 in^4. SI: 152.4 mm; 74.9×10^6 mm^4.
9.26 $\bar{I}_x = 1.5 \times 10^6$ mm^4, $\bar{I}_y = 3 \times 10^6$ mm^4.
9.28 60.7×10^6 mm^4.
9.30 $\bar{I}_x = \bar{I}_y = 2.08 \times 10^6$ mm^4.
9.32 $\bar{I}_x = 313$ in^4, $\bar{I}_y = 371$ in^4; $\bar{k}_x = 3.80$ in., $\bar{k}_y = 4.13$ in. SI: $\bar{I}_x = 130.3 \times 10^6$ mm^4, $\bar{I}_y = 154.4 \times 10^6$ mm^4; $\bar{k}_x = 96.5$ mm, $\bar{k}_y = 104.9$ mm.
9.34 6123 in^4, 1361 in^4; 7.91 in., 3.73 in. SI: 2550×10^6 mm^4, 567×10^6 mm^4; 201 mm, 94.7 mm.
9.36 $3\pi r/16$.
9.38 $(a + 3b)h/(2a + 4b)$.
9.40 (a) $\frac{1}{2}\gamma\pi r^2 h$. (b) $\frac{1}{4}\gamma\pi r^4$.
9.44 $b^2h^2/8$.
9.46 $a^2b^2/4(n + 1)$.
9.48 0.80×10^6 mm^4.
9.50 +16 in^4. SI: $+6.66 \times 10^6$ mm^4.
9.52 1.627×10^6 mm^4, 0.881×10^6 mm^4; -0.954×10^6 mm^4.
9.54 $I_u = I_v = 5a^4/24$; $P_{uv} = -a^4/8$.
9.56 −25.7°; 2.28×10^6 mm^4, 0.230×10^6 mm^4.
9.58 +28.5°; 98.3 in^4, 21.3 in^4. SI: +28.5°; 40.9×10^6 mm^4, 8.87×10^6 mm^4.
9.62 1.627×10^6 mm^4, 0.881×10^6 mm^4; -0.954×10^6 mm^4.
9.66 −25.7°; 2.28×10^6 mm^4, 0.230×10^6 mm^4.
9.72 (a) $I_{AA'} = mb^2/24$; $I_{BB'} = mh^2/18$. (b) $I_{CC'} = m(3b^2 + 4h^2)/72$.
9.74 (a) $\frac{1}{4}m(r_1^2 + r_2^2)$. (b) $\frac{1}{2}m(r_1^2 + r_2^2)$.
9.76 $ma^2/20$.
9.78 $\frac{1}{3}ma^2$; $a/\sqrt{3}$.
9.80 $\frac{1}{6}m(a^2 + 3h^2)$; $[(a^2 + 3h^2)/6]^{1/2}$.
9.82 $\frac{1}{20}m(a^2 + 12h^2)$.
9.84 $\frac{3}{10}m\dfrac{r_2^5 - r_1^5}{r_2^3 - r_1^3}$.
9.86 1.514 kg · m^2; 155.8 mm.
9.88 (a) $m(\frac{2}{5}a^2 + R^2)$. (b) $(a/R)^2/[(a/R)^2 + 2.5]$. (c) $9.98a$.
9.90 $5ml^2/24$, $ml^2/12$, $7ml^2/24$.
9.92 0.000283 lb · ft · s^2; 1.031 in. SI: 0.384×10^{-3} kg · m^2; 26.2 mm.
9.94 (a) 1.575 lb · ft · s^2. (b) 2.714 lb · ft · s^2. (c) 1.202 lb · ft · s^2. SI: (a) 2.13 kg · m^2. (b) 3.68 kg · m^2. (c) 1.629 kg · m^2.
9.96 $\frac{1}{6}m(3a^2 + 2L^2)$.
9.98 $3a^2h^2/16$.
9.100 $a\sqrt{2}/4$.

9.102 69.6 in^4, 11.64 in^4; 2.52 in., 1.029 in.
SI: 29.0 × 10^6 mm^4, 4.84 × 10^6 mm^4;
64.0 mm, 26.1 mm.
9.104 (*a*) $\frac{1}{4}ma^2$, $\frac{1}{4}mb^2$. (*b*) $\frac{1}{4}m(a^2 + b^2)$.
9.106 1.305 × 10^{-3} kg · m^2.

CHAPTER 10

10.2 30 lb. SI: 133.4 N.
10.4 900 N. ←.
10.6 $\frac{3}{2}P \tan \theta$.
10.8 $P \cos \theta / \sin \frac{1}{2}\theta$.
10.10 $M = 2Wl \sin \theta$.
10.12 50 lb. SI: 222 N.
10.14 541 N.
10.16 69.4°.
10.18 50.4°.
10.20 335 mm.
10.22 51.3°.
10.24 3.49 *W*.
10.26 78.7°; 323.8°; 379.1°.
10.30 $\frac{1}{2}Pl \tan \theta / (1 - \mu \tan \theta)$.
10.32 250 N ↑; 450 N · m ↺.
10.34 600 N.
10.36 0.625 in. ↑. SI: 15.88 mm ↑.
10.44 45°, stable; −135°, unstable.
10.46 0°, stable; 60°, unstable.
10.48 78.7° and 379.1°, stable; 323.8°, unstable.
10.50 10.3°, stable; 33.1°, unstable; 90°, stable.
10.52 (*b*) 65.1°.
10.54 10.9°; 39.5°.
10.56 $k > 4.91$ kN/m.
10.58 $P < kl/2$.
10.60 $k \leq 3P/4l$.
10.62 (*a*) $P < ka^2/2l$. (*b*) $P < ka^2/2l$.
10.64 $P < kl/3$.
10.66 23.9 kN.
10.68 $Q = P \cos \theta / \cos (\beta - \theta)$.
10.70 (*a*) 200 lb. (*b*) 10 in.
10.72 $Q = M \cos^2 \theta / r \sin \theta$.
10.74 27.0°; 11.56 in. SI: 27.0°; 294 mm.
10.76 63.4°.

CHAPTER 11

11.2 $t = 0$, $x = 15$ in., $a = -12$ in./s^2;
$t = 2$ s, $x = 7$ in., $a = 12$ in./s^2.
SI: $t = 0$; $x = 0.381$ m, $a = -0.305$ m/s^2;
$t = 2$ s, $x = 0.1778$ m, $a = 0.305$ m/s^2.
11.4 (*a*) 2 s, 4 s. (*b*) 8 m, 7.33 m.
11.6 −4 m/s; 12 m; 20 m.
11.8 (*a*) 3 ft/s^4. (*b*) $a = 3t^2$; $v = t^3 - 32$;
$x = \frac{1}{4}t^4 - 32t + 64$. SI: (*a*) 0.914 m/s^4.
11.10 25 s^{-2}.
11.12 (*a*) ±3.34 ft/s. (*b*) 7.59 ft.
SI: (*a*) ±1.018 m/s. (*b*) 2.31 m.
11.14 (*a*) 55.5 m. (*b*) Infinite.
11.16 142.7 ft/s. SI: 43.5 m/s.
11.20 $\sqrt{2gR}$.
11.22 (*a*) 6.90 m/s. (*b*) Infinite.
11.24 (*a*) 29.1 m. (*b*) 24.4 m/s.
11.26 (*a*) −3.20 ft/s^2. (*b*) 34.0 ft/s.
SI: (*a*) −0.975 m/s^2. (*b*) 10.36 m/s.
11.28 (*b*) 1.242 s; 24.8 ft. SI: (*b*) 1.242 s; 7.56 m.
11.30 1.200 m.
11.32 (*a*) 3 ft/s^2 ↑, 6 ft/s^2 ↓. (*b*) 12 ft/s^2 ↑.
SI: (*a*) 0.914 m/s^2 ↑, 1.829 m/s^2 ↓.
(*b*) 3.66 m/s^2 ↑.
11.34 (*a*) 45 mm/s^2 →; 30 mm/s^2 →.
(*b*) 90 mm/s →; 135 mm →.
11.36 (*a*) $v_A = -\frac{1}{2}(v_B + v_C)$. (*b*) 40.5 in. ↑;
27 in./s ↑. SI: (*b*) 1.029 m ↑; 0.686 m/s ↑.
11.38 $v_A = 60$ mm/s ↓; $v_B = 100$ mm/s ↑;
$v_C = 80$ mm/s ↓.
11.40 (*b*) 20 ft/s; 74 ft/s; 176 ft.
SI: (*b*) 6.10 m/s; 22.6 m/s; 53.6 m.
11.42 (*a*) 72 m. (*b*) 4 s and 15 s.
11.44 Accelerate for 14.94 s, 50.9 mi/h.
SI: 82.0 km/h.
11.46 (*a*) 270 m/s^2, −19.57 m/s^2. (*b*) 54 mm.
(*c*) 0.296 s.
11.48 (*a*) 16 s. (*b*) 310 m.
11.50 9.39 s.
11.52 22.5 ft/s^2. SI: 6.86 m/s^2.
11.54 (*a*) Local. (*b*) 122 s.
11.56 8 s.
11.58 (*a*) 123 ft/s. (*b*) 227 ft.
SI: (*a*) 37.5 m. (*b*) 69.2 m.
11.60 (*a*) 9.71 s. (*b*) 130 m.
11.62 (*a*) 90 ft/s; 180 ft. (*b*) 120 ft/s; 225 ft.
SI: (*a*) 27.4 m/s; 54.9 m.
(*b*) 36.6 m/s; 68.6 m.
11.64 104 ft. SI: 31.7 m.
11.66 (*a*) **v** = 2.69 m/s ↙ 21.8°;
a = 1.414 m/s^2 ↖ 45°.
(*b*) **v** = 1.803 m/s ↗ 33.7°;
a = 5.10 m/s^2 ↗ 11.3°.

11.68 (*a*) $\mathbf{v} = 2.83$ ft/s ⦩45°;
$\mathbf{a} = 6.32$ ft/s^2 ∠71.6°.
(*b*) $\mathbf{v} = 4.01$ ft/s ⦩3.6°;
$\mathbf{a} = 2.03$ ft/s^2 ∠10.6°.
SI: (*a*) $\mathbf{v} = 0.863$ m/s ⦩45°;
$\mathbf{a} = 1.926$ m/s^2 ∠71.6°.
(*b*) $\mathbf{v} = 1.222$ m/s ⦩3.6°;
$\mathbf{a} = 0.619$ m/s^2 ∠10.6°.

11.70 (*a*) $\mathbf{v} = 100\,\pi$ mm/s ←;
$\mathbf{a} = 100\,\pi^2$ mm/s^2 ↓.

11.74 (*a*) 14.86 m/s. (*b*) 26.0 m.

11.76 14.54 ft from *B*. SI: 4.43 m from *B*.

11.78 (*a*) 4.79 m/s ∠50°. (*b*) 0.975 s.

11.80 6.1° or 32.5°.

11.82 (*a*) 30.09 m. (*b*) 181.9 m.

11.84 15° or 75°.

11.88 76.0 mi/h, 55.2° south of west.
SI: 122.3 km/h, 55.2° south of west.

11.90 (*a*) 187.4 km/h ⦩63.7°.
(*b*) 208 m ⦩63.7°. (*c*) 353 m.

11.92 (*a*) 8.62 knots, 55.1° east of north.
(*b*) 8.62 knots, 55.1° west of south.

11.94 623 m/s ⦪36.7°; 9.81 m/s^2 ↓.

11.96 124.7 km/h.

11.98 89.7 ft/s. SI: 27.3 m/s.

11.100 9.18 ft/s^2. SI: 2.80 m/s^2.

11.102 4.56 m/s^2.

11.104 (*a*) 79.6 m. (*b*) 40.8 m.

11.106 3810 m.

11.108 592 mi. SI: 953 km.

11.110 2280 mi/h. SI: 3670 km/h.

11.112 (*a*) $v = 0$; $a = 120$ mm/s^2, $\gamma = 0$.
(*b*) $v = 170.8$ mm/s, $\beta = 69.4°$;
$a = 905$ mm/s^2, $\gamma = 135°$. Where β and γ are measured counterclockwise from the axis *OA*.

11.114 (*a*) $v = 2b\omega$; $a = 4b\omega^2$. (*b*) b. (Particle describes a circle.)

11.116 $-(v_0/b)\sin^2\theta$.

11.118 $a = d\sec^2\theta\,(\ddot{\theta} + 2\,\dot{\theta}^2\tan\theta)$.

11.120 4.28 m/s; 0.188 m.

11.122 (*a*) $v = \sqrt{2}\pi b$, $\beta = 135°$; $a = \frac{1}{2}\sqrt{5}\pi^2 b$,
$\gamma = 243.4°$. (*b*) $v = 0$; $a = \frac{1}{2}\pi^2 b$, $\gamma = 0$.
Where β and γ are measured counterclockwise from the axis $\theta = 0$.

11.124 (*a*) 21.1°. (*b*) 1.02°.

11.126 78.8 m/s.

11.128 (*a*) 1286 ft. (*b*) 1159 ft.
SI: (*a*) 392 m. (*b*) 353 m.

11.130 (*a*) 120 mm/s ↑. (*b*) 120 mm/s ↓.

CHAPTER 12

12.2 At all points: mass = m = 10 kg.
Weight: 0°, 97.81 N; 45°, 98.07 N; 90°, 98.33 N.

12.4 3.22 ft/s^2 ↓; 20 lb.
SI: 0.981 m/s^2 ↓; 89 N.

12.6 (*a*) 4.55 s. (*b*) 0.45.

12.8 0.294 m/s^2.

12.10 23.9 N.

12.12 $F_{AB} = 500$ lb *C*; $F_{BC} = 500$ lb *T*.
SI: $F_{AB} = 2220$ N *C*; $F_{BC} = 2220$ N *T*.

12.14 (*a*) 29.2 kg. (*b*) Impossible, since a_B cannot be greater than g.

12.16 (*a*) 16.1 ft/s^2 ↓; 21.5 ft/s^2 ↓; 0.767 ft/s^2 ↓.
(*b*) 64.4 ft/s ↓; 86.0 ft/s ↓; 3.07 ft/s ↓.
(*c*) 17.94 ft/s ↓; 20.7 ft/s ↓; 3.92 ft/s ↓.
SI: (*a*) 4.91 m/s^2 ↓; 6.55 m/s^2 ↓;
0.234 m/s^2 ↓. (*b*) 19.63 m/s ↓; 26.2 m/s ↓;
0.936 m/s ↓. (*c*) 5.47 m/s ↓; 6.31 m/s ↓;
1.195 m/s ↓.

12.18 (*a*) 50.2 kg. (*b*) 357 N.

12.20 (*a*) $\mathbf{a}_A = 11.15$ ft/s^2 ←;
$\mathbf{a}_B = 7.43$ ft/s^2 ←. (*b*) 7.69 lb.
SI: (*a*) $\mathbf{a}_A = 3.40$ m/s^2 ←;
$\mathbf{a}_B = 2.26$ m/s^2 ←. (*b*) 34.2 N.

12.22 (*a*) $\mathbf{a}_A = 3.72$ ft/s^2 ←; $\mathbf{a}_B = 2.48$ ft/s^2 ←.
(*b*) 9.23 lb. SI: (*a*) $\mathbf{a}_A = 1.134$ m/s^2 ←;
$\mathbf{a}_B = 0.756$ m/s^2 ←. (*b*) 41.1 N.

12.24 24.6 m.

12.26 19.58 m.

12.28 (*a*) 14.49 ft/s^2 ←. (*b*) 6.44 ft/s^2 ←.
(*c*) 8.05 ft/s^2 →. SI: (*a*) 4.42 m/s^2 ←.
(*b*) 1.962 m/s^2 ←. (*c*) 2.45 m/s^2 →.

12.30 $v = \sqrt{2gh(1 - h/l)}$; $v_{max} = \sqrt{gl/2}$.

12.32 $(v_0\,m/T_0)\ln 2$.

12.34 $\mathbf{a}_A = \mathbf{a}_B = 3.58$ ft/s^2 ↑; $\mathbf{a}_C = 14.32$ ft/s^2 ↓.
SI: $\mathbf{a}_A = \mathbf{a}_B = 1.090$ m/s^2 ↑;
$\mathbf{a}_C = 4.36$ m/s^2 ↓.

12.36 8 lb. SI: 3.63 kg (mass).

12.38 $\mathbf{a}_A = 4.76$ m/s^2 →; $\mathbf{a}_B = 3.08$ m/s^2 ↓;
$\mathbf{a}_C = 1.401$ m/s^2 ←.

12.40 (*a*) 64.0 N. (*b*) 3.25 m/s.

12.42 8.02 ft/s $< v <$ 13.90 ft/s.
SI: 2.44 m/s $< v <$ 4.24 m/s.

12.44 1.656 m/s; 9.16 m/s^2.

12.46 (*a*) $T_{CD} = W$. (*b*) $T_{CD} = \frac{1}{2}W$;
$\mathbf{a} = 0.866g$ ⦩60°.

12.48 $v_A = 67.0$ mi/h; $v_B = 104.5$ mi/h.
SI: $v_A = 107.8$ km/h; $v_B = 168.2$ km/h.

12.50 0.974 m/s.
12.52 0.815.
12.54 31.6 mi/h. SI: 50.9 km/h.
12.56 $\delta = (eVlL)/(mv_0^2 d)$
12.58 The length of the plates must be doubled.
12.60 (*a*) $F_r = 4$ N, $F_\theta = 0$. (*b*) $F_r = -21.3$ N, $F_\theta = 21.3$ N.
12.62 (*a*) $F_r = -2.30$ lb, $F_\theta = 0$. (*b*) $F_r = 0$, $F_\theta = -1.839$ lb. SI: (*a*) $F_r = 10.23$ N, $F_\theta = 0$. (*b*) $F_r = 0$, $F_\theta = -8.18$ N.
12.64 (*a*) 8.39 lb. (*b*) 1.118 lb. SI: (*a*) 37.3 N. (*b*) 4.97 N.
12.66 12.96 N.
12.70 (*a*) 24 in./s. (*b*) $\rho_A = \frac{2}{3}$ in., $\rho_B = 18$ in. SI: (*a*) 0.610 m/s. (*b*) $\rho_A = 16.93$ mm, $\rho_B = 457$ mm.
12.72 238,000 mi. SI: 383 000 km.
12.76 (*a*) 5880 km/h. (*b*) 5630 km/h
12.78 2060 mi/h. SI: 3315 km/h.
12.80 12,160 mi. SI: 19 570 km.
12.82 32 800 km/h; 14 770 km/h.
12.84 (*a*) 5590 ft/s. (*b*) 94 ft/s. SI: (*a*) 1704 m/s. (*b*) 28.7 m/s.
12.86 (*a*) 35 200 km/h. (*b*) 5150 km/h.
12.88 4 h 29 min.
12.90 3 h 45 min.
12.92 4560 m/s.
12.94 Angle $BOC = 44.8°$.
12.96 (*a*) $\frac{1}{3}v_0$. (*b*) Vehicle must increase its speed by $0.1869\ v_0$. (*c*) 75.9°.
12.98 9.56 ft/s². SI: 2.91 m/s²
12.100 1.141 m.
12.102 (*a*) 342 lb. (*b*) Impossible. SI: (*a*) 1521 N.
12.104 $-62.7° < \theta < 62.7°$.
12.106 10.36 ft/s. SI: 3.16 m/s.
12.108 $r_{max}/r_0 = \alpha^2/(2 - \alpha^2)$.

CHAPTER 13

13.2 (*a*) 699 ft·lb; 87.3 ft. (*b*) 699 ft·lb; 530 ft. SI: (*a*) 948 J; 26.6 m. (*b*) 948 J; 161.5 m.
13.4 (*a*) 15.20 m. (*b*) 7.74 m/s.
13.6 60.1 ft. SI: 18.32 m.
13.8 14.40 N.
13.10 14.16 ft/s. SI: 4.32 m/s.
13.12 1.981 m/s.
13.14 1.180 m/s.
13.16 2.45 ft. SI: 0.747 m.
13.22 (*a*) 43 m. (*b*) 44 m/s².
13.24 $\mu_s = 0.268$; $\mu_k = 0.151$.
13.26 (*a*) 5.67 ft/s. (*b*) 8.66 in. SI: (*a*) 1.728 m/s. (*b*) 0.220 m.
13.28 $\mathbf{a}_n = 2g \uparrow$, $\mathbf{a}_t = \frac{3}{4}g \leftarrow$.
13.30 (*b*) 5.25 m/s.
13.32 (*a*) 39.1 ft/s. (*b*) 27.1 in. SI: (*a*) 11.92 m/s. (*b*) 0.688 m.
13.34 1718 km.
13.36 (*a*) 10,980 ft/s. (*b*) 16,480 ft/s. SI: (*a*) 3350 m/s. (*b*) 5020 m/s.
13.38 (*a*) $\frac{1}{2}kl^2(\sin\theta + \cos\theta - 1)^2$. (*b*) $-mgl\sin\theta$.
13.40 4.83 lb SI: 21.5 N.
13.42 (*a*) 9.35 m/s. (*b*) 4.46 m.
13.44 56.7 ft/s. SI: 17.28 m/s.
13.46 (*a*) 2.64 m/s. (*b*) 0.
13.48 (*a*) 2.43 m/s. (*b*) 2.71 m/s.
13.50 (*a*) 4.91 ft/s. (*b*) 0.250 lb. SI: (*a*) 1.497 m/s. (*b*) 1.112 N.
13.52 (*a*) 22.7 ft/s. (*b*) 7.75 ft. SI: (*a*) 6.92 m/s. (*b*) 2.36 m.
13.54 (*a*) 2.67 m. (*b*) 2.02 m.
13.56 (*a*) 1.448 in.; 17.52 lb. (*b*) 2.00 in.; 12.00 lb. SI: (*a*) 36.8 mm; 77.9 N. (*b*) 50.8 mm; 53.4 N.
13.58 (*a*) 33.5×10^6 J/kg. (*b*) 45.0×10^6 J/kg.
13.60 (*a*) $mgR[1 - (R/r)]$. (*b*) $\frac{1}{2}mgR^2/r$. (*c*) $mgR[1 - (R/2r)]$.
13.62 68.9%.
13.64 (*a*) 25.0 kW. (*b*) 6.13 kW.
13.66 2.86 ft/s². SI: 0.872 m/s².
13.68 (*a*) 26.6 kW. (*b*) 26.6 kW.
13.70 (*a*) 27.3 s; 1523 ft. (*b*) 38.3 s; 2970 ft. SI: (*a*) 27.3 s; 464 m. (*b*) 38.3 s; 905 m.
13.72 3.89 km/h.
13.74 9.33 ft. SI: 2.84 m.
13.76 (*a*) 1.716 m/s ←. (*b*) 4.20 m/s ⦣65.9°.
13.78 6 mg.
13.80 5.29 m/s →.
13.82 4.47 in. SI: 113.6 mm.
13.84 233 N/m.

CHAPTER 14

14.2 219 s.
14.4 (*a*) 2.80 s. (*b*) 5.61 s.
14.8 (*a*) 38.9 s. (*b*) 10.71 kN *T*.

14.10 (*a*) 19.75 ft/s; 5.20 s. (*b*) 9.17 s.
SI: (*a*) 6.02 m/s; 5.20 s.
14.12 (*a*) 9.03 m/s. (*b*) 0.
14.14 (*a*) and (*b*) 111.1 kN.
14.16 50.2 lb ←, 204 lb ↓.
SI: 223 N ←, 907 N ↓.
14.18 18.46 ft/s. SI: 5.63 m/s.
14.20 (*a*) 5.2 km/h →. (*b*) 3.9 km/h →.
14.22 (*a*) 0.375 mi/h →. (*b*) 4270 lb.
SI: (*a*) 0.603 km/h →. (*b*) 18.99 kN.
14.24 0.742 m/s →.
14.26 (*a*) 233 lb · s. (*b*) 159.1 lb · s.
SI: (*a*) 1036 N · s. (*b*) 708 N · s.
14.28 (*a*) $\mathbf{v}_A = 9.45$ ft/s →; $\mathbf{v}_B = 12.25$ ft/s →.
(*b*) 0.24 ft · lb. SI: $\mathbf{v}_A = 2.88$ m/s →;
$\mathbf{v}_B = 3.73$ m/s→. (*b*) 0.32 J.
14.30 0.875.
14.32 $\mathbf{v}_A = 0.948$ m/s →; $\mathbf{v}_B = 1.092$ m/s →;
$\mathbf{v}_C = 1.960$ m/s →.
14.34 (*a*) $\tan\theta = \sqrt{e}$. (*b*) $\sqrt{e}\,v_0$.
14.36 (*a*) 20.9 in. (*b*) 10.20 in.
SI: (*a*) 0.531 m. (*b*) 0.259 m.
14.38 (*a*) 0.913. (*b*) 33.3 in.; 14.61 in.
SI: (*b*) 0.846 m; 0.371 m.
14.40 $\mathbf{v}_A = 0.721\,v_0 \measuredangle 16.1°$; $\mathbf{v}_B = 0.693\,v_0$ ←.
14.42 (*a*) 76.2°. (*b*) 37.1 N.
14.44 (*a*) 0.80. (*b*) 0.32.
14.46 10,140 lb. SI: 45.1 kN.
14.50 (*a*) 6670 J. (*b*) 24 km/h.
14.52 (*a*) 1.288 lb. (*b*) 1.479 lb.
SI: (mass) (*a*) 0.584 kg. (*b*) 0.671 kg.
14.54 (*a*) 1.155 m. (*b*) 5.20 m/s.
14.56 2.84 ft/s. SI: 0.866 m/s.
14.58 $v_r = 9.05$ ft/s, $v_\theta = 9.14$ ft/s.
SI: $v_r = 2.76$ m/s, $v_\theta = 2.79$ m/s.
14.64 5160 ft/s; 79.9°. SI: 1573 m/s; 79.9°.
14.66 2280 m/s.
14.68 $v_{max} = 27{,}200$ ft/s, $\phi_0 = 90°$;
$v_{min} = 16{,}330$ ft/s, $\phi_0 = 0$.
SI: $v_{max} = 8290$ m/s, $\phi_0 = 90°$;
$v_{min} = 4980$ m/s, $\phi_0 = 0$.
14.70 $65.7° \leqslant \phi_0 \leqslant 114.3°$.
14.76 (*b*) $\frac{1}{3}\sqrt{6}\,v_{esc}$, $\frac{1}{2}\sqrt{2}\,v_{esc}$.
14.78 218 lb. SI: 970 N.
14.80 $Q_1 = \frac{1}{2}Q(1 - \sin\theta)$; $Q_2 = \frac{1}{2}Q(1 + \sin\theta)$.
14.82 $\mathbf{A}_x = A\rho v_A^2(1 - \cos\theta)$ ←,
$\mathbf{A}_y = A\rho v_A^2 \sin\theta$ ↑,
$\mathbf{M}_A = A\rho v_A^2 R(1 - \cos\theta)$ ↺.
14.84 $\mathbf{C} = 149.7$ N ↓; $\mathbf{D}_x = 250$ N →,
$\mathbf{D}_y = 283$ N. ↓.
14.86 $\mathbf{C}_x = 67.1$ lb →, $\mathbf{C}_y = 367$ lb ↑;
$\mathbf{D} = 433$ lb ↑. SI: $\mathbf{C}_x = 298$ N →,
$\mathbf{C}_y = 1632$ N ↑; $\mathbf{D} = 1926$ N ↑.
14.88 49.7 kg/s.
14.90 461 mi/h. SI: 742 km/h.
14.92 4310 lb. SI: 19.17 kN.
14.94 1143 lb. SI: 5.08 kN.
14.96 (*a*) 32.4 m/s. (*b*) 101.9 m³/s.
(*c*) 64 600 J/s.
14.98 170.0 rpm.
14.100 (*a*) $\eta = 2(V/v_A)[1 - (V/v_A)]^2(1 - \cos\theta)$.
(*b*) $\frac{16}{27}$.
14.102 (*a*) qv.
14.104 (*a*) $mgvt + mv^2$. (*b*) $(l - vt)mg$.
14.106 qv.
14.108 $v = m_0v_0/(m_0 + qt)$;
$a = -m_0v_0q/(m_0 + qt)^2$.
14.110 (*a*) 53.7 lb/s. (*b*) 13.42 lb/s.
SI: (*a*) 24.4 kg/s. (*b*) 6.09 kg/s.
14.112 (*a*) 11.19 m/s²; 0. (*b*) 28.4 m/s²; 814 m/s.
(*c*) 200.2 m/s²; 3952 m/s.
14.114 (*a*) 814 m/s. (*b*) 5210 m/s.
14.116 522,000 ft. SI: 159.2 km.
14.118 (*a*) Five. (*b*) 2 m/s →. (*c*) Same as original.
14.120 (*a*) 7220 lb, 2.47 ft below *B*. (*b*) 5240 lb, 6.80 ft below *B*. SI: (*a*) 32.1 kN, 0.753 m below *B*. (*b*) 23.3 kN, 2.07 m below *B*.
14.124 Case *1*: (*a*) $\frac{1}{3}g$. (*b*) $v = \sqrt{2gl/3}$.
Case *2*: (*a*) gy/l. (*b*) $v = \sqrt{gl}$.
14.126 (*a*) 5.54 ft/s. (*b*) 0.642 ft from *B*.
SI: (*a*) 1.689 m/s. (*b*) 0.1957 m from *B*.
14.128 (*a*) 135.6 N · s. (*b*) 108.5 N · s. (*c*) 368 J; 294 J.

CHAPTER 15

15.2 (*a*) 2 s. (*b*) 2 rad; 8 rad/s².
15.4 3.93 rad/s; 20.6 rad/s².
15.6 (*a*) 60 rev. (*b*) 750 rev.
15.8 $\alpha = 8.38\,t$, $\omega = 4.19\,t^2$, $\theta = 1.396\,t^3$;
16.76 rad/s.
15.10 (*a*) 1524 ft/s; 0.1111 ft/s².
(*b*) 1168 ft/s; 0.0851 ft/s². (*c*) 0.
SI: (*a*) 465 m/s; 0.0339 m/s².
(*b*) 356 m/s; 0.0259 m/s². (*c*) 0.
15.12 1.174 s; 7.05 rad/s.
15.14 (*a*) 6.93 rad/s ↺. (*b*) 2.09 s.
15.16 (*a*) 50 rpm. (*b*) 3.29 m/s² ↑; 13.16 m/s² ↑.
15.18 (*a*) 24 rad/s² ↻. (*b*) 30.6 rev.

15.20 (*a*) 8.92 s. (*b*) $\boldsymbol{\omega}_A = 26.8$ rad/s ↺; $\boldsymbol{\omega}_B = 44.6$ rad/s ↻.

15.22 $\alpha = bv^2/2\pi r^3$.

15.24 (*a*) 2.22 rad/s ↺. (*b*) 18.77 ft/s $\angle 40°$. SI: (*b*) 5.72 m/s² $\angle 40°$.

15.26 (*a*) 3.46 rad/s ↻. (*b*) 0.693 m/s →.

15.28 (*a*) 3.93 rad/s ↻. (*b*) 0.578 m/s →.

15.30 (*a*) 1.500. (*b*) $\frac{1}{3}\omega_A$ ↺.

15.32 210 rpm ↺.

15.34 (*a*) $\boldsymbol{\omega} = 0$; $\mathbf{v}_D = 84$ in./s →. (*b*) $\boldsymbol{\omega} = 3.5$ rad/s ↺; $\mathbf{v}_D = 24.5$ in./s →. (*c*) $\boldsymbol{\omega} = 0$; $\mathbf{v}_D = 84$ in./s ←. SI: (*a*) $\boldsymbol{\omega} = 0$; $\mathbf{v}_D = 2.13$ m/s →. (*b*) $\boldsymbol{\omega} = 3.5$ rad/s ↺; $\mathbf{v}_D = 0.622$ m/s →. (*c*) $\boldsymbol{\omega} = 0$; $\mathbf{v}_D = 2.13$ m/s ←.

15.36 6.52 m/s ↓; $\boldsymbol{\omega}_{BD} = 20.8$ rad/s ↺.

15.38 $\boldsymbol{\omega}_{BD} = 1.5$ rad/s ↻; $\boldsymbol{\omega}_{DE} = 3$ rad/s ↺.

15.40 $\boldsymbol{\omega}_{BD} = 6$ rad/s ↺; $\boldsymbol{\omega}_{DE} = 2$ rad/s ↺.

15.42 (*a*) $\mathbf{v}_A = 10$ in./s →; $\boldsymbol{\omega}_{AC} = 0$. (*b*) $\mathbf{v}_A = 38.1$ in./s →; $\boldsymbol{\omega}_{AC} = 0.968$ rad/s ↻. SI: (*a*) $\mathbf{v}_A = 0.254$ m/s →; $\boldsymbol{\omega}_{AC} = 0$. (*b*) $\mathbf{v}_A = 0.968$ m/s →; $\boldsymbol{\omega}_{AC} = 0.968$ rad/s ↻.

15.44 (*a*) 1.5 rad/s ↻. (*b*) 0.520 m/s $\angle 30°$.

15.46 Vertical line intersecting zx plane at $x = 0$, $z = 5.25$ ft. SI: $z = 1.600$ m.

15.48 (*a*) 12 rad/s ↻. (*b*) $\mathbf{v}_R = 2.4$ m/s →; $\mathbf{v}_D = 2.16$ m/s $\angle 56.3°$.

15.50 (*a*) 3.08 rad/s ↻. (*b*) 83.3 in./s $\nwarrow$ 73.9°. SI: (*b*) 2.12 m/s $\nwarrow$ 73.9°.

15.52 (*a*) 5.33 rad/s ↻. (*b*) 105.8 in./s $\nwarrow$ 49.1°. SI: (*b*) 2.69 m/s $\nwarrow$ 49.1°.

15.54 (*a*) $\boldsymbol{\omega}_{AB} = 2.25$ rad/s ↺; $\boldsymbol{\omega}_{BD} = 5.00$ rad/s ↺. (*b*) 0.938 m/s $\angle 36.9°$.

15.56 (*a*) $\boldsymbol{\omega}_{AB} = 1.039$ rad/s ↺; $\boldsymbol{\omega}_{BD} = 0.346$ rad/s ↻. (*b*) 69.3 mm/s →.

15.58 (*a*) 3 rad/s ↺. (*b*) 5.78 in./s ←. (*c*) 15.12 in./s $\nwarrow$ 52.5°. SI: (*a*) 3 rad/s ↺. (*b*) 0.1468 m/s ←. (*c*) 0.384 m/s $\nwarrow$ 52.5°.

15.60 (*a*) 0.250 m/s $\angle 36.9°$. (*b*) 0.500 m/s ↑.

15.74 $\mathbf{a}_A = 20$ ft/s² ↑; $\mathbf{a}_B = 4$ ft/s² ↓. SI: $\mathbf{a}_A = 6.10$ m/s² ↑; $\mathbf{a}_B = 1.219$ m/s² ↓.

15.76 (*a*) 200 mm from *B*. (*b*) 225 mm from *A*.

15.78 $\mathbf{a}_B = 632$ m/s² ↓. $\mathbf{a}_E = 632$ m/s² ←.

15.80 (*a*) $\boldsymbol{\alpha}_A = 40$ rad/s² ↻; $\boldsymbol{\alpha}_B = 20$ rad/s² ↻; $\boldsymbol{\alpha}_C = 40$ rad/s² ↻. (*b*) $\mathbf{a}_A = 10$ in./s² →; $\mathbf{a}_B = 5$ in./s² →; $\mathbf{a}_C = 0$. SI: (*b*) $\mathbf{a}_A = 0.254$ m/s² →; $\mathbf{a}_B = 0.1270$ m/s² →; $\mathbf{a}_C = 0$.

15.82 $\boldsymbol{\omega}_{AB} = \frac{1}{2}\omega_0$ ↺, $\boldsymbol{\alpha}_{AB} = 0$; $\boldsymbol{\omega}_{BC} = 0$, $\boldsymbol{\alpha}_{BC} = \frac{3}{2}\omega_0^2$ ↺.

15.84 (*a*) 296 m/s² ↑. (*b*) 157.0 m/s² ↑.

15.86 (*a*) 2.7 rad/s² ↻. (*b*) 5.4 rad/s² ↺.

15.88 (*a*) 1.848 rad/s² ↺. (*b*) 9.24 ft/s² $\searrow$ 60°. SI: (*b*) 2.82 m/s² $\searrow$ 60°.

15.90 (*a*) 1.848 rad/s² ↺. (*b*) 8.12 ft/s² $\angle 80.2°$. SI: (*b*) 2.47 m/s² $\angle 80.2°$.

15.92 (*a*) 12 rad/s² ↺. (*b*) 1.875 m/s² ↑.

15.94 $\omega = (v_B \sin \beta)/(l \cos \theta)$

15.96 (*a*) $v_B = r\omega \cos \theta$. (*b*) $a_B = r\alpha \cos \theta - r\omega^2 \sin \theta$.

15.98 $a_x = (v^2/r) \sin (vt/r)$; $a_y = (v^2/r) \cos (vt/r)$.

15.100 $\omega = -v_0/(2l \sin \theta)$; $\alpha = -(v_0^2/4l^2)(\cos \theta)/\sin^3 \theta$.

15.102 (*a*) $\omega = (v/R) \cos^2 \theta$. (*b*) $\alpha = -2(v/R)^2 \sin \theta \cos^3 \theta$.

15.104 (*a*) $\omega = (v/r)(\cos \theta)/\tan \theta$. (*b*) $\alpha = -(v/r)^2(1 + \sin^2 \theta)/\tan^3 \theta$.

15.106 $\boldsymbol{\alpha} = 2(v_A/b)^2 \sin \theta \cos^3 \theta$ ↺.

15.108 (*a*) 2.73 rad/s ↻. (*b*) 0.616 m/s $\swarrow$ 70°.

15.110 $\boldsymbol{\omega}_{BP} = 4.38$ rad/s ↺; $\boldsymbol{\omega}_{AH} = 1.327$ rad/s.

15.112 (*a*) $\boldsymbol{\omega}_{BD} = \boldsymbol{\omega}_{AH}$; $\mathbf{v}_{P/AH} = l\omega$ ↓; $\mathbf{v}_{P/BD} = 0$. (*b*) $\mathbf{v}_{P/AH} = l\omega/\sqrt{3}$ $\swarrow$ 60°; $\mathbf{v}_{P/BD} = l\omega/\sqrt{3}$ $\searrow$ 60°.

15.114 $\mathbf{a}_1 = [r\omega^2 \rightarrow] + [2u\omega \downarrow]$, $\mathbf{a}_2 = [r\omega^2 \downarrow] + [2u\omega \rightarrow]$, $\mathbf{a}_3 = [r\omega^2 \leftarrow] + [u^2/r \leftarrow] + [2u\omega \leftarrow]$, $\mathbf{a}_4 = [r\omega^2 \uparrow] + [2u\omega \uparrow]$.

15.116 $a_P = \omega^2(r^2 + 4b^2)^{\frac{1}{2}}$

15.118 (*a*) 1.3 m/s² →. (*b*) 22.9 m/s² ←. (*c*) 16.22 m/s² $\searrow$ 48.2°.

15.120 61.2 mi/h. SI: 98.5 km/h.

15.122 8.08 rad/s² ↺.

15.124 (*a*) 0. (*b*) 2.31 $l\omega^2$ $\nwarrow$ 30°.

15.126 95.1 m/s² $\swarrow$ 41.9°.

15.128 (*a*) 6.91 rad/s² ↻. (*b*) 1.671 m/s² ←.

15.130 0.731 ft/s² ↙. SI: 0.223 m/s² ↙.

15.132 7.85 ft/s ←; 92.7 ft/s² →. SI: 2.39 m/s ←; 28.2 m/s² →.

15.134 (*a*) 0.110 m/s ↑. (*b*) 0.1667 rad/s ↻.

15.136 (*a*) $\boldsymbol{\alpha}_{BC} = 7.5$ rad/s² ↻; $\boldsymbol{\alpha}_{CD} = 3.75$ rad/s² ↺. (*b*) 97.5 in./s² $\swarrow$ 67.4°. SI: (*b*) 2.48 m/s² $\swarrow$ 67.4°.

15.138 2.25 rad/s ↺; 23.3 rad/s² ↻.

CHAPTER 16

16.2 (*a*) $5 \text{ m/s}^2 \leftarrow$. (*b*) $\mathbf{B} = 41.6 \text{ N} \uparrow$; $\mathbf{C} = 36.9 \text{ N} \uparrow$.

16.4 (*a*) $\frac{1}{2}W$. (*b*) $2r/5$.

16.6 (*a*) $0.612g$. (*b*) 5.00.

16.8 (*a*) $2.55 \text{ m/s}^2 \rightarrow$. (*b*) $h \leq 1.047$ m.

16.10 17.36 ft. SI: 5.29 m.

16.12 42.3 ft. SI: 12.89 m.

16.14 5.44 m.

16.16 (*a*) $5.37 \text{ ft/s}^2 \leftarrow$. (*b*) $\mathbf{B} = 300 \text{ lb} \uparrow$; $\mathbf{C} = 200 \text{ lb} \uparrow$. SI: (*a*) $1.637 \text{ m/s}^2 \leftarrow$. (*b*) $\mathbf{B} = 1334 \text{ N} \uparrow$; $\mathbf{C} = 890 \text{ N} \uparrow$.

16.18 $F_{AD} = 5.39$ N *C*; $F_{BE} = 45.6$ N *C*.

16.20 (*a*) $27.9 \text{ ft/s}^2 \measuredangle 60^\circ$. (*b*) $1.443 \text{ lb} \uparrow$. SI: (*a*) $8.50 \text{ m/s}^2 \measuredangle 60^\circ$. (*b*) $6.42 \text{ N} \uparrow$.

16.22 (*a*) $0.500g \measuredangle 30^\circ$. (*b*) Platform: $1.250g \measuredangle 30^\circ$; block: $0.625g \downarrow$.

16.24 $M_B = 6.37 \text{ N} \cdot \text{m}$.

16.26 $89.7 \text{ N} \cdot \text{m}$.

16.28 $61.8 \text{ rad/s}^2 \circlearrowright$.

16.30 (1) $19.62 \text{ rad/s}^2 \circlearrowleft$; $39.2 \text{ rad/s} \circlearrowleft$; $19.81 \text{ rad/s} \circlearrowleft$. (2) $14.01 \text{ rad/s}^2 \circlearrowleft$; $28.0 \text{ rad/s} \circlearrowleft$; $16.74 \text{ rad/s} \circlearrowleft$. (3) $6.54 \text{ rad/s}^2 \circlearrowleft$; $13.08 \text{ rad/s} \circlearrowleft$; $11.44 \text{ rad/s} \circlearrowleft$. (4) $10.90 \text{ rad/s}^2 \circlearrowleft$; $21.8 \text{ rad/s} \circlearrowleft$; $10.44 \text{ rad/s} \circlearrowleft$.

16.32 (*a*) $5.66 \text{ ft/s}^2 \downarrow$. (*b*) $8.24 \text{ ft/s} \downarrow$. SI: (*a*) $1.725 \text{ m/s}^2 \downarrow$. (*b*) $2.51 \text{ m/s} \downarrow$.

16.34 $112.2 \text{ kg} \cdot \text{m}^2$.

16.36 153.1 lb. SI: 681 N.

16.38 $\boldsymbol{\alpha} = \dfrac{2g}{r}\dfrac{\mu}{1+\mu} \circlearrowright$.

16.40 (*a*) $10.44 \text{ rad/s}^2 \circlearrowleft$. (*b*) 4.01 s.

16.42 (*a*) $\boldsymbol{\omega}_A = 221 \text{ rpm} \circlearrowright$; $\boldsymbol{\omega}_B = 368 \text{ rpm} \circlearrowleft$.

16.44 $I_R = \left(n + \dfrac{1}{n}\right)^2 I_0 + n^4 I_C$.

16.46 $\boldsymbol{\alpha} = (2g/r) \sin \phi \circlearrowleft$.

16.48 (*a*) $16.10 \text{ rad/s}^2 \circlearrowleft$. (*b*) $8.05 \text{ ft/s}^2 \rightarrow$. (*c*) 12 in. from *B*. SI: (*a*) $16.10 \text{ rad/s}^2 \circlearrowleft$. (*b*) $2.45 \text{ m/s}^2 \rightarrow$. (*c*) 0.305 m from *B*.

16.50 (*a*) $3.10 \text{ m/s}^2 \uparrow$. (*b*) $0.206 \text{ rad/s}^2 \circlearrowleft$.

16.52 0.593 rad/s^2; $a_x = a_y = 0$, $a_z = 0.200 \text{ m/s}^2$.

16.54 $\mathbf{a}_A = 21.5 \text{ ft/s}^2 \uparrow$; $\mathbf{a}_B = 10.73 \text{ ft/s}^2 \uparrow$. SI: $\mathbf{a}_A = 6.55 \text{ m/s}^2 \uparrow$; $\mathbf{a}_B = 3.27 \text{ m/s}^2 \uparrow$.

16.56 $\mathbf{a}_A = 8.46 \text{ m/s}^2 \uparrow$; $\mathbf{a}_B = 1.362 \text{ m/s}^2 \uparrow$.

16.58 (*a*) $10.00 \text{ rad/s}^2 \circlearrowleft$. (*b*) $1.990 \text{ m/s}^2 \uparrow$.

16.60 (*a*) $1.098g/a \circlearrowleft$. (*b*) $1.761g \measuredangle 36.6^\circ$. (*c*) $1.416\, g \measuredangle 2.0^\circ$.

16.62 (*a*) $\overline{\mathbf{a}} = \mu g \leftarrow$; $\boldsymbol{\alpha} = 2\dfrac{\mu g}{r} \circlearrowright$. (*b*) $t = \dfrac{1}{3}\dfrac{v_0}{\mu g}$. (*c*) $x = \dfrac{5}{18}\dfrac{v_0^2}{\mu g}$. (*d*) $\overline{\mathbf{v}} = \dfrac{2}{3}v_0 \rightarrow$; $\boldsymbol{\omega} = \dfrac{2}{3}\dfrac{v_0}{r} \circlearrowright$.

16.64 75.5 mm.

16.66 (*a*) $107.1 \text{ rad/s}^2 \circlearrowright$. (*b*) $21.4 \text{ N} \leftarrow$, $39.2 \text{ N} \uparrow$.

16.68 (*a*) $\boldsymbol{\alpha} = 3Pg/WL \circlearrowright$. (*b*) $\mathbf{A}_x = \frac{1}{2}P \leftarrow$, $\mathbf{A}_y = W \uparrow$.

16.70 $13.64 \text{ kN} \rightarrow$.

16.72 $490 \text{ lb} \rightarrow$. SI: $2180 \text{ N} \rightarrow$.

16.74 (*a*) $\frac{1}{4}W \uparrow$. (*b*) $\frac{3}{2}g \downarrow$.

16.76 (*a*) $27.8 \text{ ft/s}^2 \downarrow$. (*b*) $32.2 \text{ ft/s}^2 \downarrow$. SI: (*a*) $8.47 \text{ m/s}^2 \downarrow$. (*b*) $9.81 \text{ m/s}^2 \downarrow$.

16.78 (*a*) $25.3 \text{ N} \cdot \text{m} \circlearrowleft$. (*b*) $52.4 \text{ N} \leftarrow$, $112.2 \text{ N} \uparrow$.

16.80 (*a*) $35.3 \text{ lb} \leftarrow$. (*b*) $2.46 \text{ lb} \rightarrow$, $26.3 \text{ lb} \uparrow$. SI: (*a*) $157.0 \text{ N} \leftarrow$. (*b*) $10.94 \text{ N} \rightarrow$, $117.0 \text{ N} \uparrow$.

16.84 4.19 ft. SI: 1.278 m.

16.86 1.266 m.

16.88 (*a*) $24 \text{ rad/s}^2 \circlearrowright$; $3.84 \text{ m/s}^2 \rightarrow$. (*b*) 0.016.

16.90 (*a*) $8 \text{ rad/s}^2 \circlearrowleft$; $1.280 \text{ m/s}^2 \leftarrow$. (*b*) 0.220.

16.92 (*a*) Does not slide. (*b*) $23.2 \text{ rad/s}^2 \circlearrowright$; $15.46 \text{ ft/s}^2 \rightarrow$. SI: (*b*) $23.2 \text{ rad/s}^2 \circlearrowright$; $4.71 \text{ m/s}^2 \rightarrow$.

16.94 (*a*) Slides. (*b*) $12.88 \text{ rad/s}^2 \circlearrowleft$; $3.22 \text{ ft/s}^2 \leftarrow$. SI: (*b*) $12.88 \text{ rad/s}^2 \circlearrowleft$; $0.981 \text{ m/s}^2 \leftarrow$.

16.96 $7.16 \text{ ft/s}^2 \leftarrow$. SI: $2.18 \text{ m/s}^2 \leftarrow$.

16.98 (*a*) $\mathbf{a}_A = 2g/5 \leftarrow$; $\mathbf{a}_B = 2g/5 \downarrow$. (*b*) $\mathbf{a}_A = 2g/7 \leftarrow$; $(\mathbf{a}_B)_x = 2g/7 \leftarrow$, $(\mathbf{a}_B)_y = 2g/7 \downarrow$.

16.100 (*a*) $\boldsymbol{\alpha} = \dfrac{P}{2\,mr}\dfrac{\pi}{(\pi - 2)} \circlearrowright$. (*b*) $\frac{1}{2}P/(mg + P)$.

16.102 $51.0 \text{ rad/s}^2 \circlearrowleft$.

16.104 (*a*) $23.6 \text{ N} \cdot \text{m} \circlearrowleft$. (*b*) $\mathbf{A} = 51.2 \text{ N} \uparrow$; $\mathbf{B} = 6.93 \text{ N} \leftarrow$.

16.106 (*a*) 29.9 lb. (*b*) 76.6 lb. SI: (*a*) 133.0 N. (*b*) 341 N.

16.108 (*a*) $3.53 \text{ rad/s}^2 \circlearrowright$. (*b*) 176.6 N. (*c*) 358 N.

16.110 $T_{BD} = 14.00$ lb; $T_{CF} = 6.36$ lb. SI: $T_{BD} = 62.3$ N; $T_{CF} = 28.3$ N.

16.112 $\mathbf{A}_x = 218 \text{ lb} \leftarrow$, $\mathbf{A}_y = 2.5 \text{ lb} \uparrow$; $\mathbf{B}_x = 313 \text{ lb} \rightarrow$; $\mathbf{B}_y = 2.5 \text{ lb} \uparrow$. SI: $\mathbf{A}_x = 970 \text{ N} \leftarrow$, $\mathbf{A}_y = 11.12 \text{ N} \uparrow$; $\mathbf{B}_x = 1392 \text{ N} \rightarrow$, $\mathbf{B}_y = 11.12 \text{ N} \uparrow$.

16.114 $38.1 \text{ N} \rightarrow$.

16.116 $\mathbf{A}_x = 3mr^2\omega_0^2 \leftarrow$, $\mathbf{A}_y = mgr \uparrow$.
$\mathbf{B}_x = \frac{1}{2}mr^2\omega_0^2 \rightarrow$, $\mathbf{B}_y = mgr \uparrow$.

16.118 (*a*) $\frac{1}{2}a \rightarrow$. (*b*) $2d \rightarrow$.

16.120 (*a*) 74.4 rad/s^2 ↻. (*b*) 24.8 ft/s^2 $\downarrow$.
SI: (*b*) 7.56 m/s^2 $\downarrow$.

16.122 (*a*) 63.5 rad/s^2 ↻. (*b*) 20.5 lb.
SI: (*b*) 91.2 N.

16.124 (*a*) 12.14 rad/s^2 ↻. (*b*) 11.21 m/s^2 $\nwarrow$30°.
(*c*) 14.56 N $\measuredangle$60°.

16.126 $\boldsymbol{\alpha}_{AB} = 8.28$ rad/s^2 ↻;
$\boldsymbol{\alpha}_{BC} = 41.4$ rad/s^2 ↺.

16.128 (*a*) $\frac{1}{3}g \uparrow$. (*b*) $\frac{5}{3}g \downarrow$.

16.130 $\mathbf{A}_x = \frac{9}{20}W \leftarrow$, $\mathbf{A}_y = \frac{13}{40}W \uparrow$; $M_B = \frac{3}{10}WL$.

16.132 (*a*) $M_{max} = PL/3\sqrt{3}$ at $L/\sqrt{3}$ below A.

16.134 $V_{max} = \frac{1}{3}mg$ at A; $M_{max} = 4\,mgL/81$ at $\frac{1}{3}L$ to right of A.

16.136 101.4 mi/h. SI: 163.2 km/h.

16.138 178.9 km/h.

16.140 (*a*) 45.0 km/h. (*b*) $\mathbf{A}_n = 3470$ N $\leftarrow$,
$\mathbf{B}_n = 1434$ N $\leftarrow$.

16.142 (*a*) 33.7°. (*b*) 0.67.

16.144 $\mathbf{C} = 630$ N$\uparrow$; $\mathbf{D} = 1680$ N $\downarrow$.

16.146 $A_x = -B_x = -46.0$ lb,
$A_y = -B_y = 15.33$ lb.
SI: $A_x = -B_x = -205$ N,
$A_y = -B_y = 68.2$ N.

16.148 (*a*) 48.3 rad/s^2. (*b*) $A_x = -B_x = 0.75$ lb,
$A_y = -B_y = 2.25$ lb.
SI: (*b*) $A_x = -B_x = 3.34$ N,
$A_y = -B_y = 10.01$ N.

16.150 (*a*) 4.00 N · m.
(*b*) $A_y = -B_y = -19.49$ N,
$A_z = -B_z = 8.66$ N.

16.152 $x = 2.67$ ft, $y = 4$ ft.
SI: $x = 0.813$ m, $y = 1.219$ m.

16.154 $A_x = A_y = 0$, $A_z = \frac{1}{4}\sqrt{2}\,ma\omega^2$;
$M_x = \frac{1}{6}ma^2\omega^2$, $M_y = M_z = 0$.

16.156 (*a*) $\cos\beta = 3g/2l\omega^2$. (*b*) $\sqrt{3g/2l}$.

16.158 (*a*) $4M_0/ma^2$. (*b*) $A_x = -\sqrt{2}\,M_0/a$,
$A_y = A_z = 0$; $M_x = M_y = 0$, $M_z = \frac{2}{3}M_0$.

16.160 (*a*) 21.5 ft/s^2 $\rightarrow$. (*b*) 15.95 ft/s^2 $\rightarrow$.
SI: (*a*) 6.55 m/s^2 $\rightarrow$. (*b*) 4.86 m/s^2 $\rightarrow$.

16.162 (*a*) $\mathbf{a}_x = 0.3g \leftarrow$, $\mathbf{a}_y = 0.6g \downarrow$.
(*b*) $\mathbf{a} = 0.630g \downarrow$.

16.164 (*a*) 22.4 N $\uparrow$. (*b*) 30.8 N $\uparrow$.

16.166 5.57 rad/s^2 ↻; 27.2 ft/s^2 $\nwarrow$65.8°.
SI: 5.57 rad/s^2 ↻; 8.29 m/s^2 $\nwarrow$65.8°.

16.168 (*a*) $\bar{r} = \bar{k}$. (*b*) $\frac{1}{2}W \uparrow$.

16.170 (*a*) 6.54 m/s^2 $\downarrow$. (*b*) 5.19 m/s $\downarrow$.

CHAPTER 17

17.2 71.6 N · m.

17.4 14.06 in. SI: 0.357 m.

17.6 (*a*) 8.00 lb · ft · s^2. (*b*) 3.98 rev.
SI: (*a*) 10.85 kg · m^2. (*b*) 3.98 rev.

17.8 $\mathbf{v}_A = 1.293$ m/s $\uparrow$; $\mathbf{v}_B = 2.59$ m/s $\downarrow$.

17.12 71.2 rev.

17.14 18.06 N · m.

17.16 200 lb. SI: 890 N.

17.18 9.27 rev.

17.20 9.27 ft/s $\downarrow$. SI: 2.82 m/s $\downarrow$.

17.22 (*a*) 2.40 rev. (*b*) 21.4 N $\swarrow$.

17.24 $\sqrt{4gh/3}$.

17.26 (*a*) $\sqrt{4gh/7}$. (*b*) $2W/7$.

17.28 (*a*) $\sqrt{\frac{10}{7}g(R - r)(1 - \cos\beta)}$.
(*b*) $W(17 - 10\cos\beta)/7$.

17.30 (*a*) $\boldsymbol{\omega} = \sqrt{3g/l}$; $\mathbf{R} = 2.5\,W \uparrow$.
(*b*) 5.72 rad/s ↻; 36.8 N $\uparrow$.

17.32 (*a*) 1.420 m/s $\rightarrow$. (*b*) 1.529 m/s $\rightarrow$.

17.34 5.75 ft/s $\leftarrow$. SI: 1.753 m/s $\leftarrow$.

17.38 (*a*) $\mathbf{v}_A = 1.332$ m/s $\rightarrow$;
$\mathbf{v}_B = 0.769$ m/s $\downarrow$. (*b*) $\mathbf{v}_A = 0$;
$\mathbf{v}_B = 4.20$ m/s $\downarrow$.

17.40 3.78 ft/s $\leftarrow$. SI: 1.152 m/s $\leftarrow$.

17.42 14.63 rad/s ↻.

17.44 $\mathbf{v}_B = 0$; $\mathbf{v}_D = 34.0$ ft/s $\downarrow$.
SI: $\mathbf{v}_B = 0$; $\mathbf{v}_D = 10.36$ m/s $\downarrow$.

17.46 4.37 m/s $\downarrow$.

17.48 36.4°.

17.50 (*a*) Zero. (*b*) 188.5 W.

17.52 (*a*) 0.365 lb · ft. (*b*) 1.824 lb · ft.
SI: (*a*) 0.495 N · m. (*b*) 2.47 N · m.

17.54 (*a*) 3.76 m/s $\measuredangle$45°. (*b*) 3.18 m/s $\downarrow$.

17.56 $\omega_A = \dfrac{2n}{n^2 + 1}\sqrt{\dfrac{\pi M_0}{I_0}}$.

17.58 (*a*) 16.61 ft/s $\rightarrow$. (*b*) $\mathbf{A} = \mathbf{C} = 418$ lb. $\uparrow$.
SI: (*a*) 5.06 m/s $\rightarrow$.
(*b*) $\mathbf{A} = \mathbf{C} = 1859$ N $\uparrow$.

17.60 $\sqrt{g/3r}$.

17.62 6.47 ft/s $\downarrow$. SI: 1.972 m/s $\downarrow$.

CHAPTER 18

18.2 128.4 lb · ft. SI: 174.1 N · m.

18.4 0.904 r.

18.6 $t = (r\omega_0)/(2\mu g)$.

18.10 (*a*) 3.33 N · m. (*b*) $\boldsymbol{\omega}_A = 23.5$ rad/s ↻;
$\boldsymbol{\omega}_B = 39.2$ rad/s ↺.

18.12 2.72 s.
18.18 (*a*) 12 m/s →. (*b*) 100 N ←.
18.20 21.5 ft/s ←. SI: 6.55 m/s ←.
18.22 (*a*) $5\bar{v}_0/7$ →. (*b*) $2\bar{v}_0/7\mu g$.
18.24 (*a*) 3.09 rad/s ↺. (*b*) −2.40 ft · lb.
SI: (*b*) −3.25 J.
18.26 (*a*) 334 rpm. (*b*) −6.51 J.
18.28 (*a*) and (*b*) 5.71 rad/s.
18.30 41.1 rpm.
18.32 $\omega_A = \omega_B = 159.1$ rpm ↻;
$\omega_P = 20.9$ rpm ↺.
18.34 $v_r = 3.97$ m/s; $v_\theta = 2.86$ m/s.
18.36 (*a*) 1.333 m/s →. (*b*) 6 kN ←.
18.38 $\mathbf{v}_A = 4$ ft/s ←; $\mathbf{v}_B = 20$ ft/s →.
SI: $\mathbf{v}_A = 1.219$ m/s ←;
$\mathbf{v}_B = 6.10$ m/s →.
18.40 $2r/5$.
18.44 $L/\sqrt{3}$.
18.46 $\omega_2 = 6\bar{v}_1/7L$ ↺;
$\bar{\mathbf{v}}_2 = 3\sqrt{2}\bar{v}_1/7$ ⦣ 45°.
18.48 $\omega_2 = 3\bar{v}_1/2b$ ↻; $\mathbf{v}_x = \frac{3}{4}\bar{v}_1$ →;
$\mathbf{v}_y = \frac{1}{4}\bar{v}_1$ ↑.
18.50 $\omega_2 = (2 + 5\cos\beta)\omega_1/7$ ↺;
$\bar{\mathbf{v}}_2 = (2 + 5\cos\beta)\bar{v}_1/7$ ←.
18.52 (*a*) $\mathbf{v}_A = 0$, $\boldsymbol{\omega}_A = \omega_0$ ↻; $\mathbf{v}_B = v_0$ →,
$\boldsymbol{\omega}_B = 0$. (*b*) $\mathbf{v}_A = 2v_0/7$ →,
$\boldsymbol{\omega}_A = 2\omega_0/7$ ↻; $\mathbf{v}_B = 5v_0/7$ →,
$\boldsymbol{\omega}_B = 5\omega_0/7$ ↻. (*c*) The motion of part *a* is the final motion.
18.54 $\omega_2 = \dfrac{\bar{v}_1}{l}\dfrac{6\sin\beta}{3\sin^2\beta + 1}$ ↻.
18.56 59.0° ⦣.
18.58 $\omega_{AB} = \frac{1}{4}\omega_0$ ↻, $\bar{\mathbf{v}}_{AB} = \frac{1}{8}\omega_0 L$ ↑;
$\omega_{CD} = \frac{3}{4}\omega_0$ ↺, $\bar{\mathbf{v}}_{CD} = \frac{1}{8}\omega_0 L$ ↓.
18.60 (*a*) Upward. (*b*) To the left.
18.62 7950 N · m.
18.64 22.7 lb · ft. SI: 30.7 N · m.
18.66 $A_y = 0$, $A_z = +813$ lb, $B_y = 0$,
$B_z = -813$ lb. SI: $A_y = 0$,
$A_z = +3620$ N, $B_y = 0$, $B_z = -3620$ N.
18.68 0.210 N.
18.70 1677 rpm.
18.72 6260 rpm.
18.74 (*a*) Slides. (*b*) *Pipe:* $\bar{\mathbf{v}} = 0.981$ m/s →,
$\boldsymbol{\omega} = 6.54$ rad/s ↺; *Plate:* $\mathbf{v} = 6.17$ m/s →.
18.76 (*a*) 0.522 ft. (*b*) 4.47 rad/s.
SI: (*a*) 0.1591 m.
18.78 (*a*) $\boldsymbol{\omega}_2 = 3\bar{v}_1/L$ ↻; $\bar{\mathbf{v}}_2 = \frac{1}{2}\bar{v}_1$ ↓.
(*b*) $\boldsymbol{\omega}_3 = 3\bar{v}_1/L$ ↺; $\bar{\mathbf{v}}_3 = \frac{1}{2}\bar{v}_1$ ↑.
(*c*) $\omega_4 = 0$; $\bar{\mathbf{v}}_4 = \bar{v}_1$↑.
18.80 $h_A = h\left(\dfrac{3m}{6m + m_P}\right)^2$.
18.82 $\bar{\mathbf{v}}_x = mv_0/M$ →, $\bar{\mathbf{v}}_y = 3mv_0/5M$ ↑.
18.84 $\bar{\mathbf{v}} = 6mv_0/5M$ →.

CHAPTER 19

19.2 (*a*) 0.1900 m. (*b*) 2.39 m/s.
19.4 (*a*) 2.49 mm; 0.0979 in. (*b*) 0.621 mm; 0.0245 in.
19.6 (*a*) 0.497 s. (*b*) 0.632 m/s. (*c*) 8.00 m/s².
19.8 (*a*) 0.1348 s. (*b*) 2.24 ft/s ↑; 20.1 ft/s² ↓.
SI: (*b*) 0.683 m/s ↑; 6.13 m/s² ↓.
19.10 (*a*) 0.679 s; 1.473 Hz. (*b*) 0.1852 m/s; 1.714 m/s².
19.12 4.55 lb. SI: 2.06 kg (mass).
19.14 (*a*) 0.994 m. (*b*) 3.67°.
19.16 0.971 s.
19.18 84.4 min.
19.20 1.904 Hz.
19.22 (*a*) 1.794 s. (*b*) 1.825 s. (*c*) 2.12 s.
19.24 1.850 s.
19.26 (*a*) 0.297 s. (*b*) 2.64 ft/s.
SI: (*b*) 0.805 m/s.
19.28 (*a*) 1.103 Hz. (*b*) $T_B = 79.1$ N;
$T_C = 61.1$ N.
19.30 $\tau = 2\pi\sqrt{2m/3k}$.
19.32 (*a*) $f = (1/2\pi)\sqrt{4k/m}$.
(*b*) $f = (1/2\pi)\sqrt{12k/m}$.
19.36 (*a*) $f = (1/2\pi)\sqrt{3g/2l}$. (*b*) $\frac{1}{6}l$.
19.38 (*a*) $\tau = 2\pi\sqrt{7l/6g}$. (*b*) $\tau = 2\pi\sqrt{5l/6g}$.
19.40 (*a*) 1.737 s. (*b*) 150 mm.
19.42 $\bar{k}_x = 6.08$ ft; $\bar{k}_z = 6.74$ ft.
SI: $\bar{k}_x = 1.853$ m; $\bar{k}_z = 2.05$ m.
19.44 (*a*) $\tau = (2\pi l/b)\sqrt{h/3g}$. (*b*) $\tau = 2\pi\sqrt{h/g}$.
19.46 (*a*) 0.288 s. (*b*) 0.458 m/s.
19.48 0.456 in. SI: 11.58 mm.
19.54 498 lb · ft · s². SI: 675 kg · m².
19.56 $\tau = 2\pi\sqrt{0.866l/g}$.
19.60 $\tau = 2\pi\sqrt{2l_1l_2/g(l_1 + l_2)}$.
19.62 0.926.
19.66 $\tau = 2\pi\sqrt{m/3k\cos^2\beta}$.
19.68 $\tau = 2\pi\sqrt{m_c/k\cos^2\beta}$.
19.70 (*a*) 1.429 s. (*b*) 1.845 s.
19.72 2.48 s.

19.74 $\sqrt{k/2m} < \omega < \sqrt{3k/2m}$.
19.76 $\sqrt{g/2l} < \omega < \sqrt{3g/2l}$.
19.78 (*a*) 400 rpm. (*b*) 0.00167 in.
SI: (*b*) 42.3 μm.
19.80 0.703 mm.
19.82 0.750 in. or 0.1875 in.
SI: 19.05 mm or 4.76 mm.
19.86 0.794 mm.
19.88 245 rpm.
19.90 0.509 Hz; 0.657 Hz.
19.92 70.1 km/h.
19.100 (*a*) $x = x_0 e^{-pt}(1 + pt)$. (*b*) 0.1108 s.
19.102 $\sqrt{1 - 2(c/c_c)^2}$.
19.104 (*a*) 1.509 mm. (*b*) 0.583 mm.
19.106 0.1194 in. SI: 3.03 mm.
19.108 (*a*) 270 rpm. (*b*) 234 rpm. (*c*) 8.84 mm; 9.45 mm.
19.110 $m\ddot{x}_A + 5kx_A - 2kx_B = 0$;
$m\ddot{x}_B - 2kx_A + 2kx_B = P_m \sin \omega t$.
19.112 (*a*) E/R. (*b*) L/R.
19.118 (*a*) $m\ddot{x}_m + k_2 x_m + c(\dot{x}_m - \dot{x}_A) = 0$;
$c(\dot{x}_A - \dot{x}_m) + k_1 x_A = 0$
(*b*) $L\ddot{q}_m + \frac{q_m}{C_2} + R(\dot{q}_m - \dot{q}_A) = 0$;
$R(\dot{q}_A - \dot{q}_m) + \frac{q_A}{C_1} = 0$.
19.120 (*a*) $m_1\ddot{x}_1 + c_1\dot{x}_1 + c_2(\dot{x}_1 - \dot{x}_2) + k_1x_1 + k_2(x_1 - x_2) = 0$;
$m_2\ddot{x}_2 + c_2(\dot{x}_2 - \dot{x}_1) + c_3\dot{x}_2 + k_2(x_2 - x_1) + k_3x_2 = P_m \sin \omega t$.
(*b*) $L_1\ddot{q}_1 + R_1\dot{q}_1 + R_2(\dot{q}_1 - \dot{q}_2) + \frac{q_1}{C_1} + \frac{q_1 - q_2}{C_2} = 0$;
$L_2\ddot{q}_2 + R_2(\dot{q}_2 - \dot{q}_1) + R_3\dot{q}_2 + \frac{(q_2 - q_1)}{C_2} + \frac{q_2}{C_3} = E_m \sin \omega t$.
19.122 1.363 s.
19.124 5.98 in. SI: 151.8 mm.
19.126 (*a*) 5.81 Hz; 4.91 mm; 179.2 mm/s.
(*b*) 491 N. (*c*) 159.2 mm/s ↑.
19.128 4.8 mm.
19.132 2.48 s.

Centroids of Common Shapes of Areas and Lines

Shape		$\bar{x}$	$\bar{y}$	Area
Triangular area	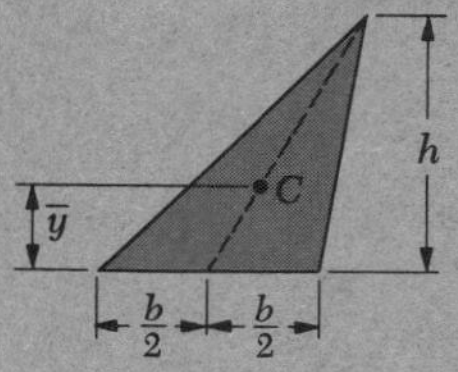		$\frac{h}{3}$	$\frac{bh}{2}$
Quarter-circular area	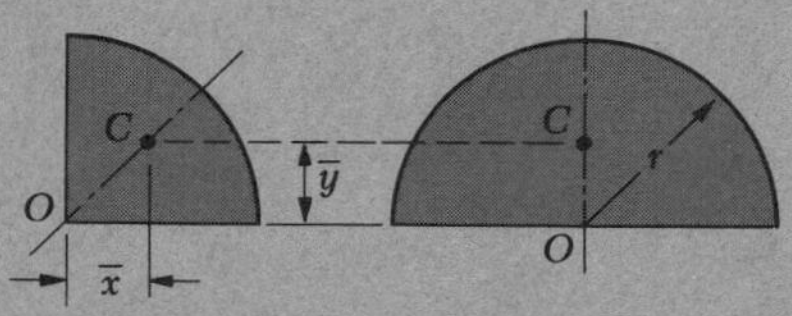	$\frac{4r}{3\pi}$	$\frac{4r}{3\pi}$	$\frac{\pi r^2}{4}$
Semicircular area		0	$\frac{4r}{3\pi}$	$\frac{\pi r^2}{2}$
Semiparabolic area	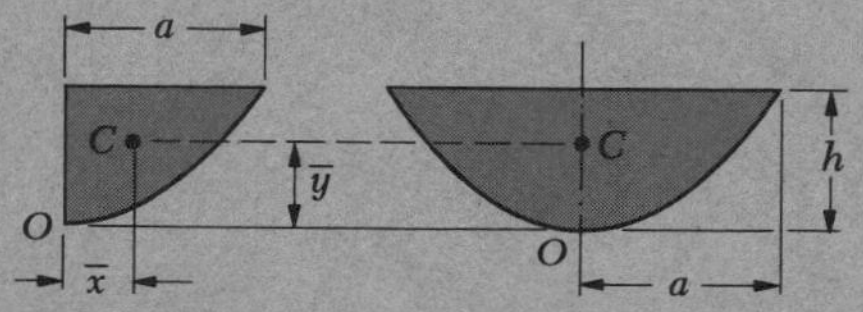	$\frac{3a}{8}$	$\frac{3h}{5}$	$\frac{2ah}{3}$
Parabolic area		0	$\frac{3h}{5}$	$\frac{4ah}{3}$
Parabolic spandrel	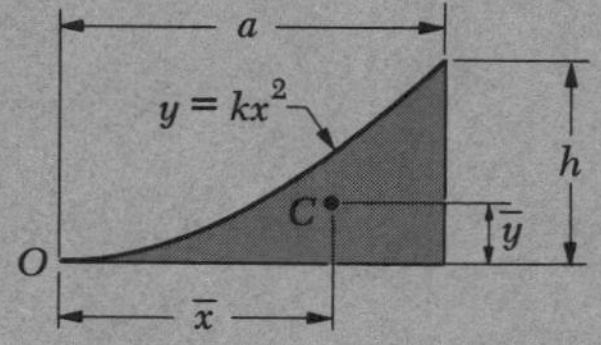	$\frac{3a}{4}$	$\frac{3h}{10}$	$\frac{ah}{3}$
Circular sector	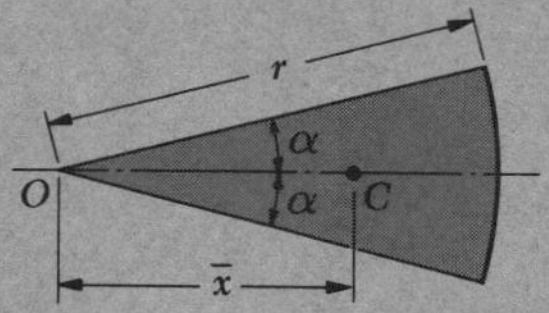	$\frac{2r \sin \alpha}{3\alpha}$	0	αr^2
Quarter-circular arc	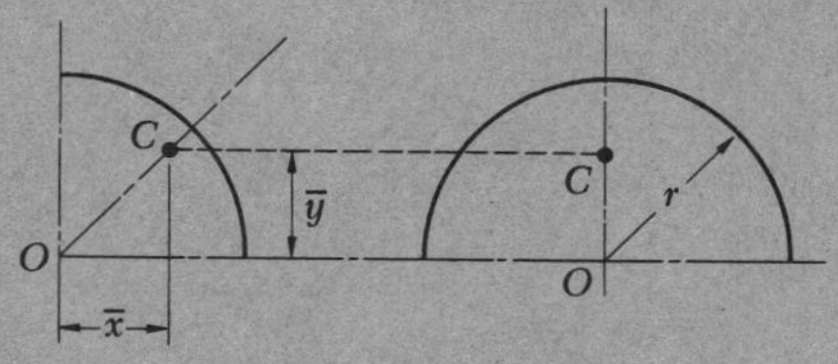	$\frac{2r}{\pi}$	$\frac{2r}{\pi}$	$\frac{\pi r}{2}$
Semicircular arc		0	$\frac{2r}{\pi}$	πr
Arc of circle	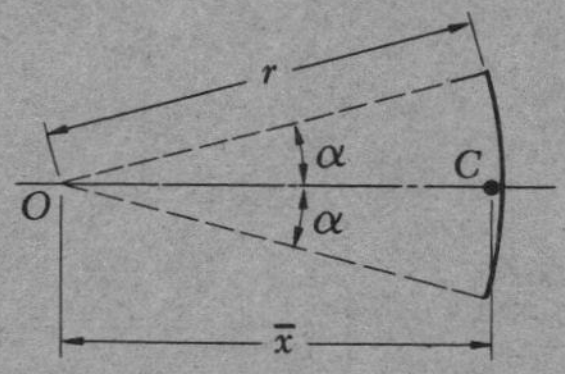	$\frac{r \sin \alpha}{\alpha}$	0	$2\alpha r$